# Finite Elements in Water Resources

Proceedings of the 4th International Conference,
Hannover, Germany, June 1982

Editors:
K.P. Holz
U. Meissner
W. Zielke
C.A. Brebbia
G. Pinder
W. Gray

Sponsors: Deutsche Forschungsgemeinschaft
International Society for Computational Methods in Engineering
International Association for Hydraulic Research

Springer-Verlag Berlin Heidelberg New York 1982
A Computational Mechanics Centre Publication

K.P. Holz
U. Meissner
W. Zielke
C.A. Brebbia
G. Pinder
W. Gray

Computational Mechanics Centre,
Ashurst Lodge
Ashurst,
Southampton,
Hampshire, SO4 2AA,
UK



ISBN 3-540-11522-6 Springer-Verlag Berlin Heidelberg NewYork
ISBN 0-387-11522-6 Springer-Verlag NewYork Heidelberg Berlin
ISBN 0-905451-09-0 Computational Mechanics Centre Southampton



Printed by Hobbs the Printers of Southampton
2061/3020/543210

## PREFACE

These proceedings contain the papers presented at the Fourth International Conference on Finite Elements in Water Resources, held in June, 1982, at the University of Hannover, Federal Republic of Germany. This Conference continued the successful series of previous conferences held at Princeton University in 1976, at Imperial College in 1978, and at the University of Mississippi in 1980.

Since Finite Elements have proved to be a powerful means for analysing water resource problems, the principal objective of the Conference was to provide an exchange of experiences in practical applications of the finite element method and to establish a forum for discussion regarding accuracy, economy, limitations and improvements. Related discretization methods were included within the scope of the Conference. New developments in numerical and computational techniques, basic mathematical formulations, and soft- and hardware aspects were considered to be equally important topics for an exchange of ideas between both theoretically and practically oriented researchers.

The Conference Organizing Committee is very grateful to the many distinguished scientists who attended the Conference, and for their contributions towards the proceedings. This collection of papers in being made available to a wider audience of engineers and scientists by CML Publications in Southampton, U.K.

We are particularly grateful for the enlightening lectures given by the guest speakers, namely, J. H. Argyris, L. Bengtsson, L. Collatz, D. Haidvogel, C. Taylor, and P. A. Witherspoon, and we would also like to express our gratitude to those speakers who were invited to present papers at individual sessions.

Hannover, June 1982

K.-P. Holz U. Meissner W. Zielke

(Local Organizing Committee of the Conference)

Please note that the editors are not responsible for the contents of the papers which were reproduced directly from the original material submitted by the authors.

ACKNOWLEDGMENT

The Conference Organizing Committee gratefully acknowledges the sponsorship given by the following organisations:

Deutsche Forschungsgemeinschaft
(German Research Society)

Niedersächsisches Ministerium
für Wissenschaft und Kunst
(Ministry of Science and Art of the State of Lower Saxony)

International Association for Hydraulic Research (IAHR)

International Society for Computational Methods in Engineering (ISCME)

The local organization of the Conference was undertaken by the following institutions of the University of Hannover:

Institut für Strömungsmechanik
und Elektronisches Rechnen im Bauwesen
(Institute of Fluid Mechanics and Electronic Computations in Civil Engineering)

Sonderforschungsbereich Wasserforschung
im Küstenbereich - SFB 79
(Interinstitutional Research Group for Water Research in Coastal Areas)

# CONTENTS

## 1. GENERAL LECTURES



## 2. MATHEMATICS UND NUMERICAL TECHNIQUES

## 3. SOLUTION OF THE NAVIER-STOKES EQUATIONS



## 4. OPEN CHANNEL FLOW CALCULATIONS



## 5. NUMERICAL TECHNIQUES FOR TIDAL AND OCEAN FLOW

10. SIMULATION OF GROUNDWATER FLOW



11. GROUNDWATER, SOIL AND INTERACTION

## 12. FLOW IN ROCKY MATERIAL



## 13. TRANSPORT PROCESSES IN GROUNDWATER FLOW



## 14. NUMERICAL TECHNIQUES FOR THE SOLUTION OF THE TRANSPORT EQUATION

## 17. FLOW INDUCED FORCES IN STRUCTURES



## 18. PARAMETER ESTIMATION AND OPTIMIZATION



## 19. BOUNDARY ELEMENT METHOD

SUPPLEMENTS

# 1. GENERAL LECTURES

As an introduction to each day of the Conference, the following general lectures, covering specific topics in numerical modelling, were presented:

J. H. Argyris:

Finite Element Methods

L. Bengtsson:

Numerical Modelling of the Global Atmosphere

L. Collatz:

Numerical Methods and Mathematical Aspects

D. Haidvogel:

Modelling of Ocean Circulation

C. Taylor:

Solving Turbulent Flow Problems Using the FEM

P. A. Witherspoon:

Overview of the Finite Element Method in Groundwater Hydrology

# NUMERICAL MODELLING OF THE GLOBAL ATMOSPHERE

Lennart Bengtsson

European Centre for Medium Range Weather Forecasts

## 1. INTRODUCTION

Among fluid systems, it is surely the atmosphere whose behaviour has attracted the greatest interest of physicists as well as scientists in other related disciplines. The understanding and predicting of the atmosphere is also of utmost importance for society, so that warnings can be given of serious weather events and planning for all kinds of weather-dependent activities (agriculture, transportation, building work, etc.) can be improved. It is therefore not surprising that the changes of the atmosphere has been more thoroughly measured than those of any other fluid system and that the prediction of these changes remains one of our most challenging scientific and technical problems. This paper will describe the problems in general of numerical modelling of the atmosphere and its application to weather prediction and climate simulation.

From a technical point of view, the problem of weather prediction is especially intriguing in that the atmospheric system is essentially non-linear and cannot be decomposed into independently acting modes. The inevitable error in observing the smallest scales of motion must then contaminate larger scales and finally destroy the accuracy of any prediction. The weather prediction can therefore be seen as an unstable problem in the sense that small initial differences have large final effects.

Although the problem as such is deterministic, it is, for practical reasons non-deterministic since the initial state can never be perfectly known. Theoretically, the weather prediction has therefore much in common with more general non-deterministic prediction problems such as economical and social systems. Atmospheric prediction models may, therefore, also serve as a useful prototype of better understanding of a more general class of problem, where the dynamical laws are not yet so well understood.

Although the basic physical laws governing the atmosphere have been known since the last century, no real progress took place before the advent of computers. The first integration of a simple 2-dimensional atmospheric model was done by Charney, Fjörtoft and von Neumann (1950). Since then there has been a rapid development of successively more realistic models following the very fast development of computers. These models have been used for a large range of atmospheric problems from short range weather prediction to the simulation of the climate of the earth, Smagorinsky (1974) Mason (1979).

The problem of predicting the atmosphere on time scales from a few days to a few weeks is particularly challenging and important. It is regarded as perhaps the most difficult prediction problem because we have to rely on accurate treatment of the atmospheric observation as well as on accurate modelling. In the appreciation of the importance of this problem, the western European countries decided to jointly set up and finance a special European Centre dedicated to medium range prediction, European Centre for Medium Range Weather Forecasts (ECMWF). ECMWF was established in 1975. It started to make daily 7-days forecasts from August 1979. The Centre has been quickly recognized and its forecasts are considered to be among the best available. Results of forecasts and numerical experiments presented in this paper have been obtained by the ECMWF model.

## 2. THE PHYSICAL AND MATHEMATICAL BASIS FOR NUMERICAL MODELS

### 2.1 The predictive equations

The behaviour of the atmosphere is governed by fundamental physical laws and their boundary conditions. The macrostructure of these laws has been known for over a century. What has been lacking, is the interaction of the macroscales (greater than a few hundred kilometers) with processes of much smaller dimensions such as radiative transfer, turbulence fluxes, cloud and precipitation processes.

The physical laws governing the atmosphere are:

The gas law
$$p = \rho RT \tag{1}$$

The continuity equation for dry air
$$\frac{d\rho_d}{dt} = - \rho_d \nabla \cdot \underline{v} \tag{2}$$

The continuity equation for moist air
$$\frac{d\rho_w}{dt} = - \rho_w \nabla \cdot \underline{v} + S \tag{3}$$

The first law of thermodynamics
$$c_p \frac{dT}{dt} - R \frac{dp}{dt} = Q \tag{4}$$

The equation of motion
$$\frac{d\underline{v}}{dt} = \frac{1}{\rho} \nabla p - g - 2\underline{\Omega} x \underline{v} + \underline{F} \tag{5}$$

For a definition of symbols and expressions see Table I. The physical parameters entering these laws are: $\underline{v}$, the three-dimensional velocity vector relative to the earth; T, the temperature; p, the pressure; $\rho_d$, the density of air; and $\rho_w$ the density of water vapour. Some atmospheric models have additional conservation laws for liquid water (cloud water) and for ozone.

The equations (1) to (5) constitutes a closed system which can be solved at all future times from a given initial state and with the necessary prescribed boundary conditions. However, the equations still contain some unspecified source and sink terms and it is necessary to provide a second set of expressions where we can specify these in terms of known physical quantities and/or in the basic parameters $\underline{v}$, T, p, $\rho_d$ and $\rho_w$.

In the case where the source and sink terms, $\underline{F}$, Q and S are zero the system is energetically closed and can only describe adiabatic processes. It is also reversible and can be integrated backward in time as well as forward. The diabatic term, $\underline{F}$ representing dissipation of momentum and Q and S representing sources and sinks for heat and water vapour, will be discussed later.

The complete atmospheric equations in (1) to (5) have not so far been used for operational forecasting. They are very general and represent in principle all scales of motion from the microscale to the largest planetary scale.

For practical purposes, we can only resolve the scale of motion which can be analyzed by standard data and it has therefore become common practise to simplify the equations by omitting small terms and filtering out unwanted motion. The approximative equations are derived by considering the temporal and spatial scales of interest which are from about 1 hour and 100 km respectively. With these assumptions, which are not only dictated by the data distribution but also as we will show later by computational considerations, the basic equations (1) to (5) can be simplified. It is found for instance that the vertical equation of motion can be reduced to a diagnostic relation where the vertical pressure force is balanced by the gravitational force:

$$\frac{1}{\rho}\frac{\partial p}{\partial z} + g = 0 \qquad (6)$$

The hydrostatic equation (6) leaves us without a prognostic equation for the vertical motion, w, but the adoption of (6) will make it possible to determine w diagnostically from the remaining equations. An additional consequence of (6) is that other approximations must be made to guarantee that the resulting system, in the absence of sources and sinks, conserves energy and momentum. These approximations lead to that w is eliminated from the expression of kinetic energy and the radial distance R is replaced by the average radius of the earth. This is justified because the atmosphere is a fairly thin envelope, like the skin of an apple, surrounding the earth. The resulting equations are generally called the "primitive equations" by the meteorological community.

An interesting consequence arising from the adoption of the hydrostatic relation is the removal of vertical travelling sound waves. As we will see later, this will eliminate a severe restriction on the numerical time-integration schemes.

## 2.2 Parameterization of subgrid scale processes

Observational and computational restrictions has made it necessary to confine the atmospheric model to the description of phenomena larger than a certain given scale. Present computers put this limit around 100 km in the horizontal and around one kilometre in the vertical. The dimension of such a volume is a measure of the computational resolution. What happens on scales smaller than that of the volume is known as subgrid scale processes and the simplified prescription of how they are related to the macroscale is known as the parameterization. Finally, the empirical constants as well as the dependent variables of the macroscale that enter into these relations are known as parameters.

The physical processes which go into the description of $\underline{F}$, Q and S consist of a manyfold of subgrid processes which must be parameterized. The major sink for momentum and the sources for heat and moisture is the earth surface. The vertical fluxes of heat are greatest in a thin layer near the ground, the planetary boundary layer, PBL. In large modelling we are not concerned with the detailed processes in the boundary layer *per se*, but rather the overall effect of the PBL on the rest of the atmosphere. If we regard for instance the grid point values, u, of the horizontal velocity as a mean value for the grid box, then within this grid box the velocity is u + u' where u' is a perturbation from the mean. This leads to Reynold stress terms or correlations such as $\overline{u'w'}$ and similarly $\overline{T'w'}$ which are the vertical fluxes of momentum and heat by subgrid scale processes. A method of relating these fluxes to the large scale parameters is by using the concept of mixing length, Prandtl (1925). Hereby $\overline{u'w'} \approx - \ell^2 \left|\frac{\partial \bar{u}}{\partial z}\right| \frac{\partial \bar{u}}{\partial z}$ that is proportional to the square of the vertical gradient of the horizontal wind multiplied by an unspecified empirical parameter, $\ell$. This is an example of a first order closure which is common in simpler models. Recently more sophisticated relations have been calculated and prognostic equations have been derived from $\overline{u'w'}$ in terms of the mean flow and high correlation with different closure assumptions about these correlations. The disadvantage with these higher order schemes is that they normally imply the use of more empirical parameters.

When large quantities of heat and moisture are supplied to the lower layers of the atmosphere (this occurs frequently when cold air masses are heated from below in particular over a relatively warm sea), the atmosphere tends to be de-stabilized. This leads to convective storms which transport large amounts of heat and moisture and momentum vertically into the atmosphere. The horizontal scale of these storms is of the order of only a few kilometres. In spite of this they play a dominating role in heating the atmosphere (through the release of latent heat) and are the driving mechanism for large tropical storms such as hurricanes. Parameterization of convection is therefore of prime importance and consists of relating the vertical fluxes to the moist static stability and to the horizontal convergence of heat and moisture.

Precipitation and evaporation do involve complex microphysical processes which must also be parameterized. So far they are treated in a very simplified way with little or no incorporation of cloud physics. In most models precipitation takes place if the model atmosphere is supersaturated. Sufficient precipitation is produced to maintain saturation and latent heat is released which modifies the temperature field.

The sun's radiation and the long wave radiation from the earth and from the atmosphere are affected by the presence of clouds, water vapour, carbon dioxide, ozone, aerosols, etc. Although the radiative processes can be treated in great detail, because the general absorption and transmission properties of different gases in the atmosphere are well known, the computational cost of such calculations turns out to be excessive. Furthermore, the clouds are essential for the radiation calculation. They must, because of their scale and their stochastic behaviour be parameterized and consequently radiation is parameterized as well. The most sophisticated radiation schemes consider the effect of water vapour, clouds, ozone and carbon dioxide. The current approach is to derive a parameterization from detailed calculations based on the latest experimental data and theory by some fitting procedure.

Some basic data about the ECMWF model is given in Table II. The parameterization scheme is typical for models used in long and medium range prediction and simulation. For more information see JSC (1980). In short range prediction, 1 - 2 days, much simpler schemes can be used.

## 3. NUMERICAL ASPECTS

We will next consider the numerical and computational aspects of atmospheric modelling. If A is the integration domain then the number of grid points in an horizontal area is $A/(\Delta s)^2$ where $\Delta s$ is the average horizontal grid size. For K vertical levels the total number of grid points is $KA/(\Delta s)^2$. If there are n variables per point, then the total number of variables at any one time are $nKA/(\Delta s)^2$ which is a measure of the number of degrees of freedom in the model. The maximum time increment $\Delta t$, to guarantee stability depends on the integration scheme. If we for instance use an explicit integration scheme it is required that

$$\Delta t < \frac{\Delta s}{\sqrt{2}(c+U_{max})} \tag{7}$$

where $c \approx 300$ m sec$^{-1}$ is the speed of the fastest gravity waves and $U_{max}$ is the maximum horizontal wind speed. In the atmosphere $U_{max} \approx 150$ m sec$^{-1}$. Since we have eliminated vertically propagating sound waves by using the hydrostatic relation the vertical condition is less severe than the horizontal one. However, with more economical integration techniques such as semi-implicit Kwizak and Robert (1971) a longer time step can be used. Hereby gravity waves are treated implicitly, while the slower Rossby waves are treated explicitly. For the ECMWF model which uses a semi-implicit scheme and a staggered grid (Figure 1) the following relation holds

$$\Delta t < \frac{\Delta s}{U_{max}} \tag{8}$$

For a total integration time T at least the following number of time steps are needed

$$\frac{T(U_{max})}{\Delta s}$$

Finally, if we need N number of operations per variable/time step the following number of arithmetic operations are needed to make a forecast with the length of time T for an area A:

$$\frac{A\ K\ nT\ (U_{max})\ N}{(\Delta s)^3} \tag{9}$$

Using a computer with the calculation time for a computer capable of I operations/sec we obtain the following expression for the time, t, it takes to make a forecast

$$t = \frac{A\ K\ n\ T(U_{max})\ N}{(\Delta s)^3\ I} \tag{10}$$

Table III gives the values for the operational ECMWF model. It can be seen that the Cray-1 model needs somewhat more than 2 hours of central processing time for a 10-day forecast. Of special significance is that the grid size occurs to the third power in the denominator. Generally the vertical resolution is also increasing as we increase the horizontal resolution so that resolution becomes by far the dominant factor in determining the computational load. Hence if we want to increase resolution by a factor of 10 in all the three space dimensions we need an increase of computer speed by a factor of $10^4$.

In addition to satisfying linear stability criteria, numerical models must have long term stability. The possibility and occurrence of other types of computational instability of numerical schemes arises because of the non-linearity of advection processes. When two fields are multiplied together, the interactions involving the shortest waves cannot be correctly represented, as these interactions produce scales which are too fine to be resolved on the original grid and are instead misinterpreted as contribution towards larger scales. These spurious interactions are usually called aliasing error and can lead to instability. These errors can be overcome by dissipation and smoothing or by designing finite difference schemes which conserve higher order moments such as energy. In the ECMWF model the square of the vorticity of the horizontal wind is conserved. Recently spectral transform models have come into use. All sub-grid scale processes are calculated in the grid point space while the advective terms are calculated spectrally. Using efficient programs for Fourier transforms, these models are generally slightly more accurate than the grid point models for the same computational cost.

They are now successfully replacing the grid point models in numerical integrations. Finite element techniques have also been applied, Cullen (1979). The most promising application seems to use finite element in the vertical dimension.

## 4. OBSERVATIONS

Numerical models need observation for the initial condition. This is a tremendous technical and economical problem because of the difficulties to obtain observation in remote areas. The total cost is of the order of 1000 M$/year. Surface observations are the easiest to acquire, and these provide, at three-hour intervals, values of pressure $p_s$ (0.1 mb), temperature $T_s$ ($1^oC$), velocity $\underline{v}_s$ ($1\ m\ sec^{-1}$), and relative humidity $r_s$ (5%), with typical errors given in parentheses. The densest spatial coverage is provided in the mid-latitudes of the northern hemisphere where a combination of fixed land stations and moving ships provides surface observations with a spacing of 200-300 km. Surface observations alone, however, are not sufficient to define the three-dimensional structure of the atmosphere; instead, more expensive observations are required.

A basic component of the 3-dimensional observing system is the set of upper-air soundings made twice daily from about 700 stations in the northern hemisphere and 100 stations in the southern hemisphere. For economic reasons, these stations tend to be concentrated in mid-latitude land areas where, in the northern hemisphere, they are spaced about 500 km apart. Soundings are made by releasing balloon-borne sensors at midnight and at noon Greenwich time. The sondes rise through the atmosphere telemetering values of temperature, pressure, and humidity back to the ground. They are also tracked so as to determine wind velocities. In accordance with the hydrostatic assumption, Equation 6, pressure p is taken as the independent variable, with other variables recorded as functions of pressure. In order to determine the height z corresponding to the measured pressure, the hydrostatic relation is integrated upward from the surface at pressure $p_s$. The measurement errors in the troposphere are $1^{\circ}C$ for temperature T(p), 5% for relative humidity r(p), and 1 m $sec^{-1}$ for velocity $\underline{v}(p)$. In the stratosphere, these errors become greater by a factor of two or more. In those regions of the globe where the spacing between observations is greater than about 500 km, errors in interpolation become more serious than instrumental errors for limiting the accuracy with which the state of the atmosphere can be determined.

To obtain more uniform spatial coverage, especially of the oceanic regions where data are sparse, satellite-based sensors have been developed to provide temperature soundings. These measure infrared radiances at a set of sharply defined wavelengths sensitive to the temperature in different layers of the atmosphere. For present sensors and inversion algorithms, a temperature profile can be determined with an error of about $2^{\circ}C$. Although the satellite sensor can provide a horizontal resolution of about 50 km, its vertical resolution and accuracy are still considerably less than that of the upper-air soundings made with sondes.

Velocity information has also been derived from satellites through the tracking of cloud elements. There remain some uncertainties about the reliability of clouds as tracers of air motion and the accuracy with which satellites determine their vertical location. The estimated accuracy is 5-10 m $sec^{-1}$.

The observing systems provide observations at locations that are distributed irregularly over the globe. The definition of an initial state of the atmosphere for use by forecasting models requires that the observations be interpolated to a regular grid by a procedure of analysis. The way to analyse the data by minimising the interpolation error is now well understood as is the assimilation of a sequence of analyses into a numerical model. For further information see Bengtsson, Ghil and Källén (1981).

5. ATMOSPHERIC PREDICTABILITY

Global atmospheric models are used for weather prediction and for climate simulation. One can, in principle, as for the ECMWF model, use the same model for both these applications. Following the continuing improvements of computers and associated progress in modelling techniques, remarkable achievements have taken place in the last 20 years.

As mentioned in the introduction numerical weather prediction can never be exact owing to the inevitable errors in the determination of the initial state coupled with an inherent tendency for errors to grow. This error growth is not an artifact of a numerical model but a consequence of the non-linearity and instability of the dynamics of the atmosphere. Considerable research has taken place to assess the predictability of the atmosphere both from data and by numerical experiments. It has been found that the doubling time for small errors is of the order of 2 - 3 days and successively decreasing as the error is growing towards some given statistical value. For practical purposes this growth seems to limit weather forecasting to something in the order of 2 - 3 weeks. However, the error growth is much slower for the largest scales of motion and it may be possible to predict large scale anomalies even beyond this time. A further improvement of the observations will extend this limit with another few days for each successive reduction of the initial error by half. So far the error growth observed in real model predictions is substantially larger than the theoretical estimates and there is consequently scope for improving the forecasts by building more realistic models. The potential improvement for the best available models today has been estimated by Lorenz (1982) to 3 - 4 days. Fig. 2 which has been taken from Lorenz (1982) shows the average error growth for the global ECMWF model. If the error is denoted by E and normalized so that it approaches unity as $t \to \infty$ the error growth curve can be approximately described by a curve of the form

$$\frac{dE}{dt} = aE - aE^2 \qquad (11)$$

where the initial exponential growth is gradually reduced by a quadratic term representing non-linear effects.

## 6. NUMERICAL PREDICTION AND SIMULATION

The error of a numerical forecast is highly scale dependent and the large scale weather systems are more predictable than the small ones. Intense cyclone developments for instance can at best only be accurately predicted up to about 4 days in advance, Fig.3, while large scale changes in the circulation pattern often can be predicted well beyond one week, Fig.4. The forecasting skill also varies from area to area and from season to season. It is generally highest over land areas of the Northern Hemisphere and it is also normally higher in winter than in summer.

Although the predictability of the atmosphere is limited *a priori* extended integrations by the same models can be used to simulate or reproduce climate. Climate here should be taken as the ensemble of a very large set of computer produced flow patterns. Such experiments started already 25 years ago but with very simple models. As the models become more realistic, they have been able to reproduce in great details the climate of the earth without any prior knowledge, except a prescription of some basic surface boundary conditions (ocean surface conditions), the solar constant and basic geophysical data (earth's radius and rotation). Examples of such simulation are given by Fig.5 and Fig.6 showing the simulated precipitation over South America and Africa in January and July.

The need for global atmospheric models in climate research is fundamental. Due to the complexity of atmospheric flow and its interaction with oceans, ice and the land surface, numerical simulation with high resolution models is probably the only feasible avenue for the understanding of climate. Of particular importance is to find out to what extent climate will possibly be affected following man-made or natural changes of the environment. $CO_2$ in the atmosphere has increased steadily since the beginning of the industrial evolution (with about 25% since 1800). $CO_2$ is an efficient absorber of terrestrial long wave radiation and $CO_2$ is enhancing the green-house effect of the atmosphere. Generally an increase of $CO_2$ will therefore give rise to an increase in the temperature of the atmosphere but the final result is related to the role of complex feedback processes between the atmosphere and the surface and in the atmosphere itself. Preliminary results indicate that the heating effect will be larger at high latitudes and during the summer.

The effect of volcanic dust can be tested in a similar way. These experiments suggest in the average a cooling although these results are not so unique as the result of the $CO_2$ simulations.

The growing concern about the climate and its possible change has lead the World Meteorological Organization (WMO), The International Commission of Scientific Unions (ICSU) and the United Nation Environmental Programme (UNEP) jointly to establish a world climate programme, WCP. Simulation experiments by global numerical models will here play a fundamental role.

## 7. REMAINING PROBLEMS

In spite of great progress there are still many remaining problems before we have a fully satisfactory modelling of the atmosphere. The most typical and serious deficiencies are systematic deviations between the simulated climate and the real one. These deficiencies differ between models but many defects are in common. Of particular importance is to understand the feed-back between the atmosphere and the ocean, between the atmosphere and the land surface and between cloud and radiation.

A common deficiency of the models is a tendency to produce an atmosphere which is colder than the real one. This is due to a generally difficulty in maintaining a heat balance in good agreement with the observed. For the ECMWF model this is related to weaker convective processes than the observed ones. In the tropics the models have general difficulties in simulating the large scale quasi-stationary circulation pattern. The low level easterlies are usually too weak and the stratospheric easterlies too strong. At higher latitudes the agreement with observations is quite good but with a tendency to favour a more zonal type of circulation. It is not clear if these errors are mainly caused by inaccuracies and inconsistencies in the numerical treatment of the equations or to errors in the parameterization of sub-grid scale processes. Only an intense future research programme can provide an answer to such questions. The international co-operation within the Climate Research Programme will hopefully contribute towards this end.

## REFERENCES

Bengtsson, L., Ghil, M. and Källén, E. 1981 Dynamic Meteorology, Data assimilation methods. Applied Mathematical Science, 36, Springer-Verlag, New York, Heidelberg, Berlin.

Charney, J.G., Fjörtoft, R. and von Neumann, J. 1950 Numerical integration of the barotropic vorticity equation. Tellus, 2 No.4, 237-254.

Cullen, J.P. 1979 The finite element method. Numerical methods used in atmospheric models. GARP Publication Series No.17, 302-339, ICSU/WMO, Joint Scientific Committee, Geneva.

JSC 1980 Catalogue of numerical atmospheric models for the First GARP Global Experiment, Part I 287-340, ICSU/WMO, Joint Scientific Committee, Geneva.

Kwizak, M. and Robert, A.J. 1971 A semi-implicit scheme for grid point atmospheric models of the primitive equations. Monthly Weather Review. Vol.99, 33-36.

Lorenz, E. 1982 Atmospheric predictability experiment with a large numerical model. Submitted to Tellus.

Mason, B.J. 1979 Some results of climate experiments with numerical models. Proceedings of the World Climate Conference. WMO - No. 537, 210-242.

Prandtl,L. 1925 Bericht über Untersuchungen zur ausgebildeten Turbulenz, Zeitschr. Augew. Math.Mech., 5, 146-139.

Smagorinsky, J. 1974 Global atmospheric modeling and the numerical simulation of climate. Weather and Climate Modification. Edited by Wilmot N. Hess, John Wiley and Sons, 633-686.

TABLE I Symbols and expressions used in the article

|  | |  |
|---|---|---|
| $\underline{v}$ | = | (u,v,w) - 3-dimensional velocity vector relative to the earth |
| T | = | temperature |
| p | = | pressure |
| $\rho_d$ | = | density of dry air |
| $\rho_w$ | = | density of water vapour |
| $\rho$ | = | density of moist air, $\rho = \rho_d + \rho_w$ |
| $\underline{F}$ | = | dissipation of momentum |
| S | = | water vapour source/sink term |
| Q | = | diabatic heating |
| R | = | gas constant for moist air |
| $c_p$ | = | specific heat at constant pressure |
| $g_a$ | = | acceleration due to gravity |
| $\underline{\Omega}$ | = | earth angular velocity |
| $\underline{R}$ | = | radial co-ordinate measure from the centre of the earth |
| g | = | $g_a - \underline{\Omega} x \underline{\Omega} x \underline{R}$ (local gravity acceleration) |
| x,y,z | = | spatial co-ordinate |
| t | = | time |
| $\frac{d}{dt}$ | = | individual derivative = $\frac{\partial}{\partial t} + \underline{v} \cdot \underline{\nabla}$ |
| $\underline{\nabla}$ | = | horizontal gradient operator |

TABLE II Characteristics of the ECMWF model

| | |
|---|---|
| Total number of levels | 15 |
| Levels in the boundary layer | 4 |
| Levels in the stratosphere | 4 |
| Top level (mb) | 25 |
| Horizontal resolution | $1.875^{o}$lat/lon |
| Insolation specified by | latitude and day |
| Absorbing gases | $H_2O$, $CO_2$, $O_3$ |
| Cloudiness | possible at all levels, time-dependent |
| Sea-surface temperature | prescribed climatologically |
| Snow cover | computed |
| Sea ice | computed |
| Soil moisture | computed |
| Land albedo | fixed |
| Land surface temperature | computed |

TABLE III Basic computational parameters for the operational ECMWF global grid point model. The <u>independent parameters</u> can be reduced to 4/level plus surface pressure by a combination of the continuity equation (2) and the hydrostatic equation (6). The average number of <u>variables</u> is 7/level.

| | | |
|---|---|---|
| A | $= 5.10^{14} m^2$ | (Area of the globe) |
| K | = 15 | (number of vertical levels) |
| n | = 4 | (number of independent parameters) |
| T | $= 10$ days $= 8.64.10^5$ sec | (length of a forecast in sec) |
| $U_{max}$ | $= 150$ m $sec^{-1}$ | (estimated maximum wind speed) |
| N | = 500 | (number of operations/grid point/parameter) |
| $\Delta s$ | $= 1.7 \cdot 10^5 m$ | (average horizontal grid length) |
| I | $= 5.10^7 \, sec^{-1}$ (CRAY-1) gives | (number of arithmetic operations/sec) |
| t | = 2.20 h | (processing time for a 10-day forecast) |

Figure Captions

Figure 1. (Top) Horizontal mesh used for the ECMWF global model $p_*$ is surface pressure and $q = {}^{\rho}w/\rho_d$. $\Delta\varphi$ and $\Delta\lambda$ is the grid length in the latitude and longitude respectively.
(Bottom) Vertical grid using $\sigma = {}^{p}/p_*$ as the vertical co-ordinate. $\Psi$ represent geopotential height gz. K = 15 and $\dot{\sigma} = \frac{d\sigma}{dt}$.

Figure 2. Global root-mean square (rms) 500 mb height error as a function of time (full line). Global rms differences between consecutive forecast days valid at the same time (dashed line). This eliminates bias between forecasts and analysis.

Figure 3. Example of 3 day forecast of surface pressure from 12 Feb 1982 successfully predicting the development of an intense depression east of New Foundland. This storm caused severe casualties on an oil rig platform. (a) Show the initial state, (b) the forecast and (c) verification. Full lines are isobars at the surface, dashed lines show isoterms of 850 mb.

Figure 4. Example of a 10 day forecast of 500 mb flow over Scandinavia for 2 May successfully predicting a change of weather for excessively cold to very warm spring weather. Average midday temperature rose during the period by about $20^{\circ}$C.

Figure 5. Total precipitation (mm) over much of South America from 50-day integrations for the periods from 21 January to 12 March (upper) and from 11 June to 31 July (lower). Regions of no rainfall are shaded.

Figure 6. As Figure 5, but for Africa.

Figure 1.

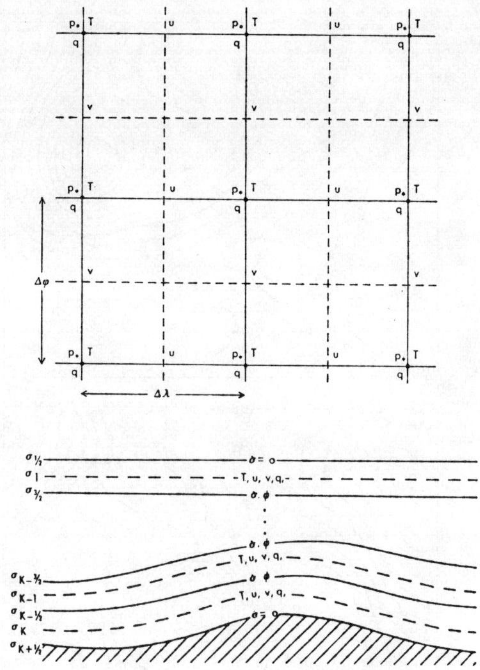


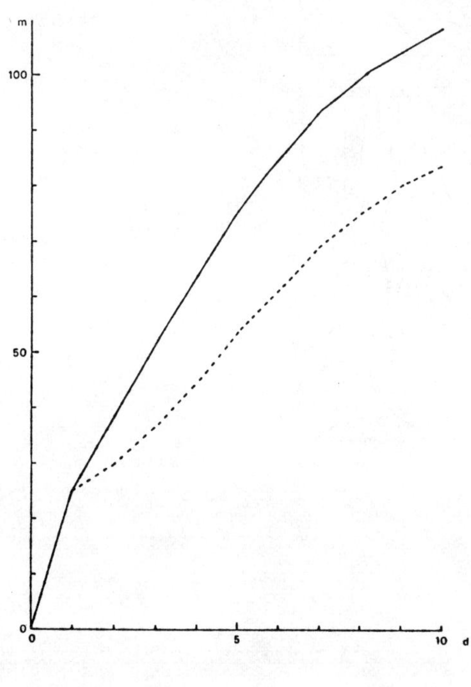


Figure 2.

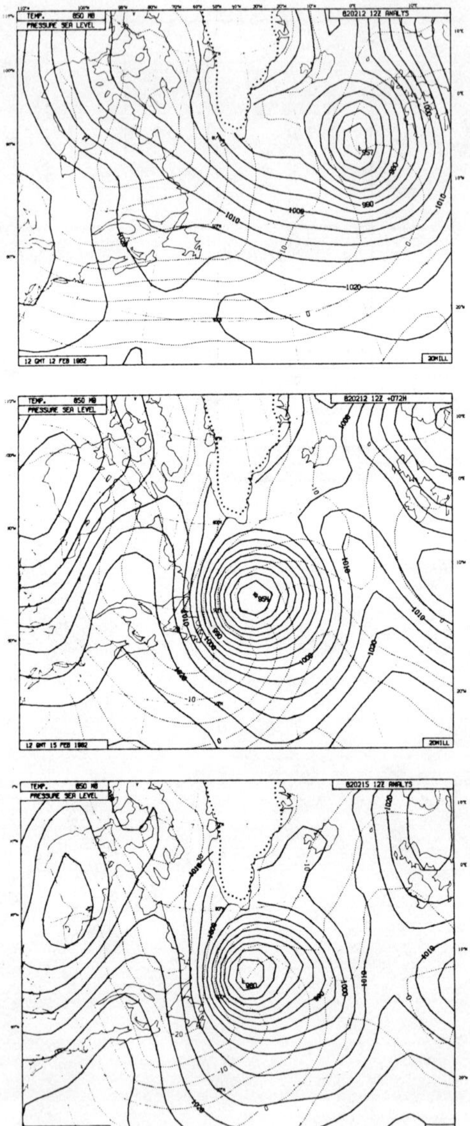
TEMP. 850 MB
PRESSURE SEA LEVEL
820212 12Z ANALYS
12 GMT 12 FEB 1982
TEMP. 850 MB
PRESSURE SEA LEVEL
820212 12Z +072H
12 GMT 15 FEB 1982
TEMP. 850 MB
PRESSURE SEA LEVEL
820215 12Z ANALYS
12 GMT 15 FEB 1982

Figure 3.

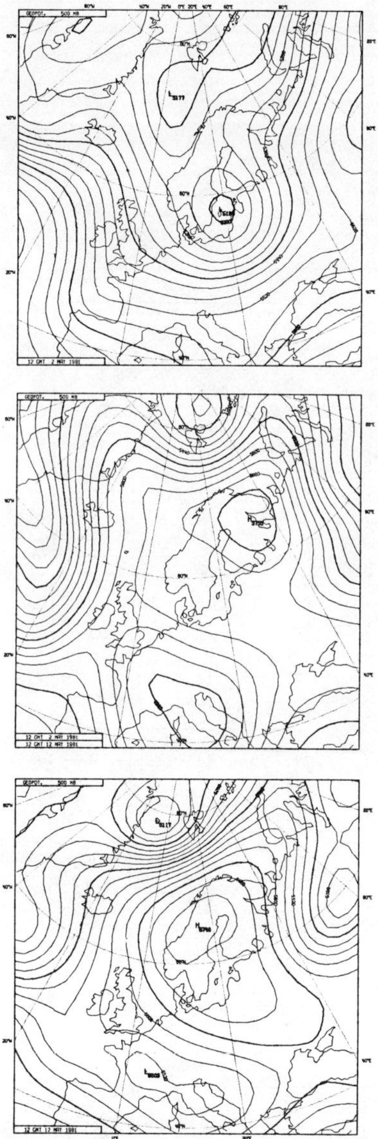
GEOPOT. 500 MB
12 GMT 2 MAY 1981
GEOPOT. 500 MB
12 GMT 2 MAY 1981
12 GMT 12 MAY 1981
GEOPOT. 500 MB
12 GMT 12 MAY 1981

Figure 4.

Figure 5.

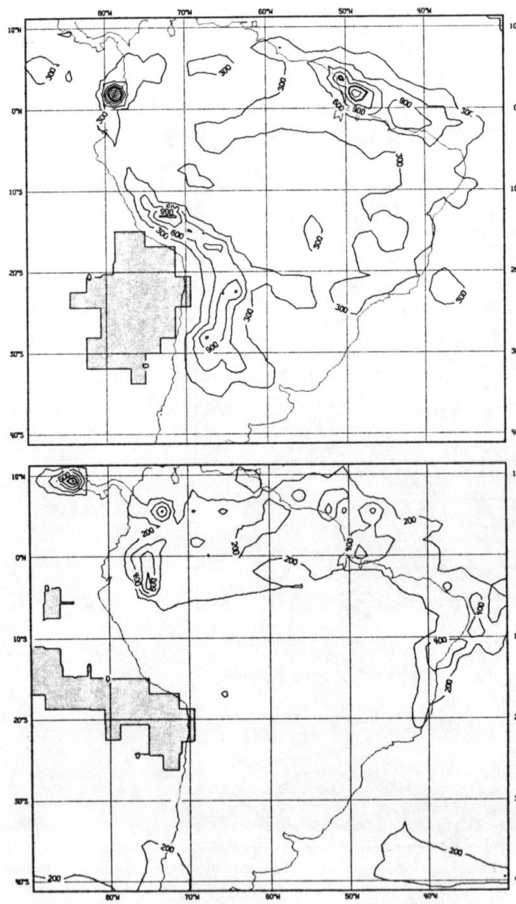

Figure 6.

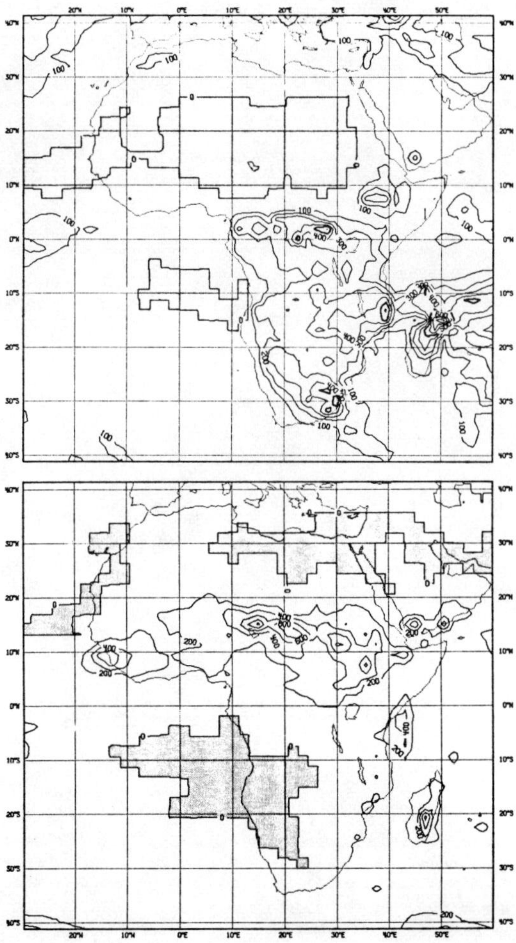

# OVERVIEW OF THE FINITE ELEMENT METHOD IN GROUNDWATER HYDROLOGY

T.N. Narasimhan and P.A. Witherspoon

Lawrence Berkeley Laboratory, Berkeley, California 94720 USA

## ABSTRACT

Since its introduction into the groundwater literature during the mid 1960's, the finite element method has developed into a very powerful numerical tool for analyzing a variety of groundwater flow problems. Applications of the method cover flow in multi-aquifer systems, flow with a free surface, saturated-unsaturated flow, land subsidence, fractured-porous systems, and large groundwater basins under steady or nonsteady conditions. The method derives its power from the fact that it uses a very general technique for the evaluation of spatial gradients in any direction at any point within the flow domain. This advantage is complemented in the method by an integral statement of the conservation equation at the point of interest. The algorithms stemming from this approach permit relatively simple geometric inputs, even when the problem of interest has complex geometries. From a conceptual perspective there is reason to suspect that alternate formulations of the finite element method may be possible in which the weighted integration technique is dispensed with in favor of an explicit definition of the subdomains of integration. The flexibility of existing finite element algorithms may be enhanced by having options for inputting preprocessed geometric inputs in addition to nodal point coordinates and element lists. Direct formulation of the finite element equations from conservation integrals may provide an alternative that deserves attention. With the advent of mini computers, the finite element method promises to become an every day tool for the practising engineer during the 1980's.

## INTRODUCTION

The power of the finite element method was first revealed early in the 1960's in the field of structural engineering. It was soon recognized that the method was sufficiently general to permit application to a much wider class of problems. The earliest attempts to apply the method in groundwater hydrology commenced during the mid 1960's and by the early 1970's the method was well established as a valuable tool for solving groundwater flow problems in complex domains. Indeed, the impact of the method in the field of quantified hydrogeology can rightly be gauged from the very organization of the present international conference which convenes every two years and continues to attract papers with many new results and novel applications to water resource problems. It is appropriate in a conference such as this for one to stop for a moment and review the finite element method, identifying its strengths as well as its weaknesses. The purpose of this paper is to attempt such an overview in so far as it relates to problems in groundwater hydrology.

Over the past decade literature on the finite element method in groundwater hydrology and related disciplines has grown at a phenomenal rate. It is therefore not realistic to attempt a comprehensive review of all the important papers that have been published on the subject. Instead we shall take the more limited approach of presenting an outline of the more important applications that have been made and to analyze some of the conceptual features of the method as they are currently understood. In developing this approach we shall rely on physical arguments rather than drawing upon the theories of mathematical operations.

A majority of groundwater problems relate to the movement of water under isothermal conditions in the subsurface. More recently, the phenomenon of chemical transport in groundwater systems has assumed a major role of practical importance. In the present work, we shall restrict our attention to the diffusion process, which not only governs the movement of water in variably saturated deformable media but also treats the transport of solutes by real or apparent diffusive (or dispersive) mechanisms. We shall not concern ourselves with the convective or advective transport of solutes by flowing water.

## SOME KEY APPLICATIONS IN GROUNDWATER HYDROLOGY

In as much as stress analysis problems deal with systems under equilibrium, it was natural that the earliest application of the method to groundwater flow problems related to steady state seepage. Zienkiewicz and his various colleagues first applied the finite element method to steady state heat flow (Zienkiewicz and Cheung, 1965), and then to steady state seepage problems in anisotropic and heterogeneous media (Zienkiewicz et al, 1966). Taylor and Brown (1967) followed quickly with applications to seepage with a free surface. At the same time, the method was being extended to the transient diffusion problem based on physical considerations by Wilson and Nickell (1966), who successfully simulated transient conduction in concrete structures. Borrowing from their work, Javandel and Witherspoon (1968, 1969) were able to extend the finite element methods to problems of transient flow of fluids in porous media. Unlike the physical approach taken by Wilson and Nickell (1966), Javandel and Witherspoon used the mathematically formal technique of Gurtin (1965) of employing variational principles to develop the finite element equations. They applied the method to the problem of transient flow of water to a well in multi-layered aquifer systems.

The phenomenon of transient seepage in deformable materials is of fundamental importance in the field of geotechnical engineering, which shares many field problems with hydrogeology. Sandhu and Wilson (1970) addressed themselves to the problem of fluid flow in saturated, deformable porous media and successfully applied the finite element method to the simulation of land subsidence due to fluid withdrawal. The work of Sandhu and Wilson paved the way for the coupled stress-strain and fluid flow analysis of saturated soil masses, which forms the basis for a wide class of problems in the field of soil mechanics. Later, during the mid 1970's other workers solved the subsidence problem using the finite element method, notable among them being Gambolati et al, (1974) and Lewis and Schrefler (1978).

Contemporaneous with this work in porous systems, an increasing interest on the problems of fluid flow in fractured rocks also developed, both by hydrogeologists and by geotechnical engineers. It was therefore natural that researchers sought to apply the newly found tool to the study of problems related to fluid flow in fractures. Wilson and Witherspoon (1970), who investigated the laminar flow of water in fractured porous rocks, applied the finite element method to simulate steady flow of water in a network of rigid fractures, representing fractures as line elements. Noorishad et al, (1971) extended Wilson's work to solve the problem of steady flow in a network of deformable fractures using the finite element method. Noorishad and Witherspoon (1981) have extended this approach even further and can now handle coupled thermo-hydro-mechanical transient behavior for non-isothermal fluid flow in fractured rocks.

The problem of groundwater flow with a free surface has interested civil engineers for nearly a hundred years since the early investigations of Dupuit, Boussinesq and Forschheimer during the late nineteenth century. The free surface problem is characterized by the fact that one segment of the boundary of the flow region, namely, the free surface, is either unknown (in the case of the steady state problem) or changes with time (in the case of the transient problem). The free surface problem is especially cumbersome to investigate using conventional numerical techniques such as the finite difference technique. Indeed, it was soon discovered that the very nature of the finite element method makes it admirably suited to manipulate problems in which the geometry needs to be changed frequently. Neuman and Witherspoon (1970) demonstrated that by using the finite element techniques in conjunction with an iterative, deformable mesh, one could efficiently solve steady free surface problems. Neuman and Witherspoon (1971) and Ehlers (1971) extended the method to non-steady problems. Later, France (1974) handled three dimensional free surface problems using the finite element method. Gureghian and Youngs (1975) also used the method to calculate steady state water table heights in drained soils.

The next important class of problems that needed attention in subsurface fluid flow involved consideration of the effects of variable water saturation, that is, the phenomenon of saturated-unsaturated flow, which is of importance in civil engineering, hydrogeology and soil physics. Neuman (1973) was among the earliest workers to apply the finite element method to analyse fluid flow in saturated-unsaturated porous media. Later, Neuman et al,(1975) extended the method to include such other features as water uptake by roots, seepage faces, and an evaporation boundary. The same problem was also solved by other workers using the finite element method and these include, Reeves and Duguid (1975), Segol (1976), Narasimhan et al, (1978), Frind and Verge (1978), and so on.

An ingeneous concept was proposed by Barenblatt et al (1960) to simulate the flow of water in fractured porous media. Noting that in a fractured porous medium the diffusivities of the fractures and the porous matrix may vary widely, Barenblatt et al reasoned that the fractures could all be lumped into one continuum while the porous materials could all be lumped into another continuum. The two continua could then interact by introducing appropriately defined coupling terms. Although this approach has certain limitations, it has provided an economical way of analysing certain field problems too difficult to simulate otherwise. This concept is familiarly known as the "double porosity" approach in the petroleum engineering literature. Duguid and Lee (1977) modified Barenblatt's concept slightly and successfully incorporated the double porosity approach into a finite element model.

From a phenomenological point of view, the investigations cited above more or less cover the spectrum of problems that are commonly encountered in groundwater hydrology. We will not attempt here to cite specific references that deal with the application of the method to site-specific problems.

## DEVELOPMENTS RELATED TO METHODOLOGY

Although the early work of Wilson and Nickell (1966) used a physically based approach to the finite element method, a desire for mathematical formalism and rigor soon established itself. Accordingly, variational calculus was introduced to formulate the finite element equations. The variational principle for the steady state problem originally proposed by Euler formed the basis for solving the elliptic partial differential equation using finite elements. It also provided the starting point for deriving extended variational principles for other steady state problems such as those involving the free surface (Neuman and Witherspoon, 1971). The variational approach, however, initially led to one difficulty, namely that a true variational principle is defined only for a system under equilibrium. To overcome this shortcoming, Gurtin (1965) introduced the novel concept of defining a variational principle in the Laplace domain, after initially eliminating the time dimension from the partial differential equation using the Laplace transform. The variational principle in such a transformed domain involves convolution integrals. These variational principles can be minimized using the Rayleigh-Ritz technique to arrive at the discretized finite element equations.

Meanwhile, yet another technique was discovered that facilitated formulation of the finite element equations: this was the method of weighted volume integration, proposed at the turn of the century by the Russian engineer, Galerkin (1915), to solve problems related to the vibration of elastic plates. Finlayson and Scriven (1967) have compared the variational technique with the weighted integration technique. They concluded that it is easier to formulate the finite element equations using the Galerkin technique than by the variational technique, although the latter has certain advantages in providing some fundamental insights into the nature of the physical problem. It is probably reasonable to state that at present the Galerkin approach is the most widely used to formulate finite element equations for fluid flow in subsurface systems.

At the root of the finite element methodology is the concept that one uses a set of non-collinear points to define the spatial variation of a given piecewise-continuous function. In the simplest and primitive case, the number of points required is three in two dimensions and four in three dimensions. These considerations lead to the simple "triangular element" and the "tetrahedral element" in two and three dimensions, respectively. Both of these simple elements permit only first degree approximations to the variation of the desired spatial variable, just as is the case with classical finite differences. Ergatoudis et al (1968) correctly reasoned that by increasing the number of non-collinear points used to define the "finite element", one could introduce higher order approximations to define the spatial variability of the function of interest. Also by taking advantage of the role of the weighting function in the volume integration process, they made the higher order approximation serve the added purpose of handling complex geometrical shapes within the elemental regions. This contribution introduced the now well-known, isoparametric elements. An added flexibility to the isoparametric elements was proposed by Pinder and Frind (1972) who devised a technique for combining different orders of approximations along different sides of the finite element; the so called mixed isoparametric elements. Unlike the triangular and tetrahedral elements, the higher order elements necessitated the numerical evaluation of the integrals before setting up the global matrix for final solution. The efficiency and accuracy of integration in this regard was greatly facilitated by the use of Gaussian quadrature (Zienkiewicz, 1971).

In its simplest form, the finite element method assumes that the permeability of the material inside an element is constant. Thus, material discontinuities coincide with interfaces of finite elements. In order to improve the accuracy of handling non-linear problems (problems where permeability changes with potential), Pinder et al, (1973) have devised the technique of functional coefficients.

A key task in the application of the finite element method is that of setting up a matrix of algebraic equations to be solved simultaneously for the various unknowns. Even with the fast computers that are increasingly becoming available, the solution of very large matrices is still not an easy process. Matrix solution techniques can be broadly divided into two classes: direct and iterative solvers. Perhaps the most widely used variant of the former is the Gaussian elimination scheme. The chief difficulty with the direct solver techniques when applied to large problems is that of computer storage. In a poorly ordered, large, sparse matrix, considerable time and space are wasted in storing and operating on the zeroes. Hence, many workers continue to develop improved direct solution techniques to go with the finite element algorithms. Among these one should mention the LU decomposition scheme of Cholesky (see Weaver, 1967) and the frontal solution algorithm of Irons (1970). In order to handle very large, sparse matrices, Gupta and Tanji (1976) developed an algorithm in which they operated on portions of the matrix at a time. They chose to store and retrieve all the intermediate information from auxiliary devices such as tapes and discs.

As opposed to direct solvers, iterative methods have the advantage of storing and operating only on non-zero quantities so that the storage requirements are always minimal. For this reason iterative schemes are continuing to be developed and applied to large problems involving the finite element method. Three such techniques that have enjoyed wide usage since the mid 50's are the Alternating Direction Implicit Procedure (Peaceman and Rachford, 1955), the Line Successive Over Relaxation Scheme (Young, 1962) and the Strongly Implicit Procedure (Stone, 1968).

There are several shortcomings associated with the iterative techniques. The first of these is of course the need to provide a good initial guess for the solution and the need to establish an acceptable estimate of error tolerance. Additionally, the iterative scheme may fail to converge or may converge extremely slowly, if at all, should the matrix be too stiff or should it lack diagonal dominance. When one seeks a solution for a large heterogeneous flow region using the finite element method, it may become very difficult to avoid these difficulties. To lessen the problems arising from storage requirements associated with large sparse matrices, Narasimhan et al, (1978) borrowed the concept of mixing explicit and implicit methods of solution during a time step. This was first developed by Edwards (1969) for his integral finite difference scheme. In a stiff matrix, the time constants associated with nodal points are widely variable; nodes with small time constants react very quickly to perturbations while others react more slowly. It is therefore possible during a given time step, to use the forward differencing approximation to decouple those groups of nodes that can be solved explicitly from those that can be solved implicitly to assure stability. In other words, the mixed explicit-implicit scheme provides a means of replacing a large matrix with a few small submatrices, leading to smaller storage requirements and faster execution times.

## CURRENT STATUS

There is little doubt that the finite element method has developed into a valuable tool for investigating complex groundwater systems. In this regard, its strengths include the ability to handle complex geometries, heterogeneities and anisotropy. Additional factors are the ease of input organization and the generality of the formulation. Many computer algorithms currently exist that incorporate the finite element method to solve a variety of groundwater problems. Some of these are general purpose codes while others have been written for a specific class of problems. Clearing houses such as the Holcomb Research Institute of Butler University in Indiana help to distribute these programs or their documentation to interested users.

However, there still exist differences of opinion about the advantages of the finite element scheme over other numerical schemes such as the classical finite differences or the more recent integral finite difference method (Narasimhan and Witherspoon, 1976). Claims of improved accuracy and speed are often made and disputed. Because of wide variations in: (a) the programming efficiencies of individual codes, and (b) the nature of the subcomponents of the different algorithms, it is often extremely difficult to prove or disprove these claims. Familiarity or personal preferences often dictate the approaches taken by many users.

Perhaps it is more relevant to attempt to synthesize the various methods rather than evaluating their individual superiorities. Such a synthesis may lead to a successful blending of the particular advantages of each method and thus generate a more powerful tool not hitherto available. The next section of this paper will attempt such a synthesis.

## A DISCUSSION OF THE FINITE ELEMENT METHOD

The physics of the transient diffusion problem requires, in a macroscopic sense, that intensive quantities such as potential be defined as averages over suitably defined domains around each point of interest in the flow region. Secondly, the physics also requires that gradients of potential be evaluated at any surface segment of interest in order that fluid flux across that segment may be calculated.

The finite element method seeks to satisfy the first requirement by associating with each nodal point, a subdomain which is some weighted fraction of the subregion formed by the union of all finite elements that contain the given nodal point as a corner point. Potentials and changes in potentials are averages integrated over these subdomains. The second requirement, namely, that of evaluating gradients of potential, leads to a technique which gave the finite element method its special identity. In this technique one first defines a "finite element", a small region formed by a set of non-collinear points, within which the potential function of interest is assumed to be piecewise continuous and is approximated in terms of the "point" values at the corner points. Once this is done, it is easy to compute the gradient of potential at any point and in any direction within the finite element by simply differentiating the piecewise continuous, approximating function at that point.

This method of gradient evaluation is obviously far more general and powerful than the classical finite differencing technique. First, the finite element approach can provide gradients in more than one direction, an ability which is vital for handling general non-isotropic conditions. Secondly, this approach can provide higher order approximations to the spatial variation of the potential function by merely increasing the number of non-collinear points that define the finite element. Both these features can only be attained using finite differences with a great deal of difficulty, if at all.

The two requirements mentioned above, that of averaging potentials over subdomains and that of evaluating gradients at interfaces, are mutually independent in nature. The uniqueness of the finite element concept is inherent in the latter requirement but not in the former. Indeed, it is possible to have other approaches such as integrated finite differences in which potentials are defined as averages over finite subdomains so that complex geometries can be handled. Yet, the finite element method, as it is commonly understood, is treated as being synonymous with defining the subdomain by a process of weighted volume integration. How relevant is it to uniquely associate the finite element method with weighted volume integration? Is it possible to use finite element gradients with alternate ways of defining the subdomain over which the accumulation term is evaluated in space?

Let us return to the problem of the physical basis for the concept of fluid movement. In order to evaluate the average change in potential over any domain in space bounded by a closed surface, we first evaluate the net accumulation of mass by summing up the fluxes crossing the bounding surface. This accumulation is then divided by the capacity of the domain to obtain the change in potential. A fundamental task therefore is to define the geometry of the subdomain over which the accumulation is evaluated by surface integration. The surface integral, thus, is primitive and is of fundamental importance. In practice, however, one could replace the surface integral by an equivalent volume integral, provided that the equivalence can be established at all times. Indeed one can show (e.g. Narasimhan and Witherspoon, 1976) that, in so far as triangular elements in two dimensions are concerned, the weighted volume integral of the finite element method is exactly equivalent to an appropriate surface integral and this equivalence is in dependent of time in the transient problem. This reasoning can be also extended to tetrahedral elements in three dimensions.

An important consequence of the weighted volume integration approach is that a geometrical description of the subdomain associated with a nodal point is only implied in the integral statement. The location or the nature of the bounding surface is never explicitly stated. While one can give a reasonable physical interpretation to such an approach for triangular and tetrahedral elements, such an interpretation becomes extremely tenuous when applied to higher order elements. If such a bounding surface exists, does it change with time? In the case of higher order elements, what can one say about the equivalence between surface integrals and their corresponding volume integrals? The equivalence is not a factor in steady state problems in which the accumulation is zero over any region surrounding a nodal point. However, the transient problem is of special interest because one must evaluate volume averages for the time derivative. There is a suspicion that if one cannot assure the exact equivalence of surface and volume integrals at all times, then, the weighted volume procedure can in fact prove to be less than desirable. It may be of interest to explore a technique in which gradients are evaluated using the finite element method, while integration of fluxes is carried out over surfaces of explicitly defined subdomains.

One of the most attactive features of the finite element method is the ease with which the geometric information to define the discretized problem can be provided as input. To solve the diffusion problem in general, three types of geometric information must be provided a priori. These are: (a) a set of one dimensional data (lengths between discrete points) to enable the evaluation of gradients of potential; (b) a set of two dimensional data (areas of interfaces between adjoining volume elements) to evaluate fluxes across surface segments; and (c) a set of three dimensional data (bulk volumes of elements) to calculate the "capacity" of each element.

The need to calculate these quantities beforehand can be bypassed by merely providing as input, the coordinates of the nodal points, supplemented by a node list identifying the corner points of each finite element. From these input data, lengths, areas and volumes are implicitly generated through the process of weighted integration. For many practical problems of interest, this ability to avoid calculating areas and volumes beforehand is extremely convenient and desirable.

Although one should take full advantage of this remarkable feature of the finite element method, it is worthwhile to consider whether or not the flexibility of the method can be enhanced by having an option to input preprocessed geometric data directly. One of the drawbacks inherent in choosing to work with nodal point coordinates is that the domain of volume integration must have a relatively simple shape (triangle, square, tetrahedron, cube) either in the real domain or in the transformed domain. Not only does this force one to write the algorithms for specific coordinate systems, but also the chances for errors of integration are increased because all actual geometric shapes must be approximated in terms of predetermined elemental shapes. When complex geometries are involved, such errors can be minimized by preprocessing the required data as accurately as one may wish.

Another situation in which the preprocessing option would be beneficial is the problem of providing for mesh refinement in certain locations. There often are situations in which a single large volume element may have to communicate with several smaller neighboring ones. Such a situation arises when a wellbore or a fracture element may communicate with several elements in the porous medium. Although one can develop the conductances between the different sized finite elements using published procedures, (e.g. Narasimhan et al, 1978), it is far simpler to handle the situation using preprocessed geometric inputs. We believe the finite element method can benefit from the added flexibility of having an option to provide geometric input information as needed.

A question that is occasionally raised for the finite element method concerns the manner in which the time derivative is handled in the discretized equations. Depending on how one chooses to handle this derivative, one obtains either a "distributed" capacity matrix or a "lumped" capacity matrix. Let us attempt to analyze this question from a conceptual perspective.

The equation of motion, namely, the Darcy-Buckingham law, relates gradient of potential to fluid flux under steady-state conditions. In order to make use of this law, the potential is assumed to be a discrete piecewise continuous function over each finite element such that the potentials at nodal points are functions only of time while the linear coefficients, which help to express the potential within an element in terms of those at the corner points, are only functions of space. Furthermore, the value at a given nodal point is an appropriate average over a simply connected subdomain immediately surrounding the point. It follows, therefore, that the approximating function in the finite element method is to be restricted solely to the spatial domain and cannot be used in the time domain. Hence, the accumulation term evaluated for a given point using the weighted volume integration procedure can only be converted into an average change in potential at that point by dividing it (the accumulation) by the capacity of the simply connected subdomain associated with that point. This manner of treating the time derivative in which the accumulation at each point is converted to a single average time derivative using a single capacity term of the point in question, gives rise to the so called "lumped" capacity matrix. From the basic physical postulates that govern the problem, it appears that the lumped capacity matrix is the most appropriate way of handling the time derivative.

However, a different situation arises when one derives the finite element equations by minimizing variational principles or by integrating the partial differential equation according to the method of weighted residuals. In the differential equation, one treats the dependent variable, the potential, to be a continuous function of space and time. In formulating finite element equations from the differential equation, one starts by approximating the dependent variable as a piecewise continuous function over a finite element and substitutes this approximating function both in the spatial derivative and in the time derivative. Then, one proceeds to integrate the differential equation over the union of finite elements by appropriately weighting the integrals with respect to each point of interest. Under these conditions, the accumulation at a point is converted into a sum of time derivatives, not only at the point of accumulation but also at its neighbors. The capacity matrix, thus, includes the capacity of the accumulation point as well as that of its neighbors. Such an operation is often referred to as a "distributed" capacity matrix. It is known that in certain problems (e.g. saturated-unsaturated flow, Neuman, 1973) better solutions are obtained by associating all accumulation at a point with a single node rather than distributing it over several nodes.

Which of the two procedures is correct; lumping or distributing? One way of judging would be to apply the procedures to many typical problems with known analytical solutions and observing which of the two methods provides a better comparison with analytic solutions. This trial and error approach, however, is not very satisfying since one cannot generalize the conclusions to other problems for which no analytical solutions are available. Therefore a choice must be made on conceptual grounds, utilizing the fundamental postulates of the diffusion process and the theoretical foundations of integration.

The concept of a "point" quantity is easily suited for setting up a differential equation and obtaining analytical solutions to it. However, if we seek to obtain a solution by a process of integration, the notion of a value at a point is of little use. Instead, the concept of an average value over a subdomain becomes critical. Therefore the question arises, how appropriate is it to substitute the approximating function in the time derivative before integration is carried out? It appears that the formulation of the finite element method has been very much restricted to one particular perspective. In as much as the numerical problem is ultimately one of integration, there must be an independent way of formulating the equations, apart from that of starting with the differential equation. One such attempt was made by Narasimhan (1978) who developed the necessary equations directly by defining the domains of integration and the operators of averaging. His analysis suggests that the "lumped" capacity matrix is logically more consistent with the physics than the "distributed" capacity matrix.

The foregoing discussion points out the need for a more critical analysis of the foundations of the finite element method from different perspectives than has hitherto been done. The modern digital computer is primarily an integrating device. The power of these devices is likely to be far better utilized by directly formulating groundwater problems as integral systems rather than relying heavily on the integration of differential equations as is done now. A basic change in perspective is clearly needed.

If the 1970's proved the decade of large digital computers, the 1980's promise to belong to the desk-top, mini-computers. The qualifier "mini" remarkably misleads one about the power of these devices. The capacities of the first generation mini-computers already exceed that of the large computers of only a decade ago. With such tools available, solutions to complex problems will come within the everyday reach of practising engineers and hydrologists. Efforts must already be underway to modify existing finite element programs to the desk top computer. The finite element method will thus find more and more applications in problems of groundwater hydrology in the new decade.

## CONCLUDING REMARKS

Since its introduction some twenty years ago, the finite element method has come to be recognized as a powerful tool of analysis in the field of groundwater hydrology. The method has been used to solve problems related to large groundwater basins, multiple aquifer systems, flow to wells, flow with a free surface and a seepage face, flow in the unsaturated zone, water uptake by plant roots and so on. In order to satisfy the needs of efficiency and rigor, the methodology of the scheme has also continued to improve. New concepts such as isoparametric elements and functional coefficients were introduced. Advantage has also been taken of better quadrature schemes and efficient matrix solvers. The method will continue to be a highly desired tool for quantitative analysis in groundwater hydrology.

How the finite element method can be further improved is a question of more than passing interest. We have attempted to answer this question from a conceptual perspective. The uniqueness of the finite element notion lies in its ability to provide a means of evaluating gradients of potential in any direction at a given location. It is this ability that sets the finite element method distinctly apart from the finite difference method. The integral aspect of the finite element method, that is, expressing potentials and changes in potentials as averages over small subdomains through a process of weighted integration, is not necessarily unique to the method. It is possible to have integral methods (such as the integrated finite difference method) which do not use finite element gradients. It will be of interest to change the traditional formulations and evaluate the advantage of the finite element method, independent of weighted volume integration. Perhaps the flexibility of the method can be enhanced by providing options to input geometric data in a form other than that of a set of nodal point coordinates. It will also be of interest to formulate the finite element method directly from the theory of integration by defining domains of integration and operators of averaging. This should provide an alternate perspective not available in the traditional approach of integrating point equations.

The mini-computer revolution is now here. The finite element method will soon become an everyday tool, well within the reach of the practising engineer and hydrologist, and will greatly enhance their ability to solve complex problems in groundwater hydrology.

REFERENCES

Barenblatt, G.E., Zheltov, I.P. and Kochina, I.N. (1960) Basic Concepts in the Theory of Homogeneous Liquids. J. Applied Math, USSR, 24:5, 1286-1303.

Duguid, J.O. and Lee, P.C.Y. (1977) Flow in Fractured Porous Media. Water Resources Res., 13:3, 558-566.

Edwards, A.L. (1969) TRUMP: A Computer Program for Transient and Steady State Temperature Distributions in Multidimensional Systems. National Technical Information Service, Springfield, VA., 265.

Ehlers, K.D. (1971) Berechnung Instationärer Grund - und Sickerwasserströmungen mit Freier Oberfläche Nach der Methode Finiter Elemente. Dissertation, Technical University of Hannover.

Ergatoudis, I., Irons, B.M. and Zienkiewicz, O.C. (1968) Curved Isoparametric, "Quadrilateral" Elements for Finite Element Analysis. Int. J. Solids Structures, 4, 31-42.

Finlayson, B.A. and Scriven, L.E. (1967) On the Search for Variational Principles. Int. Jour. Heat Transfer, 10, 799-821.

France, P.W. (1974) Finite Element Analysis of Three Dimensional Groundwater Flow Problems. J. Hydrology, 21, 381-398.

Frind, E.O. and Verge, M.J. (1978) Three Dimensional Modeling of Groundwater Flow Systems. Water Resources Res., 14:5, 844-856.

Galerkin, B.G. (1915) Rods and Plates. Series in Some Problems of Elastic Equilibrium of Rods and Plates. Vestn. Inzh. Tech., (USSR), Translation 63-18924, Clearinghouse, Federal Sci. Tech. Info., Springfield, VA.

Gambolati, G., Gatto, P. and Freeze, R.A. (1974) Mathematical Simulation of the Subsidence of Venice. Water Resources Res., 10, 563-577.

Gupta, S.R. and Tanji, K.K. (1976) A Three Dimensional Galerkin Finite Element Solution of Flow Through Multi-Aquifers in Sutter Basin, California. Water Resources Res., 12, 155-162.

Gureghian, A.B. and Youngs, E.G. (1975) The Calculation of Steady State Water Table Heights in Drained Soils by Means of the Finite Element Method. J. Hydrology, 27, 15-32.

Gurtin, M.E. (1965) Variational Principles for Linear Initial Value Problems. Quarterly Applied Math, 22, 252-256.

Irons, B.M. (1970) A Frontal Solution Program for Finite Element Analysis. Int. J. Num. Meth. Eng., 2, 5-32.

Javandel, I. and Witherspoon, P.A. (1968) Application of the Finite Element Method to Transient Flow in Porous Media. Soc. Pet. Eng. J., 8, 241-252.

Javandel I. and Witherspoon, P.A. (1969) A Method of Analyzing Transient Fluid Flow in Multilayered Aquifers. Water Resource Res., 5, 856-869.

Lewis, R.W. and Schrefler, B. (1978) A Fully Coupled Consolidation Model of the Subsidence of Venice. Water Resources Res., 14:2, 223-230.

Narasimhan, T.N. (1978) A Perspective on Numerical Analysis of the Diffusion Equation. Adv. in Water Resources, 1:3, 147-155.

Narasimhan, T.N. and Witherspoon, P.A. (1976) An Integrated Finite Difference Method for Analyzing Fluid Flow in Porous Media. Water Resources Res., 12, 57-64.

Narasimhan, T.N., Neuman, S.P. and Witherspoon, P.A. (1978) Finite Element Method for Subsurface Hydrology Using a Mixed Explicit-Implicit Scheme. Water Resources Res., 14:5, 863-877.

Neuman, S.P. (1973) Saturated-Unsaturated Seepage by Finite Elements. Proc. Am. Soc. Civil Eng., J. Hydraulics Div., 99:HY12, 2233-2250.

Neuman, S.P., Feddes, R.A. and Bresler, E. (1975) Finite Element Analysis of Two Dimensional Flow in Soils Considering Water Uptake by Roots: I Theory. Proc. Soil Sci. Soc. Am., 39:2, 224-230.

Neuman, S.P. and Witherspoon, P.A. (1970) Finite Element Method of Analyzing Steady Flow With a Free Surface. Water Resources Res., 6, 889-897.

Neuman, S.P. and Witherspoon, P.A. (1971) Analysis of Non-Steady Flow With a Free Surface Using the Finite Element Method. Water Resources Res., 7, 611-623.

Neuman, S.P. and Witherspoon, P.A. (1971) Variational Principles for Fluid Flow in Porous Media. Am. Soc. Civil Eng., J. Eng. Mech. Div., 97, 359-379.

Noorishad, J., Witherspoon, and Brekke, T. (1971) A Method of Coupled Stress and Flow Analysis of Fractured Rock Masses. Geotech. Eng. Pub. 71-6, University of California, Berkeley, CA.

Noorishad, J. and Witherspoon, P.A. (1981) Coupled Hydro Thermoelasticity: Formulation and Analysis Approach. Rept. No. LBL-12354, Lawrence Berkeley Laboratory, Berkeley, CA.

Peaceman, D. and Rachford, H.H. (1955) The Numerical Solution of Parabolic and Elliptic Differential Equations. J. Soc. Ind. Appl. Math, 3, 28-41.

Pinder, G.F. and Frind, E.O. (1972) Application of Galerkin's Procedure to Aquifer Analysis. Water Resources Res., 8:1, 108-120.

Reeves, M. and Duguid, J.O. (1975) Water Movement Through Saturated-Unsaturated Porous Media: A Finite Element Galerkin Model. Rept. ORNL-4327, Oakridge National Laboratory, Oakridge, TN.

Sandhu, R.S. and Wilson, E.L. (1970) Finite Element Analysis of Land Subsidence. In Land Subsidence, Proc. Tokyo Symposium, Int. Assoc. Sci. Hydrology-Unesco, Gentbrugge, Belgium and Paris, France, 2, 393-400.

Segol, G., (1976) A Three Dimensional Galerkin Finite Element Model of the Analysis of Contaminant Transport in Saturated-Unsaturated Porous Media, Proc. Int. Conf. on Finite Elements, Princeton University.

Stone, H.L. (1968) Iterative Solution of Implicit Approximations of Multidimensional Partial Differential Equations. J. Soc. Ind. Appl. Math., J. Num. Anal., 5, 530-558.

Taylor, R.L. and Brown, C.B. (1967) Darcy Flow Solutions with a Free Surface. Proc. Am. Soc. Civil Eng., J. Hydraulics Div., 93:HY2, 25-33.

Weaver, W. Jr. (1967) Computer Programs for Structural Analysis. D. Van Nostrand, Princeton, NJ.

Wilson, C.R. and Witherspoon, P.A. (1970) An Investigation of Laminar Flow in Fractured Rocks. Geotechnical Rept. No. 70-6, University of California, Berkeley, CA.

Wilson, E.L. and Nickell, R.E. (1966) Application of Finite Element Method to Heat Conduction Analysis. Nuclear Engineering and Design, North Holland Publishing 6, Amersterdam.

Young, D.M. Jr. (1962) The Numerical Solution of Elliptic and Parabolic Differential Equations. In the Survey of Numerical Analysis, editor, J. Todd, McGraw Hill, New York, 380-438.

Zienkiewicz, O.C. (1971) The Finite Element Method in Engineering Science. McGraw Hill.

Zienkiewicz, O.C. and Cheung, Y.K. (1965) Finite Elements in the Solution of Field Problems. The Engineer, 507-510.

Zienkiewicz, O.C., Meyer, P. and Cheung, Y.K. (1966) Solution of Anisotropic Seepage by Finite Elements. Am. Soc. Civil Eng., J. Eng. Mech. Div , 92, 111-120.

# SOLVING TURBULENT FLOW PROBLEMS USING THE F.E.M.

C. TAYLOR

READER IN CIVIL ENGINEERING, UNIVERSITY COLLEGE OF SWANSEA, SINGLETON PARK, SWANSEA SA2 8PP, U.K.

## INTRODUCTION

Although the utilisation of the finite element method to solve turbulent flow problems is in its infancy, considerable advances have already been made resulting in the analysis of diverse problems [1,2,3]. These range from fully developed pipe flow using an analytical description of effective viscosity [4], flow with separation using the two equation model of turbulence, [3].

An important feature when using the F.E.M. is that, generally, upwinding techniques [5] are not required. This leads to a more conservative system where the rate of dissipation of energy is reduced. This has been demonstrated, in the current text, when using the one equation model of turbulence to solve confined flow with separation.

Three main concepts, relating to the definition of effective viscosity, are used. These are an analytical approach, a one equation model and two equation model systems. The first is used to investigate fully developed and developing flow in a pipe. An additional feature of this work is the utilisation of special elements in the near wall zone. The one equation model is used to solve a confined flow problem with separation and the two equation model is used to analyse flow in a pipe and channel.

## MATHEMATICAL CONCEPTS

The general equations which can be utilised to predict the turbulent flow of a fluid are usually derived directly from the classical Navier-Stokes equations [6]. For steady state incompressible turbulent flow these can be expressed, for a two dimensional domain, in the form, [7],

$$\rho\, u_i \frac{\partial u_i}{\partial x_j} = \frac{\partial}{\partial x_j}\left(\tau_{ij} - \rho\, \overline{u_i' u_j'}\right) \qquad \left.\begin{matrix} i \\ j \end{matrix}\right\} 1,2 \tag{1}$$

in which $\rho$ is the fluid density, $u_i$ the local time-averaged velocities, corresponding to an orthogonal $x_i$ (i=1,2) system, $\tau_{ij}$ represents the mean stress associated with pressure and viscous effects, and $\overline{u_i' u_j'}$ are the Reynolds shear stress. These time averaged, denoted by an overbar, Reynolds stresses are often modelled by an effective viscosity concept such that,

$$-\rho\, \overline{u_i' u_j'} = \mu_T \left[\frac{\partial u_i}{\partial x_j} + \frac{\partial u_j}{\partial x_i}\right] - \frac{2}{3}\rho\, k\, \delta_{ij} \tag{2}$$

where $\mu_T$ is the local value of the turbulent viscosity, k the mean turbulent kinetic energy and $\delta_{ij}$ the Kroneker delta. Generally,

$$\delta_{ij} \left\{\begin{matrix} 1, & i = j \\ 0, & i \neq j \end{matrix}\right.$$

and

$$k = \frac{1}{2} u'^2$$

The remaining term in equation (1), namely $\tau_{ij}$, can be written in the form,

$$\tau_{ij} = - p\, \delta_{ij} + \mu \left[\frac{\partial u_i}{\partial x_j} + \frac{\partial u_j}{\partial x_i}\right] \tag{3}$$

in which p is the local time averaged pressure, and $\mu$ the laminar molecular viscosity.

The Reynolds stresses incorporated into equation (1) can be interpreted via an expression of the form,

$$\rho\, \overline{u_i' u_j'} = \mu_t \left(\frac{\partial u_i}{\partial x_j} + \frac{\partial u_j}{\partial x_i}\right) - \frac{2}{3}\rho\, k\, \delta_{ij} \tag{4}$$

in which $\mu_t$ is called the turbulent dynamic viscosity and k the mean turbulence kinetic energy defined by,

$$k = \frac{1}{2} \overline{u_i'^2} \tag{5}$$

Introducing the new variable,

$$\hat{p} = p + \frac{2}{3} k \tag{6}$$

Equation (1) can be re-written after incorporating (3), (4) and (5) in the form,

$$u_j \frac{\partial u_i}{\partial x_j} = - \frac{1}{\rho} \frac{\partial \hat{p}}{\partial x_i} + \frac{1}{\rho} \frac{\partial}{\partial x_j} \left[ \mu_e \left( \frac{\partial u_i}{\partial x_j} + \frac{\partial u_j}{\partial x_i} \right) \right] \tag{7}$$

in which $\mu_e$ is called the effective viscosity and,

$$\mu_e = \mu_t + \mu \tag{8}$$

and the effective viscosity is defined, then equation (7) combined when with the equation of continuity,

$$\frac{\partial u_i}{\partial x_i} = 0 \tag{9}$$

form a closed set and can be solved, simultaneously, for the required primitive variables $u_i$ and p provided the spatial distribution of $\mu_e$ is known.

The magnitude of the turbulent viscosity can be defined in terms of a turbulent length scale [8], $\ell$, using,

$$\mu_t = \rho \ \ell^2 \ |J| \tag{10}$$

in which

$$J = 0.5 \left( \frac{\partial u_k}{\partial x_j} + \frac{\partial u_j}{\partial x_k} \right) \left( \frac{\partial u_j}{\partial x_k} + \frac{\partial u_k}{\partial x_j} \right) \tag{11}$$

The function represented by (11) is a flow dependent function of position which can only be evaluated from a known distribution of velocity. It becomes immediately apparent that, upon using the simple definition (10), the solution procedure is iterative. An additional complication is the definition of the turbulent length scale, $\ell$. The specific definition of this will be discussed in the text where specific problems are solved.

Another commonly used expression for the local value of the turbulent viscosity is, [9],

$$\mu_t = C_\mu \ \rho \ k^{\frac{1}{2}} \ell . \tag{12}$$

where $C_\mu$ is considered to be a constant.

Two further transport equations can be derived [10,11], which may be utilised for the evaluation of both k and $\ell$ are,

$$\rho\, u_j \frac{\partial k}{\partial x_j} = \frac{\partial}{\partial x_j}\left[\left(\mu + \frac{\mu_t}{\sigma_k}\right)\frac{\partial k}{\partial x_i}\right] + \mu_t \frac{\partial u_i}{\partial x_j}\left(\frac{\partial u_i}{\partial x_j} + \frac{\partial u_j}{\partial x_i}\right)$$

$$- C_D\, \rho\, \frac{k^{3/2}}{\ell} \qquad (13)$$

for k and,

$$\rho\, u_j \frac{\partial \varepsilon}{\partial x_j} = \frac{\partial}{\partial x_j}\left[\left(\mu + \frac{\mu_t}{\sigma_\varepsilon}\right)\frac{\partial \varepsilon}{\partial x_j}\right] + C_1\, \rho\, k \frac{\partial u_i}{\partial x_j}\left(\frac{\partial u_i}{\partial x_j} + \frac{\partial u_j}{\partial x_i}\right)$$

$$- C_2\, \rho\, \frac{\varepsilon^2}{k} \qquad (14)$$

for $\ell$ where, $\varepsilon = \frac{k^{3/2}}{\ell}$ and $\sigma_k$, $\sigma_\varepsilon$, $C_D$, $C_1$ and $C_2$ are usually considered to be constant.

Specific values are usually allocated to these 'constants' which are, in the main, problem dependent and must be evaluated by using experimental data.

The so-called one equation model of turbulence consists of equations (7), (9) and (13) which implies that the length scale, incorporated in $\varepsilon$, must be evaluated algebraically. The two equation model is when these equations and (14) is solved simultaneously,

## FINITE ELEMENT FORMULATION

The, now, standard method of discretising the governing equations is adopted [12-17]. Isoparametric quadrilateral elements have been used and the usual higher order shape function employed for the velocities.

Since, for the range of Reynolds numbers presented in the current text, no upwinding is required the normal Galerkin weighted residual approach has been employed with one order higher weighting functions used in (8) than in (9). All second order derivatives are reduced by utilising Green's theorem and, upon employing the finite element discretisation procedure, a matrix equation of the form

$$\underline{H}\ \underline{\beta} = \underline{f} \qquad (15)$$

is obtained.

The surface integrals resulting from Green's theorem are all contained in $\underline{f}$ and the non-linear terms retained in $\underline{H}$. The column vector operated on by $\underline{H}$ consists of the unknown nodal values,

$$\underline{\beta}^k = \begin{Bmatrix} u_1^k \\ p^k \\ u_2^k \\ k^k \\ \varepsilon^k \end{Bmatrix} \tag{16}$$

for corner nodes and,

$$\underline{\beta}^k = \begin{Bmatrix} u_1^k \\ u_2^k \\ k^k \\ \varepsilon^k \end{Bmatrix} \tag{17}$$

for mid-side nodes.

The stream function, $\psi$, can then be determined from,

$$\nabla^2\psi = \frac{\partial u_1}{\partial y} - \frac{\partial u_2}{\partial x} \tag{18}$$

and the velocity distribution is known.

METHOD OF SOLUTION

In the iterative technique adopted for present purposes, non-linear terms such as $u_j \; \partial u_i/\partial x_j$ are replaced by $\hat{u}_j \; \partial u_i/\partial x_j$, where the circumflex denotes previous iteration values. The solution procedure then consists of updating $\hat{u}_j$ values until the desired convergence criteria are satisfied.

The specific problem and associated assumptions dictate, to a certain extent, the system of equations adopted for describing the flow regime. In certain cases $\mu_t$ may be assumed to be simply a function of the velocity field and the k,ε equations become obsolete. In others the characteristic length may be defined and the one-equation model used whereas in others the two equation model is utilised. Some examples concerning each will be presented and the appropriate assumptions discussed as they appear.

## PRACTICAL APPLICATIONS

Among the more common types of flow in water resources is flow through pipes, flow in channels and free shear flows. The concepts relating to the utilisation of the basic hydrodynamic equations, semi-empirical expressions and the finite element method will be demonstrated by solving particular problems in the categories indicated above.

### Pipe Flow

The only additional feature utilised in this section is the concept of 'special' elements in an attempt to describe the large variations close to the pipe wall. This concept has been described elsewhere [18] and only the pertinent results will be presented.

The governing equations can, with some algebraic manipulation [19], be written in terms of cylindrical coordinates for axisymmetric flow, Figure 1.

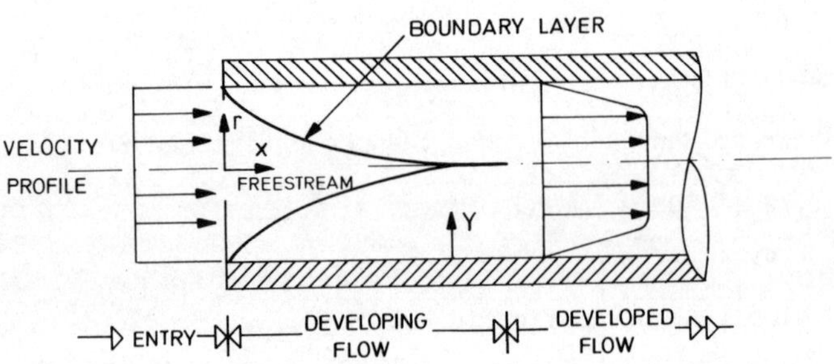


FIGURE 1 Essential features of entry problem in pipe flow

x-momentum,

$$u_1 \frac{\partial u_1}{\partial x_1} + u_2 \frac{\partial u_1}{\partial x_2} = - \frac{1}{\rho} \frac{\partial p}{\partial x_1} + \frac{1}{x_2} \left[ \frac{\partial}{\partial x_2} \left( x_2 \nu_e \frac{\partial u_1}{\partial x_2} \right) + \frac{\partial}{\partial x_1} \left( x_2 \nu_e \frac{\partial u_1}{\partial x_1} \right) \right] + \left[ \frac{\partial u_1}{\partial x_1} \frac{\partial \nu_t}{\partial x_1} + \frac{\partial u_2}{\partial x_1} \frac{\partial \nu_t}{\partial x_2} \right] \qquad (19)$$

$x_2$-momentum

$$\nu \frac{\partial u_2}{\partial x_2} + u \frac{\partial u_2}{\partial x_1} = - \frac{1}{\rho}\frac{\partial p}{\partial x_2} + \frac{1}{x_2}\left[\frac{\partial}{\partial x_2}\left(x_2 \nu_e \frac{\partial u_2}{\partial x_2}\right)\right.$$

$$\left. + \frac{\partial}{\partial x_1}\left(x_2 \nu_e \frac{u_2}{x_1}\right)\right] + \left[\frac{\partial u_2}{\partial x_2}\frac{\partial \nu_t}{\partial x_2} + \frac{\partial u_1}{\partial x_2}\frac{\partial \nu_t}{\partial x_1}\right] - \frac{\nu_e u_2}{x_2^2} \quad (20)$$

and continuity,

$$\frac{\partial u_2}{\partial x_2} + \frac{u_2}{x_2} + \frac{\partial u_1}{\partial x_1} = 0 \quad (21)$$

in which $x_1$ corresponds to the longitudinal axes and $x_2$ the radial direction in the pipe.

Typical boundary conditions are,

(i) Developing flow

Upstream boundary $\left.\begin{array}{l} u_1 = \bar{u}_1 \\ u_2 = 0 \end{array}\right\}$ for $x_1 = 0$, $0 < z_2 \leqq R$

where $\bar{u}$ is the mean velocity and R the pipe radius.

Wall of the pipe $u_1 = u_2 = 0$, for $x_2 = R$ and $x_1 \geqq 0$

Downstream boundary (assumed fully developed) $\frac{\partial u_1}{\partial x_1} = u_2 = 0.$

(ii) Developed Flow

Upstream boundary $\left.\begin{array}{l} u_1 = u_{s1} \\ u_2 = 0 \end{array}\right\}$ for $x_1 = 0$, $0 < x_2 \leqq R$

where $u_{s1}$ is the fully developed steady state velocity.

On the pipe wall $u_1 = u_2 = 0$ for $x_2 = R$ and $x_1 \geqq 0.$

For the downstream boundary,

$$\frac{\partial u_1}{\partial x_1} = u_2 = 0.$$

Modelling the effective viscosity.

Utilising a modified Van Driest Hypothesis [20], the effective kinetic viscosity can be written in the form,

$$\nu_e = \nu[1+\{1+4\kappa^2 y^{+2} \ (1 - e^{-y^+/A^+})^2\}^{\frac{1}{2}}]/2 \tag{22}$$

when $y^+ \leqq y_1^+$ and,

$$\nu_e = \nu_e(y^+ = y_1^+) \tag{23}$$

when $y^+ > y_1^+$ and $\kappa$, a universal constant.

The definition for $y^+$ is,

$$y^+ = \frac{Y}{\nu}\sqrt{\frac{\tau_s}{\rho}} \tag{24}$$

where $\tau_s$ is the wall shear stress, Y is measured normal to the wall towards the pipe centre-line and $A^+$ is a damping constant.

A restricted form of the Van Driest model [21] advocates that (22) is used when,

$$0 \leqq \frac{x_1}{2R} \leqq \ell \quad \text{and } 0 \leqq \frac{Y}{R} \leqq 0.158$$

and, $\nu_e = (\nu_e)_{Y/R = 0.158}$ for

$$0 \leqq \frac{x_1}{2R} \leqq \ell \text{ and } 0.158 \leqq \frac{Y}{R} \leqq 1.$$

Another location for truncating (22) is at, [22,23]

$$y_1^+ = \frac{\lambda \ u_k \ \delta}{K \ \nu}$$

when the truncated effective viscosity may be written

$$\nu_e = \lambda \ \rho \ u_k \ \delta.$$

The local shear can be evaluated, using laminar conditions close to the boundary,

$$\tau_s = \mu \left|\frac{\partial u}{\partial Y}\right|_{Y=0} \tag{25}$$

and therefore the shear velocity can be calculated since,

$$u_k = \sqrt{\frac{\tau_s}{\rho}}$$

If, as is usually the case, the velocity gradient at Y = 0 cannot be evaluated with sufficient accuracy then the usual law of the wall can be used,

$$\left.\begin{aligned} u_1^+ &= y^+, && 0 \leqq y^+ \leqq 5 \\ u_1^+ &= -B+C \log_n y^+, && 5 \leqq y^+ \leqq 30 \\ u_1^+ &= -D+G \log_n y^+, && 30 < y^+ \end{aligned}\right\} \tag{26}$$

in which

$$u_1^+ = \frac{\hat{u}_1}{\sqrt{\frac{\tau_s}{\rho}}}$$

where $\hat{u}_1$ is the velocity at some chosen point away from the wall surface and B, C, D and G are constants. The local boundary layer thickness, $\delta$, is taken as the distance away from the wall at which the velocity is 99% of that at the centreline [22].

The logarithmic element utilised is described fully elsewhere [18] which the interested reader can refer to quite readily.

Fully developed flow Re = $4.5 \times 10^5$.

The Reynolds number is defined in the usual manner,

$$Re = \frac{2\rho \bar{u} R}{\mu} \tag{27}$$

The element discretisation employed, the boundary conditions imposed and parameters used are as shown on Figure 2. The advantages of using the special element adjacent to the wall is immediately apparent in Figure 3. The errors are particularly apparent, and most influential, near the wall where an accurate evaluation of $_{t}$ is of importance. A more marked discrepancy in pressure is obtained as shown in Figure 4, although, again, utilisation of the logarithmic element is advantageous.

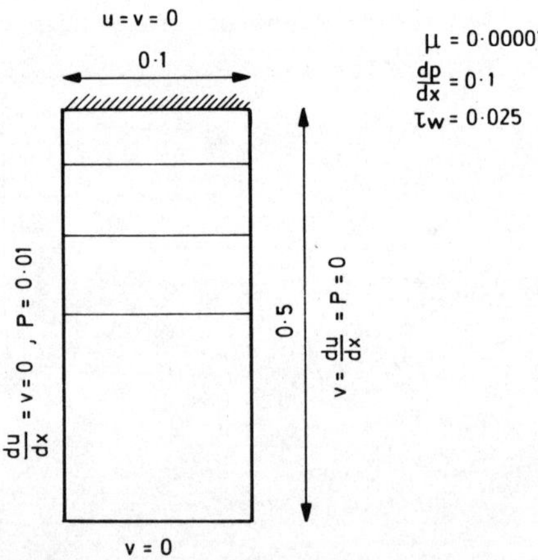


FIGURE 2 Finite element mesh and boundary conditions

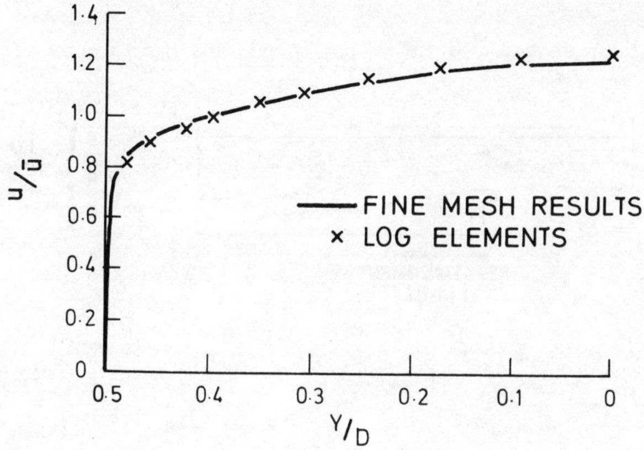


FIGURE 3 Velocity profile for fully developed flow at Re = 4.5 x $10^5$. Location x/D = 40

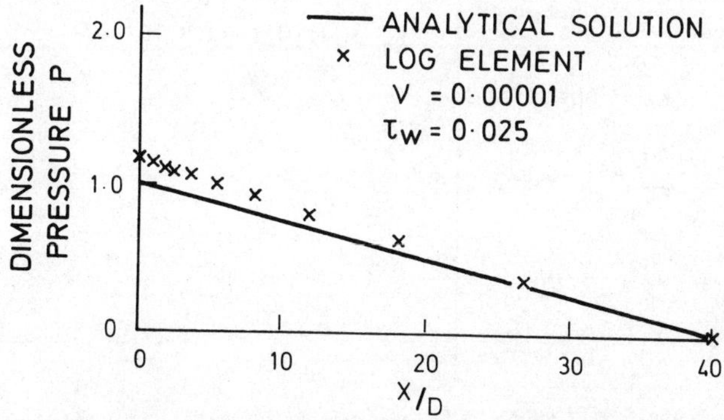


FIGURE 4 Pressure plot for fully developed flow

## CONFINED FLOW WITH SEPARATION

As an example of the utilisation of the one equation model is used to analyse a confined flow problem, namely flow over a backward facing step, Figure 5.

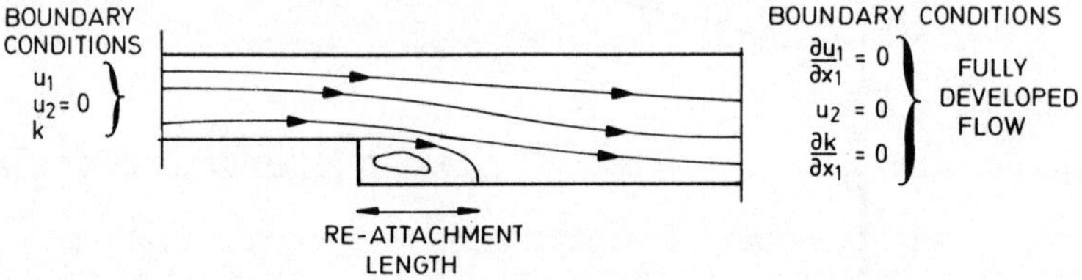


FIGURE 5 Backward facing step. Boundary conditions and schematic representation of flow.

The boundary conditions utilised in the analysis of this problem are,

Upstream

$$\left.\begin{array}{l} u_1 \ - \text{specified} \\ u_2 = 0 \\ k \ - \text{specified} \end{array}\right\} \quad x_1 = 0,\ 0 < x_2 < R.$$

Downstream

$$\left.\begin{array}{l} \dfrac{\partial u_1}{\partial x_1} = 0 \\ u_2 = 0 \\ \dfrac{\partial k}{\partial x_1} = 0 \end{array}\right\} \quad 0 < x_2 < R.$$

Pressure was nominated at one point since relative values are a permissible solution.

The Reynolds number, based on step height, solved were in the range 3,025 - 200,000., which is the same range as that covered by Atkins et. al. [24], using a finite difference approach. The inlet boundary conditions obtained, experimentally by Denham et al. [25] were used and the results are as shown on Figures 6 and 7.

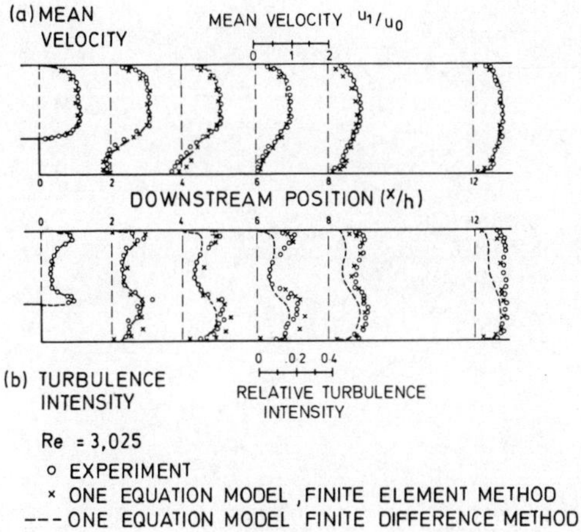


FIGURE 6 Comparison of numerical and experimental mean velocity and turbulence intensity profiles.

The agreement between the velocity profiles, obtained using the finite element method, finite difference method and experiment is particularly good. However, a discrepancy exists between the turbulence kinetic energy levels obtained using the F.E.M. and F.D. methods although the former agrees with experiment.

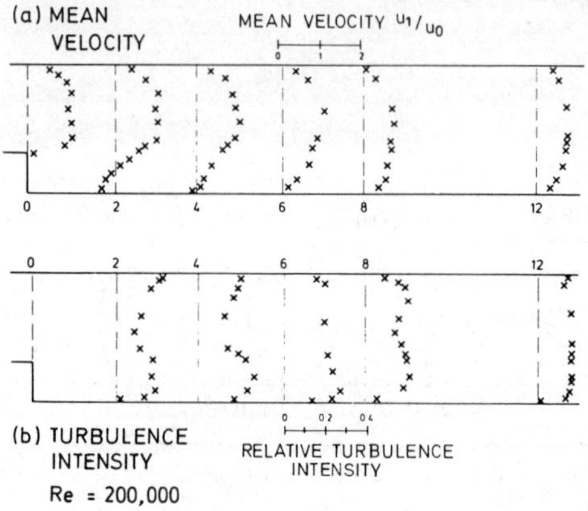


FIGURE 7 Finite element velocity and turbulence intensity profiles

Pipe and Channel Flow

The utilisation of the two equation model of turbulence is demonstrated by application to fully developed flow in a pipe and channel.

For present purposes, as previously advocated [23], the following relation, c.f. (12), is used:

$$\mu_t = \rho \ell \, k^{\frac{1}{2}} \tag{28}$$

The generalised equations used to predict the motion can be simplified, [16] since both the mean flow and level of turbulence remain the same at successive sections, leading to

Momentum

$$\frac{1}{x_2^{\alpha}} \frac{\partial}{\partial x_2} [\mu + \rho \ell \, k^{\frac{1}{2}}) \, x_2^{\alpha} \frac{\partial u_1}{\partial x_2} ] = - \frac{1}{\rho} \frac{\partial p}{\partial x_1} \tag{29}$$

Turbulence kinetic energy

$$\frac{1}{x_2^{\alpha}} \frac{\partial}{\partial x_2} \left[ \left( \mu + \rho \frac{\ell \, k^{\frac{1}{2}}}{\sigma_k} \right) x_2^{\alpha} \frac{\partial k}{\partial x_2} \right] + \rho \ell \, k^{\frac{1}{2}} \left( \frac{\partial u_1}{\partial x_2} \right)^2 - \rho \, C_D \frac{k^{3/2}}{\ell} = 0 \qquad (30)$$

Mixing length

$$\frac{1}{x_2^{\alpha}} \frac{\partial}{\partial x_2} \left( \frac{\ell \, k^{3/2}}{\sigma_1} x_2^{\alpha} \frac{\partial u_1}{\partial x_2} + \frac{\ell^2 \, k^{1/2}}{\sigma_2} x_2^{\alpha} \frac{\partial \ell}{\partial x_2} \right) + C_1 \, \ell^2 \, k^{1/2} \left( \frac{\partial u_1}{\partial x_2} \right)^2 - C_2' \, k^{3/2} = 0 \qquad (31)$$

where $\sigma_1$, $\sigma_2$ and $C_1$ are considered to be constant and,

$$C_2' = C_2 + C_w \left( \frac{\ell}{y} \right)^q$$

in which $C_2$, $C_w$ and $q$ are again considered constant and $y$ denotes distance measured normal to the wall.

The flow in the immediate vicinity of the wall exhibits a transition from laminar flow to fully turbulent flow some distance away from the wall. In the above equations, the quantities which have been assumed to be constant for fully turbulent flows are found to vary with local Reynolds number. For this reason the solution domain is not taken up to the wall but is terminated some distance away and the following expression used to determine the velocity distribution, [26]

$$u = \left( \frac{\tau_w}{\rho} \right)^{\frac{1}{2}} \left[ 2.5 \, \ell_n \frac{y}{\mu} \left( \frac{\tau_w}{\rho} \right)^{\frac{1}{2}} + 5.5 \right] \qquad (32)$$

$$k = \frac{\tau_w}{C_D^{\frac{1}{2}}}$$

$$\ell = 0.4 \, C_D^{1/4} . \, y.$$

where $C_D$ is assumed to be constant and the $x$ axis is taken parallel to the wall.

A comparison between experimental results [27] and numerical values, for the fairly coarse finite element mesh of Figure 8, is shown on Figure 9. This shows distribution of both velocity and turbulence kinetic energy in a pipe at a Reynolds number of $5 \times 10^5$. A plot of the distribution of turbulent viscosity is also presented.

Another series of results is shown on Figure 10, which effects a similar comparison for fully developed flow in a channel at a Reynolds number of $3.08 \times 10^4$. Both sets of results exhibit a good comparison with measured quantities.

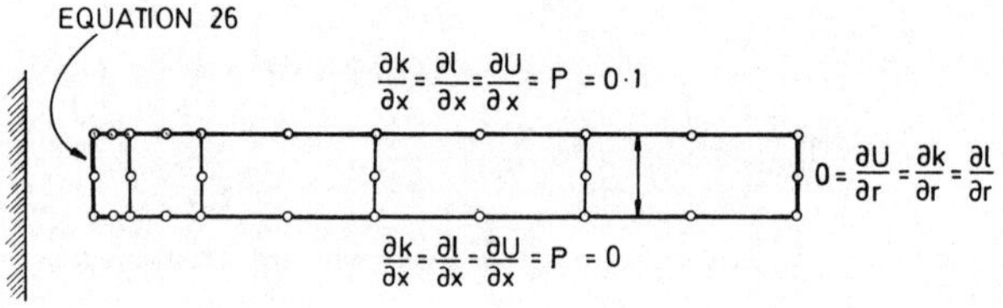


FIGURE 8 Element distribution used and corresponding boundary conditions

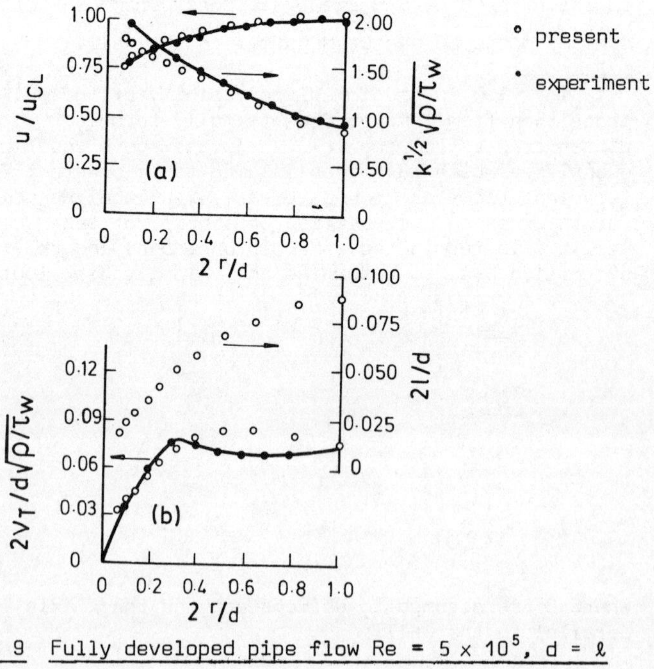


FIGURE 9 Fully developed pipe flow Re = $5 \times 10^5$, d = $\ell$

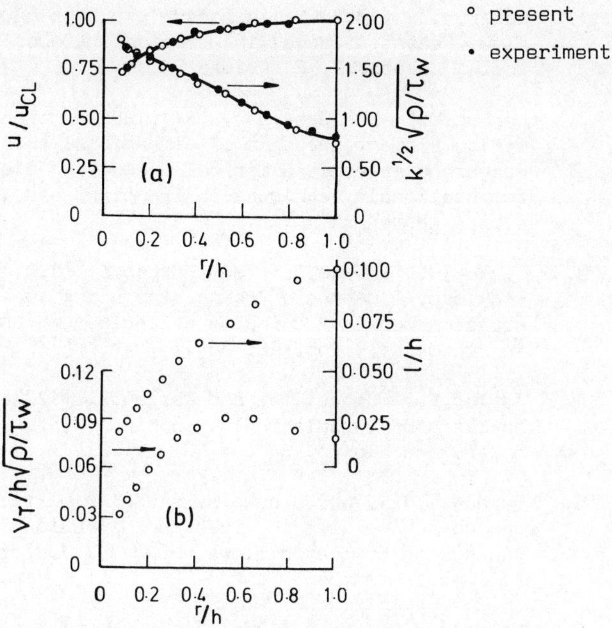


FIGURE 10 Fully developed channel flow $R_N$ = 30,800

CONCLUSIONS

It has been shown that the application of the finite element method to solve turbulent flow problems results in accurate predictions when utilising few elements. The introduction of special elements does not present undue difficulty and the utilisation of a range of turbulence models, from algebraic to the two equation model system, can be readily handled.

REFERENCES

1. Baker,A.J. (1980) Laminar and Turbulent Boundary Layer Flow, Recent Advances in Numerical Methods in Fluids, Vol.1, Chapter 9, Pineridge Press.

2. Larock,B.E. and Schamber,D.R. (1981) Approaches to the finite element solution of dimensional turbulent flow. Recent Advances in Numerical Methods in Fluids, Vol.2 - Computational Techniques in Transient and Turbulent Flow, Chapter 9.

3. Taylor,C., Thomas,C.E. and Morgan,K. (1981) Modelling flow over a backward facing step using the F.E.M. two equation model of turbulence. Int. Journ. Num. Meth. in Fluids, 1,4, pp. 295-304.

4. Taylor,C., Hughes,T.G. and Morgan,K. (1977) A numerical analysis of turbulent flow in pipes. Int. Journ. Computers and Fluids, 5, pp. 191-204.

5. Hughes,T.J.R. and Brooks,A. (1979) A multidimensional upwind scheme with no crosswind diffusion. F.E.M. for Convection Dominated Flows, Hughes,T.J.R. (ed.), A.S.M.E., Arad 34.

6. Goldstein,S. (Ed.) (1965) Modern Developments in Fluid Dynamics, Vol.1, Dover Publications.

7. Schlichting,H. (1960) Boundary Layer. McGraw-Hill.

8. Prandtl,L. (1925) Bericht uber Untersuchangen zur ausgebildeten Turbulenz, Zamm, 5.

9. Prandtl,L. (1945) Uber ein neues Formelsystem fur die ausgebildeten Turbulenz, Nachr. Akad. der Wissenschafft in Gottingen.

10. Wolfshtein,M. (1970) Some solutions of the plane turbulent impinging jet. Trans. A.S.M.E., J. Basic Engineering.

11. Wolfshtein,M. (1967) Convection processes in turbulent impinging jet. Ph.D. Thesis, Univ. London.

12. Taylor,C., Hughes,T.G. and Morgan,K. (1978) Finite element solution of one equation models of turbulent flow. Journ. Comp. Physics, 29, 2, pp. 163-172.

13. Morgan,K., Hughes,T.G. and Taylor,C. (1979) The analysis of turbulent free shear and channel flow by the finite element method. Comp. Meth. Appl. Mech. Eng. 19, pp. 117-125.

14. Taylor,C., Hughes,T.G. and Morgan,K. (1979) Analysis of turbulent free shear flow and heat transfer using the mixing length hypothesis. Mafelap Conf., Brunel, Academic Press, pp. 247-254.

15. Larock,B.E. and Schamber,D.R. (1981) Approaches to the finite element solution of two dimensional turbulent flow. Computational Techniques in Transient and Turbulent Flow, Pineridge Press, Chapter 10.

16. Thomas,C.E., Morgan,K. and Taylor,C. (1981) A finite element analysis of flow over a backward facing step. Journ. Comp. & Fluids, 9, 3, pp. 265-278.

17. Taylor,C., Thomas,C.E. and Morgan,K. (1981) Modelling flow over a backward facing step using the F.E.M. and the two equation model of turbulence. Journ. Num. Meth. in Fluids, 1, 4, pp. 295-304.

18. Taylor,C., Hughes,T.G. and Morgan,K. (1977) A numerical analysis of turbulent flow in pipes. Int. Journ. Comp. and Fluids, 5, pp. 191-204.

19. Hinze,J.O. (1959) Turbulence, McGraw-Hill, New York.

20. Van Driest,E.R. (1956) On turbulent flow near a wall. Journ. Aero. Science, 23, pp. 1007-1011.

21. Richman,J.W. and Azad,R.S. (1973) Developing turbulent flow in smooth pipes. Appl. Science Res., 28, pp. 419-441.

22. Hutton,A.G. (1971) A finite difference method for analysing the flow in the entrance section of a duct. M.Sc. Thesis, Univ. College of Swansea.

23. Oliver,A.J. (1972) An analysis of developing turbulent flow in a tube - a comparison of two turbulent models. C.E.G.B. Report No. RD/L/N105.

24. Atkins,D.J., Maskell,S.J. and Patrick,M.A. (1980) Numerical prediction of separated flows. Int. J. Num. Meth. in Engng.

25. Denham,M.K. and Patrick,M.A. (1979) Private Communication.

26. Ng,K.H. and Spalding,D.B. (1972) Turbulence model for boundary layers near walls. Phys. Fluids, 15, 1, pp. 20-30.

27. Laufer,J. (1954) The structure of turbulence in fully developed pipe flow. NACA Rep. 1174.

# 2. MATHEMATICS AND NUMERICAL TECHNIQUES

# PETROV-GALERKIN SCHEMES FOR THE STEADY STATE CONVECTION-DIFFUSION EQUATION

**M. Ahués and M. Telias ;**
**Laboratoire IMAG ; Grenoble ; France .**

## INTRODUCTION

It is well known that the main trouble in convection dominated problems is that Galerkin and Central Finite Differences schemes often give spurious oscillations which arise specially when there are downwind boundary conditions. This is due to the fact that the differential operators involved in the mathematical model are not symmetric and the matrix of the linear system which arises from discretization does not verify the discrete maximum principle. As an elementary example we take the following one dimensional problem

$$-k\phi'' + \phi' = 1 \text{ in } ]0,1[ \quad (k > 0) \ , \ \phi(0) = \phi(1) = 0 \qquad (1)$$

which when tackled by Central Finite Differences with uniform mesh size h gives :

$$\begin{aligned} &-\left(1 + \frac{h}{2k}\right)\phi_{i-1} + 2\phi_i - \left(1 - \frac{h}{2k}\right)\phi_{i+1} = \frac{h^2}{k} \ , \ 1 \leq i \leq n \\ &\phi_0 = \phi_{n+1} = 0 \end{aligned} \qquad (2)$$

It is seen that for $h > 2k$ the discrete maximum principle will be not respected. For instance if $h=1/80$ and $k=1/100$, strong numerical oscillations appear (Figure 1).
The purpose of this work is to develop a symmetric positive formulation of the steady state convection diffusion problem which when solved by a suitable Finite Element Method gives a numerical solution without spurious oscillations.
We recall the following definition and property :

**Definition 1.** An $n \times n$ matrix $A = (a_{ij})$ is said to be an M-matrix if the following conditions hold :

$$a_{ij} \leq 0 \text{ if } i \neq j \text{ and } a_{ii} > 0 \text{ for all } i . \qquad (3)$$

A is invertible and its inverse has no negative entries. (4)

We remark that condition (4) ensures that the inverse of A is monotone and hence no numerical oscillations will appear if the matrix of the linear system is an M-matrix.

**Proposition 1.** If an $n \times n$ matrix $A = (a_{ij})$ is such that

Condition (3) holds.

$$\sum_{j=1}^{n} a_{ij} \geq 0 \quad \text{for all } i \tag{5}$$

$$\sum_{j=1}^{n} a_{ij} > 0 \quad \text{for at least one } i \tag{6}$$

then A is an M-matrix.

**Proof.** (cf. Varga, 1962).

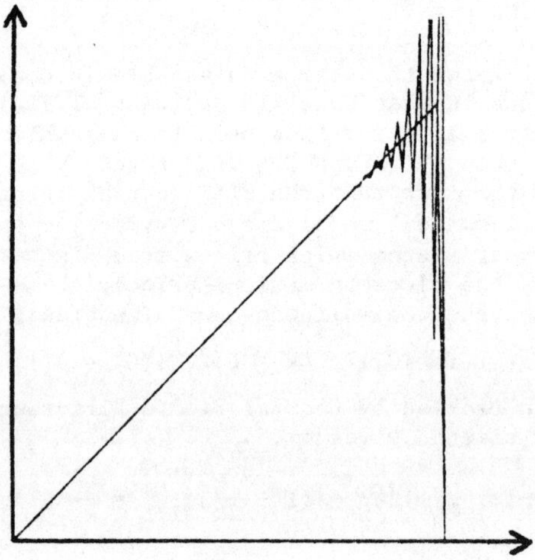

Figure 1

## THE MATHEMATICAL MODEL

We consider first the one dimensional problem

$$-k\phi'' + u\phi' = f \text{ in } ]0,1[ \ , \ \phi(0) = \phi(1) = 0 \tag{7}$$

where $k$ and $u$ are positive constants and $f$ is a given function in $X = L^2[0,1]$ .

**Proposition 2.** If $L:X\to X$ is the operator with domain $D(L) = H_0^2[0,1]$ defined by $L\phi = -\phi'' + (\frac{u}{2k})^2\phi$ and $T:X\to X$ is the operator defined by $(T\phi)(x) = \phi(x)\exp(-\frac{ux}{2k})$

i) There exists an operator $K:X\to X$ such that $L=K^*K$

ii) Problem (7) is equivalent to :
Find $\phi \in D(S)$ such that $\|S\phi-\Theta\|_X = \underset{\psi \, D(S)}{\mathrm{Min}} \|S\psi-\Theta\|_X$ (8)

where $S = KT$ and $\Theta \in D(K^*)$ satisfies $k K^* \Theta = Tf$ .

**Proof.** i) We check easily that $(L\psi,\psi)_X = \int_0^1 (\psi' + \frac{u}{2k}\psi)^2$

ii) The normal equation of Problem (8) is

$$(KT)^*(KT)\phi = \frac{1}{k} T^2 f \quad , \quad \phi \in D(KT) \tag{9}$$

which in turn is equivalent to Problem (7).

**Remark 1.** For the bidimensional problem

$$-k\Delta\phi + \vec{u} \cdot \nabla\phi = f \quad \text{in } \Omega \ , \ \phi=0 \text{ on } \partial\Omega \tag{10}$$

where $\Omega$ is an open bounded region in $R^2$ with sufficiently smooth boundary $\partial\Omega$ and $\vec{u} = (u,v)$ is a constant velocity field, we reproduce Proposition 2 setting in $X=L^2(\Omega)$ :

$$L\psi = -\Delta\psi + \frac{u^2+v^2}{4k^2}\psi \qquad \psi \in H_0^2(\Omega)$$

$$(T\phi)(x,y) = \phi(x,y)\exp(-\frac{ux+vy}{2k}) \qquad (x,y) \in \Omega \ , \ \phi \in X \ .$$

**NUMERICAL APPROACH**

For the one dimensional problem (7) we handle the following Galerkin scheme (weak formulation) on Equation (9). Let $W_h$ be the finite dimensional subspace of $H_0^1[0,1]$ generated by the following basis :

$$\omega_x(\xi) = \begin{cases} (\xi-x+h)/h & \text{if } x-h \leqslant \xi \leqslant x \\ (x+h-\xi)/h & \text{if } x \leqslant \xi \leqslant x+h \\ 0 & \text{otherwise} \end{cases} \tag{11}$$

$x \in \{x_1,\dots,x_n\}$ $x_i = ih$ , $0 \leqslant i \leqslant n+1$ , $x_{n+1} = 1$ .

We shall call **NG1 scheme** to solve Equation (9) by the Galerkin method based on $W_h$ .

**Proposition 3.** The NG1 scheme leads to a linear system with a tridiagonal symmetric M-matrix.

**Proof.** Let $\phi_h$ be the numerical solution :

$$\phi_h = \sum_{j=1}^{n} \phi_j \omega_{x_j} \tag{12}$$

Then the scalars $\phi_j$ must satisfy

$b_{i-1}\phi_{i-1} + a_i\phi_i + b_i\phi_{i+1} = d_i$ where

$$a_i = \frac{1}{h^2}\int_{x_{i-1}}^{x_{i+1}} \exp(-\frac{ux}{k})dx \ ; \quad b_i = -\frac{1}{h^2}\int_{x_i}^{x_{i+1}} \exp(-\frac{ux}{k})dx$$

$d_i = \frac{1}{k} \int_{x_{i-1}}^{x_{i+1}} f(x)\exp(-\frac{ux}{k})dx$ . We apply Proposition 1.

**Remark 2.** For the bidimensional case the situation is similar. Let $\omega_i$ be the triangular Lagrange's basis function of degree one associated to the node $(x_i,y_i)$ and let $K_i$ be its support. The Galerkin scheme reads as follows

$$\int_{K_i} T^2(\nabla\phi_h \cdot \nabla\omega_i) = \frac{1}{k} \int_{K_i} T^2 \, f \, \omega_i \tag{13}$$

For a regular triangular mesh we get a linear system with a banded matrix A such that conditions (3), (5) and (6) hold. In particular, for each interior node the corresponding row satisfies (5) as an equality.

We next show that this numerical approach can be viewed as a Petrov–Galerkin scheme on the original problem, say (7) or (10).

**Proposition 4.** Let $\tilde{W}_h$ be the finite dimensional linear space generated by the following basis

$$\tilde{\omega}_i(x) = \omega_{x_i}(x) \, \frac{1}{k} \exp(-\frac{u}{k}(x-x_i)) \ , \ 1 \leqslant i \leqslant n \tag{14}$$

where $\omega_{x_i}$ is defined in (11).

Then the NG1 method is equivalent to solve (7) by a Petrov – Galerkin scheme with $W_h$ = basis space and $\tilde{W}_h$ = test space.

**Proof.** The linear system to be solved in both cases is :

$$\sum_{j=1}^{n} \phi_j \int_{x_{i-1}}^{x_{i+1}} \omega'_{x_j}(x)\omega'_{x_i}(x) \exp(-\frac{u}{k}(x-x_i))dx =$$

$$= \frac{1}{k} \int_{x_{i-1}}^{x_{i+1}} f(x)\omega_{x_i}(x) \exp(-\frac{u}{k}(x-x_i))dx \tag{15}$$

where the unknowns $\phi_j$ are defined by (12).

**Remark 3.** The same fact stands for the bidimensional case taking as test functions

$$\tilde{\omega}_i(x,y) = \omega_i(x,y) \, \frac{1}{k} \exp(-\frac{u(x-x_i)+v(y-y_i)}{k}) \tag{16}$$

where $\omega_i$ is the Lagrange's element defined in Remark 2 .

**Proposition 5.** The NG1 scheme for the one dimensional problem (7) is equivalent to the following finite differences equation

$$\frac{u}{2h}(\phi_{i+1}-\phi_{i-1})-(\frac{k+k'}{h^2})(\phi_{i+1}-2\phi_i+\phi_{i-1}) = g(x_i)\ ,\ 1 \leqslant i \leqslant n$$

$$\phi_0 = \phi_{n+1} = 0 \tag{17}$$

$$k' = \frac{uh}{2}(\coth(\frac{uh}{2k}) - \frac{2k}{uh}) \tag{18}$$

$$g(x) = \frac{u^2 h}{4k^2 \sinh^2(\frac{uh}{2k})} \int_{x-h}^{x+h} f(\xi)\omega_x(\xi)\exp(-\frac{u}{k}(\xi-x))d\xi \tag{19}$$

**Proof.** We multiply Equation (15) by $\dfrac{u^2h}{4k\ \sinh^2(\frac{uh}{2k})}$ .

**Remark 4.** k' can be viewed as an h-dependant numerical diffusion. This finite differences equation is similar to Ill in's scheme in which $g(x_i)$ is replaced by $f(x_i)$ . It is easy to check that $g(x_i) = f(x_i) + O(h^2)$ . It is important to note that in the bidimensional case there is no analogous to (17) (unless u=0 or v=0). In this sense we can say that the NG1 scheme generalizes the Ill in scheme to $R^2$. Error bounds are available in the $H^1$-norm and at the nodes.

**Proposition 6.** Let $\phi$ denote the exact solution.

i) Both for Problems (7) and (10) :

$\|\phi_h-\phi\|_{H^1} \leqslant C(k)\|\tau_h-\phi\|_{H^1}$ for all $\tau_h \in W_h$

ii) For the one dimensional problem we have

$$|\phi_i-\phi(x_i)| \leqslant C\ h^2(\frac{1}{h+k} + \varepsilon(k))\ ,\quad 1 \leqslant i \leqslant n\ ,$$

where $0 < \varepsilon(k) \to 0$ as $k \to 0+$.

**Proof.** i) It is a classical result. C(k) is the quotient between the continuity and coercitivity constants of the associated bilinear form of the weak formulation. We note that C(k) grows exponentially in 1/k .

ii) This bound follows from the equivalence stated in Proposition 5. For details see Kellog and Tsan (1978).

**THE LIMIT CASE**

Since the diffusivity constant k is supposed to be small we are interested in the limit case $k \to 0+$ . It is well known that the exact solution is pointwise convergent to the exact solution of the pure convection problem (k=0) with suitable modifications on the boundary conditions. For the numerical scheme NG1 the situation is quite similar.

**Proposition 7.** If the function f in Equation (7) has bounded first derivative on [0,1] then, when $k \to 0+$ , Equation (15) becomes

$$\frac{u}{h}(\phi_{i+1}-\phi_i) = f(x_i) \quad , \quad 0 \leqslant i \leqslant n \quad ; \quad \phi_0 = 0 \qquad (20)$$

**Proof.** Cf. Ahués and Telias (1981).

**Remark 5.** For the bidimensional case we find the same result. For instance if $u > 0$ and $v > 0$ then, when $k \to 0+$ , the limit of Equation (13) has coefficients which are illustrated in Figure 2.

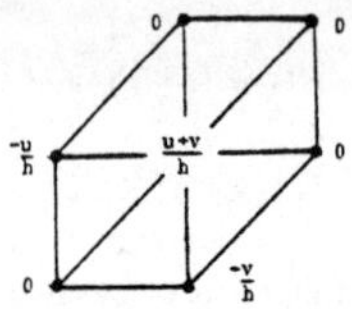


Figure 2

**Remark 6.** It is to be noted that Equation (20) leads to an approximation in $O(h)$ of the exact solution of the pure convection problem $u\phi' = f$ , $\phi(0) = 0$ .
The loss of one order in $h$ is clearly seen at nodal points letting $k$ go to 0 in Proposition 6, ii). In fact, as it is proved in Kellog and Tsan (1978), any discretization which amounts to a tridiagonal linear system with properties (3), (5) and (6) will share this property. Nevertheless, special test functions $\omega_i^*$ may be constructed to solve in order $O(h^2)$ the pure convection problem. For this it suffices that the test functions $\omega_i^*$ verify

$$\int_{x_{i-1}}^{x_i} f(x)dx = \int_{x_{i-1}}^{x_{i+1}} \omega_i^*(x)f(x)dx \quad \text{for any one degree}$$

piecewise polynomial f. For instance the test functions

$$\omega_i^* = \omega_{x_i} + \frac{h}{2}\omega'_{x_i} \quad , \quad 1 \leqslant i \leqslant n$$

proposed in Brooks and Hughes (1980) satisfy this requirement. A detailed discussion of this point and suitable examples can be found in Ahués and Telias (1981).
We present now a modification on NG1 scheme which will allow us to treat the case of non constant velocity.

## THE VARIABLE VELOCITY CASE

When $u$ in Equation (7) (resp. $\vec{u}$ in Equation (10)) is a function of x (resp. of (x,y)) we can preserve the M-matrix property of the linear system which arises from discretization by introducing the following discontinuous test functions instead of those defined in (14) (resp. (16)) :

$$\bar{\omega}_i(x) = \begin{cases} \omega_{x_i}(x)\,\frac{1}{k}\exp(-\frac{u_i}{k}(x-x_i)) & \text{if } x_i - \frac{h}{2} \leq x \leq x_i + \frac{h}{2} \\ 0 & \text{otherwise} \end{cases}$$

where $u_i = u(x_i)$ (resp.

$$\bar{\omega}_i(x,y) = \begin{cases} \omega_i(x,y)\frac{1}{k}\exp(\frac{u_i(x-x_i)+v_i(y-y_i)}{-k} & \text{if } (x,y) \in S_i \\ 0 & \text{otherwise} \end{cases}$$

where $u_i = u(x_i,y_i)$ , $v_i = v(x_i,y_i)$ and $S_i$ is the "quasi-baricentric" domain associated to node $(x_i,y_i)$ and defined in Figure 3).

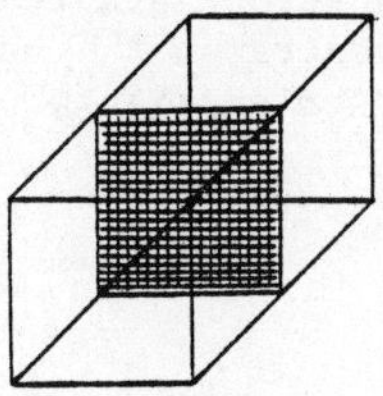

Figure 3

**Proposition 8.** Both in the one and bidimensional cases the Petrov-Galerkin scheme based on the basis functions of NG1 and on the test functions $\bar{\omega}_i$ defined above leads to an M-matrix linear system. Moreover, when $k \to 0+$ , the limit scheme coincides with that of NG1.
**Proof.** Cf. Ahués and Telias (1981).

**NUMERICAL EXAMPLES**

**One dimensional problems**
1) We take first the introductory example
$-k\phi'' + \phi' = 1$ in $]0,1[$ , $\phi(0) = \phi(1) = 0$
NG1 produces the exact solution at nodal points for any value of k and h . This situation is in fact more general.

**Proposition 9.** If in Equation (7) u and f are constant then the NG1 scheme produces the exact solution at nodal points for any value of k, u, f and h .
**Proof.** Cf. Ahués and Telias (1981).

2) The following is an example with variable velocity :
$-\phi''(x) + \frac{60}{x}\phi'(x) = x^2$ , $x \in ]1,2[$ , $\phi(1) = 1$ , $\phi(2) = 0$

With $h = 1/10$ the maximum absolute error at nodal points is seen to be $6/1000$. The exact and numerical solutions are exhibited in Figure 4.

**Bidimensional problems**

1) An interesting example with constant velocity is

$$-k\Delta\phi + \frac{\partial\phi}{\partial y} = 1 \text{ in } \Omega = ]0,1[^2 - ]1/2,1[\times]0,1/2[ \quad \phi=0 \text{ on } \partial\Omega$$

where $k=1/1000$. We set $\psi(x,y) = \phi(x,y) - y$ and we solve $-\Delta\psi + \frac{\partial\psi}{\partial y} = 0$ in $\Omega$ ; $\psi = -y$ on $\partial\Omega$. We see that for k small enough and not near the boundary, the solution looks like

$$\psi(x,y) \simeq \begin{cases} 0 & \text{if } 0 < x < 1/2 \\ -1/2 & \text{if } 1/2 < x < 1 \end{cases}$$

The exact solution has strong gradients near $y = 1$ and near $x = 1/2$. Figure 5 illustrates the numerical solution for $y = 33/50$ with a regular triangular mesh of size $h=1/50$.

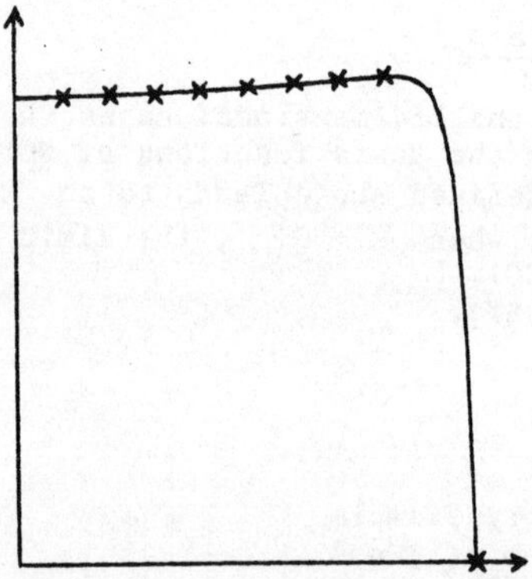

Figure 4

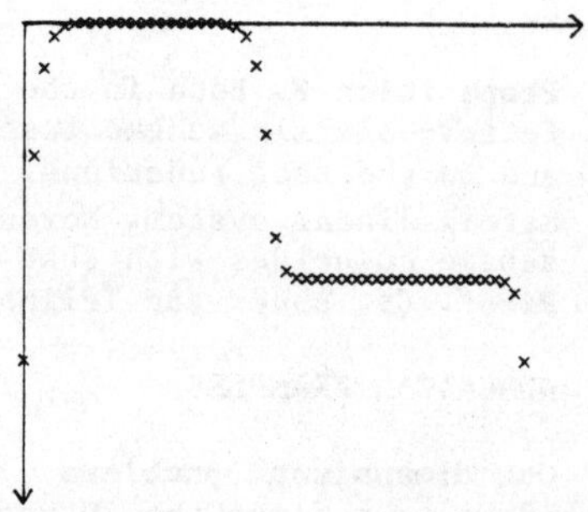

Figure 5

2) As a case of variable velocity we present the following

$$-k\Delta\phi + (x+1/2)^{-1} \frac{\partial\phi}{\partial x} - (y+1/2)^{-1} \frac{\partial\phi}{\partial y} = 0 \text{ in } \Omega = ]0,1[^2$$

$$\phi(x,y) = (x+1/2)^2+(y+1/2)^2 \quad \text{on} \quad \partial\Omega$$

With a regular triangular mesh of size $h = 1/50$ and $k = 1/1000$ the maximum absolute error at nodal points with respect to the exact solution of the pure convection problem ($k=0$) is 3/100. Figure 6 shows both solutions at the ordinates $y = 8/25$ , $y = 1/2$ and $y = 33/50$.

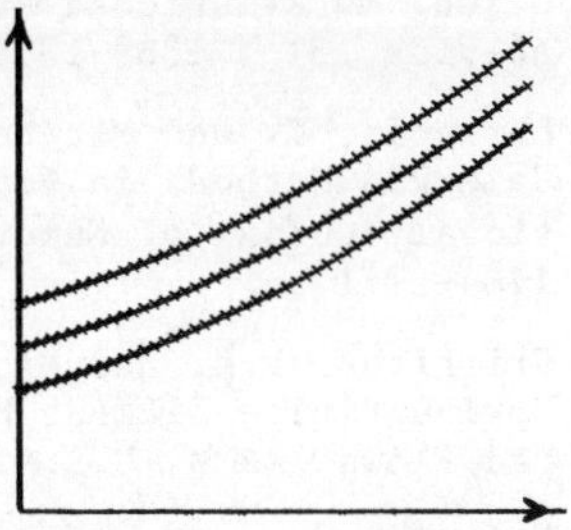

Figure 6

**CONCLUDING REMARKS**

The main feature of this work is that we are able to produce a non oscillatory numerical solution of Equation (7) or (10) for any value of the diffusivity constant k, for both constant and variable velocity fields, and without restrictions on the mesh size of discretization. Nevertheless with the NG1 procedure, and its variant for non constant velocity, the error bounds in norm and at nodal points depend on the convexity of the exact solution since we are using one degree elements. From a computational point of view we remark that negative exponent exponentials are to be carefully handled. In the numerical examples the integrals appearing in the NG1 procedure (and its variant) have been performed analytically (even in the bidimensional situation). Numerical quadrature should be carefully developed in order to use non regular triangulations and accurate computation of the exponentials.

**REFERENCES**

Ahués, M. and Telias, M. (1981). A Finite Element Method for the Steady State Convection Diffusion Equation. Rapport de Recherche IMAG n° 278. Grenoble, France.

Axelsson, O. (1980). On the Numerical Solution of Convection Diffusion Equations. Lecture Notes E.P.F.L. Switzerland.

Brooks, A. and Hughes, T.J.R. (1980). Streamline-Upwind/Petrov-Galerkin Methods for Advection Dominated Flows. 3rd. International Conference on F.E.M. in Flow Problems. Canada, 283-292.

Chatelin, F. (1982). Spectral Approximation of Linear Operators. Academic Press. U.S.A. (to appear).

Christie, I., Griffiths, D.F., Mitchell, A.R. and Zienkiewicz, O.C. (1976). Finite Element Methods for Second Order Differential Equations with Significant First Derivatives. International Journal of Numerical Methods in Engineering. 31 : 1389-1396.

Christie, I. and Mitchell, A.R. (1978). Upwind of High Order Galerkin Methods in Conduction-Convection Problems. International Journal of Numerical Methods in Engineering 12 : 1764-1771.

Griffiths, D.F. and Mitchell, A.R. (1979). On Generating Upwind Finite Element Methods. F.E.M. for Convection Dominated Flows. ASME. U.S.A. 91-104.

Heinrich J.C., Huyarkorn, P.S., Mitchell, A.R. and Zienkiewicz, O.C.(1977). An Upwind Finite Element Scheme for Two Dimensional Convective Transport Equation. International Journal of Numerical Methods in Engineering. 11 : 131-143.

Heinrich, J.C. and Zienkiewicz, O.C. (1979). The Finite Element Mehod and Upwinding Techniques in the Numerical Solution of Convection Dominated Flow Problems. F.E.M. for Convection Dominated Flows. ASME. U.S.A. 105-136.

Hughes, T.J.R. and Brooks, A. (1979). A Multidimensional Upwind Scheme with No Crosswind Diffusion. F.E.M. for Convection Dominated Flows. ASME. U.S.A. 19-35.

Kellog, R.B. and Tsan, A. (1978). Analysis of Some Difference Approximations for a Singular Perturbation Problem without Turning Points. Mathematics for Computation 32, 144 : 1025-1039.

Leonard, B.P. (1979). A Survey of Finite Differences of Opinion on Numerical Muddling of the Incomprehensible Defective Confusion Equation. F.E.M. for Convection Dominated Flows. ASME. U.S.A. 1-18.

Roache, P.J. (1976). Computational Fluid Dynamics. Hermosa Publishers. U.S.A.

Thomasset, F. (1981). Finite Element Methods for Navier Stokes Equations. Cours INRIA. France. 1-125.

Varga, R. (1962). Matrix Iterative Analysis. Prentice Hall Series in Automatic Computation. U.S.A.

# HYBRID AND MIXED FINITE ELEMENTS IN FLUID MECHANICS*

S.N. Atluri**

Center for the Advancement of Computational Mechanics
School of Civil Engineering
Georgia Institute of Technology, Atlanta, GA 30332, USA

## SUMMARY

In this paper certain new concepts in the solution of Navier-Stokes equations, for an incompressible fluid flow problem, using mixed and hybrid finite elements, are presented. These new methods are compared and contrasted with the currently popular methods such as those based on primitive variable formulations and on penalty function formulations to treat incompressibility.

## INTRODUCTION

Finite Element methods for solving the Navier-Stokes equations, governing viscous incompressible fluid flow, written in terms of the primitive variables, viz, the velocity vector and the hydrostatic pressure, may be described as "primal methods". Of late, certain expedient numerical "tricks" such as eliminating the hydrostatic pressure ab initio from the solution system through a penalty function term, wherein, further, the penalty term is under-integrated, have also met with some success and attained popularity. Representative of such methods are those discussed in [1-3]. Theoretical investigations [4] of the so-called LBB-conditions governing the penalty-under-integration schemes have shown that some of these methods may be unstable.

It has been recognized in solid mechanics that hybrid and mixed finite element methods are often advantageous in treating problems with constraint such as incompressibility, etc. These highly salutory features of these methods yet remain to be explored in fluid mechanics. Hybrid methods are those wherein certain constraints at the inter-element boundaries are introduced into the finite element formulation through Lagrange

* For presentation at 4th Int. Conf. on Finite Elements in Water Resources, June 21-25, 1982, Hannover, Fed. Rep. Germany

** Regents' Professor of Mechanics

multipliers assumed at the element boundaries. Mixed methods, broadly speaking, are those which involve more than one field variable as a directly assumed variable within each element. Some of these additional variables in each element may or may not play the role of Lagrange multipliers to enforce certain constraints within each element. For more details on these concepts, see [5].

What appears to be the first application of hybrid finite elements in fluid mechanics was presented in [6,7] in the context of "Stokes" flow of a viscous, incompressible flow wherein the convective acceleration term is zero. In this method, deviatoric fluid stresses within each element and velocities at the interelement boundaries are treated as the basic variables. The hydrostatic pressure is determined, to within an arbitrary constant, from the deviatoric stresses through the satisfaction of momentum balance conditions a priori. The final finite element system in this method involves only the nodal velocities, and the "constant" pressure term in each element. In several test cases, the method was found to be stable, and yields superior accuracies in both velocities and pressure. The so-called L.B.B. conditions for this method have been studied in [7].

In this paper, we present a new mixed method, based on a simultaneous approximation of both fluid stresses and velocities in each element, for the solution of steady Navier-Stokes equations governing the flow of a viscous incompressible fluid wherein the convective acceleration term is accounted for. This mixed method is shown to lead to the hybrid method [6,7] described above, for Stokes flow.

## MIXED FINITE ELEMENTS FOR NAVIER-STOKES FLOWS

We consider the laminar flow of an incompressible viscous Newtonian fluid. Let: $x_i$ denote spatial cartesian coordinates in a spatial domain V, $\rho$ the fluid mass density; $\overline{F}_i$ are body forces unit mass excluding inertia; $\tau_{ij}$ the fluid stress, $\tau'_{ij}$ the deviatoric stress; p the hydrostatic pressure; $v_i$ the velocity; $V_{ij}$ the velocity strains, $\overline{t}_i$ the prescribed tractions on a surface $S_t$, and $\overline{v}_i$ the prescribed velocities at $S_v$; and $(\ )_{,i}$ denotes the partial differentiation w.r.t. $x_i$. The well-known Navier-Stokes equations are:

(incompressibility): $v_{i,i}=0$ in V (1)

(momentum balance): $\tau_{ij,j}+\rho\overline{F}_i=\rho v_j v_{i,j}$ in V (2a)

$$\tau_{ij}=\tau_{ji} \text{ in } V \tag{2b}$$

(compatibility) : $V_{ij}=\frac{1}{2}(v_{i,j}+v_{j,i}) = v_{(i,j)}$ (3)

(traction B.C.) : $\tau_{ij}n_j=\bar{t}_i$ at $S_t$ (4)

(velocity B.C.) : $v_i=\bar{v}_i$ at $S_v$ . (5)

(constitutive law): $\tau_{ij}=\partial A/\partial V_{ij}$ (6)

where,

$$A = A(p,V_{\ell m}) = -pV_{kk} + \mu V_{\ell m}V_{\ell m} \tag{7}$$

$\mu$ = coefft. of viscosity

Thus,

$$\tau_{ij} = \frac{\partial A}{\partial V_{ij}} = -p\delta_{ij} + 2\mu V_{ij} \tag{8}$$

We note that:

$$\tau_{ij} = -p\delta_{ij} + \tau'_{ij} \tag{9}$$

thus, from (8) and (9) we have:

$$\tau'_{ij} = 2\mu V_{ij} \tag{10}$$

We now establish a contact transformation to express the stress-working density of the fluid in terms of stresses. Thus, let,

$$-B(\tau_{km}) = A(p,V_{km}) - \tau_{km}V_{km} \tag{11}$$

such that

$$V_{ij} = \partial B/\partial \tau_{ij} \tag{12}$$

Using (9) and (10) in (11) we obtain:

$$B(\tau_{km}) = (1/4\mu)\tau'_{km}\tau'_{km} \equiv B(\tau'_{km}) \tag{13}$$

We now consider the following mixed variational theorem, wherein $\tau'_{km}$, $p$, and $v_i$ are variables, and $\tau'_{km}=\tau'_{mk}$ a priori. Consider the functional

$$M(\tau'_{km},p,v_i) = \int_V [-B(\tau'_{km}) + \tau_{km}v_{k,m} - \rho\overline{F}_i v_i]dV - \int_{S_t} \overline{t}_i v_i ds - \int_{S_v} (v_i-\overline{v}_i)\tau_{ji}n_j ds \quad (14)$$

The term $\tau_{km}v_{k,m}$ in the r.h.s. of (14) can be rewritten as:

$$\tau_{km}v_{k,m} = -pv_{k,k} + \tau'_{km}v_{k,m} \quad (15)$$

We now show that the variational statement,

$$\delta M(\delta\tau'_{km},\delta p,\delta v_i) + \delta N(v_i) = 0 \quad (16)$$

where

$$\delta N = \int_V \rho v_j v_{i,j}\delta v_i dV \quad (17)$$

leads as the Euler equations to (i) incompressibility, Eq. (1); (ii) momentum balance (2a,b); (iii) compatibility, (3) and (iv) the appropriate b.c., (4 and 5). In order to show this, we carry out the variation in M and write:

$$\delta M + \delta N = 0 = \int_V [(-\frac{\partial B}{\tau'_{km}} + v_{(k,m)})\delta\tau'_{km} - \delta pv_{k,k} - (\tau'_{km,k}-p,_m+\rho\overline{F}_m-\rho v_k v_{m,k})\delta v_m]dV - \int_{S_t} (\overline{t}_i-n_j\tau_{ji})\delta v_i ds - \int_{S_v} (v_i-\overline{v}_i)\delta\tau_{ji}n_j ds \quad (18)$$

The Euler-Lagrange Equations from (18) are:

$$\frac{\partial B}{\partial\tau'_{km}} \equiv \frac{1}{2\mu}\tau'_{km} = v_{(k,m)} \quad (19)$$

$$\tau'_{km,k} - p,_m + \rho\overline{F}_m = \rho v_k v_{m,k} \quad (20)$$

$$v_{k,k} = 0 \quad (21)$$

$$\overline{t}_i = n_j\tau_{ji} \text{ at } S_t \quad (22)$$

$$v_i = \overline{v}_i \text{ at } S_v \quad (23)$$

It is seen that in order to be admissible, the fields in the functionals M and $\delta N$ should be such that: (a) $\tau'_{km}$ be symmetric but otherwise arbitrary; (b) p be arbitrary, and (c) $v_i$ be continuous and differentiable.

Now consider a finite element discretization. Let V be dissected into P elements, $V_m$, m=1,2,...,P. Let $\partial V_m$ be the boundary of $V_m$, such that, in general,

$$\partial V_m = \rho_m + S_{tm} + S_{vm} \tag{24}$$

where $\rho_m$ is an interelement boundary, and $S_{tm}$, $S_{vm}$ are, respectively, the segments of $\partial V_m$ where tractions, and velocities are prescribed. Let superscripts (+) and (-) denote, arbitrarily, the two immediate vicinities of $\rho_m$ in the limit as $\rho_m$ is approached from the "left" and the "right". Then, for a valid finite element formulation, it is necessary that $\tau'_{km}$, p, and $v_m$ in each $V_m$ be such that they not only satisfy Eqs. (20-24) but also the inter-element constraints that:

$$[(\tau'_{km}-\delta_{km}p)n_k]^+ + [(\tau'_{km}-\delta_{km}p)n_k]^- = 0 \text{ at } \rho_m \tag{25}$$

and

$$v_i^+ = v_i^- \text{ at } \rho_m \tag{26}$$

Eq. (25) is the Cauchy's traction reciprocity law at an interface, and (26) is the velocity continuity condition.

We now show that if: (i) $\tau'_{km}=\tau'_{mk}$ in each element, but Eq. (25) is <u>not</u> obeyed a priori, (ii) p is arbitrary in each element, (iii) $v_i$ is continuous in each element, and <u>in addition</u>, $v_i$ also obeys Eq. (26), then the statement

$$\delta(FM) + \delta(FN) = 0 \tag{27}$$

leads as Euler equations, to <u>not only</u> Eqs. (19-23) but also to Eq. (25).

In Eq. (27), it is noted that

$$(FM) = \sum_{m=1}^{P} \{\int_{V_m} [-B(\tau'_{km}) + \tau'_{km}v_{k,m} - pv_{k,k} - \rho\overline{F}_i v_i]dv$$

$$- \int_{S_{tm}} \bar{t}_i v_i ds - \int_{S_{vm}} (v_i - \bar{v}_i) \tau_{ji} n_j dS\} \tag{28}$$

and

$$\delta(FN) = \sum_{m=1}^{P} \{ \int_{V_m} \rho v_k v_{m,k} \delta v_m dV \tag{29}$$

Eq. (27) can now be expressed as:

$$\delta(FM)+\delta(FN)=0= \sum_{m=1}^{P} \{ \int_{V_m} [(- \frac{\partial B}{\partial \tau'_{km}} + v_{(k,m)}) \delta\tau'_{km} - \delta p v_{k,k}$$

$$- (\tau'_{km,k} - p_{,m} + \rho \bar{F}_m - \rho v_k v_{m,k}) \delta v_m ] dV$$

$$- \int_{S_{tm}} (\bar{t}_i - n_j \tau_{ji}) \delta v_i ds - \int_{S_{vm}} (v_i - \bar{v}_i) \delta \tau_{ij} n_j ds$$

$$+ \int_{\rho_m} (\tau'_{ij} - p\delta_{ij}) n_j \delta v_i dS\} \tag{30}$$

Since, in the last term in the r.h.s. of Eq. (30), $\delta v_i^+ = \delta v_i^-$ at $\rho_m$ a priori, and noting that the integral over $\rho_m$ occurs twice for each segment of $\rho_m$ (since it is the common boundary of two elements), it is seen that the vanishing of this term leads to Eq. (25).

The details for the finite element implementation, shape functions, and numerical results, which are omitted here for space reasons, will be presented orally at the conference.

HYBRID FINITE ELEMENTS FOR STOKES FLOW

Here we neglect the convective acceleration term. In view of this, $\delta(FN)$ of (29), can be set to zero. Thus, the momentum balance condition becomes simplified, as:

$$\tau'_{km,k} = p_{,m} + \rho \bar{F}_m = 0; \ \tau'_{km} = \tau'_{mk} \tag{31}$$

Since the momentum balance condition (31) does ont involve the coupling with $v_i$, is is easy to satisfy a priori, as was done in [6,7]. Thus, assuming that Eq. (31) is satisfied a priori, the functional (FM) of Eq. (28) can be simplified to be:

$$(FH) = \sum_{m=1}^{P} \{ \int_{V_m} -B(\tau'_{km}) + \int_{\rho_m} (\tau'_{km} - p\delta_{km}) n_k v_m$$

$$+ \int_{S_{vm}} (\tau'_{km} - p\delta_{km}) n_k \overline{v}_m dS - \int_{S_{tm}} <\overline{t}_k - (\tau'_{km} - p\delta_{km}) n_m > v_m dS \} \quad (31)$$

Thus the pressure term p occurs in the boundary integrals. Also, from Eq. (31) it is seen that p can be determined, from the chosen $\tau'_{km}$, to within an arbitrary constant. Thus, within each element, one assumes: (i) $\tau'_{km}$ with undetermined parameters $\beta$, (ii) p which involves an arbitrary constant $\alpha$ in addition to some of the $\beta$'s, and (iii) the element-boundary velocity field $v_m$ which involves noded velocities q. It can be shown [6,7] that $\delta(FH)=0$ leads as its Euler Equations, to Eq. (19), Eq. (21) and (25). Since the chosen boundary velocity field is unique at an inter-element boundary. Since $\tau'_{km}$ is arbitrary for each element, and not subject to Eq. (25) a priori, the $\beta$'s in $\tau'_{km}$ in each element can be eliminated at the element level and expressed in terms of the respective elements q's. Thus, it is seen that the finite element solution system of equations involves only q's and $\alpha$'s.

An alternative interpretation of Eq. (32) is that of a modified complementary energy principle, wherein: (i) the fields $\tau'_{km}$ and p are subject, a priori, to the momentum balance constraint, Eq. (31), (ii) but the interelement traction reciprocity constraint, Eq. (25) is enforced, a posteriori, through Lagrange multipliers $v_m$ at $\partial V_m$.

Details of implementation of the hybrid method of Eq. (31), shape functions, and numerical results will be presented orally at the conference.

Our experience so far indicates that these methods are stable, and provide better accuracies for both velocity and pressure. Some of these results are documented elsewhere [6,7].

ACKNOWLEDGEMENTS

The above results were obtained during the course of investigations supported by U.S.A.F.O.S.R. under grant 81-0057b, with Dr. A. Amos as the responsible program official. The author expresses his sincere thanks to Ms. M. Eiteman for her assistance in the preparation of this manuscript.

REFERENCES


1 Olson, M. and Tuann, S., "Primitive Variables Versus Stream Function Finite Element Solutions of Navier-Stokes Equations", Finite Elements in Fluids, Vol. 3, (R.H. Gallagher, et. al., eds), J. Wiley, 1978, pp. 73-89.

2 Lee, R.L., Gresho, P.M. and Sani, R.L., "Smoothing Techniques for Certain Primitive Variable Solutions of Navier-Stokes Equations", Intl. Jnl. of Num. Methods in Engg., 14, 1979, pp. 1785-1804.

3 Hughes, T.J.R., Liu, W.K. and Brooks, A., "Finite Element Analysis of Incompressible Fluid Flow by the Penalty Function Formulation", J. Computational Physics, 30, 1979, 1-60.

4 Oden, J.T., Kikuchi, N. and Song, S.W., "Penalty-Finite Element Methods for the Analysis of Stokesian Flows", Computer Methods in Applied Mech. and Engg., (In Press).

5 Atluri, S.N., Tong, P. and Murakawa, H., "Recent Studies in Hybrid and Mixed Finite Element Methods in Mechanics" in Hybrid and Mixed Finite Element Methods, (S.N. Atluri, R.H. Gallagher and O.C. Zienkiewicz, eds), J. Wiley, 1982, (To Appear).

6 Bratianu, C. and Atluri, S.N., "A Hybrid Finite Element Method for Stokes Flow; Part I - Formulation & Numerical Studies", Computer Methods in Applied Mechanics and Engg., (In Press).

7 Ying, L.-a. and Atluri, S.N., "A Hybrid Finite Element Method for Stokes Flow: Part II - Convergence & Stability, Computer Methods in Applied Mechanics and Engineering, (In Press).

# THE RELIABILITY ESTIMATES AND ADAPTIVITY IN FINITE ELEMENT ENGINEERING COMPUTATIONS.

Ivo Babuška

Institute for Physical Science and Technology and Mathematics Department
University of Maryland, College Park, MD 20742

## INTRODUCTION

In general the objective of computational solution of engineering problem is to obtain an acceptable, accurate and reliable prediction of the behavior of the physical phenomena under study.

The reliability depends
a) on reliability of the mathematical model and its parameters.
b) on reliability of the numerical treatment of the mathemtically formulated problem.

The mathematical model is mostly checked by various sensitivity studies, assuming that the reliability of the numerical treatment is sufficient. In addition, in many fields the certification computations are used more and more often and are sometimes unavoidable as in nuclear industry, etc.

It is essential that the reliability of the computation will be under control. By reliability we mean here, the accuracy of the approximate solution measured in a particular norm which reflects the purpose of the computation and is specified by the user.

Today's development of computer technology rapidly changes the ratio of computer to human cost. This situation should be reflected in the computational analysis of engineering problems.

From this point of view the computational analysis has to be such that:
(a) The computational analysis provides also a reliable estimation of the error in the norm specified by the user.
(b) The computation is fully adaptive using minimal physical

data as input and is very "user" friendly. For example, the meshes, etc. are adaptively constructed and the computation is adaptively terminated when achieving the set goals or exceeding the cost or computer size limitation.

It is beginning to be a widely accepted observation that for realistic problems it is rarely feasible to design numerical processes which reliably and effectively achieve the designed accuracy in the prescribed norm and yet which do not utilize some form of adaptivity. We remark that the adaptively constructed meshes strongly depends on the required accuracy and the used norm.

The error estimation should be highly reliable. By this we mean that the effectivity index

$$\theta = \frac{\text{estimated error}}{\text{true error}}$$

has to be close to 1 as the error converges to zero. More specifically we wish that $|\theta.1| \leq .1-.2$ when the relative error is on the order of 10% and $|\theta-1| \leq .1-.05$ if the relative error is on the order of 3%. (We assume of course, that round offs can be neglected.)

During the past years the effort at the University of Maryland was directed to achieve the goals formulated briefly above. The mathematical theory and the implementation were in the focus of the research. Specifically, we concentrated on

(a) Solution of linear elliptic PDE's.
(b) Solution of linear parabolic equations.
(c) Solution of linear convection diffusion equations.
(d) Solution of nonlinear problems oriented to the stability computations of structures.

In this talk we will present some illustrative results and some comments to them, omitting precise mathematical statements and details referring to other publications.

## THE FEARS PROGRAM

FEARS [Finite Element Adaptive Research Solver] is a solver dealing with a system of linear partial differential elliptic equations in 2 dimensions. It uses curved quadrilateral bilinear elements with refinement features which are adaptively constructed. FEARS provides the user with error estimation in various norms. It constructs adaptively the meshes in a successive way. The adaptively constructed meshes depends on the norm required for error measurement. The input is very simple. The domain is characterized by a set of two dimensional quadrilateral subdomains given in a most simple way (e.g. by coordinates of the vertices if if domains are polygons). For more about FEARS, a-posteriori error estimation and adaptivity, we refer the reader to [2][5] [6][7][8][9][10][11][12][17][18][20][21].

Now we will illustrate the use and performance of FEARS on a typical problem.

The steady seepage problem.
Consider here the seepage problem with known underground water level on complicated geographical domain shown in Fig. 1. The

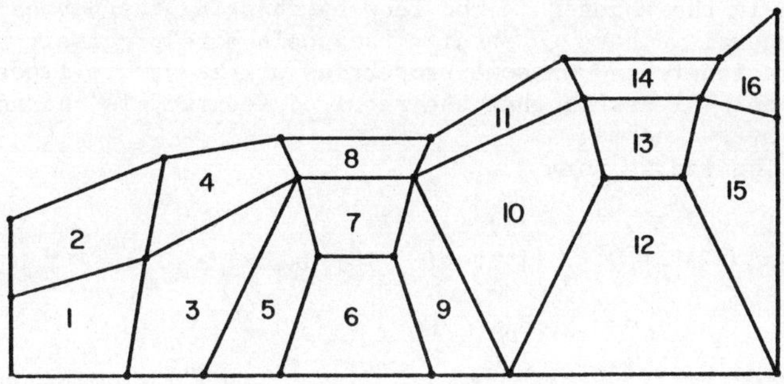


Fig. 1. The scheme of the seepage problem

domain under consideration consists of 16 subdomains with various permeabilities in particular subdomains. For simplicity we assume here isotropic media. The equation for the piezometric head $u$ is given by the equation

$$\sum_{i=1}^{2} \frac{\partial}{\partial x_i} a \frac{\partial u}{\partial x_i} = 0$$

where $a$ is a constant, but has different values in particular subdomains. The seepage velocities are then

$$v_1 = -a\frac{\partial u}{\partial x_i}$$

$$v_2 = -a\frac{\partial u}{\partial x_2} \quad .$$

We assume the usual boundary conditions. On vertical and bottom boundary we have

$$\frac{\partial u}{\partial n} = 0$$

and on the surface

$$u = x_2 \ .$$

Let the goal of computation be the analysis of the flow field characterized by the velocities. In addition, let us be interested in the changes in the flow by changing the permeability. In general we have to specify the goals more precisely when we are interested in some properties of the stream lines. Here we will assume that the required accuracy is characterized by two norms:

(a) The energy norm

$$||u||_V^2 = \int_\Omega [\sum_{j=1}^{2} a(\frac{\partial u}{\partial x_i})^2]\, dx_1 dx_2$$

(b) The $L_2$ velocity norm

$$||u||_V^2 = \int_\Omega [\sum_{j=1}^{2} v_j^{\,2}]\, dx_1 dx_2$$

Let us consider two sets A, B of permeabilities given in Table 1.

| Domain No. | A | B | Domain No. | A | B |
|---|---|---|---|---|---|
| 1 | 5.0 | 2.0 | 9 | 1.0 | 1.0 |
| 2 | 5.0 | 2.0 | 10 | 5.0 | 2.0 |
| 3 | 5.0 | 2.0 | 11 | 5.0 | 2.0 |
| 4 | 5.0 | 2.0 | 12 | 5.0 | 2.0 |
| 5 | 1.0 | 1.0 | 13 | 60.0 | 70.0 |
| 6 | 2.0 | 1.0 | 14 | 1000.0 | 1000.0 |
| 7 | 70.0 | 70.0 | 15 | 5.0 | 2.0 |
| 8 | 1000.0 | 1000.0 | 16 | 5.0 | 2.0 |

Table 1. The permeabilities coefficients.

FEARS now constructs adaptively the sequence of meshes. There are various adaptive modes available. One of them is to construct the meshes with a minimal number of elements, another one minimizes in some way the computer cost.

Tables 2-6 below relate to Set A.
Table 2 shows the sequence of meshes based on energy norm, when minimal element number is required. In addition the relative error in the energy norm $\frac{||\text{error}||_E}{||u||_E}$ is given together with the efficiency index

$$\theta = \frac{\text{estimator}}{||\text{true error}||_E} .$$

Although the exact solution is not known, we estimated by a larger scale of computation the exact energy $E_o$ of the solution. Thus of course

$$||\text{true error}||_E^2 = E-E_o$$

where $E$ is the energy of the approximate solution which is also computed.

| No. of Elements | Relative error in % | Effectivity Index |
|---|---|---|
| 64 | 19.48 | .82 |
| 82 | 18.01 | .86 |
| 94 | 15.48 | .90 |
| 112 | 13.85 | .91 |
| 160 | 13.37 | .96 |
| 226 | 11.91 | 1.01 |
| 247 | 10.95 | 1.01 |
| 349 | 9.16 | .99 |

Table 2. The adaptively constructed sequence of meshes (energy norm).

We see excellent quality of computed estimate of the error.

Table 3 shows the number of elements in the particular subdomains. We see various interesting features, e.g. that in the subdomains 8, 14 where the permeability is the largest no refinement (as expected) is taking place.

| No. of Domain | Number of Elements | | | | | | | |
|---|---|---|---|---|---|---|---|---|
| | 64 | 82 | 94 | 112 | 160 | 226 | 247 | 349 |
| 1 | 4 | 4 | 4 | 4 | 7 | 13 | 13 | 16 |
| 2 | 4 | 10 | 10 | 16 | 25 | 31 | 31 | 46 |
| 3 | 3 | 3 | 3 | 3 | 10 | 10 | 13 | 16 |
| 4 | 4 | 4 | 4 | 4 | 4 | 7 | 10 | 10 |
| 5 | 4 | 4 | 4 | 4 | 4 | 4 | 4 | 10 |
| 6 | 4 | 4 | 4 | 4 | 4 | 4 | 7 | 7 |
| 7 | 4 | 4 | 4 | 4 | 4 | 4 | 4 | 4 |
| 8 | 4 | 4 | 4 | 4 | 4 | 4 | 4 | 4 |
| 9 | 4 | 4 | 7 | 7 | 10 | 16 | 16 | 19 |
| 10 | 4 | 13 | 13 | 16 | 22 | 34 | 40 | 64 |
| 11 | 4 | 4 | 4 | 4 | 10 | 16 | 19 | 31 |
| 12 | 4 | 4 | 10 | 16 | 25 | 40 | 40 | 49 |
| 13 | 4 | 4 | 4 | 4 | 4 | 7 | 7 | 10 |
| 14 | 4 | 4 | 4 | 4 | 4 | 4 | 4 | 4 |
| 15 | 4 | 4 | 4 | 4 | 7 | 10 | 10 | 16 |
| 16 | 4 | 7 | 10 | 13 | 16 | 22 | 25 | 43 |

Table 3. Distribution of the elements in particular subdomains.

In Table 4a, b we show the results related to the meshes constructed in the adaptive mode respecting the optimization of the machine time and based on the energy and the velocity norm.

| No. of Elements | Energy | $\frac{\|\|e\|\|_E}{\|\|u\|\|_E}$ % |
|---|---|---|
| 451 | 6.7863 | 8.16 |
| 850 | 6.7706 | 6.92 |

Table 4a. The adaptive meshes based on energy norm.

| No. of Elements | Energy | $\frac{\|\|e\|\|_V}{\|\|u\|\|_V}$ % |
|---|---|---|
| 478 | 6.8035 | 7.73 |
| 886 | 6.7776 | 6.16 |

Table 4b. The adaptive meshes based on velocity norm.

Comparing tables 4a and 4b we see that the adaptivity based on energy norm yields better error in the energy (as it should) than the mesh based on the velocity norm. (We see it by comparing the values of energies in these two cases.)

In every domain the meshes consist of elements of various sizes. Table 5 shows the distribution of the various sizes of elements in the particular subdomains for meshes based on the velocity norm. The size 0 is the size of the basic elements when the domain is divided in four elements. The size j is the size $2^{-j}$ times the basic element.

| Domain | No. of Elem. | Size of the Elements | | | | | |
|---|---|---|---|---|---|---|---|
| | | 0 | 1 | 2 | 3 | 4 | 5 |
| 1 | 46 | 0 | 6 | 40 | 0 | 0 | 0 |
| 2 | 79 | 0 | 6 | 34 | 21 | 10 | 8 |
| 3 | 28 | 0 | 12 | 16 | 0 | 0 | 0 |
| 4 | 34 | 2 | 0 | 32 | 0 | 0 | 0 |
| 5 | 22 | 2 | 4 | 16 | 0 | 0 | 0 |
| 6 | 70 | 2 | 4 | 0 | 64 | 0 | 0 |
| 7 | 25 | 1 | 1 | 16 | 4 | 0 | 0 |
| 8 | 52 | 0 | 8 | 28 | 16 | 0 | 0 |
| 9 | 25 | 1 | 8 | 16 | 0 | 0 | 0 |
| 10 | 100 | 0 | 0 | 57 | 23 | 20 | 0 |
| 11 | 58 | 0 | 6 | 36 | 16 | 0 | 0 |
| 12 | 130 | 0 | 0 | 58 | 8 | 0 | 64 |
| 13 | 55 | 0 | 6 | 37 | 12 | 0 | 0 |
| 14 | 73 | 0 | 10 | 22 | 5 | 4 | 32 |
| 15 | 19 | 0 | 15 | 4 | 0 | 0 | 0 |
| 16 | 70 | 0 | 1 | 57 | 12 | 0 | 0 |
| Total | 886 | | | | | | |

Table 5. Distribution of the sizes of the elements for the meshes based on velocity norms.

We see interesting distribution of the refinements in various parts respecting among others the singularity of the solution occurring in the vertices of the interfaces where permeability changes.

In table 6 we show the number of elements in particular domains based on energy and velocity norms for the number of elements shown in table 4. We see clearly that the distribution of the elements is very dependent on which norm is used.

| Domain | Energy Norm | | Velocity Norm | |
|---|---|---|---|---|
| | 451 | 850 | 478 | 886 |
| 1 | 19 | 28 | 40 | 46 |
| 2 | 55 | 100 | 55 | 79 |
| 3 | 16 | 37 | 55 | 79 |
| 4 | 10 | 19 | 34 | 34 |
| 5 | 10 | 25 | 4 | 22 |
| 6 | 10 | 34 | 4 | 70 |
| 7 | 4 | 4 | 7 | 25 |
| 8 | 7 | 7 | 31 | 52 |
| 9 | 31 | 97 | 22 | 25 |
| 10 | 76 | 115 | 55 | 100 |
| 11 | 55 | 91 | 40 | 58 |
| 12 | 64 | 103 | 40 | 130 |
| 13 | 13 | 19 | 31 | 55 |
| 14 | 4 | 19 | 31 | 73 |
| 15 | 19 | 28 | 16 | 19 |
| 16 | 58 | 124 | 52 | 70 |

Table 6. Distribution of elements based on the energy and velocity norm adaptivity.

Sensitivity studies are based on comparison of the solutions for various input data.
Table 7 shows the velocity norm and error in the velocity norm for set A and B of the permeabilities.

| | Numbe El | Relative Error % | Absolute Error | Norm of the Solution |
|---|---|---|---|---|
| A | 478 | 7.73 | .5185 | 6.707 |
| B | 523 | 5.90 | .1731 | 2.933 |

Table 7. Comparison of the solution for sets A, B.

Computations by FEARS have various features. They consist of various parts. The main ones are:

a) Subdivision, miscellaneous.
b) Assemblage and elimination on subdomain level, (essentially in a wave front fashion based on rooted tree ordering) back substitution.
c) Elimination of the unknown located on the boundaries of the subdomains (essentially made by a sparse solver technique).
d) Error estimation calculations, design of the mesh (the final output computation is counted additionally).

The optimal ordering for elimination is achieved by a rooted tree structure of the data.
Table 8 gives relative distribution of the computer time of above mentioned parts for computation of the four cases mentioned in table 4.

| | a % | b % | c % | d % | Total |
|---|---|---|---|---|---|
| 451 | 2.55 | 50.70 | 24.84 | 21.91 | 100% |
| 478 | 3.07 | 50.39 | 25.48 | 21.06 | 100% |
| 850 | 2.74 | 51.81 | 26.14 | 19.31 | 100% |
| 886 | 2.20 | 51.31 | 28.65 | 17.84 | 100% |

Table 8. The distribution of the used machine time.

We see remarkable stability of the relative distribution of the machine time for the four basic phases.

FEARS makes essential computations in the loops of complete solutions when a mesh is constructed. In the example the machine time is of the order $O(N^{\alpha})$ with $\alpha < 1.5$ (mostly $< 1.3$) with $N$ being the number of elements. In general $\alpha$ depends on a number of subdomains and distribution of the elements. By our experience $\alpha \leq 1.8$-$1.9$ in the majority of cases. Although FEARS is a research code it has various features of today's modern computer science ideas, is covered well by mathematical theory, and was tested in a large scale engineering problem. It shows that the ideas of reliable a-posteriori estimates and adaptive approaches can be implemented practically on a large scale.

THE PARABOLIC PROBLEM

The time variant seepage problem is described by a parabolic partial differential equation. In the most simple model the equation is

$$\frac{du}{dt} = \Delta u + f$$

with proper boundary conditions. One of the best ways to solve a problem is to describe the equation in the space by finite element method and then obtain a system of ordinary differential equations. Such a method is often called method of lines. One can use today's sophisticated adaptive solvers for initial values of ODE's.

The ideas mentioned in the introduction and elaborated on in the section about the FEARS project could be applied here. New features related to this problem arise.

a) Combination of the accuracies (tolerances) for the space and time error. The error control in time has to be combined with the error control in space.

b) The design of the total (time and space) error estimation.

c) Adaptive space mesh changing during time. (The adaptive time stepping is provided by the ODE solver.)

Once more we cannot go either into mathematical or other details. We refer here to [1][14][15][16] and we will bring here only a simple illustrative example of the effectivity of the error estimation. We will restrict ourselves to the case where the space mesh is not changing in time.

<u>The time dependent problem.</u>
Let us consider the problem in one dimension

$$\frac{\partial u}{\partial t} = f(t,x)+\frac{\partial}{\partial x}(a(x)\frac{\partial u}{\partial x})-b(x)u, 0<x<1,\ t>0,$$

$$u(t,x) = g(t,x),\quad x = 0,1,\ t>0,$$

$$u0,x) = u_0(x),\ 0<x<1\ .$$

We select

$$a(x) = \cosh\ (4x-2)$$
$$b(x) = \sinh\ (2x)$$

and the functions $u_0$, f, g are chosen so that the exact solution is

$$u(t,x) = 1-\exp\left\{\frac{-x^2}{(10t+.1)}\right\}$$
$$+(\frac{1}{2}+\frac{1}{2}\ \mathrm{tgh}(10t-11))(2\sin(\pi x)+2\sin(2\pi t)\sin(2\pi x))$$

The solution was chosen so that it oscilates. For various t the solution is shown in Fig. 2.

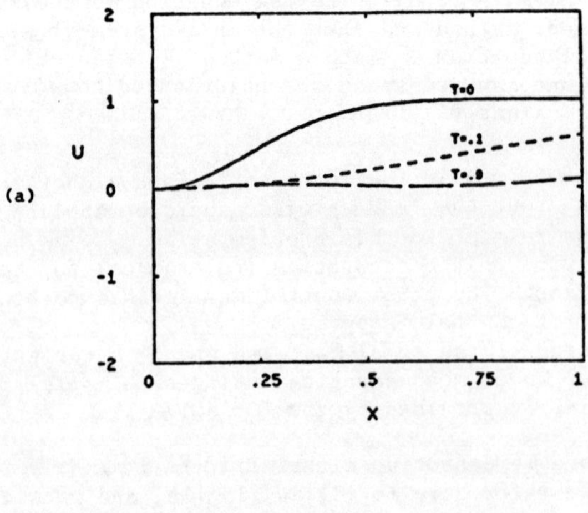

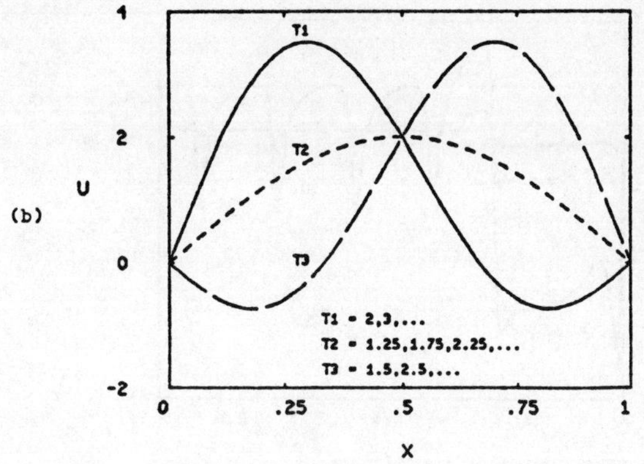


Fig. 2. Exact Solution u(t,x) for Various t .

The norm we are interested in is the energy norm

$$||u(t,x)||_E^2 = \int_0^1 \left(a\left(\frac{\partial u}{\partial x}\right)^2 + bu^2\right) dx$$

We use in our computation the uniform mesh with N intervals and piecewise linear elements. The ODE's are solved with sufficiently high accuracy.

Fig. 3 shows the relative error measured in the energy norm as the function of time.

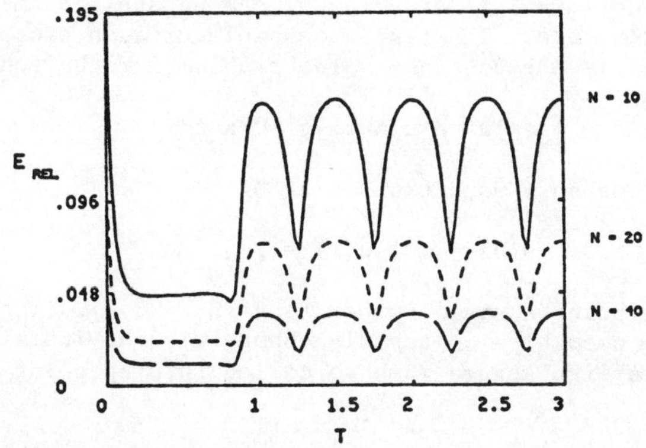


Fig. 3. Relative Error $E_{REL}(T)$ vs. t . N = Number of Uniform Finite Elements.

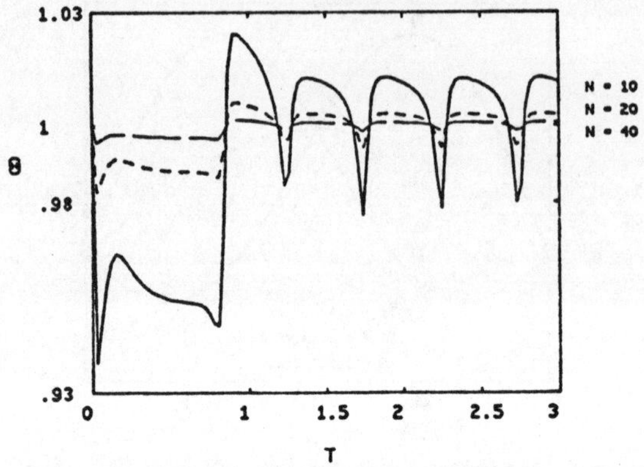


Fig. 4. Effectivity Index θ(T) vs. t .

Fig. 4 shows the effectivity index as function of the time. Once more we see the high quality of the estimation also if the accuracy is a modest one.

The theory and various tests shows that the ideas can be extended to the adaptive construction of the meshes in space and are valid also for cases with solutions of shock wave type and equations which are mildly linear.

THE CONVECTION DIFFUSION PROBLEM

Many problems related to fluids are of convection-diffusion type. The ideas explained above can be applied and implemented in this case. The typical example of such problems in one dimension is the boundary value problem for the equation

$$\varepsilon u''+a(x)u'+b(x)u=f \quad , \quad 0<x<1$$

and the boundary conditions

$$u(0) = \alpha \ , \ u(1) = \beta \ ,$$

with $\varepsilon$ being a small number (say $10^{-4}$-$10^{-5}$). The function a(x) can smoothly or rapidly change without its sign change or with a sign change when so called turning point problems occurs.

The typical and well known difficulties occurs when dealing with this problem. To achieve uniform performance of the method with respect to $\varepsilon$ it is necessary to use different trial and test functions in the finite element framework.

The solution of this problem contain boundary layers. The proper norm to measure the accuracy of an approximate solution is the $L_1$ norm. Because of theoretical reasons, the so called mesh dependent norm $||\cdot||_{L_{1,\Delta}}$ has to be used.

Here

$$||u||_{L_{1,\Delta}} = \int_0^1 |u(x)|dx + \Sigma \frac{h_i+h_{i+1}}{2} |u(x_i)| \tag{1}$$

where $h_i$ are the lengths of elements adjoint to the nodal point $x_i$. [Recall that the sum is a trapezoidal approximation of the value of the integral in the norm]. For more about the mesh dependent norms we refer to [3][4].

In [13] the finite elements with piecewise linear test functions were studied, together with the choices of trial functions spaces which guarantee the uniform quasi-optimality of the approach with respect to $\varepsilon$. These results and ideas can be extended to the construction of an adaptive approach with error estimator in various norms. We restrict ourselves here only to the norm of the space $L_{1,\Delta}$ mentioned above. For more we refer here to [19].

Once more we will restrict ourselves to an illustrative example only.

The convection diffusion problem.

Let us consider the problem:

$$\begin{aligned} -\varepsilon u'' + v(x)w' &= 0 \\ u(0) = 2 \ , \ u(1) &= 0 \end{aligned} \tag{2}$$

$$v(x) = -2 \ \mathrm{tgh}(\frac{x-1}{\varepsilon}) \tag{3}$$

This problem is a linear version of the Burger's equation:

$$-\varepsilon u'' + ww' = 0 \tag{4}$$

which expresses the character of the Navier-Stokes equation. The exact solution of equation (4) can be easily found and for $\varepsilon \leq .01$, it is practically identical with $v(x)$ defined in equation (3). This problem is in fact the problem with a turning point in $x=1$ and $w=v(x)$ is its exact solution. We see the boundary layer behavior of the solution at $x=1$.

Fig. 5 now shows the error norm given by eg (1) as a function of the number of the used elements for uniform and adaptive mesh ($\varepsilon$=.01, .0001). The effectivity of the approach is obvious. [The uniform mesh would need about 1200000 nodal points to get the same accuracy as an adaptivity connected mesh with 295 elements (for $\varepsilon$=.0001)]

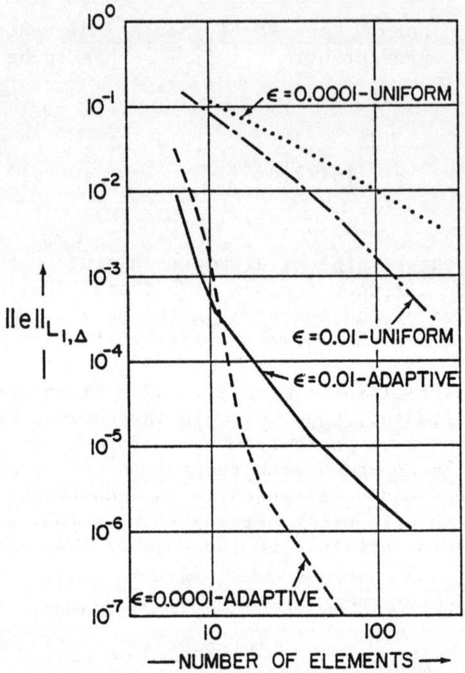


Fig. 5. Accuracy for the adaptive and uniform mesh.

Fig. 6 shows the grading function $\psi(x)$ of the adaptive meshes. The grading function of a mesh with nodal points $x_j^{(n)}$ is the function which maps uniform mesh into the given one i.e. we have $\psi(x_j^{(n)}) = \frac{j}{n}$ where $n$ is the number of the used elements and $0 \leq j \leq n$ .

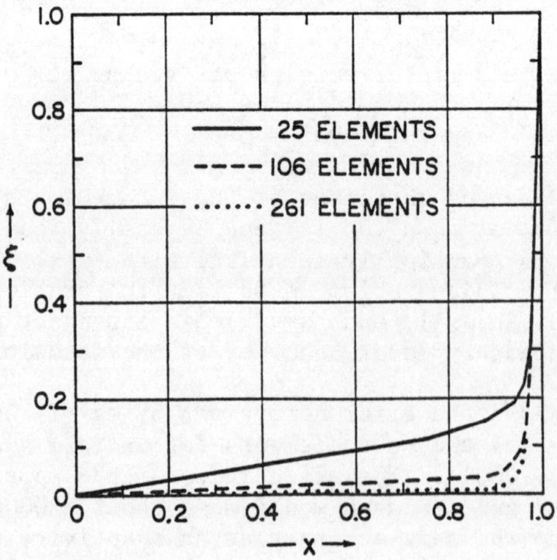


Fig, 6
The grading function

Fig. 7 shows the effectivity index in this case

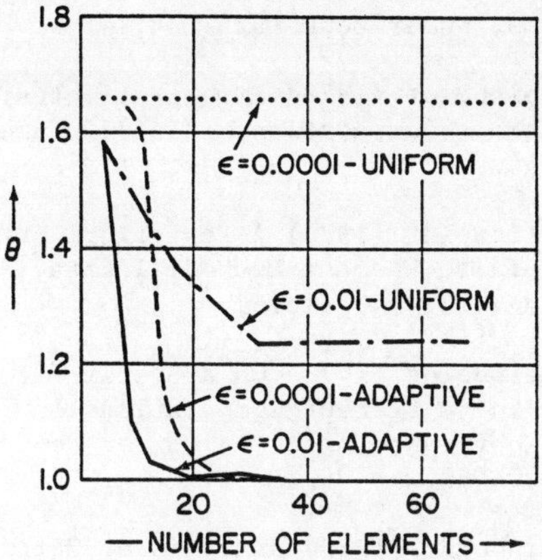


Fig. 7. The effectivity index.

We see clearly that the effectivity index is much better for adaptive meshes than the uniform one. This effect occurs also in the case of parabolic and elliptic equations mentioned in the previous chapters.

THE NONLINEAR PROBLEM

The ideas which we mentioned and elaborated on can be further extended to the analysis of nonlinear problems, the studies of the bifurcations, the equilibrium surfaces and numerical treatment of some problems related to the catastrophy theory. We will not go into details and refer to [12].

REFERENCES

Babuška, I., M. Luskin (1981) An Adaptive Time Discretization Procedure for Parabolic Problems, in "Proc. of the Fourth IMACS Int. Symp. June 30-July 2, ed. by R. Vichnevetski and R. S. Stepleman, IMACS 1981, 5-9.

Babuška, I., A. Miller, (1981 A-Posteriori Error Estimates and Adaptive Technique for the Finite Element Method, Univ. of Maryland, Institute for Physical Science and Technology Tech. Note BN-968.

Babuška, I, J. Osborn (1979) Analysis of Finite Element Methods Using Mesh Dependent Norms in Constructive Approaches to Mathematical Models. C. V. Coffman, C. J. Fix ed. Academic Press.

Babuška, I., J. Osborn (1980) Analysis of Finite Element Methods for Second Order Boundary Value Problems Using Mesh Dependent Norms, Numerische Math. 34, 41-62.

Babuška, I., W. Rheinboldt (1978) Error Estimates for Adaptive Finite Element Computations, SIAM J, Numerical Analysis 15, 736-754.

Babuška, I., W. Rheinboldt (1978) A-Posteriori Error Estimates for the Finite Element Method, Intern, Journ. of Num. Math. in Eng. 12, 1597-1615.

Babuška, I., W. Rheinboldt (1978) On a System for Adaptive Parallel Finite Element Computations in Proc. 1978 Annual Conference Association for Comp. Mach., New York, 480-489.

Babuška, I, W. Rheinboldt (1979) Analysis of Optimal Finite Element Meshes in $R^1$ Math. of Comp. 33, 435-463.

Babuška, I., W. Rheinboldt (1979) Adaptive Approaches and Reliability Estimation in Finite Element Analysis Com. Meth. in Appl. Mich. and Eng. 17/18, 519-540.

Babuška, I., W. Rheinboldt (1980) Reliable Error Estimation and Mesh Adaption for the Finite Element Method in Computational Methods in Nonlinear Mechanics ed. by J. T. Oden, North Holland Pub. Co., Amsterdam, 67-108.

Babuška, I., W. Rheinboldt (1981) A-Posteriori Error Analysis of Finite Element Solutions for One-Dimensional Problems, SIAM J Num. Anal. 18, 565-589.

Babuška, I., W. Rheinboldt, Computational Error Estimates and Adaptive Processes for Some Non Linear Structural Problems, Tech. Dep. ICMA-81-28, Inst. for Comp. Math and Applications, Dept. of Math and Statistics, University of Pittsburgh.

Babuška, I., W. G. Szymczak (1982) An Error Analysis for the Finite Element Method Applied to Convection Diffusion Problems. Tech Note BN-912, Institute for Physical Science and Technology, University of Maryland. To appear in Computer Methods in Applied Mechanics and Engineering.

Bieterman, M. (1982) Thesis, University of Maryland.

Bieterman, M., I. Babuška, The Finite Element Method for Parabolic Equations..
I A-Postiori Error Estimation Tech. Note University of Maryland, to appear.

Bieterman, M., I. Babuška, The Finite Element Method for Parabolic Elements.
II The Adaptive Mesh Selection, Tech. Note University of Maryland, to appear.

A. Miller (1981) Thesis, University of New South Wales, Brisbane, Australia.

Rheinboldt, W., C. K. Mesztenyi (1980) On a Data-Structure for Adaptive Finite Element Mesh Refinements, ACM Trans. on Math Softwear 6, 166-187.

Szymczak, W. (1982) Thesis, University of Maryland.

Zave, P., G. E. Cole, A Quantitative Evolution of the Feasibility of and Suitable Hardware Architecutre for an Adaptive, Parallel Finite Element System, University of Maryland.

Zave, P., W. Rheinboldt (1979) Design of an Adaptive Parallel Finite Element System, ACM Transaction on Math Software 5, 1-17.

# ON THE ACCURACY AND EFFICIENCY OF A FINITE ELEMENT ALGORITHM FOR HYDRODYNAMIC FLOWS

A. J. Baker and M. O. Soliman

University of Tennessee
Knoxville, TN 37996, USA

The accurate characterization of hydrodynamic flowfields in channels, streams, and river floodplanes, as well as large confined bodies of water, is an important requirement. Historically, as has occurred in almost every branch of engineering, the hydraulics laboratory served the major purpose by permitting acquisition of pertinent data on scaled models of the physical configuration. In the United States, the U.S. Army Waterways Experiment Station and TVA Norris Engineering Laboratory exemplify the concept. The emergence of the digital computer, and its phenomenal recent rate of maturation, is introducing the near-term capability to conduct numerical experiments on a mathematical model of certain engineering configurations. Aerodynamics is probably foremost in an advanced numerical analysis capability, yet even now only an elementary potential flow computer simulation can be conducted for a completely configured airplane. And while the National Laboratory wind tunnels will probably never be used to store computer output, most importantly, the cost of computer simulation is actually declining, while the physical simulation costs are steadily increasing.

These development and cost effectiveness factors have combined to initiate and nourish an ever expanding research and computer code development activity in computational methods, particularly in fluid mechanics. Pioneering progress related to numerical mathematical modeling of hydrodynamic flows were reported at the International Conference on Finite Elements in Water Resources, held at Princeton University in 1976, cf. Gray, Pinder and Brebbia, Eds (1977). The highly non-regular shorelines of natural containments provided the original impetus for application of finite element procedures to open channel flow predictions, see for example Gray (p. 4.33), Baker (p. 4.95) and King (p. 4.165) of this conference proceedings. The finite element model and code of Norton and King, originally reported by King and Norton (1975), exemplifies a procedure undergoing continued extension, as documented by King and Norton (p. 2.81) at the Second International Conference, Brebbia, Gray and Pinder, Eds. (1978). Further, other workers in the field have adopted this type numerical model concept to related configurations, cf. McAnally and Thomas (1980, p. 6.47) and Lee (1980, p. 6.3), Proceedings of the Third International Conference, Wang, Alonso, Brebbia, Gray, and Pinder, Eds. (1980). Of course, numerous numerical simulation results are reported in the open literature, as well, with the Europeans and Japanese being particularly active. In the majority of these instances, a finite element algorithm was chosen to provide the desired geometrical flexibility.

Concurrently, over the past five or so years, an extensive and comprehensive research, development, and application study of finite element numerical solution theory and methodology has been occurring in the aerodynamics. The emphasis in these studies has been keenly focused on accurate prediction of turbulent flows, and the more general theoretical areas of accuracy, convergence and efficiency. The level of mathematical sophistication has tended to be somewhat higher, since the subject areas are certainly more mature. Consequently, greater emphasis is and has been placed on critical assessment of numerical error mechanisms and their control, and in computer code efficiency and performance. The summary results of these studies have indeed confirmed the theoretical soundness of the finite element algorithm concept. In particular, its main attribute is now becoming recognized as a thoroughly formal procedure, utilizing calculus and vector field theory, for generation of optimally order-accurate numerical algorithms for classes of problem descriptions in fluid mechanics, cf. Baker (1981). In particular, the theoretical structure can yield prediction of ways to improve performance of certain finite difference algorithms. Of course, the geometry flexibility remains as an attribute, but the importance of this aspect has been rendered secondary with emergence of efficient regularizing coordinate transformation algorithms for finite difference procedures, as exemplified by the recent NASA Conference on Numerical Grid Generation Techniques, Proceedings (1981).

This paper presents the derivation, theoretical analysis, and documentary numerical results for an accurate and efficient finite element, multi-pole tensor product algorithm for prediction of free surface hydrodynamic flowfields governing by the depth-averaged shallow water equations. This equation set is highly germane to the subject. Furthermore, for flows at large Reynolds numbers, its strongly non-linear hyperbolic character places rather significant demands on algorithm construction.

## PROBLEM STATEMENT

The partial differential equation system employed for marine and certain types of estuary modeling are termed the shallow water equations. They are obtained by integrating the three-dimensional Navier-Stokes equations, for a viscous incompressible fluid, in the direction parallel to the gravity vector. Employing a time-averaged form of the Navier-Stokes equations, and permitting density variations, as might occur in saline-stratified flows for example, the governing equation system is,

Continuity:

$$L(h) = \frac{\partial h}{\partial t} + h\frac{\partial u_i}{\partial x_i} + u_i\frac{\partial h}{\partial x_i} = 0 \tag{1}$$

Momentum:

$$L(u_i) = \frac{\partial u_i}{\partial t} + u_j\frac{\partial u_i}{\partial x_j} + \frac{1}{\rho}\frac{\partial p}{\partial x_i} + \frac{\partial \sigma_{ij}}{\partial x_j} + \Omega_{ij}u_j - \tau_i\Big|_b^h = 0 \tag{2}$$

Pressure:

$$L(p) = \frac{1}{\rho}\frac{\partial p}{\partial x_i} - g\frac{\partial h}{\partial x_i} - g\frac{\partial b}{\partial x_i} - \frac{gh}{2\rho}\frac{\partial \rho}{\partial x_i} = 0 \tag{3}$$

Density Convection-Diffusion:

$$L(\rho) = \frac{\partial \rho}{\partial t} + u_i\frac{\partial \rho}{\partial x_i} - \frac{\partial}{\partial x_i}\left[k_{ij}\frac{\partial \rho}{\partial x_j}\right] = 0 \tag{4}$$

In equations 1-4, the range of the tensor index is $1 \leqslant i \leqslant 2$, $h$ is the depth of fluid above the channel bed distribution $b(x_i)$, see Figure 1, and $u_i$ is the two-dimensional depth-averaged velocity vector.

$$u_i(x_1,x_2,t) \equiv \frac{1}{h}\int_b^h \bar{u}_i(x_1,x_2,x_3,t)dx_3 \tag{5}$$

In equation 2, $\sigma_{ij} \equiv \overline{u_i'u_j'}$ is the kinematic Reynolds stress tensor, and $\Omega_{ij} = -2\omega\sin\phi\, e_{ij}$ is the Cooriolis coefficient, where $\omega$ is angular velocity, $\phi$ is latitude, and $e_{ij}$ is the alternator. Further, $\tau_i \equiv \sigma_{ij}\hat{n}_3\delta_{j3}$ is the $x_3$ normal component of shear stress, evaluated for wind shear at the free surface h, and drag at the channel bed. Representative forms for these surface stress are, cf. Brebbia and Partridge (1976),

$$\tau_i(h) = \frac{\gamma}{\rho}\frac{w_i}{h}(w_jw_j)^{\frac{1}{2}} \tag{6}$$

$$\tau_i(b) = -\frac{gu_i}{C^2h}(u_ju_j)^{\frac{1}{2}} \tag{7}$$

where $w_i$ is the wind velocity vector and C is the Chezy coefficient. Equation 4 is a species continuity equation for density, where $k_{ij}$ is the eddy viscosity tensor. (The finite element solution algorithm for equation 4 has been examined in detail cf. Baker (1980), and will not be considered further.)

For the finite element tensor product multi-pole algorithm, it is desirable to recast equations 1-3 into the conservative or divergence form. Multiplying equation 2 by h, and substituting equations 1, 3, 6, 7, and defining the new dependent variable $m_i \equiv hu_i$, the conservative form for the shallow water equation system is

$$L(h) = \frac{\partial h}{\partial t} + \frac{\partial}{\partial x_j}[m_j] = 0 \tag{8}$$

$$L(m_i) = \frac{\partial m_i}{\partial t} + \frac{\partial}{\partial x_j}[u_j m_i + \tfrac{1}{2}gh^2\delta_{ij} + h\sigma_{ij}] + hg\varepsilon_i - \sigma_{ij}\frac{\partial h}{\partial x_j} + \Omega_{ij}m_j - \frac{\gamma w_i}{\rho}(w_j w_j)^{\frac{1}{2}} + \frac{gu_i}{C^2}(u_j u_j)^{\frac{1}{2}} = 0 \tag{9}$$

In equation 9, $\varepsilon_i(\vec{x},t) \equiv \frac{\partial b}{\partial x_i}$ is the slope distribution of the channel bed. The density derivative term in equation 3 has been discarded, consistent with neglect of equation 4.

## FINITE ELEMENT ALGORITHM

Equations 8-9 are three coupled non-linear, principally hyperbolic partial differential equations written on the vector-valued dependent variable $q_\alpha(\vec{x},t) \equiv \{h, m_i, \sigma_{ij}\} = \{h, hu_1, hu_2, \sigma_{11}, \sigma_{22}, \sigma_{12}\}$. Each equation is of the form,

$$L(q_\alpha) = \frac{\partial q_\alpha}{\partial t} + \frac{\partial}{\partial x_j}[u_j q_\alpha + f_{j\alpha}] + s_\alpha = 0 \tag{10}$$

where $f_{j\alpha} = \frac{1}{2}gh + h\sigma_{ij}$, and $s_\alpha$ contains the remaining terms in equation 9. The domain of definition of equation 10 is $\Omega \equiv R^2 \times t = \{(\vec{x},t): x \in R^2 \text{ and } t \in [t_0,t)\}$. On the boundary $\partial\Omega = \partial R \times t$, the boundary condition statement for equation 10 is

$$\ell(q_\alpha) = a_1^\alpha q_\alpha + a_2^\alpha \frac{\partial q_\alpha}{\partial x_j}\hat{n}_j + a_3^\alpha = 0 \tag{11}$$

For example, on an inflow boundary, either h or $m_i$ is specified by setting $a_\alpha^2 = 0$ in equation 11. At a downstream outflow boundary, the typical boundary condition is vanishing derivative, which is generated by setting $a_1^\alpha \equiv 0 \equiv a_3^\alpha$. The differential equation statement is completed by an initial condition specification

$$q_\alpha(\vec{x},0) = q_\alpha^0(\vec{x}) \tag{12}$$

For the finite element solution algorithm for equations 10-12, the semi-discrete approximation $q_\alpha^h(\vec{x},t)$ to the dependent variable set $q_\alpha$ is constructed from members of a finite dimensional subspace of $H_0^1(R)$, the Hilbert space of all functions possessing square integrable first derivatives and satisfying equation 11. Hence,

$$q_\alpha(\vec{x},t) \approx q_\alpha^h(\vec{x},t) \equiv \sum_{e=1}^{M} q_\alpha^e(\vec{x},t) \tag{13}$$

The finite dimensional subspace chosen is typically polynomials in the cardinal basis form,

$$q_\alpha^e(\vec{x},t) \equiv \{N_k(\vec{\eta})\}^T\{QI(t)\}_e \tag{14}$$

In equation 14, the semi-discrete free index "I" denotes $q_\alpha^h$ evaluated at the nodal coordinates $\vec{x}_i$ of the discretization $UR^2 = R^2$ at anytime t. The sub- and super-script e denotes pertaining to the finite element domain $\Omega_e \equiv R_e^2 \times t$. The elements of the row matrix $\{N_k(\vec{\eta})\}$ are polynomials, complete to degree k, and written on the tensor product coordinate system $\vec{\eta} = \vec{\eta}(x_i,\bar{x}_i)$ spanning $R_e^2$.

To render the error in $q_\alpha^h$ extremum, the finite element algorithm for equations 10-11 requires the semi-discrete approximation error, $L(q_\alpha^h)$ and $\ell(q_\alpha^h)$, to be orthogonal to the cardinal basis employed to construct $q_\alpha^h$. Further, a multi-pole expansion is invoked to constrain the derivative of the error, ie., $\nabla L(q_\alpha^h)$, since high frequency dispersion is a dominant error mechanism and $L(q_\alpha^h)$ is therefore locally non-smooth. These constraints are combined, on the assembly $(S_e)$ of the finite element algorithm statement over $UR_e^2 = R^2$, by identification of the (Lagrange) multiplier set $\beta_i$, yielding

$$S_e\left[\int_{R_e^2}\{N_k\}L(q_\alpha^h)d\vec{x} + \vec{\beta}_1\cdot\int_{R_e^2}\{N_k\}\nabla L(q_\alpha^h)d\vec{x} + \beta_2\int_{\partial R_e \cap \partial R}\{N_k\}\ell(q_\alpha^h)d\vec{x}\right] \equiv \{0\} \tag{15}$$

Upon definition of k in equation 14, equation 15 is a coupled non-linear system of ordinary differential equations of the form

$$[C]\{QI\}' + [U]\{QI\} + [FIJ]\{QJ\} + \{SI\} = \{0\} \tag{16}$$

Since the [C] matrix is non-diagonal, the Θ-implicit finite difference algorithm, where $\Theta = \frac{1}{2}$ yields the trapezoidal rule, is

$$\{FI\} = \{QI\}_{j+1} - \{QI\}_j - \Delta t\left[\Theta\{QI\}'_{j+1} + (1-\Theta)\{QI\}'_j\right] = 0 \quad (17)$$

Equation 16 provides the definition of the derivative $\{QI\}'$. Upon substitution, equation 17 becomes a non-linear algebraic equation system written on $\{QI\}_{j+1}$. The Newton matrix iteration algorithm is

$$[J(FI)]^p_{j+1} \ \{\delta QI\}^{p+1}_{j+1} = - \{FI\}^p_{j+1} \quad (18)$$

The dependent variable $\{\delta QI\}$ yields the solution

$$\{QI\}^{p+1}_{j+1} = \{QI\}^p_{j+1} + \{\delta QI\}^{p+1}_{j+1} \quad (19)$$

The Jacobian for equation 18 is defined as,

$$[J(FI)] \equiv \frac{\partial\{FI\}}{\partial\{QJ\}} \quad (20)$$

Generalized Coordinates

A principal requirement for free surface flow prediction is to accurately interpolate global domain boundaries $\partial\Omega$ that are non-regular. The versatility of the triangular finite element for forming such discretization of $R^2$, was an early impetus to generation of finite element algorithms. An alternative procedure, that can yield significant solution economies when employing an implicit solution algorithm, is definition and use of "boundary fitted" coordinate transformations. The use of such procedures is commonplace in computational aerodynamics, cf., the Proceeding of the NASA Conference (1981).

The basic concept is to generate a coordinate transformation,

$$x_i = x_i(\eta_j) \quad (21)$$

wherein coordinate surfaces of $\eta_j$ are defined coincident with appropriate segments of $\partial R$. For example, Figure 2 from Johnson and Thompson (1978), is a graph in physical space of the coordinate system $\eta_j$ that interpolates the boundary $\partial R$ of a segment of the Charleston, N.C., harbor, and the illustrated discretization. If this discretization were plotted in the transform space spanned by $\eta_j$, it would appear as the assembly of regular rectangular domains illustrated in Figure 3. In this domain, the $\eta_i$ coordinate system is (ideally) rectangular Cartesian, and the discretization uniform. In comparison, the $\eta_1$ coordinate runs nominally parallel to the river flow direction, $\eta_2$ spans across the channels and the river mouth, and the island is a "hole" in the mesh.

The concept of the body-fitted coordinate system fits in a natural way into the finite element algorithm statement. Each quadrilateral domain $R_e^2$ on $R^2$, Figure 2, can be mapped to the unit square using an isoparametric interpolation, equation 14. Specifically, the local form of equation 21 on $R_e^2$ is

$$x_i = \{N_k(\vec{\eta})\}^T\{XI\}_e \ , \quad x_i \in R_e^2 \tag{22}$$

The elements of the column matrix $\{XI\}_e$ are the coordinates $\bar{x}_i$ of the nodes of $R_e^2$, and for example, the bilinear cardinal basis is

$$\{N_1^+(\vec{\eta})\} = \frac{1}{4}\begin{Bmatrix}(1-\eta_1)(1-\eta_2)\\(1+\eta_1)(1-\eta_2)\\(1+\eta_1)(1+\eta_2)\\(1-\eta_1)(1+\eta_2)\end{Bmatrix} \tag{23}$$

The finite element algorithm statement, equation 15, is the assembly of integrals over $R_e^2$, each of which requires formation of the divergence of a vector field, see equations 8-9. The computational requirement therefore is to transform the local divergence operator as

$$\frac{\partial}{\partial x_j} = \frac{\partial}{\partial \eta_k}(\cdot)\left[\frac{\partial \eta_k}{\partial x_j}\right]_e \tag{24}$$

The transformation matrix $[\cdot]_e$ is readily determined, using equation 22-23, as

$$\left[\frac{\partial \eta_k}{\partial x_j}\right]_e = \frac{1}{\det J_e}\ [\text{Cofactor } J]_e^T \tag{25}$$

where

$$[J]_e = \left[\frac{\partial x_i}{\partial \eta_j}\right]_e \tag{26}$$

Also, using equation 26, the differential element in equation 15 becomes

$$d\vec{x} = \det[J]_e\ d\vec{\eta} \tag{27}$$

For the bilinear cardinal basis, equation 23, and assuming the elements of $[\partial\eta_k/\partial x_j]_e$ are interpolated on $R_e^2$ using equation 14, yields

$$\left[\frac{\partial \eta_k}{\partial x_j}\right]_e \equiv \frac{1}{\det[J]_e}\ \{N_1^+(\vec{\eta})\}^T\{ETAKJ\}_e \tag{28}$$

It is easy to show that

$$\{ETAKJ\}_e = \frac{1}{2}\left\{\begin{array}{l}\{Y4-Y1,\ Y3-Y2,\ Y3-Y2,\ Y4-Y1\}_e,\ (1,1)\\ \{X1-X4,\ X2-X3,\ X2-X3,\ X1-X4\}_e,\ (1,2)\\ \{Y1-Y2,\ Y1-Y2,\ Y4-Y3,\ Y4-Y3\}_e,\ (2,1)\\ \{X2-X1,\ X2-X1,\ X3-X4,\ X3-X4\}_e,\ (2,2)\end{array}\right\} \qquad (29)$$

In equation 29, the indices in parenthesis indicate (K,J), and the entries XI, YI are the nodal coordinates $\bar{x}_i$ of $R^2_e$ for the numbering sequence shown in Figure 4. By the same token, for

$$\det[J]_e \equiv \{N_1^+(\vec{\eta})\}^T\{DET\}_e \qquad (30)$$

the elements of $\{DET\}_e$ are

$$\{DET\}_e = \frac{1}{4}\left\{\begin{array}{l}(X2-X1)(Y4-Y1)\ -\ (X4-X1)(Y2-Y1)\\ (X2-X1)(Y3-Y2)\ -\ (X3-X2)(Y2-Y1)\\ (X3-X4)(Y3-Y2)\ -\ (X3-X2)(Y3-Y4)\\ (X3-X4)(Y4-Y1)\ -\ (X4-X1)(Y3-Y4)\end{array}\right\}_e \qquad (31)$$

Therefore, the local coordinate transformation is uniquely defined by the node coordinates of $R^2_e$.

The next step is to transform the derivative $\partial/\partial x_i$ in equations 8-9, after an integration by parts, and to perform the integrals in equation 15. Using equation 8 as the example, and defining the bi-linear element formulation, k=1 in equation 14, the first term in equation 15 becomes

$$\int_{R^2_e}\{N_1^+\}L(q_1^h)d\vec{x} = \int_{R^2_e}\{N_1\}\left(\frac{\partial h}{\partial t} + \frac{\partial}{\partial x_j}m_j^h\right)d\vec{x}$$

$$= \int_{R^2_e}\{N_1\}\frac{\partial h^h}{\partial t}dx - \int_{R^2_e}\frac{\partial\{N_1\}}{\partial x_j}m_j^h\,d\vec{x} + \oint_{\partial R_e}\{N_1\}m_j^h\cdot\hat{n}_j d\vec{\sigma} \qquad (32)$$

The last term vanishes identically, upon assembly over all interior boundaries $\partial R_e$ of the discretization. Substituting equations 14, 24, 28, and 30 yields,

$$\int_{R^2_e}\{N_1\}L(q_1^h)d\vec{x} = \int_{R^2_e}\{N\}\{N\}^T\{H\}'_e\{DET\}^T_e\{N\}d\vec{\eta}$$

$$- \int_{R^2_e}\frac{\partial}{\partial\eta_k}\{N\}\{MJ\}^T_e\{N\}\{N\}^T\{ETAKJ\}_e\,d\vec{\eta}$$

$$+ \int_{\partial R_e\cap\partial R}\{N\}\{N\}^T\{MJ\}_e\,\hat{n}_j d\vec{\sigma} \qquad (33)$$

Rearranging the order of scalar terms, and extracting the element dependent matrices $\{\cdot\}_e$ from the integrals, since they possess no $\vec{\eta}$ dependence, yields

$$\int_{R_e^2} \{N_1^+\} L(q_1^h) d\vec{x} = \{DET\}_e^T \int_{R_e^2} \{N\}\{N\}\{N\}^T d\vec{\eta}\{H\}'$$

$$- \{ETAKJ\}_e^T \int_{R_e^2} \{N\} \frac{\partial}{\partial \eta_k} \{N\}\{N\}^T d\vec{\eta}\{MJ\}_e$$

$$+ \int_{\partial R_e \cap \partial R} \{N\}\{N\}^T d\vec{\sigma}\,\{MJ\}_e \cdot \hat{n}_j \tag{34}$$

Each of the integrals in equation 34 now involves products (only) of elements of the cardinal basis $\{N_1^+\}$ and the first derivative. These are easily evaluated using numerical quadrature, and are universally valid for each element $R_e^2$ of the discretization. Noting that $d\vec{\sigma} = \Delta_e d\eta$ on the boundary intersection $\partial R_e \cap \partial R$, and using hypermatrix notation, cf. Baker (1981), the final form for equation 34 is

$$\int_{R_e^2} \{N_1^+\} L(q_1^h) d\vec{x} = \{DET\}_e^T [B3000]\{H\}_e' - \{ETAKJ\}_e^T [B30K0]\{MJ\}_e$$

$$+ \hat{n}_J \, \Delta_e \, [A200]\{MJ\}_e \tag{35}$$

In equation 35, [B3000] is the two-dimensional matrix equivalent of interpolation of two variables on $R_e^2$, [B30K0] is the equivalent of the directional derivative (gradient) parallel to $\eta_k$, and [A200] is a one-dimensional interpolation on the global boundary $\partial R$ with unit normal vector $\hat{n}_J$. The range of the embedded tensor indices is $1 \leqslant (K,J) \leqslant 2$.

Inserting equation 24 in the momentum flux equation 9, leads to definition of the contravariant components $\bar{u}_k$ of the convection velocity field $u_j$ as,

$$\bar{u}_k \equiv \det[J]_e \left[\frac{\partial \eta_k}{\partial x_j}\right]_e u_j \tag{36}$$

Using equations 25, 28-29, the finite element nodal approximation for equation 36 on $R_e^2$ becomes,

$$\{UBARK\}_e = u_j \{ETAKJ\}_e \tag{37}$$

where $u_j = m_j/h_j$ is the convection velocity at the nodes of $R_e^2$. With equation 37, and intergration by parts, the first term of the finite element algorithm statement for equation 9, becomes

$$\int_{R_e^2} \{N_1\} L(m_i^h) d\tau = \int_{R_e^2} \{N_1\} \left( \frac{\partial m_i^h}{\partial t} + \frac{\partial}{\partial x_j} (u_j^h m_j^h + \tfrac{1}{2} g h^2 \delta_{ij} + h\sigma_{ij}^h) + s_i^h \right) d\vec{x}$$

$$= \{DET\}_e^T [B3000] \{MI\}_e' - \{UBARK\}_e^T [B30K0] \{MI\}_e$$

$$- \frac{g}{2} \{ETAKI\}_e^T \left( [B40K00] \{H\}_e \right) \{H\}_e$$

$$+ \{ETAKJ\}_e^T \left( [B40K00] \{H\}_e \right) \{SIGIJ\}_e$$

$$+ \{DET\}_e^T [B3000] \{SI\}_e$$

$$+ \int_{\partial R_e \cap \partial R} \{N_1\} \left( u_j^h m_i + \tfrac{1}{2} g h^2 \delta_{ij} \right) \hat{n}_j \, d\vec{\sigma} \tag{38}$$

In equation 38, [B40K00] is a hypermatrix of degree two, which accounts for the added interpolation of $\{H\}_e$ within the directional derivative equivalent of $\partial/\partial\eta_k$. As occurred in equation 35, the closed surface integral on $\partial R_e$ vanishes identically upon assembly except for the occurrence $\partial R_e \cap \partial R \neq 0$, whereupon the last term in equation 38 may contain non-zero entries dependent upon the boundary conditions.

Multi-pole Tensor Product Algorithm Statement

The basic construction aspects of the generalized coordinates formulation are introduced. The second term requires definition of the multi-pole expansion coefficient $\vec{\beta}_1$. Based upon the analysis of Baker (1981), for the compressible flow Navier-Stokes equations, the form is

$$\vec{\beta}_{1e}^{\alpha} = \nu \det[J]_e \; [\vec{\nu}_{\alpha}^{t} + \vec{\nu}_{\alpha}^{x}] \tag{39}$$

Here, $\det[J]_e$ is the element characteristic measure, and $\nu \equiv 1/\sqrt{15}$ yields six-order phase accuracy for the $\{N_1^\dagger\}$ algorithm statement for the free-surface equation 8. The non-dimensional coefficients $\vec{\nu}_{\alpha}^{\gamma}$ are bounded about unity, eg., $0 < \nu_{\alpha}^{t} < 2$, and permit fine tuning of the dissipation mechanisms. The superscripts (t,x) constrain use to the time and space derivatives in equations 8-9, respectively.

With equation 39, and assuming for simplicity that the stress tensor terms in equation 9 are higher order, hence negligible, the multi-pole finite element algorithm statement for equations 8-9 inserted in equation 15 can be completed. Noting that for equation 17, $\{FI\} \equiv S_e\{FI\}_e$, $1 < I < 3$, and letting $\{QI\} \equiv \{H, M1, M2\} = \{H, HU1, HU2\}$, and defining $\Theta \equiv \tfrac{1}{2}$, the finite element algorithm statements on $R_e^2$ are

$$\{FH\}_e = \left(\{DET\}_e^T[B3000] + \nu\nu_{HJ}^t\{ETAKJ\}_e^T[B40K00]\{DET\}_e\right)\{H\}'_{j+1}$$

$$+ \frac{\Delta t}{2}\left(-\{ETAKJ\}_e^T[B30K0]\{MI\}_e\right.$$

$$\left. + \nu\nu_{HJ}^x\{ETAKJ\}_e^T[B40KL0]\{UBARL\}_e\{H\}_e\right)_{j+1,j} \qquad (40)$$

$$\{FMI\}_e = \left(\{DET\}_e^T[B3000] + \nu\nu_{IJ}^t\{ETAKJ\}_e^T[B40K00]\{DET\}_e\right)\{MI\}'_{j+1}$$

$$+ \frac{\Delta t}{2}\left(-\{UBARK\}_e^T[B30K0]\{MI\}_e - \tfrac{1}{2}g\{ETAKI\}_e^T([B40K00]\{H\}_e)\{H\}_e\right.$$

$$+ g\{ETAKI\}_e^T([B400K0]\{H\}_e)\{B\}_e + \{DET\}_e^T[B3000]\{SI\}_e$$

$$\left. + \nu\nu_{IJ}^x\{ETAKJ\}_e^T([B40KL0]\{UBARL\}_e)\{MI\}\right)_{j+1,j} \qquad (41)$$

The second and last terms in equations 40 and 41 are the multipole terms, and the tensor indices J and (I,J) appearing throughout have range $1 \leq (I,J) \leq 2$. For equation 41, the fifth term is explicitly the channel bed gradient term, $\varepsilon_i = \partial b/\partial x_i$, and the sixth contains the remainder of the source terms in equation 9. The notation $\{\cdot\}'_{j+1}$ denotes $\{QI\}^p_{j+1} - \{QI\}_j$, and $(\cdot)_{j+1,j}$ defines the evaluation $(\cdot)^p_{j+1} + (\cdot)_j$. Additional detail on notational structure is given by Baker (1981).

The final formulation step is to construct the Newton algorithm Jacobian, equation 20. Equations 40-41 can be analytically differentiated with respect to {QI}, since all functional dependence is explicitly detailed. For the considerable solution economy it provides, the analytical Jacobian is replaced by its matrix tensor product, ie.,

$$[J(QI)] \Rightarrow [J_1] \otimes [J_2] \qquad (42)$$

Each of the matrices $[J_\eta]$ exhibits an $\alpha$-block tridiagonal structure for the $\{N_1^+\}$ basis form of the algorithm, and Baker (1981) presents complete derivational detail. For the shallow water equation statement 40-41, and recalling by definition,

$$[J_\eta] \equiv S_e[J_\eta]_e \qquad (43)$$

the self-coupling and cross-coupling terms in the typical tensor matrix product Jacobian are

$$\frac{\partial\{FI\}_e}{\partial\{QI\}_e} = \frac{\partial\{FH\}_e}{\partial\{H\}_e} \equiv [JHH]_e = \Big[\{DETK\}_e^T[A3000] + \nu\nu_{HJ}^t\{ETAKJ\}_e^T[A40K00]\{DETK\}_e + \frac{\Delta t}{2}\,\nu\nu_{HJ}^x\{ETAKJ\}_e^T[A40KL0]\{UBARL\}\Big] \quad (44)$$

$$\frac{\partial\{FH\}_e}{\partial\{MI\}_e} \equiv [JHMI]_e = \frac{\Delta t}{2}\Big[-\{ETAKI\}_e^T[A30K0]\Big] \quad (45)$$

$$\frac{\partial\{FMI\}_e}{\partial\{H\}_e} \equiv [JMIH]_e = -\frac{\Delta t}{2}\Big[\left(\frac{\bar{m}}{\bar{h}^2}\right)\{MI\}_e^T[A30K0] + g\{ETAKJ\}_e^T([A40K00]\{H\}_e) - g\{ETAKI\}_e^T([A4000K]\{B\}_e)\Big] \quad (46)$$

$$\frac{\partial\{FMI\}_e}{\partial\{MI\}_e} \equiv [JMIMI]_e = \{DETK\}_e^T[A3000] + \nu\nu_{HJ}^t\{ETAKJ\}_e^T[A40K00]\{DETK\}_e + \frac{\Delta t}{2}\Big[-\{UBARK\}_e^T[A30K0] + \nu\nu_{IJ}^x\{ETAKJ\}_e^T([A40KL0]\{UBARL\}_e\Big] \quad (47)$$

$$\frac{\partial\{FMI\}}{\partial\{MJ\}} \equiv [JMIMJ]_e = \frac{\Delta t}{2}\left(\frac{1}{\bar{h}}\right)\Big[-\{MJ\}_e^T[A30K0] + \nu\nu_{IJ}^x\{MJ\}_e^T[A40KK00]\{ETAKL\}\Big]\bar{D}_{KJ} \quad (48)$$

The entirety of the algorithm standard matrices [A•••] are presented in Baker (1981, Appendix A). The tensor index K in equations 44-48 is not eligible for summation, but instead now denotes the matrix component parallel to the grid sweep direction $\eta_k$. The components of $\{ETAKJ\}_e$ are the direction cosines of the affine transformation $\eta_i = a_{ij}x_j$ at the node coordinates of $R_e^1$.

The operation of the tensor matrix form of the Newton algorithm is as follows. For the orientation in transform space, see Figure 3, the grid lines $\eta_k$ form a Cartesian network. The tensor product Newton algorithm sweeps the mesh parallel to lines $\eta_k$ = constant. Hence, two grid sweeps are required to generate $\{\delta QI\}_{j+1}^{p+1}$, equation 18. Defining the intermediate solution matrix {PI}, the solution sequence is

$$[J_1]\{PI\} = -\{FI\} \quad (49)$$

$$[J_2]\{\delta QI\} \equiv \{PI\} \quad (50)$$

The three-block triadiagonal Jacobian for equation 49 is

$$S_e \begin{bmatrix} [JHH]_e & [JHM1]_e & 0 \\ [JM1H]_e & [JM1M1]_e & 0 \\ [JM1M2]_e & [JM2M1]_e & [JM2M2]_e \end{bmatrix} \{PI\} = - \{FI\} \tag{51}$$

The second algorithm step, equation 50, is

$$S_e \begin{bmatrix} [JHH]_e & 0 & [JMH2]_e \\ [JM1H]_e & [JM1M1]_e & [JM1M2]_e \\ [JM2H]_e & 0 & [JM2M2]_e \end{bmatrix} \{\delta QI\} = \{PI\} \tag{52}$$

## DISCUSSION AND RESULTS

The principal requirement is to assess the accuracy and stability of the finite element multi-pole algorithm for the basic inviscid shallow water equation system. An elementary test is to determine the steady-state solution for flow in a uniform cross-section channel with an abrupt change in bed slope. For a step of positive slope, and dependent upon whether the upstream flow is super- or sub-critical, where $h_{cr} = \sqrt[3]{m_1^2/g}$, the velocity and free surface height will accordingly change abruptly. Figure 5 summarizes the algorithm steady-state prediction for h and $u_1 = m_1/h$, for both the subcritical and supercritical upstream flow specification. The bed profile is also graphed for comparison. Both solution sets are crisp and devoid of excess artificial diffusion using the theoretical specification for $\vec{\nu}_\alpha = \nu\{1,1\}$ where $\nu = 1/\sqrt{15}$. A very modest "2Δx" wave is evidenced in the upstream free surface height for the subcritical specification. Both solutions specified $m_1$ at inlet, and employed vanishing normal derivatives for h at inlet and both h and $m_1$ at the outlet.

The second test is subcritical flow from a reservoir onto a broad-crested weir and down a spillway of sufficient slope to cause the flow to go super critical. The foot of the spillway is given a smooth transition to horizontal, whereupon a geometry change is imposed to induce a hydraulic jump to subcritical flow again. The flow from the reservoir is assumed to occur through a given height opening; therefore, h is specified and $m_1$ is unknown, hence given a vanishing derivative boundary condition. The downstream boundary condition is vanishing normal derivative for both h and $m_1$. Figure 6a graphs the steady-state free surface height distribution without hydraulic jump. Figure 6b is the comparison case with hydraulic jump, which is computed to occur over about 3-4 mesh intervals. A modest "2Δx" wave occurs on the downstream side of the jump, as obtained using the theoretical values for $\vec{\nu}_\alpha$. Figure 6c-d are the corresponding computed velocity distributions, which clearly illustrate the sharp decrease in speed occurring over about 3-4 mesh intervals for the super-critical case. A rather pronounced velocity overshoot is predicted on the upstream side, for $\vec{v}_\alpha$ at theoretical levels, but overall the solution appears crisp.

The final test case is subcritical flow in a two-dimensional channel with bed profile. Figure 7 illustrates the computed free-surface, and channel velocity distributions, occurring for the illustrated saw-tooth bed profile. This two-dimensional case employed about 250 finite elements with three-dependent variables and seven pieces of metric data per node yielding about 2500 nodal variables. The solution required only 40 K words of central memory for execution, and during the transient the Newton algorithm iterative convergence rate was just lower than quadratic.

## CONCLUDING REMARKS

A finite element numerical solution algorithm for the shallow water equations has been established. The theoretical framework is a multi-pole expansion of the basic Galerkin-Weighted Rsiduals statement, yielding highly selective phase error control for the limit of large Reynolds number flows. Using an implicit integration technique, the resultant non-linear algebraic equation system is solved using the tensor matrix product approximation to the Newton iteration algorithm Jacobian. This approximation is readily formed using the concept of boundary fitted coordinate transformations, yielding a completely general geometry capability. Example numerical solutions are discussed documenting aspects of accuracy, efficiency and versatility of the finite element algorithm.

## ACKNOWLEDGEMENTS

The authors wish to acknowledge the excellent research atmosphere provided by the University of Tennessee, Department of Engineering Science and Mechanics, and the generosity of the University of Tennessee Computing Center in providing computer resources.

## REFERENCES

—— (1981) Proceedings, Numerical Grid Generation Techniques, NASA Conference Publication 2166.

Baker, A. J. (1981) Research on Numerical Algorithms For The Three-Dimensional Navier-Stokes Equations, II. Dissipative Finite Element. USAF Technical Report AFWAL-TR-80--3157.

Baker, A. J. and Soliman, M. O. (1980) Analysis of a Finite Element Algorithm For Numerical Predictions In Water Resources Resources Research. Proceedings, Finite Elements In Water Resources III, Rose Printing, Tallahassee, 1.41.

Baker, A. J. (1978) The Finite Element Method Applied to Flow Field Predictions in Environmental Hydrodynamics. Proceedings, Finite Elements In Water Resources I, Pentech Press, London, 4.95.

Brebbia, C. A. and Partridge, P. W. (1976) Finite Element Simulation of Water Circulation in the North Sea. J. Appl. Math Modeling, V. 1, 101.

Gray, W. G. (1977) An Efficient Finite Element Scheme For Two-Dimensional Surface Water Computation. Proceedings, Finite Elements In Water Resources I, Pentech Press, London, 4.33.

Johnson, B. H. and Thompson, J. F. (1978) A Discussion of Boundary-Fitted Coordinate Systems and Their Applicability to the Numerical Modeling of Hydraulic Problems. U.S. Army Misc. Paper H-78-9, WES, Vicksburg, MS.

King, I. P. and Norton, W. R. (1978) Recent Application of RMA's Finite Elemnt Models For Two-Dimensional Hydrodynamics and Water Quality. Proceedings, Finite Elements In Water Resources II, Pentech Press, London, 2.81.

King, I. P. (1977) Finite Element Models For Unsteady Flow Routing Through Irregular Channels. Proceedings, Finite Elements In Water Resources I, Pentech Press, London, 4.165

King, I. P., Norton, W. R. and Iceman, K. R. (1975) A Finite Element Solution For Two-Dimensional Stratified Flow Problems. Chapter 7 in Finite Elements In Fluids, Wiley, London, V. I, 133.

Lee, J. K. (1980) Two-Dimensional Finite Element Analysis of the Hydraulic Effect of Highway Bridge Fills in a Complex Flood Plane. Proceedings, Finite Elements In Water Resources III, Rose Printing, Tallahassee, 6.3

McAnally, W. H. and Thomas, W. A. (1980) Finite Element Models In a Hybrid Model Study of Estuarine Sedimentation. Proceedings, Finite Element In Water Resources III, Rose Printing, Tallahassee, 6.47.

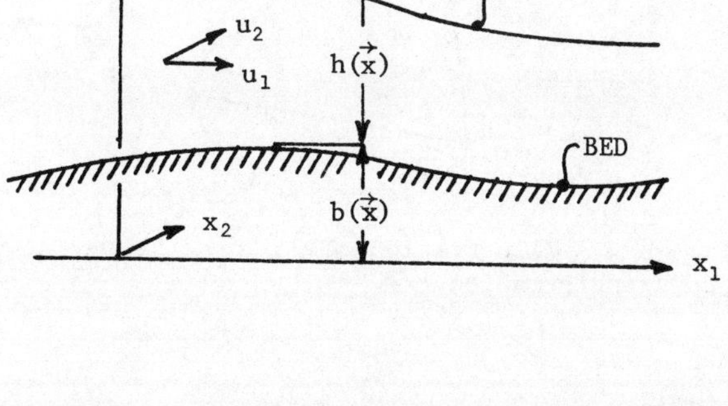


Figure 1. Geometry and Coordinate System

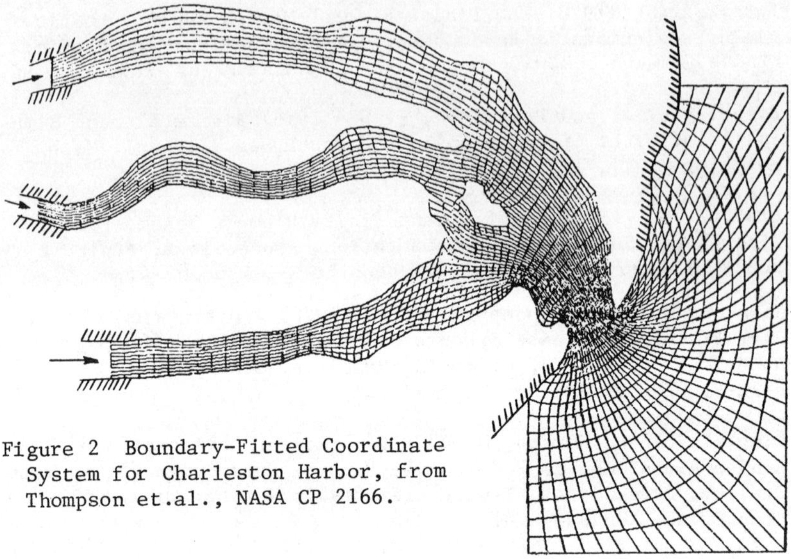

Figure 2 Boundary-Fitted Coordinate System for Charleston Harbor, from Thompson et.al., NASA CP 2166.

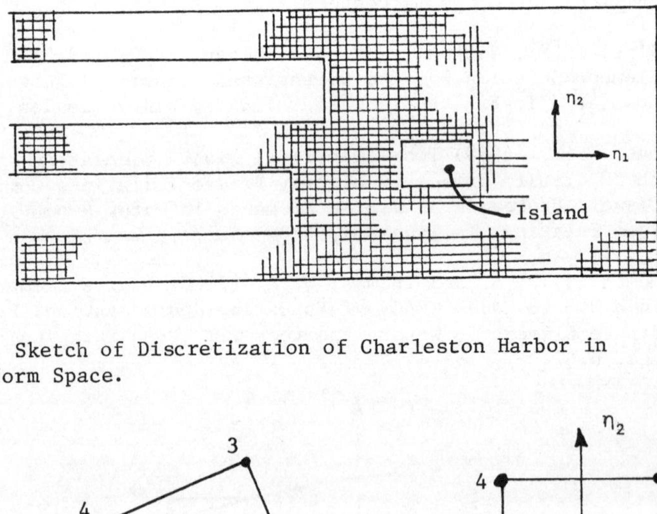


Figure 3. Sketch of Discretization of Charleston Harbor in Transform Space.

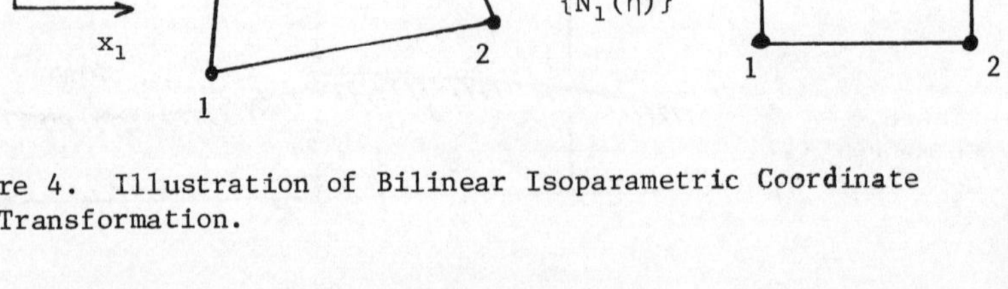


Figure 4. Illustration of Bilinear Isoparametric Coordinate Transformation.

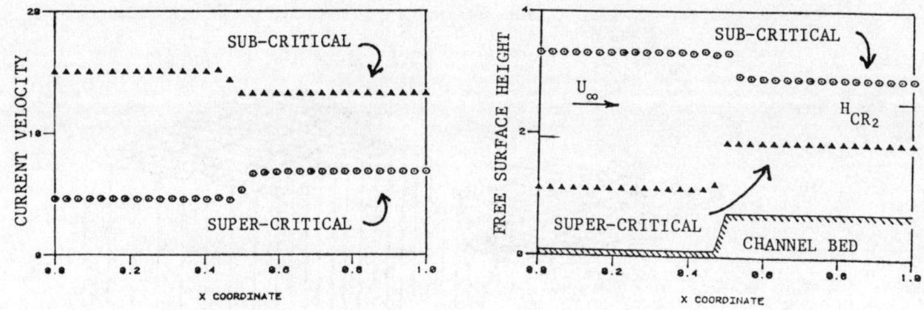


Figure 5 Computed Steady State Velocity and Free Surface Distribution, One-Dimensional Channel Flow With Step.

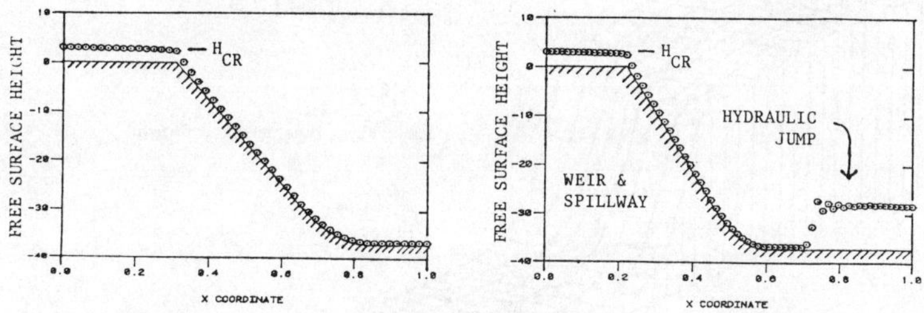


a) Free Surface, Without Hydraulic Jump

b) Free Surface With Hydraulic Jump

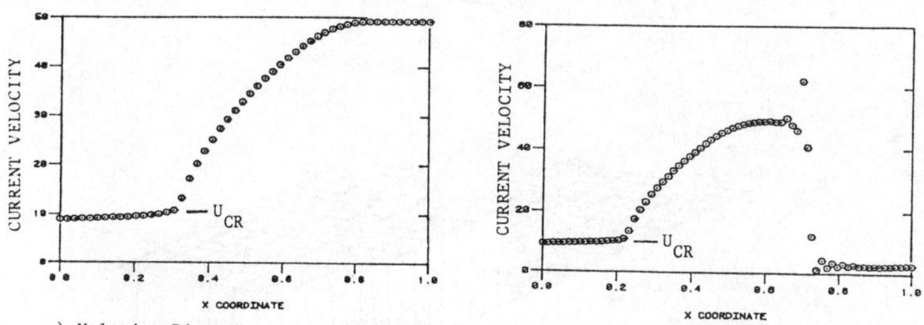


c) Velocity Distribution, No Hydraulic Jump

d) Velocity Distribution With Hydraulic Jump

Figure 6 Computed Steady-State Free Surface and Velocity Distribution, Broad Crested Weir With Spillway.

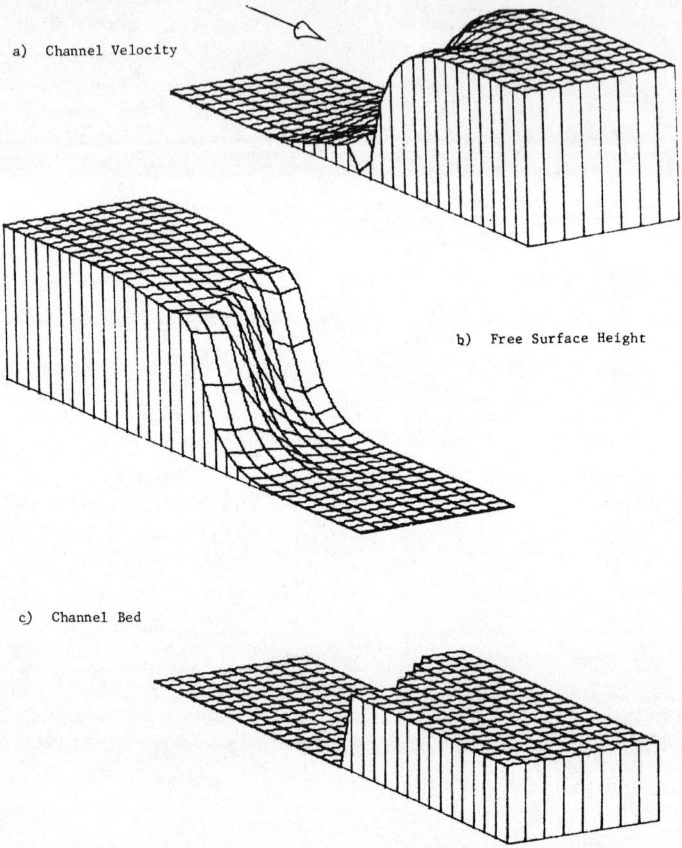


Figure 7 Computed Velocity and Free Surface Distributions, Two-Dimensional Channel Flow with Bed Profile.

# IMPLEMENTATION OF A DIRECT METHOD FOR THE BIHARMONIC PROBLEM IN THE CONTEXT OF A CODE USING SUBSTRUCTURATION

A. Bossavit

Electricité de France, Clamart, France.

## INTRODUCTION

The biharmonic problem

$$\Delta^2\psi = 0 \quad \text{in } \Omega \qquad \psi_{|\Gamma} = h, \quad \partial\psi/\partial n = g \text{ on } \Gamma, \tag{1}$$

(for given g and h on the boundary Γ of a domain Ω of the plane) is familiar in fluid dynamics. Solving it is often part of the more general problem of solving Navier-Stokes equations in two dimensions.

Many investigators have been tempted by the idea of solving Equation 1 with tools originally designed for the Poisson problem (operator Δ). See for instance Pal'tsev (1966), Smith (1968), Bossavit (1971).

This paper is concerned with implementation problems, with the "substructure" approach in mind. Its purpose is to show that once the basic pieces of a finite element code for ordinary Poisson problems have been elaborated, they can be combined together to yield a code for Problem 1. More, the cost of solving Equation 1, in terms of computer time, is only marginally superior to the cost of solving a Poisson problem.

The basic idea is stated in Section 1. Its discrete counterpart is given in Section 2. Section 3 deals with substructuration. The same symbols (R, K, ...) will often be used for operators or functions and for their discrete approximations. This confusion, when consciously done, is feeled to be more helpful than harmful.

1 THE BIHARMONIC ADMITTANCE

To solve Equation 1, let us set $\Delta\psi = \omega$, thence

$$\left|\begin{array}{l} -\Delta\psi = \omega, \\ \psi_{|\Gamma} = h, \\ \partial\psi/\partial n = g. \end{array}\right. \qquad \left|\begin{array}{l} -\Delta\omega = 0, \\ \\ \\ \end{array}\right. \tag{2}$$

This system is unbalanced : the $\omega$ problem lacks a boundary condition, the $\psi$ problem has too many. If the trace u of $\omega$ on $\Gamma$ were known, one could solve sequentially, first

$$-\Delta\omega = 0, \quad \omega_{|\Gamma} = u, \tag{3}$$

then

$$-\Delta\psi = \omega, \quad \psi_{|\Gamma} = h. \tag{4}$$

The unknown u should be found using the remaining data, i.e. the Neumann condition,

$$\partial\psi/\partial n = g. \tag{5}$$

A first idea is to find u iteratively, by a loop for which Equation 5 is taken as exit condition (when satisfied to some prescribed accuracy), the loop-invariant being equalities 3 and 4. The structure of the code would be :

*$h \leftarrow h_0$ {affect to h some starting value $h_0$}.*

*<u>while</u> (5) is not achieved <u>do</u>*

> *use $\partial\psi/\partial n - g$ to alter u ;*
>
> *solve the system (3)(4)*

There are many ways to "alter u". Whatever the choice made (this is where the proposed methods differ), the net result is that Equation 1 can be solved through a succession of Poisson problems.

Another idea leads to a direct method. Let us introduce two boundary operators as follows. Let $\phi$ be the solution of

$$-\Delta\phi = 0, \quad \phi_{|\Gamma} = w, \tag{7}$$

and $\psi$ the solution of

$$-\Delta\psi = \omega, \quad \psi_{|\Gamma} = 0, \qquad -\Delta\omega = 0, \quad \omega_{|\Gamma} = u. \tag{8}$$

Define

$$R\,w = \partial\phi/\partial n \qquad (9)$$

$$K\,u = -\,\partial\psi/\partial n \qquad (10)$$

Now the relation between the data g and h and the unknown u is

$$Ku + g = Rh \qquad (11)$$

So, by first solving Equation 11, then System 3-4, one achieves the desired result. Again two Poisson problems are to be solved.

Notice that $R \in \mathcal{L}(H^{1/2}(\Gamma), H^{-1/2}(\Gamma))$ and $K \in \mathcal{L}(H^{-1/2}, H^{1/2})$. These are isomorphisms (for R, up to constants). Useful properties of R and K may be derived from the equalities

$$(Rw, w) = \int_\Omega \mathrm{grad}^2\psi \qquad (12)$$

$$(Ku, u) = \int_\Omega \omega^2 \qquad (13)$$

where ( , ) is the scalar product in $L^2(\Gamma)$. Operator R will be referred to as the "harmonic stiffness" and K is the "biharmonic admittance". (A close analog to K is used in Glowinski-Pironneau (1978).)

Let us now show that the same discretizing elements which are used to solve Equation 3 or 4 allow one to compute R and K without extra computational effort.

## 2 DISCRETIZATION

<u>The concept of "approximation scheme" revisited</u>

Whatever the approximation scheme one chooses, it establishes a correspondence between vectors in $R^n$ and functions in $H^1(\Omega)$, and between some bilinear forms on $H^1$ and bilinear forms on $R^n$. One can always suppose (changing the basis if necessary) that functions of trace zero on $\Gamma$ correspond to vectors in the subspace of the n-m first components, for some $m < n$. (This renumbering is assumed here for clarity, but is not compulsory in practice.)

The correspondence between bilinear forms and matrices is as follows (the partition corresponds to the n-m inner nodes and to the m boundary nodes, distinguished by the use of subscripts 1 and 2 respectively):

$$\int_\Omega \mathrm{grad}\,u \,.\, \mathrm{grad}\,v \quad \text{-----} \quad \begin{vmatrix} A & B^t \\ B & C \end{vmatrix}$$

$$\int_\Omega u\,v \quad \text{-----} \quad \begin{vmatrix} D & 0 \\ 0 & E \end{vmatrix}$$

$$\int_\Omega uv \qquad \text{-----} \qquad \begin{vmatrix} 0 & 0 \\ 0 & F \end{vmatrix}$$

Most often, D, E and F are diagonal matrices (to be completely general, extra-diagonal blocks should appear in the second matrix, but this would complicate the calculations without changing our conclusions in any interesting way).

This correspondence is used as follows. If a Dirichlet problem like Equation 4 is to be solved, one starts from the variational formulation

$$\begin{aligned} &\int_\Omega \operatorname{grad}\psi \cdot \operatorname{grad}\psi' = \int_\Omega \omega\psi' \qquad \forall\, \psi' \in H_0^1(\Omega) \\ &\int_\Gamma \psi\psi' = \int_\Gamma h\psi' \qquad \forall\, \psi' \in H^1(\Omega) \end{aligned} \tag{14}$$

and substitutes vectors and bilinear forms. This gives

$$\{{}^t\psi_1', 0\}\left(\begin{vmatrix} A & B^t \\ B & C \end{vmatrix} \begin{vmatrix} \psi_1 \\ \psi_2 \end{vmatrix} - \begin{vmatrix} D & 0 \\ 0 & E \end{vmatrix} \begin{vmatrix} \omega_1 \\ \omega_2 \end{vmatrix}\right) = 0$$

$${}^t\psi_2' F(\psi_2 - h) = 0$$

for all $\psi'$, so

$$A\,\psi_1 + B^t h = D\omega_1$$

is the discretized form of Equation 14.

We so insist on this to emphasize what follows : given u in $R^n$, to be interpreted as the approximation of some function in $H^1(\Omega)$, there is no canonical mean to define the dicrete analogs of either $\Delta u$ or $\partial u/\partial n$. Using the Green formula,

$$\int_\Omega \operatorname{grad} u \cdot \operatorname{grad} v = -\int_\Omega \Delta u\ v \qquad \forall\, v \in H_0^1(\Omega),$$

the above correspondence only gives

$$\{{}^t v_1, 0\} \begin{vmatrix} A & B^t \\ B & C \end{vmatrix} \begin{vmatrix} u_1 \\ u_2 \end{vmatrix} = \{{}^t v_1, 0\} \begin{vmatrix} D & 0 \\ 0 & E \end{vmatrix} \begin{vmatrix} (-u)_1 \\ (-u)_2 \end{vmatrix}$$

for all v, thus

$$A\,u_1 + B^t u_2 = D\,(-\Delta u)_1$$

and this leaves $(-\Delta u)_2$ undefined. If all v in $R^n$ are taken into account, the approximation of the Green formula

$$\int_\Omega \text{grad } u.\ \text{grad} v = \int_\Omega -\Delta u\, v + \int_\Gamma \partial u/\partial n\ v$$

yields

$$Bu_1 + Cu_2 + E\,(\Delta u)_2 = F\,(\partial u/\partial n)_2 \qquad (15)$$

Let us denote by $\partial u$ the right-hand side of Equation 15 ("discrete boundary flux"). Neither $\partial u$ nor $\Delta u$ can be defined independently. (This is no surprise, since $\partial$ and $\Delta$ are unbounded operators on $H^1(\Omega)$.)

The discrete analogs of R and K

Now we are ready to define the discrete analogs of R and K. First solve $\Delta u = 0$, $u = w$ on $\Gamma$, i.e., in discrete terms,

$$Au_1 + B^t w = 0, \qquad u_2 = w,$$

then use Equation 15, knowing that $\Delta u = 0$. One gets

$$R = C - B\,A^{-1}B^t \qquad (16)$$

As for K, carry the same discretization process on System 8. One first obtains

$$A\omega_1 + B^t\omega_2 = 0, \quad \omega_2 = u,$$

then

$$A\psi_1 = D\omega_1$$

and finally $\partial\psi = B\psi_1 - E\omega_2$, so

$$K = B\,A^{-1}\,D\,A^{-1}\,B^t + E \qquad (17)$$

How to compute them

We call "selective Cholesky factorization" the process by which S and T are found such that

$$M = \begin{vmatrix} A & B^t \\ B & C \end{vmatrix} = \begin{vmatrix} S & 0 \\ T & 1 \end{vmatrix} \begin{vmatrix} S^t & T^t \\ 0 & U \end{vmatrix}$$

and S is lower triangular. If the unknowns are renumbered the way we assumed above, this is just a Cholesky factorization interrupted just after column n - m. In the general case, as one wants to avoid any rearrangement of data that would only be justified by the necessity to call a library Cholesky routine, it is better to write a special piece of code for the purpose of selective factorization. (We have listed one in Bossavit-Meyer (1979).)

Remark that S, T and C (or, equivalently, U) constitute all the information necessary to solve any Dirichlet or Neumann problem in $\Omega$, and in the right form (the Cholesky root of A is the right thing to have if one wants to solve Ax = b).

Now our main point. Easy substitution shows that

$$R = C - T\,T^t \qquad (= U) \tag{18}$$

$$K = E + T\,S^{-1}\,D\,S^{-t}\,T^t \tag{19}$$

so R and K are easily obtained from the data just mentioned. More, the extra cost is marginal, since factorization of M requires $(n-m)^3/6 + m(n-m)^2/2$ operations, as $m(n-m)^2/2 + (n-m)m^2/2$ are necessary to get K.

Putting together all the steps we have explored, the algorithm to solve Equation 1 (the biharmonic problem) can be described like this:

*form a mesh on* $\Omega$ ;
*obtain {by any method}* $A, B, C, D, E$;
*obtain {by selective factorization of M}* $S$ *and* $T$ ;
*compute* $K$ *by (19);*
$u \leftarrow K^{-1} g$ ; *{*$m^3/6$ *operations}*
*compute* $\psi$ :
| $\omega_1 \leftarrow - S^{-t}\, T^t u$;
| $\psi_1 \leftarrow S^{-t}\, S^{-1}\, D\, \omega_1$ ;
| $\psi_2 \leftarrow 0$

"Any method" will generally mean finite element assembly, but finite differences or Galerkine global approximations can be considered as well.

## 3 SUBSTRUCTURATION

Let us recall briefly the substructuration process for the Laplace problem with Dirichlet boundary conditions. Let $u_1$, $u_2$, u, be given traces on $\Gamma_1$, $\Gamma_2$, (Fig. 1). If the harmonic stiffness $R^1$ relative to $\Omega_1$ is known, one can get the boundary normal derivative of $\phi_1$, harmonic in $\Omega_1$, by using $R^1$. In block form :

$$\begin{vmatrix} \partial\phi_1/\partial n_{|\Gamma_1} \\ \partial\phi_2/\partial n_{|\Sigma} \end{vmatrix} = \begin{vmatrix} R^1_{\Gamma\Gamma} & R^1_{\Gamma\Sigma} \\ R^1_{\Sigma\Gamma} & R^1_{\Sigma\Sigma} \end{vmatrix} \begin{vmatrix} u_1 \\ u \end{vmatrix}$$

The same for $R^2$. We say that $\Gamma_1 \cup \Sigma$ is the "support" of $R^1$. We define the "assembly" of $R^1$ and $R^2$ as the following operator, in block form :

$$\mathcal{R} = \begin{vmatrix} R^1_{\Gamma\Gamma} & R^1_{\Gamma\Sigma} & 0 \\ R^1_{\Sigma\Gamma} & R^1_{\Sigma\Sigma} + R^2_{\Sigma\Sigma} & R^2_{\Gamma\Sigma} \\ 0 & R^2_{\Sigma\Gamma} & R^2_{\Sigma\Sigma} \end{vmatrix}$$

Its support is $\Gamma_1 \cup \Sigma \cup \Gamma_2$. From $u_1$, $u_2$, u, R generates the triple

$$\{\partial\phi_1/\partial n_{|\Gamma_1},\ \partial\phi_1/\partial n + \partial\phi_2/\partial n_{|\Sigma},\ \partial\phi_2/\partial n_{|\Gamma_2}\}$$

By equating the middle term to zero, one obtains an equation for u :

$$(R^1_{\Sigma\Sigma} + R^2_{\Sigma\Sigma})u = -(R^1_{\Sigma\Gamma}\ u_1 + R^2_{\Gamma\Sigma}\ u_2) \qquad (20)$$

For this value of u, $\phi = \phi_1$ on $\Omega_1$ and $\phi_2$ on $\Omega_2$ is harmonic in $\Omega$. Using Equation 20, we can compute $\partial\phi/\partial n$ on $\Gamma_1$ and $\Gamma_2$, and obtain the stiffness associated with $\Omega$ as a whole :

$$R = \begin{vmatrix} R^1_\Gamma - R^1(R_\Sigma)^{-1}\ {}^tR^1 & R^1_\Gamma - R^1\ R_\Sigma^{-1}\ R^2 \\ R^2_\Gamma - {}^tR^2 R_\Sigma^{-1}\ {}^tR^1 & R^2_\Gamma - R^2(R_\Sigma)^{-1}\ {}^tR^2 \end{vmatrix}$$

(To simplify the notations, we put $R_\Sigma$ for $R^1_{\Sigma\Sigma} + R^2_{\Sigma\Sigma}$, $R^1_\Gamma$ for $R^1_{\Gamma\Gamma}$, $R^1$ and ${}^tR^1$ for $R_{\Gamma\Sigma}$ and its transpose $R_{\Sigma\Gamma}$.)

The process just described is well known as "static condensation". The two steps are <u>assembly</u>, by which $\mathcal{R}$ is formed from $R^1$ and $R^2$, and <u>condensation</u>, which gives R from $\mathcal{R}$. Assembly and condensation also act on supports: the support of an assembly is the union of the supports of the factors, and the condensation amputates the support of a part (  in the example above). One conceives easily how the assembly-condensation process can be generalized to more than two sub-domains (it is then referred to as the frontal method (Irons (1970), Bossavit-Frémond (1976)). Since subdivision of the domain has to be prescribed by the programmer, as part of the description of the problem, the programmer may as well control the assembly-condensation process at data-input time, if suitable syntactic facilities exist. We have made an attempt in Bossavit (1980) to specify such facilities by introducing abstract program objects, called SUPPORTS and RIGIDITIES, and shown how this formalization can help in the design of a library of Fortran subroutines to manipulate supports and rigidities.

The functionalities proposed there are also convenient for the biharmonic problem. To make this point, we shall only show how the biharmonic admittance K is found in the union of two sub-domains (Fig. 1) on which $K^1$ and $K^2$ are known. Consider the sub-problem

$$\left|\begin{array}{l} -\Delta\psi_i = \omega_i \\ \psi_{i|\Gamma_i} = h_i \\ \psi_{i|\Gamma} = h \end{array}\right. \qquad \left|\begin{array}{l} -\Delta\omega_i = 0 \\ \omega_{i|\Gamma_i} = u_i \\ \omega_{i|\Sigma} = u \end{array}\right.$$

in $\Omega_i$ and set $g_i = -\partial\psi_i/\partial n$ on $\Gamma_i$ (i = 1 or 2).

Then, by the definitions given above (Equations 9 and 10), u and h should be such that

$$K^i \begin{pmatrix} u_i \\ u \end{pmatrix} + \begin{pmatrix} g_i \\ 0 \end{pmatrix} = R^i \begin{pmatrix} h_i \\ h \end{pmatrix}$$

for i = 1, 2. Solving for u and h, we obtain the global K and R such that

$$K \begin{pmatrix} u_1 \\ u_2 \end{pmatrix} + \begin{pmatrix} g_1 \\ g_2 \end{pmatrix} = R \begin{pmatrix} h_1 \\ h_2 \end{pmatrix}$$

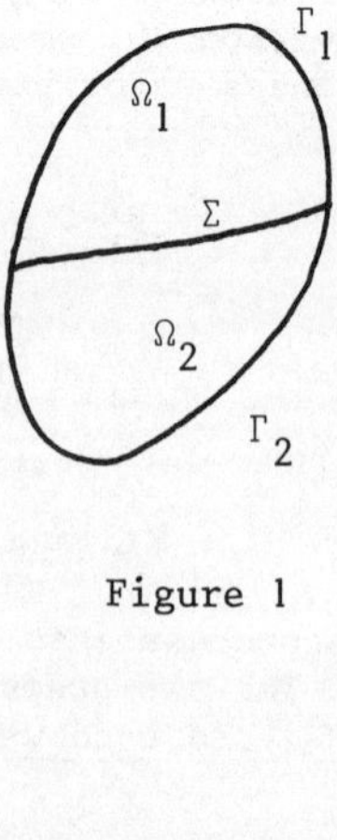


Figure 1

The computation has already been done for R. It is similar for K, and rather tedious. The result is (in the simplified notation already used)

$$K = \begin{vmatrix} K_\Gamma^1 - ({}^tK^1 R_\Sigma^{-1} R^1 + {}^tR^1 R_\Sigma^{-1} K^1) + {}^tR^1 R_\Sigma^{-1} K_\Sigma R_\Sigma^{-1} R^1 & K^{12} \\ K^{21} & K_\Gamma^2 - ({}^tK^2 R_\Sigma^{-1} R^2 + {}^tR^2 R_\Sigma^{-1} K^2) + {}^tR^2 R_\Sigma^{-1} K_\Sigma R_\Sigma^{-1} R^2 \end{vmatrix}$$

where

$$K^{21} = {}^tK^{12} = {}^tR^2 R_\Sigma^{-1} K_\Sigma R_\Sigma^{-1} R^1 - ({}^tR^2 R_\Sigma^{-1} K^1 + {}^tK^2 R_\Sigma^{-1} R^1)$$

CONCLUSION

The popular idea of substructuration by static condensation is feasible for the biharmonic problem, just like for the Poisson problem. The computational cost is not significantly greater. The programming effort is also comparable, once the basic functionalities (assembly and condensation in sub-domains) have been implemented.

REFERENCES

Bossavit, A. (1971) Une méthode de décomposition de l'opérateur biharmonique. Internal report HI585/2, EDF, 92141 Clamart.

Bossavit, A. (1980) On the Implementation of Mixed FEM-BEM methods: Rigidities as a New "Type" in Programming Languages. Proc. Southampton, New Developments in Boundary Element Methods, CML, 380-392.

Bossavit, A. and Frémond, M. (1976) The Frontal Method Based on Mechanics and Dynamic Programming, Comp. Meth. in Appl. Mech. Engng., 8:153-178.

Bossavit, A. and Meyer, B. (1979) On the Constructive Approach to Programming (The Case of Selective Cholesky Factorization). IMACS Symposium, Bethleem, Pa, USA, in Advances in Computing Methods for PDE's, III, 287-291.

Glowinski, R. and Pironneau, O. (1978) Approximation par éléments finis du problème de Stokes. CRAS, 286:225-228.

Irons, B.M. (1970) A Frontal Solution Program for Finite Element Analysis. Int. J. Numer. Meths. Engng. 2:5-32.

Pal'tsev, B.V. (1966) The expansions of Solutions of Dirichlet's problem. Zh. Vycisl. Mat. mat. Fiz., 1:43-51.

Smith, J. (1968) SIAM J. Numer. Anal., 2:323-339.

# DISCRETE OPTIMIZATION OF FINITE-ELEMENT SOLUTIONS: A CONCEPT DISCUSSED FOR THE DIFFUSION-ADVECTION PROBLEM

S. Gärtner

Hannover University

## GENERAL CONCEPT

We consider the equation

$$Lu = f \quad \text{in } \Omega , \tag{1}$$

where $L:D(L) \subset E \to F$ is a linear differential-operator. Equation (1) represents the pointwise comparison

$$(Lu)(X) = f(X) \quad \forall\, X \in \Omega . \tag{2}$$

We denote by F' the space of all bounded linear functionals $m:F \to \mathbb{R}$ which we call mean-functional, if $m(1)=1$. The functional-equation

$$m(Lu) = m(f) \quad \forall\, m \in F' \tag{3}$$

and the differential-equation (1) are equivalent mathematical models which may be used to find a solution $u \in D(L)$.

Now let $P_i^h : E \longrightarrow S_i^h$ be a locally defined projector, where $i=1(1)N$. The N local approximations

$$(P_i^h u)(X) = \sum_{X_k \in M_i^h} \varphi_k^i(X) u(X_k) \tag{4}$$

are connected over the discrete values $u(X_k)$, whereby

$$M^h = \{ X_1, X_2, \ldots, X_N \} = \bigcup_{i=1}^{N} M_i^h \tag{5}$$

represents a set of N points or nodes $X_k \in \overline{\Omega}$. If we are concerned with difference-schemes (FDM), the basis functions $\varphi_k^i$ are defined as Lagrange-type polynomials with the biorthonormality-relation

$$\varphi^{i}_{k}(X_j) = \delta_{jk} \quad \forall\, X_j \in M^h_i. \tag{6}$$

In contrast, the finite-element technique (FEM) leads to piecewise constructed polynomials having the same property (6). Furthermore, a global FEM-projector $P^h: E \rightarrow S^h$ is defined as

$$(P^h u)(X) = \sum_{k=1}^{N} \varphi_k(X)u(X_k), \tag{7}$$

so that

$$(P^h u)(X) = (P^h_i u)(X) \quad \forall\, X \in \bar{\Omega}_i\,, \tag{8}$$

where the patch $\bar{\Omega}_i$ is the union of all finite elements $\bar{\Omega}^e$ which contain the common node $X_i$.

Using the projection (4) in the sense of FDM or FEM, we define the following <u>finite model of equation (3)</u>:

$$m_i(LP^h_i u^h) = m_i(f) \qquad (i=1(1)N,\ m_i \in F'), \tag{9}$$

where $u^h$ denotes an approximation of u. This system of difference equations can be written as

$$\sum_{X_k \in M^h_i} m_i(L\varphi^i_k)u^h(X_k) = m_i(f) \tag{10}$$

or

$$\sum_{k=1}^{N} a_{ik}u^h(X_k) = m_i(f). \tag{11}$$

Subtracting $\Sigma\, a_{ik}u(X_k) = m_i(LP^h_i u)$ from equation (11) we obtain (with f=Lu) the relation

$$\sum_{k=1}^{N} a_{ik}(u^h(X_k)-u(X_k)) = m_i(Lu-LP^h_i u) \tag{12}$$

between the discrete errors and the local defects, given by

$$E^h_i(u) = m_i(Lu-LP^h_i u). \tag{13}$$

<u>From the foregoing, it is seen that there is no difference between FDM and FEM.</u>

Let $\mathbb{P}^k$ denote the space of all polynomials $w: \Omega \rightarrow \mathbb{R}$ with degree less or equal to k. If

$$\mathbb{P}^{k_o} \subset S^h_i \qquad \text{and} \qquad \mathbb{P}^{k_o+1} \not\subset S^h_i, \tag{14}$$

then we define the ordinary order $q_o$ of consistency of equation (9) in the asymptotic sense by

$$E^h_i(u) = m_i(L(u-P^h_i u)) = O(h^{q_o}) \quad \forall\, m_i \in F' \tag{15}$$

which is a result of the relation

$$P_i^h w = w \quad \forall\, w \in S_i^h \implies P_i^h w = w \quad \forall\, w \in \mathbb{P}^{k_o} \implies \tag{16}$$

$$E_i^h(w) = 0 \quad \forall\, w \in \mathbb{P}^{k_o}. \tag{17}$$

We do not repeat the standard arguments of the convergence theory. It is clear, that an improved order $q > q_o$ of consistency induces an acceleration of the discrete convergence based upon the usual well-known assumptions. Therefore, <u>we shall attempt to construct local defects (13) with certain mean-functionals $m_i$, such that the estimate</u>

$$E_i^h(u) = O(h^q) \tag{18}$$

<u>holds for an order as high as possible</u>. In order to achieve this, it is necessary to replace $k_o$ by k in condition (17). This leads to the condition

$$E_i^h(w) = 0 \qquad \forall\, w \in \mathbb{P}^k \quad (k > k_o) \tag{19}$$

or to the weaker condition

$$E_i^h(w) = O(h^q) \quad \forall\, w \in \mathbb{P}^k \quad (k > k_o). \tag{20}$$

The foregoing conditions lead to a system of equations which may be solved using a mean-functional $m_i$, corresponding to the maximum possible value of k.

<u>In the FDM-case</u>, we can define the Hermitian-method (Mehrstellenverfahren) of COLLATZ [4] using the following mean-functional

$$m_i(f) = \sum_{X_k \in M_i^h} b_{ik} f(X_k), \tag{21}$$

which has the special property

$$m_i(P_i^h f) = m_i(f). \tag{22}$$

Now, condition (19) leads to an algebraic system for the coefficients $b_{ik}$ introduced in definition (21). But it is not necessary to work with this special mean-functional in certain cases, and in others it is not even possible. In particular, unsymmetric problems can be solved with more general mean-formulas which do not have the property (22); [5].

<u>In the FEM-case</u>, which is dealt with here, it is usual to distinguish between strong and weak solutions. Considering L be a second-order elliptic operator, the approximation (7) of strong solutions must be of the Hermite-type. The condition (20) generates linear independent mean-functionals. This is useful since we require more than one difference equation for each node. But, for the purpose of introducing our concept we prefer Lagrange-

type approximations for weak solutions initially. Therefore we define functionals $m_i$ as follows:

$$m_i(f) = (f, \psi_i) = \int_\Omega \psi_i(X) f(X) d\Omega \quad , \tag{23}$$

where $\psi_i$ represent certain test functions with support

$$\text{supp } \psi_i = \bar{\Omega}_i \tag{24}$$

with $\bar{\Omega}_i$ as in equation (8). Integrating the functional-equation

$$(Lu, \psi_i) = (f, \psi_i) \quad (\psi_i \in H_0^1(\Omega)) \tag{25}$$

by parts, we obtain the weak form

$$B(u, \psi_i) = (f, \psi_i) \quad \forall \psi_i \in H_0^1(\Omega), \tag{26}$$

where

$$H_0^1(\Omega) = \{ w \in H^1(\Omega); w=0 \text{ on } \partial\Omega \}, \tag{27}$$

assuming the boundary condition u=g on $\partial\Omega$. Let us suppose that the set $\{\varphi_i\}$ of trial functions - which appear in definition (7) - spans a subspace $S^h$ of $H^1(\Omega)$, and that $S_0^h$ is defined as follows:

$$S_0^h = \{ w \in S^h; w=0 \text{ on } \partial\Omega \} . \tag{28}$$

Assuming g can be spanned by $S^h$ on $\partial\Omega$, the Bubnov-Galerkin approximation $\bar{u}^h$ is obtained by solving the equations

$$B(P^h\bar{u}^h, \varphi_i) = (f, \varphi_i) \quad \forall \varphi_i \in S_0^h \quad (\bar{u}^h - u_0^h \in S_0^h), \tag{29}$$

where $u_0^h \in S^h$ solves the boundary condition.
Suppose in a Galerkin-Petrov method the test functions $\{\psi_i\}$ span a subspace $T_0^h$ of $H_0^1(\Omega)$ and the corresponding approximation $u^h$ is obtained from

$$B(P^h u^h, \psi_i) = (f, \psi_i) \quad \forall \psi_i \in T_0^h \quad (u^h - u_0^h \in S_0^h). \tag{30}$$

Then the problem is to choose $T_0^h$ corresponding to a given finite-element space $S^h$.

BARRETT and MORTON [1] introduced a transformation of the trial functions,

$$\psi_j = \rho\, \varphi_j \tag{31}$$

which (approximately) leads to a symmetric bilinear form $B_\rho( \, , )$

$$B_\rho (v,w) = B(v, \rho w), \quad B_\rho (v,w) = B_\rho (w,v). \tag{32}$$

Then, by using test functions $\rho\varphi_i$ in the Galerkin-Petrov method,

$$B(u^h_\rho, \rho\varphi_i) = (f, \rho\varphi_i) \quad \forall\, \varphi_i \in S^h_o \quad (u^h_\rho - u^h_o \in S^h_o), \tag{33}$$

we obtain an approximation $u^h_\rho$ which is the best approximation from $S^h$ to the unknown u in the norm induced by the bilinear form $B_\rho$ (,). That is,

$$\|u^h_\rho - u\| \leq \|u^h - u\| \qquad \forall\, u^h \in u^h_o + S^h_o, \tag{34}$$

where

$$\|v\|^2 = B_\rho\,(v,v), \qquad v \in H^1_o(\Omega).$$

<u>Our approach is to choose test functions in the Galerkin-Petrov method which improve the asymptotic rate of convergence in the discrete norm,</u>

$$|u| = \max_j \left|u(X_j)\right|. \tag{35}$$

Therefore, following the same arguments as before, equation (26) leads to the local defects

$$E^h_i(u) = B(u - P^h u, \psi_i), \tag{36}$$

in which certain test functions $\psi_i$ must be determined when applied to condition (20).

A more subtle estimation of linear functionals on Hilbert and Sobolev spaces is given in [2] or [3]. In the following heuristic considerations, however, we only require the asymptotic rate (18). It should be noted, however, that we are only interested here in the relation $q > q_o$ and that q does not represent the correct order of discrete convergence (which is smaller).

TEST FUNCTIONS

As a consequence of definition (36), condition (19,20) represents a restriction for the test function $\psi_i$. By introducing continuous test functions of the form

$$\psi_i(X) = \rho_i(X)\,\varphi_i(X), \quad \text{supp}\,\psi_i = \overline{\Omega}_i, \tag{37}$$

we obtain the Galerkin-Petrov equation

$$B(u^h, \rho_i\varphi_i) \equiv \sum_{k=1}^{N} B(\varphi_k, \rho_i\varphi_i)\,u^h(X_k) = (f, \rho_i\varphi_i) \tag{38}$$
$$\forall\, \varphi_i \in S^h_o,\ u^h - u^h_o \in S^h_o.$$

For the case of $\rho_i \equiv 1$, this equation represents the Bubnov-Galerkin method.

Equation (38) requires the determination of a continuous weighting function $\rho_i$ which has the properties

$$\rho_i(X_i) = 1, \tag{39}$$

$$E_i^h(w) \equiv B(w-P^h w,\ \rho_i \varphi_i) = O(h^q) \quad \forall\ w \in \mathbb{P}^k\ (k > k_0). \tag{40}$$

A possible means of representing the above weighting function is to use the truncated Taylor expansion

$$\rho_i(X) = 1 + \sum_{k=1}^{n} (x_k - x_k^i)\left(\frac{\partial \rho}{\partial x_k}\right)(X_i) \tag{41}$$

$$= 1 + \sum_{k=1}^{n} (x_k - x_k^i)\,\alpha_k$$

or any other function which seems to be appropriate. Condition (39,40) defines a proplem $P(B,P^h;\rho_i)$ which depends upon the projector $P^h$ and the bilinearform B( , ). It should be noted that a weighting function $\rho_i$ with $q > q_0$ does not exist in each (multidimensional) case. We thus have to identify certain projectors $P^h$ which allow q to exceed $q_0$. Even though we may be unable to obtain a solution to $P(B,P^h;\rho_i)$, it is nonetheless possible to obtain a number of improvements to the Galerkin method in most cases.

SYMMETRICAL PROBLEM

In order to compare symmetrical and unsymmetrical problems, let us consider the symmetrical eigenvalue-problem

$$Lu \equiv \frac{d^2u}{dx^2} + \lambda u = f \quad \text{in } (0,1) \quad (u(0)=u(1)=0)$$

which, in the weak form is

$$B(u,\psi) \equiv -\left(\frac{du}{dx},\frac{d\psi}{dx}\right) + \lambda\,(u,\psi) = (f,\psi) \quad \forall\ \psi \in H_0^1(0,1). \tag{42}$$

Equation (42) may be solved numerically using the Galerkin equations (38) with the following trial functions

$$\varphi_i(x) = \begin{cases} (x-x_{i-1})/a = \xi & \text{in } [x_{i-1},x_i] \\ (x_{i+1}-x)/b = 1-\eta & \text{in } [x_i,x_{i+1}] \\ 0 & \text{otherwise,} \end{cases} \tag{43}$$

where $a = x_i - x_{i-1}$ and $b = x_{i+1} - x_i$.

In order to estimate the local defect (36), it is necessary to evaluate

$$E_i^h(w) = E_i^{i-1}(w) + E_{i+1}^i(w) \tag{44}$$

with the element-wise defined terms

$$E_i^{i-1}(w) = {}_0\!\int^1 \Big[-\frac{1}{a}\frac{d}{d\xi}(w-P^h w)\frac{d}{d\xi}(\rho_i\varphi_i) + \lambda a\,(w-P^h w)\rho_i\varphi_i\Big]d\xi,$$

$$E_{i+1}^i(w) = {}_0\!\int^1 \Big[-\frac{1}{b}\frac{d}{d\eta}(w-P^h w)\frac{d}{d\eta}(\rho_i\varphi_i) + \lambda b\,(w-P^h w)\rho_i\varphi_i\Big]d\eta.$$

For the case of $\rho_i \equiv 1$ with $w = 1,\ x-x_i,\ (x-x_i)^2$ we obtain the following key results:

$$E_i^{i-1}(\ 1) = E_{i+1}^i(\ 1) = 0, \tag{45}$$

$$E_i^{i-1}(x-x_i) = E_{i+1}^i(x-x_i) = 0,$$

$$E_i^{i-1}((x-x_i)^2) = -\lambda a^3/12,$$

$$E_{i+1}^i((x-x_i)^2) = -\lambda b^3/12.$$

Finally we obtain the asymptotic estimate

$$E_i^h(u) = O(h^3) \qquad (a,b \le h), \tag{46}$$

which arises from the Bubnov-Galerkin method.

UNSYMMETRICAL PROBLEM

Let us consider the steady-state one-dimensional advection-diffusion equation

$$Lu \equiv \frac{d}{dx}(-k\frac{du}{dx} + vu) = 0 \quad \text{in } (0,1) \quad (u(0)=0,\ u(1)=1)$$

with constant diffusion-coefficient $k>0$ and constant velocity $v$. The above may be expressed in the following weak form

$$B(u,\psi) \equiv (\frac{du}{dx} - \frac{v}{k}u,\ \frac{d\psi}{dx}) = 0 \quad \forall\psi \in H_0^1(0,1) \tag{47}$$

which is generated by partial integration and dividing by k. We apply the Galerkin-Petrov equation (38) with the trial functions (43) and with the weighting function

$$\rho_i(x) = 1 + \alpha\,(x-x_i) = \begin{cases} 1 + \alpha\, a(\xi-1) & \text{in } [x_{i-1},x_i] \\ 1 + \alpha\, b\,\eta & \text{in } [x_i,x_{i+1}], \end{cases} \tag{48}$$

in which the same notation is used as in (43) with an unknown parameter $\alpha$ as in (41). The test function and its derivative are given by

$$\psi_i = \rho_i \varphi_i = \begin{cases} \xi + \alpha a\,(\xi^2-\xi) & \text{in } [x_{i-1}, x_i] \\ 1-\eta + \alpha b\,(\eta-\eta^2) & \text{in } [x_i, x_{i+1}], \end{cases} \tag{49}$$

$$\frac{d\psi_i}{dx} = \begin{cases} (1+\alpha a\,(2\xi-1))/a & \text{in } [x_{i-1}, x_i] \\ -(1+\alpha b\,(2\eta-1))/b & \text{in } [x_i, x_{i+1}]. \end{cases} \tag{50}$$

As before we calculate the local defect (44) applied to equation (47),

$$E_i^{i-1}((x-x_i)^2) = a\int_0^1 \left[\frac{d}{d\xi}(\xi^2-\xi) - \frac{va}{k}(\xi^2-\xi)\right]\left[1+\alpha a(2\xi-1)\right]d\xi$$

$$= \left(\frac{v}{6k} + \frac{\alpha}{3}\right)a^2, \tag{51}$$

$$E_{i+1}^{i}((x-x_i)^2) = -\left(\frac{v}{6k} + \frac{\alpha}{3}\right)b^2.$$

Setting $\alpha = -v/(2k)$ we obtain the weighting function

$$\rho_i = 1 - \frac{1}{2}\frac{v}{k}(x-x_i) = \begin{cases} 1 - \frac{1}{2} Pe_{i-1,i}\,(\xi-1) & \text{in } [x_{i-1}, x_i] \\ 1 - \frac{1}{2} Pe_{i,i+1}\,\eta & \text{in } [x_i, x_{i+1}] \end{cases} \tag{52}$$

with the Peclet-numbers $Pe_{ij} = v(x_j - x_i)/k$. This yields

$$\begin{aligned} E_i^{i-1}((x-x_i)^3) &= -(Pe_{i-1,i})^2 a^2/120, \\ E_{i+1}^{i}((x-x_i)^3) &= +(Pe_{i,i+1})^2 b^2/120. \end{aligned} \tag{53}$$

Defining $Pe = vh/k$ $(a,b \le h)$ we obtain the final result

$$E_i^h(u) = O(|Pe|^2 h^2) = O((|v|/k)^2 h^4). \tag{54}$$

If $\alpha = 0$ or $\rho_i \equiv 1$, then estimate (51) only leads to

$$E_i^h(u) = O(|Pe|\,h) = O((|v|/k)\,h^2) \tag{55}$$

which arises from the Bubnov-Galerkin or from the central difference scheme (a=b=h). Both these schemes generate the well-known oscillations for $|Pe| > 2$. The improved scheme can be written as

$$-(1+\frac{Pe^2}{12})(u_{i-1}-2u_i+u_{i+1})/h^2 + Pe(u_{i+1}-u_{i-1})/(2h) = 0, \qquad (56)$$

if we set a=b=h and devide by h. Compared to Il'ins scheme

$$-\frac{Pe}{2}\coth\frac{Pe}{2}(u_{i-1}-2u_i+u_{i+1})/h^2 + Pe(u_{i+1}-u_{i-1})/(2h) = 0, \qquad (57)$$

which has no discrete error, we can interprete the improved Galerkin-Petrov scheme using the following Taylor series,

$$\frac{Pe}{2}\coth\frac{Pe}{2} = (1+\frac{Pe^2}{12}) - (\frac{Pe^4}{720} - \frac{Pe^6}{30240}) - \dots \quad |Pe| < 2\pi . \qquad (58)$$

The development of scheme (57) with a truncated series (58) is free of oszillations, if an even number of terms is employed. Numerical solutions to problem (47) are shown in figure 1.

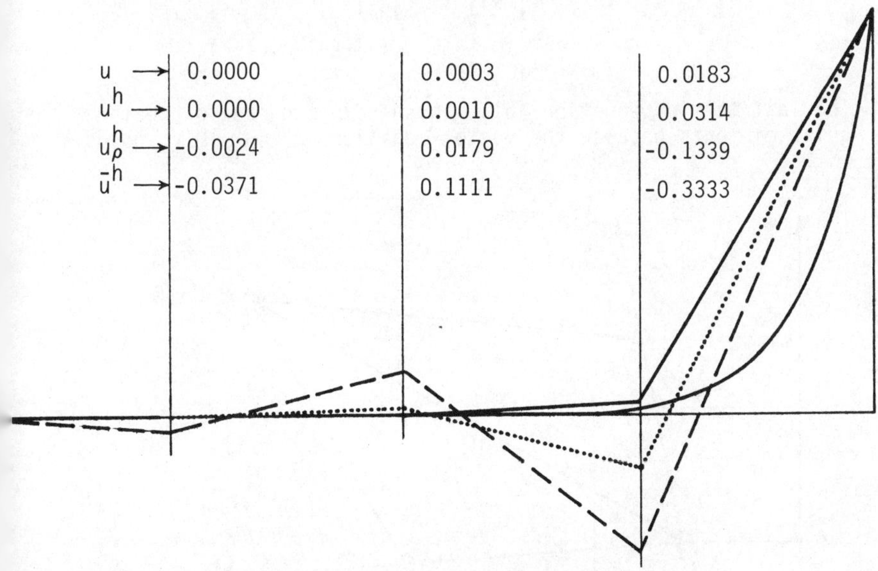


Figure 1: Solutions to the problem (47) for Pe=4:
the exact solution u is shown as a full curve,
the Bubnov-Galerkin solution $\bar{u}^h$ as dashed lines,
the Galerkin-Petrov solution of BARRETT/MORTON $u_\rho^h$ as dotted lines,
and the present (dicrete improved) Galerkin-Petrov solution $u^h$ as full lines.

## HEURISTICAL ARGUMENTS IN THE MULTI-DIMENSIONAL CASE

The Lagrange-type interpolation (7) is piecewise constructed by form functions $\varphi_i^e: \overline{\Omega}^e \rightarrow \mathbb{R}$ which are polynomials in certain local coordinates. The following examples for these special coordinates $\xi_j^i \in [0,1]$ are considered:

The 1-simplex $\overline{\Omega}^e \subset \mathbb{R}$ with two nodes $X_i$, $X_j \in \mathbb{R}$ induces the relation

$$X = X_i + \xi_j^i H_{ij} \qquad (H_{ij} = X_j - X_i) \quad \text{(i-j-cyclic)}. \tag{59}$$

The 2-simplex $\overline{\Omega}^e \subset \mathbb{R}^2$ (triangle) with three nodes $X_i$, $X_j$ and $X_k$ and with vectors such as $H_{ij}=X_j-X_i$ induces the relation

$$X = X_i + \xi_j^i H_{ij} + \xi_k^i H_{ik} \qquad \text{(i-j-k-cyclic)}. \tag{60}$$

The isoparametric finite element $\overline{\Omega}^e \subset \mathbb{R}^2$ with four nodes $X_i$, $X_j$, $X_k$ and $X_l$ and with vectors such as $H_{ij}=X_j-X_i$ induces the relation

$$X = X_i + \xi_j^i H_{ij} + \xi_l^i H_{il} + \xi_j^i \xi_l^i (H_{ik}-H_{ij}-H_{il}) \tag{61}$$

$$\text{(i-j-k-l-cyclic)}.$$

The last term of equation (61) vanishes in the case of rectangular elements because the vector-addition $H_{ik}=H_{ij}+H_{il}$ holds.

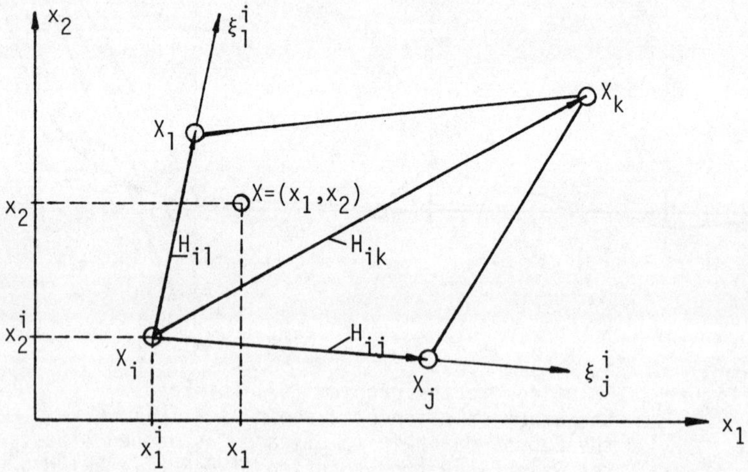


Figure 2: Local coordinates of the four-node isoparametric element with origin node $X_i$.

In the following deductions, the local coordinates are node-wise defined, for simplicity. By this means we obtain individual systems for each node belonging to the finite element. Once the following node- or patch-wise considerations have been dealt with, further calculations will be carried out with one uniform system of (natural, barycentric) coordinates.

As we have seen in the one-dimensional case, the Bubnov-Galerkin method applied to the symmetrical problem (42) automatically has an improved order (46). Excluding a number of special cases, such as rectangular elements, we only obtain ordinary estimates in multi-dimensional problems. Therefore an improvement does not exist, if multi-dimensional unsymmetrical problems are considered. Nevertheless, it seems to be appropriate to generalize the one-dimensional improvement.

The weighting function (52) complies with the Taylor expansion (41), where n=1 and

$$(\frac{\partial \rho_i}{\partial x})(x_i) = -\frac{1}{2}\frac{v}{k}\,.$$

The generalization for $n \geq 1$ would appear to be given by

$$(\frac{\partial \rho_i}{\partial x_l})(X_i) = -\frac{1}{2}\frac{v_l^i}{k} \qquad (l=1(1)n). \tag{62}$$

This assumption leads to the n-dimensional weighting function

$$\rho_i(X) = 1 - \tfrac{1}{2}(X-X_i)V_i' \tag{63}$$

in which the scalar product of the two vectors $X-X_i$ and $V_i'=(v_1^i, v_2^i, \ldots, v_n^i)/k$ must be calculated for each finite element which is included in the patch $\bar{\Omega}_i$. It should be noted that $V_i'$ must be calculated either in the patch $\bar{\Omega}_i$ as a mean value or at the origin node $X_i$, in order that a continuous function $\rho_i$ results.

With the element-wise introduced transformations (59) - (61) of $X-X_i$ we obtain a natural definition of Peclet numbers:

For the triangle we have

$$(X-X_i)V_i' = \xi_j^i H_{ij} V_i' + \xi_k^i H_{ik} V_i' = \xi_j^i Pe_{ij} + \xi_k^i Pe_{ik}, \tag{64}$$

whilst, using the four-node isoparametric finite element, we obtain the formulation

$$(X-X_i)V_i' = \xi_j^i Pe_{ij} + \xi_l^i Pe_{il} + \xi_j^i \xi_l^i (Pe_{ik} - Pe_{ij} - Pe_{il}). \tag{65}$$

In the one-dimensional case considered before, the weighting function (52) is as follows,

$$\rho_i(x) = 1 - \frac{1}{2}\,\xi_j^i\,Pe_{ij} \qquad \text{in} \quad \begin{cases} [x_j,x_i] = [x_{i-1},x_i] \\ [x_i,x_j] = [x_i,x_{i+1}]. \end{cases} \tag{66}$$

Following the arguments of HEINRICH et al. [6] we should construct the bilinear form

$$\rho_i(X) = (1-\tfrac{1}{2}\,\xi_j^i Pe_{ij})(1-\tfrac{1}{2}\,\xi_l^i Pe_{il}). \tag{67}$$

However, using definition (63) and transformation (65) our result should be

$$\rho_i(X) = 1 - \frac{1}{2}\left[\,\xi_j^i Pe_{ij} + \xi_l^i Pe_{il} + \xi_j^i\,\xi_l^i(Pe_{ik} - Pe_{ij} - Pe_{il})\right]. \tag{68}$$

It would appear that in formulation (68) the third additional Peclet-number $Pe_{ik}$ will be relevant if the distance $|H_{ik}|$ between the opposite node $X_k$ and the origin node $X_i$ involves greater rates; see figure 2. However, if a rectangular finite element is given, then the whole non-linear term does not appear, and therefore $P_{ik}$ vanishes.
In contrast, formulation (67) includes a non-linear term with the coefficient $Pe_{ij}Pe_{il}$.

Let us now consider an interesting comparison with a modified Hermitian method which has been investigated by the author; [5]. We are aware that the central differencing method and the Bubnov-Galerkin method with regular grids generate similar schemes. However, our improved schemes of FDM and FEM are also similar. We consider only that result which is given by a generalized FDM. Working with 3-point (9-point) schemes we have to employ the quadratic (biquadratic) interpolation (4), as defined for FDM. By improving the order of consistency, we obtain a mean functional $m_i$ with the property

$$m_i\begin{pmatrix} 1 & \eta & \eta^2 \\ \xi & \xi\eta & \xi\eta^2 \\ \xi^2 & \xi^2\eta & \xi^2\eta^2 \end{pmatrix} = \begin{pmatrix} 1 & -Pe_y/12 & 1/6 \\ -Pe_x/12 & 0 & -Pe_y/24 \\ 1/6 & -Pe_x/24 & 1/36 \end{pmatrix}.$$

If $Pe_x = Pe_y = 0$, our result is the well-known Hermitian formula applied to the biharmonic equation $-\Delta u + \lambda u = f$. It is now evident that the rectangular case does not induce terms with the product $Pe_{ij}Pe_{il}$ as was seen in definition (67).

## CONCLUSIONS

The major objective of this paper is to improve the discrete convergency within a wide class of difference schemes of the FDM and FEM type. The mathematical tools which are used consist in local approximation (interpolation of discrete values) and the mean-functional $m_i$, which needs to be calculated. This method has been succesfully applied to initial-, boundary- and initial-boundary-problems. This paper provides an initial insight into a common foundation which combines FDM and FEM. Arising from this general approach, various schemes are obtained which are similar to those successfully developed by other authors for more specific cases.
In addition, we should investigate the numerical problems associated with the Bubnov-Galerkin method applied to the well-known cases. Some revealing facts might also be obtained by solving the problem $P(B,P^h;\rho_i)$ or $P(\mathbf{L},P^h_i;m_i)$ with greater flexibility.

## REFERENCES

1. J.W.Barrett and K.W.Morton, 'Optimal finite element solutions to diffusion-convection problems in one dimension', Int.J. num.Meth.Engng.,15,1457-1474(1980).
2. J.H.Bramble and S.R.Hilbert, 'Bounds for a class of linear functionals with applications to Hermite interpolation', Num.Math.,16,362-369(1971).
3. P.G.Ciarlet and P.A.Raviart, 'General Lagrange and Hermite interpolation in $\mathbb{R}^n$ with applications to finite element methods', Arch.Ration.Mech.Anal.,46,177-199(1972).
4. L.Collatz, 'Hermitean methods for initial value problems in partial differential equations', in J.J.H.Miller (ed.) Topics in numerical analysis, Proceedings of the Royal Irish Academy Conference on Numerical Analysis, 1972, Academic Press, London and New York.
5. S.Gärtner, 'Eine Verallgemeinerung des Mehrstellenverfahrens mit Anwendungen auf unsymmetrische und instationäre Probleme', to appear, ZAMM.
6. J.C.Heinrich, P.S.Huyakorn, A.R.Mitchell and O.C.Zienkiewicz, 'An 'upwind' finite element scheme for two-dimensional convective transport equation', Int.J.num.Meth.Engng.,11,131-143 (1977).

# ON A DIRECT APPROACH FOR THE SOLUTION OF LINEAR SPACE-INVARIANT 2-D DIFFERENTIAL CONVOLUTION MODELS

F.C. Incertis

IBM Madrid Scientific Center

ABSTRACT

2-D differential convolution operators in a rectangular discrete domain are modelled as linear differential matrix equations of the form $\dot{X}(t) = \Sigma A_j X(t) B_i + F(t)$, where the coefficient matrices $A_i \in \mathbb{R}(MxM), B_i \in \mathbb{R}(NxN)$ and $F(t) \in \mathbb{R}(MxN)$ is a piecewise continuous function matrix of t. Under some periodicity assumptions an analytical solution method is presented making use of the 2-D discrete Fourier transform (DFT). For most general problems an iterative computational method founded in the Peano-Baker series is proposed.

## INTRODUCTION

The study of 2-D differential convolution models has been a subject of considerable interest in many fields, such as boundary problems in partial differential equations; Greenspan 1974, two-dimensional systems, Rabiner and Rader 1972; Fourier transforms, Goodman 1968; particle dynamics, Berger and Davids 1965; fluid flow dynamics, Fromm 1969; digital filtering, Hunt 1973 and image restoration, Andrews and Hunt 1977. In the framework of linear and space-invariant convolution kernels (impulse response, diffusion law, point spread function, etc.), 2-D differential convolution processes are modelled by the superposition or Fredholm integral-differential equation of the first kind. The discrete formulation of this equation in a rectangular domain is given by the following system of ordinary differential equations

$$\dot{X}(m,n,t) = \sum_{a=-U}^{U} \sum_{b=-V}^{V} H(a,b)X(m-a,\ n-b,\ t) + F(m,n,t) \qquad (1)$$

where the set of initial conditions $X(m,n,t_o)$ is known at advance, $X(m,n,t)$ and $F(m,n,t)$ are defined in a rectangular lattice $m = 0,1,\ldots, M-1$; $n = 0,1,\ldots,N-1$ and, for all $t \geq t_o$ $F(m,n,t)$ are piecewise continuous functions of t. Moreover, we shall view matrices a rectangular arrays of complex numbers, like "rectangular windows" into an infinite array and, in the absence of further information, all values outside these "windows" are assumed to be zeroes. Thus, we shall be mainly concerned with zero boundary value problems for all $t \geq t_o$.

Centering our attention on the solutions of system Equations (1), they can be viewed as a coupled system of MxN scalar differential equations for the state variables X(m,n,t); hence, we can write in the compact matrix-vector notation, Dunford and Schwartz 1966

$$\dot{\bar{x}}(t) = H^*\bar{x}(t) + \bar{f}(t) \qquad (2)$$

where $\bar{x}(t)$ and $\bar{f}(t)$ denote MxN-dimensional column vectors formed by stacking the rows of X(t), $F(t) \in \mathbb{R}(MxN)$, and $H^*$ is of dimension (MN)x(MN). This matrix-vector form is the usual notation utilized for 2-D differential convolutions and is based on the concept of tensor or Kronecker's product of two matrices and that of the vector form of a matrix. The main drawback with this approach comes from the considerable increase in the matrix space dimensions: A problem formely formulated in $\mathbb{R}(MxN)$ is transformed into a problem in $\mathbb{R}(MNxMN)$. This "dimensionality course" gives rise to serious computational troubles when solving Equation (2) by using available software for linear systems. Moreover, from the point of view of theoretical analysis use of the matrix vector form obscures the determination of the eigensystem of matrix $H^*$ and then the fundamental modes computation. Thus, application of this formulation, while quite general, limits theoretical and computational developments for specific problems. The approach in this paper is different in the sense of determining the class of differential matrix equations that models linear, space-invariant 2-D differential convolutions defined in rectangular discrete domains, and then analyzing these equations by using the formal and computational tools today available for the solution of such problems. We demonstrate the equivalence of the $H^*\bar{x}$ operator and a matrix operator of the form $\Sigma A_i X B_i$, from which differential convolution processes will be represented by the general linear matrix equation

$$\dot{X}(t) = \Sigma\, A_i X(t) B_i + F(t) \qquad (3)$$

where $A_i \in \mathbb{R}(MxM)$ and $B_i \in \mathbb{R}(NxN)$. This equation is then investigated in order to devise adequate solution algorithms. The problem in its full generality is far from tractable in an analytical way, although, by assuming periodicity in the convolution kernel and input functions, we shall be able to solve the algorithmic problem in a closed form. For the periodicity case the solution is derived in terms of the eigensystem of particular circulant matrices $A_i$ and $B_i$. The complete set of eigenfunctions is shown to be given by exponentials of the discrete Fourier transform coefficients of kernel H(a,b) and FFT techniques are shown to be useful to solve very large 2-D differential deconvolutions. For the most general (aperiodic) deconvolution problem a computational numerical method based on matrix series iterations is presented jointly with an analysis of the convergence behaviour; this method leads to a natural extension of the Peano-Baker series method to solve homogeneous linear state equations.

To save notation, in the following sections we shall neglect the explicit representation of time variable t in most equations. Also, without any loss of generality we shall use square matrices instead of rectangular ones, being possible to extend the results to rectangular domains in a trivial manner.

The theory in this paper can be applied to solve many problems arising in physics and engineering of the form

$$\frac{\partial \Psi}{\partial t} = \mathcal{D}(\Psi) + F; \; \mathcal{D}(.): \text{ linear differential operator}$$

in an infinite or finite rectangular domain. This is the case of the heat equation and the diffusion equation in infinite porous media. Moreover, the results in this paper endeavours to enlarge previous results, Incertis 1981, on the algebraic formulation for solving 2-D differential convolution equations which are closely linked with many initial free-boundary value problems in partial differential equations.

THE DIFFERENTIAL CONVOLUTION MATRIX EQUATIONS

Let $X \in \mathbb{R}(MxM)$ be a square matrix with elements X(m,n); m,n=0,1, ...M-1. Let H be a real matrix with elements H(a,b) and where a,b are integers in the intervals $a\in[-U,U]$, $b\in[-V,V]$. The axis conventions used for the representation of matrices X and H are shown in Figure 1

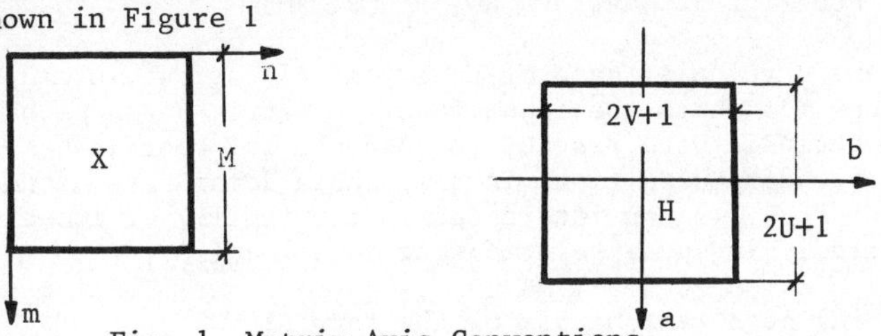


Fig. 1. Matrix Axis Conventions

We now introduce a class of matrix operators from which a linear differential matrix equation for (1) will be derived. Consider a real matrix $L \in \mathbb{R}(MxN)$ with elements

$$L(m, m+1) = 1 \;, \qquad M=0,1,...,M-2 \tag{4}$$

and L(m,n) = 0 otherwise. Given $X \in \mathbb{R}(MxM)$ we can analize the two-dimensional transformations induced by the operations $Z = L^a X L^b$, $L^a X L^{Tb}$, $L^{Ta} X L^b$, $L^{Ta} X L^{Tb}$, where the symbol T denotes transpose and a, b are positive integers. Lexicographical computation with those operators leads to the following results

$$L^a X L^b \quad : \quad Z(m,n) = X(m+a, \; n-b) \tag{5.1}$$

$$L^a X L^{Tb} \quad : \quad Z(m,n) = X(m+a, \; n+b) \tag{5.2}$$

$$L^{Ta}XL^{b} \quad : \quad Z(m,n) = X(m-a,\ n-b) \tag{5.3}$$

$$L^{Ta}XL^{Tb} \quad : \quad Z(m,n) = X(m-a,\ n+b) \tag{5.4}$$

for all $m,n \in [0,\ M-1]$ and where $X(m,n) = 0$ for all $m,n \notin [0,\ M-1]$. Now, it is possible to rearrange the summation terms of Equation (1), to give

$$\dot{X}(m,n)-F(m,n) = \sum_{a=o}^{U} \sum_{b=o}^{V} H(a,b)X(m-a,n-b) +$$

$$+ \sum_{a=1}^{U} \sum_{b=o}^{V} H(-a,b)X(m+a,n-b) + \sum_{a=1}^{U} \sum_{b=1}^{V} H(-a,-b)X(m+a,n+b) +$$

$$+ \sum_{a=o}^{U} \sum_{b=1}^{V} H(a,-b)X(m-a,n+b) \tag{6}$$

Hence, by forming $(\dot{X} - F) \in \mathbb{R}(MxM)$ from the above equations and by substituting the relationships given by Equations (5), we obtain

$$\dot{X} - F = \sum_{a=o}^{U} \sum_{b=o}^{V} H(a,b)L^{Ta}XL^{b} + \sum_{a=1}^{U} \sum_{b=o}^{V} H(-a,b)L^{a}XL^{b} +$$

$$+ \sum_{a=1}^{U} \sum_{b=1}^{V} H(-a,-b)L^{a}XL^{Tb} + \sum_{a=o}^{U} \sum_{b=1}^{V} H(a,-b)L^{Ta}XL^{Tb} \tag{7}$$

which is a differential linear matrix equation of the general form

$$\dot{X} - F = \sum_{a=-U}^{U} \sum_{b=-V}^{V} A_a X B_b \tag{8}$$

where the coefficient matrices $A_a$, $B_b \in \mathbb{R}(MxM)$.

An important particular case of Equations (7) arises in many applications for which the input function matrix F can be assumed to be periodic with respect to both spatial coordinates and where the boundary conditions in the square domain are assumed to be free. Thus, we are interested in the familly of two-dimensional periodic in space solutions of Equation (7).

Let $P \in \mathbb{R}(MxM)$ be a real matrix of the form

$$P(M-1,0) = 1,\ P(m,m+1) = 1 \text{ for all } m = 0,1,\ldots,M-2 \tag{9}$$

and where $P(m,n) = 0$ otherwise. This circulant matrix fulfils the orthonormality condition $P^T = P^{-1}$. Now, if matrices X,F, and H are supposed to be periodic, with period M in both dimensions we have that

$$X(m,n) = X(m+\alpha M,\ n+\beta N) \tag{10}$$

for all $m,\ n \in [0,\ M-1]$ and all integers $\alpha, \beta \in [-\infty, \infty]$. By using lexicographical computations it is easily proved that the 2-D transformation induced by operator $Z = P^{Ta}XP^{b}$ is given by

$$Z(m,n) = X(m-a,\ n-b) \tag{11}$$

for all $a \in [-U,U]$, $b \in [-V,V]$ and all $m,n \in [0,M-1]$. Thus, by forming $(d/dt\ X - F) \in \mathbb{R}(MxM)$ by substituting the above relationship into equation (1), we obtain

$$\dot{X} - F = \sum_{a=-U}^{U} \sum_{b=-V}^{V} H(a,b)P^{Ta} XP^{b} \tag{12}$$

which is a differential linear matrix equation of the form of Equation (8) and where all the pre-and post-multiplying coefficient matrices are integer powers of the same basic circulant matrices $P^T$ and P. In the following section we demonstrate the usefulness of this matrix formulation to solve 2-D differential deconvolution problems.

ANALYTICAL SOLUTION METHODS

For the general Equation (7) the best we can do to compute an analytical solution is to formulate the problem in terms of the equivalent matrix-vector differential equation: Let by definition A⊗B be the direct (tensor or Kronecker) product of matrices A and B, Neudecker 1969 and let $\bar{x},\bar{f}$ be the column vectors associated with matrices X and F, by stacking its row vectors in ordered form. It is a routine matter to verify that if

$$H^* = \sum_{a=o}^{U} \sum_{b=o}^{V} H(a,b)L^{Ta} \otimes L^{Tb} + \sum_{a=1}^{U} \sum_{b=o}^{V} H(-a,b)L^{a} \otimes L^{Tb} + \\ + \sum_{a=1}^{U} \sum_{b=1}^{V} H(-a,-b)L^{a} \otimes L^{b} + \sum_{a=0}^{U} \sum_{b=1}^{V} H(a,-b)L^{Ta} \otimes L^{b} \tag{13}$$

then Equation (7) is equivalent to

$$\dot{\bar{x}} = H^* \bar{x} + \bar{f} \tag{14}$$

where $H^* \in \mathbb{R}(M^2 x M^2)$. Thus, if $\bar{x}_o$ is the column vector associated with $X(t_o) = X_o$ then Equation (14) can be integrated, Brokett 1970 to give

$$\bar{x}(t) = \Phi(t,t_o)\bar{x}_o + \int_{t_o}^{t} \Phi(t,r)\bar{f}(r)dr \tag{15}$$

where

$$\Phi(t,t_o) = \exp H^*(t-t_o) \tag{16}$$

is the transition matrix of $H^*$. Let $\sigma(H^*)$ be the spectrum of $H^*$. In the general case, referred to Equation (7) it is difficult to determine $\sigma(H^*)$ as an analytical function of H and L. The first special case which allows us to determine $\sigma(H^*)$ is the following, Lancaster 1970: Consider Equation (12) with $P \in \mathbb{R}(MxM)$ and where H(a,b) are real numbers. If we define the polynomial in two scalar variables

$$\phi(z_1, z_2) = \sum_{a=-U}^{U} \sum_{b=-V}^{V} H(a,b) z_1^a z_2^b \tag{17}$$

and then define the polynomial with matrix arguments

$$\phi(P^T, P) = \sum_{a=-U}^{U} \sum_{b=-V}^{V} H(a,b)P^{Ta} \otimes P^b \tag{18}$$

it is then a routine matter to prove that Equation (12) is

equivalent to Equation (14) with $H^* = \phi(P^T, P^T)$. The remarkable fact about this class of problems is the following result, which is a particular case of a theorem proved in Lancaster 1970:

Theorem 1: The spectrum of matrix $H^* = \phi(P^T, P^T)$, associated to Equation (12), is the set of all complex numbers of the form $\phi(\lambda_m, \lambda_n)$, where $\lambda_m, \lambda_n \epsilon \sigma(P)$.

The application of the above results on the eigensystem analysis of matrix $H^*$ require the computation of the complete set of eigenvalues of matrix P. Now, from González and Wints 1977 the spectrum of this circulant is given by the formula

$$\lambda_k = \exp \theta k \quad ; \quad k = 0,1,\ldots,M-1 \tag{19}$$

with $\theta = j\,2\pi/M$, $j = \sqrt{-1}$. Moreover, for every $\lambda_k \epsilon \sigma(P)$ the associa̲ted eigenvector $\vec{w}_k$ is given by

$$\bar{w}_k^T = [1, \exp \theta k, \exp 2\theta k, \ldots, \exp(M-1)\theta k]; \; k=0,1,\ldots,M-1 \tag{20}$$

An analytical approach to the solution of the periodic differential deconvolution Equation (12) by using matrix operators is obtained by reducing matrix P to the Jordan normal form. Because P is a real matrix with only linear elementaty divisors, it can always be diagonalized, so that

$$P = W \Lambda W^{-1} \tag{21}$$

where

$$W = [\bar{w}_o, \bar{w}_1, \ldots, \bar{w}_{M-1}] \text{ and } \Lambda = \text{Diag}[\lambda_o, \lambda_1, \ldots, \lambda_{M-1}] \tag{22}$$

Now, since the eigenvectors structure, Equation (20), the components of matrices W and $W^{-1}$ are given by

$$W(m,n) = \exp \theta mn \; ; \; W^{-1}(m,n) = 1/M \exp - \theta mn \tag{23}$$

for all $m,n = 0,1,\ldots,M-1$. Now, from the above equations W is a symmetric matrix; thus, being $P^T = P^{-1}$, we can write

$$P^T = (W \Lambda W^{-1})^T = W^{-1} \Lambda W = P^{-1} = W \; \Lambda^{-1} W^{-1} \tag{24}$$

Substituting Equation (21) into (12), this equation takes the form

$$\frac{d}{dt} X - F = \sum_{a=-U}^{U} \sum_{b=-V}^{V} H(a,b)\, W^{-1} \Lambda^a W X \Lambda^b W^{-1} \tag{25}$$

and, by using the notations

$$X_w = W X W \quad \text{and} \quad F_w = W F W \tag{26}$$

it is then easily shown that Equation (12) is equivalent to

$$\dot{X}_w - F_w = \sum_{a=-U}^{U} \sum_{b=-V}^{V} \Lambda^a X_w \Lambda^b \tag{27}$$

where $X_w(t_o) = X_{wo} = W X_o W$. Now, since $\Lambda$ is a diagonal matrix, we can write

$$H_w^* = \phi(\Lambda,\Lambda) = \sum_{a=-U}^{U} \sum_{b=-V}^{V} H(a,b)\Lambda^a \circledast \Lambda^b \tag{28}$$

From which $H_w^*$ is a diagonal matrix, with diagonal elements $\phi(\lambda_m, \lambda_n)$, for all $m,n = 0,1,\ldots,M-1$. Furthermore, from Equations (17) and (19)

$$\phi(\lambda_m, \lambda_n) = \sum_{a=-U}^{U} \sum_{b=-V}^{V} H(a,b) \exp\theta\,(ma+nb) \tag{29}$$

which clearly coincides with the 2-D DFT (Discrete Fourier Transform) of the sequence $H(a,b)$; $a\epsilon[-U,U]$, $b\epsilon[-V,V]$ for all points $m,n = 0,1,\ldots,M-1$. Thus, Equation (12) is decoupled as

$$\dot{X}_w(m,n) - F_w(m,n) = \phi(\lambda_m, \lambda_n)\, X_w(m,n) \tag{30}$$

and by using the initial condition $X_{wo}(m,n)$ at $t_o$, we get

$$X_w(m,n,t) = [\exp\phi(\lambda_m, \lambda_n)(t-t_o)]X_{wo}(m,n) + + \int_{t_o}^{t} [\exp\phi(\lambda_m, \lambda_n)(t-r)]F_w(m,n,r)dr \tag{31}$$

where the arguments in the exponentials are given by Equation (29). Finally, the required solution $X(t)$ is obtained by inverting transformation (26), by means of

$$X(t) = W^{-1} X_w(t)\, W^{-1} \tag{32}$$

Now, since Equation (26) by using lexicographical computations it is a routine matter to verify that

$$F_w(m,n,t) = \sum_{k=o}^{M-1} \sum_{\ell=o}^{M-1} F(m,n,t) \exp\theta(mk+n\ell) \tag{33}$$

$$X_{wo}(m,n) = \sum_{k=o}^{M-1} \sum_{\ell=o}^{M-1} X_o(m,n) \exp\theta(mk+n\ell) \tag{34}$$

which are, be definition, the 2-D DFT of matrices $F(t)$ and $X_o$. Moreover, by means of Equation (32), lexicographical computation gives

$$X(m,n,t) = 1/N^2 \sum_{k=o}^{M-1} \sum_{\ell=o}^{M-1} X_w(k,\ell,t) \exp -\theta(mk+n\ell) \tag{35}$$

which is, by definition, the 2-D inverse DFT of the time function matrix $X_w(t)$. Equations (29),(31),(33 - 35) formally characterize the computational algorithm to solve the periodic 2-D differential deconvolution problem (Equations (1)) or, what is equivalent, differential linear matrix Equation (12). Moreover the usefulness of the discrete Fourier Analysis (frequency domain analysis) to solve 2-D differential deconvolutions has been demonstrated. The algorithm steps can be summarized as follows:

i) Compute the 2-D DFT of the kernel $H(a,b)$; $a\epsilon[-U,U]$, $b\epsilon[-V,V]$ for all points $m,n = 0,1,\ldots,M-1$, to obtain $\phi(\lambda_m, \lambda_n)$.

ii) Compute the 2-D DFT of the forcing function matrix $F(t)$

and the initial state matrix $X_o$, to give $F_w(t)$ and $X_{wo}$

iii) Compute, by using Equation (31) $X_w(m,n,t)$ for all $m,n = 0,1,\dots,M-1$, to obtain $X_w(t)$.

iiii) Compute the 2-D inverse DFT of $X_w(t)$ to determine solution $X(t)$.

Now, consider the situation in which the state and input matrices are sampled periodically in time, so that their values are available every T seconds, or $t = kT$ where $k \geq 0$. Under these circumstances Equation (31) can be integrated to give

$$X_w(m,m(k+1)T) = C(m,n)X_w(m,n,kT) + D(m,n)F_w(m,n,kT) \tag{36}$$

with $X_w(m,n,0) = X_{wo}(m,n)$, and where

$$C(m,n) = \exp \phi(\lambda_m, \lambda_n)T \tag{37}$$

$$D(m,n) = \int_o^T [\exp \phi(\lambda_m, \lambda_n)r]dr \; ; \; m,n = 0,1,\dots,M-1 \tag{38}$$

These equations are readily programmed for a digital computer to provide the digitally simulated 2-D system response. By using FFT techniques to implement 2-D direct and inverse DFT, Cooley and Tuckey 1972, we can give a roughly estimation of the computational cost (number of floating-point real additions and products) required to implement the above analytical procedure. For a complete simulation with a total of S time steps the operations count is about $N_{op} = 4M^2(1+2\log_2 M)S$. In this formula the cost of computing matrices C and D has not been included since it represents an small percentage of the total cost.

## NUMERICAL SOLUTIONS METHODS

The homogeneous differential matrix equation associated to Equation (7) can be written in the general equivalent form

$$\dot{X}(t) = \sum_i A_i X(t) B_i \quad , \quad X(t_o) = X_o \tag{39}$$

Now, by utilizing an extension of the Peano-Baker series method, Brockett 1970, we can formulate without proof the following

Theorem 2. If the sequence of matrices $Y_k$ is defined recursively by

$$Y_k = X_o + \sum_i A_i \int_{to}^t Y_{k-1}(r)dr \, B_i \; ; \quad Y_o = X_o \tag{40}$$

then, the sequence of matrices $Y_o, Y_1,\dots,Y_k,\dots$ converges uniformly on the interval $t_o \leq t \leq t_1$. Moreover, if $\|.\|$ denotes any kind of matrix norm, such as the euclidean matrix norm, and if $\eta = \|\Sigma A_i B_i\|$ then $\|Y_k - Y_{k-1}\| \leq [\eta^k (t-t_o)^k/k!].\|X_o\|$ which gives a bound for the k-th iteration error norm. Moreover, observe that, since the band structure of matrix L (Equation (4)) computation of the recurrence formulas (40) involve, as elementary operations, row and column shifts (rotations) of matrix $Y_{k-1}(t)$, matrix additions and scalings (multiplication of a matrix by scalar). Now, since matrices $A_i$ and $B_i$ are constant the Peano-Baker series converges uniformly and absolutely on any finite interval. The explicit representation of the solution is

$$X(t) = X_o + \sum_i A_i X_o B_i (t - t_o) + \sum_i \sum_i A_i A_k X_o B_k B_i (t-t_o)^2/2! + \ldots \quad (41)$$

The computational cost to determine S+1 terms in Equation (44) is approximately given by $N_{pb} = \nu M^2 S$ basic operations, where $\nu$ is the number of non-zero coefficients in the convolution Kernel H(a,b). Note that the convergence rate of this iterative method strongly depends on the value of $\eta$. Thus, it is possible to improve the preformance of the analytical FFT based methods in some circumstances for which $\nu$ and $\eta$ are small, with the aditional advantage that all computations are implemented by using only arithmetical packages for real numbers.

EXAMPLE

The following example illustrate the method and some possible extensions and applications. Example: we consider the hyperbolic differential equation (wave equation) written in normalized form $\Psi_{tt} = \Psi_{xx} + \Psi_{yy}$. By discretizing this equation in a rectangular lattice of points $\mathbb{R}(MxN)$ by using $\varepsilon$ discretization step the 2-D second order differential convolution equation is given by

$$\ddot{X} = [-4X + A_M X + X A_N]/\varepsilon^2$$

where $A_M = L_M + L_M^T$ and $A_N = L_N + L_N^T$. Now, from Incertis 1981, the complete eigensystem of $A_M$ and $A_N$ is known. The eigenvalues are given by $\lambda_{Mk} = 2\cos(\pi k/M+1)$; k=0,1,...,M-1 and $\lambda_{Nk} = 2\cos(\pi k/N+1)$; k=0,1,...,N-1. Hence, if $W_M$ and $W_N$ are the respective eigenvector matrices, following the basic main results in this paper, we introduce the new variable $X_w(t) = W_M X(t) W_N$ and then, the wave differential equation, with zero boundary conditions, can be decoupled into the set of MxN scalar equations

$$\ddot{X}_w(m,n,t) = \phi(\lambda_{Mm}, \lambda_{Nn})\, X_w(m,n,t)$$

with

$$\phi(\lambda_{Mm}, \lambda_{Nn}) = (2/\varepsilon^2)[-2 + \cos(\pi m/M+1) + \cos(\pi n/N+1)]$$

Now, this decoupled system can be inmediately integrated in closed analytical form by using as initial conditions $X_w(t_o)$ and $\dot{X}_w(t_o) = W_M \dot{X}(t_o) W_N$, to obtain $X_w(t)$ and, by reversing the transformation we finally determine $X(t) = W_M^{-1} X_w(t) W_N^{-1}$.

CONCLUDING REMARKS

We have discussed in considerable detail some new theoretical results leading to closed analytical solutions for 2-D differential convolution equations arising in dynamical distributed parameter systems analysis and in some, free-boundary, initial value problems in partial differential equations. The new compact matrix notation introduced in this paper appears to be very useful for theoretical and computational developments making possible to apply powerful results from matrix analysis to solve partial differential equations.

## REFERENCES

Andrews, H.C. and B.R. Hunt (1977) Digital Image Restoration, Prentice Hall, Englewood Cliffs, New Jersey.

Berger, R.L. and N. Davids (1965) General computer method analysis of condition and diffusion in biological systems with distributive sources, Rev. Sci. Instr. 36, 88-93.

Brockett, R.W. (1970) Finite Dimensional Linear Systems, John Wiley ans Sons, New York.

Cooley J.W. and J.W. Tuckey (1972) An algorithm for the machine computation of complex Fourier series, IEEE Press, Compiled in Digital Signal Processing, 223-228.

Dunford, N, and J.T. Schwartz (1966) Linear Operators, Part 1 Interscience, New York.

Fromm, J.E. (1969) Lectures on large scale finite difference computation of incomprensible fluid flows, Rpt. RJ 617, IBM Research, San Jose, Calif.

Gonzãlez, R.G. and P. Wints (1977) Digital Image Processing, Addison Wesley, Advanced Book Program, New York.

Goodman, Y.W. (1968) Introduction to Furier Optics, Mac Graw-Hill, New York.

Greenspan, D. (1974) Discrete Numerical Methods in Physics and Engineering, Academic Press, New York.

Hunt, B.R. (1973) The application of constrained least-squares estimation to image restoration by digital computer, IEEE Trans. Comput. vol. C-22, 9:805-812.

Incertis, F.C. (1981) On a novel matrix approach for linear space invariant 2-D deconvolutions, (in press) IEEE Trans. Automat. Contr. vol. AC-27, 4.

Lancaster, P. (1970) Explicit solutions of linear matrix equations, SIAM J. Appl. Math., 12, 4:554-556.

Neudecker, H. (1969) A note on Kronecker matrix product and matrix equation systems, SIAM J. Appl. Math. 17: 603-606.

Rabiner, L.R. and C.M. Rader (1972) Digital Signal Processing, IEEE Press, New York.

# MIXED FINITE ELEMENT METHODS FOR NONLINEAR PROBLEMS

M. Aslam Noor and Kh. Z. Elahi
Mathematics Department, King Saud University, Riyadh, S.Arabia

K. Inayat Noor
Mathematics Department, Science College of Education for Girls, Riyadh, Saudi Arabia

## 1. Introduction and Formulations

Finite element methods based on standard variational principle of minimum potential energy have been extensively studied from both practical and theoretical point of view. The mathematical theory of such methods is firmly established and is a part of the numerical analysis. In recent years, considerable efforts have been made to give a complete mathematical analysis of other finite element methods including mixed methods. All these works are related to the linear elleptic problems except that of Noor [1981a] which provides general ideas to obtain error estimates for the mixed finite element approximation of mildly nonlinear elliptic problems. It should be remarked that the nonlinear cases are much complicated, since each problem has almost to be treated individually. This is one of the reason that there is no unified and general technique to obtain the error estimates for the nonlinear problems. The mathematical model, we consider in this paper is highly nonlinear and has been the subject of recent investigations by many research workers including, Glowinski and Marroco [1975], Pelissier [1975], Noor [1981], Babuska [1976] and Bermudez and Moreno [1981].

For simplicity, we consider the problem of finding the velocity $u$ of the glacier which is required to satisfy the nonlinear problem of the type:

$$\left.\begin{aligned} -\nabla(|\nabla u|^{p-2}\nabla u) &= f(u) \quad \text{in } \Omega \\ u &= o \quad \text{on } \partial\Omega \end{aligned}\right\} \qquad (1)$$

where $\Omega$ is the cross-section of the glacier and $f(u)$ is the Coulomb friction. The presence of $f(u)$ may be interpreted as a body heating term, it arises

from resistivity and is known as local Joule heating effect. For the mathematical formulation, see Pelissier [1975].
The problem(1) is a generalization of the nonlinear problem of finding u such that

$$\begin{aligned} Tu &= f \quad \text{in } \Omega \\ u &= o \quad \text{on } \Omega \end{aligned}$$

where $Tu = -\nabla(|\nabla u|^{p-2}\nabla u)$, for which the error estimates have been derived by Noor [1982] by using the mixed finite element approximation. The presence of the nonlinear term f(u) needs a different approach for obtaining the error estimates and this is the main motivation of this paper.
The usual variational form of (1), see Noor [1981] consists in finding $u\varepsilon V$ such that

$$I[u] = \inf_{u \varepsilon V} I[v]$$

where

$$I[v] = \frac{1}{p}\int_\Omega |\nabla v|^p d\Omega - 2\int_\Omega\int_0^v f(\eta)\, d\eta\, d\Omega$$

$$\equiv J(v) - 2F(v),$$

is the energy functional associated with (1). Standard finite element methods have been extensively studied and convergence results are now classical, see Noor [1981].
On the other hand, it has been noticed that one could weaken the requirement of interelement continuity for the function of the space and still obtain a convergent finite element method. This type of methods have been first introduced by the engineers, see Pian and Tong [1969] and can be viewed as a generalization of nonconforming methods. In this paper, we use a more general approach, which is based on extended variational principle in which the constraint of interelement continuity has been removed at the expense of introducing a Lagrange multiplier. By this method, the unknown u and some of its derivatives are simultaneously approximated. It should be pointed out that the mixed formulation will appear to be more complicated than the minimum energy formulation because there are more independent variables. However, it has been discovered that the mixed method is actually much more efficient. An outline of the paper is as follows. Section 2 will be devoted to an abstract construction of approximate methods which can be derived from a mixed variational principle. In section 3, using

the abstract results of the previous section, we derive the error estimates for our model nonlinear problem with homogeneous boundary conditions arising in the study of field of glaciology.
Due to the space limitation, we only consider the special case for p=2 and give few details of the results obtained. The complete results and proofs will appear elsewhere. We like to remark that the approach adopted in this paper can be applied to a wider class of nonlinear problems having more complicated boundary conditions.

## 2. Abstruct Results.

Let V,W and H be three Hilbert spaces with their duals V',W',H' and norms $||.||_V$, $||.||_W$, and $||.||_H$ respectively. Let a(.,.) and b(.,.) be continuous bilinear forms on VxV and VxW respectively and F be a continuous functional on V and $g\varepsilon W'$. We denote by <.,.>, the pairing between V and V' or W and W'. We now consider the problem of finding the saddle point of the functional

$$I[v,\phi]=a(v,v)+2b(v,\phi)-2F(v)-2<g,\phi>. \qquad (2)$$

It has been shown by Noor [1980], that under suitable conditions, $(u,\psi)\varepsilon$ VxW is the saddle point of the functional $I[v,\phi]$ defined by (2) if and only if $(u,\psi)\varepsilon$ VxW satisfies the variational equations.

$$a(u,v)+b(v,\psi) = <F'(u),v>, \text{ for all } v\varepsilon V \qquad (3)$$

$$b(u,\phi) = <g,\phi> \quad \text{for all } \phi\varepsilon W, \qquad (4)$$

where F'(u) is the frechet differential of a nonlinear functional F on V. It is obvious that the variational problem (2) and the weak formulation (3) and (4) are equivalent. This equivalence will play an important part in deriving the error estimates for the mixed finite element technique.
In order to derive the abstract results, we consider the following problem, which we refer to as problem (P):
Find $(u,\psi)\varepsilon$VxW satisfying

$$a(u,v) + b(v,\psi) = <F'(u),v>, \text{for all } v\varepsilon V \quad (5)$$

$$b(u,\phi) = <g, \phi > \quad \text{for all } \phi\varepsilon W \quad (6)$$

We consider this problem for a subclass of data, i.e. for $(F'(u),g)\varepsilon D$, where D is a subclass of V'xW'. We assume that: (H1). For $(F'(u),g)\varepsilon D$,(P) has a unique solution.
In the analysis of (P),we also consider the adjoint problem,

given $d\epsilon G'$, where G is a Hilbert space satisfying $W\subseteq G$ with a continuous embedding, find $(y,\lambda)=(y_d, \lambda_d)$ $\epsilon V \times W$ satisfying

$$a(v,y)+b(v,\lambda)= o \quad \text{for all } v\epsilon V \quad (7)$$
$$b(y,\phi)= \langle d,\phi\rangle, \text{for all } \phi\epsilon W \quad (8)$$

We shall assume that:
(H2). Problem (7) and (8) has a unique solution for each $d\epsilon G'$.
Let $V_h$ and $W_h$ be given finite dimensional subspaces of V and W respectively. We now consider the following approximate problem $(P_h)$:
Find $(u_h,\psi_h)\epsilon V_h x W_h$ such that

$$a(u_h,v_h)+b(v_h,\psi_h)=\langle F'(u_h),v_h\rangle, \text{for all } v_h\epsilon V_h \quad (9)$$
$$b(u_h,\phi_h)=\langle g,\phi_h\rangle, \quad \text{for all } \phi_h\epsilon W \quad (10)$$

We are interesting in deriving error estimates for $u-u_h$ and $\psi-\psi_h$. We now state several assumptions which are required in the proof of our main results.
(H3). There is a constant $\alpha>o$ such that

$$a(v,v)\geq\alpha||v||^2 \qquad \text{for all } v\epsilon Z_h,$$

where

$$Z_h =\{v_h\epsilon V_h: b(v_h,\phi_h)=o, \quad \text{for all } \phi_h\epsilon W_h\}$$

(H4). There exists a number S(h) satisfying

$$||v_h||_V\leq S(h)||v_h||_H, \quad \text{for all } v_h\epsilon V_h.$$

(H5). There is an operator $\Pi_h: Y\longrightarrow V$ satisfying

$$b(y-\Pi_h y,\phi_h)=o, \text{ for all } y\epsilon Y \text{ and } \phi_h\epsilon W,$$

where

$Y =\text{span}(\{y_d\}_{d\epsilon G'}, u)$, $(u,\psi)$ is the solution of (P) and $(y_d,\lambda_d)$ is the solution of (7) and (8) corresponding to $d\epsilon G'$.

(H6). F'(u) is required to satisfy the Lipschitz condition, that is there exists a constant $\gamma>o$ such that

$$||F'(u)-F'(v)||_{H'}\leq\gamma||u-v||_H \quad \text{for all } u,v\epsilon H.$$

(H7). F'(u) is antimonotone on V, i.e.,

$$\langle F'(u)-F'(v), u-v\rangle\leq o, \text{ for all } u, v\epsilon V$$

For (5), we note that

$$a(\pi_h u,v_h)+b(v_h,\psi)=\langle F'(u),v_h\rangle+a(\pi_h u-u,v_h), \text{ for all } v\varepsilon V_h \qquad (10)$$

Subtracting (5) from (10), we get for all $v_h \varepsilon V_h$,

$$a(\pi_h u-u,v_h)+b(\psi-\psi_h,v_h)=\langle F'(u)-F'(u_h),\ v_h\rangle+a(\pi_h u-u,v_h). \qquad (11)$$

This shows that in the nonlinear case, we have the term $\langle F'(u)-F'(u_h),v_h\rangle$, which is not present in the linear problems, i.e., $F'(u) = F$, see Falk and Osborn [1981]. We therefore, introduce the concept of generalized pseudo-projection to obtain the inequalities bounding the error $u-u_h$ and $\psi-\psi_h$. This idea is basically due to Noor and Whiteman (1976) and Noor [1981] for deriving the error bounds for $u-u_h$ in the standard finite element methods for the mildly nonlinear elliptic boundary value problems.

Define $\bar{u}_h \varepsilon V_h$ to be pseudo-projection of $u\varepsilon V$ by the conditions.

$$a(\pi_h u-\bar{u}_h,v_h)+b(v_h,\psi-\psi_h)=a(\pi_h u-u,v_h) \text{ for all } v_h\varepsilon V_h. \qquad (12)$$

$$b(\pi_h u-\bar{u}_h,\phi_h)=o, \qquad \text{for all } \phi_h\varepsilon W_h. \qquad (13)$$

These conditions play an important part in deriving the error estimates. It should be notedthat in the final analysis, $\bar{u}_h\varepsilon V_h$ does not appear.

We now state the main results of this section, which are the abstract error bounds for $u-u_h$ and $\psi-\psi_h$.

<u>Theorem 1</u>
Suppose that the hypothesis (H1)-(H7) hold. If $(u,\psi)$ and $(u_h,\psi_h)$ are solutions of problems (P) and $(P_h)$ respectively, then

$$\|u-u_h\|_H\leq C_1\|u-\pi_h u\|_H+C_2\|\psi-\phi_h\|_W, \text{ for all } \phi_h\varepsilon W_h. \qquad (14)$$

$$\|u-u_h\|_V\leq\|u-\pi_h u\|_V+C_3\|\pi_h u-u\|_H+C_4\|\psi-\phi_h\|_W \qquad (15)$$

$$\|\psi-\psi_h\|_G=\sup_{d\varepsilon G'}\{b(y_d-\pi_h y_d,\psi-\phi_h)+a(u_h-u,\pi_h y_d-y_d) \qquad (16)$$

$$+b(u-u_h,\lambda_d-\eta)+\langle F'(u)-F'(u_h),y_d\rangle\}\ /\|d\|_G,$$

$$\text{for all } \eta\varepsilon W_h. \qquad (17)$$

If in addition, $Z_h\subset Z=\{v\varepsilon V, b(v,\phi)=o \text{ for all } \phi\varepsilon W\}$ then

$$\|u-u_h\|_H\leq C_5\|\pi_h u-u\|_H \qquad (18)$$

$$\|u-u_h\|_V \leq \|u-\pi_h u\|_V + C_6 \|u-\pi_h u\|_H,$$

where C's are constants independent of u and $\psi$ .

Remark:

If F' is independent of u, that is F'(u)=f, then the Lipschitz constant is zero. Consequently our results reduce to the ones proved by Falk and Osborn [1981 ] for the linear elliptic problems. It is obvious that our results not only generalize their results, but also includes them as special cases. Theorem 1 allows us to consider the nonlinear elliptic boundary value problems of type (1)

## 3. APPLICATIONS

We consider our model problem of finding u such that

$$\left.\begin{aligned} -\nabla(|\nabla u|^{p-2}\nabla u) &= f(u), && \text{in } \Omega \\ u &= o && \text{on } \partial\Omega \end{aligned}\right\} \tag{19}$$

where the given function f(u) is Lipschitz continuous and antimonotone. Throughout, we shall use the classical Sobolev spaces, see Ciarlet [1978], for notations and definitions.

Following Raviart and Thomas [1977], let

$H(\text{div};\Omega)=\{v\varepsilon(L_p(\Omega))^p:\text{div } v\varepsilon L_p(\Omega)\}$ with the norm

$$\|v\|^p_{H(\text{div};\Omega)} = \|v\|^p_o + \|\text{div } v\|^p_o$$

In this setting, the energy functional associated with (19) can be written in the form:

$$I[v,\phi]=\frac{1}{p}\int_\Omega (v)^p d\Omega + \int_\Omega \phi \text{div } v d\Omega - \int_\Omega\int_0^v f(\eta)d\eta d\Omega. \tag{20}$$

If follows that $(u,\psi)\varepsilon H(\text{div};\Omega)xL_p(\Omega)$ is the saddle point of the functional $I[v,\psi]$, defined by (20) if and only if $(u,\psi)\varepsilon H(\text{div};\Omega)xL_p(\Omega)$ such that

$$\int_\Omega u.vd\Omega + \int_\Omega \psi \text{div } vd\Omega = \int_\Omega f(u).vd\Omega, \text{ for all } v\varepsilon H(\text{div};\Omega) \tag{21}$$

and

$$\int_\Omega \phi \text{div } ud\Omega = o \quad \text{for all } \phi\varepsilon L_p(\Omega) \tag{22}$$

One can easily see that (21) and (22) is an example of problem (P) with $V=H(\text{div};\Omega)$, $W=L_p(\Omega)$, $H=\{L_p(\Omega)\}^p$,

$a(u,v) = \int_\Omega u.v d\Omega$, $b(u,\psi) = \int_\Omega \psi \text{div } u d\Omega$, $g=o$, $\langle F'(u), v\rangle = \int_\Omega f(u) v d\Omega$, see Noor and Whiteman [1976].

For simplicity, we consider the case p=2, only. The subclass D of data for which (H1) is satisfied is given by $D \equiv O x W'$. Since the bilinear form a(u,v) is symmetric, it can be easily shown that (H2) and (H3) are satisfied where $\alpha=1$.

Let $\{T_h\}_{h>o}$ be a regular family, see Ciarlet [1978], of triangulations of $\Omega$ and defind.

$$V_h = \{v_h \varepsilon H(\text{div};\Omega),\text{ for } T\varepsilon T_h,\ v_h|_T \varepsilon Q_T\},$$

where

$$Q_T = \{v\varepsilon H(\text{div};T): v\varepsilon \hat{Q}\},\text{see Falk and Osborn [1981]}$$

and

$$W_h = \{\phi_h \varepsilon L_2(\Omega),\text{ for all } T\varepsilon T_h, \phi_h|_T \varepsilon P_k\},$$

where $P_k$ is apolynomial of degree k.

To apply the results of section 2, we must verify that the appropriate hypothesis are satisfied. Actually (H4) and (H5) are shown by Falk and Osborn [1981] to hold in this case. Furthermore, it has been shown that for $v\varepsilon\{H^{r-1}(\Omega)\}^2, r\geq 2$, the following results hold

$$\|v-\P_h v\|_o \leq Ch^\ell \|v\|_\ell, 1\leq \ell \leq \min(r-1,k+1) \tag{23}$$

and

$$\|\text{div}(v-\P_h v)\|_o \leq Ch^m \|\text{div } v\|_m,\ o\leq m\leq \min(r-2,k+1). \tag{24}$$

From the definitions of $V_h$ and $W_h$, it follow that for all $v_h \varepsilon V_h$, $\text{div } v_h|_T \varepsilon P_k$. Thus $v_h \varepsilon Z_h$ implies that $\text{div } v_h = 0$ and so $v_h \varepsilon Z$. Thus $Z_h \subset Z$ and we are in special cases of theorem 1. In order to apply our results, it remains to show that (H6) and (H7) are satisfied. Actually these have been proved in Noor and Whiteman [1976]. Thus from the above discussion we gather that all the hypotheses (H1)-(H7) are satisfied for our model problem (19) and so we can apply the results of theorem 1.

Furthermore, we now derive the error estimates. Assuming $\psi\varepsilon H^r(\Omega)$, $r\geq 2$, from (18) and (23), we obtain

$$\|u-u_h\|_o \leqslant C\|u-\Pi_h u\|_o, \quad \text{for } k>o.$$

$$\leqslant Ch^t\|u\|_t \leqslant Ch^t\|\psi\|_{t+1}, \text{ if } u = \text{grad}\psi, \quad (25)$$

where $t = \min(r-1,k+1)$

Again using the technique and the results of Falk and Osborn [1981], we can also show that in our case the following estimates are true.

$$\|\psi-\psi_h\|_o \leqslant h^\mu\|\psi\|_\mu \text{ , } \mu = \min(r,k+1)$$

and

$$\|\psi-\psi_h\|_o \leqslant Ch\|\psi\|_2, \quad k = 2.$$

## 4. Acknowledgement

The research of Dr. M. Aslam Noor, was in part supported by the Research Centre, Faculty of Engineering, King Saud University, under grant No.1402/11.

## References

Babuska, I. (1976) Singularity problems in the finite element method in Formulations and Computational algorithms in Finite Element Analysis (ed.) K. Bathe, J. Oden and W. Wunderlich, M.I.T. Press U.S.A., 748-792.

Bermudez, A. and Moreno, C. (1981) Duality methods for solving variational inequalities, Comp. & Math. Appl. 7, 43-58.

Ciarlet, P. (1978) The finite element method for elliptic problems, North-Holland.

Falk, R. and Osborn, J. (1981) Error estimates for mixed methods, RAIRO, Numerical Analysis.(in press).

Glowinski, R. and Marroco, A. (1975) Sur l'Approximation, par Elements Finis d'Ordre un et la Resolution par Penalisationdualite d'Une Classe de Problems de Dirichlet non Lineaires, RAIRO, R-2, 41-76.

Noor, M.A. (1980) Error bounds for the mixed finite element method,C.R.Math.Rep.Acad.Sc.Canada,2:227-230.

Noor, M.A. (1981a) Mixed finite element methods for mildly nonlinear elliptic problems, Conf. Math. Fin.El.& Appl. MAFELAP IV. Brunel University,London.

Noor, M.A. (1981) Finite element approximation theory for strongly nonlinear problems, Comm. Math. Univ. St. Pauli. (in press).

Noor, M.A. (1982) Mixed finite element analysis of nonlinear elliptic problems, 4th Int. Conf. Fin.El. Methods, Australia.

Noor, M.A. and Whiteman, J.R. (1976), Error bounds for finite element solutions of mildly nonlinear elliptic boundary value problems, Num. Math. 26, 107-116.

Pellissier, M.C. (1975) Sur quelques problems non-lineaires en glaciologie, Universite Paris XI, U.E.R. Math. Orsay, France, T/R 110-75-24.

Pian, T.H.H. and Tong, P. (1969) Basis of finite element methods for solid continua, Int. J. Num. Meth. Eng. 1:3-28.

Raviart, R. and Thomas, J. (1977). A mixed finite element method for second order elliptic problems, Lecture Notes in Mathematics, No.606, Springer-Verlag, Berline, New York, London, 292-315.

# CERTAIN VARIATIONAL PRINCIPLES WITH NO CONSTRAINTS FOR FLUID-SOLID STRONG INTERACTION

Nesrin Sarıgül and M. Cengiz Dökmeci

The University of Arizona, Tucson, Arizona, and Istanbul Technical University, P.K. 9, Istanbul

## ABSTRACT

The classical variational principles of fluid-solid interaction impose certain constraint as well as stringent interface conditions. The conditions which are often undesirable in obtaining approximate direct solutions are relaxed through Friedrichs's transformation. Hence, the unconstrained variational principles are formulated which yield, as the Euler equations, all the fundamental equations of a viscous incompressible fluid and an anisotropic elastic solid immersed within the fluid of finite extent, and their interface continuity conditions.

## 1. INTRODUCTION

With their familiar important features in formulating finite element models, variational principles, though only a few, were also constructed for the problems of fluid-solid dynamic interaction. The strong dynamic interaction occurs wherever the relative motions of fluid and solid continuum take place, and it requires a simultaneous treatment of both the two continua at high frequencies. Early developments of variational principles and their use for related interaction problems were due to Gladwell (1966) and his colleagues (1966, 1971) in acousto-elasticity, and Zienkiewicz and Newton (1969), Craggs (1971) and Tong (1971) in hydro-elasticity. Then, among others, Zarda, Chien and Skalak (1977), Dökmeci(1977), Hamdi, Ousset and Verchery (1978), Morand and Ohayon (1978), and Dökmeci and Sarıgül(1982) contributed to this area and exhibited certain applications as well. Besides, Nickell (1975), Oden and Reddy (1976), and Geers and Belytschko (1977) supplied comprehensive surveys, as did Dökmeci (1977). All the existing variational principles exclude the initial conditions and they impose certain constraint conditions which are often burdensome in the choice of trial functions to satisfy them. Accordingly, variational principles with as few constraints as possible have computational advantages over the others in obtaining certain approximate direct solutions (for instance, Tiersten (1968), Wu (1977) and Simkins (1981) ). Thus to formulate the unconstrained variational principles for the fluid-solid strong interaction is taken up in the present paper.

Besides the method of the mirror or adjoint equation, the quasi-variational method and the restricted variational methods (for instance, Dökmeci (1977,80), two main lines of attack, that is, Friedrichs's transformation (Courant and Hilbert (1953) ) and Gurtin's (1963) method of convolution, were followed in formulating variational principles with as few constraints as desirable. Of these methods, due to its simplicity and versatility, Friedrichs's method is employed herein in relaxing the constraint conditions. The inclusion of non-holonomic constraint conditions into a variational principle denies an explicit functional (Lanczos (1952) ) as in the cases of certain fundamental principles of mechanics (for instance, d'Alembert's principle). In fact, a true variational principle, that is, one with an explicit functional, is known to be inexistent for viscous fluids (Millikan (1929) and Finlayson (1972) ) and clearly for the fluid-solid interaction. However, a constrained variational principle, deduced from a general principle of physics or given by a trial-and-error method, can be always transformed into an unconstrained variational principle through Friedrichs's transformation; this will be illustrated below for the problems of fluid-solid interaction.

In Section 2, stating the principle of virtual power for a viscous incompressible fluid of finite extent, a one-field variational principle is derived for the fluid field. Similarly, in Section 3, the principle of virtual work is stated for an anisotropic elastic solid immersed within the fluid, and a one-field variational principle is formulated for the solid field. In Section 4, the two variational principles are combined and a constrained but unified, two-field variational principle is obtained. Then, the rest of constraint conditions is removed through the dislocation potentials and Lagrange multipliers (Friedrichs's transformation). That is to say, the holonomic and non-holonomic constraint conditions are incorporated into the unified variational principle by adding each one multiplied by an undetermined factor to the principle and then treating all variations as free. Thus a generalized variational principle is formulated which leads, as the appropriate Euler equations, to all the fundamental equations of the fluid and solid fields as well as to the natural continuity conditions of traction and velocity at the fluid-solid interface. Further, certain special cases of unconstrained variational principles are pointed out.

## 2. THE PRINCIPLE OF VIRTUAL POWER FOR VISCOUS INCOMPRESSIBLE FLUIDS. A VARIATIONAL PRINCIPLE

The principle of virtual power (Truesdell and Toupin (1960), and Connor and Brebbia (1977) ) is stated as

$$\int_B (t^{kl}-g^{kl}p)\delta d_{kl}\,dV = \int_B \rho(f^k-a^k)\delta v_k\,dV + \int_{\partial B} t^k_* \delta v_k\,dS \quad (1)$$

for a regular, finite and bounded region of viscous fluid, $B+\partial B$, with its boundary surface $\partial B$. In this equation, $t^{kl}$, $g^{kl}$, $d^{kl}$, $f^k$, $v^k$, $a^k=\dot{v}^k$, $p$, $\rho$, $t^l = n_k(t^{kl}-g^{kl}p)$, in this order, denote the symmetric viscous stress tensor, the metric tensor, the velocity strain tensor, the body force vector, the velocity and acceleration vectors, the hydrodynamic pressure, the mass density and the traction vector. By using

$$d_{kl} = 1/2\ (v_{k;l} + v_{l;k}) \quad (2)$$

in Equation (1) and integrating it over T, and then applying the general version of Green-Gauss integral transformation for the fluid region, $B+\partial B$, with a surface of discontinuity, S (the entire boundary surface of the solid region), one finally obtains a variational principle of viscous fluids in the form

$$\begin{aligned}\delta J_f\langle v_k\rangle = \int_T dt \Big\{ &\int_B [t^{kl}{}_{;k} - g^{kl}p_{,k} + \rho(f^l-a^l)]\delta v_l\,dV \\ &+ \int_{\partial B-S} [t^l_* - n_k(-g^{kl}p+t^{kl})]\delta v_l\,dS \\ &+ \int_S \nu_k [\![ (-g^{kl}p+t^{kl})\delta v_l ]\!]\,dS \Big\} = 0 \end{aligned} \quad (3)$$

where $\nu^k$ and $n^k$ stand for the unit outward vectors normal to S and $\partial B$, respectively.

In the equations above and henceforth, standard vector and tensor notation (Synge and Schild (1949) ) is employed. Accordingly, Einstein's summation convention is in operation for repeated indices. Superposed dots stand for time differentiation, and commas and semicolons for partial and covariant differentiations with respect to space coordinates, respectively. Stars denote prescribed quantities, and boldface brackets indicate a jump of enclosed quantity across S. Also $B(t_0)$ represents B at time $t_0$, $T=[t_0,t_1)$ and $x^k$ a fixed, right-handed system of curvilinear coordinates in a 3-D Euclidean space E.

The one-field variational principle (3) has, as its constraints, the kinematic relations (2) and the rest of the linear fundamental equations of viscous incompressible fluid which undergoes only small amplitude vibrations (Landau and Lifshitz (1959) ), namely,

$$g^{kl}v_{l;k} = 0 \quad , \quad t^{kl} = \mu^{klmn}d_{mn} \quad \text{in } (B\cup\partial B)XT \tag{4}$$

$$v_k - v_k^* = 0 \text{ on } \partial_v BXT, \; v_k(x^l,t_o) - w_k^*(x^l) = 0 \text{ in } B(t_o) \tag{5}$$

in the Euclidean coordinate system. The region B is filled by a viscous incompressible fluid with no surface waves, no radiation condition at infinity and no cavitation on S. Equations(4) and (5) denote the equation of continuity, the constitutive relations of Newtonian incompressible fluid with its isothermal viscosity coefficients $\mu_{klmn}$, and the velocity boundary and initial conditions. Further, the variational principle (3) recovers that given by Truesdell and Toupin (1960) in a constrained body with no surface of discontinuity, and that derived by Sarıgül and Dökmeci (1980) through Hamilton's principle.

## 3. THE PRINCIPLE OF VIRTUAL WORK. AN ASSOCIATED VARIATIONAL PRINCIPLE FOR SOLIDS

In the $x^k$-coordinate system of E, consider a regular, finite and bounded region, D+∂D, with its boundary surface ∂D=S, immersed within the viscous fluid, B+∂B. For the linear elastic, anisotropic, non-polar region, the principle of virtual work asserts that

$$\int_D \sigma^{kl}\delta e_{kl}dV = \int_D q(h^k - b^k)\delta u_k dV + \int_{\partial D} \sigma_*^k \delta u_k dS \tag{6}$$

Here, $\sigma^{kl}$, $e^{kl}$, $h^k$, q, $u^k$, $b^k=\ddot{u}^k$, and $\sigma^l = n_k\sigma^{kl}$ denote the symmetric stress tensor, the linear strain tensor, the body force per unit mass of elastic solid, with its mass density q, the displacement and acceleration vectors, and the traction vector, respectively. By substituting the linear strain tensor in the form

$$e_{kl} = 1/2\,(u_{k;l} + u_{l;k}) \tag{7}$$

into Equation (6) and, as before, integrating it over T and then applying the Green-Gauss integral transformation, a one-field variational principle of elastic solid is obtained. This variational principle is expressed in the form

$$\delta J_s \langle u_k \rangle = \int_T dt \Big\{ \int_D [\sigma^{kl}{}_{;k} + q(h^l - b^l)] \delta u_l dV + \int_S (\sigma^l_* - \nu_k \sigma^{kl}) \delta u_l dS \Big\} = 0 \tag{8}$$

which is a familiar one (for instance, Dökmeci (1979) and it can be deduced from Hamilton's principle (Dökmeci (1973) ) as well. The variational principle is to hold all variations consistent with the constraints, that is, the strain-displacement relations (7), the remaining fundamental equations of elastic solid (Mason (1966), and Green and Zerna (1954) ) as

$$\sigma^{kl} = C^{klmn} e_{mn} \quad \text{in } (D \cup S) \times T \tag{9}$$

$$u_k(x^l, t_o) - u^*_k(x^l) = 0 \ , \ \mathring{u}_k(x^l, t_o) - y^*_k(x^l) = 0 \ \text{in } D(t_o) \tag{10}$$

where $C^{klmn}$ are the isothermal elastic stiffenesses, and the interface conditions of the form

$$\nu_k(t^{kl} - \sigma^{kl}) = 0 \ , \quad v_k - \mathring{u}_k = 0 \quad \text{on } S \times T \tag{11}$$

These equations express the continuity of traction, which results from Newton's third law and that of velocity due to no cavitation on S.

## 4. VARIATIONAL PRINCIPLES FOR FLUID-SOLID STRONG INTERACTION

By use of the fundamental lemma of the calculus of variations, the variational principles (3) and (8) yield the equations of motion and the associated natural traction boundary conditions of viscous incompressible fluid and those of linear elastic solid, respectively. By summing up Equations (3) and (8), a unified two-field variational principle is found for the fluid and solid regions, namely,

$$\delta J_{fs} \langle u_k, v_k \rangle = \delta (J_f + J_s) = 0 \tag{12}$$

which has Equations (2), (4), (5), (7) and (9)-(11) as its constraints. The constraint conditions are now incorporated into the variational principle (12) through Friedrichs's transformation.

To begin with, the dislocation potential $J_i$ is introduced so as to remove the interface constraint conditions (11) from the variational principle (12) as

$$\delta J = \delta J_f + \delta J_{se} + \delta J_i = 0 \tag{13a}$$

with

$$\delta J_{se} = \int_T dt \int_D [\sigma^{kl}{}_{;k} + q(h^l - b^l)] \delta u_l dV \tag{13b}$$

$$\delta J_i = \int_T dt \int_S \lambda^k (v_k - \mathring{u}_k) dS \tag{13c}$$

Here, $\lambda^k$ denotes Lagrange undetermined multipliers, and, in view of Equations (11), the traction boundary conditions of elastic solid are dropped out. By executing the variations of $J_i$ and then combining the terms on the surface S, one reads

$$\delta J = \delta(J_{fe}+J_{ft})+\delta J_{se}+\delta J_{in} = 0 \tag{14a}$$

with the definitions

$$\begin{aligned}\delta J_{fe} &= \int_T dt \int_B [t^{kl}{}_{;k} - g^{kl}p_{,k} + \rho(f^l - a^l)]\,\delta v_l\, dV \\ \delta J_{ft} &= \int_T dt \int_{\partial B-S} [t^l_* - n_k(-g^{kl}p + t^{kl})]\,\delta v_l\, dS\end{aligned} \tag{14b}$$

and

$$\begin{aligned}\delta J_{in} = \int_T dt \int_S \{\nu_k [(-g^{kl}p+t^{kl})\delta v_l - \sigma^{kl}\delta\mathring{u}_l] \\ +\delta\lambda^k (v_k - \mathring{u}_k) + \lambda^k(\delta v_k - \delta\mathring{u}_k)\} dS\end{aligned} \tag{14c}$$

It thus follows from Equation (14) the conditions

$$v_k - \mathring{u}_k = 0$$

$$\lambda^l + \nu_k \sigma^{kl} = 0\ , \quad \lambda^l + \nu_k(-g^{kl}p+t^{kl}) = 0$$

for arbitrary and independent variations $\delta\lambda^k$, $\delta\mathring{u}_k$ and $\delta v_k$. From the last two equations, the multipliers $\lambda^k$ are appropriately solved in the form

$$\lambda^l = -1/2\ \nu_k(-g^{kl}p + t^{kl} + \sigma^{kl})$$

By susbtituting the multipliers into Equation (14a) and then rearranging terms, a variational principle of fluid-solid interaction is obtained as follows

$$\delta J = \delta(J_{fe}+J_{ft})+\delta J_{se}+\delta(J_{it}+J_{iv}) = 0 \tag{15a}$$

with

$$\begin{aligned}\delta J_{it} &= \int_T dt \int_S 1/2\ \nu_k(-g^{kl}\delta p + \delta t^{kl} + \delta\sigma^{kl})(\mathring{u}_l - v_l) dS \\ \delta J_{iv} &= \int_T dt \int_S 1/2\ \nu_k [(-g^{kl}p + t^{kl}) - \sigma^{kl}](\delta\mathring{u}_l + \delta v_l) dS\end{aligned} \tag{15b}$$

The variational principle (15) is in agreement with the constrained variational principle derived recently by Dökmeci and Sarıgül (1982b).

It thus remains to incorporate the rest of the constraint conditions into the variational principle (15), as in the case of the constraint conditions (11) by the dislocation potentials and Lagrange multipliers. Following Tiersten (1968) for the initial conditions, Dökmeci and Sarıgül (1982a) for the constraint conditions of viscous fluid and Dökmeci (1979) for those of elastic solid, all the constraint conditions are relaxed and a variational principle with no constraints is formulated in a systematic way. This variational principle is expressed in the form

$$\delta J_{FS}\langle \Lambda_{FS}\rangle = \delta(J_F+J_S+J_I) = 0 \tag{16a}$$

with

$$\Lambda_{FS} = \Lambda_F \cup \Lambda_S \tag{16b}$$

$$\Lambda_F = \{v_k,\ t^{kl},\ d_{kl},\ p,\ t^k\},\ \Lambda_S = \{u_k,\ \sigma^{kl},\ e_{kl},\ \sigma^k\}$$

and

$$\begin{aligned}
\delta J_F &= \delta(J_{fe}+J_{ft}+J_{fc}+J_{fd}+J_{fp}+J_{fi})\\
\delta J_S &= \delta(J_{se}+J_{sc}+J_{su}+J_{si})\\
\delta J_I &= \delta(J_{it}+J_{iv})
\end{aligned} \tag{16c}$$

Here, the first variations of the form

$$\begin{aligned}
\delta J_{fc} &= \int_T dt\int_B (t^{kl}-\mu^{klmn}d_{mn})\delta d_{kl}\,dV\\
\delta J_{fd} &= \int_T dt\int_B [d_{kl}-1/2(v_{k;l}+v_{l;k})]\delta t^{kl}dV\\
\delta J_{fp} &= \int_T dt\int_B g^{kl}v_{l;k}\delta p\,dV\\
\delta J_{fi} &= \int_B [v^k(x^m,t_o)-w_*^k(x^m)]\rho\delta v_k(x^n,t_o)dV
\end{aligned}$$

and

$$\begin{aligned}
\delta J_{sc} &= \int_T dt\int_D (\sigma^{kl}-C^{klmn}e_{mn})\delta e_{kl}\,dV\\
\delta J_{su} &= \int_T dt\int_D [e_{kl}-1/2(u_{k;l}+u_{l;k})]\delta\sigma^{kl}dV\\
\delta J_{si} &= \int_D \{[u^k(x^m,t_o)-u_*^k(x^m)]\delta\dot{u}_k(x^n,t_o)\\
&\quad + [\dot{u}^k(x^m,t_o)-y_*^k(x^m)]\delta u_k(x^n,t_o)\}\rho\,dV
\end{aligned} \tag{17b}$$

are defined, and $\delta J_{se}$, $\delta J_{fe}$ and $\delta J_{ft}$, $\delta J_{it}$ and $\delta J_{in}$ are already given in Equations (13b), (14b) and (15b) respectively. The unified variational principle (16) evidently yields, as its Euler equations, the equations of motion, the natural traction boundary conditions, the equation of continuity, the constitutive equations, the kinematic relations and the natural initial conditions for the region of viscous incompressible fluid; the equations of motion, the constitutive equations, the strain-displacement relations and the natural initial conditions for the region of linear elastic solid; and the natural continuity conditions of traction and velocity at the interface of the two regions. Conversely, if all the fundamental equations of viscous fluid and those of elastic solid as well as the interface continuity conditions are met, the variational principle (16) is obviously satisfied. This unclassical variational principle with no constraints contains certain earlier variational principles of viscous fluids, those of elastic solids and those of fluid-solid interaction (e.g., Oden and Reddy (1976), Dökmeci (1979), Dökmeci and Sarıgül (1982), and references therein), as special cases.

## 5. CONCLUDING REMARKS

This analysis contains the derivation of a non- classical variational principle of fluid-solid strong interaction. First, a one-field variational principle of viscous incompressible fluid is deduced from the principle of virtual power and that of elastic solid from the principle of virtual work. Next, the two principles are combined and then augmented by means of the dislocation potentials and Lagrange multipliers so as to construct the unconstrained variational principle of fluid-solid strong interaction. The variational principle is completely equivalent to all the field equations, the boundary and initial conditions, and the interface continuity conditions of the viscous fluid and the linear elastic anisotropic solid immersed within the fluid of finite extent. Though it is non-classical, the unconstrained variational principle presents a unified basis for the finite element models in the fluid-solid strong interaction. This type of interaction which occurs at high frequencies requires a simultaneous treatment of both the fluid and solid fields. Further, the variational principle provides a great deal of flexibility that can be taken advantage of when choosing the trial functions of direct methods of solution.

The present variational principle of fluid-solid interaction does agree, among others, with the analogous but constrained variational principles due to Tong(1971), Zarda, Chien and Skalak (1977), and Dökmeci and Sarıgül (1982). This variational principle allows a simultaneous approximation on all the field variables, that is, the trial functions need not to satisfy any requirements. However, if these functions satisfy some of the constraint conditions at the outset, the approximation will be obviously more accurate. The latter is also more restrictive and in certain cases unusable, and hence the former is, in general, desirable in applications. This point of importance was discussed very thoroughly by Lanczos (1952) and Tiersten (1968).

In closing, some applications of the unconstrained variational principle, its special cases, and its extensions through Friedrichs's transformation will be taken up in a forthcoming memoir.

REFERENCES


Connor, J.J. and Brebbia, C.A. (1977) Finite Element Techniques for Fluid Flow. London: Newness-Butterwh.

Courant, R. and Hilbert, D. (1953) Methods of Mathematical Physics. New York: Interscience Publishers

Craggs, A. (1971) The Transient Response of a Coupled Plate-Acoustic System Using Plate and Acoustic Finite Elements. J. Sound Vibration, 15, 509-528.

Dökmeci, M.C. (1973) A Generalized Variational Theorem in Elastodynamics, with Application to Shell Theory. Meccanica, 8, 4:252-260.

Dökmeci, M.C. (1977) Hydroelastodynamics. Lecture notes at Istanbul Technical University.

Dökmeci, M.C. (1979) Dynamic Variational Principles for Discontinuous Elastic Fields. J. Ship Research, 23, 2:115-122.

Dökmeci, M.C. (1980) Recent Advances: Vibrations of Piezoelectric Crystals. Int. J. Engng. Science, 18, 3:431-448.

Dökmeci, M.C. and Sarıgül, N. (1982) Certain Variational Principles for Viscous Incompressible Fluids. To appear.

Dökmeci, M.C. and Sarıgül, N. (1982b) A Variational Principle for Fluid-Solid Interaction. To appear.

Finlayson, B.A. (1972) Existence of Variational Principles for the Navier-Stokes Equation. Physics of Fluids, 15, 963-967.

Geers, T.L. and Belytschko, T., Edt. (1977) Computational Methods for Fluid-Structure Interaction Problems. New York: ASME.

Gladwell, G.M.L. (1966) A Variational Formulation of Damped Acousto-structural Vibration Problems. J. Sound Vibration, 4, 172-186.

Gladwell, G.M.L. and Mason, V. (1971) Variational Finite Element Calculation of the Acoustic Response of a Rectangular Panel. J. Sound Vibration,14,115-35.

Gladwell, G.M.L. and Zimmermann, G. (1966) On Energy and Complementary Energy Formulations of Acoustic and Structural Vibration Problems. J. Sound Vibration, 3, 233-241.

Green, A.E. and Zerna, W. (1954) Theoretical Elasticity. Oxford: Clarendon Press.

Gurtin, M.E. (1963) Variational Principles in the Linear Theory of Viscoelasticity. Arch. Rational Mech. Anal., 13, 179-191.

Gurtin, M.E. (1972) The Linear Theory of Elasticity. Handbuch der Physik, VIa/2, 1-295. Berlin: S. Verlag.

Hamdi, M.A., Ousset, Y. and Verchery, G. (1978) Displacement Method for the Analysis of Vibrations of Coupled Fluid-Structure Systems. Int. J. Num. Meth. Engng., 13, 139-150.
Lanczos, C. (1952) The Variational Principles of Mechanics. Toronto: University of Toronto Press.
Landau, L.D. and Lifshitz, E.M. (1959) Fluid Mechanics. London: Pergamon Press.
Mason, W.P. (1966) Crystal Physics of Interaction Processes. New York: Academic Press.
Millikan, C.B. (1929) On the Steady Motion of Viscous Incompressible Fluids; with Particular Reference to a Variation Principle. Phil Mag. 7, 4:641-662.
Morand, H. and Ohayon, R. (1978) Substructure Variational Analysis of the Vibrations of Coupled Fluid-Structure Systems. Finite Element Results. Int. J. Num. Meth. Engng., 14, 741-755.
Nickell, R.E. (1975) Applications of the Finite Element Method in Solid Mechanics, Fluid Mechanics, and Heat Transfer. Developments in Mechanics, 8, 599-626. Norman, Oklahoma: University of Oklahoma Press.
Oden, J.T. and Reddy, J.N. (1976) Variational Methods in Theoretical Mechanics. Berlin: Springer-Verl.
Sarıgül, N. and Dökmeci, M.C. (1980) Variational Principles for Fluid-Solid Interactions, with Application to Ship Beams. J. Acoust. Soc. Amer., 68,S116
Simkins, T.E. (1981) Finite Elements for Initial Value Problems in Dynamics. AIAA J., 19, 10:1357-62.
Synge, J.L. and Schild, A. (1949) Tensor Calculus. Toronto: University of Toronto Press.
Tiersten, H.F. (1968) Natural Boundary and Initial Conditions from a Modification of Hamilton's Principle. J. Math. Phys., 9, 9:1445-1451.
Tong, P. (1971) The Finite Element Method for Fluid Flow. Advances in Matrix Methods of Structural Analysis and Design, 787-808. Huntsville, Alabama: University of Alabama Press.
Truesdell, C. and Toupin, R.A. (1960) The Classical Field Theories. Handbuch der Physik, III/1, 226-793. Berlin: Springer-Verlag.
Wu, J.J. (1977) Solutions to Initial Value Problems by Use of Finite Elements-Unconstrained Variational Formulations. J. Sound Vibration, 53, 341-356.
Zarda, P.R., Chien, S. and Skalak, R. (1977) Interaction of Viscous Incompressible Fluid with an Elastic Body. Computational Methods for Fluid-Structure Interaction Problems, 65-82. New York: ASME.
Zienkiewicz, O.C. and Newton, R.E. (1969) Coupled Vibrations of a Structure Submerged in a Compressible Fluid. Finite Element Techniques, 359-379. Stuttgart : Inst. Statik Dynamik.

# 3. SOLUTION OF THE NAVIER-STOKES EQUATIONS

# CONVECTIVE AND RADIATIVE TRANSFER PAST A POROUS FLAT PLATE IN COMPRESSIBLE FLOW

A. R. Bestman
Mathematics Department, University of Science and Technology, Port Harcourt, Nigeria.

## INTRODUCTION

A principal difficulty in the study of radiating gas flow centers around the radiation flux term which in general is a complicated integral expression. By and large one is called upon to solve a non-linear integro-differential equation which is a formidable task. To circumvent this difficulty, differential approximations have been used extensively in the literature on the assumption that the mean free path of radiation is very small. Briefly we shall only mention the recent work of Helliwell and Mosa (1979) who carried out an analytical - numerical study of an MHD channel flow with temperature gradient.

However in many high temperature phenomena the mean free path may not be too small and the differential approximation is not strictly valid. This paper considers the class of flows where the mean free path could be fairly large or of order O (1).

The model comprises of an infinite porous flat plate, maintained at a very high temperature To, and moving in its own plane with a very large velocity $U_o$. The porosity of the plate makes it an opague boundary at which black body radiation properties are fairly well approximated. Also only variations along the y-axis are permissible (see Fig. 1), so that the problem is a strictly one-dimensional one. Finally fluid is sucked from the porous plate with velocity $V_o$.

This model reduces to a set of ordinary integro-differential equation. The method of weighted residual is then deployed to derive the finite element equations. An iterative method, which is particularly suitable for this type of problem, is devised to solve the non-linear global finite element system.

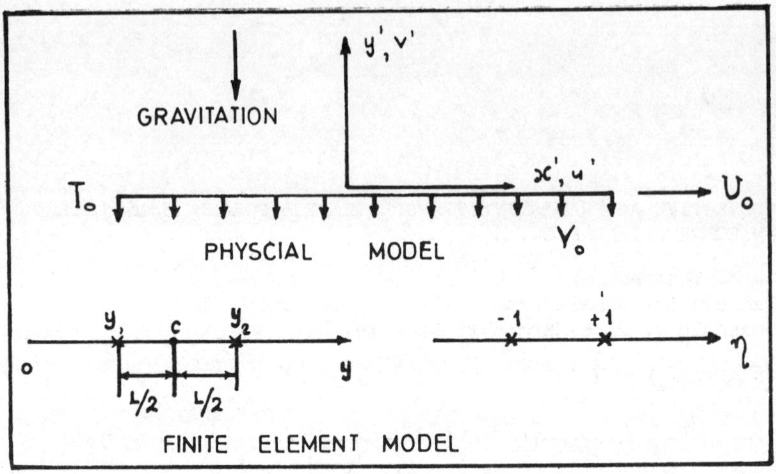


Figure 1

GOVERNING EQUATIONS

Let the velocity field in the (x', y') cartesian co-ordinate system be given by (u', v'). If we employ the following non-dimensional quantities

$$y = y^1 U_o/\nu_\infty \; ; \; (u,v) = (u^1, v^1)/U_o, \; \rho = \rho'/\rho_\infty, \; \Theta = T^1/T_\infty$$

$$q = q^1/\rho_\infty C_p T_\infty U_o, \; \alpha_o = V_o/U_o, \; \beta = T_w/T_\infty,$$

$$M_o = U_o/a_\infty, \; P_r = \mu C_p/k, \; F_r = U_o^3/\nu_\infty g,$$

$$B_o = \rho_\infty C_p U_o/\sigma T_\infty^3, \; \lambda = \alpha \nu_\infty/U_o \tag{1}$$

the equations of motion could be written in the form

$$\rho v = \alpha_\infty ,$$

$$\alpha_\infty \frac{du}{dy} = \frac{d}{dy}\left(\Theta^\omega \frac{du}{dy}\right) ,$$

$$\alpha_\infty \frac{dv}{dy} - \frac{1}{F_r}\rho = -\frac{1}{\gamma M_o^2}\frac{d}{dy}(\rho\Theta) + \frac{4}{3}\frac{d}{dy}\left(\Theta^\omega \frac{dv}{dy}\right)$$

$$\alpha_\infty \frac{d\Theta}{dy} - \frac{\gamma-1}{\gamma} v \frac{d}{dy}(\rho\Theta) = \frac{1}{P_r}\frac{d}{dy}\left(\Theta^\omega \frac{d\Theta}{dy}\right) + (\gamma-1) M_o^2 \Theta^\omega\left\{\left(\frac{du}{dy}\right)^2 + \frac{4}{3}\left(\frac{dv}{dy}\right)^2\right\} - \frac{dq}{dy} \; . \qquad (2)$$

In deriving equation (2) we have assumed the perfect gas law so that $a_\infty$ is the isentropic sound speed and the Prandtl number $P_r$ is assumed constant. Also we assume that the viscosity variation is given by

$$\mu / \mu_\infty = T'^{\omega} / T_\infty^{\omega} \; ,$$

where supscrit $\infty$ refers to conditions at infinity. The Mach number ($M_o$) and the buoyancy term or Froude number ($F_r$) are assumed large. The radiation term is given by $dq/dy$. For the strictly one-dimensional flow under consideration here, this radiation term is given by Vincenti and Kruger (1965) as

$$-\frac{dq}{dy} = \frac{2\lambda}{B_o}\left[\beta^4 E_2\{\lambda(y-y_w)\} + \lambda\int_0^{y-y_w} \Theta^4 E_1\{\lambda(y-\tilde{y})\}\, d\tilde{y} + \int_{y-y_w}^{\infty} \Theta^4 E_1\{\lambda(\tilde{y}-y)\}\, d\tilde{y} - 2\Theta^4\right]. \qquad (3)$$

$E_1$ and $E_2$ are exponential integrals, $B_o$ is the Boltzmann number and $\lambda$ is a reciprocal of a Reynolds number based on the photon mean free path $1/\alpha$. Subscript w refers to the conditions at the flat plate.

The boundary conditions of the problem are

$$u = 1, \quad v = \alpha_o, \quad \Theta = \beta \qquad \text{on} \qquad y = o \; ,$$

$$q = o, \quad v < \infty, \quad \Theta = 1 \qquad \text{as} \qquad y \to \infty \qquad (4)$$

Equations (2) to (4) constitute a set of ordinary non-linear integro-differential equations. No analytical methods are envisaged for this problem. A numerical approach is in order.

FINITE ELEMENT FORMALISM

In the numerical simulation, the Galerkin integral approximation within the Method of Weighted Residuals (MWR) are employed to derive the finite element equations. Using line elements and linear isoparametric interpolation functions, we write

$$v^{(e)} = V_m N_m(\eta) \text{ etc.}, \quad N_m(\eta) = \tfrac{1}{2}(1+\eta_m \eta), \; m=1,2, \tag{5}$$

(see Fig. 1). Substituting (5) in (2), the residuals in the Galerkin formalism become

$$\int_{Y_1}^{Y_2} \left(\rho^{(e)} v^{(e)} - \alpha_\infty\right) N_r \, dy = 0 \qquad \text{etc.}$$

Thus the finite element equations, on employing the simple transformation

$$\eta = \frac{2}{L}(y - y_c),$$

could be written as

$$A^{(e)}_{mnr}\rho^{(e)}_m v^{(e)}_n - \alpha_\infty \hat{i}_r = 0,$$

$$\alpha_\infty I^{(e)}_{mr} u_m + B^{(e)}_{mnr} \Theta_m u_n = S^{(1)(e)}_r$$

$$\alpha_\infty I^{(e)}_{mr} v^{(e)}_m - \frac{1}{F_r} C^{(e)}_{mr} \rho^{(e)}_m + \frac{1}{\gamma M_o^2} D^{(e)}_{mnr} \rho^{(e)}_m \Theta^{(e)}_n + E^{(e)}_{mnr} \Theta^{(e)}_m v^{(e)}_n = S^{(2)(e)}_r$$

$$\alpha_\infty I^{(e)}_{mr} \Theta^{(e)}_m - \frac{\gamma-1}{\gamma} F^{(e)}_{mnpr} v^{(e)} \rho^{(e)}_n \Theta^{(e)}_p + \frac{3}{4P_m} E^{(e)}_{mnr} \Theta^{(e)}_m \Theta^{(e)}_n - \tag{6}$$

$$(\gamma-1) M_o^2 G^{(e)}_{mnpr} \Theta^{(e)}_p \left(u_m u^{(e)}_n + \frac{4}{3} v^{(e)}_m v^{(e)}_n\right) + \int_{-1}^{1} \frac{dq^{(e)}}{dy} d\eta = H^{(e)}_r,$$

where $\hat{i}_r$ is a unit vector. The other coefficients are given by

$$A^{(e)}_{mnr} = \int_{-1}^{1} N_m N_n N_r \, d\eta = \frac{1}{4}\left[1 + \frac{1}{3}\left(\eta_m \eta_n + \eta_n \eta_r + \eta_r \eta_m\right)\right]$$

$$B^{(e)}_{mnr} = \frac{2}{L^{(e)}}\int_{-1}^{1} N_m \frac{dN_n}{d\eta}\frac{dN_r}{d\eta}\, d\eta = \frac{1}{2L^{(e)}}\eta_n\eta_r$$

$$G^{(e)}_{mr} = \int_{-1}^{1} N_m N_r\, d\eta = \tfrac{1}{2}\left(1+\frac{1}{3}\eta_m\eta_r\right),\quad D^{(e)}_{mnr} = \int_{-1}^{1}\frac{d}{dy}(N_m N_n)N_r\, d\eta = \frac{1}{2L^{(e)}}\left(\eta_m+\eta_n+\frac{2}{3}\eta_m\eta_n\eta_r\right),$$

$$E^{(e)}_{mnr} = \frac{16}{3L^{(e)2}}\int_{-1}^{1} N_m \frac{dN_n}{d\eta}\frac{dN_r}{d\eta}\, d\eta = \frac{8}{3L^{(e)}}B^{(e)}_{mnr},\quad I^{(e)}_{mr} = \int_{-1}^{1}\frac{dN_m}{dy}N_r\, dy = \frac{L^{(e)}}{4}\eta_m,$$

$$F^{(e)}_{mnpr} = \frac{2}{L^{(e)}}\int_{-1}^{1} N_m \frac{d}{d\eta}(N_n N_p) N_r\, d\eta = \frac{1}{4L^{(e)}}\left[\eta_n+\eta_p+\frac{1}{3}\,2\eta_m\eta_n\eta_p + 2\eta_n\eta_p\eta_r+\eta_m\eta_n\eta_r+\eta_m\eta_p\eta_r)\right],$$

$$G^{(e)}_{mnpr} = \int_{-1}^{1}\frac{dN_m}{dy}\frac{dN_n}{dy}N_p N_r\, d\eta = \frac{1}{2L^{(e)}}\eta_m\eta_n\left(1+\frac{1}{3}\eta_p\eta_r\right),\quad H^{(e)}_r = -\frac{4}{P_r L^{(e)2}}\left\{N_r\Theta^{(e)}\frac{d\Theta^{(e)}}{d\eta}\right\}_{-1}^{1}, \tag{7}$$

$$S^{(1)(e)}_r = \frac{2}{L^{(e)}}\left\{N_r\Theta^{(e)}\frac{du^{(e)}}{d\eta}\right\}_{-1}^{1},\quad S^{(2)(e)}_r = -\frac{16}{3L^{(e)2}}\left\{N_r\Theta^{(e)}\frac{dv^{(e)}}{d\eta}\right\}_{-1}^{1}$$

In deriving (6) and (7) we have taken $\omega = 1$, an approximation which is often made in boundary layer analysis ($\omega = 0.76$ gives a good approximation for air). $S_r^{(1),(2)}$ and $H_r$ are in general unknown surface reactions. What remains now is to calculate the heat flux term in equation (6). Unfortunately we cannot use equation (3) for this calculation since this equation assumes that the flow domain extends from $y = o$ to $y = \infty$. The finite element method discretizes this domain, so that equation (3) should be replaced appropriately. We assume that each element is bounded by two opaque perpendicularly surfaces. The gray gas approximation holds and further the absorptivity coefficient is independent of temperature. This latter assumption, which is also made in the non-dimensional equation (3), is more valid in this case of small though finite elements. Following Pai (1966), equation (3) may be replaced for each finite element by

$$-\frac{dq}{dy} = \frac{2\lambda}{B_o}\left\{\frac{d}{d\eta}\left(\int_{-1}^{\eta}\Theta^4\,d\eta - \int_{\eta}^{1}\Theta^4\,d\eta\right) - \Theta_1^4 - \Theta_{-1}^4\right\}. \tag{8}$$

Equation (8) assumes that the mean free path is fairly large. If it is not, a modified version of (8) could be used in which exponential integrals are approximated by appropriate exponentials. For simplicity we shall use (8).

By virtue of (8), we can show that

$$-\int_{-1}^{1}\frac{dq}{dy}N_r\,d\eta = \frac{\lambda}{4B_o}\left\{J^{(e)}_{mnpqr}\,\Theta^{(e)}_m\Theta^{(e)}_n\Theta^{(e)}_p\Theta^{(e)}_q - \tfrac{1}{2}\left(\Theta_{-1}^4+\Theta_1^4\right)\hat{i}_r\right\} \tag{9}$$

where

$$J^{(e)}_{mnpqr} = 1 + \frac{1}{3}\left[(\eta_m+\eta_n+\eta_p+\eta_q)\eta_r + (\eta_m\eta_n+\eta_m\eta_p+\eta_m\eta_q+\eta_n\eta_p+\eta_n\eta_q+\eta_p\eta_q) + \frac{1}{5}\left[(\eta_m\eta_n\eta_p+\eta_m\eta_n\eta_q+\eta_m\eta_p\eta_q+\eta_n\eta_p\eta_q + \eta_m\eta_n\eta_p\eta_q\right]. \tag{10}$$

Equations (9) and (10) complete the element equations of (6) and (7).

It is now routine to combine the element equations into a global system. We obtain

$$A_{ijs}\rho_i v_j - \alpha_\infty \hat{i}_s = 0\,; \tag{11}$$

$$\alpha_\infty I_{is}u_i + B_{ijs}\Theta_i\Theta_j = S^{(1)}_s\,, \tag{12}$$

$$\alpha_\infty I_{is}v_i - \frac{1}{F_r}C_{is}\rho_i + \frac{1}{\gamma M_o^2}D_{ijs}\rho_i\Theta_j + E_{ijs}\Theta_i v_j = S^{(2)}_s\,, \tag{13}$$

$$\alpha_\infty I_{is}\Theta_i - \frac{\gamma-1}{\gamma} F_{ijks} v_i \rho_i \Theta_k + \frac{3}{4P_r} E_{ijs}\Theta_m \Theta_n - (\gamma-1) M_o^2 G_{ijks} \times$$

$$\Theta_k\left(u_i u_j + \frac{4}{3} v_i v_j\right) + \frac{\lambda}{4B_o} J_{ijkls}\Theta_i \Theta_j \Theta_k \Theta_l = H_r + \frac{\lambda}{8B_o}\Theta_w^4 \hat{i}_r . \tag{14}$$

We note also that $S_s^{(1)}$, $^{(2)}$ and $H_s$ are different from zero except at the wall and at infinity.

METHOD OF SOLUTION

First of all in the high temperature phenomenon involved in hypersonic flow, $F_r$ and $M_o$ are expected to be large. On the other hand since the mean free path is of order O(1) or fairly large one would expect $\lambda$ and $B_o$ to be small. A simple iterative process can therefore be devised for solving equations (11) - (14) subject to equation (4). We start by neglecting the quadratic terms in (12) and (13) and since the Froude number is large the linear density term may be neglected in equation (13) in comparison with the velocity term term. Thus we get

$$\alpha_\infty I_{is} u_i^{(o)} = S_s^{(1)}, \quad \alpha_\infty v_i^{(o)} = S_s^{(1)} \tag{15}$$

Next, with the help of equation (15) we can solve for $\rho_i^{(o)}$ from equation (11). Finally in equation (14) we neglect quadratic terms except the term involving $M_o^2$ which is expected to be large. That is

$$\alpha_\infty I_{is}\Theta_i^{(o)} - (\gamma-1) M_o^2 G_{ijks}\Theta_k^{(o)}\left(u_i^{(o)} u_j^{(o)} + \frac{4}{3} v_i^{(o)} v_j^{(o)}\right) = H_s + \frac{\lambda}{8B_o}\Theta_w^4 \hat{i}_s \tag{16}$$

where $\Theta_w = \beta$ on $y=o$, $\Theta_w = 1$ on $y=\infty$ and $\Theta_w = o$ else where respectively.

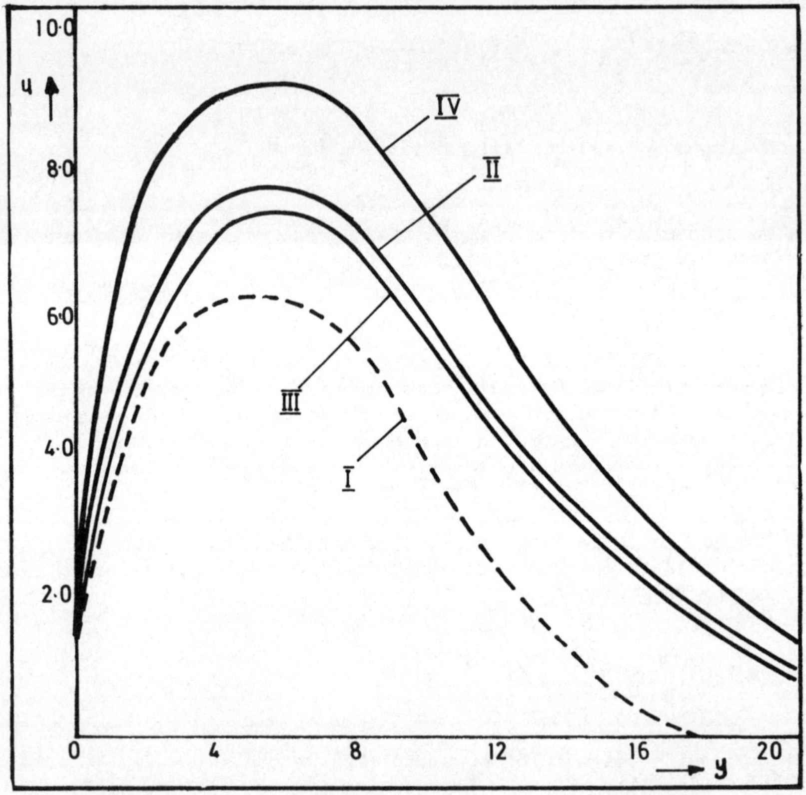


Figure 2

| | $M_o$ | $F_r$ | |
|---|---|---|---|
| I | 10 | 50 | ••• no radiation |
| II | 10 | 50 | —— radiation |
| III | 10 | 100 | |
| IV | 20 | 50 | |

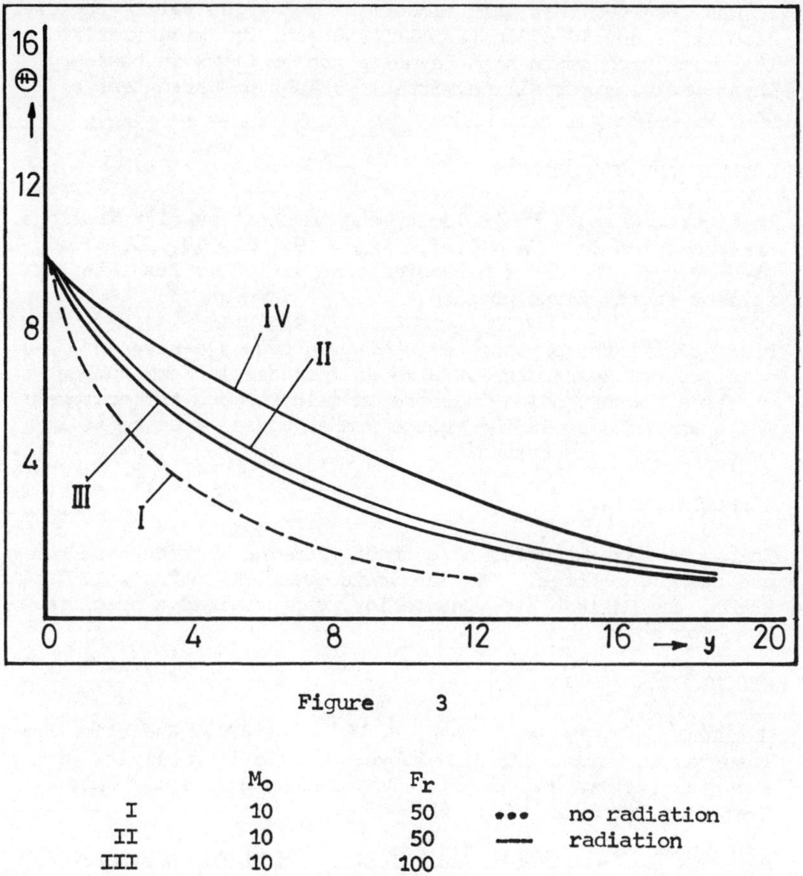


Figure 3

| | $M_o$ | $F_r$ | | |
|---|---|---|---|---|
| I | 10 | 50 | ••• | no radiation |
| II | 10 | 50 | —— | radiation |
| III | 10 | 100 | | |
| IV | 20 | 50 | | |

With the known values of $u_i^{(o)}$ and $v_i^{(o)}$ equation (16) can now be solved for $\Theta_i^{(o)}$. These give the basic approximation.

We iterate on this by replacing the quadratic terms and the neglected linear density term involving the Froude number in equations (12) and (13) by the basic approximations. The solutions $u_i^{(1)}$ and $v_i^{(1)}$ obtained by this process enables $\rho_i^{(1)}$ to be calculated from equation (11), which in turn could be utilized to solve for $\Theta_i^{(1)}$ noting that the quadratic terms involving $M_o^2$ are replaced by $u_i^{(1)}$ and $v_i^{(1)}$ while the rest are replaced by $u_i^{(o)}$, $v_i^{(o)}$, $\rho_i^{(1)}$; and so on.

It was observed that this alogarithm converged fairly rapidly. Also it is devoid of instabilities associated with similar iterative process in high Reynolds number flows in boundary layer analysis for all permissible values of parameters of this problem.

RESULTS AND DISCUSSIONS

In figures 2 and 3 the velocity and temperature distributions are presented for $\alpha_o = 0.01$, $\alpha_\infty = 10$, $\beta = 10$, $\lambda = 0.1$, $B_o = 0.01$. The other parameters are varied to simulate the salient facts of the problem.

First of all the presence of radiation causes a rise in velocity and temperature. Also an increase in Mach number causes a corresponding increase in velocity and temperature while an increase in the Froude number causes a decrease in velocity and temperature.

ACKNOWLEDGEMENT

This work was motivated by a grant from the Unitech Research and Grant Committee. The author is grateful to Prof. T. T. Isoun, the Unitech Vice-Chancellor, who provided a specific fund for editing.

REFERENCES

Helliwel, J. B. and Mosa, M. F. - (1979) Radiative Heat Transfer in Horizontal Magnetohydrodynamic Channel flow with Buoyancy Effects and an Axial Temperature Gradient. Int. J. Heat Mass Transfer, 22: 657 - 668.

Pai, S. I. - (1966) Radiation Gasdynamics. Springer-Verlag, New York.

Vincenti, W. G. and Kruger, C. H. - (1965) Introduction to Physical Gasdynamics. John Wiley, New York.

# EFFECTIVE TRANSIENT COMPRESSIBLE FLOW ANALYSIS IN PIPE NETWORKS

ORTWIN OHTMER

MESSERSCHMITT-BÖLKOW-BLOHM GMBH, MUNICH, GERMANY

## 1. Summary of Analysis

The publication is based on [1] and the literature named there.

Pipe networks are computed in an analoque manner as frame-works in structural mechanics loaded only by moments. The mesh method (force method) is applied. Due to the boundary layer effects in special pipe members the flow problem is in general nonlinear. Therefore the NEWTON-RAPHSON Interation Procedure is used to solve the nonlinear system of equations, see Appendix 1. Using the computed flow rates in the TREE-Structure (determined by the graph theory) as initial values the iteration procedure converges rapidly to a user specified tolerance value. The loss coefficients of pressure for different pipe members (TUBE, VALVE, BOW, TEE, PUMP, KNEE, ± CONTR, ± DIFSR) must only be given in diagrams, see Appendix 2. These diagrams are used in digitalized form. In the back-substitution phase with known flow rates in all members the pressure at the joints is computed.

The main advantages of the analysis as outlined are that no initial values for the member flow rates must be known, the iteration procedure converges rapidly, and within each iteration step only small systems of linear equations must be solved. Due to the fact that the loss coefficients of pressure must only be given in diagrams arbitrary nonlinear networks can be analysed by the unchanged programsystem. A flow rate assumption may be specified in the input for a member of a mesh.

The outlined procedures in [1] are limited to in-transient in-compressible flows. In this paper the analysis is extended to transient compressible flows.
Due to time-dependent flow rates the NEWMARK-Method published in [2] is used to approximate the accelerations in the time-integration scheme. The iteration procedure outlined in [1] to fulfil the equilibrium conditions must now be applied at each time step.

## 2. The governing Equation of the Mesh-Method for transient Pipe-Flows in a gravitational field

In fig. 1 the same example is given as in [1] to illustrate the procedure. Equation (1) is an extension of (3) in [1] . The first term ( $[B_G] \{\dot{q}^*\}$ ) is derived from the time-dependent term in the well-known EULER DIFFERENTIAL EQUATION. The vector $\{S\}$ on the right hand side of (1) takes gravitational effects into account. (For illustration see fig. 2)

The vectors $\{P\}(t)$ and $\{Q(t)\}$ are now time-dependent.

$$[B_G] \{\dot{q}^*\} + [F_G] \{q^*\} = [C]^T \cdot \Big\{ \{P(t)\} - [F] \{Q(t)\} + \{S\} \Big\} \tag{1}$$

with

$$[B_G] = [C]^T \cdot [B][C] \; ; \; [F_G] = [C]^T [F] [C]$$

$$[B] = \begin{bmatrix} B_1 & & & 0 \\ & B_2 & & \\ & & \ddots & \\ 0 & & & B_n \end{bmatrix} ; \qquad B_i = \frac{4 \cdot Li}{\pi \cdot D_i^2} ;$$

$$[F] = \begin{bmatrix} F_1 & & & 0 \\ & F_2 & & \\ & & \ddots & \\ 0 & & & F_n \end{bmatrix} ; \qquad F_i = \frac{8}{\pi^2} \cdot \frac{\left(\frac{Li \cdot \lambda_i}{Di} + K_v^i\right)}{\varrho_i \cdot Di^4} \cdot Q_i$$

$$\{S\} = \begin{Bmatrix} \rho_1 \ L_1 \ \sin \beta_1 \\ \rho_2 \ L_2 \ \sin \beta_2 \\ \cdots\cdots\cdots\cdots \\ \rho_n \ L_n \cdot \sin \beta_n \end{Bmatrix} .g; \quad \begin{matrix} \text{n is the number of all members} \\ i = 1,2,3,\ldots\ldots\ldots n \end{matrix}$$

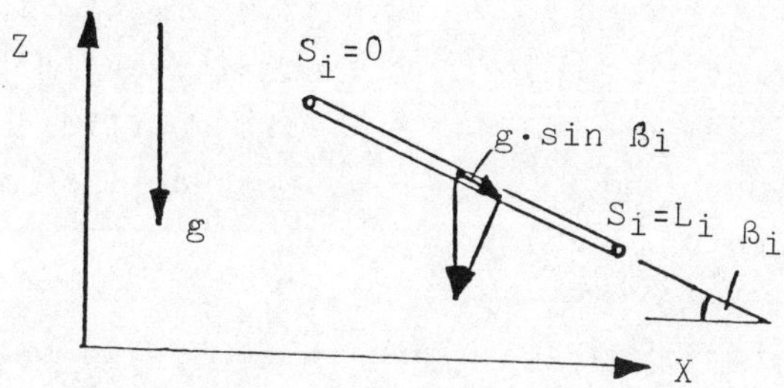


fig. 2

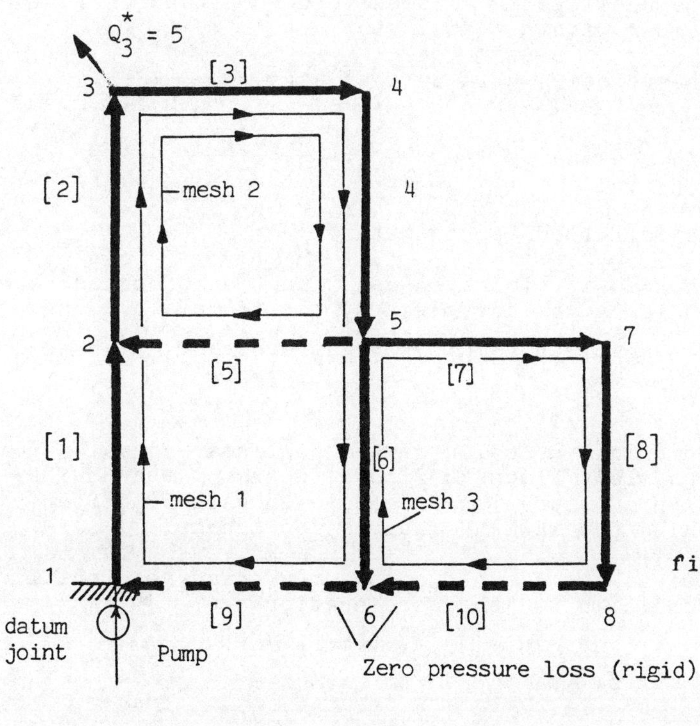


fig. 1

$[C]$ is the transformation matrix cont-aining the topology of the network

$\{q^*\}$ is the mesh flow rate vector; $\left(\frac{\partial q^*}{\partial t} = \dot{q}^*\right)$

$\{P\}$ is the applied member pressure loss vector

$\{Q\}$ is the applied member flow rate vector

$\lambda$ = tube resistance coefficient;

L = tube length [m]

D = tube diamter [m]

p = pressure [Pa]

Q = flow rate [kg/sec]

g = gravitational constant [m/sec$^2$]

$\rho$ = density [kg/m$^3$]

$K_v$ = loss coefficient

ß = angle of flow member to horizontal

The computation of pressure losses $\Delta p$ in the "flow members" is described in [1] and the literature mentioned there.

The density is assumed to be constant in a flow member within a time step $\Delta t$.

The velocity v is defined by $v = \frac{4 \cdot Q}{\rho \cdot \pi \cdot D^2}$

The continuity law $\frac{\partial \rho}{\partial t} = -\frac{\partial (v \cdot \rho)}{\partial x}$

is fulfilled approximately.

The density is given as a function of pressure in a diagram or for gases by the formula $\rho = \frac{p}{R\,T}$ ;

T = absolute temp.; R = gas constant;

3. Numerical Solution of the nonlinear Equation for transient Flows in Pipe-Networks by the NEWTON-RAPHSON Iteration Procedure in each Time Step using the NEWMARK Integration Scheme.

Due to the NEWMARK constant-average-acceleration scheme in (2) the acceleration

$$\dot{q}^*_{t+\Delta t} \text{ in (1) at time } t+\Delta t$$

can be replaced by $\dot{q}^*_t$ at time t and the flow rates $q^*_{t+\Delta t}$ and $q^*_t$ at the times $t+\Delta t$ and t.

(2) is illustrated in fig. 3.

$$\delta \dot{q}^*_{t+\Delta t} = -\dot{q}^*_t (1-\delta) + \frac{1}{\Delta t}\cdot(q^*_{t+\Delta t} - q^*_t) \qquad (2)$$

Linear acceleration method: $\delta = \frac{1}{2}$

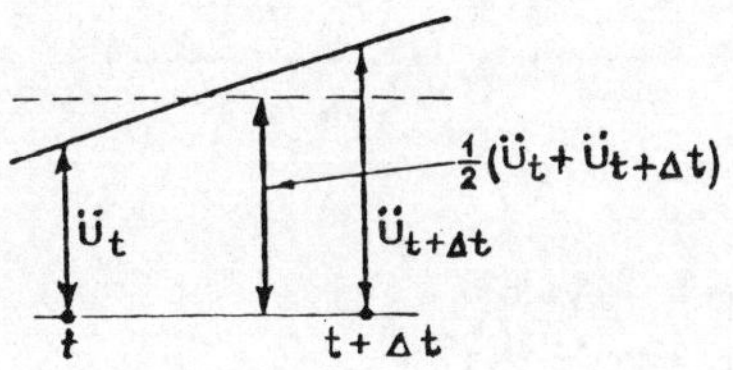


fig. 3

Applying (2) in (1) we obtain (3).

$$\left[[F_G] + \frac{1}{\delta\Delta t}[B_G]\right]\left\{q^*_{t+\Delta t}\right\} = [C]^T\left\{\left\{P_{t+\Delta t}\right\} + \left\{S\right\}\right\} -$$

$$[C]^T [F]\left\{Q_{t+\Delta t}\right\} + [B_G]\left\{\left(\frac{1-\delta}{\delta}\right)\left\{\dot{q}^*_t\right\} + \frac{1}{\delta\Delta t}\left\{q^*_t\right\}\right\} \qquad (3)$$

The nonlinear equation in (3) is solved for the unknown vector $\{q^*_{t+\Delta t}\}$ applying the NEWTON-RAPHSON Procedure as described in [1].

With the computed mesh flow rate $\{q^*_{t+\Delta t}\}$, the derivation of the mesh flow rate $\{\dot{q}^*_{t+\Delta t}\}$ is obtained applying the NEWMARK formula in (2).

As mentioned in [1] no initial values are necessary to start the iteration procedure.

Applying the EULER-DIFF. EQUATION in (4) for the TREE-Structure the initial values $\dot{q}^*$ (t=0) can be computed ($q^*$ (t=0) = 0).

$$\frac{\partial v}{\partial t} + \frac{1}{\rho} \cdot \frac{\partial p}{\partial s} + g \cdot \frac{\partial z}{\partial s} = 0 \tag{4}$$

With $\frac{\rho \cdot \pi \cdot D^2}{4} \cdot \frac{\partial v}{\partial t} (t=0) = \dot{q}^* (t=0)$

an approximation of the initial values $\dot{q}^*$ (t=0) is given in (5) being illustrated in fig. 4.

$$\dot{q}(t=0) = \frac{\pi D^2 m}{4} \left( \frac{P_o - P_u}{L_s} + g \rho \frac{(H_o - H_u)}{L_s} \right) \tag{5}$$

$$\left\{ \dot{q}^*(t=0) \right\}_{TREE} = \left[ C \right]^T \cdot \left\{ \dot{q}(t=0) \right\}_{TREE} \tag{5a}$$

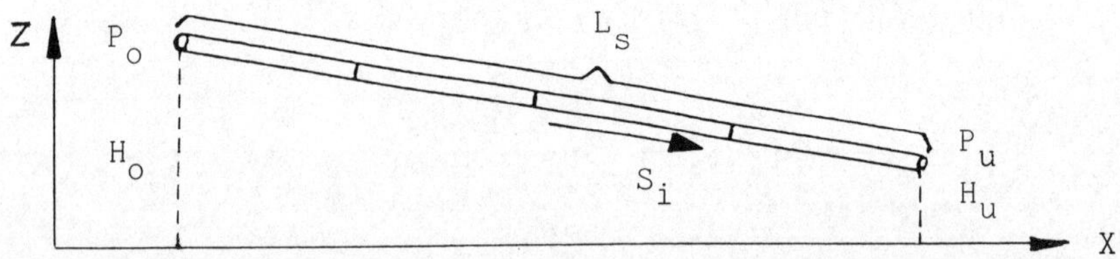


fig. 4

$D_m$ = average tube diameter [m]

$L_s$ = TREE - length [m]

$P_o - P_u$ = pressure difference [Pa]

$H_o - H_u$ = highness difference [m]

Applying the mesh-method (see [1] ) given flow rates can easily be taken into account on the right hand side of (3). To take into account special boundary conditions for the pressure (constant pressure between two joints) the mesh flow rates $q_i^*$ must fulfil special kinematic constraint conditions.

(In statics, applying the displacement-method those kinematic constraint conditions for the displacements are well known, e.g. rigig body motion within an elastic Finite Element Idealization)

## 4. Conclusion

The main advantages of the analysis as outlined are that no initial values for the member flow rate must be known, the iteration procedures converge rapidly and within each iteration step, only small systems of linear equations must be solved. Due to the fact that the loss coefficients of pressure must only be given in diagrams, arbitrary nonlinear networks can be analysed by the unchanged program system. A flow rate assumption may be specified in the input for a member of a mesh. Elastic pipes or damping effects can easily be taken into account.

## Literature:

[1] O. Ohtmer

Convergence of Iteration Procedures in the Nonlinear Analysis of Pipe Networks
International Journal Num. Methods in Eng. 1982

[2] K.J. Bathe, E.L. Wilson

Numerical Methods in Finite Element Analysis,
Prentice Hall, Inc. 1976

Ortwin Ohtmer, Isegrimstraße 10, D-8000 München 83

Appendix 1

Example:
INPUT FILE 10

```
STRUDL

JOINT COORD
1 0. 0. S
2 0. 2.
3 0. 4.
4 2. 4.
5 2. 2.
6 2. 0.
7 4. 0.
8 4. 2.
```

TREE-STRUCTURE

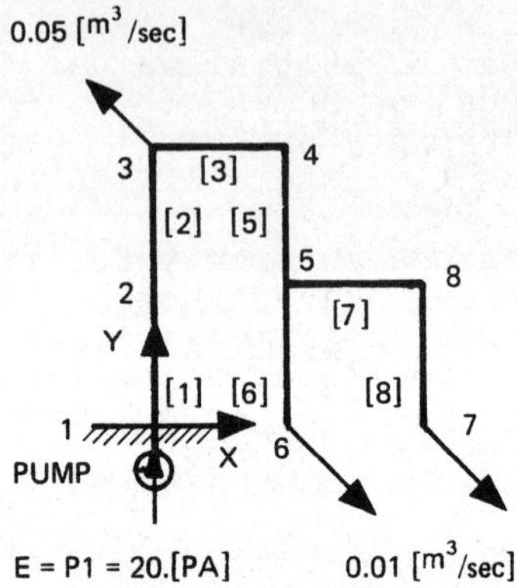


```
TYPE PLANE FRAME
MEMBER INCIDENCES
1 1 2
2 2 3
3 3 4
5 4 5
4 5 2
6 5 6
7 5 8
8 8 7
```

ORIGINAL STRUCTURE
(having a closed mesh)

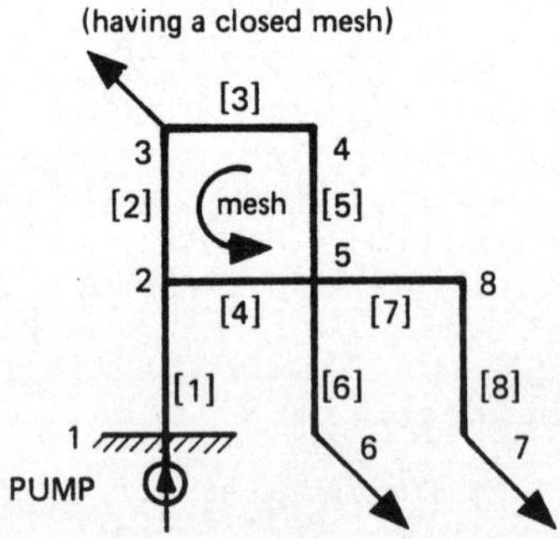


```
   CONSTANT E 20.0 ALL
   MEM 1 TO 8 PROP NON STA TYPE 'TUBE' '    ' .06 845. 1.E-5 1.E-3 1.E+1
C  MEM 4         PROP NON STA TYPE '+CONTR-1' TABLE 'DIAGRAM' .06 845. A10 0.03 -
   A11 0.06 A12 4
   CONVERGENCE TOLERANCE 0.0001
   MAXIMUM NUMBER OF CYCLES 10
   LOADING 1
   JOINT LOADS
   3 MOMENT Z .05
   6 7 MOMENT Z .01
   LOADING LIST ALL
   NETWORK ANALYSIS
```

```
OUTPUT
***-CONVERGENCE ACHIEVED IN CYCLE 9

DATUM JOINT: 1
GIVEN PRESSURE AT DATUM JOINT: 20.
```

Appendix 1 *cont.*

MEMBER FLOW INTENSITIES

| MEMBER | LOADING | JOINT | /---FLOW INTENSITY---/ |
|---|---|---|---|
| 1 | | | |
| | 1 | 1 | −0.0700000 |
| | | 2 | 0.0700000 |
| 2 | | | |
| | 1 | 2 | −0.0406960 |
| | | 3 | 0.0406960 |
| 3 | | | |
| | 1 | 3 | 0.0093040 |
| | | 4 | −0.0093040 |
| 5 | | | |
| | 1 | 4 | 0.0093040 |
| | | 5 | −0.0093040 |
| 4 | | | |
| | 1 | 5 | −0.0293040 |
| | | 2 | 0.0293040 |
| 6 | | | |
| | 1 | 5 | −0.0100000 |
| | | 6 | 0.0100000 |
| 7 | | | |
| | 1 | 5 | −0.0100000 |
| | | 8 | 0.0100000 |
| 8 | | | |
| | 1 | 8 | −0.0100000 |
| | | 7 | 0.0100000 |

JOINT PRESSURES – SUPPORTS

| JOINT | | LOADING | /-----PRESSURE-----/ |
|---|---|---|---|
| 1 | GLOBAL | | |
| | | 1 | 20.0000000 |

JOINT PRESSURES – FREE JOINTS

| JOINT | | LOADING | /-----PRESSURE-----/ |
|---|---|---|---|
| 2 | GLOBAL | | |
| | | 1 | 11.9718828 |
| 3 | GLOBAL | | |
| | | 1 | 8.1872520 |
| 4 | GLOBAL | | |
| | | 1 | 8.8363250 |
| 5 | GLOBAL | | |
| | | 1 | 9.4853992 |
| 6 | GLOBAL | | |
| | | 1 | 8.7826223 |
| 7 | GLOBAL | | |
| | | 1 | 8.0798454 |
| 8 | GLOBAL | | |
| | | 1 | 8.7826223 |

**The following flow branches known as "flow members" were implemented:**

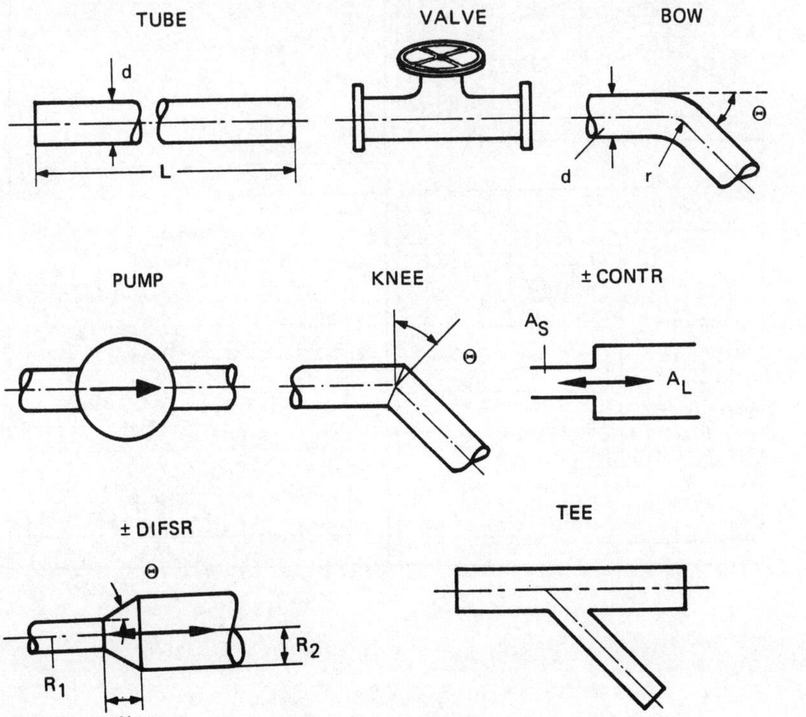


$K_v$(TEE) is computed for $\lambda = 0$; L = 0 with the GARDEL EQUATIONS given in [1].

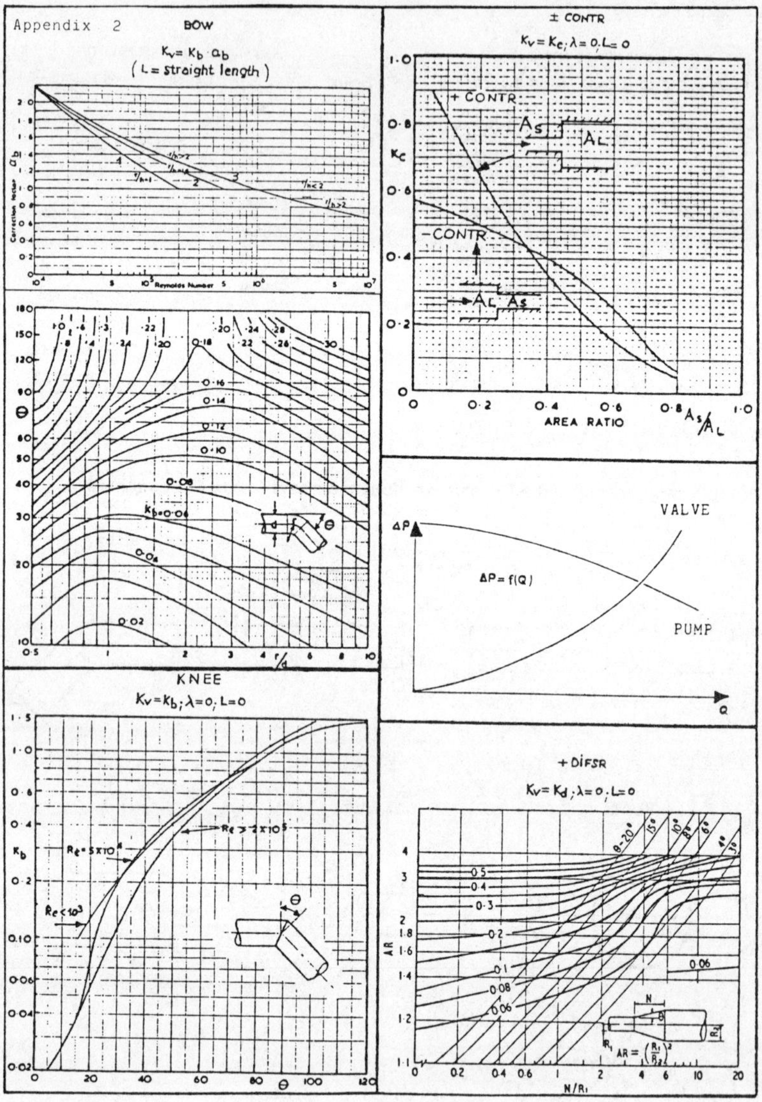

Appendix 2
BOW
$K_v = K_b \cdot a_b$
( L = straight length )
Correction factor $a_b$
Reynolds Number
KNEE
$K_v = K_b; \lambda = 0; L = 0$
$K_b$
$R_e < 10^3$
$R_e = 5 \times 10^4$
$R_e > 2 \times 10^5$
± CONTR
$K_v = K_c; \lambda = 0; L = 0$
+ CONTR
- CONTR
$K_c$
AREA RATIO
$A_s/A_L$
VALVE
PUMP
ΔP
ΔP = f(Q)
Q
+ DIFSR
$K_v = K_d; \lambda = 0; L = 0$
AR
N/R₁
$AR = \left(\frac{R_1}{R_2}\right)^2$

# CONTINUATION METHODS FOR THE FINITE ELEMENT SOLUTION OF TURBULENT FLOW

D. R. Schamber, University of Utah
B. E. Larock, B. A. DeVantier, University of California, Davis

## INTRODUCTION

This paper concentrates on the solution of some turbulent flow problems by the combined use of the Galerkin finite element method and a continuation method; turbulence will be simulated by the k-ε model. This turbulence model is adopted here because it currently appears to represent correctly most major features of many relatively complex flows while not overburdening the user with an unduly large number of unknowns per computational node (Rodi, 1980a). Application of the finite element method to the governing equations produces a nonlinear set of algebraic equations. Obtaining solutions to these equations is the primary problem facing the investigator.

Schamber and Larock (1980, 1981) have noted the difficulty of creating an appropriate initial estimate of the solution vector which is the starting point for a convergent computation. The continuation approach avoids these initialization problems by constructing a mathematical expression which effects an incremental transition from a problem with a known solution to the problem one wants to solve. This mathematical expression, or embedding function, is constructed in terms of a parameter t in such a way that t=0 corresponds to the system of equations with a known solution and t=1 corresponds to the final system of equations to be solved. The solution is sought in a controlled manner by incrementing t from zero towards unity.

This article presents an algorithm which heuristically selects an appropriate increment δt on the basis of convergence criteria that are evaluated at the current step. A predictor-corrector scheme is employed; Lagrangian extrapolation may be used for the predictor, and Newton's method is used for the corrector. Three one-dimensional flow examples, each of which are highly nonlinear, illustrate the features of the method. Both stratified and nonstratified flows are considered. Application of the method to two-dimensional flows is discussed briefly.

## FLOW EQUATIONS

The equations for one-dimensional, steady, turbulent, density-stratified flow are (DeVantier and Larock, 1981)

$$0 = gS_o - \frac{1}{\rho}\frac{\partial P}{\partial x} + \frac{\partial}{\partial z}(\nu_t \frac{\partial U}{\partial z}) \quad \text{(x-momentum transport)} \quad (1)$$

$$0 = P_r - \varepsilon + \frac{\partial}{\partial z}(\frac{\nu_t}{\sigma_k}\frac{\partial k}{\partial z}) + G \quad \text{(k transport)} \quad (2)$$

$$0 = c_{\varepsilon 1}\frac{\varepsilon}{k}[P_r + (1-c_{\varepsilon 3})G] - c_{\varepsilon 2}\frac{\varepsilon^2}{k} + \frac{\partial}{\partial z}(\frac{\nu_t}{\sigma_\varepsilon}\frac{\partial \varepsilon}{\partial z}) \quad (\varepsilon \text{ transport}) \quad (3)$$

$$0 = \frac{\partial}{\partial z}[\Lambda(1-\Lambda)V_s] + \frac{\partial}{\partial z}(E_s \frac{\partial \Lambda}{\partial z}) \quad (\Lambda \text{ transport}) \quad (4)$$

with x=streamwise coordinate, z=transverse coordinate, g=gravity, $S_o$=channel slope, $\rho$=bulk fluid density, P=mean pressure, U=mean x-velocity, $\Lambda$=mean sediment volume fraction, $V_s$=suspension settling velocity, $E_s=\gamma\nu_t$=sediment diffusion coefficient. Turbulence production by shear, bouyancy-induced production, and the kinematic turbulent viscosity are

$$P_r = \nu_t (\frac{\partial U}{\partial z})^2 \quad (5)$$

$$G = \eta\gamma\nu_t \frac{\partial \Lambda}{\partial z} g \quad (6)$$

$$\nu_t = c_\mu k^2/\varepsilon \quad (7)$$

The constant $\gamma$=1.2 (DeVantier and Larock, 1981) while the k-$\varepsilon$ turbulence model constants $c_\mu$=0.09, $c_{\varepsilon 1}$=1.45 (1.44), $c_{\varepsilon 2}$=1.92 (1.90), $c_{\varepsilon 3}$=0.8, $\sigma_k$=1.0 and $\sigma_\varepsilon$=1.3 (Rodi, 1980b). (Values in parentheses are used for only the nonstratified flow examples.) Finally, $\eta=\rho_{sed}/\rho_{fluid}-1$.

For stratified flow all terms in Equations 1-4 are retained, except the pressure gradient term which is deleted for free surface flows. For nonstratified flow the $\Lambda$ transport equation is omitted from the equation set, and $gS_o$ and G are deleted in Equations 1-3.

## FINITE ELEMENT - CONTINUATION FORMULATION

The following finite element forms of Equations 1-4 are written for an element domain $\Omega_e$ bounded by the closed contour $\Gamma_e$. Embedding functions for Equations 2 and 3 are included in the expressions. Mathematically,

$$f_U = \int_{\Omega_e}[N_i gS_o - N_i \frac{\partial P}{\partial x} - \nu_t \frac{\partial N_i}{\partial z}\frac{\partial U}{\partial z}]d\Omega + \int_{\Gamma_e} N_i \nu_t \frac{\partial U}{\partial z} \ell_z d\Gamma \quad (8)$$

$$f_k = t \int_{\Omega_e} [N_i (P_r - \varepsilon + G) - \frac{\nu_t}{\sigma_k} \frac{\partial N_i}{\partial z} \frac{\partial k}{\partial z}] d\Omega + t \int_{\Gamma_e} N_i \frac{\nu_t}{\sigma_k} \frac{\partial k}{\partial z} \ell_z d\Gamma + (1-t) H_k \quad (9)$$

$$f_\varepsilon = t \int_{\Omega_e} N_i \{c_{\varepsilon 1} \frac{\varepsilon}{k} [P_r + (1-c_{\varepsilon 3})G] - c_{\varepsilon 2} \frac{\varepsilon^2}{k}\} d\Omega - t \int_{\Omega_e} \frac{\nu_t}{\sigma_\varepsilon} \frac{\partial N_i}{\partial z} \frac{\partial \varepsilon}{\partial z} d\Omega + t \int_{\Gamma_e} N_i \frac{\nu_t}{\sigma_\varepsilon} \frac{\partial \varepsilon}{\partial z} \ell_z d\Gamma + (1-t) H_\varepsilon \quad (10)$$

$$f_\Lambda = \int_{\Omega_e} \frac{\partial N_i}{\partial z} \{\Lambda(1-\Lambda)V_s + E_s \frac{\partial \Lambda}{\partial z}\} d\Omega - \int_{\Gamma_e} N_i \{\Lambda(1-\Lambda)V_s + E_s \frac{\partial \Lambda}{\partial z}\} \ell_z d\Gamma \quad (11)$$

The Galerkin weighting functions are chosen to be the 8-node biquadratic basis functions $N_i$, and $\ell_z$ is the z component of the outward unit normal to the flow boundary $\Gamma$. Boundary integral terms are evaluated only when $\Gamma_e$ coincides with $\Gamma$.

For the case of nonstratified, symmetric channel flow $H_k$ and $H_\varepsilon$ are

$$H_k = \int_\Omega N_i \, (k_o - k) \, (U_o/L_o) \, d\Omega \quad (12)$$

$$H_\varepsilon = \int_\Omega N_i \, (\varepsilon_o - \varepsilon) \, (U_o/L_o) \, d\Omega \quad (13)$$

in which $U_o$=reference velocity and $L_o$=reference length. These reference scales are used along with a reference concentration $\Lambda_o$ to nondimensionalize the variables in Equations 1-13.

The particular form of $H_k$ and $H_\varepsilon$ is suggested by a constant viscosity simulation. For $t_o$=0 the solution to Equations 9 and 10 yields $k=k_o$ and $\varepsilon=\varepsilon_o$ throughout the solution domain. Specifically, $k_o$ and $\varepsilon_o$ are set to the appropriate wall boundary conditions for symmetric plane channel flow. This yields, via Equation 7, $\nu_t = c_\mu k_o^2/\varepsilon_o$=constant. The result is a parabolic velocity profile for a specified pressure gradient or channel slope $S_o$, according to Equation 1. The form of $H_k$ and $H_\varepsilon$ are problem dependent. Appropriate functions for these terms are presented with the examples.

## SOLUTION STRATEGY

The assembled forms of Equations 8-11 may be written as

$$\mathbf{F}(\mathbf{x}, t) = 0 \quad (14)$$

in which **F**=vector representing the system of nonlinear equations and **x**=vector of nodal unknowns for the system. For $0 \leq t_j \leq 1$ a basic Newton or corrector scheme solves Equation 14 by writing

$$\mathbf{A}(\mathbf{x}^k)\ \delta\mathbf{x}^k = -\ \mathbf{F}(\mathbf{x}^k,\ t_j) \tag{15}$$

in which k=iteration counter, **A**=Jacobian of the equation matrix, and $\delta\mathbf{x}^k$=set of incremental corrections to the vector $\mathbf{x}^k$. Solving Equation 15, one then obtains the new estimate of the solution as

$$\mathbf{x}^{k+1} = \mathbf{x}^k + \delta\mathbf{x}^k\ . \tag{16}$$

The solution strategy begins by obtaining the constant viscosity simulation from the solution of Equation 14 for $t_o$=0. Divergence of the corrector routine is rare for this simple case. The $t_o$-solution is then used to initialize the corrector scheme at $t_1=t_o+\delta t_o$. Once three or more successive solutions have been computed (corresponding to $t_{j-3}$, $t_{j-2}$, $t_{j-1}$ for $j \geq 3$), a Lagrangian predictor formula may be used to generate a new solution estimate corresponding to $t=t_j$. If the nonlinearity of the problem is severe and $\delta t_{j-1}=t_j-t_{j-1}$ is too large, the Lagrangian predictor may yield values which are a poorer approximation to $\mathbf{x}_j$ than $\mathbf{x}_{j-1}$ itself. To monitor this situation, the ratio $r=|x_j-x_{j-1}|/|\Delta x|$, where $\Delta x$ is the maximum difference between $x_{j-1}$, $x_{j-2}$, and $x_{j-3}$ for each unknown, is required to be less than unity. If $r<1$, then the predicted value of $\mathbf{x}_j$ is used to initialize the corrector at the new station $t_j$. If $r \geq 1$, then $\mathbf{x}_{j+1}=\mathbf{x}_j$ is used as the predictor. Rheinbolt (1977) presents other predictor methods which depend on formulating a differential equation in t from Equation 14. His procedure requires the solution of an additional set of equations at each step to determine d**x**/dt, which is then used in a predictor equation. This type of procedure was not pursued because of the additional cost and effort needed to determine d**x**/dt. If the method were to give a more accurate prediction of $\mathbf{x}_j$ than does the Lagrangian predictor, however, then the method may reduce the overall number of iterations.

Step size selection $\delta t$ is the key to the performance of the continuation algorithm. An adaptive step-size selection scheme (Schmidt, 1978) is adopted here. The procedure is heuristic in nature but is based on the contractive mapping theorem; it is briefly described now.

Each succeeding step is found from the current $\delta t_j$ via

$$\delta t_{j+1} = g\ \delta t_j \tag{17}$$

in which $\delta t_j=t_j-t_{j-1}$. The function $g$ is

$$g = g_2 + (g_1 - g_2)\,\alpha_L\,(1/\alpha - 1)/(1 - \alpha_L) \qquad \text{for } \alpha_L < \alpha < 1 \qquad (18)$$

$$g = g_1 \qquad \text{for } \alpha \leq \alpha_L \qquad (19)$$

$$g = g_2 + (g_2 - g_3)(1/\alpha - 1) \qquad \text{for } \alpha > 1 \qquad (20)$$

in which $\alpha$ is an estimate for the contractive constant given by

$$\beta_k = \|\mathbf{x}^{k+1} - \mathbf{x}^k\| \,/\, \|\mathbf{x}^k - \mathbf{x}^{k-1}\| \qquad (21)$$

where $\alpha = \max(\beta_k)$ for $k=1,2,3,\ldots$ Herein the Euclidian norm is used exclusively. Equation 21 follows directly from the definition of a contractive mapping (Schmidt, 1978).

An iterative algorithm, e.g. Newton's method, is a contractive mapping if $\alpha \leq 1$. Furthermore, the convergence of a contractive iteration is assured. Equation 18 increases the step size whenever $\alpha$ falls below unity, while Equation 20 decreases the step size when $\alpha > 1$. A transition value $\alpha_T < 1$ is selected to produce $g_T = 1$. For $\alpha > \alpha_T$ the step size is reduced. Values $\alpha < \alpha_T$ increase the step. The lower limit $\alpha_L$ is determined from Equation 18 by specifying $g_1$, $g_2$, and $\alpha_T$. The minimum reduction in step size is controlled by $g_3$.

The value of $\delta t_{j+1}$ from Equation 17 may be too large to produce a convergent solution at $t_{j+1}$. For $\alpha > 1$ with the step $\delta t_{j+1}$ the contractive properties of the algorithm no longer hold, and a reduction in $\delta t_{j+1}$ and hence a new value of $t_{j+1}$ appear to be required. Schmidt (1978) has found this criterion to be overly conservative because the algorithm may in fact converge for $\alpha > 1$. To control step size reduction the following conditions are checked at the kth iteration and jth step: (1) $\beta_k > 1$ and (2) $\|\mathbf{x}^k - \mathbf{x}^{k-1}\| / \max(|\mathbf{x}^{k-1}|) > 0.1$. Condition (2) limits the maximum change in the variables (norm sense) to some fraction of the magnitude of the largest variable. If conditions (1) and (2) are both met, then the current step size is reduced via Equation 17 using the current value of $\beta_k$. If ever $\delta t_{j+1} < \delta t_{min}$ (specified), the program is terminated. In addition, computations stop if the number of iterations k exceeds a specified value $k_{max}$.

Convergence is checked by requiring

$$[\beta_k/(1-\beta_k)]\ \|\mathbf{x}^{k+1} - \mathbf{x}^k\| \leq \theta \qquad (22)$$

in which $\theta$ is a specified tolerance. Equation 22 follows from the contractive theorem (Schmidt, 1978). Once Equation 22 is satisfied, the current $t_{j+1}$ cycle is completed, and a new increment $\delta t_{j+2}$ is determined from Equation 17.

## EXAMPLES

### Symmetric Channel Flow

Turbulent flow between smooth parallel walls is now examined. The computational domain of six elements starts a small distance from the solid boundary and ends at the channel center. At the solid boundary law-of-the-wall functions (Schamber and Larock, 1980) determine the essential boundary conditions for k, $\varepsilon$ and U. At midchannel symmetry requires the normal derivatives of k, $\varepsilon$, and U to be zero. These derivatives are also zero in the streamwise direction for fully developed flow. The streamwise pressure gradient is specified. Also specified are $g_1=2.0$, $g_2=0.5$, $g_3=0.1$, $g_T=1.0$, $\alpha_T=0.25$, $\delta t_{min}=0.0001$, and $\theta=0.0001$. From Equation 18 $\alpha_L=0.10$. For $\delta t_o=0.05$ algorithm performance is presented in Table I, where $k_n$=number of iterations at $t_j$ to satisfy Equation 22. The starred entries indicate

Table I. Computation Schedule, Plane Channel Flow

| t | 0 | 0.05 | 0.133 | 0.260* | 0.431* | 0.642* | 0.802* | 0.952* | 1.0 |
|---|---|---|---|---|---|---|---|---|---|
| $\alpha$ | | 0.126 | 0.139 | 0.164 | 0.185 | 0.392 | 0.277 | 0.293 | 0.963 |
| $k_n$ | 2 | 4 | 4 | 4 | 4 | 5 | 3 | 5 | 5 |

steps for which $r<1$ and the Lagrangian predictor is used to initialize the next step. The total number of iterations is 36. The selection of $\delta t_o$ appears reasonable considering the size of $\alpha$ at each step. Increasing $\theta$ to 0.001 reduces the total number of iterations by 3. Profiles of k, $\varepsilon$ and U are shown in Figures 1, 2 and 3, respectively, for several t values. Quantities presented have been nondimensionalized on the half channel width d and shear velocity $u_\tau$ at the wall. The Reynolds number is based on d and the centerline channel velocity.

### Asymmetric Channel Flow

This flow has been experimentally studied by Hanjalic and Launder (1972). An asymmetric flow between parallel walls is generated by roughening one wall with square ribs. The other wall is smooth. Nine elements were placed across the channel width with the edge nodes displaced from the wall to accommodate the law-of-the-wall boundary conditions. Essential boundary conditions at the smooth and rough surfaces are determined from appropriate wall functions (c.f. Schamber and Larock, 1980). Streamwise derivatives for k, $\varepsilon$ and U are zero for fully developed flow. Specifying $u_{\tau R}/u_{\tau S}$ and $u_{\tau S}d/\nu$, where $u_{\tau R}$ and $u_{\tau S}$ are the rough and smooth wall shear velocities, d=channel halfwidth, $\nu$=fluid kinematic viscosity, fixes the streamwise pressure gradient and the wall boundary conditions. Because both boundaries require essential conditions, Equations 12 and 13 are modified to

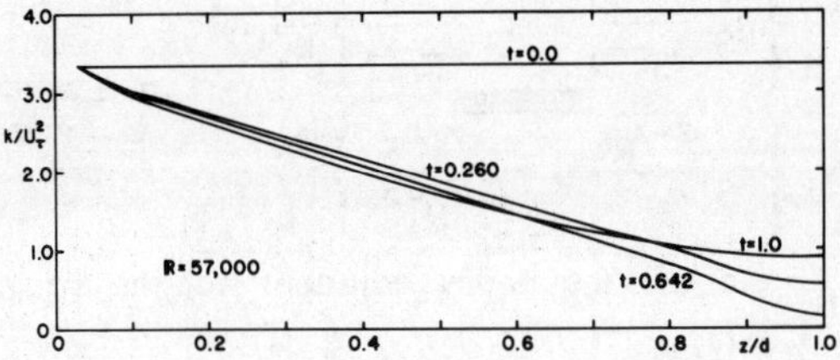


Figure 1. Symmetric Channel, $k/u_\tau^2$ vs. z/d.

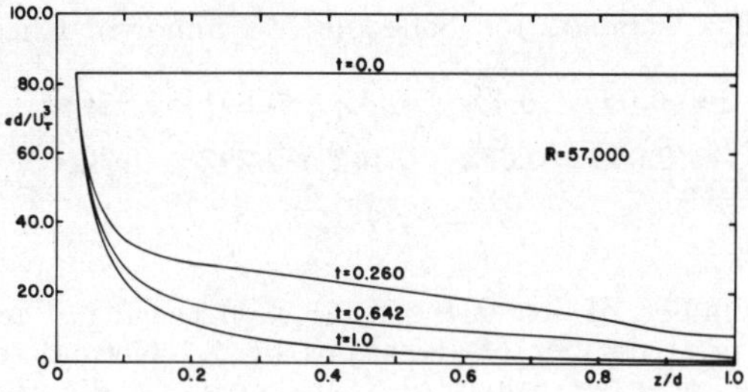


Figure 2. Symmetric Channel, $\varepsilon d/u_\tau^3$ vs. z/d.

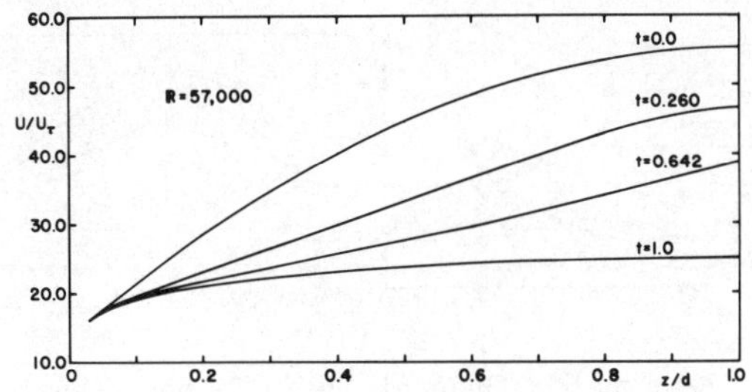


Figure 3. Symmetric Channel, $U/u_\tau$ vs. z/d.

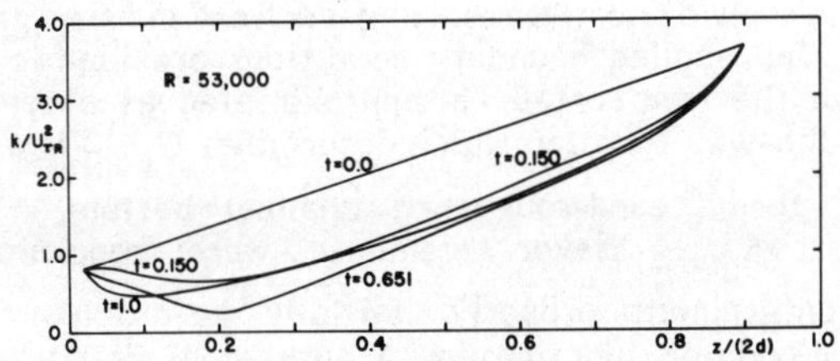


Figure 4. Asymmetric Channel, $k/u_{\tau R}^2$ vs. z/(2d).

$$H_k = \int_\Omega N_i \left\{ k_S \frac{z-z_R}{z_S-z_R} + k_R \frac{z-z_S}{z_R-z_S} \right\} \frac{U_o}{L_o}\, d\Omega \tag{23}$$

$$H_\varepsilon = \int_\Omega N_i \left\{ \varepsilon_S \frac{z-z_R}{z_S-z_R} + \varepsilon_R \frac{z-z_S}{z_R-z_S} \right\} \frac{U_o}{L_o}\, d\Omega \tag{24}$$

in which $z_R$ and $z_S$ are nodes adjacent to the rough and smooth walls, respectively.

The input parameters for this example are identical to the symmetric channel problem. The computational schedule is presented in Table II.

Table II. Computation Schedule, Asymmetric Channel Flow

| $t$ | 0 | 0.05 | 0.15 | 0.35 | 0.651 | 0.924* | 1.0 |
|---|---|---|---|---|---|---|---|
| $\alpha$ | | 0.055 | 0.078 | 0.142 | 0.292 | 0.682 | 0.252 |
| $k_n$ | 2 | 4 | 4 | 4 | 5 | 6 | 4 |

The total number of iterations is 29. Increasing $\theta$ to 0.001 would reduce the total number of iterations by 5. Normalized plots of k, $\varepsilon$ and U for selected values of t are given in Figures 4, 5 and 6, respectively. The reported Reynolds number is based on d and the maximum channel velocity.

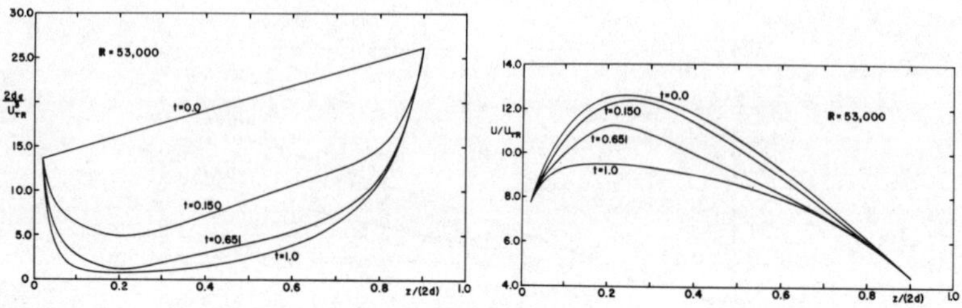


Figures 5,6. Asymmetric Channel, $2d\varepsilon/u_{\tau R}^3$ and $U/u_{\tau R}$ vs. $z/(2d)$.

## Stratified Channel Flow

Steady sediment transport in open channel flow results in a stably stratified flow owing to the resulting vertical density gradient of the bulk fluid. The applied boundary conditions are identical to example one, because the free surface is approximated as a symmetry plane. The law-of-the-wall function which determines $U_{wall}$ is altered slightly to account for a sand-roughened channel bottom. The pressure gradient is $gS_o$. Eleven elements were required to resolve concentration gradients properly. Initially the problem was solved by use of an embedding function for $\Lambda$ similar to Equations 12 and 13. The solution required over 200 iterations because the free surface value of $\Lambda$ was 5 orders of magnitude smaller than $\Lambda_{wall}$. Since $\Lambda$

at the free surface plays an important role in determining the distribution of $\Lambda$, the $\Lambda$ profile did not begin to change significantly until t exceeded 0.99.

To get a solution in a more reasonable number of iterations, no embedding function was used (as indicated in Equation 11). First $E_S$ was held at $10E_{So}$ $(E_{So}=\gamma\, c_\mu k_o^{\,2}/\varepsilon_o)$ until t=1, because allowing $E_s$ to vary with $\nu_t$ produced gradients in $\Lambda$ far too steep for the element mesh to resolve. Thus the $\Lambda$ profile remained constant for all values of t until t=1, when $E_S$ assumed its proper value.

Table III. Computation Schedule, Stratified Flow

| | | | | | | | |
|---|---|---|---|---|---|---|---|
| t | 0 | 0.2 | 0.311 | 0.432 | 0.568 | 0.712 | 0.851 |
| $\alpha$ | | 0.754 | 0.220 | 0.210 | 0.228 | 0.266 | 0.344 |
| $k_n$ | 2 | 3 | 3 | 3 | 3 | 3 | 3 |

| | | | | | | |
|---|---|---|---|---|---|---|
| t | 0.964 | $0.977^R$ | 0.996 | $0.9993^R$ | 0.9999 | 1.0 |
| $\alpha$ | 0.570 | 0.143 | 0.469 | 5.152 | 0.451 | 0.108 |
| $k_n$ | 4 | 5+3 | 4 | 5+7 | 4 | 2 |

Table III details progress to a solution, obtained in a total of 54 iterations. The superscript R indicates t stations at which the original step size was reduced and the computation restarted using the predictor from the previous t-station. The greater number of iterations for t > 0.9 is a result of steeper gradients in k and $\varepsilon$ compared to the symmetric channel flow. The variation of velocity and concentration with t are shown in Figures 7 and 8. Plots of k and $\varepsilon$ are very similar to Figures 1 and 2 and thus are not shown. The Reynolds number is much larger in this case than in the other examples so that a significant sediment volume fraction could be suspended. The larger Reynolds number also appears to be a good indicator that the problem should be more difficult to solve.

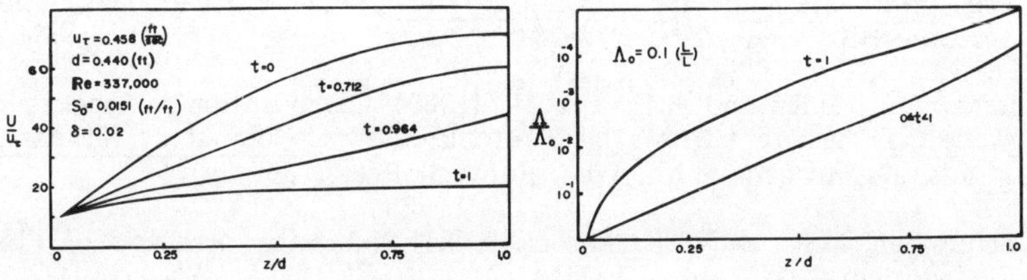


Figures 7,8. Stratified Channel Flow, log $\Lambda/\Lambda_o$ and $U/u_\tau$ vs. z/d.

## SUMMARY AND OUTLOOK

The Galerkin finite element formulation of the k-ε turbulence model has been presented for solution by the continuation method. Several examples are presented for nonstratified and density-stratified one-dimensional flow. The method works well for all cases reported.

A modified continuation algorithm may be promising for two-dimensional turbulent flow simulations. An embedding function in the spirit of Equations 12, 13, 23 and 24 needs to be developed which would properly interpolate essential boundary conditions in two dimensions. Furthermore, partitioning the solution between mean flow variables (velocities and pressure) and turbulence quantities (k and ε) and solving the set sequentially rather than simultaneously may be more cost effective.

## ACKNOWLEDGEMENT

Support of this work by National Science Foundation Grants CME-7914762 and CME-8006743 is sincerely appreciated.

## REFERENCES

DeVantier, B.A. and B.E. Larock (1981) Sediment Transport in Stratified Turbulent Flow, Third Symposium on Turbulent Shear Flows, L.J.S. Bradbury et al., eds., University of California, Davis, Sept. 9-11, 12.12-12.17.

Hanjalic, K. and B.E. Launder (1972) Fully Developed Asymmetric Flow in a Plane Channel, J. Fluid Mech., 5, 301-335.

Rheinboldt, W.C. (1977) Numerical Continuation Methods for Finite Element Applications, Formulations and Computational Algorithms in Finite Element Analysis: U.S.-Germany Symposium, K.J. Bathe et al., eds., August 1976, M.I.T. Press.

Rodi, W. (1980a) Turbulence Models and Their Application in Hydraulics, State-of-the-Art Paper, IAHR.

Rodi, W. (1980b) Turbulence Models for Environmental Problems, Prediction Methods for Turbulent Flow, von Karman Inst. Pub., McGraw-Hill, New York, 276-281.

Schamber, D.R. and B.E. Larock (1980) Computational Aspects of Modeling Turbulent Flows by Finite Elements, Computer Methods in Fluids, K. Morgan et al., eds., Pentech Press, 339-361.

Schamber, D.R. and B.E. Larock (1981) Numerical Analysis of Flow in Sedimentation Basins, J. Hyd. Div., ASCE, HY5, 575-591.

Schmidt, W.R. (1978) Adaptive Step Size Selection for use with the Continuation Method, Int. J. Num. Meth. Engrg., 12, 677-694.

# THE APPLICATION OF FEM AND FDM TO FLOW SEPARATION PATTERN ; A COMPARISON STUDY

Nabil A. Zaghloul

Assistant Professor, Univ. of Petroleum & Minerals, Saudi Arabia

## INTRODUCTION

With the advent of high speed computer, it has become possible to develop methods for studying theoretically many of the unsteady incompressible fluid flow problems which previously had been hopelessly complicated for analysis. The purpose of this paper is to describe in detail two popular numerical methods namely; the Finite Element Method (FEM) and the Finite Difference Method (FDM). Both methods are applied to solve the flow separation pattern past an obstruction in a two-dimensional flow field. The differential equations that govern the phenomenon are the Helmholtz type vorticity equation and the Poisson type stream function equation. The solution of the two equations will provide the variation of the vorticity and stream function fields with respect to time. The Finite Element Model is developed using two extremum functionals for the Poisson and Helmholtz vorticity equations (Zienckiewicz 1977). A triangular elements are utilized in the solution with 6-nodal elements used for the Poisson solution and 3-nodal elements for the Helmholtz equation. The Finite Difference Model is developed using a square mesh at which the stream function and vorticity values are evaluated at the mesh points. The derivatives of the Helmholtz and the Poisson equations are replaced with central differences both in space and time (Smith 1965). The FEM and FDM are formulated, developed and tested. The solutions for the separation pattern are compared using both models. The results predicted by both models are compared to an experimental separation pattern using pitot-tube measurements.

## THE GOVERNING DIFFERENTIAL EQUATIONS

The governing equation of motion for incompressible turbulent flow, is described by the Reynolds equation as

$$\rho\left[\frac{\partial \overline{U}_i}{\partial t} + \overline{U}_j \frac{\partial \overline{U}_i}{\partial x_j}\right] = - \frac{\partial \overline{P}}{\partial x_i} + \frac{\partial}{\partial x_j}\left(\mu \frac{\partial \overline{U}_i}{\partial x_j} - \rho \overline{u_i u_j}\right) + \overline{F}_i \qquad (1)$$

where $\overline{U}$ = the ensemble average velocity at a point; $\overline{P}$ = the ensemble average pressure at a point; $\rho$ = fluid mass density; $\overline{F_i}$ = external body force; $\rho\overline{u_i u_j}$ = the turbulent or Reynolds stress
The Reynolds stresses were later assumed by Boussinesq (Hinze 1959), to act like viscous stresses. Using the proposed eddy viscosity, Equation 1 can be transformed to a Helmholtz type vorticity equation as

$$\frac{\partial \vec{\omega}}{\partial t} + \nabla x \vec{\omega} x \vec{U} = (\nu+\xi) \nabla^2 \vec{\omega} \tag{2}$$

where $\vec{U}$ = the ensemble average velocity vector; $\vec{\omega}$ = the ensemble average vorticity vector; $\nu$ = the kinematic viscosity; $\xi$ = the kinematic eddy viscosity.
In a two-dimensional flow field using Cartesian coordinates, the vorticity is given as

$$\omega = -\frac{\partial u}{\partial y} + \frac{\partial v}{\partial x} \tag{3}$$

where u and v are the velocity components in the x and y directions respectively and can be defined in terms of the stream function $\psi$ as,

$$u = \frac{\partial \psi}{\partial y} \quad \text{and} \quad v = -\frac{\partial \psi}{\partial x} \tag{4}$$

Substituting Equation 4 into Equation 3, yields the following Poisson type equation.

$$\frac{\partial^2 \psi}{\partial x^2} + \frac{\partial^2 \psi}{\partial y^2} = -\omega \tag{5}$$

Also, the time dependent Helmholtz vorticity Equation 2, becomes

$$\frac{\partial \omega}{\partial t} + \frac{\partial}{\partial x}\left[u\omega - (\nu+\xi)\frac{\partial \omega}{\partial x}\right] + \frac{\partial}{\partial y}\left[v\omega - (\nu+\xi)\frac{\partial \omega}{\partial y}\right] = 0 \tag{6}$$

which is a non-linear second order partial differential equation similar to the diffusion convection equation. Equations 5 and 6 describe the flow phenomenon in a two-dimensional field. The numerical solution for these equations can be achieved either by Finite Element or Finite Difference methods.

## THE FINITE ELEMENT MODEL

The variational form of the Poisson equation is given as

$$\chi = \iint\left[\tfrac{1}{2}\left\{\left(\frac{\partial \psi}{\partial x}\right)^2 + \left(\frac{\partial \psi}{\partial y}\right)^2\right\} - \omega\psi\right]dx\, dy \tag{7}$$

To obtain the variational form for the Helmholtz equation two assumptions have to be utilized : 1) The convective term in the equation is assumed to be constant over the element; 2) The value of $(\nu+\xi)$ is assumed as a constant average value within the element.
With these two assumptions the equivalent variational

formulation is given as

$$\overline{\chi} = \iint [\tfrac{1}{2} \{(\frac{\partial \omega}{\partial x})^2 + (\frac{\partial \omega}{\partial y})^2\} + \frac{1}{(\nu+\xi)} (K + \frac{\partial \omega}{\partial t})\omega] \, dx \, dy \qquad (8)$$

where $\frac{\partial \omega}{\partial t}$ is considered invariant, and K is the convective term.

The Poisson equation

A typical triangular element allowing a quadratic variation of a function is shown in Figure 1. The stream function $\psi$ can be expressed as a complete 2nd order polynomial in the local x and y co-ordinates. Six nodal triangular element is used, 3-corner nodes and 3-mid side nodes. Zienkiewicz (1977) and Desai (1972) present a detailed description of this element. The value of stream function $\psi$ in the mth element may be expressed as

$$\psi^m = N_i \, \psi_i \qquad (9)$$

where $\psi_i$ is the value of $\psi$ at the ith node and $N_i$ is the shape function. The contribution to $(\frac{\partial \chi}{\partial \psi})$ from the mth element may be expressed as

$$\{\frac{\partial \chi}{\partial \psi}\}^m = [S]^m \{\psi\}^m + \{F\}^m \qquad (10)$$

where $[S]$ is (6 x 6) element stiffness matrix and $\{F\}$ is (6 x 1) element vector

$$S_{ij} = \iint (\frac{\partial N_i}{\partial x} \frac{\partial N_j}{\partial x} + \frac{\partial N_i}{\partial y} \frac{\partial N_j}{\partial y}) \, dx \, dy \qquad (11)$$

$$F_i = - \iint \omega \, N_i \, dx \, dy \qquad (12)$$

The minimization of $\chi$ is achieved by successively partial differentiating $\chi$ with respect to each of the unknown values of $\psi$ and equating each differentiation to zero.

Hence

$$\frac{\partial \chi}{\partial \psi_i} = \Sigma \frac{\partial \chi^m}{\partial \psi_i} = 0 \qquad (13)$$

Assembling the element stiffness matrix into a global stiffness matrix by Equation 13 with appropriate modifications for boundary conditions leads to

$$[S^*] \{\psi\} + \{F^*\} = 0 \qquad (14)$$

where $[S^*]$ is (N x N) global stiffness matrix and $\{F^*\}$ is (N x 1) global vector. N represents the number of unknown nodes in the stream function field.

The Helmholtz equation

The same triangular elements are used for the Helmholtz equation. A linear vorticity field is assumed and the $\omega$ value within the mth element may be expressed as

$$\omega^m = M_i \, \omega_i \qquad (15)$$

where $\omega_i$ is the value of $\omega$ at the ith node and $M_i$ is the shape function for linear triangular elements. Similar to Equation 10, the element contributions to $(\overline{\frac{\partial \chi}{\partial \omega}})$ is given as

$$\{\overline{\frac{\partial \chi}{\partial \omega}}\}^m = [H]^m \{\omega\}^m + [p]^m \{\frac{\partial \omega}{\partial t}\}^m + \{Q\}^m \tag{16}$$

where [H] and [p] are (3 x 3) element stiffness matrices and {Q} is (3 x 1) element vector.

$$H_{ij} = \iint (\frac{\partial M_i}{\partial x} \frac{\partial M_j}{\partial x} + \frac{\partial M_i}{\partial y} \frac{\partial M_j}{\partial y})\, dx\, dy \tag{17}$$

$$P_{ij} = \frac{1}{\nu+\xi} \iint M_i M_j\, dx\, dy \tag{18}$$

$$Q_i = \frac{K}{\nu+\xi} \iint M_i\, dx\, dy \tag{19}$$

The minimization of the whole region can be achieved by assembling the contribution of each element into a global stiffness matrices given as

$$[H^*]\{\omega\} + [p^*] \{\frac{\partial \omega}{\partial t}\} + \{Q^*\} = 0 \tag{20}$$

where [H*] and [p*] are (N1 x N1) global stiffness matrices and {Q*} is (N1 x 1) adjusted global vector. N1 represents the number of unknown nodes in the vorticity field.

Solution in time domain Using an average value of a backward and forward differences' approximations for the term $\{\frac{\partial \omega}{\partial t}\}$, Equation 20 yields the following recursion formula.

$$([H^*] + \frac{2}{\Delta t} [p^*])\{\omega\}_t = [p^*] (\frac{2}{\Delta t} \{\omega\}_{t-t\Delta} + \{\frac{\partial \omega}{\partial t}\}_{t-\Delta t}) - \{Q^*\}_t \tag{21}$$

with the initial condition given as

$$- [p^*] \{\frac{\partial \omega}{\partial t}\}_{t=0} = [H^*]\{\omega\}_{t=0} + \{Q^*\}_{t=0} \tag{22}$$

where $\Delta t$ is the time interval

## BOUNDARY CONDITIONS

The boundary conditions pertinent to the present problem are :

a) Inflow and outflow boundaries
These boundaries were chosen at considerable distances upstream and downstream of the obstacle where a steady, uniform irrotational flow condition can be assumed. Along these boundaries, the following two conditions were met :
i) $\omega = 0$ at all times, and ii) $\psi$ varies linearly with y.

b) Upper and lower wall boundaries
The wall boundaries were treated as streamlines; it follows that a natural condition on the wall boundary is that of a constant $\psi$. Theoretically, a no slip condition applies at the wall, i.e. a vorticity of $\omega = -\partial u/\partial y$ exists at the walls. However, for the problem under consideration, the main vorticities affecting the phenomenon are caused by the presence of the

obstruction while the wall vorticities are strictly confined to the wall zone and do not play a significant role in shaping the flow separation pattern. Thus, the zero vorticity boundary condition was imposed on the walls. Hence, on the upper wall,

$\psi = \overline{Q}$, & $\omega = 0$, and on the lower wall $\psi = 0$ & $\omega = 0$

c) The obstacle boundary conditions

In treating flow around obstructions, an essential requirement is that the periphery of the obstacle must be a streamline. Therefore, $\psi$ must be everywhere the same on that periphery and is equal to $\psi$ on the upper wall. In addtion, a no-slip condition was assumed along the obstacle. Thus, the boundary conditions for the dike are

$\psi = \overline{Q}$, and $v = 0$

The presence of the obstacle, with the no-slip condition,introduces vorticity that causes separation. Experimentation with the model showed that the best representation of the singularity of the obstacle's tip could be obtained by the use of a thin plate obstacle and neglecting the velocity gradient $\partial u/\partial y$, indicating that the separation line emerges, from the tip, tangential to the upstream face of the obstacle.

## THE NUMERICAL SOLUTION OF THE FINITE ELEMENT METHOD

To apply the Finite Element Method, the flow field is discretized into 72 triangular elements as shown in Figure 2. The boundary conditions used for $\psi$ and $\omega$ are also shown in this figure. The total number of nodal points used for the Poisson Solution is 171 with N = 119 interior points. The total number of nodal points for the Helmholtz Solution is 50 with N1 = 24 interior points. The nodes' numbers are arranged, as suggested by Finn (1967) in such a way as to minimize the band width of the Matrix of solution. The interior nodes are numbered first and the boundary points are numbered last. Equations 14 and 21 are solved alternately for each time step to determine the variation in $\psi$ and $\omega$ with time. The solution of the Poisson equation is initiated by assuming a zero vorticity field and a Laplace solution is obtained as given in Figure 3. The final solution is achieved when the variation in the value of $\psi$ and $\omega$ for any two successive solutions is negligible and a relatively steady separation pattern is obtained as shown in Figure 4. The stability of the solution depends on the choice of time increment $\Delta t$.Cheng (1972) reports that the solution procedure is always stable for sufficiently small $\Delta t$. As a guideline $\Delta t$ should be chosen so that

$\Delta t < 0.1 L^2 R_N$

where L is the characteristic length of an element; $R_N$ is the Reynolds Number.

## THE FINITE DIFFERENCE MODEL

The Poisson Equation 5 and the Helmholtz Equation 6 are solved numerically by the Finite Difference Scheme. The derivatives are approximated with central differences both in space and time. In order to save computational time, successive over relaxation formulation (Smith 1965) is utilized to solve the discretized Poisson equation and is given as

$$\psi_{ij}^{K+1} = (1-W)\,\psi_{ij}^{K} + \frac{W}{4}\left[\psi_{i-1,j}^{K+1} + \psi_{i+1,j}^{K} + \psi_{i,j-1}^{K+1} + \psi_{i,j+1}^{K} + a^2\omega_{ij}\right] \tag{23}$$

where a is the mesh size and W is the over relaxation factor. The Helmholtz equation is represented in Finite Difference notation along the x and y axes by an implicit scheme. The method used is anologous to the technique proposed by Peaceman and Rachford (1955) and is given as

In the x-direction

$$-r\omega_{i-1,j}^{n+1} + (1+3r)\,\omega_{ij}^{n+1} - r\omega_{i+1,j}^{n+1} = r(\omega_{i,j-1}^{n} + \omega_{i,j+1}^{n}) + (1-r)$$
$$\omega_{ij}^{n-1} + r1\left[(u\omega)_{i-\frac{1}{2},j}^{n} - (u\omega)_{i+\frac{1}{2},j}^{n} + (v\omega)_{i,j-\frac{1}{2}}^{n} - (v\omega)_{i,j+\frac{1}{2}}^{n}\right] \tag{24a}$$

In the y-direction

$$-r\omega_{i,j-1}^{n+2} + (1+3r)\,\omega_{ij}^{n+2} - r\omega_{i,j+1}^{n+2} = r(\omega_{i-1,j}^{n+1} + \omega_{i+1,j}^{n+1}) + (1-r)$$
$$\omega_{ij}^{n} + r1\left[(u\omega)_{i-\frac{1}{2},j}^{n+1} - (u\omega)_{i+\frac{1}{2},j}^{n+1} + (v\omega)_{i,j-\frac{1}{2}}^{n+1} - (v\omega)_{i,j+\frac{1}{2}}^{n+1}\right] \tag{24b}$$

where $r = 2(\nu+\xi)\,\Delta t/a^2$ and $r_1 = 2\Delta t/a$

## THE NUMERICAL SOLUTION OF THE FINITE DIFFERENCE METHOD

A finite difference scheme with a regular computational grid consisting of about 1400 nodes is utilized in the solution Figure 5. The optimum over relaxation factor W, based on the least number of iterations is found to equal 1.75. The convective term in the Helmholtz equation is linearized by defining it in terms of the velocities and vorticities at beginning of each computational step. To obtain the different values for the convective terms, which are not positioned at the node points, the averages of the adjacent values are used. Equations 23 and 24 are solved alternately to obtain the variations in $\psi$ and $\omega$ with time. The initial Laplace Solution is found to be exactly similar to the solution obtained by the Finite Element Model Figure 3. The final fully developed separation pattern is shown in Figure 6.The stability of the alternating direction implicit scheme adopted was studied by Zaghloul and McCorquodale (1975), and was found to be stable for appropriate boundary conditions.

## EXPERIMENTAL VERIFICATION

A pitot-tube and depth measurements are conducted to determine the stream function distribution around an obstruction. A 1.5 ft. wide by 2 ft. deep and 15 ft. long rectangular channel is used. The obstruction is represented by a steel plate fixed against the channel wall and protruded 6 inches into the channel normal to the flow direction. The contracted width of the channel at the obstruction site is 12 inches and the opening ratio is 0.67. The flow depth and velocity values are measured at different locations upstream and downstream of the obstruction. The velocities are recorded at 0.2 and 0.8 of the depth. Along each section, measurements are made at different points. The spacings between these points are closer, in the separation zone than in the other areas, to achieve better accuracy of the separation line. In the wake zone, near the wall, the velocity of the reversed flow is also recorded. The reattachment of the separation line to the wall is detected by introducing a dye on the water surface, near the wall, and determining the zone at which the dye starts to flow in the forward direction. The depths, at all the points where the velocity was recorded, are measured using the piezometric reading of the manometer connected to the pitot-tube. Also, the depths at some selected points are measured by means of a point gauge. The velocity and depth values are integrated over the appropriate cross-sections to yield the dimensionless stream function $\psi$ . Figure 7 shows the shape and location of the separation line. The section of maximum contraction is found to be 12 inches downstream the obstruction. The coefficient of contraction is calculated based on the average pitot-tube depths and the depths obtained by means of a point gauge and found to have an average value of $0.66 \pm 0.02$. The reattachment of the separation line to the channel wall is observed, from Figure 7, to be at a distance ranging between 39 to 42 inches downstream from the obstruction.

The dimensionless separation pattern computed by the Finite Element Method, Figure 4, produce a value of $C_c = 0.66$ at about 13 inches downstream the obstruction. The reattachment of the separation line occurs at about 40 inches downstream the obstruction.

The dimensionsless separation pattern computed by the Finite Difference Method Figure 6, produce a value of $C_c = 0.66$ at about 13.5 inches downstream. The reattachment of the separation line occurs at about 35 inches downstream the obstruction.

CONCLUSIONS

1. The equations governing the phenomenon of flow separation can be solved numerically by either a Finite Element Method or a Finite Difference Scheme. The Poisson and Helmholtz equations are solved alternately to produce the flow separation pattern.
2. Both the Finite Element and the Finite Difference Models are found to be numerically stable with the proper selection of the time increment $\Delta t$.
3. The computed dimensionless plots of the fully developed separation pattern as obtained from the Finite Element Model Figure 4 and from the Finite Difference Model Figure 6 agreed fairly well with the experimental pattern using pitot-tube measurements.

ACKNOWLEDGEMENT

The author gratefully acknowledges the use of the computer facilities at the University of Petroleum and Minerals.

REFERENCES

Cheng, R.T. (1972), "Numerical Solution of the Navier-Stokes Equations by the Finite Element Method", Phys. Fluids, Vol.15, No.12

Desai, C.S. and Abel, J.F. (1972), Introduction to Finite Element Method, Van Nostrand Reinhold Co.

Finn, W.D. (1967), "Finite Element Analysis of Seepage Through A Dam", Journal of Soil Mechanics Division, ASCE, Vol.93, No. SM6, Proc. paper 5552.

Hinze, J.O. (1959), Turbulence, McGraw-Hill Book Co., Inc.,New York, N.Y.

Peacemen, D.W., and Rachford, H.H. (1955), "The Numerical Solution of Parabolic and Elliptic Differential Equations", The American Mathematical Society, Vol.3, No.7.

Smith, G.D. (1965), Numerical Solution of Partial Differential Equations, Oxford University Press, London.

Zaghloul, N.A. and McCorquodale, J.A. (1975), "A Stable Numerical Model for Local Scour", I.A.H.R., Journal of Hydraulic Research 13, No.4.

Zienckiewicz, O.C. (1977), The Finite Element Method, Third Edition, McGraw-Hill Book Co., Inc., New York, N.Y.

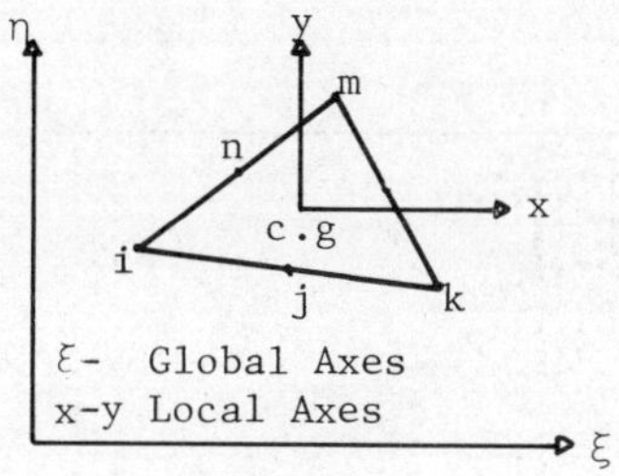


Figure (1) The Characteristics of a 6-nodal triangular element

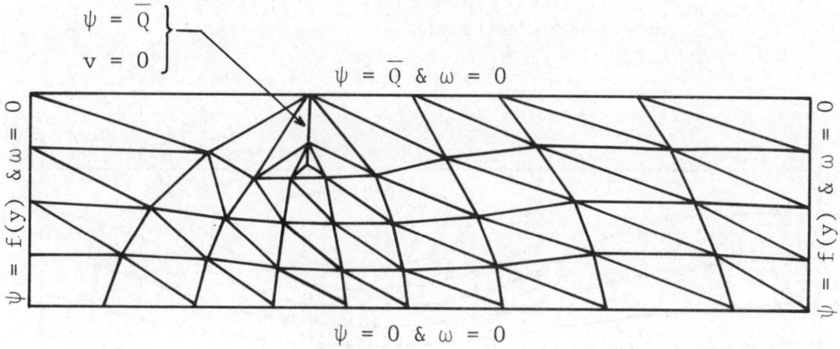


Figure (2) Finite Element Space discretization and boundary conditions

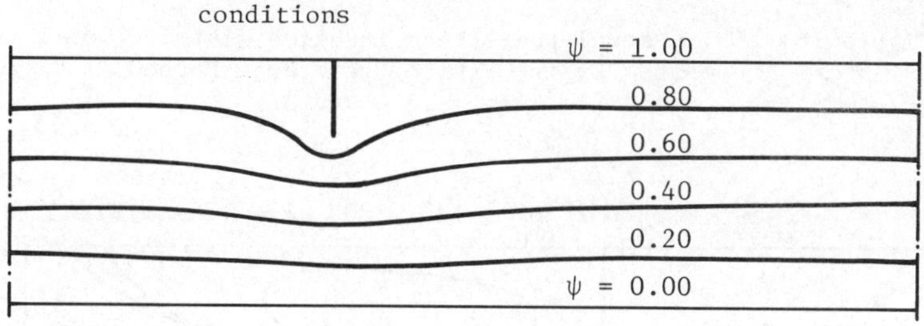


Figure (3) Stream function distribution (Initial Laplace Solution)

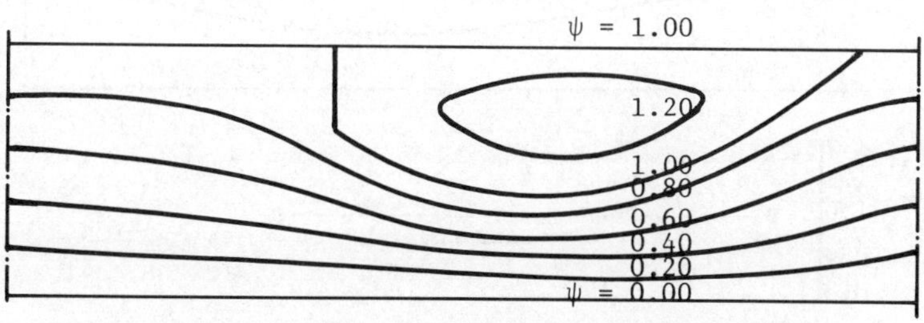


Figure (4) Fully developed stream function distribution computed by the Finite Element Method

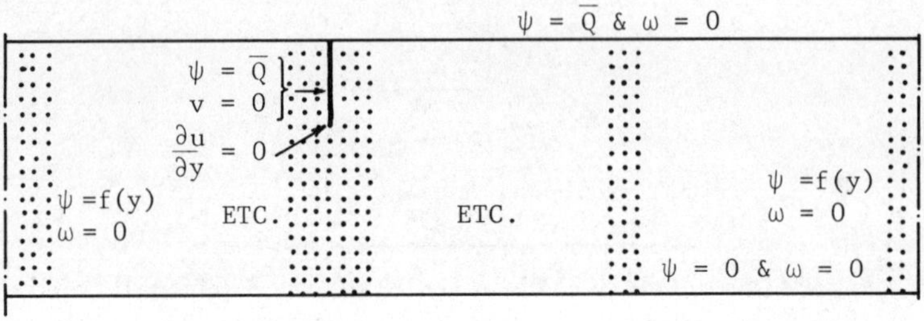


Figure (5) Finite Difference Space discretization and boundary conditions

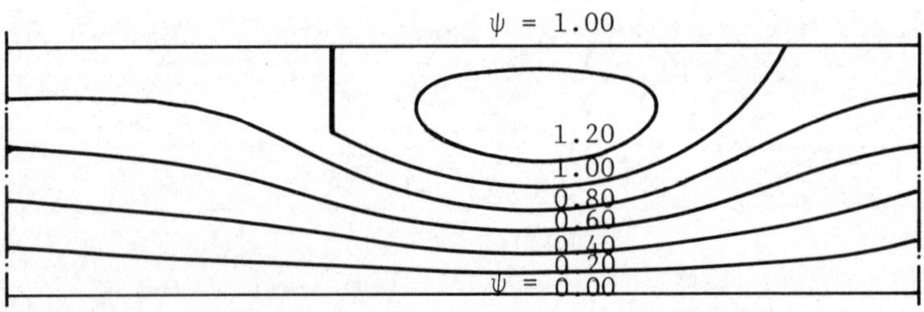


Figure (6) Fully developed stream function distribution computed by the Finite Difference Method

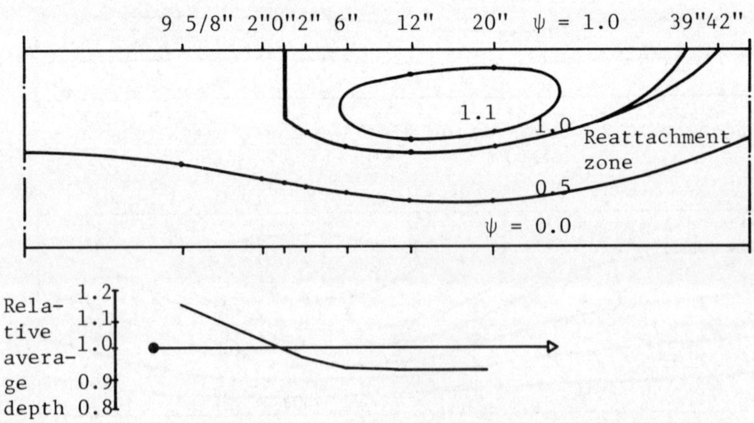


Figure (7) Dimensionless Experimental Stream function distribution

# CURRENT PROGRESS IN SOLVING THE TIME-DEPENDENT, INCOMPRESSIBLE NAVIER-STOKES EQUATIONS IN THREE-DIMENSIONS BY (ALMOST) THE FEM

P. M. Gresho, and C. D. Upson

Lawrence Livermore National Laboratory, University of California, Livermore, California 94550 USA

## INTRODUCTION

We have been pursuing the development of numerical methods which, while originally derived via the conventional Galerkin Finite Element Method (GFEM), have been significantly modified in the interest of cost-effectiveness to the point where we now refer to them as a useful blend of finite element and finite difference methods. Starting with the simplest 3-D isoparametric element and the simplest time integration method (explicit Euler), the principal modifications of GFEM are the following: (1) the "mass" matrix is diagonalized via lumping, (2) all Galerkin integrals are evaluated approximately using one-point quadrature. Additional gains in computational speed have been achieved by two modifications of the time integration method: (1) we have introduced a procedure called subcycling, which permits less frequent updates of the pressure relative to the stability-limited processes of advection and diffusion, and (2) the often expensively small time-step restriction required to assure numerical stability has been eased via the introduction of an additional anisotropic viscosity. This paper summarizes the current status of our hybrid method and presents sample numerical results.

## GOVERNING EQUATIONS AND SPATIAL DISCRETIZATION

The equations of interest here are the Navier-Stokes (NS) equations for an incompressible, constant property fluid in the Boussinesq approximation:

$$\partial \underline{u}/\partial t + \underline{u} \cdot \nabla \underline{u} = -\nabla P + Re^{-1}\nabla^2 \underline{u} + Fr^{-1}\underline{k}T\,, \qquad \nabla \cdot \underline{u} = 0\,, \qquad (1a,b)$$

and $$\partial T/\partial t + \underline{u} \cdot \nabla T = (RePr)^{-1}\nabla^2 T\,, \qquad (1c)$$

where $\underline{u} \equiv (u,v,w)$ is the velocity, P is the pressure, T is the temperature, $Re = u_o L/\nu$ is the Reynolds number, $Fr = u_o{}^2/(\gamma \Delta T g L)$ is the Froude number, [or inverse bulk Richardson number], and Pr is the Prandtl number. Given an initial temperature field, an initial velocity field $\underline{u}_0(\underline{x})$ which satisfies $\nabla \cdot \underline{u}_0=0$, and appropriate boundary conditions for $\underline{u}$ and T, Eqs. (1) can be solved (in principle) for $\underline{u}$, P, and T as functions of space and time.

If Eqs. (1) are spatially discretized using the GFEM, the corresponding matrix system of coupled ordinary differential equations (ODE's) with concomitant algebraic constraints, is obtained:

$$M\dot{u} + K(u)u + CP = f \;; \quad u(0) = u_o \quad \text{where} \quad C^T u_o = 0 \,, \tag{2a}$$

$$C^T u = 0 \,, \quad \text{and } M_s\dot{T} + K_s(u)T = f_s \,, \tag{2b,c}$$

where now u is a global vector containing nodal values of u, v, and w, P is a global pressure vector, and f is a "forcing" vector which incorporates the buoyancy term and boundary conditions on velocity (either essential or natural). M is the mass matrix (which we henceforth simplify to a "lumped mass" approximation; e.g., via row-sum at element level), K(u) is the "advection + diffusion" matrix, C is the gradient matrix and its transpose, $C^T$, is the divergence matrix. The subscript s refers to similar, but smaller, sets of coefficients and T is a global temperature vector.

In order to integrate (2) with an explicit method, we will need the consistent Poisson equation for the pressure, which is easily derived from (2):

$$AP = C^T M^{-1} [f - K(u)u] \,, \tag{2d}$$

where $A \equiv C^T M^{-1} C$ is a discretized approximation to the Laplacian operator into which the velocity boundary conditions have been (automatically) incorporated. This equation is an approximation to the continuum Poisson equation

$$\nabla^2 P = Fr^{-1} \nabla \cdot (\underline{k}T) - \nabla \cdot (\underline{u} \cdot \nabla \underline{u}) \,, \tag{1d}$$

which is implied by (1a) and (1b). Since (2a) and (2d) imply (2b), we henceforth consider (mainly) these; the resulting velocity field will always satisfy mass continuity, (2b).

While Eqs. (2) describe a general FEM approximation to velocity, temperature, and pressure, we are currently interested only in the simplest and least-expensive combination on the 8-node isoparametric brick, in which $\underline{u}$ and T are approximated via trilinear basis functions and pressure is piecewise constant ("defined" at the centroid of each element). In order to reduce costs further, we also employ approximate integration to generate the element matrices; in fact, all coefficient matrices (except element volume) are usually generated via 1-point quadrature, in which each integral is approximated by its value at the element centroid multiplied by the element volume. For further discussion of 1-point quadrature and its effects, see Chan and Gresho (1982).

## TIME INTEGRATION

In the same spirit of simplicity expressed in the spatial approximation, we employ the simplest ODE integration scheme:

explicit (or forward) Euler. Applied to Eqs. (2) this first-order scheme is

$$u_{n+1} = u_n + \Delta t\, M^{-1}[f_n - K(u_n)u_n - CP_n] \quad , \text{ where} \tag{3a}$$

$$AP_n = C^T M^{-1}[f_n - K(u_n)u_n]; \text{ and } T_{n+1} = T_n + \Delta t M_s^{-1}[f_s - K_s(u_n)T_n]. \tag{3b,c}$$

Thus, given $u_n$ satisfying $C^T u_n = 0$, (3b) is first solved for $P_n$ (which assures that $C^T u_{n+1}=0$) and then (3a,c) are simply "marched forward" from $t_n$ to $t_{n+1}$. For reasons previously enumerated (Upson et al. (1982)), we currently solve the linear algebraic system for $P_n$ using a skyline version of Gaussian elimination; the symmetric and constant matrix A is factored into $LDL^T$ and stored on disk (or in memory, if the problem is small enough). Each pressure update thus consists of one forward reduction and one back substitution.

Although (3) represents a very simple algorithm, which is easily vectorized when 1-point quadrature is employed, its cost-effectiveness warrants improvement for the following reasons:

1. The forward Euler integration scheme exhibits (at least) two undesirable properties: (i) it (like almost any explicit method) requires the use of rather small time steps to preclude numerical instability and (ii) it reduces the effective diffusivity in the streamline direction, especially when the flow is advection-dominated (large Reynolds number).
2. For large problems (which, of course, describes most three-dimensional simulations), the I/O cost associated with reading the factored A matrix from disk is relatively high on the CRAY-1 computer.

While we have considered (and are still considering) alternate methods for solving the discretized Poisson equation such as pre-conditioned conjugate gradient or multi-grid, and while alternate time integration schemes <u>may</u> be more cost-effective, our principal efforts thus far have been directed toward improving the speed of the given algorithm, which we now address (with respect to items (1) and (2) above).

<u>Introduction of tensor viscosity</u>

Since the NS equations are, in some sense, a (highly glorified) generalization of the simpler (and scalar) advection-diffusion (AD) equation, the latter is commonly used to help understand the former. If the simple forward-time, centered-space (FTCS) finite difference approximation is used to solve the constant-velocity AD Eq., (1c), on a regular grid, the time step is limited by the following stability requirements (see Hindmarsh and Gresho (1982)):

$$\Delta t \leq \frac{Pe/2}{1/\Delta x^2 + 1/\Delta y^2 + 1/\Delta z^2}, \quad \Delta t \leq \frac{2/Pe}{u^2 + v^2 + w^2}, \tag{4a,b}$$

where Pe = RePr is the Peclet number. We have found that these results (when applied fairly conservatively) are still reasonably accurate when solving the variable-velocity, variable-grid NS equations and that (4b) is especially restrictive when Re is large (it tends to require the Courant number to be less than the reciprocal of the largest grid Reynolds number which is usually >> 1), thus often forcing the use of a much smaller $\Delta t$ than might be reasonably required based on accuracy.

In order to elucidate the cause of the advective-diffusive stability limit (4b) and to suggest a cure, we next present a summary of a Taylor-series analysis (in time) when the forward Euler method is used to integrate (1c). Given the solution of (1c) at time $t_n$, the exact solution at time $t_{n+1}$ is given (formally) by

$$T(t_{n+1}) = T_n + \Delta t \left.\frac{\partial T}{\partial t}\right|_n + \frac{\Delta t^2}{2} \left.\frac{\partial^2 T}{\partial t^2}\right|_n + O(\Delta t^3) \tag{5}$$

which, using (1c) with $L \equiv Pe^{-1}\nabla^2 - u\cdot\nabla$, can be written

$$T(t_{n+1}) = T_n + \Delta t \left. LT\right|_n + \frac{\Delta t^2}{2}\left(\left.L^2 T\right|_n + \frac{\partial L}{\partial t}\left. T\right|_n\right) + O(\Delta t^3)\,. \tag{6}$$

Next, we ask: "For what approximation to L, say $\tilde{L}$, is the forward Euler integration of (1c) equivalent, in some sense, to the exact result given by (6)?". The approximate problem is thus

$$T_{n+1} = T_n + \Delta t\, \tilde{L} T_n\,, \tag{7}$$

from which $\tilde{L}$ can be obtained by equating $T_{n+1}$ and $T(t_{n+1})$:

$$\tilde{L} = L + \frac{\Delta t}{2}\left(L^2 + \frac{\partial L}{\partial t}\right) + O(\Delta t^2)\,. \tag{8}$$

Restricting attention to that subset of first-order terms which resembles diffusion gives, for $\nabla\cdot\underline{u} = 0$,

$$\tilde{L} = L + \nabla\cdot(\underline{\underline{\tau}}\cdot\nabla\ ) = [Pe^{-1}\nabla^2 + \nabla\cdot(\underline{\underline{\tau}}\cdot\nabla\ ) - \underline{u}\cdot\nabla]\,,\quad \text{where } \tau_{ij} = u_i u_j\,\Delta t/2 \tag{9a,b}$$

and $\nabla\cdot(\underline{\underline{\tau}}\cdot\nabla\ ) = (\Delta t/2)(\underline{u}\cdot\nabla)^2$; i.e., artificial diffusion is (as usual) caused by the advection term. It then follows that if the forward Euler method is employed to integrate (1c), i.e., using the <u>true</u> spatial operator, the result is equivalent to solving the modified continuum equation

$$\partial T/\partial t + \underline{u}\cdot\nabla T = \nabla\cdot[(Pe^{-1}\underline{\underline{I}} - \underline{\underline{\tau}})\cdot\nabla T]\,, \tag{10}$$

at least through diffusion-like terms to $O(\Delta t)$. Finally, it can be shown that the RHS of (10) can only exhibit "positive" diffusion if the total diffusion tensor is positive-definite which in turn leads to (4b), a stability requirement completely independent of spatial discretization.

We have thus shown that the straightforward application of the forward Euler method will introduce a negative (tensor) diffusion effect owing to time truncation error and that this destabilizing term is apparently independent of the spatial discretization scheme. To compensate for this effect, we simply replace $Pe^{-1}\nabla^2 T$ (or its appropriate tensor counterpart if diffusion is anisotropic) by $\nabla\cdot[(Pe^{-1}\underline{\underline{I}} + \underline{\underline{\tau}})\cdot\nabla T]$ in (1c) and, replacing T by $\underline{u}$ and Pe by Re, the equivalent substitution is made in (1a). (<u>Remarks</u>: (1) the tensor viscosity concept has been previously discussed by Dukowicz and Ramshaw (1979) and demonstrated in a finite difference context, (2) the same analysis applied to the backward (implicit) Euler scheme shows that the same tensor viscosity, $\underline{\underline{\tau}}$ , is effectively added to the true viscosity, making the scheme stable but overly diffusive.)

In addition to improving the accuracy of the time integration, the use of the added tensor viscosity term also (and importantly) improves the stability and phase speed of the scheme. The often restrictive advective-diffusive stability limit (4b) is relaxed to one which, for Pe>>1, is closer to the usual "Courant limit"; it is

$$u\Delta t/\Delta x \leq (L/\Delta x)Pe^{-1}[\sqrt{(\Delta x Pe/L)^2+1}-1] \,. \tag{11}$$

Although this stability limit was derived in one dimension (Hindmarsh and Gresho (1982)), numerical results suggest that it is also reasonable for the two- and three-dimensional NS equations. Turning now to the effect of balancing diffusion on the phase speed error for the 1-D AD equation and, in particular, the limiting case of pure advection for which the forward Euler scheme is unconditionally unstable in the absence of balancing diffusion, we have

$$\partial T/\partial t + u\partial T/\partial x = u^2\Delta t/2 \; \partial^2 T/\partial x^2 \,, \tag{12}$$

where u is constant and the physical diffusivity is zero. Application of the FTCS discretization to (12) on a uniform mesh (or FEM using linear elements and mass lumping) leads to the following equation for the numerical phase speed, $u_p$:

$$u_p = \frac{u}{ck\Delta x} \tan^{-1}\left[\frac{c \sin k\Delta x}{1 - c^2(1-\cos k\Delta x)}\right] \,, \tag{13}$$

where $c = u\Delta t/\Delta x$ is the Courant number and k is the wave number. Fig. 1 shows how $u_p/u$ varies with c and $k\Delta x$, the key observation being that time integration error ($c > 0$) increases the phase speed which helps compensate for the lagging error caused by the spatial discretization (sometimes too much so - for $1/\sqrt{2}<c<1$ and short waves). Thus, not only does balancing diffusion cause a previously unstable method to be stable (for $c\leq 1$), it also enhances the accuracy as $\Delta t$ is increased toward the (Courant) stability limit.

These desirable effects are further demonstrated in Fig. 2 which shows some results from the A-D equation solved on the unit span with 100 (lumped) linear elements (FTCS) and periodic boundary conditions. The initial condition is a Gaussian wave with $\sigma = 2\Delta x = .02$ centered at x = 0.5. The velocity is 1.0 and the diffusion coefficient is .001 which gives a Peclet number, $Pe \equiv u\sigma/k$, of 20 and a grid Peclet number of 10. The exact solution, $T(x,t) = e^{-(x-x_o-ut)/(2\sigma^2+4kt)}/\sqrt{1+2kt/\sigma^2}$, is shown at t = 1.0 (dotted) along with three numerical results: (1) FTCS using a small $\Delta t$ to (nearly) eliminate time truncation error ($\Delta t$ = .0002 which is 10% of the stability limit, giving c = .02), (2) FTCS at its stability limit ($\Delta t$ = .002, c = 0.2) and (3) FTCS plus balancing diffusion at its larger stability limit ($\Delta t$ = .00905, c = .905). Clearly the result using balancing diffusion is superior and at reduced cost owing to the larger time step. The result in curve (2) shows the typical oscillatory

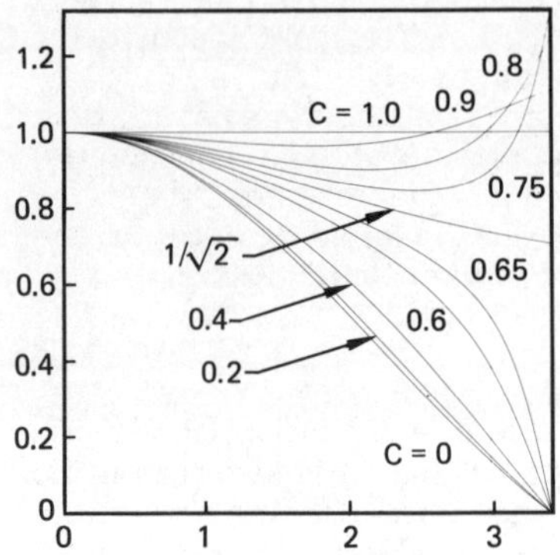


Fig. 1. Relative phase speed vs dimensionless wave number.

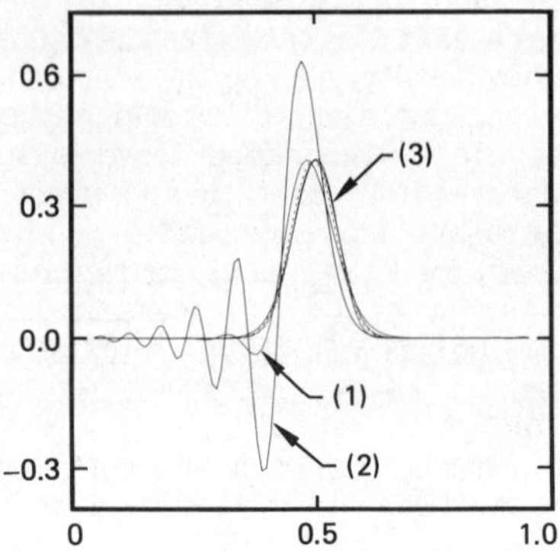


Fig. 2. Advection and diffusion of a Gaussian wave.

solution associated with pure advection on a coarse mesh, thus corroborating the Taylor series analysis which predicts an effective diffusivity of zero at the stability limit.

These pure advection results have also been extended to 2-D via the "rotating cone problem" (see, e.g. Orzag (1971)) and the results were much the same: The 4-node bilinear element with lumped mass and 1-point quadrature yielded stable results which displayed accuracy equivalent to a typical centered second-order finite difference method (e.g. Arakawa's); i.e. noticeable phase error and dispersion, but no obvious artificial diffusion, crosswind (see below) or otherwise. Thus, we are well-convinced that the combination of forward Euler and added balancing diffusion (or tensor viscosity for NS) is natural, useful, and quite cost-effective.

It is also noteworthy that the artificial tensor viscosity of (9b) acts in such a way that it only affects (i.e., modifies) the true viscosity for the streamline component of the velocity (with a magnitude given by $(u^2+v^2+w^2)\Delta t/2$) since the streamline direction is the only principal direction of the tensor $\underline{\underline{\tau}}$ with a non-zero principal value; i.e., there is no "crosswind diffusion" effect. In this regard, the tensor-viscosity correction shares the attributes of the method advocated by Hughes (see Brooks and Hughes (1982)) as a means of controlling the wiggles sometimes caused by the advection terms in GFEM models, quite independently of time integration effects.

Since, however, the tensor viscosity correction was derived using a Taylor-series expansion in time, it follows that neither the correction nor the underlying analysis applys to steady state flows. Thus, it is unfortunately true that the use of added tensor viscosity endows the fluid with an additional artificial streamline viscosity for flows which approach (and/or attain) steady state. Denoting the "magnitude" of $\underline{\underline{\tau}}$ by $|\underline{u}|^2\,\Delta t/2$, it is clear that the effective Re at steady state is reduced to $Re/(1+Re|\underline{u}|^2\Delta t/2)$, but only in the streamline direction (cf. streamline upwinding). Actually, this "wiggle suppressant" (see Gresho and Lee (1981)) is even present during time-dependent simulations for any portions of the flow which are only slowly changing. It is of course possible, but based on our experience not very fruitful, to remove $\underline{\underline{\tau}}$ (<u>and</u> reduce $\Delta t$, to retain stability) as a steady state is approached.

<u>Subcycling</u>

For many cases of interest, the stability limits associated with an explicit time integration method are quite restrictive, even when the tensor viscosity correction is applied (i.e., there is nothing really sacrosanct that requires the Courant number to be < 1 <u>everywhere</u> in the flow at <u>all</u> times -- cf. implicit methods). The cost of using small time steps is especially burdensome for the NS equations when a pressure update is (in principle) required at each time step. To ease this problem we devised a cost-effective shortcut called subcycling, which is based on the following premises: (i) stability usually dictates a $\Delta t$ which is much smaller than would be necessary to resolve with sufficient accuracy the solution to the ODE's and (ii) the pressure gradient and associated continuity equation do not affect the stability of the time integration scheme. Both premises have been justified a posteriori from the results of many experiments.

A four-step summary of the subcycling strategy is as follows: (i) The minor (smaller, and fixed) time step is based on the stability estimates discussed above and is used to (accurately) compute advection and diffusion <u>only</u> (these processes are "subcycled"); the pressure gradient is approximated via simple (linear) extrapolation and the continuity equation is completely ignored during subcycling, (ii) a constrained least-squares method is used at the conclusion of the sub-cycle process in order to re-enforce the satisfaction of the continuity equation (2b), (iii) the pressure field which is consistent with the now "solenoidal" velocity field is updated in the usual manner, and finally (iv) the next major (larger) time step is dynamically computed based on the desired temporal accuracy by using a local error estimate. Unfortunately, space restrictions preclude a thorough description and demonstration of this multi-faceted strategy and only a brief description of the four steps can be presented here. A more thorough exposition is planned as a future publication since the method has proven to be quite cost-effective and is applicable to other discretization methods than those employed herein. The overall scheme is represented graphically in Fig. 3, which shows the velocity and pressure at three times and depicts the subcycle process between $t_n$ and $t_{n+1}$. Here $\tilde{u}$ and $\tilde{P}$ are the approximate velocity and pressure during subcycling and u,P are their mass consistent analogs (at major time steps only).

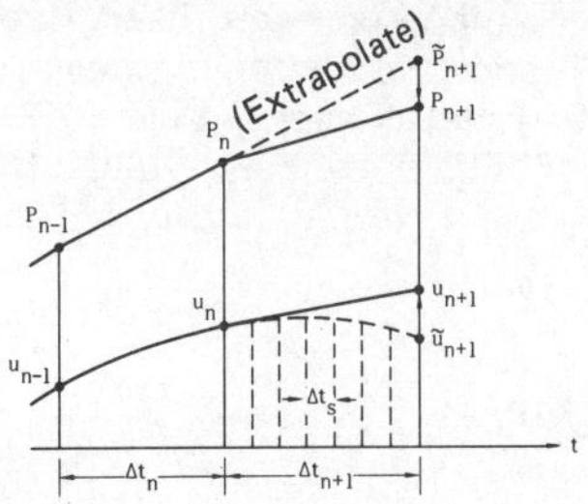


<u>Fig. 3.</u> Schematic of the sub-cycling process.

<u>The subcycling process</u> Between $t_n$ and $t_{n+1}$, the following ODE's:

$$M\dot{\tilde{u}} + K(\tilde{u})\tilde{u} + C\tilde{P} = f \quad ; \quad \tilde{u}(t_n) = u_n \,, \tag{14}$$

are solved using the forward Euler method,

$$\tilde{u}_{m+1} = \tilde{u}_m + \Delta t_s \dot{\tilde{u}}_m = \tilde{u}_m + \Delta t_s\, M^{-1}[f_m - K(\tilde{u}_m)\tilde{u}_m - C\tilde{P}_m]\,;\ \tilde{u}_o = u_n \,, \tag{15}$$

where $\Delta t_s$ is the subcycle time step and m is the subcycle index, $\tilde{P}(t)$ is known via linear extrapolation, $K(\tilde{u})$ includes the augmented tensor viscosity in the viscous portion, and $u_n$ is available. This simple marching scheme is employed for a previously-determined number, $s \simeq \Delta t_{n+1}/\Delta t_s$, of (stable) subcycle steps (the computation of s will be discussed later) at which time we have $u_s$ as an approximation to the desired velocity, $u_{n+1}$.

The return to mass consistency Given a velocity field $\tilde{u}$ which satisfies the desired boundary conditions and nearly satisfies the equations of mass and momentum conservation, we seek a nearby field, u, which is "as close as possible to $\tilde{u}$" in some sense, and which satisfies (2b). The appropriate way (we believe) to perform this velocity adjustment is to find the extremum of the functional

$$F(u,\lambda) = \frac{1}{2}(u - \tilde{u})^T M(u - \tilde{u}) + \lambda^T C^T \tilde{u}\,, \tag{16}$$

over all vectors u which satisfy the same boundary conditions as $\tilde{u}$. Here $\lambda$ is a vector of Lagrange multipliers (one for each discretized continuity equation -- like the pressure) and the distance measure between u and $\tilde{u}$ is seen to be related to kinetic energy since $KE \equiv (1/2)u^T Mu$. Setting the first variations of $F(u,\lambda)$ to zero yields the coupled system

$$M(u - \tilde{u}) + C\lambda = 0\,, \quad C^T u = 0\,, \tag{17a,b}$$

which is more conveniently solved sequentially as

$$(C^T M^{-1} C)\,\lambda = A\lambda = C^T \tilde{u} \quad \text{and} \quad u = \tilde{u} - M^{-1} C\lambda\,, \tag{18a,b}$$

since A is already available in factored form. Application of (18) to $u_s$ gives the minimally-adjusted mass-consistent field at $t_{n+1}$, viz. $u_{n+1}$; the kinetic energy of the difference between u and $u_s$ is minimal.

The pressure update Once a mass-consistent velocity field is available, the compatible pressure field, $P_{n+1}$, is obtained by solving (3b), with n replaced by n+1; i.e., the factored A matrix is "hauled out" one more time, a procedure which indicates (properly) that subcycling is only cost-effective when s is significantly larger than 2.

The next major step-size The last process in the subcycle strategy is the estimation of the next step-size, $\Delta t_{n+2}$ or, what is equivalent, the new subcycle ratio, $s = \Delta t_{n+2}/\Delta t_s$. In the actual algorithm, the step-size estimate is the first process in the sequence rather than the last; the order of discussion was selected to provide (hopefully) greater clarity.

Given u, $\dot{u}$, and P at $t_n$ and $t_{n+1}$, the local time truncation error vector, $d_{n+1}$, can be estimated from

$$d_{n+1} \equiv u_n - u(t_{n+1}) \cong -(\Delta t_{n+1})^2\, \ddot{u}_{n+1}/2 = -\Delta t_{n+1}(\dot{u}_{n+1} - \dot{u}_n)/2\,, \tag{19}$$

where $u(t_{n+1})$ is the (unknown) exact solution, $\dot{u}$ is computed directly from (2a), and we currently neglect the effects of subcycling, which is probably conservative, and temperature (which may not be). After computing a relative RMS norm of $d_{n+1}$, say $e_{n+1}$, (19) is used again to provide the following equation for the next time step:

$$\Delta t_{n+2} = \Delta t_{n+1} (\varepsilon/e_{n+1})^{1/2}, \tag{20}$$

where ε is the user-specified (typically $10^{-3}$) tolerance parameter on the local error.

NUMERICAL RESULTS

While our intended applications are in the atmospheric boundary layer, the example selected for this paper is ostensibly much simpler: the transient development of the flow and temperature in a three-dimensional lid-driven cavity in which the density stratification is initially linear (and stable). The incentive for this simulation was provided by current laboratory research (at Stanford University; see Koseff and Street (1982)) which seemed to be ideal for further code "verification." The cavity is 15 x 15 cm in cross section and 45 cm long. For this exploratory study, we have assumed that the flow will be symmetric, and thus Fig. 4 shows a mesh describing a cavity which is 22.5 cm long. For our main simulation we selected Re = 1000 (about the lowest attainable in the laboratory) at an initial temperature difference in the fluid (water, Pr = 7.1) of 1°C, giving Fr ≃ .168 and a nominal buoyancy period ($2\pi\sqrt{L/\gamma g \Delta T} = 2\pi L\sqrt{Fr}/u_0$; see Turner, 1973) of ∿ 58 sec, which is more relevant for our purposes. The graded mesh was designed by a simple iterative method: trial and error. Seeking an affordable simulation while retaining some credibility finally led to the 18 (vertical) by 22 (moving lid direction) x 16 (transverse) mesh shown. After several 2-D simulations on the 18 x 22 mesh, however, we became somewhat suspicious of the results and therefore halved the mesh (36 x 44), which did two things: (i) it gave nicer looking results and (ii) it showed that the coarse mesh was not all that bad with respect to the overall flow features. Thus, we proceeded with more confidence on the 3-D mesh (although it is probably still a bit too crude). Since the 3-D simulation generated results which were (disappointingly, considering the cost) little different from 2-D except near the side wall, we first present a summary of the latter which is quite interesting and very different from an isothermal

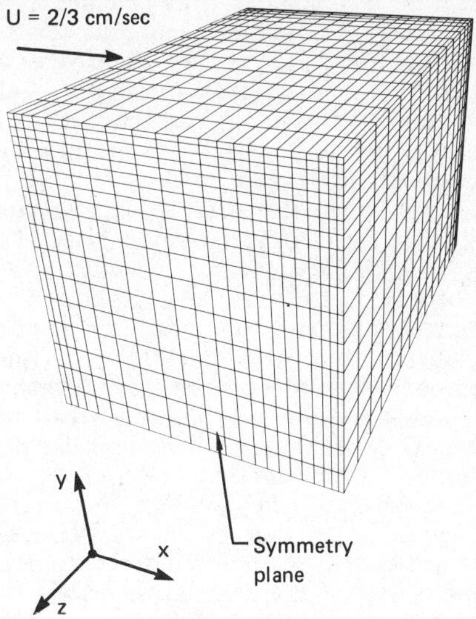


Fig. 4. Graded mesh of 6336 elements.

cavity, involving multiple time scales, internal gravity waves, and even a hint of shear instability (Kelvin-Helmholtz).

The boundary conditions are: u = 0 at the top corners, rising linearly over the first two elements to obtain the lid speed of 2/3 cm/sec which pays due respect to the "spurious CB constraint" explained by Sani et al. (1981); u = 0 elsewhere, v = 0, T = 0, 1 respectively at the cavity bottom and top, and $\partial T/\partial n = 0$ on the side walls. The initial temperature is linear in y (heat conduction profile) and the initial velocity is "zero" except that such a flow field is not mass consistent with the "non-leaky" boundary conditions. Hence, our actual initial condition is that given by the least squares velocity adjustment scheme explained earlier. Brief comparison with results using a "leaky cavity" (u = 2/3 everywhere on the top) showed only small differences; however, we still believe that the non-leaky cavity (with near-zero initial velocity) models the corner singularities somewhat better than the leaky version (with zero initial velocity).

An overview of our interpretation of the phenomena is as follows: the initial impulsively-generated lid motion rather quickly (< 30 sec) sets up a viscous flow which is only moderately affected by buoyancy (the stable stratification does limit the vertical penetration of the developing vortex); see Fig. 5. This primary eddy basically remains in place thereafter (in marked contrast to the isothermal case where it migrates at about 45° toward the center of the cavity - see Gresho et al., 1982) and gains in strength. This initially "mechanically-generated flow" tilts and distorts the isotherms (Fig. 5b), which sets the stage for the next phenomenon: internal gravity waves. These waves are most intense for about 5 min or so after which they subside and are much smaller after about 10 min, although they never really disappear, even after an hour or so. The next phase (1-2 hr) involves the generation of locally sharp temperature gradients at the bottom of the main eddy followed by subsequent overturning (heavy fluid above) and the growth of a near isothermal or "mixed layer" in the top third of the cavity. The final phase seems to be a gradual approach to a steady state over "many" hours, leading to a mixed layer above and a nearly linearly stratified and essentially uncoupled (stagnant) layer below. We next discuss portions of this development in somewhat more detail.

Figs. 6 through 9 show snapshots of the solution at several interesting times and Fig. 10 shows a portion (1 hr) of the time histories of velocity and temperature at the two nodes indicated in the previous figures. Fig. 6 shows the effect of the internal waves via the streamlines and the tilting of the isotherms as well as a very tight temperature gradient below the main vortex and the beginning of a rollup of the T = 0.8 isotherm. By 8 min (Fig. 7) the isotherm tilting and the associated flow in the lower portion have largely ceased; also, the T = 0.8 isotherm is in the process of overturning (it is pinched off within two more minutes). Fig. 8, at t = 80 min, shows the nearly-developed mixed layer in the upper 30% of the cavity and the overturning of the T = 0.9 isotherm. Apparently the nearly isothermal hot core at a temperature close to 1.0 (it is actually between 0.99 and 1.0), when impinging on the stably stratified lower region, provides a mechanism for wiggle generation since the temperature slightly exceeds 1.0 ($T_{max}$ is 1.016); tensor viscosity does not totally suppress the wiggles. The rollup and overturning appear to be caused,

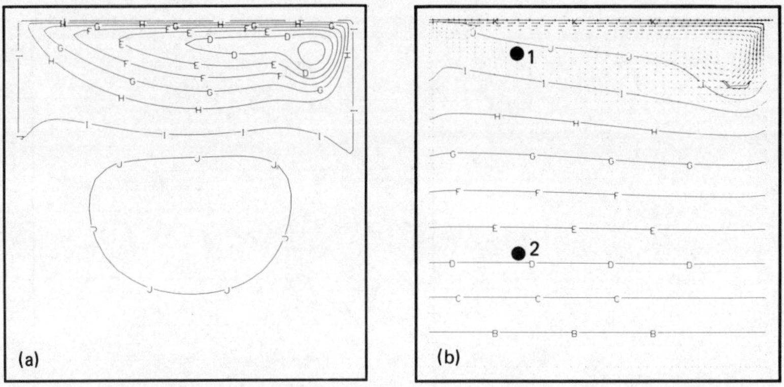


Fig. 5. Solution at t = 30 sec: (a) Streamlines, (b) Vectors and isotherms. Streamline contour levels: A = -0.4, B = -0.35, C = -0.3, D = -0.25, E = -0.2, F = -0.15, G = -.10, H = -.05, I = 0, J = 0.025, K = 0.05. Temperature contour levels: B = 0.10, $\Delta T$ = 0.10, K = 1.0.

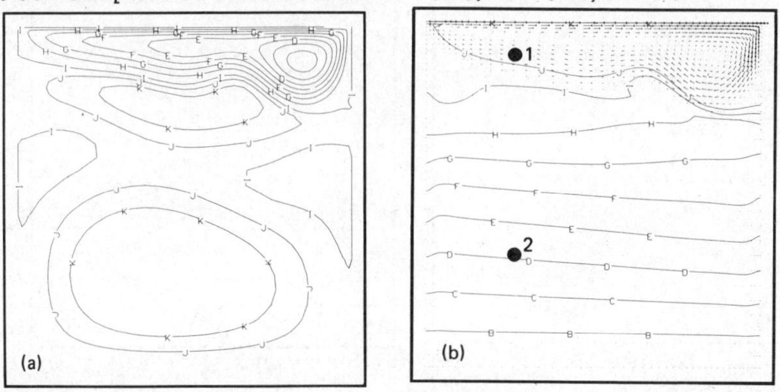


Fig. 6. Solution at t = 2 min: (a) Streamlines, (b) Vectors and isotherms.

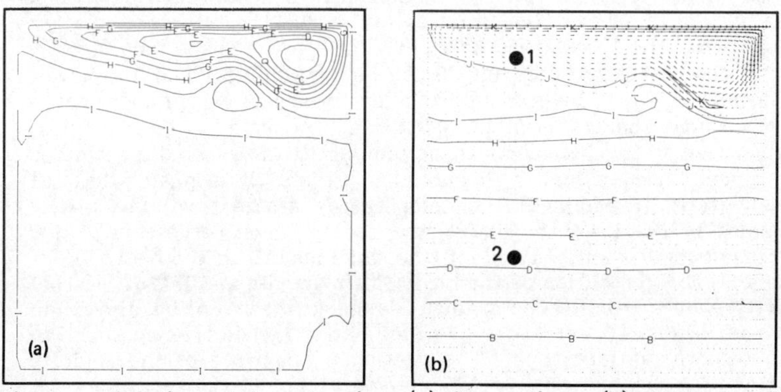


Fig. 7. Solution at t = 8 min: (a) Streamlines, (b) Vectors and isotherms.

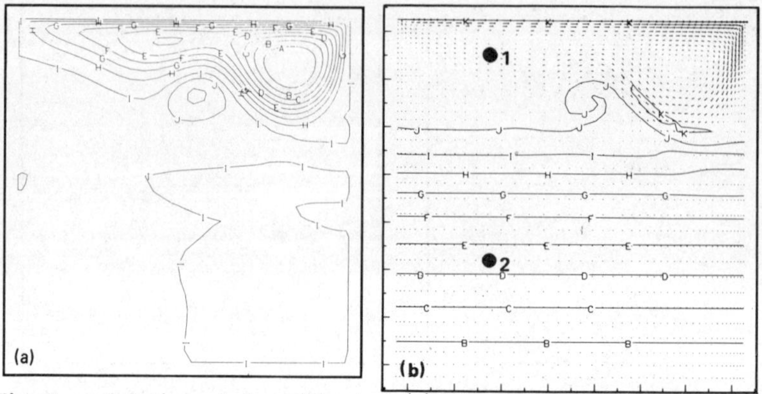


Fig. 8. Solution at t = 80 min: (a) Streamlines, (b) Vectors and isotherms.

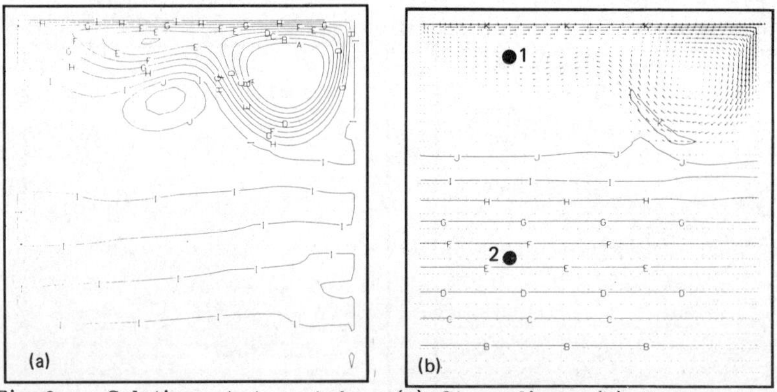


Fig. 9. Solution at t = 4 hr: (a) Streamlines, (b) Vectors and isotherms.

at least in part, by the Kelvin-Helmholtz shear instability mechanism. Finally, Fig. 9 shows the solution at t = 4 hr, the end of the simulation. The T = 0.9 isotherm has changed its shape only slightly during the last two hours (as have the streamlines) and the lower isotherms appear to be gradually tending toward a nearly linear profile in y, via heat conduction in the essentially uncoupled stagnant region. In some respects, this seems to be similar to the shear-driven mixed layer and thermocline in lakes and oceans. It appears that at steady-state (if attained) the temperature gradient will be nearly constant in the lower $\sim$ 2/3 of the cavity which represents a 50% increase over the initial gradient (the same applies to the heat flux of course). The generation of the basic flow, the rise and fall of internal gravity waves, and their gradual damping is shown by the time histories in Fig. 10, the latter of which (Node 2) shows the approach to $\sim$ 0 velocity at late time. The maximum amplitude of the velocity at Node 2 appears to be $\sim$ .025 cm/sec at t = 90-100 sec; about 4% of the lid speed. The dominant period of the internal waves was about 77 sec, which seems to agree with laboratory data and is about 75% of

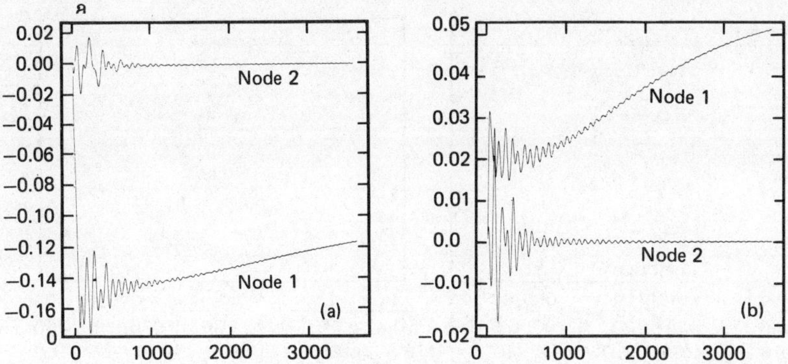


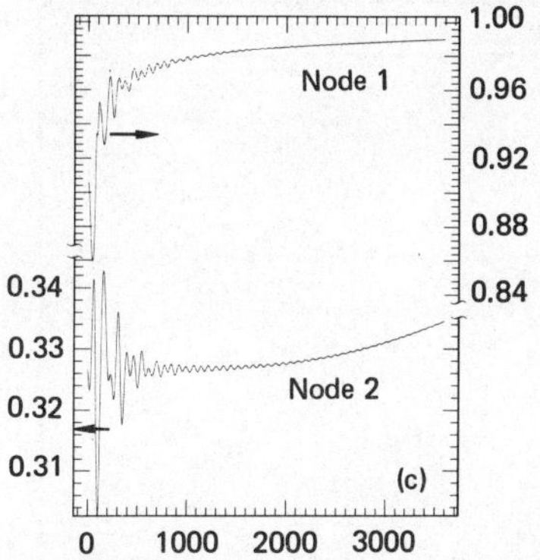


Fig. 10. Time history of two nodes: (a) Horizontal velocity, (b) Vertical velocity, (c) Temperature.

the minimum possible period (58 sec) based on 1° C/15 cm, but about 90% of that based on the later gradient (1° C/10 cm). The decay rate of these internal waves seems to yield a 7-8 min. time constant which, if analyzed via the transient heat equation using Fourier modes, gives characteristic wavelengths on the order of the cavity size; a result which seems appropriate and also correlates fairly well with the normal modes of the linear inviscid theory.

Turning briefly to the 3-D results, we frankly find little additional interesting behavior. The solution appears to be largely two-dimensional except close to the end wall, where the no-slip boundary condition necessarily induces three-dimensionality. This effect is not especially strong, however, and the various 2-D cross sections look much like one another, except those within 2-3 cm of the end wall. Figs. 11 and 12 show two cross-sections through the primary vortex center at $\sim$ 1 hr (the end of the simulation). Fig. 11, viewed from below (1.85 cm below the lid in the isothermal region) with the lid moving upward, shows some 3-D effects near the end wall. In Fig. 12, the lid is at the top, moving normal to the paper (toward you), the end wall is at the right, and the plane is 2.45 cm from the back wall; again there is a noticeable 3-D effect in the velocity projection (maximum speed $\simeq$ 1/6 the lid speed) but the nearly level isotherms are consistent with a basically 2-D behavior.

Perhaps of equal interest to the 3-D results are those at a different Reynolds number. Fig. 13 shows the markedly different solution at Re = 2000 and t = 3 hours (not steady state) with the same initial stratification-on a 32 x 36 mesh. Here the mixed region includes about 70% of the cavity and the isotherms are compressed to the lower 20-30%. If this mesh is fine enough (which we have not yet verified), this flow appears to be the victim of a repetitive

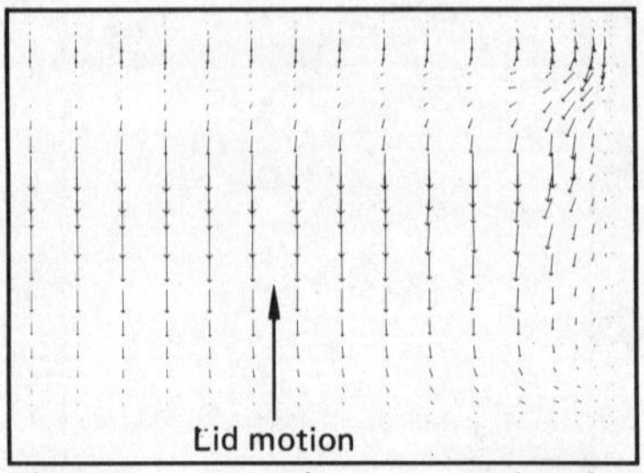


Fig. 11. Velocity projection ($u_{max}$ = .26) on a horizontal plane 1.85 cm below the moving lid.

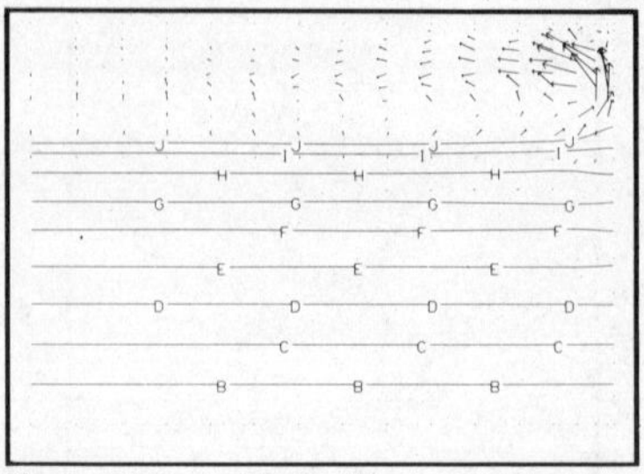

Fig. 12. Velocity projection ($u_{max}$ = .11) and isotherms on a vertical plane (yz) located 2.45 cm from the back wall.

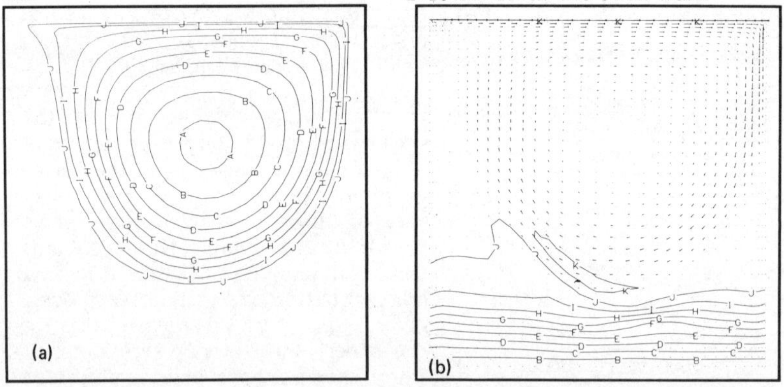


Fig. 13. Solution at t = 3 hr for Re = 2000: (a) Streamlines, (b) Vectors and isotherms. Streamline contour levels: A = -1.8, $\Delta\psi$ = 0.2, I = -0.2, J = -0.1. Same temperature contours as before.

Kelvin-Helmholtz instability. The temperature "spikes" are alternately formed and ripped off by the flow, and appear to generate significant gravity waves affecting the entire flow field. The overall process appears to be verging on a turbulence mechanism via the creation of "patches of turbulence" (Turner (1973)).

## DISCUSSION

These were some of the most difficult (and interesting, and expensive) calculations we have performed with this code, partly because the physics is complex and partly (less so, we believe) because our solution method is not yet sufficiently robust.

The fine mesh 2-D simulation using a minor $\Delta t$ of 0.25 sec required about 3.5 min CPU and ∿ 1.5 min I/O (CRAY-1) per hour of real time, with an average of 26 subcycle steps. On the other hand the (coarse mesh) 3-D simulation, using the same minor step size, required about 1 hour CPU and 2.5 hr I/O per hour of real time - here the average subcycle ratio was 34. The high cost of I/O for the 3-D simulation, which would have been ∿ 35 hr (!) with no subcycling, is directly related to the relatively low rate of data transfer from the disk: ∿ 0.5 million words/sec. The factored A matrix requires about 1.8 million words of storage and must be read four times per major time step (the CPU time associated with the forward reductions and back substitutions is negligible by comparison). Smaller problems (typically 2-D) permit A to be stored in memory and are thus much less expensive (the I/O cost for the 2D simulation above was due primarily to the large amount of data output to disk for post-processing; e.g. movie); see also Gresho et al. (1982).

These simulations revealed two "problem areas" related to our new algorithms: (i) the forward Euler method is unconditionally unstable with respect to internal gravity waves according to linear inviscid theory (diffusion restores stability, conditionally). When we tried to solve the same problem on the coarse mesh with a $\Delta t_S$ of 1 sec (vs 0.25), the internal gravity waves, whose phase speeds are of the same order as the lid speed, gradually became unstable (caused in part by (ii), below) and grew, causing violent distortion of the isotherms and large velocities throughout. Not even tensor viscosity is effective against this instability and we are currently considering certain modifications of the time integration scheme to better cope with these and other wave-like phenomena. (ii) The subcycling strategy causes some difficulty when dealing with stratified flows containing "normal modes" (internal gravity waves) - the mass adjustment process excites these modes. The amplitude of these spurious internal waves is usually, but not always small ($O(\varepsilon)$), where $\varepsilon$ is the tolerance parameter); it depends on both the subcycle ratio and mesh resolution.

To end on a happier note, we now briefly turn to experimental verification. While there are little quantitative data yet available, the qualitative comparisons, in the form of dye streaklines, indicate that our results are very good-at both Reynolds numbers.

## SUMMARY AND CONCLUSIONS

We have described and demonstrated our current multi-dimensional, time-dependent method for solving the incompressible Navier-Stokes equations. The various shortcuts employed (mass lumping, 1-point quadrature, added tensor viscosity, subcycling) appear to be viable and generally lead to rather cost-effective computations, although stratified flows offer additional complications that warrant further algorithm improvement.

## ACKNOWLEDGMENTS

The authors are indebted to Dr. A. C. Hindmarsh (LLNL) and Prof. R. L. Sani (Univ. of Colorado) for numerous enlightening discussions. The computer code described herein was largely written by Dr. S. T. Chan of LLNL. This work was performed under the auspices of the U. S. Department of Energy by the Lawrence Livermore National Laboratory under contract No. W-7405-Eng-48.

## REFERENCES

Brooks, A. and T. Hughes (1982) Streamline Upwind/ Petrov-Galerkin Formulation for Convection-Dominated Flows with Particular Emphasis on the Incompressible Navier-Stokes Equations. Comp. Meth. Appl. Mech. Eng., 1982 (to appear).

Chan, S. and P. Gresho (1982) Solution of the Multi-Dimensional, Incompressible Navier-Stokes Equations Using Low-Order Finite Elements and One-Point Quadrature. In Proceedings of the Fourth International Symposium on Finite Elements in Flow Problems, Tokyo, Japan, July 26-29, 1982.

Dukowicz, J. and J. Ramshaw (1979) Tensor Viscosity Method for Convection in Numerical Fluid Dynamics. J. Comp. Phys., 32, 71.

Gresho, P., and R. Lee (1981) Don't Suppress the Wiggles — They're Telling You Something. Comp. and Fluids, 9, 223.

Gresho, P., C. Upson, and S. Chan (1982) Current Progress in Solving the Time-Dependent, Incompressible Navier-Stokes Equations in Three-Dimensions by (Almost) the FEM. In Proceedings of the 4th International Conference on Finite Elements in Water Resources, Hannover, Germany, June 21-25.

Hindmarsh, A. and P. Gresho (1982) The Stability of Explicit Euler Time-Integration for Certain Finite Difference Approximations of the Multi-Dimensional Advection-Diffusion Equation. To be submitted to Int. J. Num. Meth. Fluids.

Koseff, J. and R. Street (1982) Visualization Studies of a Shear-Driven Three-Dimensional Recirculating Flow. In Proceedings of the ASME Symposium on Three-Dimensional Turbulent Shear Flows, St. Louis, Missouri.

Orzag, S. (1971) Numerical Simulation of Incompressible Flows Within Simple Boundaries: Accuracy. J. Fl. Mech., 49, 75.

Sani, R., P. Gresho, R. Lee and D. Griffiths (1981) The Cause and Cure(?) of the Spurious Pressures Generated by Certain FEM Solutions of the Incompressible Navier-Stokes Equations: Part 1. Int. J. Num. Meth. Fluids, 1, 17.

Turner, J. (1973) Buoyancy Effects in Fluids, Cambridge at the University Press.

Upson, C., P. Gresho, R. Sani, S. Chan, and R. Lee (1982) A Thermal Convection Simulation in Three-Dimensions by a Modified Finite Element Method. In Numerical Properties and Methodologies in Heat Transfer — Proceedings of the Second International Symposium, Hemisphere Publishing Co., Washington, D. C., 1982 (to appear).

# 4. OPEN CHANNEL FLOW CALCULATIONS

# A DISTRIBUTED RAINFALL-RUNOFF MODEL

F.J. Aparicio and M. Berezowsky

Instituto de Ingeniería, National University of Mexico

## INTRODUCTION

Design, construction and operation of hydraulic works always require a previous determination of a maximum flood for a given frequency. When it is desired to forecast changes on the output hydrographs due to alterations in the morphology or land use of the watersheds, the methods at present available are, in general, of limited applicability.

In order to avoid such limitations, models based on physical characteristics of the watershed must be developed, so that input parameters for their calibration and operation can be measured or easily estimated in advance .

A model of this kind is described in this paper. As a first step,the model transforms the total rainfall in effective rainfall using Chu's method (Chu, 1978). The input parameters for this method are, besides total rainfall, the retention capacity, the hydraulic conductivity, the capillary potential at the wetting front and the moisture content of the soil. The effective rainfall is computed simulating the change of infiltration capacity with time. Afterwards, this effective rainfall is routed throughout the watershed, first as overland flow and then as channel flow. The main subject of this paper deals with this routing,which is simulated using the kinematic wave theory and the Galerkin formulation of the finite element method (FEM) on both overland and channel flow. In trying to solve the whole Saint-Venant equations for supercritical flow, several difficulties were found and are mentioned below; finally, the calibration of the model on a small watershed located in the Mexico City area is described.

## OVERLAND FLOW SIMULATION

The continuity equation for the kinematic wave approach to shallow water flow is

$$\frac{\partial A}{\partial t} + \frac{\partial Q}{\partial x} = q \tag{1}$$

where A = area ($m^2$), Q = discharge ($m^3/s$), t = time (s), x = space coordinate (m) and q = lateral inflow ($m^2/s$); it is assumed that friction forces balance gravitational forces, so that the dynamic equation becomes

$$S_f \approx S_o \tag{2}$$

where $S_f$ = friction slope and $S_o$ = bottom slope.

Kinematic wave theory can be applied when the following condition is satisfied:

$$K = \frac{S_o L}{Fh} > 10 \tag{3}$$

where K = kinematic flow number (dimensionless), L = lenght of domain (m) and h and F are the depth (m) and Froude number respectively, both computed for steady state conditions. Woolhiser and Liggett (1967) stated that, for many overland flow problems, K has a magnitude of well over 1000, so the kinematic approximation can be taken as very accurate for this kind of flow.

For the operation of the model, the zones of the watershed where runoff does not take place in well-defined channels are discretized in finite elements. These elements are ordered in rows oriented in the direction of the main path of the drainage, so that the flow in each row can be considered one-dimensional.

Equation (1) is integrated in each row by FEM and Galerkin's procedure using linear basis funtions, which lead to the following system of equations

$$M \, \overline{\dot{A}} = \overline{H} + \overline{C} \tag{4}$$

where

$$M=1/6 \begin{bmatrix} 2(\ell_1+\ell_2) & \ell_2 & 0 & \cdots & 0 & 0 \\ \ell_2 & 2(\ell_2+\ell_3) & \ell_3 & \cdots & 0 & 0 \\ \vdots & \vdots & \vdots & & \vdots & \vdots \\ 0 & 0 & 0 & \cdots & 2(\ell_{n-2}+\ell_{n-1}) & \ell_{n-1} \\ 0 & 0 & 0 & \cdots & \ell_{n-1} & 2\ell_{n-1} \end{bmatrix}$$

$$\overline{\dot{A}} = \begin{bmatrix} \dot{A}_2 \\ \dot{A}_3 \\ \vdots \\ \dot{A}_{n-1} \\ \dot{A}_n \end{bmatrix} ; \quad \overline{H}=1/2 \begin{bmatrix} Q_1-Q_3 \\ Q_2-Q_4 \\ \vdots \\ Q_{n-2}-Q_n \\ Q_{n-1}-Q_n \end{bmatrix} ; \quad \overline{C}=1/2 \begin{bmatrix} q_1\ell_1+q_2\ell_2-\ell_1\dot{A}_1/6 \\ q_2\ell_2+q_3\ell_3 \\ \vdots \\ q_{n-2}\ell_{n-2}+q_{n-1}\ell_{n-1} \\ q_{n-1}\ell_{n-1} \end{bmatrix}$$

In the above equations, the point over the variable denotes time derivation, the subindex denotes the number of the node, $\ell_i$ is the lenght of the element within nodes i and i+1 and n is the total number of nodes. The resulting system of ordinary differential equations is then solved iteratively with a Crank-Nicolson finite difference scheme; the discharges $Q_i$ are calculated through Manning's formula. Note that, being matrix M tridiagonal and symmetrical, it does not depend of time, so the solution procedure becomes extremely fast.

Initial discharge and depth are assumed zero in all the overland flow elements. As boundary conditions, discharge and depth are zero in the first node of each row, for all times.

A test of the above procedure on a plane of overland flow is shown in figure 1 and is compared with experimental results reported by Yu and Mc Nown (1964).

CHANNEL FLOW SIMULATION

Flow in channels during floods can be either subcritical or supercritical and it can even alternatively shift from one regime to another. Thus, it is desirable to simulate channel flow solving the Saint-Venant equations without simplifying assumptions, regardless of regime. However, as Zovne and Martin (1979) have pointed out, transient supercritical simulation is a largely unexplored domain which urgently needs further research. These authors found severe difficulties in trying to solve Saint-Venant equations by the method of characteristics. Similar difficulties were also experienced here when several methods were tested; none of these was found to have a satisfactory behavior for both regimes. It is believed that such experiences should be published, but, for reasons of space, only a summary of the tests carried out is given here.

Saint-Venant equations are the continuity equation (1) and the momentum equation:

$$V \frac{\partial V}{\partial x} + g\frac{\partial h}{\partial x} + g\frac{\partial z}{\partial x} + \frac{\partial V}{\partial t} + gn^2 V|V|/R^{4/3} = 0 \qquad (5)$$

where g = gravitational acceleration ($m/s^2$), z = elevation of channel bed (m), V = velocity (m/s), n = Manning's roughness coefficient and R = hydraulic radius (m).

Finite-difference schemes

Two explicit and two implicit finite difference schemes were tested. The explicit schemes were only tested for supercritical flow, while the implicit schemes were tested for sub- and supercritical conditions.

Explicit diffusive scheme This scheme (Liggett and Cunge, 1975) uses the following approximation of derivatives:

$$\frac{\partial f}{\partial t} = \frac{f_j^{k+1} - \left\{\alpha f_j^k + (f_{j+1}^k - f_{j-1}^k)(1-\alpha)/2\right\}}{\Delta t}; \quad \frac{\partial f}{\partial x} = \frac{f_{j+1}^k - f_{j-1}^k}{2\Delta x} \tag{6}$$

where f is any function, $\alpha$ is a diffusion coefficient varying from 0 to 1 and indexes k and j denote a step in time and space respectively. It can be shown that the stability condition for this scheme ($\alpha$=0) is the Courant condition: $Cr = c\Delta t/\Delta x < 1$, where $c = V + \sqrt{gh}$

Tests were carried out for a rectangular, 6.1m wide and 3200m long channel, divided in ten equal reaches with n=0.02 and $S_o$=0.05. Velocities were such that flow was always supercritical. Boundary conditions were fixed by calculating depth and velocity with Manning's formula at the upstream end of the channel for a known discharge; the boundary conditions so established are independent of the equations governing the flow inside the domain, and were the same for all methods of solution tested for supercritical flow. Three kinds of tests were used: a) steady state flow, b) routing a hydrograph entering at the upstream boundary and c) routing a constant lateral inflow entering along the channel. For steady state conditions ($Q = 23m^3/s$), instabilities were observed when $\alpha$=0 and $\Delta t$=5s ($\Delta t_{max} \approx$ 30s, according to Courant condition) and, for values of $\alpha$ between 0.5 and 0.7, the scheme was stable. Similar results were obtained for type b test. For test c, stability was attained at low values of $\Delta t$ (about 5s) and $\alpha$=0.5. Nevertheless, a loss of mass was observed, as equilibrium conditions were not reached. For higher $\Delta t$ (about 10s), results such as figure 2 were observed: there are not instabilities, but the achieved conditions are quite different from the correct ones. Although it could not be demonstrated, the authors believe that this kind of results are due to the influence of ignored higher order terms.

<u>"Fractioned step" scheme</u> In this scheme, the finite difference equations are as follows (Berezowsky, 1980)

$$B_j^k \frac{H_j^{k+1} - H_j^k}{\Delta t} + \frac{V_j^k A_{j+1/2}^k - V_{j-1}^k A_{j-1/2}^k}{\Delta x} = q_j^k \tag{7}$$

$$\frac{V_j^{k+1} - V_j^k}{\Delta t} + V_j^{k+1} \frac{V_{j+1}^k - V_{j-1}^k}{2\Delta x} + g\psi \frac{H_j^{k+1} - H_{j+1}^{k+1}}{\Delta x} + $$

$$+ g(1-\psi) \frac{H_j^k - H_{j+1}^k}{\Delta x} + g \frac{n^2 V_j^{k+1} |V_j^k|}{R_{j+1/2}^{k\;4/3}} = 0 \tag{8}$$

where H = h+z, $\psi$= weighting factor ($0<\psi<1$) and Bj = mean width of the free surface in the time interval. This scheme showed a very stable behavior for steady state supercritical flow for $5s \leq \Delta t \leq 40s$ and all values of $\psi$, but for lateral inflow it fails (see figure 3). Here, and in the following two schemes, velocities are computed between two successive heads.

Abbott type implicit scheme A scheme similar to Abbott's (Abbott, 1979) was also tested. It works quite well in subcritical flow, where setting of boundary conditions have been well established. Unfortunately, this is not the case for supercritical flow (see Cunge et al, 1980, for a general discussion about boundary conditions in this type of schemes) where instabilities were observed even for uniform flow.

Cruickshank-Berezowsky implicit scheme The following implicit scheme (Cruickshank & Zumaya, 1974, and Berezowsky, 1981) works also very well for subcritical flow. However, its behavior for supercritical flow is similar to the Abbott-type scheme. It is believed that, besides the possible effect of higher order terms, at least part of the observed instabilities are due to the existence of a small error in the computation of the friction slope in the first nodes, which is propagated downstream and amplified without control with time. It was intended to counteract this effect with modifications on both schemes but the problem always persisted.

$$B_j^k \frac{H_j^{k+1} - H_j^k}{\Delta t} + \frac{V_j^{k+1} A_{j+1/2}^k - V_{j-1}^{k+1} A_{j-1/2}^k}{\Delta x} = q_j^k \tag{9}$$

$$\frac{V_j^{k+1} - V_j^k}{\Delta t} + V_j^{k+1} \frac{V_{j+1}^k - V_{j-1}^k}{2\Delta x} + g\psi \frac{H_j^{k+1} - H_{j+1}^{k+1}}{\Delta x} +$$

$$+g(1-\psi) \frac{H_j^k - H_{j+1}^k}{\Delta x} + g \frac{n^2 V_j^{k+1} |V_j^k|}{\left[R_{j+1/2}^k\right]^{4/3}} = 0 \tag{10}$$

Finite element simulation Another method tested to solve Saint Venant equations was FEM and Galerkin's procedure using linear basis functions.

Due to its high nonlinearity, several forms of approximation were tested for the friction term $gn^2V|V|/R^{4/3}$ in momentum equation. However, various difficulties arouse in the use of the method for subcritical flow, even when simulating steady state conditions. In the authors' opinion, these problems can be originated either at the management of the boundaries or during the integration with respect to time which was also done by means of a Crank-Nicolson scheme. Further research is being done in order to improve the formulation, but, at the moment, the complications found prevent its use in the model.

Kinematic simulation Although condition 3 is not always satisfied for channel flow, due to the difficulties pictured above the problem was worked out applying the kinematic wave approach already described. In this case, care has to be taken to assure that the Froude number does not grow too much; Henderson (1966) considers 2 as a safe upper Froude number limit for the applicability of kinematic wave theory. A type c test made with this method is shown in figure 4. It was observed that equilibrium discharge is exactly correct and that the behavior of the method is stable up to values of Courant number slightly greater than 1.

INTEGRATION OF THE MODEL AND CALIBRATION

For application of the model, the finite elements in which overland flow zones have been divided (see figures 5 and 6) must be of rectangular shape. As this sometimes cannot be done,"equivalent" rectangular elements can be considered, such that, for the same lenght, the original area is conserved. Each row can have any number of elements (2 as a minimum), but this number should be set having in mind that the larger the number of elements, the longer the time of computation. Elements of the order of 1 $km^2$ have been tested successfully. Even though any shape of cross-section of the river reaches could be taken into account,due to the above simplifications it is not necessary to consider shapes other than rectangular. Discharge of overland flow constitutes lateral inflow to channels; therefore, the lenght of each channel reach must be equal to the width of the overland flow rows which discharge into it. Once established the idealized configuration of the watershed (figure 6), it is decided which elements fall within the influence area of each rain gage. Afterwards, the following parameters must be estimated for each element: Manning's coefficient, slope, saturated conductivity, initial and saturation moisture content and capillary potential at the wetting front. Manning's coefficient is estimated from the type of soil and vegetation; slope can be evaluated from topographic data and the hydraulic properties of the soil can be measured *in situ* or in laboratory. All these parameters are constant, except initial moisture content which can be measured for each single storm by means of tensiometers. If there are not available methods to measure hydraulic properties of soil,they should be calibrated changing their values until the volume of runoff of three or four measured hydrographs is properly simulated. If, once the hydraulic properties of soil have been determined, considerable differences remain between measured and simulated hydrographs, the only parameter to be calibrated would be Manning's coefficient.

Figures 7a and b show two simulated hydrographs in the Río Mixcoac watershed near Mexico City. Obviously, all the input parameters, except initial moisture content and rainfall, are the same for both storms. For almost all the elements, Manning's n equals 0.015 and the hydraulic properties of the soil were obtained by calibration.

CONCLUSIONS

A distributed rainfall-runoff model has been developed, which uses only parameters with physical significance. Once programmed,the model is easy to use and, if a proper estimation of the input parameters is made, hydrographs resulting from a given storm are fairly well simulated. In most cases, kinematic wave approach can be considered sufficient for simulation of overland flow. Several problems were found in supercritical flow on river channels simulation, so the need of an additional effort in this direction is pointed out. Due to this problems, the kinematic wave approach had to be used also for simulation of channel flow.

Although the model seems to be somewhat complicated, its features make it advantageous with respect to other methods in the determination of design floods, in real-time simulation and in forecasting of changes in the runoff pattern of the watersheds due to alterations on the morphology or land use of the basin or parts of it.

REFERENCES

Abbott,M. B.(1979) Computational Hydraulics. Pitman, London

Berezowsky, M. (1980) Escurrimiento a Superficie Libre Chap. A.2.9 of *Manual de Diseño de Obras Civiles*,C.F.E.,Mexico

Chu, S.T. (1978) Infiltration During an Unsteady Rain. Water Resources Research, V.14, no.3

Cruickshank, C. and Zumaya, E. (1974) Simulación Matemática del Flujo de Avenidas sobre Llanuras de Inundación. Proceedings, VI Latin -American Hydraulics Congress (IAHR), Colombia

Cunge, J.A., Holly, F.M., Verwey, A. (1980) Practical Aspects of Computational Hydraulics. Pitman, London

Henderson, F.M. (1966) Open Channel Flow. McMillan, New York

Liggett,J.A. and Cunge, J.A. (1975) Numerical Methods of Solution of the Unsteady Flow Equations. Chap. 4 of *Unsteady Flow in Open Channels*, Water Res. Publications, Colorado

Woolhiser, D.A. and Ligget, J.A. (1967) Unsteady One Dimensional Flow Over a Plane - The Rising Hydrograph, Water Resources Research, V. 3, no.3

Yu, Y.S. and McNown, J.S. (1964) Runoff from Impervious Surfaces. J. Hyd. Res.,V. 2, no. 1

Zovne, J.J. and Martin,C.S. (1979) Simulation of Transient Supercritical Channel Flow. J. Hyd. Div. ASCE, HY7

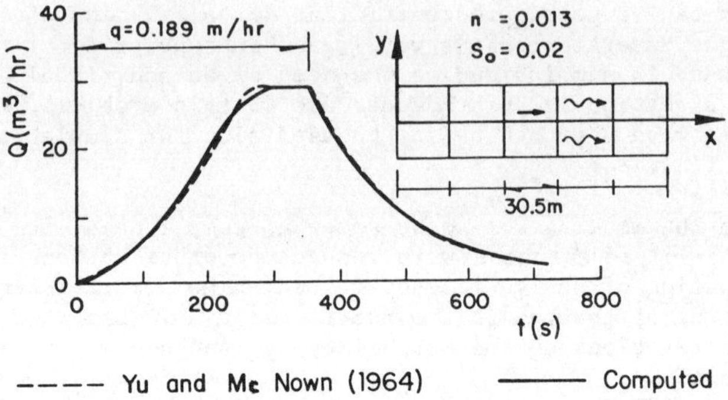


Figure 1. Measured and computed overland flow, simulated with the kinematic wave approach and FEM

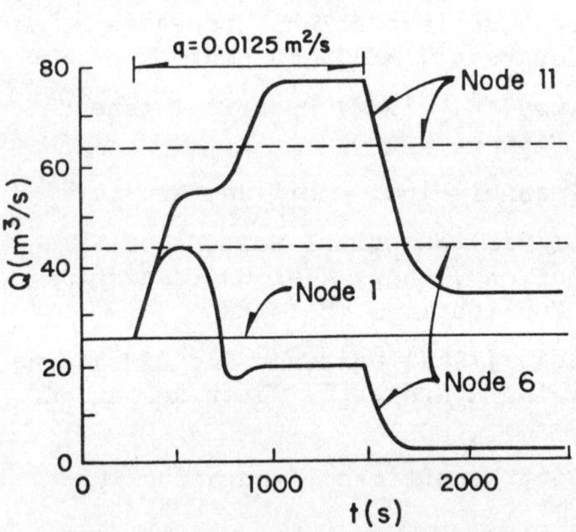


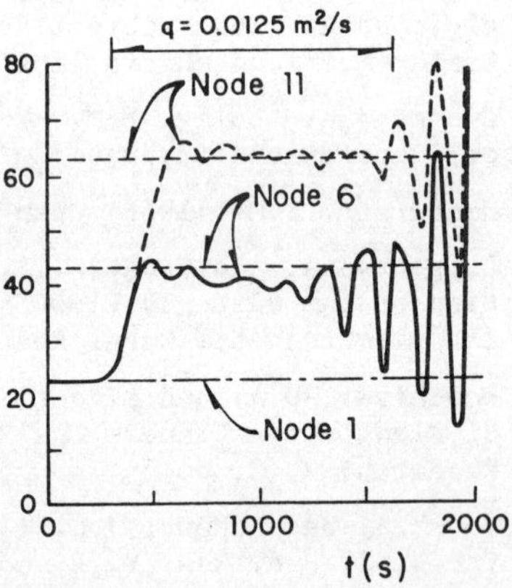


----- Equilibrium discharge

Figure 2. Explicit diffusive scheme with constant lateral flow $\Delta t$=10s, $\alpha$=0.5

Figure 3. "Fractioned step" scheme with constant lateral inflow $\Delta t$ = 5s, $\psi$ = 0.5

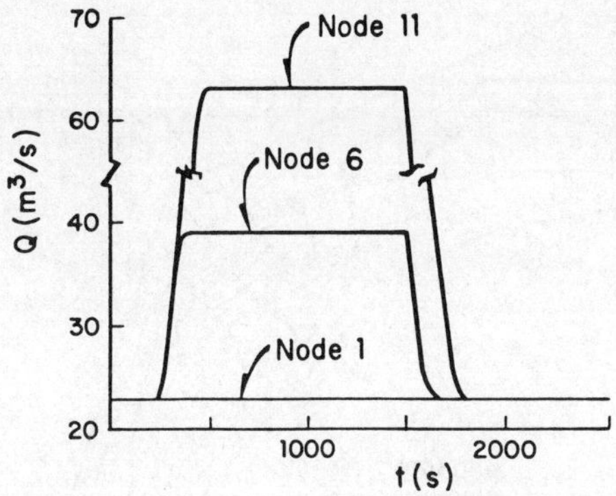


Figure 4. Kinematic wave approach by FEM for the same conditions as figures 2 and 3

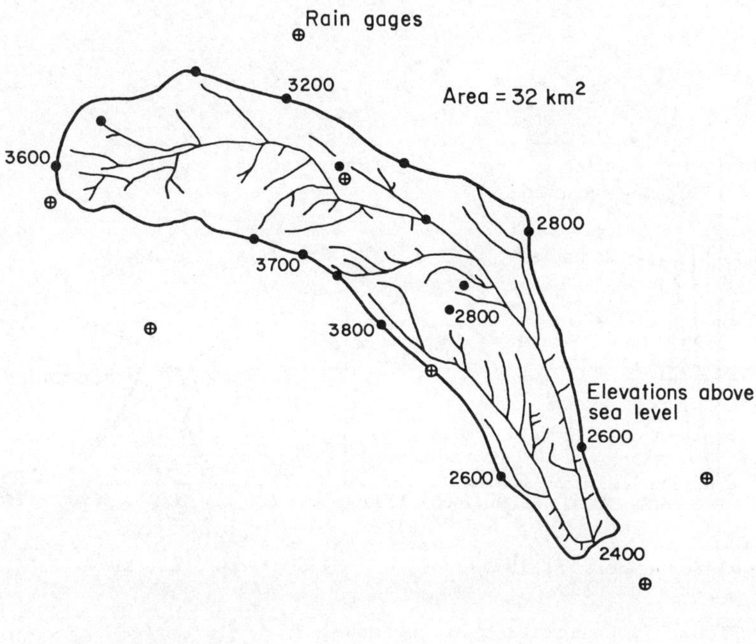


Figure 5. Río Mixcoac watershed

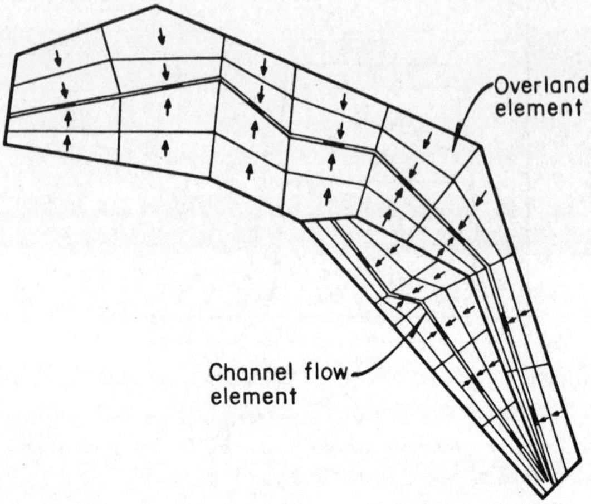


Figure 6. Finite element discretization for Río Mixcoac watershed

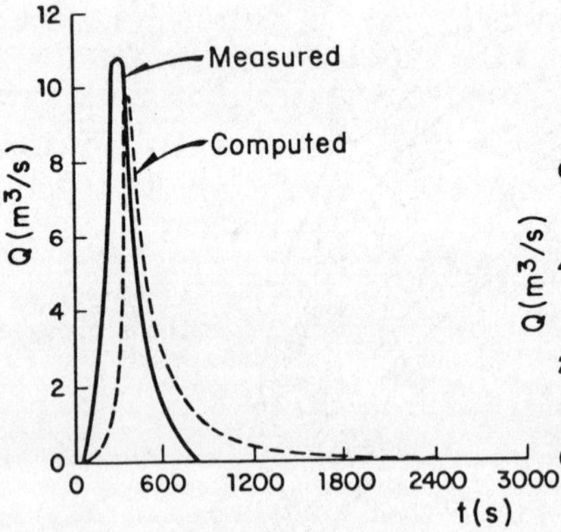


a) September 18, 1977

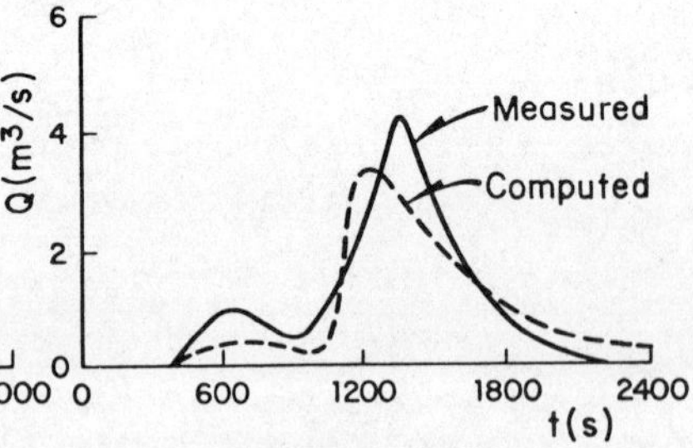


b) September 21, 1977

Figure 7. Calibration of the model

# NUMBERICAL COMPUTATION OF SUPERCRITICAL AND SUBCRITICAL FLOWS IN OPEN CHANNELS WITH VARYING CROSS-SECTIONS

A.O. Demuren
Sonderforschungsbereich 80
Universität Karlsruhe

## ABSTRACT

A two-dimensional numerical procedure has been developed for the calculation of both supercritical and subcritical, steady flows in open channels with curvature or varying cross-sections. The flow is assumed to be fully mixed in the vertical direction so that vertical variations may be neglected. The mixing length hypothesis of Prandtl is used to specify the turbulent diffusion coefficients in the calculation of subcritical, turbulent flows. The present predictions of the wave heights and the wave angles of the supercritical flow in various channels agree very well with experimental data, except at high Froude numbers. The present predictions are equally good or better than those calculated using the method of characteristics.

## INTRODUCTION

The flow of water in open channels with varying cross-section or curvature would in general require a three-dimensional analysis, however, by assuming that the flow is fully-mixed in the vertical direction a two-dimensional analysis may be employed. Analysis of supercritical flows in open channel was first carried out in detail by Preiswerk (1940), using the method of characteristics developed by Buseman (1931) for supersonic gas flow computations. The analogy between free surface water flows and compressible gas flows was pointed out by Riabouchinsky (1932). It may be summarised as illustrated in Table 1. Buseman's method was applied by Ippen et al. (1949) to calculate the hydraulic jumps produced in supercritical water flows in channels with varying cross-sections. In the presence of pressure forces

Table 1: Analogy between free surface water flow and compressible gas flow

| | Two-dimensional compressible gas flow | Free-surface water flow |
|---|---|---|
| Medium | Gas with $\gamma (\equiv C_p/C_v) = 2$ | Incompressible fluid |
| Equivalent quantities | Velocity ratio, $U/U_{IN}$ | Velocity ratio, $U/U_{IN}$ |
| | Temperature ratio, $T/T_{IN}$ | Water depth ratio, $Z/Z_{IN}$ |
| | Density ratio, $\rho/\rho_{IN}$ | Water depth ratio, $Z/Z_{IN}$ |
| | Pressure ratio, $P/P_{IN}$ | Square of water depth ratio, $(Z/Z_{IN})^2$ |
| | Pressure ratio, $P/P_{IN}$ | Ratio of driving force per unit area |
| | Velocity of sound, c | Wave velocity, $(gZ)^{1/2}$ |
| | Mach number, U/c | Froude number, $U/(gZ)^{1/2}$ |
| | Supersonic flow | Supercritical flow |
| | Subsonic flow | Subcritical flow |
| | Shock waves | Hydraulic jumps or cross waves |

due to surface elevation or recirculation zones, subcritical flows in open channels are elliptic, otherwise they are parabolic. Two-dimensional finite-difference calculation methods have been developed for the latter by McGuirk and Rodi (1978) and for the former by Rastogi and Rodi (1978). Pavlović (1981) subsequently applied a combination of both methods to calculate the dispersion of heated discharges into various rivers.

In the present paper, a two-dimensional numerical procedure is presented for calculating both supercritical and subcritical flows in open channels with varying cross-sections. The procedure takes advantage of the fact that the equations governing the supercritical flow are hyperbolic, so that influences only travel in the downstream direction, hence, the solution is obtained by a single integration sweep. The subcritical flow equations, on the other hand, may be parabolic or elliptic (as discussed above). When they are parabolic, the pressure effects are either unimportant or can be determined a priori, so that the solution may also be obtained by a single integration sweep. When the equations are elliptic the pressure field is important and unknown, thus many integration sweeps of flow field are required in obtaining the solution. The present procedure is restricted to the calculation of those elliptic flows without recirculation zones; i.e., the flows which are elliptic only in the pressure field, and hence called semi-elliptic flows. In the present paper, both the theoretical and the practical differences between these classes of open-channel flows are illustrated.

## MATHEMATICAL MODEL

### Governing equations

The time-averaged and depth-averaged partial differential equations governing the steady flow of water in open channels may be written in cartesian coordinates as:

$$ZU\frac{\partial U}{\partial x} + Z\frac{V\partial U}{\partial y} = \frac{\partial}{\partial y}\left(Z\nu_{eff}\frac{\partial U}{\partial y}\right) - \frac{Z}{\rho}\frac{\partial P}{\partial x} \quad (1)$$

$$ZU\frac{\partial V}{\partial x} + ZV\frac{\partial V}{\partial y} = \frac{\partial}{\partial y}\left(Z\nu_{eff}\frac{\partial V}{\partial y}\right) - \frac{Z}{\rho}\frac{\partial P}{\partial y} \quad (2)$$

and the continuity equation as:

$$\frac{\partial UZ}{\partial x} + \frac{\partial VZ}{\partial y} = 0 \quad (3)$$

where, the symbols have their usual meaning and are also defined in the Nomenclature. $_{eff}$ represents the effective viscosity. This is of significant magnitude only in subcritical flows, and is usually calculated as the sum of the molecular and the turbulent viscosities. The turbulent viscosity is calculated from Prandtl's mixing length formula as:

$$\nu_t = \ell_m^2 \left|\frac{\partial u}{\partial y}\right| \quad (4)$$

where, $l_m$, the mixing length is prescribed as:

$$\ell_m = \kappa y \quad ; \quad \text{when} \quad \frac{y}{b} \leq \frac{\lambda}{\kappa}$$

and (5)

$$\ell_m = \kappa b \quad ; \quad \text{when} \quad \frac{y}{b} > \frac{\lambda}{\kappa}$$

( $\lambda$ and $\kappa$ are empirical constants having values of 0.09 and 0.4, respectively.)

In equations (1) and (2), the pressure gradient terms are derived from first principles as:

and, (6)

$$\frac{\partial P}{\partial x} = g\left\{\frac{Z}{2}\frac{\partial \rho}{\partial x} + \rho\left(\frac{\partial Z}{\partial x} + \frac{\partial Z_o}{\partial x}\right)\right\}$$

$$\frac{\partial P}{\partial y} = g\left\{\frac{Z}{2}\frac{\partial \rho}{\partial y} + \rho\left(\frac{\partial Z}{\partial y} + \frac{\partial Z_o}{\partial y}\right)\right\} ,$$

or, if density gradients are neglected (in the absence of buoyancy forces):

$$\frac{\partial P}{\partial x} = \rho g\left(\frac{\partial Z}{\partial x} + \frac{\partial Z_o}{\partial x}\right)$$

and

$$\frac{\partial P}{\partial y} = \rho g\left(\frac{\partial Z}{\partial y} + \frac{\partial Z_o}{\partial y}\right) \quad . \tag{7}$$

In each case, the first term represents pressure variations due to differences in the water depth, and the second, those due to channel bottom surface variations.

Solution procedure

To enable their solution, the partial-differential equations (1) and (2) are discretised by integrating them over a finite-difference grid such as is illustrated in figure 1. The resulting finite-difference equation has the form, for the node i:

$$A_i \phi_{i,d} = A_{i+1}\phi_{i+1,d} + A_{i-1}\phi_{i-1,d} + C_i \quad , \tag{8}$$

which is subsequently solved by a Tri-Diagonal Matrix Algorithm (TDMA), after the coefficients A and C have been calculated. The solution procedure depends on whether the flow is classified as parabolic, hyperbolic or semi-elliptic.

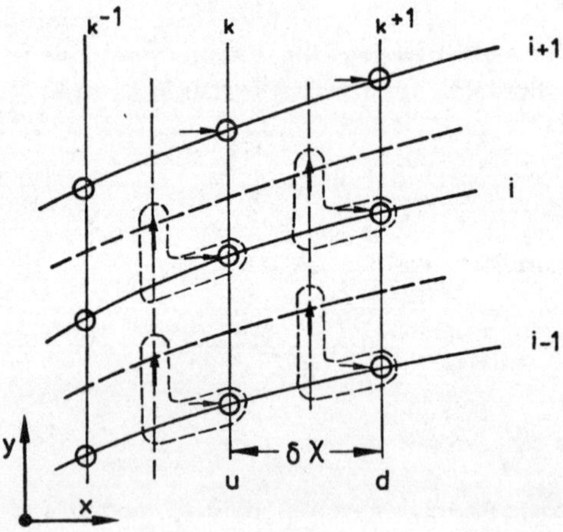


| Location | Variable Stored |
| --- | --- |
| → | U |
| ↑ | V |
| O | P, Z |

Fig. 1: The finite-difference grid

Uniform flow may, in general, be classified as parabolic, since pressure influences are usually unimportant. In solving this class of flows, the velocities are prescribed at the inlet plane as initial conditions, the coefficients of the finite-difference equations are calculated from known quantities, and the resulting set of equations is solved by a TDMA. The solution sequence proceeds in a line by line march from the inlet plane to the end of the calculation domain.

Non-uniform flow may be classified as hyperbolic, if supercritical, or semi-elliptic, if subcritical. In these classes of flows, the pressure influences are important so it is necessary to solve an extra-equation to determine the pressure field. This is the so-called pressure-correction equation, proposed by Patankar and Spalding (1972), which is a simplified, finite-difference representation of the Poisson equation. Its main task is to ensure that, the continuity equation (3) is satisfied. A full derivation of the particular equation, applicable to the flows under discussion here is given in Demuren (1979). It may be expressed, for the $i^{th}$ node as:

$$(A_{i+1}+A_{i-1}+A_{i,u})Z'_{i,u}+(A_{i+1}+A_{i-1}+A_{i,d})Z'_{i,d}$$

$$= A_{i+1}(Z'_{i+1,u}+Z'_{i+1,d})+A_{i-1}(Z'_{i-1,u}+Z'_{i-1,d})+C_i \quad , \quad (9)$$

where,

$$A_{i-1}=\left(\frac{g\Delta x^2}{2U^2\Delta y}\right)_{i-j} \quad ; \quad A_{i+1}=\left(\frac{g\Delta x^2}{2U^2\Delta y}\right)_{i+1} \quad ;$$

$$A_{i,d}=\left(\frac{\Delta y}{Z}\right)\left(1-\frac{1}{F^2}\right). \quad ; \quad A_{i,u}=-\left(\frac{\Delta y}{Z}\right)\left(1-\frac{1}{F^2}\right) \quad . \quad (10)$$

In equation (9), the pressures have been replaced by the flow depths using relations given by equation (7); and $C_i$ is the mass source error due to continuity imbalance. Similar equations are obtained at the other grid nodes. Inspection of equation (9) shows that, unless all the equations for the whole-pressure field are solved simultaneously, with boundary conditions prescribed at the initial and final planes, there would be more unknowns than equations. In order to keep the line by line approach of the solution scheme, physical reasoning is used to reduce the number of unknowns to equal the number of equations. If the flow is supercritical, pressure effects cannot travel upstream, so only the corrections for the downstream plane are included; the upstream ones being set to zero. If the flow is subcritical, pressure effects may travel in all directions. To enable the pressure influences to travel upstream, upstream corrections are employed; the downstream ones being set to zero. The correction to the downstream depths would take place during the next integration step. This practice ensures that the coefficients of equation (9) are always positive; which is a necessity for the use of the TDMA. The solution sequence for non-uniform flow computations, thus, consists of: calculating an approximate velocity field, with a guessed pressure field, and then subsequently correcting the pressure and velocity fields with calculated pressure corrections, and the corresponding velocity corrections. Only a single sweep (from initial to final plane) of the marching integration sequence is necessary to compute the supercritical flow, but the computations of subcritical flows require many such sweeps: the pressure need to be stored for the whole field and are updated during each sweep.

Boundary conditions

At the inlet plane, the axial velocity and the flow depth are prescribed as uniform, while the lateral velocity is set to zero. At symmetry planes, the boundary conditions are prescribed as: zero lateral gradients for the axial velocity and flow depth, and the lateral velocity is set to zero. No-slip conditions are assumed at the walls for subcritical flows and inviscid conditions for supercritical flows, the axial velocity at the nearest grid node to the wall is assigned a value dictated by the logarithmic law of the wall. Along the bottom surface, the wall shear stress is prescribed as:

$$\tau_W = 0.03\rho U^2 \quad . \tag{11}$$

## RESULTS AND DISCUSSION

Computational details

A computer program based on the GENMIX code of Spalding (1978) is employed in all the calculations. The same program is used for the calculation of both supercritical and subcritical flows; with the switch from one type to the other being made internally, depending on whether the Froude number of the flow is greater or less than unity. A supercritical flow calculation with 60 cross-stream and 200 axial grid nodes requires about 25 secs CPU time on the CDC 6600 Computer. The subcritical flow calculations are carried out with about four times less grid nodes in each direction, but with 20 sweeps of the whole field. The CPU time required is about twice that of the former.

Comparison of calculations with experiments

In figures 2 - 7, computations of various supercritical flows are compared with experimental data and calculations using the method of characteristics. Figures 2 and 3 show that the present calculation method predicts the water depth variation near the outer wall of curved channel flows quite well. Figures 4 and 5 compare the predictions of the flow in channels with straight wall contractions

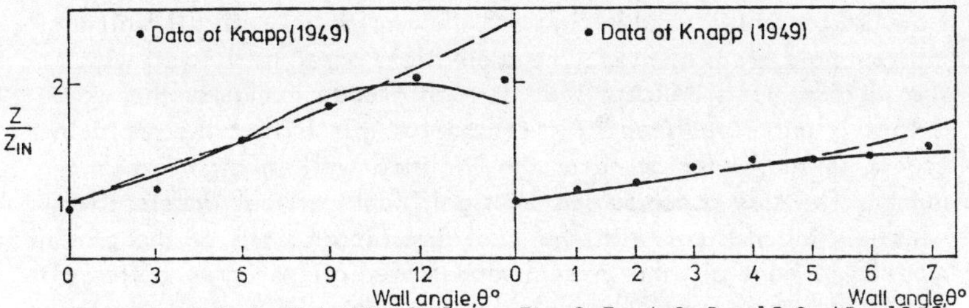


Fig. 2: F = 4·6, $R_I$=7·4, AR = 9·4 Fig. 3: F = 4·2, $R_I$ = 15·0, AR = 12·5

Prediction of water surface height along the outer wall of circular channels; --- method of characteristics, —— present predictions

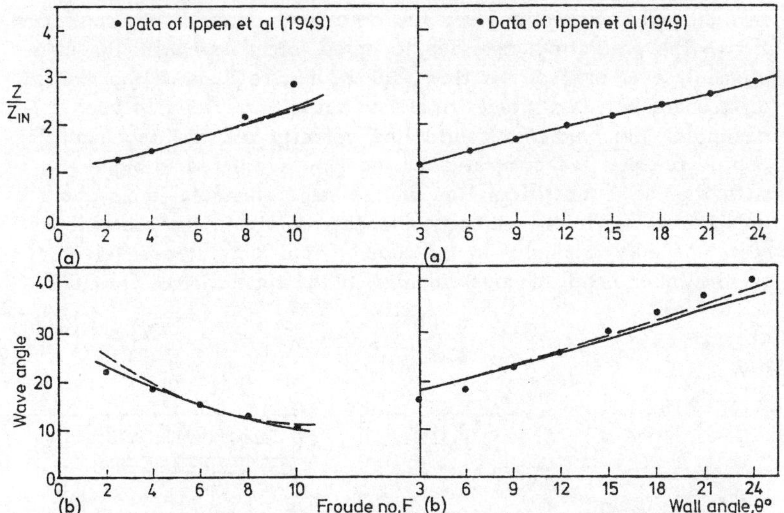


Fig. 4: Variation with Froude no. at θ=6° Fig. 5: Variation with θ, at F = 3·9

Prediction of the height and angle of the first wave produced in channels with straight wall contraction; --- method of characteristics, —— present prediction

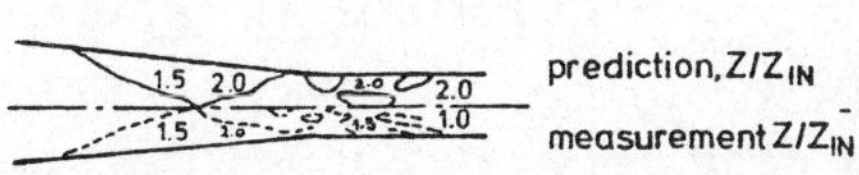


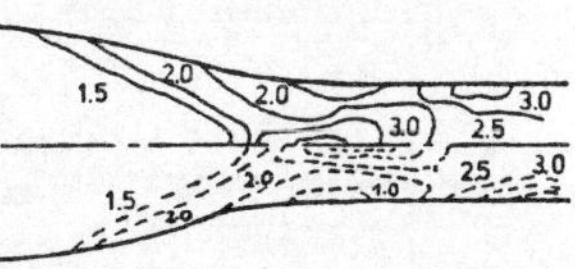

Fig. 6: Straight wall contraction Fig. 7: Circular wall contraction

Contours of water surface height in channel contractions at Froude number F = 4 with measurements. In figure 4, the variations of the predicted wave heights and wave angles with Froude number of the oncoming flow are presented, and in figure 5, the variations with the wall inclination angle are presented. The predictions agree fairly well with the measurements, except that at high Froude numbers, the wave height is underpredicted. The reason for this is that the rapid increase in the water depth at the inclined wall in high Froude number flows is accompanied by significant vertical acceleration which is inconsistent with the two-dimensional basis of the present model. In general, the present model predictions agree better with the measurements than the calculations using the method of characteristics. Figures 6 and 7 show comparison of the predicted water depth contours with the measurements in two channels: one with straight wall contractions and the other with circular arc contractions. The predictions are seen to agree fairly well with the measurements over most of the flow regime. Figure 8 shows the main differences between supercritical and subcritical flows in open channels. The predicted centre line velocity and the near wall depth variation are compared for the flow presented in figure 6 with that of a subcritical flow in the same channel. It may be noted that, while the velocity changes rapidly in the subcritical flow and only gradually in the supercritical flow, the water depth, on the other hand, changes rapidly in the supercritical flow and

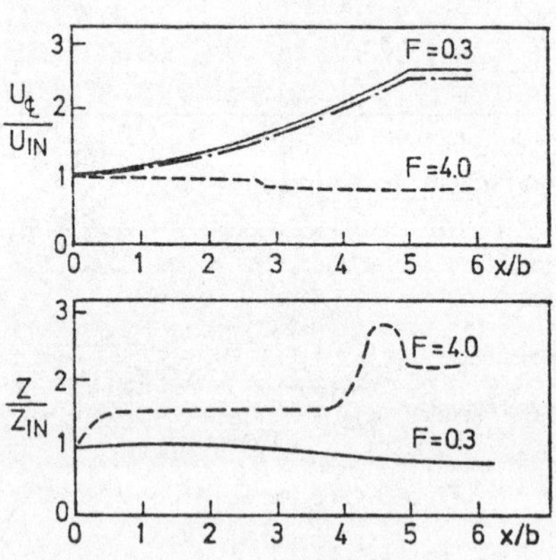


Fig. 8: Predictions of flow in channels with straight wall contraction

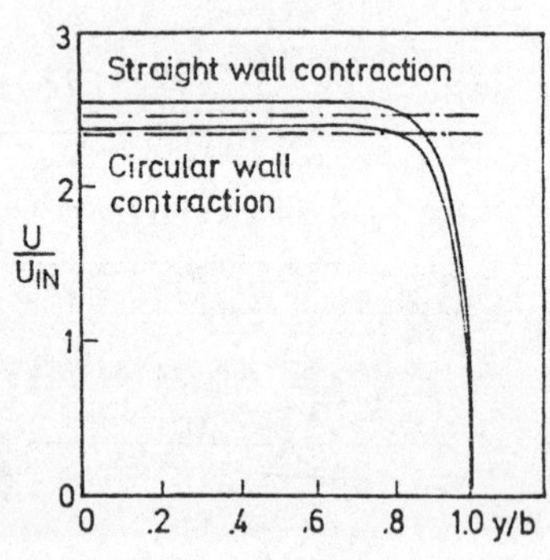


Fig. 9: Predictions of the velocity profiles in the channels, at F = 0·28, x/b = 5.0;
-·-·- Inviscid, —— no slip

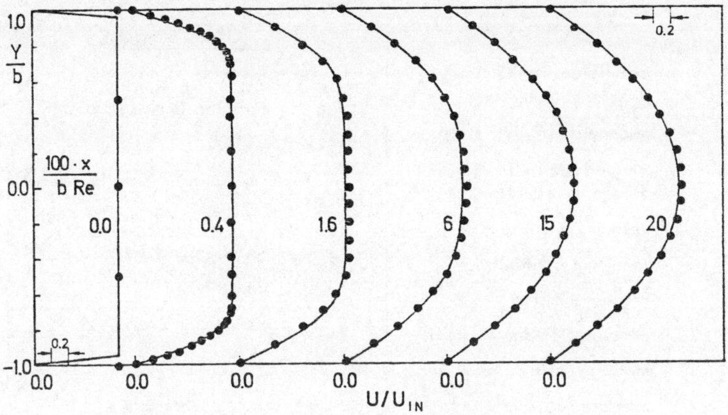


Fig. 10: Prediction of the laminar developing flow in a straight open channel: • analytical solution ; — present prediction

only gradually in the subcritical flow. The directions of the changes are also opposite to one another in the two flow types. The effect of assuming that the turbulent, subcritical flow is inviscid is only noticeable in the velocity. The predicted lateral profiles of the axial velocity presented in figure 9 show that these deviate from a constant value only near the wall, and an inviscid flow calculation may be acceptable in most cases. The last comment is, however, not applicable to laminar flows. Figure 10 shows comparisons of the present predictions of the velocity profiles of the developing flow in a straight open channel with the analytical results of Schlichting (1968). The agreement is seen to be quite good. This flow is parabolic in the context of the present calculations, since the pressure variations are negligible.

## CONCLUSIONS

A finite-difference numerical procedure for calculating steady supercritical and subcritical flows in open channels with wall curvature or with varying cross-section. The procedure has been shown to produce predictions which agree well with experimental data over a wide range of cases, so long as three-dimensional effects do not become very important. In most predictions of subcritical turbulent flows, viscous flow calculations would produce results

which do not differ much from inviscid flow ones.

## NOMENCLATURE

|  |  |
|---|---|
| A | coefficient in the finite-difference equation |
| b | channel half width |
| C | source term in the finite-difference equation |
| c | velocity of sound |
| $C_p, C_v$ | ideal gas constants |
| F | Froude number |
| g | acceleration due to gravity |
| $l_m$ | mixing length of turbulence |
| P | pressure |
| Re | Reynolds number |
| T | absolute temperature |
| x | longitudinal dimension |
| y | lateral dimension |
| Z | channel flow depth |
| $Z_o$ | channel bottom level |
| U | longitudinal velocity |
| V | lateral velocity |

**Greek Symbols**

|  |  |
|---|---|
| $\| \ \|$ | Modulus |
| $\kappa, \lambda$ | turbulence model constants |
| $\gamma$ | $C_p/C_v$ |
| $\nu$ | molecular kinematic viscosity |
| $\nu_{eff}$ | $= \nu_t + \nu$ , effective viscosity |
| $\nu_t$ | turbulent viscosity |
| $\rho$ | density |
| $\tau_w$ | wall shear stress |
| $\phi$ | velocity U or V |

**Subscripts**

|  |  |
|---|---|
| ℄ | centre-line |
| d, u | downstream, upstream |
| i | $i^{th}$ nodal value |
| IN | initial condition |

## REFERENCES


Busemann, A. (1931) Gasdynamik. Handbuch der Experimentalphysik, vol. IV.

Demuren, A.O. (1979) Prediction of steady surface-layer flows. University of London, phd. thesis.

Ippen, A.T., Dawson, J.H. (1949) Design of channel contractions. Trans.ASCE. 116, 326.

Knapp, R.T. (1949) Design of channel curves for supercritical flow. Trans.ASCE. 116,296.

McGuirk, J.J., Rodi, W. (1978) A depth-averaged mathematical model for the near-field of side discharges into open channel flow. J. Fluid Mech. 86, 761.

Patankar, S.V., Spalding, D.B. (1972) A calculation procedure for heat mass and momentum transfer in three-dimensional parabolic flows. IJHMT. 15, 1787.

Pavlović, R.N. (1981) Numerische Berechnung der Wärme- und Stoffausbreitung in Flüssen mit einem tiefengemittelten Modell. Universität Karlsruhe, Doktor-Ingenieur-Disseration.

Preiswerk, E. (1940) Application of the methods of gas dynamics to water flows with a free surface. Naca tech. Memo Nos. 934 and 935.

Rastogi, A.K., Rodi, W. (1978) Two and three-dimensional calculations of heat and mass transfer in open channel flows. ASCE, J. hyd. Div. HY3, 397.

Riabouchinsky, D. (1932) Sur l'analogie hydraulique des mouvements d'un fluide compressible. Compte Rendus de l'Acad. de Sci. 195, 998.

Schlichting, H. (1968) Boundary-layer theory. McGraw Hill .

Spalding, D.B. (1978) A general computer program for two dimensional parabolic phenomena. HMT series 1, Pergamon Press.

# FINITE ELEMENT SOLUTION OF THE LAGRANGIAN EQUATIONS OF UNSTEADY FREE-SURFACE FLOWS ON DRY RIVER BEDS

Di Monaco, A. and Molinaro, P.

ENEL - DSR - Centro Ricerca Idraulica e Strutturale
Milano, Italy

## SUMMARY

A finite element model that solves the governing equations of a one-dimensional free-surface flow in the lagrangian form is presented. It offers the advantage of operating on a grid relating only to the initial configuration of the fluid, thus keeping both trial functions and mass matrix time-independent. This technique is applied to the problem of propagation of a submersion wave on a dry river-bed. Numerical solutions are compared with analytical ones to be found in literature.
A comparison is also made, in terms of accuracy and computational effort, with a finite element model based on the eulerian formulation of the same problem.

## INTRODUCTION

In some hydraulic problems one has to tackle the study of the free-surface motion of a mass of water not bounded by vertical walls, that as a consequence covers bottom-surfaces continuously changing in time. The mathematical modelling of such flows is rather complicated, and therefore it seems not possible to directly apply the usual algorithms that make use of a fixed spatial discretization.
Among moving-boundary problems the dam-break wave problem is certainly of special interest. For solving such a problem, several unidimensional models have been proposed. The algorithms based upon the method of characteristics generally use a fixed mesh (Hartree scheme; see Mahmood K. and Yevjevich V., 1975), and apply the shock equations (Chen C.L., 1980) at the downstream positive wave front. These algorithms need an initial condition corresponding to a con-

stant "small" discharge. Such a condition may be difficult to achieve if the river-bed geometry causes the formation of hydraulic jumps. Also finite difference algorithms generally use a fixed grid, and utilize special techniques, not easily generalizable, to apply the appropriate conditions at the moving boundary (Richtmyer R.D. and Morton K.W., 1967; Chervet A. and Dallowes P., 1970; Rajar R., 1978). A finite difference algorithm which allows the grid-points to move has been developed by Vasiliev (1970) and applied by Barr D.I. and Dass M.M. (1980). Anyway use is made of a fixed grid on which the starting values of the dependent variables must be interpolated at the beginning of each time-step. All the above mentioned models solve numerically the shallow water equations expressed in terms of eulerian independent variables.

In this paper we propose a one-dimensional model, for the simulation of the propagation of submersion waves on dry beds. It starts from the lagrangian form of the De Saint Venant equations and makes use of a finite element method for the spatial discretization. A particular feature of the proposed method is that the trial functions are time-independent as they are defined on a fixed grid which refers only to the initial position of the water body. This fact produces a constant mass matrix and therefore makes convenient the use of an explicit time integration algorithm.

Besides, the application of proper boundary conditions at the advancing and withdrawing fronts turns out to be very straightforward.

Numerical results of this technique are compared with well known analytical solutions.

Finally a comparison is made, in terms of accuracy and computational effort, with a F.E. model based on an eulerian formulation.

## PROPOSED MODEL

According to the lagrangian description of the motion, the physical variables which define the state of the flow are related to the initial position of the water particles and to time. Following such a description of the motion, if "a" denotes the initial position of the particles and "t" the time, the governing equations of a one-dimensional shallow, free-surface flow are :

$$\frac{\partial z}{\partial t} = u \frac{dz_b}{dx} - \frac{A}{B_s}\frac{\partial u}{\partial x} - u \frac{1}{B_s} \frac{\partial A}{\partial x}\bigg|_{h=c} \qquad (1\,a)$$

$$\frac{\partial u}{\partial t} = - g \frac{\partial z}{\partial x} - g S_f \qquad (1\,b)$$

where:

$x = x(a, t)$ : current position of the water particles,
$z = z(a, t)$ : free-surface elevation above a reference datum,
$u = u(a, t)$ : mean velocity in the cross-section,
$h = h(a, t)$ : flow depth,
$A = A(x, h)$ : wetted cross-sectional area,
$z_b = z_b(x)$ : bottom elevation,
$B_s = B_s(x, h)$: surface width,
$g$ : gravity acceleration,
$S_f$ : friction slope.

In Equation (1 a) $\frac{\partial A}{\partial x}\Big|_{h=c}$ represents the rate of change of cross-sectional area A with respect to longitudinal distance x, with depth h held constant. Equations (1) can be formally derived from the motion equations in the eulerian form by simply deleting the advective terms.

The principle of conservation of mass can also be expressed through the equation :

$$\left(\frac{\partial x}{\partial a}\right)^{-1} = \frac{A}{A_o} \qquad (2)$$

where $A_o$ denotes the initial value of the cross-sectional area A, i.e. : $A_o = A\ [x(a, 0),\ h(a, 0)]$.

## Discretization of the governing equations

The governing equations (1) are transformed in a system of ordinary differential equations by a semi-discrete finite element procedure. The domain initially occupied by the fluid is subdivided into elements (fig. 1) and the dependent variables are approximated through the equations :

$$\hat{z}\,(a, t) = \underline{F}\,(a) \cdot \underline{Z}\,(t) \qquad (3\,a)$$

$$\hat{u}\,(a, t) = \underline{F}\,(a) \cdot \underline{U}\,(t) \qquad (3\,b)$$

where $\underline{F}(a)$ is the row matrix of the trial functions, and $\underline{Z}(t)$, $\underline{U}(t)$ are the vectors containing the nodal values of z and u :

$$\underline{F}(a) = [\,f_1\,(a)\ f_2\,(a)\ \ldots\ldots f_N\,(a)\,] \qquad (4\,a)$$

$$\underline{Z}^T\,(t) = [\,z_1\,(t)\ z_2\,(t)\ \ldots\ldots z_N\,(t)\,] \qquad (4\,b)$$

$$\underline{U}^T\,(t) = [\,u_1\,(t)\ u_2\,(t)\ \ldots\ldots u_N\,(t)\,] \qquad (4\,c)$$

In equations (4) N is the number of nodes.

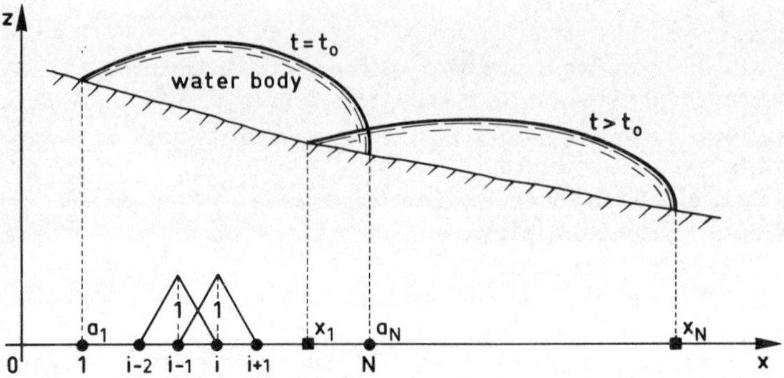


Figure 1 - Discretization of the domain initially occupied by the fluid and linear trial functions

The position of the particles $\hat{x} = \hat{x}\,(a, t)$ is computed by the equation :

$$\hat{x}\,(a, t) = a + \int_0^t \hat{u}\,(a, t')\,dt' = \underline{F}\,(a)\,\underline{X}\,(t) \tag{5}$$

where $\underline{X}\,(t)$ is the vector containing the current position of the nodes. Application of the Galerkin method leads to the system of ordinary differential equations :

$$\left\{\begin{aligned} \underline{M}\,\dot{\underline{Z}} &= -\,\underline{R}\,(\underline{Z}, \underline{U}) \\ \underline{M}\,\dot{\underline{U}} &= -\,\underline{S}\,(\underline{Z}, \underline{U}) \\ \dot{\underline{X}} &= \underline{U} \end{aligned}\right. \tag{6}$$

where

$$\underline{M} = \int_{a_1}^{a_N} \underline{F}^T\,\underline{F}\,da \tag{7 a}$$

$$\underline{R} = \int_{a_1}^{a_N} \underline{F}^T\left(-\,\hat{u}\,\frac{dz_b}{dx} + \frac{A}{B_s}\,\frac{\partial \hat{u}}{\partial x} + \hat{u}\,\frac{1}{B_s}\,\frac{\partial A}{\partial x}\,\Big|_{h=c}\right) da \tag{7 b}$$

$$\underline{S} = \int_{a_1}^{a_N} \underline{F}^T\left(g\,\frac{\partial \hat{z}}{\partial x} + g\,S_f\right) da \tag{7 c}$$

The dots over the variables denote time differentiation.

## Boundary conditions

Boundary conditions only applied on the extreme moving fluid sections are examined. Conditions imposed with x held constant are not considered; anyway they are not natural for the present model.

The simplest boundary conditions applicable to a problem governed by equations (1) are :

$$z(a, t) = \varphi_1 (t) \tag{8}$$

$$u(a, t) = \varphi_2 (t) \tag{9}$$

where $a = a_1$ or $a = a_N$ , and $\varphi_1 (t)$, $\varphi_2 (t)$ are specified functions of time.

A condition which turns out to be very simple to apply concerns waves travelling on dry river-beds. At the wave front the depth is always zero since the pressure vanishes there; therefore the following equation is valid at the front :

$$z (a, t) = z_b [x (a, t)] \tag{10}$$

The present model is particularly devoted to the study of problems where it is necessary to apply condition (10).

Dam-break wave problem belongs to this class of problems.

## Adjustment of eulerian F.E. models for solving the governing equations in the lagrangian form

The following operations are necessary to enable an eulerian F.E. model (E.M.) to solve the lagrangian governing equations (1) :

a) introduction of a vector $\underline{X}$ containing the current nodal positions;

b) introduction of a vector $\underline{\mathcal{A}}$ containing the initial positions: these are utilized to compute the trial functions $\underline{F}$ and their derivatives $\underline{F}_{,a}$ ;

c) erasure of the advective terms, i.e.: $u \frac{\partial z}{\partial x}$ , $u \frac{\partial u}{\partial x}$ ;

d) the spatial derivative of any variable (for example $\hat{z}$ ) must be evaluated as follows :

$$\frac{\partial \hat{z}}{\partial x} = \frac{\partial \hat{z}}{\partial a} \left(\frac{\partial x}{\partial a}\right)^{-1} = \underline{F}_{,a} \underline{Z} (\underline{F}_{,a} \underline{X})^{-1} ;$$

e) description of the river-bed geometry (cross-sections and bottom elevations) free from nodal positions; in particular, denoting $G = G (x)$ any geometric variable, its value related to any fluid section is $G = G [x (a, t)]$ .

To define $G = G (x)$ the introduction of a fixed-grid is needed for the necessary interpolations.

## Time integration of the semi-discrete equations

The solution of the system of ordinary differential equations (6) is carried out using an explicit three-stage Runge-Kutta method of order three, owing to the shape of its stability domain that makes this method suited for the solution of hyperbolic problems (Lambert , 1973).

According to the hypotheses usually assumed for the stability analysis of time integration methods, we found that the following condition holds for the chosen method:

$$\Delta t \leqslant \frac{\sqrt{3}}{\lambda} \frac{\Delta x}{\sqrt{gh}} \tag{11}$$

where: $\Delta x$ is the current lenght of the finite elements (supposed constant in space and in time): $h = A/B_s$ is the hydraulic depth (constant); employing linear shape functions, $\lambda = \sqrt{3}$ for consistent mass matrix and $\lambda = 1$ for lumped mass matrix.

APPLICATIONS

For the verification of the validity of the present lagrangian model (L.M.), reference was made to the dam-break wave problem for which analytic solutions are available. In particular in the hypotheses of sudden failure of the dam, horizontal bottom, rectangular and infinitely wide cross-section and no hydraulic resistance, the analytic solution of the governing equations (1) (Ritter, A., 1892; see Stoker, J.J., 1957) is :

$$h = \frac{1}{9g}(2\sqrt{gh_o} - \frac{x}{t})^2 \tag{12a}$$

$$u = \frac{2}{3} ( \sqrt{gh_o} + \frac{x}{t}) \tag{12b}$$

where $h_o$ denotes the initial water depth.
When a hydraulic resistance is present on the channel bottom, for relatively small values of the resistance itself and of the time t, the analytic solution due to Dressler (1952) is valid (see also Su S.T., 1970).
In this solution the friction slope $S_f$ is computed through the Chezy formula $S_f = u^2/(C_z^2\ h)$ ($C_z$ : Chezy resistance parameter). In the case of zero hydraulic resistance the starting initial conditions were derived from the Ritter's solution at a time $t_o > 0$; in fact the strong discontinuity in z at the dam-site prevents starting of the numerical solution at $t = 0$.
For $h_o = 10$ m and $t_o = 20.193$ s the analytic solution gives a dam-break wave 600 m long. The wave was discretized with 20 elements, each of them 30 m long; another 30 elements of the same length were set upstream. At the boundary the following conditions were assigned: $u_1 = 0$, $z_{51} = 0$.
In the model linear trial functions and lumped mass-matrix were used. In fig. 2 numerical results against analytic solution are represented.
In the case of non-zero hydraulic resistance a constant Chezy parameter $C_z = 140.1\ m^{1/2}/s$ was assigned.

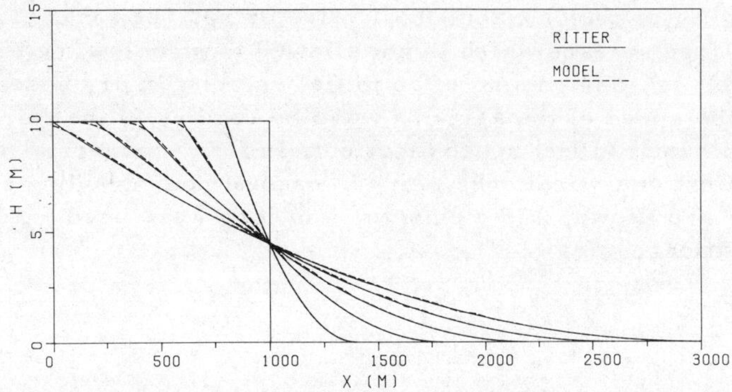


Figure 2 - Free surface elevations in the case of absence of hydraulic resistance. Comparison between Ritter's analytic solution and L.M. results

The starting initial conditions at $t_o$=20,193 s were derived from Dressler's analytic solution. At $t_o$ the wave is 502 m long; it was discretized with 17 elements (the first 16 of which 30 m long); another 30 elements were set upstream, as in the case of zero resistance. The model results are shown in fig. 3 together with the analytic solution.

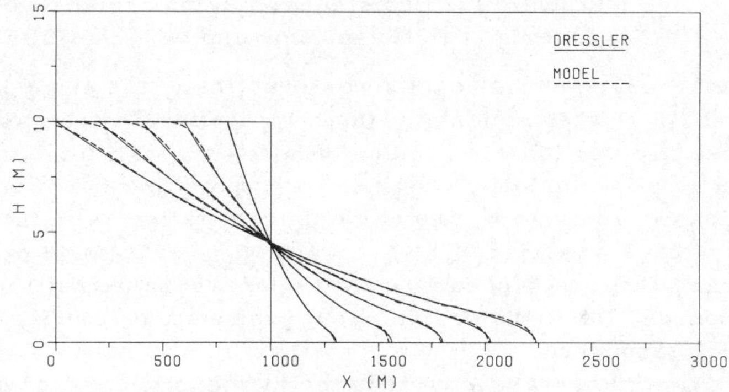


Figure 3 - Free surface elevations in the case of Chezy resistance parameter equal to 140.1 $m^{1/2}/s$. Comparison between Dressler's analytic solution and L.M. results

To make a comparison in terms of accuracy of results and computational effort between the present L.M. and an E.M. the same previous two cases were studied also with an E.M. To enable this model to deal with waves travelling on dry beds the addition of the following particular procedures was necessary:

a) distinction between wetted elements and dry elements, which give no contribution, depending on whether the flow depth is greater than a fixed small value or not; b) constraint on the free surface which is not allowed to go below the bottom level; c) assignment of an initial velocity in dry nodes equal to the speed of the arriving wave (according to the spirit of Whitham's (1955) approximation). In fig. 4 numerical results against analytical ones, for the case of no hydraulic resistance, are shown. 105 elements 30 m long were used for the calculations.

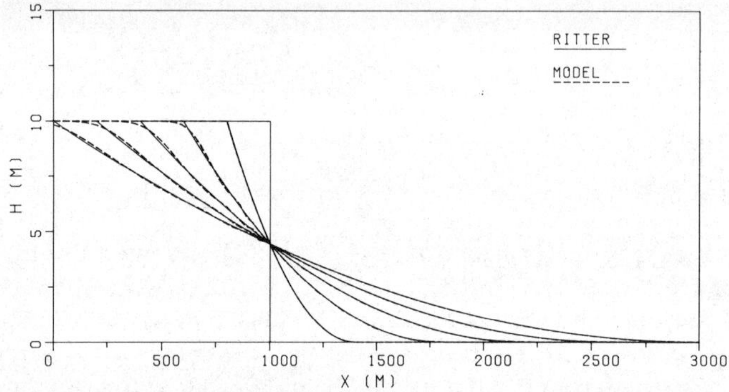


Figure 4 - Free-surface elevations in the case of absence of hydraulic resistance. Comparison between Ritter's analytic solution and E.M. results

The accuracy of the results was practically the same as for the L.M. (50 elements) but the computational time was four times greater (double number of nodes and half time integration step for stability requirement). For the case of non-zero hydraulic resistance, the same discretization as in the previous case was used. The accuracy was of the same order as for the L.M. except for the tip region where parasitic waves appeared. Nevertheless the front propagation velocity was well reproduced.

To damp the parasitic waves, the introduction of a proper viscosity term in the momentum equation is needed. However, the accuracy in the tip region is less than that given by the L.M. due to the constant size of the elements. In the L.M. the length of the elements automatically suits the gradients of velocity and free-surface level, so that it decreases in the tip region and increases in the back-wave region. In fig. 5 numerical results of the E.M. against analytical ones are represented; use was made of a suitable pseudo-viscosity.

Note that in figs. 2, 3, 4, 5 the time interval between two successive free-surface profiles is about 20,2 s.

Finally a hypothetic case was simulated of a dam-break wave propagation, making use of the L.M. The computation was carried out by discretizing the mass of water with 70 elements, initially 30 m long. The initial condition was derived from Dressler's analytical solution.

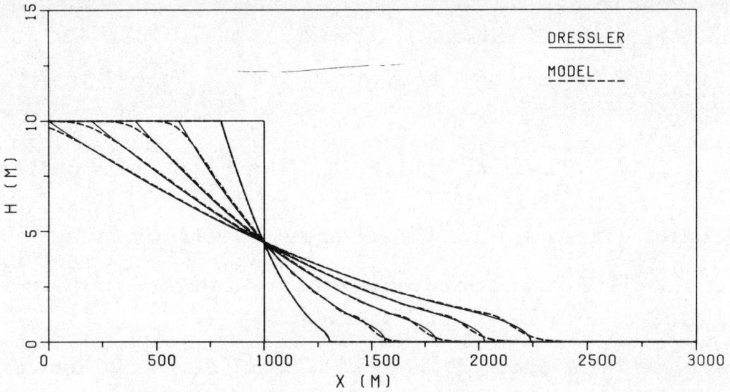


Figure 5 - Free surface elevation in the case of Chezy resistance parameter equal to 140.1 $m^{1/2}s^{-1}$. Comparison between Dressler's analytic solution and E.M. results

In this case we found necessary to introduce in the momentum equation a suitable viscosity term, acting only in the neighbourhood of the upstream trailing edge. In our opinion this is likely to depend on the development of hydraulic bores in that zone. This point, however, has to be further investigated.

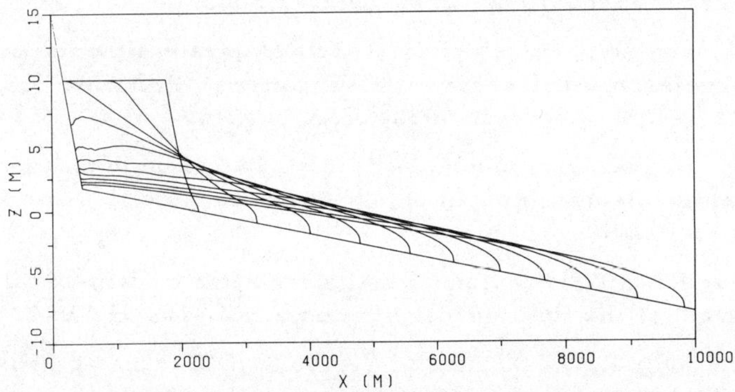


Figure 6 - Hypothetic case of dam-break wave. Rectangular cross-section. Bottom with two different slopes: upstream slope ~0,03, downstream slope 0,001. Chezy coefficient 140.1 $m^{1/2}s^{-1}$. Time interval between two successive free-surface profiles ~80 s.

ACKNOWLEDGMENTS

The Authors wish to express their sincere gratitude to Prof. M.Fanelli for his valuable suggestions. Recognition is also due to E. Piazzardi for his expert help in the computational aspects of this work.

REFERENCES

Barr D.I.H., Dass M.M. (1980). Numerical simulation of dam-burst and reflections, with verification against laboratory data. Proc. Instn. Civ. Engrs. Part 2, 69 June:359-373.

Chen C.L.(1980). Laboratory verification of a dam-break flood model, ASCE, 106, HY4: 534-556

Chervet A., Dallowes P.(1970). Calcul de l'onde de submersion consecutive a la rupture d'un barrage, Schweizerische Baureitung, 7 Mai

Dressler R.F. (1952). Hydraulic resistance effect upon dam--break functions. Journal of Research of the National Bureau of Standards, 49, 3: 217-225.

Dressler R.F. (1954). Comparison of theories and experiments for the hydraulic dam-break wave. International Ass. of Scientific Hydrology, 3, 38: 319-328.

Lambert J.D. (1973). Computational methods in ordinary differential equations, John Wiley & Sons, London.

Mahmood K., Yevjevich V.(1975), Unsteady flow in open channels. Water Resources Publications, P.O. Box 303, Fort Collins, Colorado 80522, USA, pag. 100 Vol.I

Preissmann A., Cunge J.A.(1961). Calcul du mascaret sur machine electronique, Intumescences, Societé Hydrotechnique de France, 15 juin.

Rajar R. (1978). Mathematical simulation of dam-break flow, Journal of the Hydraulics Division, ASCE, 104, HY 7:1011-1025.

Richtmyer R.D., Morton K.W.(1967). Difference methods for initial value problems, Interscience, New York.

Sakkas G.J., Strelkoff T.(1973). Dam-break flood in aprismatic dry channel, ASCE, 99, HY12: 2195-2216.

Stoker J.J.(1957). Water Waves, Interscience publishers, Inc., New York.

Strelkoff T. (1969). One-dimensional equations of open-channel flow. ASCE, 95, HY3: 861-876

Su S.T. (1970). Geometric and frictional effects on sudden releases. ASCE, 96, HY11: 2185-2200

Vasiliev O.F. (1970). Numerical solution of non-linear problems of unsteady flows in open channels, Proc. 2nd International Conference on numerical methods in fluid dynamics, Berkeley, Calif., September.

Whitham G.B. (1955). The effects of hydraulic resistance in the dam-break problem. Royal Society of London, Proceedings Series A, Mathematical and Physical Science Vol. 226-227.

# A HYDRODYNAMIC MODEL FOR OPEN CHANNEL FLOW PROBLEMS

G. A. Keramidas

Fluid Dynamics Branch
Naval Research Laboratory

## ABSTRACT

A variational formulation of the vertically-integrated differential equations for open channel flow is presented, and based on this formulation a finite element model is derived for solving the non-linear system of equations. The time integration scheme employed is also discussed and the obtained results show its good stability and accuracy.

Two types of problems are considered for applying the derived hydrodynamic model. The first problem deals with an open channel of uniform depth and the second one with a channel of variable depth. The inflow boundary condition is prescribed in terms of the velocity and the outflow boundary condition is specified in terms of the water elevation. The outflow condition is specified for two cases, the first one for a reflecting boundary and the second one for a non-reflecting boundary. Emphasis is placed on the latter boundary condition and results show that the numerical model can simulate the non-reflecting boundary in a satisfactory manner.

The simplicity of the derived model and its good performance suggests that one can simulate open channel flows without using high order approximations or complicated numerical schemes. Furthermore, the ability of the computational model to correctly simulate outflow boundary conditions is of special interest since such conditions usually create problems for many numerical algorithms.

## INTRODUCTION

Time-dependent flow phenomena in open channels are usually described by a set of non-linear partial differential equations known as the shallow water equations. The non-linearities of these equations usually preclude analytical solutions for general classes of problems. Many efforts have been made to obtain approximate or numerical solutions of these equations and a wide variety of techniques have been implemented for this purpose. For example, the finite difference method has been used extensively by many researchers. However, the applications of this method are generally restricted to problems of simple geometry where the boundaries conform to the finite difference grid in some way. On the other hand, the finite element method has been more successful in dealing with problems involving complex boundaries, Grotkop (1973), Lynch and Gray (1978). A large and growing body of literature is available on the relative merit of these two methods as well as other numerical schemes and approximate methods, Katopodes and Strelkoff (1979).

The objective of this paper is to examine one difficulty common to most numerical approaches and to certain approximate methods as well. This difficulty is associated with problems where wave propagation in an infinite physical domain must be modeled by a finite computational domain. At the point of domain truncation a boundary or outflow condition must be imposed which behaves like the classical analytical radiation condition for waves at infinity. An improper outflow condition at the end of the computational domain will result in reflections of wave energy which may have no counterpart in the physical problem. Certain techniques have been developed to address this issue either by stretching the computational grid or by using hybrid approaches where the last point in the computational domain is matched to a farfield analytical solution. In this study a simple but effective outflow condition is utilized that not only can provide the proper radiation condition but also it can be adjusted to produce partial wave reflections.

The analytical development of the computational model is presented in the first part of this paper. A generalized variational approach, based on the principle of virtual work, is utilized to derive a set of equations from which numerical or approximate methods can be readily obtained. One such method is the finite element and is implemented here in its simplest form by using linear approximations for the space and time coordinates. The computational model is applied to solve two problems dealing with the water wave propagation in an open channel.

In the first problem the propagation of waves in a channel of uniform depth is simulated by assuming two types of outflow boundary conditions. The first type corresponds to a finite length channel and the second type to an infinite one. In the second problem the channel depth is assumed to be non-uniform and a partial reflection is assured at the outflow boundary.

The time dependent solution is based on a linear finite element approximations and an artificial viscosity term has been implemented into the model, Keramidas, (1982). This artificial viscosity eliminates any oscillations present in the numerical solution which are generated by the non-linear character of the equations. By using such an artificial damping one can avoid higher order approximations or other time consuming numerical schemes.

The obtained results, for the case studies considered in this paper, not only demonstrate the ability of the computational model to correctly simulate outflow boundary conditions, but also the results show the range of applicability of the simple depth-averaged hydrodynamic model.

MODEL DEVELOPMENT

The governing equations describing flow in shallow water are obtained by vertically integrating the continuity and momentum equations. If one neglects effects due to viscocity and turbulence, then the equations are given by

$$\frac{\partial h}{\partial t} + \frac{\partial}{\partial x_i}(HV_i) = 0 \tag{1}$$

$$\frac{\partial V_i}{\partial t} + V_j \frac{\partial V_i}{\partial x_j} + g \frac{\partial h}{\partial x_i} = 0 \tag{2}$$

Here summation is assumed over repeated indices. In the above equations, h is the water surface elevation from the mean surface level, H is the total depth of the fluid, which can be a function of $x_i$, $V_i$ is the depth-averaged velocity in the $x_i$ direction, t is time and g is the gravitational acceleration.

In order to derive the computational model from the above equations a generalized variational formulation is introduced. For this purpose it is convenient to rewrite Eq. (1) in a form similar to Eq. (2) or

$$\frac{\partial h}{\partial t} + V_i \frac{\partial h}{\partial x_i} + h \frac{\partial V_i}{\partial x_i} + \frac{\partial}{\partial x_i}(V_i d) = 0 \tag{3}$$

where $H(x_i,t)$ is given by

$$H(x_i,t) = h(x_i,t) + d(x_i) \tag{4}$$

and $d(x_i)$ is the water depth in the channel. The variational formulation is derived by implementing the principle of virtual work which for Eqs. (2) and (3) is expressed as follows

$$\int_v \left[\frac{\partial h}{\partial t} + V_i \frac{\partial h}{\partial x_i} + h\frac{\partial V_i}{\partial x_i} + \frac{\partial}{\partial x_i}(V_i d)\right]\delta h\, dv = 0 \qquad (5)$$

$$\int_v \left[\frac{\partial V_i}{\partial t} + V_j \frac{\partial V_i}{\partial x_j} + g\frac{\partial h}{\partial x_i}\right]\delta U_i\, dv = 0 \qquad (6)$$

where $\delta h$ and $\delta U_i$ are virtual displacements corresponding to the elevation h and velocity $V_i$ and the integration extends over a volume v of the fluid. Integrating Eq. (6) by parts one obtains

$$\int_v \left[\frac{\partial V_i}{\partial t}+V_j \frac{\partial V_i}{\partial V_j}\right]\delta U_i dv+g\int_v \frac{\partial}{\partial x_i}(h\,\delta U_i)dv-g\int_v h\frac{\partial}{\partial x_i}(\delta U_i)dv = 0 \qquad (7)$$

or

$$\int_v \left[\frac{\partial V_i}{\partial t}+V_j \frac{\partial V_i}{\partial x_j}\right]\delta U_i dv-g\int_v h\frac{\partial}{\partial x_i}(\delta U_i)dv=-g\int_s \eta_i h\,\delta U_i ds=0 \qquad (8)$$

The surface integral is extended over the boundary surface s of the volume v and $\eta_i$ is the unit vector normal to the boundary pointing outward. Equations (7) and (8) can be considered respectively as the variational statements for the continuity and momentum equations. These equations must be verified for arbitrary variations of the displacement fields h and $U_i$ and these variational equations also must be understood in a broad sense as they correspond to the principle of virtual work.

A physical system can be described by a set of generalized coordinates by assuming that the state variables are functions of the coordinates. For the problem considered here the displacement fields are assumed to be functions of the form.

$$h(x_j,t) = h(p_k,x_j,t)$$

$$(9)$$

$$U_i(x_j,t) = U_i(q_k,x_j,t)$$

where $p_k$ and $q_k$ (k=1,..,n) are the generalized coordinates used to describe the physical system. The concept of generalized coordinates has been widely used in classified mechanics and by introducing such coordinates into the variational equations one can derive a set of equations which are similar to the Lagrangian type of equations in mechanics.

If one assumes some arbitrary variations $\delta p_k$ and $\delta q_k$ of the generalized coordinates, the corresponding displacement variations are due entirely to the variations $\delta p_k$ and $\delta q_k$

$$\delta h = \frac{\partial h}{\partial p_k} \delta p_k, \quad \delta U_i = \frac{\partial U_i}{\partial q_k} \delta q_k \tag{10}$$

Taking into account the above definitions the variational equations, Eqs. (7) and (8), are written in the following compact forms

$$\frac{\partial D}{\partial \dot{p}_k} + \frac{\partial E_p}{\partial p_k} = 0 \tag{11}$$

$$\frac{d}{dt}\left(\frac{\partial K}{\partial \dot{q}_k}\right) + \frac{\partial E_q}{\partial q_k} = Q_k \tag{12}$$

where D represents a dissipation function similar to Rayleigh's dissipation function, and K is the kinetic energy $E_p$ and $E_q$ correspond to potential energy functions and $Q_k$ are boundary forces. The complete derivations for the two equations, Eqs. (11) and (12), can be found in Keramidas, (1982). The expressions for D, K, $E_p$, $E_q$ and $Q_k$ are given by

$$D = 1/2 \int_V \left(\frac{dh}{dt}\right)^2 dv, \quad K = 1/2 \int_V V_i V_i \, dv$$

$$E_p = 1/2 \int_V \frac{\partial V_i}{\partial x_i} (h)^2 dv + \int_V h \frac{\partial}{\partial x_i} (V_i d) dv \tag{13}$$

$$E_q = -g \int_V h \frac{\partial V_i}{\partial x_i} dv, \quad Q_k = -g \int_S h n_i \frac{\partial V_i}{\partial q_k} ds$$

The computational model is obtained from the above formulation by assuming a specific representation of the elevation and velocity in terms of the generalized coordinates. Since we are interested in deriving a linear finite element model, h and $V_i$ are expressed as linear combinations of the generalized coordinates. Furthermore, since the model will be restricted to one dimension and to first order approximations with respect to the space coordinate the elevation h and velocity V are expressed as

$$\begin{aligned} h(x,t) &= f_1 h_1 + f_2 h_2 \\ V(x,t) &= f_1 q_1 + f_2 q_2 \end{aligned} \tag{14}$$

where

$$f_1 = 1 - \frac{x}{\Delta x}, \quad f_2 = \frac{x}{\Delta x} \tag{15}$$

Here $\Delta x$ is the element length for the space discretization. The generalized coordinates $(p_1, p_2)$ and $(q_1, q_2)$ are functions of time and correspond to the unknown elevation and velocity at the nodes of the element, or $p_1 = h_1$, $p_2 = h_2$, $q_1 = V_1$, $q_2 = V_2$, etc.

If a linear approximation in time is also assumed, then Eqs. (14) are expressed as

$$h(x,t) = f_1 g_1 p_1^1 + f_1 g_2 p_1^2 + f_2 g_1 p_2^1 + f_2 g_2 p_2^2$$
$$V(x,t) = f_1 g_1 q_1^1 + f_1 g_2 q_1^2 + f_2 g_1 q_2^1 + f_2 g_2 q_2^2 \tag{16}$$

where the functions $f_i$ (i = 1,2) are given by Eqs. (15). The terms $p_i^n$ and $q_i^n$ (i = 1,2, n = 1,2) are the values of the generalized coordinates at the $i^{th}$ point in space and at the $n^{th}$ point in time, and the functions $g_1$ are given by

$$g_1 = 1 - \frac{t}{\Delta t}, \quad g_2 = \frac{t}{\Delta t} \tag{17}$$

with Δt the time step size. By substituting Eqs. (15) and (16) into Eqs. (13) and the results into Eqs. (11) and (12) the finite element equation is obtained and it can be represented by the following matrix equation

$$[A]\ \{q^{n+1}\} = \{Q\} - [B]\ \{q^n\} \tag{18}$$

where the superscripts denote time level and

$$[A] = [M] + [K]$$
$$[B] = [M] - [K] \tag{19}$$
$$\{q\} = [h_i, V_1, h_{i+1}, V_{i+1}]^T$$

with

$$[M] = \frac{1}{\Delta t}\begin{bmatrix} 2 & 0 & 1 & 0 \\ 0 & 2 & 0 & 1 \\ 1 & 0 & 2 & 0 \\ 0 & 1 & 0 & 2 \end{bmatrix}, \quad \{Q\} = \frac{3g}{\Delta x}\begin{Bmatrix} 0 \\ h_i^{n+1} + h_i^n \\ 0 \\ h_{i+1}^{n+1} + h_{i+1}^n \end{Bmatrix} \tag{20}$$

and

$$[K] = \frac{1}{2\Delta x}\begin{bmatrix} -V_1 + 2\Delta V & -H_1 + 2\Delta H & V_1 + \Delta V & H_1 + \Delta H \\ 3g & -V_1 + D_o & 3g & V_1 - D_o \\ -V_2 + \Delta V & -H_2 + \Delta H & V_2 + 2\Delta V & H_2 + 2\Delta H \\ 3g & -V_2 - D_o & -3g & V_2 + D_o \end{bmatrix} \tag{21}$$

The quantities in the [K] matrix are defined as follows

$$V_1 = 2\overline{V}_i + \overline{V}_{i+1}, \quad V_2 = \overline{V}_i + 2\overline{V}_{i+1}$$

$$H_1 = 2d_i + d_{i+1}, \quad H_2 = d_i + 2d_{i+1} \tag{22}$$

$$\Delta V = \overline{V}_{i+1} - \overline{V}_i, \quad \Delta H = d_{i+1} - d_i$$

with

$$\overline{V}_i = \tfrac{1}{2}(V_i^n + V_i^{n+1})$$

where $V_i$ is the velocity and $d_i$ is the water depth at the $i^{th}$ nodal point. The quantity $D_o$ in Eq. (21) is the artificial viscosity term with a value depending on the time step size, Keramidas, (1982).

## MODEL APPLICATIONS

The computational model presented in the previous section is implemented here to simulate the propagation of water waves in an open channel. The required computational domain has a length L = 100, the water depth is d = 10 and the domain is divided in elements of length $\Delta x = 1.0$. The time step used is $\Delta t = 0.02$ and the value for the artificial viscosity is $D_o = 2.0$.

The two boundary conditions which are required for the solution of the matrix equations are one for the velocity and one for the water elevation. At the left boundary (inlet of the channel) the velocity is specified as

$$V(0,t) = V_o \sin \omega t \tag{23}$$

where $\omega = 2\pi/T$ with T the wave period and $V_o$ the velocity amplitude. At the right boundary (outlet of channel) the elevation is specified as a function of the velocity at that point given by

$$h(L,t) = V(L,t)\,\frac{d(L)}{C_o} \tag{24}$$

where d(L) is the water depth and $C_o = \sqrt{gd}$ is the wave velocity at the same point. The condition given by Eq. (24) results in a non-reflecting outflow boundary or a boundary with partial reflection depending on the value of the water depth d(L). Equation (24) is derived from continuity considerations which require that the volumetric flow in the channel is constant at any x location. This conservation requirement is expressed as a constant mass of fluid per unit surface and unit time. In order to demonstrate the use of the outflow condition together with the computational model three example problems are presented.

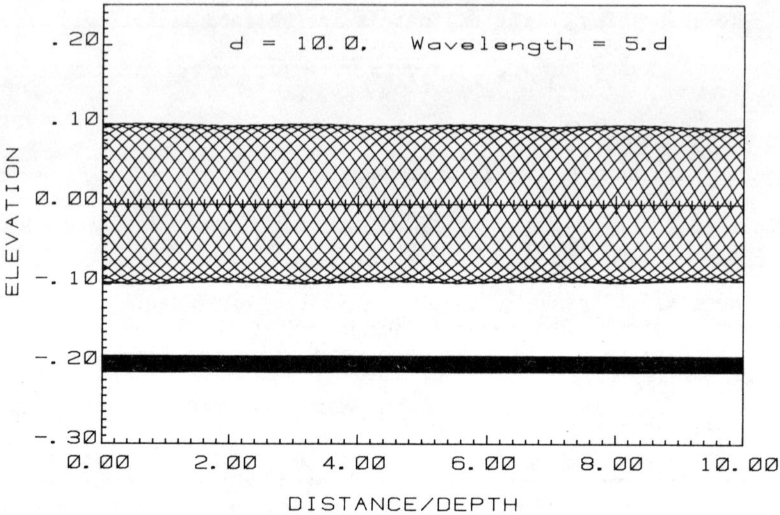


Figure 1. Surface elevation along channel during 25 wave cycles with non-reflecting boundary on the right and a wave generator on the left.

In the first example the computational model simulates the propagation of water surface waves in a semi-infinite channel with uniform depth. The wave period used is T = 5.0 and the velocity amplitude is $V_o = 0.1$. Results are presented in Figs. (1), (2) and (3). In Fig. (1) a series of wave profiles at sequential time steps are superimposed over 25 cycles to give a simulated time exposure of the propagated waves. Since there should be no reflections at the outflow boundary an envelope of constant height forms along the channel and the waves propagate unchanged. The lower portion of Fig. (1) shows the same data in scale with the channel geometry. The non-reflecting boundary is further demonstrated in Figs. (2) and (3) where results for the elevation and velocity are presented at the inlet and outlet of the channel. From both figures one can observe that the end of the computational domain has no effect on the incoming waves since the outflow boundary condition simulates the radiation condition correctly.

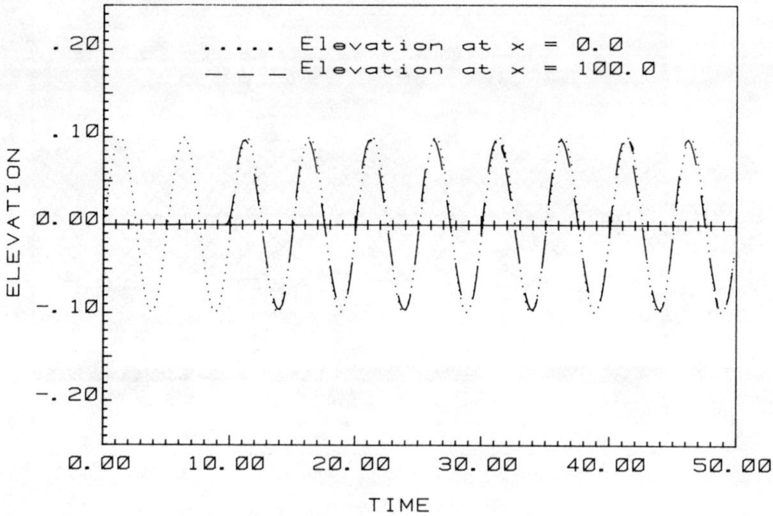


Figure 2. Time history of wave elevation at the left and right boundaries for case shown in Figure 1.

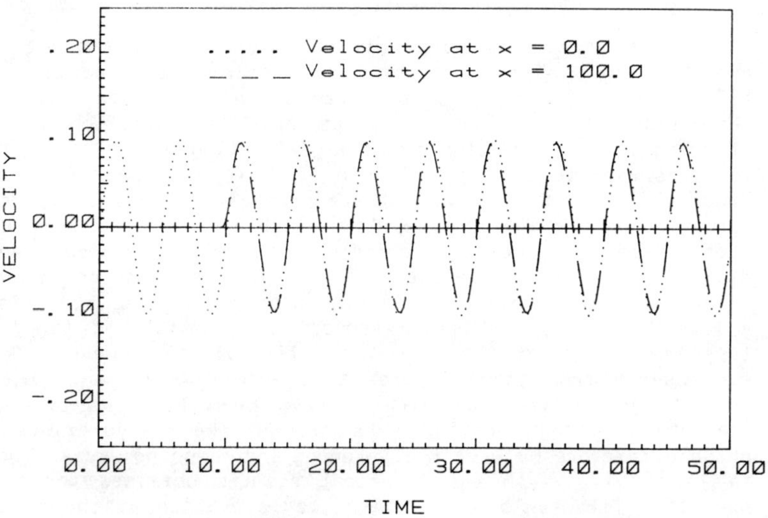


Figure 3. Time history of velocity at the left and right boundaries for the case shown in Figure 1.

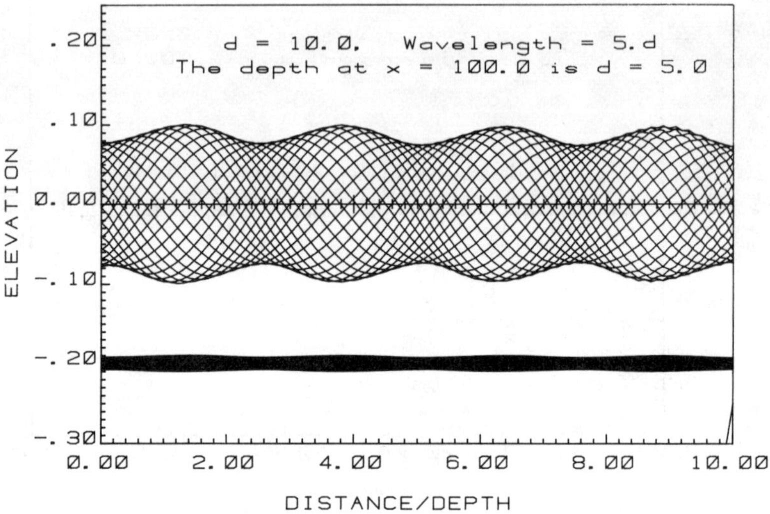


Figure 4. Surface elevation along the channel during 25 wave cycles with reflecting boundary on the right. Waves generated on the left.

In the second example only the depth of the channel at the last point of the computational domain is changed to d = 5.0. The effect of a different depth will be partial reflection of the incoming waves and the formation of a standing wave by the incoming and reflecting waves. Results for this example are presented in Fig. (4), (5) and (6). In Fig. (4) the time exposure of the propagated waves shows the formation of a standing wave, from which one can calculate the approximate reflection coefficient from the maximum and minimum values of the envelope. In order to demonstrate the effect of the partially reflecting boundary on the elevation and velocity, results are presented in Figs. (5) and (6) for these quantities at the inlet and outlet of the channel. One can observe from these figures that not only the amplitudes for the elevation and velocity have been affected by the presence of the reflecting boundary, but there also exists a phase difference between the incoming and outgoing waves shown in Fig. (5). From the numerical results obtained for the specified velocity and the calculated elevation at the inlet of the channel the validity of Eq. (24) can be verified.

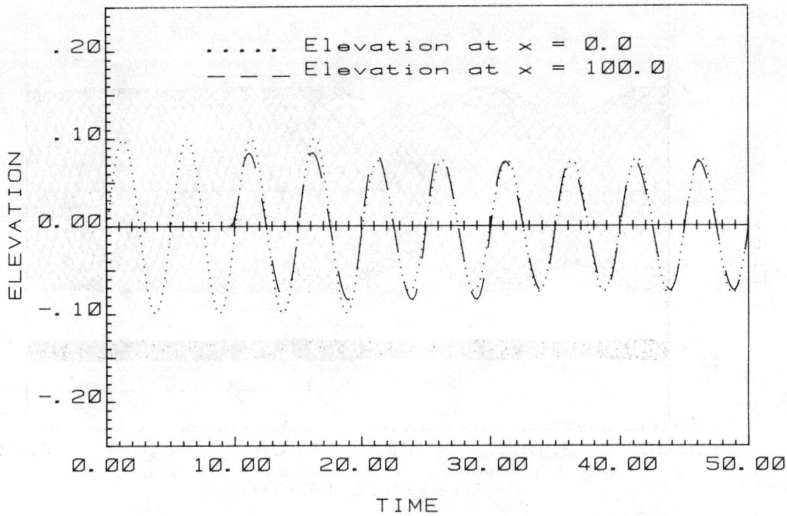


Figure 5. Time history of wave elevation at the left and right boundaries for case shown in Figure 4.

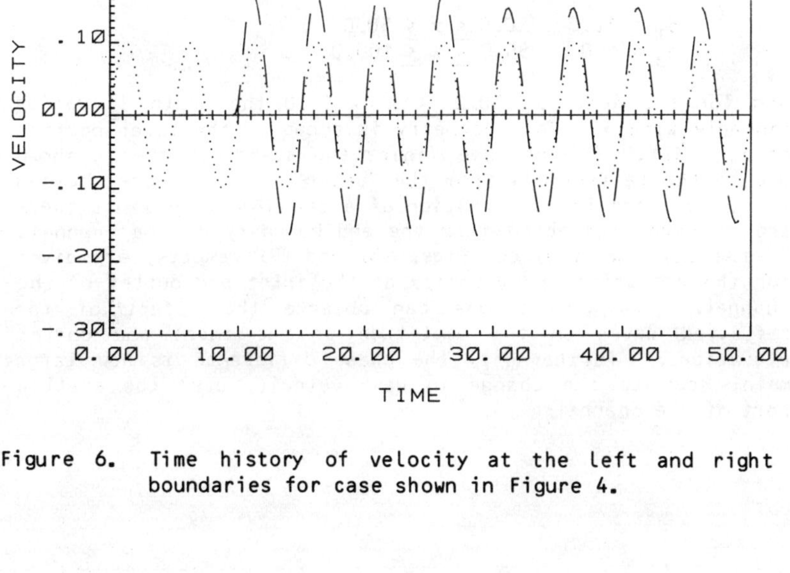


Figure 6. Time history of velocity at the left and right boundaries for case shown in Figure 4.

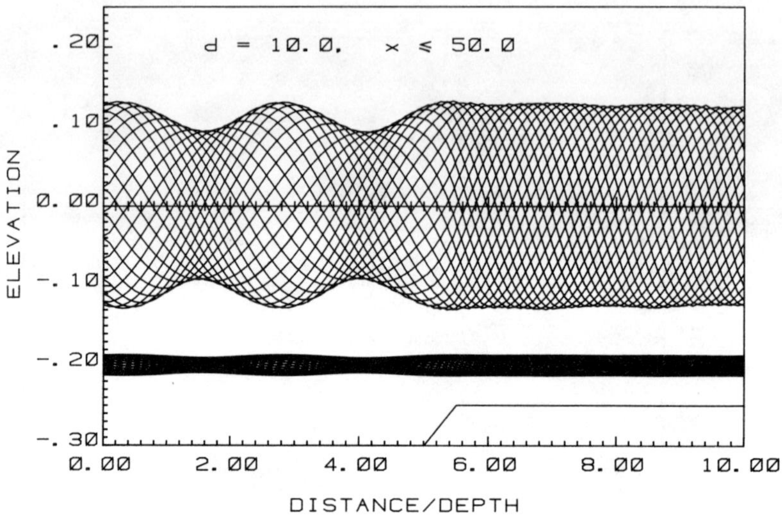


Figure 7. Surface elevation along channel during 25 wave cycles with non-uniform depth. Non-reflecting boundary on the right.

To further demonstrate the flexibility of the computational model a third example is presented for a different channel geometry. The water depth is non-uniform with

$$d_1 = 10.0 \qquad 0.0 \leq x \leq 50.0$$
$$d_3 = 5.0 \qquad 55.0 \leq x \leq 100.0$$

and for the interval $50.0 \leq x \leq 55.0$ the depth $d_2$ varies linearly with x. This geometry is shown in the lower portion of Fig. (7). On the same figure the standing wave is shown due to the reflections from the "beach". Furthermore, beyond x = 55.0 there is no formation of a standing wave since there are no waves reflected from the end boundary of the channel. In the last two figures, Figs. (8) and (9) results, are given for the elevation and velocity at the inlet and outlet of the channel. As before one can observe the effect of the reflected waves on the amplitude of the inlet and outlet elevations. Furthermore, the phase difference is now large mainly due to the change in wave velocity over the shallow part of the channel.

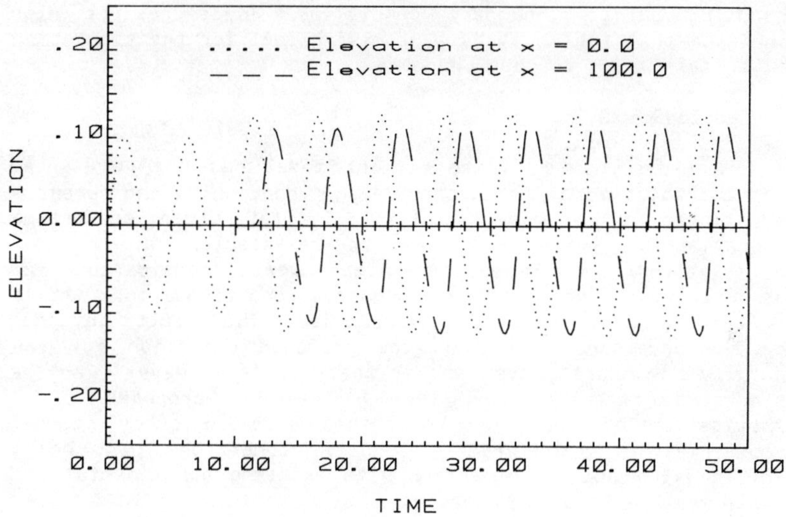


Figure 8. Time history of wave elevation at the left and right boundaries for case shown in Figure 7.

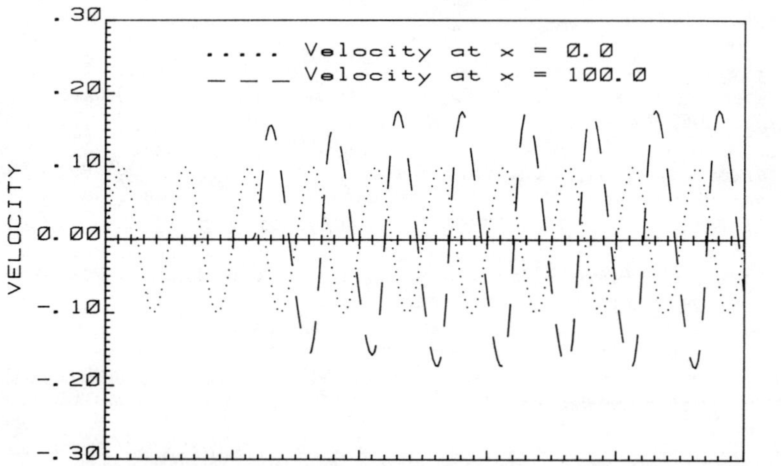


Figure 9. Time history of velocity at the left and right boundaries for the case shown in Figure 7.

The results obtained for the above three problems demonstrate the efficiency of the computational model in handling outflow boundary conditions. These results also indicate that the model is a promising tool for the simulation of surface waves in open channels.

## CONCLUSIONS

A hydrodynamic finite element model has been presented for the solution of the depth-averaged continuity and momentum equations. The model employs a first order space-time approximation which is shown to be stable and free of oscillations. Three examples were considered for demonstrating the outflow boundary conditions for either reflecting or non-reflecting boundaries. The results obtained by the developed model and the proposed condition for the outflow boundary have shown that surface waves can be simulated correctly by the formulation proposed here. Application of the model described here to other channel geometries or different boundary conditions is being investigated and a comparison with existing experimental and analytical results is underway.

## REFERENCES

Grotkop, G., "Finite Element Analysis of Long-Period Water Waves", Computer Methods in Applied Mechanics and Engineering, Vol. 2, 1973.

Katopodes, N. and Strelkoff, T., "Two-Dimensional Shallow Water Wave Models", ASCE, Journal of Engineering Mechanics, Vol. 105, 1979.

Keramidas, G.A., "One and Two-Step Time Integration Methods for Non-Linear Partial Differential Equations", to be presented at the 10th IMACS World Congress, Aug. 1982.

Keramidas, G.A., "Variational Formulations and Approximate Solutions for Transport Phenomena", NRL Memorandum Report, 1982.

Keramidas, G.A., "A Computational Model for Surface Waves", NRL Memorandum Report, 1982.

Lynch, D.R., and Gray, W.G., "Finite Element Simulation of Shallow Water Problems With Moving Boundaries", 2nd Int. Conf. on Finite Elements in Water Resources, 1978.

# ON THE USE OF EXPLICIT METHODS OF SOLUTION OF THE SAINT VENANT EQUATIONS FOR OPEN CHANNEL FLOW

K.Sivaloganathan

Senior Lecturer in Civil Engineering
Coventry (Lanchester) Polytechnic, Coventry, England

## INTRODUCTION

Explicit methods of solution, of the Saint Venant equations for open channel flow, have not been favoured for use in practice by some engineers. The reason, given for this disfavour, is that the observance of the so called Courant condition for stability imposes a severe restriction on the size of the time step used at each stage of the computation, even though the acceptable accuracy may not warrant the use of such small time steps, and that this in turn results in large computation times. The limitation, on the size of the time steps seems, to the writer, to stem indirectly from an assumption that is implicit in the procedure normally used. This assumption seems to be that in the computation of the dependent variables at a grid point on the new time level, its domain of dependence should lie within the adjoining cells of the grid used. This paper shows how, when this assumption is relaxed, it is possible to use arbitrary time steps without violating the essence of the Courant condition.
Sivaloganathan (1979) has shown how arbitrary time steps may be used with the Rectangular Grid Broken Characteristic method. This paper extends this facility to all explicit methods. The paper details the proposed techniques using the Diffusing method. Their adoption to the Lax-Wendroff method and the Leap Frog method is similar and presents no problems (Sivaloganathan, 1980).

## SAINT VENANT EQUATIONS

The Saint Venant equations exist in various forms (Liggett,1968; Strelkoff,1970; Sivaloganathan,1980). The following general form is used in this paper :-

$$\frac{\partial A}{\partial t} + \frac{\partial Q}{\partial x} = q \qquad (1)$$

$$\frac{\partial Q}{\partial t} + \frac{\partial F}{\partial x} = E \tag{2}$$

where $F = Q^2/A + gA\bar{z}$ , $E = gA(S_o - S)$
A = area of the water section at distance x at time t ; B = width of section at the water surface at distance x at time t ; Q = discharge at distance x at time t ; q = lateral inflow per unit length of channel at distance x at time t ; S = friction slope at distance x at time t ; $S_o$ = bed slope at distance x ; $\bar{z}$ = the depth of the centroid of the water section below the water surface and g = gravitational acceleration.

Characteristic Form :
Equations 1 and 2 may be expressed in characteristic form as follows (Sivaloganathan,1980) :-

$$\frac{dx}{dt} = \frac{Q}{A} + \sqrt{\frac{gA}{B}} \tag{3}$$

$$\frac{dQ}{dt} - \left(\frac{Q}{A} - \sqrt{\frac{gA}{B}}\right)\frac{dA}{dt} = G_f \tag{4}$$

$$\frac{dx}{dt} = \frac{Q}{A} + \sqrt{\frac{gA}{B}} \tag{5}$$

$$\frac{dQ}{dt} - \left(\frac{Q}{A} + \sqrt{\frac{gA}{B}}\right)\frac{dA}{dt} = G_b \tag{6}$$

where $G_f = E - q\left(\frac{Q}{A} - \sqrt{\frac{gA}{B}}\right) + \frac{gA}{B}\left(\frac{\partial A}{\partial x}\right)_y$ and

$G_b = E - q\left(\frac{Q}{A} + \sqrt{\frac{gA}{B}}\right) + \frac{gA}{B}\left(\frac{\partial A}{\partial x}\right)_y$ and $\left(\frac{\partial A}{\partial x}\right)_y$ is the rate of increase of A with respect to x when y is held constant. The compatibility equation 4 holds along positive characteristics given by equation 3 and the compatibility equation 6 holds along negative characteristics given by equation 5.

DIFFUSING METHOD

Traditional Furmulation :
Reference is made to figure 1. In advancing the solution from a time level $t^j$, $\Delta t$ is first determined such that the Courant condition $\Delta t \leq \Delta x/(Q/A + \sqrt{gA/B})$ is satisfied at all grid points on time level $t^j$. The values of the dependent variables at grid points on time level $t^{j+1} = t^j + \Delta t$ are then determined by point by point calculation using the finite difference equations below, for the interior points :-

$$\frac{A_i^{j+1} - \frac{A_{i+1}^j + A_{i-1}^j}{2}}{\Delta t} + \frac{Q_{i+1}^j - Q_{i-1}^j}{2\Delta x} = q_i^j \tag{7}$$

$$\frac{Q_i^{j+1} - \frac{Q_{i+1}^j + Q_{i-1}^j}{2}}{\Delta t} + \frac{F_{i+1}^j - F_{i-1}^j}{2\Delta x} = E_i^j \tag{8}$$

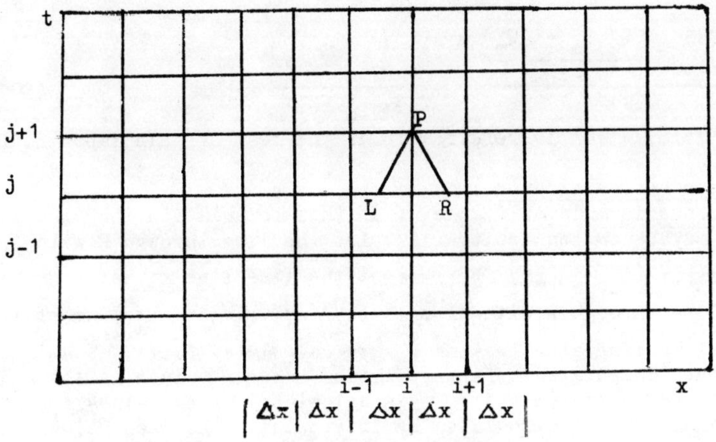


FIG.1

Proposed Formulation 1 :
Reference is made to figure 2. In this formulation, $\Delta t$ may be chosen arbitrarily. Let the line through P with slope equal to $(Q_i^j / A_i^j + \sqrt{gA_i^j / B_i^j})$ meet the line $t = t^j$ at L and let the grid point to the left of L be $(m,j)$. The difference equations that follow are used for the calculation of the dependent variables at the interior grid points such as P on time level $t^{j+1}$.

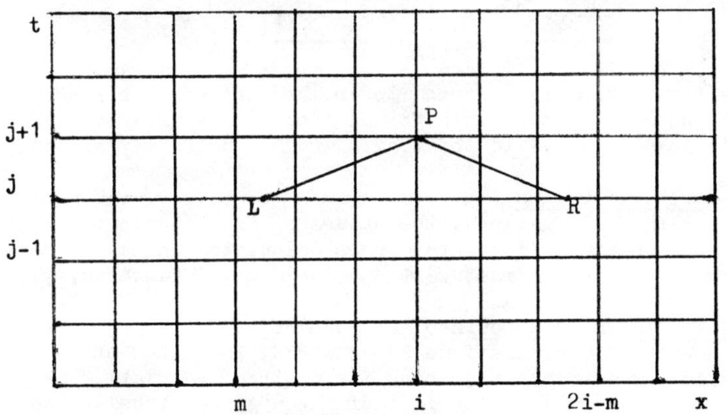


FIG.2

$$\frac{A_i^{j+1} - \dfrac{A_{2i-m}^j + A_m^j}{2}}{\Delta t} + \frac{Q_{2i-m}^j - Q_m^j}{x_{2i-m} - x_m} = q_i^j \qquad (9)$$

$$\frac{Q_i^{j+1} - \frac{Q_{2i-m}^j + Q_m^j}{2}}{\Delta t} + \frac{F_{2i-m}^j - F_m^j}{x_{2i-m} - x_m} = E_i^j \qquad (10)$$

This formulation is referred to in the rest of this paper as PF1.

Proposed Formulation 2 :
Reference is made to figure 2. In this formulation too, the time step may be chosen arbitrarily. Let the line through P with slope $(Q_i^j/A_i^j + \sqrt{gA_i^j/B_i^j})$ meet the time line $t^j$ at L and let the line through P with slope $-(Q_i^j/A_i^j + \sqrt{gA_i^j/B_i^j})$ meet the same time line at R. $x_L$ and $x_R$ are calculated first. $A_L$ and $Q_L$ are then calculated by interpolation between values at the grid points (m,j) and (m-1,j). $F_L$ is calculated using values $A_L$ and $Q_L$. $A_R$ and $Q_R$ are calculated by interpolation between values st the grid points (2i-m-1,j) and (2i-m,j). $F_R$ is calculated using values $A_R$ and $Q_R$. (The use of non-linear interpolation for the calculation of $A_L$, $Q_L$, $A_R$, $Q_R$ is under investigation). The values of $A_i^{j+1}$ and $Q_i^{j+1}$ are then determined using the difference schemes that follow.

$$\frac{A_i^{j+1} - \frac{A_L + A_R}{2}}{\Delta t} + \frac{Q_R - Q_L}{x_R - x_L} = q_i^j \qquad (11)$$

$$\frac{Q_i^{j+1} - \frac{Q_L + Q_R}{2}}{\Delta t} + \frac{F_R - F_L}{x_R \quad x_L} = E_i^j \qquad (12)$$

This formulation is referred to in the rest of this paper as PF2. It is noteworthy that this formulation can be used even if the grid lines parallel to the t axis are not uniformly spaced.

Boundary Grid Points :
For all the formulations, the values of the dependent variables at the boundary grid points may be calculated using the Rectangular Grid Broken Characteristic method (Sivaloganathan,1979).

Grid Points in Neighbourhood of Boundaries :
In the case of formulations PF1 and PF2, for a few of the interior grid points adjoining the upstream boundary, the point L (Fig.2) may fall to the left of the upstream boundary and for a few of the interior grid points in the neighbourhood of the downstream boundary, the point R (Fig.2) may fall to the right of the downstream boundary. The calculations for such points are best done by the Rectangular Grid Broken Characteristic method as detailed by Sivaloganathan (1979).

STABILITY, ACCURACY AND COMPUTATION TIME

Stability :
The essence of the Courant condition for stability has been adhered to in the formulations PF1 and PF2. The necessary condition for stability has thus been satisfied in these formulations. Nevertheless, numerical experiments were done by the writer to test the stability of these formulations, by introducing perturbations on steady flow profiles. No signs of any instability were observed at any time, in any of the related computations.

Accuracy and Computation Time :
The accuracies, of formulations PF1 and PF2, for different sizes of time steps, were assessed by comparison of the computed depths using them with the depths computed using the traditional formulation of the Diffusing method with the same distance steps. The solution using the traditional formulation is referred to, as the standard solution, in the rest of this paper. The percentage differences between solutions using each of the formulations PF1 and PF2 and the standard solution were were computed and examined. To highlight the differences between solutions, fairly steep input discharge hydrographs were used. As should be expected, the use of larger time steps does result in less accurate solutions. However less accurate solutions are acceptable for practical purposes. Depending on the acceptable accuracy, computation times can be reduced by the use of larger time steps. Formulation PF2 has been found to produce more accurate solutions than formulation PF1 for same time step while it takes more computation time. For the larger time steps, the depth hydrographs of solutions by PF2 are found to be smoother than the depth hydrographs of solutions by PF1 for a given time step. The writer expects that the use of quadratic interpolation for determining the values of the dependent variables at points L and R (Fig.2) would improve the accuracy of solutions by PF2. This is under investigation. A repeat calculation during each time step, using $q_i^{j+\frac{1}{2}}$ and $E_i^{j+\frac{1}{2}}$ on the right hand sides of equations 11 and 12 respectively is also under investigation and is expected to improve accuracy.

## APPLICATION

An example of flow routing is dealt with in this section. Computed values of depths, for different time steps, using formulations PF1 and PF2 are compared with the depths of the standard solution. The computations were done on a digital computer.

Example : The channel in which the flow routing is done is circular in section with radius 1 m and length 60 m. It is sufficiently open at the top to permit lateral inflow, and discharges freely at its downstream end. The channel has a Manning's constant of 0.012 and a bed slope of 0.0006 .
A trapezoidal discharge hydrograph is superimposed at the

upstream end, on a steady flow of 0.8 $m^3/s$ to give the following discharges in $m^3/s$ :-

$Q(0,t) = 0.8 \quad 0 \leq t \leq 10$ ; $Q(0,t) = 0.8+0.04(t-10) \quad 10 < t \leq 35$ ;
$Q(0,t) = 1.8 \quad 35 < t \leq 60$ ; $Q(0,t) = 1.8-0.04(t-10) \quad 60 < t \leq 85$ ;
$Q(0,t) = 0.8 \quad 85 < t$

In addition, the following lateral inflow hydrograph, in $m^3/s/m$ length of channel is also superimposed over the whole length of the channel:-

$q = 0.0 \quad 0 \leq t \leq 10$ ; $q = 0.0024(t-10) \quad 10 < t \leq 35$ ;
$q = 0.06 \quad 35 < t \leq 37$ ; $q = 0.06-0.0024(t-37) \quad 37 < t \leq 62$ ;
$q = 0.0 \quad 62 < t$

The computations presented are up to t=140 s. Table 1 gives comparisons between solutions by the formulations PF1 and PF2 using different time steps and the standard solution. The means, standard deviations, maximum and minimum values have been calculated for values of the percentage differences calculated at 2 second intervals of time. Table 2 gives a comparison of the CPU times as a percentage of the CPU time taken for the standard solution. Figures 3 and 4 show the depth hydrographs of the standard solution and PF2 at x= 12 m and x= 48 m when time steps of 2 seconds are used for the latter. Figure 5 shows the depth hydrographs of the standard solution and PF1 at x= 36 m when time steps of 2 seconds are used for the latter.

TABLE 1 : Percentage Differences of Depths from Depths of the Standard Solution

| Formulation No. | Mean Value | Standard Deviation | Maximum Value | Minimum Value | At Distance in m |
|---|---|---|---|---|---|
| | | ($\Delta t$ = 0.5 s) | | | |
| PF1 | -1.7183 | 0.8493 | -0.0069 | -2.5575 | x = 12 m |
| PF2 | -1.2537 | 0.5900 | -0.0047 | -1.9569 | |
| PF1 | -1.8620 | 0.8166 | -0.0098 | -2.6439 | x = 24 m |
| PF2 | -1.3459 | 0.5812 | -0.0067 | -1.9941 | |
| PF1 | -2.0403 | 0.7585 | -0.0151 | -2.9438 | x = 36 m |
| PF2 | -1.4734 | 0.5343 | -0.0104 | -2.1711 | |
| PF1 | -2.3442 | 0.9350 | -0.0270 | -4.0902 | x = 48 m |
| PF2 | -1.5561 | 0.3786 | -0.0189 | -2.1365 | |
| | | ($\Delta t$ = 1.0 s) | | | |
| PF1 | -3.0142 | 1.3457 | -0.0170 | -4.6758 | x = 12 m |
| PF2 | -2.5784 | 1.0992 | -0.0121 | -3.7680 | |
| PF1 | -3.1320 | 1.1974 | -0.0241 | -4.4249 | x = 24 m |
| PF2 | -2.6938 | 1.0700 | -0.0172 | -3.7461 | |
| PF1 | -3.3562 | 1.2411 | -0.0372 | -5.0845 | x = 36 m |
| PF2 | -2.8659 | 0.9805 | -0.0268 | -4.1300 | |
| PF1 | -3.2914 | 0.9912 | -0.0667 | -5.5065 | x = 48 m |
| PF2 | -2.7878 | 0.8987 | -0.0488 | -4.4806 | |

| Formulation No. | Mean Value | Standard Deviation | Maximum Value | Minimum Value | At Distance in m |
|---|---|---|---|---|---|
| | | ($\Delta t = 2.0$ s) | | | |
| PF1 | -5.3368 | 2.1438 | -0.0307 | -7.4098 | x = 12 m |
| PF2 | -4.9616 | 1.9702 | -0.0272 | -7.1866 | |
| PF1 | -5.3267 | 2.0456 | -0.0434 | -8.6584 | x = 24 m |
| PF2 | -4.9866 | 1.9371 | -0.0387 | -7.7719 | |
| PF1 | -4.9446 | 1.7769 | -0.0670 | -7.2946 | x = 36 m |
| PF2 | -4.5853 | 1.6738 | -0.0602 | -7.1032 | |
| PF1 | -4.3455 | 1.6775 | -0.4867 | -6.8219 | x = 48 m |
| PF2 | -4.0908 | 1.5246 | -0.4216 | -6.5667 | |

TABLE 2 : CPU Times as % of CPU Time of Standard Solution

| $\Delta t$ = | 0.5 | 1.0 | 2.0 |
|---|---|---|---|
| PF1 | 61 | 36 | 22 |
| PF2 | 79 | 45 | 27 |

## DISCUSSION AND CONCLUSION

The computed results above are illustrative of the comments made under 'Accuracy and Computation Time'. For practical purposes the accuracies achieved above with time steps of 2 s may be acceptable and would result in considerable savings in computation time. The savings in computation times would be even more dramatic when less steep input discharge hydrographs are involved such as in river flow routing. Formulation PF2 gives more accurate solutions than formulation PF1 for the same time steps. This paper has shown how arbitrary time steps may be used with explicit methods, without violating the stability condition. Considerable savings in computation time are now possible when using explicit methods, the extent of savings depending on the acceptable accuracy of the solution.

## REFERENCES

Liggett J.A., (Aug 1968) Mathematical Flow Determination in Open Channels, Journal of the Eng. Mech. Div. ASCE, EM4, 947-963.

Sivaloganathan K., (Jun 1980) Numerical Computation of Flows in Storm Water Sewers, PhD Thesis, University of London, 53-58.

Sivaloganathan K., (Jul 1979) Channel Flow Computations Using Characteristics, Journal of Hydr. Div. ASCE, HY7, Vol 105, 899-910.

Strelkoff T., (May 1969) One Dimensional Equations of Open Channel Flow, Journal of the Hydr. Div. ASCE, HY3, Vol 95, 861-876.

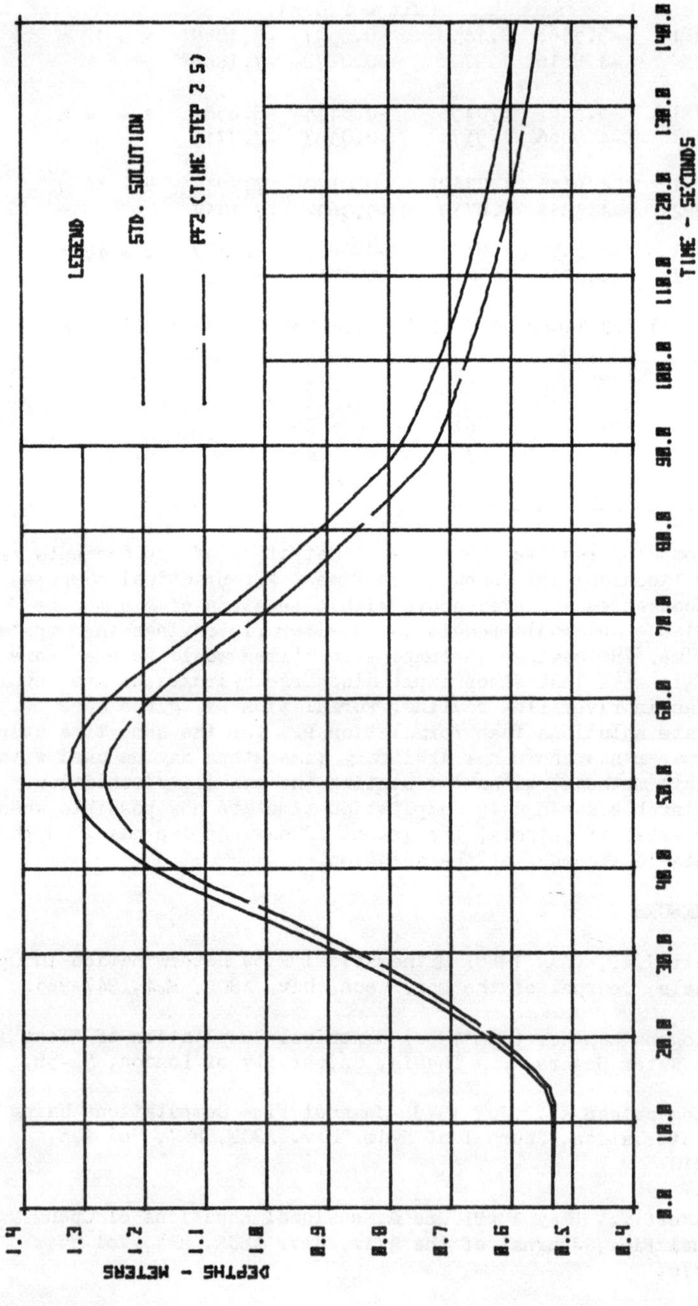


FIG.3 - DEPTH HYDROGRAPHS AT X= 12.0 M

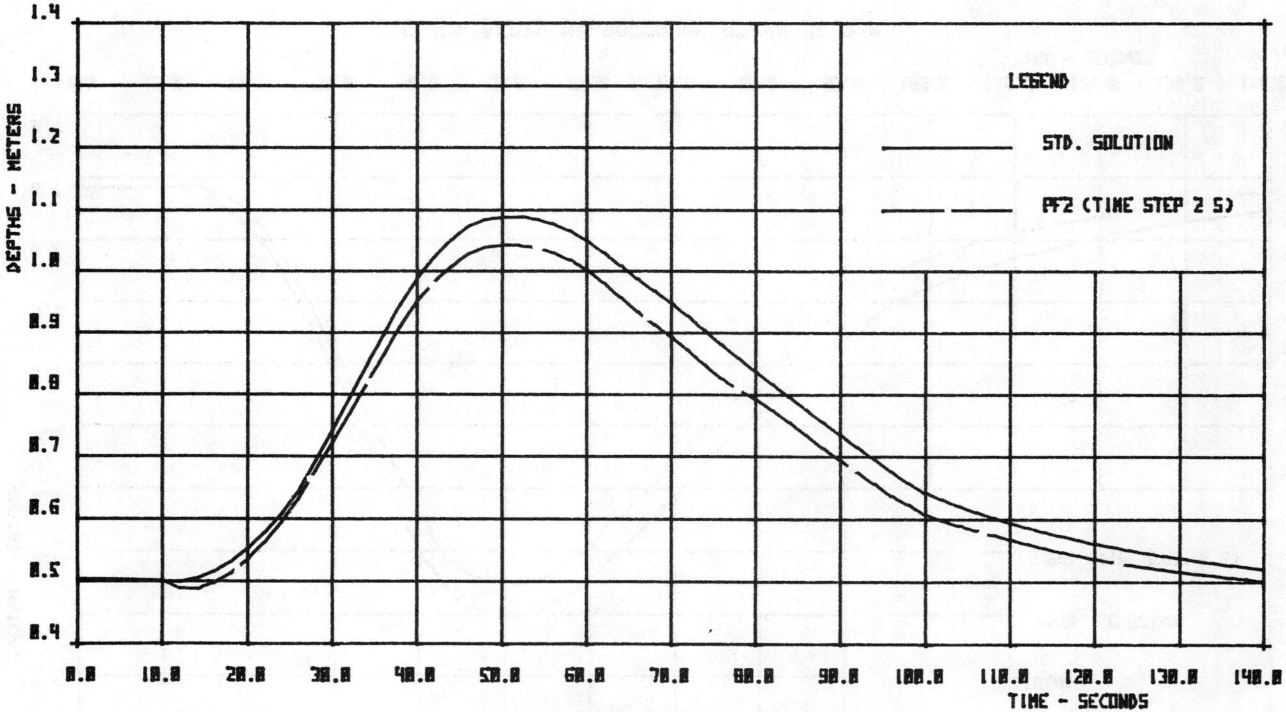


FIG.4 - DEPTH HYDROGRAPHS AT X= 48.0 M

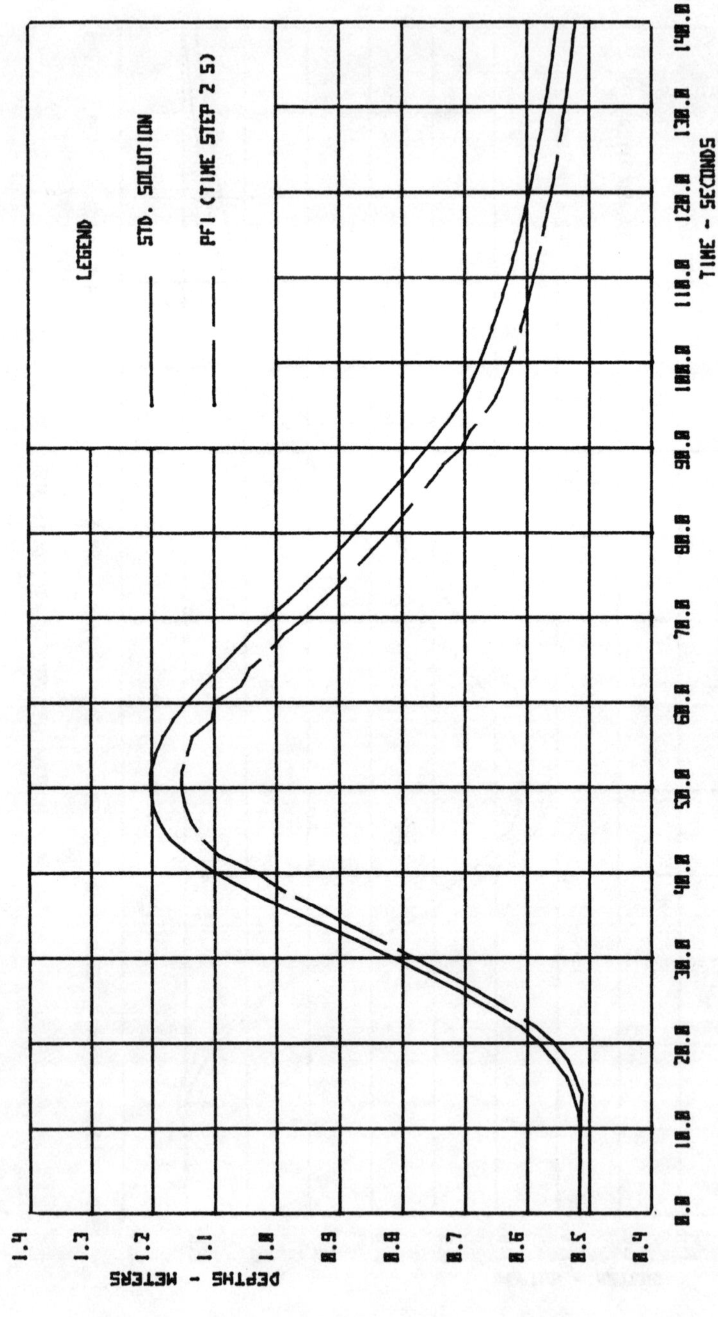


FIG.5 - DEPTH HYDROGRAPHS AT X= 36.0 M

# 5. NUMERICAL TECHNIQUES FOR TIDAL AND OCEAN FLOW

# SOME RECENT DEVELOPMENTS AND PROSPECTS IN FINITE DIFFERENCE METHODS.

M.B. Abbott

International Institute for Hydraulic and Environmental Engineering, Delft, The Netherlands

## 1. INTRODUCTION

By "finite differences" we may understand finite methods for modelling differential equations. In the classical theory of finite differences we seek (strong and generalised) consistency with differential equations,(weak and strong) convergence to solutions of differential equations and stability of the corresponding numerical processes (e.g. Richtmyer and Morton, 1967; Abbott, 1979; Rosinger, 1980) while striving to make optimal trade-offs between investment costs and running costs. This classical structure is currently undergoing an extensive "rethinking" process. The present contribution is intended only to outline some of the more immediately practical developments that have already arisen from this rethinking.

The main themes are the following:
1. The use of higher accuracy (in classical terms, third/ fourth-order) methods,
2. The introduction of de-averaging (de-filtering) processes at the level of the differential equations,
3. The introduction of control components in system frames.

These developments are illustrated by examples.

## 2. THE USE OF METHODS OF HIGH-ORDER ACCURACY

The principle of generating high-order accuracy is simple and, if algorithmic constraints and their centring problems are put to one side, almost trivial. One works out the Taylor-series truncation error generated by any proposed difference scheme with a given centring (usually one that has some good accuracy/ speed characteristics in the first place) and one then modifies the scheme so as to cancel-out the highest-order part of this truncation error. One then calculates the truncation error of the corrected scheme and modifies it again in order to cancel-out further errors, and so on. The realisation is complicated by the wish to maintain those algorithmic structures that provided the original accuracy/speed advantages, so that the subtracting-out becomes correspondingly constrained, and by the need to carry through this programme in the presence of truncation errors that involve a very large number of terms. Thus the main two-dimensional production systems used in hydraulic engineering practice (e.g. those of RAND, DHI, EDF/SOGREAH) all make use of fast ADI algorithms which involve, in principle, between 100 and 200 correction terms, all of which must be cancelled out, one against the other, in order to attain third/fourth-order accuracy.

The systematisation of this process in hydraulics (coastal and offshore engineering, oceanology etc.) has occupied a considerable number of researchers over the last decade. A very readable description of one such systematisation and a review of earlier work is provided by Navon and Riphagen (1979). These authors also give references to studies of boundary conditions applicable to such high-order schemes, but they provide only a restricted stability analysis. The von Neumann stability of the linearised two-dimensional schemes and of the high-order one-dimensional schemes consistent with the Boussinesq equations has been discussed by Abbott, McCowan and Warren (1981). These last authors also provide an analysis of the numerical vorticity generated by such schemes while Leendertse _et al_ (1981) provide an example whereby correction terms are introduced precisely in order to cancel-out the numerical vorticity (and thence estrophy) generation. The principles of this procedure date back, of course, to Arakawa (1966), who indeed treated the problem of conserving any integral invariant of the fluid motion. Since it is known, since Benney (1974), that the number of these invariants is infinite, there is an infinite scope for improvement in this direction. In the real world, _sub species necessitatis_, the choice of integral invariant is made on the basis of the requirements of the application.

A high-order accuracy is needed whenever the truncation errors of the scheme give rise to properties of their solutions that have a significant unrealistic physical interpretation. In order to make this statement less tautological, some examples are required. In the case of the propagation of dispersive wave trains, it was shown by Lighthill and Whitham (1954) that small perturbations in the energy balance of the train would cause considerable differences in the wave lengths and thence in the celerities of the individual waves. This as well appears from a truncation-error analysis of second-order mass and momentum difference equations, where the third order terms are seen to have the same form as and a comparable magnitude to the terms introduced by those vertical accelerations that close the equation system. The truncation error then provides an additional, pseudo-dispersive mechanism. Thus the influence of the third-order error is to change, unacceptably, the wave lengths and celerities observed in the solution. The correction process and its results are described by Abbott, Petersen and Skovgaard (1978). An example of a recent engineering application of these methods is provided in Fig. 1.

A further example arises in simulations of two-phase flows, where errors introduced in the mass and momentum equations are translated, through the energy equation, into spurious evaporations and condensations. A further example again, is found in dynamic structural analysis, when material properties change rapidly through the structure. The truncation errors give rise, locally, to quite unphysical relations between the stress resultants, and these errors then propagate further through the numerically modelled structure.

Some current difficulties in these applications should be described. In hydraulics the problems are increasingly those of shear complexity. An industrial modelling system such as the DHI's S21 HD (HD = Hydrodynamic stage) is already too complicated to be completely understood by any one person. Thus the open-boundary conditions alone can be entered through five structurally-different procedures, so that each of the eight descriptive parameters appearing at a boundary grid point can be constructed in five different ways. The closed boundary conditions could, in principle, have a similar complexity, but in practice (at low Froude numbers) the problems arising at closed boundaries are not usually so severe as those experienced at open boundaries. The problem of extending high-accuracy methods to these boundary conditions, internal as well as external, then becomes one of almost overwhelming complexity and a solution is rather sought by changing the nature of the internal boundary conditions, at least, through the use of curvilinear coordinate systems. This shifts the problem back into the more general one of optimising for accuracy of schemes built upon curvilinear grids, and reopens the problem of stability. These questions are considered by G.S. Rodenhuis, elsewhere in this conference.

Another method for improving the resolution properties of a scheme is to use local refinements of the computational grid. In the case of the S21 HD, this was first done for second-order accuracy schemes (Mark 6-3 version) and later extended to third/fourth order schemes (Mark 8-3 version). Some results obtained with the Mark 6-3 version are shown in Fig. 2. The problems of maintaining accuracy from one scale of resolution to another, without creating spurious (numerical) reflections and refractions, is again one of considerable algebraic complexity.

The problems of hydrodynamics are continued in the problems of superstructures to hydrodynamic schemes, and in particular to the transport-dispersion (TD) superstructures. The S21 TD uses output from the S21 HD to transport and disperse indicators while working interactively with an appropriate water quality stage (so one or another subset of S21 WQ). In the S21 TD, a third - order scheme is used, in the form of a 12 point operator that is selected from out of a 21 point potential operator by the indication of the direction quadrant of the velocity vector (Hinstrup, Kej and Kroszynski, 1977). Here the problems of internal and external boundary conditions do appear to have been "solved enough" for most present applications. However, a recent development is that of "multi-solution" methods, in which more than one solution is generated during a computation. The different solutions are then so linked as to augment the overall accuracy. The augmentation can then be higher than can be obtained by simple refinement of the distance and time discretizations. A recent but already classical example is the scheme of Holly and Preissmann (1977, see also Glass and Rodi, 1981). A quite other approach to multi-solution working has been followed at DHI, based on fundamental research dating from 1976-77 but so far unpublished. This approach is currently being developed for both transport stages and hydrodynamic stages (Mark 9) of the S21.

Behind all this work on higher-order schemes lies the assumption of analytic continuation in Taylor series. However, many problems are hyperbolic in character, so possessing real characteristics, so that analytic continuation cannot be uniformly justified (Abbott, 1979, p.140). A classical example is provided by the propagation of numerical information in a pressurised flow (in a full pipe, under a ship, etc.) where the highest time-level boundary conditions must dominate the solution, implying the use of foreward (and sideways) weighted operators, despite the result from the Taylor series expansion that centred schemes should be the most accurate. Examples of such effects in pressurised flows are given by Abbott et al (1982).Another example of the inadequacy of local analysis is provided by the analysis of convective terms, as described in the next section.

Even when there is no discontinuity in the Taylor series expansion, it has been realised, at least since Hademard (1923), that this expansion must be regarded critically. The Taylor series is equivalent (and even term-by-term equivalent) to an expansion in the Weierstrass basis $\{x^m\}$, which is only one choice of basis, albeit the only one that translates directly into numerical approximations. We read that in recent finite-element work there is a feeding-back of first-order error in order to smooth (=filter) the solutions of second-order schemes (Leonard, 1979). This practice died-out in professional finite-difference modelling following Cunge (1969). However, finite-difference modelling is still not averse to feeding back some third-order error in order to smooth-out fourth-order schemes. The practice may have "moved up-market", but it is still with us.

## 3. THE INTRODUCTION OF FILTERING AND DEFILTERING PROCESSES

The use of finite methods implies the existence of more or less defined limits of resolution. One method of introducing these limits while keeping the structure of a differential equation, sub species aeternitatis, is to filter that differential equation (or some parts of such an equation) which would be provided by an infinite capacity to resolve. Thus the integral form of the differential transport equation

$$\frac{\partial f}{\partial t} + u\,\frac{\partial f}{\partial x} = 0 \tag{1}$$

is

$$u(x,t_2) = u(x-\bar{u}(t_2-t_1),t_1) \tag{2}$$

where $\bar{u}$ is the filtered velocity given by

$$\bar{u} = \frac{1}{t_2-t_1}\int_{t_1}^{t_2} u\,dt$$

Expanding (2) in Taylor series provides the differential equation

$$\frac{\partial f}{\partial t} + \bar{u}\,\frac{\partial f}{\partial x} + \frac{\partial}{\partial t}\left(\frac{\partial f}{\partial t} + \bar{u}\,\frac{\partial f}{\partial x}\right)\frac{\Delta t}{2} + \text{H.O.T.} = 0 \tag{3}$$

The terms in second and higher derivatives are then correcting for the introduction of the filtered velocity $\bar{u}$ in place of the unfiltered velocity u. They thus describe a "defiltering process". Equations (3) reduced to equation (1) when $\Delta t \to 0$, so when the filter function converges to a $\delta$-function, which is the neutral element of the convolution operator and corresponds to an infinite capacity to resolve. Equations of the type of (3) were discovered by C.H. Rassmusen (DHI unpublished report) in 1977.

Defiltering has become popular for subgrid scale modelling, where the Navier-Stokes equations

$$\frac{\partial u_i}{\partial t} + \frac{\partial u_i u_j}{\partial x_j} = -\frac{1}{\rho}\frac{\partial p}{\partial x_i} - \nu\nabla^2 u_i \tag{4}$$

are filtered to provide

$$\frac{\partial \bar{u}_i}{\partial t} + \frac{\partial \overline{u_i u_j}}{\partial x_j} = -\frac{1}{\rho}\frac{\partial \bar{p}}{\partial x_i} - \nu\nabla^2 \bar{u}_i \tag{5}$$

Then

$$\overline{u_i u_j} = \overline{\bar{u}_i \bar{u}_j} + \overline{\bar{u}_i u'_j} + \overline{u'_i \bar{u}_j} + \overline{u'_i u'_j}$$

and, by Taylor series expansion,

$$\overline{\bar{u}_i \bar{u}_j} = \bar{u}_i \bar{u}_j + \frac{\Delta^2}{4\gamma} \frac{\partial^2(\bar{u}_i \bar{u}_j)}{\partial x_k \partial x_k} + \text{H.O.T.} \tag{6}$$

$$\bar{u}_i u'_j = - \bar{u}_i \frac{\Delta^2}{4\gamma} \frac{\partial^2 u_j}{\partial x_k \partial x_k} + \text{H.O.T.} \tag{7}$$

$$\overline{u'_i u'_j} = \frac{\Delta^2}{96\gamma} \left( \frac{\partial^2 u_i}{\partial x_k \partial x_k} \cdot \frac{\partial^2 u_j}{\partial x_l \partial x_l} \right) + \text{H.O.T.} \tag{8}$$

where $\gamma$ is the shape parameter of the filter and $\Delta$ is its effective width ($\approx$ 2 x the grid size). Equation (6) was proposed by Leonard in 1974, equation (7) by Clark in 1977 and equation (8) by Abbott, McCowan and Warren in 1981. As it is necessary to replace u by $\bar{u}$ as approximations in (7) and (8), the accuracy of these expressions is uncertain. However, they also illustrate how the introduction of filtering, or averaging, leads to de-filtering, or de-averaging, terms additional to the original differential equation.

It has been observed that the introduction of such terms is necessary in order to obtain spectral consistency in large-eddy simulations, while the use of averaged coefficients in difference schemes is also found to lead to difficulties unless such corrections are introduced. (However, the problems of spectral preservation is still scarcely solved by these devices: see Bedford, 1981).

## 4. THE INTRODUCTION OF CONTROL COMPONENTS IN SYSTEM FRAMES

The level of resolution required at any stage in a computation often depends upon the results of the computation, so that it changes during the course of the computation. One attempt to solve this problem, through the use of differencing in the independent variables, to provide "inverse" schemes, has been described by Abbott and Vium (1977. See also Abbott et al, 1982B). Another approach is to use a grid that is refined automatically as the need for additional resolution is expressed through the solution. This approach is most easily realised in the time coordinate by setting the time step to vary as the rate of variation of the solution. Thus, when variations occur slowly, a larger time step is used, while when these variations occur more rapidly a shorter time step is introduced. This facility has been incorporated into the controlling "frame" of the SHE system (Système Hydrologique

Européen). In this case,very short time steps (a few minutes) are automatically selected in order to resolve canopy and upper soil processes occuring during an intense storm, while, at the other extreme, time steps of the order of a day are selected in order to describe accurately long periods with little hydrological activity. Even as the time step of processes occuring at the soil surface level are reduced,the time steps used for simulations at greater depths (saturated-zone processes) are still maintained at an appropriate scale. The matching of data-exchanges between components through the varying time steps is one of the major functions of the frame codes. The technique used in the SHE is schematized in Fig. 3.

## 5. SOME CONCLUDING REMARKS

The examples selected here illustrate the beginnings of some radical changes in the practise of modelling with difference methods. The very accomplishments of this modelling provide the impetus for a thorough re-evaluation of its foundations and provide, at the same time, the material resources to realise this re-evaluation. It is remarked of work in this area that each new development necessitates an increasingly heavy investment, often carried back into fundamental-research areas. The cost of the "entry ticket" into the different areas of professional mathematical modelling continues to increase correspondingly. However, the next generation of modelling systems in hydraulics, for example, appear to be unrealisable without drastic re-formulations of hydrodynamic theory and the cost of this re-formulation must be associated with the cost of these new systems.

## 6. REFERENCES

Abbott, M.B. (1979) Computational Hydraulics, Pitman, London.

Abbott, M.B., Havnø, K., Hoff-Clausen, N.E. and Kej, A. (1982). A Modelling System For the Design and Operation of Storm Sewer Networks. In Abbott, M.B. and Cunge, J.A. (Eds.) Engineering Applications of Computational Hydraulics: Homage to Alexandre Preissmann, Vol. I. Pitman, London.

Abbott, M.B. Kastberg Andersen, J., Havnø, K, Høgh Jensen, K., Kroszynski, U.I., and Warren, I.R. (1982) Research and Development for the Unsaturated Zone Component of the European Hydrologic System - Système Hydrologique Européen (SHE). In Abbott, M.B. and Cunge, J.A. (Eds.) Engineering Applications of Computational Hydraulics, Homage to Alexandre Preissmann, Vol. I., Pitman, London.

Abbott, M.B., McCowan, A. and Warren, I.R. (1981). Numerical Modelling of Free-Surface Flows that are Two-Dimensional in Plan. In Fischer, H.B. (Ed.) Transport Models for Inland and Coastal Waters, Academic Press, New York.

Abbott, M.B., Petersen, H.M. and Skovgaard, O. (1978). On the Numerical Modelling of Short Waves in Shallow Water, J. Hyd. Res. 16 pp. 173-203.

Arakawa, A. (1966). Computational Design for Long-Term Numerical Integration of the Equations of Fluid Motions etc. J. Comp. Phys. 1, pp. 119-143.

Benny, D.J. (1974). Long Waves, Lectures in Applied Math. 15. Am. Math. Soc. pp. 49-60.

Bedford, K.W. (1981). Spectral Preservation Capabilities of Great Lakes Transport Models, Fischer, H.B. (Ed.) Transport Models for Inland and Coastal Waters, Academic Press, New York.

Clark, R.A., Ferziger J.H. and Reynolds, W.C. (1977). Evaluation of Subgrid Scale Turbulence Models etc., Dept. of Mech. Eng'g, Stanford Univ. Rep. No. TF9.

Cunge, J.A. (1969). Au Sujet d'une Méthode de Calcul de Propagation de Crues (Méthod Muskingum). J. Hyd. Res. 7, 2, pp. 205-229.

Hadamard, J. (1923). Lectures on Cauchy's Problem in Linear Partial Differential Equations, Dover (reprinted 1952), New York.

Hinstrup, P.I., Kej, A and Kroszynski, U. (1977). A High-Accuracy Two-Dimensional Transport-Dispersion Model for Environmental Applications, Proc. 17th Congr. Int. Assoc. Hydraulic Res. B17 pp. 129-137.

Holly, F.M. and Preissmann A. (1977). Accurate Calculation of Transport in Two Dimensions, JHYD ASCE 103 No. HY11.

Leendertse, J.J., Langerak, A. and de Ras, M.A.M. (1981). Two-Dimensional Tidal Models of the Delta Works, in Fischer H.B. (Ed.) Transport Models for Inland and Coastal Waters, Academic Press, New York.

Leonard, A. (1974) Energy Cascade in Large-Eddy Simulations of Turbulent Fluid Flows, Adv. in Geophys. A 18, pp. 237-248.

Leonard, B.P. (1979). A Survey of Finite Differences of Opinion on Numerical Muddling of the Incomprehensible Defective-Confusion Equation. Proc. Am. Soc. Mech. Eng's Publ. No. AMD-34.

Lighthill, M.J. and Whitham, G.B. (1955). On kinetic Waves, I. Flood Waves in Rivers, Proc. Roy. Soc. London A 229, pp. 281-316.

Navone, I.M. and Riphagen, H.A. (1979). Compact Fourth-Order Algorithm for Solving the Shallow-Water Equations in Conservation

Form. Monthly Weather Review, 107, 9, pp. 1107-1127.

Richtmyer, R.D. and Morton, K.W. (1967). Difference Methods for Initial Value Problems, Interscience, New York.

Rossinger, E.E. (1980). Non-linear Partial Differential Equations, North Holland, Amsterdam.

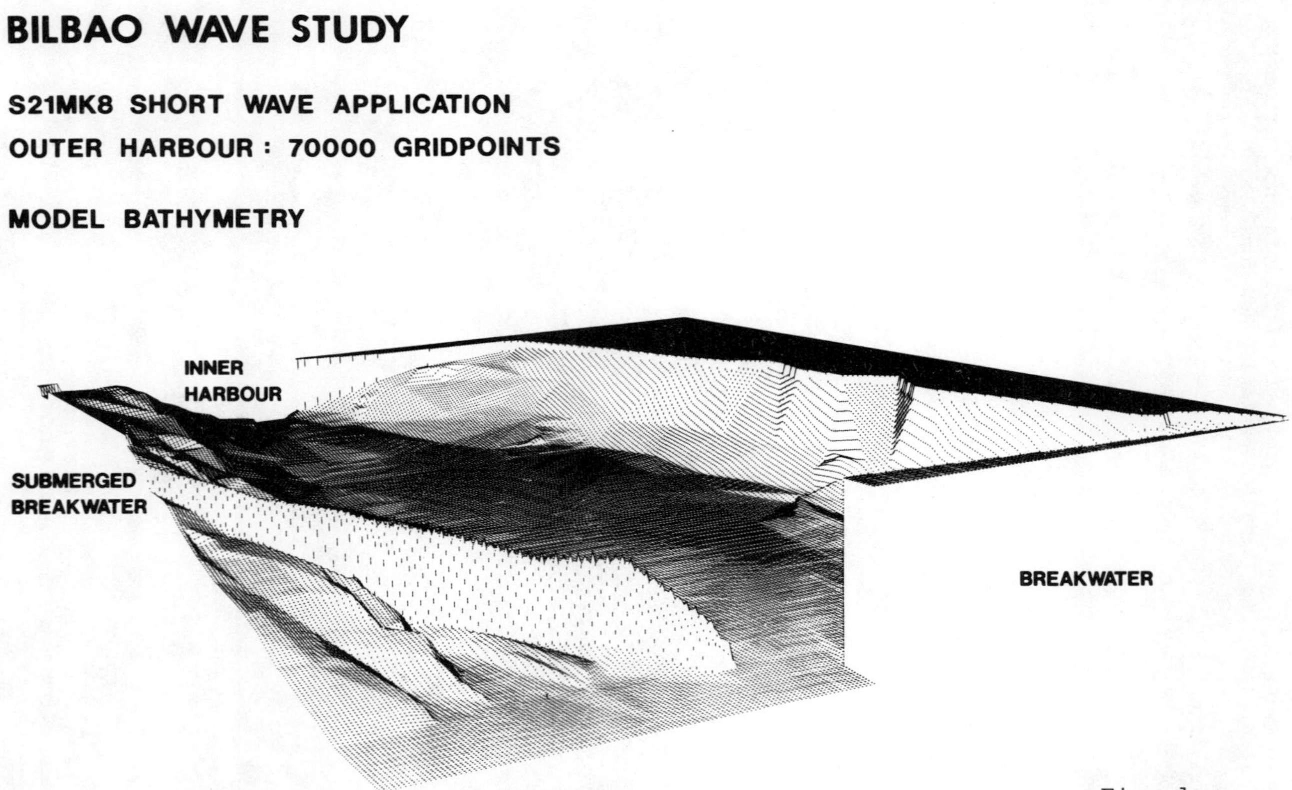
BILBAO WAVE STUDY
S21MK8 SHORT WAVE APPLICATION
OUTER HARBOUR : 70000 GRIDPOINTS
MODEL BATHYMETRY
INNER HARBOUR
SUBMERGED BREAKWATER
BREAKWATER
BAY OF BISCAY

Fig. 1 a

BILBAO WAVE STUDY

S21MK8 SHORT WAVE APPLICATION

OUTER HARBOUR : 70000 GRIDPOINTS

IRREGULAR UNIDIRECTIONAL WAVE INPUT

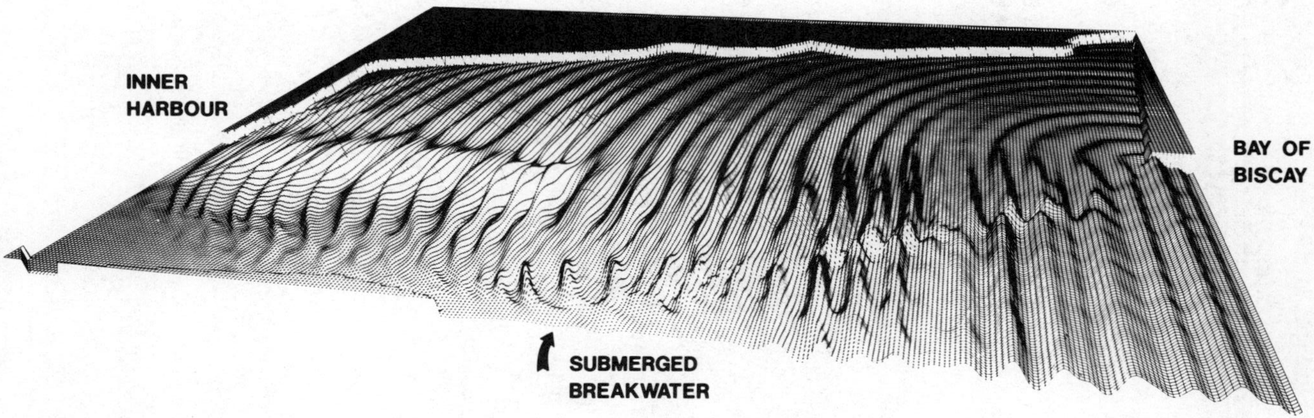


Fig. 1 b

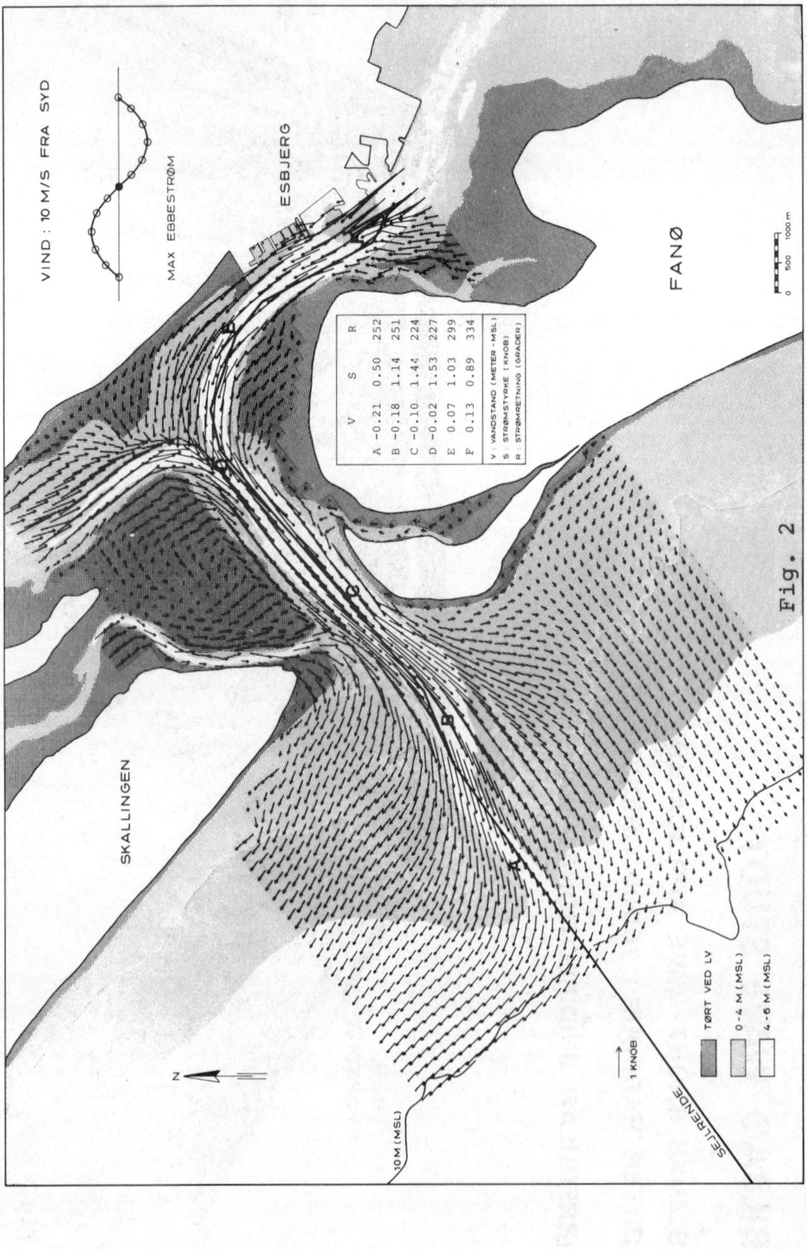
VIND : 10 M/S FRA SYD
MAX EBBESTRØM
ESBJERG
FANØ
SKALLINGEN
V S R
A -0.21 0.50 252
B -0.18 1.14 251
C -0.10 1.44 224
D -0.02 1.53 227
E 0.07 1.03 299
F 0.13 0.89 334
V : VANDSTAND (METER - MSL)
S : STRØMSTYRKE (KNOB)
R : STRØMRETNING (GRADER)
0 500 1000 m
N
1 KNOB
TØRT VED LV
0-4 M (MSL)
4-6 M (MSL)
10 M (MSL)
SEJLRENDE

Fig. 2

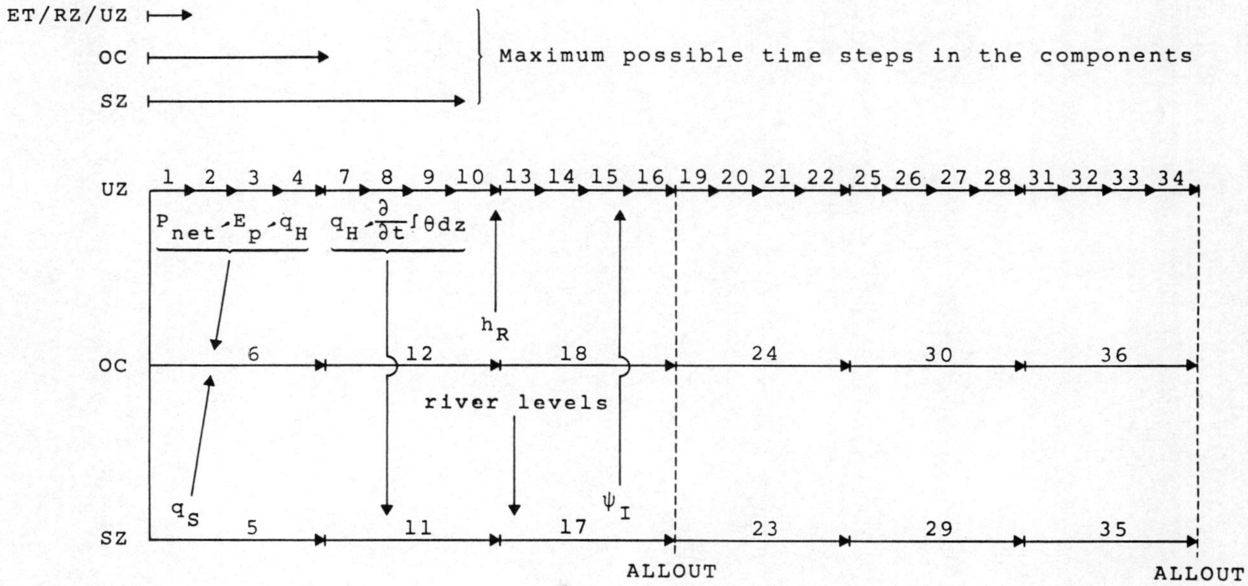


Fig. 3 Example of the sequence of calculation created in SHE by the Chef d'Orchestre also showing the transfer of data between components.

# A TIME INTEGRATION TECHNIQUE FOR MODELLING OF SMALL AMPLITUDE TIDAL WAVES

V. Argintaru-Hydraulics Lab. National Research Council, Canada
L. Spraggs - McGill University, Montreal Canada

## INTRODUCTION

The development of new techniques for modelling of tidal estuaries, such as hybrid modelling (the real time coupling between numerical and physical models) has redefined the criteria for efficient numerical models. Whereas traditionally a numerical tidal model was judged on the basis of cost per computer run, the requirement for a hybrid model now becomes computational time per physical time step. [Funke 79][Holtz 77]

In general, either implicit or explicit discretizations were chosen to represent the time domain, the choice always being made between the inherent stability of implicit schemes versus the ease of computation of explicit schemes [Rahman'80]. The time step size imposed by an explicit formulation, however, is too small for hybrid modelling applications, while the long computational times of implicit formulations render difficult the task of real time synchronization with the physical model.

The problem to be modelled was a tidal estuary subjected to highly nonlinear flow. The mathematical model would be used to simulate the tidal action up to the complex geometries of the estuary and a physical model would then be used.

This present work was primarily concerned with determining the feasibility of using a finite element model to interface with the physical model. The object was to find or develop a stable implicit scheme which approached the computational speed of the explicit formulation.

## MATHEMATICAL MODEL

The two dimensional, vertically integrated equations for propagation of long waves in shallow water were discretized to give the following matrix equation [Argintaru,1980]

$$[M]\{\dot{\phi}\} + [K(\phi)]\{\phi\} = \{B\} \tag{1}$$

where:

$$[M] = \sum_{e=1}^{L} \int_{A_e} dA \begin{bmatrix} [N]^T[N] & [0] & [0] \\ [0] & [N]^T[N] & [0] \\ [0] & [0] & [N]^T[N] \end{bmatrix} \tag{2}$$

$$[K] = \sum_{e=1}^{L} \int_{A_e} dA \bullet$$

$$\begin{bmatrix} [0] & -\frac{\partial [N]^T}{\partial x}[N] & -\frac{\partial [N]^T}{\partial y}[N] \\ g[N](h+\eta)[N]^T\frac{\partial [N]}{\partial x} & \frac{g}{C^2}[N]^T[N]\left(\frac{(q_x{}^2+q_y{}^2)^{\frac{1}{2}}}{(h+\eta)^2}\right) & -\Omega[N]^T[N] \\ g[N](h+\eta)[N]^T\frac{\partial [N]}{\partial y} & \Omega[N]^T[N] & \frac{g}{C^2}[N]^T[N]\left(\frac{(q_x{}^2+q_y{}^2)^{\frac{1}{2}}}{(h+\eta)^2}\right) \end{bmatrix} \tag{3}$$

$$\{B\} = \sum_{e=1}^{L} \begin{Bmatrix} \{0\} \\ \{0\} \\ \{0\} \end{Bmatrix} \tag{4}$$

and $\{\phi\}$ is the vector of nodal variables:

$$\{\phi\} = \sum_{e=1}^{L} \begin{Bmatrix} \eta \\ q_x \\ q_y \end{Bmatrix} \tag{5}$$

where the following notation was used:

- $\eta$ = water level above mean sea level
- $h$ = water depth below sea level
- $q_x, q_y$ = flow components per unit width in the horizontal cartezian plane (vertically averaged)
- $C$ = empirical Chezy coefficient
- $\Omega$ = coriolis acceleration
- $g$ = acceleration of gravity
- $N$ = element shape functions
- $A$ = elemental area
- $L$ = number of elements in the domain

As the objective was to develop a computationally expediant scheme, linear triangular elements were considered to be sufficiently accurate, thus allowing the required synchronization with the physical model.

## TIME STEPPING SCHEME AND EQUATION LINEARIZATION

The nonlinear character of equation 1 is due to the time varying terms in the coefficient matrix $[K(\phi)]$. Therefore some means of handling the nonlinearity must be found and this is usually achieved together with the discretization of the time domain.

It is customary to treat the time integration via finite difference approximations,a more economical approach to approximating equal sized time steps than the use of finite elements in time. The discretization is performed by letting:

$$\{\phi\} = \frac{\{\phi\}^{t+\Delta t} - \{\phi\}^{t}}{\Delta t} \tag{6}$$

where t and t + $\Delta t$ represent the known and the unkown time levels respectively and $\Delta t$ is the magnitude of the numerical time step.

If $[K(\phi)]\{\phi\}$ in equation 1 is taken at the known time level the problem can be formulated explicitly:

$$[M]\{\phi\}^{t+\Delta t} - [M]\{\phi\}^{t} + \Delta t[K(\phi)]^{t}\{\phi\}^{t} = 0 \tag{7}$$

or:

$$\{\phi\}^{t+\Delta t} = [M]^{-1}[[M]-\Delta t[k(\phi)]^{t}]\{\phi\}^{t} \tag{8}$$

The explicit formulation requires minimum computational effort, however, experimentation with this model showed the method to remain stable only for smaller time steps than would be required by a real time application.

Alternatively, taking $[K(\phi)]\{\phi\}$ at the unknown time level yields the following fully implicit scheme:

$$[M]\{\phi\}^{t+t\Delta} - [M]\{\phi\}^{t} + \Delta t[K(\phi)]^{t+\Delta t}\{\phi\}^{t+\Delta t} = 0 \tag{9}$$

In spite of its strong stability characteristics, the implicit scheme presents the disadvantage that a solution of the implicit system of equations is required. This becomes particularly costly for unsteady problems.

A different approach was adopted here by isolating the constant terms of [K] into a separate matrix $[K_1]$ and the time varying terms into $[K_2(\phi)]$ where:

$$[K_1] = \sum_{e=1}^{L} \int_{A_e} \begin{bmatrix} [0] & -\frac{\partial [N]^T}{\partial x}[N] & -\frac{\partial [N]^T}{\partial y}[N] \\ g[N](h)[N]^T\frac{\partial [N]}{\partial x} & [0] & -\Omega[N]^T[N] \\ g[N](h)[N]^T\frac{\partial [N]}{\partial y} & \Omega[N]^T[N] & [0] \end{bmatrix} \quad (10)$$

and

$$[K_2] = \sum_{e=1}^{L} \int_{A_e} \cdot \quad (11)$$

$$\begin{bmatrix} [0] & [0] & [0] \\ g[N](\eta)[N]^T\frac{\partial [N]}{\partial x} & \frac{g}{C^2}[N]^T[N](\frac{(q_x{}^2+q_y{}^2)^{\frac{1}{2}}}{(h+\eta)^2}) & [0] \\ g[N](\eta)[N]^T\frac{\partial [N]}{\partial y} & [0] & \frac{g}{C^2}[N]^T[N](\frac{(q_x{}^2+q_y{}^2)^{\frac{1}{2}}}{(h+\eta)^2}) \end{bmatrix}$$

The implicit formulation (Equation 9 ) now becomes:

$$[M[\{\phi\}^{t+\Delta t} - [M]\{\phi\}^t + \Delta t[K_1]\{\phi\}^{t+\Delta t} + \Delta t[K_2(\phi)]^{t+\Delta t}\{\phi\}^{t+\Delta t} = 0 \quad (12)$$

Since the terms of $[K_2(\phi)]$ (proportional to $\eta$) are small in comparison with the terms of $[K_1]$ (proportional to h). the time varying contribution $[K_2(\phi)]\{\phi\}$ is now treated explicitly:

$$[M]\{\phi\}^{t+\Delta t} - [M]\{\phi\}^t + \Delta t[K_1]\{\phi\}^{t+\Delta t} + \Delta t[K_2(\phi)]^t\{\phi\}^t = 0 \quad (13)$$

or:

$$[[M] + \Delta t[K_1]]\{\phi\}^{t+\Delta t} = [[M] - \Delta t[K_2(\phi)]^t\{\phi\}^t \quad (14)$$

and since the coefficient matrix of $\{\phi\}^{t+\Delta t}$ is now constant $\{\phi\}^{t+\Delta t}$ can be explicitly expressed in terms of $\{\phi\}^t$:

$$\{\phi\}^{t+\Delta t} = [[M] + \Delta t[K_1]]^{-1}[[M] - \Delta t[K_2\{\phi\}]^t\{\phi\}^t \quad (15)$$

The idea of splitting the coefficinet matrix, while treating the constant contribution implicitly and the time-varying contribution explicitly is not new [King, 1973][Argintaru,1980]. In fact, for general flow problems this method was found to have relatively poor stability [King,1973]. For problems of tidal flow however, where $\eta/h$ is small and hence the terms of $[K_2]$ are much smaller than those of $[K_1]$, the strong stability of the fully implicit scheme is preserved (15 was obtained as an extension from 9), while the computing times are reduced to those of explicit formulations.

## MODEL VERIFICATION

### Comparison with the Implicit Scheme

In order to verify the validity of the above scheme, a simple regular grid was defined (see Figure 1), 40 km on each side with a water depth of 10 m and constant throughout. A uniform tide of 1 m amplitude and 12 hrs. period was specified at the open boundary.

A comparison between the scheme developed here and the fully implicit scheme is shown in Figure 2. All the results represent relative differences between the closed end of the model and the open boundary. Figure 2.a compares the amplitude variations for different values of the numerical time step, while Figure 2.b compares the phase lags. Good agreement is obtained between the two methods, particularly in the range where a reasonable time step size is used, the error introduced at larger time steps representing an inherent feature of implicit schemes when full advantage of their stability is taken [Argintaru,1980]. On the other hand the computing time of the above method was 1/30 that of the implicit method.

### Comparison with Theory

Furthermore, the above method was compared to the analytical solution given by Ippen(1966):

$$\frac{\eta_o}{\eta_{o_b}} = \frac{1}{\cos \frac{\pi \ell}{\lambda}} \tag{16}$$

where:

$\eta_o$ = tidal amplitude at the closed end

$\eta_{o_b}$ = tidal amplitude at the open boundary

$\ell$ = length of the channel

$\lambda$ = wavelength

Figure 3 illustrates the comparison for varying channel lengths. Excellent agreement is obtained away from a resonant solution. (near resonance, higher harmonics are generated by the numerical scheme due to the presence of the nonlinear terms).

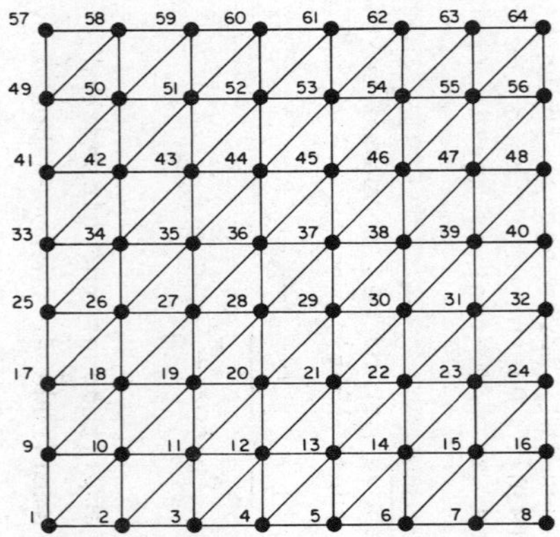


**FIG.1 SIMPLE MODEL - DOMAIN DISCRETIZATION**

**NODES 1 TO 8 REPRESENT THE OPEN-SEA BOUNDARY**

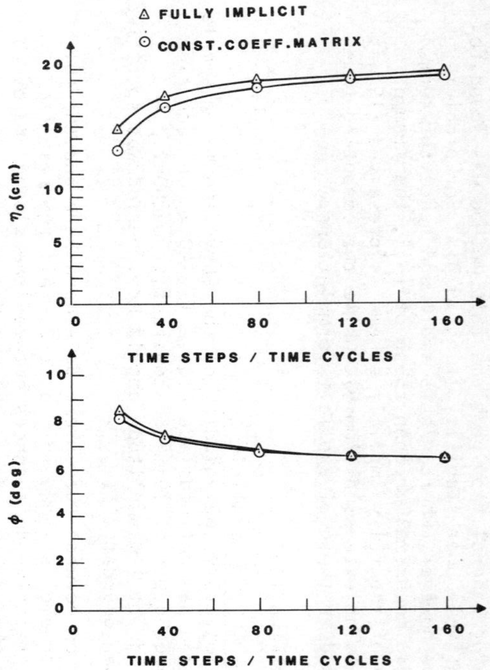


**FIG.2 AMPLITUDE AND PHASE AT THE CLOSED END OF THE MODEL, COMPARISON BETWEEN THE TWO METHODS**

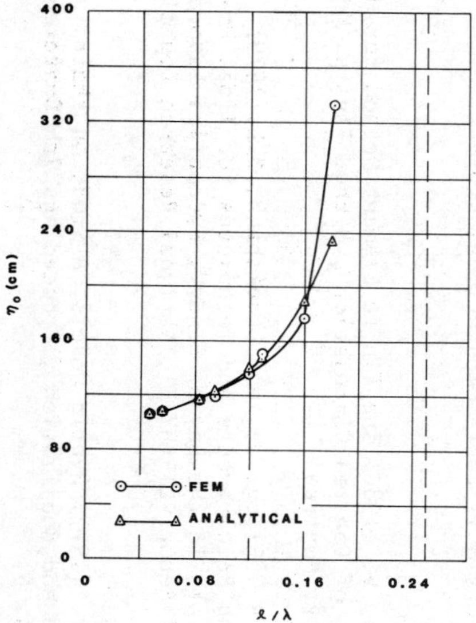


FIG.3 PREDICTED AMPLITUDE FOR VARYING CHANNEL LENGTH TO WAVELENGTH RATIOS. THE BROKEN LINE REPRESENTS RESONANCE

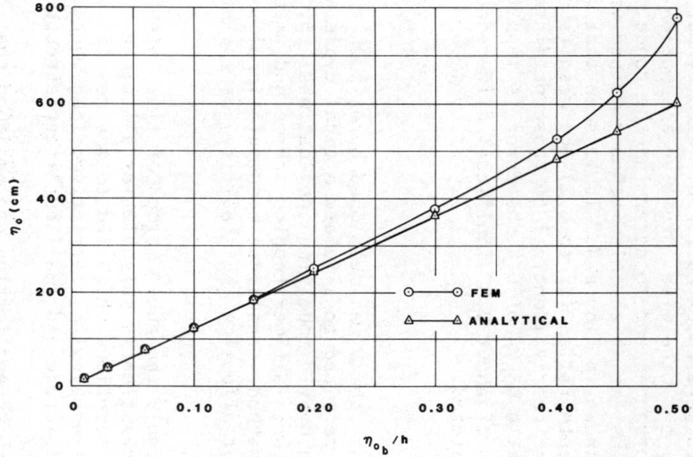


FIG.4 PREDICTED AMPLITUDE FOR VARYING OPEN BOUNDARY AMPLITUDE TO WATER DEPTH RATIOS

Next, the results were compared for different input tides, in order to verify the otherwise intuitive assumption of small amplitude disturbances and to determine the range of amplitude over depth ratios for which this method remained valid. The comparison is shown in Figure 4. It is seen that for small amplitude to depth ratios (up to 0.2) the results obtained with the FEM are nearly identical to those predicted analytically. Even at a $\eta_{ob}/h$ of 0.4 the error is less than 10%. Only at a $\eta_{ob}/h$ of about 0.5 significant differences (about 30%) are encountered.

## CONCLUSIONS

A finite element scheme was developed here, such that the execution time required to simulate a tidal cycle would be compatible with hybrid modelling applications. By a suitable combination between a time-stepping scheme and an equation linearization method, considerable computational savings were achieved without affecting the stability, or the accuracy of the solution.

This does not imply the above method of time integration and equation linearization is suitable for all fluid problems. In fact, the assumptions which led to the results of the numerical tests were quite restrictive; long waves, shallow water, small amplitude and small nonlinear effects.

However, if such an application can be found that this method can remain stable and reproduce particular flows with relative accuracy, the inherent efficiency of solution of the equations should lead to significant computational savings.

## REFERENCES

1. Argintaru, V. (1980)"The Finite Element Method for Hybrid Modelling of Coastal Circulation", M.Eng. Thesis, McGill University, Montreal.

2. Funke, E.R., Ploeg, J., and Crookshank, N.L.(1979) "Modelling Requirements and Techniques for Tidal Power Developments in the Bay of Fundy", Hydraulics Laboratory Technical Report LTR-HY-71, National Research Council Canada.

3. Holtz, K.P. (1977)"Hybrid Models, A Study on their Principle and Realization", Proceedings 7th Conference IAHR, Vol. 6.

4. Ippen, A.T. (1966)" Estuary and Coastline Hydrodynamics", McGraw Hill

5. King, I.P., Norton, W.R. and Orlob, G.T. (1973) " A Finite Element Solution for Two-Dimensional Density Stratified Flow", Water Resources Engineers Inc.

6. Rahman, M., Prandle, D., Spraggs, L., and Argintaru,V., (1980) "Numerical Methods for Tidal Propogation Applied to a Hybrid Model of the Bay of Fundy", Reprints of Papers from the 3rd International Conference on Finite Elements in Water Resources, Oxford, Miss.

# FORMULATING A THREE DIMENSIONAL HYDRODYNAMIC SEA MODEL USING A MIXED GALERKIN-FINITE DIFFERENCE METHOD.

A.M. Davies

I.O.S., Bidston Observatory, Birkenhead, Merseyside, U.K.

## INTRODUCTION

The classic method of solving the two dimensional hydrodynamic equations in the horizontal, which describe flow in a sea area, has been to use a rectangular finite difference grid. Recently this method has been supplemented by the finite element technique in which linear or curved two-dimensional elements are used in the horizontal space domain.

The numerical solution of the three-dimensional equations has primarily been developed using a three-dimensional fixed grid. (For a review of 3D models, see Cheng et al (1976)). The use of three-dimensional elements is also possible (for a review of finite elements see Fix (1975)). A mixed analytical numerical approach was proposed by Heaps (1972).

An alternative to the use of a three-dimensional grid, or three-dimensional finite elements is the method formulated by Davies (1977), in which a conventional finite difference grid is used in the horizontal and the Galerkin method with a basis set of piecewise functions (a finite element method) or continuous functions (a spectral method) is employed in the vertical.

In the first section of this paper a mixed Galerkin-finite difference method is developed for the solution of the three-dimensional equations which described flow in a homogeneous sea. A brief review of the various types of basis function which have been used is given. With certain types of basis function the equations describing the mean flow (the external mode) can be separated from the equations describing the deviation from the mean flow (the internal mode). This deviation is associated with shears due to the external stresses applied at surface and bottom. In this case the equations can be economically integrated using a time splitting algorithm in which the two flows are integrated with different time steps.

This paper shows that the separation of the solution into

an external mode and internal modes can also be accomplished when an arbitrary basis set is used.

In the final section of the paper the application of the Galerkin method to stratified seas is briefly described. Again a time splitting method can be used and some results for an idealized sea are presented.

HYDRODYNAMIC EQUATIONS

The linear three-dimensional hydrodynamic equations which described flow in a homogeneous sea, using Cartesian Co-ordinates are given by,

$$\frac{\partial \zeta}{\partial t} = -\frac{\partial}{\partial x}\int_0^h U dz - \frac{\partial}{\partial y}\int_0^h V dz \tag{1}$$

$$\frac{\partial U}{\partial t} = \gamma V - g\frac{\partial \zeta}{\partial x} + \frac{\partial}{\partial z}\left(\mu \frac{\partial U}{\partial z}\right) \tag{2}$$

$$\frac{\partial V}{\partial t} = -\gamma U - g\frac{\partial \zeta}{\partial y} + \frac{\partial}{\partial z}\left(\mu \frac{\partial V}{\partial z}\right) \tag{3}$$

where in these equations : t denotes time, x, y, z a left handed set of co-ordinates, with z the depth below the undisturbed surface, h the undisturbed depth, of water, $\zeta$ elevation, U, V the x and y components of current, $\gamma$ geostrophic coefficient, g acceleration due to gravity and $\mu$ vertical eddy viscosity.

In order to solve equations (1) to (3), sea surface and sea bed boundary conditions have to be specified. Surface boundary conditions evaluated at z = 0 are

$$-\rho\left(\mu \frac{\partial U}{\partial z}\right)_0 = F_S \;, \quad -\rho\left(\mu \frac{\partial V}{\partial z}\right)_0 = G_S \tag{4}$$

where $F_s$, $G_s$ denote components of wind stress acting on the water surface in the x and y directions; $\rho$, density of sea water assumed constant.

Similarly at the sea bed

$$-\rho\left(\mu \frac{\partial U}{\partial z}\right)_h = F_B \;, \quad -\rho\left(\mu \frac{\partial V}{\partial z}\right)_h = G_B \tag{5}$$

where $F_B$, $G_B$ denote the components of bottom friction.

Using a linear slip condition at the sea bed :

$$F_B = k\rho U_h \;, \quad G_B = k\rho V_h \tag{6}$$

where k is the coefficient of bottom friction.

Before applying the Galerkin method to the solution of

equations (1) to (3), it is convenient to transform them from the interval $0 \leq z \leq h$ which varies with horizontal positions into a depth-following set of co-ordinates from 0 to 1, namely sigma co-ordinates, using

$$\sigma = z/h \tag{7}$$

Transforming equations (1), (2) and (3) using (7) gives,

$$\frac{\partial \zeta}{\partial t} = -\frac{\partial}{\partial x}\left\{h\int_0^1 u\,d\sigma\right\} - \frac{\partial}{\partial y}\left\{h\int_0^1 v\,d\sigma\right\} \tag{8}$$

$$\frac{\partial u}{\partial t} = \gamma v - g\frac{\partial \zeta}{\partial x} + \frac{1}{h^2}\frac{\partial}{\partial \sigma}\left(\mu\frac{\partial u}{\partial \sigma}\right) \tag{9}$$

$$\frac{\partial v}{\partial t} = -\gamma u - g\frac{\partial \zeta}{\partial y} + \frac{1}{h^2}\frac{\partial}{\partial \sigma}\left(\mu\frac{\partial v}{\partial \sigma}\right) \tag{10}$$

Boundary conditions (4) and (5) are transformed into sigma co-ordinates in an analogous manner.

We now consider the solution of equations (8), (9) and (10) using the Galerkin method. Expanding the two components of velocity in terms of m depth-dependent functions $f_r(\sigma)$ (the basis functions) and coefficients $A_r(x,y,t)$ and $B_r(x,y,t)$, gives

$$u = \sum_{r=1}^{m} A_r f_r \tag{11}$$

$$v = \sum_{r=1}^{m} B_r f_r \tag{12}$$

Substituting (11) and (12) into (8) gives

$$\frac{\partial \zeta}{\partial t} = -\sum_{r=1}^{m}\left[\left(\frac{\partial}{\partial x}(A_r h) + \frac{\partial}{\partial y}(B_r h)\right)\cdot\int_0^1 f_r\,d\sigma\right] \tag{13}$$

Applying the Galerkin method to equation (9), we multiply it by $f_k$ and integrate with respect to $\sigma$. By integrating the term involving the eddy viscosity the surface and bottom boundary conditions can be included, (see Davies and Owen 1979 for details). Also, using expansions (11) and (12) we obtain,

$$\sum_{r=1}^{m} \frac{\partial A_r}{\partial t} \int_0^1 f_r f_k \, d\sigma = \gamma \sum_{r=1}^{m} B_r \int_0^1 f_r f_k \, d\sigma - g \frac{\partial \zeta}{\partial x} \int_0^1 f_k \, d\sigma$$

$$- \frac{F_B}{\rho h} f_k(1) + \frac{F_s}{\rho h} f_k(0) - \frac{1}{h^2} \sum_{r=1}^{m} A_r \int_0^1 \mu \frac{df_r}{d\sigma} \frac{df_k}{d\sigma} d\sigma \qquad (14)$$

where $k=1,2,\ldots,m$

In a similar manner we can obtain from (10) the Galerkin form of the v equation of motion.

Before equation (14) can be solved, it is necessary to compute integrals involving $\mu$. For the general case in which $\mu$ varies with x, y, z and t, it is advantageous to expand it in terms of coefficients $E_j(x,y,t)$ and vertically varying functions $\Phi_j(\sigma)$, thus

$$\mu = \sum_{j=1}^{m'} E_j \Phi_j(\sigma) \qquad (15)$$

where $m'$ is the number of terms, not necessarily equal to m.

Although equation (15) allows for a general form of eddy viscosity; in many practical computations the vertical variation of $\mu$ is fixed although its magnitude varies with horizontal position and time (see for example Davies and Furnes 1980). When the vertical variation of $\mu$ is fixed we have

$$\mu = \psi(x,y,t)\,\Phi(\sigma) \qquad (16)$$

where $\Phi$ is a fixed function representing the vertical variation of $\mu$. This special case is considered in detail in the next section.

Having evaluated the integrals involving the basis functions and $\mu$, equations (13), (14) and the corresponding equation for the v component of current, can be used to advance $\zeta$ and the coefficients $A_r$, $B_r$ through time. A forward time stepping method is used and a staggered finite difference grid in the horizontal (see Davies (1980), Davies and Owen (1979) for details).

## FUNCTIONAL FORM OF THE BASIS SET

The rate of convergence of the Galerkin method with a basis set of either Chebyshev or Legendre polynomials (a Galerkin-spectral model) has been examined by Davies and Owen (1979). They found that expansions of Chebyshev and Legendre polynomials converged very rapidly (accurate current profiles could be computed using only four functions).

An alternative to using a basis set of continuous functions is to use piecewise functions in the vertical (a Galerkin-finite element method).

Davies (1977) used a basis set of B-splines, generated

using a recurrence algorithm due to Cox (1972). By this means B-splines of an arbitrary order could be generated.

For the case in which the eddy viscosity takes the form given in (16), an alternative to using a basis set of polynomials or B-splines to solve the hydrodynamic equations is to first solve an eigenvalue problem of the form

$$\frac{d}{d\sigma}\left[\Phi \frac{df}{d\sigma}\right] = -\varepsilon f \tag{17}$$

In general $\mu$ will vary through the vertical in an arbitrary manner, and hence $\Phi$ is arbitrary and it is necessary to determine numerically the set of eigenfunctions f and eigenvalues $\varepsilon$ associated with equation (17).

Equation (17) can be readily solved using the Galerkin method. Thus, expanding the r'th eigenfunction $f_r$ in terms of a set of $\bar{m}$ coefficients $d_{ir}$ and B-splines $M_i$, gives

$$f_r = \sum_{i=1}^{\bar{m}} d_{ir} M_i(\sigma) \tag{18}$$

Applying the Galerkin method to (17), the r'th eigen equation is multiplied by $f_k$ and integrated over the region 0 to 1. Also by integrating the term involving $\Phi$ by parts we obtain

$$\Phi \frac{df_r}{d\sigma} f_k \Big|_1 - \Phi \frac{df_r}{d\sigma} f_k \Big|_0 - \int_0^1 \Phi \frac{df_r}{d\sigma}\frac{df_k}{d\sigma} d\sigma = -\varepsilon_r \int_0^1 f_r f_k \, d\sigma \tag{19}$$

where k=1,2,...,m

The eigenvalues and eigenvectors determined by the solution of (19) will be real, provided

$$\Phi \frac{df_r}{d\sigma} f_k \Big|_1 = 0 \quad , \quad \Phi \frac{df_r}{d\sigma} f_k \Big|_0 = 0 \tag{20}$$

In which case (19) becomes

$$\int_0^1 \Phi \frac{df_r}{d\sigma}\frac{df_k}{d\sigma} d\sigma = \varepsilon_r \int_0^1 f_r f_k \, d\sigma \tag{21}$$

where k=1,2,...,m

For convenience the eigenfunctions are normalised by requiring that $f_r(0) = 1$ for all r. Substituting equation (18) into equation (21) gives a standard eigenfunction problem which may be readily solved using conventional numerical methods (see Davies (1982a) for details of the method used to solve (21)).

The eigenfunctions $f_r$ can be accurately computed using a basis set of B-splines (Davies 1982a). B-splines are piecewise polynomials which are non zero only over a finite interval of space (see figure 1). Points along the $\sigma$ axis at which the

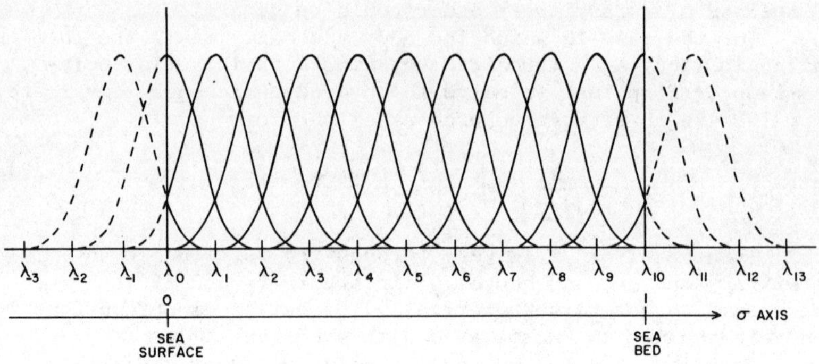


Figure 1. Distribution of B-splines and associated knots with depth.

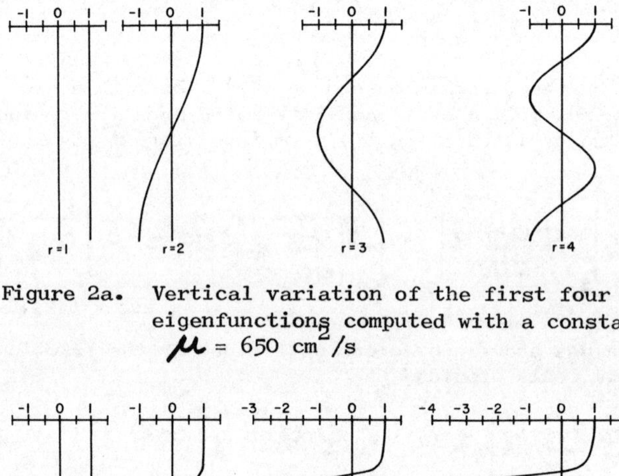


Figure 2a. Vertical variation of the first four eigenfunctions computed with a constant $\mu = 650\ cm^2/s$

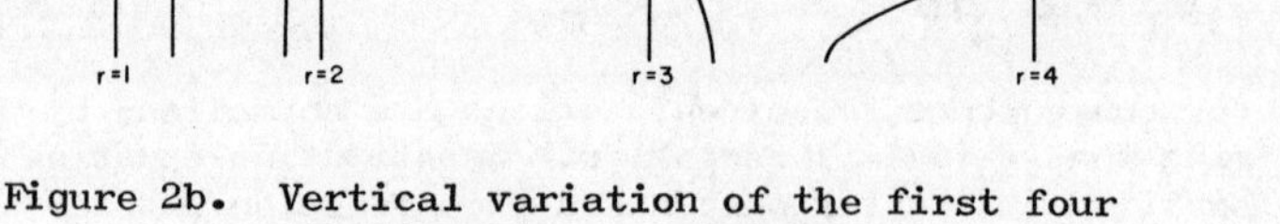


Figure 2b. Vertical variation of the first four eigenfunctions computed with $\mu_1 = 500\ cm^2/s$, $\mu_2 = 200\ cm^2/s$, $\mu_3 = 10\ cm^2/s$ and $\mu_4 = 100\ cm^2/s$. (see Figure 3 for vertical variation of $\mu$ )

B-spline changes from a zero to a non-zero function are termed knots, denoted by $\lambda_i$. The position of the knots and their separation is arbitrary, and hence resolution can be increased by increasing the number of knots in the region where higher resolution is required.

For the case in which the basis functions are eigenfunctions, using (16) and (21), equation (14) reduces to

$$\frac{\partial A_k}{\partial t}\int_0^1 f_k f_k \, d\sigma = \gamma B_k \int_0^1 f_k f_k \, d\sigma - g\frac{\partial \xi}{\partial x}\int_0^1 f_k \, d\sigma$$

$$-\frac{F_B}{\rho h} f_k(1) + \frac{F_S}{\rho h} f_k(0) - \psi \frac{\varepsilon_k A_k}{h^2}\int_0^1 f_k f_k \, d\sigma \tag{22}$$

where k=1,2,...,m

In deriving (22) advantage has been taken of the orthogonality property of eigenfunctions, namely

$$\int_0^1 f_r f_k \, d\sigma = 0 \qquad r \neq k \tag{23}$$

It should be noted that (22) is of the form of k partial differential equations which are only coupled together through the bottom stress term $F_B$ which is given by (6), with $U_h$, $V_h$ computed from (11) and (12).

## FORM OF EDDY VISCOSITY AND SEPARATION INTO EXTERNAL AND INTERNAL MODES

The first four eigenfunctions of (17), computed for the case in which the eddy viscosity is constant ($\mu = 650\ cm^2/s$) are shown in Figure 2(a). In this particular example (21) can be solved analytically, giving

$$f_r = \cos \alpha_r \sigma, \quad \varepsilon_r = \alpha_r^2 \quad \text{where } \alpha_r = (r-1)\pi, \; r=1,2,\ldots,m \tag{24}$$

It is evident from (24) that

$$\int_0^1 f_1 \, d\sigma = 1, \quad \int_0^1 f_1^2 \, d\sigma = 1, \quad f_1(0) = 1, \quad f_1(1) = 1$$

$$\int_0^1 f_r \, d\sigma = 0, \quad r = 2,3,\ldots,m \tag{25}$$

A consequence of (24), (25) and the orthogonality property (23) is that equation (22) can be separated into one equation describing the mean flow (the external mode), given by

$$\frac{\partial A_1}{\partial t} = \gamma B_1 - g\frac{\partial \zeta}{\partial x} - \frac{F_B}{\rho h} + \frac{F_s}{\rho h} \tag{26}$$

and the set of equations

$$\frac{\partial A_k}{\partial t}\int_0^1 f_k f_k \, d\sigma = \gamma B_k \int_0^1 f_k f_k \, d\sigma$$

$$-\frac{F_B}{\rho h} f_k(1) + \frac{F_s}{\rho h} f_k(0) - \psi \frac{\varepsilon_k A_k}{h^2}\int_0^1 f_k f_k \, d\sigma \tag{27}$$

where k=1,2,...,m

describing the deviations from the mean flow (the internal mode). The two sets are coupled through the bottom friction term $F_B$, which is given by (6).

It is evident that the free surface elevation $\zeta$ only enters into equation (26) and not into equation (27). If an explicit time stepping method is used to evaluate (26) and (27) then the time step $\tau$ used to integrate (26) is restricted by the speed of propagation of free surface waves. Then, using a staggered grid (Davies 1980) with grid spacing $\Delta$, $\tau < \Delta/\sqrt{2gh}$. Equation (27) however does not involve surface waves and can therefore be integrated with a much longer time step $\Delta t$, where for convenience we choose $\Delta t = n\tau$, with n in theory arbitrary. However, since the two equations are coupled through the bottom stress $F_B$ and driven by a surface stress $F_s$, it is necessary to ensure that time variations in $F_B$ and $F_s$ can be adequately described by the time step $\Delta t$. Consequently when rapid fluctuations in wind stress $F_s$ occur, or bottom currents change rapidly, $\Delta t$ will have to be comparable with $\tau$ (e.g. $n$ = 2 to 4). However with steady winds, particularly in stratified seas, where bottom currents are shielded from the surface wind stress and change slowly, $\Delta t$ can be much larger than $\tau$ (e.g. $n$ = 6 to 8).

In practice the continuity equation (13) and the U and V equations describing the external mode are integrated with a time step $\tau$ for n steps. Then the two equations describing the internal part of U and V are integrated through a time step $\Delta t = n\tau$ with the bottom friction used in the external mode calculation.

In stratified seas $\mu$ is not constant but varies significantly through the vertical (Figure 3). In this case eigenfunctions and eigenvalues cannot be derived analytically but can be readily computed from (21). Figure 2b shows the first four eigenfunctions computed using an expansion of twenty five B-splines with the eddy viscosity distribution shown in Figure 3, and $\mu_1 = 500$ cm$^2$/s, $\mu_2 = 200$ cm$^2$/s, $\mu_3 = 10$ cm$^2$/s, $\mu_4 = 100$ cm$^2$/s.

Comparing Figures 2a and 2b it is evident that for a given r, the eigenfunctions show the same characteristics. However

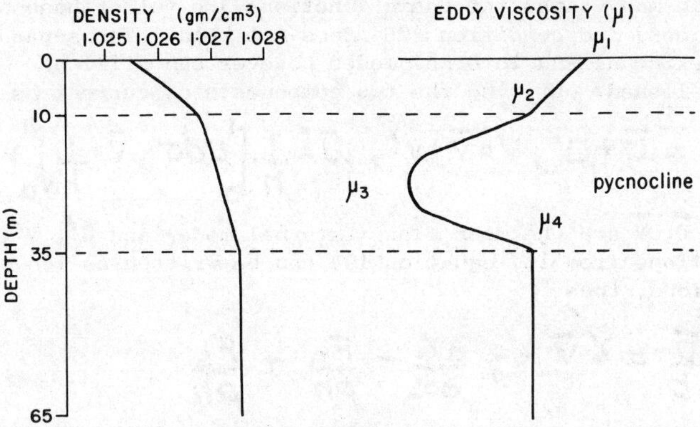


Figure 3. Schematic diagram showing the depth distributions of density and eddy viscosity used in the calculation.

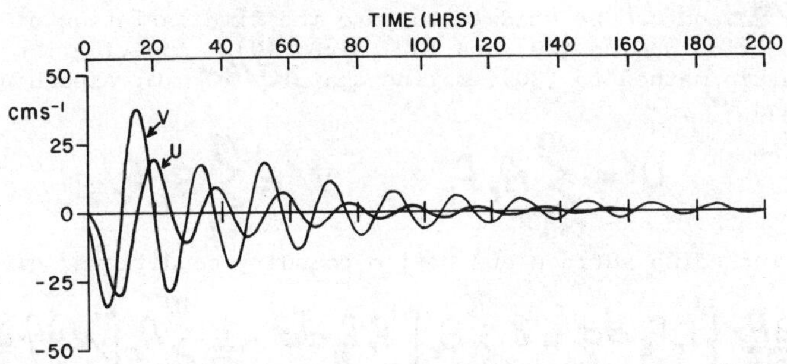


Figure 4a. Time variation of the U and V components of surface current.

the higher eigenfunctions in Figure 2b change abruptly in the region of the pycnocline, reflecting the uncoupling of surface and bottom flows produced by the pycnocline. (This point will be discussed in more detail in the next section).

In many cases the basis functions are not orthogonal (e.g. B-splines) and condition (25) does not hold. The separation into external and internal modes however can still be accomplished. Writing the two components of current as

$$U = \bar{U} + U', \quad V = \bar{V} + V', \quad \bar{U} = \frac{1}{h}\int_0^1 U \, d\sigma, \quad \bar{V} = \frac{1}{h}\int_0^1 V \, d\sigma \tag{28}$$

where $\bar{U}$, $\bar{V}$ are the mean flow (external mode) and $U'$, $V'$, deviations from it, equation (9) can be written as two equations, thus

$$\frac{\partial \bar{U}}{\partial t} = \gamma \bar{V} - g\frac{\partial \zeta}{\partial x} - \frac{F_B}{\rho h} + \frac{F_S}{\rho h} \tag{29}$$

$$\frac{\partial U'}{\partial t} = \gamma V' + \frac{1}{h^2}\frac{\partial}{\partial \sigma}\left(\mu \frac{\partial U}{\partial \sigma}\right) + \frac{F_B}{\rho h} - \frac{F_S}{\rho h} \tag{30}$$

Equation (29) is the conventional two-dimensional form of (9), obtained by vertically integrating (9) and using surface and bottom boundary conditions. Equation (30) does not involve $\partial\zeta/\partial x$ and can be used to compute the time variation of $U'$. By inspection adding (29) and (30) gives (9). Applying the Galerkin method to (30), noting that $\partial \bar{U}/\partial\sigma = 0$, expanding $U'$ and $V'$ as

$$U' = \sum_{r=1}^{m} A_r f_r, \quad V' = \sum_{r=1}^{m} B_r f_r, \tag{31}$$

and including surface and bottom boundary conditions, gives

$$\sum_{r=1}^{m} \frac{\partial A_r}{\partial t}\int_0^1 f_r f_k \, d\sigma = \gamma \sum_{r=1}^{m} B_r \int_0^1 f_r f_k \, d\sigma - \frac{1}{h^2}\sum_{r=1}^{m} A_r \int_0^1 \mu \frac{\partial f_r}{\partial \sigma}\frac{\partial f_k}{\partial \sigma} d\sigma$$

$$- \frac{F_B}{\rho h} f_k(1) + \frac{F_S}{\rho h} f_k(0) + \frac{F_B}{\rho h}\int_0^1 f_k \, d\sigma - \frac{F_S}{\rho h}\int_0^1 f_k \, d\sigma \tag{32}$$

where $k=1,2,\ldots,m$

Equations (29) and (32) form a set of coupled equations describing respectively the external and internal modes and can be integrated using the time splitting algorithm described previously. For the case in which the f in (32) satisfy conditions (25), then it can be readily seen that (29) and (32) (with r, k = 2, 3...., m) are identical to (26) and (27) with $\bar{U} = A_1$, $\bar{V} = B_1$.

STRATIFIED FLOW

The linear equations which describe motion in a stratified sea are given by

$$\frac{\partial u}{\partial x} + \frac{\partial v}{\partial y} + \frac{\partial w}{\partial z} = 0 \tag{33}$$

$$\frac{\partial \rho}{\partial t} + w\frac{\partial \rho}{\partial z} = 0 \tag{34}$$

$$\frac{\partial u}{\partial t} - \gamma v = -g\frac{\partial \zeta}{\partial x} - \frac{g}{\bar{\rho}}\frac{\partial}{\partial x}\left[\int_0^z \rho\, dz\right] + \frac{\partial}{\partial z}\left(\mu \frac{\partial u}{\partial z}\right) \tag{35}$$

$$\frac{\partial v}{\partial t} + \gamma u = -g\frac{\partial \zeta}{\partial y} - \frac{g}{\bar{\rho}}\frac{\partial}{\partial y}\left[\int_0^z \rho\, dz\right] + \frac{\partial}{\partial z}\left(\mu \frac{\partial v}{\partial z}\right) \tag{36}$$

where the notation is as used previously, except that the density $\rho$ is now a function of x, y, z and t, $\bar{\rho}$ is the mean density, and w the vertical velocity.

The solution of these equations using the Galerkin method has been described by Davies (1981a, 1982b), and details will not be given here. Davies (1981a, 1982b) developed a solution, in terms of an internal displacement, and replaced w by the time derivative of this displacement. The internal displacement together with U, V and $\rho$ were then expressed in terms of an expansion of basis functions.

This method is particularly suitable for studying internal waves, but is not applicable if mixing occurs. An alternative is to vertically integrate (33), and incorporate the free surface boundary condition to give equation (1), which is a prognostic equation from which $\zeta$ can be calculated. Using equation (33), w can be eliminated from (34) to form a prognostic equation for $\rho$. By vertically integrating (35) and (36), these equations can be separated into external and internal modes as described in the previous section. The solution then proceeds as before using the Galerkin method but with $\rho$, $U'$ and $V'$ expanded in terms of basis functions. In a stratified sea, particularly in the absence of external stresses, the external mode is associated with barotropic flow, and the internal mode with baroclinic flow, and this terminology will be used when describing stratified seas.

The separation into a barotropic mode and a baroclinic mode and the use of the time splitting algorithm is particularly applicable in stratified flows. In this case the physical nature of the problem produces a natural separation into short period, long wave surface flows (the barotropic mode), and long period, short wave internal flows (the baroclinic mode). Since the barotropic mode has a short period, it requires a short time step, although because of its long wavelength, only moderate grid resolution is required. The baroclinic mode in a strongly stratified sea, on the other hand due to its short wavelength requires a fine grid, but a long time step can be used to integrate the solution.

In practice it is computationally economic to use both time and space splitting particularly in strongly stratified seas. The barotropic flow is computed on a coarser grid than the baroclinic flow which is computed on a fine grid. A linear interpolation method which is consistent with the horizontal finite differencing used to solve the equations, is employed to interpolate bottom currents from the coarse to the fine grid.

The Galerkin method has been applied previously in stratified seas to examine wind induced internal displacements at the Norwegian coast (Davies 1981a) and meteorologically induced currents in a lake (Davies 1982b).

Preliminary calculations using a rectangular closed North Sea basin of constant depth h = 65m are presently in progress. The basin is identical to that used by Davies and Owen (1979) and details of the basin can be found in that paper.

The vertical distribution of density and eddy viscosity used in the model are shown in Figure 3. A linear variation of viscosity with $\mu_1 = 500\ cm^2/s$ and $\mu_2 = 200\ cm^2/s$ corresponding to a wind generated turbulent layer was employed in the model. Below this layer there is a pycnocline in which turbulence is supressed and an appropriate value of $\mu_3$ here is $10\ cm^2/s$ (Davies 1981a). In the bottom layer turbulence was assumed to be of tidal origin with $\mu_4 = 100\ cm^2/s$. A linear law of bottom friction was used in the model with k = 0.2 cm/s.

In the calculation described here, a basis set of five B-splines was used in the vertical, although a basis set of eigenfunctions (Figure 2b) has also been used.

The water in the basin was initially at rest and motion was generated by a suddenly imposed sinusoidal wind stress from the North of magnitude 5 $dyn/cm^2$ and duration of 12 hours.

Wind induced currents in the basin are generated in response to this wind stress and Figure 4a shows a typical time series of surface currents. The initial response of the surface layer to the wind field is a current in the direction of the wind (Figure 4b). However once such a flow is generated, rotational effects become important, and in an infinite sea a purely rotational current (an inertial current) is generated with a period $T = 2\pi/\gamma$. In an inviscid fluid this circular flow would be maintained indefinitely, however in the real sea and in the model viscous effects are important and damping occurs (see Figures 4a, 4b).

In an enclosed basin, lateral boundaries modify the flow. Also, particularly in a homogeneous sea, the wind excites the natural periods of oscillation of the basin and currents associated with these will be present together with the inertial currents.

Although in nature $\gamma$ is determined by the angular speed of the earth's rotation and latitude of the basin, in the model any value of $\gamma$ can be used. A range of values of $\gamma$ were employed in the model and the period of the surface current changed with $\gamma$, confirming the presence of inertial currents.

Currents shown in Figures 4a, 4b were computed with $\gamma = 0.0009\ s^{-1}$, and it is evident from the figure that these currents have a period of approximately 19 hours, which agrees well with the theoretical period of 19.39 hours. However these currents are not purely inertial currents but also contain currents due to boundary effects.

Due to viscous and boundary effects these currents do not show purely circular motion, but exhibit an elliptical spiral motion decaying rapidly with time (Figure 4b). Stratification in this model is fairly weak and consequently the inertial current is quickly damped.

After the wind impulse had decayed, a $180^{o}$ phase difference in currents above and below the pycnocline was evident, which is characteristic of inertial currents (Pollard 1970). The vertical variation of current was characterized by a surface layer through which the velocity changed very little; a rapid nearly linear variation within the pycnocline, through which the flow reversed. Below the pycnocline the current varied little in the vertical except for a layer of frictionally retarded flow near the sea bed. The main characteristics of this flow could be adequately described using the second eigenfunction ($r = 2$) shown in Figure 2b.

## CONCLUDING REMARKS

Previous papers (Davies and Furnes 1980, Davies 1981a,b) have shown that the Galerkin method can be applied to computing currents in physically realistic models (results from which will be presented during the lecture).

In this paper the mathematical development of the Galerkin method for homogeneous or stratified flow has been described. Also, by separating the barotropic and baroclinic motion a computationally efficient time splitting algorithm has been developed.

Preliminary calculations of flow in a stratified basin look very encouraging and further calculations using higher vertical resolution and finer grids are in progress and will be reported subsequently.

## ACKNOWLEDGEMENTS

The author is indebted to Dr. N.S. Heaps for helpful

discussions and comments on this work. Thanks are due to Mr. J.E. Jones for some plotting programs, Mr. R. Smith for drawing the diagrams and Mrs L. Parry for typing the paper.

This work was funded by the Natural Environment Research Council and the Department of Energy.

REFERENCES

Cheng, R.T.; Powell, T.M. and Dillon, T.M. (1976) Numerical models of wind-driven circulation in lakes, Applied Mathematical Modelling 1, 141-159.

Cox (1972) The numerical evaluation of B-splines, J. Inst. Math. Appl. 10, 134-149.

Davies, A.M. (1977) The numerical solution of the three-dimensional hydrodynamic equations using a B-spline representation of the vertical current profile, in : J.C.J. Nihoul (ed), Bottom Turbulence, Proceedings of the 8th Liège Colloquim on Ocean Hydrodynamics (Elsevier 1977) pg 1-25.

Davies, A.M. (1980) Application of the Galerkin method to the formulation of a three-dimensional non-linear hydrodynamic numerical sea model. Applied Mathematical Modelling 4, 245-255.

Davies, A.M. (1981a) Three-dimensional hydrodynamic numerical models. Part 1. A homogeneous ocean-shelf model. Part 2. A stratified model of the northern North Sea, pg 370-426 in Proceedings of the Norwegian Coastal Current Symposium, Geilo, Norway. Published Bergen University.

Davies, A.M. (1981b) Three-dimensional modelling of surges, pg 45-74, in Floods due to High Winds and Tides, ed D.H. Peregrine, published Academic Press.

Davies, A.M. (1982a) Formulation of a linear three-dimensional hydrodynamic sea model using a Galerkin-Eigenfunction method. (in press) International Journal of Numerical Methods in Fluids.

Davies, A.M. (1982b) On computing the three-dimensional flow in a stratified sea (in press) Applied Mathematical Modelling.

Davies, A.M. and Owen, A. (1979) Three-dimensional numerical sea model using the Galerkin method with a polynomial basis set, Applied Mathematical Modelling 3, 421-428.

Davies, A.M. and Furnes, G.K. (1980) Observed and Computed $M_2$ tidal currents in the North Sea. Journal of Physical Oceanography, 10, 237-257.

Fix, G.J. (1975) A survey of numerical methods for selected problems in continuum mechanics, pg 268-284 in Numerical Models of Ocean Circulation ed R.O. Reid, published National Academy of Sciences.

Heaps, N.S. (1972) On the numerical solution of the three-dimensional hydrodynamic equations for tides and storm surges, Mem. Soc. r. Sci Liège Ser 6, 2, 143-180.

Pollard, R.T. (1970) On the generation by wind of inertial waves in the ocean, Deep Sea Research 17, 785-812.

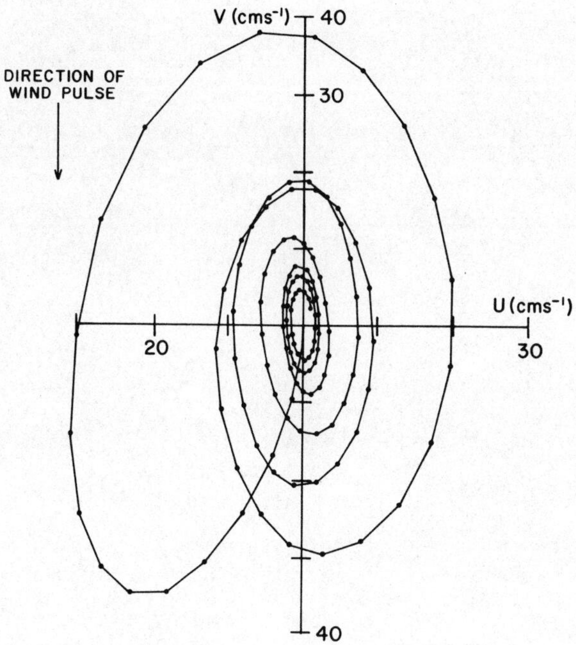


Figure 4b. Surface U and V components of current plotted at hourly intervals in polar form.

# FEASABILITY OF FINITE ELEMENT METHODS FOR OCEANIC GENERAL CIRCULATION MODELLING

Eric Dumas and Christian Le Provost, Institute of Fluid Mechanics
Alain Poncet, Institute of Applied Mathematics,
Grenoble, France

## INTRODUCTION

### The problem considered

The response of the ocean to the action of a wind-stress at the surface has been investigated by many scientists since the middle of this century. The pioneering investigations were carried by Sverdrup (1947), Stommel (1948) and Munk (1950) : they approximate the real ocean flow induced by the wind at the mid-latitudes in terms of total mass transport of a homogeneous layer of fluid of uniform depth on a beta plane (a projection of the sphere on a tangent plane where the Coriolis parameter is taken to be $fo + \beta y$, with $fo$ and $\beta$ constant, and $y$ the north south coordinate). The fluid is set in motion by a torque which models the effects of the zonally averaged large scale wind pattern of the westerlies and the trades. Thus, the circulation appears as a gyre in which the Coriolis forces balance the pressure gradients and the driving force of the wind stress, in the greater part of the basin (Sverdrup balance) ; every streamline is going through a boundary layer region, in the western part of the basin, where an important dissipation of vorticity arises : this sink of vorticity is produced by bottom friction in Stommel's model, and by lateral diffusion of momentum in Munk's model ; some non linear analytical models were then developped to include the inertial effects which are of evidence very important : Munk, Groves and Carrier (1950), Carrier and Robinson (1962), Moore (1963), Il'in and Kamenkovich (1963), Niiler (1966) ; but these models were questionable on many points : for example they assumed a priori the existence of steady state solutions, without proof. Thus numerical models have been developped to investigate the general time dependent non linear case : Bryan (1963) extended in the non linear and time dependent domain the Munk's model, Veronis (1966) did the same for the Stommel's model, Brandford (1971) realized a synthesis of the two. Since these early works, numerous contributions by numerical modellers improved our understanding of the wind induced circulation of the ocean which is impossible to review in this introduction. But, within this paper, it must be pointed out that all these numerical models have been developped on the basis of finite difference technics.

The aim of the present contribution

Some years ago, Fix (1975) has investigated the properties of the Finite Element technics for modelling ocean circulation problems in their quasi geostrophic approximation (see latter on for the corresponding formulation) and shown their interests on several areas : precision, conservative formulation, natural treatment of the boundary conditions, flexibility of the triangulation for the resolution on complex areas. He has proved the stability and convergence of the semi-discrete finite element formulation of such a problem, and established the conservative properties of this numerical approximation for energy, vorticity, and enstrophy, independently of the irregularity of the grid used for the spatial integration.

To our knowledge, only one model can be found in the litterature, based on the finite element technic for solving that class of problem : Haidvogel, Robinson and Schulman (1980) have compared the precision of a finite difference model, a finite element model, and a spectral model for applications to open ocean problems, and shown the interest of finite element methods. However, this model has been applied on a regular grid in the domain of the stability and precision. In this paper, a finite element model is presented for the resolution of the classical wind induced general oceanic circulation problem, briefly introduced in the preceding chapter. The performances of the method are investigated by comparison to the analytical solutions of the linear case, and to the classical finite difference solutions presented in the litterature for non linear cases. As the corresponding solutions include a western boundary layer with intense velocity gradients, a particular interest is devoted to the use of the Finite Element technic flexibility to increase the resolution of the model along the western wall of the basin.

## MATHEMATICAL FORMULATION

### The primitive equations

If the fluid is assumed to be in hydrostatic balance and of uniform density, the Navier-Stokes Equations, written on a beta plane, are reduced to :

$$qt + q.\nabla q + (f_o + \beta y)k \times q = - g\nabla\eta + F \qquad (1)$$

where k is the vertical unity vector and F is the body force per unit of mass, which included the vertically averaged effects of wind stress and friction.

q is the depth averaged horizontal velocity vector.

$\eta$ is the elevation of the sea surface.

t is the time and g the earth gravity.

If h is the uniform depth when the fluid is at rest, the continuity equation is then :

$$\eta_t = -\nabla [(h + \eta) q] \tag{2}$$

Thus, we have to solve a problem of three unknowns : $\eta$ and the two components of q in x and y , of three variables x, y and t .

The vorticity equation

Let us consider that the body force F includes the wind driving stress $\tau$, the bottom drag - Kq, and the lateral friction $A\nabla^2 q$, and introduce the stream function $\psi$ as :

$$q = k \times \nabla\psi \tag{3}$$

If we take the "rigid lid" approximation, which is equivalent to neglecting the terms involving $\eta$ in (2), it is possible to eliminate $\eta$ from (1) by substitution of (3) into (1) and cross differentiation : thus we obtain the vorticity equation :

$$\zeta_t + J(\psi,\zeta) + \beta\psi_x = \frac{1}{h}\,\mathrm{curl}\,\tau + A\nabla^2\zeta - K\zeta \tag{4}$$

where $\zeta = \nabla^2\psi$ and $J(\psi,\zeta) = \psi_x\zeta_y - \psi_y\zeta_x$

It is convenient to non dimensionalize the system so as to introduce a minimum number of parameters. This is done by taking :

$(x,y) = (x'L,\ y'L),\ t = t'T,\ \psi = \psi' LU,\ \zeta = \zeta' U/L,\ \tau = \tau_o z'$

where L is a typical size of the domain $\Omega$ of integration, and U a typical velocity scale. Thus we reduce the vorticity equation (4) into :

$$\zeta'_{t'} + RJ(\psi',\zeta') + \psi'_{x} = \frac{1}{\pi}\,\mathrm{curl}\,\tau' + \frac{R}{Re}\nabla^2\zeta' - \varepsilon\zeta \tag{5}$$

if we choose the time scale and the velocity scale as :

$$T = (\beta L)^{-1} \qquad U = \pi\tau_o(\rho h\beta L)^{-1}$$

the parameters R, Re and $\varepsilon$ are defined as :

$$R = U(\beta L^2)^{-1},\ Re = ULA^{-1},\ \varepsilon = K(\beta L)^{-1}$$

R is representative of the non linear processes, compared to the beta effect of the Coriolis force and can be considered as a Rossby number ; Re can be interpreted as the Reynolds number of the western boundary layer, the ratio R/Re being representative of the lateral diffusion effects, compared to the beta effect ; $\varepsilon$ characterises the magnitude of the frictional processes. The primer will be dropped in the following.

Typical order of magnitude of the parameters

If L = 2000 km, $\beta = 2.10^{-11}$ $(ms)^{-1}$, h = 4000 m, $\tau_o = 0{,}3\ N/m^2$, A = 2,5 $10^{-3}$ $m^2/s$, it follows that T = 7h, R = 7 $10^{-5}$, Re = 5 and = 0,016.

Boundary conditions

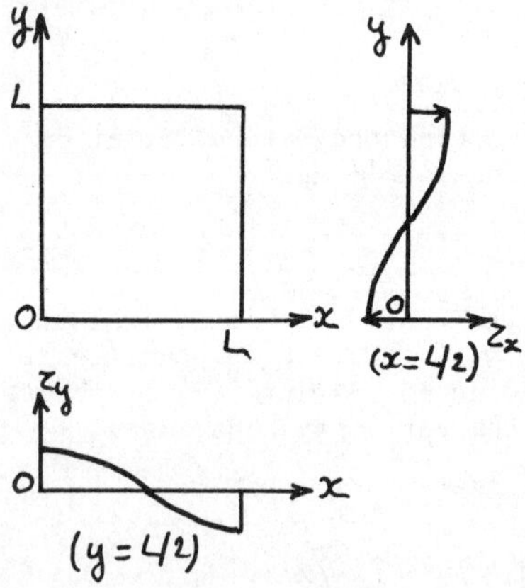


The domain L of integration is a square of size L . The depth of the basin is h = $C^{te}$. The wind field is taken as :

$$\tau_x = -\frac{\tau_o}{2} \sin \frac{\pi x}{L} \cos \frac{\pi y}{L}$$

$$\tau_y = \frac{\tau_o}{2} \cos \frac{\pi x}{L} \sin \frac{\pi y}{L}$$

such that

$$\text{curl } \tau = -\frac{\pi}{L} \tau_o \sin \frac{\pi x}{L} \sin \frac{\pi y}{L} \tag{6}$$

The boundary $(\Omega)$ of the basin is supposed to be a coast ; thus, $\psi = 0$ and $\zeta = \zeta_o$ (data), for x = 0, L and y = 0, L.

NUMERICAL METHOD

The time discretization

Time is divided into increments dt and for t = ndt we note f(t) as $f^n$. Thus, a traditional discrete formulation of the vorticity equation (5) is obtained by using second order centered finite differences (Crank - Nicholson).

$$\frac{\zeta^{n+1} - \zeta^{n-1}}{2dt} + RJ(\psi^n, \zeta^n) + \psi_x^n = -\sin \pi x \sin \pi y - \varepsilon \frac{\zeta^{n+1} + \zeta^{n-1}}{2} + \frac{R}{Re} \nabla^2 \frac{\zeta^{n+1} + \zeta^{n-1}}{2} \tag{7a}$$

$$\nabla^2 \psi^{n+1} = \zeta^{n+1} \tag{7b}$$

The dissipative terms are evaluated as the average of their values at steps n - 1 and n + 1, because that implicit evaluation leads to a numerical integration which is stable, with respect to these terms.

Centered differencing will possibly lead to computational instability ; a maximum limit time step is in fact empirically found for each simulation, which appears to be independent of the horizontal grid size, but connected to the stability of the Coriolis term and the Jacobian convective term.

## The variational formulation

For each time step, we have to solve equation (7a) and (7b), two elliptic equations of second order : one of the unknown $\zeta^{n+1}$ and the other of $\psi^{n+1}$. Our purpose is to search a solution of (7a,b) in a weak sense, i.e. in the space $H_o^1(\Omega)$ for stream function $\psi$ and vorticity $\zeta$.

Integrating (7a) with a testing function $\zeta'$ over the domain $\Omega$, we have :

$$\forall \zeta' \in H_o^1(\Omega) : \left\{ (1 + \varepsilon dt - \frac{R}{Re} dt \nabla^2) \zeta^{n+1}, \zeta' \right\} = \left\{ f(\psi^n, \zeta^n), \zeta' \right\} \quad (8)$$

with the notation $\{a, a'\} = \iint_\Omega a.a' \, d\Omega$

and $f(\psi^n, \zeta^n) = (1 - \varepsilon dt + \frac{R}{Re} dt \nabla^2) \zeta^{n-1} - 2R \, dt \, J(\psi^n, \zeta^n)$

$$- 2dt(\psi_x^n - \sin \pi x \sin \pi y)$$

Using the Green formula :

$$\{\nabla^2 \zeta, \zeta'\} = -\{\nabla \zeta, \nabla \zeta'\} + \oint_\Gamma \frac{\partial \zeta}{\partial n} \zeta' \, d\Gamma$$

and noting that the curvilinear integral is nul because $\zeta'/_\Gamma = 0$, we deduce the variational formulation of (7a) :-

$$(1 + \varepsilon dt) \{ \zeta^{n+1}, \zeta' \} + \frac{R}{Re} dt \{ \nabla \zeta^{n+1}, \nabla \zeta' \} = \{ f(\psi^n, \zeta^n), \zeta' \} \quad (9a)$$

we can deduce the same kind of formulation for the stream function equation

$$\{ \nabla \psi^{n+1}, \nabla \psi' \} = - \{ \zeta^{n+1}, \psi' \} \quad (9b)$$

## The finite element approximation

The domain $\Omega$ is subdivided into triangles and we consider a finite element space on this grid, $S^\Delta$, in which any function $\phi$ can be written :

$$\phi^\Delta(x,y) = \sum_{j=1}^{N} \phi^\Delta (z_j) \varphi_j(x,y)$$

$z_j$, j = 1, N, are the nodes of the elements
$\varphi_j$ are the basic functions.

Thus, the approximation for the vorticity and the stream function are of the form :

$$\zeta^{\Delta}(x,y,t) = \sum_{j=1}^{N} \zeta_j^{\Delta}(t)\, \varphi_j(x,y)$$
$$\psi^{\Delta}(x,y,t) = \sum_{j=1}^{N} \psi_j^{\Delta}(t)\, \varphi_j(x,y) \qquad (10)$$

and the problem is to determine the weights $\zeta_j^{\Delta}(t)$ and $\psi_j^{\Delta}(t)$ for each time step, which is done by using the variational formulations (9a,b). Let us take the matrix notations :

$M = \{\varphi_i, \varphi_j\}$ , the "mass matrix",

$K = \{\nabla\varphi_i, \nabla\varphi_j\}$ , the "stiffness matrix",

the equations (9a) and (9b) can be written consequently :

$$\left[(1 + \varepsilon\, dt)M + \frac{R}{Re}\, dt\, K\right] \zeta^{\Delta,n+1} = B^n \qquad (11a)$$

$$K\, \psi^{\Delta,n+1} = C^n \qquad (11b)$$

with $B^n = \left\{ f(\psi^{\Delta,n}, \zeta^{\Delta,n}),\ \zeta' \right\}$

$C^n = -\left\{ \zeta^{\Delta,n+1},\ \psi' \right\}$

The matrices M and K are built at the beginning of the iteration, and don't vary while the iteration occurs. Thus, the two matrices on the left hand side of (11a) and (11b) are inverted only once, at the beginning of the computation, and the problem reduces at each time step to the computation of the second members of the equations (11a) and (11b), and the resolution of these linear systems by direct substitutions. As we shall see later, this procedure requires very reasonable computational work per time step.

Following a demonstration similar to the one given by Fix (1975), it is possible to proof the stability and the convergence of the semi discrete finite element system (11a - 11b), at least when the dissipative terms of friction and lateral diffusion, and the forcing term of wind stress are set to zero. It is also possible to show that within this model, vorticity, energy and entrophy are conserved, independently of the irregularity of the grid dividing $\Omega$ .

<u>Eddy viscosity and boundary conditions on $\zeta$</u>

If the term $\frac{R}{Re}$ is neglected (related to the mesh-size), the condition $\zeta = \zeta_0$ has no longer to be imposed at the nodes on the boundary : equation (8) becomes of zero order and the mass matrix M can be lumped, producing an explicite scheme for $\zeta^{n+1}$.

In the other case, the calculation of significant values for $\zeta_o$ on the boundary increases computational work (Ciarlet, Glovinski, 1975).

NUMERICAL TESTS

The linearized problem

As we noticed when introducing the problem, analytical steady state solutions exist when the problem (5) is linearized. In this paper, we limit our application to the non viscous case : for R = 0 (linearized model), A = 0 (no lateral eddy viscosity), and $\frac{\partial}{t} \equiv 0$ equation (5) reduces to :

$$\psi_x + \varepsilon \nabla^2 \psi = - \sin \pi x \sin \pi y \qquad (12)$$

with $\psi = 0$ on $(\Gamma)$
and the analytical solution (Stommel, 1948) is :

$$\psi_a(x,y) = \frac{\sin \pi y}{\pi(1 + 4\pi^2 \varepsilon^2)} \left[ 2\pi\varepsilon \sin \pi x + \cos \pi x + \frac{1}{e^{R1} - e^{R2}} \times \left[ (1 + e^{R2})\, e^{R1x} - (1 + e^{R1})\, e^{R2x} \right] \right] \qquad (13)$$

with $\quad R1(2) = \left[ - 1 + (-)\, (1 + 4\pi^2 \varepsilon^2)^{1/2} \right] / 2\varepsilon$

Thus, as a first series of tests, we apply our finite element model to the time dependent problems (5) restricted to the non viscous case R/Re = 0. Starting from an initial situation in which the fluid is at rest in the basin, we apply at t = 0 the wind curl (6) and look at the time evolving solution of the stream functions until a steady state is obtained. The corresponding solution in then compared to the analytical solution (13). As we know from this Stommel's solution, the steady state solution (if it exists) is an anticyclonic gyre presenting an intense boundary layer along the western coast of the basin. As an illustration, we present on figure 1 the steady state solution obtained for $\varepsilon = 0.05$ and on figure 2 the total kinetic energy in the basin as a function of time, for two different time steps (dt = 4 and 8).

The typical width of the western boundary layer in the Stommel's model is $d_F = \varepsilon L$, i.e. related to the intensity of the friction parameter $\varepsilon$ : the less intensive the bottom friction is, the more intense the circulation is and the closer the stream function lines are near the western boundary. Consequently, the precision of the numerical solutions is particularly sensitive to the size of the triangulation along the western wall. A series of computations were carried over a large range of values for $\varepsilon$, and two triangulations of the domain $\Omega$ , which are presented on figures 3. The steady state solutions obtained are compared to the analytical solutions through the RMS errors on the streamfunction.

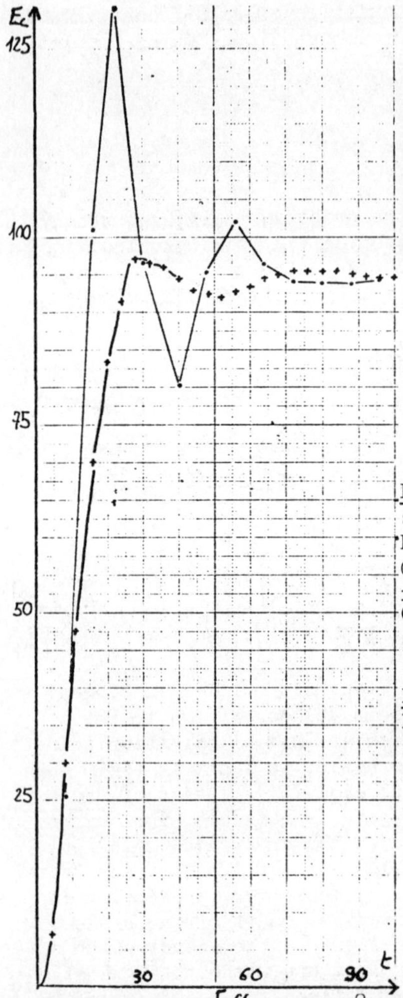


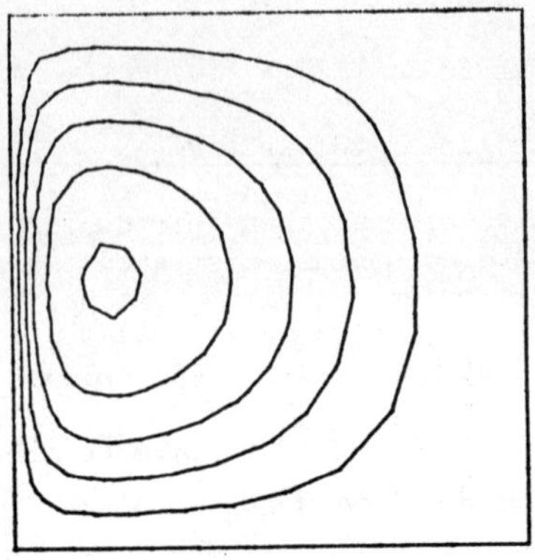

Fig. 1 : linear steady state solution for = 0.05 and R/Re = 0, after t = 160 T. Contours of $\psi$ normalized with respect to its maximum value. $\psi_{max}$ = 0.48 in non dimensionalized unit.

Fig. 2 : plot of the total kinetic energy of the basin as a function of time, in non dimensionalized unit ($\varepsilon$ = 0.05).

+ —— + time step dt = 4

. —— . time step dt = 8

$$\mathrm{RMS}(\psi) = \left[\iint_{\Omega} (\psi - \psi_a)^2 \, d\Omega \Big/ \iint_{\Omega} \psi_a^2 \, d\Omega\right]^{1/2}$$

Thus, for the case presented on figure 1 ($\varepsilon$ = 0.05) with the triangulation $T_1$, the RMS($\psi$) is $1.6 \times 10^{-2}$. Figure 4 summarizes the corresponding results : the RMS of the stream functions computed for different values of $\varepsilon$, and with the two triangulations are plotted on the same graph ; it can be seen that, with a given triangulation, the computations are valid until the width of the boundary layer becomes of the order of magnitude of the size of the triangles along the western wall. All these computations have been done by using piecewise linear elements ($P_1$) and the corresponding RMS errors on $\psi$ are of the order of $10^{-2}$, when using a correct triangulation.

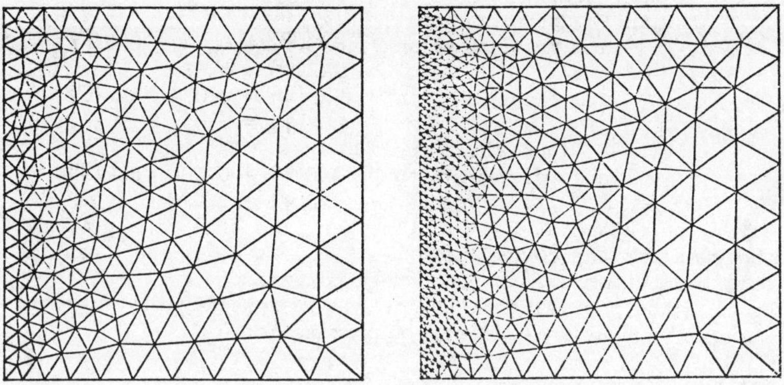

Fig. 3

Triangulation $T_1$
391 triangles - 220 nodes
typical size of the mesh
0.05 on the left
0.15 on the right

Triangulation $T_2$
937 triangles - 510 nodes
typical size of the mesh
0.02 on the left
0.15 on the right

Fig. 4

RMS of the stream function as a function of $\varepsilon$, with triangulation $T_1$ and $T_2$ and approximation $P_1$ and $P_2$. The typical width of the western boundary layer corresponding to the size of the meshes along the western wall, for triangulation $T_1$ and $T_2$ is indicated by an arrow.

With the same triangulation, the precision is improved when using quadratic elements in the FEM ; as an example, for the case $\varepsilon = 0.5$, we obtain :

with $P_1$ approximation

and triangulation $T_1$ $RMS(\Psi) = 0{,}017$ $RMS(\nabla\Psi) = 0{,}200$
$RMS(\psi) = 0{,}130$

and triangulation $T_2$ $RMS(\psi) = 0{,}011$ $RMS(\nabla\psi) = 0{,}103$
$RMS(\psi) = 0{,}043$

with $P_2$ approximation

and triangulation $T_1$ $RMS(\psi) = 0{,}008$ $RMS(\nabla\psi) = 0{,}021$
$RMS(\psi) = 0{,}020$

Given the fact that the CPU time per time step is approximately the same for the computation $(T_2 - P_1)$ and $(T_1 - P_2)$, an important conclusion arises from these comparisons :

the precision of the method is considerably improved when using quadratic approximation.

## The non linear problem

For the non linear problem, we have no more analytical solution for comparison, and the only way to control our results is to refer to the previous results on the same problem, obtained by using a finite difference method. We know from Veronis (1966) that the non linearities introduce an alternating northward and southward circulation near the north west corner of the basin ; in term of vorticity, this corresponds to a shift toward the north of the point of maximum negative vorticity, which is situated at $(x = 0, y = L/2)$ in the linear case, and to the production of a cell of positive vorticity near the north west corner. Results obtained with the present model are shown on figures 5 for the linear case and on figures 6 in the case where the inertial boundary layer thickness, which can be demonstrated analytically to be $d_I = \sqrt{RL}$, is six tenths of the frictional boundary width : i.e. $R = 9.10^{-5}$. These results are in good agreement with those of Veronis, though they have been obtained in these preliminary tests with $P_1$ elements.

## CONCLUSION

A model based on finite element technic is presented for solving the classical barotropic wind driven oceanic circulation problem formulated under the non linear quasi geostrophic approximation.

Several typical problems have been investigated. For the simpler linearized case, the results of the simulations are checked to an analytical solution : the comparison is very satisfying ; the interest of refinement of the triangulation along the western boundary and the superiority of quadratic elements are clearly shown. For the non linear cases, the results are favourably compared to previous results obtained by finite difference technics. The method appears thus to be very attractive and will be applied in the future to more realistic situations.

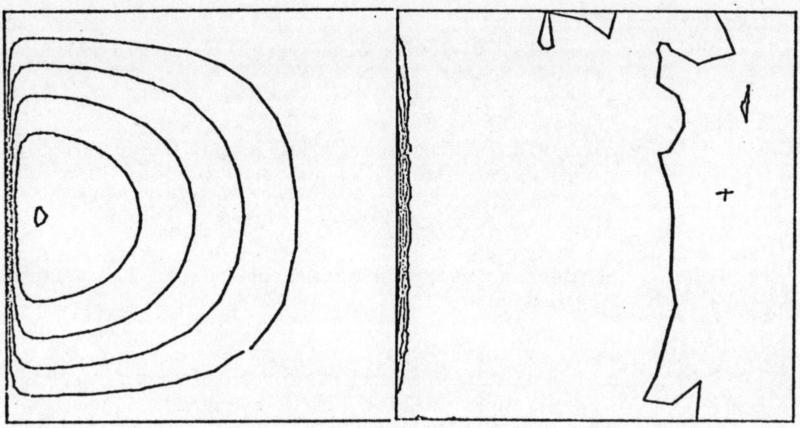

linear case
$\varepsilon = 0.016 - R = 0$

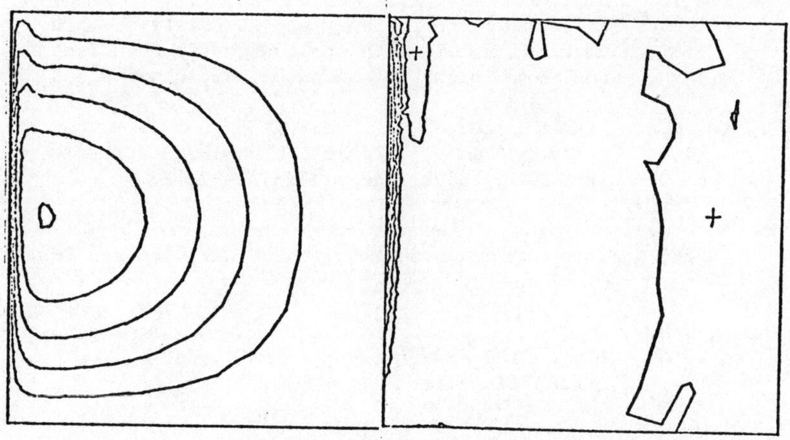

non linear case
$\varepsilon = 0.016 - R = 9.10^{-5}$

Fig. 5 : contour lines of $\psi$ and $\zeta$ normalized with respect to their maximum value.

REFERENCES

Blandford, R.R.
1971. Boundary conditions in homogeneous ocean models. Deep Sea Research, 18 : 739-751.

Bryan, K.
1963 A numerical investigation of a non linear model of a wind driven ocean. J. Atmos. Sciences, 20 : 594-606.

Carrier, G.F. ; ROBINSON, A.R.
1962 On the theory of the wind driven ocean circulation. J. Fluid Mech. 12 : 49-80.

Ciarlet, P.G. ; Glovinsky, R.
1975 Dual iterative techniques for solving a finite element approximation of the bi-harmonic equation. Comp. Math in Applied Mech. in Eng., 5 : 277-295.

Fix, G.J.
1975 Finite element models for ocean circulation problems. SIAM J. Appl. Math. 29-3 : 371-387.

Haidvogel, D.B. ; Robinson, A.R. ; Schulman, E.E.
1980 The accuracy, Efficiency and Stability of three Numerical Models with application to Open Ocean Problems. Journal of Computational Physics, 34-1-53.

Il'in, A.M. ; Kamenkovich, V.M.
1963 On the influence of friction on ocean currents. Dakl. Akad. Nauk, SSSR : 150-1274-1277.

Moore, D.W.
1963 Rossby waves in ocean circulation. Deep Sea Research, 10 : 735-747.

Munk, W.H.
1950 On the wind driven ocean circulation. Journal of Meteor. 7 : 79-93.

Munk, W.H. ; Groves, G. and Carrier, G.F.
1950 Note on the dynamics of the Gulf Stream. J. of Marine Res. 9 : 218-238.

Niiler, P.P.
1966 On the theory of the wind driven ocean circulation. Deep Sea Research, 13 : 597-606.

Stommel, H.M.
1948 The westward intensification of wind driven ocean currents. Trans. Amer. Geoph. Un. 29 : 202-206.

Sverdrup, H.U.
1947 Wind driven currents in a barotropic ocean, with application to the equatorial currents of the Eastern Pacific. Proc. U.S. Nat. Acad. Sc. 33 : 318-326.

Veronis, G.
1966 Wind driven ocean circulation. Deep Sea Research, 13 : 17-29, 31-55.

# THE USE OF KALMAN-BUCY FILTERS IN FORECASTING THE WATERLEVELS IN THE DUTCH COASTAL AREA.

A.W. Heemink.
Applied mathematician, Data Processing Division, Rijkswaterstaat, The Netherlands.

B. de Jong.
Applied mathematician, Twente University of Technology, The Netherlands.

## SUMMARY.

A discrete Kalman-Bucy filter has been derived on the basis of the one dimensional shallow water equations. The filter gives estimates of the waterlevel and watervelocity, together with the parameters in the model which essentially have a random character, e.g. bottom friction and wind stress. Using on-line information the estimates are adapted automatically by the filter when the physical circumstances change, e.g. during the rising of a storm surge. The Kalman-Bucy filter is applied to field data gathered in the Dutch coastal area and it will be shown that the filter gives satisfactory results in forecasting the waterlevels during storm surge periods.

## INTRODUCTION.

Most of the southwestern part of The Netherlands lies below or just above sealevel. Therefore,special precautions have to be taken with respect to the dikes and the coastline in general. After the disaster of February 1953, when during an extreme stormperiod the dikes in this area broke and approximately 1800 people were drowned, the Dutch government planned to close all the estuaries except the free waterway to Antwerp, the Western Scheldt (figure 1). However, under pressure exerted by ecologists, who wanted to preserve the salt-water milieu, plans had to be changed. For the Eastern Scheldt it has now been decided to insert a storm surge barrier, which will be closed only in cases of extreme high water. The procedure to close the barrier takes a few hours; consequently accurate short time predictions of the waterlevel are needed to ensure that the storm surge barrier is closed in time.
Previously, tidal prediction models have been either statistical or deterministic (Dronkers 1964, 1975).
Statistical methods (e.g. the harmonic method, the response method) are based on the stationarity of the observed process, so that they are unable to forecast the waterlevel during a

storm surge period. Deterministic methods include analytical or numerical solutions to hydrodynamic equations. In these models, the parameters which represent the influences of physical phenomena, such as bottom friction or wind stress, are inserted as constants. Because these parameters have essentially a random character and are sensitive to changing conditions deterministic methods too are unable to forecast the water-level during a storm surge period, when conditions can change rapidly. Moreover, when using a deterministic method, it is impossible to incorporate on-line information directly to improve the predictions. These problems suggest the use of a Kalman-Bucy filter with which it becomes possible to adapt the model to changing conditions by using on-line information (Jazwinsky 1970).

DISCRETE KALMAN-BUCY FILTER THEORY.

Assume that modeling techniques have produced an adequate description in the form of a linear stochastic system to describe the propagation in time of a state vector $\underline{X}_{t_k}$

$$\underline{X}_{t_k} = \Phi(t_{k-1},t_k)\ \underline{X}_{t_{k-1}} + \underline{W}_{t_k}, \quad k = 1,2,3,.... \tag{1}$$

$$\underline{X}_{t_o} = \underline{X}_o$$

where $\Phi$ is the system dynamics matrix and $\underline{W}_{t_k}$ is a white Gaussian noise process with statistics:

$$E\{\underline{W}_{t_k}\} = 0$$

$$E\{\underline{W}_{t_k}\ \underline{W}^T_{t_l}\} = \begin{cases} Q(k) & k = l, \quad k,l = 1,2,3,... \\ 0 & k \neq l \end{cases} \tag{2}$$

The initial condition $\underline{X}_o$ is also assumed to be normally distributed with mean $\hat{\underline{X}}_o$ and covariance matrix $P_o$. It can be proved that under these conditions the state vector $\underline{X}_{t_k}$, $k = 1,2,3,...,$ is also normally distributed (Jazwinsky 1970).
Measurements are available at discrete points $t_1$, $t_2$, .... and are modeled by the relation

$$\underline{Z}_{t_k} = M(t_k)\ \underline{X}_{t_k} + \underline{V}_{t_k} \tag{3}$$

where $\underline{Z}_{t_k}$ is the measurement process, M is the measurement matrix and $\underline{V}_{t_k}$ is a white Gaussian noise process with statistics:

$$E\{\underline{V}_{t_k}\} = 0$$

$$E\{\underline{V}_{t_k}\,\underline{V}^T_{t_l}\} = R(k) \quad k = 1, \quad k,l = 1,2,3,.... \qquad (4)$$
$$\qquad\qquad 0 \quad k \neq l$$

It is further assumed that $\underline{X}_o$, $\underline{W}_k$ and $\underline{V}_k$ are mutually independent.
It is desired to combine the measurement $\underline{Z}_{t_k}$, taken from the actual system, with the information provided by the system model in order to obtain an optimal (least squares) estimate of the system state $\underline{X}_{t_k}$. The recursive Kalman-Bucy filter equations to accomplish this task are:

$$\underline{\hat{X}}(k|k-1) = \Phi(t_{k-1},t_k)\underline{\hat{X}}(k-1|k-1) \qquad (5)$$

$$P(k|k-1) = \Phi(t_{k-1},t_k)P(k-1|k-1)\Phi(t_{k-1},t_k)^T + Q(k) \qquad (6)$$

$$\underline{\hat{X}}(k|k) = \underline{\hat{X}}(k|k-1) + K(k)\,[\underline{Z}_{t_k} - M(t_k)\underline{\hat{X}}(k|k-1)] \qquad (7)$$

$$P(k|k) = P(k|k-1) - K(k)M(t_k)P(k|k-1) \qquad (8)$$

with:

$$K(k) = P(k|k-1)M(t_k)^T\,[M(t_k)P(k|k-1)M(t_k)^T + R(k)]^{-1} \qquad (9)$$

as the Kalman gain and

$\underline{\hat{X}}(k|l)$ = the least squares estimate of $\underline{X}_{t_k}$ based on the observations taken at $t_1, t_2, ...., t_l$

$P(k|l)$ = the covariance of the estimation error $\underline{X}_{t_k} - \underline{\hat{X}}(k|l)$

The initial condition for the recursion is given by $\underline{\hat{X}}(0|0) = \underline{X}_o$ and $P(0|0) = P_o$. The Kalman-Bucy filter has a predictor-corrector structure. Based on all previous information, a prediction of the value that the state vector will have at the next measurement time is made by means of the equations (5) and (6). Once the prediction of the state vector is known, it is possible to calculate the prediction of the next measurement by means of the measurement equation (3). Then, after the next measurement has been taken, the difference between this measurement and its predicted value is used to correct the prediction of the state vector by using the equations (7), (8) and (9).
Most problems in practical applications are nonlinear and linearization methods have been developed to extend the

Kalman-Bucy filter to nonlinear applications (extended Kalman filter, Jazwinsky 1970). In applications these methods have turned out to be useful in dealing with nonlinear problems. A detailed review of the Kalman-Bucy filter theory and its application to both linear and nonlinear problems can be found in the excellent work of Jazwinsky (1970).

THE DETERMINISTIC MODEL.

The tidal movement in the Dutch coastal area is approximately one dimensional (figure 1), so to describe this phenomenon the one dimensional shallow water equations (Dronkers 1964, 1975) can be used. These equations consist of the momentum equation:

$$\frac{\partial u}{\partial t} + u\frac{\partial u}{\partial x} + g\frac{\partial h}{\partial x} + \mu\frac{u|u|}{D+h} + \gamma\frac{v_x^2}{D+h} = 0 \qquad (10)$$

and the continuity equation:

$$\frac{\partial h}{\partial t} + \frac{\partial[u(D+h)]}{\partial x} = 0 \qquad (11)$$

where: $t$ = time
$x$ = direction of the tidal wave propagation
$h$ = water level
$u$ = water velocity
$D$ = depth
$g$ = gravitational acceleration
$\mu$ = bottom friction coefficient
$\gamma$ = wind stress coefficient
$v_x$ = wind speed in the x-direction

The boundary conditions consist of specified waterlevels at $x = 0$ and free outflow at $x = x_L$.

SYSTEM REPRESENTATION.

In order to apply the discrete Kalman-Bucy filter theory it is necessary to discretisize the equations (10) and (11). The finite difference scheme of Lax has been chosen to approximate these equations so that, if $F(x,t)$ is some function:

$$F(x,t) \simeq \frac{\Delta x_m}{\Delta x_{m-1} + \Delta x_m} F^k_{m-1} + \frac{\Delta x_{m-1}}{\Delta x_{m-1} + \Delta x_m} F^k_{m+1} \qquad (12)$$

$$\frac{\partial F(x,t)}{\partial x} \simeq \frac{F^k_{m+1} - F^k_{m-1}}{\Delta x_{m-1} + \Delta x_m} \qquad (13)$$

$$\frac{\partial F(x,t)}{\partial t} \simeq \frac{F_m^k - [\frac{\Delta x_m}{\Delta x_{m-1} + \Delta x_m} F_{m-1}^k + \frac{\Delta x_{m-1}}{\Delta x_{m-1} + \Delta x_m} F_{m+1}^k]}{\Delta t} \quad (14)$$

where: $\Delta t = t_{n+1} - t_n$ is the time step

$\Delta x_m = x_{m+1} - x_m$, $m = 1,\ldots, L-1$ are the dinstance steps

m,k are indices for space and time respectively.

The spatial distribution of $x_m$, $m = 1,\ldots, L$ is given by the location of the tide gauges (figure 1). For stability the Friedrichs-Courant-Lewy condition must be applied. The characteristic equations have been used to discretisize the boundary conditions (Roach 1972).
Applying the difference scheme leads to the following system equation:

$$\underline{x}_{t_{k+1}} = \Phi(\underline{x}_{t_k}) \quad (15)$$

where $\underline{x}_{t_k}$ is the state vector at $t_k$ consisting of the waterlevels $h_m^k$ and watervelocities $u_m^k$ at $x_m$, $m = 1,2,\ldots, L$.

THE STOCHASTIC MODEL.

The stochastic model is represented by

$$\underline{X}_{t_{k+1}} = \Phi(\underline{X}_{t_k}) + \underline{W}_{t_k} \quad (16)$$

in which $\Phi$ is defined by equation (15), $\underline{W}_{t_k}$ is the system noise with statistics given by equation (2). The unknown value of the convariance matrix Q(k) of $W_{t_k}$, which represents the uncertainties in the deterministic model (15), had to be estimated by trial and error: One simulates the filter for various values of Q(k) until a statisfactory filter performance is obtained. Observations at $t_k$ consist of the measurement of the waterlevels $h_m^k$ at $x = x_m$, $m = 1, \ldots, L$, so that the observation equation becomes:

$$\underline{Z}_{t_k} = \begin{bmatrix} (1+a^1)\,h_1^k \\ (1+a^2)\,h_2^k \\ \cdot \\ \cdot \\ (1+a^{L-1})\,h_{L-1}^k \\ h_L^k \end{bmatrix} + \underline{V}_{t_k} = \underline{m}(a^1,..,a^{L-1},\underline{X}_{t_k}) + \underline{V}_{t_k} \quad (17)$$

where $\underline{V}_{t_k}$ is the measurement noise with statistics given by equation (4). The parameters $a^i$, $i = 1,\ldots, L-1$ ($|a|<<1$), are introduced to account for the fact that most tide gauges are not located exactly on the x-axis (figure 1) and owing to wind effects, perpendicular to the x-axis, and the Coriolis force corrections had to be made.

PARAMETER ESTIMATION.

As mentioned in the introduction, the parameters $\mu$ and $\gamma$ in the system equation (16) and the parameters $a^i$ introduced in the observation equation (17) are not constant, but have essentially a random character, therefore, its values have to be estimated by the Kalman-Bucy filter. In order to obtain these estimates the parameters are considered as components of the state vector with system equations:

$$\mu_{t_{k+1}} = \mu_{t_k} \tag{18}$$

$$\gamma_{t_{k+1}} = \gamma_{t_k} \tag{19}$$

$$a^i_{t_{k+1}} = a^i_{t_k}, \; i = 1,\ldots, L-1 \tag{20}$$

By adding system noise in these equations the random character of the parameters can be taken into account. The observation equation (17) does not change.

APPLICATION.

Having derived the system and observation equations in the previous sections the extended Kalman filter can be applied to estimate the waterlevels, and the watervelocities together with the parameters in the model. Predictions at $t_k$ are made by using the nonlinear system equations (16), (18), (19) and (20) without system noise. As initial condition the estimates of the waterlevels and watervelocities at time $t_k$ in the points $x_i$, $i = 1,2, \ldots L$, are used. Since these estimates have been filtered by using the measurements at $t_k$, they are adapted to changing weather conditions. Since the parameters in the model have also been estimated by using these measurements, the prediction model is adapted to the changing conditions as well.
The extended Kalman filter has been applied to predict the waterlevel in the mouth of the Eastern Scheldt during a storm surge period (september 11-13, 1975). The predictions were made 1.5 hours ahead by using 8 tide gauges in the Dutch coastal area (figure 1). The results are plotted in figure 2. The high and low waterlevels of the astronomical tide, i.e.

the tide without the effects caused by the changing weather conditions, are plotted in figure 2 as well. It can be seen that the filter gives satisfactory results in forecasting the waterlevel during this storm surge.

## CONCLUSIONS.

In the study described in this paper a discrete extended Kalman filter for nonlinear systems has been derived. By using on-line information the model and the predictions can be adapted by the filter to changing physical circumstances. The filter is based on the one dimensional shallow water equations. Inaccuracies due to this one dimensional approximation are corrected by the filter by adapting the parameters $a^i$ in the model. Some other errors in the model are corrected by estimating the parameters $\mu$ and $\gamma$, e.g. the numerical damping is partly corrected by the estimation of $\mu$. The filter has been applied to field data gathered in the Dutch coastal area and has turned out to be a promising method of forecasting the waterlevel at short notice, especially during storm surge periods.

## REFERENCES.

Abbott, M.B. (1979) Conputational Hydraulics, Pitman, London.

Dronkers, J.J. (1964) Tidal Computations in Rivers and Coastal Waters, North Holland Publ. Amsterdam.

Dronkers, J.J. (1975) Tidal Theory and Computation, Advances in Hydrosciences, Vol. 10, ed. V.T. Chow, Academic Press, New York pp. 145-230.

Forsythe, G.E. and W.R. Wasow (1960) Finite Difference Methods for Partial Differential Equations, John Wiley and Sons Inc., New York, 1960.

Jazwinski, A.H. Stochastic Processes and Filtering Theory, Academic Press, New York and London, 1970.

Langerak, A., M.A.M. de Ras and J.J. Leendertse, Adjustment and Verification of the Randdelta II Model, Proc. 16th Int.Conf. on Coastal Engineering, Hamburg, 1978.

Maybeck, P.S. Stochastic Models, Estimation and Control, Vol.1, Academic Press, New York and London, 1979.

Meditch, J.S. Stochastic Optimal Linear Estimation and Control, McGraw-Hill, New York, 1969.

Ree, W.J. van de, J. Voogt and J.J. Leendertse, A Tidal Survey for a Model of an Offshore Area, Proc. 16th Int. Conf. on Coastal Engineering, Hamburg 1978.

Roach, P.J. Computational Fluid Dynamics, Hermosa Publ., Abuquerque, N.M., 1972.

Sawaragi, Y., T. Soeda and S. Omatu, Modeling Estimation, and their Applications For Distributed Parameter Systems, Springer Verlag, Berlin, 1978.

Tzafestas, S.G. Distributed Parameter State Estimation, Distributed Parameter Systems, ed. W.H. Ray and D.G. Lainiotis, New York and Basel, 1978.

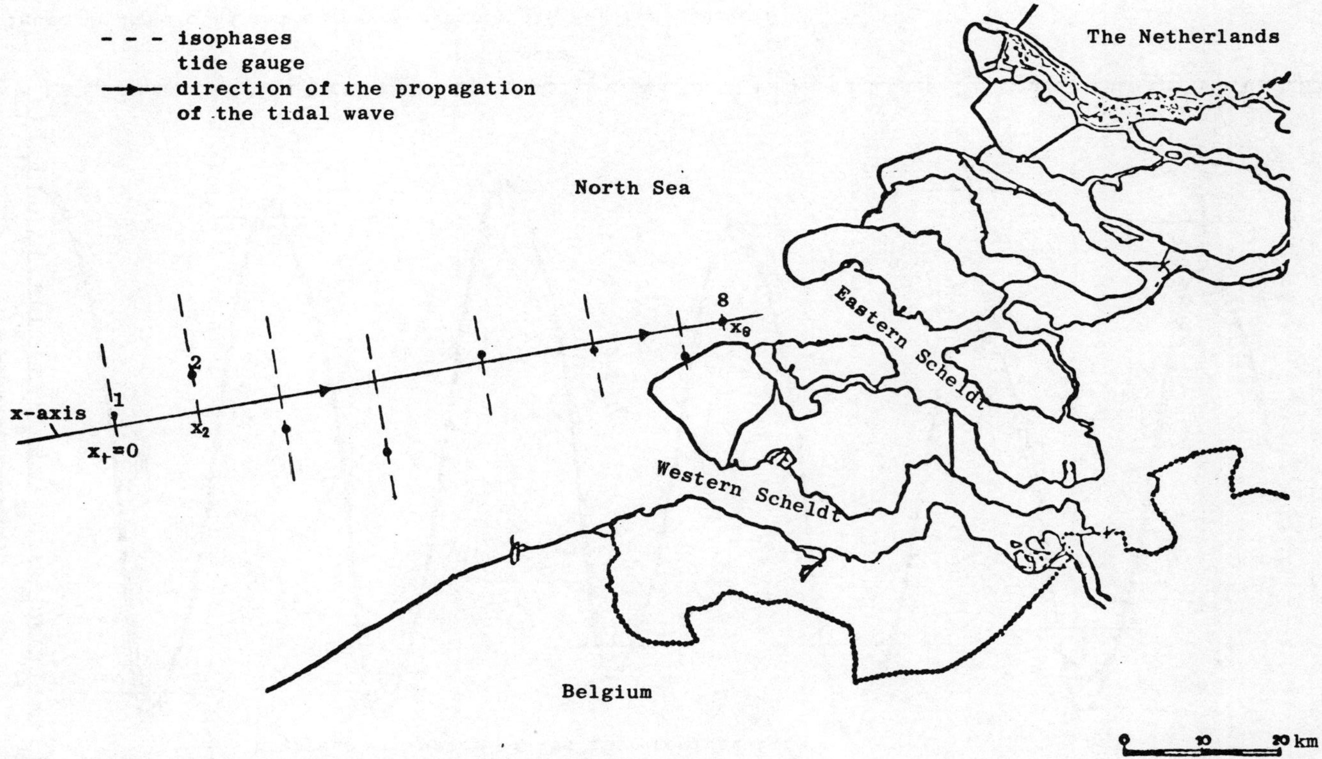


Figure 1 The tidal movement in the Dutch coastal area.

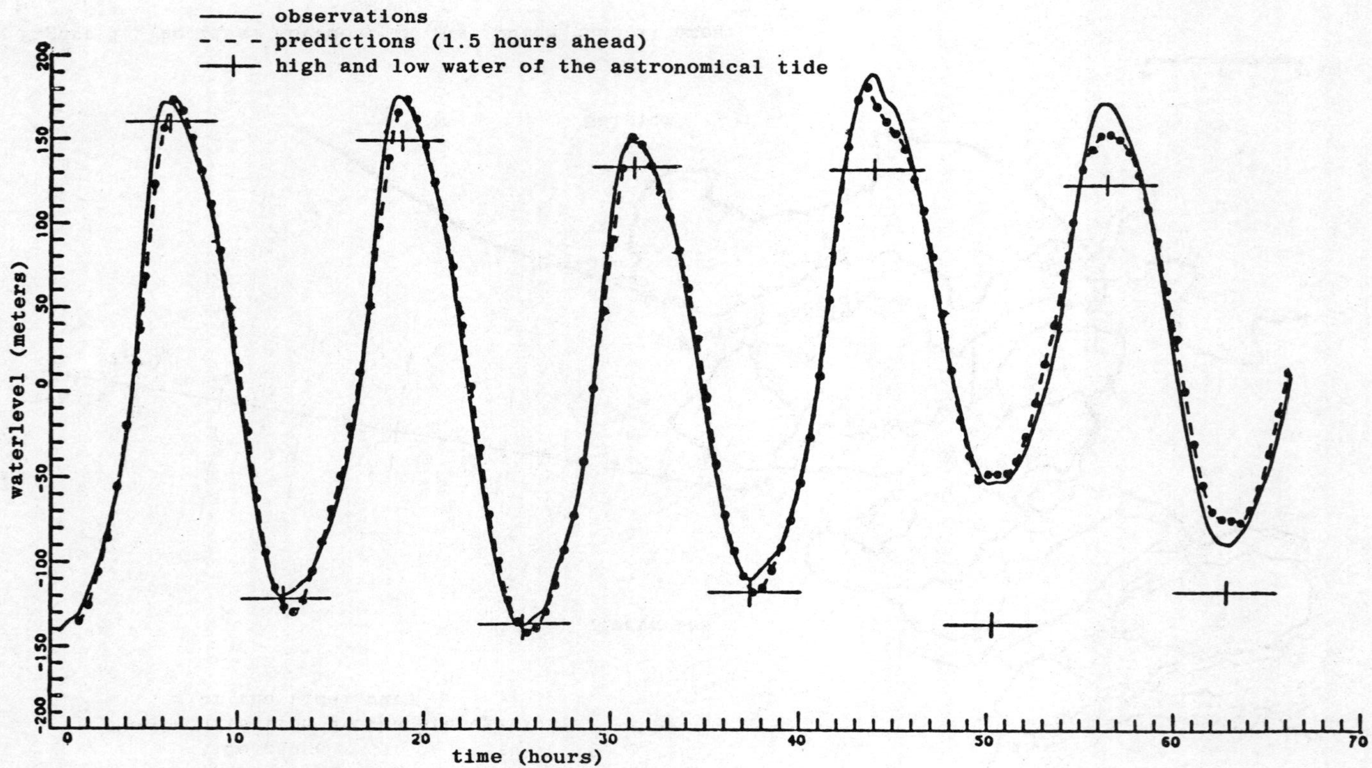


Figure 2 Predicted and observed waterlevels for tide gauge 8.

# TIME-WEIGHTING OF THE MOMENTUM EQUATION IN EXPLICIT WAVE EQUATION MODELS OF SURFACE WATER FLOW

Ingemar P.E. Kinnmark
William G. Gray
Princeton University

## INTRODUCTION

Much research effort has been directed toward finding numerical solutions to the shallow water equations (see Pritchard, 1971, for a clear derivation of these equations). Although finite difference modelers have enjoyed much success in solving these equations, finite element modelers have experienced many difficulties in obtaining stable, accurate solutions (Gray, 1980). Recently Lynch and Gray (1979) have reported on a wave equation formulation which seems to provide significant numerical advantages for finite element simulations.

However Lynch and Gray (1980) have pointed out that the numerical analysis of the linearized wave equation-momentum equation system is complicated by the uncoupling of discharge from the wave equation. They point out that the standard propagation factor analysis does not indicate the total numerical character of the solution scheme because it is independent of the numerical algorithm for the momentum equation. Lynch and Gray (1980) point out that the distribution factor, which is determined by considering solution schemes for both the wave and momentum equations, does complement the propagation factor and allow for complete numerical analysis.

In the current study, a few inherent properties of the wave equation-momentum equation scheme are shown. Numerical artifacts which arise from discrete approximation to the differential momentum equations and their effect on the solution are considered. A scheme is proposed which solves the equation set while suppressing the numerical artifacts.

## BASIC EQUATIONS

The equations analyzed in this study are the linearized, one-dimensional continuity and momentum equations and the wave equation, which is derived from these two. The linearized continuity equation is

$$\frac{\partial \phi}{\partial t} + \frac{\partial Q}{\partial x} = 0 \tag{1}$$

where
- $\phi$ is the free surface elevation above some datum,
- Q is the volumetric discharge per unit width,
- x is the spatial dimension, and
- t is time.

The momentum equation in linearized form with bottom friction is

$$\frac{\partial Q}{\partial t} + \tau Q + c^2 \frac{\partial \phi}{\partial x} = 0 \tag{2}$$

where
- $\tau$ is the linearized bottom friction parameter, and
- c is the wave speed in the absence of friction.

These two equations will be referred to as the primitive equations and their nonlinear forms are most commonly solved in computer models. Note that both equations contain terms in $\zeta$ and Q. The wave equation, which is solved in lieu of continuity equation (1), is obtained from equations (1) and (2).

First differentiate equation (1) with respect to time to obtain

$$\frac{\partial^2 \phi}{\partial t^2} + \frac{\partial}{\partial x}\left(\frac{\partial Q}{\partial t}\right) = 0 \tag{3}$$

Now substitution of equation (2) into equation (3) for $\frac{\partial Q}{\partial t}$ yields

$$\frac{\partial^2 \phi}{\partial t^2} - \tau \frac{\partial Q}{\partial x} - c^2 \frac{\partial^2 \phi}{\partial x^2} = 0 \tag{4}$$

Finally equation (1) is back substituted into equation (4) to eliminate $\frac{\partial Q}{\partial x}$ and give the wave equation:

$$\frac{\partial^2 \phi}{\partial t^2} + \tau \frac{\partial \phi}{\partial t} - c^2 \frac{\partial^2 \phi}{\partial x^2} = 0 \tag{5}$$

This equation does not have any terms in the discharge Q. Thus, in the linearized form, the surface elevations may be solved for without obtaining Q. Final solution for Q requires that momentum equation (2) be solved after equation (5).

## NUMERICAL ARTIFACTS

Gray and Lynch (1977) and Lynch and Gray (1980) have performed Fourier analyses which demonstrate the virtue of using the wave equation in lieu of the continuity equation for tidal simulations. Their linear analyses indicate that the numerical coupling of the momentum equation to the wave equation is complex. These analyses have concentrated on proper formulation of the wave equation. It will be shown here that although the wave equation does significantly reduce oscillations in the stage solution, the numerical formulation of the momentum equation is important in reducing oscillations in the velocity solution.

As a simple example for analysis, let us consider the following finite difference approximation to equation (2)

$$\theta \frac{Q_i^{k+1} - Q_i^k}{\Delta t} + (1 - \theta) \frac{Q_i^k - Q_i^{k-1}}{\Delta t}$$

$$+ \tau(\theta Q_i^{k+1} + (1 - 2\theta)Q_i^k + \theta Q_i^{k-1}) = - c^2 \frac{\phi_{i+1}^k - \phi_{i-1}^k}{2\Delta x} \tag{6}$$

where k is an index in the time coordinate,
i is a spatial index, and
$\theta$ is a parameter selected to define the solution scheme which typically ranges from 0 to 1.

The actual solution to equation (2), and its discrete counterpart equation (6), can be considered to be composed of two superposed solutions. The first part is the non-homogeneous part which results when the distribution of $\phi$, the surface elevation, forces the flow. Because the solution for $\phi$ is obtained from the wave equation, the excellent properties of the wave equation solution impact on this part of the solution. The second part of the solution is due to the homogeneous counterpart of equation (2), or equation (6), when the surface elevation term is neglected. Thus this part of the solution is independent of the wave equation solution scheme. The numerical approximation to this part of solution thus depends strictly on the discretization of equation (2) given by equation (6) with the right hand side neglected.

To perform the Fourier analysis of equation (6), let

$$Q = Q_o e^{\hat{i}\beta' t + \hat{i}\sigma x} \tag{7}$$

where $\beta'$ is the numerically obtained approximation to the frequency,
$\sigma$ is a spatial wave number equal to $2\pi$/wavelength, and
$\hat{i}$ is $\sqrt{-1}$.

Substitution of equation (7) into equation (6) with the right hand side neglected yields

$$\frac{\theta}{\Delta t} Q_o(e^{\hat{i}\beta'\Delta t} - 1) + \frac{(1 - \theta)}{\Delta t} Q_o(1 - e^{-\hat{i}\beta'\Delta t})$$

$$+ \tau Q_o[\theta e^{\hat{i}\beta'\Delta t} + (1 - 2\theta) + \theta e^{-\hat{i}\beta'\Delta t}] = 0 \tag{8}$$

If we denote the amplification at each time step, which is $e^{\hat{i}\beta'\Delta t}$, by $\lambda'$, equation (8) becomes

$$\theta(1 + \tau\Delta t)\lambda'^2 + (1-2\theta)(1 + \tau\Delta t)\lambda' + \theta(1 + \tau\Delta t) - 1 = 0 \qquad (9)$$

Solution of this equation for $\lambda'$ yields

$$\lambda' = 1 - \frac{1}{2\theta} \pm \frac{1}{2\theta}\sqrt{\frac{1 + \tau\Delta t(1 - 4\theta)}{1 + \tau\Delta t}} \qquad (10)$$

Numerical stability requires the $|\lambda'|$ be less than or equal to 1. Thus for stability

$$\theta \geq \frac{2 + \tau\Delta t}{4 + 4\tau\Delta t} \simeq \frac{1}{2} \qquad (11)$$

Thus regardless of the time stepping scheme selected for the wave equation, the momentum equation will carry its own separate stability constraint.

To get an approximate polynomial expression which indicates the dependence of $\lambda'$ on $\theta$, we expand the term under the radical in equation (1) to obtain

$$\lambda' = 1 - \tau\Delta t + \tau^2\Delta t^2(1 - \theta) - \tau^3\Delta t^3(1 - 2\theta + 2\theta^2) + \ldots \qquad (12a)$$

and

$$\lambda' = 1 - \frac{1}{\theta} + \tau\Delta t - \tau^2\Delta t^2(1 - \theta) + \tau^3\Delta t^3(1 - 2\theta + 2\theta^2) + \ldots \qquad (12b)$$

where it has been assumed that $\tau\Delta t < 1$ as is usually the case. These numerical approximations to the amplification must be compared to the exact expression, $\lambda$, obtained by solving the homogeneous version of equation (2)

$$\lambda = e^{-\tau\Delta t} \simeq 1 - \tau\Delta t + \frac{1}{2}\tau^2\Delta t^2 - \frac{1}{6}\tau^3\Delta t^3 + \ldots \qquad (13)$$

Immediately we see that there are two numerically obtained expressions for the amplification, whereas from the analytical analysis we would expect only one root. A comparison of equations (12) and (13) clearly indicates that the numerical root given by equation (12a) is the analogue to the physical root given by equation (13). The second numerical root, equation (12b) has no physical counterpart but is a numerical artifact which should be suppressed to avoid computing a temporal oscillations in the velocity field. Further inspection reveals that the highest order of accuracy for equation (12a) is obtained when $\theta$ equals 0.5. However because $\tau\Delta t$ is small, selection of $\theta$ greater than 0.5 will not seriously effect the accuracy of the numerical approximation to the physical root. On the other hand, $\theta$ dramatically influences the numerical artifact in equation (12b).

For purposes of discussion of the numerical artifact, we will assume that $\tau\Delta t$ in equation (12b) is small. Thus by selection of $\theta = 0.5$, the optimal value for the physical root, $\lambda'$ for the artificial root will approximately equal -1. This root would produce a spurious velocity oscillation from time step to time step which does not damp. In contrast, if a value for $\theta$ of 1.0 is selected, the artificial $\lambda'$ will approximately equal $\tau\Delta t$ which corresponds to a highly damped artifact. Thus a $\theta$ value of 1.0 essentially eliminates the numerical artifact without seriously distorting the root corresponding to the physical solution.

DISTRIBUTION FACTOR ANALYSIS

Lynch and Gray (1980) have demonstrated that the distribution factor is an important measure of the quality of the numerical solution for a coupled set of equations such as (2) and (5) where one of the dependent variables appears in only one of the equations. This factor is obtained first by defining a factor $\gamma$ for the analytic solution where

$$\gamma = \frac{Q}{\phi} \tag{14}$$

A similar factor, $\gamma'$, is defined for the numerical solution such that

$$\gamma' = \frac{Q_o}{\phi_o} \tag{15}$$

where $\phi_o$ is analogous to $Q_o$ defined in equation (7).

If the numerical scheme were a perfect representation of the differential equation, the ratio $\gamma'/\gamma$ would equal 1.0. This ratio is called the distribution factor D such that

$$D = \frac{\gamma'}{\gamma} \tag{16}$$

where its magnitude and phase are indicators of the quality of the solution.

From equation (2) with

$$Q = Q_o e^{\hat{i}\beta t + \hat{i}\sigma x} \tag{17a}$$

$$\phi = \phi_o e^{\hat{i}\beta t + \hat{i}\sigma x} \tag{17b}$$

$\gamma$ may be readily obtained as

$$\gamma = -\frac{\hat{i}\, c^2 \sigma}{\tau + \hat{i}\beta} = \frac{-\hat{i}\, c^2\, \sigma}{\frac{\tau}{2} \pm \hat{i}\sigma\sqrt{c^2 - (\frac{\tau}{2\sigma})^2}} \tag{18}$$

We may compute $\gamma'$ from equation (8) after including an approximation to $\frac{\partial\phi}{\partial x}$ using a second order expression at time level t and obtain

$$\gamma' = \frac{-\hat{i}\lambda' \frac{c^2 \Delta t}{\Delta x} \sin(\sigma\Delta x)}{\theta(\lambda'^2 - \lambda') + (1-\theta)(\lambda' - 1) + \tau\Delta t[\theta(\lambda'^2 + 1) + (1 - 2\theta)\lambda']} \tag{19}$$

where $\lambda' = e^{\hat{i}\beta'\Delta t}$ as computed from a numerical approximation to equation (5). For the current study equation (5) is approximated by the explicit formula

$$\frac{\phi_i^{k+1} - 2\phi_i^k + \phi_i^{k-1}}{\Delta t^2} + \tau \frac{\phi_i^{k+1} - \phi_i^{k-1}}{2\Delta t} - c^2 \frac{\phi_{i+1}^k - 2\phi_i^k + \phi_{i-1}^k}{\Delta x^2} = 0 \tag{20}$$

Note that the distribution factor in a sense measures the quality of the inhomogeneous part of the solution for Q because $\lambda'$ is obtained from the wave quation approximation and not from the momentum equation. The magnitude and phase of the distribution factor as a function of $\theta$ and the friction are presented in figures 1 and 2. It can be seen that the magnitude of the distribution factor is best for $\theta$ of 1.0 while the phasing is best for $\theta$ of 0.5. Perhaps the intermediate value for $\theta$ of $1/(1 + \tau\Delta t)$ is the best compromise value because it provides good damping of the numerical artifact in equation (12b) while improving the phasing of the distribution factor over the $\theta = 1.0$ case.

## NUMERICAL CALCULATIONS

To illustrate the importance of the above considerations, numerical calculations were performed using the hypothetical basin modeled by Lynch and Gray (1980) using linear triangular finite elements and by Gray et. al. (1982) using nine-node quadratic finite elements. The physical parameters used in the current study are the same as in Gray et. al. and the grid is shown in figure 3. Boundary conditions imposed were a sinusoidal stage with a period of 12.4 hours along the southwest edge, specified constant normal inflow on the two other open boundaries, and zero normal flow along the solid walls. Bottom friction was modeled using a Manning's n of 0.022. Bathymetry varies from 0.36 to 10.67 m in a rather irregular fashion.

The finite element model used in this study is WAVEQ which is described in Gray et. al. This model has the attractive feature that no matrices need be solved because all integrations on the nine-node quadrilaterals are performed using Simpson's rule. The equations solved are as follows wave equation

$$\frac{\partial^2 H}{\partial t^2} + \tau \frac{\partial H}{\partial t} - \frac{\partial}{\partial x}\left[\frac{\partial HUU}{\partial x} + \frac{\partial HUV}{\partial y} + gH \frac{\partial}{\partial x}(H - h) - fHV - A_x\right]$$

$$- \frac{\partial}{\partial y}\left[\frac{\partial HUV}{\partial x} + \frac{\partial HVV}{\partial y} + gH \frac{\partial}{\partial y}(H - h) + fHU - A_y\right]$$

$$- HU \frac{\partial \tau}{\partial x} - HV \frac{\partial \tau}{\partial y} = 0 \qquad (21)$$

x-momentum

$$\frac{\partial U}{\partial t} + U \frac{\partial U}{\partial x} + V \frac{\partial U}{\partial y} + g \frac{\partial}{\partial x}(H - h) - fV + \tau U = \frac{A_x}{H} \qquad (22)$$

y-momentum

$$\frac{\partial V}{\partial t} + U \frac{\partial V}{\partial x} + V \frac{\partial V}{\partial y} + g \frac{\partial}{\partial y}(H - h) + fU + \tau V = \frac{A_y}{H} \qquad 23)$$

where x is positive eastward,
y is positive northward,
U is the vertically averaged eastward velocity,
V is the vertically averaged northward velocity,
H is the total fluid depth,
h is the bathymetry and varies in space,
f is the Coriolis' parameter,
$\tau$ is the non-linear bottom friction function, and
$A_x, A_y$ are atmospheric forcing in the x- and y-directions, respectively.

The first set of computations used the wave and momentum equations without convective terms. Computations were run for 2100 time steps with a step size of 111.75 sec., or a little more than five cycles with the imposed sinusoidal forcing. Using values for $\theta$ of 0.5, 0.75, and 1.0, we were able to obtain stable solutions. The eastward velocity computed at node 43 for time steps 400 to 420 are plotted in figure 4. It is apparent that for $\theta$ of 0.5 there is a significant oscillation from time step to time step due to the computational root. When $\theta$ is increased above 0.5, this undesirable root is damped out while the fundamental solution is not severely affected.

A similar set of calculations were performed on the full equations with the convective terms included. For approximately the first 1000 time steps, computed stage and flow fields were very similar. However at this point the $\theta$ = 0.5 results began to diverge until at step 1340 a negative total depth of water was computed at one node causing the model to become unstable. For the $\theta$ values of 0.75 and 1.0, this phenomenon did not occur and, in fact, with computations continued until step 3300, a periodic system was achieved.

Figure 5 shows the computed eastward velocity at node 43 for a few consecutive time steps. It is clear that for $\theta$ = 0.5 there is an undesirable oscillation. This oscillation may feed back into the computations through the convective terms and be ultimately responsible for the demise of the $\theta = 0.5$ simulation. However this has not been precisely determined.

## CONCLUSIONS

The results presented here demonstrate that for a wave equation model, the proper temporal weighting of the momentum equation is important to ensure that numerical artifacts are suppressed. The weighting presented here seems effective in suppressing these oscillations while impacting the physical solution to a minimal degree. It has been demonstrated that in analyzing numerical schemes, it is important to consider both the homogeneous and inhomogeneous equations.

## ACKNOWLEDGMENT

This work has been supported by grant number CME-7921076-01 obtained from the United States National Science Foundation.

## REFERENCES

Gray, W.G. (1980) Do finite element models simulate surface flow? Proceedings of the Third International Conference on Finite Elements in Water Resources, pp. 1.122-1.136.

Gray, W.G. and D.R. Lynch (1977) Time-stepping schemes for finite element tidal model computations, Advances in Water Resources, 1, 2, 83-95.

Gray, W.G., D.R. Lynch, I. Kinnmark (1982) An explicit model for two-dimensional tidal circulation using isoparametric finite elements: WAVEQ User's Manual, Princeton University Water Resources Program Report 82-WR-1, 91 pp.

Lynch, D.R. and W.G. Gray (1979) A wave equation model for finite element tidal computations, Computers and Fluids 7, 3, 207-228.

Lynch, D.R. and W.G. Gray (1980) On the analysis of accuracy for two-equation transient problems, Intl. J. Numer. Meth. in Engrg., 15, 55-62.

Pritchard, D.W. (1971) Two-dimensional models, in G.H. Ward and W.H. Espery, eds., Estuarine Modeling: An Assessment, #PB206-807, Water Quality Office, U.S. Environmental Protection Agency.

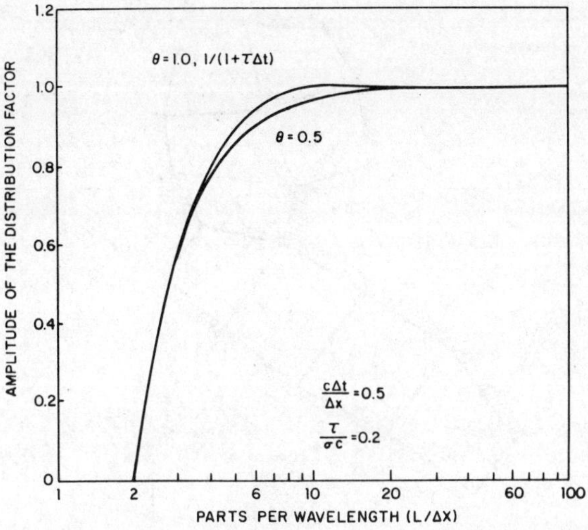


Figure 1. Amplitude of the Distribution Factor as a Function of Wavelength Divided by Grid Spacing for Different Values of the Time Weighting Parameter.

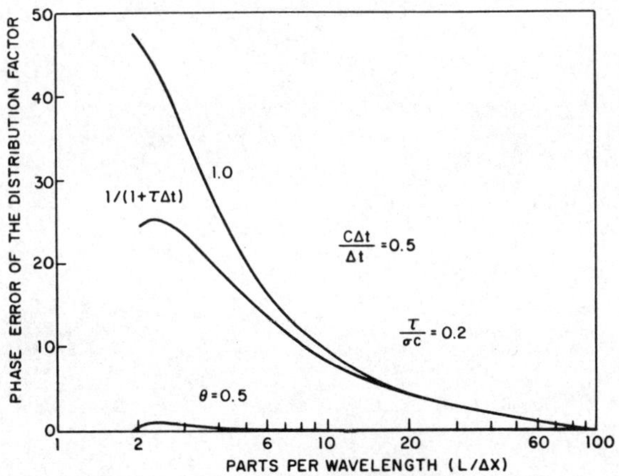


Figure 2. Phase Error of the Distribution factor as a Function of Wavelength Divided by Grid Spacing for Different Values of the Time Weighting Parameter.

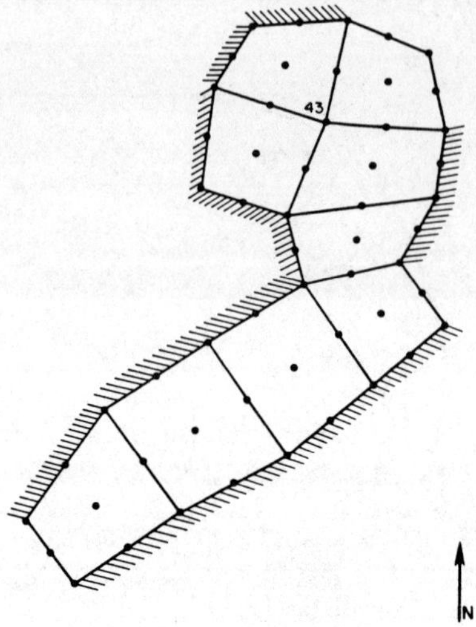


Figure 3. Finite Element Mesh for the Hypothetical Estuary Used in Numerical Calculations.

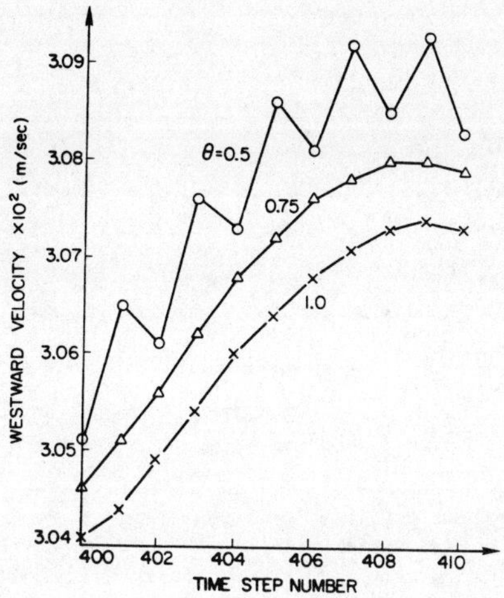


Figure 4. Westward Velocity as a Function of Time for Node Number 43 for Different Values of Time Weighting. Convective Terms Neglected.

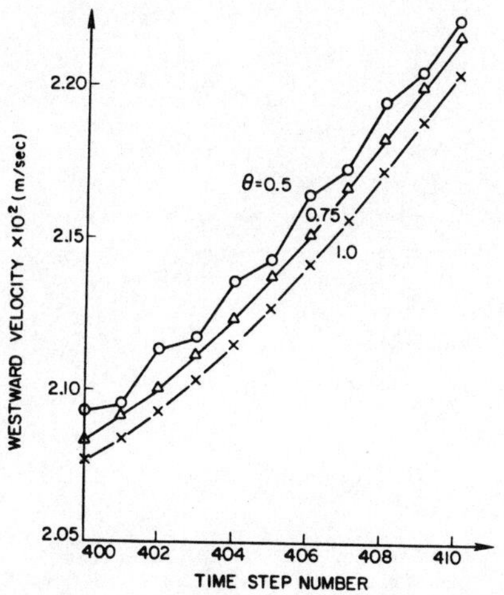


Figure 5. Westward Velocity as a Function of Time for Node Number 43 for Differential Values of Time Weighting. Convective Terms Included.

# FORMULATING LAYERED AND SEMI-ANALYTIC FINITE ELEMENT FLOW PROGRAMS WITH VARIABLE EDDY VISCOSITY

Jeffrey P. Laible

Department of Civil Engineering University of Vermont

## INTRODUCTION

A number of different finite element programs for the analysis of flows in lakes and oceans have evolved over the past several years under an increased demand to understand the hydrodynamics of these natural bodies of water. These programs are all founded on the basic Eulerian momentum and continuity equations for incompressible flow. Different analysis formulations have evolved depending on the importance of different flow velocities. Three dimensional transient flow programs are generally required to provide a reasonable description of the major flow characteristics. Although there are a number of ways to classify finite element programs of this type one feature which can be used to distinguish two groups is the means by which the vertical structure of the flow is represented. The two general possibilities are to use a semi-analytic formulation for the vertical structure or use a Galerkin based finite element or spectral formulation. Semi-analytic numerical formulations incorporating finite elements (Cheng, 1977; Liggett,1973) have been successfully used in the analysis of circulations in lakes. Fully three dimensional Galerkin based finite element models are generally of the multi-layered type (Kawahara, 1980; Wang and Connor, 1975; King and Norton, 1973; Koutitas, 1980). Single layered finite element models have also been formulated using higher degree polynomial bases functions (Laible, 1980) or Fourier series (Koutitas, 1978) to model the vertical structure of the flow.

The purely finite element models are generally capable of including a depth varying eddy viscosity coefficient, $e_\nu(z)$. The semi-analytic numerical models on the other hand have generally assumed the eddy viscosity coefficient to be constant. There are however a number of purely analytic studies which have investigated the effect of variable eddy viscosity for the linearized form of the governing equations and for problems of simple geometry (Thomas, 1975; Witton and Thomas, 1975; Madsen, 1977; Jordan and Baker, 1980). These studies, incorporating various forms of the eddy viscosity coefficient i.e., linear, parabolic, exponential, etc. have demonstrated that the variation of $e_\nu(z)$ can have a significant effect on the flow characteristics. It follows that a desirable feature of a purely finite element or semi-analytic numerical model would be the capability to include this variation.

The objective here is to describe the formulation of a two layered finite element program which includes a quadratic law variation of the eddy viscosity coefficient. With a very minor effort the program can be modified to model any form of $e_\nu$ distribution. As a means to check the accuracy of the method a comparison is made with the exact steady state solution of an x-z problem with variable eddy viscosity. The two layered model is also compared with a semi-analytic transient solution and the single layered version of the two layered model.

## THE GOVERNING EQUATIONS

The equations of momentum and continuity for nearly horizontal flow conditions can be derived from the Navier Stokes equations using the usual shallow water approximations. The major restrictions are that the vertical momentum is negligible as compared with the horizontal momentum and the pressure in the vertical direction is hydrostatic. Vertical velocities can however be determined based on continuity. With these restrictions the governing equations of momentum and continuity are:

$$\frac{\partial u}{\partial t} + u\frac{\partial u}{\partial x} + v\frac{\partial u}{\partial y} - fv - g\frac{\partial \zeta}{\partial x} - \frac{\partial}{\partial z}\left(e_\nu\frac{\partial u}{\partial z}\right) - e_h\left(\frac{\partial^2 u}{\partial x^2} + \frac{\partial^2 u}{\partial y^2}\right) = 0 \quad (1)$$

$$\frac{\partial v}{\partial t} + u\frac{\partial v}{\partial x} + v\frac{\partial v}{\partial y} + fu - g\frac{\partial \zeta}{\partial y} - \frac{\partial}{\partial z}\left(e_\nu\frac{\partial v}{\partial z}\right) - e_h\left(\frac{\partial^2 v}{\partial x^2} + \frac{\partial^2 v}{\partial y^2}\right) = 0 \quad (2)$$

$$\frac{\partial \zeta}{\partial z} - \frac{\partial}{\partial x}\int_o^H u dx - \frac{\partial}{\partial y}\int_o^H v dy = 0 \quad (3)$$

in which f is the Coriolis parameter, g is gravity, $e_\nu$ and $e_h$ are eddy viscosity representations of the tubulent Reynolds stresses and t is time. The remaining terms are defined in figure 1.

## BASIS FUNCTIONS FOR TWO LAYERED GALERKIN MODEL

The usual procedures of the Galerkin method may be applied to equations 1-3. The weighting functions are a subclass of the basis function chosen to approximate the dependent variables u, v and $\zeta$. In this study the basis functions for horizontal and vertical discretization are of two types. The velocities are approximated by the product series of the form:

$$N_1 = 2(S)^2 - 3(S)^2 + 1$$

$$N_2 = -2(S)^3 + 3(S)^2$$

$$N_3 = h((S)^3 - 2(S)^2 + S)$$

$$N_4 = h((S)^3 - (S)^2)$$

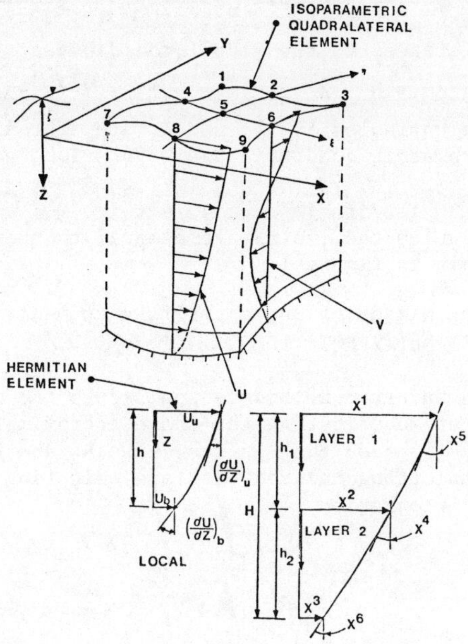


Figure 1. Coordinates, Variable Definition and Finite Elements

$$u \simeq N^i(S) \quad \phi^j(x,y)\ u^{ij}(t) \tag{4}$$

$$v \simeq N^i(S) \quad \phi^j(x,y)\ v^{ij}(t) \tag{5}$$

$$\zeta \simeq \quad \phi^j(x,y) \cdot \zeta^j(t) \tag{6}$$

$$S = z/h \qquad h = f(x,y)$$

Superscript indices refers to global functions or variables while subscripts refer to local functions or variables. The Einstein summation convention of summing over repeated indices is valid unless otherwise indicated by summation symbols. To avoid confusion when a quantity is raised to a power it will be enclosed in brackets ( ). The quantities $N_i$ are the local hermetian polynomial basis functions, the quantities $\phi_j$ are the local isoparametric quadratic basis functions for a nine node quadralateral. The hermitian basis functions are included in figure 1. The distribution of u and v over the vertical direction referenced to a layers local coordinates are given by

$$u = N_i X_i = N_1(S)\ u_u + N_2(S)\ u_L + N_3(S)\left(\frac{\partial u}{\partial z}\right)_u + N_4(S)\left(\frac{\partial u}{\partial z}\right)_L \tag{7}$$

$$v = N_i Y_i = N_1(S)\; v_u + N_2(S)\; v_L + N_3(S)(\frac{\partial v}{\partial z})_u + N_4(S)(\frac{\partial v}{\partial z})_L \quad (8)$$

Referenced to the global coordinates

$$u = N^i \;\; X^i(x,y) \quad v = N^i \;\; Y^i(x,y) \quad (9)$$

The variables $X_j$, $Y_j$ and $\zeta_j$ are approximated by the isoparametric quadratic basis functions as

$$X_j = \phi_r X_{jr} \qquad Y_j = \phi_r Y_{jr} \qquad \zeta = \phi_r \zeta_r \quad (10)$$

in which the $\phi$'s are referenced to the element numbering shown in figure 1.

## APPLICATION OF THE GALERKIN METHOD TO THE MOMENTUM AND CONTINUITY EQUATIONS

The Galerkin method is applied to the momentum equations by first substituting the above approximations into equation (1) and (2) resulting in residuals $R_x$ and $R_y$. The residuals are then orthogonalized with the weighting function in two stages of integration

$$0 = \int_A \phi^m \int_o^H N^N R_x \; dz dA \quad (11)$$

$$0 = \int_A \phi^m \int_o^H N^N R_y \; dz dA \quad (12)$$

In this formulation the surface and bottom velocity gradients will be prescribed by the relations

$$-e_\nu(H)(\frac{\partial u}{\partial z})_H = C_b \; u(H) \qquad -e_\nu(H)(\frac{\partial v}{\partial z})_H = C_b \; v(H) \quad (13)$$

$$-e_\nu(0)(\frac{\partial u}{\partial z})_o = C_w W_x((W_x)^2+(W_y)^2)^{\frac{1}{2}} \quad (14)$$

$$-e_\nu(0)(\frac{\partial v}{\partial z})_o = C_w W_y((W_x)^2+(W_y))^{\frac{1}{2}}$$

in which $W_x$, $W_y$ = wind velocity components, and $C_w$, $C_b$ = wind and bottom shear coefficients. By directly prescribing these values the boundary residuals are not required.

In the following the diffusion term will be omitted. These terms are rather lengthy and can be found in detail in Laible (1980). The inner integral over the depths in equations 11 and 12 are carried out first followed by a numerical integration over the surface area. The surface integration is accomplished numerically in the curvilinear coordinates using the coordinate transformations

$$\begin{Bmatrix} \frac{\partial \phi}{\partial \varepsilon} \\ \frac{\partial \phi}{\partial \eta} \end{Bmatrix} = [J] \begin{Bmatrix} \frac{\partial \phi}{\partial x} \\ \frac{\partial \phi}{\partial y} \end{Bmatrix} \qquad [J] = \begin{bmatrix} \frac{\partial x}{\partial \xi} & \frac{\partial y}{\partial \xi} \\ \frac{\partial x}{\partial \eta} & \frac{\partial y}{\partial \eta} \end{bmatrix}, \; dA = |J| d\xi d\eta \quad (15)$$

in which [J] is the Jacobian matrix and |J| is the determinant of [J]. The numerical integration scheme used is the two dimensional trapezoidal rule developed by Gray (1976) for this element. The approach effectively reduces the procedure to a one point integration scheme. The final equation for conservation of momentum in the x direction for a node $\ell$ is:

$$\sum_{e=1}^{NE} \sum_{L=1}^{2} (|J|_e W_e (\alpha_j^i \frac{\partial X_j}{\partial t})_L)^{\ell} = F_x \tag{16}$$

$$F_x = - \sum_{e=1}^{NE} (|J|_e W_e \sum_{L=1}^{2} [\beta_{jk}^i (X_j (\frac{\partial \phi_s}{\partial x})_e X_{ks} + Y_i (\frac{\partial \phi_s}{\partial y})_e Y_{ks}]_L$$

$$+ [\gamma_{jk}^i (X_j (\frac{\partial \phi_s}{\partial x} h_s)_e X_k + Y_j (\frac{\partial \phi_s}{\partial y} h_s) X_k]_L$$

$$+ [-\alpha_j^i Y_i f - K_j^i X_j - \lambda^i (\frac{\partial \phi_s}{\partial x} h_s)_e]_L)^{\ell}$$

in which $W_e$ = numerical integration factor (trapezoidal rule) i=1,6 j=1,4 k=1,4 $\ell$=all non variables evaluated at node $\ell$, NE = number of elements attached to node $\ell$, L=layer number, ' = $\partial/\partial z$ and

$$\alpha_j^i = \int_0^{H_L} N^i N_j dz, \quad \beta_{jk}^i = \int_0^{H_L} N^i N_j N_k dz, \quad \gamma_{jk}^i = \int_0^{H_L} N^i N_j \frac{\partial N_k}{\partial h} dz \tag{17}$$

$$\lambda^i = \int_0^{H_L} N^i dz \qquad K_j^i = \int_0^{H_L} N^i (N_j' e_\nu'(z) + e_\nu(z) N_j'') dz$$

Starting with equation 2 an analogous y momentum equation may also be formed.

Using the same quadratic basis functions the Galerkin method applied to the continuity equation results in

$$\sum_{e=1}^{NE} (|J|_e W_e)^{\ell} \frac{\partial \zeta}{\partial t} = F_\zeta$$

$$F_\zeta = \sum_{e=1}^{NE} |J|_e W_e [(\frac{\partial \phi_r}{\partial x} q_{xr})_e^{\ell} + (\frac{\partial \phi_r}{\partial x} q_{yr})_e^{\ell}] \tag{18}$$

$$q_x = \int_0^H u dz \qquad q_y = \int_0^H v dz$$

In a matrix format equation 16, its y analogy and the continuity equation may be written as:

$$[\alpha]_{6x6} \left\{\frac{\partial X}{\partial t}\right\} = \{F_x\} \qquad [\alpha]_{6x6} \left\{\frac{\partial Y}{\partial t}\right\} = \{F_y\} \tag{19}$$

$$m \, \partial\zeta/\partial t = F_\zeta \tag{20}$$

In terms $\partial X_5/\partial t$, $\partial X_6/\partial t$, $\partial Y_5/\partial t$, $\partial Y_6/\partial t$ are treated as prescribed variables. The [α] matrices are partitioned resulting in a 4x4 matrix for each node in the grid. These matrixes are inverted and stored in the program once per analysis. The right hand sides of equation (19) are $\{F_x\}$, $\{F_y\}$,

$F_y$ plus the contributions from the prescribed values noted above. An explicit split time integration technique, developed by Wang and Connor (1975), is used to solve equation (19). In this procedure $\partial\zeta/\partial t$ is used to compute $\zeta$ at $t+\Delta t/2$ while $\{\partial x/\partial t\}$ is used to compute $\{X\}_{t+\Delta t}$ for the x direction. The variables $\{X\}$ and $\{Y\}$ are therefore computed using $\zeta$ values which are centered in time. The backwards difference operator is used to advance each step.

SEMI-ANALYTIC FINITE ELEMENT MODEL

The semi-analytic model is developed by first rearranging equations (1) and (2) as:

$$\frac{\partial}{\partial t}\left(e_\nu \frac{\partial u}{\partial z}\right) - \frac{\partial u}{\partial t} + fv = g\frac{\partial \zeta}{\partial x} + C_x \tag{21}$$

$$\frac{\partial}{\partial t}\left(e_\nu \frac{\partial v}{\partial z}\right) - \frac{\partial v}{\partial t} - fu = g\frac{\partial \zeta}{\partial y} + C_y \tag{22}$$

in which $C_x = u\partial u/\partial x + v\partial u/\partial y$, $C_y = u\partial v/\partial x + v\partial v/\partial y$ and z is now positive upward and zero at the bottom.

Two different time integration schemes have been investigated.

1) a weighted method with a predictor corrector step for computation of the surface elevation
2) a wave equation type scheme

In the weighted difference approach the following operators are used

$$\partial F/\partial t = (F_{k+1} - F_k)/\Delta t \tag{23}$$

$$F = \varepsilon F_{k+1} + (1-\varepsilon)F_k \tag{24}$$

in which F is any function of time, k is a time level and $\varepsilon$ is a weighting factor. Using 21 as an example substitution of 23 and 24 eventually leads to

$$(D_x)_{k+1} = \left(\frac{1-\varepsilon}{\varepsilon}\right)(D_x)_k - \left(\frac{u}{\varepsilon^2\Delta t}\right)_k + \left(\frac{1-\varepsilon}{\varepsilon}\right)(F_x)_k + (F_x)_{k+1} \tag{25}$$

$$D_x = \frac{\partial}{\partial z}\left(e_\nu \frac{\partial u}{\partial z}\right) - \frac{u}{\varepsilon\Delta t} + fv \tag{26}$$

$$F_x = q\frac{\partial \zeta}{\partial x} + C_x \tag{27}$$

$D_x$ is treated as an ordinary differential equation along with its y analogy. The values of $\zeta$ as required for $F_x$ are computed from equations (18) and (20). A backwards difference operator is used to advance $\zeta$, however a corrector step is required when $\varepsilon<1$. The correction step may be expressed as

$$\zeta^*_{k+1} = ((1-\varepsilon_1)(\partial\zeta/\partial t)_{k+1} + \varepsilon_1(\partial\zeta/\partial t)_k)\Delta t + \zeta_k \tag{28}$$

in which $\varepsilon_1$ is a corrector weighting factor $\varepsilon_1 < \varepsilon$. The case of $\varepsilon=1$ was used by Cheng (1977) in his semi-analytic model with constant $e_\nu$. Both methods compute $C_x$ at level k.

In the wave equation approach 21 and 22 are differentiated with respect to time. The continuity equation is substituted eliminating $\zeta$. The following temporal operators are then used

$$\frac{\partial^2 F}{\partial t^2} \simeq \frac{F_{k+1} - 2F_k + F_{k-1}}{(\Delta t)^2} \tag{29}$$

$$\frac{\partial F}{\partial t} \simeq \frac{F_{k+1} - F_{k-1}}{\Delta t} \tag{30}$$

The momentum equation becomes

$$(D_x)_{k+1} = (D_x)_{k-1} - 2\left(\frac{u}{(\frac{1}{2}\Delta t)}\right)_k + 2\left(\frac{u}{\frac{1}{2}\Delta t}\right)_{k-1} + (F_x)_k \tag{31}$$

$$F_x = 2\Delta t\left(-g\,\frac{\partial G}{\partial x} + \dot{C}_x\right) \tag{32}$$

$$G = \partial q_x/\partial x + \partial q_y/\partial y \tag{33}$$

Gray and Lynch (1977) have demonstrated the attractiveness of this approach for the vertically integrated flow equation. The surface elevation is no longer computed (except as required for output) and no correction scheme is required. The wave equation scheme is centered in time. $\dot{C}_x$ however is computed by using a backwards difference $\dot{C}_x = ((Cx)u - (Cx)_{k-1})\Delta t$.

Regardless of the time integration scheme used the equations to be solved are

$$D_x = \frac{\partial}{\partial z}\left(e_\nu \frac{\partial u}{\partial z}\right) - \frac{u}{\varepsilon \Delta t} + fv = R_x \tag{34}$$

$$D_y = \frac{\partial}{\partial z}\left(e_\nu \frac{\partial v}{\partial z}\right) - \frac{v}{\varepsilon \Delta t} - fu = R_y \tag{35}$$

Rx and Ry are the right hand sides of either 25 and 31 (and the y analogies). At this point complex notation is introduced.

$$W = u + vi \qquad i = \sqrt{-1} \tag{36}$$

resulting in

$$D(z) = \frac{\partial}{\partial z}\left(e_\nu \frac{\partial w}{\partial z}\right) - \left(\frac{1}{\Delta t \cdot \varepsilon} + fi\right)W = 0 \tag{37}$$

In this work the eddy viscosity coefficient $e_\nu(z)$ will be taken to be of the quadratic form used by Swift (1979).

$$e_\nu(z) = -az^2 + bz + c \tag{38}$$

Introducing $e_\nu(z)$ and the transformation:

$$\xi = (z/h - b/2ha)/\sqrt{c/ah^2 + (b/2ah)^2} \tag{39}$$

into 37 yields

$$D(\xi) = (1-\xi^2)\frac{\partial^2 W}{\partial \xi^2} - 2\xi\frac{\partial W}{\partial \xi} + \lambda W = R = \overline{f}(\xi) + \overline{q} \tag{40}$$

in which $\lambda = -(1/a\Delta t\varepsilon + (f/a)i)$ and R is the complex right hand side of equations (34) and (35). $D(\xi)$ is the well known

Legendre equation. Using the previously defined surface and bottom boundary conditions the solution to (40) may be found in terms of the even and odd Legendre series P and Q. The final solution is quite lengthy but may be expressed as:

$$W = F_g \bar{g}/\lambda + F_w w^2 + F_p + F_q \tag{41}$$

in which $F_g$, $F_w$, $F_p$ and $F_q$ are complex functions of P & Q and $\bar{g} = (\partial\zeta/\partial x + \partial\zeta/\partial y\ i)$; $w^2 = w_x^2 + w^2_y\ i$ (complex wind velocity). The terms Fg and Fw represent the slope and drift current respectively. Fp and Fq are the tranient effects and the contribution from the convective term.

The finite element procedure is used in this method to compute spatial derivatives $\partial/\partial x$, $\partial/\partial y$, etc. Using the same isoparametric quadralateral element as in the layered model the Galerkin method may be applied to the simple relation $H(x,y) = \partial L(x,y)/\partial x$ to obtain

$$\int \phi_i\ \phi_j\ H_j\ dxdy = \int \phi_i \frac{\partial\phi_j}{\partial x} L_j\ dxdy \tag{42}$$

Using the nodal numerical integration for the above integrals yields

$$H_\ell = W(\ell)^e_j L^e_j,\quad W(\ell)^e_j = \left[\sum_{e=1}^{NE} (\partial\phi_j/\partial x)^\ell |J|^\ell_e\right] \Big/ \sum_{e=1}^{NE} |J|^\ell_e$$

in which $H_\ell$ is the x derivative of function L(x,y) at node $\ell$. To obtain y derivatives $\partial\phi/\partial x$ is replaced by $\partial\phi/\partial y$. A similar approach was used by Cheng (1977) and Forristall (1980) for a four node quadralateral element.

## NUMERICAL RESULTS

### x-z Problem

The numerical schemes were tested on an x-z basin measuring 200 m by 1000 m and 5 m deep. The straight basin was modeled by 5 elements with $\Delta x = 100$ m. A constant x direction wind was applied to the surface. To compare the different numerical schemes for the semi-analytic model, the time history of the surface velocity at the center node of the basin is shown in figure 2. In this test $e_\nu$ = constant = .1 m$^2$/s, $\Delta t$=10 sec, g = 10 m/s$^2$, $w^2{}_x$ = 10 m$^2$/s$^2$, $C_w$ = .0001, $C_b$ = .05 m/s. The plot shows that the wave equation solution demonstrates the greatest damping. For the weighted difference scheme with $\varepsilon$=1 or $\varepsilon$=.5 the peak values lag behind those of the wave equation solution. The response for $\varepsilon$=.5 and the wave equation show the greatest agreement.

Tests using a constant $e_\nu$ were also carried out using the two layer, semi-analytic and single layer version of the two layer model. The time history compares well although the semi-analytic model using $\varepsilon$=1, $\varepsilon_1$=.5 shows the greatest damping. All flow profiles showed excellent agreement with the exact steady solution for constant $e_\nu$.

The problem of a quadratic $e_\nu$ distribution was also investigated and compared with the exact solution. In this test all parameters were the same as above except the $e_\nu$ distribution was prescribed by $e_\nu(H) = .02\ m^2/s$, $e_\nu(H/2) = .05\ m^2/s$ and $e_\nu(o) = .02\ m^2/s$. The flow profiles as the steady state is reached are shown in figure 3 for all models and the exact steady solution. The steady solution was found to be given by:

$$u = C_w W_x^2(C_1 - R(z)) + G(C_2 + F(z))$$

in which $R(z) = (2/\sqrt{Q}) \tanh^{-1} (2az + b)/\sqrt{Q}$

$$F(z) = (H + b/2a)\ R(z) + (1/2a)\ell n(e_\nu(z))$$

$$G = g\frac{\partial\zeta}{\partial x} = -C_w W_x^2 \left[\frac{C_1 H - \int_0^H R(z)dz}{C_2 H - \int_0^H F(z)dz}\right]$$

$$\int_0^H R(z)dz = \left(\frac{1}{a}\right)\Big|_{\xi(o)}^{\xi(H)} \tanh^{-1}\xi + \frac{1}{2}\ \ell n(1-\xi^2)$$

$$\int_0^H F(z)dz = \left[\left(\frac{1}{2a}\right)\Big|_0^H\left(z + \frac{b}{2a}\right)\ell n\ e(z) - 2z + \frac{Q}{2a}\ R(z)\right] + \left(H + \frac{b}{2a}\right)\int_0^H R(z)dz$$

$$\xi = (2az + b)/\sqrt{Q}, \quad C_1 = \frac{1}{C_b} + R;$$

$$C_2 = -\left(HC_1 + \frac{1}{2a}\ell n(c) + \frac{b}{2a}\ R\right)$$

$$R = \frac{2}{\sqrt{Q}}\tanh^{-1}\frac{b}{\sqrt{Q}}\ ; \quad \sqrt{Q} = \sqrt{b^2 - 4\ ac}$$

The two layered model and the semi-analytic model show excellent agreement with the exact solution. The single layer model appears to minimize the error between the exact and a cubic distribution, but not without significant error. The two layer model actually gives a closer fit to the exact solution than the semi-analytic model.

Three dimensional problem

The nearly circular lake with variable depths was analyzed for a constant x direction wind from the west. The maximum depth at the center is 50 meters. Boundary depths are 15m with all other depths varying according to a parabolic distribution. Six isoparametric elements (35 nodes) were used to model the basin which measures 120 km x 120 km along the major axes. Boundary conditions required flow normal to the boundary to be zero. Other parameters were $C_w = 0.001$, $C_b = .5$, $w^2 = 10\ m^2/s^2$ $\Delta t = 240$ sec. Constant and variable $e_\nu$ analyses were carried out. For a constant analysis $e_\nu = .1\ m^2/s$. For the variable $e_\nu(h) = .07\ m^2/s$ $e_\nu(max) = .1\ m^2/s$ $e(0) = .02\ m^2/s$. The flow patterns at $z = H$ (surface) and $z = 0$ (bottom) as the steady state is approached are shown in figure 4.

Both analyses produce the characteristic two-celled circulation pattern below the surface with a clockwise gyre in the south. An s-shaped flow pattern along the central axis is evident for both the surface and return currents. The variable $e_\nu$ has the general effect of increasing velocities and accentuating the s-shaped pattern at all depths. The centers of the gyres are also deflected closer to the center of the basin with the northern gyre moving slightly to the east and the southern gyre moving to the west.

The effect of the variable $e_\nu$ is more vividly revealed by a comparison of the vertical structure at the center node. The flow profiles for this node reveals the increased complexity of the flow for a variable $e_\nu$ (figure 5). Figure 5 also illustrates the significant increase in the reverse current as $e_\nu$ near the bottom is reduced. These results were obtained using the semi-analytic model.

The two layered program was used to analyze the same basin but with a more sharply decreasing $e_\nu$. A linear variation of $e_\nu$ was used with $e_\nu$ surface = .06 $m^2/s$ and $e_\nu$ bottom = .01 $m^2/s$. The circulations at the lower depths were similar to those of figure 4, however the gyres now extend to the surface as shown in figure 6.

RUN TIME AND BEHAVIOR

The semi-analytic, two layer and single layer programs were all run on a DEC 20-60. To compare run times the ratio CPU/(NNxNS) is used. CPU is in seconds of run time, NN is the number of nodes in the grid and NS is the number of time steps required to reach a steady state. The ratios for the single layer, two layer and semi-analytic model are 5 $10^{-3}$ 9 $10^{-3}$ 30 $10^{-3}$ respectively. The single layer model is the most efficient but cannot adequately account for a variable eddy viscosity. The semi-analytic model requires more than three times the run time as the two-layered model. Overall the two layered model appears to be the most versatile and computationally efficient procedure of those considered here for analyses with depth varying eddy viscosity.

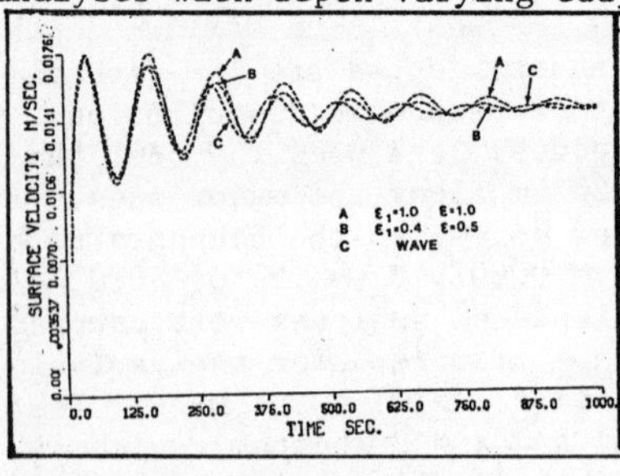


Figure 2. Comparison of Integration Schemes - Semi-Analytic Model

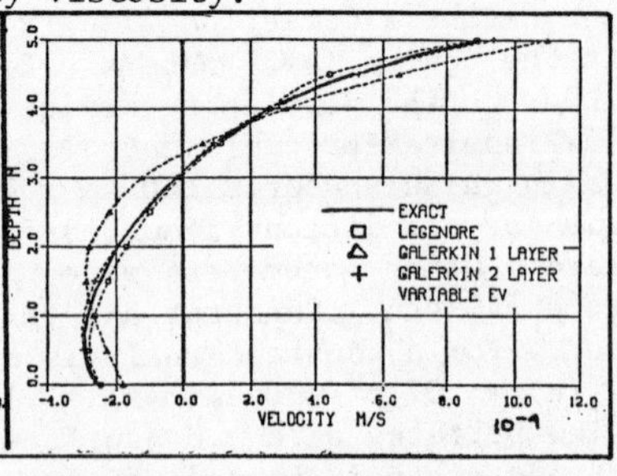


Figure 3. Flow Profiles at Steady State Variable $e_\nu(z)$

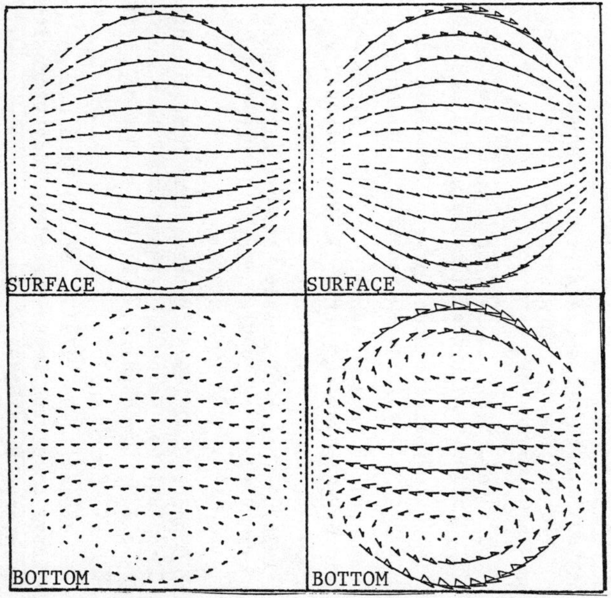


Constant $e_\nu$ Wind → Variable $e_\nu$

Figure 4. Flow patterns in circular lake with parabolic depth profile.

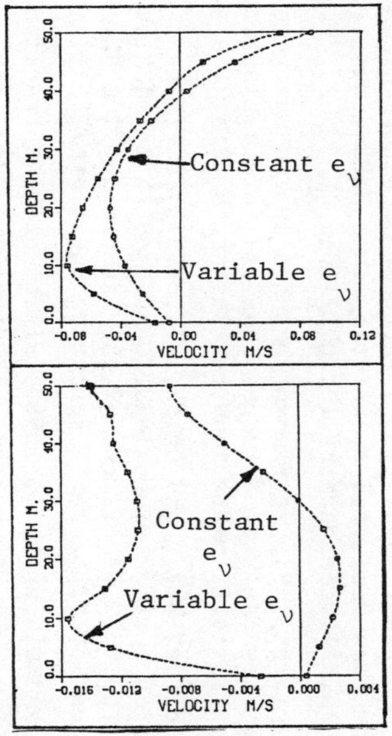


Figure 5. Flow profiles at center node.

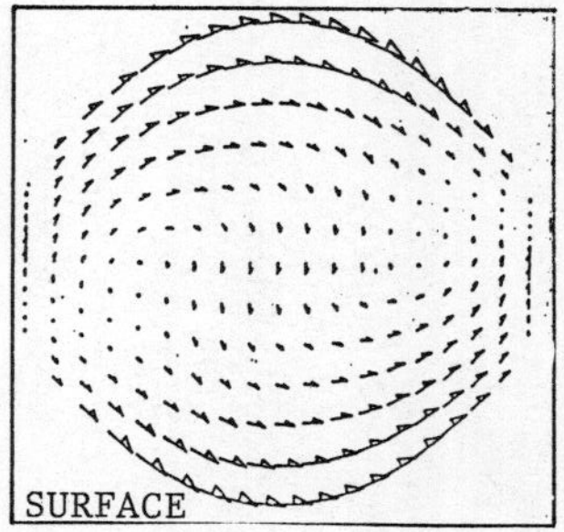


Figure 6. Surface flow for linear varying $e_\nu$.

## REFERENCES

Cheng, R.T. (1977) Transient Three-Dimensional Circulation of Lakes. J. Engr. Mechanics Div., 103, EM1:17.

Forristall, G.Z. (1980) A Two Layer Model for Hurricane-Driven Currents in a Large Lake with Depth-Dependent Eddy Viscosity. 7:248-255.

Gray, W.G. (1976) An Efficient Finite Element Scheme for Two-Dimensional Surface Water Computations. First International Conference on Finite Elements in Water Resources.

Gray, W.G. and Lynch, D.R. (1977) Time Stepping Schemes for Finite Element Tidal Model Computations. Adv. in Water Resources, 1, 2.

Jordan, T.F. and Baker, J.R. (1980) Vertical Structure of Time-Dependent Flow Dominated by Friction in a Well-Mixed Fluid. J. Physical Oceanography 10:1091-1103.

Kawahara, M. (1980) Finite Element Methods in Two-Layer and Multi-leveled Flow Analysis. Proceedings 3rd International Conference on Finite Elements in Water Resources. 2:5.3.

King, I.P., Norton, W.R., and Orlob, G.T. (1973) A Finite Element Solution for Two-Dimensional Density Stratified Flow. Prepared for the U.S. Department of the Interior Office of Water Resources Research, Water Resources Engineers.

Koutitas, C. and O'Connor, B. (1980) Finite Element-Fractional Steps Solution of the 3-D Coastal Circulation Model. Proceedings 3rd International Conference on Finite Elements in Water Resources, 2:5.64.

Laible, J.P. (1980) Finite Element Analysis of Depth Varying Flow in Lakes. Proceedings 3rd International Conference on Finite Elements in Water Resources. 2:5.72.

Liggett, J.A. and Young, D.L. (1973) Transient Finite Element Shallow Lake Circulation. J. Hydraulics Div., 103, HY 2:109.

Madsen, O.S. (1977) A Realistic Model of the Wind-Induced Ekman Boundary Layer. J. Physical Oceanography 7:248-255.

Swift, M.R., Reichard, R. and Celibbol, B. (1979) Stree and Tidal Current in a Well-mixed Estuary. J. Hydraulics Div., 7:785-799.

Thomas, J.H. (1975) A Theory of Steady Wind-Driven Currents in Shallow Water with Variable Eddy Viscosity. J. Physical Oceanography. 5:136-142.

Wang, J.D. and Connor, J.J. (1975) Mathematical Modeling of Near Coastal Circulations, Ralph M. Parsons Laboratory for Water Resources and Hydromechanics, MIT, Department of Civil Engineering, Report No, 200.

Witton, A.J. and Thomas, J.H. (1975) Steady Wind Driven Currents in a Large Lake with Depth-Dependent Eddy Viscosity. J. Physical Oceanography, 6:85-92.

# TIDAL PROPAGATION IN A RECTANGULAR BASIN OF THE BAY OF FUNDY

M. Rahman, Department of Applied Mathematics, Technical University of Nova Scotia, Halifax, Nova Scotia

## ABSTRACT

Two dimensional tidal propagation in a rectangular basin of the Bay of Fundy has been studied in this paper. The dynamic response of the tidal currents and tidal elevation inside the basin has been obtained by the finite element method using triangular elements with three nodes. Analytical solutions to the linearized tidal equations are also presented to test the accuracy of the numerical solutions. The initial tidal elevations has been found to amplify near the closed end of the basin or at the tidal barrier site.

## INTRODUCTION

It is well known that the tides in the Bay of Fundy are among the highest in the world. For instance, at the head of the Minas Basin, the spring tide attains a height of 52.9 feet (16.12 m) and in Chignecto Bay, the tidal range is 46 feet (14.02 m). The Bay of Fundy tidal behaviour is semidiurnal, with two high tides and two low tides of approximately the same height each day. For this reason, the Federal Government of Canada and the Provincial Governments of New Brunswick and Nova Scotia have looked at various schemes for the generation of tidal power throughout the Bay of Fundy.

The Minas Basin and Cumberland Basin provide the greatest promise of economic advantage. An estimated 4000 megawatts of power could be developed by a dam across the Cumberland Basin using existing techniques. The construction of such a dam or barrier calls for much experience and expertise in the specialty field of tidal power development.

A knowledge of tidal elevations and tidal currents at the tidal power development sites is very important, both from the standpoint of construction operations and also from that of build-up of sediment after completion of the site works. Up to the present, relatively little work has been done in this area, except that a few previous researchers including Greenberg (1976) have provided some information relating to some parts of the Bay of Fundy.

The problem of tidal response in estuaries which was first considered by Taylor (1920) concerns the reflection of Kelvin waves from the closed end of a semi-infinite rectangular canal. He considered two Kelvin waves of equal amplitude travelling in opposite directions in the canal; the origin of the coordinates was chosen such that their phases matched at x = 0. Taylor used Fourier expansion method. Brown (1973) demonstrated to find the solution of Taylor's problem by using the method of collocation.

Numerical models of tidal response in two dimensions have been studied by many investigators including Liggett and Woolhiser (1967). Comparison of various numerical schemes has been described by them.

In this study, we have developed a simple mathematical model for the region of the Bay of Fundy shown in Figure 1. A hypothetical barrier or dam is created marked with dashed lines on the east side of the Bay and an open boundary is assumed (marked with dashed lines) joining Digby, N.S. and Saint John, N.B. This simple geometrical configuration is chosen to obtain some physical insight from the linearized two-dimensional mathematical model. The straight shore lines provide enormous advantages to obtain the analytical solutions easily from the linearized model. In this paper, Coriolis effect and bottom friction have been taken into consideration. Finite element method has been used to obtain the tidal response and currents from the non-linear mathematical model. Then these results were checked by the analytical solutions.

FINITE ELEMENT FORMULATION

Galerkin's method on finite elements is used to solve the basic governing equations given in the work of Rahman et al (1980). The flow domain is first discretized into a number of triangular elements. Then basic functions are chosen which interpolate the dependent variables and parameters over the elements. Thus for each element the following equations result:

Continuity:

$$\sum_{n_e} \left( \iint \left( [N]^T [N] \frac{\partial}{\partial t} \{\eta\} - \frac{\partial}{\partial x} [N]^T [N] \{U\} - \frac{\partial}{\partial y} [N]^T [N] \{V\} \, dV \right) \right) = \int [N]^T q_n \, dS \tag{1}$$

x-Momentum:

$$\sum_{n_e} \iint \left( [N]^T [N] \frac{\partial}{\partial t} \{U\} + [N]^T \{C_1\} \frac{\partial}{\partial x} [N] \{\eta\} + [N]^T \{F_1\} [N] \{U\} - [N]^T \Omega [N] \{V\} \right) dV = 0 \tag{2}$$

y-Momentum:

$$\sum_{n_e} \iint ([N]^T[N] \frac{\partial}{\partial t} \{V\} + [N]^T \{C_1\} \frac{\partial}{\partial y} [N] \{\eta\} +$$

$$+ [N]^T \{F_1\} [N]\{V\} + [N]^T \Omega[N] \{U\}) \, dV = 0$$

where
$$\{C_1\} = g(D+\eta) \tag{3}$$

$$\{F_1\} = \frac{g\sqrt{U^2+V^2}}{C^2(D+\eta)^2}$$

are the column vectors of the previous time step, and dS represents an elementary boundary. The term $q_n$ represents the inflow per unit width normal to the boundary due to tributaries flow, etc. N represents the shape function used, $N^T$ the transpose of N, and V is the area of each grid element. Application of initial and boundary conditions to these sets of equations yields:

$$[M] \{\dot{P}\} + [K]\{P\} = \{F_0\} \tag{4}$$

where [M] is the mass matrix
[K] is the system matrix
and $\{F_0\}$ contains the forcing terms.

To solve these equations, appropriate boundary conditions must be applied. It is to be noted that at solid boundaries, the surface integral of Eq. (1) will necessarily be zero, while at the open boundaries they may be assumed zero if the boundary is sufficiently far from the source that mixing may be considered complete. Physically, the boundary condition at a solid boundary is the no slip requirement, such that the tangential and normal flows are zero. In this problem, however the imposition of no slip condition is not realistic and the zero normal flow at a solid boundary is employed.

ANALYTICAL SOLUTION

Considering only one tidal constituent with period $T = \frac{2\pi}{\sigma}$, we may assume solutions of the linearized equations in the form

$$\eta = \text{Re} \{\eta(x,y) \, e^{-i\sigma t}\} \tag{5}$$

$$U = \text{Re} \{U(x,y) \, e^{-i\sigma t}\} \tag{6}$$

$$V = \text{Re} \{V(x,y) \, e^{-i\sigma t}\} \tag{7}$$

Introducing the following boundary conditions:

$$U = 0 \quad \text{at} \quad x = 0$$

$$V = 0 \quad \text{at} \quad x = x_1 \tag{8}$$

$$V = 0 \quad \text{at} \quad y = y_1$$

The solutions of the equations may be written as:

<u>The Water Elevation:</u>

$$\eta = R\,H_0 \exp\{\frac{\Omega m_2 x}{\sigma+if} + i\,m_2\,(y_1-y) - i\sigma t\}$$

$$+ H_0 \exp\{\frac{\Omega m_2 (x_1-x)}{\sigma+if} - i\,m_2\,(y_1-y) - i\sigma t\}$$

$$+ \sum_{m=1}^{\infty} C_m\,[\cos\,(k_{1m}x) + \frac{\Omega k_{2m}}{k_{1m}(\sigma+if)} \sin\,(k_{1m}x)]$$

$$\cdot \exp\{-\,ik_{2m}\,(y_1-y) - i\sigma t\} \tag{9}$$

The horizontal velocity components take the form

$$U = \frac{ic^2\,(\sigma+if)}{(\sigma^2-f^2-\Omega^2+2i\sigma f)} \sum_{m=1}^{\infty} C_m\,k_{1m}\,[1 + \frac{\Omega^2\,k_{2m}^2}{k_{1m}^2\,(\sigma+if)^2}]\,\sin\,(k_{1m}x)$$

$$\cdot \exp\{-\,ik_{2m}\,(y_1-y) - i\sigma t\} \tag{10}$$

and

$$\frac{(\sigma+if)V}{c^2} = H_0 m_2\,[R \exp\{\frac{\Omega m_2 x}{\sigma+if} + i\,m_2\,(y_1-y) - i\sigma t\}$$

$$- \exp\{\frac{\Omega m_2 (x_1-x)}{\sigma+if} - i\,m_2\,(y_1-y) - i\sigma t\}\ ]$$

$$+ \sum_{m=1}^{\infty} C_m\,[k_{2m}\,\cos\,(k_{1m}x) + \frac{\Omega\sigma}{k_{1m}c^2}\,\sin\,(k_{1m}\,x)]$$

$$\cdot \exp\{-\,ik_{2m}\,(y_1-y) - i\sigma t\} \tag{11}$$

The unknown constants R, $C_1$, $C_2$, ... $C_m$ can be obtained satisfying the condition $V(x,y_1,t) = 0$ at $y = y_1$ giving

$$H_0 m_2\,[R \exp\{\frac{\Omega m_2 x}{\sigma+if}\} - \exp\{\frac{\Omega m_2 (x_1-x)}{\sigma+if}\}]$$

$$+\sum_{m=1}^{\infty} C_m \left[k_{2m} \cos (k_{1m}x) + \frac{\Omega\sigma}{k_{1m}c^2} \sin (k_{1m}x)\right] = 0 \tag{12}$$

The numerical values of R, $C_1$, $C_2$, ..., $C_m$ are obtained by the matrix method. Here the R represents the reflection coefficient of the Kelvin wave.

The example considered is for a tide of 12 hour period in a region of the Bay of Fundy. The dimensions of the region were taken as 40 km square with a depth of 10 m, a friction coefficient f = 0 to .0001 and a coriolis parameter $\Omega$ = .0001 to .001 $s^{-1}$. The amplitudes specified along the open boundary (y = 0) of the rectangular basin have been varied from 1 meter to 15 meters. The results show the distribution of the tidal elevation along the closed end of the basin and are summarized in terms of the amplitudes and phases of the specified constituent.

DISPERSION RELATIONS OF KELVIN WAVE AND POINCARE WAVE

Some discussions regarding the dispersion relation of the Kelvin wave and that of Poincare's wave are necessary when we apply this result to prototype problems. The dispersion relation due to Kelvin waves may be written as

$$m_2 = \pm \left(\frac{\sigma}{c}\right) \left[1 + \frac{f^2}{\sigma^2}\right]^{1/4} \left[\cos\left(\frac{1}{2} \tan^{-1} \frac{f}{\sigma}\right) + i \sin\left(\frac{1}{2} \tan^{-1} \frac{f}{\sigma}\right)\right] \tag{13}$$

The special case of f = 0 suggests that $m_2$ must be negative in such a way that the exponential terms in equations ( 9 ) and (11) do not blow up for large values of x. Thus in obtaining the numerical results one has to take into consideration the negative value of $m_2$ in the relation (13). In this study, we have used this information in our calculations. Next, it is extremely important to consider the dispersion relation due to the Poincare waves which may be written as

$$k_{2m}^2 = \frac{\sigma^3 + 2if\sigma^2 - \sigma(f^2+\Omega^2)}{c^2(\sigma+if)} - \frac{m^2\pi^2}{x_1^2} \tag{14}$$

and which may be subsequently rewritten as

$$k_{2m}^2 = -\left[\left\{\frac{m^2\pi^2}{x_1^2} - \frac{\sigma^2(\sigma^2+f^2-\Omega^2)}{c^2(\sigma^2+f^2)}\right\} - i\left\{\frac{\sigma f\,(f^2+\Omega^2)}{c^2(\sigma^2+f^2)}\right\}\right] \tag{15}$$

Again, the particular case f = 0 suggests that (15) reduces to

$$k_{2m}^2 = -\left[\frac{m^2\pi^2}{x_1^2} - \frac{\sigma^2-\Omega^2}{c^2}\right]$$

It can be easily verified that the particular problem considered here, the term $(m^2\pi^2/x_1^2)$ always greater than the term $(\sigma^2-\Omega^2/c^2)$ Thus $k_{2m}$ is always $\pm$ imaginary quantity. In this problem we have chosen a negative imaginary value of $k_{2m}$ such that the exponential term in (9), (10) and (11) do not blow up for large values of $(y_1-y)$, where $(y_1-y)$ is always positive within the basin. In this manner, the numerical calculations have been performed in this study.

## RESULTS

The concepts of Kelvin waves and Poincare waves have been used to obtain the analytical solutions. Triangular finite element methods has been used for numerical computations. The region has been divided into 128 triangular elements with 81 nodal points.

Amplification factors of tidal elevations along the Basin width from the open end to the closed boundary (barrier site) have been computed from the analytical solutions for the friction factor f = 0.0001 and the coriolis term. $\Omega$ = 0.0001 and is shown in Figure 2.

The finite element solutions for the amplitude and phase along the closed boundary are compared with the corresponding results of analytical solution in Figure 3 for f = 0.0 and $\Omega$ = 0.001 given an initial tidal amplitude of 100 cms at the open boundary of the basin. The comparison shows good agreement between the results of these two methods.

## CONCLUSION

It appears that the mathematical model considered here predicts the effects of the tidal barrier on the response of tidal constituents. The model predicts that the amplification of the tidal amplitudes is dependent on two physical parameters; namely, coriolis term and the bottom friction, in addition to many other factors including the estuarine shape, as pointed out by Prandle and Rahman (1980). This investigation reveals that the initial tidal elevations are amplified at the barrier site and this amplification is larger for large values of the friction factors while samller for large values of coriolis force.

This study constitutes a simple mathematical model corresponding to the simple geometrical configuration of the Bay of

Fundy. A comprehensive numerical model will include the Bay of Fundy and Gulf of Maine far out to the Continential Shelf for better knowledge of the Fundy tidal behaviour. Future investigation will be directed to study this comprehensive model stated above.

REFERENCES

Brown, P. J. (1973) Kelvin Wave Reflection in a Semi-Infinite Canal. J. Mar. Res., 31, 1-10.

Greenberg, D. A. (1976) Mathematical Description of the Bay of Fundy - Gulf of Maine Numerical Model. Tech. Note 16, Marine Env. Data Service, Environment Canada, Ottawa.

Liggett, J. A. and Woolhiser, D. A. (1967) Difference Solutions of the Shallow Water Equations. Proc. ASCE, Vol. 95, E.M. 2.

Prandle, D. and Rahman, M. (1980) Tidal Response in Estuaries. J. Phy. Ocean., Vol. 10, No. 10, October, 1552-1573.

Rahman, M., Prandle, D., Spraggs, L. and Argintaru, V. (1980) Numerical Methods for Tidal Propagation Applied to a Hybrid Model of the Bay of Fundy. 3rd International Conference on Finite Elements in Water Resources, The University of Mississippi, Oxford, May 19-23.

Taylor, G. I. (1920) Tidal Oscillations in Gulfs and Rectangular Basins. Proc. Lond. Math. Soc., 20, 148-181.

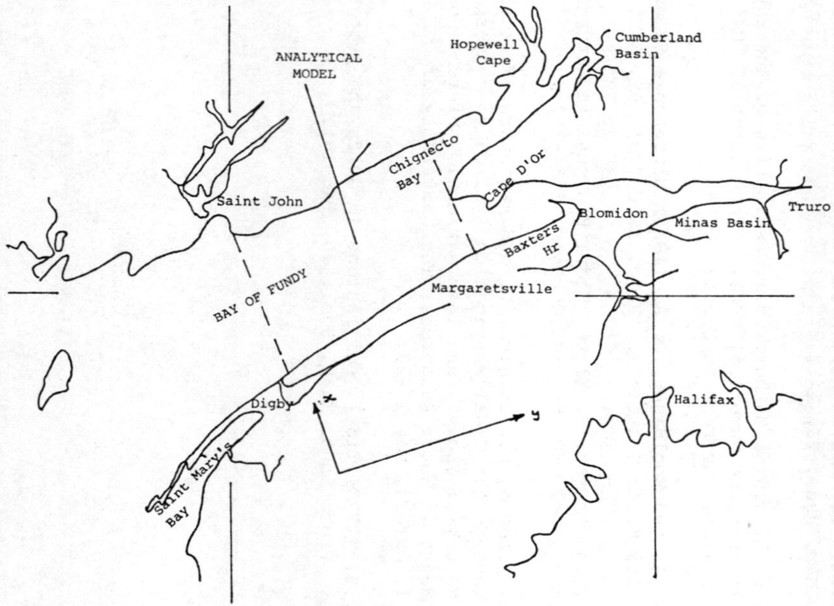


Figure 1 Schematic Diagram of the Bay of Fundy

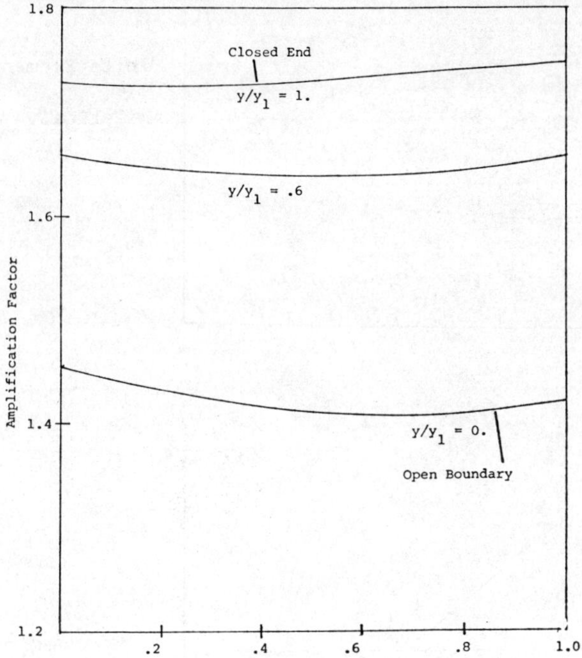


Dimensionless Basin Width

Figure 2 Amplification of Tides at $y/y_1 = 0.0$, 0.6 and 1.0 for $f = 0.0001$ and $\Omega = 0.0001$

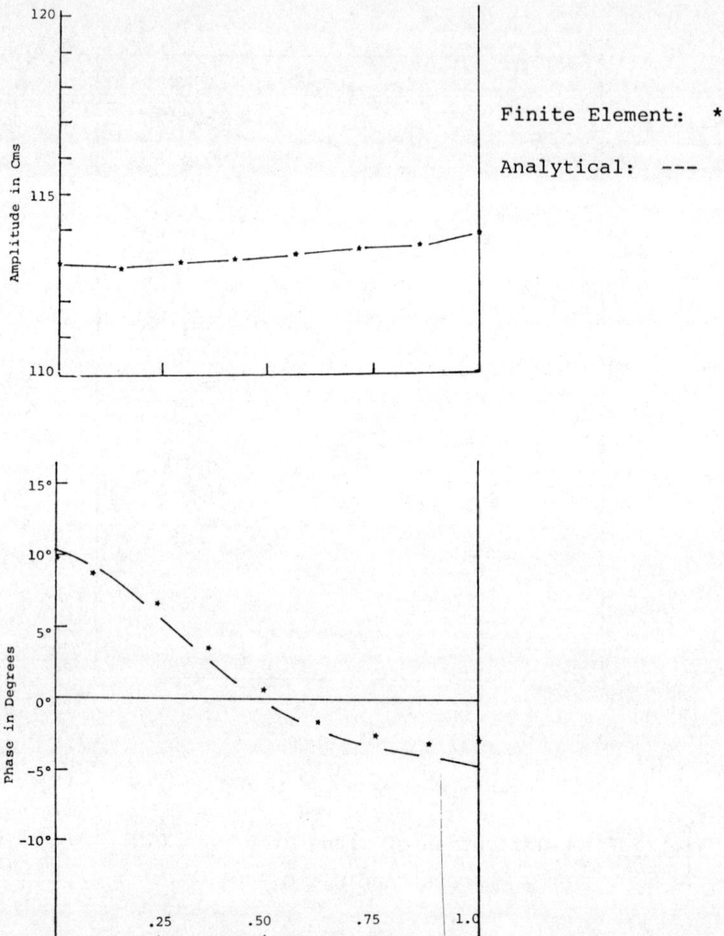


Figure 3

Comparison of FEM Solutions with Analytical Solutions for f = 0.00 and $\Omega$ = 0.0005 Along the Closed End of the Basin

# F.E. ANALYSIS OF WAVES IN A SEMI-INFINITE LAYER OF FLUID

P.Wilde, K.Szmidt

Institute of Hydroengineering
Polish Academy of Sciences, Gdańsk, Poland

## FORMULATION OF THE PROBLEM.

Solutions obtained by the finite element methods are constructed in a discrete space of chosen points while analytical solutions are constructed in a continuum. It is usually assumed that the net is sufficiently dense so that the approximation is good. It must be stressed that solutions in a countable set ( set of discrete points ) have different properties compared to solutions in continuum.

In the present paper the differences are discussed in the case of the linear problem of harmonic waves in a semi-infinite layer of fluid caused by harmonic vibration of a wall - Fig 1. The discussion leads to the formulation of radiation conditions in the discrete space of the assumed net. The steady state is considered, thus time enters as a multiplication factor $\exp(i\omega t)$ where $\omega$ is the angular frequency of vibrations. It is assumed the fluid is compressible, nonviscose, the disturbances are small so that the linear theory may be used and the potential theory is applied.

In the discussed case the potential $\Phi(x,z,t)$ may be assumed in the following form:

$$\Phi(x,z,t) = \varphi(x,z) \cdot e^{i\omega t} \tag{1}$$

and the governing equation is given by the following expression:

$$\nabla^2 \varphi + k_c^2 \cdot \varphi = 0 , \quad k_c = \omega / c \tag{2}$$

where $c$ is the speed of sound, $\nabla^2$ denotes the Laplace operator.

Three cases of boundary conditions on the upper surface are considered - Fig 1 :

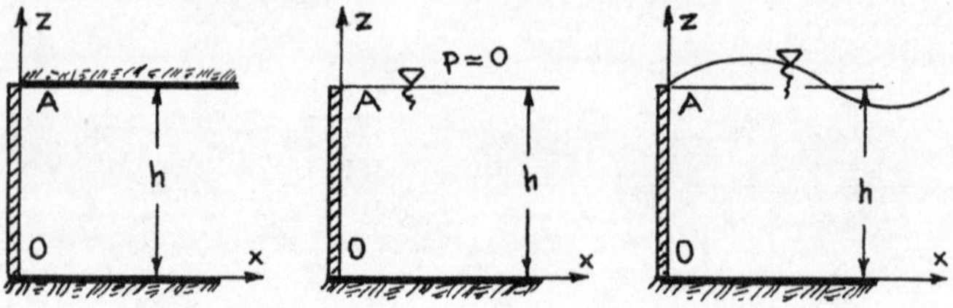


Fig 1. Semi-infinite layer of fluid.

$$\frac{\partial\varphi}{\partial z}\Big|_{z=h}=0, \quad \varphi\Big|_{z=h}=0, \quad \left(-\omega^2\varphi+g\cdot\frac{\partial\varphi}{\partial z}\right)\Big|_{z=h}=0. \tag{3}$$

In the first case there is no flow through the boundary, in the second case the pressure is zero and in the third case propagation of gravitational surface waves is possible. The common boundary conditions are: no flow through the bottom surface and the condition on the vertical wall that the horizontal velocieties of the fluid are equal to the prescribed velocieties of the wall $V_x$ . Thus it follows:

$$\frac{\partial\varphi}{\partial z}\Big|_{z=0}=0, \quad \frac{\partial\varphi}{\partial x}\Big|_{x=0}=V_x. \tag{4}$$

To obtain a unique solution the Sommerfeld condition must be added that the waves die out or propagate to the right at infinity.

The classical analytical solution is constructed by the method of separation of variables and results in a countable set of solutions corresponding to propagating waves and standing waves. In the case of no flow through the upper boundary there is only one propagating dilatational wave; when pressure is zero on the surface there are standing dilatational waves only and in the third case there is one propagating surface wave and a set of standing dilatational waves as discussed by P.Wilde and K.Szmidt(1979).

For the finite element formulation it is useful to apply the variational formulation. The functional which corresponds to the differential equation (2) and has the natural boundary conditions as stated in the problem assumes the following form:

$$T=\frac{1}{2}\iint\left[\left(\frac{\partial\varphi}{\partial x}\right)^2+\left(\frac{\partial\varphi}{\partial z}\right)^2-k_c^2\cdot\varphi^2\right]dx\,dz\;+$$
$$+\int_0^h\varphi(0,z)\cdot f(z)\,dz\;-\;\frac{1}{2}\,\frac{\omega^2}{g}\cdot\int_0^\infty\varphi^2(x,h)\,dx \tag{5}$$

where $f(z)$ is defined by $V_x = i\cdot\omega\cdot f(z)$.

In the case of no flow through the boundary and zero pressure on the surface the last line integral must be omitted. The terms which appear in the Eq (5)have physical meaning. The surface integral corresponds to the kinetic energy and strain energy of the fluid. The line integral along the wall corresponds to the power introduced by the moving wall and the line integral on the free surface is connected with the potential energy of the gravitational waves.

Rectangular and linear elements are assumed as shown on Fig 2. It is assumed the size of elements may not be uniform in the vertical direction.

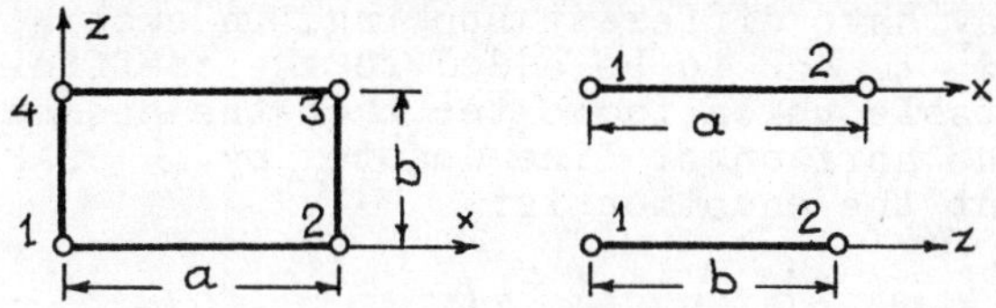


Fig 2. Finite elements.

The following shape functions are assumed for the rectangular elements and for the line elements respectively:

$$\varphi(x,z)=\frac{1}{a\cdot b}\left[\varphi_1\cdot(x-a)\cdot(z-b)-\varphi_2\cdot x\cdot(z-b)+\varphi_3\cdot x\cdot z-\varphi_4\cdot(x-a)\cdot z\right],$$

$$\varphi(x)=\frac{1}{a}\left[-\varphi_1\cdot(x-a)+\varphi_2\cdot x\right],\quad \varphi(z)=\left[-\varphi_1\cdot(z-b)+\varphi_2\cdot z\right]. \qquad (6)$$

Substitution into the functional (5) and integration yields a polynomial of second degree with respect to the values of $\varphi_{i,j}$ at the points of the net. Differentiation with respect to $\varphi_{i,j}$ yields the linear set of equations of the problem:

$$\mathbf{K}\cdot\boldsymbol{\delta}=\mathbf{f} \qquad (7)$$

where $\boldsymbol{\delta}^T=(\varphi_{11},\varphi_{21},\dots,\varphi_{12},\varphi_{22}\dots)$, $\mathbf{f}$ is the column matrix of the right side given by the velocities of the wall, $\mathbf{K}$ is the matrix of equations.

For a typical point in the region only the elements in the neighbourhood enter into the equation for the considered point. The coefficients for one rectangular element are shown in table 1.

Table 1.

$\mathbf{K}=\frac{a\cdot b}{36}\cdot$

| | | | |
|---|---|---|---|
| $\frac{12}{a^2}+\frac{12}{b^2}-4k_c^2$ | $\frac{-12}{a^2}+\frac{6}{b^2}-2\cdot k_c^2$ | $-(\frac{6}{a^2}+\frac{6}{b^2}+k_c^2)$ | $-(\frac{-6}{a^2}+\frac{12}{b^2}+2k_c^2)$ |
| | $k_{11}$ | $k_{14}$ | $k_{13}$ |
| | | $k_{11}$ | $k_{12}$ |
| SYMMETRY | | | $k_{11}$ |

For convenience in the following expression $k_{13}$ and $k_{14}$ will be replaced by $-k_{13}^*$ and $-k_{14}^*$. For the line elements on the free surface a square (2 x 2) matrix results with the following elements:

$$k_{11} = k_{22} = -\frac{\omega^2}{g}\cdot\frac{a}{3}, \quad k_{12} = k_{21} = -\frac{\omega^2}{g}\cdot\frac{a}{6}. \tag{8}$$

Let us consider a typical point denoted by the index $i$ and $j+s$. Let us assume the spacing of all the vertical lines is equal to $a$ but the horizontal lines may have different spacing. In such a situation an index $i$ has to be added to the coefficients $k$ in the table which indicates that the element is just above the horizontal line denoted by $i$. For typical point the equation is:

$$\begin{aligned}
&-\underset{i-1}{k_{13}^*}\cdot\varphi_{i-1,j+s-1} + (\underset{i-1}{k_{12}} + \underset{i}{k_{12}})\cdot\varphi_{i,j+s-1} - \underset{i}{k_{13}^*}\cdot\varphi_{i+1,j+s-1} + \\
&-2\underset{i-1}{k_{14}^*}\cdot\varphi_{i-1,j+s} + 2\cdot(\underset{i-1}{k_{11}} + \underset{i}{k_{11}})\cdot\varphi_{i,j+s} - 2\cdot\underset{i}{k_{14}^*}\cdot\varphi_{i+1,j+s} + \\
&-\underset{i-1}{k_{13}^*}\cdot\varphi_{i-1,j+s+1} + (\underset{i-1}{k_{12}} + \underset{i}{k_{12}})\cdot\varphi_{i,j+s+1} - \underset{i}{k_{13}^*}\,\varphi_{i+1,j+s+1} = 0.
\end{aligned} \tag{9}$$

Now let us introduce the separation of variables by the following formula:

$$\varphi_{i,j+s} = \varphi_{i,j}\cdot X_s . \tag{10}$$

It means we express all the values in terms of the values on the $j$-vertical line. Substitution into the Eq (9) and division by $X_s$ yields the following result:

$$\begin{aligned}
&\Big[-\underset{i-1}{k_{13}^*}\cdot\underset{j}{\varphi_{i-1}} + (\underset{i-1}{k_{12}} + \underset{i}{k_{12}})\underset{j}{\varphi_i} - \underset{i}{k_{13}^*}\cdot\underset{j}{\varphi_{i+1}}\Big]\cdot\frac{X_{s-1}+X_{s+1}}{X_s} + \\
&+2\cdot\Big[-\underset{i-1}{k_{14}^*}\,\underset{j}{\varphi_{i-1}} + (\underset{i-1}{k_{11}} + \underset{i}{k_{11}})\cdot\underset{j}{\varphi_i} - \underset{i}{k_{14}^*}\cdot\underset{j}{\varphi_{i+1}}\Big] = 0.
\end{aligned} \tag{11}$$

The set of equations should be satisfied for each value of $i$ and $s$, it means $(X_{s-1}+X_{s+1})/X_s$ should be equal to a number say $2\lambda$. Thus the following set of difference equations is obtained:

$$X_{s-1} - 2\lambda X_s + X_{s+1} = 0. \tag{12}$$

Two sets of general solutions are obtained:

$$X_s = \exp(-s\cdot r\cdot a), \quad X_s = \exp(s\cdot r\cdot a) \tag{13}$$

when the following notation is used:

$$\lambda = \cosh ra . \tag{14}$$

If equations of the type given by Eq (11) are written for all $i$ including the bottom and the upper boundary the result is a set of equations with $n$ unknowns corresponding to the $n$ points along the vertical line of the type:

$$(A + \lambda \cdot B) \cdot \varphi = O \tag{15}$$

where $A$ and $B$ are square matrices, $\varphi$ and $O$ are column matrices. For the cases considered in this paper by appropriate transformation the set of equations (15) will be transformed to the standard form for eigenvalues and eigenvectors with a symmetric matrix thus it may be shown that the eigenvalues are real numbers. For real values of $\lambda$ it is useful to substitute:

$$\lambda = \begin{cases} \cosh ra & \text{if} \quad \lambda \geqslant 1 \\ \cos ra & \text{if} \quad -1 \leqslant \lambda \leqslant 1 \\ -\cosh ra & \text{if} \quad \lambda \leqslant -1 \end{cases} \tag{16}$$

and the solutions are:

$$\begin{aligned} X_s &= \exp(-s \cdot r \cdot a) \quad , & \lambda \geqslant 1 \\ X_s &= \exp(-i \cdot s \cdot r \cdot a) \quad , & -1 \leqslant \lambda \leqslant 1 \\ X_s &= (-1)^s \cdot \exp(-s \cdot r \cdot a) \quad , & \lambda \leqslant -1 \end{aligned} \tag{17}$$

The first solution corresponds to a standing wave; the second to a propagating wave and the third does not correspond to a solution in continuum and is a property of the discrete set. In the last solution the sign changes when one moves from one point to the next along the horizontal lines.

## THE SOLUTIONS FOR A DIVISION IN EQUAL ELEMENTS.

In the case the spacing is uniform in the vertical direction all the finite elements are equal and in the expression for the separation of variables Eq (10) it is useful to assume $i = p+m$ and to write:

$$\varphi_{p+m,\,j+s} = \varphi_{p,j} \cdot Z_m \cdot X_s, \tag{18}$$

where $\varphi_{p,j}$ is the value at a chosen point $(p,j)$ .

Substitution of the notation given by Eq (18) transforms Eq (11) to the following form:

$$-Z_{m-1} + 2\varrho \cdot Z_m - Z_{m+1} = 0 \tag{19}$$

where:

$$k^*_{14} + \lambda \cdot k^*_{13} \neq 0, \quad \varrho = \frac{k_{11} + \lambda k_{12}}{k^*_{14} + \lambda \cdot k^*_{13}} \, , \quad \lambda = \frac{\varrho \cdot k^*_{14} - k_{11}}{k_{12} - \varrho \cdot k^*_{13}} \, . \tag{20}$$

In the case $k^*_{14} + \lambda \cdot k^*_{13}$ is equal to zero the matrix reduces to a diagonal form. Thus a nontrivial solution is only possible if at least one of the elements of the diagonal matrix is equal to zero.

If $\lambda$ is a real value $\varrho$ is a real value too

and to solve the difference equations (19) it is convenient to substitute:

$$\rho = \begin{cases} \cosh kb & \text{if} \quad \rho \geq 1 \\ \cos kb & \text{if} \quad -1 \leq \rho \leq 1 \\ -\cosh kb & \text{if} \quad \rho \leq -1 \end{cases} \tag{21}$$

and to obtain the following two set of solutions in real numbers:

$$\begin{aligned} Z_m &= \cosh kb \quad , \quad Z_m = \sinh kb \\ Z_m &= \cos kb \quad , \quad Z_m = \sin kb \\ Z_m &= (-1)^m \cdot \cosh kb \quad , \quad Z_m = (-1)^m \cdot \sinh kb . \end{aligned} \tag{22}$$

For the case of no flow through the upper boundary the set of equations assumes the following form:

$$A \cdot \varphi = 0 \tag{23}$$

where $\varphi^T = (\varphi_1, \varphi_2, \ldots, \varphi_n)$ and the matrix $A$ is given in table 2.

Table 2.

$$A = \begin{bmatrix} \rho & -1 & & & & \\ -1 & 2\rho & -1 & & & \\ & & \ddots & & & \\ & & & -1 & 2\rho & -1 \\ & & & & -1 & \rho \end{bmatrix}$$

The matrix $A$ is symmetric. The equation (23) may be transformed to the standard form for eigenvalues with the help of nonsingular matrix $Q$ according to the formula:

$$Q \cdot A \cdot Q \cdot Q^{-1} \cdot \varphi = 0 \tag{24}$$

The elements of the diagonal matrix $Q$ are given as follows:

$$Q_{diag} = (1, 1/\sqrt{2}, \ldots, 1/\sqrt{2}, 1) \tag{25}$$

The transformed set of equations assumes the following form:

$$A^* \cdot \varphi^* = 0 \tag{26}$$

where $A^* = Q \cdot A \cdot Q$ and $\varphi^* = Q^{-1} \cdot \varphi$

The matrix $\varphi^*$ is expressed in terms of $\varphi$ as follows:

$$\varphi^{*T} = (\varphi_1, \sqrt{2}\varphi_2, \ldots, \sqrt{2}\varphi_{n-1}, \varphi_n) \tag{27}$$

For a symmetric matrix $A^*$ the eigenvalues $\rho_v$ and eigenvectors $\varphi^*_{iv}$ are real numbers. There are $n$ eigenvalues and if they are distinct all the eigenvectors are mutually orthogonal.

Let us express the solution in terms of the value $\varphi_{1,j}$ . A suitable expression which satisfies the first equation in Eqs (23) is:

$$\varphi_{j,1+m} = \varphi_{1,j} \cdot \cos mb \qquad (28)$$

Substitution into the last equation in (23) leads to the following equation:

$$\sin kb \cdot \sin(n-1)kb = 0, \rightarrow k_{\nu} b = \frac{\nu \cdot \pi}{n-1}, \quad \nu = 0, 1, 2, \ldots \qquad (29)$$

If $kb = 0$ , thus $\varrho = 1$ and from Eq (20) it follows:

$$\lambda_0 = 1 - \frac{9(a \cdot k_c)^2}{18 + 3(a \cdot k_c)^2}. \qquad (30)$$

It is easy to see that $-1 < \lambda_0 < 1$ for $ak_c < 1$ and thus this is a progressive wave. If $ak_c \ll 1$ it may be easily seen that the wave velocity is very close to the speed of sound. In such a case the following relation results:

$$\cos ra \sim 1 - \tfrac{1}{2} \cdot (ak_c)^2, \rightarrow r \sim k_c \qquad (31)$$

From Eq (20) it follows:

$$\lambda_{\nu} = 1 + \frac{[18\eta^2 - 6(ak_c)^2] - [18\eta^2 + 3(a \cdot k_c)^2]\varrho_{\nu}}{[12 - 6\eta^2 + 2(ak_c)^2] + [6 + 6 \cdot \eta^2 + (ak_c)^2] \cdot \varrho_{\nu}}, \qquad (32)$$

where $\eta = a/b$

In this equation for $\nu = 1$ it may happen that $\lambda_1$ is less than one and we have a progressive wave which has no physical meaning. This corresponds to the case :

$$\cos \frac{\pi}{n-1} > \frac{18\eta^2 - 6(ak_c)^2}{18\eta^2 + 3(a \cdot k_c)^2} \qquad (33)$$

and the choice of $a, b, n$ should be taken in such a way that this case is excluded.

For a special value of $\varrho_{\nu}$ it may happen that the denominator goes to zero and for value over it $\lambda_{\nu} < -1$ . This case happens for practical proportion. The influence may be important if higher harmonics are important. To avoid this case one should take:

$$\eta^2 < \tfrac{1}{2} - \tfrac{1}{12} \cdot (ak_c)^2 \qquad (34)$$

If this case is not excluded solutions of the following type occur:

$$\varphi_{1+m, j+s} = (-1)^s \cdot \cos mkb \cdot e^{-sra} \qquad (35)$$

It may be easily seen that if one takes $v = 1,2,\dots,n-1$ $n-1$ distinct eigenvalues are obtained and with $\varrho_o$, $\lambda_o$ they form a complete set of eigenvalues. If $v$ goes beyond $n-1$ the eigenvectors are repeated. It may be easily verified that the other cases $\varrho = \cosh kb$ and $\varrho = -\cosh kb$ give no other solutions.

Let us consider the case of zero pressure on the surface. A corresponding matrix $\mathbf{A}$ may be obtained by crossing the last row and the last column in matrix $\mathbf{A}$ . Thus the matrix may be reduced as before to the standard form with symmetric matrix. The last equation (23) reduces to the following relations:

$$\varphi_1 \cdot \cos nkb = 0, \rightarrow k_v b = (2v-1)\cdot\pi/2n, \quad v = 1,2,\dots,n-1. \tag{36}$$

For this case the inequality (33) assumes the following form:

$$\cos\frac{\pi}{2n} > \frac{18\eta^2 - 6\cdot(ak_c)^2}{18\eta^2 + 3(a\cdot k_c)^2} \tag{37}$$

It may be easily seen that the set given by Eq (36) completes all the solutions.

In the case of gravitational waves on the upper surface all equations are the same as in the first case only the last equation assumes the following form:

$$-(k_{14}^* + \lambda\cdot k_{13}^*)\cdot Z_{m-1} + \left[k_{11} + \lambda\cdot k_{12} - \frac{6\omega^2}{b\cdot g}\cdot(2+\lambda)\right]\cdot Z_m = 0. \tag{38}$$

It may be shown that the set of equations may be reduced to a standard form for eigenvalues but in this case the elements of the matrix are real numbers only if

$$\alpha = \frac{\omega^2\cdot b}{3g} < 1 \tag{39}$$

If the first type of solutions according to Eqs (21) is considered the last equation reduces to the relation:

$$\varphi_1\cdot[\sinh kb\cdot\tanh(n-1)kb - \alpha\cdot(2+\cosh kb)] = 0. \tag{40}$$

Nontrivial solution is obtained in real numbers when the inequality (39) is satisfied. The corresponding dispersion equation in the analytic solution is:

$$\omega^2 = g\cdot k_o\cdot\tanh k_o h \tag{41}$$

The numerical solution is close to the analytical solution for small values of $kb$ such that:

$$\sinh k_o b \sim k_o b, \quad \cosh k_o b \sim 1 \tag{42}$$

In the discussed case it is possible that $k_{14}^* + \lambda\cdot k_{13}^*$ is equal to zero. This case happens only if $\alpha$ is equal to one. It may be shown that in this case:

$$\lambda = 1 - \frac{18 + 3\cdot(b\cdot k_c)^2}{6 + 6(b/a)^2 + (b\cdot k_c)^2} \tag{43}$$

and if $-1 < \lambda < 1$ the solution represent a progressive wave. If $\lambda < -1$ a standing wave results with signs changing when one moves from point to point along the horizontal line.

Now let us consider the case of trigonometric functions. In this case the last equation reduces to the following one:

$$\varphi_1 \cdot [\sin kb \cdot \tan(n-1)kb + \alpha\cdot(2 + \cos kb)] = 0 \tag{44}$$

The Eq (44) has an infinite number of roots. In this case the analytical solution leads to the following dispersion relation:

$$\omega^2 + g\cdot k_v \tan k_v h = 0 \tag{45}$$

The numerical solution is a good approximation of the analytical solution only for such small values of $k_v b$ that:

$$\sin k_v b \sim k_v b \;, \quad \cos k_v b \sim 1 \tag{46}$$

It may be shown that in the interval $0 \div \pi$ there are $n-1$ roots of the Eq (44) . The values $k_v h$ are in the intervals:

$$\frac{2v-1}{2}\cdot\pi < k_v h < v\cdot\pi \;, \quad v = 1, 2, \cdots, n-1 \tag{47}$$

The continuation over the interval $0 \div \pi$ leads to values of $kh$ which have the same eigenvectors. If $\alpha \gg 1$ than the eigenvalues approach the values $k_v h = (2v-1)\pi/2$ which are the eigenvalues for the case of zero pressure at the upper surface. Thus for large values of $\alpha$ it is useful to neglect the gravity surface waves. It may be shown that for the values $\varsigma_v = \cos k_v h$ the corresponding values of $\lambda_v$ are greater than one and dicussion of Eq (32) leads to Eq (34) and to appropriate inequality of the type given by (33) .

It may happen that in numerical calculations it is difficult to observe the condition $\alpha < 1$ . In such a situation in a general case the values $\lambda$ and $\varsigma$ are complex numbers. The analysis in this case is very tedious. It may be easily shown that in the case the imaginary part of $k_o b$ is equal to $\pi$ Eq (40) assumes the following form:

$$\varphi_1 \cdot [\sinh k_o b \cdot \tanh(n-1)k_o b + \alpha\cdot(2 - \cosh k_o b)] = 0 \tag{48}$$

This case corresponds to the third case $\rho = -\cosh k_o b$. It may be seen that there is no solution for $\alpha < 1$. A numerical solution was established for such a case. For $a/b = 0.5$, $\alpha = 2.446$ the resulting values are : $k_o b = 1.855$, $\lambda = -0.3682$ . The solution was obtained by tedious analytical considerations and directly from the numerical solution. It represents a gravitational wave which changes sign when one moves along the vertical line from point to the next one.

For illustration of the obtained results on Fig 3. eigenvectors for different values of $\alpha$ are plotted.

- progressive waves

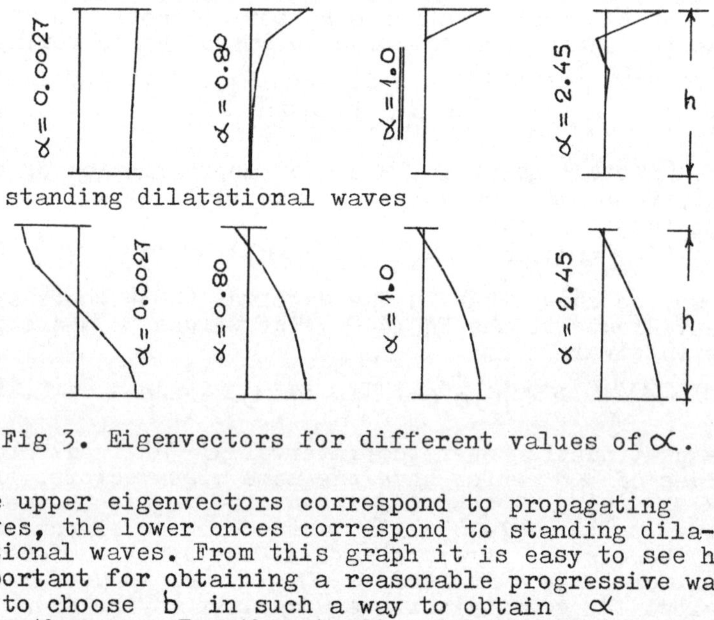


Fig 3. Eigenvectors for different values of $\alpha$.

The upper eigenvectors correspond to propagating waves, the lower onces correspond to standing dilatational waves. From this graph it is easy to see how important for obtaining a reasonable progressive wave is to choose $b$ in such a way to obtain $\alpha$ less than one. For the standing dilatational wave there is no such necessity. If the value of $\alpha$ is greater than one the solution for the standing wave approaches the solution for zero pressure on the upper surface of the layer.

REFERENCES


Lysmer,J. Waas,G. (1972) Shear waves in plane infinite structures. J.Eng.Mech.Div.,Proc.ASCE,Febr.1972.

Wilde,P.Szmidt,K. (1979) Hydrodynamical loads on a vibrating rigid plate in water of finite depth. Arch. Hydrotechniki. tom XXVII,z.3, 357-370.

# 6. SIMULATION OF TIDAL AND OCEAN FLOW PROCESSES

# NUMERICAL MODELING OF STORM SURGE GENERATION PHENOMENA

S.A. Dendrou, Camp Dresser & McKee, Annandale, VA
C.I. Moore, Camp Dresser & McKee, Annandale, VA
B.A. Dendrou, Agbabian Associates, El Segundo, CA

## INTRODUCTION

Storm surges are long ocean waves produced by atmospheric anomalies (hurricanes, cyclonic storms and others) acting on the ocean surface in the form of atmospheric pressure gradients and wind fields. Total observed surges near a coastline represent an amalgam of effects: the local effect of the atmospheric forcing, the effect of surges produced at earlier positions of the storm and propagating as long-gravity waves and their interaction with the astronomic tides. Customarily, storm surges are simulated by distributed parameter finite difference and finite element models that solve numerically the Navier-Stokes equations with appropriate atmospheric forcing terms. In the context of Monte Carlo experiments for flood-frequency studies, these models may be exercised hundreds of times. Therefore, computational economy and efficiency are two highly desirable properties for the numerical models used to simulate storm surge generation phenomena.

The objective of this paper is to present a systematic review of the physical phenomena that lead to coastal flooding conditions and to justify the selection of appropriate simulation models. The emphasis is placed on determining the numerical characteristics of the models as a function of the modeled physical phenomena. Theoretical excursions are also made, in an attempt to cover a wider range of phenomena, based on a real case, coastal flood-frequency study for the City of New York (Dendrou, et.al., 1982). First, a statistical formulation of flood-frequency studies is presented.

## STATISTICAL FORMULATION

Meteorologic phenomena are predominantly stochastic in nature. Consequently, the determination of the frequency of high waters is also a problem of statistical inference. Two types of approaches exist: those based on a statistical analysis of the historical record at tidal gages, and those based on infering surge frequencies from meteorological statistics by means of a Monte Carlo experiment. The work presented here deals with the second approach exclusively. The frequency of storm surges depends both on the frequency of occurrence of the storms and on the probability that the storm will achieve a certain intensity.

Assuming that there is a one-to-one correspondence between the meteorologic causes (storms) and their associated high water elevations, their joint occurrence may be viewed as a Poisson point stochastic process. Denoting by $F_s$ the average frequency of storm occurrence, the average frequency of exceedance events is given by:

$$F_H(h) = F_s \cdot (1-P_H(h)) \tag{1}$$

where $P_H(h)$ is the probability distribution function for high water events. In spite of the many assumptions involved in its development, Equation (1) represents the fundamental relationship for the evaluation of frequency-elevation information. The Joint Probability Monte Carlo approach provides a way to evaluate the term in parenthesis in Equation (1). Assuming statistical independence between the various meteorological parameters, the term $P_H(h)$ is calculated as the product of the corresponding probabilities of the statistical parameters. The Monte Carlo method consists of simulating storm surges from a sample of synthetic storms of adequate sample size and evaluating their frequencies through Equation (1). This "mapping" is conceptually illustrated in Figure 1, where $\Delta P$ denotes the atmospheric gradient and all other meteorologic parameters are grouped in the term P (surge/ P).

## PHYSICAL PROCESSES

The period, wave length, and amplitude characteristics of long ocean waves depend on two sets of conditions: (1) the geometric properties of the water body in which the waves propagate, i.e., bathymetry and boundary conditions, and (2) the characteristics of the forcing excitation. The New York City study area illustrated in Figure 2 incorporates a variety of embayments that cover a range of conditions. The most important are: the coastal plain (continental shelf) which has a mild bottom slope with an average depth of 50 to 100 feet, and which is connected to the deep ocean at the continental break 100 miles from the shore; and the inland embayments, namely, the Upper and Lower New York Harbor at the mouth of the Hudson River, Harlem River, and East River extending to Long Island Sound, Jamaica Bay and the waterways around Staten Island, Raritan Bay, and Newark Bay.

Storm surges in the latitudes of New York City are generally caused by North Atlantic meteorologic phenomena, namely, hurricanes or tropical storms, and northeasters or winter storms. Both are atmospheric low pressure disturbances moving over an area covering several hundreds of square miles. They differ in size (aerial extent) and kinematic characteristics. Hurricanes are smaller in size, 10 to 20 miles in diameter, and move very fast, while winter (extratropical) storms are much larger and often quasi-stationary over periods of 2 to 3 days. The atmospheric pressure gradients cause counter-clockwise (in the northern hemisphere) winds, varying in direction and magnitude across the lower "friction" layer of the atmosphere, the so-called Planetary Boundary Layer (PBL), as illustrated in Figure 2. Atmospheric pressure gradients and wind provide the forcing mechanism that generates surges. Pressure gradients produce the "inverted barometer" effect, i.e., a swell proportional to the local pressure deficit as compared to the far-field background atmospheric pressure. The amplitude of the swell is also proportional to the water depth. This mechanism is dominant in hurricanes over deep waters. The quasi-instantaneous formation of swells follows the movement of the "eye" of the hurricanes. These swells, in turn, propagate as long gravity waves.

The effect of the wind on the ocean is produced by the transfer of momentum between the Planetary Boundary Layer and the surface of the ocean. This transfer is usually parameterized through the wind shear stress, or more accurately, through the wind drag force, which is proportional to the square of the wind velocity. The coefficient of proportionality is called the wind drag coefficient, which depends, among other things, on the magnitude of the wind vector and on the presence of air-sea temperature gradients. Being a surface phenomenon, it is predominant in shallow waters. The coastal embayment response to wind is much slower than the inverted barometer effect and depends primarily on the natural frequency of the body of water. Coastal land forms are shaped over thousands of years in response to a variety of forces. It is not unreasonable then that they be tuned to the periodic astronomic tide, and exhibit a natural period of 6 hours. Wind is the primary forcing mechanism in winter storms, which tend to last long enough to fully excite the embayments. Surges generated a distance from the shore are combined with astromic tides and propagated through the inland embayments.

## NUMERICAL MODELS

The generation and propagation of surges is governed by the 2-D, vertically integrated, Navier-Stokes equations. The x-component of the momentum equation is written as:

$$\frac{\partial u}{\partial t} + u\frac{\partial u}{\partial x} + v\frac{\partial u}{\partial y} = -g\frac{\partial H}{\partial x} + f.v. - \frac{\tau bx}{\rho H} + \frac{\tau wx}{\rho H} - \frac{1}{\rho}\frac{\partial p}{\partial x} \qquad (2)$$

The continuity equation is written as:

$$\frac{\partial H}{\partial t} + \frac{\partial uH}{\partial x} + \frac{\partial vH}{\partial y} = 0 \qquad (3)$$

where u, v are the horizontal velocity components, H the water depth, f is the Coriolis parameter, p is the atmospheric pressure and $^{\tau}bx$, and $^{\tau}wx$, respectively the bottom friction and the wind stress. The wind stress and the pressure gradient are the forcing terms in Equ. (2) and may be represented as a function of the storm parameters, namely the wind velocity and central pressure depression $\Delta p$. Equations (2) and (3) govern most tidal circulation situations. However, it is numerically advantageous and more economic to use different numerical solutions for the two phases, namely the surge generation, and the surge propagation and transformation in the zone of near coastal embayments.

Strictly speaking, the forcing terms of Equation (2) are present in the entire domain, i.e., regions 1, 2, and 3, in Figure 2. However, they affect mostly the deep ocean and the continental shelf where surges are created, while in the inland embayments they only tend to modify the surges propagating from off-shore. For hurricanes and extratropical storms, the continental shelf is the primary medium in which surges are generated. Moreover, observation of historical surges reveals that the free-body dynamics of this body of water play a fundamental role. For Northeasters, under constant wind forcing the continental shelf becomes fully arisen and exhibits a characteristic period of 6 hours. Interestingly, the same period of 6 hours is observed in resurgences occuring after the passage of hurricanes, indicating a free-body response of the medium to the relatively fast (3-4 hours) excitation from the hurricane forcing. Lumped mass models are known to simulate accurately dynamic phenomena. The applicability of Boundary Integral models was investigated, but they proved to be impractical because of the distributed nature of the meteorologic forcing. An explicit space staggered finite-difference scheme was used instead with a 5 x 5 n. mile spatial resolution covering all the continental shelf and portion of the deep ocean, as illustrated in Figure 3. The sharp interface between shallow continental shelf and deep ocean provides the natural reflecting boundary that is at the origin of the formation of resurgences after the passage of the storm.

The atmospheric forcing from hurricanes was modeled in terms of the geometric and kinematic characteristics of the storm. These are the central pressure depression, indicative of the intensity of the storm; the radius to maximum winds, indicative of the size of the storm, and the forward velocity. Figure 4 shows their respective effect on the resulting surge for a range of values around the base case (curve 1) of 1.5 in. Hg, 36 n. miles and 19 knots respectively. Curves 2 and 3 correspond to a central pressure depression of 1 and 2 in. Hg respectively; curves 4 and 5 to a radius of 20 and 50 n. miles respectively; and curves 6 and 7 to a velocity of 10 and 27 knots respectively. It is observed that pressure depression affects the surge amplitude, and storm size affects the arrival time, while forward velocity affects both the magnitude and the arrival time of the surge. A particularly important boundary condition is that of the open ocean. Hurricanes do affect the deep ocean, region 1 in Figure 2, and that effect, in turn, is compounded with the forcing over the continental shelf. Dispersive and non-linear effects in deep water are important and can account for as much as 20 percent of variation over the basic inverted barometer effect. Deep ocean data however are scanty at the present time and hence the subject is matter for speculation. This is especially true since numerous other effects come into play before the surge reaches a gage usually located near the shore.

For extratropical storms, a geometric description of the forcing proved inadequate. Wind is the predominant forcing mechanism and the wind drag force on the surface of the ocean depends intimately on the structure of the Planetary Boundary Layer, particularly on the presence of air-sea temperature gradients. A geostrophic atmospheric circulation model combined with a Planetary Boundary Layer (Powell, 1980) model were used to simulate the wind drag on the ocean as a function of observed atmospheric pressure fields over the area.

Finally, the effects of the inland waterways (region 3 in Figure 2), reflection refraction and shoaling, were simulated by a 1-D, explicit, space-staggered finite element model appropriately interfaced with the off-shore model as illustrated in Figure 3. Among the major effects encountered here is the nonlinearity or non-validity of the superposition between surges and astronomic tides. Figure 5 illustrates that this effect is negligible near the open shore (Fort Hamilton) but significant inland (Kearny Point, Newark Bay) in the presence of tidal flats. The combined surge in Figure 5 was simulated by superposing astronomic tide and surge at the interfacing boundary. An example of application of the complete set of models, namely the atmospheric wind circulation model, the off-shore finite-difference model, and the inland 1-D finite-element model, is shown in Figure 6, for the northeaster of November 1950. Considering the variety of phenomena that affect the water levels at the Battery and Fort Hamilton, the match between observed and simulated marigrams is remarkably close.

## CONCLUDING REMARKS

This study dealt with the modeling and simulation of coastal storm surge phenomena. Therefore, the following remarks on their modeling apply only to surge phenomena at the exclusion of all other coastal circulation processes. The same fundamental hydrodynamic equations govern a wide range of hydrodynamic processes. Nevertheless, it must be recognized that different numerical solution schemes are better suited for different phenomena, and that the idea of one general purpose versatile code may not be a viable one. From the set of models that were selected for the New York City Flood-Frequency case study, the following remarks are formulated.

Distributed parameter models, finite difference and finite element, were used because of the distributed nature of the forcing mechanism. Lumped models such as the Boundary Integral models do not seem to be practical in this regard. Finite-difference, fixed $\Delta x$, $\Delta y$ models proved more handy over areas with little spatial variation. A 1-D model is sufficient to model 1-D/2-D inland areas so long as some precautions are taken. Constant shape functions can prove adequate when combined with a space-staggering scheme whereby velocities and elevations are solved for at different nodes of the network. Proper selection of the simulation domain and boundary conditions is by far the most important task: (1) modeling of the natural dynamics of the coastal plain requires incorporating in the domain the transition from shallow to deep ocean; (2) non-linearities are most often due to geometric rather than dynamic causes. In the presence of so many interacting effects, it is difficult to secure acceptable calibration and validation of the model; e.g., it is easy to bias a model by arbitrarily adjusting the friction term. There is a necessity to adequately model the free-body dynamics, unless the model is only meant to be used as a glorified interpolator. Improvement of the understanding of the physical phenomena and governing equations should be pursued in parallel with the selection of numerical schemes; for example, dispersive effects are negligible in shallow waters (continental shelf) especially in the presence of distributed forcing, but are important, and may contribute to as much as 10 to 20 percent of the total surge in deep ocean.

## REFERENCES

Dendrou, S.A., et. al. (1982) New York City Flood Insurance Study. Reports No. 1 - 8. Camp Dresser & McKee Inc., Annandale, Virginia.

Powell, M.D. (1980) "Evaluations of Diagnostic Marine Boundary Layer Models Applied to Hurricanes, "Monthly Weather Review, Vol. 108.

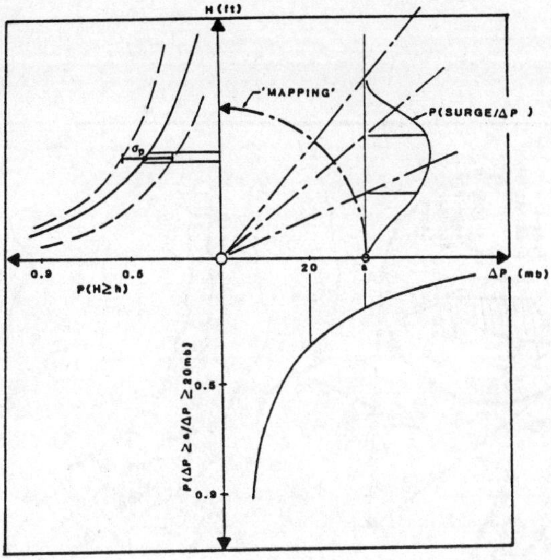


Figure 1: "Mapping" of Meteorologic Frequencies into Surge Frequencies

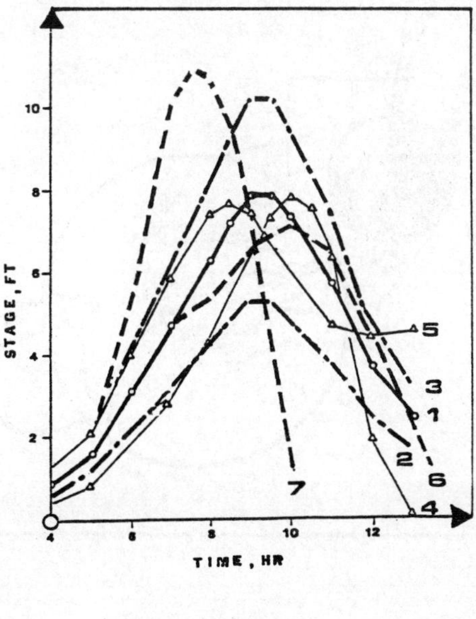


Figure 4: Effect of Various Hurricane Parameters

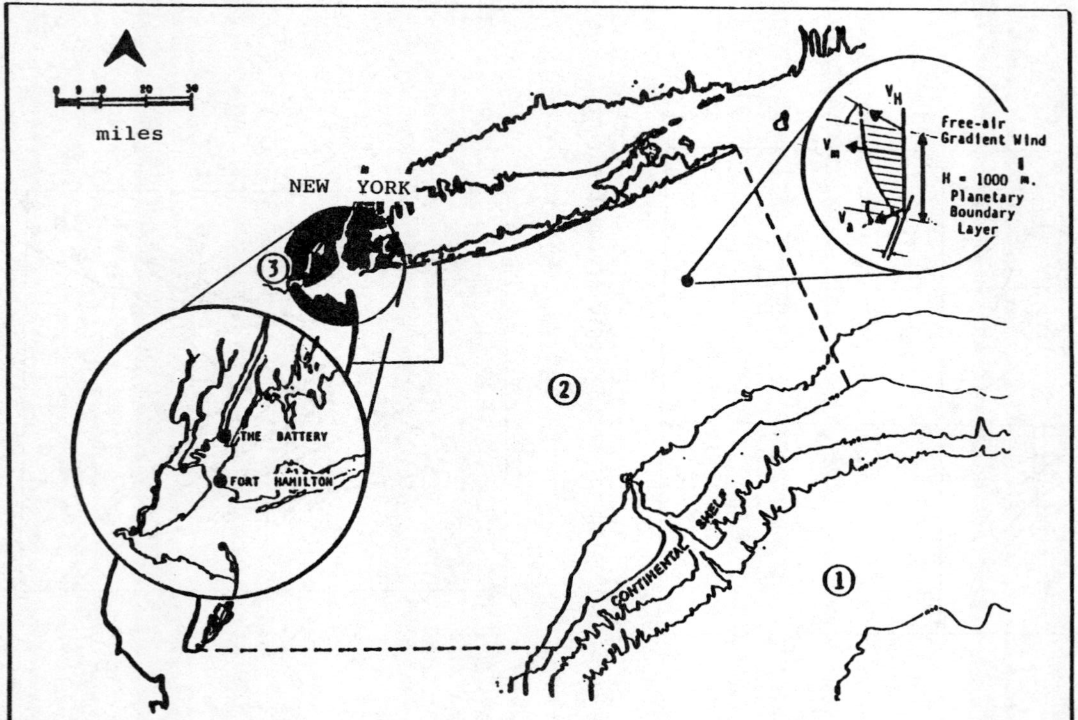


Figure 2: New York City Study Area

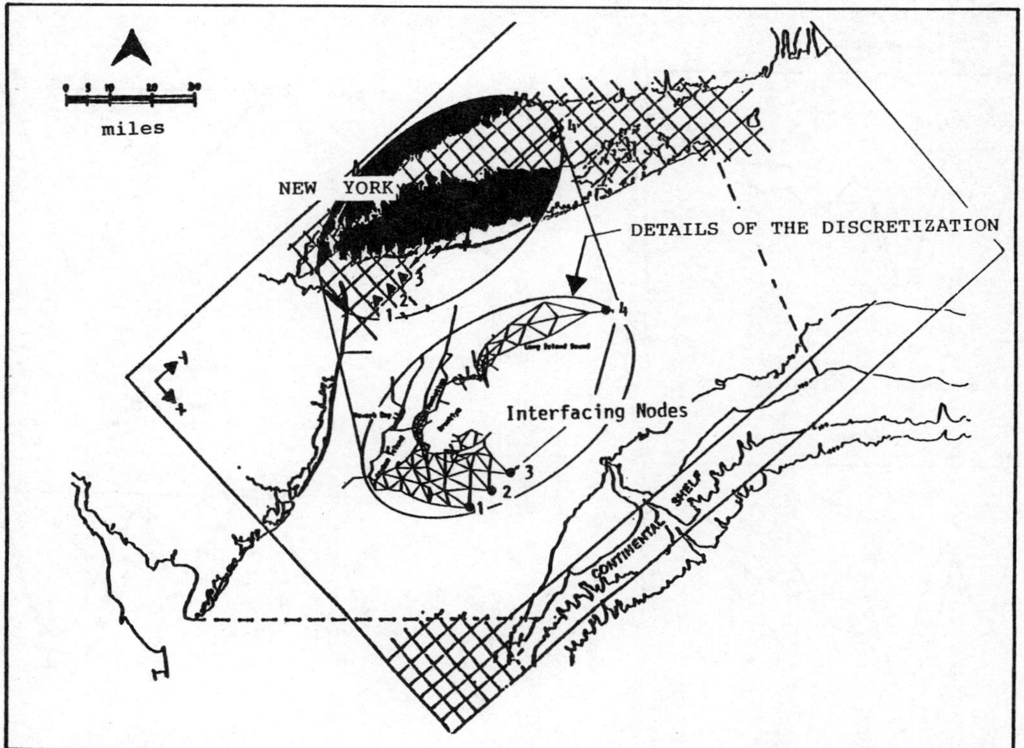


Figure 3 Spatial Discretizations

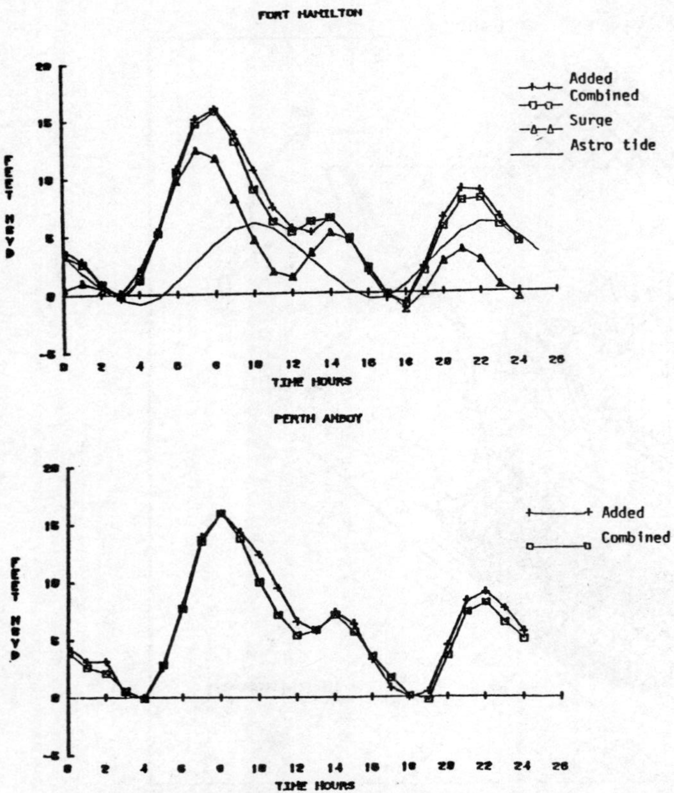


Figure 5: Nearshore Non-linearity Effects

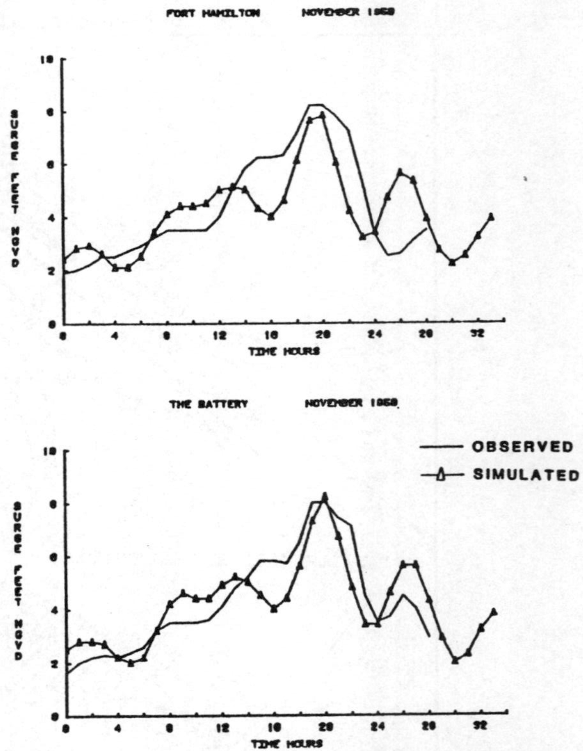


Figure 6: Observed and Simulated Marigrams from Northeaster Storm of November 1950.

# TYPHOON SURGE ANALYSIS BY SELECTIVE LUMPING TWO STEP EXPLICIT FINITE ELEMENT METHOD

Mutsuto Kawahara

Department of Civil Engineering, Chuo University
Kasuga, Bunkyo-ku, Tokyo, Japan

## INTRODUCTION

Since Japan is an island country of which frontier is all surrounded by the ocean, tremendous numbers of typhoons sweep down on coastal area every year. One of main human damages is caused by the typhoon surge, which is a raise of the water surface by the atmospheric pressure drop due to the typhoon. The elevation of the water surface is amplified because of resonance, which is fundamentally related to the configuration of the bay and the water bottom topography. Along the coastline of the Tokyo Bay, there are a population of about 22,000,000, a half of which remains in the city of Tokyo. The area is extremely endangered by the damages of the typhoon surge because the characteristic period of the Tokyo Bay is likely coincident with that of the typhoon surge. Thus, there have been great demands to predict the typhoon surge considering the approximation as exact as possible for the configuration of the coastline and bottom topography of the Tokyo Bay. For this purpose, the finite element method is the most suitable tool for the analysis.

To predict the typhoon surge problem, many numerical analyses based on the finite difference method have been presented, e.g., O'Brien and Reid (1966), Heaps (1971), Overland and Myers (1976), Gjevik and Røed (1976), Reid et al. (1977), Reid (1977) and Yeh and Chou (1979). Recently, Tanaka and Ono (1978) have analyzed the typhoon surge problem by using the implicit finite element method. Kawahara et al. (1978-a, b, c) have applied the explicit finite element method to the tsunami wave propagation analysis. The typhoon surge is also analyzed by Kawahara et al. (1980). Contrary to these former analyses, this paper presents the finite element method based on the selective lumping two step explicit scheme, which is originally argued by Kawahara et al. (1979, 1982). The typhoon surge of the Tokyo Bay has been precisely analyzed. It is noted that the trajectory rather than the intensity of the typhoon is more significant for the surge and the water elevation will reach to about 145 cm at the most interior part of the Tokyo Bay in case of the typhoon, a scale of which corresponds to the one most frequently attacks the coastal area of the bay.

## BASIC EQUATIONS

A typhoon surge is a sort of long wave propagating through a surface of the sea induced by the pressure drop of the typhoon. According to the long wave theory, the basic equation is expressed in the following form.

$$\frac{\partial u_i}{\partial t} + u_j u_{i,j} + g(\zeta - \zeta_a) + \tau_i^B = \tau_i^S \tag{1}$$

$$\frac{\partial \zeta}{\partial t} + \{(h+\zeta)u_i\}_{,i} = 0 \tag{2}$$

where $u_i$ and $\zeta$ represent water velocity and water elevation measured from the undisturbed water level and g, h are gravity acceleration and water depth of the sea, respectively. Here and in the followings the indecial notation and the usual summation convention with repeated indices are employed. For the bottom friction $\tau_i^B$, the Chezy formula is introduced as

$$\tau_i^B = \frac{u_i}{C^2 h} \cdot \sqrt{u_k u_k} \tag{3}$$

where C is the Chezy coefficient. The surface friction $\tau_i^S$ is assumed to be given by a function of wind velocity,

$$\tau_i^S = DW_i \sqrt{W_k W_k} \tag{4}$$

where $W_i$ denote wind velocity and D is a constant. The atmospheric pressure head $\zeta_a$ is derived from the atmospheric pressure difference $P_a$ as follows.

$$\zeta_a = \frac{P_a}{\rho_a g} \tag{5}$$

where $\rho_a$ is density of atmosphere.

For the boundary conditions, the velocity condition,

$$u_i = \hat{u}_i \qquad \text{on } S_1 \tag{6}$$

and the water elevation condition

$$\zeta = \hat{\zeta} \qquad \text{on } S_2 \tag{7}$$

are introduced, where $S_1$ or $S_2$ is the boundary on which velocity or pressure is prescribed and $\hat{u}_i$ and $\hat{\zeta}$ are the prescribed values on $S_1$ and $S_2$ respectively. For the initial condition, the appropriate initial state is assumed as

$$u_i = \hat{u}_i^0 \qquad \text{at} \quad t = 0 \tag{8}$$

$$\zeta = \hat{\zeta}^0 \qquad \text{at} \quad t = 0 \tag{9}$$

where $\hat{u}_i^0$ and $\hat{\zeta}^0$ represent the initial water velocity and elevation respectively, which are normally assumed to be zero.

The standard Galerkin procedure leads to the weighted residual equation multiplying both sides of equations (1) and (2) by weighting functions $u_i^*$ and $\zeta^*$, respectively, integrating the resulted equations over the domain V and rearranging the terms:

$$\int_V \left(u_i^* \frac{\partial u_i}{\partial t}\right) dV + \int_V (u_i^* u_j u_{i,j}) dV + \int_V \{u_i^* g(\zeta - \zeta_a)_{,i}\} dV + \int_V (u_i^* \tau_i^B) dV = \int_V (u_i^* \tau_i^S) dV \tag{10}$$

$$\int_V \left(\zeta^* \frac{\partial \zeta}{\partial t}\right) dV + \int_V \zeta^* \{(h+\zeta) u_i\}_{,i} dV = 0 \tag{11}$$

The natural boundary conditions associated with equations (10) and (11) have been precisely discussed by Kawahara et al. (1980).

## FINITE ELEMENT EQUATIONS

The flow field to be analyzed is assumed to be divided into small triangular domains called finite elements. Let the trial and weighting functions for velocity and water elevation in each finite element be expressed as:

$$u_i = \Phi_\alpha u_{\alpha i} \qquad u_i^* = \Phi_\alpha u_{\alpha i}^* \tag{12), (13}$$

$$\zeta = \Phi_\alpha \zeta_\alpha \qquad \zeta^* = \Phi_\alpha \zeta_\alpha^* \tag{14), (15}$$

where $\Phi_\alpha$ represents the interpolation function, $u_{\alpha i}$ is the nodal value of the velocity in each finite element at $\alpha$th node in the ith direction, $\zeta_\alpha$ is the nodal value of water elevation at $\alpha$th node, respectively, and $u_{\alpha i}^*$ and $\zeta_\alpha^*$ are the corresponding nodal values of the weighting functions. In the numerical computation in this paper, linear interpolation function based on three-node triangular finite element is used for $\Phi_\alpha$. By introducing equations (12) ~ (15) into the weighted residual equations (10) and (11), rearranging the terms and considering the arbitrariness of $u_{\alpha i}^*$ and $\zeta_\alpha^*$, the finite element equations for each finite element is derived as follows.

$$M_{\alpha\beta}\dot{u}_{\beta i} + K_{\alpha\beta\gamma j}u_{\beta j}u_{\gamma i} + H_{\alpha i\beta}\zeta_{\beta} + R_{\alpha i} = \hat{\Omega}_{\alpha i} \quad (16)$$

$$M_{\alpha\beta}\dot{\zeta}_{\beta} + C_{\alpha\beta j\gamma}(h_{\gamma}+\zeta_{\gamma})u_{\beta j} = 0 \quad (17)$$

where

$$M_{\alpha\beta} = \int_V(\Phi_\alpha\Phi_\beta)dV$$

$$K_{\alpha\beta\gamma j} = \int_V(\Phi_\alpha\Phi_{\beta,j}\Phi_\gamma)dV$$

$$H_{\alpha i\beta} = g\int_V(\Phi_\alpha\Phi_{\beta,i})dV$$

$$C_{\alpha\beta j\gamma} = \int_V\{\Phi_\alpha(\Phi_\beta\Phi_\gamma)_{,j}\}dV$$

$$R_{\alpha i} = \int_V(\Phi_\alpha\tau_i^B)dV$$

$$\hat{\Omega}_{\alpha i} = \int_V(\Phi_\alpha\tau_i^S)dV + g\int_V(\Phi_\alpha\zeta^a_{,i})dV$$

in which water depth is also interpolated in each finite element as follows.

$$h = \Phi_\alpha h_\alpha \quad (18)$$

where $h_\alpha$ denotes water depth at nodal points of each finite element.

Using the wellknown superposition procedures leads to the finite element equation for the whole flow field in the following form.

$$M_{\alpha\beta}\dot{v}_{\beta} + K_{\alpha\beta\gamma}v_{\beta}v_{\gamma} + H_{\alpha\beta}Z_{\beta} + R_{\alpha} = \hat{\Omega}_{\alpha} \quad (19)$$

$$M_{\alpha\beta}\dot{Z}_{\beta} + C_{\alpha\beta\gamma}(h_{\gamma}+Z_{\gamma})v_{\beta} = 0 \quad (20)$$

where $v_\beta$ and $Z_\beta$ represent velocity and water elevation at whole nodal points of the flow field respectively. All coefficientsaare derived from equations (16) and (17).

The selective lumping two step explicit scheme is effectively used for the numerical integration in time as:

$$\bar{M}_{\alpha\beta}\cdot v_\beta^{n+1/2} = \tilde{M}_{\alpha\beta}\cdot v_\beta^{n} - \frac{\Delta t}{2}(K_{\alpha\beta\gamma}v_\beta^n v_\gamma^n + H_{\alpha\beta}Z_\beta^n + R_\alpha^n - \hat{\Omega}_\alpha^n) \quad (21)$$

$$\bar{M}_{\alpha\beta}\cdot Z_\beta^{n+1/2} = \tilde{M}_{\alpha\beta}\cdot Z_\beta^{n} - \frac{\Delta t}{2}\{C_{\alpha\beta\gamma}(h_\gamma+Z_\gamma^n)v_\beta^n\} \quad (22)$$

$$\bar{M}_{\alpha\beta}\cdot v_\beta^{n+1} = \tilde{M}_{\alpha\beta}\cdot v_\beta^{n} - \frac{\Delta t}{2}\,(K_{\alpha\beta\gamma} v_\beta^{n+1/2} v_\gamma^{n+1/2} + H_{\alpha\beta} Z_\beta^{n+1/2} + R_\alpha^n - \hat{\Omega}_\alpha^n) \qquad (23)$$

$$\bar{M}_{\alpha\beta}\cdot Z_\beta^{n+1} = \tilde{M}_{\alpha\beta}\cdot Z_\beta^{n} - \frac{\Delta t}{2}\,\{C_{\alpha\beta\gamma}(h_\gamma + Z_\beta^{n+1/2}) v_\beta^{n+1/2}\} \qquad (24)$$

where $\bar{M}_{\alpha\beta}$ denotes the lumped coefficient of $M_{\alpha\beta}$ and $\tilde{M}_{\alpha\beta}$ expresses the combination of $\bar{M}_{\alpha\beta}$ and $M_{\alpha\beta}$ as:

$$\tilde{M}_{\alpha\beta} = e\bar{M}_{\alpha\beta} + (1-e)M_{\alpha\beta} \qquad (25)$$

in which e is the selective lumping parameter, which controls the numerical stability and artificial damping. In equations (21) ~ (24), $v_\beta^n$ and $Z_\beta^n$ denote velocity and water elevation at nth time point and $\Delta t$ is short time interval between n+1th and nth time points. Precise discussions on the selective lumping two step scheme have been presented by Kawahara et al. (1982).

## MODELLING OF TYPHOON

The typhoon surge of the Tokyo Bay has been analyzed by the present finite element method. The domain to be analyzed is set to be within a coastal sea area of the Pacific Ocean including the Tokyo Bay. The typhoon in this area can be modelled as follows.

First, the central coordinate $(x_c, y_c)$ of the typhoon is expressed by:

$$x_c = V_x \times (\text{ellapsed time}) + x_0\,[\text{km}] \qquad (26)$$

$$y_c = V_y \times (\text{ellapsed time}) + y_0\,[\text{km}] \qquad (27)$$

where $V_x$ and $V_y$ [Km/hr.] denote the velocity of typhoon and $(x_0, y_0)$ is the initial position of typhoon. Second, atmospheric pressure distribution is modelled after Fujita (1952).

Let pressure distribution at position r be described by p, it follows that

$$p = p_\infty - \frac{a}{\sqrt{1+(\frac{r}{r_0})^2}} \qquad [\text{mb}] \qquad (28)$$

$$r = \sqrt{(x-x_c)^2 + (y-y_c)^2} \qquad [\text{km}] \qquad (29)$$

where $p_\infty$ represents the atmospheric pressure distribution outside the typhoon, $r_0$ is the radius of typhoon and a is the maximum pressure difference. Third, wind velocity $W_x$, $W_y$

[m/sec] is assumed to be given by:

$$W_x = \frac{1}{3.6} c_1 V_x \exp\left(-\frac{r}{500}\Pi\right)$$
$$+ 500c_2|f|(1-\sqrt{1+G})\{0.5(x-x_c) + 0.866(y-y_c)\} \quad (30)$$

$$W_y = \frac{1}{3.6} c_1 V_y \exp\left(-\frac{r}{500}\Pi\right)$$
$$+ 500c_2|f|(1-\sqrt{1+G})\{0.5(x-x_c) + 0.866(y-y_c)\} \quad (31)$$

where

$$G = \frac{4\times10^{-7}a}{\rho_a f^2 r_0^2}\{1+(\frac{r}{r_0})^2\}^{-3/2}$$

in which $\rho_a$, f denote density of air and Coriolis parameter and the following values were employed.

$$\rho_a = 1.22 \times 10^{-3} \qquad [g/m^3]$$

$$f = -8.3304 \times 10^{-5} \qquad [rad/sec]$$

Empirical constants $c_1$ and $c_2$ will be written in table (1). Fourth, the equivalent upset of the water elevation $\zeta_a$ can be converted from equation (28).

$$\zeta_a = \frac{1}{10\rho_\omega g}\cdot\frac{a}{\sqrt{1+(\frac{r}{r_0})^2}} \quad (32)$$

where $\rho_\omega$ and g represent density of water and gravity acceleration. This is the driving force of the storm surge propagation and will be substituted into equation (1). Finally, the surface friction is assumed as a function of wind velocity:

$$\tau_x^s = 3.26 \times 10^{-6} W_x\sqrt{W_x^2 + W_y^2} \qquad [m/sec^2] \quad (33)$$

$$\tau_y^s = 3.26 \times 10^{-6} W_y\sqrt{W_x^2 + W_y^2} \qquad [m/sec^2] \quad (34)$$

Three modelling cases for the typhoon computation have been carried out. The first two cases are concerned with the comparison for the different trajectories and the last case is chosen to be the comparison for the different power of the typhoon. All constants used are summarized in table (1).

table (1)

| | a (mb) | $r_0$ (km) | $c_1$ | $c_2$ |
|---|---|---|---|---|
| Case 1<br>Case 2 | 50 | 45 | 0.714 | 0.6 |
| Case 3 | 70 | 75 | 0.68 | 0.571 |

TYPHOON SURGE ANALYSIS

In the plainland along the coastline of the Tokyo Bay, a plenty of cities including Tokyo are located, in which there are about 20% of the Japanese population. Therefore, it is extremely important to predict the human damage by the typhoon surge along the Tokyo Bay area. Figure (1) shows the finite element idealization of the coastal area of the Pacific Ocean including the Tokyo Bay. The north - south and east - west distances are 192.0 km and 120.0 km respectively. The total numbers of nodal points and finite elements are 1862 and 3422 respectively. Along the coastline, the reflection condition is imposed, i.e., the normal velocity to the coastline is assumed to be zero. The rest of boundary is the artificial one, on which the water elevation is postulated to be zero. This means that the relative water upset from the still water elevation can be computed by the present computation.

Case 1 is a numerical simulation of typhoon surge by a moderate size typhoon. The typhoon modelled in Case 1 is a modelling of the most frequently attacked one to the coastal area of the Tokyo Bay. Figure (2) shows the computed water elevation versus ellapsed time at various cities along the coastline. At Makuhari City, which is located at the most interior part of the bay, the maximum amount of the water elevation has been computed. The value is 140cm. The second largest water elevation is observed at Chiba City, the amount is 122cm. On both cities, the time interval between the beginning of the surge and the time at which the surge reached to the maximum are extremely short, for instance, 2 hours. The time interval will be referred to as the arrival interval. The shortness of the arrival interval effects dangerous for the refuge of the public population. At Tokyo City, the maximum water elevation is about 85cm and the arrival time interval is about 7 hours. At cities outside the Tokyo Bay, the surge is not comparably strong, say, less than 50cm.

Case 2 is a modelling that the power of the typhoon is the same as in Case 1 but the trajectory is shifted to 30km west of the one in Case 1. The computed water elevation is plotted in figure (3). The maximum water elevation is observed at Makuhari City, but, the amount is 138cm, which is less than that of Case 1. Contrary to this, the maximum water elevation at Tokyo City is computed as high as 100cm. This shows that the maximum water elevation is not always observed in case that typhoon passes just on the observation station. The maximum amount and behavior of the water elevation depends on the trajectory of typhoon. Case 3 is the computation on the larger size typhoon. The trajectory is assumed to be the same as in Case 1. The water elevation is illustrated in figure (4). The maximum water elevations computed are 145cm at Makuhari City, 138cm at Chiba City and 98.8cm at Tokyo City.

Comparing the numerical results obtained by the above three computations, the following issues have been drived.
I) In the Tokyo Bay, the maximum amount of water elevation computed is 145cm at Makuhari City, which is located at the most interior part of the bay.
II) The trajectory of the typhoon is more significant for the water elevation rather than the intensity of the typhoon is.
III) The arrival time interval, which is the difference between the beginning of the surge and the time at which the maximum surge occurred, is about 2 hours at Makuhari City. This is important for public refuge.
IV) At various cities along the west coast of the Tokyo Bay, e.g., Misaki, Yokohama, Kawasaki and Tokyo, the arrival time interval is rather long, for instance, 5 hours to 7 hours.
V) From the mouth to the interior of the bay, the maximum values of the water elevation are gradually increasing.
VI) The surge computed outside the Tokyo Bay is smaller than that inside the bay.
The information derived above are useful for the prediction of the human damage in the coastal area along the Tokyo Bay.

REFERENCES


1. Fujita, T. (1952): "Pressure distribution in typhoon," Geophs. Mag. Vol. 23

2. Gjevik, B. and P. Røed (1976): "Storm surge along the western coast of Norway," Tellus, XXVIII (2) pp. 166 ~ 182

3. Heaps, N.S. (1971): "On the numerical solution of the three-dimensional hydrodynamical equations for tides and storm surges," Mem. Soc. Sci., Liege, Vol. 1, pp. 143 ~180

4. Kawahara, M., N. Takeuchi and T. Yoshida (1978-a): "Two step explicit finite element method for tsunami wave propagation analysis," Int. J. num. Meth. Engng. Vol. 12 (2), pp. 331 ~ 351

5. Kawahara, M., S. Nakazawa, S. Ohmori and K. Hasegawa (1978-b): "Tsunami wave propagation analysis by the finite element method," 2nd int. conf. on Finite Element Method in Water Resources (Eds. C.A. Brebbia et al.) pp. 2.131 ~ 2.150

6. Kawahara, M. and S. Nakazawa (1978-c): "Finite element method for unsteady shallow water equation," U.S. Japan seminar on Interdisciplinary Finite Element Methods (Eds. Abel et al.) Cornell University, pp. 267 ~ 283

7. Kawahara, M. (1979): "On the finite element methods in shallow water long wave flow analysis," Proc. 2nd TICOM Conf. on Computational Methods in Nonlinear Mechanics, the University of Texas at Austin, pp. 261 ~ 287

8. Kawahara, M., S. Nakazawa, S. Ohmori and T. Takagi (1980): "Two-step explicit finite element method for storm surge propagation analysis," Int. J. num. Meth. Engng. Vol. 15, pp. 1129 ~ 1148

9. Kawahara, M., H. Hirano, K. Tsubota and K. Inagaki (1982): "Selective Lumping Finite Element Method for Shallow Water Flow," Int. J. num. Meth. Fluid., Vol. 2 (In press)

10. O'Brien, J.J. and R.O. Reid (1966): "The non-linear response of a two-layer, baroclinic ocean to a stationary, axially symmetric hurricane," J. Atmos. Sci., Vol. 24, 197 ~ 207

11. Overland, J.E. and V.A. Myers (1976): "Models of hurricane tide in Cape Fear estuary," J. Waterways, Harbours and Coastal Engng. Div., Proc. A.S.C.E., Vol. 102, No. WW4, pp. 407 ~ 424

12. Reid, R.O. and B.R. Bodine (1968): "Numerical model for storm surges in Galveston Bay," J. Waterways and Harbors Div., Proc. A.S.C.E., Vol. 94, No. WW1, pp. 33 ~ 57

13. Reid, R.O. and A.C. Vastano, R.E. Whittaker and J.J. Wanstrath (1977): "Experiments in storm surge simulation," The Sea (Ed. E.D. Goldberg), John Wiley & Sons, pp. 145 ~ 168

14. Tanaka, T. and Y. Ono (1978): "Finite element analysis of typhoon surge in Ise Bay," U.S. Japan Seminar on Interdisciplinary Finite Element Methods, (Eds. Abel et al.) Cornell University, pp. 307 ~ 332

Figure (1)

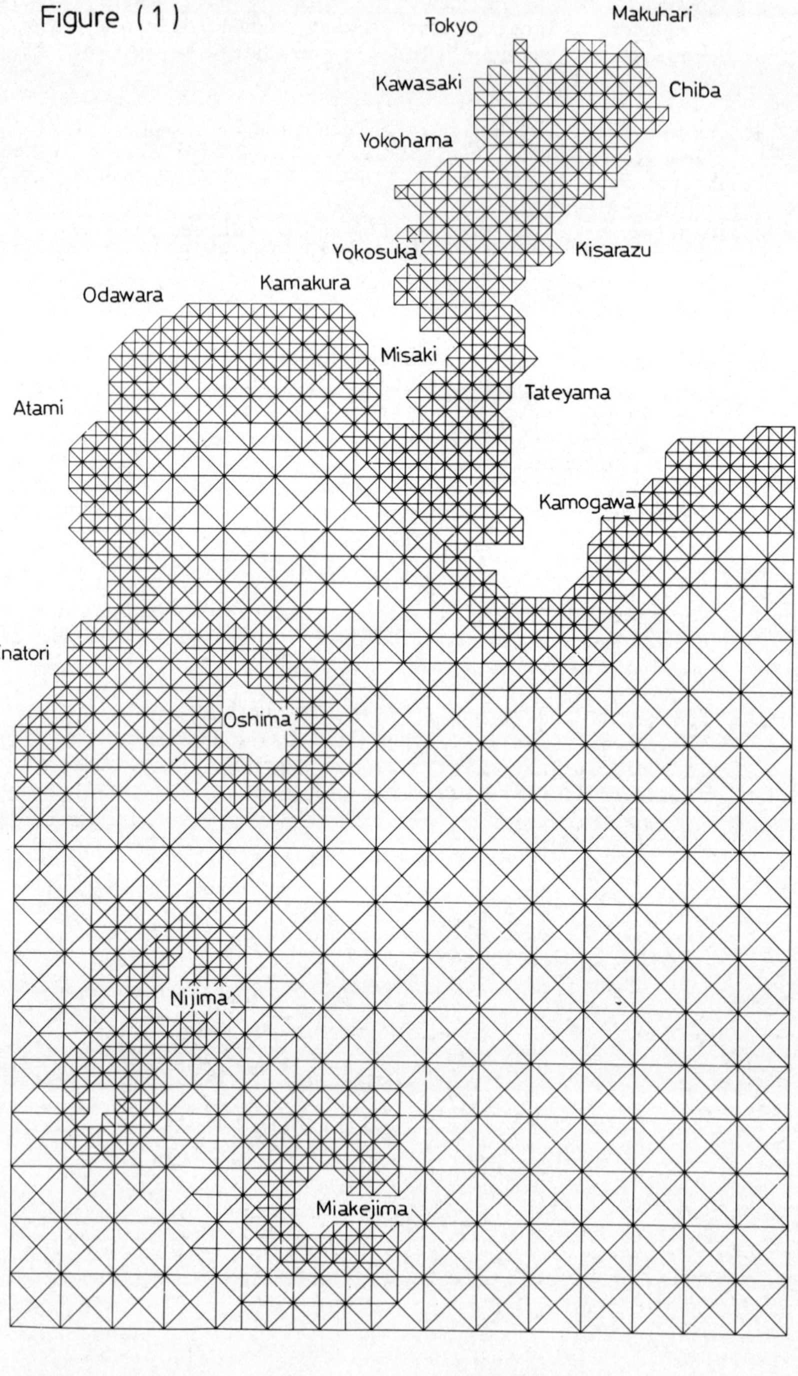
Tokyo
Makuhari
Kawasaki
Chiba
Yokohama
Yokosuka
Kisarazu
Odawara
Kamakura
Misaki
Tateyama
Atami
Kamogawa
Inatori
Oshima
Nijima
Miakejima

CASE 1 $\Delta t = 12.0$ sec

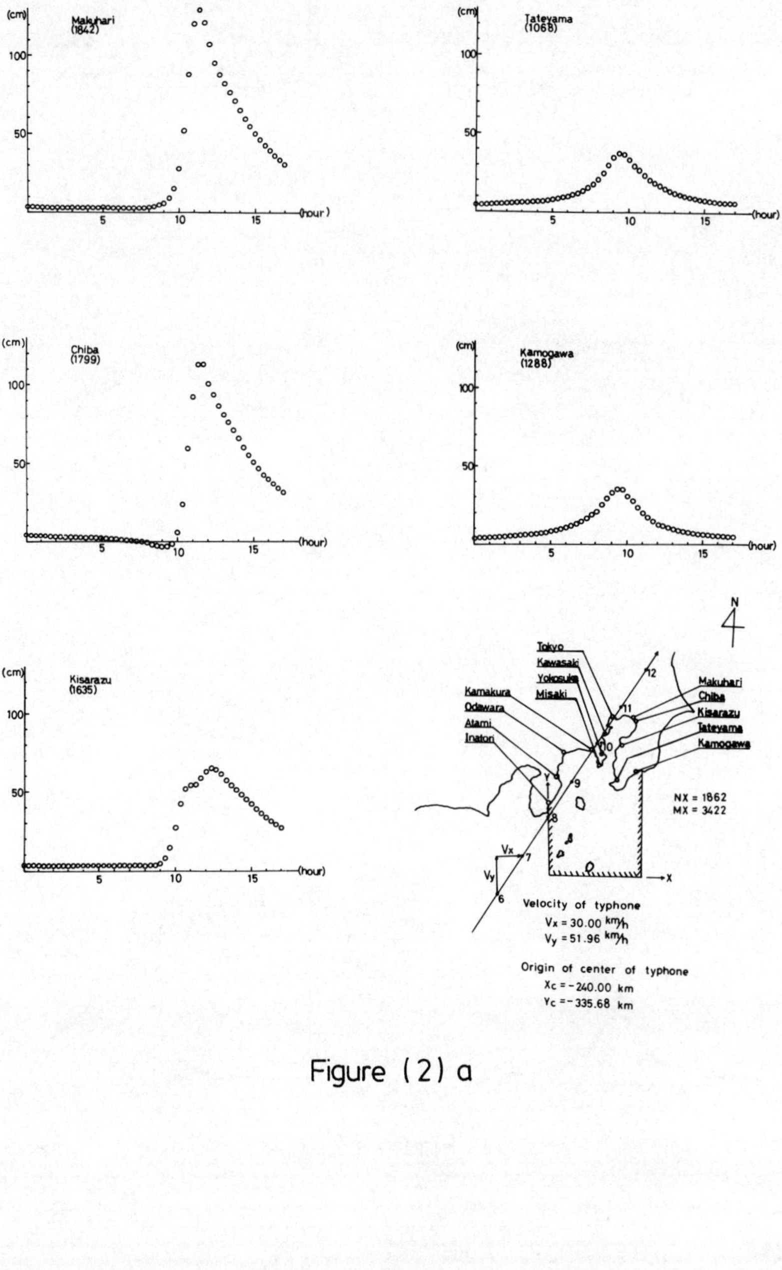


Figure (2) a

CASE 1 $\Delta t = 12{,}0$ sec

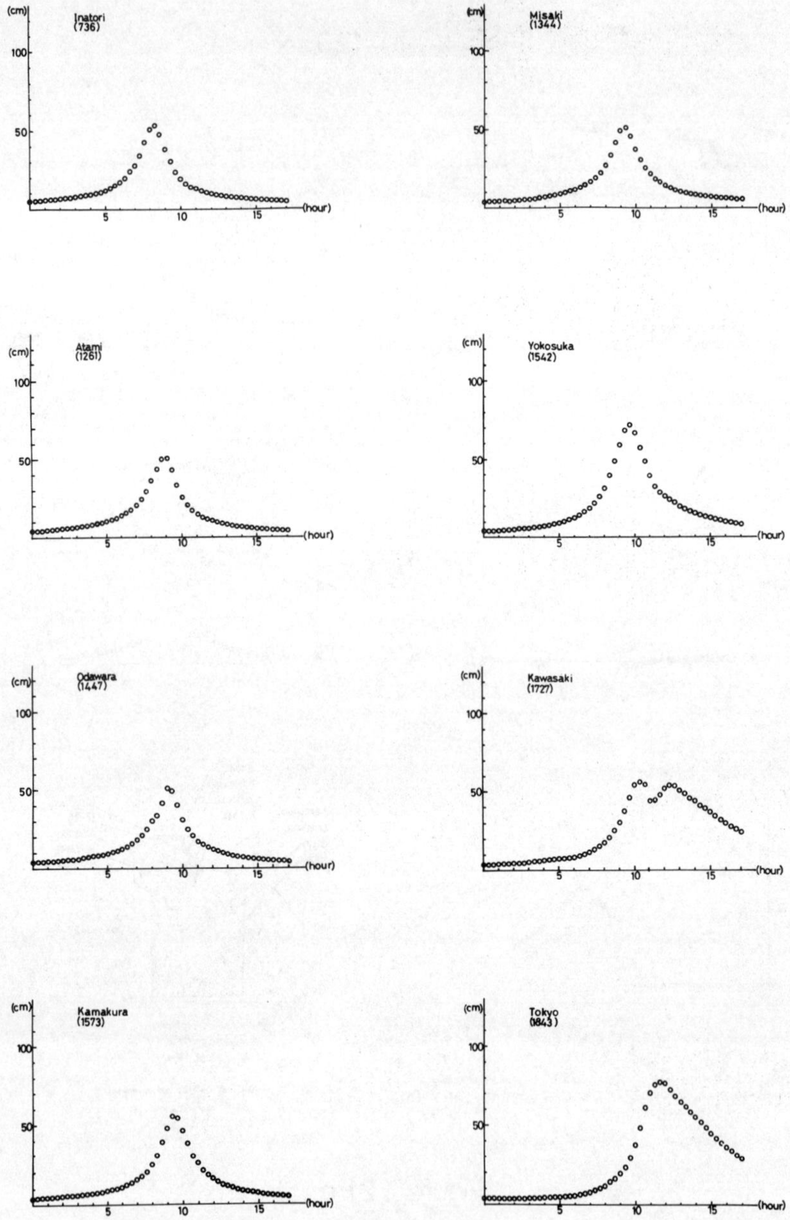


Figure (2) b

CASE 2 $\Delta t=12.0$ sec

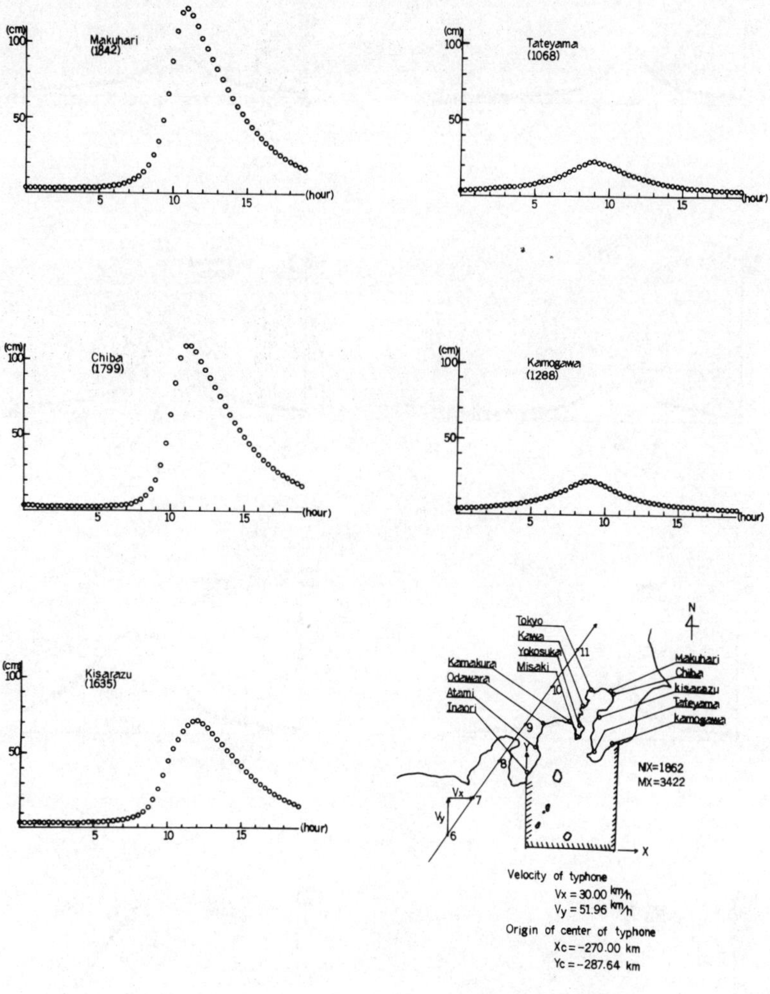


Figure (3) a

CASE 2 $\Delta t$=12.0 sec

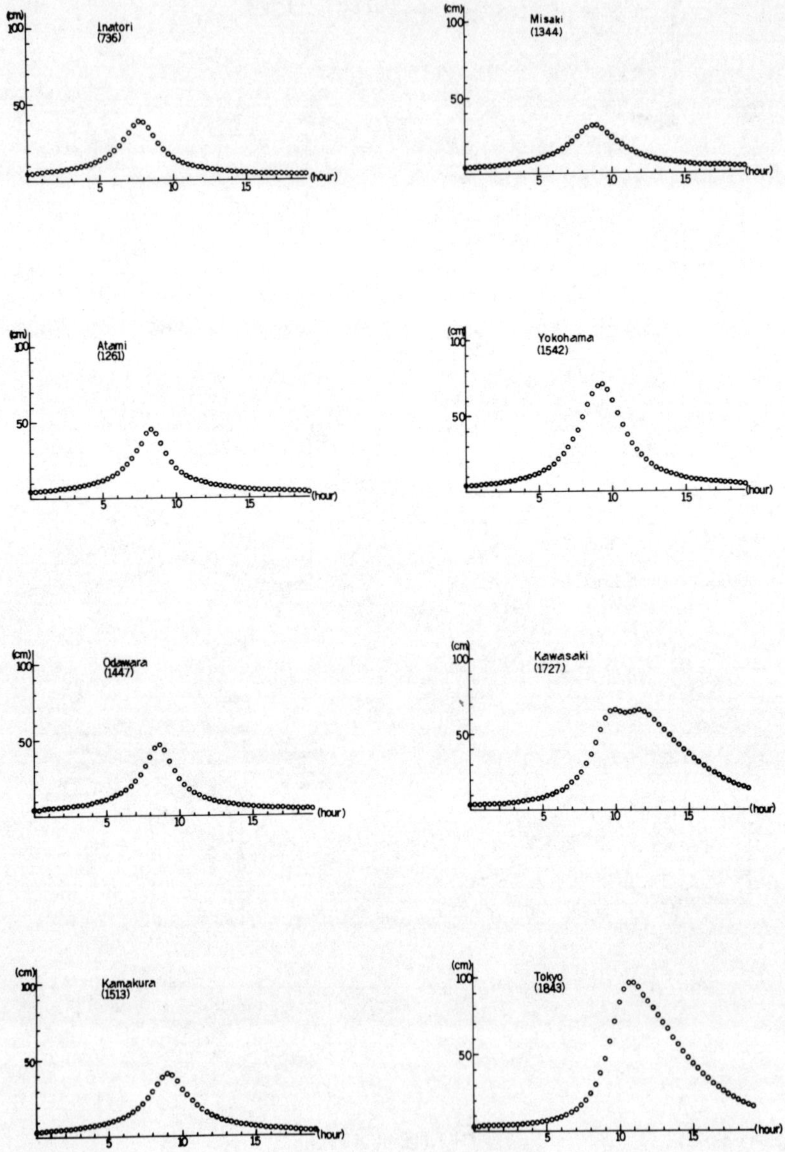


Figure (3) b

CASE 3 $\Delta t = 12.0$ sec

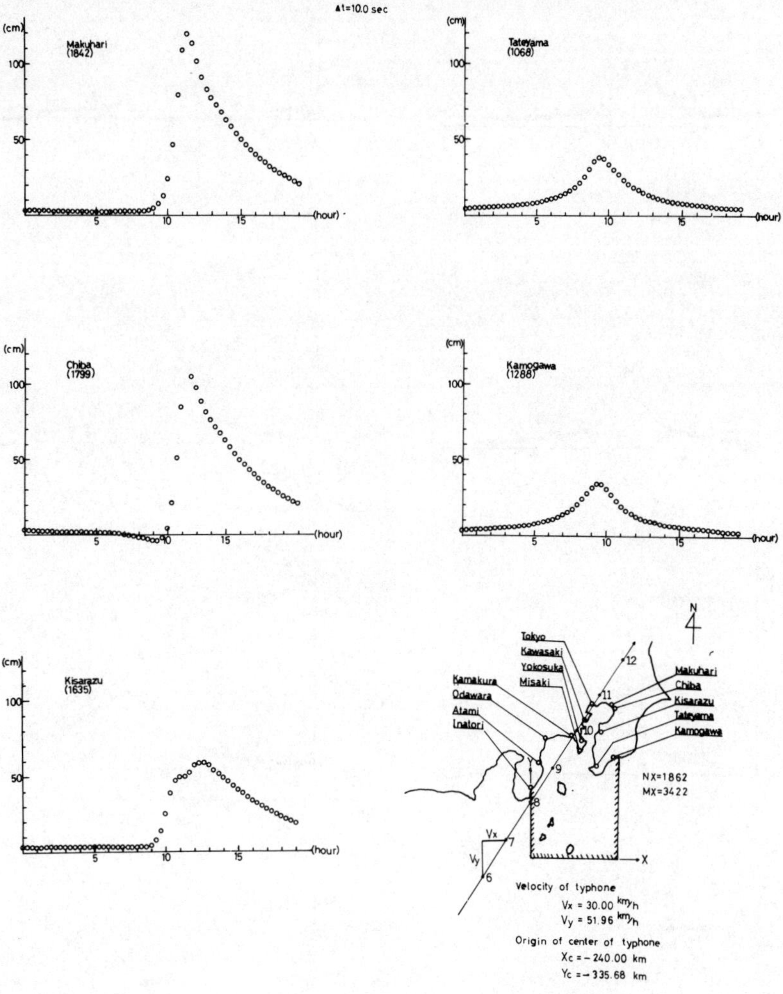


Figure (4) a

CASE 3 $\Delta t = 12.0 sec$

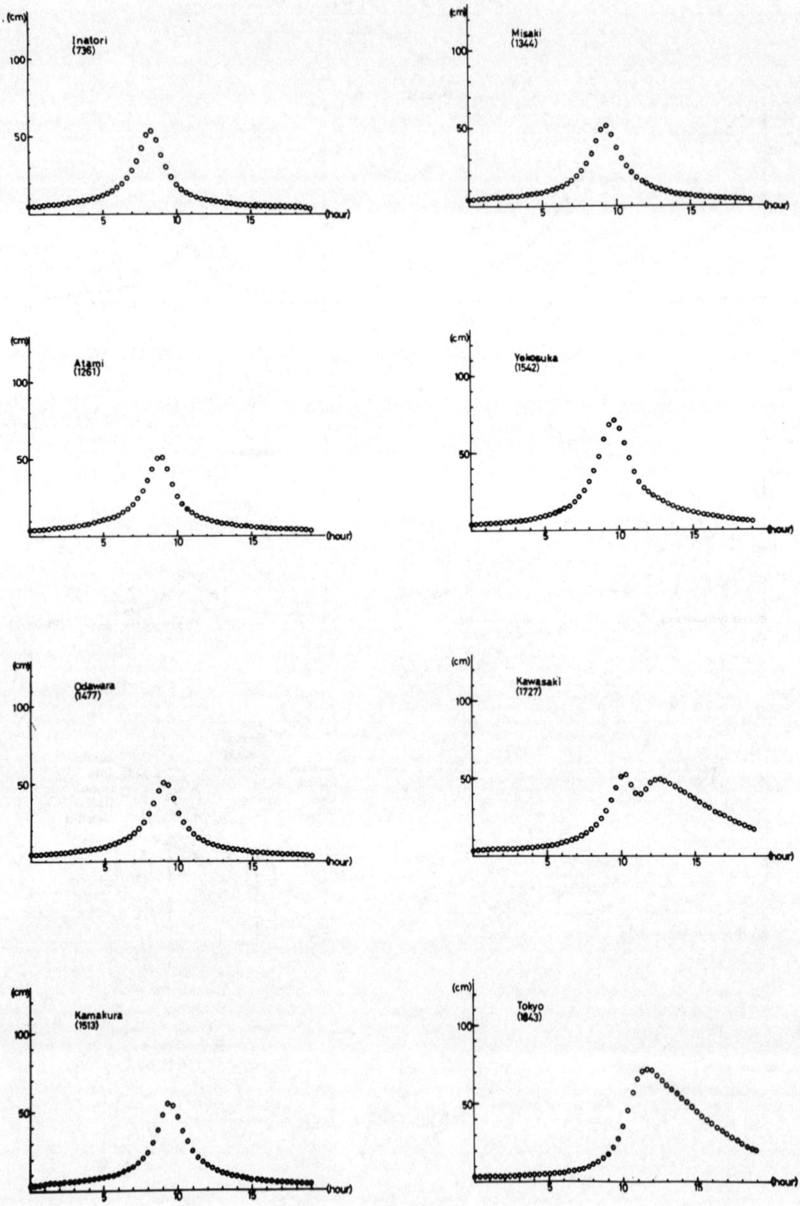


Figure (4) b

# TIDAL RIVER FLOW CALCULATIONS WITH MEASURED VELOCITIES ON THE OPEN BOUNDARIES

A. Nehlsen, W. Michaelis, A. Müller

Institut für Physik, GKSS-Forschungszentrum
2054 Geesthacht, Germany

ABSTRACT

A two-dimensional tidal model for shallow waters using measured velocity cross-profiles on the open boundaries is described. Comparisons of numerical results with measured field data are performed. The influence of model parameters on the calculated flow fields is discussed.

## INTRODUCTION

Since sufficient flow data are not available depth-averaged models for tidal river flow calculations in general use measured surface elevations as open boundary conditions. GKSS developed a new technique for fast flow measurements (Nehlsen, 1982) based on the moving boat method (Smoot, Novak, 1969). The system essentially consists of a two-axis ultrasonic current meter in connection with an electromagnetic distance measuring system to determine continuously position and velocity of the boat. Measurements were performed on the Lower Weser river in September 1980. The flow on the cross-profiles at km 39.5 and km 41.5 as well as in the interior field were measured over one tidal period in time intervals of one hour. The cross-profile data are used as open boundary values for the numerical calculation of the flow field between km 39.5 and 41.5.

## NUMERICAL MODEL

The model is based on the horizontal momentum equations

$$\frac{\partial u}{\partial t} + u\frac{\partial u}{\partial x} + v\frac{\partial u}{\partial y} + \frac{\tau_B^{(x)}}{H} - \frac{\tau_S^{(x)}}{H} - A_h \Delta u - fv + g\frac{\partial \xi}{\partial x} = 0 \quad (1)$$

$$\frac{\partial v}{\partial t} + u\frac{\partial v}{\partial x} + v\frac{\partial v}{\partial y} + \frac{\tau_B^{(y)}}{H} - \frac{\tau_S^{(y)}}{H} - A_h \Delta v + fu + g\frac{\partial \xi}{\partial y} = 0 \quad (2)$$

and the continuity equation

$$\frac{\partial(Hu)}{\partial x} + \frac{\partial(Hv)}{\partial y} + \frac{\partial \xi}{\partial t} = 0 \qquad (3)$$

where $\Delta$ is the horizontal Laplacian operator; $\xi$ is the height of water surface above mean sea level; u and v are the components of the horizontal velocity vector $\vec{v}$ with Cartesian coordinates x and y; $H = \xi + h$ is the total height of the water column, where h is the depth of the bottom below mean sea level; g is the gravity; and f is the Coriolis parameter. Bottom friction $\tau_B^{(x)} = r|\vec{v}|u$, $\tau_B^{(y)} = r|\vec{v}|v$ and wind stress on the surface $\tau_S^{(x)} = \lambda|\vec{v}_w|u_w$, $\tau_S^{(y)} = \lambda|\vec{v}_w|v_w$ are represented by quadratic terms with coefficients r and $\lambda$, where $\vec{v}_w = (u_w, v_w)$ is the wind velocity. The horizontal mixing processes are represented by a constant eddy viscosity coefficient $A_h$.

Eqs. (1), (2) and (3) are solved numerically by the finite difference scheme due to Hansen (1956) using modified open boundary conditions. Instead of surface elevation, measured velocities are specified. The grid notation and the topography of the Lower Weser river between kam 39.5 and km 41.5 are shown in Fig. 1. Each cell element is of the size 125 m x 40 m and characterized by a central elevation point while the velocity components are evaluated at the midpoints of the sides.

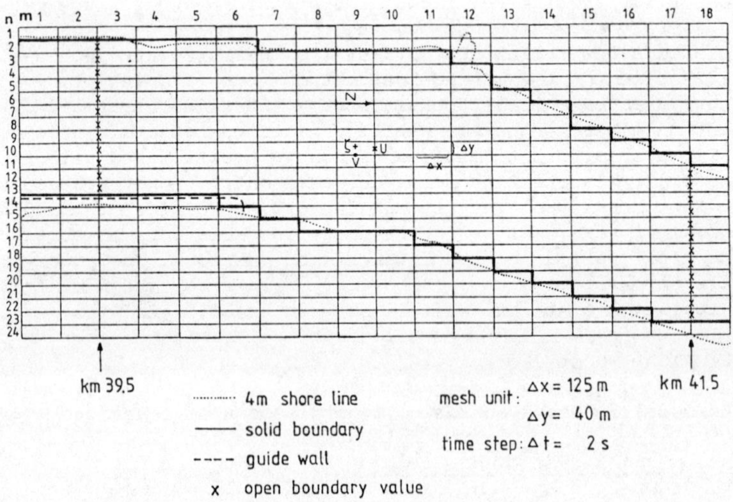


Figure 1: Solution area

## TIME AND SPATIAL INTERPOLATION

Since the shallow domains of the Lower Weser river near the shore line were not accessible for the boat, the corresponding boundary values remain undefinite. These values are calculated by a spatial interpolation based on the assumption that the velocities in the shallow region are mainly determined by the actual depth. The unknown velocity components are set proportional to the neighbouring measured velocity component weighted by the ratio at the corresponding depths. Measurement and spatial interpolation determine the boundary values at time intervals of about one hour. The numerical calculation is performed with a time step of two seconds. The necessary time interpolation is done by a cubic spline routine.

## MASS BALANCE CALCULATION

To guarantee continuity a mass balance algorithm was developed which changes the measured velocities slightly in such a way that the calculated surface elevation always fits the empirical value. The principle of the algorithm is shown in Fig. 2.

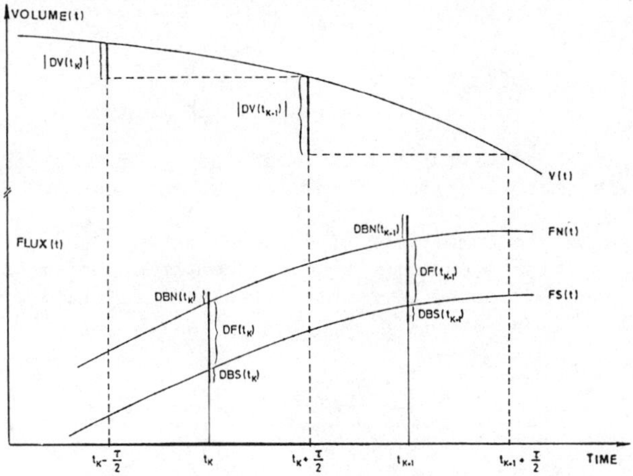


Figure 2: Principle of mass balance calculation

The time scale is divided into equidistant intervals $[t_K, t_{K+1}]$ of length $\tau = 200$ s. At $t_K - \tau/2$, $t_K$ and $t_K + \tau/2$ the fluxes $FN(t_K)$ and $FS(t_K)$ through the northern and southern open boundaries are determined and the mean net flux $\overline{DF}(t_K)$ is calculated by integration. $\overline{DF}(t_K)$ is compared with the change of water volume $DV(t_K)$ in the solution area obtained from the measured surface elevation.

The difference gives the amount of mass balance deviation $DB(t_K)$ for the time interval concerned:

$$DB(t_K) = DV(t_K) - \overline{DF}(t_K) \tag{4}$$

The mass balance deviation $DB(t_K)$ is divided in terms $DBN(t_K)$ and $DBS(t_K)$ which gives the necessary amount of flux correction on the northern and southern open boundaries, such that the percental deviations between measured and balance corrected flux are the same for both open boundaries:

$$DBN(t_K) = \frac{DB(t_K)}{1 + |FS(t_K)/FN(t_K)|} \tag{5}$$

$$DBS(t_K) = DB(t_K) - DBN(t_K) \tag{6}$$

The mass balance corrected boundary velocities $u_N^c(t_K)$ and $u_S^c(t_K)$ are given by

$$u_N^c(t_K) = u_N(t_K) + \frac{DBN(t_K)}{QN(t_K)} \tag{7}$$

$$u_S^c(t_K) = u_S(t_K) - \frac{DBS(t_K)}{QS(t_K)} \tag{8}$$

where $u_N(t_K)$ and $u_S(t_K)$ are the measured velocities and QS and QN are the cross-sections of the open boundaries.

Fig. 3 shows the comparison of measured and calculated surface elevations, Fig. 4 and Fig. 5 give a comparison of measured and mass balance corrected mean velocities as a function of time. In addition high water (HW) and low water (LW) are indicated.

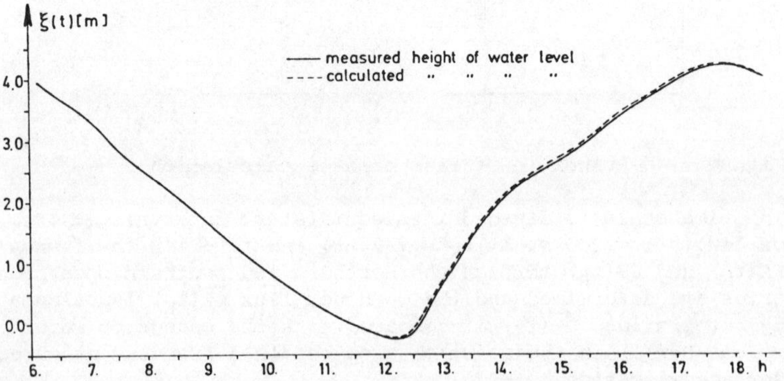


Figure 3: Comparison of measured and calculated surface elevation

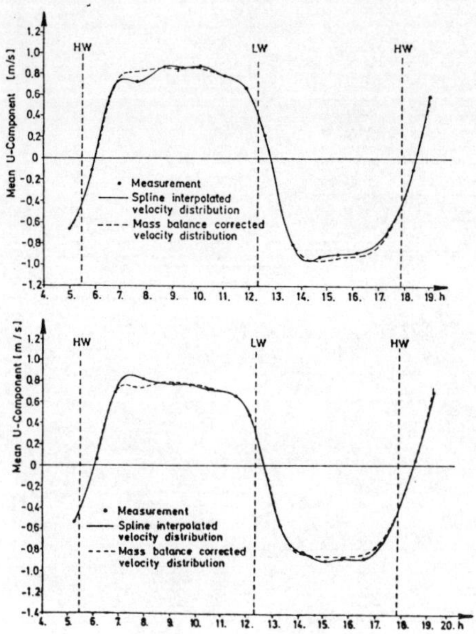


Figure 4 and 5:

Measured and mass balance corrected velocities on the cross-sections at km 39.5 and 41.5, resp.

RESULTS

Fig. 6 to Fig. 9 sketch the calculated flow fields at four states of the tidal cycle. The dashed line marks dry regions in the eastern parts of the solution area. The arrows on the open boundaries correspond to the experimental velocities.

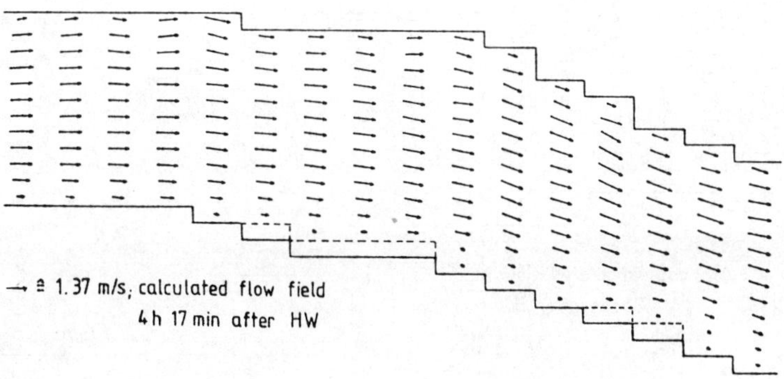


Figure 6

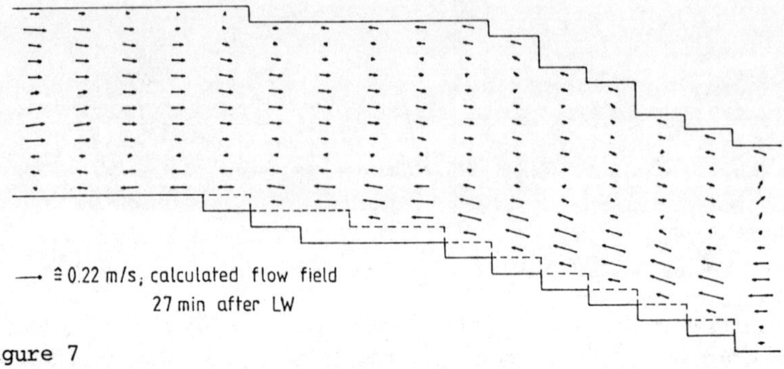


Figure 7

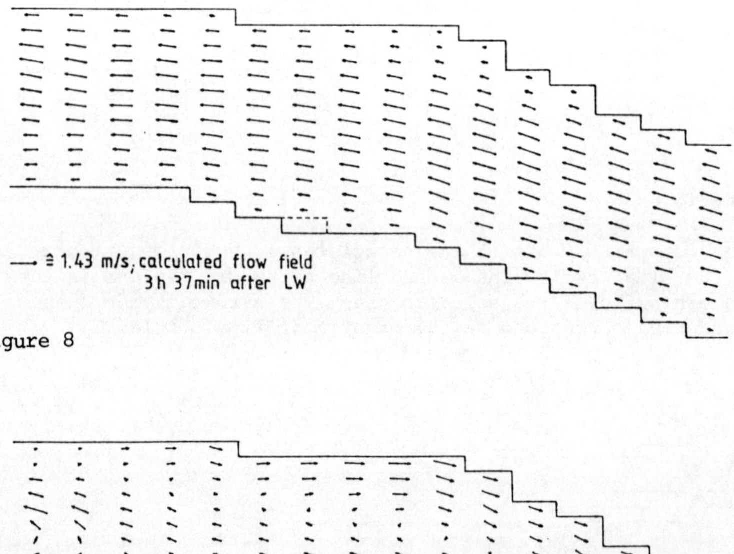


Figure 8

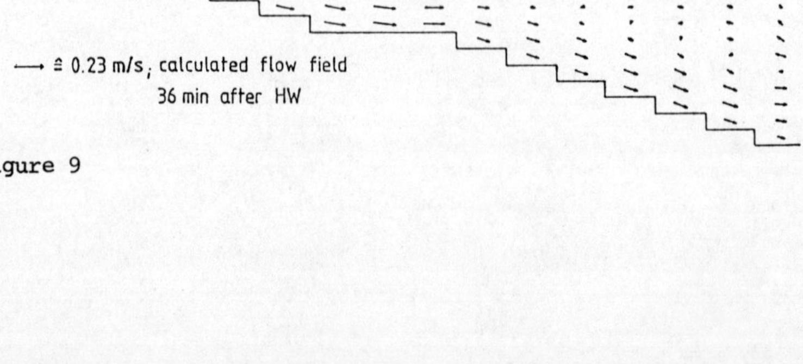


Figure 9

Fig. 10 and Fig. 11 demonstrate the influence of different bottom friction parameters r and lateral and longitudinal exchange coefficients $A_{hx}$ and $A_{hy}$ on the calculated distribution of the velocity component u. A variation of these parameters modifies the structure of the u profile and has remarkable effects on the calculated surface elevation gradient between the open boundaries (Fig. 12 and Fig. 13).

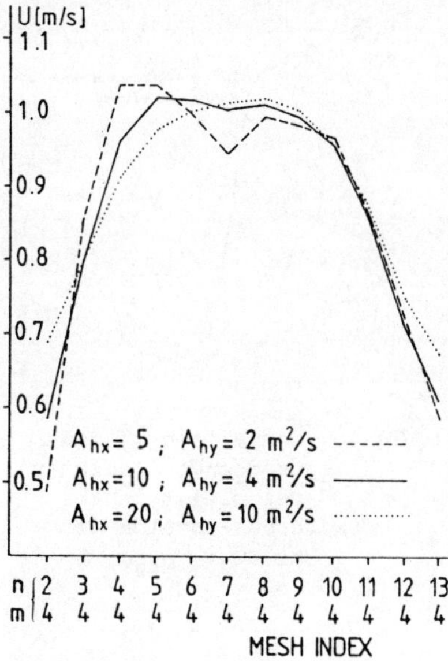


Figure 10:
Variation of exchange coefficients

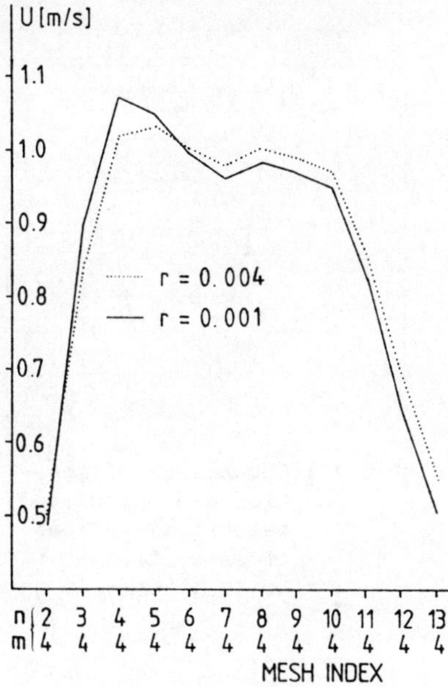


Figure 11:
Variation of bottom friction parameter

Figure 12 and 13:
Calculated surface elevations on the open boundaries for different exchange coefficients $A_{hx}$ and $A_{hy}$

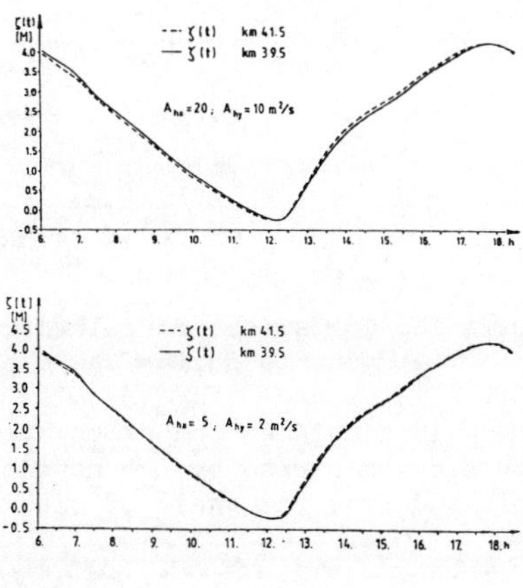

Fig. 14 and Fig. 15 show the calculated velocities for one tidal period at two places on the cross-section at km 40.5 in comparison with flow data, as obtained from a stationary measurement by the Wasser- und Schiffahrtsamt Bremerhaven (1980).

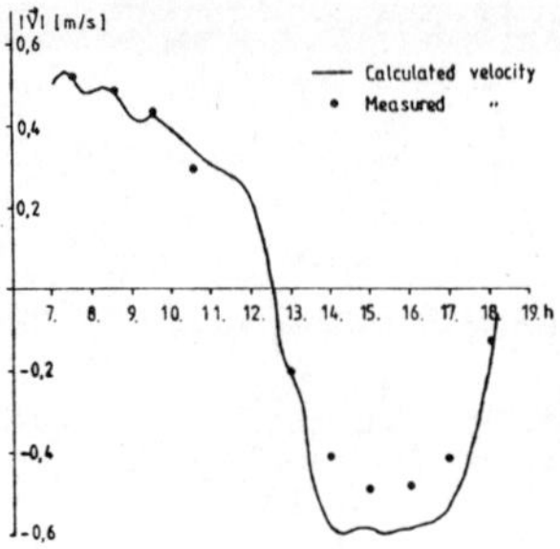


Figure 14: Comparison of measured and calculated velocities on the cross-section at km 40.5, place 1

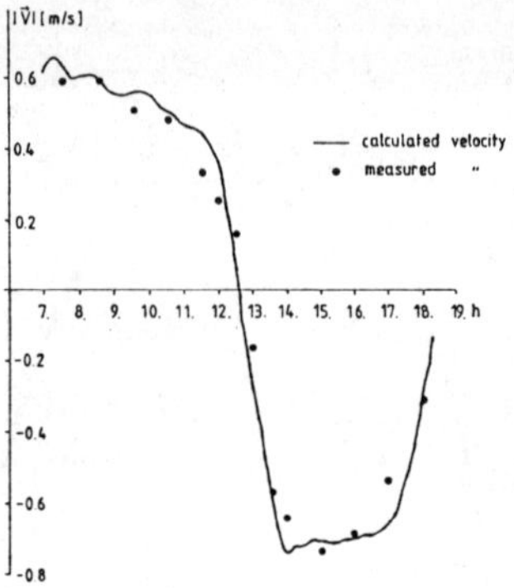


Figure 15: Comparison of measured and calculated velocities on the cross-section at km 40.5, place 2

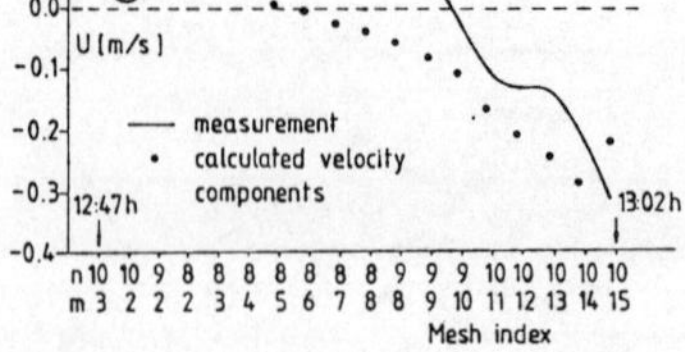


Figure 16: Comparison of calculated and measured velocity components during the phase of slack water

Figure 16 displaces the calculated velocities in comparison with flow data, measured by the moving-boat-method on a longitudinal profile during the phase of slack water.

## CONCLUSIONS

The model version described permits the calculation of horizontal flow fields in tidal rivers from measured velocities on the open boundaries. The flow data are adjusted to a simultaneously measured water level to guarantee continuity and are completed by time and spatial interpolation. Variation of model parameters results in modification of the lateral structure of the calculated velocities and has a remarkable effect on the surface elevation gradient between the open boundaries. The verification of the model will be continued by future measurements and calculations.

## REFERENCES

Hansen, W. (1956) Theorie zur Errechnung des Wasserstandes und der Strömungen in Randmeeren nebst Anwendungen. Tellus, 8.

Nehlsen, A. (1982) Berechnung und Messung horizontaler Strömungsfelder in Tideflüssen. Dissertation, Universität Hamburg (GKSS external report, to be published).

Smoot, H.F. Novak, C.E. (1969) Measurement of discharge by the moving boat method, US Geological Survey Techniques, Water Resources Inv., book 3, chap. A 11.

Wasser- und Schiffahrtsamt Bremerhaven (1980) Current measurement data (unpublished).

# RESIDUAL CURRENTS DURING CHANGING METEOROLOGICAL CIRCUMSTANCES

by
Niek Praagman
Delft University of Technology
Mathematical Department
Julianalaan 132
2628 BL Delft
The Netherlands

## ABSTRACT

The influence of a weak wind field on residual velocities is studied. It is shown that the influence on the residual currents is small but considerable larger than the influence on the depth mean velocities. Reduction of computing time is considered for the case that in connection with several windfields residual velocity fields have to be computed. It turns out that the residuals behave linearly with respect to the wind.

## INTRODUCTION

During the second part of the seventies the interest in mathematical models that treat the behaviour of pollutants in coastal seas increased considerably. These models, often called water quality models, are based on a system of partial differential equations which represent conservation of mass and momentum. (see Praagman, [1981]). In many practical applications one is interested in the behaviour of the pollutants for periods that cover several days or even weeks. Considering the speed of modern computers it is not attainable to compute tidal currents and related processes using the original mathematical equations as given in Dronkers, [1964] and Ippen, [1966]. In order to overcome this problem time averaging techniques are used. (see for instance Hinze, [1959]).

Application of a moving time average to the equations that represent the conservation of N-quantities $c_i$ results in the system:

$$\frac{\partial c_i^0}{\partial t} + \frac{U^0}{h}\frac{\partial c_i^0}{\partial x} + \frac{V^0}{h}\frac{\partial c_i^0}{\partial y} = F(c_i^0,\underline{x},t) \qquad i=1,2,\ldots,N. \qquad (1)$$

The variables $c_i^0$, $U^0$ and $V^0$ are defined by

$$c_i^0 = \frac{1}{2T} \int_{-T}^{+T} c_i(\underline{x},\tau)\, d\tau \tag{2}$$

$$U^0 = \frac{1}{2T} \int_{-T}^{+T} (h+\zeta)u\, d\tau \tag{3}$$

and

$$V^0 = \frac{1}{2T} \int_{-T}^{+T} (h+\zeta)v\, d\tau \tag{4}$$

while:

| | |
|---|---|
| $c_i(\underline{x},t)$ | : concentration of quantity i at position $\underline{x}$ and time t |
| $(u,v)$ | : depth mean velocity vector |
| $h(\underline{x})$ | : depth at position $\underline{x}$ |
| $F(c_i,\underline{x},t)$ | : term in which several processes are taken into account (connected with sources, sinks, interactions, dispersion) |

The computation of the vector field

$$\left(\frac{U^0}{h}, \frac{V^0}{h}\right) \tag{5}$$

has been a topic of interest since many years. (See for instance Tee, [1976], Nihoul and Ronday [1975], Alfrink and Vreugdenhil [1981]).
Several models for the computation of the residuals

$$\left(\frac{U^0}{h}, \frac{V^0}{h}\right)$$

have been proposed but for actual computations the straightforward method as given in Flather [1976] seems to be preferable, (See also Praagman [1982]) if windinfluence is not taken into account. In the present study the influence of a weak wind both on the depth mean current and on the depth mean residual current is considered. Computations show that the influence of such a weak wind on the residual currents is small but considerable larger than the influence on the velocity. Attention is paid to the possibility to reduce computing time in the case that the residual velocity field has to be computed for several wind-fields.

## THE MATHEMATICAL EQUATIONS

In this study two systems of equations play a role. In the following merely the mathematical formulation is given. (For a derivation see Dronkers [1964], Nihoul and Ronday [1975], Praagman [1979], [1981].)

The first system is the system of the shallow water equations:

$$\frac{\partial H}{\partial t} + \frac{\partial}{\partial x}(Hu) + \frac{\partial}{\partial y}(Hv) = 0 \tag{6}$$

$$\frac{\partial}{\partial t}(Hu) + \frac{\partial}{\partial x}(Hu^2) + \frac{\partial}{\partial y}(Huv) - fHv + gH\frac{\partial \zeta}{\partial x} + F_x = D_x + W_x$$

$$\frac{\partial}{\partial t}(Hv) + \frac{\partial}{\partial x}(Huv) + \frac{\partial}{\partial y}(Hv^2) + fHu + gH\frac{\partial \zeta}{\partial y} + F_y = D_y + W_y$$

with

- $H$ : total waterdepth : +h
- $f$ : parameter of Coriolis
- $g$ : acceleration of gravity
- $D_x, D_y$ : diffusion terms
- $F_x, F_y$ : bottom friction terms
- $W_x, W_y$ : surface friction terms, i.e. windstress.

Utilizing a time averaging process and a stream function $\psi$ such that

$$U^0 = -\frac{\partial \psi}{\partial y}, \quad V^0 = \frac{\partial \psi}{\partial x} \tag{7}$$

the second system, an elliptical equation in $\psi$, can be obtained:

$$K\Delta\psi + f\left(\frac{\partial h}{\partial x}\frac{\partial \psi}{\partial y} - \frac{\partial h}{\partial y}\frac{\partial \psi}{\partial x}\right) - \frac{2K}{h}\left(\frac{\partial h}{\partial x}\frac{\partial \psi}{\partial x} + \frac{\partial h}{\partial y}\frac{\partial \psi}{\partial y}\right) =$$
$$= h^2\left\{\frac{\partial}{\partial y}\left(\frac{\tau_x - (w_x)^0}{h}\right) - \frac{\partial}{\partial x}\left(\frac{\tau_y - (w_y)^0}{h}\right)\right\} \tag{8}$$

with:

$$\tau_x = \frac{\partial}{\partial x}\left(\left(\frac{U^2}{H}\right)^0\right) + \frac{\partial}{\partial y}\left(\left(\frac{UV}{H}\right)^0\right) + g\frac{\partial}{\partial x}\left(\left(\frac{1}{2}\zeta^2\right)^0\right) \tag{9}$$

$$\tau_y = \frac{\partial}{\partial x}\left(\left(\frac{UV}{H}\right)^0\right) + \frac{\partial}{\partial y}\left(\left(\frac{V^2}{H}\right)^0\right) + g\frac{\partial}{\partial y}\left(\left(\frac{1}{2}\zeta^2\right)^0\right) \tag{10}$$

For the derivation of (8) it has been assumed that

$$(F_x)^0 = \frac{KU^0}{h}, \quad (F_y)^0 = \frac{KV^0}{h} \tag{11}$$

with K a constant, moreover the terms $D_x$ and $D_y$ representing the dispersive processes have been taken zero. Mark that, contrary to (8), in system (6) the bottom friction term still can be specified. (linear, quadratic or an intermediate form). As shown in Praagman [1982], use of different formula's does not necessarily change the solution procedure but does have influence on the residual currents. In the present study a quadratic bottom friction formula is used:

$$F_x = \frac{g}{C^2 H} u\sqrt{u^2+v^2} \;, \quad F_y = \frac{g}{C^2 H} v\sqrt{u^2+v^2} \tag{12}$$

in system (6).

THE VELOCITY FIELDS (u,v) AND $(\frac{U^0}{h},\frac{V^0}{h})$.

In order to obtain a numerical solution system (6) has been discretized in space using the Galerkin finite element technique. The resulting system of ordinary differential equations has been solved in time using a Runge-Kutta method. (see Praagman [1979]). Numerical results for the residuals have been obtained using the trapezoid rule and equations (3) and (4). (See Praagman [1982].

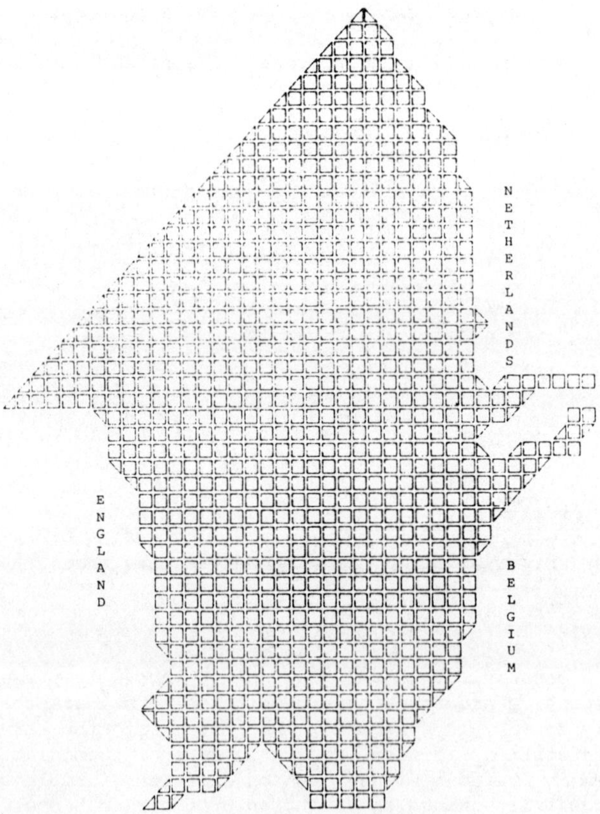


Fig. 1 : Finite element discretization of the Southern Part of the North Sea

Velocity fields are shown in figures 2-5. For $W_x$ and $W_y$ the formula's

$$W_x = \gamma \bar{u} \sqrt{\bar{u}^2 + \bar{v}^2}$$

$$W_y = \gamma \bar{u} \sqrt{\bar{u}^2 + \bar{v}^2}$$

with

$$\gamma = 3.2 * 10^{-6} \text{ (see Dronkers [1964])}$$

and

$(\bar{u},\bar{v})$ : wind velocity near to the water surface

have been used.
In the first run, (fig. 2 and fig. 4)
$(\bar{u},\bar{v}) = (0,0)$ has been used, while the second run: (fig. 3 and fig. 5) corresponds to
$(\bar{u},\bar{v}) = (5 \text{ m/sec}, -5 \text{ m/sec})$.

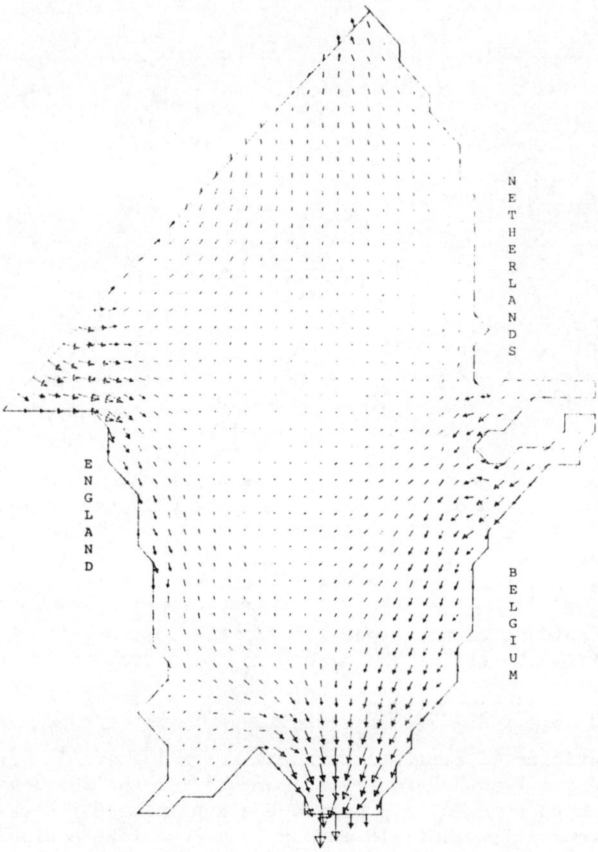


Fig. 2 : Velocity field at time $6^h 15^{min}$, no wind

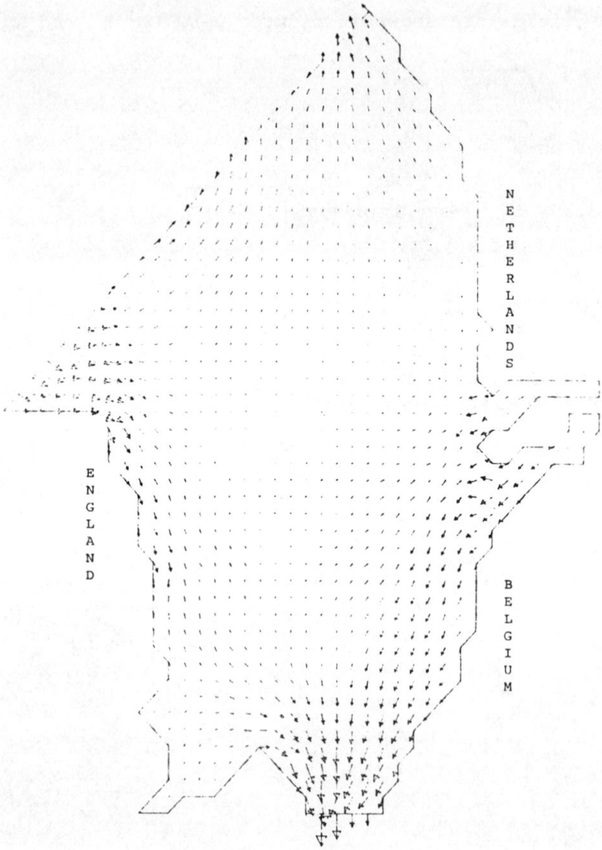


Fig. 3 : Velocity field at time $6^h 15^{min}$, weak S-W wind

Comparing fig. 2 and fig. 3, i.e. the depth mean velocities, differences are hardly visible, while the differences between figure 4 and figure 5, the figures that represent the residual velocities, are obvious. Once it has been shown that the influence of a weak wind is visible the problem of computing time arises, since the computation of residual currents for each new windfield is rather expensive. In the next section two possibilities to reduce the computing costs are considered.

## REDUCTION OF COMPUTING TIME

The computation of residual currents for all possible windfields is very expensive when for every computation the above mentioned method has to be used, i.e. solve the shallow water equations for each specific windfield and apply a numerical averaging process. In order to save computing time two possibilities are considered in the following.

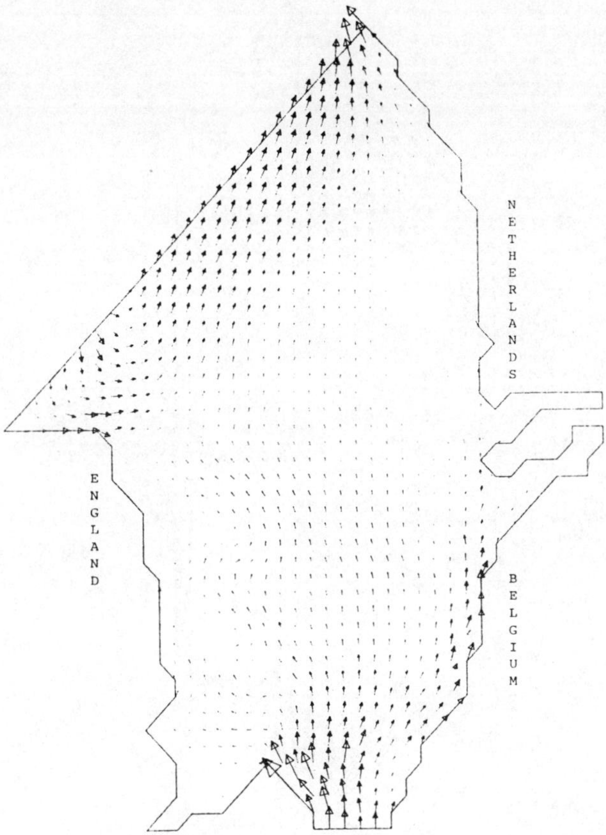


Fig. 4 : Residual velocity $(\frac{U^0}{h},\frac{V^0}{h})$, no wind

System (8) and in the influence of wind

A special advantage of system (8) could be the fact that only the right-hand side terms are winddependent. Hence only these terms are influenced by a changing windfield. In order to study the importance of this influence two series of values were computed. In the first series the quantity

$$h^2\{\frac{\partial}{\partial y}\ (\frac{\tau_x-(W_x)^0}{h}) - \frac{\partial}{\partial x}\ (\frac{\tau_y-(W_y)^0}{h})\}$$

was computed for each nodal point of fig. 1 while $\tau_x$ and $\tau_y$ obtained in a computation without wind, were used. In the second series $\tau_x$ and $\tau_y$ obtained in a computation with $(\tilde{u},\tilde{v})$ = (5 m/s, -5 m/s) were used. The difference between the values, was less than two percent for every internal nodal point. So, using this approach almost no difference will be found

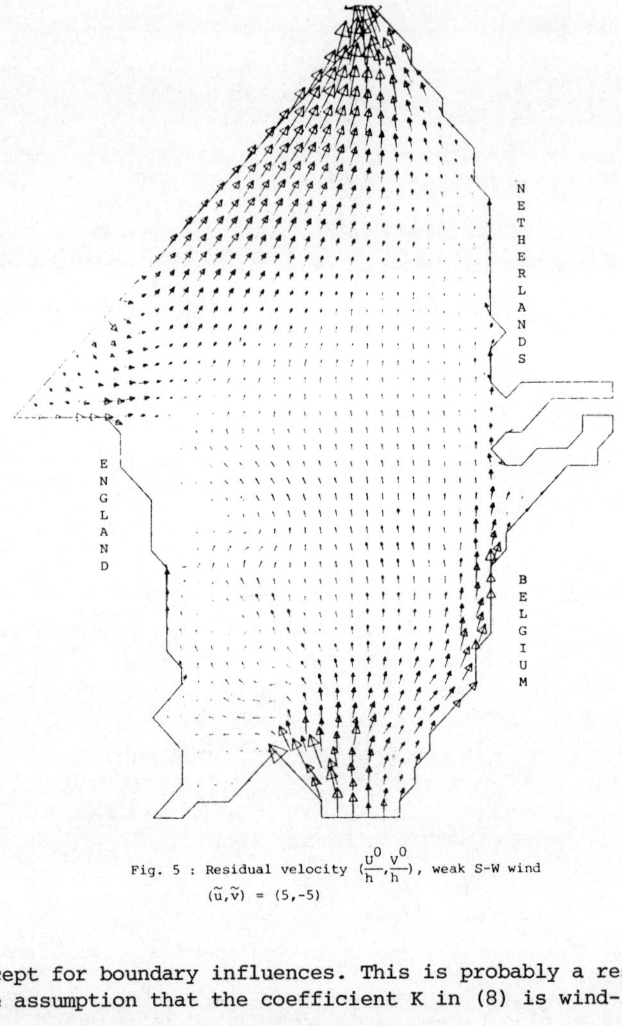


Fig. 5 : Residual velocity $(\frac{U^0}{h},\frac{V^0}{h})$, weak S-W wind $(\tilde{u},\tilde{v}) = (5,-5)$

except for boundary influences. This is probably a result of the assumption that the coefficient K in (8) is wind-independent.

<u>System (6) and the influence of the wind</u>

Also the following approach was studied:

(i) compute the residual current in the case of no wind. Call this velocity field $(\frac{\bar{U}^0}{h},\frac{\bar{V}^0}{h})$.

(ii) compute the residual current in case of a windfield $(\tilde{u},\tilde{v})$ and call the residual field $(\frac{\tilde{U}^0}{h},\frac{\tilde{V}^0}{h})$.

(iii) for every wind field $(\alpha\tilde{u},\alpha\tilde{v})$, $\alpha$ real, compute the related residual velocity field $(\frac{U^0}{h},\frac{V^0}{h})$ using the formula:

$$\left(\frac{U^0}{h},\frac{V^0}{h}\right) = \left(\frac{\bar{U}^0}{h},\frac{\bar{V}^0}{h}\right) + \alpha\left(\frac{\tilde{U}^0-\bar{U}^0}{h},\frac{\tilde{V}^0-\bar{V}^0}{h}\right).$$

Comparing the results of three computations:

a) $(\tilde{u},\tilde{v}) = (\ 0,\ 0)$ m/s

b) $(\tilde{u},\tilde{v}) = (\ 5,-5)$ m/s

c) $(\tilde{u},\tilde{v}) = (10,-10)$ m/s

see figure 4, figure 5 and figure 6, it appeared that this approach is applicable here.

A generalization of this idea to general windfields is of course straightforward. Evidently this is only permitted as long as the amplitude of the wind, given by $\alpha$, and the used friction parameter $\gamma$ are in accordance with each other. (Dronkers [1964]).

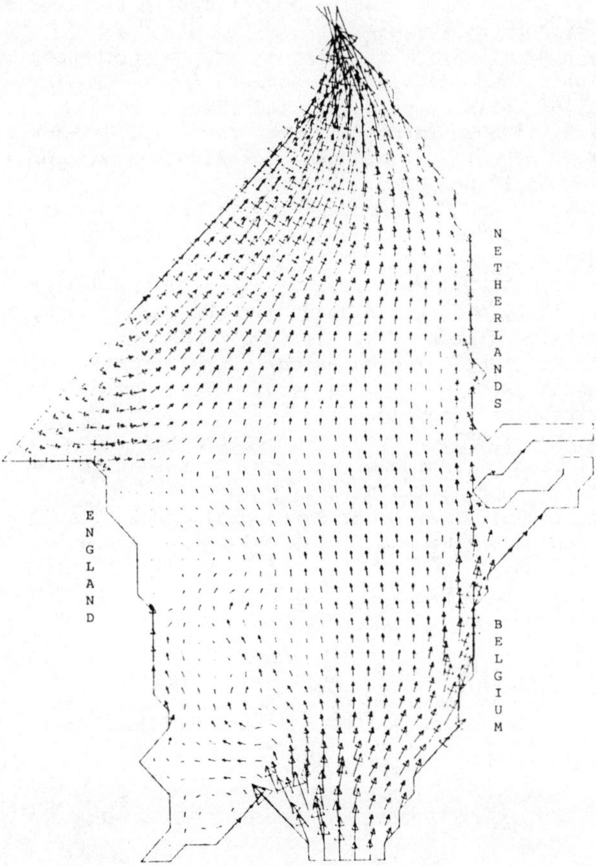


Fig. 6 : Residual velocity $\left(\frac{U^0}{h},\frac{V^0}{h}\right)$, S-W wind, $(\tilde{u},\tilde{\sigma}) = (10,-10)$ m/s.

Hence for strong winds additional computations have to be performed.

CONCLUSION

It has been shown that the influence of even weak winds on residual currents is visible. Reduction of computing time can be obtained, utilizing interpolation, when residual velocity fields corresponding to different windfields have to be computed.

REFERENCES

Alfrink, B.J. and C.B. Vreugdenhil (1981): Residual Currents-Analysis of Mechanisms and Model types, Delft Hydraulics Laboratory, R 1469-11.

Dronkers, J.J. (1964): Tidal computations in rivers and coastal waters, John Wiley, New York.

Flather, R.A. (1976): A tidal model of the north-west European continental shelf. Memoires Societé Royale des sciences de Liége, 6e serie, tome X, pp. 141-164.

Hinze, J.O. (1959): Turbulence, Mc Graw-Hill, New York.

Ippen, A.T. (1966): Estuary and coastline hydrodynamics, Mc Graw-Hill, New York.

Nihoul, J.C.J. and F.C. Ronday (1975): The influence of the "tidal stress" on the residual circulation. Tellus XXVII, pp. 484-489.

Praagman, N. (1979): Numerical Solution of the Shallow water equations by a finite element method. Ph.D.Thesis, Delft University of Technology.

Praagman, N. (1981): The numerical computation of residual currents. Report 81-16, Mathematical Department, Delft University of Technology.

Praagman, N. (1982): A comparison of numerical models for the computation of residual currents. To appear in: Deutsche Hydrographische Zeitschrift.

Tee, K.T. (1976): Tide induced residual current, a 2D non-linear numerical tidal model. Journal of Marine Research, 34, pp. 603-628.

# SOME COMPUTATIONAL PROBLEMS OF OCEANOGRAPHY

William Carlisle Thacker

Atlantic Oceanographic and Meteorological Laboratories
4301 Rickenbacker Causeway, Miami, Florida 33149 (USA)

## ABSTRACT

Three computational problems will be discussed. The first is numerical forecasting of hurricane-induced storm surges. Because population densities are highest around bays and estuaries, forecast models must be capable of representing the irregular shorelines of these natural basins. The computational domain must extend from the shallow inland waters out into the deeper waters of the continental shelf. The distance between adjacent computational points should vary inversely with the square root of the depth, if the solutions are to be uniformly accurate throughout the entire basin. And a storm which takes twelve hours in crossing the basin must be simulated in only a few minutes due to operational constraints involved with issuing warnings to evacuate vulnerable coastal areas. Finite element methods are too slow to be used for this problem, and coordinate transformation methods are incompatible with the required distribution of spatial points. However, a finite difference method can represent the irregular coastline while providing the proper interior resolution and computational efficiency.

The other two problems are encountered in numerically forecasting wave conditions in the ocean. Problem two is a new twist to the well-known problem of solving the advection equation. For this problem there are a family of distributions, one for each frequency and direction of the waves making up the spectrum of the chaotic sea. Thus, a family of advection equations must be solved simultaneously, each with a different advecting velocity. As a result, after long times numerical solutions must be characterized by anomalous dispersion. Because the finite-element method provides a means of interpolating the advecting velocity between discrete wave components of the numerical model, it provides a direction in which to look for a solution to this problem.

The third problem concerns the nonlinear interaction among the components of the wave spectrum. This involves the evaluation of a six-dimensional integral at each time-step of a wave forecast model. The finite element method can be used to replace this integral equation by a discrete analog, thus allowing the nonlinear interactions to be explicitly included in wave models.

## INTRODUCTION

Perhaps the most widely hailed feature of finite element methods is their ability to accommodate irregularly shaped domains. This feature is particularly attractive within the context of water resources because it is often necessary to compute flows in rivers, lakes, estuaries, or oceans, all of which have naturally irregular boundaries. For this reason, finite element methods might appear to be the obvious choice for computations of this sort. Before jumping to this conclusion, it is important to recognize that finite difference methods can offer this same flexibility, sometimes for considerably less computational expense.

Finite difference methods for irregular domains fall into two distinct categories. In the first, the irregular domain is mapped into a geometrically regular region for which a conventional, uniform computational grid is appropriate. This mapping is specified as a coordinate transformation, and transformed equations are solved using standard finite difference methods on the uniform grid. Finally, the inverse transfomation is used to obtain the physically significant results. This approach has been used to compute flows in irregularly shaped basins by Reid and Vastano (1966), Birchfield and Murty (1974), and Wanstrath, *et al.* (1976). Other examples of this approach are given by Chu (1971), Potter and Tuttle (1973), Meyder (1975), Thompson, *et al.* (1977), and Haussling (1979).

The second category of finite difference methods for irregular domains is quite like finite element methods in that the computational points need not form a geometrically regular pattern. In fact, a finite element mesh might serve equally well as a finite difference mesh. This approach dates back to the work of MacNeal (1953), if not farther. The computation of flows having moving boundaries using finite difference methods in conjunction with time dependent, irregular meshes have been discussed by Noh (1964), Crowley (1971) and Fritts and Boris (1979). The computation of atmospheric flow over the spherical surface of the earth by Sadourny, *et al.* (1968) also falls into this category.

The problem of numerically forecasting storm tides, the first of the three computational problems of oceanography discussed here, is presented as an example of the economic advantage that might be realized by using finite difference methods on irregular grids. A corresponding Galerkin finite element method having similar accuracy would be at least an order of magnitude more expensive.

What distinguishes this category of finite difference methods from finite element methods is the way in which the discrete equations are obtained. The finite element method starts with the assumption that the solutions to the partial differential equations can be approximated as piecewise polynomial functions, and the criterion used to specify the best approximation determines the form of the discrete equations. On the other hand, the finite difference equations are obtained simply by replacing partial derivatives with finite difference approximations constructed from functional values at neighboring grid points. The great similarity between the discrete equations that are obtained has been clearly pointed out by Gray and Pinder (1976) and by Thacker (1978). The finite element equations can be written so that terms can be identified as averaged finite diffence forms.

In this sense, finite element methods can be considered to be systematic procedures for generating finite difference equations. For example, Jespersen (1974) showed that the finite difference form for the Jacobian suggested by Arakawa (1966), which conserved mean vorticity, mean-square vorticity, and kinetic energy, can be obtained using finite elements. In fact, Shen (1977) suggests that finite element methods might be regarded as schemes that provide difference formulas for near optimal distributions of computational points. Thus, it is really not so important to distinguish between finite difference and finite element methods as it is to recognize the scope and variety of methods that are available.

The other two computational problems to be discussed are encountered in numerical models for forecasting wave conditions in the ocean. The first is the problem of anomalous dispersion, which can result in computed wave distributions appearing patchy when they should be smooth. This problem should not be confused with the well-known problem of numerical dispersion due to finite spatial and temporal resolution. Anomalous dispersion is a necessary consequence of the discretization of the wave spectrum and the only way to eliminate the problem is to increase spectral resolution. Unfortunately, this type of model already strains available computing facilities to their limits, so increasing resolution is not presently feasible.

The unwanted patchiness can be eliminated at the expense of lowering peak values, if some sort of smoothing is incorporated into the model. The anomalous dispersion results when spectral bands become too narrow to be resolved by the grid of wave frequencies and wave directions, which suggests that spectral averaging should be appropriate. A finite element model would have some sort of spectral averaging built-in. However, this does not mean that a finite element model is the answer to the problem. Here, as for the previous problem, it is best to examine all of the possibilities and to choose the one that is best for this particular case.

The last computational problem to be discussed involves the evaluation of an integral. An essential part of wave models is the exchange of energy between waves due to their nonlinear interaction. The term in the governing equations which describes the wave-wave interaction takes the form of a multi-dimensional integral. Because this integral is so expensive to evaluate, it has been impossible to obtain updated values at each time step. Instead, estimates based on tabulated values are used. Finite element methods suggest that much of the expense of evaluating the integral can be done once as preprocessing, and the updated values at each time step can be obtained by matrix multiplications.

## FORECASTING STORM SURGES

When a hurricane is approaching a coastline, a great deal of attention is focused on the problems of forecasting the storm's size, strength, and track. Because of these meteorological uncertainties, the numerical model for forecasting the hurricane-driven tides has to be run several times in order to bracket the entire range of possibilities. For this reason, it is most important that the model be fast. Only a few minutes of computational time can be spent to simulate approximately twelve hours of oceanic response to the passage of the hurricane.

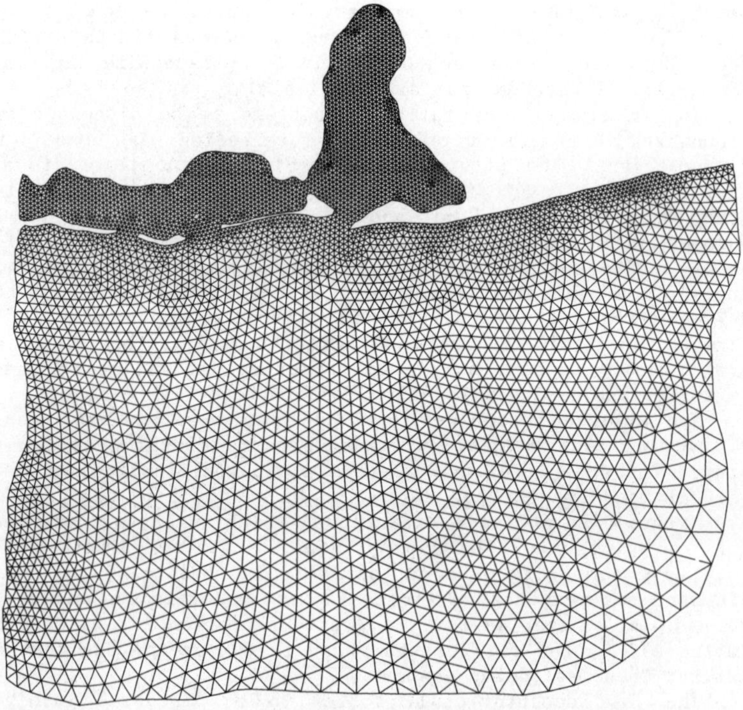

Figure 1. Grid for computing storm surges in Mobile Bay.

The storm surge is governed by the shallow water equations,

$$\frac{\partial \vec{U}}{\partial t} + \vec{\nabla}\cdot\frac{\vec{U}\vec{U}}{D} = -gD\vec{\nabla}H - f\hat{k}\times\vec{U} + \vec{T} - \vec{B} \qquad (1)$$

$$\frac{\partial H}{\partial t} + \vec{\nabla}\cdot\vec{U} = 0, \qquad (2)$$

where $\vec{U}$ is the vertical integral of the horizontal velocity, H is the deviation of the sea surface from its mean, D is the total depth, g and f are, respectively, the gravitational and Coriolis parameters, $\hat{k}$ is the unit vector in the vertical, and $\vec{\nabla}$ is the horizontal gradient operator. A simple spatially and temporally varying functional form is used to account for the atmospheric forcing, represented here by $\vec{T}$. The bottom friction, $\vec{B}$, is a function of $\vec{U}$ and D.

The computational grid must be able to resolve curving coastlines and the flow through narrow inlets and must extend well out into the sea. The grid should be designed so that the distance separating neighboring computational points is roughly proportional to the square root of the local depth (Thacker, *et al.*, 1980). This does more than simply concentrate the points in the important nearshore region. Because the phase speed of shallow water waves is proportional to the square root of the depth, it guarantees that the same number of grid points per wave length are used to resolve the wave in shallow water, where the wave is shorter, as in deep water, where the wave is longer. The Courant number, and thus the resolution, is approximately the same everywhere on the grid.

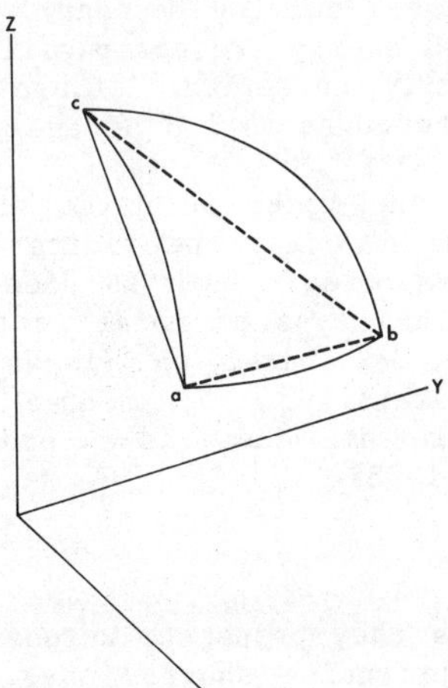


Figure 2. The slope of the plane approximates the slope of the curved surface.

Although such an irregular grid might suggest that the model is based on the finite element method, that is not the case. A finite difference formula is used to approximate the derivatives directly. The basic idea is that the slope of a plane passing through three nearby points on a curved surface approximates the slope of the tangent plane in that vicinity (Thacker, 1977). Partial derivatives at a grid point are simply averages of the slopes computed using the triangular facets approximating the surface about that point. This can be compactly summarized with a finite difference formula for the gradient of a function, $\phi$,

$$(\vec{\nabla}\phi)_j \doteq \frac{\sum_{i=1}^{N} \phi_i(\vec{r}_{i+1} - \vec{r}_{i-1}) \times \hat{k}}{\sum_{i=1}^{N} (\vec{r}_i \times \vec{r}_{i+1}) \cdot \hat{k}} . \tag{3}$$

The subscript, j, indicates that the gradient is approximated at the $j^{th}$ grid point. The index, i, is assigned to the N neighboring points in a consecutive, cyclic, counterclockwise order. The vectors, $\vec{r}_i$, are the positions of the neighboring points. This formula can be generalized for use with curved domains (Thacker, 1980) by letting the vertical unit vary from point to point. On a uniform rectangular grid, this formula reduces to the familiar central difference approximation.

Because this formula satisfies a discrete counterpart of Green's theorem, it is easy to guarantee that mass will be conserved. In fact, both mass and energy can be rigorously conserved for an irregular grid, so long as the boundary is rectangular. However, for a curving boundary, the appropriate flux vectors for mass and energy are not parallel, so in general both cannot be exactly conserved. Conservation of mass does guarantee approximate conservation of energy, so there is no practical problem.

The formulation of the model is quite simple. Spatial derivatives are approximated using the finite difference formula, and an explicit leap-frog scheme is used to extrapolate forward in time. When this type of model is compared with a corresponding Galerkin model based on linear elements, the economic advantages are startling. To produce results of comparable accuracy, the Galerkin scheme is an order of magnitude more expensive (Thacker, 1978).

## ANOMALOUS DISPERSION

Gravity waves disperse as they propagate across the surface of the sea, longer waves outrunning shorter ones. As the waves propagate and disperse, energy that is initially continuously distributed over a localized area and also continuously distributed among waves of varying frequency and direction later is spread over a much wider area, but both the spatial and spectral distributions remain smooth and continuous. However, numerical models for forecasting wave spectra may yield distributions that are patchy rather than continuous and smoothly varying.

Ignoring the complications of generation of waves by the wind, dissipation by breaking, and non-linear interactions, the computational problem reduces to solving

$$\frac{\partial E}{\partial t} + \vec{v} \cdot \vec{\nabla} E = 0, \tag{4}$$

where E is the wave spectrum and $\vec{v}$ is the group velocity. At first sight this can be recognized to be the advection equation, perhaps the most studied of all partial differential equations, at least from the point of view of numerical methods. However, both E and $\vec{v}$ are functions of wave vector, or equivalently of wave frequency and direction, $\omega$ and $\theta$, as well as functions of the usual space and time variables. In this sense, (4) is actually a family of equations rather than a single equation. Numerical methods necessarily replace this infinite family of equations with a finite number of equations corresponding to the discrete representation of the continuous spectrum. The anomalous dispersion is entirely a result of this limited spectral resolution.

Suppose that the solution were known exactly for all times, and consider how it would appear if transferred to a discrete grid. The grid can be chosen so that the solution is initially well resolved and contours of the initial state are quite smooth. But after a sufficiently long time this grid will no longer provide adequate resolution. As the waves propagate and disperse and energy is spread over a larger spatial region, at any given point in that region the energy is confined to a narrower band of frequencies and directions. This clearly must be the case for the total energy to be conserved. Perhaps the best way to visualize this is to imagine the ripples resulting from a rock tossed into a pond. When they arrive at a point far away across the pond, the energy is contained in a tight directional band. Similarly, the longer waves arrive first, so the energy is also in a narrow band of frequencies. Because the spectral bands are too narrow to be resolved by the discrete grid for exact solutions, it should be clear that no computational method can eliminate the problem.

Still, it might be desirable to avoid the patchiness. Highly diffusive schemes, such as upwind differencing, produce less patchy results than some other schemes. This amounts to a form of spatial smoothing, which really is not appropriate because spatial resolution is always adequate. Another possibility is spectral smoothing, which can be used to guarantee that the spectral bands never become too narrow to be resolved. Of course, peak values will be reduced as the energy is spread over a broad band.

Finite element methods have not yet been used for models of wave spectra. If they had been, the idea of spectral smoothing would have been immediately recognized. For example, if Equation (4) were reduced to one dimension, then a Galerkin method with linear elements would yield,

$$\frac{1}{6}\frac{\partial}{\partial t}(E_{j-1}+4E_j+E_{j+1}) + \frac{1}{12}\Big[(v_{j-1}+v_j)\frac{\partial E_{j-1}}{\partial x}$$
$$+(v_{j-1}+6v_j+v_{j+1})\frac{\partial E_j}{\partial x} + (v_j+v_{j+1})\frac{\partial E_{j+1}}{\partial x}\Big] = 0 \qquad (5)$$

As with the the previous problem, it should be emphasized that finite elements should not be used without first ascertaining that no better approach exists. For this problem, smoothing spectra after every few time steps is likely to give good results at minimal expense.

NONLINEAR WAVE INTERACTIONS

Nonlinear interactions of surface waves are governed by the equation,

$$\frac{\partial n_4}{\partial t} + \int C[n_1n_2(n_3+n_4) - n_3n_4(n_1+n_2)]d^2k_1d^2k_2d^2k_3 = 0, \qquad (6)$$

where $n_i \equiv n(\vec{k}_i)$, $i = 1,2,3,4$. The function, $n(\vec{k}) = E(\vec{k})/\omega(\vec{k})$, is the wave action, the spectrum divided by the frequency. The coefficient, C, is a complicated function of four wave vectors $\vec{k}_1$, $\vec{k}_2$, $\vec{k}_3$, and $\vec{k}_4$, containing delta function factors that vanish unless the conditions, $\vec{k}_1+\vec{k}_2=\vec{k}_3+\vec{k}_4$ and $\omega_1+\omega_2=\omega_3+\omega_4$, are satisfied. The integral appearing in Equation (6) should be evaluated at each step forward in time. Even though the delta functions reduce the dimensionality of the integral from six to three, it has nevertheless been prohibitively expensive to evaluate. A finite element approach is suggested here, which might reduce the cost substantially.

Suppose that $n(\vec{k})$ can be approximated by a piecewise polynomial function $\tilde{n}(\vec{k}) = \Sigma_i n_i \phi_i(\vec{k})$. Substitution of $\tilde{n}$ for n in the left side of Equation (6) results in a residual. Requiring that integrals of products of the residual with each of the $\phi_j$ vanish results in model equations of the form,

$$\sum_a \frac{dn_a}{dt} A_{ad} + \sum_{a,b,c} n_a n_b n_c B_{abcd} = 0. \qquad (7)$$

The arrays A and B are independent of the spectral shape, and so they can be computed once and stored:

$$A_{ad} = \int \phi_a(\vec{k})\phi_d(\vec{k})d^2k \qquad (8)$$

$$B_{abcd} = \int C(\vec{k}_1,\vec{k}_2,\vec{k}_3,\vec{k}_4,)\{\phi_a(\vec{k}_1)\phi_b(\vec{k}_2)[\phi_c(\vec{k}_3)+\phi_c(\vec{k}_4)]$$

$$-\phi_a(k_3)\phi_b(k_4)[\phi_c(k_1)+\phi_c(k_2)]\}\phi(k_4)d^2k_1d^2k_2d^2k_3d^2k_4. \quad (9)$$

If piecewise constant functions are used, then $A_{ad}$ would be diagonal.

One problem that can be anticipated is the size of the B array. If the spectrum were represented by ten frequencies and ten directions, then each subscript would range form 1 to 100, so B would have $10^8$ entries. Because of the delta functions contained in C, most of the elements of B should be identically zero. By only storing the non-zero elements, the dimensionality of B should be dramatically reduced. Symmetries of C under exchanges, $\vec{k}_1 \leftrightarrow \vec{k}_2$, $\vec{k}_3 \leftrightarrow \vec{k}_4$, and $\vec{k}_1,\vec{k}_2 \leftrightarrow \vec{k}_3,\vec{k}_4$, result in further reduction in the required dimensionality. Because the delta functions reduce the six-dimensional integral to a three-dimensional one, it is reasonable to expect that the dimensionality of B would be reduced to the order of $10^4$. In other words, resolutions consistent with models currently in use would require tens to hundreds of thousands of additions and multiplications at each time-step for evaluating the wave-interaction integral. Although this is an improvement over direct evaluation at each time-step, it is still quite time consuming. However, this is entirely feasible with the large, fast computers that now exist.

## REFERENCES


Arakawa, A. (1966) Computational Design for Long Term Numerical Integration of the Equations of Motion: Two-Dimensional Incompressible Flow. Part I. J. Comp. Phys., 1:119-143.

Birchfield, G. E. and T. S. Murty (1974) A Numerical Model for Wind-Driven Circulation in Lakes Michigan and Huron. Mon. Wea. Rev., 102:157-165.

Chu, Wen-Hwa (1971) Development of a General Finite Difference Approximation for a General Domain. Part I: Machine Transformation. J. Comp. Phys., 8:392-408.

Crowley, W. P. (1971) FLAG: A Free-Lagrange Method for Numerically Simulating Hydrodynamic Flows in Two Dimensions. Proc., Second Int. Conf. Numerical Methods in Fluid Dynamics (M. Holt, Ed.), Springer-Verlag, 37-43.

Fritts, M. J. and J. P. Boris (1979) The Lagrangian Solution of Transient Problems in Hydrodynamics using a Triangular Mesh. J. Comp. Phys., 31:173-215.

Gray, W. G. and G. F. Pinder (1976) On the Relationship between the Finite Element and Finite Difference Methods. Int. J. Num. Math. Engng., 10:893-923.

Haussling, H. J. (1979) Boundary-Fitted Coordinates for Accurate Numerical Solution of Multi-Body Flow Problems. J. Comp. Phys., 30:107-124.

Jespersen, D. C. (1974) Arakawa's Method is a Finite-Element Method. J. Comp. Phys., 16:383-390.

MacNeal, R. J. (1953) An Asymmetrical Finite-Difference Network. Q. Appl. Math., 11:285-310.

Meyder, R. (1975) Solving the Conservation Equations in Fuel Rod Bundles Exposed to Parallel Flow by Means of Curvilinear-Orthogonal Coordinates. J. Comp. Phys., 17: 53-67.

Noh, W. F. (1964) CEL: A Time-Dependent Two-Space-Dimensional Coupled Eulerian-Lagrange Code. Meth. Comp. Phys., 3: 117-179.

Potter, D. E. and G. H. Tuttle (1973) The Construction of Discrete Orthogonal Coordinates. J. Comp. Phys., 13: 483-501.

Reid, R. D. and A. C. Vastano (1966) Orthogonal Coordinates for the Analysis of Long Gravity Waves Near Islands. Coastal Engineering, Proc., Santa Barbara Specialty Conf., American Soc. Civil Eng., 1-20.

Sadourny, R., A. Arakawa and Y. Mintz (1968) Integration of the Nondivergent Barotropic Vorticity Equation with an Icosahedral-Hexagonal Grid for the Sphere. Mon. Wea. Rev., 96:351-356.

Shen, Shan-fu (1977) Finite-Element Methods in Fluid Mechanics. Ann. Rev. Fluid Mech., 9:421-445.

Thacker, W. C. (1977) Irregular Grid Finite-Difference Techniques: Simulations of Oscillations in Shallow Circular Basins. J. Phys. Oceanogr., 7:284-292.

Thacker, W. C. (1978) Comparison of Finite-Element and Finite Difference Schemes. Part II: Two-Dimensional Gravity Wave Motion. J. Phys. Oceanogr., 8:680-689.

Thacker, W. C. (1980) A Geodesic Finite-Difference Method for Curved Domains: Simulation of Tidal Motion on a Sphere. J. Comp. Phys., 37:355-369.

Thacker, W. C., A. Gonzalez and G. E. Putland (1980) A Method for Automating the Construction of Irregular Computational Grids for Storm Surge Forecast Models. J. Comp. Phys., 37: 371-387.

Thompson, J. F., F. C. Thames and C. W. Mastin (1977) TOMCAT: A Code for Numerical Generation of Boundary-Fitted Curvilinear Coordinate Systems on Fields Containg any Number of Arbitrary Two-Dimensional Bodies. J. Comp. Phys., 24: 274-302.

Wanstrath, J. J., R. E. Whitaker, R. O. Reid and A. C. Vastano (1976) Storm Surge Simulation in Transformed Coordinates. Vol I: Theory and Application. U.S. Army Corps of Engineers Tech. Rept. No. 76-3.

# 7. LAKE CIRCULATION

# COMPUTATION OF WIND-DRIVEN CIRCULATION IN SHALLOW LAKES

Bogusław L. Jackowski

Polish Academy of Sciences, Institute of Hydroengineering, Gdańsk, POLAND.

ABSTRACT

A three-dimensional, non-stationary, shallow-water model of wind-induced circulation in lakes is formulated. The finite element approach is used for spatial discretization. Uniform elements /triangular prisms/ with the simplest possible shape functions are used. Since the model is linear and has an evolutionary form, spatial discretization yields a linear system of ordinary differential equations. An implicit, one-step scheme is applied for time integration. The obtained system of algebraic equations is solved iteratively.

Partition into elements is supported by a heuristic algorithm of surface triangulation.

## PREFACE

Lake circulation seems to be rather a well-recognized problem from the theoretical point of view, i.e. from the point of view of mathematical equations. A shallow-water approximation is commonly used for the purpose of modelling of circulation. The problem, however, turns out to be fairly hard from the computational point of view. The aim of this paper is to present an attempt of a new effective numerical and computational approach. The paper, therefore, should be regarded rather as a technological than a theoretical contribution.

The detailed description of the computer program for IBM360 /including listing in PASCAL and sample results/ will be available during the conference.

NOTATION

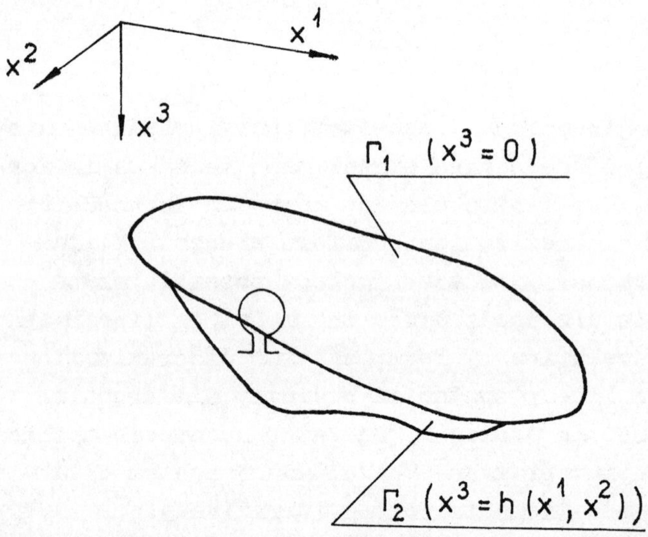


Figure 1. Site of a lake.

A computer-oriented /i.e. avoiding greece letters, dashes, asterisks, etc./ notation has been chosen to facilitate transition from mathematical formulation to a computer program.

Underlined symbols denote vectors, twice underlined symbols denote matrices. Upperscripts 1,2,3 denote respective vector components. Upperscript 0 denotes planar /two-dimensional/ object, e.g. if $\underline{x}=(x^1,x^2,x^3)$ is a point in $\mathbb{R}^3$ then $\underline{x}^0=(x^1,x^2,0)$ or, simply, $\underline{x}^0=(x^1,x^2)$ is a point in $\mathbb{R}^2$. Upperscript T denotes transposition of vectors and matrices. Comma-upperscript denotes a derivative /or Jacobi matrix, in general/.

$\Omega$ - considered region,

$\Gamma=\Gamma_1\cup\Gamma_2$ - boundary of $\Omega$,

$\gamma=\Gamma_1\cap\Gamma_2$ - boundary of $\Gamma_1$,

$\partial_t, \partial_\alpha\equiv\partial_{x^\alpha}$ - partial derivatives,

$\mathcal{L}=\sum_{\alpha=1}^{3}\partial_\alpha(l_\alpha\partial_\alpha)$ - elliptic operator, where $l_1,l_2,l_3$ are eddy-viscosity coefficients /in the simplest case $l_1=l_2=0\neq l_3$/. For conviniency reason the viscosity will be treated as a diagonal tensor $\underline{\underline{l}}$ throughout the text. In general $\underline{\underline{l}}=\underline{\underline{l}}(\underline{x},\underline{w},h)$,

grad $=(\partial_1,\partial_2)$ - horizontal differential operator,

$\underline{u}=(u^1,u^2)=\underline{u}(\underline{x},t)$ - horizontal water velocity,

$\underline{v}=\int_0^h\underline{u}\,dx^3$ - vertically integrated water velocity,

$d=d(\underline{x}^0,t)$ - small depression of the surface,

f - Coriolis' parameter /$\underline{\underline{f}}\,\underline{u}$ is the Coriolis acceleration, where $\underline{\underline{f}}=\begin{bmatrix}0 & f\\ -f & 0\end{bmatrix}$,

g - gravity acceleration,

$\underline{w}=(w^1,w^2)=\underline{w}(\underline{x}^0,t)$ - wind velocity,

$\underline{s}=(s^1,s^2)=\underline{s}(\underline{w})$ - wind extortion at the surface /$\underline{w}$ and $\underline{s}$ are given functions/.

The remaining symbols are displayed partially on Figure 1 and partially in the text.

## MODEL OF CIRCULATION

### Governing equations

The model is described by well-known shallow-water equations /cf. VOLCIGER et al. /1977//:

$$\partial_t \underline{u} = g \text{ grad}(d) + \mathcal{L}(\underline{u}) + \underline{\underline{f}}\, \underline{u} \qquad /1/$$

$$\partial_t d = \partial_1 v^1 + \partial_2 v^2 \qquad /2/$$

Initial and boundary conditions are as follows:

for t=0:

$$\underline{u} = \underline{u}_0(\underline{x}) \qquad /3/$$

$$d = d_0(\underline{x}^0) \qquad /4/$$

for t≥0:

$$\underline{u}|_{\Gamma_2} = \underline{0} \qquad /5/$$

$$\partial_3 \underline{u}|_{\Gamma_1} = \underline{s} \qquad /6/$$

$$\nu^1 \partial_1 v^1|_\gamma + \nu^2 \partial_2 v^2|_\gamma = 0 \qquad /7/$$

where $\underline{u}_0, d_0$ - given functions, $\underline{\nu} = (\nu^1, \nu^2)$ - vector normal to $\gamma$. Note, that these equations do not contain the vertical water velocity. It is a characteristic feature of shallow-water approximation. The vertical velocity can be computed, if necessary, from the continuity equation.

Variational formulation

Let $q: \mathbb{R}^3 \rightarrow \mathbb{R}$ and $q^0: \mathbb{R}^2 \rightarrow \mathbb{R}$ be time-independent trial functions. The standard procedure /see e.g. CIARLET /1978/ or CONNOR et al. /1977// applied to Equations /1/ and /2/, and utilizing Equations /6/ and /7/ leads to

$$\int_\Omega q\, \partial_t \underline{u}\, d\Omega = \int_\Omega q \{\text{grad}(d) + \underline{\underline{f}}\, \underline{u}\}\, d\Omega - \int_\Omega q' \underline{\underline{l}}\, \underline{u}'^T\, d\Omega - \int_{\Gamma_1} q\, l_3\, \underline{s}\, d\Gamma_1 \qquad /8/$$

$$\int_{\Gamma_1} q^0 \partial_t d\, d\Gamma_1 = -\int_{\Gamma_1} v^1 \partial_1 q^0 + v^2 \partial_2 q^0\, d\Gamma_1 \qquad /9/$$

/here ′ denotes spatial derivatives, i.e. $\underline{u}'$ is a 2 × 3 matrix/. Equations /8/ and /9/ hold for every appropriate functions q and $q^0$, provided $\underline{u}|_{\Gamma_2} = \underline{0}$.

Semidiscrete formulation

Some comments on partitioning of a region into elements will be perhaps relevant here. Let $\Gamma_1$ be partitioned into $e_1^o, e_2^o, \ldots, e_{n^o}^o$. It is assumed that if $e_\alpha^o \cap \gamma \neq \emptyset$ then element $e_\alpha^o$ is tetragonal and two of its nodes belong to $\gamma$, else $e_\alpha^o$ is a triangle. Partition of $\Omega$ into elements $e_1, e_2, \ldots, e_n$ is determined by that of $\Gamma_1$, namely, there are equidistantly placed $n^v$ nodes below every node $b_\alpha^o \in \Gamma_1 \setminus \gamma$, where $n^v$ is fixed for a given partition - see Figure 2.

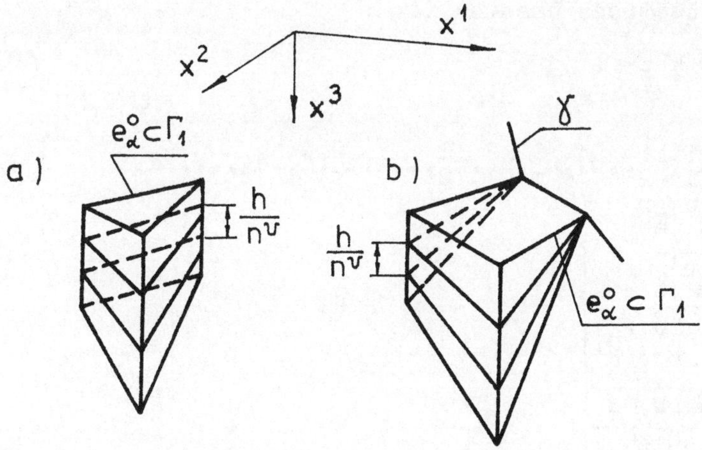


Figure 2. Partition of $\Omega$ into elements:
a/ for inner surface elements,
b/ for coastal surface elements.

Therefore the only three-dimensional elements used are prisms. Moreover, it suffices to keep information about a structure of $\Gamma_1$, as the information about a structure of partition of $\Omega$ can hence be easily computed.

Let $b_1^o, b_2^o, \ldots, b_{m^o}^o$ and $b_1, b_2, \ldots, b_m$ be the sets of nodes of two- and three-dimensional elements, respectively. To every node there is associated a function

/$q^o_\alpha$ to $b^o_\alpha$ , $q_\alpha$ to $b_\alpha$/, so called shape function , defined in a rutine way, cf. CONNOR et al. /1977/. The simplest possible polynomials are used for this purpose /such that functions $q^o_\alpha$ and $q_\alpha$ are linear over each edge/. The solution is approximated by a linear combination of shape functions:

$$\underline{u} = \sum_{\alpha=1}^{m} \underline{u}_\alpha\, q_\alpha , \qquad d = \sum_{\alpha=1} d_\alpha\, q^o_\alpha ,$$

where $\underline{u}_\alpha$ and $d_\alpha$ depend on time only /note, that while $\underline{u}|_{\Gamma_2} = \underline{0}$, $d|_\gamma \neq 0$, in general/.

The obtained system of ordinary linear differential equations has the form:

$$\underline{\underline{m}}\, \underline{z}' = \underline{\underline{k}}\, \underline{z} + \underline{c} , \qquad /10/$$

where

$$\underline{z} = (u^1_1, u^1_2, \ldots, u^1_m, u^2_1, u^2_2, \ldots, u^2_m, d_1, d_2, \ldots, d_m) ,$$

$$\underline{\underline{k}} = \left[\begin{array}{cc|c} \underline{\underline{k}}^u & \underline{\underline{m}}^u & \underline{\underline{k}}^d \\ -\underline{\underline{m}}^u & \underline{\underline{k}}^u & \\ \hline \multicolumn{2}{c|}{\underline{\underline{k}}^v} & \underline{\underline{0}} \end{array}\right] ,$$

$$\underline{\underline{m}} = \left[\begin{array}{c|c|c} \underline{\underline{m}}^u & \underline{\underline{0}} & \underline{\underline{0}} \\ \hline \underline{\underline{0}} & \underline{\underline{m}}^u & \underline{\underline{0}} \\ \hline \underline{\underline{0}} & \underline{\underline{0}} & \underline{\underline{m}}^d \end{array}\right] ,$$

$\underline{c}$ is an extortion vector / $c_\alpha \neq 0 \Rightarrow b_\alpha \in \Gamma_1$ /,

where, in turn,

$\underline{\underline{k}}^u$ is $m \times m$ matrix, involving integrals $\int_\Omega q'_\alpha \underline{\underline{1}}\, q'^T_\beta\, d\Omega$ ,

$\underline{\underline{m}}^u$ is $m \times m$ matrix, involving integrals $\int_\Omega q_\alpha\, q_\beta\, d\Omega$ ,

$\underline{\underline{k}}^d$ and $\underline{\underline{k}}^v$ are $2m \times m^o$ and $m^o \times 2m$ matrices, respectively, involving integrals $\int_{\Gamma_1} q^o_\alpha\, \partial_1 q^o_\beta\, d\Gamma_1$, $\int_{\Gamma_1} q^o_\alpha\, \partial_2 q^o_\beta\, d\Gamma_1$,

$\underline{\underline{m}}^d$ is $m^o \times m^o$ matrix, involving integrals $\int_{\Gamma_1} q^o_\alpha\, q^o_\beta\, d\Gamma_1$.

All matrices occuring here are of course sparse. Note, that for numerical quadrature utilizing nodal values only $\underline{\underline{m}}^u$, $\underline{\underline{m}}^d$ and, therefore, $\underline{\underline{m}}$ become diagonal. As Nedoma has proved /NEDOMA /1978//, it suffices to apply a quadrature of $\frac{n}{2}-1$ order for n-dimensional space. This means that the simplest quadrature, i.e. the average of nodal values times the volume of an element, is sufficient here.

Time integration scheme

For non-trivial cases $\underline{\underline{m}}^{-1}\ \underline{\underline{k}}$ is a stiff matrix, i.e. it has a large spectrum of eigenvalues. For this reason special time integration schemes are recommended /cf. NEDOMA /1978/,/1979/ or DAHLQUIST et al. /1971//.

The well-known "$\Theta$- method" has been applied:

$$\Delta t^{-1}\ \underline{\underline{m}}\left\{\underline{z}(t+\Delta t)-\underline{z}(t)\right\} = \underline{\underline{k}}\left\{\Theta\,\underline{z}(t+\Delta t)-(1-\Theta)\,\underline{z}(t)\right\} + \underline{c} \qquad /11/$$

The value of $\Theta$ should be chosen to provide possibly high numerical stability of the method. The necessary condition for $A_0$-stability of the method is $\Theta > \frac{1}{2}$ /cf. CRYER /1973//. It seems that Liniger's proposal /LINIGER /1968// to choose $\Theta$ to minimize

$$\max_{0\leq \tau\leq\infty} \left| e^{-\tau} - (1-(1-\Theta)\tau)/(1+\Theta\tau)\right| ,$$

which gives $\Theta = 0.878$, may suits well this purpose. The $\Theta$- method is of course implicit if $\Theta \neq 0$. Therefore the method involves solving of a system of algebraic equations at each stage. Not very sophisticated, though quite efficient is Gauss-Seidel overrelaxation method. Young has pointed out /YOUNG /1971// that appropriately selected overrelaxation factor can speed-up the iterative process by an order of magnitude. The proper value of this factor can be estimated iteratively at the begining of computation.

A HEURISTIC TRIANGULATION ALGORITHM

It is commonly acknowledged fact that inputing of a mesh is an important part of a computer program. Any attempt to facilitate mesh generation seems therefore to be relevant. The main idea of the proposed algorithm of mesh generation is illustrated on Figure 3.

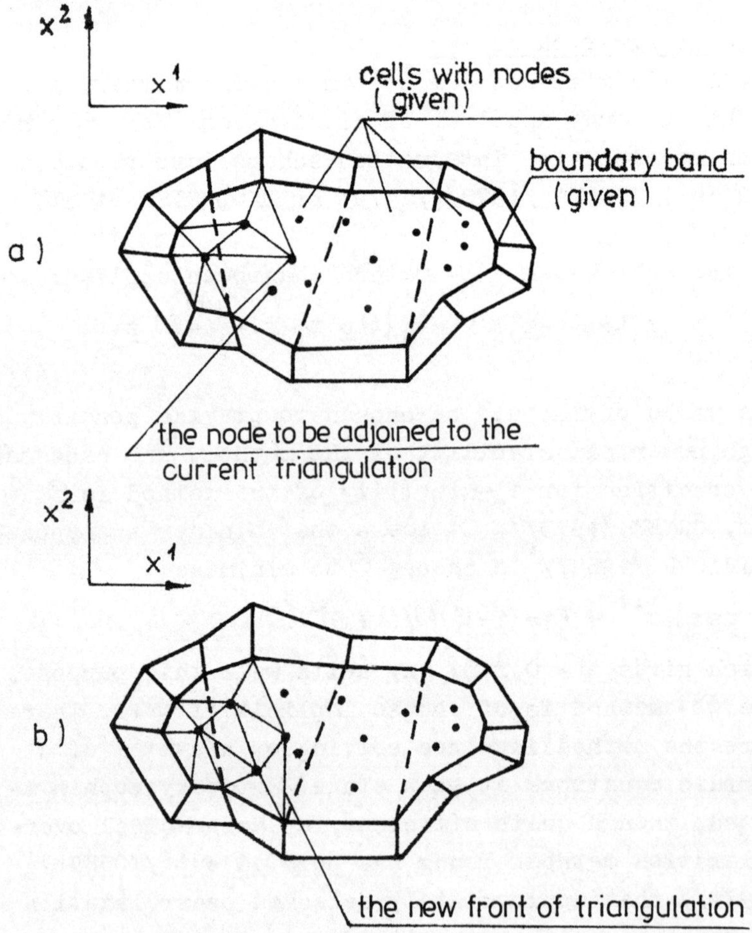


Figure 3. Adjoining a new node to a partially generated mesh.

input data comprise the geometry of a shore-line and distribution of nodes /the depth of a lake is given in nodes only/. Moreover, since computers hardly recognize patterns, some further information may significantly optimize the algorithm; here it is assumed that also a crude partition of the surface into cells is given. The main invariant of the algorithm is: current triangulation does not contain any "free" node inside. After addition of a new node the invariant have to hold. Then a local improvement of the mesh, based on a heuristic criterion /keep the minimal angle possibly large, but, in the same time, keep an order of a node possibly low/ is performed. Next, before proceeding to another node, the front of triangulation is smoothed, where possible.

Of course, one can apply another algorithm instead of the proposed one / see SIMPSON /1979/ for a survey/.

## FINAL REMARKS

The presented model seems to be a handy tool for investigation of wind-driven circulation. The applied finite element approach suits very well this purpose, as the method manages well irregular regions. Moreover, a grid is constructed semi-automatically, what brings the problem of data input significantly down to size and facilitates mesh size variations. The computed results can serve as a basis for further computations, e.g. for computation of sediment transport or propagation of pollutants.

## ACKNOWLEDGMENT

The author expresses his deep gratitude to dr. Jerzy

Kołodko for his patience, scepticism and for many helpful discussions.

REFERENCES

1. CIARLET, P.G. /1978/ - The finite element method for elliptic problems. North Holland Publ. Co., Amsterdam.
2. CONNOR, J.J., BREBBIA, C.A. /1977/ - Finite element techniques for fluid flow. Newnes - Butterworth, London.
3. CRYER, C.W. /1973/ - A new class of highly-stable methods. $A_0$-stable methods. BIT 13,153-159.
4. DAHLQUIST, G. et al. /1971/ - Survey of stiff ordinary differential equations. Roy. inst. Tech. Stockholm, Dept. of Inf. Proc., Report NA 70.
5. LINIGER, W. /1968/ - Optimization of a numerical integration method for stiff systems of ordinary differential equations. IBM Res. Rep. RC2198.
6. NEDOMA, J. /1978/ - The finite element solution of parabolic equations. Aplikace matematiky, svazek, 23, č. 6, 408-438.
7. NEDOMA, J. /1979/ - the finite element solution of elliptic and parabolic equations using simplicial isoparametric elements. R.A.I.R.O. Numerical Analysis, 3, 3, 257-289.
8. SIMPSON, R.B. /1979/ - A survey of two dimensional finite element mesh generation. Proc. of the IX Manitoba Conf. on Num. Math. and Comp.
9. VOLCIGER, N.E., PIASKOVSKIJ, R.V. /1977/ - Teoria mielkoi vody. Gidrometeoizdat, Leningrad.
10. YOUNG, D.M. /1971/ - Iterative solution of large linear systems. Academic Press, N.Y.

# A THREE DIMENSIONAL FINITE ELEMENT MODEL FOR STRATIFIED FLOW

Ian P. King

Resource Management Associates
Lafayette, California

## INTRODUCTION

The purpose of this paper is to describe a three dimensional model suitable for the analysis of stratified flow in hydrodynamic systems. The model (RMA10) uses a finite element solution to a three dimensional set of nonlinear equations. The equations are transformed to a uniform vertical dimension, and the assumption of hydrostatic pressure is applied. The working model is demonstrated with a test case from an experimental flume and a 180 degree river bend. Comparisons are made with both experimental data and results from earlier finite difference studies.

## BASIC EQUATIONS

The Reynolds form of the Navier-Stokes Equations are used as the governing equations for flow. Additional equations describe continuity, convection-diffusion of heat (or salinity) and the equation of state for density. In order to simplify the problem the momentum equation defining flow in the vertical direction is simplified to state the assumption that pressure is hydrostatic. The equations may thus be written as:

$$\rho\frac{\partial u}{\partial t} + \rho u\frac{\partial u}{\partial x} + \rho u\frac{\partial u}{\partial y} + \rho w\frac{\partial u}{\partial z} -$$

$$\frac{\partial}{\partial x}\left(\varepsilon_{xx}\frac{\partial u}{\partial x}\right) - \frac{\partial}{\partial y}\left(\varepsilon_{xy}\frac{\partial u}{\partial y}\right) - \frac{\partial}{\partial z}\left(\varepsilon_{xz}\frac{\partial u}{\partial z}\right) + \frac{\partial p}{\partial x} - \tau_x = 0 \qquad (1)$$

$$\rho\frac{\partial v}{\partial t} + \rho u\frac{\partial v}{\partial x} + \rho v\frac{\partial v}{\partial y} + \rho w\frac{\partial v}{\partial z} -$$

$$\frac{\partial}{\partial x}(\varepsilon_{yx}\frac{\partial v}{\partial x}) - \frac{\partial}{\partial y}(\varepsilon_{yy}\frac{\partial v}{\partial y}) - \frac{\partial}{\partial z}(\varepsilon_{yz}\frac{\partial v}{\partial z}) + \frac{\partial p}{\partial y} - \tau_y = 0 \quad (2)$$

$$\frac{\partial p}{\partial z} - \tau_z = 0 \quad (3)$$

$$\frac{\partial u}{\partial x} + \frac{\partial v}{\partial y} + \frac{\partial w}{\partial z} = 0 \quad (4)$$

$$\rho C_T[\frac{\partial T}{\partial t} + u\frac{\partial T}{\partial x} + v\frac{\partial T}{\partial y} + w\frac{\partial T}{\partial z}$$

$$- \frac{\partial}{\partial x}(D_x\frac{\partial T}{\partial x}) - \frac{\partial}{\partial y}(D_y\frac{\partial T}{\partial y}) - \frac{\partial}{\partial z}(D_z\frac{\partial T}{\partial z})] - \theta_s = 0 \quad (5)$$

$$\rho - F(T) = 0 \quad (6)$$

where the following definitions apply:

$x,y,z$ = cartesian coordinates

$\rho$ = density

$\varepsilon_{xx},\varepsilon_{xy}$,etc = turbulent eddy coefficients

$T$ = temperature

$\tau_x,\tau_y,\tau_z$ = external tractions

$u,v,w$ = velocities

$p$ = pressure

$D_x,D_y$ = heat diffusion coefficients

$f(T)$ = equation of state

$\theta_s$ = external heat source

Because of the assumption of hydrostatic pressure it is convenient to reestablish the continuity equation in terms of an integral over the depth, i.e.

$$\int_a^{a+h} (\frac{\partial u}{\partial x} + \frac{\partial u}{\partial y})dz + (u(x,y,e)\frac{\partial e}{\partial x})\Big|_a^{a+h} + v(x,y,e)\frac{\partial e}{\partial y}\Big|_a^{a+h} = 0 \quad (7)$$

where a = elevation of bottom, h = depth of water

The representation of velocity along the boundary of three dimensional elements presents difficulties. If flow is to be continuous in the network without crossing boundaries then element shapes must have conforming slopes where they join. This may be achieved for horizontal boundaries with the methods used for two-dimensional models but extension to the free surface in the vertical elevation cannot be generally achieved. In fact such a definition of the free surface or of the bottom profile with linear shape functions (assuming a mixed interpolation approach) is impossible.

In order to eliminate this difficulty the governing equations have been geometrically transformed so that the modified system will have uniform depth over the entire domain.

The necessary transformation may be simply stated as:

$$z' = (z - a(x,y))/h(x,y) \quad , \quad x'=x \quad y'=y$$

If the above transformations are applied and the primes dropped for convenience, the equations can be restated as:

$$\rho h\frac{\partial u}{\partial t} + \rho hu\frac{\partial u}{\partial x} + \rho hv\frac{\partial u}{\partial y} + \rho w\frac{\partial u}{\partial z} - \rho(us_x + vs_y)\frac{\partial u}{\partial z} - h\frac{\partial}{\partial x}(\varepsilon_{xx}\frac{\partial u}{\partial x})$$

$$- h\frac{\partial}{\partial y}(\varepsilon_{xy}\frac{\partial u}{\partial y}) - \frac{\partial}{\partial z}(\varepsilon_{xz}\frac{\partial u}{\partial z}) + h\frac{\partial p}{\partial x} - S_x\frac{\partial p}{\partial z} - h\tau_x = 0 \tag{8}$$

$$h\frac{\partial v}{\partial t} + hu\frac{\partial v}{\partial x} + hv\frac{\partial v}{\partial y} + w\frac{\partial v}{\partial z} - (us_x + vs_y)\frac{\partial v}{\partial z} - h\frac{\partial}{\partial x}(\varepsilon_{yx}\frac{\partial v}{\partial x})$$

$$- h\frac{\partial}{\partial y}(\varepsilon_{yy}\frac{\partial v}{\partial y}) - \frac{\partial}{\partial z}(\varepsilon_{yz}\frac{\partial v}{\partial z}) + h\frac{\partial p}{\partial y} - S_y\frac{\partial p}{\partial z} - h\tau_y = 0 \tag{9}$$

$$\int_z^1 \rho gh dz - p = 0 \tag{10}$$

$$\int_0^1 \{\frac{\partial}{\partial x}(hu) + \frac{\partial}{\partial y}(hv)\}dz = 0 \tag{11}$$

$$\rho C\{h\frac{\partial T}{\partial t} + hu\frac{\partial T}{\partial x} + hv\frac{\partial T}{\partial y} + w\frac{\partial T}{\partial z} - (us_x + vs_x)\frac{\partial T}{\partial z}$$

$$- h\frac{\partial}{\partial x}(D_x\frac{\partial T}{\partial x}) + h\frac{\partial}{\partial y}(D_y\frac{\partial T}{\partial y}) + \frac{\partial}{\partial z}(D_z\frac{\partial T}{\partial z})\} - h\theta_s = 0 \tag{12}$$

where $s_x = (\frac{\partial h}{\partial x}z + \frac{\partial a}{\partial x})$ $\qquad s_y = (\frac{\partial h}{\partial y}z + \frac{\partial a}{\partial y})$

Note that in the restated form the eddy and diffusion coefficients have a slightly different definition. Since these terms are approximate this is not considered a significant modification. Equations (6), and (8) through (12) form the basic set of equations that are solved in the primary model. Dependent variables are u, v, T, and h. For modeling purposes it is assumed that the influence of vertical velocity is small and although its influence is incorporated in Equations (8), (9) and (12) it may be treated as semi-uncoupled and directly computed from the original continuity equation (4).

## FINITE ELEMENT FORMULATION AND MODEL

The model formulation follows the Newton-Raphson iterative approach with Galerkin weighted residuals that has been used in earlier two-dimensional models (Norton et al., 1973). In fact, the program has been constructed so that in areas where two-dimensional analysis will suffice, the models marry together and provide a consistent solution at the interface. Because the depth (h) is a function only of x and y, all nodes below the surface of the system must line up vertically with nodes in the surface plan view. Nodes in any vertical line are thus associated with a unique depth and element coefficients are contributed for all nodes in one vertical line. The model uses mixed interpolation with quadratic definition for u, v, and T, and a linear description for h. Four types of elements are constructed from an input stream that describes the plan view and the number of nodes in a vertical line. The element types are defined in Table 1.

TABLE 1. ELEMENT DEFINITION

| | BRICK | PRISM | PYRAMID | TETRAHEDRON |
|---|---|---|---|---|
| Nodes | 20 | 15 | 13 | 10 |
| Total degrees of freedom | 68 | 51 | 44 | 34 |

The model operates in a two step mode. In step 1, the horizontal velocities and head are developed. The pressure is derived by integration of the density term over depth and that in order to achieve consistency (solution remaining steady for tank with horizontal layers) constant values for density are used to develop the pressure distribution inside an element.

In step 2 the continuity equation is used to solve for vertical velocities over the entire system. In order to convert it to a boundary value problem the continuity equation (4) is differentiated with respect to the vertical dimension and boundary conditions are applied at both the top and bottom from the known horizontal components.

EXAMPLE PROBLEM 1

In order to demonstrate the model and compare performance with a prototype case, experimental measurements from the Generalized Reservoir Hydrodynamics (GRH) flume, maintained by U.S. Corps of Engineers in Vicksburg, were compared with simulation results from RMA10. This flume has been used as a comparison case for a number of models which are described in detail in Johnson (1981).

The basic flume dimensions are shown in Figure 1. It is designed to simulate reservoir hydrodynamics. In the particular case studied, the flume was started with water at a uniform initial temperature of 21.44°C. Water at 16.67°C was introduced under a baffle and allowed to flow into the system, over the lower 0.15m. Outflow was removed from a 0.0254m diameter port located on the outer line 0.15m from the bottom. The inflow was 0.00063 $m^3$/sec with the outflow rate the same.

The cold water was dyed and observed to demonstrate classical flow with the most dense fluid showing a wedge of approximately 0.075m thickness. Observation included travel time of the wedge, and temperature at the exit and some measurements of velocity in a cross section. Figure 1 also shows cross section views of the network used in the simulation. The network has 458 nodes and 55 elements. The system was simulated using 2 minute time intervals for 12 steps. (note that these times represent intervals at the model scale and are not truly representative of the intervals necessary to model an actual reservoir. Figure 2 presents a set of vectors showing (to a distorted scale) velocities along the centerline of the system after 4 minutes and 14 minutes. Overlying the vectors is a contour line showing the progress of the dense wedge. The major recirculating eddy is clearly demonstrated in this figure. At earlier times the eddy is confined to the upper end but as time progresses the eddy grows in extent and steadily moves downstream. The figure also demonstrates the shape of the cold water wedge that develops and shows how it progresses along the bottom of the flume with very little vertical mixing. This shows good agreement with photographic evidence from the flume.

A comparison of computed and measured horizontal velocities at the centerline for section A of Figure 1 is presented in Figure 3. Because of interpolation and location difficulties the finite difference model results are shown at a slightly different location and time. However, the general form of the velocity vectors is relatively constant. The figure indicates how the finite element model succeeds in describing the eddy particularly with respect to the thickness of the wedge. The finite difference model that uses a stair

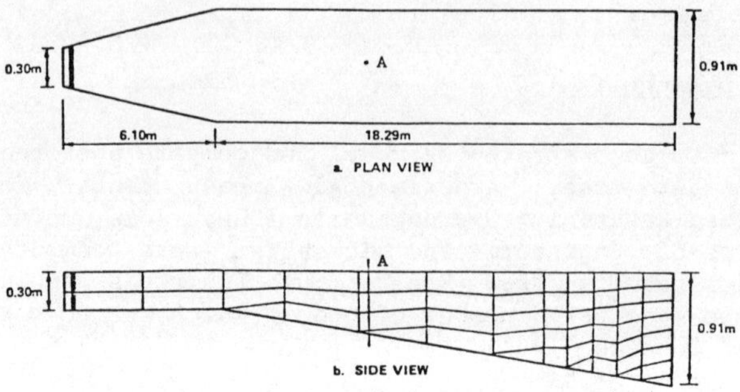


FIGURE 1. SCHEMATIC OF GRH FLUME

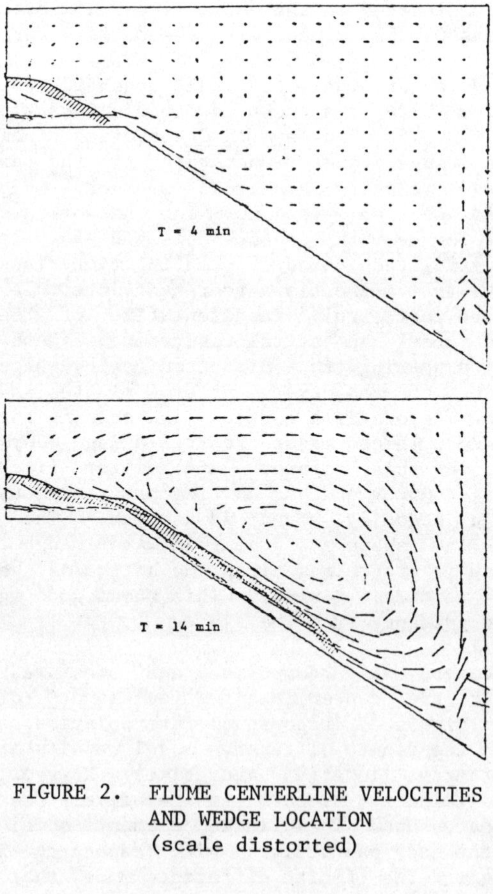


FIGURE 2. FLUME CENTERLINE VELOCITIES AND WEDGE LOCATION (scale distorted)

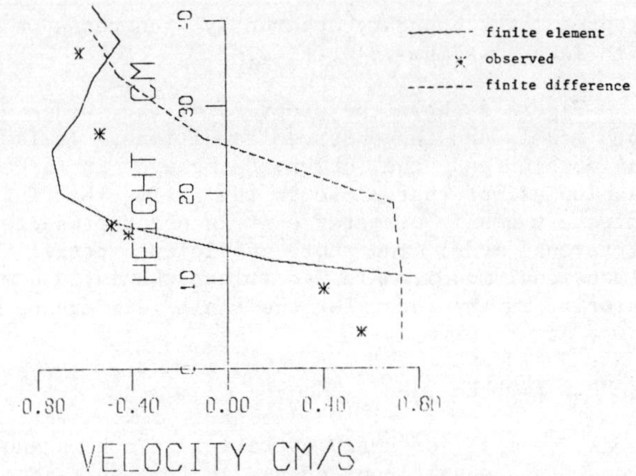


FIGURE 3. COMPARISON OF COMPUTED AND RECORDED HORIZONTAL VELOCITIES AT SECTION A

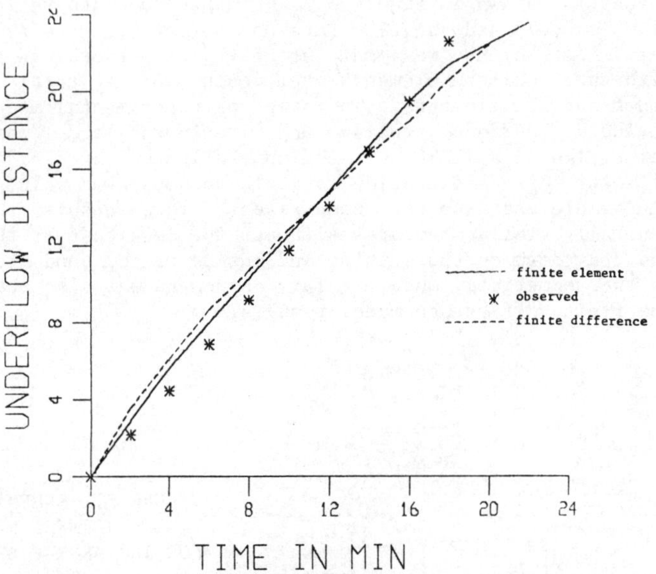


FIGURE 4. COMPARISON OF COMPUTED AND RECORDED UNDERFLOW SPEED

stepped lower boundary apparently generates a thicker wedge with lower maximum velocities.

Figure 4 shows the progress of the wedge downstream for both models and the observed conditions. It is noteworthy that the models and the observed behave in a generally similar fashion except that close to the exit the influence of the extremely small diameter exit is not represented by the finite difference model and not sufficient detail by the three dimensional model. In addition the influence of the reduced velocity in the wedge for the finite difference model shows in the flatter curve.

EXAMPLE PROBLEM 2

As an example of RMA10 operating in a homogeneous flow situation with a significant change in free surface, a problem of flow around a 180° bend originally described by Rosovskii (1957) and more recently utilized as a test case for a finite difference model by Leschziner and Rodi (1979) will be presented.

The basic horizontal plan together with the finite element layout is shown in Figure 5. The depth to width ratio for the flow is 0.075 and the flow rate was set at 0.012 $m^3$/sec. The Chezy coefficient was set at 60. The network was generated with three uniform elements over depth for a total of 1263 nodes and 210 elements. The original experimental measurements included surface profile and transverse and longitudinal velocities at a number of sections. Figure 6 presents contours of water surface elevation for both the observed values and the finite element results. Comparison of these results shows the generally similar response although the magnitude of the set up and draw down on the outside and inside of the bend seem offset by the numerical model. This phenomena was also observed in the finite difference model results.

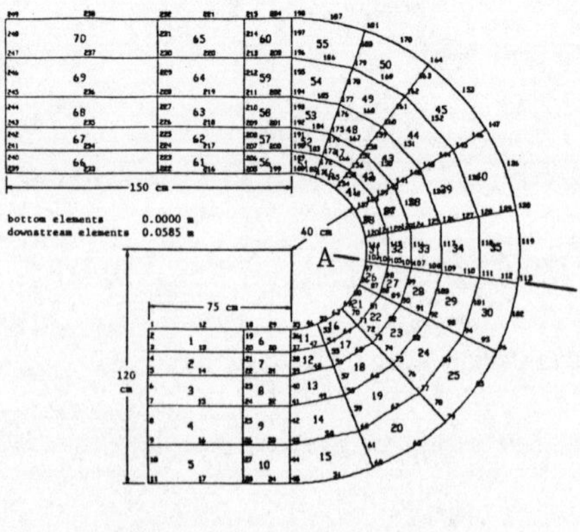


FIGURE 5. SCHEMATIC AND FINITE ELEMENT NETWORK FOR 180 DEGREE BEND

A

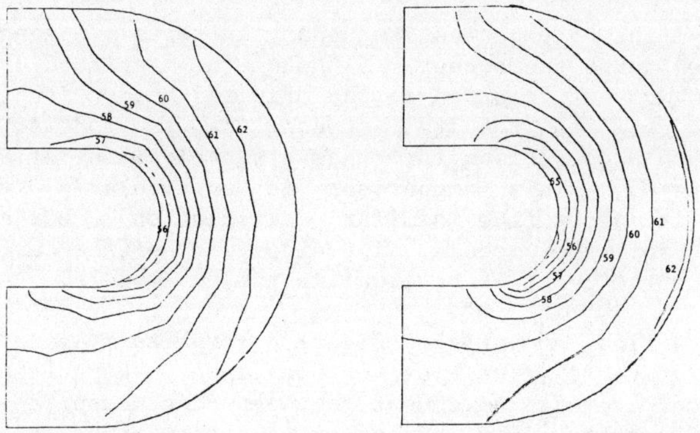


a) measured b) finite element

FIGURE 6. CONTOURS OF WATER SURFACE (in mm)

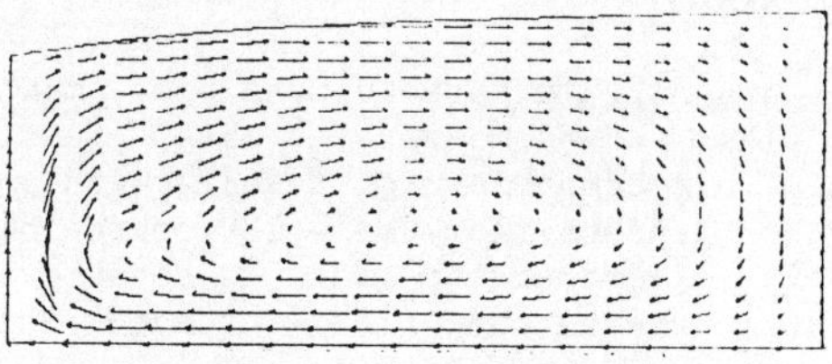

a) finite element model

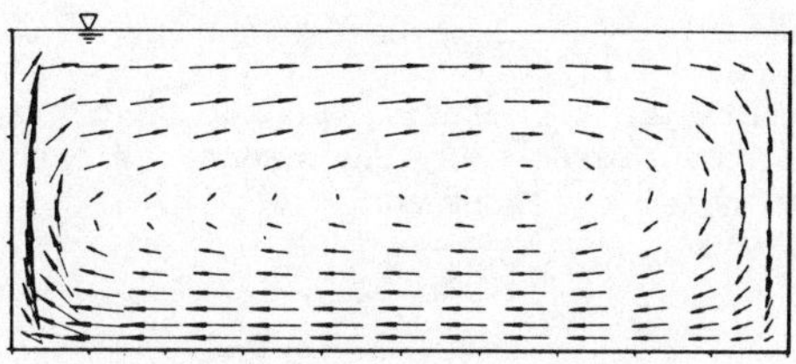

b) finite difference model

FIGURE 7. TRANSVERSE CIRCULATION AT SECTION A

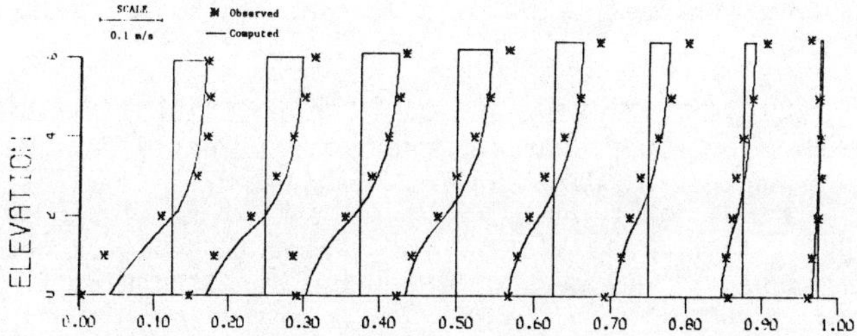


FIGURE 8. TRANSVERSE VELOCITIES AT SECTION A

As might be expected, secondary currents are generated as the flow turns the corner. Although the magnitude of vertical currents was not recorded in the original study it is worthwhile to compare results between the two models. Figure 7 shows such a comparison for transverse velocities at Section A of Figure 5. This comparison is particularly interesting because it shows the similar circulation characteristics derived from the models. The slope of the free surface seems to affect the point of reversal in the two models.

As a final comparison, Figure 8 compares measured and computed transverse velocities for Section A. Here the results show good overall agreement between the measured values and computed results. Similar agreement occurs at other locations around the bend. Although not shown here the longitudinal velocities also showed the same degree of consistency.

## CONCLUDING REMARKS

This paper has presented a three dimensional model for three dimensional flow and demonstrated its feasibility for stratified flow in a comparison of results with an experimental flume and for homogeneous flow for a 180 degree bend. Although not presented in this paper the model is capable of simulating a tidal system where the free surface influences are significant.

## ACKNOWLEDGEMENT

The author would like to acknowledge the Waterways Experiment Station, U.S. Army Corps of Engineers, Vicksburg, for their support of the model development.

## REFERENCES

Johnson, B. H. (1981) A Review of Numerical Reservoir Hydrodynamic Modeling, Technical Report E-81-2, prepared for Office, Chief of Engineers, U.S. Army, Washington, D.C.

Leschziner, M. A. and W. Rodi, (1979) Calculation of Strongly Curved Open Channel Flow, in Journal of the Hydraulics Division, ASCE.

Norton, W. R., I. P. King and G. T. Orlob, (1973) A Finite Element Model for Lower Granite Reservoir, prepared for U.S. Army Corps of Engineers, Walla Walla, Washington.

Rozovskii, I. L. (1957) Flow of Water in Bends of Open Channels, Academy of Sciences of the Ukrainian SSR, Kiev.

# FINITE ELEMENT ANALYSIS OF WATER QUALITY IN LAKE ERIE

**D.C.L. Lam**

National Water Research Institute
Canada Centre for Inland Waters, Burlington, Ontario, Canada

## ABSTRACT

Thermal stratification dominates the physical circulation and biochemical variable distribution in Lake Erie during the summer period. The Central Basin hypolimnion is often depleted of oxygen because of strong sediment oxygen demand. The water circulation and temperature can be computed by using the two dimensional x-z momentum and energy conservation equations. Results from a finite element model using mixed interpolation basis functions showed that oxygen replenishment from the Eastern Basin depends on favourable westerly winds, but not so much on the earth's rotational effect.

## INTRODUCTION

Lake Erie is the shallowest lake in the Great Lakes system in North America. It can be divided, along its 390 km longitudinal (x) axis, into three basins: Western, Central and Eastern, according to the mean depths ($z$ = 10 m, 25 m and 60 m, respectively). Because of the topographical differences, the magnitude and direction of the transports across these basins are sensitive to the changes in the wind forcing, heat input and environmental turbulence. There are many approaches to modelling the interbasin transports, using finite-difference methods on the x-z and y-z planes (e.g. Hollan and Simons, 1978; Heinrich et al., 1981). In the present study, a finite element x-z model is proposed, in which the vertical eddy diffusivity formulation is based on an environmental turbulence submodel. The model results should then provide the basis for analysing the distribution of water quality variables in the basins. Thus, the computed current velocities and water temperature will be compared to the observational values and the discussion be made with reference to observed oxygen and soluble reactive phosphorus distributions.

## TWO-DIMENSIONAL X-Z MODEL EQUATIONS

The governing equations of the two-dimensional model are based on the continuity equation, the momentum equation and the thermodynamic energy equation. Hydrostatic flow is assumed and hence the equation of motion in the vertical reduces to a pressure distribution condition. Non-linear advection terms in the other equations of motion can also be neglected because their effects have been shown to be small in models of this type with small Rossby numbers, Hollan and Simons (1978). Thus, although the following equations are three-dimensional, they will be further simplified:

$$\frac{\partial u}{\partial x} + \frac{\partial v}{\partial y} + \frac{\partial w}{\partial z} = 0 \tag{1}$$

$$\frac{\partial u}{\partial t} = \frac{\partial}{\partial x} A_h \frac{\partial u}{\partial x} + \frac{\partial}{\partial y} A_h \frac{\partial u}{\partial y} + \frac{\partial}{\partial z} A_v \frac{\partial u}{\partial z} + fv - \frac{\partial}{\partial x}(P+Q) \tag{2}$$

$$\frac{\partial v}{\partial t} = \frac{\partial}{\partial x} A_h \frac{\partial v}{\partial x} + \frac{\partial}{\partial y} A_h \frac{\partial v}{\partial y} + \frac{\partial}{\partial z} A_v \frac{\partial v}{\partial z} - fu - \frac{\partial}{\partial y}(P+Q) \tag{3}$$

$$\frac{\partial T}{\partial t} = -\frac{\partial uT}{\partial x} - \frac{\partial vT}{\partial y} + \frac{\partial}{\partial x} K_h \frac{\partial T}{\partial x} + \frac{\partial}{\partial y} K_h \frac{\partial T}{\partial y} - \frac{\partial wT}{\partial z} + \frac{\partial}{\partial z} K_v \frac{\partial T}{\partial z} \tag{4}$$

where u, v and w are the velocities along the three coordinates x, y and z, respectively; $A_h$ and $A_v$ are the horizontal and vertical eddy viscosities; $K_h$ and $K_v$ are the horizontal and vertical thermal eddy diffusitives; f is the Coriolis parameter; P and Q represent the barotropic and baroclinic pressure components as defined below:

$$P = g\eta + P_s/\rho; \quad Q = \int_{-z}^{\eta} \frac{\Delta\rho}{\rho} dz; \quad \frac{\Delta\rho}{\rho} = \alpha_T(4-T)^2 \tag{5}$$

where $\eta$ denotes the free surface elevation; g is the gravitational acceleration; $P_s$ is atmospheric pressure; $\alpha_T$ is the thermal expansion coefficient.

The boundary conditions at surface (s) and bottom (b) are:

$$\rho A_v \frac{\partial u}{\partial z}\Big|_s = \tau_{sx} \qquad \rho A_v \frac{\partial v}{\partial z}\Big|_s = \tau_{sy} \qquad \rho C_p K_v \frac{\partial T}{\partial z}\Big|_s = q \tag{6}$$

$$\rho A_v \frac{\partial u}{\partial z}\Big|_b = \tau_{bx} \qquad \rho A_v \frac{\partial v}{\partial z}\Big|_s = \tau_{by} \qquad \frac{\partial T}{\partial z}\Big|_b = 0 \tag{7}$$

where $\tau_s$ is the wind stress and $\tau_b$ is the bottom stress formulated as a quadric function of the bottom velocities.

By assuming no variations in the y direction, the following equations can be obtained from Equations (1), (2) and (4):

$$\frac{\partial u}{\partial x} + \frac{\partial w}{\partial z} = 0 \tag{8}$$

$$\frac{\partial u}{\partial t} = \frac{\partial}{\partial x} A_h \frac{\partial u}{\partial x} + \frac{\partial}{\partial z} A_v \frac{\partial u}{\partial z} - \frac{\partial}{\partial x} (P + Q) \tag{9}$$

$$\frac{\partial T}{\partial t} = - \frac{\partial uT}{\partial x} + \frac{\partial}{\partial x} K_h \frac{\partial T}{\partial x} - \frac{\partial wT}{\partial z} + \frac{\partial}{\partial z} K_v \frac{\partial T}{\partial z} \tag{10}$$

Notice that there is only one momentum equation (Equation (9)) in the two-dimensional case, which mainly consists of the frictional terms and pressure gradient terms. However, in the energy conservation equation, Equation (10), the advective and convective transports of heat is included, as well as horizontal and vertical turbulent mixing of heat. The boundary conditions are essentially the same as given by Equations (6) and (7).

## VERTICAL THERMAL EDDY DIFFUSIVITY

There are a number of models using the turbulent diffusivity concept to bring the heat conservation equation into closure form. The main difference in these models is the definition of $K_v$ which can be functions of wind mixing, Coriolis force, buoyance effects, internal-wave frequencies and turbulent boundary layer thicknesses. For example, in Mellor and Durbin (1975), the diffusivity is parameterized through the balance of kinetic energy and energy dissipation in a two-equation system. Others, e.g. MacArthur and Norton (1980), use constant eddy viscosity and dispersion coefficients.

The following formulation, however, has been tested with Great Lakes data and has produced satisfactory results (Lam, 1980). Briefly, the vertical turbulent thermal diffusity is defined over three layers:

$$\text{Epilimnion:} \quad K_v = K_o(1 + \sigma R_i)^{-1} - \gamma \partial\rho / \partial z \cdot \frac{g}{\rho} \tag{11}$$

$$\text{Mesolimnion:} \quad K_v = K_{TC}(N_{TC}^2 / N^2)^{\frac{1}{2}} \tag{12}$$

$$\text{Hypolimnion:} \quad K_v = K_B - \gamma \partial\rho / \partial z \cdot \frac{g}{\rho}; \; K_B \leq \beta K_{TC} \tag{13}$$

where $K_o$ is the air-water diffusion parameter; $K_B$ is the bottom diffusion parameter; $\sigma$, $\gamma$, $\beta$ are model coefficients; $\rho$ is water density; $R_i$ is the Richardson number; N is the Brunt-Väisälä frequency; TC denotes the value at the thermocline where the maximum temperature gradient occurs. Mathematically, matching of the three regions of formulations given in Equations (11) to (13) is required to ensure the continuity in the diffusivity. For example, in the middle layer (mesolimnion), the diffusivity decreases to its smallest value at the thermocline and then increases to a constant value, $K_B$, as the lower layer (hypolimnion) is reached.

There are two approaches of computing the Richardson number, $R_i$, which essentially compares the thermal buoyancy effect to the current shear mixing effect:

$$R_i = \frac{-g \frac{\partial \rho}{\partial z}}{\rho (\frac{\partial u}{\partial z})^2} \tag{14}$$

The first approach is to assume certain semi-empirical velocity profiles, such as the von Karman's logarithmic profile (Walters et al., 1978). The second approach is to solve the velocity gradient from the momentum equation directly, which is used in the present model.

Note that there are three adjustable constants: $\sigma$, $\gamma$ and $\beta$, which assume the values of 0.03, 0.0046 and 5 respectively. To define the values of $K_v$ in Equations (11) to (13) requires the knowledge of the water density which is related to the computed temperature through Equation (5). Thus, an iterative algorithm is needed between the thermal eddy diffusivity and the computed water temperature. To obtain the vertical momentum eddy viscosity, a multiplicative factor of 5 is used on the thermal eddy diffusivity at high Richardson numbers, otherwise the two are assumed equal. The horizontal counterparts, $A_h$ and $K_h$, are assumed to have constant values of $10^7$ and $5\text{x}10^5$ $cm^2/s$, respectively.

**FINITE ELEMENT METHOD**

Briefly, two types of basis functions, Hood and Taylor (1974) are used: the quadratic basis function $[\phi]$ is used for u and w, and the linear basis function $[\psi]$ is used for T. The Galerkin method is then applied to the Equations (8) to (10) resulting in:

$$\int_e \left(\frac{\partial \phi}{\partial x} \hat{u} + \frac{\partial \phi}{\partial z} \hat{w}\right) dxdz = 0 \tag{15}$$

$$\int_e \left(\phi \frac{d\hat{u}}{dt} + A_h \frac{\partial \hat{\phi}}{\partial x} \frac{\partial \phi}{\partial x} \hat{u} + (\psi \hat{A}_v) \frac{\partial \hat{\phi}}{\partial z} \frac{\partial \phi}{\partial z} \hat{u} + \frac{\partial \psi}{\partial x} \hat{Q}\right) dxdz$$

$$- \int_s \phi \, \hat{\tau}_{sx} dx + \int_b \phi \, \hat{\tau}_{bl} dl = 0 \tag{16}$$

Fig. 1(a) Finite element mesh of the Lake Erie x-z model.

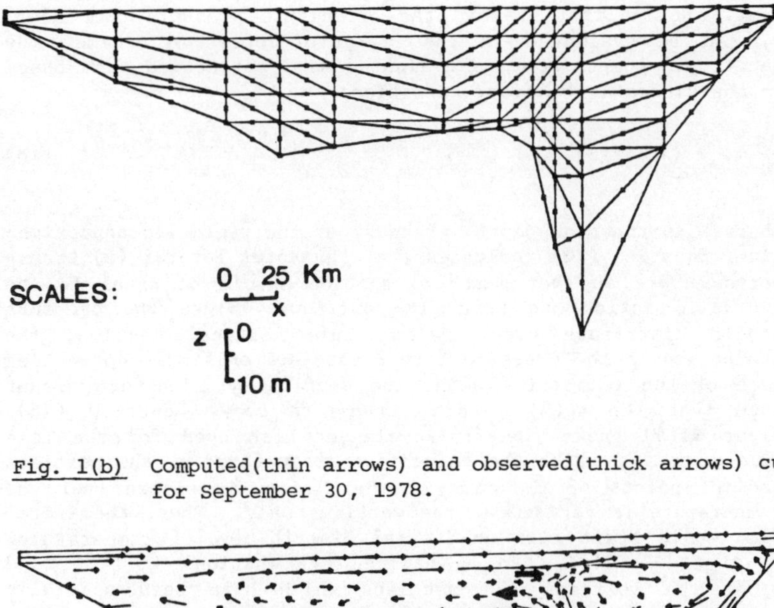


Fig. 1(b) Computed(thin arrows) and observed(thick arrows) currents for September 30, 1978.

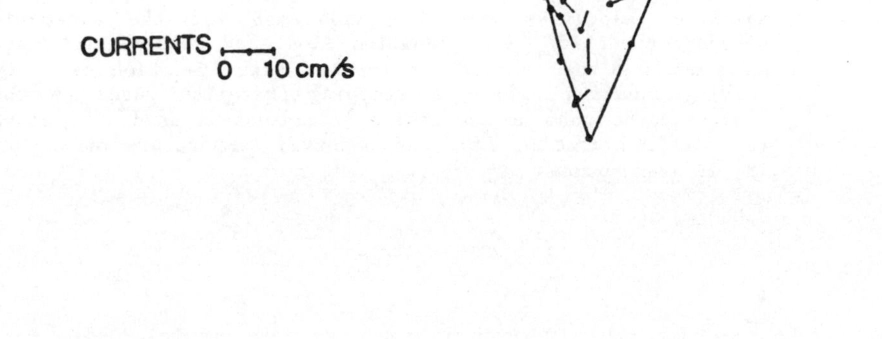

$$\int_e \left(\frac{d\hat{T}}{dt} + \frac{\partial(\phi\hat{u})(\psi\hat{T})}{\partial x} + K_h \frac{\partial\psi}{\partial x}\frac{\partial\psi}{\partial x}\hat{T} + \frac{\partial(\phi\hat{w})(\psi\hat{T})}{\partial z}\right.$$

$$\left. + (\psi\hat{K}_v)\frac{\partial\psi}{\partial z}\frac{\partial\psi}{\partial z}\hat{T}\right) dxdz - \int_s (\psi\frac{\hat{q}}{C_p\rho})\, dx = 0 \qquad (17)$$

where $\hat{\phi}$ is the tranpose of $\phi$ etc; e indicates the domain of the elements; s and b indicate surface and bottom boundaries; and the surface integrals are obtained from integration by parts. A predictor-corrector time integration method (Lam, 1980) is used to advance the solution from the one time level to the next. Notice that whereas the baroclinic pressure component, Q, can be evaluated in the predictor-corrector scheme, the barotropic component P is evaluated as a correction term based on the kinematic boundary condition:

$$\frac{\partial\eta}{\partial t} + \frac{\partial}{\partial x}\left\{\int_{-D}^{\eta} udz\right\} = 0 \qquad (18)$$

where D is the mean depth. Because of the rigid lid approximation, Equation (18) indicates that the total lateral (x) transports across any vertical (z) section should be equal to the hydraulic inflow and hence the outflow. Since the pressure should distribute evenly along this vertical section, the evaluation of the pressure term P consists of simply correcting each of the velocities along the section by a uniform amount such that the total lateral transport obeys Equation (18). Figure 1(a) shows the finite-element mesh used for the Lake Erie model, in which the velocities are defined at the vertices and mid-points of the side of the traingular element but the temperature is defined at the vertices only. Thus, the iteration would start with an initial distribution of temperature, so that $K_v$ and $A_v$ can be defined by Equations (11) to (14) along the vertical (z) sections which have been specially aligned in the mesh (Figure 1(a)). The derivatives of variables and the variables themselves can be obtained from the interpolating basis functions, if necessary. Using these guesses, the momentum and continuity equations (Equations (15) and (16)) are solved by the predictor-correction scheme. The computed velocities are then corrected for the kinematic boundary condition, i.e. Equation (18), and subsequently used in computing the temperature for the next iteration cycle by solving Equation (17). A stopping criterion based on the least-squares norm of successive iterations is used. To start the model, hydraulic flow and observed temperature on August 18, 1978 are used.

Fig. 2(a) Observed temperature (September 30, 1978)

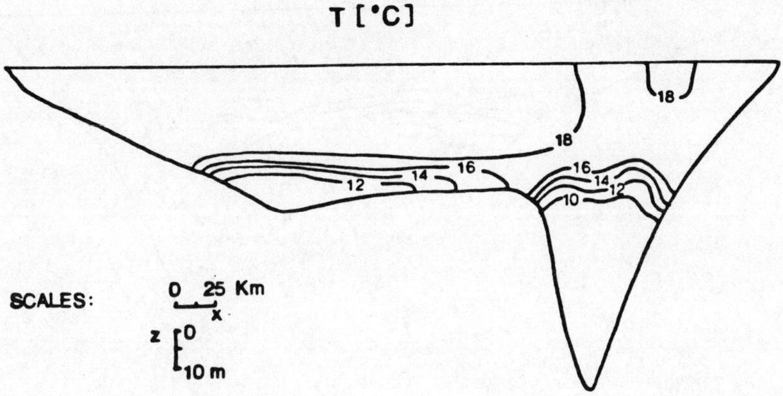


Fig. 2(b) Computed temperature (September 30, 1978)

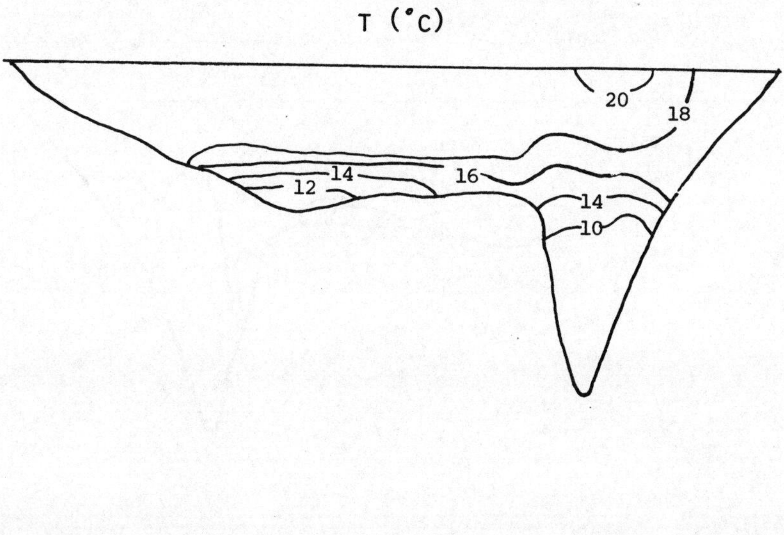

Fig. 3(a) Observed dissolved oxygen contours (September 30, 1978)

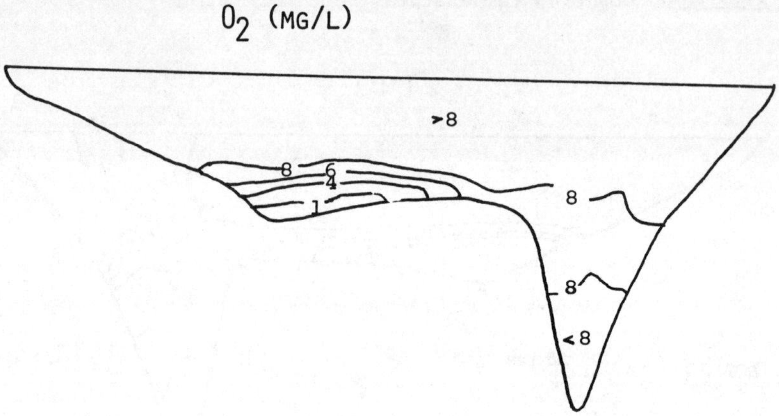


Fig. 3(b) Observed soluble reactive phosphorus contours (September 30, 1978)

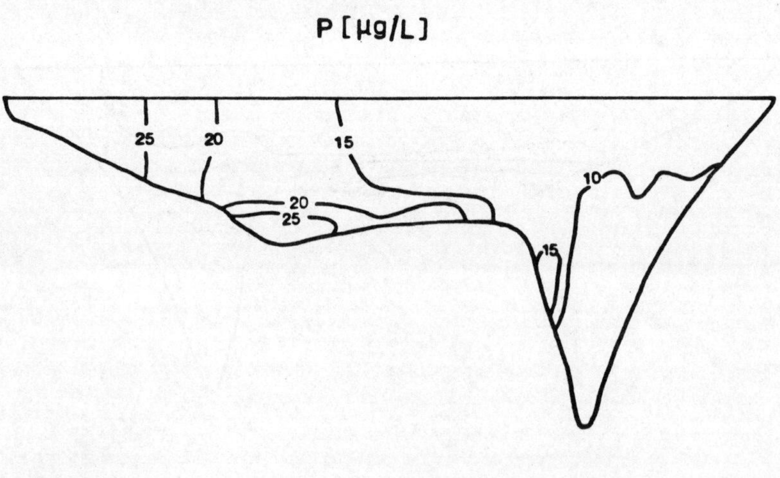

## COMPUTED AND OBSERVED RESULTS

Figure 1(b) shows the computed and observed currents for September 30, 1978. The epilimnetic currents are within a 7 m layer beneath the surface, predominantly westerly flows which are in line with the prevailing wind. The return flow occupies almost the remaining portion of the water column. The thermocline is at about 20 m from the surface, indicating that vertical gyres exist in the epilimnia of both Central and Eastern Basins. Of particular interest is the strong east-to-west return flow between these two Basins, because, with the 20 m thermocline, the Central Basin hypolimnion becomes anoxic, but then this interbasin transport will bring new supply of oxygen from Eastern Basin into the anoxic areas.

Figure 2(a) shows the observed temperature and Figure 2(b) the computed temperature for September 30, 1978. The isotherms compare satisfactorily, except for the epilimnion of the Eastern Basin where the temperature is overpredicted. Because of the mesh spacing, the computed temperature did not show as sharp a vertical gradient as in the observed temperature profile. However, in general, the thermal stratification is simulated and the transport of warmer water from the Eastern Basin into the Central Basin hypolimnion well established. Figures 3(a) and (b) further demonstrate these physical effects on the distribution of observed dissolved oxygen and soluble reactive phosphorus.

## CONCLUSION

The interbasin transport is found to be important in affecting the distribution of water quality variables. Anoxia, as evidenced by low oxygen content and high levels of regenerated phosphorus is confined to the western portion of the Central Basin hypolimnion on September 30, 1978 because of a strong westward hypolimnetic current. This return flow, in turn, is due to the prevailing westerly winds blowing at the lake surface. In the absence of the Coriolis effect, the finite element x-z model appears to be able to simulate both the flow field and the thermal regimes. The implementation of isoparametric elements are being currently considered, together with better ways to incorporate the kinematic boundary condition.

## ACKNOWLEDGEMENT

The author wishes to thank T.J. Simons for helpful discussions, W.M. Schertzer for the meteorological inputs, A.S. Fraser for the observed water quality data, F. Chiocchio for the observed currents, and J. Hodson for the programming.

**REFERENCES**

Heinrich, J., W. Lick and J. Paul (1981) Temperatures and Currents in a Stratified Lake: A Two-dimensional Analysis. J. Great Lakes Res., 7:264-275.

Hollan, E. and T.J. Simons (1978) Wind Induced Changes of Temperature and Currents in Lake Constance. Arch. Met. Geophys. Bioclimatology, Series A, 27(3-4):333-373.

Hood, P. and C. Taylor (1974) Navier - Stokes Equations Using Mixed Interpolation, Finite Element Methods in Flow Problems, Ed. Oden, J.T., UAH Press, Huntsville, 121-132.

Lam, D.C.L. (1980) Finite Element Simulation of Anisotropic Turbulent Diffusion in Lakes. Proc. 3rd Int. Conf. on Finite Elements in Water Resources, Ed. Wang et al., University of Mississippi, 3.20-3.35.

Mellor, G.L. and P.A. Durbin (1975) The Structure and Dynamics of the Ocean Surface Mixed Layer. J. Phys. Ocean., 5:718-728.

MacArthur, R.C. and W.R. Norton (1980) Application of the Finite Element Method to Vertically Stratified Hydrodynamic Flow and Water Quality. Proc. 3rd Int. Conf. Finite Elements in Water Resources, Wang et al. ed., University of Mississippi, 5.92-5.102.

Walters, R.Z., G.F. Carey and D.F. Winter (1978) Temperature Computation for Temperate Lakes. Appl. Math. Modelling, (2):41-48.

# CALCULATION OF LAKE-CIRCULATION WITH THREEDIMENSIONAL FINITE ELEMENTS

J. Trösch

Laboratory of Hydraulics, Hydrology and Glaciology,
Federal Institute of Technology, Zurich, Switzerland.

## INTRODUCTION

Limnologists, studying biological and chemical processes in lakes know about the importance of physical transport processes, induced by wind, temperature differences, in- and outflows. For these purposes, a lot of models are developed which calculate the currents and the temperature in lakes. Normally, twodimensional vertically-integrated models are used, with the assumption of hydrostatic pressure distribution. In certain cases, when vertical acceleration cannot be neglected, threedimensional models should be used. An example is the artificial aeration of lakes by blowing air in the anaerobic layers near the bottom, which causes strong upward velocities.

## BASIC EQUATIONS

The governing equations for the turbulent meanflow of an incompressible fluid in a rotating coordinate system are the continuity equation

$$v_{i,i} = 0 \tag{1}$$

and conservation of linear momentum

$$\rho \frac{\partial v_i}{\partial t} + \rho\, v_j\, v_{i,j} - \rho\, k_i + p_{,i} - t_{ij,j} = 0, \quad (i = 1,2,3). \tag{2}$$

Here, $v_i$ are the velocity components, $\rho$ the density of water, t the time, p the pressure and $t_{ij}$ the stress tensor, containing the turbulent Reynoldsstresses. $k_i$ is the vector of the external forces with gravity- and Coriolisterms

$$k_i = (f\,v_2, -f\,v_1, -g) \tag{3}$$

with f the Coriolisparameter and g the gravitational acceleration.

The boundary conditions are the no-slip condition

$$v_i = 0 \tag{4}$$

at the bottom and normal and tangential stress conditions at the free surface, predominantely the wind stress

$$\tau_s = \rho_a \, c_D \, u|u|. \tag{5}$$

Here, $\tau_s$ is the horizontal windstress, $\rho_a$ the air density, $c_D$ the drag coefficient and u the wind velocity.

As the surface is not a fixed boundary, the kinematic boundary condition

$$\frac{\partial\zeta}{\partial t} + v_1 \frac{\partial\zeta}{\partial x} + v_2 \frac{\partial\zeta}{\partial y} - v_3 = 0 \tag{6}$$

holds, where $\zeta(x,y,t)$ is the surface elevation measured positive upwards from a surface at rest.

Likewise initial conditions are prescribed

$$\begin{aligned} v_i &= v_i\Big|_{t=0} \\ p &= p\Big|_{t=0} \end{aligned} \tag{7}$$

Nonlinear terms as convection terms in equations (2) and (6) are linearized by taking the velocity from the past time step. This is justified, as the velocities are slow and slowly changing.

The closure problem of the turbulent Reynoldsstresses is solved by a simple eddy viscosity approximation. Although the coefficients of viscosity may vary in space, this turbulence model is not satisfactory, a better model will be introduced as next step.

The free surface given by equation (6) leads to a solutiondependent domain for the integration of the fieldequations. This nonlinear problem could be solved by an uneconomic iteration. As the surface fluctuations are very small compared with the depth of a lake, the solution domain is approximated as independent of time. As consequence we have to consider the elevation in the pressure term by

$$p^{eff} = p + \rho g \eta. \tag{8}$$

Equation (8) is then introduced in equation (2).

## FINITE ELEMENT APPROXIMATION

The unknowns $\{x\} = \{{v_i \atop p}\}$ and the surface elevation $\zeta$ are described by

$$v_i(x,y,z,t) = \sum_{\alpha=1}^{n} N_\alpha(x,y,z) \cdot v_{\alpha i}(t), \quad (i = 1,2,3) \tag{9}$$

$$p(x,y,z) = \sum_{\alpha=1}^{m} M_\alpha(x,y,z)\cdot p_\alpha \qquad (10)$$

$$\zeta(x,y,t) = \sum_{\alpha=1}^{\ell} L_\alpha(x,y)\cdot\zeta_\alpha(t) \qquad (11)$$

where $N_\alpha$, $M_\alpha$ and $L_\alpha$ are shape functions. The finite element analog of the equations (1) and (2) is obtained by the Galerkin-method (Finlayson (1972)). The incompressibility condition (1) is weighted by $M_\alpha$, equations (2) by $N_\alpha$. Thus equation (1) is transformed in

$$\int_V (M_\alpha \sum_\beta N_{\beta,i}\, v_{\beta i})\, dV = 0 \quad (\alpha = 1,2,\ldots,m) \qquad (12)$$

and the equations (2)

$$\int_V N_\alpha\Big(\rho\, N_\beta \frac{\partial v_{\beta i}}{\partial t} + \rho\, N_\beta\, \overset{o}{v}_{\beta j}\, N_{\gamma,i}\, v_{\gamma i} - \rho\, k_i + M_{\beta,i}\, p_\beta + \rho\, g\, L_{\beta,i}\, \zeta_\beta - t_{ij,j}\Big)\, dV = 0 \qquad (13)$$

$$(\alpha = 1,2,\ldots,n;\;\; i = 1,2,3)$$

where the summation sign for $\beta$ and $\gamma$ has been omitted. Integration by parts of the last three terms yields

$$\int_V\Big(\rho N_\alpha N_\beta \frac{\partial v_{\beta i}}{\partial t} + \rho\, N_\alpha N_\beta N_{\gamma,i}\, \overset{o}{v}_{\beta j}\, u_{\gamma i} - \rho\, N_\alpha\, k_i - N_{\alpha,i}\, M_\beta\, p_\beta - \rho\, g\, N_{\alpha,i}\, L_\beta\, \zeta_\beta + N_{\alpha,j}\, t_{ij}\Big)\, dV - \int_F N_\alpha\Big((M_\beta\, p_\beta + \rho g\, L_\beta\, \zeta_\beta)\delta_{ij} - t_{ij}\Big) n_j\, dF = 0 \qquad (14)$$

$$(\alpha = 1,2,\ldots,n;\;\; i = 1,2,3)$$

with

$$k_i = (+ f N_\beta\, v_{\beta 2}, - f N_\beta\, v_{\beta 1}, - g) \qquad (15)$$

It is important, that the surface integral in equation (13) does not vanish and has to be evaluated at the free surface.

The finite element analog of equation (6) becomes

$$\int_V\Big(L_\alpha L_\beta \frac{\partial \zeta_\beta}{\partial t} + L_\alpha\Big(N_\gamma\, \overset{o}{v}_{\gamma 1}\, L_{\beta,1} + N_\gamma\, \overset{o}{v}_{\gamma 1}\, L_{\beta,2}\Big)\zeta_\beta - L_\alpha\, N_\beta\, V_{\beta 3}\Big)\, dF = 0 \qquad (16)$$

Equations (12), (14) and (16) form a system of linear differential equations for the time t, it can be written in matrix form as

$$[A]\,\{\frac{\partial x}{\partial t}\} + [B]\,\{x\} + \{C\} = 0 \qquad (17)$$

where the coefficients of the matrices [A], [B] and {C} are given by appropriate parts of the equations (12), (14) and (16), and the vector {x} contains the unknowns $v_i$, p and $\zeta$.

The differential equations (17) are integrated in time by an implicit finite difference scheme with time step $\Delta t$

$$\{\frac{\partial x}{\partial t}\} = \left(\{x\}^{t+\Delta t} - \{x\}^{t}\right)/\Delta t \tag{18}$$

and

$$\{x\}^{t+\Delta t/2} = \left(\{x\}^{t+\Delta t} + \{x\}^{t}\right)/2. \tag{19}$$

Introducing equations (18) and (19) in (17) leads to a system of linear equations

$$\left([B] + 2[A]/\Delta t\right)^{-1} \{x\}^{t+\Delta t/2} - \left(2[A]/\Delta t \ \{x\}^{t} - \{C\}\right) = 0 \tag{20}$$

When solving these equations, two problems are encountered: The matrix is unsymmetric and rather huge. The only equationsolver for unsymmetric matrices was the program from Hood (1976), but in a computer without virtual memory a frontwidth of several hundred nonzero coefficients could not be handled. For this reason, the version of Hood (1976) was enlarged. The space consuming working matrix is horizontally divided in several blocks, each block containing a certain number of rows. The blocks are stored out-of-core only the space of one block is exploited in the central unit. Although a lot of bookkeeping has to be done, computer time did not augment due to improvements in coding. For 4000 unknowns and a frontwidth of 250 the time consumed on a CDC Cyber 176 is 2400 seconds.

Shape functions

In the presented work, isoparametric quadrilateral elements are used. As is well known, shape functions of the same degree for velocities and pressure are not consistent, the problem is overconstrained, as was shown for twodimensional elements (Stummel (1979)). A reduction of the degree of the shape function for pressure is a remedy. We choose triquadratic functions for the velocity ($N_\alpha$ in equation (9)) and trilinear functions for pressure ($M_\alpha$ in equation (10)). The element used consists of eight corner nodes for velocity and pressure and twelve midside nodes for velocity. For surface elements four supplementary elevation points are added.

## RESULTS

Some preliminary calculations showed, that the element with triquadratic and trilinear shape functions was consistent, but not unconditionally. If more than one side of an element lies on a boundary with a no-slip condition then the system becomes overconstrained. This is not a severe restriction, as natural lakes have smooth bottoms without corners and edges. In testbasins with rectangular cross-sections some care is needed when generating the finite element mesh.

Results from a winddriven basin with rectangular cross-sections were compared to other threedimensional models (Spraggs (1975)). The results were in good agreement.

## CONCLUSIONS

For the tested simple configurations the threedimensional finite element model is consistent, if not more than one side of an element lies on a boundary with a no-slip condition. Further calculations are needed to show if complicated domains as natural lakes can be treated.

## ACKNOWLEDGEMENTS

The author wishes to thank Prof. Dr. D. Vischer, Laboratory of Hydraulics, Hydrology and Glaciology at the Federal Institute of Technology, Zurich, for his support in this investigation.

## REFERENCES

Finlayson B.A. (1972) The Method of Weigthed Residuals and Variational Principles. Academic Press, New York.

Hood P. (1976) Frontal Solution Program for Unsymmetric Matrices. Int. J. Num. Meth. Eng., 10, 1976.

Spraggs L.D. and R.L. Street (1975) Threedimensional Simulation of Thermally-Influenced Hydrodynamic Flows. Stanford University, Technical Report No. 190.

Stummel F. (1979) The Generalized Patch Test. SIAM J. Numerical Analysis, 16, 3: 449-471.

# 8. MODERN SOFTWARE DEVELOPMENTS

# A MICROCOMPUTER PROGRAM FOR DESIGN OF THE SELF RETIMING MECHANICS OF LARGE TIDAL POWER SITES

H. Kinno

Faculty of Engineering, The University of Tokushima
Tokushima-shi 770, Japan

## INTRODUCTION

It is very difficult to find a computer program in the papers of tidal power generation inspite of their long year research progress. In order to push the tidal power technology, the author presents a newly developed microcomputer program written by BASIC for design of the self retiming mechanics of large tidal power sites of which system structure and analysis used mathematical models was reported by Kinno(1981).

The program can be used in a range of Z≧5 m, in where famous tidal power sites as Severn Estuary, La Rance and Fundy Bay at A8 and B9 are included. System similarity could be obtained when sea level, machine head and low pond area are shown by ratios to Z and A.

## SYSTEM, SYMBOL AND PROGRAM

Figure 1 shows the system mechanism in which upper part is top view and under part is side view. Turbine T1 generates 5 hr at each ebb tide only and pump is driven by the turbine T1 at nighttime. Nighttime is taken here from 20.00 hr to 08.00hr and daytime is 08.00 hr to 20.00 hr. Turbines T2 and T3 generate power at daytime only. Table 1 is symbols appeared in Program 1. If (head of T1)<(minimum limit H4), T1 has to stop even in the 5 hr at nighttime and to be switched to T2 or T3 at daytime. If (head of T2 or T3)<H4, the system is defined to instability, and calculation in the computer stop and jump to next input data. Value of H4 is determined as a function of Z which is shown in the input data of program as H4=K2*Z. Initial water levels of X and Y are determined as X=0.8Z and Y=-0.8Z which can be obtained easily from spring tide.

As examples of input data, Severn Estuary, La Rance, Fundy Bay at A8 and B9 having each two data of rated pump head H5 are used in Program 1. Table 2 is output data for La Rance and Table 3 is for Fundy Bay at A8.

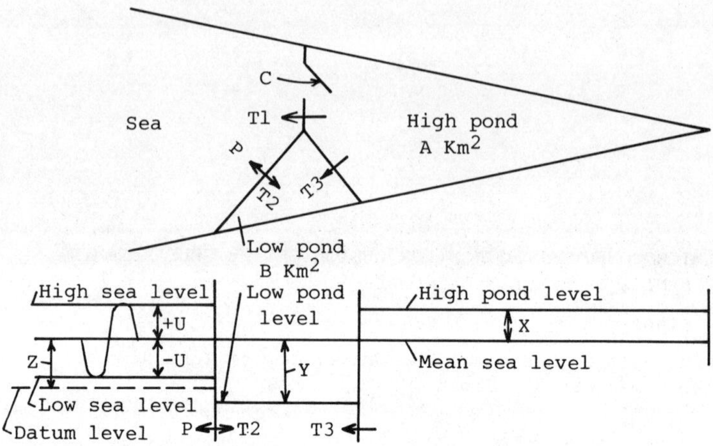


Figure 1 System mechanism and variables

Table 1 Principal variables

| Symbol | Dimension | Description |
|---|---|---|
| A | $Km^2$ | area of high pond |
| B | $Km^2$ | area of low pond |
| C | $m^{2.5}$/sec | gate characteristic |
| D | day | computing period |
| D1 | day | number of day |
| E0 | MWH | energy generated at a nighttime |
| E1 | MWH | energy generated at a daytime |
| E2 | MWH, GWH | total energy of E0 during 31 days |
| E3 | MWH, GWH | total energy of E1 during 31 days |
| E5 | GWH | annual no retime output energy, see 830 |
| E6 | GWH | annual retimed output energy, see 840 |
| F | | retiming efficiency, see 850 |
| G | hour | time used in the line printer |
| H0 | m | head of turbine T1 at nighttime |
| H1 | m | head of turbine T1 at daytime |
| H2 | m | head of turbine T2 at daytime |
| H3 | m | head of turbine T3 at daytime |
| H4=K2*Z | m | mimimum head limit of turbine T1,T2 & T3 |
| H5 | m | rated pump head |
| H6=H5*2 | m | maximum head limit of pump |
| H7 | m | pump head |
| I | | loop control variable |
| J, J1 | | printout pitch control variables |
| K, K1 | | time coefficients obtaining G |
| K0 | | total steps of small time during 31 days |
| K2 | | coefficient of Z obtaining H4 |
| L$ | | site name |

| | | |
|---|---|---|
| M0,M1,M2,M3,M4,M5,M6 | | initial value obtaining maximum values of machine head and water level, see 970 |
| N | | loop control variable |
| N0,N1,N2,N3,N4,N5,N6 | | initial value obtaining minimum values of machine head and water level, see 980 |
| M7,M8 | | initial values obtaining maximum tide level in a nighttime and a daytime |
| N7,N8 | | initial values obtaining minimum tide level in a nighttime and a daytime |
| P1 | MW | power generated at a nighttime |
| P2 | MW | power generated at a daytime |
| R1 | m | tidal range at a nighttime |
| R2 | m | tidal range at a daytime |
| S | hour | time delay along coast line at a site |
| T | hour | time |
| T0 | hour | system start time from the tidal 0 hour |
| T1,T2,T3 | | name of turbine |
| T5 | hour | operating time of turbine T1 and pump in a nighttime |
| T6 | hour | operating time of turbine T1 in a daytime |
| U,U1 | m | tide level used for computation |
| U2,U3 | m | tide level used for output data and graph |
| V,W | | medium variables obtaining U and U1 |
| X | m | water level of high pond used for computation |
| X2,X3 | m | water level of high pond used for output data and graph |
| X0 | m | sum of each X in a small time in order to obtain mean water level of X |
| X1 | m | small change of X in a small time, 0.1 hr |
| X4 | m | mean water level of X |
| Y | m | water level of low pond used for computation |
| Y2,Y3 | m | water level of low pond used for output data and graph |
| Y1 | m | small change of Y in a small time, 0.1 hr |
| Z | m | datum depth at mean sea level |
| PHAS | | machine operating phase |
| SUB 00 | | all machine stop at nighttime |
| SUB RF | | refilling condition at nighttime but all machine stop |
| SUB TP | | turbine drive the pump at nighttime |
| SUB T1 | | turbine T1 is generating condition |
| SUB T2 | | turbine T2 is generating condition while tide is refilling condition |
| SUB T3 | | turbine T3 is generating condition |

The microcomputer used in this program is SHARP MZ-80B which has 64 KB in the main memory, and line printer is SHARP MZ-80P4 which has 136 column. In the program,

120 PRINT is print to display

130 PRINT/P is print to line printer.

Program 1 The microcomputer program written by BASIC

```
10 REM PROGRAM FOR DESIGN OF THE SELF RETIMING MECHANICS OF
   LARGE TIDAL POWER SITES
20 DIM I(240), L$(25)
30 READ D, J
40 READ L$,A,B,C,H5,T0,S,Z,K2
50 IF A=0 THEN 1140
60 X=0.8*Z
70 Y=-X
80 X2=INT(X*100+0.5)/100
90 Y2=INT(Y*100-0.5)/100
100 H4=K2*Z
110 H6=H5*2
120 PRINT L$,A,B,C,H5,T0,S,Z,K2
130 PRINT/P L$
140 PRINT/P:PRINT/P
150 PRINT/P "COMPUTE PERIOD,";D;"DAY";TAB(26);"PRINTOUT PITCH,"
    ;0.1*J;"HOUR"
160 PRINT/P
170 PRINT/P "A=";A;TAB(13);"B=";B;TAB(26);"C=";C;TAB(40);"H5=";
    H5
180 PRINT/P "T0=";T0;TAB(13);"S=";S;TAB(26);"Z=";Z;TAB(40);"H4=
    ";K2;"*Z"
190 PRINT/P:PRINT/P
200 PRINT/P "DAY";TAB(5);"TIME";TAB(12);"MW";TAB(18);"U";TAB(25
    );"X";TAB(32);"Y";TAB(38);"THD";TAB(45);"PHD";TAB(53);"PHAS
    ";
210 PRINT/P ;TAB(59);"-20M";TAB(78);"-10M";TAB(100);"0";TAB(119
    );"10M"
220 E2=0:E3=0
230 M0=-50:N0=50:M1=-50:N1=50:M2=-50:N2=50:M3=-50:N3=50:M4=-50:
    N4=50:M5=-50:N5=50:M6=-50:N6=50
240 X0=0:K0=0
250 FOR N=0 TO D-1
260 M7=0:N7=0:M8=0:N8=0
270 PRINT/P TAB(60);"|---------|---------|---------|---------I
    ---------|---------|---------|"
280 FOR I=0 to 119
290 IF I>=120 THEN 340
300 GOSUB 1160
310 IF U1>M7 THEN M7=U1
320 IF U1<N7 THEN N7=U1
330 NEXT I
340 FOR I=120 TO 239
350 IF I>=240 THEN 780
360 GOSUB 1160
370 IF U1>M8 THEN M8=U1
380 IF U1<N8 THEN N8=U1
390 NEXT I
400 R1=M7-N7
410 P1=INT(0.1206*A*R1^2+0.5)
```

```
420 R2=M8-N8
430 P2=INT(0.05025*A*(R1^2+R2^2)+0.5)
440 E1=P2*12
450 E3=E3+E1
460 REM CALCULATION OF U,X AND Y ******************************
470 T5=0:T6=0
480 FOR I=0 TO 239
490 T=N*24+I/10
500 D1=INT(T/24)+1
510 K=T+20
520 K1=INT(K/24)*24
530 G=INT((K-K1)*10+0.5)/10
540 V=SIN(0.01689*(T0+T))
550 W=SIN(0.5067*(T0+T-S))
560 U=(0.675+0.225*V)*W*Z
570 U2=INT(U*100)/100
580 IF I>=120 THEN 620
590 IF U>X GOSUB 1480:GOTO 660
600 IF (U<=0)*(T5<5)*((X-U)>H4)*(H6>(U-Y)) GOSUB 1240:GOTO 660
610 GOSUB 1590:GOTO 660
620 IF I>=240 THEN 780
630 IF U>X GOSUB 1910:GOTO 660
640 IF (U<=0)*(T6<5)*((X-U)>H4) GOSUB 1710:GOTO 660
650 GOSUB 2070
660 IF X2>M5 THEN M5=X2     REM In 600 & 640, IF (   )*(   ) is
670 IF X2<N5 THEN N5=X2                       IF (   )AND(   )
680 IF Y2>M6 THEN M6=Y2
690 IF Y2<N6 THEN N6=Y2
700 X0=X0+X
710 K0=K0+1
720 X=X+X1
730 Y=Y+Y1
740 NEXT I
750 IF N>=D THEN 40
760 E0=P1*T5
770 E2=E2+E0
780 NEXT N
790 E3=E3/1000
800 E2=E2/1000
810 X4=X0/K0
820 X4=INT(X4*100+0.5)/100
830 E5=INT(E2*2*12*0.98+0.5)
840 E6=INT(E3*12*0.98+0.5)
850 F=E6/E5
860 F=INT(F*100+0.5)/100
870 PRINT/P:PRINT/P
880 PRINT/P L$
890 PRINT/P
900 PRINT/P "A=";A;TAB(12);"B=";B;TAB(24);"C=";C;TAB(36);"H5=";
    H5
910 PRINT/P "T0=";T0;TAB(12);"S=";S;TAB(24);"Z=";Z;TAB(36);"H4=
    ";K2;"*Z"
920 PRINT/P
```

```
930 PRINT/P "ANNUAL NO RETIME OUTPUT ENERGY =";E5;TAB(40);"GWH"
940 PRINT/P "ANNUAL RETIMED OUTPUT ENERGY   =";E6;TAB(40);"GWH"
950 PRINT/P "AVERAGE RETIMING EFFICIENCY    =";F
960 PRINT/P:PRINT/P
970 PRINT/P "MAX HEAD OF T1 AT NIGHTTIME=";M0;TAB(36);"M"
980 PRINT/P "MIN HEAD OF T1 AT NIGHTTIME=";N0;TAB(36);"M"
990 PRINT/P "MAX PUMP HEAD              =";M4;TAB(36);"M"
1000 PRINT/P "MIN PUMP HEAD              =";N4;TAB(36);"M"
1010 PRINT/P "MAX HEAD OF T1 AT DAYTIME  =";M1;TAB(36);"M"
1020 PRINT/P "MIN HEAD OF T1 AT DAYTIME  =";N1;TAB(36);"M"
1030 PRINT/P "MAX HEAD OF T2             =";M2;TAB(36);"M"
1040 PRINT/P "MIN HEAD OF T2             =";N2;TAB(36);"M"
1050 PRINT/P "MAX HEAD OF T3             =";M3;TAB(36);"M"
1060 PRINT/P "MIN HEAD OF T3             =";N3;TAB(36);"M"
1070 PRINT/P "MAX WATER LEVEL  OF X      =";M5;TAB(36);"M"
1080 PRINT/P "MIN WATER LEVEL  OF X      =";N5;TAB(36);"M"
1090 PRINT/P "MEAN WATER LEVEL OF X      =";X4;TAB(36);"M"
1100 PRINT/P "MAX WATER LEVEL  OF Y      =";M6;TAB(36);"M"
1110 PRINT/P "MIN WATER LEVEL  OF Y      =";N6;TAB(36);"M"
1120 PRINT/P:PRINT/P:PRINT/P
1130 GOTO 40
1140 END
1150 REM *****************************************************
1160 REM SUBROUTINE U1
1170 T=N*24+I/10
1180 V=SIN(0.01689*(T0+T))
1190 W=SIN(0.5067*(T0+T-S))
1200 U1=(0.675+0.225*V)*W*Z
1210 RETURN
1220 END
1230 REM *****************************************************
1240 REM SUBROUTINE TP
1250 H0=X-U
1260 H7=U-Y
1270 X1=(-0.04455)*P1/(A*H0)
1280 Y1=(-0.0285)*(H6-H7)*P1/(B*H5^2)
1290 H0=INT(H0*100+0.5)/100
1300 H7=INT(H7*100+0.5)/100
1310 IF H0>M0 THEN M0=H0
1320 IF H0<N0 THEN N0=H0
1330 IF H7>M4 THEN M4=H7
1340 IF H7<N4 THEN N4=H7
1350 X2=INT(X*100+0.5)/100
1360 Y2=INT(Y*100-0.5)/100
1370 T5=T5+0.1
1380 GOSUB 2230
1390 IF J1>0 THEN RETURN
1400 PRINT/P D1;TAB(4);G;TAB(10);P1;TAB(16);U2;TAB(23);X2;TAB(3
    0);Y2;TAB(37);H0;TAB(45);H7;TAB(53);"TTPP";
1410 IF U3<=Y3 THEN 1440
1420 PRINT/P ;TAB(Y3);"Y";TAB(U3);"*";TAB(X3);"X"
1430 RETURN
```

```
1440 PRINT/P ;TAB(U3);"*";TAB(Y3);"Y";TAB(X3);"X"
1450 RETURN
1460 END
1470 REM ******************************************************
1480 REM SUBROUTINE RF
1490 X1=(0.00036*C*SQR(U-X))/A
1500 Y1=0
1510 X2=INT(X*100+0.5)/100
1520 Y2=INT(Y*100-0.5)/100
1530 GOSUB 2230
1540 IF J1>0 THEN RETURN
1550 PRINT/P D1;TAB(4);G;TAB(10);"  ";TAB(16);U2;TAB(23);X2;TAB
     (30);Y2;TAB(37);"  ";TAB(45);"  ";TAB(53);"RRFF";TAB(Y3);
     "Y";TAB(X3);"X";TAB(U3);"*"
1560 RETURN
1570 END
1580 REM ******************************************************
1590 REM SUBROUTINE OO
1600 X1=0
1610 Y1=0
1620 X2=INT(X*100+0.5)/100
1630 Y2=INT(Y*100-0.5)/100
1640 GOSUB 2230
1650 IF J1>0 THEN RETURN
1660 PRINT/P D1;TAB(4);G;TAB(10);"  ";TAB(16);U2;TAB(23);X2;TAB
     (30);Y2;TAB(37);"  ";TAB(45);"  ";TAB(53);"OOOO";TAB(Y3);
     "Y";TAB(U3);"*";TAB(X3);"X"
1670 RETURN
1680 END
1690 REM ******************************************************
1700 REM DAYTIME OPERATION
1710 REM SUBROUTINE T1
1720 H1=X-U
1730 X1=(-0.04455)*P2/(A*H1)
1740 Y1=0
1750 H1=INT(H1*100+0.5)/100
1760 IF H1>M1 THEN M1=H1
1770 IF H1<N1 THEN N1=H1
1780 X2=INT(X*100+0.5)/100
1790 Y2=INT(Y*100-0.5)/100
1800 T6=T6+0.1
1810 GOSUB 2230
1820 IF J1>0 THEN RETURN
1830 PRINT/P D1;TAB(4);G;TAB(10);P2;TAB(16);U2;TAB(23);X2;TAB(3
     0);Y2;TAB(37);H1;TAB(45);"  ";TAB(53);"T  1";
1840 IF U3<=Y3 THEN 1870
1850 PRINT/P ;TAB(Y3);"Y";TAB(U3);"*";TAB(X3);"X"
1860 RETURN
1870 PRINT/P ;TAB(U3);"*";TAB(Y3);"Y";TAB(X3);"X"
1880 RETURN
1890 END
1900 REM ******************************************************
1910 REM SUBROUTINE T2
```

```
1920 H2=U-Y
1930 X1=(0.00036*C*SQR(U-X))/A
1940 Y1=0.04455*P2/(B*H2)
1950 H2=INT(H2*100+0.5)/100
1960 IF H2<H4 THEN 40
1970 IF H2>M2 THEN M2=H2
1980 IF H2<N2 THEN N2=H2
1990 X2=INT(X*100+0.5)/100
2000 Y2=INT(Y*100-0.5)/100
2010 GOSUB 2230
2020 IF J1>0 THEN RETURN
2030 PRINT/P D1;TAB(4);G;TAB(10);P2;TAB(16);U2;TAB(23);X2;TAB(3
     0);Y2;TAB(37);H2;TAB(45);"  ";TAB(53);"T  2";TAB(Y3);"Y";
     TAB(X3);"X";TAB(U3);"*"
2040 RETURN
2050 END
2060 REM ***********************************************************
2070 REM SUBROUTINE T3
2080 H3=X-Y
2090 X1=(-0.04455)*P2/(A*H3)
2100 Y1=(A/B)*(-X1)
2110 H3=INT(H3*100+0.5)/100
2120 IF H3<H4 THEN 40
2130 IF H3>M3 THEN M3=H3
2140 IF H3<N3 THEN N3=H3
2150 X2=INT(X*100+0.5)/100
2160 Y2=INT(Y*100-0.5)/100
2170 GOSUB 2230
2180 IF J1>0 THEN RETURN
2190 PRINT/P D1;TAB(4);G;TAB(10);P2;TAB(16);U2;TAB(23);X2;TAB(3
     0);Y2;TAB(37);H3;TAB(45);"  ";TAB(53);"T  3";TAB(Y3);"Y";
     TAB(U3);"*";TAB(X3);"X"
2200 RETURN
2210 END
2220 REM ***********************************************************
2230 REM SUBROUTINE FOR GRAPH OF U,X  AND Y
2240 U3=INT(U*2)+100
2250 X3=INT(X*2)+100
2260 Y3=INT(Y*2)+100
2270 J1=I-INT(I/J)*J
2280 RETURN
2290 END
3000 REM INPUT DATA ********************************************
3010 DATA 31,10
3020 DATA "*SEVERN ESTUARY*",400,200,161000,6.0,92,0,6.7,0.45
3030 DATA "*SEVERN ESTUARY*",400,200,161000,4.0,92,0,6.7,0.45
3040 DATA "****LA RANCE****", 16,  8,  6400,6.0,92,0,6.7,0.45
3050 DATA "****LA RANCE****", 16,  8,  6400,4.0,92,0,6.7,0.45
3060 DATA "**FUNDY BAY A8**",100, 50, 46800,7.0,92,0,7.8,0.45
3070 DATA "**FUNDY BAY A8**",100, 50, 46800,4.7,92,0,7.8,0.45
3080 DATA "**FUNDY BAY B9**",280,140,136000,7.3,92,0,8.1,0.45
3090 DATA "**FUNDY BAY B9**",280,140,136000,4.9,92,0,8.1,0.45
5000 DATA "**END OF DATA***",0,0,0,0,0,0,0,0,0
```

Table 2 Output data for La Rance tidal power site

```
***LA RANCE****
COMPUTE PERIOD, 31DAY         PRINTOUT PITCH, 1HOUR

A= 16          B= 8           C= 6400        H5= 6
T0= 92         S= 0           Z= 6.7         H4= 0.45*Z
```

| DAY | TIME | MW | U | X | Y | THD | PHD | PHAS |
|---|---|---|---|---|---|---|---|---|
| 1 | 20 | | 2.93 | 5.36 | -5.37 | | | OOOO |
| 1 | 21 | | 0 | 5.36 | -5.37 | | | OOOO |
| 1 | 22 | 280 | -2.93 | 4.25 | -7.2 | 7.17 | 4.27 | TTPP |
| 1 | 23 | 280 | -5.12 | 3.24 | -9.37 | 8.36 | 4.25 | TTPP |
| 1 | 0 | 280 | -6.03 | 2.32 | -11.4 | 8.34 | 5.37 | TTPP |
| 1 | 1 | 280 | -5.42 | 1.31 | -12.99 | 6.73 | 7.57 | TTPP |
| 1 | 2 | 280 | -3.45 | -0.2 | -13.87 | 3.25 | 10.41 | TTPP |
| 1 | 3 | | -0.62 | -0.44 | -13.91 | | | OOOO |
| 1 | 4 | | 2.36 | 0.75 | -13.91 | | | RRFF |
| 1 | 5 | | 4.75 | 2.73 | -13.91 | | | RRFF |
| 1 | 6 | | 5.94 | 4.65 | -13.91 | | | RRFF |
| 1 | 7 | | 5.63 | 5.74 | -13.91 | | | OOOO |
| 1 | 8 | 231 | 3.91 | 5.74 | -13.91 | 19.64 | | T 3 |
| 1 | 9 | 231 | 1.21 | 5.4 | -13.24 | 18.64 | | T 3 |
| 1 | 10 | 231 | -1.79 | 4.68 | -12.89 | 6.47 | | T 1 |
| 1 | 11 | 231 | -4.34 | 3.79 | -12.89 | 8.13 | | T 1 |
| 1 | 12 | 231 | -5.79 | 3.04 | -12.89 | 8.83 | | T 1 |
| 1 | 13 | 231 | -5.79 | 2.29 | -12.89 | 8.08 | | T 1 |
| 1 | 14 | 231 | -4.33 | 1.39 | -12.89 | 5.72 | | T 1 |
| 1 | 15 | 231 | -1.79 | 0.38 | -12.51 | 12.88 | | T 3 |
| 1 | 16 | 231 | 1.19 | 0.32 | -11.47 | 12.65 | | T 2 |
| 1 | 17 | 231 | 3.86 | 2.02 | -10.51 | 14.37 | | T 2 |
| 1 | 18 | 231 | 5.56 | 3.97 | -9.64 | 15.2 | | T 2 |
| 1 | 19 | 231 | 5.85 | 5.47 | -8.79 | 14.64 | | T 2 |
| 2 | 20 | | 4.68 | 5.35 | -7.87 | | | OOOO |

continue till 31DAY

```
ANNUAL NO RETIME OUTPUT ENERGY = 561    GWH
ANNUAL RETIMED OUTPUT ENERGY   = 608    GWH
AVERAGE RETIMING EFFICIENCY    = 1.08

MAX HEAD OF T1 AT NIGHTTIME= 8.6     M
MIN HEAD OF T1 AT NIGHTTIME= 3.02    M
MAX PUMP HEAD              = 11.39   M
MIN PUMP HEAD              =-0.73    M
MAX HEAD OF T1 AT DAYTIME  = 9.17    M
MIN HEAD OF T1 AT DAYTIME  = 3.02    M
MAX HEAD OF T2             = 19.6    M
MIN HEAD OF T2             = 7.85    M
MAX HEAD OF T3             = 20.07   M
MIN HEAD OF T3             = 6.53    M
MAX WATER LEVEL  OF X      = 5.77    M
MIN WATER LEVEL  OF X      =-1.64    M
MEAN WATER LEVEL OF X      = 2.47    M
MAX WATER LEVEL  OF Y      =-2.93    M
MIN WATER LEVEL  OF Y      =-14.42   M
```

THD = turbine head
PHD = pump head

Table 3 Output data for Fundy Bay A8 site

```
**FUNDY BAY A8**

A= 100        B= 50        C= 46800        H5= 7
TO= 92        S= 0         Z= 7.8          H4= 0.45*Z

ANNUAL NO RETIME OUTPUT ENERGY = 4771  GWH
ANNUAL RETIMED OUTPUT ENERGY   = 5149  GWH
AVERAGE RETIMING EFFICIENCY    = 1.08

MAX HEAD OF T1 AT NIGHTTIME= 10.08  M
MIN HEAD OF T1 AT NIGHTTIME= 3.52   M
MAX PUMP HEAD              = 13.38  M
MIN PUMP HEAD              =-0.59   M
MAX HEAD OF T1 AT DAYTIME  = 10.82  M
MIN HEAD OF T1 AT DAYTIME  = 3.52   M
MAX HEAD OF T2             = 22.38  M
MIN HEAD OF T2             = 9.28   M
MAX HEAD OF T3             = 23.53  M
MIN HEAD OF T3             = 7.9    M
MAX WATER LEVEL  OF X      = 6.81   M
MIN WATER LEVEL  OF X      =-1.92   M
MEAN WATER LEVEL OF X      = 2.95   M
MAX WATER LEVEL  OF Y      =-3.71   M
MIN WATER LEVEL  OF Y      =-16.85  M
```

It is necessary to use a line printer with 136 column for Program 1. Because output data shown in Table 2 are printed on the left half space of the line printer and transients of U,X and Y are printed on the remained right half space as rough graph. Annual energy, average retiming efficiency, maximum and minimum head of machine and water level of each pond are listed at the end of data as Tables 2 and 3.

Take care, however, the last listed data are correct only when D=31. The value of J is taken to 1 or 5 for detailed output and to 10 or 20 for rough output, and J=240 is used for a main purpose obtaining the listed data as Table 3. The calculations are made always with 0.1 hr pitch, neverthless change of the value of J.

## CONCLUSIONS

It is revealed that the stored tidal energy generated at nighttime into the system can be received next day with 100 percent efficiency and plus 8 percent interest as same as money deposit in a bank without any help of energy from grid.

## REFERENCE

Kinno, H. (1981) CAD of the Self-Retiming Mechanism at Large Tidal Power Sites. Proc. of the Second International Conf. on Engineering Software, London, 496-511.

# AUTOMATIC GENERATION OF FINITE ELEMENT NETWORK FOR BIDIMENSIONAL AND QUASI-THREE-DIMENSIONAL EQUATIONS GOVERNING THE GROUNDWATER FLOW

L. Sartori, V. Aurelio, M. Riccioni, G. Svarca.

Aquater S.p.A. - San Lorenzo in Campo (PS) - Italy

## INTRODUCTION

The FE network generation method is based on the use of digitalized thematic maps and is particularly adapted for groundwater modelling. The generated network follows properly both the area contour and the internal lines such as rivers, faults and geological boundaries, even in very complex configurations.The mesh areas can vary within a large range, in different zones of the grid. The meshes are always sufficiently regularly-shaped to avoid numerical criticity during the solution of the finite element equations. Transition zones between areas of different density, with connection elements of a gradually varying area, are automatically generated.
The attribution of the different physical parameters to the elements and nodes of the grid is done automatically by means of digitalized thematic maps. A practical application of this methodology is shown by Pizzi G. and Sartori L. (1982) with the model of the western Libya regional aquifer system.

## DIGITALIZED THEMATIC MAPS

The digitalized thematic maps are defined on the basis of generation units, arranged hierarchically by means of a hierarchy graph.Every generation unit is formed by a simply connected area, whose boundary is represented by a closed line of straight segments. To each generation unit is given a value of the variable. In the region between two areas, is given the value pertaining to the outermost one.
Fig. 1 gives an example of a digitalized thematic map; Fig. 2 shows the hierarchy graph which organizes the generation units.

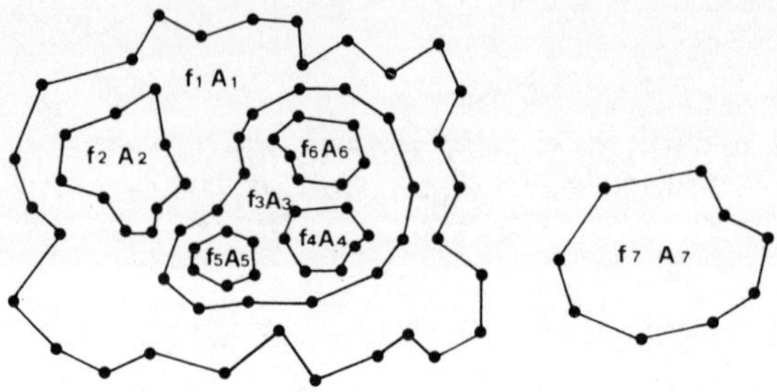


Fig. 1-Example of a digitalized thematic map. $A_i$ indicates various generation units; $f_i$ the pertaining values.

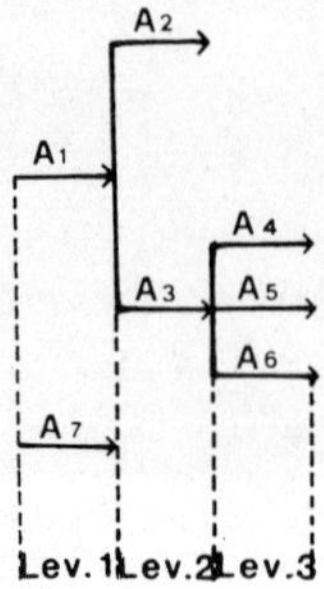


Fig. 2-Hierarchy graph of the generation units related to the thematic map

BASIC ALGORITHMS

The methodology used for the automatic enquiring of a digitalized thematic map is based on three algorithms.

Bresenham algorithm

This algorithm (Bresenham 1965) allows the approximation of a straight segment which connects any two points of a regular square grid, by means of a sequence of vectors selected among the 8 vectors which form the base.

Fig. 3 illustrates how this algorithm consents the approximation of the segment connecting the origin of the axes with a point P = (5h, 3h) of the network(h is the network step).

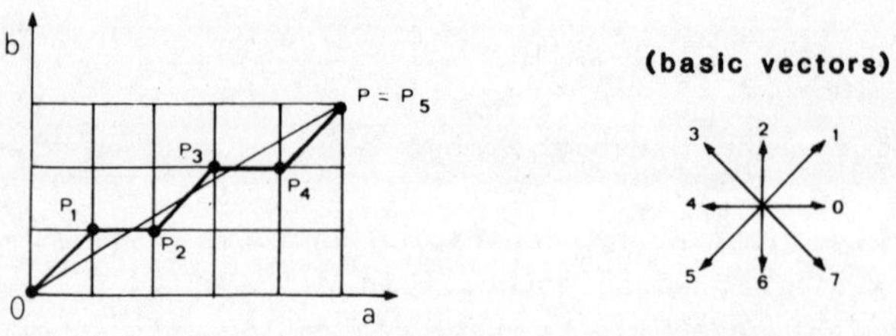


Fig. 3 - Approximation of a segment with the Bresenham algorithm.

Merrill Algorithm

This algorithm (Merrill 1973) tests the location (inside, autside or on the boundary) of a square grid vertex point, in relation to a simply connected area, whose boundary is defined on the same square grid, by a sequence of elementary vectors as considered by Bresenham (fig. 4).

This is done by analyzing the "y-partitions", defined as the list of boundary nodes intersected by a straight line parallel to the x axis, arranged according to the increasing value of the x coordinate.

In order to eliminate same critical aspect related to the method, same boundary points are doubled in the "y-partitions" (the larger points of fig. 4).

Now given on the square grid a point P of integer coordinates (a, b) we determine in the "y-partition" y=b two consecutive points $a_i$ and $a_{i+1}$ such as $a_i \leqslant a < a_{i+1}$ (defining $a_o = -\infty$, $a_{s+1} = +\infty$, s the number of points in the "y-partition" y=b).

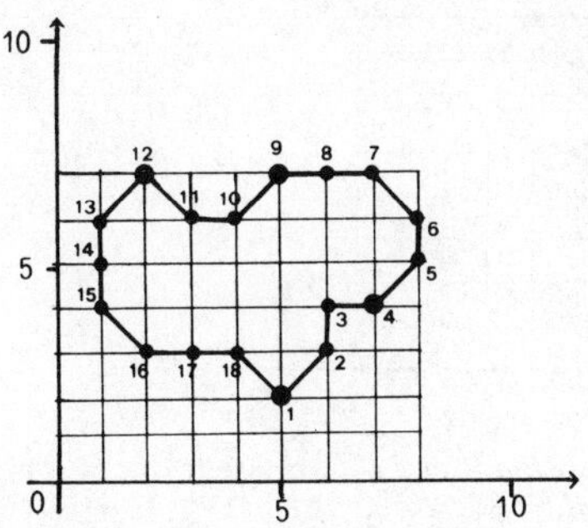


Fig. 4 - Simply connected area on a square grid.

Follows:

- Point P is on the boundary if $a = a_i$
- Point P is internal if i equals an odd number
- Point P is external if i is even.

In addition Merrill gave a criteria to determine the boundary point of the area, closest to a given point P located on the square grid.

Algorithm for the intersection of segments with the area's boundaries

This algorithm tests if a straight segment parallel to axes or to the axes bisectors, intersects with the boundary of a simply connected area, analyzing the x and y partitions and the

diagonal ascending and descending ones.
As for the y partition the x partition is defined in relation to the straight line parallel to the y axis while the ascending and descending diagonal partitions are related to the respective bisector axes.
We define the partition interval as the interval determined by 2 subsequent points of the same partition. Considering a segment parallel to the x axis, this segment intersects the area's boundary, both if one of the two end points belongs to the boundary, or if the two end points are located on different partition intervals of the same y or x partition.
The application of the above criteria to segments parallel to diagonals, is not valid because there is a phenomenon of transparency which conceals the intersection with the boundary.

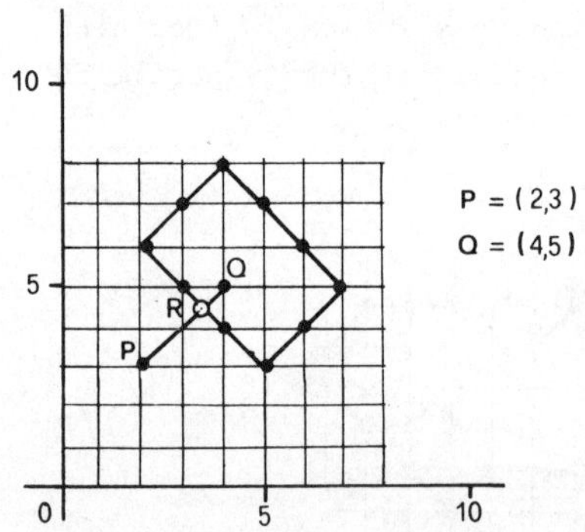


Fig. 5 - Boundary transparency for segments parallel to the bisector of the first quadrant.

The transparency phenomenon as shown in fig. 5 is due to the intersection of the PQ segment with the area's boundary at a point R which is not a node of the square grid.
The transparent boundary point P relative to the ascending diagonal $y = x + \alpha_o$ ($\alpha_o$ entire) is present if and only if the ascending diagonal partitions $\alpha' = \alpha_o + 1$ and $\alpha'' = \alpha_o - 1$ exist and in them two points of x coordinates $a'_i$ and $a''_j$ respectively, with $a''_j = a'_i + 1$, are present.
For the descending diagonal $y = -x + \beta_o$ the situation is similar with $\beta' = \beta_o + 1$ and $\beta'' = \beta_o - 1$. For the points in the diagonal partitions, it must be now $a'_i = a''_j + 1$. In both cases the x coordinate of the transparency point is given by $x = (a'_i + a''_j)/2$.

In order to re-establish the validity of the segment's intersection criteria, it is necessary to carry out a preliminary examination of the boundary's transparent points, for the ascending and descending diagonals, and to enlarge the ascending and descending diagonal partition by including the transparent points so found.

## AUTOMATIC CONSULTATION OF A DIGITALIZED THEMATIC MAP

Thematic maps as previously defined must be treated so as to prepare them for the automatic consultation algorithm. The generation units' boundaries are processed with the Bresenham algorithm. Subsequently, the y-partitions are prepared.

At the same time, for each generation unit it is necessary to define a containing rectangle (the minimum rectangle with sides parallel to the axes, which contain the unit). In the thematic map utilized for the two-dimensional network generation, the most external generation units, which represent the boundaries of the network generation area, are destined to be examined even in relation to a possible intersection with the segments. Even the x-partition and the enlarged diagonal partitions both ascending and descending, are predisposed for these units.

To evaluate if a point P is located within a generation unit of the thematic map, we examine first the containing rectangle of the units, and if the point P is inside it; we also examine the unit's y-partition with the Merrill criteria.

To point out the assigned value in P of the function, it is necessary to first locate the outermost generation units containing the point. Subsequently the units included in it should be pointed out in order to give to point P the value pertaining to the function, by means of the criteria originating from the hierarchic organization of the generation units.

## GENERATION OF A BI-DIMENSIONAL NETWORK

Generation of a bi-dimensional network occurs starting from the network generation map (which can be obtained with the aid of a digimeter).

The function represented by this map is the network generation parameter $\delta$, defined as the number of elements contained inside a reference area assumed as base element. Said parameter can take on, as a value, every power of 2. The network generation map is joined to the no-crossing lines, which represent the

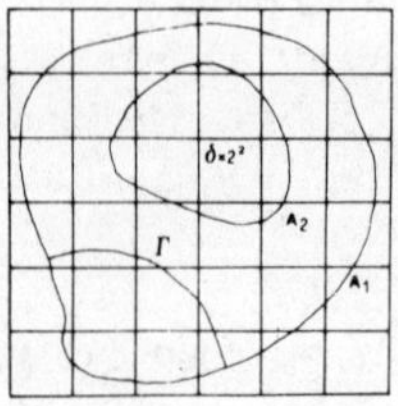


Fig. 6 - Network generation map

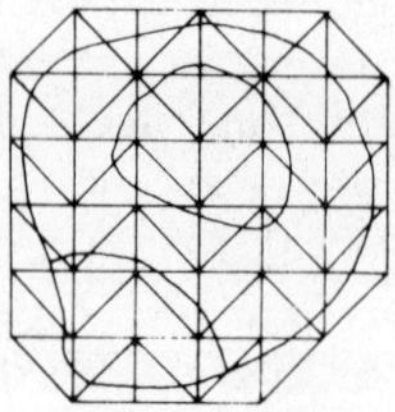

Fig. 7 - Basic network

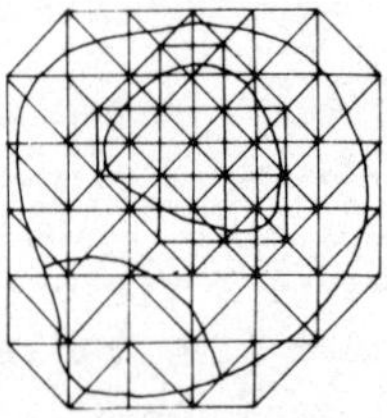

Fig. 8 - Thickened network

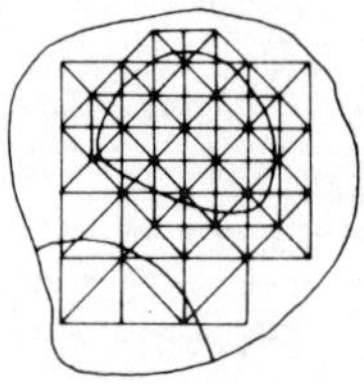

Fig. 9 - Reduced network

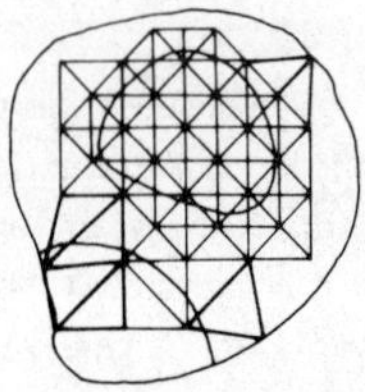

Fig. 10 - Node transportation

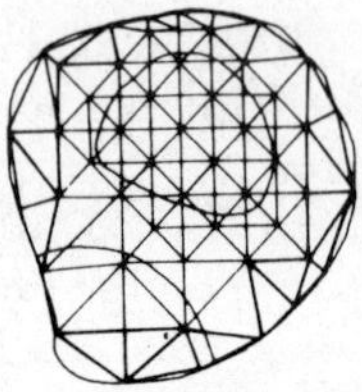

Fig. 11 - Boundary completation

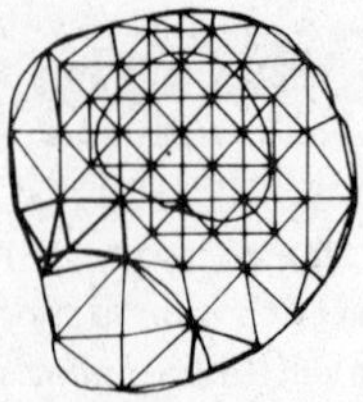

Fig. 12 - No-crossing line

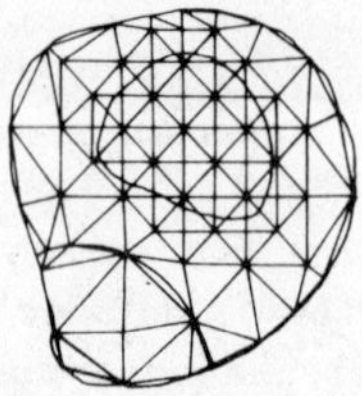

Fig. 13 - Boundary refinement

lines which the network must respect and reproduce.
An example of a network generation map is given in Fig. 6. This esample gives the real boundaries of the problem. Said boundaries are at first approximate with straight segments lines and then treated with the Bresenham algorithm. In this figure $\Gamma$ represents a no-crossing line. The square grid defines the reference mesh.
Generation of a bi-dimensional network occurs in three phases.

1st phase - Generation of a basic network

The network generation map is initially predisposed, for automatic interrogation, based on the Bresenham step equal to one-eighth of the smallest side of the triangle to be generated. One by one, all the basic square meshes (fig. 5) will be examined to point out those within the generation area, or those which intersect the boundary. The analysis is carried out by utilizing the Merrill algorithm and the algorithm for the analysis of segment intersection. The internal squares (with the vertexes inside the area and no side intersecting the boundary) generate two internal triangles, selecting the diagonal every time, by employing the alteration criteria. The boundary squares (having at least one side which crosses the generation area boundary) generate at least one boundary triangle. The other triangle could be external (and is thus excluded from the network), or internal or on the boundary.

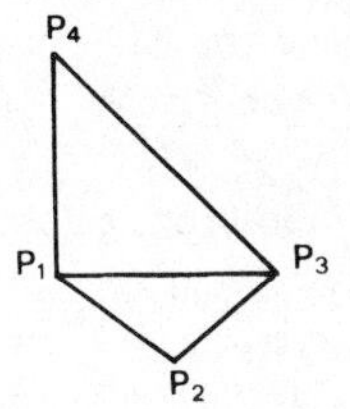


Fig. 14 - Adjacent triangles with a hypothenuse-cathetus contact.

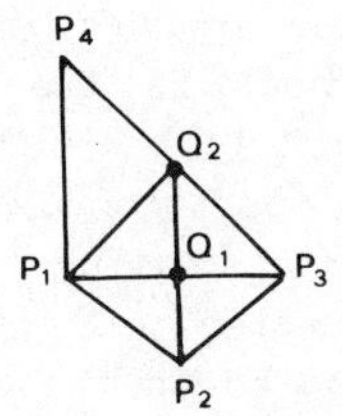


Fig. 15 - Their subdivision

Fig. 7 illustrates the entire set of triangles of the basic network so obtained for the example of Fig. 6.
Triangles generated in this manner are processed for subsequent subdivisions until the density requirements of the network generation map have been satisfied. A triangle is declared

suitable for subdivision when at least one of its nodes is inside an area which requires this subdivision.
In the subsequent subdivision of triangles, situations like those of Fig. 14 and 15 can occur.
All triangles of the same form are generated.
This subdivision procedure brings to the generation, in a highly dense area, of a halo of connecting elements having a fairly variable area. Each triangle generated in this manner is adjacent only to elements whose area is half, the double or the same. Fig. 8 illustrates the evolution of Fig. 7's network, after the two subsequent thickenings required by the network generation map of Fig. 6. All the elements generated by these subdivisions are classified as internal or on the boundary. Those which are external are excluded from the network.
The subdivision of internal triangles gives internal triangles. The subdivision of boundary triangles can generate internal, external or boundary triangles, whose classification requires utilization of the Merrill algorithm and of that for the intersection of segments with the area's boundary.

<u>2nd phase : adapting the network to the area's boundary</u>
During the second phase all the so-called boundary triangles are excluded from the network. Thus a reduced network is obtained (Fig. 9). At this point the boundary nodes of the reduced network are examined, and for each of them the closest boundary point is located (Merrill algorithm), and the proper distance is estimated. If this distance is less than one-third of the shortest side which converges onto the boundary node, the node is transported on the real boundary, at the point nearest to the node (Fig. 10).
This operation involves a slight deformation of the triangles without causing critical situations. At this point on the generation unit's real boundary, a number of new nodes are generated on the network. This occurs in correspondance to the real boundary point closest to the boundary nodes of the reduced network, which has not been transported on the real boundary. These new nodes determine triangles or trapezoids. The latter are divided into two triangles, choosing between the two diagonals the one which is shortest (Fig. 11).

<u>3rd phase : Attribution of the no-crossing lines</u>
As a first step, it is necessary to point out the triangles crossed by the no-crossing lines, defined by the vertexes of the straight segment line $P_1 \ldots P_m$, which approximate the real

line.
A preliminary analysis is made to locate a triangle of the network containing point $P_1$, the line's origin. This triangle is pointed out by locating first the triangles which contain the point inside its containing rectangle.
Subsequently the equations of the oriented sides of these triangles are examined to find the one that includes the point. Triangle $T_1$, pointed out in this manner, forms the first triangle of the sequence of triangles crossed by the no-crossing lines. Pointing out the subsequent triangles of the sequence occurs after having pointed out the side of the triangle crossed by the line.
In this manner all the crossed triangles and the cross points on their sides are pointed out. If one side is crossed in more than one point because the line is idented, only one of these points is taken into consideration (the last one). At this point it is necessary to examine the sides of the triangle crossed by the line. If the distance from the cross point to the end points is less than one-fourth of the side's length, the nearest node of the crossed side in trasported to the cross point. The remaining crossed tringles are then subdivided on the basis of the cross-point (Fig. 12).
At the end the network boundary sides are analyzed, and for each boundary side it is necessary to locate the point farthest from the real boundary. If this distance is more than one-fourth of the side's lenght, another triangle is created by determining a point of the real boundary farthest from it (Fig. 13).

## ATTRIBUTION OF PHYSICAL PARAMETERS TO THE NODES AND TO THE NETWORK'S ELEMENTS

The physical parameters (transmissivity, thicknesses, distributed source terms, etc) are attributed on the basis of digitalized thematic maps, still using the Merril and Bresenham algorithms.

## GENERATION OF A QUASI-THREE-DIMENSIONAL NETWORK AND AUTOMATIC ATTRIBUTION OF VARIABLES

The generation of a quasi-three-dimensional network occurs in two phases. In the first one, using the methods above illustrated a reference two-dimensional network is generated on the area covered by the aquifer system, starting from the network generation map.

Subsequently the quasi-three-dimensional network is created. The aquifers are composed by substructures. The elements which define each substructure are pointed out starting from every substructure boundary (which must define a simply connected area). These substructure boundaries are expressed by means of a closed line formed by reference two-dimensional network sides.
The boundary of each substructure is processed with the Bresenham algorithm. The elements of the reference two-dimensional network are analyzed one by one. By employing the Merrill algorithm it is possible to point out those elements whose barycentres are located inside the substructures. The element is thus attributed to these substructures. Subsequently, the aquitards are generated.
Generation of aquitards could occur automatically, based on the aquifer's topology and by assigning an aquitard macroelement to each couple of aquifer nodes, located on the same vertical but on different levels. Other aquifer nodes must not be located between them.
As an alternative, the generation could occur on the basis of thematic maps which define its field of existance by employing the Merrill algorithm.
Attribution of the physical constant to the quasi-three-dimensional network occurs using the same procedure illustrated for the bi-dimensional case. This procedure operates level by level on the aquifers and, on the aquitards, by thematic maps.

## ELEMENT SEQUENCING AND RESEQUENCING

Substructures originate from the reference plan network, maintaining for the elements, the same sequence which they present within the above-said network.
This sequence is obtained by SORT on the barycenter's coordinate of the elements.
This method usually provides for the substructures, an arrangement which is suitable for the frontal method. An ulterior improvement of the front is obtained by employing an algorithm for resequencing the substructure's elements. This algorithm originates from the Akin Pardue algorithm(1976) and takes into account the substructures within the frontal method.
When the substructured frontal method is employed, the nodes of the substructure boundary, common also the other substructures, remain until the end of the assembly process, once they have appeared on the active variables front. To these nodes is given

a negative number.
We overcome this difficulty by delaying the appearance of negative node in the first fase of the Akin methods (node renumeration) taking into account the positive ones, first.
The method obtained in this manner, as for the original Akin Pardue one, is a semi-empirical method and does not provide the optimal sequence. Considering this fact, after having applied this algorithm, the average linear front is evaluated and compared with that of the previous situation. Resequencing of the substructures is carried out only if it has really decreased the front.

## REFERENCES

Akin J.F., Pardue R.M. (1976). Elemental resequencing for frontal solutions.
The mathematics of finite elements and applications II MAFELAP 1975. Proceedings of the Brunel University Conference of the Institute of Mathematics and its applications held in April 1975. Academic Press.

Bresenham J.E. (1965) Algorithm for computer control of a digital plotter.
IBM System Journal Vol. 4 n° 1.

Irons M.B. (1970) A frontal solution program for finite element analysis.
Int. J. for numerical methods in engineering. Vol. 3. 5-32.

Merrill R.D. (1973) Representation of contour and regions for efficient computer search.
Com. Of A.C.M. Vol. 16, February.

Pizzi G., Sartori L. (1982) Interconnected GROundwater System Simulations (IGROSS). Description of the system and its applications to the Western Libya regional aquifer.
4th Int. Conf. on Finite Elements in Water Resources. Hannover.

Sartori L., Peverieri G. (1982) A frontal method based solution of the quasi-three-dimensional finite element model for interconnected aquifer systems and fluid mass balance evaluation. Steady and unsteady equations.
4th Int. Conf. on Finite Elements in Water Resources. Hannover.

# PROCEDURE OF MESH DATA GENERATION IN CODES FOR THE ANALYSIS OF VISCOUS FLOW

Sebastián, M.A.; Pérez, J.M.; Sánchez, A.M.

Dpto. Tecnología Mecánica. E.T.S. Ing. Industriales.
Universidad Politécnica de Madrid.
José Gutierrez Abascal, 2. Madrid-6. Spain.

## SUMMARY

This paper describes the main characteristics of a code for finite element mesh generation. The objective of the code exposed is to generate every coordinates of each nodal point and the conectivity among these points with minimun effort by the user.

The philosophy of the method is that the only data it requires are contour data, which in some single cases consists in a very few parameters. In the discretization of a plane shape with symmetries respect to two axis, it is begining with a mesh that have a rough likeness with the original shape.

It is given a motions to the outside nodes in such a way that they go to the positions that define the discretization of the shape. This plane results, are generalized to spatial forms in which there is a line along which the sections varies as it is usual in fluid flow.

## INTRODUCTION

During last years, several studies about metalforming processes with large plastic deformations has been carried out in the Department before quoted. One of the most powerful ways for these studies is the finite element method, which gives information about forces, stresses and deformations or speeds simultaneously.

It have been implemented a code for bidimensional purposes, in which metallic materials that suffers plastic flow are treated like a viscous non-newtonian fluid under incompressibility conditions.

The procedures are based on the analogy between the mathematical models of viscous flow and non-linear elasticity. There were several published works, such as those of Zienckiewicz (1977) and (1978); and Owen (1980); which could be contrasted in their application to hot forming processes with predominant compresive stresses.

One difficult in the use of a FEM code is the complexity of data preparation, it is due to the high number of routinish operations. The input data of a FEM code may be grouped in the following terms:

- identification and control data
- geometric data
- boundary conditions data
- material behaviour data

The first and fourth groups, do not present important problems since the number of materials used in the field here concerned, is generally only one; and similar reasons can be exposed about general data and control data. However, the second and third groups, relating to geometric data and boundary data can be very large if the discretization employed is close.

The procedure presented for the automated generation of geometric and boundary data, is based on two steps. First step deals with a number of subprograms for solving certain problems. Second step has a more complicated planning that allows to solve these problems with more generality from the results obtained in the first step. Inmediately afterwards it will be exposed both steps, and it will be shown some results of each one.

## FIRST STEP OF DISCRETIZATION

As it have been mentioned previously, the specific application for which it have been required an automated procedure of discretization, was the analysis of the viscous flow present in the strained zone of plastic deformation forming processes. Among these, extrusion was the process studied due to the very large deformations in it concerned.

The fact of treat with discretization problems in deformation zones of extrusion processes, lead to shapes like that showed in figure 1. In this figure are presented the two basic shapes, which cover an high percentaje of the possible morphologies that is used to employ. Shadow zones in figure 1, represents areas to divide in elements of the type chosen for the resolution by FEM.

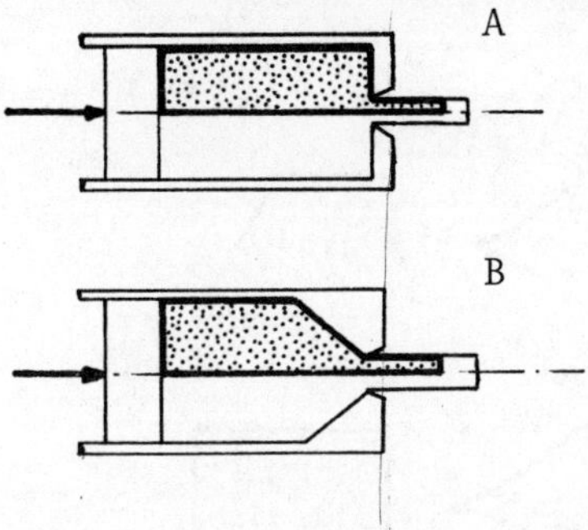


Figure 1. Types of configurations

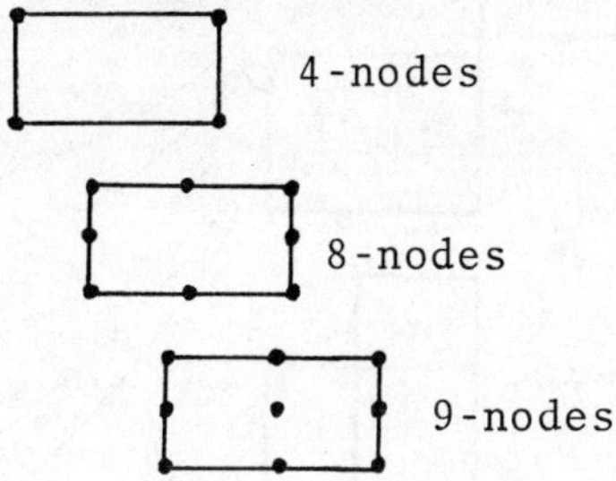


Figure 2. Types of elements

In bidimensional problems, the element used has been the nine-node quadrilateral, nevertheless the procedure can be adapted to other types of elements such as those shown in figure 2, that is, four-nodes element, eight-nodes element, and nine-nodes element. The local numbering of the element nodes has been given with the anticlockwise criterion.

Specific programs for the shapes shown in figure 1 has been adopted due to the reduced number of types that present the domains involved. These programs for automatic discretization should accomplish the following specifications:

- easy preparing of input data
- capability to modify some process parameters
- to have several levels or steps of discretization
- automated generation of nodal coordinates
- automated generation of element conectivities
- complete or partial, automated generation of boundary conditions
- a relatively low incidence on execution time

Within these specifications, it has been developed a program with two options named A for configurations such as that showed in part A of figure 1, and B for that lettered in this way in figure 1.

In figure 3 it can be seen the domain to divide in elements, and the zones I, II, III and I', I", III' considered for the discretization process.

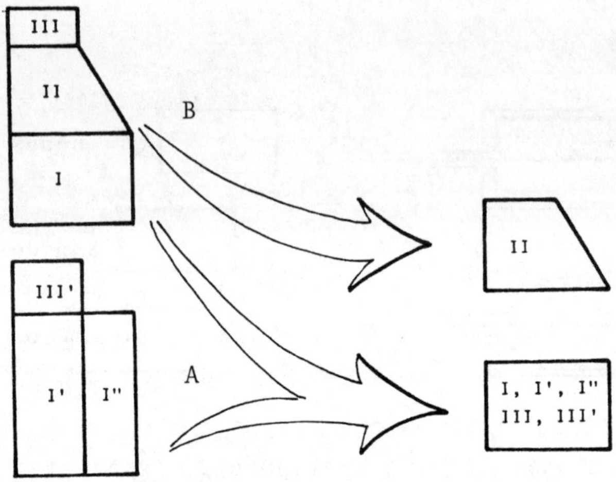


Figure 3. Configurations type A and B.
Method of discretization.

The sole difference between form A and B, is the presence of a trapezoidal area in the latter, and their connection configuration.

In order to have a more flexible use of this program, it had been developed three types of discretization. The discretization type is chosen by means a control variable. The first type (figure 4) is based on the use of equal elements in every rectangular area, and similar elements on the trapezial area. This is a very simple discretization and it is not capable of reproduce the strong variations in deformations that occur in a lot of cases, which requires the use of a more close division.

This problem is solved in part with the use of discretization type 2 (figure 4), since its elements are smaller and can accomplish such variations. The element size increase as they are away from the zones of strong gradients, the dimensions of elements varies with a geometric series whose base can be chosen in the input data preparation.

The use of base values close to 1.5 has produced distributions of element sizes very suitable to the configurations studied, according to the results obtained.

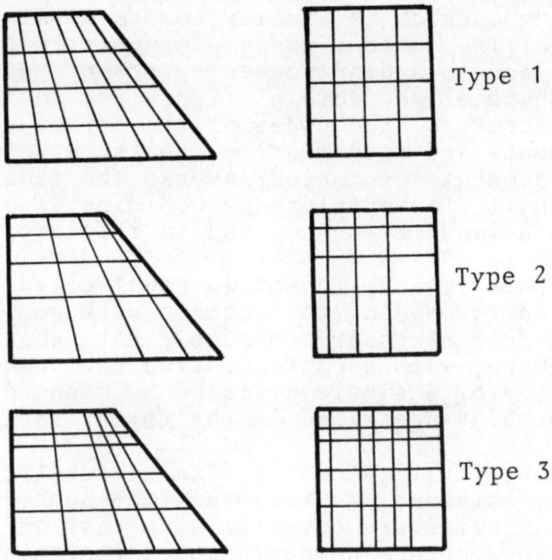


Figure 4. Distributions of element sizes in the deformation zone.

But in several cases, it is necessary to create a specific distribution of element sizes, from what it have been implemented the third type of discretization (figure 4) in which the user defines the relative sizes of elements.

SECOND STEP OF DISCRETIZATION

Let us consider the example presented in figure 5.

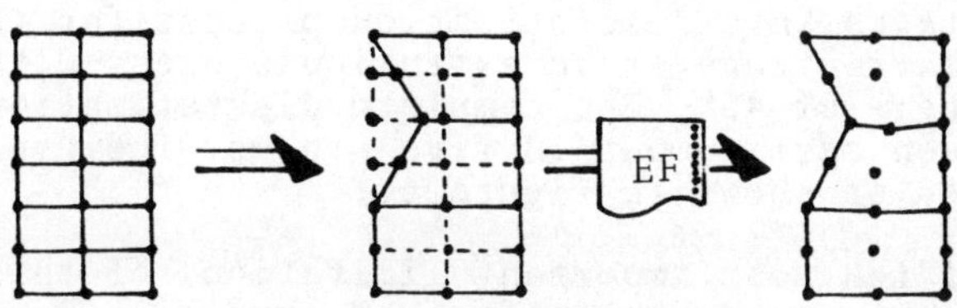


Figure 5. Diferent stages in the discretization.

Part "c" represents the domain to divide into elements, obtained from a structure like shown in part "a", which is similar to the domain but is more simple. The contour nodes of such structure are subjected to a displacements set in order to adopt the shape of the domain (figure 5b). Owing to these displacements, the nodes of the structure will suffer movements and they must be consistent with the displacements of contour nodes. The final location give us the internal nodes coordinates of bidimensional elements used in this discretization.

With this method, to obtain nodal coordinates of the mesh corresponding to a domain with complex shape, it is only necessary to supply data about a bar structure, with a configuration that can be generated directly by a simple division by means of some of the methods described in the former section.

The conectivities of the final discretization are easily obtained provided that the number of nodes of the bar structure coincide with that of the domain and the secuence numbering of the structure nodes be in consistency with the criterion chosen for numbering nodal points. It is necesary to prepare an algorithm for each element type.

The procedure used to pass from the situation represented in figure 5b to 5c, it is based on a finite element program prepared for such purpose.

When the values of the displacements given to the contour nodes are of the same order that dimensions fo elements, it is necessary to give such displacements by means of consecutives increments. This complicates the procedure since the finite element program must be executed several times, using as starting data the results obtained from the previous execution.

In figure 6 they are shown the results obtained from odd stages in a discretization process for the deformation zone of an extrusion process with semiangle of 45°. The complete discretization process has been carried out in five stages. The final state reached is shown in figure 6c.

One of the most important limitations of this method is that imposed by the element type considered, since elements can be distortioned, what can be harmful in the use of finite element codes for elasto-plastic analysis, and can arise situations in which there is a lack in the unicity of the change from natural coordinates to global coordinates.

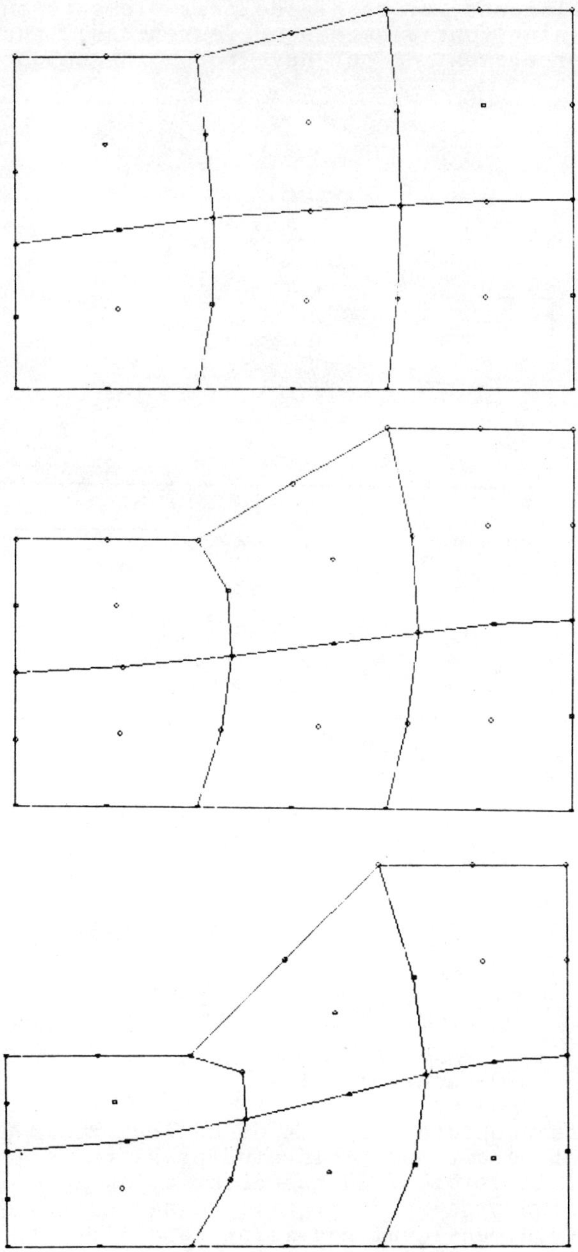

(a)

(b)

(c)

Figure 6. Discretization after stages 1, 3 and 5.

An example may be the element stood out in figure 7. However, it is foreseen adopt corrected solutions for that elements, or to divide that elements in four, (obtaining four elements of four nodes), though this type of element do not have to be the unique. (Figure 7).

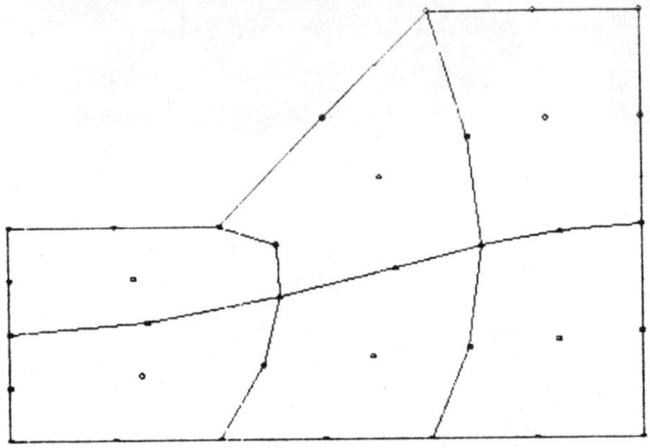

(a)

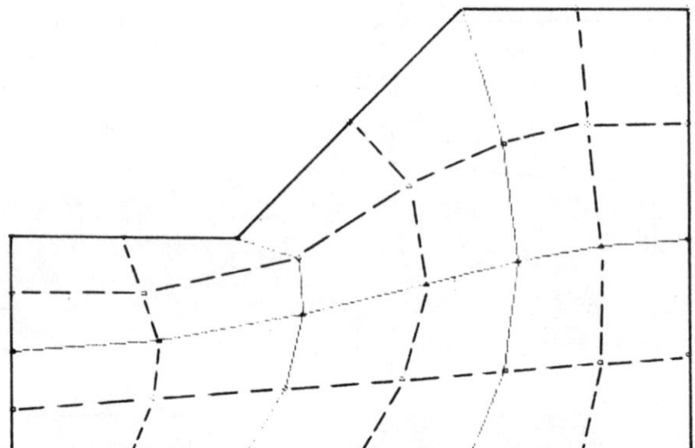

(b)

Figure 7. 45° extrusion. Case of element deformed (above) and subdivision in elements.

TRIDIMENSIONAL APPLICATIONS

For the complete analysis of the behaviour of viscous flow of metals under elevated pressures in convergent ducts, as it occur in the deformation zone of extrusion processes, it will be necessary to deal with tridimensional modelling capabilities. By this reason it has been undertaken the elaboration of a computer code for tridimensional analysis by finite

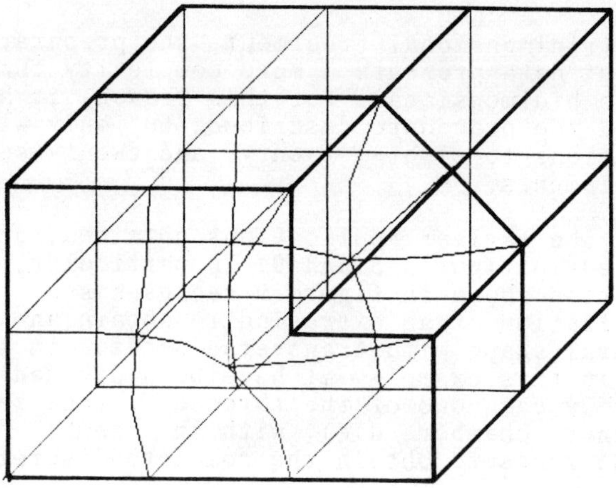

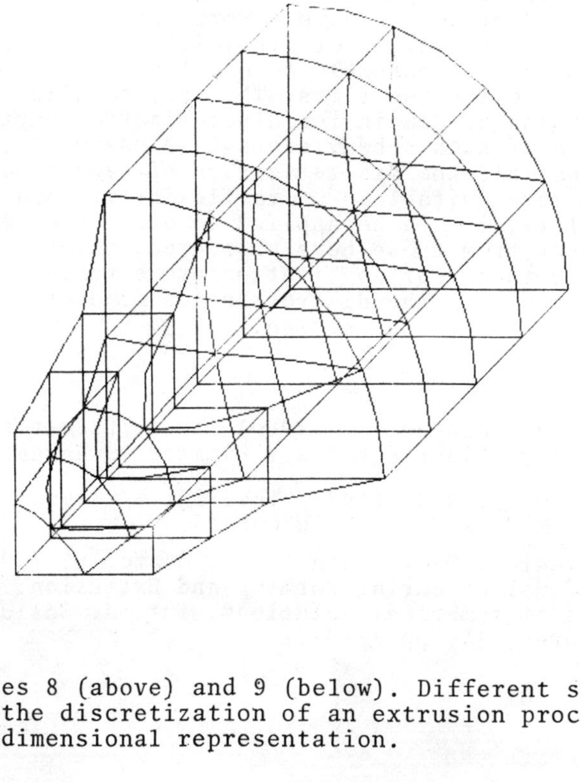

Figures 8 (above) and 9 (below). Different stages in the discretization of an extrusion process. Tridimensional representation.

elements, based in the code developed for bidimensional analysis.

In the tridimensional treatment, the preparation of the input data presents a more complexity than in the case bidimensional. For this reason, it have been employed the procedure described, but only with applications to eight-, twenty- and twenty-seven nodes elements.

Some of the earlier applications obtained, are presented in figures 8 and 9. In particular, the application shown in figure 9 represents a discretization of an extrusion to obtain an structural shape with transverse section in form of cross. In this example, it have been divided separately each one of the three different zones (container, chamber, die), with an assembly process of these zones to obtain the complete discretization.

## CONCLUSIONS

It is evident the necessity to have automated procedures for input data in finite element codes. The method presented have several steps of complexity with different levels of flexibility, since in simple applications is possible to use the direct steps with saving of execution times. The fact to allow subdivision of domain for discretization partial in each one of such subdivisions with posterior assembly, make possible the discretization of each zone by the method more suitable. Due to flexibility and generality, it can be applied to other problems different from those here presented; however it is possible that your application prove to be adventageous in the discretization of ducts with variable transverse section.

## REFERENCES

Owen, R.J.; Hinton, E. (1980) Finite Elements in Plasticity. Pineridge Press Limited. Swansea.

Zienckiewicz, O.C. (1977) The Finite Element Method. 3rd ed. McGraw-Hill. Maindenhead.

Zienckiewicz, O.C.; Jain, P.C.; Oñate, E. (1978). Flow of Solids during Forming and Extrusion: Some aspects of numerical solutions. Int. J. Solids Structures, 14, pp 15-38.

# 9. NUMERICAL TECHNIQUES IN GROUNDWATER FLOW

# SOME EXPERIENCES USING FINITE ELEMENT METHODS FOR FLUID FLOW PROBLEMS

K.J. Bathe, V. Sonnad and P. Domigan
Department of Mechanical Engineering
Massachusetts Institute of Technology
Cambridge, Massachusetts

## ABSTRACT

In this paper we report some of our experiences recently gained in the finite element solution of fluid flow problems. Firstly, an algorithm for transient free surface seepage problems has been developed. The advantage of this procedure is that the mesh is stationary with the free surface passing through the elements. Secondly, the solution of Navier-Stokes equations governing compressible flow with heat transfer has been researched.

## 1. INTRODUCTION

During the recent years a considerable research effort has been spent on the development of finite element methods for the solution of fluid flow problems. The success in using finite element techniques for various fluid flow problems has been demonstrated, but there remain a large number of research tasks, both with regard to the development of new and more effective procedures, as well as the testing and evaluation of methods that are already available.

The objective in this paper is to present some of our experiences that we have recently gained in the solution of fluid flows. Two kinds of problems are considered. First, we consider the solution of transient seepage problems. For these problems we have extended the algorithm that was previously proposed by Bathe and Khoshgoftaar for steady-state conditions[1]. The essence of this algorithm is that the seepage analysis is carried out without updating the finite element mesh. The approach used has in general a great deal to offer for free-surface flow problems.

The second kind of problems considered are those of Newtonian fluids in laminar flow with heat transfer. For the solution of these problems a more general finite element scheme has been developed, and in particular we have investigated the effect of operating on the compressible flow equations for the analysis of problems in which the fluid physically behaves as incompressible.

All the developments and experiences briefly reported in the following are part of our overall objective to develop a somewhat general finite element solution capability for fluid flow problems.

## 2. TRANSIENT FREE-SURFACE SEEPAGE ANALYSIS

The major problem in free surface transient analysis is the satisfaction of the mixed boundary condition at the free surface, whose location is unknown. Assuming that the flow is governed by the usual Laplace equation these boundary conditions along the free surface can in two-dimensional analysis be written as [2]

$$\phi(y,z,t) = h^{FS}(y,z,t) \tag{1}$$

and

$$-k_n \frac{\partial \phi}{\partial n} = S_y \frac{\partial h^{FS}}{\partial t} \cos\theta \tag{2}$$

where

$\phi$ = total potential,

$h^{FS}$ = height of free surface from reference level,

$k_n$ = permeability across free surface,

$S_y$ = specific yield,

$\theta$ = angle between differential element of free surface and horizontal axis,

and y,z,t are the coordinates and time variable, respectively. In steady-state analysis, the right hand size of Eq. (2) is zero, and for the solution of such problems an effective algorithm was presented in ref. [1].

The essence of our solution procedure is that the mesh is not changed or updated but instead the free surface is allowed to pass through the (stationary) elements. The free surface is located by an iteration which operates on Eqs. (1)

and (2) until--within the finite element spatial discretization and the time integration scheme used--both equations are satisfied for a certain convergence tolerance. The relation in Eq. (1) is imposed by testing where within the solution domain the value of total potential is equal to the height from the datum reference level, i.e. where the pressure is zero. This test is conveniently carried out at the Gauss integration points of the isoparametric elements used. If an integration point is above the free surface, a zero (or rather very small) material permeability is assumed so that no flow takes place. Along the free surface, the actual position of which is located as described in Section 2.1, the flow condition in Eq. (2) is imposed. The difference to the algorithm presented in ref. [1] is thus that we now need to impose the right-hand-side of flow given in Eq. (2), which was zero for the steady-state case.

### 2.1 Location of Free Surface for Imposition of Eq. (2)

The solution of the governing finite element flow equilibrium equations yields the total potential at the finite element nodal points. The free surface location is then evaluated using Eq. (1). Namely, writing Eq. (1) for a finite element as (with z measuring the height of a material point)

$$\phi - z = \sum_{i=1}^{n} h_i(r,s)(\phi_i - z_i) \tag{3}$$

where the $h_i(r,s)$ are the element interpolation functions and the element has n nodes, it follows that the condition to locate the free surface is

$$\sum_{i=1}^{n} h_i(r,s)\,(\phi_i - z_i) = 0 \tag{4}$$

This equation could be solved for the actual r and s values of the free surface within an element. However, in our work we simply assumeda linear variation of height of free surface between the points of zero pressure on the element edges.

### 2.2 Free Surface Flow Condition

To satisfy Eq. (2) the following flow is imposed along the free surface

$$\underline{Q}^{FS} = \int_{FS} S_y\, \underline{H}^T\, \underline{H} \cos\theta \, ds\, \dot{\underline{\phi}} \tag{5}$$

where the matrix $\underline{H}$ stores the finite element interpolation functions. We should note that the free surface (denoted as FS) over which the integration is performed passes through the element and is located as described in Section 2.1.

## 2.2 Total Element Equilibrium

Including the contribution from Eq. (5) the flow equilibrium equations for a typical finite element are

$$\left( {}^{t}\underline{K}^{k} + \underline{K}_{b} + \frac{1}{\Delta t}\; {}^{t}\underline{C}^{FS} \right) \Delta\underline{\phi}^{(i)} = {}^{t+\Delta t}\underline{Q}^{k(i-1)} - {}^{t+\Delta t}\underline{Q}^{FS(i-1)} - \underline{K}_{b}\; {}^{t+\Delta t}\underline{\phi}^{(i-1)} \qquad (6)$$

where we employ our usual notation [2] and the Euler backward time integration is used. Note that the matrix $\underline{K}_b$ is a result of imposing the essential boundary conditions around the domain of interest as described in ref. [1], and that a modified Newton-Raphson scheme is used to ensure flow equilibrium.

In Eq. (6) only the material below the free surface is considered, and when flow equilibrium for the whole finite element mesh is reached, all material and boundary flow conditions are satisfied. Hence, provided an adequate mesh and time step are employed, an accurate solution to the transient free surface problem is obtained.

## 3. LAMINAR NEWTONIAN FLOW WITH HEAT TRANSFER

In recent years much emphasis has been directed to the finite element solution of the Navier-Stokes equations corresponding to inviscid incompressible, inviscid compressible and viscous incompressible fluid flow, see e.g. [3]. Although our ultimate aim is to provide a formulation and solution scheme that can be employed for general flow conditions, so far we have considered only low Reynolds number fluid flow including heat transfer conditions. In the solution we have solved the compressible flow equations, but for the physical situations considered the fluid behaves as almost incompressible. Therefore, a specific objective of our work was to evaluate our solution results obtained with the more general formulation when compared with those obtained assuming incompressible flow conditions.

## 3.1 Governing Continuum Mechanics Equations

Using standard indicial notation, and a Cartesian reference frame, $x_1$, $x_2$, $x_3$, the motion of the fluid we consider is governed by the following equations:

Continuity:

$$\frac{\partial \rho}{\partial t} + (\rho v_i)_{,i} = 0 \tag{7}$$

Momentum:

$$\rho\left(\frac{\partial v_i}{\partial t} + v_j v_{i,j}\right) - \rho f_i^B - \tau_{ji,j} = 0 \tag{8}$$

Energy:

$$\rho c_p\left(\frac{\partial \theta}{\partial t} + v_i \, \theta_{,i}\right) + q_{i,i} = q^B \tag{9}$$

Constitutive:

$$\tau_{ij} = -p\delta_{ij} + \mu(v_{i,j} + v_{j,i} - 2/3 \, v_{r,r} \, \delta_{ij}) \tag{10}$$

State:

$$p = \rho R \Theta \tag{11}$$

in which:

$v_i$ = velocity component in direction $x_i$

$\rho$ = mass density

$f_i^B$ = component of body force per unit mass

$\tau_{ij}$ = component of stress tensor

$q^B$ = heat generated per unit volume

$\theta$ = temperature

$c_p$ = specific heat at constant pressure

$q_i$ = heat flow in direction $x_i$

$p$ = pressure

$\delta_{ij}$ = Kronecker delta

$\mu$ = fluid viscosity

k = thermal conductivity

R = ideal gas constant

Using the above equations we consider a compressible fluid in laminar flow subjected to heat transfer. We assume that the fluid is governed by the ideal gas law in Eq. (11). In the solution of Eqs. (7) to (11), we also use the following dimensionless numbers:

$$\text{Reynolds number} = \frac{\rho\, v_i h}{\mu}$$

$$\text{Prandtl number } P_r = \frac{c_p \mu}{k}$$

$$\text{Rayleigh number } R_a = \frac{g\alpha\ \Delta\theta\ h^3\ \rho^2\ \ c_p}{\mu k}$$

in which

g = constant of gravity

$\alpha$ = coefficient of thermal expansion

$\Delta\theta$ = characteristic temperature difference

h = characteristic length

## 3.2 Finite Element Formulation

To obtain the finite element governing equations we use in our formulation the primitive variables $\rho$, $v_i$, $\theta$ and the Galerkin method on Eqs. (7) to (9). We have considered only two-dimensional planar flow conditions although the formulation can directly be applied to axisymmetric and three-dimensional flows.

For solution we employ the 9-node element with biquadratic interpolations on coordinates, velocities and temperature,

$$v_l(r,s,t) = \sum_{k=1}^{9} h_k(r,s)\ v_l^k(t) \qquad (12)$$

$$v_2(r,s,t) = \sum_{k=1}^{9} h_k(r,s)\, v_2^k(t) \tag{13}$$

$$\theta(r,s,t) = \sum_{k=1}^{9} h_k(r,s)\, \theta_k(t) \tag{14}$$

and bilinear interpolation on mass density,

$$\rho(r,s,t) = \sum_{k=1}^{4} \tilde{h}_k(r,s)\, \rho_k(t) \tag{15}$$

This element and the interpolations have been selected based on the experiences available with the various finite elements currently in use. We note that the element can have a curved boundary.

In the Galerkin formulation we weigh the continuity, momentum, and energy equations with the respective $h_k$'s from Eqs. (12-15). Finally, we substitute for p from Eq. (11) to obtain the governing equations for a typical finite element

$$\underline{C}\,\dot{\underline{x}} + \underline{K}\,\underline{x} = \underline{R} \tag{16}$$

in which

$$\underline{C} = \begin{bmatrix} \underline{C}_{\rho} & & \text{zeros} & \\ & \underline{C}_{v_1} & & \\ & & \underline{C}_{v_2} & \\ & & & \underline{C}_{\theta} \end{bmatrix} \tag{17}$$

$$\underline{K} = \begin{bmatrix} \underline{K}_{\rho\rho} & \underline{K}_{\rho v_1} & \underline{K}_{\rho v_2} & \underline{0} \\ \underline{K}_{v_1\rho} & \underline{K}_{v_1 v_1} & \underline{K}_{v_1 v_2} & \underline{0} \\ \underline{K}_{v_2\rho} & \underline{K}_{v_2 v_1} & \underline{K}_{v_2 v_2} & \underline{0} \\ \underline{0} & \underline{K}_{\theta v_1} & \underline{K}_{\theta v_2} & \underline{K}_{\theta\theta} \end{bmatrix} \tag{18}$$

$$\underline{x} = \begin{bmatrix} \underline{\rho} \\ \underline{v}_1 \\ \underline{v}_2 \\ \underline{\theta} \end{bmatrix} \quad ; \quad \underline{R} = \begin{bmatrix} \underline{0} \\ \underline{R}_{x_1} \\ \underline{R}_{x_2} \\ \underline{Q} \end{bmatrix} \tag{19}$$

and the entries in the matrices $\underline{C}$ and $\underline{K}$ and $\underline{R}$ are defined in ref. [4].

## 4. SAMPLE SOLUTIONS

In this section we present the experiences of some analyses that have been performed using the above described solution capabilities. In the first two analyses we present solutions obtained using our algorithm for free surface calculations, and in the last problem we present an application of the solution of the Navier-Stokes equations with heat transfer.

### 4.1 Square Dam with Instantaneous Drop in Head

The square dam shown in Fig. 1 is modeled using two finite element meshes. The first mesh consists of 16 x 16 four-node linear interpolation elements and the second mesh consists of 8 x 8 eight-node quadratic interpolation elements. The dam is initially saturated and the upstream and downstream water levels are maintained at sixteen feet. The permeability to specific yield ratio is $\frac{k}{S_y} = 1$.

At time $t = 0^+$ the water level on the down-stream face falls to zero. Five time steps are used to reach the steady-state conditions. The results are shown in Fig. 2 and are compared with those obtained by Herbert using an electrical resistance network technique [5]. Good agreement is noted despite the fundamentally different methods employed to predict the movement of the free surface.

### 4.2 Transient Axisymmetric Flow to a Well

Figure 3 shows the two-dimensional axisymmetric well problem analyzed. The model consists of 6 by 6 four-node linear interpolation elements. The initial conditions are $\phi$ = 48 feet everywhere, corresponding to an initially full well with zero discharge. The permeability is $k = 0.1$ ft/hr and specific yield is $S_y = 0.3$. At time $t=0^+$ the water level drops at a prescribed rate as shown in Fig. 4. A similar problem was analyzed by Neuman and Witherspoon [6] who determined the rate of discharge from the well to be 39.045 $ft^3/hr$. Five time steps are employed, half the number used by Neuman and Witherspoon.

The solution results given in Fig. 4 are compared to those obtained by Neuman and Witherspoon. It is observed that the two response predictions agree quite closely.

### 4.3 Analysis of Natural Convection in a Square Cavity

In this solution we consider a square cavity filled with fluid. As shown in Fig. 5 the fluid is subjected to gravity and the vertical walls are at different temperatures causing a natural convection. The response of the fluid depends on the Rayleigh and Prandtl numbers in the incompressible case. In our analysis we assumed $P_r = 1000$ and $R_A = 1000$ and 10000. An 8 x 8 mesh of the 9-node finite element was used.

Figure 6 shows the response predicted in our analyses assuming an incompressible and a compressible fluid. It is noted that the temperature and velocity distributions predicted with the two assumptions are very similar.

A comparison of the results with those of Taylor and Ijam [7] is also given in Fig. 6. It is seen that these solutions compare very well with the response predicted in our study.

## REFERENCES


[1] Bathe, K.J., and Khoshgoftaar, M., "Finite Element Free Surface Seepage Analysis without Mesh Iteration", Int. J. Num. and Anal. Meth. in Geomechanics, Vol. 3, pp. 13-22, 1979.

[2] Bathe, K.J., Finite Element Procedures in Engineering Analysis, Prentice-Hall, Inc., 1982.

[3] T. J. Chung, Finite Element Analysis in Fluid Dynamics, McGraw-Hill Book Co., 1978, New York.

[4] Bathe, K.J., and Sonnad, V., "On Finite Element Analysis of Compressible and Incompressible Temperature-Dependent Fluid Flow", Proceedings, AIAA Conference on Computational Fluid Dynamics, Palo Alto, California, June 1981.

[5] Herbert, R., "Time Variant Ground Water Flow by Resistance Network Analogues", J. of Hydrology, Vol. 6, 1968, pp. 237-264.

[6] Neuman, S.P., and Witherspoon, P.A., Analysis of Nonsteady Flow with a Free Surface Using the Finite Element Method", Water Resources Research, Vol. 7, No. 3, June 1971, pp. 611-623.

[7] C. Taylor and A. Z. Ijam, "A Finite Element Numerical Solution of Natural Convection in Enclosed Cavities", J. Comp. Meth. in Appl. Mech. and Eng., Vol. 19, 1979, pp. 429-446.

ACKNOWLEDGEMENTS

We greatly appreciate the financial support of the ADINA users group and the Forschungszentrum GKSS, Geesthacht, W. Germany for this work.

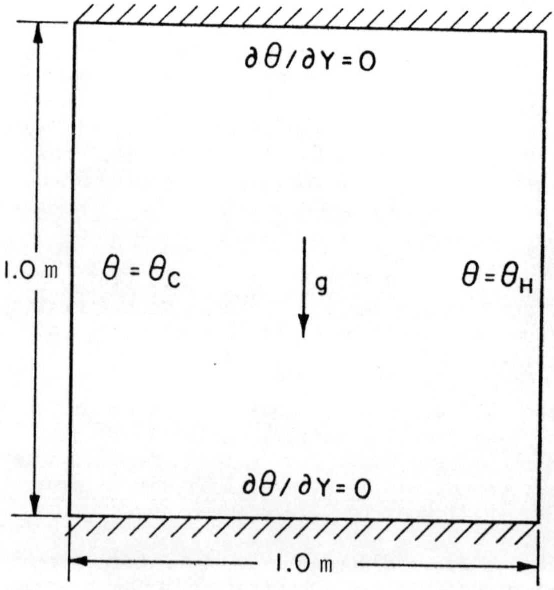


Fig. 5 Natural convection in a square cavity;
$P_r = 1000$, $\Delta\theta = 20°K$

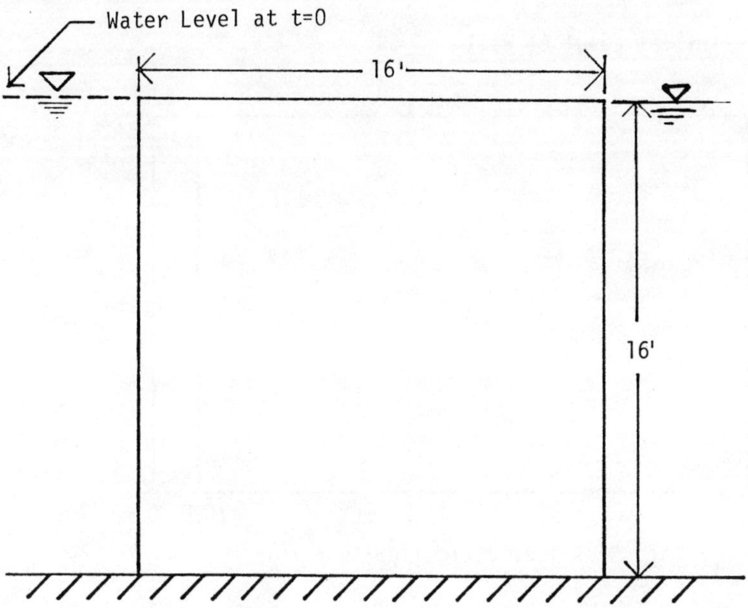


Fig. 1 A square dam with instantaneous drop in head

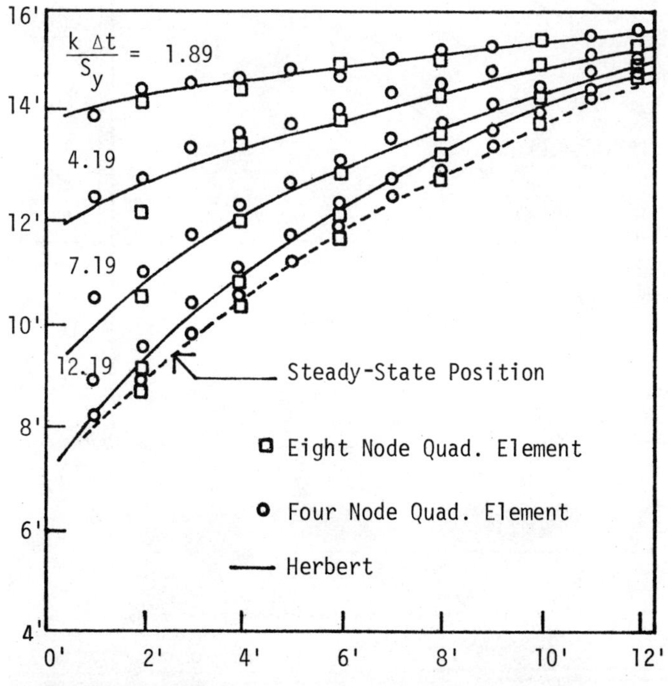


Fig. 2 A square dam solution: instantaneous drawdown

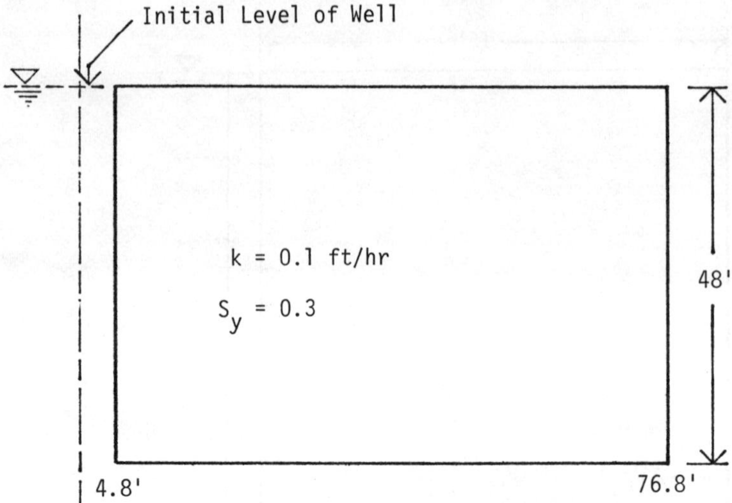


Fig. 3 Axisymmetric flow to a well

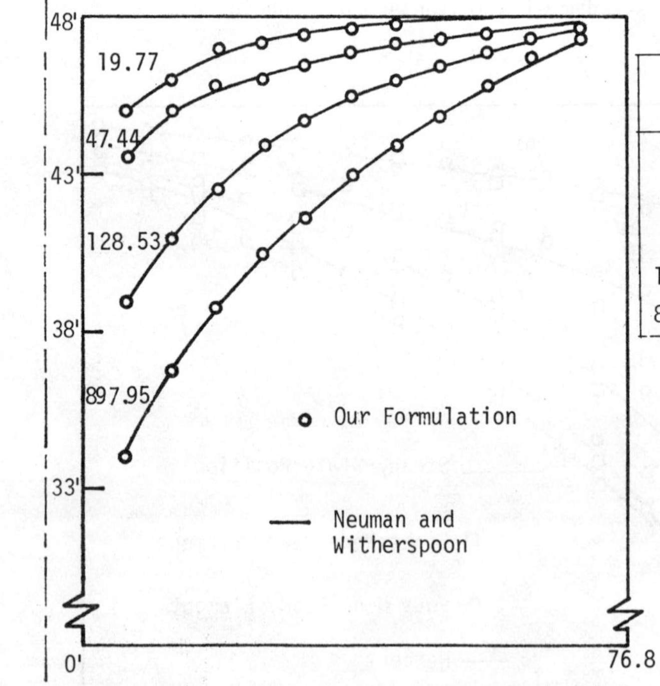


| t Hrs. | Level ft |
|---|---|
| 0 | 48' |
| 19.77 | 40' |
| 47.44 | 33' |
| 128.53 | 24' |
| 897.95 | 12' |

Fig. 4 A well solution: axisymmetric flow

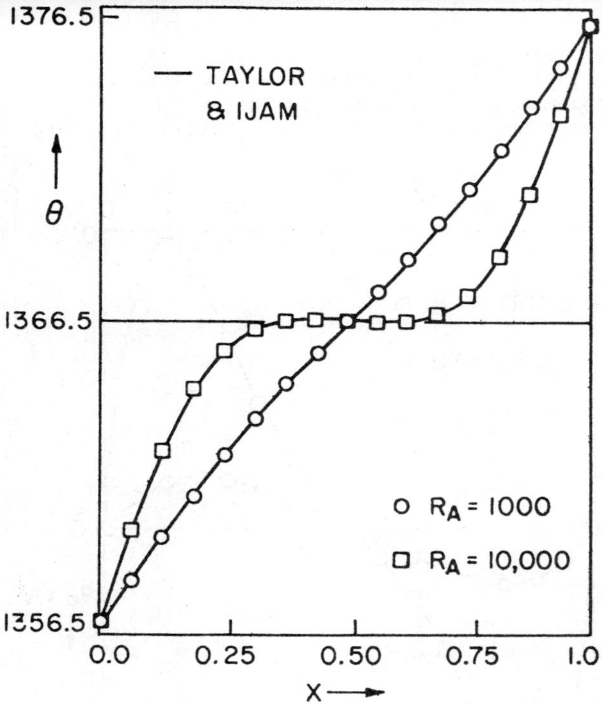


Fig. 6a Temperature along mid-height of cavity using incompressible flow equations. The solid lines represent the solutions of Taylor and Ijam.

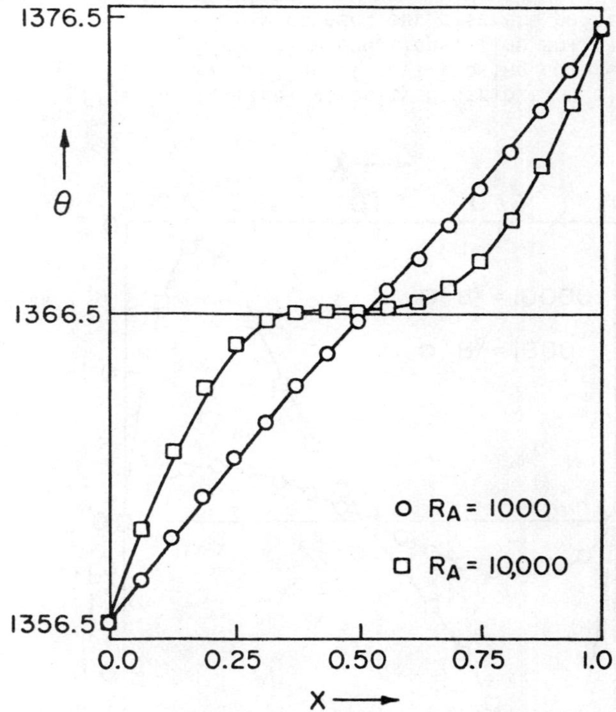


Fig. 6b Temperature along mid-height of cavity using compressible flow equations. The solid lines simply connect our discrete points of solution.

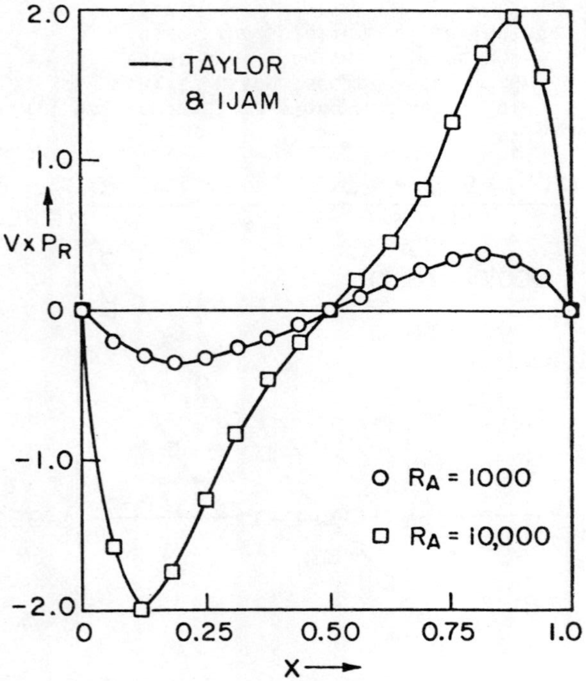


Fig. 6c Vertical velocity distribution along mid-height of cavity using incompressible flow equations. The solid lines represent the solutions of Taylor and Ijam.

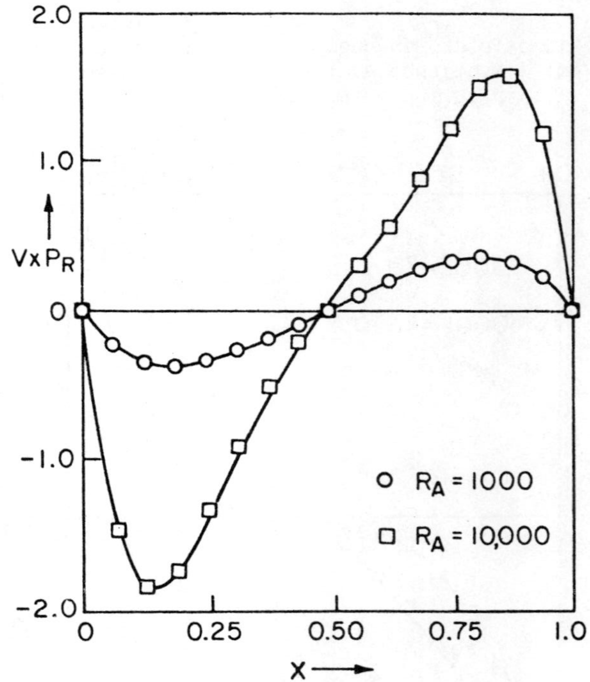


Fig. 6d Vertical velocity distribution along centerline of cavity using compressible flow equations. The solid lines simply connect our discrete points of solution.

# SOFTWARE DEVELOPMENTS FOR FINITE ELEMENT APPLICATIONS

R. Damrath

Technische Universität Berlin, Germany

### Abstract

An efficient iterative procedure for the analysis of unconfined three dimensional seepage flow is presented. The method is based on an invariant net of finite elements. The system equations are assembled and decomposed only once for the entire domain. Variations in the location of the free surface and the extent of the seepage surface are taken into account by slight modifications of the decomposed equations. The implementation of the method is described in the paper.

## 1. Introduction

The analysis of the unconfined seepage flow poses two particular problems: the location of the free surface and the extent of the seepage surface are not known a priori. If the problem is treated as a stationary boundary value problem, the boundaries of the seepage zone have to be determined within the analysis procedure. The location of the free surface and the extent of the seepage surface must be assumed and corrected iteratively.

Several iterative procedures for the unconfined seepage problem have been published. In [5–8] it is proposed to modify the finite element net in each iteration step. This method requires the assembly and the solution of the system equations in each iteration step and leads to a large amount of computational effort. The procedure for the modifications of the net are difficult if the seepage zone contains domains with different material properties. In some analyses, for instance stability analyses for embankments and earth dams, the finite element net for the static analysis comprises the entire body including the volume above the free surface. It is desirable to use identical nets for the static and the seepage analysis. This implies that the geometry of the net should not be depended on the location of the free surface.

In this paper, a consistent method for the analysis of unconfined seepage flow, based on an invariant finite element net, is presented. Contrary to the method proposed in [3,4], the errors in the assumed location of the free surface are not corrected directly by nodal fluxes. Instead, the permeability matrix of the complete domain is treated as a superposition of the matrix of the seepage zone and the matrix representing that part of the domain which does not form a part of the seepage zone. It is shown that the changes in the assumed location of the free surface and the assumed extent of the seepage surface are readily accounted for by small modifications in the decomposed system equations. The iterative solution can therefore be achieved with reasonable effort.

## 2. Finite Element Approximation

### 2.1 Physical Laws

The physical laws of the seepage problem are described. The primary unknown is the piezometric head.

$$h = z + \frac{p}{\gamma} \qquad (1)$$

$h$ = piezometric head
$z$ = elevation head
$p$ = fluid pressure
$\gamma$ = specific fluid weight

The fluid velocity is computed by Darcy's law.

$$\mathbf{v} = \mathbf{K}\,\mathbf{g} \qquad (2)$$

$$\mathbf{v}^T = [\; v_x \quad v_y \quad x_z \;]$$

$\mathbf{v}$ = vector of fluid velocity

$$\mathbf{g}^T = \begin{bmatrix} \frac{\partial h}{\partial x} & \frac{\partial h}{\partial y} & \frac{\partial h}{\partial z} \end{bmatrix}$$

$\mathbf{g}$ = piezometric gradient vector

$$\mathbf{K} = \begin{bmatrix} k_{xx} & k_{xy} & k_{xz} \\ k_{yx} & k_{yy} & k_{yz} \\ k_{zx} & k_{zy} & k_{zz} \end{bmatrix}$$

$\mathbf{K}$ = permeability matrix
$\mathbf{K} = \mathbf{K}^T$

The continuity equation is the governing differential equation for the seepage zone.

$$\frac{\partial v_x}{\partial x} + \frac{\partial v_y}{\partial y} + \frac{\partial v_z}{\partial z} = 0 \tag{3}$$

The boundary conditions are defined by specifying the piezometric head or the velocity normal to the surface.

$$v_n = \mathbf{n}^T \mathbf{v} \qquad \mathbf{n} = \text{outer normal vector of the surface} \tag{4}$$

Fig. 1 shows four different types of boundary surfaces:

| | | | | |
|---|---|---|---|---|
| Potential surface | P : | $h = h_o$ | | (5) |
| Impermeable surface | I : | | $v_n = 0$ | (6) |
| Seepage surface | S : | $h = z$ | $v_n \geqslant 0$ | (7) |
| Free surface | F : | $h = z$ | $v_n = 0$ | (8) |

While the potential surface and the impermeable surface are given, the seepage surface and the free surface are not known a priori. Therefore they are described by two conditions and are determined iteratively.

## 2.2 Variational Approach

Fig.2 shows a finite element discretization of an earth dam. The entire domain of the porous medium is subdivided into finite elements. An iterative algorithm for the computation of the free surface and the seepage surface is developed. The finite element net is not modified.

If the free surface and the seepage surface are known, the piezometric head and the flux can be computed on the basis of the following variational problem.

$$F = \frac{1}{2} \int_{V_s} \mathbf{g}^T \mathbf{K} \mathbf{g} \, dv + \sum_m q_m \, (h_m - z_m) \rightarrow \quad \text{stationary} \tag{9}$$

$h = h_p$ = prescribed piezometric heads on the potential surface (10)

$V_s$ = volume of the seepage zone

$q_m$ = flux across the seepage surface at node m

m = node on the seepage surface

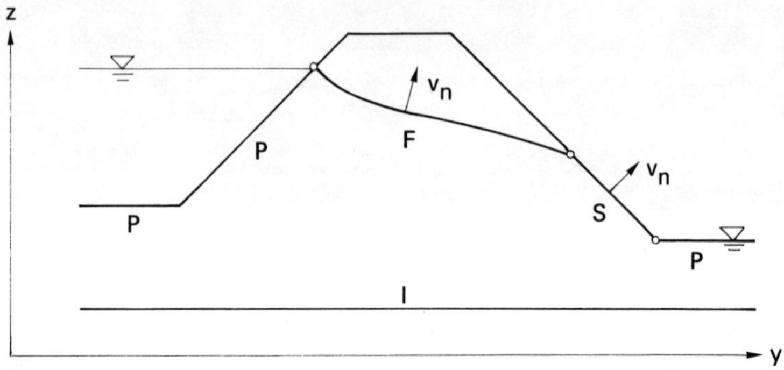


P = Potential surface

I = Impermeable surface

S = Seepage surface

F = Free surface

$v_n$ = velocity normal to the surface

Fig.1 Surface definitions for the unconfined seepage problem

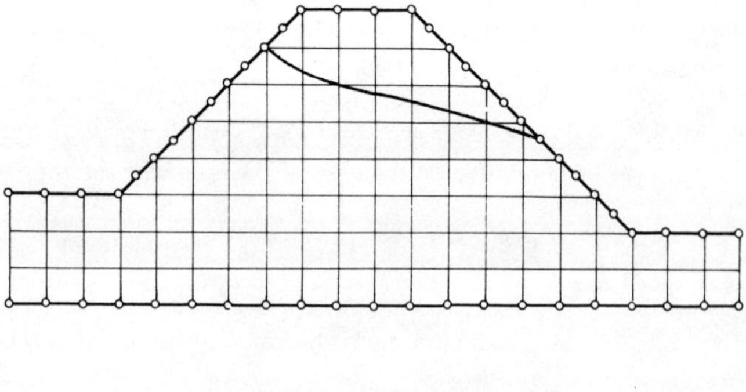

Fig.2 Finite element discretization with isoparametric elements

The boundary conditions $v_n = 0$ for the impermeable surface and the free surface are implicitly satisfied by the variational calculus. The first variation of the functional F with respect to the piezometric head and the flux must be zero.

$$\delta F = \int_{V_s} \delta \mathbf{g}^T \mathbf{K} \mathbf{g} \, dV + \sum_m \delta h_m q_m + \sum_m \delta q (h_m - z_m) = 0 \qquad (11)$$

$$\delta h = 0 \quad \text{on the potential surface} \qquad (12)$$

The finite element approximation accounting for the prescribed piezometric heads on the potential surface leads to the following system of linear equations:

$$\begin{aligned} \mathbf{C}_S \mathbf{h} + \mathbf{E}\,\mathbf{q} &= \mathbf{p}_o \\ \mathbf{E}^T \mathbf{h} &= \mathbf{z}_S \qquad \mathbf{h}_S = \mathbf{E}^T \mathbf{h} \end{aligned} \qquad (13)$$

$\mathbf{h}$ = vector of the nodal piezometric heads

$\mathbf{h}_S$ = vector of the nodal piezometric heads on the seepage surface

$\mathbf{z}_S$ = vector of the nodal elevation heads on the seepage surface

$\mathbf{q}$ = vector of the nodal fluxes on the seepage surface

$\mathbf{p}_o$ = vector due to the prescribed nodal piezometric heads on the potential surface

$\mathbf{C}_S$ = permeability matrix of the seepage zone

$\mathbf{E}$ = topology matrix

For the exact solution, the nodal fluxes are positiv or zero. The permeability matrix can be expressed by:

$$\mathbf{C}_S = \mathbf{C} - \mathbf{C}_F \qquad (14)$$

$\mathbf{C}$ = permeability matrix of the entire domain

$\mathbf{C}_F$ = permeability matrix of the domain above the free surface

The iterative algorithm for the determination of the free surface is constructed on the basis of equations (13) and (14).

$$\begin{aligned} \mathbf{C}\,\mathbf{h}_{i+1} + \mathbf{E}\,\mathbf{q}_{i+1} &= \mathbf{p}_o + \mathbf{r}_i \\ \mathbf{E}^T \mathbf{h}_{i+1} &= \mathbf{z}_S \end{aligned} \qquad (15)$$

$\mathbf{r}_i = \mathbf{C}_{Fi}\,\mathbf{h}_i$ = residial vector

$i, i+1$ = indices for iteration steps

The nodal piezometric heads $\mathbf{h}_i$ and the nodal fluxes $\mathbf{q}_i$ are known from the iteration step i. The free surface can be determined approximately from the condition $h = z$. The permeability matrix of the domain above the free surface $\mathbf{C}_{Fi}$ and the vector $\mathbf{r}_i$ are calculated. In the iteration step i+1, the corrected nodal piezometric heads $\mathbf{h}_{i+1}$ and nodal fluxes $\mathbf{q}_{i+1}$ are computed using equation (15). In each iteration step, the conditions on the seepage surface are checked. If the seepage surface changes during the iteration, the matrix $\mathbf{E}$ in equation (15) has to be modified. The iteration procedure is described in detail in the next sections.

## 2.3 Specifications of the Iterative Process

The following terminology is used in describing the iteration process:

Potential seepage surface

The potential seepage surface consists of all nodes through which the true seepage surface can pass. For each node on the potential seepage surface, the nodal flux is introduced as a variable.

Location of elements relative to the free surface

Fig.3 shows the different locations of the elements relative to the free surface:

- elements cut by the free surface
- adjacent elements above the free surface
- adjacent elements below the free surface
- distant elements above the free surface
- distant elements below the free surface

Location of nodes relative to the free surface

Fig.3 shows the different locations of the nodes relative to the free surface:

- adjacent nodes above the free surface
- adjacent nodes below the free surface
- distant nodes above the free surface
- distant nodes below the free surface

The adjacent nodes are the nodes of the elements which are cut by the free surface.

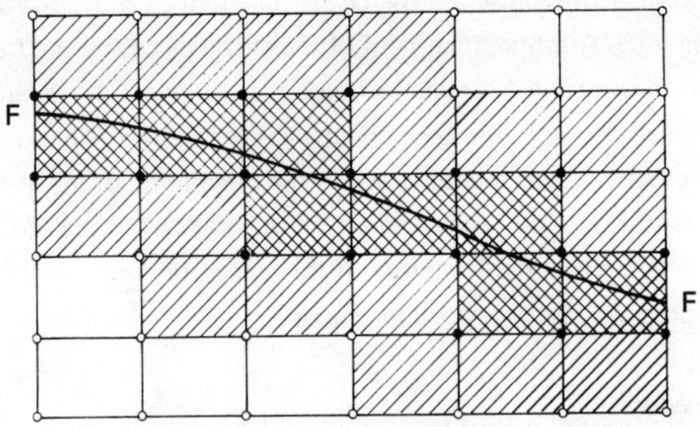


● adjacent nodes above/below the free surface F

○ distant nodes above / below the free surface F

element cut by the free surface F

adjacent element above / below the free surface F

distant element above / below the free surface F

Finite element cut by the free surface

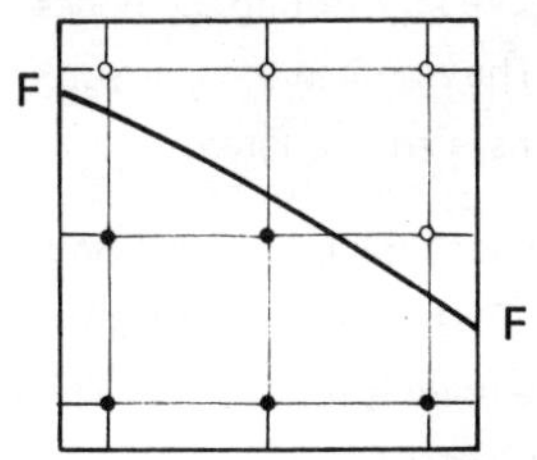


○ Gauss points above the free surface $h < z$

● Gauss points below the free surface $h > z$

Fig.3 Location of nodes and elements relative to the free surface

Numerical integration

In each iteration step, the permeability matrix $\mathbf{C}_F$ of the domain above the free surface has to be computed. A special treatment of the elements cut by the free surface is necessary. If numerical integration techniques are used for the computation of the element matrices, only the Gauss points above the free surface are significant. In order to achieve adequate accuracy in the integration of the element matrices for cut elements, it is advantageous to use an integration scheme of higher order than for the standard elements. The order of this integration scheme should not be built into the programs, but specified by the user.

## 2.4 Iteration Procedure

The iteration procedure is derived from equation (15). It consists of the following steps:

1. The initial assumed seepage surface is identical with the potential seepage surface. The system of linear equations shown in Fig.4 is initially assembled setting the residual vector $\mathbf{r}$ to zero. The symmetric coefficient matrix is decomposed by the Gaussian algorithm.

2. The system of linear equations for the nodal piezometric heads and the nodal fluxes is solved using the decomposed coefficient matrix.

3. The conditions for the seepage surface are checked by testing the nodes on the potential seepage surface in sequence. The assumed seepage surface is modified if one the following conditions is encountered:

   - The nodal flux is found to be negative and hence the boundary condition for the seepage surface is not satisfied. The node is removed from the assumed seepage surface. The flux at the node is set to zero. The associated row and column in the system of linear equations is modified as shown in Fig. 4.

   - The piezometric head at a node which is not a part of the assumed seepage surface is found to be greater than the elevation head. The node is added to the assumed seepage surface. The associated row and column in the system of linear equations is set back to the original form shown in Fig.4 (reverse modification).

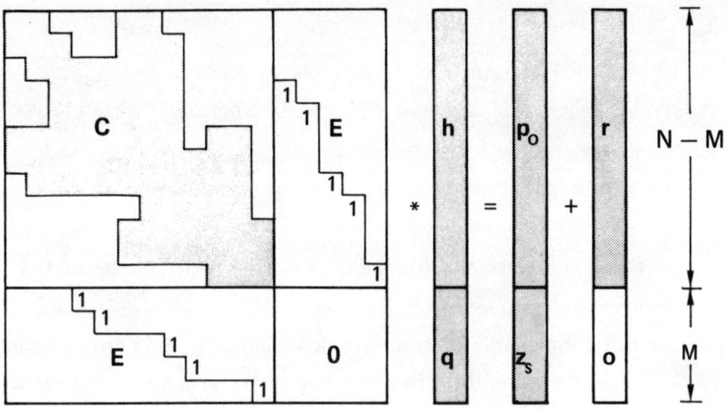


Original system of linear equations

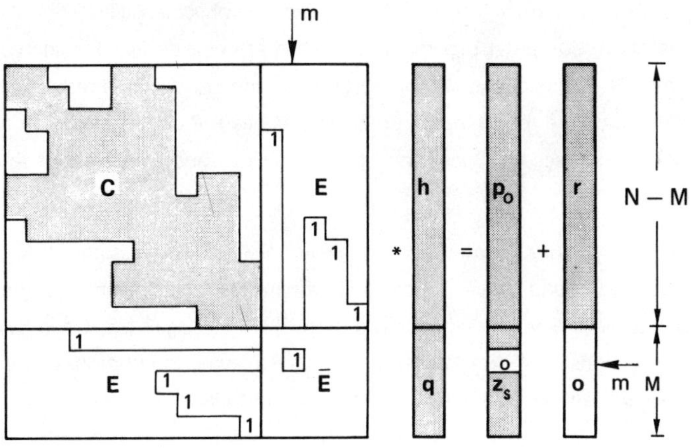


Modified system of linear equations

Fig.4 Skyline structure of the system of linear equations

If the assumed seepage surface is changed, the coefficient matrix and its decomposition have to be modified. The computational effort for the modification of the decomposed matrix is not large. The procedure is described in chapter 5.

4. On the assumed free surface, the boundary condition $h = z$ must be satisfied. The residual vector $\mathbf{r} = \mathbf{C}_F \; \mathbf{h}$ is computed. The coefficients of the vector $\mathbf{r}$ are zero with the exception of those which are associated with the nodes adjacent to the free surface. The computation of the permeability matrix $\mathbf{C}_F$ requires the evaluation of the element matrices for the elements cut by the free surface and for the adjacent elements above the free surface. It is not necessary to assemble the matrix $\mathbf{C}_F$ explicitly. Instead, the contributions of the elements to the residual vector $\mathbf{r}$ are computed individually and assembled directly. The coefficients of the vector $\mathbf{r}$ corresponding to the nodes of the potential surface are set to zero.

5. The iteration steps 2, 3 and 4 are repeated until sufficient convergence is reached. It is difficult to establish a general convergence criterion for the iteration. Computer programs should therefore be designed in such a way that the user can control the iteration by inspection. Data required for the inspection are the changes of piezometric head, the changes in flux of the seepage surface, the location at the free surface and the extent of the seepage surface.

It should be noted that the method of analysis presented here assumes that the seepage surface begins and ends at a node. The fineness of the element net therefore influences the accuracy with which the extent of the seepage surface can be determined. Precautions must be taken in the analysis to avoid oscillations of the endpoint of the seepage surface between two adjacent nodes.

## 3. Command Language

The user controls the data flow and the operations of the program with the command language. Fig.5 shows a simple command language for the unconfined seepage problem. Each command is interpreted and executed directly. The rules for a more comfortable command language and the application to finite element programs are described in [1].

| | | |
|---|---|---|
| **START** | | program name ; |
| **INITIALIZE** | **FILE** | file number ; |
| **READ** | **DATA** | file number ; |
| **STORE** | **DATA** | file number ; |
| **DEFINE** | **NODE** | [* node number, coordinates, {**P** piezometric head} ]* |
| **DEFINE** | **ELEMENT** | [* element number, type number, material number, node list ]* ; |
| **DEFINE** | **SEEPAGE** | node list ; |
| **DEFINE** | **MATERIAL** | [* material number, material values ]* ; |
| **DEFINE** | **ANALYSIS** | number of iteration steps , {* element type number, number of Gauss points for standard and cut elements }* ; |
| **CHECK** | **TOPOLOGY** | ; |
| **COMPUTE** | **HEAD** | ; |
| **COMPUTE** | **VELOCITY** | ; |
| **PRINT** | **TOPOLOGY** | ; |
| **PRINT** | **HEAD** | { node list } ; |
| **PRINT** | **VELOCITY** | { element list }; |
| **END** | | program name |

Symbols

| | |
|---|---|
| [ ] | Simple mandatory selection |
| { } | Simple optional selection |
| [* ]* | Multiple mandatory selection |
| {* }* | Multiple optional selection |

Fig.5 Examples for user commands

## 4. Data structures

The global data of the unconfined seepage problem can be catergorised as follows:

– Descriptive data and computed results for nodes in tabular form

– Descriptive data and computed results for elements in tabular form

– System equations in matrix and vector form

It is assumed that the program is implemented as an in–core version. The data structures for the three categories of global data are described. For an out–of–core version, the data management in the finite element analysis is treated in [ 2,10].

### 4.1 Table for the Properties of Nodes

Fig.6 shows the table for the properties of nodes. The table is indexed with the external number of the node. Each line contains the data for the topology and the internal number of the nodal variables, the data for the actual free surface and the potential seepage surface, the coordinates and the computed results. The condition for each node on the potential seepage surface is described by an integer variable called condition code. The sign of the variable indicates whether the node is a part of the assumed seepage surface or not. The absolute value is the number of changes of the condition during the iteration. This information is used to avoid oscillations during the iteration process.

### 4.2 Table for the Properties of Elements

Fig.7 shows the table for the properties of the elements. The table is indexed with the external number of the element. Each line contains the data for the topology, for the numerical integration and for the assumed free surface.

### 4.3 Skyline Structure of the System Equations

Fig.8 shows the skyline structure of the system equations. The coefficients of the upper part of the matrix including the diagonal elements are stored columnwise, The adresses of the skyline elements and the addresses of the diagonal elements are stored in arrays. This allows a simple management of the modifications of the matrix during the iteration process. For the decomposition of the matrix by the standard Gaussian algorithm, it is advantageous to store the diagonal elements in a special array.

| 1 | 2 | 3 | 4 | 5 | 6 | 7 | 8 | 9 | 10 | 11 |
|---|---|---|---|---|---|---|---|---|---|---|
| | | | | | | | | | | |
| | | | | | | | | | | |
| | | | | | | | | | | |
| | | | | | | | | | | |
| | | | | | | | | | | |

N ↓

N = External number of the nodes

| Column | Type | Contents |
|---|---|---|
| 1 | i | Internal number of the node |
| 2 | i | Number of the piezometric head variable at the node<br>Number of the flux variable at the node<br>(only for the nodes on the potential seepage surface) |
| 3 | i | Number of elements at the node |
| 4 | i | Specification of prescribed piezometric heads<br>0 = not specified<br>1 = specified |
| 5 | i | Location of the node relative to the free surface<br>±1 = adjacent node above/below the free surface<br>±2 = distant node above/below the free surface |
| 6 | i | Condition code for nodes on the potential seepage surface |
| 7–9 | R | Coordinates x, y, z |
| 10 | R | Piezometric head h |
| 11 | R | Nodal flux q |

Fig.6 Table for the properties of nodes

| 1 | 2 | 3 | 4 | 5 | 6 | 7 | 8 | 9 | | | M |
|---|---|---|---|---|---|---|---|---|---|---|---|
| | | | | | | | | | | | |
| | | | | | | | | | | | |
| | | | | | | | | | | | |
| | | | | | | | | | | | |
| | | | | | | | | | | | |

i ↓

i = External number of the element

| Column | Type | Contents |
|---|---|---|
| 1 | i | Internal number of the element |
| 2 | i | Number of the element type |
| 3 | i | Number of the material type |
| 4 | i | Number of Gauss points for standard elements |
| 5 | i | Number of Gauss points for elements cut by the free surface |
| 6 | i | Location of the element relative to the free surface |
| | | 0 = element cut by the free surface |
| | | ±1 = adjacent element above/below the free surface |
| | | ±2 = distant element above/below the free surface |
| 7 | i | Number of nodes of the element |
| 8–M | i | Numbers of the nodes of the element |

Fig.7 Table for the properties of elements

System of linear equations **A x = b**

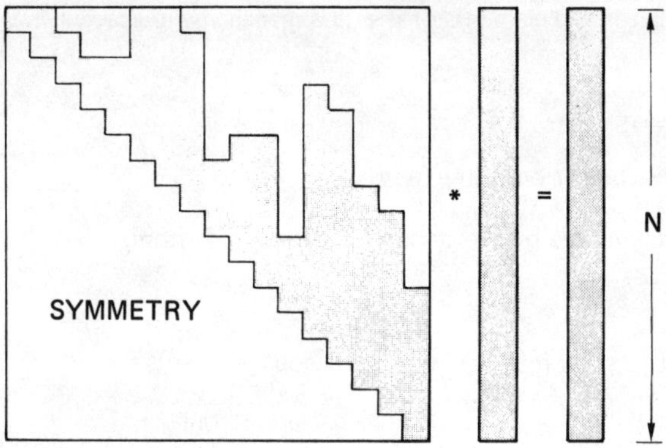


Data and storage structure

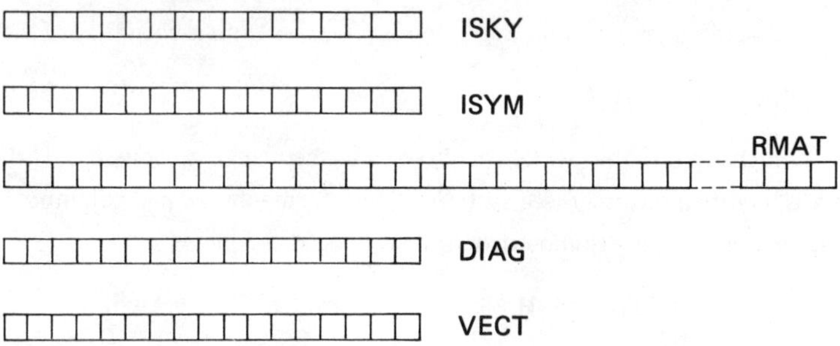


ISKY = array for the adresses of the skyline elements in RMAT

ISYM = array for the adresses of the diagonalelements in RMAT

RMAT = array for the coefficients of the matrix, stored columnwise

DIAG = array for the diagonal elements of the matrix

VECT = array for the load or solution vector

Fig.8 Data and storage structure for the system of linear equations

## 5. Procedure of the Matrix Decomposition

Before the iteration starts, the matrix **A** of the system equations is assembled. The relationsship between the elements of the data management arrays ISKY and ISYM shown in Fig.8 is:

$$\text{ISKY}(n+1) = \text{ISYM}(n) + 1 \qquad n = 1, \ldots, N-1 \tag{16}$$

$$N = \text{Dimension of the matrix}$$

The matrix **A** is decomposed by the standard Gaussian algorithm:

$$\mathbf{A} = \mathbf{L}\,\mathbf{D}\,\mathbf{L}^T \tag{17}$$

**L** = Lower triangular matrix with unit diagonal elements

**D** = Diagonal matrix

The matrix $\mathbf{L}^T$ is stored columnwise in the array RMAT and the diagonal matrix **D** in the array DIAG. If the matrix ist modified in the m–th row and column during the iteration as shown in Fig.4, the skyline structure is changed as follows:

$$\text{ISKY}(m) = \text{ISYM}(m) \tag{18}$$

The decomposition in the last (N–m+1) rows of the matrix is computed. If the matrix is modified reversely (see section 2.4) in the m–th row and column during the iteration, the skyline structure is changed as follows

$$\text{ISKY}(m) = \text{ISYM}(m-1) + 1 \tag{19}$$

The decomposition in the last (N–m+1) rows of the matrix is computed. The computational effort for the modification of the decomposed matrix is not large.

## 6. Conclusions

An efficient iterative algorithm for the solution of the unconfined seepage problem is developed. Modification of the finite element ne t are not necessary. The command language, the data structures and the procedure for the modifications of the matrix decomposition during the iteration process are described.

**Acknowledgement**

The method for the solution of the confined seepage problem described in this paper was developed in cooperation with my colleague at the Institut für Allgemeine Bauingenieurmethoden, Prof.Dr. Peter Jan Pahl.

REFERENCES

1. Damrath, R. and Beilschmidt, L.
Entwicklung und Einsatz von Kommandosprachen für FEM–Programme
Finite Elemente in der Baupraxis, Hannover 1978
Verlag von Wilhelm Ernst & Sohn

2. Damrath, R. and Pahl, P.J.
Data and storage structures for a system of finite elements
Formulation and Computional Algorithms in Finite Element Analysis: U.S.–Germany Symposium
Massachusetts Institute of Technology 1977

3. Desai, C.S.
Analysis and evaluation of some finite element schemes for flow through rigid and deformable media
Finite Elements in Water Resources, 1976

4. Desai, C.S.
Finite element residual schemes for unconfined flow
International Methods in Engineering, Volume 10, Issue No 6

5. Diersch, H.J. and Martin, H.
The numerical treatment of free surface flow by finite elements
Finite Elements in Water Resources 1978

6. France, P.W.
Finite element analysis of two and three dimensional unconfined seepage problems
Finite Elements in Water Resources, 1976

7. Guymon, G.L. and Berg, R.L.
Variable domain finite element analysis of free surface problems
Finite Elements in Water Resources 1976

8. Kazda, I.
Free surface seepage flow in earth dams
Finite Elements in Water Resources 1978

9. Meißner, U.
Die Berechnung ebener und räumlicher Grund– und Sickerströmungen mit verallgemeinerten Variationsprinzipien
Mitteilung des Sonderforschungsbereiches 79 im Küstenbereich der Technischen Universität Hannover 1973

10. Pahl, P.J.
Data management in finite element analysis
Nonlinear Finite Element Analysis in Structural Mechanics
Proceedings of the Europe–U.S. Workshop, Ruhr-Universität Bochum, 1980

# FINITE ELEMENT ANALYSIS OF INTERFACE PROBLEM IN NONSTEADY SEEPAGE

Iichiro KOHNO and Makoto NISHIGAKI

Department of Civil Engineering, Okayama University, Okayama, JAPAN

## INTRODUCTION

The saltwater intrusion into coastal aquifers and the lealcage of oil from the oil tank or pipe line into groundwater belong to the interface problem of seepage of two fluids with different densities and different viscosities. The free surface and the seepage surface of unconfined groundwater can be dealt with similarly because in that flow one liquid is groundwater and the other is air, namely it being the special case of the interface problem. Actually, freshwater and saltwater are two miscible fluids with a trasition zone between them. Across this zone dispensive mixing will occur between freshwater and saltwater. However, the width of this transition zone is usually thin enough to compare with the thickness of the aquifers (Bear, 1979). So it can be assumed that an immiscible interface is introduced between freshwater and seawater. In this paper are described on the finite element analysis of the nonsteady behavior of interface between salt- and freshwater in coastal groundwater and some experimental studies.

## GOVERNING EQUATIONS

Figure 1(a) shows the interface problem of seepage of two fluids with different densities ($\gamma_f$, $\gamma_s$) and different viscosities ($\mu_f$, $\mu_s$). A schematic sketch of the interface in coastal aquifer is shown on Figure 1(b). The well-known equation 1 and 2 are formed for two dimensional flow in (x, z) plan respectively as,

Freshwater:

$$S_f \frac{\partial h_f}{\partial t} = \frac{\partial}{\partial x}\left(k_{xf}\frac{\partial h_f}{\partial x}\right) + \frac{\partial}{\partial z}\left(k_{zf}\frac{\partial h_f}{\partial z}\right) + q_{0f} \quad (1)$$

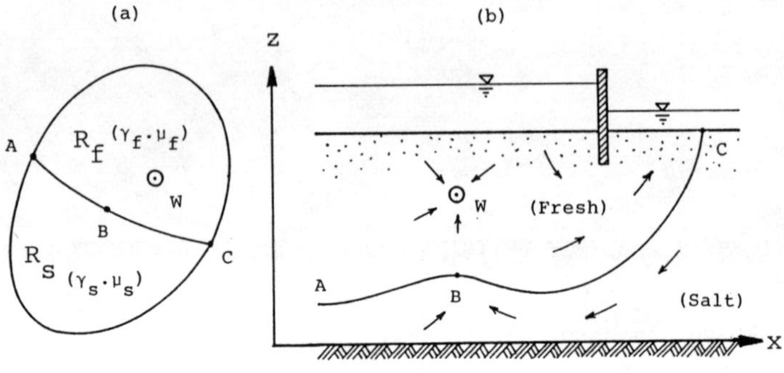


Figure 1. Domains and interface.

Saltwater:

$$S_s \frac{\partial h_s}{\partial t} = \frac{\partial}{\partial x}\left(k_{xs}\frac{\partial h_s}{\partial x}\right) + \frac{\partial}{\partial z}\left(k_{zs}\frac{\partial h_s}{\partial z}\right) + q_{0s} \tag{2}$$

where

$$h_f = z_f + p_f/\gamma_f \tag{3}$$

$$h_s = z_s + p_s/\gamma_s \tag{4}$$

In these equations the suffix f and s represent freshwater and saltwater, respectively, and

k = permeability,
S = specific yield or specific storage,
h = piezometric head,
q = sink/source term,
p = pressure,
γ = specific weight.

Based on the assumptions that the two fluids are immiscible, the pressures of two fluids $p_f$ and $p_s$ should be equal along an interface,

$$p_f = p_s \tag{5}$$

Substituting equations 3 and 4 into equation 5 gives

$$h_f = \frac{\gamma_s}{\gamma_f} h_s + z_B\left(\frac{\gamma_f - \gamma_s}{\gamma_f}\right) \tag{6}$$

or

$$z_B = \frac{\gamma_s h_s - \gamma_f h_f}{\gamma_s - \gamma_f} \tag{7}$$

where $z_B$ is the height of the interface above the reference datum. Equation 6 means that the total head of fresh groundwater at the interface is expressed as a linear function of $z_B$ and $h_s$. In the case of nonsteady state, the following conservation equation has to be derived on the interface,

$$\frac{\partial v_f}{\partial n} = \frac{\partial v_s}{\partial n} \tag{8}$$

where n shows the direction perpendicular to the interface (see Figure 2).

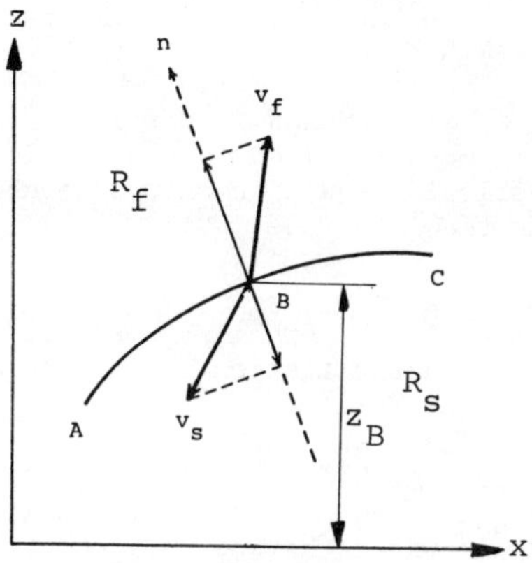


Figure 2. Conservative relationship

EXPRESSION WITH FINITE ELEMENT

To solve equations 1 and 2 in the conditions of equations 7, 8 and initial-boundary conditions, the finite element method is used. Putting $\gamma_s = 0$ into equation 7 gives

$$h_f = z_B \tag{9}$$

This result can be explained that the interface flow of saltwater and freshwater is reduced to a same problem of unconfined groundwater flow. In each subregion $R_f$, $R_s$ storage of water due to volumetric deformations of water and soil will be disregarded (S = 0 in equation 1, 2). Then the differential equation is the same as for steady flow. Time dependency will only be introduced that moving boundary conditions (for example free surface and interface) is time dependent. The domain is subdivided into a larg number (say N) of elementary sub-domains. The alternative formulation of this problem is obtained by the variational principle (Kohno, 1974 (a), (b))

$$\sum_{j=1}^{N^f} P_{ij}^f h_j^f + Q_i^f + Q_i^f{}^* = 0 \quad (i=1\sim N^f) \tag{10}$$

$$\sum_{j=1}^{N^s} P_{ij}^s h_j^s + Q_i^s + Q_i^s{}^* = 0 \quad (i=1\sim N^s) \tag{11}$$

where

$P_{ij}$ = seepage array,

$Q_i$ = discharge (or recharge) rate at prescribed flux boundary,

$Q_i{}^*$ = discharge rate at prescribed head boundary,

$N^f$, $N^s$ = nodal point number.

DETERMINATION OF INTERFACE

Along the interface as shown in Figure 3, the solution must satisfies not only equation 7 but also equation 8. Then, since by equation 8, following equation should be checked for the nodal point along the interface,

$$Q_k^f{}^* + Q_k^s{}^* = 0 \quad (k=1,2,\cdots\cdots,m) \tag{12}$$

The procedure of the determination of interface in this finite element analysis is as follow:

(1) Initially set the shape and the position of interface.

(2) Calculate values of $Q^f{}^*$ and $Q^s{}^*$

(a) Compute $h_s$ in the interior of the domain $R_s$ by use of equation 11, where the boundary condition of interface has to be dealt with as an impermiable boundary.

(b) Calculate $h_f$ along the interface from equation 6 substituting $h_s$.

(c) Compute $h_f$ values in the interior of the domain $R_f$ by using equation 10, where the boundary condition of interface is the prescribed head condition.

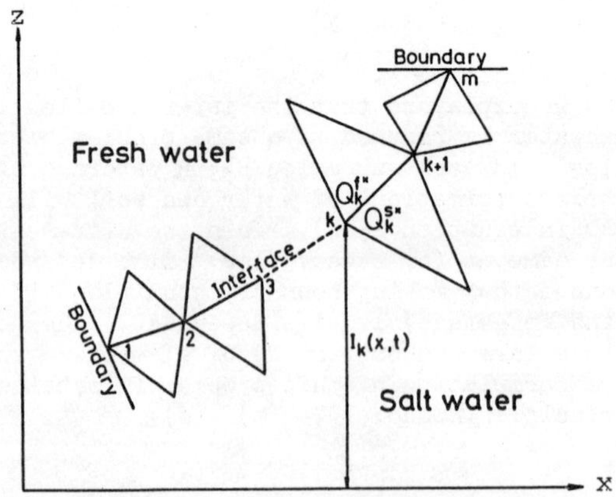


Figure 3. Finite element along interface.

(d) From these procedure, the nodal flows $Q^{f*}$, $Q^{s*}$ are calculated for the nodal points along the interface.

(3) From these values of $Q^{f*}$, $Q^{s*}$, an error criterion of $\varepsilon_i$ can be determined,

$$\varepsilon_i = Q_k^{f}* + Q_k^{s}* \tag{13}$$

This cycles are repeated until the error is within a specified tolerance.

(4) By using the values of $Q^*$, and $\Delta t$ one can get the position of interface in the next step.

The flow chart of this procedure is shown in Figure 4.

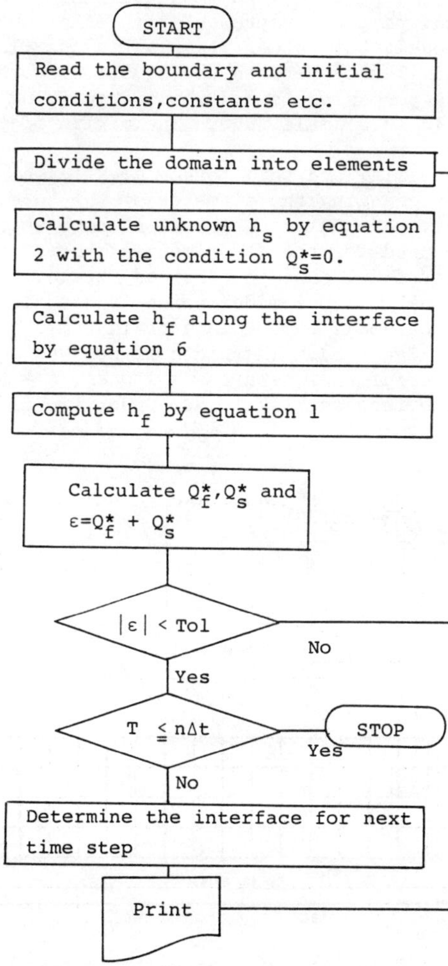


Figure 4. Flow chart of calculation.

COMPARISON OF EXPERIMENTAL AND NUMERICAL RESULTS

In order to investigate the validity and the accuracy of the numerical analysis solution, in general, two methods are used, that is, numerical results are compared with the analytical results of simple problems for which rigorous analytical solutions are available and numerical results are compared with experimental results. The flow problem in this work invalves so complex conditions that there is no analytical solution as far as we know. Then to check the validity and its applicability to practical problems of saltwater intrusion, laboratory experimental studies by use of a small tube-net viscous model were performed (Kohno 1970). A small tube-net viscous model is not an actual physical representation of porous media as is the sand-box model, but rather it is a viscous flow analog based on the similarity between the equations governing saturated flow in a porous media and those describing the flow of a viscous liquid in a small tube. The merits of this model are as follows.

(1) The permeability and the storage coefficient can be controlled by changing the diameter of the tube, then one can easilly make anisotropic and nonhomogeneous modeles.

(2) It can be used for the experiment of three dimensional groundwater flow, also in nonsteady state.

The small tube-net viscous model which is used in this research is constructed of small tube (diameter is 3 mm) and acrylic connector (see Figure 5). The width of net was equal to 10 cm, the model aquifer dimensions are $\ell$ = 180 cm, d = 40 cm. Movement of the interface can take place under two different types of flow conditions:

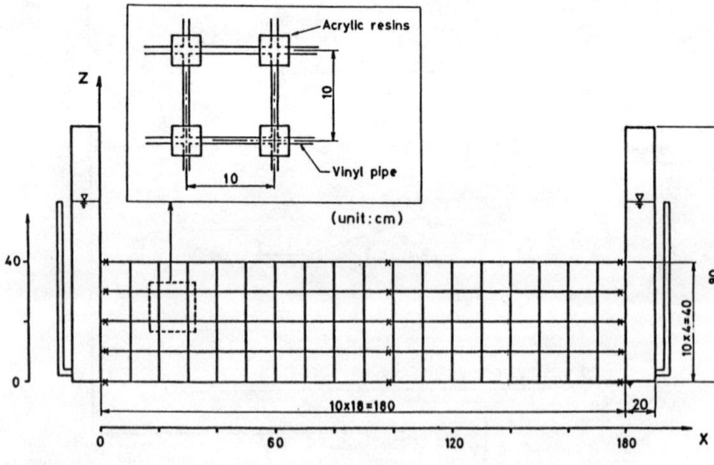


Figure 5. Schematic diagram of the experimental equipment and the dimension of small tube net model.

(1) The model aquifer is divided two region with closed tubes at x = 100 cm (see Figure 5). The freshwater and tracer solution are filled up in both region respectivelly. At time t = 0, the closed tubes are opened suddenlly. In this case, total head on right and left boundary are fixed.

(2) In the second type, initial condision is the same as the first type, but the boundary conditions on right and left hand are impermiable. At time t = 0, as the same manner, the closed tubes at the position x = 100 cm are opened suddenly.

Hence, in two cases a confined aquifer is simulated by the model. During each experiment, the behavior of the interface at various times is recorded and photographed.

Two main aspects were compared: the variation of interface toe with time and the shape of the moving interfaces. Figures 6, 7 show numerical and experimental results for these moving interfaces. Figure 8 shows similar case 2 for a motion of the interface toe. As noted in the figure, all the results are consistent. The comparisons between the two sets of results show that the numerical model is quite accurate in their description of the movement of the interface.

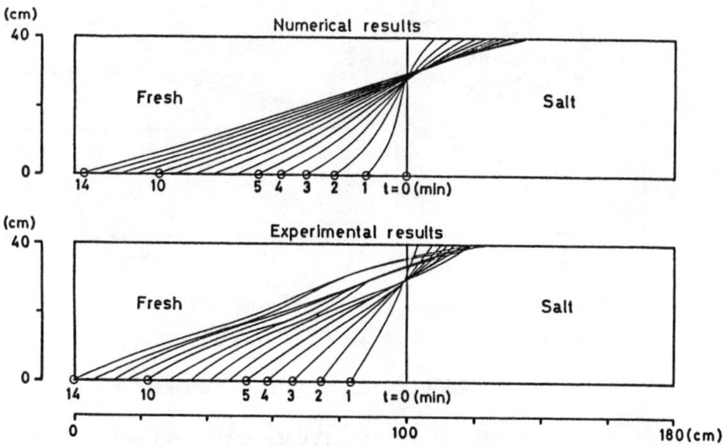


Figure 6. Comparison between the numerical and the experimental results of the interface shapes with the constant head boundary.

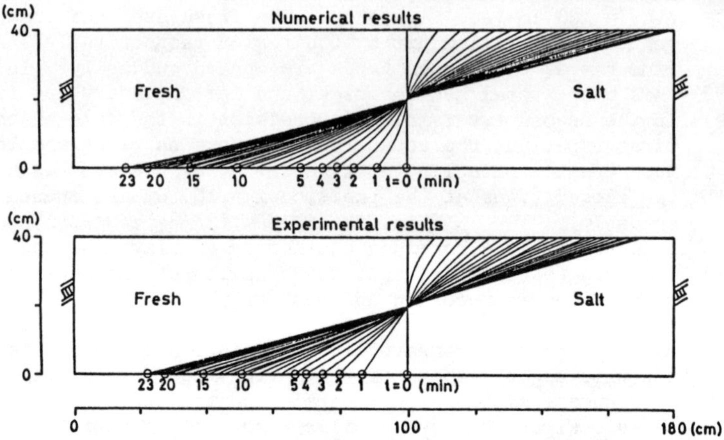


Figure 7. Comparison between the numerical and the experimental results of interface shapes with the impervious boundary.

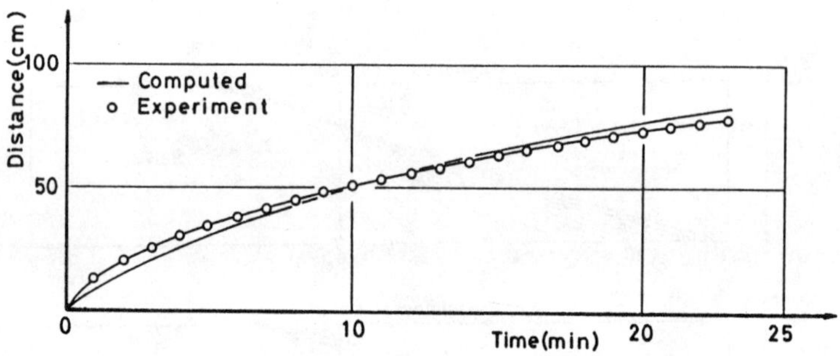


Figure 8. Motion of the interface toe with the impervious boundary.

## APPLICATION OF THIS METHOD TO PLACTICAL INTERFACE PROBLEM

When a drain is operating in the fresh groundwater zone, the phenomenon of raising the interface, which is called "upconing", appears in coastal groundwater (see Figure 1). Here, two example analyses of this phenomenon were performed,

(1) The lower boundary in Figure 9 is impervious (in the case of a shallow coastal aquifer).

(2) The lower boundary is a constant head (in the case of a deep coastal aquifer).

In both cases, the uper boundary is a constant head. Provided that the coefficient of permeability: $k_{xf} = k_{zf} = k_{xs} = k_{zs} = 2.4$ m/day $\doteqdot 3 \times 10^{-2}$ cm/s, specific yield: $S = 0.2$, and specific weight $\gamma_f = 1.00$ t/m³, $\gamma_s = 1.03$ t/m³, Figures 10, 11 show the results with the discharge rate $Q_p = \pm 100$ m³/d/m at the point P(-450 m, -50 m).

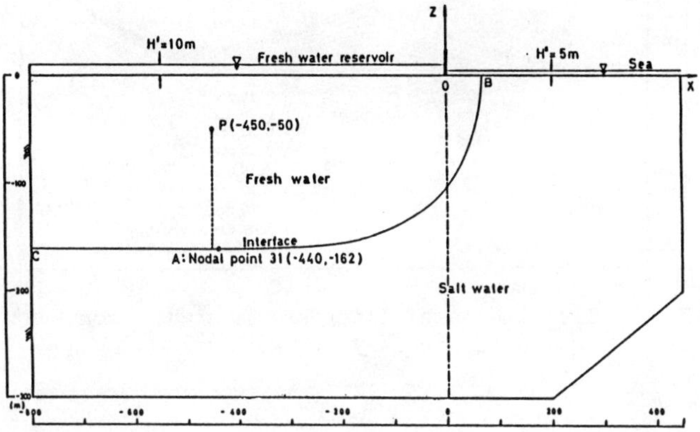


Figure 9. Initial and boundary conditions of analytical model.

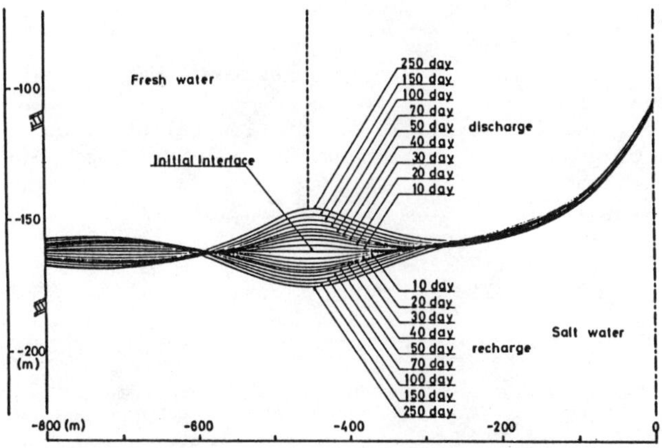


Figure 10. Shapes of interface (with the impervious boundary at the bottom).

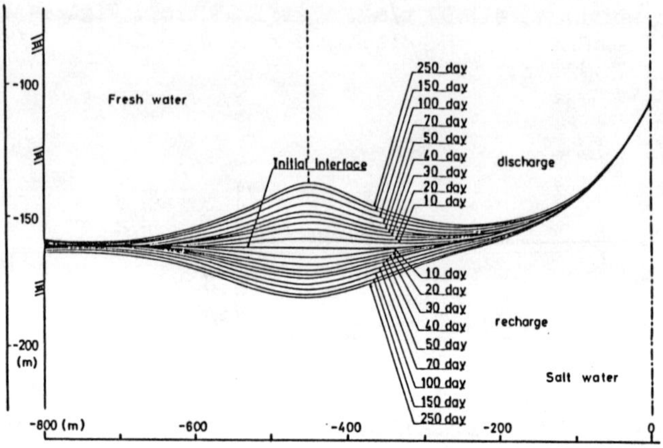


Figure 11. Shapes of interfare (with the constant head boundary at the bottom).

CONCLUSION

In this paper some attempts have been made to apply the finite element method to the interface flow problem in coastal groundwater and some examples. The results calculated by this finite element analysis have been confirmed by the laboratory model tests. The numerical results showed the very good agreement with the experimental data.

REFERENCES


Bear, J. (1979) Hydraulics of groundwater. McGraw-Hill Book Co., New York.

Matuo, S., and Kohno, I. (1970) Experimental study on the control of groundwater by impervious wall, Proc. of JSCE, No. 178, 53-61.

Kohno, I. (1974(a)) Finite element analysis of interface problem in groundwater, Proc. of JSCE, No. 228, 109-116.

Kohno, I. (1974(b)) Finite element analysis of a nonsteady seepage problem, Soils and Foundations, Vol. 14(4), 75-85.

# A FRONTAL METHOD BASED SOLUTION OF THE QUASI-THREE-DIMENSIONAL FINITE ELEMENT MODEL FOR INTERCONNECTED AQUIFER SYSTEMS AND FLUID MASS BALANCE EVALUATION. STEADY AND UNSTEADY EQUATIONS.

L. Sartori, G. Peverieri.

Aquater S.p.A. - San Lorenzo in Campo (PS) - Italy

## INTRODUCTION

The quasi-three-dimensional equations governing the groundwater flow in regional aquifer systems are the most suitable to describe the physical system.

K. Fujinawa (1977) and Don W. Chorley, Emil O. Frind (1978) describe the finite element discretization of the problem in the hypothesis of overlying aquifers with interbedded aquitards, disregarding therefore the problem of branching aquifers.

This problem is here exposed in relation to an adequate solution of the linear system deriving from it. A method for evaluation of fluid mass balances is presented.

This method allows rigorous interpretation of the leakage term estimation used by K. Fujinawa and gives an answer to some perplexities pointed out by the author. The K. Fujinawa method also uses time derivatives of the nodal piezometry to estimate the leakage terms.

## QUASI-THREE-DIMENSIONAL MODEL EQUATIONS

In the quasi-three-dimensional model the aquifer system is composed of aquifers and aquitards. It originates from coupling two-dimensional equations for aquifers with mono-dimensional equations for aquitards. Following are the water flow's two-dimensional equations for aquifers:

$$\frac{\partial}{\partial x}\left(T\frac{\partial h}{\partial x}\right)+\frac{\partial}{\partial y}\left(T\frac{\partial h}{\partial y}\right)=L_2-L_1+Q+S\frac{\partial h}{\partial t}$$

where:

$T$ = transmissivity (fixed for confined aquifers, linearly dependent on the piezometry for phreatic aquifers).

$L_2$ and $L_1$ = top and bottom leakage terms

$Q$ = source term

$S$ = storage coefficient for confined aquifers and effective porosity for phreatic aquifers

$h$ = averaged aquifer piezometry along thickness

The following mono-dimensional equation is used for aquitards:

$$\frac{\partial}{\partial z}\left(K \frac{\partial h^*}{\partial z}\right) = S_s \frac{\partial h^*}{\partial t} \qquad (2)$$

where:

$K$ = vertical permeability of aquitard

$S_s$ = specific storage coefficient

$h^*$ = piezometry along aquitard thickness

Coupling of the aquifer two-dimensional equations (1) with the aquitard mono-dimensional equations (2) takes place through leakage terms:

$$L_2 = -K\left(\frac{\partial h^*}{\partial z}\right)_{z=z_T} \quad ; \quad L_1 = -K\left(\frac{\partial h^*}{\partial z}\right)_{z=z_B} \qquad (3)$$

($z_T$ and $z_B$ are the quotas at top and bottom of the aquitard respectively) and identify the piezometry of the aquitard with that of the aquifer in correspondance to separation surface at top and bottom of the aquitard:

$$h^*(t,x,y,z_T)=h_T(t,x,y) \ ; \ h^*(t,x,y,z_B)=h_B(t,x,y) \qquad (4)$$

Finite element discretization and fluid mass balance

The Galerkin method is applied for the finite element discretization, with linear shape functions both for aquifers and aquitards.

Triangular elements are used for aquifers. On the aquitards mono-dimensional finite elements which connect two nodes located along the same vertical are used. They are all the same lenght and together form an aquitard macroelement.

Fig. 1 illustrates this discretization pointing out the branching line treatment and the connections between aquifers and aquitards.

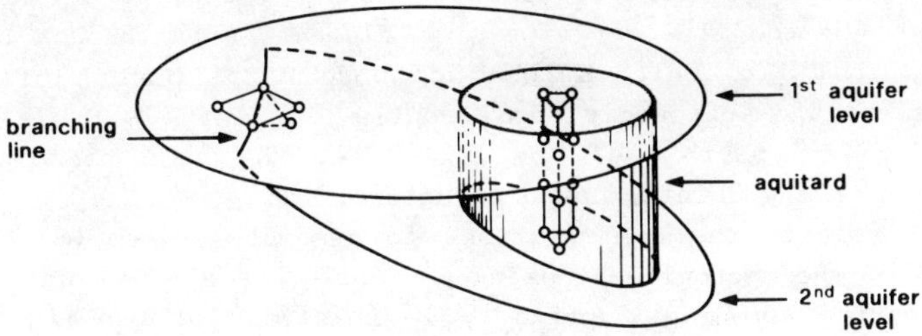


Fig. 1 - Finite element discretization of an interconnected aquifer system.

By applying the Galerkin method to equation (2) valid for aquitards, we obtain the differential equations system:

$$C^*\dot{h}^* + A^*h^* = b^* \qquad (5)$$

These equations concern only the nodes within aquitard macro-elements, or the boundary nodes which are subject to imposed flow conditions.
The right hand terms of the system are all null except those related to nodes near those subject to imposed piezometry conditions, or subject to imposed flow conditions. For the latter, the right hand term is:

$$b_i = - L \qquad (6)$$

where:
L = imposed flow (or leakage) on the node

The fluid mass balance can be estimated for each element on the basis of the following four items:
$F_i$ = imposed flow
$F_c$ = flow deriving from contiguous elements
$F_p$ = flow equivalent to imposed piezometry conditions
$F_u$ = storage depletion
This balance concerns a column of aquitards of unit section which extends over the entire element thickness. The first item is present only if a node of the element is subject to imposed flow conditions. The second item is present in correspondance to each node which is neither subject to imposed flow conditions nor to imposed piezometry conditions.
In order to define, on an element node, the flow deriving from

contiguos elements, we will divide the macroelement into two parts A and B in correspondance to node $P_i$. We will define in element ($l$) the flow deriving from element ($l$+1) through node $P_i$, as that flow which imposed on node $P_i$ at part A of the macroelement, supposedly interrupted on $P_i$, is able to reproduce on part A of the structure, the same piezometry obtained on the uninterrupted structure.

If we indicate the contributions to assembly system (5) from part A of the macroelement using $a_{is}^A$ and $c_{is}^A$ and those deriving from part B using $a_{is}^B$ and $c_{is}^B$, in relation to node $P_i$, the equation of system (5) related to node $P_i$ (internal) could be written as follows:

$$\begin{cases} \Sigma a_{is}^A h_s^* + \Sigma c_{is}^A \dot{h}_i^* = -L \\ -L = -\Sigma a_{is}^B h_s^* - \Sigma c_{is}^B \dot{h}_s^* \end{cases} \qquad (7)$$

The term L which appears in (7) represents, on the basis of given definition, the flow originating from part B on part A through node $P_i$. In order to define the flow with these equations it is necessary to previously calculate the piezometry, thus solving system (5). In this case, for the evaluation of flow L it is convenient to use the first equation of system (7). This allows determination of the flow deriving from part B on part A through node $P_i$, resorting only to part A data.

The incoming flow through the nodes i and j of the element is:

$$F_{ci} = K \frac{h_j^* - h_i^*}{\Delta z} - \frac{S_s \, \Delta z}{6} \left(2\dot{h}_i^* + \dot{h}_j^*\right) \qquad (8)$$

$$F_{cj} = -K \frac{h_j^* - h_i^*}{\Delta z} - \frac{S_s \, \Delta z}{6} \left(\dot{h}_i^* + 2\dot{h}_j^*\right) \qquad (9)$$

In a similar manner, we could define the flow equivalent to the imposed piezometry conditions as the flow imposed on a node, instead of the imposed piezometry condition, able to reproduce the same piezometric values on the aquitard macroelement.

The same equations (8) and (9) can be obtained for the terms $F_{pi}$ and $F_{pj}$. The last balance item, the flow which represents storage depletion in unsteady state, is given by the following:

$$F_u = \int_{z_i}^{z_j} S_s \dot{h}^* (z) \, dz = S_s \, \Delta z \frac{\dot{h}_i^* + \dot{h}_j^*}{2} \qquad (10)$$

for each element the following equation was verified:

$$F_i + F_c + F_p + F_u = 0 \qquad (11)$$

which shows how this procedure of evaluation of fluid mass balance preserves the mass.
Since condition (4) of coupling between aquitards and aquifers is comparable to an imposed piezometry condition for aquitards, it would be appropriate to utilize expression (8) and (9) for the definition of leakage terms $L_2$ and $L_1$ indicated in equation (1), as done by Fujinawa (1977).
For aquifers, applying the Galerkin method to equation (1), we obtain the following linear system:

$$C\dot{H} + AH = B \qquad (12)$$

$$C=\begin{bmatrix} C^{o} & C^{o*} \\ C^{*o} & C^{*} \end{bmatrix};\; A=\begin{bmatrix} A^{o} & A^{o*} \\ A^{*o} & A^{*} \end{bmatrix};\; B=\begin{bmatrix} b \\ b^{*} \end{bmatrix};\; H=\begin{bmatrix} h \\ h^{*} \end{bmatrix}$$

$C^o$, $A^o$, b and h represent the discretization of equation (1) on aquifers; and $C^*$, $A^*$, $b^*$ and $h^*$ the discretization of equations (2) on aquitards. $C^{o*}$ and $A^{o*}$ represent the coupling conditions between aquifers and aquitards and derive from leakage terms $L_2$ and $L_1$, whereas $C^{*o}$ and $A^{*o}$ represent the coupling conditions between aquitards and aquifers, by assimilating condition (4) of coupling, to an imposed piezometry condition on the aquitards.
If $Q = q_o\, \delta(x-x_o)\, \delta(y-y_o)$ represents a concentrated source, with intensity $q_o$ located on point $(x_o, y_o)$, we have for the right hand term of equations (12).

$$b_l = - q_o N_l\, (x_o, y_o) \qquad l = i, j, k \qquad (13)$$

$N_i$, $N_j$, $N_k$ indicate the linear shape functions, of the element containing the point $(x_o, y_o)$.
In particular if point $(x_o, y_o)$ coincides with one of the 3 nodes, for example l=i, we have:

$$b_i = - q_o \qquad b_j = b_k = 0 \qquad (14)$$

This observation allows interpretation of the right hand term of equation (12) related to the aquifer nodes, as for a point source, located on node $P_s$ with an intensity of $-b_s$. Once this has been established, we are able to define the balance

evaluation procedure, on the aquifers as described in the following.
The items used in this fluid mass balance are:

$q_1$ = distributed source
$q_2$ = concentrated source
$q_3$ = top leakage
$q_4$ = bottom leakage
$q_5$ = source equivalent to the imposed flow conditions
$q_6$ = source equivalent to the imposed piezometry conditions
$q_7$ = fluid deriving from contiguous elements
$q_8$ = storage depletion

$q_1$ $q_3$ $q_4$ $q_8$ are evaluated by integrating theyr corresponding function on the element. $q_3$ and $q_4$ are supposed to linearly vary $q_8$ come from term $S\partial h/\partial t$. $q_2$ is obtained summining all the element's punctual source terms. $q_5$ is obtained by integrating the flow along the element's sides subjected to the imposed flow condition.
Each of these 8 balance items related to the element can be divided into 3 terms associated to 3 element nodes.
They are obtained directly or by weighting with the shape functions pertaining to each node before integrate.
The balance terms associated to nodes allows to interpret the equation of system (12) as balance equations.
The nodal source $q_6$ equivalent to the imposed piezometry condition on a node; is defined in the same manner for aquifers as for aquitards. It is defined as a point source term, located on a node of such an intensity able to reproduce the same piezometry on the whole system, after having removed the imposed piezometry condition on the node.
Removed on system (12) the imposed piezometry conditions, pertaining to the node, on the basis of the observations related to formulas (13) and (14), this balance term can be determined:

$$-q_{6m} = \sum_s a_{ms} h_s + \sum_s c_{ms} \dot{h}_s - b_m \qquad (15)$$

To define the source equivalent to fluid deriving from the contiguous element, we will consider the elements which meet on node $P_r$, divided into 2 parts A and B, as in Fig. 2.
Indicating with $a^A_{rs}$, $c^A_{rs}$, $b^A_r$, $a^B_{rs}$, $c^B_{rs}$, $b^B_r$ the contributions to the assembly of equations in system (12) for node $P_r$, which derive from the two groups of elements meeting on node $P_r$, we have:

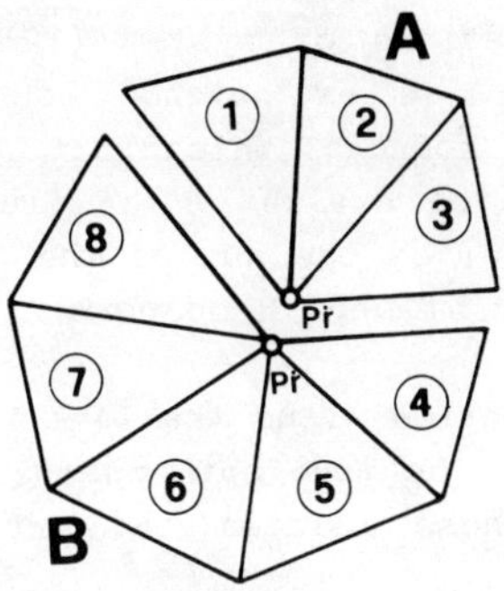


Fig. 2 - Section of a structure in correspondance to node $P_r$ for evaluation of the nodal source equivalent to the influence of part B on part A through node $P_r$.

$$\begin{cases} \Sigma a^A_{rs} h_s + \Sigma c^A_{rs} \dot{h}_s = b^A_r - q^B_r \\ -q^B_r = b^B_r - \Sigma a^B_{rs} h_s - \Sigma c^B_{rs} \dot{h}_s \end{cases} \tag{16}$$

According to the above-said observations, this interpretation allows to define $q^B_r$ as a source term equivalent to the influence of part B on part A through node $P_r$. To evaluate $q^B_r$ it would be more suitable to use the first of the equations in (16).

$$q^B_r = b^A_r - \Sigma a^A_{rs} h_s - \Sigma c^A_{rs} \dot{h}_s \tag{17}$$

This because it permits evaluation of the fluid flow from part B to part A on node $P_r$, using the only data deductible from the assembly of the elements found in part A.
It is easy to ascertain that this procedure for evaluation of fluid mass balance on the element, is conservative.
In fact, by summing the 3 expression which supply the fluid deriving from contiguous elements on 3 element nodes:

$$\begin{cases} q^B_i = b^e_i - \Sigma a^e_{is} h_s - \Sigma c^e_{is} \dot{h}_s \\ q^B_j = b^e_j - \Sigma a^e_{js} h_s - \Sigma c^e_{js} \dot{h}_s \\ q^B_k = b^e_k - \Sigma a^e_{ks} h_s - \Sigma c^e_{ks} \dot{h}_s \end{cases} \tag{18}$$

we obtain the following equation:

$$q_{1e} + q_{2e} + \dots + q_{8e} = 0 \tag{19}$$

(Symbol e refer to element data).
Moreover, the equations of system (12) on this base can be rigorously interpreted as balance equations, regarding the equivalent nodal sources which act upon eah node. This procedure permits the evaluation of fluid mass flow, through any aquifer section, as the sum of the nodal sources, equivalent to the action of part B on part A, for all the nodes of the section line.
This balance evaluation along the line is conservative, in the sense that it could be made both on the basis of data related to part A and those related to part B, and both values coincide.
Such a characteristic is also true for branching lines. Mass conservation along the branching line is rigorously verified.

## NUMERICAL SOLUTION IN STEADY STATE CASE

In the steady state case, the discretization of finite element brings to a system of ordinary linear equations.

$$Ah = b \tag{20}$$

With a uniform permeability along the z axis, the piezometry trend along this axis is linear and every aquitard macroelement can be reduced to a single element.
Matrix A, in this case is limited to part A°. The steady state equations are utilized only for calibration of the mathematical model, consisting in the solution by various attempts of an inverse problem. In this case the piezometry is known in advance and thus the wet thickness in the aquifer's phreatic zones.
For this reason, it is possible to use equations of aquifers under pressure even in phreatic zone, system (20) thus results, linear.
Application of the frontal method in our case, similarly to three-dimensional ones, sets same limits to the problems which can be treated, due to the high number of active variables which we can have in the front. This is due to 2 causes: aquifer branching and presence of aquitards. The method which we propose eliminates this problem.
The front increase caused by aquifer branching is limited resorting to the substructuring. This permits a temporary elimination, from the active variables front, of the variables corresponding to the branching lines. These appear subsequently, when only the variables of bordering nodes between

substructures are processed.
For the efficiency of this method, the branching line nodes should coincide with the bordering nodes between substructures. However, the substructuring technique does not solve the problem of front increase caused by the presence of aquitards, which relate the nodes of one level to those of the underlying level.
The variables of nodes subject to leakage remain in the active variables front, until all the elements which refer to it in every level, have been assembled.
In order to solve this other problem, we inserted an iterative process within the substructured frontal method. The above process transfers the contribution caused by leakage, to the right hand side of the equations.
Splitting matrix A of system (20) into $A_o$ and $A_1$, where part $A_1$ derives from the assembly of contributions due to leakage terms $L_2$ and $L_1$, system (20) can now be written as follows:

$$A_o h = b - A_1 h \tag{21}$$

From the system written as above we can deduce 2 iterative processes.

a - Simple iteration method

This is obtained by iteratively applying the following relation:

$$A_o h^n = b - A_1 h^{n-1} \tag{22}$$

with $h^o = 0$

This iterative scheme determines firstly solution $h^1$ assuming that the aquitards are impermeable. Subsequently, on the basis of this piezometry, the leakage terms are determined and transferred to the right hand term.
Thus a new piezometry is determined iteratively.
Indicated as:

$$\varepsilon^n = h^n - h \tag{23}$$

the difference between the n-th iteration and the solution of system (20) this error will propagate with the following law:

$$\varepsilon^n = P\, \varepsilon^{n-1} \tag{24}$$

$$P = -A_o^{-1} A_1 \tag{25}$$

the iterative process (22) is convergent if and only if all the P eigenvalues modulus are inferior to one.
The terms of the $A_1$ matrix are directly proportional to aquitard permeability and inversely proportional to the thickness.
Consequently, even the dominating eigenvalue of the matrix P is directly proportional to the permeability and inversely proportional to the aquitard thickness. Therefore, we can see that the above-said process is not always converging. In particular, it doesn't for problems with high permeability and minor thickness on aquitards.
Another important aspect for the following considerations is that the described iterative procedure, both when it converges and when it diverges, presents a fluctuating behaviour. This fact was experimented by us in all its numerous applications to practical case and hypothetical ones.

b - Method of forced iterations

The system described as in (21) brings to the second iterative scheme:

$$A_o (h^n - h^{n-1}) = \alpha(b - Ah^{n-1}) \tag{26}$$

where $\alpha$ indicates the forcing factor of iterations. The iterative scheme can be written as follows:

$$h^n = (1 - \alpha) I h^{n-1} + \alpha A_o^{-1} (b - A_1 h^{n-1}) \tag{27}$$

To study the convergence of this second iterative scheme we analize the eigenvalues of matrix $\overline{P}$ which guides the propagation of error;

$$\overline{P} = (1- \alpha) I + \alpha P \tag{28}$$

Eigenvalues of the two matrices are tied by the same linear relation:

$$\mu = (1 - \alpha) + \alpha \lambda \tag{29}$$

$\mu$ indicates the eigenvalues of $\overline{P}$ and $\lambda$ those of P.
The oscillating trend of the iterative solution technique (22) points out that the dominating eigenvalues of matrix P, which

is real (P is symmetrical), is negative.
We will consider in this iterative solution technique, that particular forcing factor $\alpha_0$ which in equation (29) changes the dominating eigenvalue of P to 0.

$$\alpha_0 = \frac{1}{1 - \lambda_0} \tag{30}$$

Where $\lambda_0$ is the dominating eigenvalue. In this case all the negative eigenvalues of matrix P are transformed into positive and less than 1 eigenvalues (in fact less than $1 - \alpha_0$).
Eventual positive eigenvalues less than 1 of P remain positive and less than 1 in $\overline{P}$. Thus we can see that this second iterative scheme is convergent even when the first scheme is not convergent, provided that the dominating eigenvalue of P is negative and P does not include positive eigenvalues higher than 1 in its spectrum.
Both these iterative schemes are used in the iterative method which we associated to the frontal method in order to avoid enlargement of the front due to the presence of aquitards.

c - The iterated frontal method

The frontal method is a gaussian method and thus for the solution of system (20) uses a factorization of matrix A in the form of:

$$A = LDL^T \tag{31}$$

where:
D = diagonal matrix
L = lower triangular matrix
In the forward elimination phase the system is written in the equivalent form:

$$Uh = b^* \tag{32}$$

with

$$U = DL^T \tag{33}$$

$$b^* = L^{-1}b \tag{34}$$

In the back substitution phase the system is solved and the following solution is obtained:

$$h = U^{-1}b^* \tag{35}$$

Eventual subsequent solutions of the same system with different right hand terms can be obtained in "resolution", by using the previously calculated matrix U.

The iterative method used with the frontal method is substantially the forced iteration method as described previously.

However, several initial iterations are carried out with the first iterative method so as to estimate the dominating eigenvalue of matrix P and thus define by (30) the forcing factor in the second iterative method.

Three iterations resulted to be suitable for this objective, in all the practical applications made by us, resorting to the Rayleigh's quotient to estimate the dominating eigenvalue of matrix P.

In the iterated frontal method the first iteration is determined by considering the leakage terms as null.

The three following simple iterations are made in resolution.

By utilizing the piezometry of the last iteration carried out, the system's right hand term is updated with the leakage terms.

By applying the "resolution" procedure to the obtained right hand term, the result is:

$$b^{**(n)} = L^{-1} (b - A_1 h^{(n-1)}) \qquad (36)$$

and one proceeds with a back substitution in equation (35), with:

$$b^{*(n)} = b^{**(n)} \qquad (37)$$

We then proceed to forcing iterations after having verified that the dominating eigenvalue of P is negative, on the basis of the iteration's oscillating behaviour.

We start from the last previous iteration if the dominating eigenvalue of P is estimated less than one in modulus. In the opposite case, we should start from the initial iteration.

The right hand term is processed in resolution according to equation (36). In the back substitution phase this method differs from the first by applying in equation (35) the following right hand term:

$$b^{*(n)} = (1 - \alpha_o)\, b^{*(n-1)} + \alpha_o b^{**(n)} \qquad (38)$$

## NUMERICAL SOLUTION FOR THE UNSTEADY CASE

The numerical solution for the unsteady case is given by a predictor-corrector procedure. Use of the Crank-Nicolson method

for the solution of the differential equations of system (12) points out the same problems for the solution of the frontal method as in the steady case.
The problem is more complicated due to the non-linearity of the system in relation to the phreatic zone and to the dependence of the right hand term to the piezometry.
This last factor occurs in relation to particular requirements of the hydrogeological models, especially in relation to problems of aquifer-river exchange.
The proposed method solves these problems retaining the performance of the Crank-Nicolson method.
The differential equation system (12) could be written in an equivalent form:

$$C_o\dot{h}+A_o h = b-C^o{*}\dot{h}^*-C_1\dot{h}-A^o{*}h^*-A_1 h \tag{39}$$

$$C^*\dot{h}^*+A^*h^* = b^*-C^{*o}\dot{h}-A^{*o}h \tag{40}$$

with

$$C^o = C_o + C_1,\ A^o = A_o + A_1 \tag{41}$$

$C_o$ and $A_o$ are obtained by $C^o$ and $A^o$ without considering the contribution due to leakage.
The leakage contributions define the matrices $C_1$ and $A_1$. The predictor-corrector procedure initially determines the predictor piezometry for all the aquifer nodes subject to leakage.
This piezometry defines the right hand term of system (40) that is solved by resorting to the Crank-Nicolson method, and Thomas method for the linear system (since the relative matrix is tridiagonal).
In this way the predictor piezometry on the aquitards is obtained and by it the predictor estimate of leakage is done. These appear on the right of system (39). It thus determines the corrector piezometry on the aquifers, discretizing initially the time part with the Crank-Nicolson method, and solving the resulting linear system with the substructured frontal method previously described.
The timestep calculation is concluded by the determination of the corrector piezometry on aquitards, which is determined using the same procedure utilized for the predictor piezometry on aquitards. The only difference is that the right hand term of equation (40) is determined on the basis of the corrector piezometry on aquifers.

The time advancement scheme of the predictor-corrector method is based on a sequence of time macrosteps. Each time macrostep is composed of an arbitrary number of timestep; all of the same lenght. This permits to have the forward elimination phase with the frontal method only at the beginning of each macrostep, operating in resolution for all the subsequent steps of the macrostep, thus saving a lot of C.P.U. time.
At the beginning of each macrostep, the matrix of the system in phreatic zones is updated.

Predictor piezometry on aquifers

The predictor piezometry on aquifers is determined for each node subject to leakage, solving the system (39) with an explicit procedure, in which $C_o$ is evaluated using lumped capacity.
This procedure uses the forward differences for the time derivatives of system (39), and gives directly the values of the piezometry at time $t+\Delta t$. (Being matrix $C_o$ diagonal now).
In 2 different circumstances this procedure is not adequate and causes a high fluctuation of the solution.
The first case occurs in correspondance to nodes near which considerable water extractions are made, with high values of the time increment $\Delta t$. The cause of this is that the previous formulation don't takes correctly into account the evolution of the fluid flowing from nearest node in that $\Delta t$ interval.
To overcome the above difficulty we have introduced an alternative method for evaluation of predictor piezometry for all nodes which, based on the previous evaluations, present a piezometric variation exceeding 10 m.
Using this method, to determine the drawdown on node $P_1$, we will imagine that all source terms acting on the elements converging into node $P_1$ are located on the element's barycentre and assimilated to punctual source terms.
Drawdown on node $P_1$ is estimated as the sum of the drawdown pertaining to each source term located on the element barycentres, defined on the basis of the Theis formula.
The second case in which the explicit method becomes critical is represented by the case of source points with piezometric control. For this case, a predictor procedure is adopted which purposely takes into account the law of dependence of punctual source terms to piezometry.
We will use q to indicate the source term value as a piezometry function:

$$q = q^* f\ (h(P)) \tag{42}$$

$q^*$ is the constant reference source term and f (h) is the law of dependence on the piezometry on point P, represented with a step function or with a straight segment function. In this case we can formulate the equations resorting to the forward difference for the temporal terms as in the previous case taking now into account the formula (42) for the source term.

Corrector piezometry on the aquifers

The corrector piezometry is determined by applying the Crank Nicolson method to system (39), where the right-hand term is evaluated by utilizing the predictor piezometry.
The system obtained is solved by using the substructured frontal method.
The solution occurs with assembly of equations and forward eliminations at the beginning of each macrostep, and in resolution for every other step of the time macrostep.

Predictor and corrector piezometry on aquitards

The predictor and corrector piezometry is determined with the same procedure on each aquitard macroelement.
Application of the Crank-Nicolson method to the equation of system (40) related to internal nodes of each aquitard macroelement, yields a system in which the matrix is tridiagonal. Therefore to solve this system, the Thomas method is applied.

## APPLICATIONS

The illustrated method has been applied in very complex practical cases. One of thus is described in detail by Pizzi G. and Sartori L. (1982).
This case with 7500 elements and 4690 nodes required on 3033 I.B.M. 260 sec of C.P.U. for the steady state case (7 iterations) and 1147 sec of C.P.U. to simulate a 50 years history using three macrosteps and a total of 24 steps.
The model cover a total surface of 800.000 $km^2$.
The minimum area element has a surface of 3 $km^2$; the maximum area is approx. 3000 $km^2$.

## AKNOWLEDGEMENTS

The authors wish to thank Mr. J. Donea and Mr. S. Giuliani of the Joint Research Center Ispra Establishement, and Mr. G. Pizzi of Aquater S.p.A. in San Lorenzo in Campo (PS) for their critical contributions and suggestions.

## REFERENCES

Chorley Don W. and Emil O. Frind (1978) An iterative quasi-three-dimensional finite element model for heterogeneous multiaquifer systems.
Water Resources Research 14.5.943-952.

Connor J.J. and Brebbia C.A. (1977) Finite Element Techniques for Fluid Flow.
Newnes-Butterworths. London.

Edelman J.H. (1972) Groundwater Hydraulics of Extensive Aquifers.
International Institute for Land Reclamation and Improvement. Wageningen.

Fujinawa K. (1977) Finite element analysis of groundwater flow in multiaquifer system, I. the behavior of hydrogeological properties in an aquitard while being pumped.
Journal of Hydrology 33.59-72.

Fujinawa K. (1977) Finite element analysis of groundwater flow in multiaquifer systems, II A quasi-three-dimensional flow model.
Journal of Hydrology 33.349-362.

Irons B.M. (1970) A frontal solution program for finite element analysis.
International Journal for Numerical Method in Engineering 2.5-32.

Neumann S.P. and Witherspoon P.A. (1969) Theory of flow in a confined two-aquifers system.
Water Resources Research 5.817.

Pizzi G., Sartori L. (1982) Interconnected GROundwater System Simulations (IGROSS). Description of the system and its applications to the Western Lybia regional aquifer.
4th Int. Conf. on Finite Elements in Water Resources. Hannover.

# SOLUTION OF IMMISCIBLE DISPLACEMENT IN POROUS MEDIA USING THE COLLOCATION FINITE ELEMENT METHOD

Allen M. Shapiro

Dept. of Water Resources Engineering
Royal Institute of Technology
Stockholm, Sweden

George F. Pinder

Dept. of Civil Engineering
Princeton University
Princeton, New Jersey, U.S.A.

## ABSTRACT

The equations governing the immiscible displacement of one fluid by another in a porous medium are solved using the collocation finite element method. Capillary forces are neglected yielding a partial differential equation which is hyperbolic and non-linear. The application of a standard orthogonal collocation formulation to this problem fails to provide acceptable solutions. However, as in other numerical schemes a method of upstream weighting is adopted which alleviates the problem of an oscillatory solution while causing a minimum amount of numerical dispersion in association with sharp changes in the solution profile. Solutions obtained using the collocation method are compared with finite difference, Galerkin finite element and subdomain finite element formulations.

## INTRODUCTION

The computationally attractive features of the collocation finite element method have been recognized by several authors in the solution of linear partial differential equations (see e.g. Finlayson, 1972; Pinder, et al., 1977). For these problems the collocation scheme has been shown to be at least comparable in accuracy to more widely used numerical techniques. However, adopting the collocation method to non-linear problems has achieved only varying degrees of success. Authors have noted difficulties in arriving at physically acceptable solutions to problems using the collocation method, where other weighted residual techniques and finite difference methods have been used successfully (see e.g. Sincovec, 1977).

An important example of this phenomena and the one addressed in this paper, involves the flow of two immiscible fluids through a porous medium. The equations governing this problem constitute a severe test to methods of solving non-linear, partial differential equations, since in the absence of capillary forces they reduce to a form which is first order and hyperbolic. Equations of this type are known to cause oscillatory numerical solutions especially when abrupt changes in the solution profile are encountered. By adopting upstream weighting techniques several numerical schemes have alleviated this difficulty. It is the intent of this paper to demonstrate that similar methods can also be applied to a collocation finite element formulation. The success in reproducing analytic solutions to the immiscible displacement problem implies that the computationally attractive features of collocation can then be adopted to more demanding areal simulations in oil recovery, as well as other non-linear problems in subsurface hydrology.

## THEORETICAL DEVELOPMENT

### Governing Equations

The equations describing the uni-directional flow of two immiscible fluids (e.g. oil (o) and water (w)) in a porous medium can be written in the following manner when capillary forces are assumed to be negligible:

$$R(S_w) = \varepsilon \frac{\partial S_w}{\partial t} + q \frac{\partial}{\partial S_w}\left(\frac{\mu_o}{k_o} K\right) \frac{\partial S_w}{\partial x} = 0 \tag{1}$$

where $\varepsilon$ is porosity,

$S_\alpha$ is the saturation of the void space occupied by $\alpha$-phase ($\sum_\alpha S_\alpha = 1$),

$q$ is the total volumetric flow rate per unit cross-section ($q = \sum_\alpha q_\alpha$),

$q_\alpha$ is the Darcy velocity of the $\alpha$-phase,

$\mu_\alpha$ is the dynamic fluid viscosity of the $\alpha$-phase,

$k_\alpha$ is the permeability of the $\alpha$-phase ($k_\alpha = k\, k_{r\alpha}$),

$k$ is the medium permeability,

$k_{r\alpha}$ is the relative permebility of the $\alpha$-phase, and

$$K = \frac{k_w k_o}{k_o \mu_w + k_w \mu_o}$$

Throughout the remainder of the text an investigation of the water-flooding of an oil saturated reservoir is to be conducted. Boundary and initial conditions for this problem are given by

$$S_w(0,t) = 1 - S_{or} \qquad t \geq 0 \tag{2}$$

$$S_w(x,0) = S_{wr} \qquad 0 \leq x \leq L \tag{3}$$

where $S_{\alpha r}$ is the residual saturation of the α-phase. For computational purposes the relative permeability functions are assumed to be the following:

$$k_{rw}(S_w) = \frac{(S_w - S_{wr})^2}{(1 - S_{or} - S_{wr})^2} \tag{4}$$

$$k_{ro}(S_w) = \frac{(1 - S_{or} - S_w)^2}{(1 - S_{or} - S_{wr})^2} \tag{5}$$

with $S_{wr} = 0.16$ and $S_{or} = 0.20$. The rate of water injection q is $1.51 \times 10^{-7}$ m/sec and the porosity is 0.20. Furthermore, the viscosities of oil and water are assumed to be equivalent.

Numerical Solution Technique

Solutions to equation (1) are to be defined by the use of the collocation finite element method. This technique is one member in the family of numerical schemes based on the method of weighted residuals, i.e.

$$\int_\ell R(\hat{S}_w)\, \omega_i(x)\, dx = 0 \qquad i = 1,2, \ldots M \tag{6}$$

where the weighting functions are denoted by $\omega_i$ and the discretized representation of the true solution is given by $\hat{S}_w$. Substituting equation (1) into (6) yields

$$\int_\ell \left( \frac{\partial \hat{S}_w}{\partial t} + V(\hat{S}_w)\frac{\partial \hat{S}_w}{\partial x} \right) \omega_i(x)\, dx = 0 \qquad i = 1,2, \ldots M \tag{7}$$

where the non-linear coefficient is now defined by $V(S_w)$,

$$V(S_w) = \frac{q}{\varepsilon} \frac{\partial}{\partial S_w} \left( \frac{\mu_o}{k_o} K \right) \tag{8}$$

The integrations noted above may also be written in a local coordinate system $(-1 \leq \xi \leq 1)$ and defined over each spatial increment, i.e.

$$\int_{-1}^{1} \left(\frac{\partial \hat{S}_w}{\partial t} + V(\hat{S}_w) \frac{\partial \hat{S}_w}{\partial \xi} \frac{\partial \xi}{\partial x}\right) \omega_i(x(\xi)) \frac{dx}{d\xi} \, d\xi = 0 \qquad i = 1,2, \ldots M \tag{9}$$

In the collocation method the weighting functions are chosen as the Dirac delta functions, which are defined in the local coordinate system by

$$\omega_i(x(\xi)) = \delta(\xi - \xi_i) \tag{10}$$

Properties of the Dirac delta function reduce the integration of equation (9) to the evaluation of the integrand at the collocation points, $\xi_i$. This feature makes the collocation method attractive in comparison with other weighted residual techniques since very little computational effort is required in matrix formulation. Here, the collocation points are chosen as the Gauss points within each spatial increment. This is referred to as orthogonal collocation, a discussion of which appears in Finlayson (1972).

The spatial functions for the trial solution are selected as Hermitian cubic polynomials, i.e.

$$\hat{S}_w = \sum_{j=1}^{N} \left(S_j(t)\phi_{oj}(\xi) + \frac{\partial S_j}{\partial x}(t) \frac{\partial x_j}{\partial \xi} \phi_{1j}(\xi)\right) \tag{11}$$

where
$$\phi_{oj} = -\left(\tfrac{1}{4}\right)\beta_j(\xi+\beta_j)^2(\xi-2\beta_j)$$
$$\phi_{1j} = \left(\tfrac{1}{4}\right)(\xi+\beta_j)^2(\xi-\beta_j)$$

and $\beta_j = \pm 1$. This provides two unknowns at each nodal location, the water saturation and its spatial gradient. Thus, solutions are continuous through the first derivative. Furnishing two collocation points in each spatial increment provides 2N-2 equations where N is the number of nodes. The two additional restrictions which are required result from conditions imposed at the boundaries of the domain.

Substituting equations (10) and (11) into equation (9), and introducing a finite difference approximation for the time derivative yields the following set of non-linear algebraic equations:

$$\sum_{j=1}^{N} \left( (S_j^{t+\Delta t} - S_j^{t}) \phi_{oj} + \left(\frac{\partial S_j^{t+\Delta t}}{\partial x} - \frac{\partial S_j^{t}}{\partial x}\right) \frac{\partial x_j}{\partial \xi} \phi_{1j}\right.$$

$$+ \gamma(\Delta t)\, V(\hat{S}_w)^{t+\Delta t}\left(S_j^{t+\Delta t} \frac{d\phi_{oj}}{d\xi} + \frac{\partial S_j^{t+\Delta t}}{\partial x} \frac{\partial x_j}{\partial \xi} \frac{d\phi_{1j}}{d\xi}\right) \frac{\partial \xi}{\partial x}$$

$$+ (1-\gamma)\, \Delta t\, V(\hat{S}_w)^t\left(S_j^t \frac{d\phi_{oj}}{d\xi} + \frac{\partial S_j^t}{\partial x} \frac{\partial x_j}{\partial \xi} \frac{d\phi_{1j}}{d\xi}\right) \frac{\partial \xi}{\partial x}\Big)\Big|_{\xi=\xi_i} = 0$$

$$i = 1,2, \ldots M \qquad (12)$$

where the value of $\gamma$ $(0 \leq \gamma \leq 1)$ defines the location of the convective term in the time domain and $\Delta t$ is the time increment. The presence of the non-linear coefficient V requires an iterative method for obtaining solutions to the above set of equations. The method adopted in this analysis is that of updating the non-linear coefficient based on the solution at the previous iteration.

The hyperbolic nature of the governing equation and the sharp interface proposed by the boundary and initial conditions cause solutions for water saturation to be similar to those of a propagating step function. With the use of standard numerical techniques, convergence to an incorrect solution or one which is oscillatory often is the result. One means of alleviating this problem has been to use one-sided representations of the spatial derivative in the convective term along with a method of evaluating the non-linear coefficient which is biased with upstream values of water saturation. These and other numerical techniques that have been used in solving the water-flooding problem are discussed by Mercer and Faust (1976).

As one may anticipate, the standard formulation of the collocation finite element method as presented above, yields unacceptable solutions to the water-flooding problem. For this reason a modification in the numerical formulation is proposed which serves as an upstream weighting technique. Here, asymmetric basis functions are applied in the evaluation of the convective term. These functions have the following form (Pinder and Shapiro, 1980):

$$\phi'_{oj} = -(\tfrac{1}{4})\beta_j(\xi+\beta_j)^2(\xi-2\beta_j) - \alpha_o\beta_j(\xi^2-1)^2 \qquad (13)$$

$$\phi'_{1j} = (\tfrac{1}{4})(\xi+\beta_j)^2(\xi-\beta_j) - \alpha_1\beta_j(\xi^2-1)^2 \qquad (14)$$

where $\beta_j = \pm 1$. The magnitude of the undetermined parameters $\alpha_o$ and $\alpha_1$ define the additional emphasis to be placed on upstream values of water saturation and its slope. The convergence and stability properties associated with this technique when applied to linear, partial differential equations has been investigated by Shapiro and Pinder (1981). In addition, upstream weighting of the non-linear coefficient is proposed using interpolation functions which are similar to equations (13) and (14).

The method of upstream weighting described above, however, is in contrast to the majority of techniques used to define one-sided approximations for the convective term in other weighted residual schemes. Authors have often defined upstream weighting with the use of an asymmetric weighting function (see e.g. Huyakorn, 1977). In a recent publication, Allen and Pinder (1981) applied this concept to collocation by evaluating the convective term at points which lie upstream of the Gauss points.

## DISCUSSION OF RESULTS

### Upstream Weighted Collocation

A comparison of the standard and upstream weighted orthogonal collocation formulations is displayed in Figure 1. Although the standard formulation conserves mass its representation of the oil-water interface is incorrect. By applying an asymmetric basis function to the convective term a mass conserving solution is also achieved, however, the sharp interface is more correctly identified. In both cases the non-linear coefficients are evaluated by first calculating water saturation at the collocation points with the use of the Hermitian cubic interpolation functions, after which equation (8) is employed.

The solution obtained using upstream weighting proves to be quite sensitive to the choice of the coefficients $\alpha_0$ and $\alpha_1$. Larger values of $\alpha_0$ than that employed do not smear the sharp front as one might anticipate. Instead, the position of the sharp interface is impeded as in the case where no upstream weighting is employed. Non-zero values of $\alpha_1$, however, have only an adverse effect on the solution profile. Oscillations in the vicinity of the sharp interface result from gradients in water saturation at the oil-water interface which are steeper than the case of $\alpha_1$ equal to zero.

One distressing fact that can be noted concerning the collocation solutions is their inability to reproduce the spatial gradients predicted by the analytic solution. In some instances this difference is an order of magnitude, which indicates the insensitivity of the solution for water saturation to its gradient. The cause of this discrepancy is judged to be the result of the initial conditions prescribed in the collocation method. Equations (2) and (3) denote a discontinuous function where the slope at $x = 0$ and $t = 0$ is undefined. For $C^0$ continuous elements, implementation of these boundary and initial conditions poses no additional problem. However, in using Hermite cubic basis functions one is forced to specify the gradient at $x = 0$. Here, an

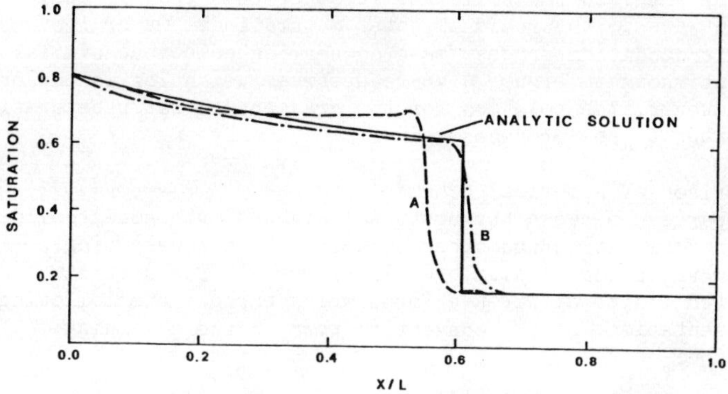


Figure 1 Comparison of standard (A) and uptsream weighted (B) orthogonal collocation formulations ($\gamma$ = 0.5, $\Delta x$ = 15.24 m, $\Delta t$ = 10. days, $\alpha_0$ = 0.10, $\alpha_1$ = 0.00) at an elapsed time of 1500 days.

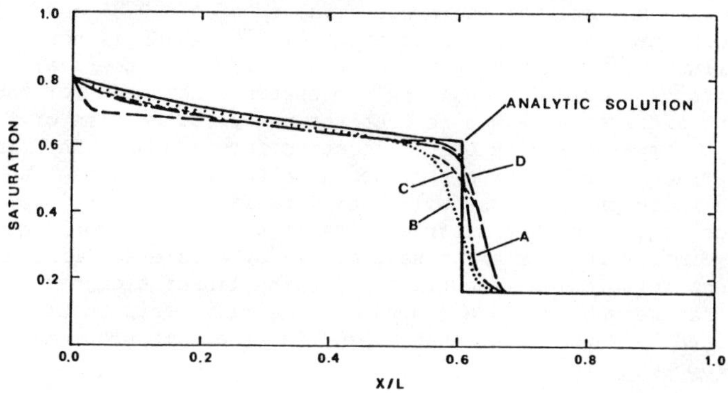


Figure 2 Comparison of upstream weighted orthogonal collocation (A) ($\gamma$ = 0.5, $\Delta x$ = 15.24 m, $\Delta t$ = 10. days, $\alpha_0$ = 0.10, $\alpha_1$ = 0.00), upstream weighted finite difference (B) ($\gamma$ = 0.5, $\Delta x$ = 7.62 m, $\Delta t$ = 10. days, $\sigma$ = 0.85, $\lambda$ = 0.80), Galerkin finite element (C) (from Mercer and Faust, 1976) and subdomain finite (D) ($\gamma$ = 0.5, $\Delta x$ = 15.24 m, $\Delta t$ = 10. days, $\alpha_0$ = 0.30, $\alpha_1$ = 0.00) formulations at an elapsed time of 1500 days.

initially steep slope is chosen which imparts no additional mass of water in the soil profile other than that already present due to the residual water saturation. Other choices of this condition produce mass conserving solutions similar to that shown in Figure 1 where upstream weighting is employed, however, the solution for the gradient of water saturation continues to be erroneous.

Comparison of Numerical Schemes

A comparison between the upstream weighted orthogonal collocation formulation and other numerical techniques using upstream weighting is displayed in Figure 2. The upstream weighted finite difference formulation employs the following representation of the convective term in the governing equation:

$$V(S_w)\frac{\partial S_w}{\partial x} \approx V(\overline{S}) \left( \frac{(1+\sigma)}{2\Delta x}(S_i - S_{i-1}) + \frac{(1+\sigma)}{2\Delta x}(S_{i+1} - S_i) \right)$$

where

$$\overline{S} = \lambda\, S_{i-1} + (1-\lambda)\, S_i$$

and $\Delta x$ is the spatial increment. Values of $\sigma = 0.85$ and $\lambda = 0.80$ provide a mass conserving solution which is non-oscillatory. A just comparison between the two numerical schemes is provided by halving the spatial increment for the finite difference method so that the number of degrees of freedom is comparable with the collocation formulation. The finite difference approximation provides a better solution over the range which precedes the oil-water interface, however, the representation of the interface itself tends to be more smeared. Similar results are also provided by a Galerkin finite element formulation (see Figure 2) using linear elements, upstream weighting and a lumped mass matrix. This result was obtained by Mercer and Faust (1976) for the same problem that is proposed here.

The subdomain finite element method also displayed in Figure 2 is another type of weighted residual scheme. The weighting functions are defined as step functions with a unit magnitude. Here two weighting functions are defined in each spatial increment, i.e.

$$\omega_i(\xi) = \begin{cases} 1 & -1 \leq \xi \leq 0 \\ 0 & 0 < \xi \leq 1 \end{cases}$$

$$\omega_i(\xi) = \begin{cases} 0 & -1 \leq \xi < 0 \\ 1 & 0 \leq \xi \leq 1 \end{cases}$$

Hermite cubic basis functions and the method of upstream weighting used for the collocation scheme are also employed. Suprisingly, the collocation formulation provides a more accurate solution. The integrations associated with the subdomain method are carried out using Gaussian integration. Thus, each equation is the result of the weighted sum of the residuals evaluated at several locations in the element. In contrast, the collocation scheme considers the residual at only one point in defining a given equation. One might then have considered the collocation formulation to be of less accuracy, however, the results here do not support this conclusion.

## CONCLUSIONS

A standard orthogonal collocation formulation to the immiscible displacement of oil by water in a porous medium yields solutions which incorrectly identify the position of the sharp interface between the two fluids. By adopting an upstream weighting technique which employs an asymmetric basis function in the evaluation of the convective term, solutions for water saturation show a greater degree of accuracy in comparison to the analytic solution. However, errors continue to exist in the numerically generated spatial gradient in water saturation. Furthermore, a comparison with other numerical techniques which also employ upstream weighting indicates that the upstream weighted orthogonal collocation scheme provides an equivalent degree of accuracy for a similar amount of computational effort.

## REFERENCES


Allen, M. and G. Pinder, (1981) Collocation simulation of multi-phase porous medium flow, Princeton University Water Resources Program Report 81-WR-6, Princeton, New Jersey.

Finlayson, B., (1972) The Method of Weighted Residuals and Variational Principles, Academic Press, New York.

Huyakorn, P., (1977) Solution of the steady-state, convective equation using an upwind finite element scheme, Applied Mathematical Modelling, Vol. 1, pp. 187-195.

Mercer, J. and C. Faust, (1976) The application of finite element techniques to immiscible flow in porous media, in Finite Elements in Water Resources, ed. W. Gray, G. Pinder and C. Brebbia, Pentech Press, London.

Pinder, G., E. Frind and M. Celia, (1978) Groundwater flow simulation using collocation finite elements, in Finite Elements in Water Resources, ed. C. Brebbia, W. Gray and G. Pinder, Pentech Press, London.

Pinder, G. and A. Shapiro, (1979) A new collocation method for the solution of the convection-dominated transport equation, Water Resources Research, Vol. 15, pp. 1177-1182.

Shapiro, A. and G. Pinder, (1981) Analysis of an upstream weighted collocation approximation to the transport equation, Journal of Computational Physics, Vol. 39, pp. 46-71.

Sincovec, R., (1977) Generalized collocation methods for time-dependent, non-linear, boundary value problems, Society of Petroleium Engineers Journal, Vol. 17, pp. 345-352.

# THE STREAMFUNCTION METHOD
# A DIRECT APPROACH TO SIMULATE GROUNDWATER FLOW

Gerhard Willms

Ruhr-Universität Bochum, W-Germany
Inst. f. Konstr. Ingenieurbau, Arbeitsgruppe IV

## INTRODUCTION

The streamfunction $\psi$ can be used to describe two dimensional groundwater flow. Because the given properties of the flow can not be applied immediately to the streamfunction formulation this method was used only rarely or indirectly up to now.

In this paper the numerical application of the $\psi$-model will be discussed. The theory of potential and streamfunction is given else where /1/, /2/ and will not be repeated here. It will be shown that all given physical data, on the boundary or in the interior, can be transformed to the $\psi$-model. In addition examples will be shown where the $\psi$-model is more adequate then the potential model which fails due to missing boundary conditions or due to numerical difficulties /2/.

The calculations were performed with the finite element program SICK-100 /4/ on a PRIME 400 with double precision accuracy.

## MATHEMATICAL DESCRIPTION

Darcy's law, conservation of mass and existence of a potential are the principles describing groundwater flow. Two boundary value problems follow:

| Potential ($\phi$-model) | Streamfunction ($\psi$-model) |
|---|---|
| differential equations | |
| $(t\, k_{ij}\, \phi,_j),_i = q$ | $e_{ij}\, e_{ks}(\frac{1}{t}\, c_{ij}\, \psi,_s),_i = 0$ |

boundary conditions

| | |
|---|---|
| $\phi = \bar{\phi}$ on $\Gamma_1$ | $u_t = \bar{u}_t$ on $\Gamma_1$ |
| $v_n = \bar{v}_n$ on $\Gamma_2$ | $\psi = \bar{\psi}$ on $\Gamma_2$ |

$\Gamma = \Gamma_1 \cup \Gamma_2$ : boundaries of the region (exterior and interior),
$q$ : source/sink term ,
$k_{ij}$ : components of the permeability,
$c_{ij}$ : components of the hydraulic resistance $c_{ij} = k_{ij}/(k_{11}\, k_{22} - k_{12}^2)$ ,
$t$ : thickness of the aquifer ,
$v_n$ : flow velocity normal to the boundary (flux),
$u_t$ : component of the hydraulic gradient tangential to the boundary,
$e_{ij}$ : permutation symbol ($e_{11}= e_{22}=0$, $e_{12}=1$, $e_{21}=-1$),
$\phi$ : hydraulic head.

The boundary values $\bar{\psi}$ and $\bar{u}_t$ are related to the physical values as follows:

$$\bar{\psi}(S) = \bar{\psi}_o + \int_o^S \bar{v}_n\, t\, ds = \bar{\psi}_o + Q(S) \quad \text{on } \Gamma_2, \tag{1}$$

where $\bar{\psi}_o = \bar{\psi}(S{=}0)$ is arbitray and $Q(S)$ is the sum of flux over $\Gamma_2$ up to the point $S$ on $\Gamma_2$,

$$\bar{u}_t = - \frac{d\bar{\phi}}{ds} \quad \text{on } \Gamma_1 \; . \tag{2}$$

Equation (2) leads to the definition of the so called "circulation" Z. One obtains for example for a boundary part $\Gamma_1$:

$$\bar{Z}(\Gamma_1) = \int_{\Gamma_1} \bar{u}_t(s)\, ds \mathrel{\hat{=}} - \int_{\Gamma_1} d\phi = - \Delta\bar{\phi} \, . \tag{3}$$

$\bar{Z}$ is an entity which defines a tangential velocity along $\Gamma_1$ compatible with the increase of the potential.

The definition of Z allows the determination of a free surface flow. For each value $\bar{\psi}$ on $\Gamma_2$ the corresponding Z will be computed. Therefore one can find all boundary potentials by the following equation:

$$\phi(S) = - \phi_o + \int_o^S u_t(s)\, ds = - \phi_o + Z(\Gamma_S) \quad \text{on } \Gamma, \tag{4}$$

where $\phi_o = \phi(S{=}0)$ has to be chosen as a known elevation of the free surface and $Z(\Gamma_S)$ is the circulation along $\Gamma$ from S=0 up to s=S.

A special iterative method with fixed element grid to determine the free surface will be presented in the next part.

Sources and sinks lead to multi-valued streamfunctions. These multi-valued functions can be avoided by arbitrary Riemannian cuts along element boundaries from the source point to the exterior boundary of the region.

NUMERICAL PROCEDURE

The finite element approximation

$$\psi(x_i) = \Omega^\alpha(x_i)\psi^\alpha \quad , \quad \alpha = 1,2,\dots,N$$

with $\Omega^\alpha$ : patch function at node $\alpha$, $\psi^\alpha$ : function value at node $\alpha$, N : number of nodes, leads to the set of equations:

$$C^{\alpha\beta}\,\psi^\beta = Z^\alpha \qquad \alpha = 1,2,\dots,N \quad ,$$

where

$$C^{\alpha\beta} = \int_D \frac{1}{t}\, c_{ij}\, e_{ik}\, \Omega^\alpha_{,k}\, e_{js}\, \Omega^\beta_{,s}\, dD \, ,$$

$$Z^\alpha = \int_\Gamma \Omega^\alpha \phi(s)_{,s}\, ds \quad .$$

D is the considered region with boundary $\Gamma$, $\bar{\psi}^\beta$ and $\bar{Z}^\alpha$ are the discrete nodal values which follow from equation (1) and (3), $c_{ij}$ is the hydraulic resistance obtained as the inverted permeability. Special care has to be taken if part $\Gamma_2$ of the boundary is not a single segment. If $\Gamma_2$ consists of several segments interrupted by $\Gamma_1$ or if the inner region is multiply connected the integration in equation (1) has to be discontinued since the discharge along $\Gamma_1$ us unknown. Two constraints have to be introduced:

<u>Constraint 1</u> :

$\psi = \psi^*$ = const, but unknown on $\Gamma_a$

where $\Gamma_a$ is a part of $\Gamma_2$, i.e. all nodes of $\Gamma_a$ have to be assigned to the same unknown value $\psi^*$.

<u>Constraint 2</u> :

$\psi(A) = \psi(B) + Q$, $\psi(B)$ unknown,

where A,B are nodes on $\Gamma_2$ and Q is a constant equal to the flux between A and B.

This leads to the concept of super nodes:
All nodes which are adressed by the same constraint form a super node. Algebraicly the equations of these nodes have to be added. In case of constraint 2 the effect of the quantity Q enters on the right hand side of the equation of the super node.

In Fig. 1 the application of constraint 1 in a multiply connected region is shown. Boundary conditions and constraints for this example are given in Table 1. Here two super nodes were defined: the bottom (C-D) with a circulation $\bar{Z}=\Delta\phi$ and the inner boundary $\Gamma_I$ without any circulation . The potential lines and streamlines, also shown in Fig. 1, are orthogonal in every point and the velocities are equal in magnitude and direction.

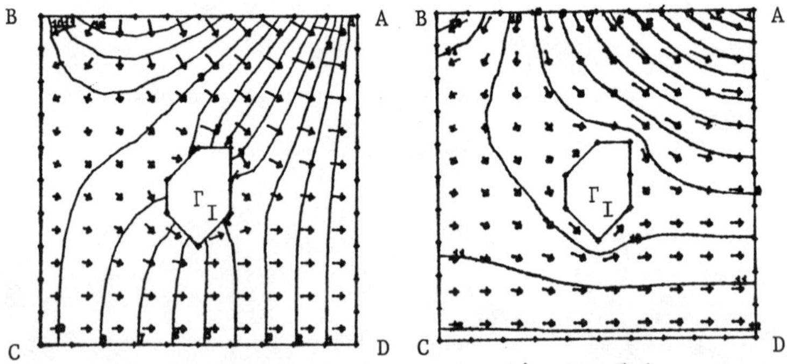


a) $\phi$-model b) $\psi$-model

Fig. 1: Multiply connected region with constraint 1

| Boundary | $\phi$-model | $\psi$-model |
|---|---|---|
| A-B | $v_n\ t > 0$ | $\psi(S) = \int_A^S v_n\ t\ ds \quad S \in \overline{AB}$ |
| B-C | $\phi = \phi_o$ | $\frac{d\phi}{ds} = 0 \rightarrow Z = 0$ |
| C-D | $v_n\ t = 0$ | $\psi = \psi^* = \text{const}\ ,\ Z = \phi_1 - \phi_o = \Delta\phi$ |
| D-A | $\phi = \phi_1$ | $\frac{d\phi}{ds} = 0 \rightarrow Z = 0$ |
| $\Gamma_I$ | $v_n\ t = 0$ | $\psi = \tilde{\psi} = \text{const}\ ,\ Z = 0$ |

Table 1: Boundary conditions and constraints for example Fig. 1

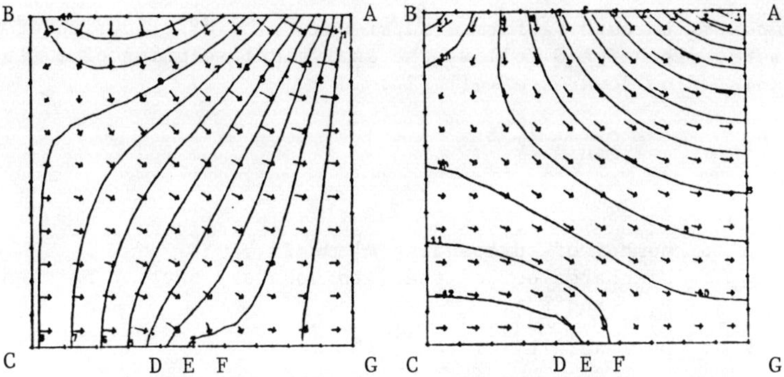


a) $\phi$-model b) $\psi$-model

Fig. 2: An application of constraint 2

| Boundary | $\phi$- model | $\psi$- model |
|---|---|---|
| A-B | $v_n\ t > 0$ | $\psi(S) = \int_A^S v_n\ t\ ds\ ,\ S \in \overline{AB}$ |
| B-C | $\phi = \phi_o$ | $\frac{d\phi}{ds} = 0 \rightarrow Z = 0$ |
| C-D | $v_n = 0$ | $\psi(\overline{CD}) = \psi^* = \text{const}\ ,\ Z = \phi_1 - \phi_o$ |
| D-E | $v_n\ t < 0$ | $\psi(D) = \psi^*\ ;\ \psi(E) = \psi^* + \int_D^E v_n\ t\ ds$ |
| E-F | $v_n\ t < 0$ | $\psi(F) = \psi(E) + \int_E^F v_n\ t\ ds$ |
| F-G | $v_n = \upsilon$ | $\psi(\overline{FG}) = \psi(F) = \text{const}$ |
| G-A | $\phi = \phi_1$ | $\frac{d\phi}{ds} = 0 \rightarrow Z = 0$ |

Table 2: Boundary conditions and constraints for example Fig. 2

In Fig. 2 the application of constraint 2 is illustrated. Table 2 gives the boundary conditions and constraints. Here the boundary from C to G was connected to one super node.

Sources and sinks

Again the concept of super nodes together with the constraint 2 can be used. In Fig. 3b the discretization of a source point, made up out of boundary sections of 5 elements,is shown. The nodes $P_0$ through $P_5$ form a super node. The distribution of the source discharge Q follows the analytical solution of a steady radial flow towards a well /3/:

$$Q_j = Q \frac{\Theta_j k_j}{\sum_{i=1}^{n} \Theta_i k_i} \qquad j = 1,2,\ldots,n$$

n : number of surrounding elements,

$Q_j$ : discharge out of the j-th boundary section between $P_{j-1}$ and $P_j$ ,

$k_j$ : permeability of the j-th element,

$\Theta_j$ : angle of the j-th element (see Fig. 3a) .

With this construction one obtains the following constraints for the source (super node $P_o$-$P_5$):

$$\psi(P_o) = \psi^*, \text{ unknown },$$

$$\psi(P_j) = \psi^* + \sum_{i=1}^{j} Q_i \quad , j = 1,2,\ldots,5 \quad . \tag{5}$$

Along the cut double nodes (super nodes as well) are defined. constraint 2 for A1, $A_2$ in Fig. 3b is as follows:

$$\psi(A_1) = \tilde{\psi} \quad , \text{unknown },$$

$$\psi(A_2) = \tilde{\psi} + Q . \tag{6}$$

Similar constraints have to be specified for the other double nodes along the cut.

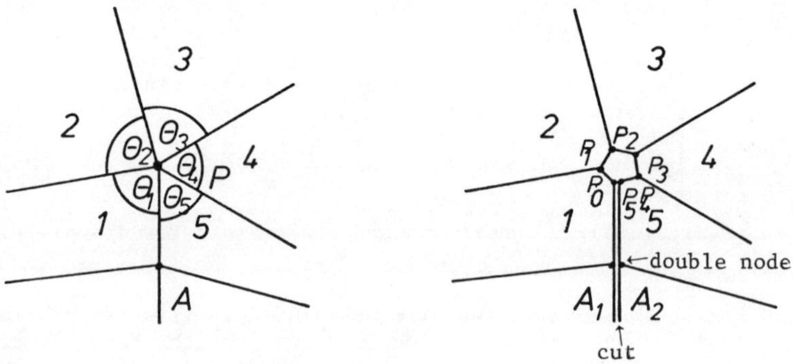


a) $\phi$-model b) $\psi$-model

Fig. 3: Discretization of a source point

In Fig. 4 the streamfunction approach for an internal sink is shown. This is a typical example where a solution exists in the $\psi$-model but not in the $\phi$-model because there is one boundary condition lacking. Table 3 gives the boundary conditions and constraints.

A dipole (a source with inflow = + Q and sink with outflow = - Q) is shown in Fig. 5. Only one cut is needed.

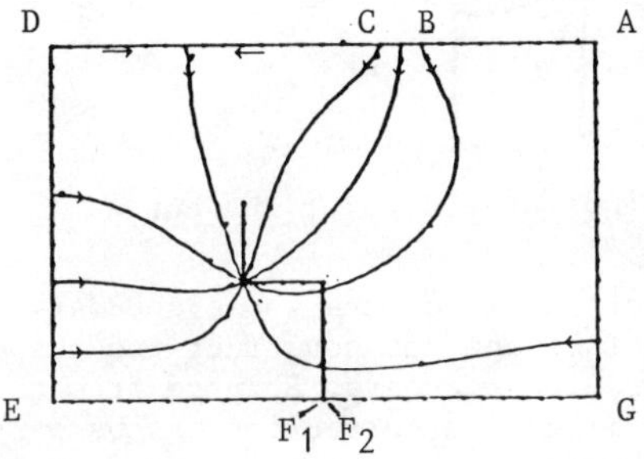


Fig. 4: Streamlines of an internal sink computed with the $\psi$-model

| Boundary | $\psi$-model | $\phi$-model |
|---|---|---|
| A-B | $\psi = 0$ | $v_n = 0$ |
| B-C | $\psi(S) = \int_B^S v_n \, t \, ds \, , \; S \in \overline{BC}$ | $v_n \, t > 0$ |
| C-D | $\psi = \psi(C)$ | $v_n = 0$ |
| D-E | $u_t = \frac{d\phi}{ds} = 0 \rightarrow Z = 0$ | $\phi = \phi_o$ |
| $E-F_1$ | $\psi = \hat{\psi} = \text{const.}$ | $v_n = 0$ |
| cut | equation (6) for each double node | ./. |
| sink | equation (5) | $q = Q$ |
| $F_2-G$ | $\psi = \hat{\psi} + Q = \text{const.}$ | $v_n = 0$ |
| G-A | $u_t = 0 \rightarrow Z = 0$ | $\phi$, $v_n$ unknown |

Table 3: Boundary conditions and constraints for example Fig.4

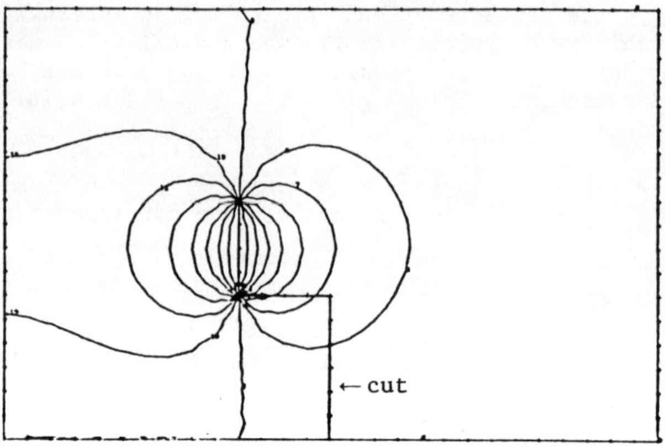


Fig. 5: Streamlines of a dipole; one cut

Free surface flow

With equation (4) the potentials of the whole boundary can be found. Hence it follows that one can construct the shape of the free surface from the assumed boundary. Seepage from the top is taken into account. The iterative correction can be performed with a fixed grid. The knwon $\psi$-values have to be carried over to the new free surface and the region above it has to be made more and more impermeable by increasing the reciprocal transmissivities. The impermeability can be simulated by the definition of factors $\alpha_E = A_S/A_T$, where $A_S$ is the saturated and $A_T$ the total area of an element. In the program one may not decrease $\alpha_E$ to zero to avoid division by zero. Streamlines will develop from the assumed boundary to the final free surface.

A corresponding method exists for the $\phi$-model: Here the seepage from the top has to be carried over to the new free surface.

A fixed grid method is especially advantageous if there are horizontal layers in the aquifers.

Fig. 6 shows the calculation of a free surface with the $\psi$- and $\phi$-model.

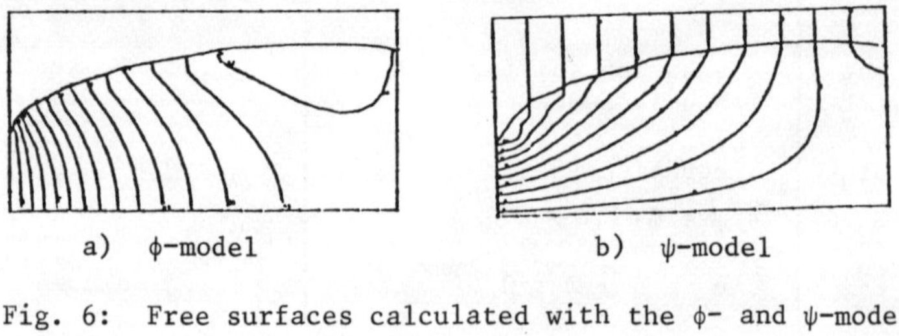

a) $\phi$-model b) $\psi$-model

Fig. 6: Free surfaces calculated with the $\phi$- and $\psi$-model

## CONCLUSION

Seepage flow can be formulated by the potential or streamfunction approach. With the indicated techniques for the $\psi$-model it is possible to compute the velocity field and the streamlines in a direct way. Only a few changes in an existing finite element program are necessary. The advantages of the streamfunction method are:

- direct flow path computation
- good numerical behaviour, where the $\phi$-model may fail
- solutions exist although the $\phi$-model fails due to missing boundary conditions.

In addition the streamfunctions are of great advantage for groundwater quality problems where travel path and travel time are of interest.

## ACKNOWLEDGEMENTS

The author is thankful to Prof. Dr. G. Schmid and Dr. P. Obermann (Ruhr-University Bochum) for helpful suggestions and critically going through the manuscript and to Mrs. Hoogterp for typing.

## REFERENCES

/1/ D. Withum, Elektronische Berechnungen ebener und räumlicher Sicker- und Grundwasserströmungen durch beliebig berandete, inhomogene anisotrope Medien, Mitteilungen des Inst. f. Wasserwirtschaft, TH Hannover, 1967.

/2/ G. Schmid, Seepage flow in extremely thin aquifers, Adv. Water Resources, 1981, Volume 4, 134 - 136.

/3/ C. van den Akker, The Calculation of streamlines in an inhomogeneous, anisotropic porous medium with two-dimensional flow, in W.v.Duijvenbooden et al (Eds.): Quality of Groundwater, Elsevier Scientific publish. Comp., Amsterdam, 1981.

/4/ SICK-100, Ein Programmsystem zur Berechnung von Grundwasserströmungen mit der Methode der finiten Elemente, Benutzerhandbuch, Mitteilungen des Inst. f. Konstr. Ingenieurbau, AG IV, Ruhr-Universität Bochum, Februar 1981.

# 10. SIMULATION OF GROUNDWATER FLOW

# INFLUENCE OF A SOIL PIPE NETWORK ON CATCHMENT HYDROLOGY

Mark D. Barcelo and John L. Nieber

Texas A&M University

## INTRODUCTION

Water flow in natural soil pipes (pipe flow) is recognized as a potentially significant mechanism in subsurface flow contributions to stream flow (Whipkey, 1969; Jones, 1975). It is a phenomenon known to occur in both humid and arid climates and is not restricted to a particular soil type. Enough measurements of pipe flow under different environmental conditions have not been made to actually quantify the significance of pipe flow in generating contributions to storm and drought hydrographs. However, there are a few reports on the limited measurements that have been made (Jones, 1975; Jones, 1978; Gilman and Newson, 1980).

Where subsurface flow (matrix flow, or throughflow, as defined by Kirkby and Chorley, 1967) has been regarded generally as a matrix phenomenon and hence a slow response system, it is now conceived that soil pipes impart a system of flow conduits to the soil matrix thereby increasing the overall matrix response to rainfall or snowmelt. In the partial area concept of runoff (Dunne and Black, 1970), subsurface flow is essential in supplying moisture to the saturated seepage zone. The mechanism of soil pipes may accelerate this rate of supply by "short circuiting" the slope between productive source areas of runoff and the valley bottom source areas (Newson, 1976). Soil pipes also have the potential to accelerate the drainage of the profile. Depending on the climate this accelerated drainage can have a beneficial or detrimental effect. For rangeland areas in the southwestern United States, Brown (1961) observed that soil pipes are detrimental to the survival of rangeland vegetation.

In previous work by the authors (Barcelo and Nieber, 1981), the simulation of a single soil pipe in the vertical cross section of a hillslope (taken parallel to the longitudinal axis of the hillslope) was performed. The presence of the soil pipe was found to influence significantly the storm hydrograph for a small hillslope (1 meter wide, 1 meter deep and 5 meters long). In view of this work and the fact that soil pipes generally occur as a network of pipes, it was felt that the significance of soil pipes was in their contribution to the catchment runoff hydrograph as a soil pipe network rather than as a single soil pipe. Therefore, the work presented in this paper emphasizes the influence that a soil pipe network exerts on the hydrology of a small catchment. A computer-based simulation of a network of soil pipes in a hillslope is used to illustrate this influence.

## BACKGROUND

The origin of soil pipes is associated with structural and biotic voids in the soil matrix. Voids of a biotic nature generally result from decayed tree roots or burrows dug by small rodents. Whipkey (1969) brings this to light in his observation of discharge occurring from root holes, decayed root channels and worm holes in field plots he studied. The initiation of soil pipes is due to water draining through these voids in saturated or unsaturated soils or is a result of "boiling" under hydrostatic and/or hydrodynamic pressure in saturated soil (Jones, 1975). Elements of both processes commonly appear together and are frequently found near the location of the wet weather water table. In his dissertation, Jones (1975) discusses the causes and the initiation of soil pipes in great detail.

In the past decade, a debate has prevailed over the relative importance of stream flow contributions from subsurface flow (matrix flow) and Hortonian overland flow (Chorley, 1978). There is now increasing interest in determining the significance of contributions of soil pipe flow relative to that of subsurface flow and Hortonian overland flow. Reports by Whipkey (1969) and Jones (1975) provide some information regarding the significance of soil pipe flow relative to other hydrograph contributions.

In his field experiments, Whipkey (1969) sought to determine the mechanics of subsurface flow from a forested watershed. In one experiment, 16 min after an application rate of 25 mm per hr had begun, Whipkey observed outflow to come from a root hole 122 cm below the soil surface. After observing discharge from his plots to occur via these biological and structural channels, Whipkey dug trenches across the slope of his plots to break up the continuity of the existing channels. Subsequent sprinkler tests yielded total stormflow equal to half of that which occurred prior to trenching. In a watershed experiment, Jones (1975) inferred the importance of the contribution of a network of soil pipes to stormflow. A single pipe Jones monitored was found to contribute 1 percent of the stream flow he measured from his experimental watershed. Visual reconnaissance of the area suggested that there were 25 pipes of similar magnitude feeding the same stream. This lead Jones to hypothesize that the relative contribution of pipe flow to stream flow could be raised to 25 percent.

Where the hydraulic conductivity value ($9.4 \times 10^{-6}$m per sec) of the soil material overlying a soil pipe indicated that a time of 12 hr was required for pipe discharge to initiate, Jones (1975) observed time to peak for the pipe hydrograph to be 1.73 hr. Thereby, Jones concluded that pipe flow was derived by means other than flow through the micropores of overlying soil material. He identified the source of pipe flow as coming from cracks and holes in the matrix which connect the soil surface directly to the soil pipe. Where the soil beneath the pipe is serviced by this same mechanism, pipe flow results from seepage into the pipe due to the saturation of the surrounding soil.

METHODS OF ANALYSIS

The computer simulation of a network of soil pipes was conceived for a hillslope with a shallow soil overlying an impermeable layer. The flow region was defined as being the two-dimensional plan view of a hillslope (Figure 1). The

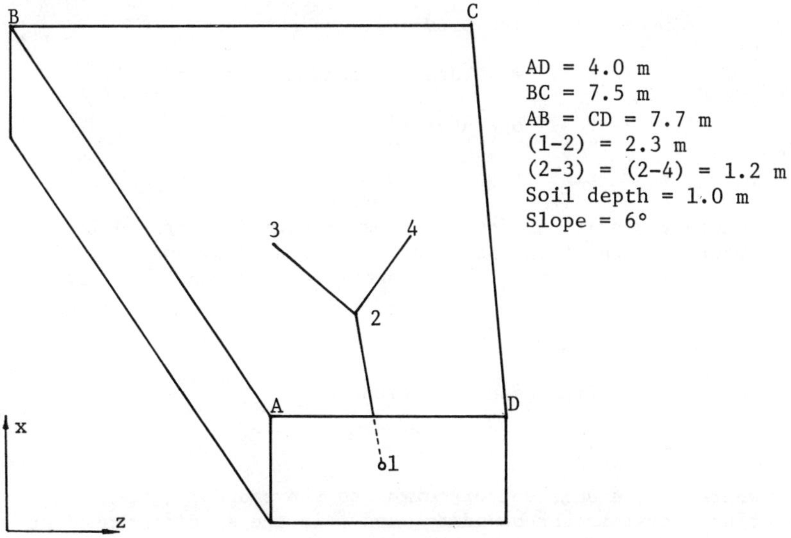


Figure 1. Illustration of the flow region for the problem considered.

depth of soil within the flow region was allowed to vary in a realistic fashion. The flow conditions normal to the impermeable layer were assumed to be hydrostatic so that a two-dimensional flow analysis could be used. No attempt was made to simulate the delay time normally associated with an advancing wetting front. Water infiltrating the soil surface was assumed to reach the saturated zone immediately.

The governing equation used to describe the saturated/unsaturated flow problem was the two-dimensional form of the Richards equation for flow through porous media. To simulate a three-dimensional flow problem in a hillslope using a two-dimensional analysis, the Richards equation was rederived to incorporate a variable depth term (Beven, 1977). This yields:

$$BC\frac{\partial\phi}{\partial t} = \nabla\cdot(BK\nabla\phi) \tag{1}$$

where $B = B(x,z) =$ depth of soil (m),

$C = C(h) = \frac{\partial\theta}{\partial h}$, = specific moisture capacity $(\frac{1}{m})$,

$\theta =$ volumetric moisture content $(\frac{m^3}{m^3})$,

$\phi = h + E =$ hydraulic head (m),

$h =$ capillary pressure head (m),

$E =$ elevation head (m),

$K = K(h) =$ hydraulic conductivity (m/sec),

$\nabla =$ "del" operator $(\frac{1}{m})$, and

$t =$ time.

Boundary conditions for the flow problem illustrated in Figure 1 were of the Neumann (specified flux) and Dirichlet (specified head) type. The specified flux condition is:

$$-K\frac{\partial\phi}{\partial n} = q_n \tag{2}$$

and the specified head condition is

$$\phi = \Phi \tag{3}$$

where n is a unit vector normal to the boundary, $q_n$ is the flux normal to the boundary, and $\Phi$ is the specified head at specified points on the boundary. The hillslope boundaries AB, BC and DC were specified as boundaries with $q_n = 0$. The seepage boundary, AD, was a specified head boundary with $\Phi = E(z)$. The required initial condition is given by

$$\phi\ (x,z,t) = \phi_o \text{ at } t = 0 \tag{4}$$

To parameterize expressions for C(h) and K(h), the following equations were used:

$$C(h) = \frac{mA\beta h^{\beta-1}}{(A+h^{\beta})^2}\ ; \tag{5}$$

$$K(h) = K_o \left(\frac{A}{A+h^{\beta}}\right)^N$$

Where m is the drainable porosity, $K_o$ is the saturated hydraulic conductivity and A, B and N are constants related to the pore size properties of a given soil.

Equations (1) to (4) were solved using finite difference and finite element techniques. The finite difference method was used to discretize the time domain and the finite element method was used to transform the space derivatives. This transformation procedure is outlined in detail by Neuman (1973). The resulting set of nonlinear equations take the form:

$$\sum_j (A_{ij} + \Delta t\ B_{ij})\ \phi_j^K = \sum_j A_{ij}\ \phi_j^{k-1} + \Delta t Q_j\ ;\ i = 1, 2, \ldots . \tag{6}$$

The terms in this equation are defined in detail by Nieber (1980).

The soil parameters C and K were assumed to vary linearly over each element. To assign values of C, K and volumetric moisture content to a particular node, an integrated average was made over the profile at that node location.

Any node within region ABCD (Figure 1) can be specified as either a source or sink node. The flux rate into or out of a node j is applied to its corresponding term, $Q_j$, in Equation (6). $Q_j$ is assigned as positive for a source and negative for a sink. For the case of infiltration at node j, the $Q_j$ term is positive and is set equal to the infiltration rate. For the case of evapotranspiration at node j, the $Q_j$ term is negative and is set equal to the evapotranspiration rate. For the case where a soil pipe passes through node j the value of $Q_j$ is negative for flow from the soil matrix into the pipe, and positive for flow from the pipe into the soil matrix.

To calculate flow into a soil pipe, a tile drain formula derived by Kirkham (1949) was employed. This equation relates the flux into the pipe (at any point along the pipe) to the elevation of the phreatic surface above the pipe (at that point). The equation is:

$$Q' = \frac{2\pi K_o(H-r)}{\ln\left(\dfrac{\tan\left(\dfrac{\pi(2H-r)}{4d}\right)}{\tan\left(\dfrac{\pi r}{4d}\right)}\right)} \tag{7}$$

where $Q'$ = flux into drain per unit length of drain ($m^3$ per sec),

$K_o$ = saturated hydraulic conductivity (m per sec),

$H$ = height of phreatic surface above center of pipe (m),

$r$ = radius of pipe (m), and

$d$ = depth of saturated flow region (m).

After determining the position of the phreatic surface from the known values of hydraulic head, given by the finite element solution, H and d were then calculated. The magnitude of the flux Q' computed at a pipe node, let us say node j, is assigned as a negative number to the term Q in Equation (6).

It is possible to simulate the case where water enters the pipe and flows into the surrounding soil matrix. This would be possible by assigning a positive $Q_j$ equal to the supply rate or by assigning a positive pressure if the supply rate exceeds the infiltration rate of node j.

For the simulations presented here, deterministic rainfall and antecedent soil moisture conditions were used. Also, the only mechanism of pipe flow to be considered here is where pipe flow results from seepage into the pipe from the surrounding saturated soil.

RESULTS AND DISCUSSION

Results of four simulations are presented to illustrate the potential influence soil pipes have on hillslope hydrology. One simulation was made for each of the following hillslope-soil pipe arrangements: 1) one each for a single soil pipe situated parallel to the soil surface a distance of 0.5 and 0.25 meters above the impermeable boundary (branch 1-2 in Figure 1); 2) a soil pipe network situated parallel to the soil surface a distance of 0.5 meters above the impermeable boundary (branch 1-2-3-4 in Figure 1); and 3) an unpiped hillslope. In each case where a soil pipe was present, the pipe outlet was located along the seepage boundary AD (Figure 1). The soil constants used for Equation (5) were: $K_o = 3.5 \times 10^{-5}$ m per sec, m = 0.318, A = 0.218, β = 1.045, and N = 6.56. Since the rainfall rate, 45.0 mm per hr, was less than the saturated hydraulic conductivity of the soil, the infiltration rate into region ABCD was set equal to the rainfall rate. The duration of the rainfall event was 150 min. The hillslope geometry was as pictured in Figure 1. The pipe radius was 0.025 m. The initial hydraulic head condition was uniform across the slope in the direction perpendicular to the longitudinal axis of the hillslope. The initial hydraulic head for a particular node was set equal to the elevation of the node plus 0.25 meters of pressure head. All nodes were located on the impermeable boundary.

The relative flow contributions of the soil pipes and the soil matrix to their respective runoff hydrographs are shown in Figure 2. From the hydrographs it is apparent that pipe flow increases with a decrease in pipe elevation. Since the phreatic surface is more easily maintained above the elevation of the lower pipe and since the only mechanism of pipe flow considered in these simulations was due to seepage into the pipe from the surrounding saturated soil, a larger volume of discharge was attributable to the lower pipe than the higher pipe. For the same rainfall and hillslope conditions it is apparent that the soil pipe network contributes more discharge to stormflow, relative to the corresponding matrix contribution, than does a single pipe.

The influence exerted by soil pipes on the total runoff hydrographs for the hillslope is illustrated in Figure 3. In each simulation, the piped hillslope exhibited a greater peak discharge than the unpiped hillslope. The depth of the pipe in the soil profile did not appreciably affect the peak discharge. Significant, though relative to the case for the other hillslopes, is the greater magnitude of peak discharge resulting from the hillslope with the soil pipe network. Corresponding with the increase in peak discharge was the enhanced ability of the piped hillslopes to remove a greater volume of water than the unpiped hillslope during the first 1000 min of the simulation.

The soil moisture distributions in the hillslope for the four hillslope cases at 800 min are illustrated in Figure 4. It is seen that the presence of the soil pipe network had a greater effect on soil moisture distribution in the hillslope than did the presence of a single pipe. The effect of the soil pipe network was to effectively drain more of the hillslope. A single pipe also had an effect on soil moisture distribution but did not affect as great an area of the hillslope as did the pipe network. Again the effect of pipe depth was evident. The deeper pipe had a more dramatic effect on the hillslope soil moisture distribution.

For the simulated hydrographs presented here, it is evident

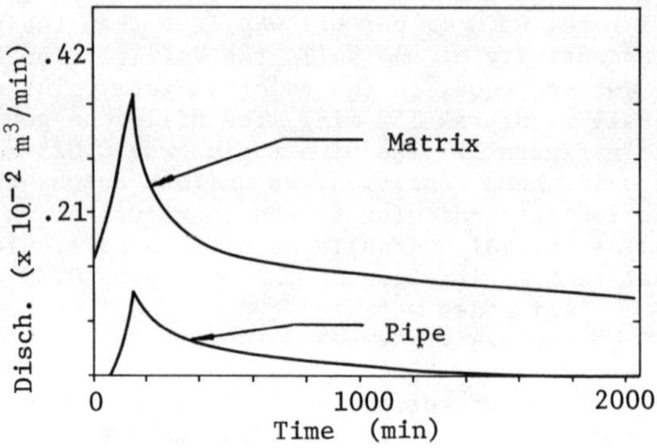


a. Single pipe at 0.5 meters.

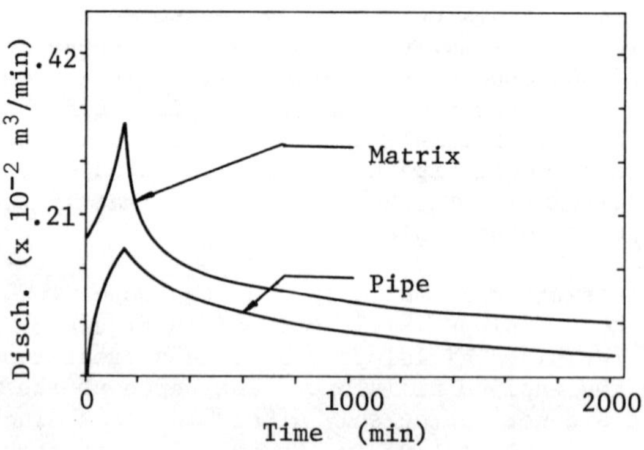


b. Single pipe at 0.25 meters.

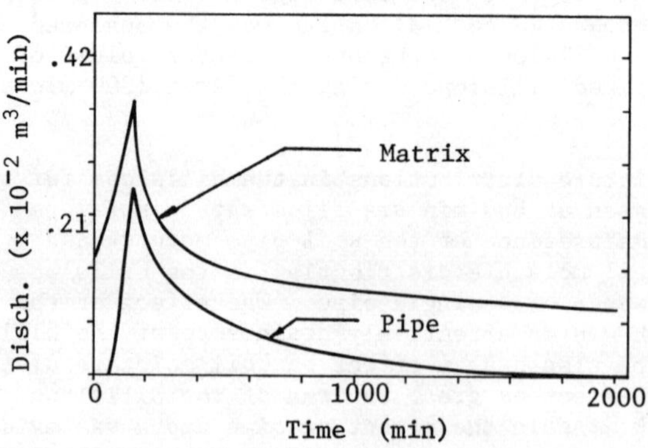


c. Pipe network at 0.5 meters.

Figure 2. Soil matrix and pipeflow contributions to the runoff hydrograph.

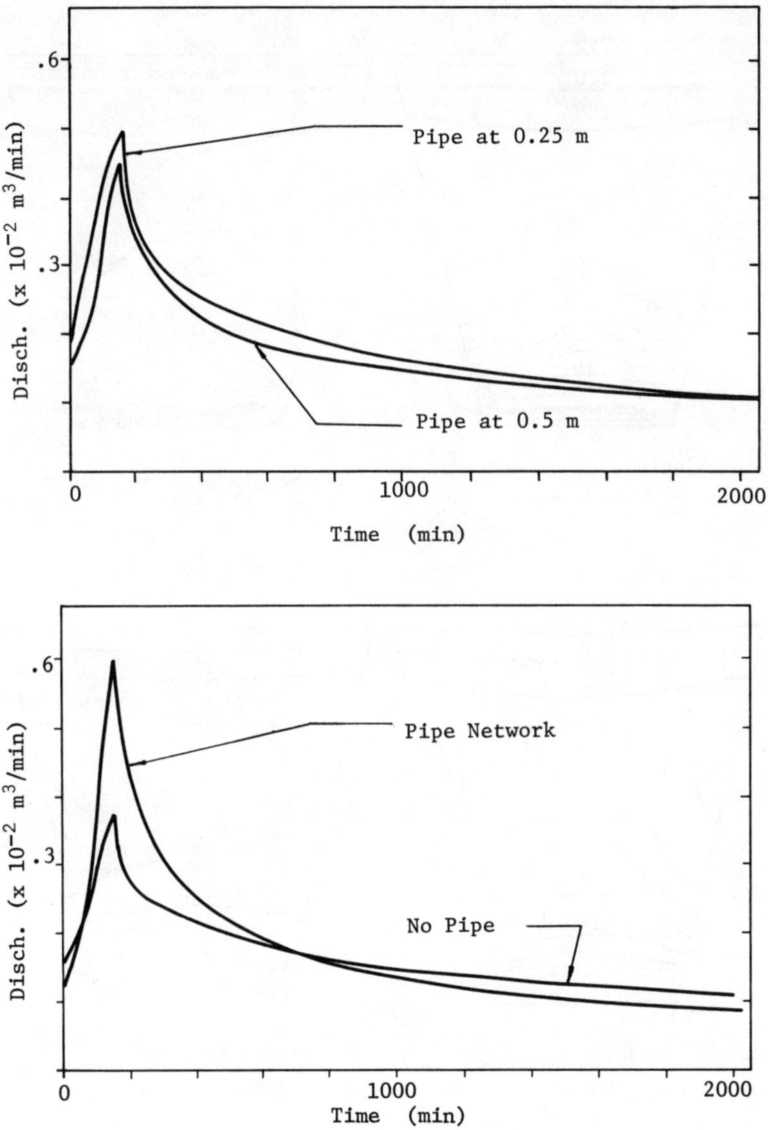


Figure 3. Total runoff hydrographs for each hillslope case.

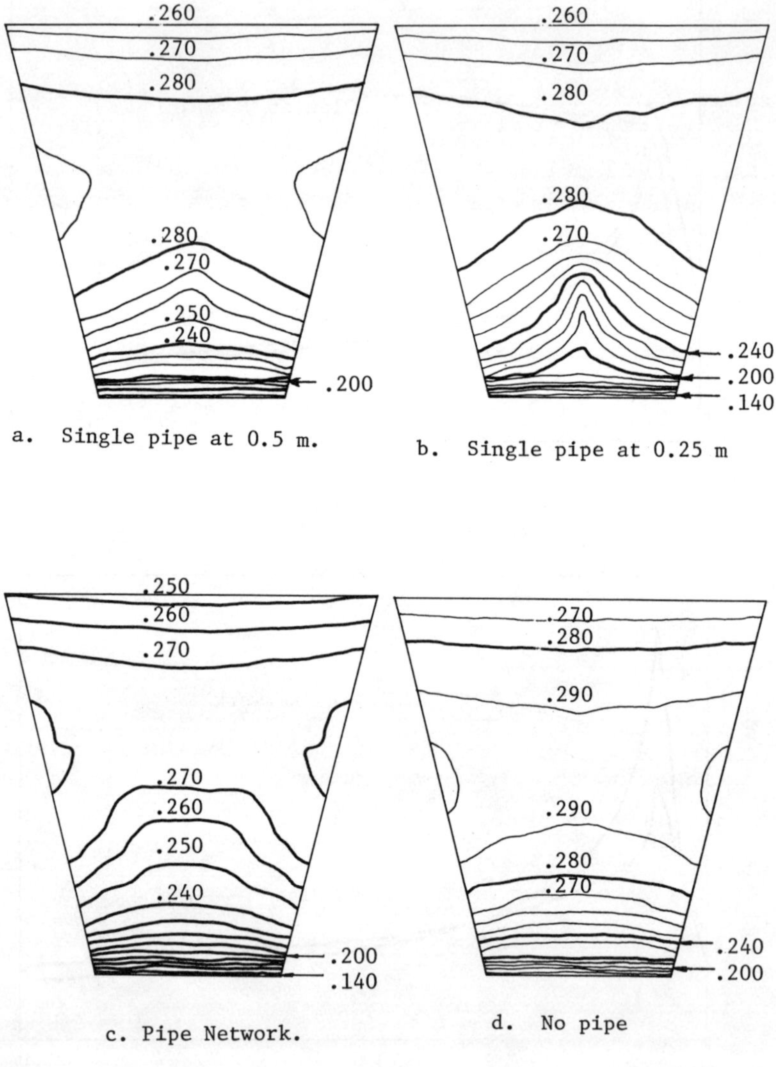


a. Single pipe at 0.5 m.

b. Single pipe at 0.25 m

c. Pipe Network.

d. No pipe

Figure 4. Lines of equal moisture content at 800 minutes.

that the presence of soil pipes in a hillslope can increase the peak discharge, increase the volume of water removed during the stormflow period and significantly affect the distribution of soil moisture in the hillslopes. Too, it is apparent that the significance of soil pipes, in their contribution to stormflow, is their existence as a network. The overall effect of soil pipes, as investigated in this paper, on the simulated hydrology of a hillslope was to provide more rapid drainage of the hillslope and thereby reduce the potential for moisture storage in the soil matrix.

The limitations of this study include the fact that only deterministic rainfall and initial conditions were used. To completely evaluate the effect soil pipes have on hillslope hydrology it would be necessary to make continuous simulations that involve stochastic rainfall and evaporative inputs. In addition, only the mechanism of pipe flow which results from seepage into the pipe from a saturated zone was considered. Flow that directly accesses soil pipes via surface cracks and interconnected channels need also be quantified to more completely evaluate the influence of soil pipes on catchment hydrology.

REFERENCES

Barcelo, M. D. and J. L. Nieber. 1981. Simulation of the Hydrology of Natural Pipes in a Soil Profile. American Society of Agricultural Engineers Paper No. 81-2028, St. Joseph, MI 49085.

Beven, K. 1977. Hillslope hydrographs by the Finite Element Method. Earth Surf. Proc. 2:13-28.

Brown, G. W. 1961. Some Physical and Chemical Soil Properties as Possible Causes of Piping Erosion. Master of Science Thesis, Colorado State University, Fort Collins. 61 pp.

Chorley, R. J. 1978. The Hillslope Hydrological Cycle. In: M. J. Kirkby (ed.) Hillslope Hydrology. John Wiley and Sons, New York. p 1-42.

Dunne, T. and R. D. Black. 1970. Partial Area Contributions to Storm Runoff in a Small New England Watershed. Water Res. Res. 6:1296-1311.

Gilman, K. and M. D. Newson. 1980. Soil Pipes and Pipe Flow. Geo. Abstracts, England.

Jones, J.A.A. 1975. Soil Piping and the Subsurface Initiation of Stream Channel Networks. Unpub. Ph.D. Thesis, University of Cambridge, England. 467 pp.

Jones, J.A.A. 1978. Soil Pipe Networks: Distribution and Discharge. Cambria. 5:1-21.

Kirkby, M. J. and R. J. Chorley. 1967. Throughflow, Overland Flow and Erosion. Bull. Intern. Assoc. Sci. Hydrol. 12: 5-21.

Kirkham, D. 1949. Flow of Ponded Water into Drain Tubes in Soil Overlying an Impervious Layer. Transactions, American Geophysical Union 30:369-385.

Neuman, S. P. (1973) Saturated-Unsaturated Seepage by Finite Elements. Journal of Hydraul. Div., ASCE 99(HY-12):2230-2250.

Newson, M. 1976. Soil Piping in Upland Wales: A Call For More Information. Cambria. 1:33-39.

Nieber, J. L. 1980. Evaluation of Coefficient Matrices of Linear Triangular Elements Involving Unsaturated Soil Moisture Flow. In: Proc. of Third International Conference on Finite Elements in Water Resources, University of Miss., p. 2.260-2.269.

Whipkey, R. Z. 1969. Storm Runoff from Forested Catchments by Subsurface Routes. Intern. Assoc. Sci. Hydrol. 85:773-779.

# ADAPTATION OF FEAP FOR SEEPAGE WITH A FREE SURFACE

A. Das Gupta and G. Mustafa

Asian Institute of Technology, P.O. Box 2754
Bangkok, Thailand

## SUMMARY

This paper deals with the modifications of the General Purpose Finite Element Program, FEAP by Taylor (1977) to convert the package into a continuous simulator for seepage problem with a free surface boundary. The FEAP was developed for direct solution of the problem with well defined boundaries for the solution domain. For seepages problems with a free surface, since the profile is unknown, computation procedure has to pass through several iterations till the specific conditions on free surface are satisfied. To extend the applicability of FEAP as a continuous simulator for problems with unknown boundary, modifications are introduced in input data module; algorithms for checking convergence of the solution, for adjustment of the free surface boundary according to the discrepancy between calculated head and specified head and for internal changing of boundary conditions in input data module due to the adjustment are incorporated. An application to a specific case of seepage from open channel is described to illustrate the capabilities.

## INTRODUCTION

Many studies have been made to analyse the problem of seepage with a free surface. Analytical studies are based on simplifying assumptions which restrict their practical applicability while analog studies have limited application to a specific problem where the effect of gravity forces is predominant. Exclusive treatment of the analytical methods are given in Muskat (1946) whereas analog studies were conducted by Bouwer (1969). On the other hand, numerical approach has more flexibility in analysing the seepage in heterogeneous and anisotropic aquifer situation and with complicated boundary conditions. The finite element method of numerical analysis has advantage over finite difference method in treating non-linear boundary conditions and in treating problems in physical plane with unknown boundary whose profile has to be adjusted through iteration. The current state-of-the-art in finite element modelling has reached a stage where the development of new models is perhaps unnecessary for application in seepage problems. However, this does not imply that improvements and modifications are unnecessary, but that it is not necessary to start from scratch when embarking on a problem involving modelling.

This paper describes a case in point. The general purpose finite element analysis program, FEAP developed by Taylor (1977) at the University of California, U.S.A. is extensively used for structural analysis. This program has the capability of solving Navier-Stokes equation in Fluid Mechanics. It has efficient input data system, matrix assembly system and associated operating system which reduce the computation time and have made its application more promising. The standard form of FEAP can be used for direct solution of seepage with flow domain well defined. For seepage with free surface, the profile of the free surface is unknown and has to be adjusted in the solution scheme through iteration till the conditions of atmospheric pressure along the profile and specified flow are satisfied simultaneously. Because of two conditions to be satisfied, each iteration is performed in two steps. In order to use FEAP as a continuous simulator to analyse seepage problem with a free surface, the data input module has been modified; automatic data set changing facility has been incorporated according to the adjustment of the free surface profile; and a convergence test is introduced for terminating the program when required accuracy has been reached. An application to a specific case of seepage from open channel is described to illustrate the capabilities.

## THEORY AND SOLUTION PROCEDURE

Combining Darcy's Law with the continuity equation, the governing equation for two-dimensional steady seepage problems in a non-homogeneous, anisotropic aquifer is

$$\frac{\partial}{\partial x}\left(K_x \frac{\partial h}{\partial x}\right) + \frac{\partial}{\partial y}\left(K_y \frac{\partial h}{\partial y}\right) = 0 \qquad (1)$$

where, h is the hydraulic head = $y + p/\gamma_w$; y is the static head in xy cartesian coordinate plane with respect to some specified datum, p is the liquid pressure and $\gamma_w$ is the specific weight of flowing liquid; and $K_x$ and $K_y$ are coefficient of permeability in respective coordinate directions.

The boundary conditions to be satisfied for a specific case of seepage from open channel (see Figure 1) are as follows

$$h[x, y_f(x)] = y_f(x) \text{ on the free surface, } B_1 \quad (2)$$

$$K_x \frac{\partial h}{\partial x} \cos\theta_1 + K_y \frac{\partial h}{\partial y} \cos\theta_2 = Wn_2 \text{ on the free surface, } B_1 \quad (3)$$

$$h[L, y(L)] = h_w \text{ on the outflow face, } B_2 \quad (4)$$

$$K_x \frac{\partial h}{\partial x} \cos\theta_1 + K_y \frac{\partial h}{\partial y} \cos\theta_2 = V \text{ on the outflow face, } B_2 \quad (5)$$

$$K_x \frac{\partial h}{\partial x} \cos\theta_1 + K_y \frac{\partial h}{\partial y} \cos\theta_2 = 0 \text{ on the impermeable boundary,} B_3 \quad (6)$$

and along the vertical axis of symmetry, $B_4$

$h(x,y) = H_w$ on boundary, $B_5$ where the groundwater is in contact with the water in the channel, which is supposed to be in hydrostatic equilibrium. (7)

In Equation (2) through (7), $y_f$ denotes coordinate of points along the free surface; $\theta_1$ and $\theta_2$ are angles between normal direction and the respective coordinate directions x and y for specified boundaries; $h_w$ represents the depth of flow at the outflow boundary; $H_w$ is the depth of water in the channel; W is the net vertical specific rate of infiltration at the free surface; $n_2$ stands for unit outer normal vector; and V represents specific flux across the outflow face. The equivalent variational formulation for this problem following the generalised principle (Mauersberger, 1965) can be expressed as follows

$$I[h, y_f(x)] = \iint_R \frac{1}{2} [K_x (\frac{\partial h}{\partial x})^2 + K_y (\frac{\partial h}{\partial y})^2] dx\, dy$$

$$- \int_C (h-A)(K_x \frac{\partial h}{\partial x} \cos\theta_1 + K_y \frac{\partial h}{\partial y} \cos\theta_2) ds - \int_C y_f W n_2 ds + \int_C Vh\, ds \quad (8)$$

where; $\iint_R$ denotes the integral over the flow domain; $\int_C$ stands for the line integral along the specified boundary; ds corresponds to an elementary length along the boundary; and A stands for the specified hydraulic head for the respective boundaries.

The required solution is obtained by following the well established sequences of operation for the minimisation of the functional as outlined in Desai (1979). The flow region is discretised by a network of quadrilateral elements. A bilinear interpolation function in terms of nodal values defines uniquely the hydraulic head distribution within each element. Since the location of the free surface is unknown, the numerical computation starts with the assumed profile. The solution is then iterated by adjusting the profile of the free surface at the end of each iteration till the two specified conditions along the profile are satisfied simultaneously. Because of the restriction of specifying only one boundary condition along the unknown profile at one time, each iteration is performed in two steps. During the first step of iteration, the free surface is considered to be known hydraulic head boundary along with the remaining boundaries of the seepage domain and accordingly the minimisation of the functional gives the hydraulic head distribution within the flow domain, from which, the specific flux V on the outflow face is determined. In the second step of iteration, the free surface and the outflow face are considered as specified flow boundaries. Subsequent solution gives the hydraulic head distribution along the free surface and if the calculated head does not satisfy the condition given in Equation (2), to an acceptable degree of accuracy, the profile of the free surface is then adjusted following the procedure suggested by Neuman and Witherspoon (1970). This process determines new profile of the unknown boundary and the entire procedure is repeated until the conditions are satisfied.

To accommodate the iterative solution procedure as outlined above in the original version of FEAP, two data sets (data set 1: specified head; data set 2: specified flux) are introduced in the input module; and algorithms for convergence test, adjustment of boundary and automatic input data changing for next iteration are incorporated.

ADAPTATION OF FEAP

The operation procedure of FEAP is schematically shown in Figure 2. Dotted line diagram represents the basic operational flow diagram for seepage problems whereas the full line diagram shows the operational mode based on which FEAP was developed for direct solution of the problem. Improvement and modifications according to the operational flow diagram for seepage problem are incorporated in the subroutines PCONTR and PMACR of the FEAP program. The flow chart of the modified form of control program is given in Figure 3.

The storage allocation for FEAP is partitioned to store all the data arrays, as well as, the global arrays. Each array is variably dimensioned to the exact size required for each program by using a set of pointers established in the control program (see Figure 3). For seepage problems, additional fortran variable names are used for data storage as given in Table 1. Accordingly additional variables are incorporated in the input data format for control information as given in Table 2. The first step in users instruction is to specify problem title and control information which are used during subsequent data input and also are used to allocate memory in the program. Once the control data is supplied the program expects the data cards for the mesh description. The flow of data to the program is controlled by set of *macro* commands given in Table 3. The basic aspect of the variable algorithm program is a macro instruction language which can be used to construct modules for specific algorithms as needed. Table 4 gives the sequence of macro pro-

Table 1 Additional Fortran Variables Used for Data Storage

| Variable Name (dimension) | Description |
|---|---|
| NCCRDS | Number of coordinate data cards |
| NFCRDS | Number of Specified head data cards |
| NFFS | Channel boundary node number on free surface |
| NBFL | Number of node on the free surface excluding boundary nodes |
| NX1CAR, NX2CAR | X-array number where x-coordinate of first and last adjustable node on the free surface are stored |
| NY1AR, NY2AR | X-array number where y-coordinate of first and last adjustable node on the free surface are stored |
| NSDR, NSLDR | DR-array number of first and last adjustable node point on the free surface for which calculated head value are stored at the end of each iteration |
| WDNN (4) | Variable used for data initialisation |
| NLM1 : NLM16 | Specific element number with defined flux |

Table 2 Control Information Formats

| CONTROL CARD ONE-FORMAT(16I5) | | | |
|---|---|---|---|
| Columns | Variable | Columns | Variable |
| 1-5 | NUMNP | 51-55 | NBFL |
| 6-10 | NUMEL | 56-60 | NXCAR |
| 11-15 | NUMMAT | 61-65 | NY1AR |
| 16-20 | NDM | 66-70 | NY2AR |
| 21-25 | NDF | 71-75 | NSDR |
| 26-30 | NEN | 76-80 | NSLDR |
| 31-35 | NAD | CONTROL CARD TWO-FORMAT(16I5) | |
| 36-40 | NCCRDS | 1-5 | NLM1 |
| 41-45 | NFCRDS | . | . |
| | | . | . |
| 46-50 | NFFS | 76-80 | NLM16 |

Table 3 Macro Control Statements (input is controlled by the value assigned to CC)

| CC Value | Data to be input |
|---|---|
| NOPR | Do not print subsequent mesh data |
| COOR | Coordinate data |
| ELEM | Element data |
| BOUN | Boundary condition data |
| FORC | Prescribed nodal force data |
| MATE | Material data |
| PRIN | Print subsequent mesh data (default mode) |
| END | Must be last card in mesh data,terminates mesh input |

Table 4 Macro Programming Commands

| Instruction | Description |
|---|---|
| MACR | Call MACRO solution module for establishing solution algorithm |
| TANG | Symmetric tangent stiffness formulation |
| FORM | Form right hand side of the equation |
| SOLV | Solve tangent equations |
| DISP | Print nodal head every step in the iteration |
| STRE | Print element variables |
| CONV | Convergence test for solution |
| END | End of macro programming instructions |

gramming commands which the user needs to supply for the solution of seepage problems.

APPLICATION

The solution to seepage from a channel embedded in a uniform soil and underlain by less permeable (considered as impermeable) material as shown in Figure 1 is considered for application of the modified version of FEAP program. The specific dimensions and estimated parameters for the problem are as follows (see Figure 1): $H_w = 0.5$ m; $W_b = 2.0$ m, $W_s = 3.0$ m; $D_i = 6.0$ m, $D_w = 4.0$, $\alpha = 45^o$, $L = 10W_b$ and $I_s/K = 0.4183$. Initial guess of the free surface is a straight line along which the hydraulic head varies linearly from the upstream head of $(D_i + H_w)$ to $(D_i + H_w - D_w)$ at the point where uniform flow condition is preassumed beyond the outflow boundary. The seepage domain is discretised by 192 quadrilateral elements with 234 node points. The flow pattern drawn from the computed head at the end of final iteration is shown in Figure 4. Figure 5 shows how solution converges to the acceptable one starting with the initial approximation. At the end of each iteration, the absolute value of the maximum difference between the computed head and specified head for nodal points on the free surface is expressed as percentage of the maximum available head in the flow domain and the iteration is continued till the percentage error goes below 0.4% which is selected as convergence criterion for this problem. The numerically estimated seepage rate (seepage per unit width of the water surface in the channel and per unit length of the channel) expressed in non-dimensional form $(I_s/K)$ is compared with the electric analog result by Bouwer (1969) and is shown in Figure 6.

CONCLUSION

Modifications needed to use the General Purpose Finite Element Program, FEAP as continuous simulator for seepage problem with unknown boundary have been presented. To treat the unknown boundary as adjustable one, additional control variables are introduced for data input module; algorithms for checking the convergence of the solution, for adjustment of the coordinates of nodal points on the free surface and for updating the input data on coordinates and specified flux along the free surface for successive iteration are incorporated. The application problem of seepage from open channel illustrates the applicability of the modified FEAP program.

## REFERENCES


Bouwer, H. (1969) Theory of Seepage from Open Channels. *in Advances in Hydroscience* (V.T. Chow,ed.), 5, 121-172.

Desai, C.S. (1979) Elementary Finite Element Method. Englewood Cliff, N.J., Prentice-Hall, 299-321.

Mauersberger, P. (1965) A Variational Principle for Steady State Groundwater Flow with a Free Surface. Pure Appl. Geophys., 60, 101-106.

Muskat, M. (1946) The Flow of Homogeneous Fluids Through Porous Media, McGraw-Hill Book Co., New York, 326-336.

Neuman, S.P. and Witherspoon, P.A. (1970) Finite Element Method of Analysing Steady Seepage with a Free Surface. Water Resources Research, 6, 3: 889-897.

Taylor, R.L. (1977) Computer Procedures for Finite Element Analysis. *in Finite Element Method* (O.C. Zienkiewicz, au.), McGraw-Hill Book Co., U.K., 677-773.

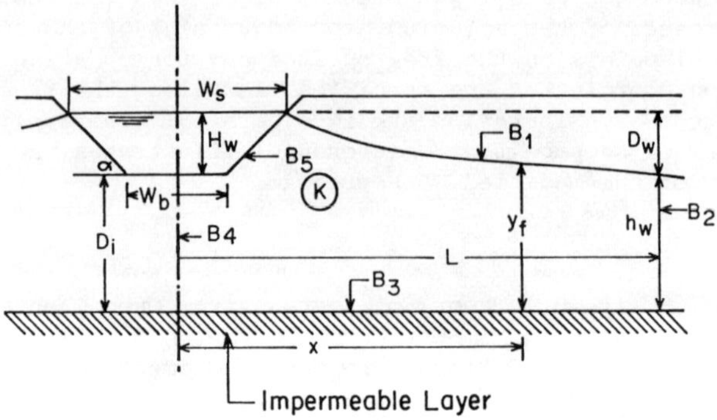


Figure 1 Geometry and Symbols for Channel in Soil Underlain by Impermeable Material

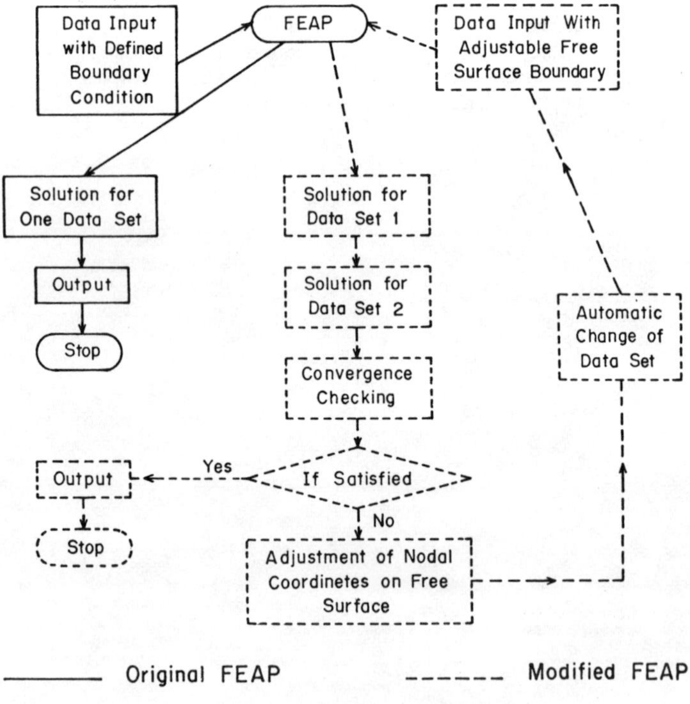


Figure 2 Schematic Representation of Original FEAP and Modified FEAP

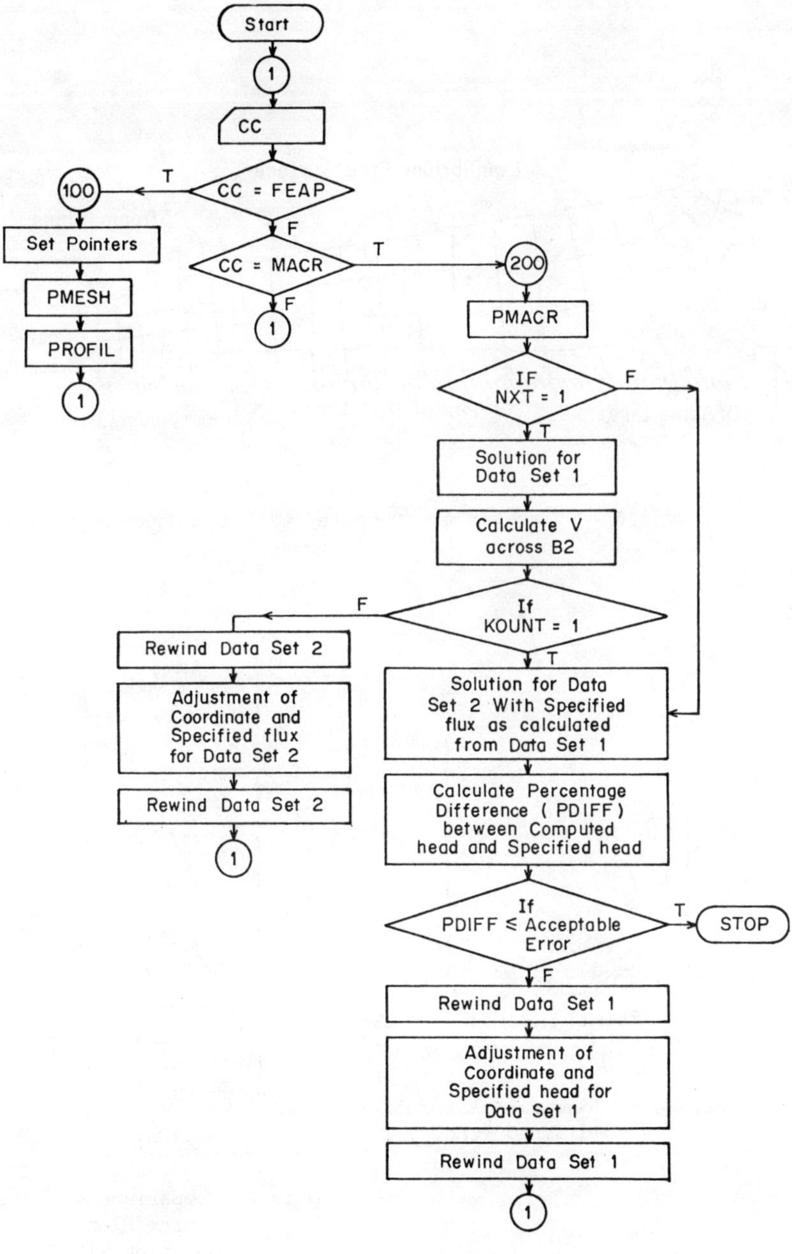


Figure 3 Control Program of Modified FEAP

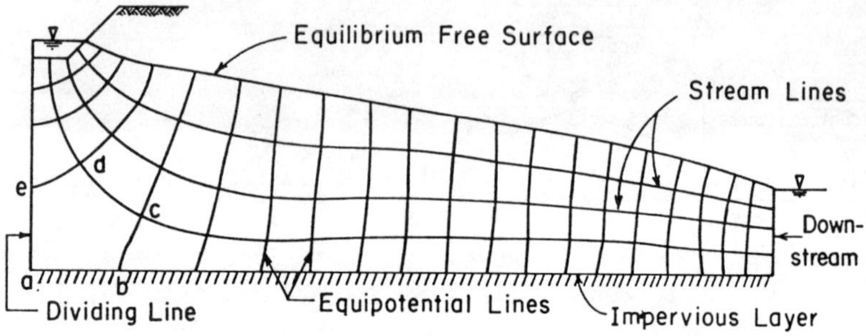


Figure 4 Flow Pattern Computed with the Model

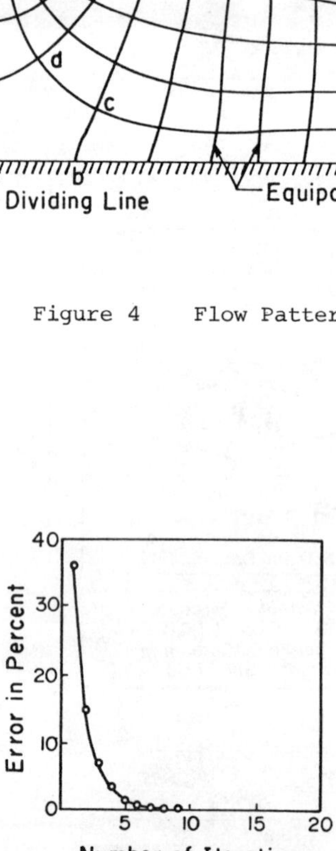


Figure 5 Convergence for Adjusted Free Surface

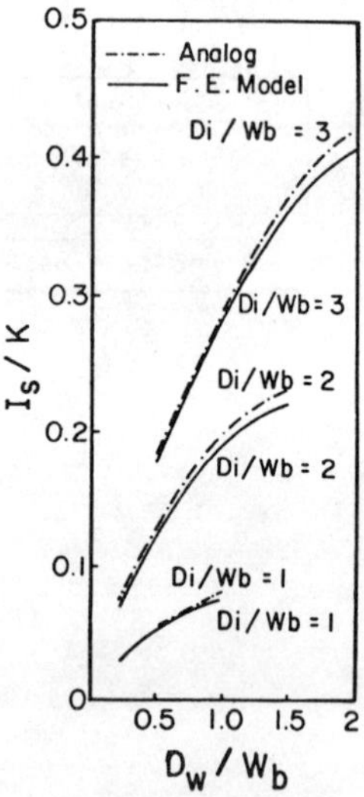


Figure 6 Comparison of Finite Element Solution with Bouwer's Analog Solution

# FINITE ELEMENT COMPUTATIONS OF HORIZONTAL GROUNDWATER FLOW WITH MOVING BOUNDARIES

Bruno Herrling

Institute of Hydromechanics, University of Karlsruhe, W.Germany

## INTRODUCTION

In the region of the Alps' foreland in West Germany numerous glacial valleys exist in which primarily quaternary sand and gravel sediments are deposited. In these basin and channel systems the impermeable groundwater bottom is often characterized by very steep slopes of the lateral periphery and thalweg slopes of the aquifer of 1% or more. The groundwater flow is mainly charged by the relative high precipitation in these areas and by the inflow from transverse valley aquifers. In some regions the distance between the groundwater table and the surface approach 20 m, whereas at other places the streams are recharged by escaping groundwater. The latter is important to increase the low water discharge during dry periods.

Periods of high precipitation and the sloping surface of the ground often make it necessary to build retarding basins as flood control reservoirs. A proposal of current interest is to set up a groundwater model to investigate the consequences of using the high-water of the streams as an artificial groundwater recharge, thus eliminating the need for expensive retarding basins. It has been proposed that this artificial groundwater recharge can be realized using existing or newly excavated gravel-pits (Werner et al. (1974)).

In order to investigate such complex aquifer systems in replenished glacial valleys, time-dependent, horizontal models used in conjunction with the finite element method (FEM) have been employed. The effective boundary of the computational domain is of particular importance in situations with partly steep lateral boundaries and large variations of the groundwater level (over 10 m) due to artificial recharging (i.e. with changing groundwater levels the lateral boundaries of the aquifer must also be altered in the model). Furthermore, the variable groundwater level leads to a variable saturated thickness of the unconfined aquifer which must be accounted for in the model via an updated local transmissivity as usual.

The use of a moving lateral boundary in the groundwater model has also proven advantageous in computations of similar valley aquifers without groundwater recharge, especially during the calibration phase of the model. In this way peripheral nodes or elements which become dry during certain time intervals do not pose any problems in the numerical computation.

In addition, the moving boundary feature of the present model is well suited to handle the horizontal propagation of recharged water in a normally dry and permeable aquifer which is confined by a thin aquitard from underneath as part of a stratified underground. An artificial groundwater recharge in the top layer can lead to highly transient conditions with large variations of the effective lateral boundaries.

## COMPUTATION OF TIME-DEPENDENT, HORIZONTAL GROUNDWATER FLOW

### Basic equations

The basic equations describing the time-dependent, two-dimensional horizontal groundwater flow in an unconfined aquifer are the continuity equation

$$S_y h,_t + q_{i,i} = q^* \tag{1}$$

and Darcy's equation of motion

$$q_i = -T_{ij} h,_j \tag{2}$$

summing over i,j = 1,2. The symbols $,_i$ and $,_t$ denote partial differentiations with respect to the coordinates $x_i$ and the time t. These equations result from a vertical integration using the Dupuit assumption. The unknown parameters are the groundwater level h (height of water table) and the horizontal flux per unit length $q_i$ defined in a Cartesian coordinate system $x_i$. An inflow in the area of computation is specified by $q^*$. In addition there is the specific yield $S_y$ for the free surface flow and the transmissivity tensor $T_{ij}$.

Setting equation (2) in the continuity equation (1) results in

$$S_y h,_t - (T_{ij} h,_j),_i = q^* \tag{3}$$

with only the groundwater level as the unknown parameter. $q_i$ can be computed in a second step when h is known.

Boundary conditions are the hydrographs of groundwater level or of the flux across the boundaries:

$$h - \bar{h} = 0 \qquad \text{on } S_h \tag{4}$$

$$q_i n_i + \bar{q} = 0 \qquad \text{on } S_q \tag{5}$$

$\bar{h}$ signifies a prescribed water level and $\bar{q}$ a given flux normal to the outer boundaries. On closed boundaries $\bar{q}$ is equal to zero. $n_i$ represents a unit vector normal to the boundary.

Finite element formulation

The method of weighted residuals is the basis of the application of the finite element analysis. In time, the equations are solved in a stepwise manner as usual, starting with a given initial condition. The domain of computation is subdivided into finite elements for the integration of the weighted basic equation (3)

$$\int_A \int_t \delta h \, S_y \, h,_t \, dt \, dA - \int_A \int_t \delta h \, (T_{ij} \, h,_j),_i \, dt \, dA =$$

$$\int_A \int_t \delta h \, q \, dt \, dA + \int_S \int_t \delta h \, \bar{q} \, dt \, dS + \int_t \delta h_K \, Q_K \, dt. \tag{6}$$

The elements are defined in space (triangles) and time, in both instances with linear shape functions. The weighting functions $\delta h$ are also functions in space and time. In space they are identical with the shape functions (Galerkin method), and in time, constant weighting functions are chosen which give the same factors as the Crank-Nicolson time integration. The previous inflow $q^*$ is replaced by an areal distributed groundwater increment q(e.g. as a consequence of the precipitation), by a line source of mass inflow $\bar{q}$ defined on element boundaries (e.g. a seepage along a line) and by a point source of mass inflow $Q_K$ at node K of the element system (e.g. withdrawal of groundwater from wells).

Equation (6) is transformed by partial integration. Using Gauss' integral theorem and employing again equation (2) this results in:

$$\int_A \int_t \delta h \, S_y \, h,_t \, dt \, dA + \int_A \int_t \delta h,_i \, T_{ij} \, h,_j \, dt \, dA$$
$$+ \int_S \int_t \delta h \, q_i \, n_i \, dt \, dS$$
$$= \int_A \int_t \delta h \, q \, dt \, dA + \int_S \int_t \delta h \, \bar{q} \, dt \, dS + \int_t \delta h_K \, Q_K \, dt. \tag{7}$$

While the boundary condition (4) is considered in the final set of equations, the condition (5) is inserted directly into equation (7) and can be combined with the second term on the right hand side.

In horizontal models it is very important to consider the water exchange with the streams. Therefore a line source term (leakage term) is formulated

$$\bar{q}' = \lambda \cdot B \cdot (h_S - h) \tag{8}$$

with the leakage parameter $\lambda$, which signifies the quotient from the conductivity and the thickness of the bottom layer of a stream, together with the width B and the water level $h_S$ of a stream. The leakage term (8) is treated in the same manner as the line source term. Equation (7) then becomes

$$\int_A \int_t \delta h\, S_y\, h_{,t}\, dt\, dA + \int_A \int_t \delta h_{,i}\, T_{ij}\, h_{,j}\, dt\, dA =$$

$$\int_A \int_t \delta h\, q\, dt\, dA + \int_S \int_t \delta h\, \bar{q}\, dt\, dS$$

$$+ \int_S \int_t \delta h\, \lambda\, B\, (h_S - h)\, dt\, dS + \int_t \delta h_K Q_K\, dt \tag{9}$$

Since the leakage term is individually formulated and integrated along element sides S it is possible to simulate the important exchange situation near weirs which are often situated in streams. At nodes located at a weir the water levels in the stream and the leakage parameters $\lambda$ are allowed to differ from one element side to the next. Furthermore two parameters for $\lambda$ representing the infiltration and exfiltration are introduced. Altogether this allows realistic modelling of the exchange between groundwater and surface water.

The integration of the weighted equation (9) is carried out in time steps of $\Delta t$ and over the whole area A of computation and the boundaries S respectively. In practice, the finite element technique reduces the integration in space to an element area $A_e$ which leads to three element equations with three unknown parameters. For the integration over the entire area A, the individual element equations are summed together to give the final system of equations. After the boundary condition (4) is inserted, the equation system can then be solved applying Cholesky's method. As the transmissivity is a function of the groundwater level h, the problem is non-linear. Therefore at each particular time step an iteration is used. This means obtaining a solution several times with an updated coefficient matrix until the maximum difference of the computed groundwater levels from two successive iterations is smaller than a prescribed limit.

## Moving boundaries

Moving boundaries are necessary in a model when the impermeable bottom of an aquifer has large gradients, for example at the lateral edges of a glacial valley filled with sediments. An artificial groundwater recharge can lead to groundwater flood waves which are much greater than in times of high precipitation. This can enlarge the lateral extension of the flooded aquifer and can result in significant changes of its hydraulic characteristics. Actually, the boundaries of an aquifer do not become dry since there always exists a small subterraneous inflow from the slope. However, to obtain a stable solution in a horizontal computation a large number of iterations are necessary because of the nonlinearities introduced by the shallowness of the saturated aquifer. Furthermore the basic assumptions for a horizontal model are violated in these sloped areas, specificly the Dupuit assumptions. To save computer time

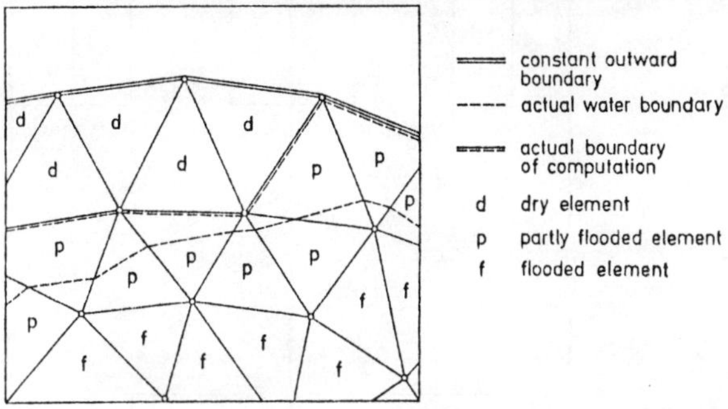


Figure 1.--Principle sketch of the area of moving boundaries

and to offer a solution which is consistent with the accuracy of the rest of the model, the following computational procedure is suggested for use with moving boundaries:

- The discretization in space remains constant, except that elements having three dry nodes at the beginning of a time step are removed from the domain of computation, i.e. they are made passive (figure 1).
- When at least one node of an element is flooded at the beginning of a time step the respective element is active (flooded or partly flooded).
- In the partly flooded elements the mean thickness of the saturated aquifer, which is used to determine the transmissivity, is roughly calculated as the quotient of the water volume (WV, see figure 3a) and the total element area. The mean transmissivity of an element changes linearly within a time step.
- Within individual elements in contact with the moving boundary the exact flow pattern is estimated only approximately, however the continuity of the existing groundwater volume is considered exactly.
- To accelerate the computational procedure an exact observance of continuity is only carried out at the beginning of each time step on the basis of the previous calculated time step. In this way inaccuracies of the water balance arising in boundary elements can be corrected at this time (see below).
- Prescribed boundary inflows and the exchange of the above mentioned water volumes take place only at flooded nodes. An inflow at dry nodes is shifted to flooded ones using a trajectory based on the fall line.

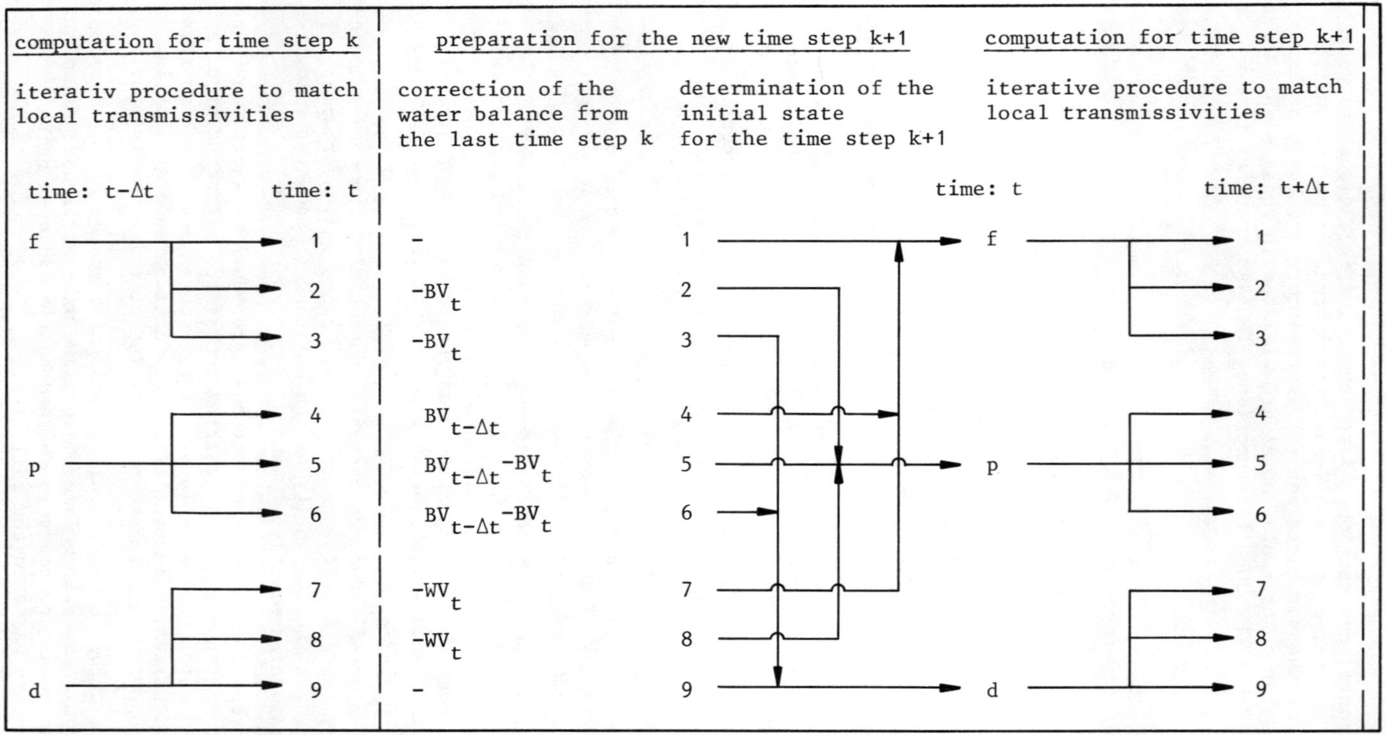


Figure 2.--Schematic representation of the computational procedure (notations see table 1)

Table 1.--Designation of initial and final states of an element within one time step

| initial state | final state |
| --- | --- |
| f<br>flooded | 1 flooded<br>2 partly flooded<br>3 dry |
| p<br>partly flooded | 4 flooded<br>5 partly flooded<br>6 dry |
| d<br>dry | 7 flooded<br>8 partly flooded<br>9 dry |

The following comments help to clarify the above-mentioned guidelines. In particular the corrections to the overall water volume balance which are made at the beginning of each time step will be illustrated using two examples - a boundary element in which the groundwater level is increasing (figure 3a) and one in which it is decreasing (figure 3b).

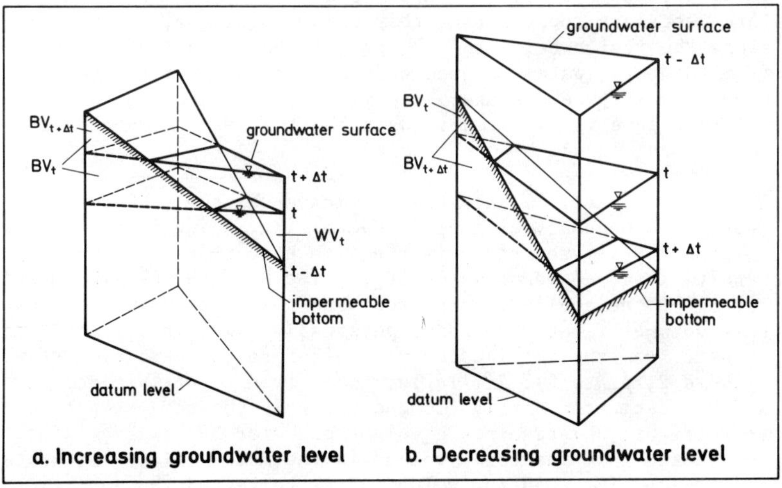


Figure 3.--Flooding and dehydrating of elements

Starting with figure 3a, a node of a hitherto dry element becomes flooded at time t when for example the groundwater surface in the adjacent partly flooded elements increases. Thus in the next time interval t to t+Δt the former dry element will be active (state 7 or 8 in table 1 and figure 2) and is flooded or partly flooded. To activate this element groundwater levels must be determined to describe the initial condition for this time step in hitherto dry nodes. This is accomplished by using a negative gradient estimated in advance to extrapolate the groundwater level from neighbouring nodes. This yields a lower level at the node in question than the average of the surrounding flooded nodes. The choice of negative gradient is especially important when the actual boundary of computation moves on a horizontal bottom. In principle it is also possible that several rows of elements may become active in the same time step and therefore the gradient should not be chosen too small for reasons of model stability. Physically the gradient describes the slope of the groundwater surface near the moving boundary.

Having defined the initial condition of the element, as described above, consideration must also be given to the additional volume of water in elements which become newly active. For instance in figure 3a an element is shown which changes from dry to partly flooded conditions in the time interval t-Δt to t. This results in a water volume $WV_t$ (figure 3a) being added to the water balance of the model which must be corrected by subtraction of the volume $WV_t$ (figure 2, state 8) at the beginning of the time interval t to t+Δt. Should the element remain partly flooded at time t+Δt, as illustrated in figure 3a, an apparent volume increase $BV_t - BV_{t+\Delta t}$ of the total groundwater will be incorrectly calculated. It is therefore necessary to substract this volume of water, which represents the region below the impermeable bottom of the aquifer, from the total water balance before proceeding to the next time step (figure 2 state 5). Similarly, a correction of the water balance is also required in the following time step if the element should become flooded (state 4, figure 2).

In a second case, illustrated in figure 3b, the groundwater level in a boundary element is decreasing. Going from a flooded to a partly flooded state in the time step t-Δt to t, which results in a volume of water $BV_t$, lying beneath the impermeable bottom of the aquifer, being added to the total volume. Since this volume of water does not physically exist it must then be removed from the total water volume before the next time step (figure 2, state 2). Similarly, going from a partly flooded to another decreased partly flooded state in the time step t to t+Δt (figure 3b) requires a volume of water $BV_{t+\Delta t}-BV_t$ to be subtracted from the total volume in the model (figure 2, state 5). At any time when the volume BV is corrected the specific yield $S_y$ of the actual element have of course to be considered.

A similar procedure has proven successful in models employed to simulate inter-tidal flats in front of coastlines (computation of shallow water waves)(Herrling 1976). A number of modifications to previously applied schemes have been used in the present investigation as described above to increase the speed and stability of the calculation without loss of accuracy.

## NUMERICAL RESULTS

This section presents computational results of the foregoing considerations with moving boundaries obtained by means of a coded FORTRAN program. Initially three computations are presented which investigate the performance of the moving boundary feature under various test conditions. Finally, a practical application is demonstrated.

### Test to examine the conservation of the water volume

The first example demonstrates the use of moving boundaries under the aspect of preservation of the water volume in an aquifer sloped at one end. Figure 4 shows the element network of the aquifer model and figure 5 the height of the impermeable bottom (the left part of the aquifer bottom is horizontal). The aquifer dimensions are 2000 m x 180 m, the hydraulic conductivity amounts to 0.01 m/s and the specific yield to 0.2. Starting in the whole model with a constant groundwater level of 30 m above datum level the groundwater volume is reduced by a withdrawal of 933120 $m^3$ (that is 49.35% of the total initial water volume) at the left boundary during a time of 24 days. After this time the withdrawal of water is stopped and the groundwater surface becomes horizontal during the following 175 days. Then a recharge of the same water volume (933120 $m^3$) is initiated at the left boundary again spread over a period of 24 days. After stopping the inflow the groundwater surface adjusts itself horizontally. Several hydrographs of calculated groundwater levels are shown in figure 6, presenting nodes marked in figure 4.

The analyticly calculated groundwater level of 16.667 m is exactly reproduced by the numerical computation in the whole flooded aquifer after the phase of withdrawal. By then four rows of elements are fallen dry. Furthermore, after the flooding period the initial groundwater level of 30 m is exactly calculated. This proves not any volume of water is lost during the whole computation. The calculation was carried out with time steps of one day during periods of prescribed withdrawal or recharge. The time steps were increased to 5, 10 and 20 days in the reassurance phase.

### Horizontal propagation of recharge water on a plane aquitard

The second and third example are used to demonstrate the horizontal propagation of a line source, caused by an artificial groundwater recharge, in the top layer of a stratified aquifer.

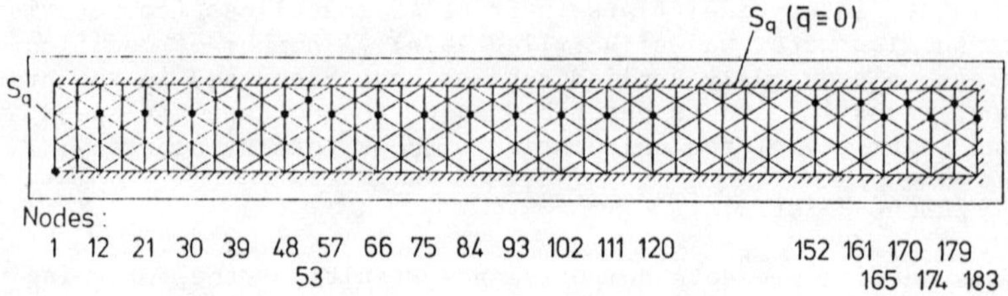


Figure 4.--Network of elements for the horizontal computation

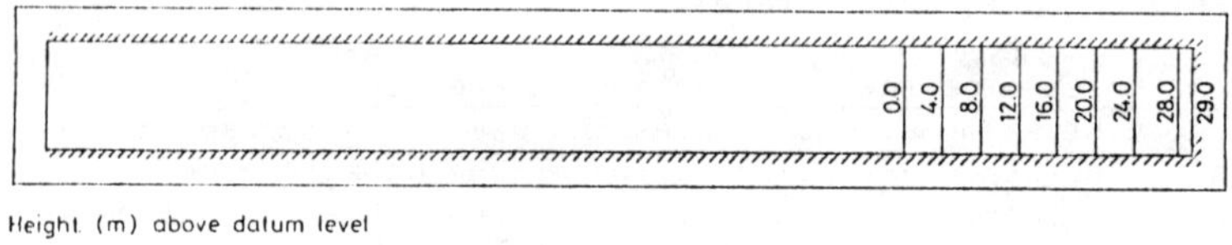


Figure 5.--Contour lines of the aquifer bottom

The recharge water enters a dry and permeable aquifer which is limited by an aquitard with a thickness of 1 m from beneath. Another permeable aquifer with a free groundwater surface is located below this aquitard. In the horizontal model the vertical flow through the aquitard is simulated using the leakage term in equation (9). As atmospheric pressure is assumed below the aquitard the value $h_s$ in equation (8) and (9) amounts to the level of the lower margin of the aquitard. For reasons of symmetry only one side of the recharge domain and furthermore scarcely a strip is considered. The element network of the horizontal computation has a configuration as shown in figure 4. The layers are assumed to be horizontal. The line source is situated at the left boundary of the model.

In the first of these two further test computations a steady flow case of the horizontal propagation is calculated using two different methods: Method one bases on a time-dependend, horizontal calculation carried out with a constant recharge of water yielding a steady solution after some time. Method two carries out the computation with a steady saturated-unsaturated model in a vertical profile. The second is based on the theory as for instance described by Neuman (1973) but using only the steady state part of the basic equation which is formulated in pressure head $\psi$ (Leismann (1981)). The nonlinearities have been solved in an iterative way. To lower the numerical effort of the vertical computation the length of the model has been limited to 200 m using a nodal distance of maximum 4 m in the horizontal and of 0.33 m in the vertical coordinate with a total model height of 3 m (triangular element network). The size of the horizontal model amounts to 200 m x 18 m (see element network

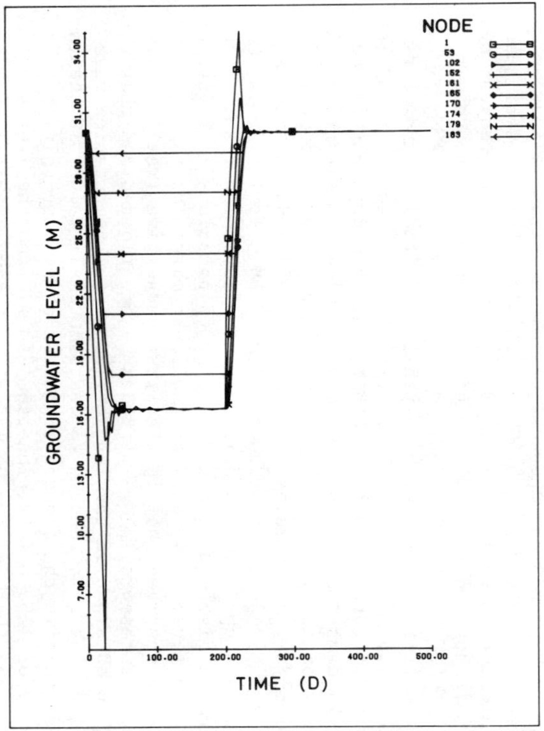


Figure 6.--Groundwater levels during periods of withdrawal and artificial recharge

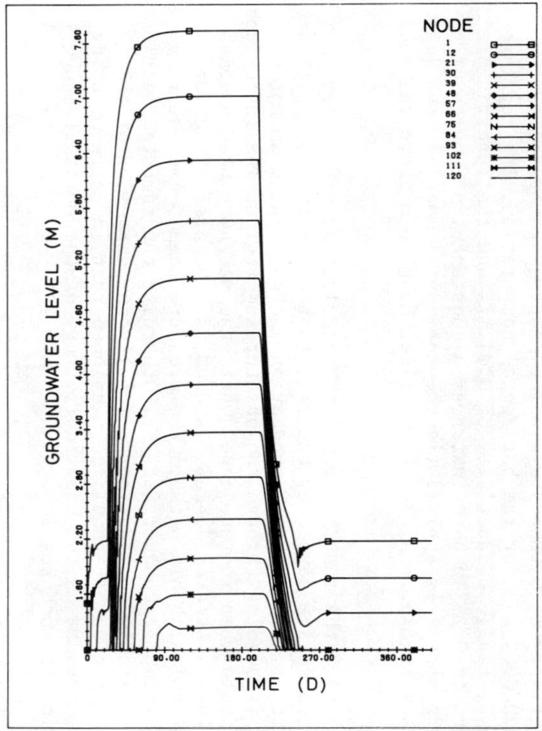


Figure 8.--Propagation of artificial recharged water on a semi-permeable horizontal aquitard

figure 4). The aquitard of thickness 1 m received a hydraulic conductivity of 0.000001 m/s which means a leakage parameter $\lambda = 0.000001\ s^{-1}$ for the horizontal computation. The top layer has a hydraulic conductivity of 0.01 m/s and the specific yield amounts to 0.2. In both instances a constant line source of 0.00025 $m^3$/s m is applied (in the horizontal model at the left boundary in figure 4, in the vertical model 2 m above the aquitard).

In figure 7 the results from both calculations are compared. Differences are caused by the fact that above the zero pressure surface of the vertical computation a considerable amount of water is seeping through the unsaturated region. The latter can be reduced when in the scope, where the pressure head becomes negative, a steeper descent is used for the prescribed functional relationship between the pressure head $\psi$ and the relative hydraulic conductivity $K_r$. Otherwise the computed results are comparable considering especially the expansion of the flooded region. It should be mentioned that in spite of using moving boundaries the numerical effort of the horizontal computation is much smaller.

The third test computation demonstrates in a similar case of artificial recharge as before the efficiency of the computational procedure under the condition of a line source changing quantitatively in time. During this flow simulation the boundary of computation moves along the horizontal semi-permeable aquitard. The size of the model totals to 2000 m x 180 m (see element network figure 4), the top layer has again a hydraulic conductivity of 0.01 m/s and a specific yield of 0.2, the leakage parameter amounts to $\lambda = 0.0000001\ s^{-1}$ (the thickness of the aquitard equals to 1 m). Further an extrapolation gradient of -0.035 is applied for newly flooded elements (see above).

The computation starts with a specified initial groundwater level of 0.5 m above the aquitard at the four nodes on the left boundary (figure 4), all the other nodes are dry. Within the first 24 days a line source of 0.00005 $m^3$/sm is prescribed at the left boundary. This inflow is suddenly increased up to 0.0005 $m^3$/sm and remains constant during 175 days. Afterwards it is again reduced to 0.00005 $m^3$/sm. The computation ran with time steps of one and in parts two days. The results are plotted in figure 8 (the nodes refer to notations in figure 4; the datum level is located at the lower margin of the aquitard). It can be observed that in the first phase five (really six, but not represented) rows of nodes become flooded and that this is again obtained in the last period of the computation when the line source is reduced to the same small value. In the intermediate time the flooded region enlarges immensely up to a distance of 1250 m (starting from 250 m) from the line source which means that 26 rows of nodes or 25 rows of elements become flooded.

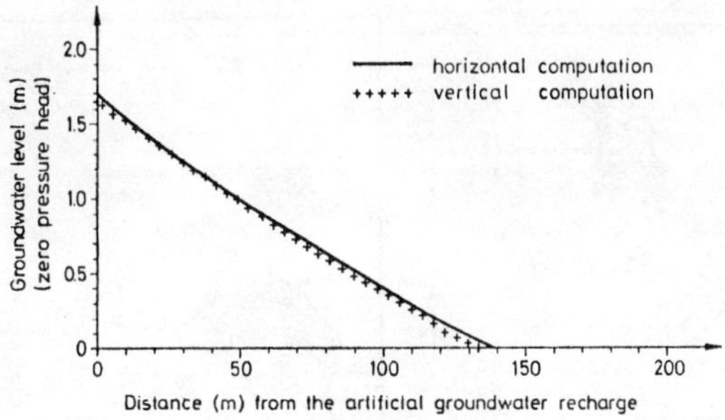


Figure 7.--Comparison of horizontally and vertically computed water levels (zero pressure head) relative to the surface of the aquitard

These results show that the algorithm of the moving boundaries works well. In each time step normally two maximum ten iterations have been used if a tolerance of 0.001 m for the greatest difference between groundwater levels of two successive iterations is prescribed.

## Practical applications

One of the practical applications which requires moving boundaries in a horizontal model is an investigation concerned with the downstream consequences of an artificial groundwater recharge in a glacial valley north of the Alps as described above. Here the stream Eschach near the town Leutkirch, West Germany, is known to be very susceptible to severe high-water conditions (average streamflow 2 $m^3/s$, but maximum values of 52 $m^3/s$ during peak periods). When the high-waters are used for an artificial groundwater recharge in regions with a groundwater table deep under the ground the possibility of waterlogging have to be investigated down in the valley.

The established model of Leutkirch encompasses an area of 18 km x 3 km. Figure 9 shows the network of elements (929 elements, 535 nodes) and figure 10 the contour lines of the aquifer bottom as it is realized in the model. The represented island within the area of computation corresponds to the region in which the impermeable bottom rises to the ground. The element network is constructed in such a way that the streams run along element sides (seepage on a line and leakage), and wells and groundwater gauges lie on nodes. The measured lysimeter data are integrally incorporated in the model as an areal distributed inflow. The nodes with prescribed water levels (boundary conditions) are marked in figure 9 by circular black symbols and the planned location of the recharge pit by a quadratic symbol.

Calibration of the model was done using field data. The here presented computations cover a period of 8.5 months from November 5th, 1979 to July 21st, 1980. This period includes several incidents of high-water conditions in the stream Eschach with

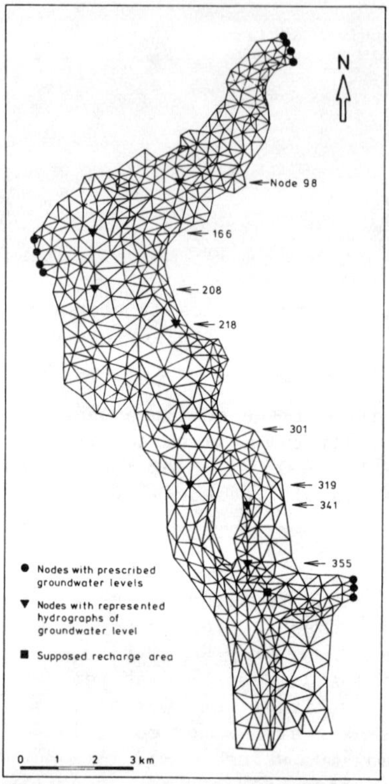


Figure 9.--Network of elements

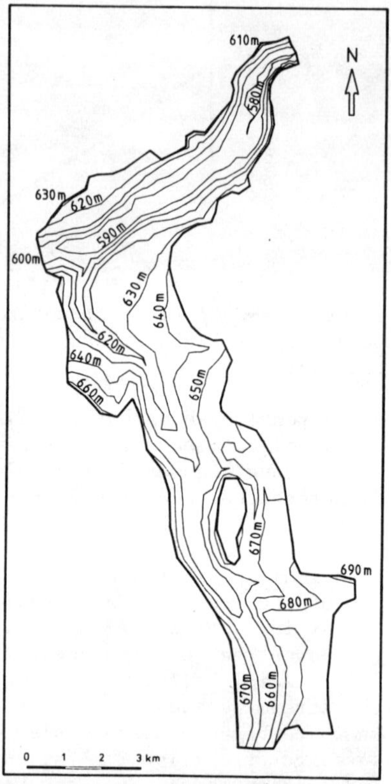


Figure 10.--Contour lines of the aquifer bottom

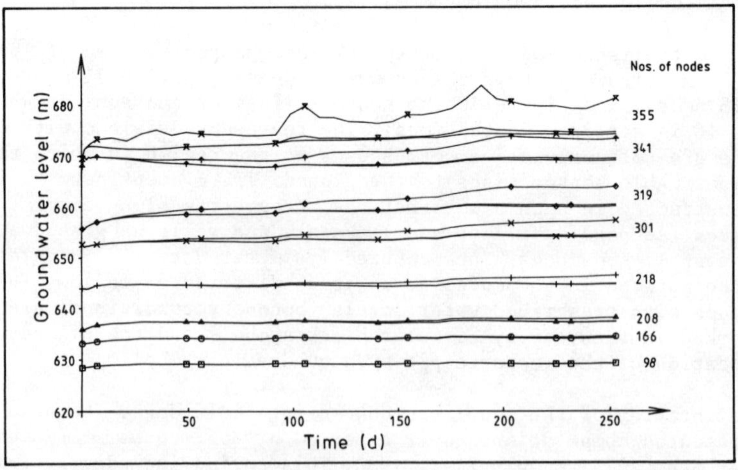


Figure 11.--Consequences of an artificial groundwater recharge

daily average stream flows up to 16 m³/s (maximum 22 m³/s). Though the model calibration has not yet been finished the hydraulic conductivities determined hitherto range from 0.022 m/s to 0.0005 m/s, further a specific yield of 0.2 is applied.

Using the previous calculated modelling constants several modifications were then made in order to investigate the influence of an artificial groundwater recharge. Altogether three runs have been carried out: The first without any additional recharge; the second with a recharge regulated to prevent the total discharge in the stream Eschach from exceeding 10 m³/s during high-water conditions. The third run investigates the case of recharging almost the entire stream discharge into the groundwater, leaving in the stream only a minimum flow of 2 m³/s. The latter is carried out under the assumption that the required recharge basin could be physically realized.

Results of the three runs of the model are presented in figure 11 at eight nodes (triangular marked nodes in figure 9). As expected the consequences of the second run are very small but those of the more powerful recharge in the third run (the top line of each node in figure 11) are more significant.

Though the computed situation does not reproduce the most unfavourable case, these first results show that the represented numerical model is a useful tool in this investigation. The moving boundary algorithm turned out to be a practicable instrument for a realistic computation of the changing lateral expansion of the horizontal model.

ACKNOWLEDGEMENTS

The author wishes to thank Dipl.-Ing. D. Willibald, State Board for Environmental Protection, and Dipl.-Geol. F. Kupsch, Geological State Board both of Baden-Württemberg, West Germany, for their hydrological and geological advice and cand. ing. H. M. Leismann for the performance of the unsaturated flow computation.

REFERENCES

Herrling, B.(1976) A finite element model for estuaries with inter-tidal flats. In:Proc. 15th Coast.Eng.Conf., Honolulu: 3396-3415.

Leismann, H.M.(1981) Erstellung und Erprobung eines Finite-Elemente-Programms zur Berechnung von wassergesättigten sowie -ungesättigten, vertikal ebenen, stationären Sickerströmungen. Diplom Thesis, University of Karlsruhe.

Neuman, S.P.(1973) Saturated-Unsaturated Seepage by Finite Elements. J.Hydr.Div.Amer.Soc.Civil Eng.,Vol.99(Hy12):2233-2250.

Werner,J., Strayle,G., Walser,M.(1974) Möglichkeiten der Grundwassererschließung und -anreicherung im Gebiet der Leutkircher Heide (Oberschwaben). Das Gas- und Wasserfach 115 (12): 525-568.

# Finite-element Investigation of the Groundwater Basin of the Szentendre Island, Hungary

G. Molnár
hydraulic
engineer

G. Popper
mathematician,
cand.of techn.sci.

## Abstract

The main water supply region for Budapest, the groundwater basin of the Szentendre Island has been investigated. Contributing to the establishment of a Safe, yet economical water supply policy, simmulation modell, have been developed.

The govering equations of groundwater flow has been solved by the Galerkin-finite element method and the influence of the both categaries of boundary conditions /the essential and the natural/ and some nonlineraty of the problem has been considered.

The stage-discharge relationships of the simulation are utilized by the Budapest, Water Works in the operational policy of withdrawal.

## 1. Introduction

The daily water consuption of the Hungarian capital, Budapest varies between 8oo-12oo thousand cubic meters, depending on season and weather conditions.

5o-7o% of this vast amount is extracted from the bank-filtered ground-water basin of Szentendre Island that is situated in the Danube river, on the north of the town. Contributing to the establishment of a safe, yet economical water supply policy, simulation modells have been developed to investigate the hydrodynamical processes of the ground-water regime.

In this paper both the simulation model, applying finite-element method, and the results of the analysis are presented.

## 2. Description of the Szentendre Island

The Szentendre Island located to the north of Budapest in the Danube river is nearly 31 km long, its maximum width is 3.5 km, its area: 55,8 km2. The surface of the island is diversified, the difference in the elevation of the lowest and highest point is about 2o meters. The surface for the most part is under cultivation. Besides tillage recreational area and forest can be found on the island.

The overlying layer is 1.5 - 2o.o m thick, consisting of low permeable silty sand. Below this a gravel layer is situated with high permeability and thickness of 5.o to 19.o meters. It can be characterised by a hydraulic conductivity varying between 5.o and 8oo,o m/day and storage coefficient of o,2o-o,24.
The water-bearing layer is followed by an impervious one, consequently the whole area can be modelled as a singleaquifer ground-water basin.

## 3. Mathematical formulation of the problem

Based on the afore-metioned, the governing differential equation of the hydrodynamic simulation may be written in the following form:

$$\frac{\partial}{\partial x}\left(km\,\frac{\partial h}{\partial x}\right)+\frac{\partial}{\partial y}\left(km\,\frac{\partial h}{\partial y}\right)=S\,\frac{\partial h}{\partial t}+I+Q+\frac{k'}{b}(h_f-h) \qquad /1/$$

where

x and y - Cartesian coordinates /m/

k - hydraulic conductivity /m/day/

m - thickness of the aquifer /m/

h - piezometric head in the aquifer /m above reference level/

S - storage coefficient of the aquifer /$m^3/m^3$/

Q - concentrated recharge /discharge to/from the aquifer /$m^3$/day/

I - vertical water flux /$m^3/m^2$/day/

k' - hydraulic couductivity of the bed material of a river or lake influencing the regime of the ground-water basin /m/day/

b - thickness of the river bed /m/

$h_f$ - stage of the river or lake /m above reference level/

Initial conditions are the h values existing at the first instance of the simulation procedure, the initial water levels:

$$h(x,y,0)=h^{\circ} \qquad /2/$$

Two types of boundary conditions may de distinguished:

a./ pervious boundary

$$T\frac{\partial h}{\partial n}=Q \quad \text{on the } \Gamma_1 \text{ section} \qquad /3/$$

if Q=0, the boundary is impervious

b./ boundary with given water level

$$h(x,y,t)=\hat{h} \text{ on the } \Gamma_2 \text{ section} \qquad /4/$$

where

$h^o$ - initial water level of the aquifer /m above reference level/

n - normal to the $\Gamma_1$ boundary section

$\Gamma_1, \Gamma_2$ - set of boundary sections representing different boundary conditions /$\Gamma_1+\Gamma_2=\Gamma$ /

The foregoing bivariate partial differential equation is of parabolic type. Applying the numerical methods available, a procedure was developed primarily based on finite element method. Solutions for the most general case namely when the transmissibility, storage coefficient and vertical water flux are also dependent on h /thus the differential equation is nonlinear /were established as well.

4. Solution of the differential equation

Equation /1/ should be solved over the spatial domain $\Omega$ , subject to the external boundary conditions /3/ and /4/ being valid on the margin $\Gamma=\Gamma_1\cup\Gamma_2$

For the spatial discretization of this boundaryvalue problem the Galerkin-type, finite element method, for temporal discretization a finite difference approximation scheme was employd.

From the application of the Galerkin equations follows, that both the differential equation and the natural boundary conditions /4/ are satisfied on the average only. After the usual partial integration, because of the homogeneous form of the boundary conditions essential, the Galerkin equations will contain first derivatives at most. Consequently it is sufficient to adopt basis functions corresponding the finite element method, that are linear.

As a result of spatial discretization one should obtain the following system of ordinary differential equations, including the time derivative:

$$\underline{\underline{A}}\,\underline{h} + \underline{\underline{B}}\,\frac{dh}{dt} + \underline{F} = \underline{0} \qquad /5/$$

Proceeding as straightforward as possible, /5/ can be approximated by an implicite finite difference scheme, may be written in the form:

$$\left(\underline{\underline{A}} + \frac{1}{\Delta t}\,\underline{\underline{B}}\right)\underline{h}\,(t+\Delta t) = \frac{1}{\Delta t}\,\underline{\underline{B}}\,h(t) - \underline{F} \qquad /6/$$

Setting out of the initial conditions $t=t_0;\underline{h}\ /t_0/$ we wish to obtain repeated solutions for the linear algebraic equations /6/, where the coefficient

$$\underline{\underline{A}} + \frac{1}{\Delta t}\,\underline{\underline{B}}$$

is a banded symmetric matrix.

The problem becomes more complex when, vector $\underline{F}$ being a nonlinear function of the potential vector $\underline{h}$ /e.g.in case of confined aquifers/, the recent time formula defined by the set of nonlinear equations /6/ has first to be solved by "posterior iteration".

Finally let's consider the essential boundary condition, effective on the boundary $\Gamma_1$, that can be given in the form:

$$h(t) = \bar{h}(t) \qquad /7/$$

Practically it means the definition of the i

indices of those nodes theat have $h_i/t/$ values specifide. In doing so, the following simple device was used: If $h_i$ /t/ and $h_i$ /t+$\Delta$t/ are specified boundary values at time t and t+$\Delta$t respectively, $h_i$ /t/ representing element i in vector $\underline{h}$ /t/, than element /i,i/ in diagonal of matrix $\underline{\underline{B}}$ should be substituted for a number, that is large enough compared to other elements. Let's also substitute element i of vector $\underline{F}$ for the product of this large number and the difference $h_i$ /t/- $h_i$ /t+$\Delta$t/. In this way, element i of the solution vector $\underline{h}$ /t+$\Delta$t/ from equation /6/ will be equal with the specified boundary value $h_i$/t+$\Delta$t/.

5. Applicatons and their results

Purpose of these investigations was to give a good approximation of the hydrodynamic processes taking place in the ground-water basin of the island. The results of the simulation were tested on the map of the ground-water levels. For sake of synopticity, only 1/3 of this map is demonstrated in Figure 1.

At first, simulation was carried out using data based on hydrogeological exposures. For the calculations a triangular grid, shown in Figure 2., was applied. Results of the simulation were compared to the measured water levels of the map /Fig.1./. Where the differences between the simulated and measured levels were significant, there permeability coefficients has been modified by personal judgement. Care has been taken not to modify more than one order of magnitude. After three modifications the measured and simulated ground-water tables coincided reasonably well. Figure 3. represents the map of simulated ground-water levels.

In the next stages of the analysis, the efforts were concentrated on the question, how the increase or dicrease of withdrawal influences the water table in the anquifer. The stage-discharge relationships thus derived are utilized by the Budapest Water Works in the operational policy of withdrawal.

The authors as a continuation of these investigations are working on the refinement of the identification of permeability and storage parameteres. They intended to do this amalgamating mathematcal methods available and engineering experience.

## References

Coonor, J.J.-Brebbia, C.A/1977/ Finite Element Techniques for Fluid Flow. Newnes-Butterworths, London, 1977.

Durbin T.J./1978/ Calibration of a Mathematical Model of the Antelope Valley Ground-Water Basin, California, Geological Survey Water-Supply Paper n° 2o46. U.S. Government Printing Office, Washington, 1978.

Kovács Gy./1972/ A szivárgás hidraulikája. Akadémiai Kiadó, Budapest,1972.

Marcsuk G.I./1976/. A gépi matematika numerikus módszerei. Parciális differenciálegyenletek. Müszaki Könyvkiadó, Budapest,1976.

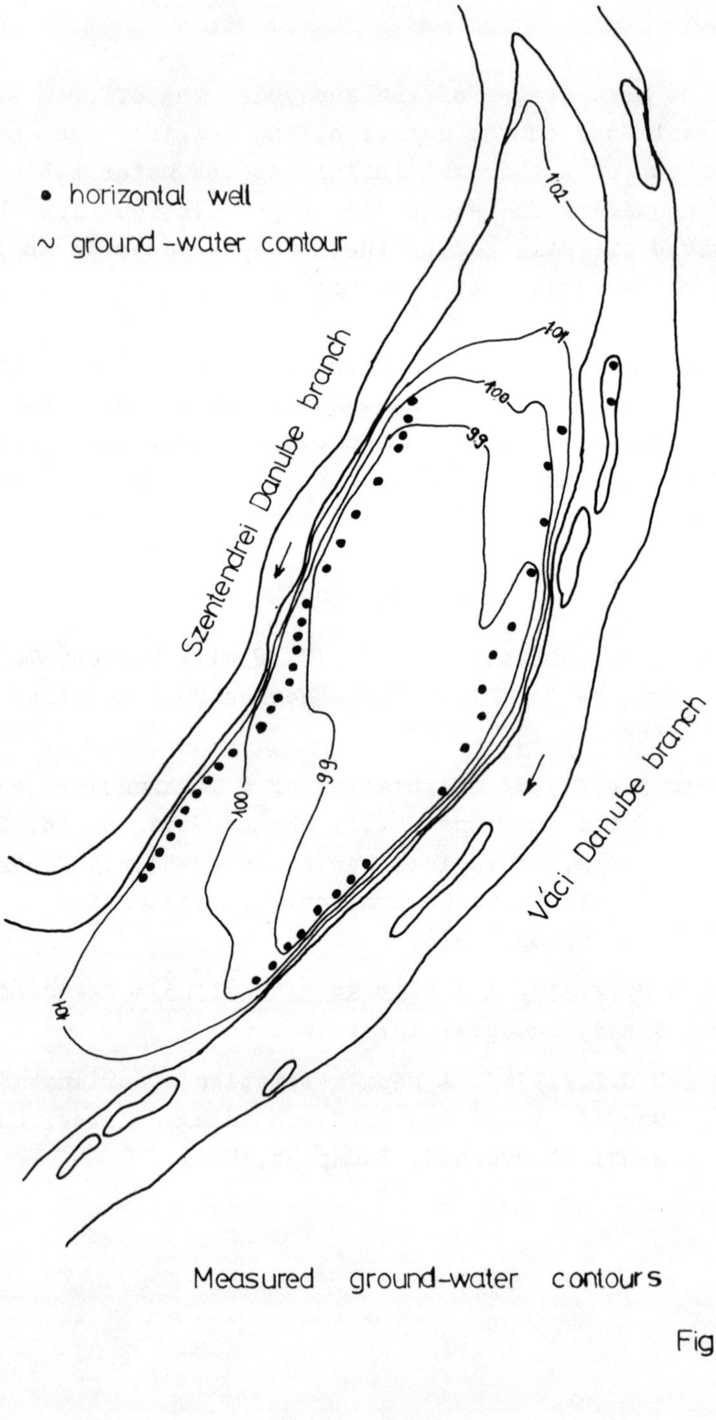


Measured ground-water contours

Fig 1.

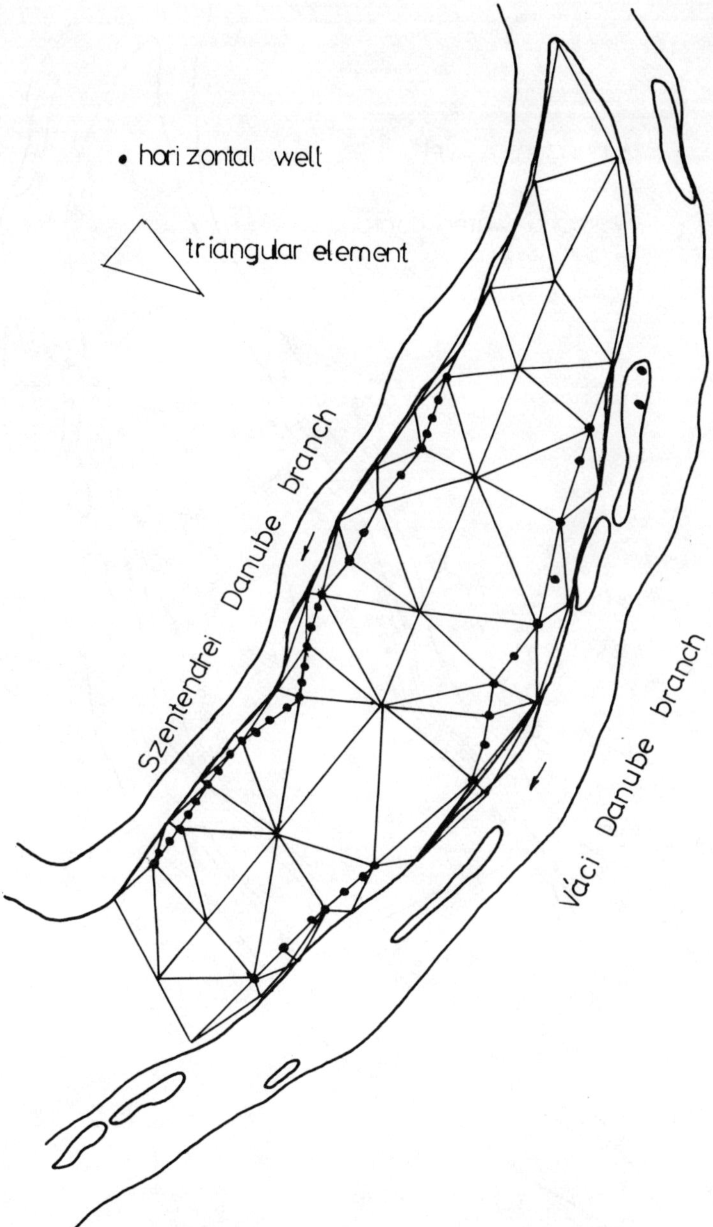


Division of the area investigated to triangular elements

Fig 2.

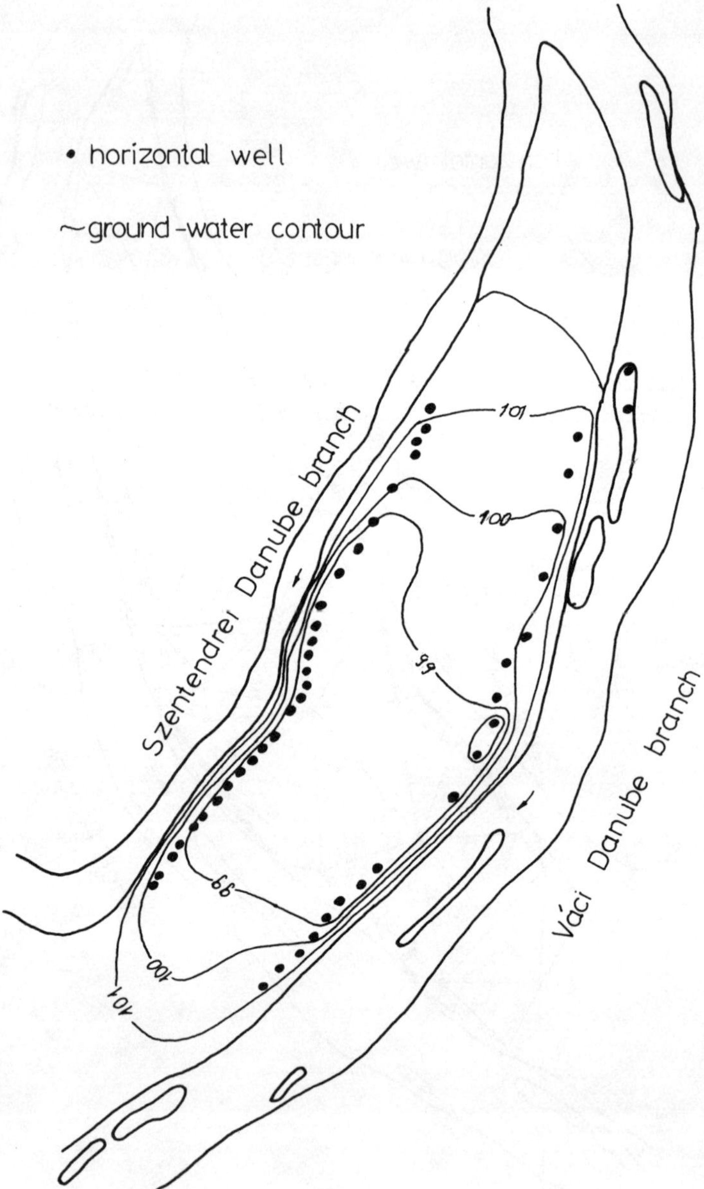


Simulated ground-water contours

Fig 3

# INTERCONNECTED GROUNDWATER SYSTEMS SIMULATION (IGROSS) DESCRIPTION OF THE SYSTEM AND ITS APPLICATION TO THE WESTERN LIBYA REGIONAL AQUIFER.

G. Pizzi, L. Sartori.

Aquater S.p.A. - San Lorenzo in Campo (PS) - Italy

## INTRODUCTION

IGROSS represents the natural evolution of the use of finite elements methods towards industrial applications of groundwater mathematical models.The same evolution occurred during the last decade for the applications of finite elements methods to stress analysis.
This evolution is characterized by the passage from experimental programs, aimed to solve the finite elements equations, to integrated systems of computer programs, aimed to guarantee the product's quality and reduce the production cost, according to an industrial outlook.
Based on the quasi-threedimensional finite element approach, IGROSS has been conceived to simulate not only the groundwater movement (flow model) but also the most important phenomena coupled with the groundwater flow: convective dispersive mass transport (quality model) energy transport (geothermal model), soil consolidation (subsidence model) and integrated surface-subsurface rainfall-runoff (Watershed model).
Presently it is dealing only with quantitative aspects of the groundwater flow.

## DESCRIPTION OF IGROSS

IGROSS is composed of four computer programs connected by five data files as shown in the figure 1. The above programs can be divided into three groups.

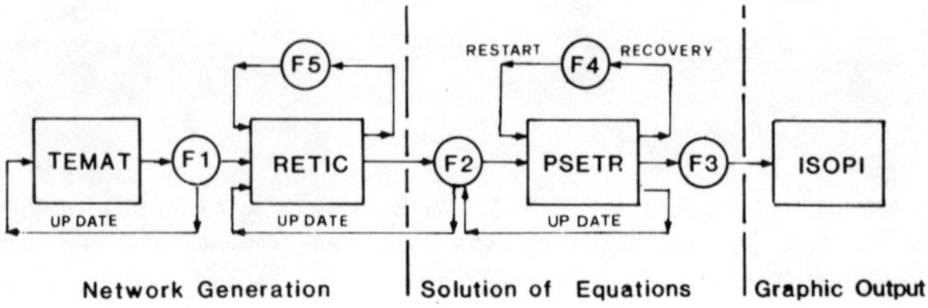


Fig. 1 - Block diagram of IGROSS.

a - Programs for the network generation

IGROSS network generation programs automatically generate the FE quasi-threedimensional network and automatically attribute the physical parameters to each network element.

The methods employed are based on the use of digitalized thematic maps and are described by L. Sartori, V. Aurelio, M. Riccioni and G. Svarca (1982).

The digitalized thematic maps are generated by TEMAT. Due to the regional extension of the models the basic data are often contained on several maps of different scales and mapping systems. TEMAT allows direct acquisition of data by means of a digimeter, from the reference maps. For this purpose the program has a routine allowing the translation from and into any one of usual mapping systems (conformal cylindrical Gauss projection, U.T.M. System, conformal conical Lambert projection, natural projection). TEMAT has the possibility of graphically reproducing thematic maps in any desired scale and mapping system.

The digitalized thematic maps, generated by TEMAT, are stored in file F1, and subsequently employed by RETIC.

RETIC has been designed to produce complex networks with unlimited possibilities of mesh variation. The generated networks respect both the external boundaries and a series of internal lines. These lines, called no-crossing lines, permit the correct reproduction of rivers, faults and geological formations' boundaries, with a succession of network sides.

The generation of the network takes place in two phases. In the first phase a two-dimensional reference network of triangular meshes is generated and stored in file F5; in the second phase the quasi-threedimensional FE network is generated, the physical parameters are attributed and finally the network is stored in file F2.

The two-dimensional reference network is generated starting from a digitalized thematic map called the network generation map containing the network density and the no-crossing lines. The network density is defined as the number of elements contained within a reference element.
The figure 2 illustrates the network generation map utilized for the construction of the model of Western Libya groundwater system.
Figure 3 presents the two-dimensional reference network automatically produced by RETIC on the basis of the above network generation map.

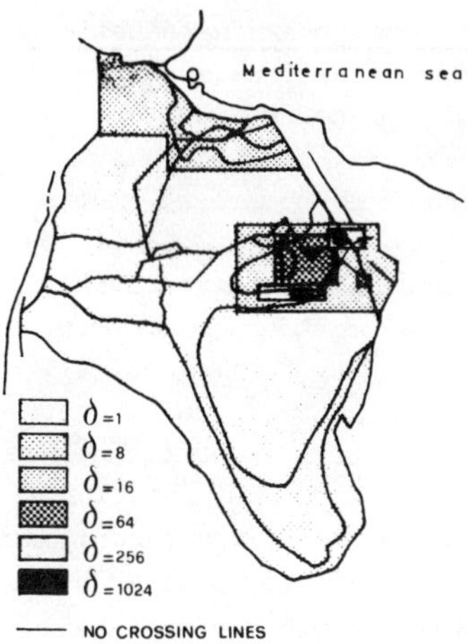


Fig. 2 - Network generation map

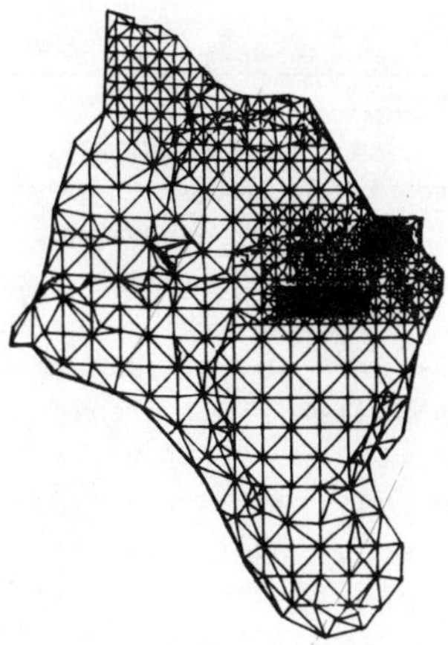

Fig. 3 - Two-dimensional reference network

The part of the quasi-threedimensional network related to aquifers is constructed utilizing two-dimensional substructures of triangular elements taking into account the lateral connections existing among the different aquifers of the groundwater system. The above connections are obtained by identifying the aquifers' branching nodes. The substructure's elements are automatically identified on the basis of their boundary on the two-dimensional reference network.
The part of the network related to aquitards is constructed connecting the nodes of superimposed aquifers located on the same vertical by means of linear macroelements, each one composed by linear elements of the same length.

The attribution of the system's physical parameters to the aquifer's and aquitard's elements is made on the basis of an automatic consultation of the digitalized thematic maps.
Similarly to the TEMAT program, RETIC has the possibility of graphically reproducing the generated network in any scale and mapping system.
As the computer time required to solve the equations depends on the order in which the elements of each substructure are numbered, RETIC has been provided with a substructure elements resequencing routine.
Moreover RETIC can interpolate the parameters associated to a network onto another one. This possibility allows to change the network during the model calibration.

b - Programs for the solution of the governing equations

The PSETR program solves finite element equations according to the quasi-threedimensional scheme in the steady and unsteady state, using linear shape functions.
The governing equations, along with the numerical methods utilized for their solution and for evaluation of fluid mass balance, are described by L. Sartori and G. Peverieri (1982).
The numerical methods employed derive from M.B. Irons frontal method (1970) and allow to overcome the two strongly limiting conditions which characterize this method in the quasi-threedimensional equations system. These limiting conditions are caused by the occurrence of very wide fronts, in relation to the aquifer branchings and to the presence of aquitards. The above disadvantages are eliminated in IGROSS numerical methods, and the frontal method operative performances are preserved.
Use of substructures limits the frontal increase in correspondance to the aquifer branching. In the steady state, the frontal method is integrated with an iterative solution technique to eliminate the frontal increase caused by the presence of aquitards.
It converges very rapidly using a forcing technique with an automatic parameter definition. In the unsteady case the same scope is achieved using a predictor-corrector procedure which employs the Crank-Nicolson method in the corrector phase.
This very stable procedure permits use of fairly long timesteps and concerns the case of source terms depending on piezometry (problem of interaction between water table and river).
A high convergence velocity has been observed in the iterative method, requiring normally 2-3 iterations and a maximum of 6-7 iterations in the most complex cases. This method has not presented any numerical criticity in all the cases which have

been dealt with up to now, even in situations which make critical the other numerical methods.
The method of equivalent nodal sources, used for the evaluation of fluid mass balance, is conservative. This means that along every aquifer section, the computed flow on both parts of the section is the same. This allows an exact calculation of the aquifer mass balance and, in particular, of the flow repartition in correspondence of aquifer branchings.
PSETR disposes of specific procedures which considerably simplifies data management (UPDATE Procedure) and reduces computer time in certain situations (RECOVERY/RESTART Procedures).
Two methods are used for the updating of physical parameters associated to network nodes or elements.
The SINGLE method assigns a new value of the variable to the nodes or elements specified in a list.
The AREA method assigns a new value of the variable to all elements or nodes contained inside a specified area defined by means of a sequence of nodes laying on its boundary.
Moreover PSETR has been provided with a user-oriented macrolanguage which allows to manage the input with a large flexibility. The macrolanguage is composed of 56 instructions structured on 4 hierarchic leveles.
The results of calculations are stored by PSETR on the File F3 to be successively utilized by the graphic post-processing programs and/or are directly printed.

c - Programs for the graphic output

Presently, in the IGROSS system, the only graphic output program used is the ISOPI program, which operates the graphic display of the results (piezometry, drawdown and leakage), in the form of isovalue lines.
This program also uses automatic mapping routines which allow to reproduce the results in any desired scale and mapping system.

APPLICATION: THE MATHEMATICAL MODEL OF THE WESTERN LIBYA REGIONAL GROUNDWATER SYSTEM.

a - The structure of the model

The Western Libya regional groundwater system extends over an area of 800,000 $km^2$ and is composed of four aquifer hydraulically interconnected, both along lateral aquifer branchings and through aquitards.
An exhaustive description of the Western Libya regional

groundwater system is given in the works of Pallas (1978), and Aquater (1981).

The quasi-threedimensional scheme utilized for the model construction consists of three aquifer levels interbedded by two aquitards.

- The first aquifer level is the Mizda aquifer (Upper Cretaceous dolomitic limestones).
- The first aquitard is the Cenomanian aquitard.
- The second aquifer level is represented by two aquifers: the Kikla and the Nubian (Triassic, Jurassic and Lower Cretaceous Sandstones).
- The second aquitard is the Carboniferous Aquitard.
- The third aquifer level is the Paleozoic Aquifer (Cambro-Ordovician, Upper Silurian and Lower Devonian continental sandstones).

Figure 4 illustrates the scheme of the lateral connections existing among the aquifers of the Western Libya groundwater system.

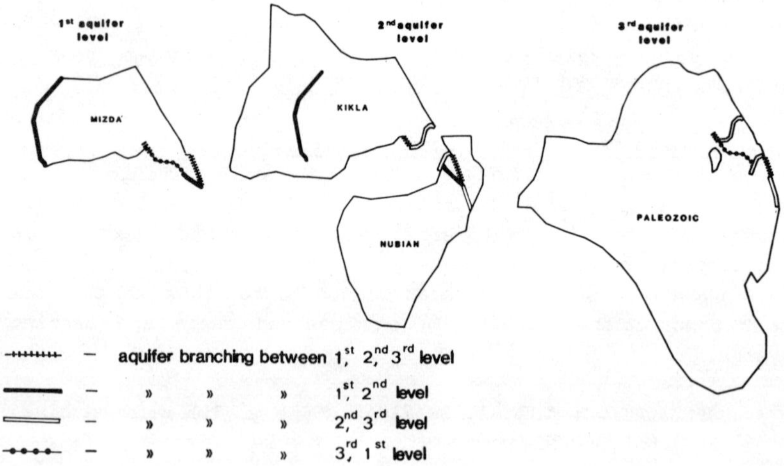


Fig. 4 - Scheme of the lateral connections existing among the aquifers of the Western Libya groundwater systems.

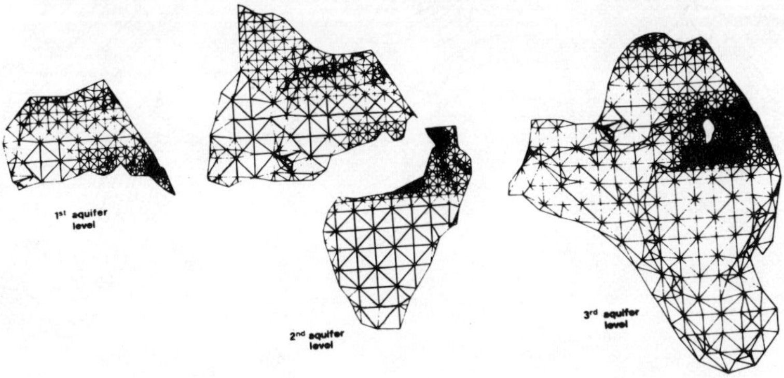


Fig. 5 - Aquifer's quasi-threedimensional network.

In figure 5 the quasi-threedimensional network corresponding to the aquifers is presented. The connections existing among the aquifers are reproduced simply by identifying the aquifers' branching nodes.
The part of the network corresponding to the aquitards was automatically generated utilizing linear macroelements, each composed of four elements, linking the nodes of superimposed aquifers within the existence field of each aquitard.
A total of 7511 elements are included in the generated network: 4407 triangular elements for the aquifers and 3104 linear elements for the aquitards; the nodes are in total 4490: 2362 on the aquifers and 2328 on the aquitards. The finite elements of the aquifers are assembled into 20 substructures: 6 related to phreatic zones, 14 to confined zones. The branching lines nodes are 62.

Steady state model calibration

The first step towards calibration of the model was the reproduction of the state of the system prior to its modern development, when it was subject only to the natural regime.
The reference piezometric configuration utilized for the steady state calibration was the one observed in the year 1960 (fig. 6).

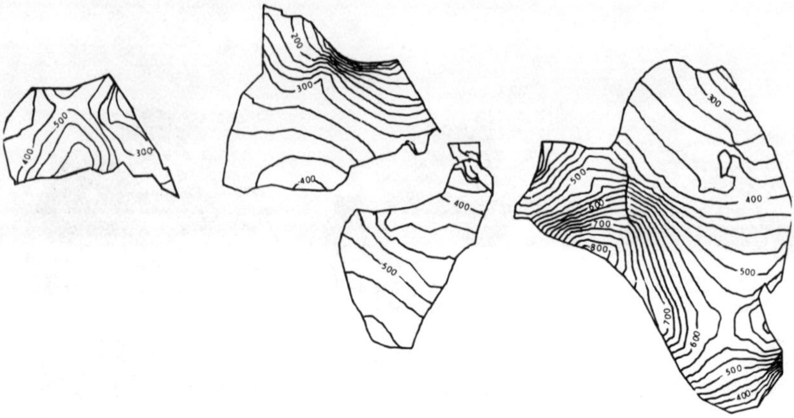


Fig. 6 - Piezometric distribution in the year 1960.

Only the natural outflow of the groundwater system have been considered, imposing for each aquifer the piezometric levels in the corresponding points (including the Sebkas and springs laying inside the study area).

The zero flow rate condition has been imposed along the impervious boundaries of each aquifer.

Let us note that in correspondance of the aquifer branchings, IGROSS automatically takes into account the aquifers' continuity and therefore no boundary conditions must be imposed.

The steady-state model calibration was conducted with the usual trial and error procedure. A total of 113 steady-state calculations were run to obtain a reasonable fitting between the computed and observed piezometric configurations.

The efficiency of the PSETR and particularly the updating procedure, allowed to conduct this calibration with limited human and computer resources.

### c - Unsteady-state model calibration

The artificial groundwater extraction in the Saharian Oasis started in the year 1960.

A typical exploitation history is given in figure 7.

In the concerned area only the Mizda Aquifer is exploited. The piezometric decline observed in the Paleozoic aquifer is due to the combined effect of the leakage through the Cenomanian Aquitard and of the lateral contact between the Mizda and the Paleozoic Aquifers along the aquifer branching.

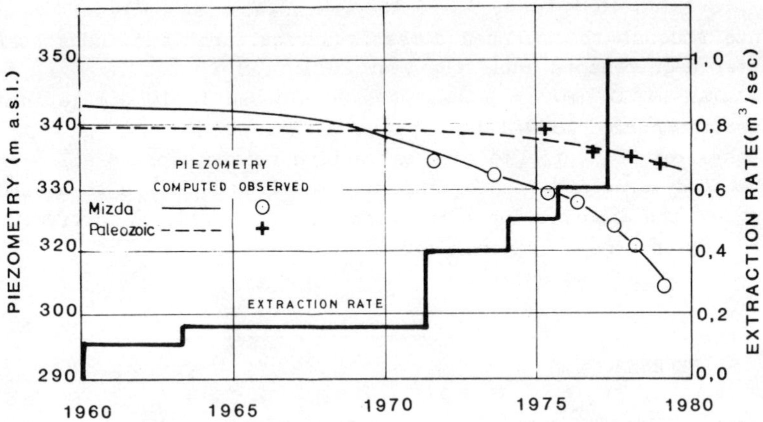


Fig. 7 - Typical groundwater production history.

The unsteady state calibration of the model consisted in the reproduction of the whole groundwater system's history during the industrial exploitation period ranging from 1960 up to 1979.

Therefore together with the natural outflow all the groundwater extractions that took place in said period were taken into account.

The initial conditions assumed have been the piezometric configuration resulting from the steady state model calibration corresponding to the year 1960.

The external boundary conditions were the same used for the steady state case.

As for the springs and the sebkas, laying inside the study area, in order to correctly take into account the decay of their outflow due to artificial extraction, a piezometry dependent discharge rate has been imposed.

A total number of 25 unsteady state computations were run to reach a good fitting between the computed and measured piezometric decline. An example of fitting is given in fig. 7.

## Forecast and optimization of the groundwater system exploitation.

After the steady and unsteady state calibration, the model has been assumed as representative of the groundwater system.

In the last phase of the study, the model was therefore used as a prevision tool to optimize the groundwater extraction taking into account the planned domestic agricultural and industrial water requirements until the year 2030.
A total of 31 new projects, were considered in 10 simulations over the period 1979-2030.
A total of 281 variable rate extraction points, spread all over the study area and tapping the four modelled aquifers were used during the simulations. For each of these points, different production schedules were tested.

## REFERENCES

Aquater (1981). Hydrogeological Study of Wadi Ash Shati, Al Jufrah and Jabal Fezzan area. Prepared for Secretariat of Agricultural Reclamation and Land Development. Tripoli. Libya.

Irons M.B. (1970). A frontal solution program for finite element analysis.
Int. J. for numerical methods in engineering. Vol. 3 5-32.

Pallas P. (1978) Water Resources of the Socialist People's Libyan Arab Jamahiriya.
Tripoli. Libya.

Sartori L., Peverieri G. (1982). A frontal method based solution of the quasi-three-dimensional finite element model for interconnected aquifer systems and fluid mass balance evaluation. Steady and unsteady equations.
4th Int. Conf. on Finite Elements in Water Resources. Hannover.

Sartori L., Aurelio V., Riccioni M., Svarca G. (1982) Automatic generation of Finite Element network for bidimensional and quasi-three-dimensional equations governing the groundwater flow.
4th Int. Conf. on Finite Elements in Water Resources. Hannover.

# FINITE ELEMENT SIMULATION OF THE TRANSIENT EXPLOITATION OF A COASTAL AQUIFER

António Sá da Costa

Inst. Superior Técnico, Universidade Técnica de Lisboa, Portugal

## INTRODUCTION

The province of Algarve, in the South of Portugal, has in the last ten to fifteen years sufered a tremendous development. This development has two main components: agriculture and tourism. Both these activities demand water, and sometimes in large quantities, but water is not easy to get in the Algarve. As usually two sources of water are possible: surface water and groundwater. Due to topographic reasons, surface water will require the construction of dams and a large number of long cannals and mains, because in many instances it is required to import surface water, this means that this type of water supply sistem will take several years to become operational and has a high cost. The other alternative, groundwater, is rigth in the place of consumption, most of the times, and in few working days could be available to both irrigation and urban water supply.

As expected groundwater is supplying many communities and irrigating many fields, however, many of the new wells were drilled quite close to the coast, and almost all of them with very little, if any, study of the local situation in respect to ambient flow in the aquifer system or seawater intrusion. Lately some problems start to rise and the municipalities, because the problems are first felt in the urban water supply, turned to the DGRAH (Direcção Geral de Recursos e Aproveitamentos Hidráulicos) which is the governement agency that supervises the use of water in Portugal, for help.

DGRAH has planned a quite large intervention in the Algarve. Due to the configuration of this province the global problem can be subdivided in small areas which are more or less isolated, from the point of view of groundwater. In 1981 a contract was celebrated betweenn the DGRAH and the Instituto Superior Técnico to study one fo these areas, the "Aquifer of the Dunes of Vila Real de Santo António". This aquifer is located at the Southeast tip of Algarve, Figure 1, and is one of the smallest areas mentioned before, but one that has more problems.

The main objective of this study is to predict how far the town of Vila Real de Santo António can go in the exploitation of the aquifer without threatning contaminate, by seawater, the existing or future wells. It is also pretended with this study to determine the relationship that must exist between the annual natural recharge and the amount of water that can be withdraw annually from the aquifer.

## DESCRIPTION OF THE MODEL

As in most practical situations, in the aquifer of Vila Real de Santo António the assumption of an immiscible interface between seawater and freshwater is appropriated. Since the aquifer is quite shallow, the depth ranges between 10 and 20 meters, the toe of the interface, which is defined by the intersection of the interface with the bottom of the aquifer, is a good indicator of the state of seawater enroachment in the aquifer. Therefore this situation called for the use of a numerical model that can represent a "moving toe" accuratelly. The model SWIM,

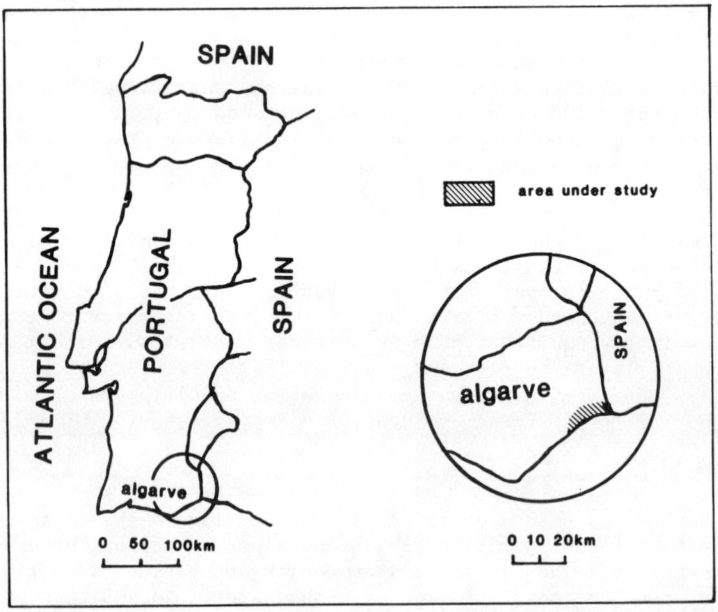


FIGURE 1 Localization of the area under study

an acronym for SeaWater Intrusion Model, developed at the Massachusetts Institute of Technology, USA, Sá da Costa and Wilson (1979) and (1980), is such a model and was selected to be used in this study.

The governing equations in tensorial form are:

FRESHWATER:

$$\frac{\partial}{\partial x_i}\left(K^f_{x_i x_j} b^f \frac{\partial \phi^f}{\partial x_j}\right) + \frac{K'^f_m}{b'^f_m}(\phi'^f_m - \phi^f) + q^f + N =$$

$$= \left(S^f + n\frac{\gamma^f}{\Delta\gamma}\right)\frac{\partial \phi^f}{\partial t} - n\frac{\gamma^s}{\Delta\gamma}\frac{\partial \phi^s}{\partial t} \qquad i,j,m=1,2 \qquad (1)$$

SEAWATER:

$$\frac{\partial}{\partial x_i}\left(K^s_{x_i x_j} b^s \frac{\partial \phi^s}{\partial x_j}\right) + \frac{K'^s_m}{b'^f_m}(\phi'^s_m - \phi^s) + q^s =$$

$$= \left(S^s + n\frac{\gamma^s}{\Delta\gamma}\right)\frac{\partial \phi^s}{\partial t} - n\frac{\gamma^f}{\Delta\gamma}\frac{\partial \phi^f}{\partial t} \qquad i,j,m=1,2 \qquad (2)$$

where i,j,m=1,2 refer to the orthogonal spatial coordinates, $x_i$, $x_j$; and m=1,2 refers to the leakage from/to the above or below aquifer layer. The interface depth, $\zeta$, is defined by continuity in pressure at both sides of the interface. In terms of piezometric heads, this leads to the expression:

$$\zeta = \frac{\gamma^s}{\Delta\gamma}\phi^s - \frac{\gamma^f}{\Delta\gamma}\phi^f \qquad (3)$$

In these equations the superscripts $^f$ and $^s$ represent a freshwater and a seawater quantity, respectively, and

$K_{x_i x_j}$ = components of the hydraulic conductivity tensor,

b = saturated thickness,

o = piezometric head,

$K'_m$ = vertical permeability in the semi-pervious leaky layer ( =0 if nonexistent), where m=1 is the top layer and m=2 the bottom layer,

$b'_m$ = thickness of the leaky layer,

$\phi'_m$ = piezometric head in the vertically adjacent aquifer on the other side of the leaky layer,

q = source/sink term,

N = natural accretion ( =0 in confined aquifers),

S = elastic storage for confined aquifers, or specific yield for phreatic aquifers,

n = effective porosity,

$\gamma$ = specific weight, $\Delta\gamma = \gamma^s - \gamma^f$ , and

t = time.

The major assumptions behind the derivation of these equations are:

- immiscible interface separating freshwater and seawater,
- Darcy's law is applicable,
- vertical variations of the aquifer properties are neglected,
- Dupuit assumption is valid,
- vertical flow is only considered for leaky effects and for accretion,
- constant specific weight for both freshwater and seawater,
- these two phases are considered homogeneous, isotropic, and they completely fill all voids in the saturated zone of the porous media.

Using Galerkin approximation to perform the space integration, the governing equations 1 and 2 can be reduced to a simpler matrix equation, Sá da Costa and Wilson (1979):

$$\underline{\underline{C}} \frac{d\underline{X}}{dt} + \underline{\underline{K}}\underline{X} = \underline{\underline{F}} \qquad (4)$$

where: $\underline{\underline{C}}$ represents the storage matrix,

$\underline{\underline{K}}$ represents the transmissivity and leakage matrices,

$\underline{F}$ is the vector of independent variables,

$\underline{X}$ is the vector of the unknown piezometric heads.

Mixed isoparametric elements can be used to integrate spatially the equations and a modofied Newton-Raphson scheme is used for the time integration.

SWIM uses a "toe tracking algorithm" with a fixed finite element mesh. This algorithm uses the Gauss points of the finite elements integration, combined with a non-linear rate of variation of the permeability of the vanishing phase at the toe (in the more common case seawater), to determine the location of the the toe. This technique gives very good results and is practically costless, may be only two or three percent increase in the cost of a simulation run.

## DESCRIPTION OF THE AREA UNDER STUDY

Geographically, Figure 1, the aquifer of the dunes of Vila Real de Santo António is located in the Southeast corner of Portugal. It is limited at the South by the Atlantic Ocean and at the East by the Guadiana River. The North and West boundaries are defined by a large contact on a clay with some sand. The aquifer is about 6 km long and has an averaged width of 1.5 to 2 km, the areal extent is of about 10 $km^2$.

There are three important urban centers supplied by wells drilled in this aquifer, Vila Real de Santo António, Monte Gordo and Castro Marim. The two first located in the aquifer zone and the third one outside. This last one it is also partially supplied by other sources besides the aquifer in study. The first two centers are touristically orientated, and this explains the great variations on the population, and its corresponding water comsumptions, along the year. The population off-season in the area is, nowadays, around 15 000 inhabitants and in the summer this number rises by 15 or 20 000 . Population estimates for the year 2 000 show a population of 40 000 off-season and 82 000 during the season.

It is shown by records that usually this consumption increases more than twice, and this is explained by the climate in the region, which is quite dry and warm. The mean annual precipitation is 460 mm, 90% of it occurs from October-April, and is more or less well distributed along this months. The averaged temperature is 17ºC, the monthly mean has a minimum in December January of 11ºC, and from June to August this value ranges between 21ºC and 24ºC. These two factors combined with warm waters in good beaches bring a lot of tourists from may to October, and in other and make the natural recharge quite low and the evapotranspiration quite high. The annual recharge varies between 30 and 80 mm, which is quite low for the amount withdraw from the aquifer.

A rough estimation shows that the annual natural recharge will vary between $3x10^5$ and $8x10^5$ $m^3$. On the other hand it is known that the consumption in 1980 was of the order of $10^6 m^3$, therefore it must have some water recharge the aquifer, or the aquifer is being mined. If this last hypothesis is true, that is some water is being withdraw from storage, and this situation can not last very long.

Geologically the aquifer is formed by a quite thin layer of sand and gravel, 10-20 m thick, which is commonly called the "Dunes of Vila Real de Santo António". These dunes are covered in a great percentage by pine trees. The aquifer is limited underneath by a silt, clay and sand layer, that is believed to be quite impervious by analogy with some outcrops at the west side of the aquifer, since there are no permeability measurements.

In this aquifer it has been drilled about 10 wells used for water supply and 7 more used as observation wells. Unfortunately the period of observation started in some wells in 1978, however, the records are more reliable only after the second half of 1979. In the Figure 2 is shown the localization of the wells. In Figure 3 is presented the precipitation in the region, the water level in some observation wells, and the evolution of consumptions.

As a general comment to the water level records, one can say that the almost steady lowering of the water level confirms the early assumption concerning the mining of the groundwater. This is particulary serious in dry years has the period 1979-1980. Regarding the consumptions, one can see that it is rising about 7% from 1979 to 1980, and that the peak season extents from May to October, with a maximum in August.

## NUMERICAL SIMULATION

The numerical simulations were carried out in a Data General computer, model Eclipse MV-8 000 . The Finite element grid used is presented in Figure 4. It has 100 nodes and 86 elements. Due to the great amount of elements, imposed by a certain degree of detail required in certain areas, it was decided to use only linear sides in the elements. This procedure is less expensive to run, both in CPU time and core requirements, even though that more linear elements are necessary that if quadratic or mixed elements were used.

The first runs were used to calibrate the model. Two sets of runs were done for calibration, one set for steady state conditions, using average values for the recharge and withdraw, and the other set for transient conditions. the variables whose changes were used to reproduce the piezometric heads observed were: permeability, aquifer thickness, natural recharge and recharge from leakage.

The field information about permeability and specific yield led to the assumption that both parameters could be assumed uniform over the entire domain. The aquifer thickness is not constant, it pinchs out toward the north and west, nevertheless, the varia-

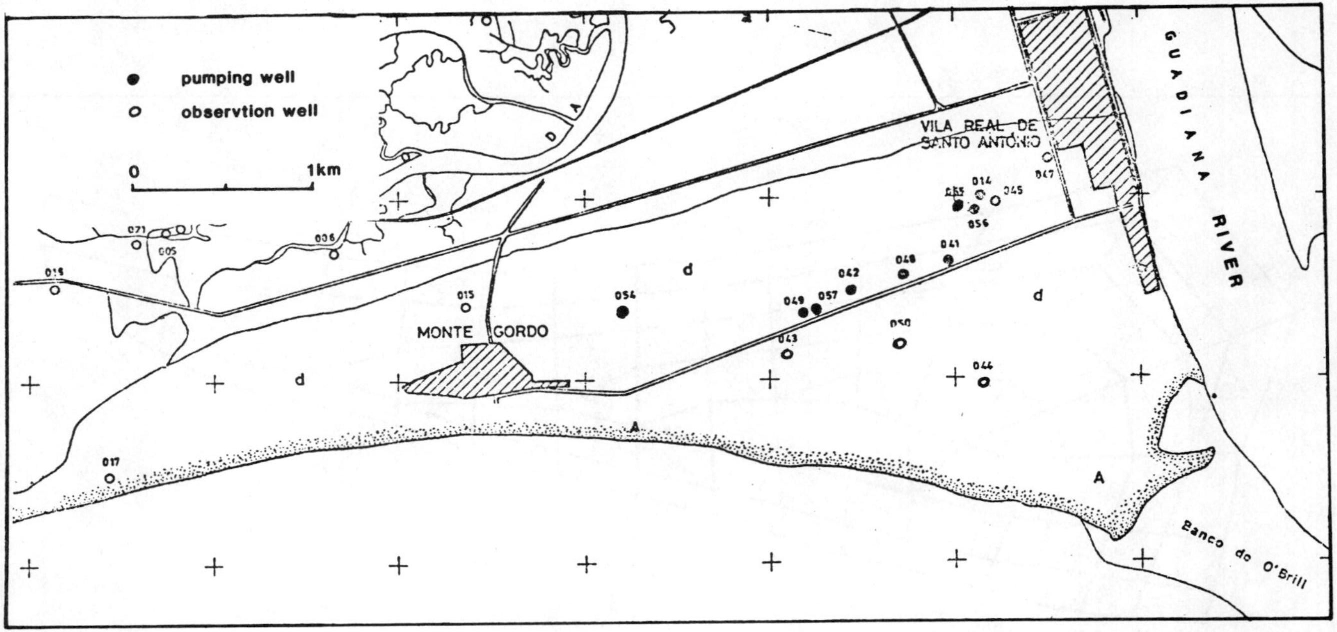


FIGURE 2 - Aquifer limits and well localization

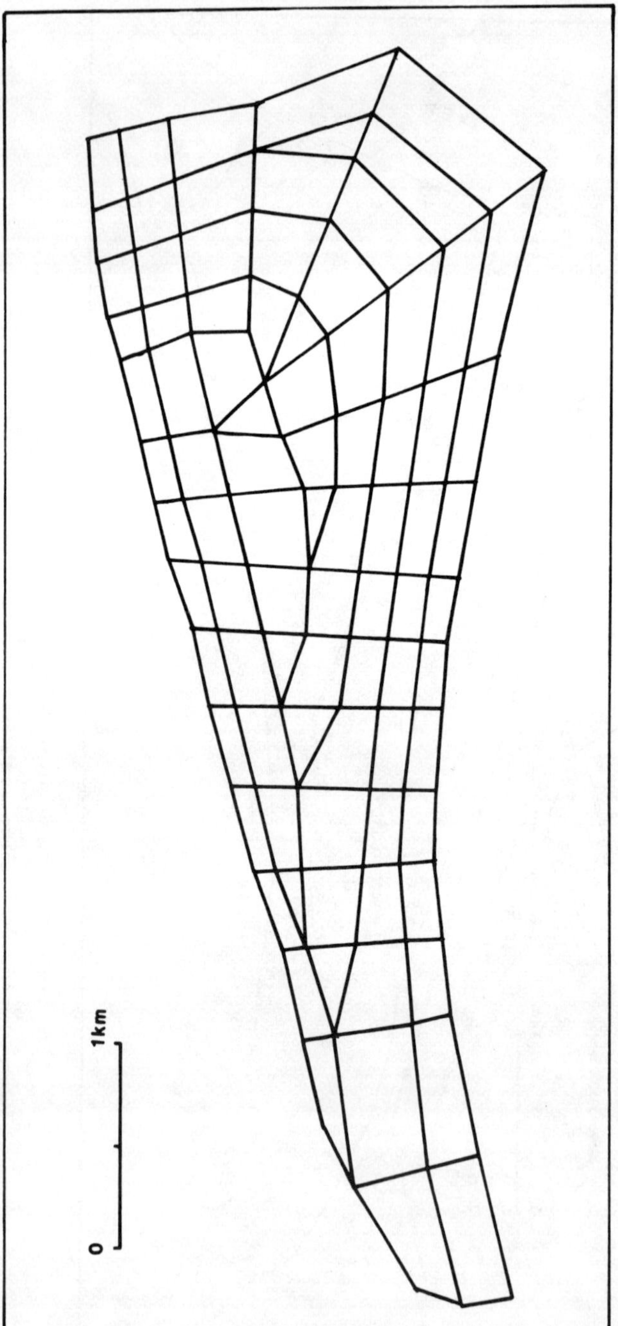


FIGURE 3 - Finite element grid

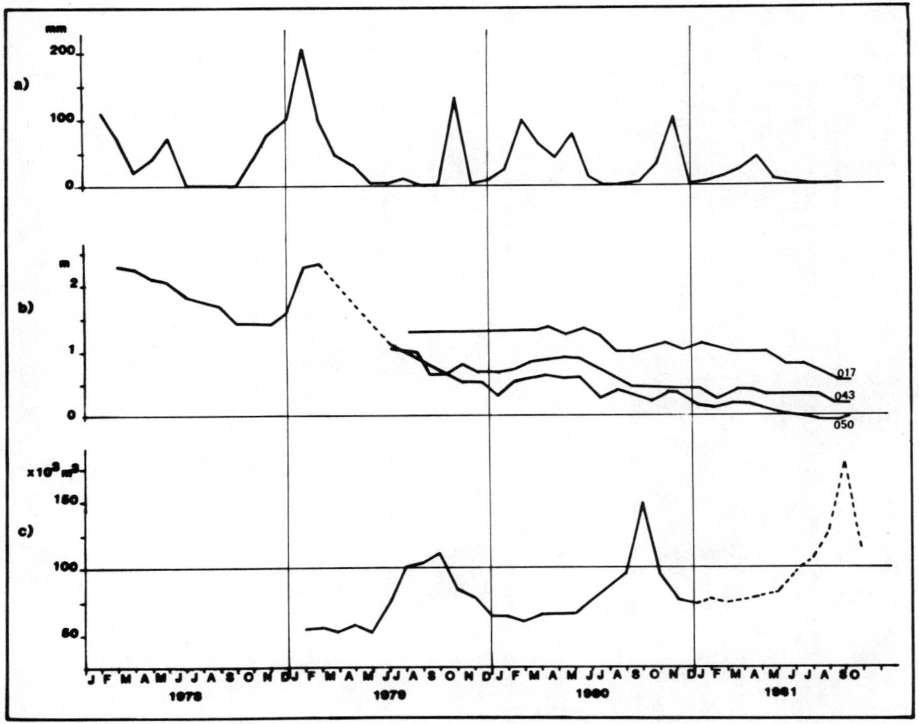


FIGURE 4 - Observed data: a) precipitation; b) water level; c) comsumption.

tions in the area under study were always relatively small. The natural recharge was more or less well defined for the average year, and quite well determine for the transient case.

The greater unknown was the recharge underneath the aquifer due to leakage. There is no information about the variables associated with this factor, like for example the extent in area of the leaky region, the piezometric head in the adjacent formation nd the permeability of the leaky layer.

Assuming no leakage and using the following parameters:

K = 40 m/day
b = 12 2 18 m
n = 0.2
Q = 2 200 $m^3$/day (withdraw distributed over 6 nodes)
N = $3x10^3$ $m^3/m^2$/day (natural recharge distributed over the entire domain)

the resulting piezometric head distribution and its comparison with the calculated averaged values are shown in Figure 5.

Based on this values, that were hold constant, and trying different combinations for the leakage parameters it was possible to match, almost exactely, the observed piezometric heads. But again, it should be refered that, only the feeling of the geologists that studied the area could give some support to this hypothesis, since no data were available to confirm it.

After the calibration phase several runs were done to evaluate the total potentialities of the aquifer and the relationship between the natural recharge, the amount that can be withdrawn and the advance of the seawater wedge. This process is not yet concluded and the results in hand can already give us some feeling about the future exploitation of the aquifer under study.

## CONCLUSIONS

Two types of conclusions can be drawn from this work, one concerning the model SWIM and the other concerning the application itself. Relatively to the computer code SWIM it was demonstrated that it is an accurate and efficient model for seawater intrusion in aquifers, when an immiscible interface can be assumed. It is also relatively simple to use, and the toe tracking algorithm simulates very well the extent of the seawater intrusion.

Concerning the Vila Real de Santo António aquifer it can be said that: the system it is quite close to its maximum production capacity; the drilling of new wells should be done the farthest from the shore line and from the existing wells; and that the influence of the annual variations of the precipitation can modify the exploitation policy, however, the establishment of a relationship between the natural recharge and the wa-

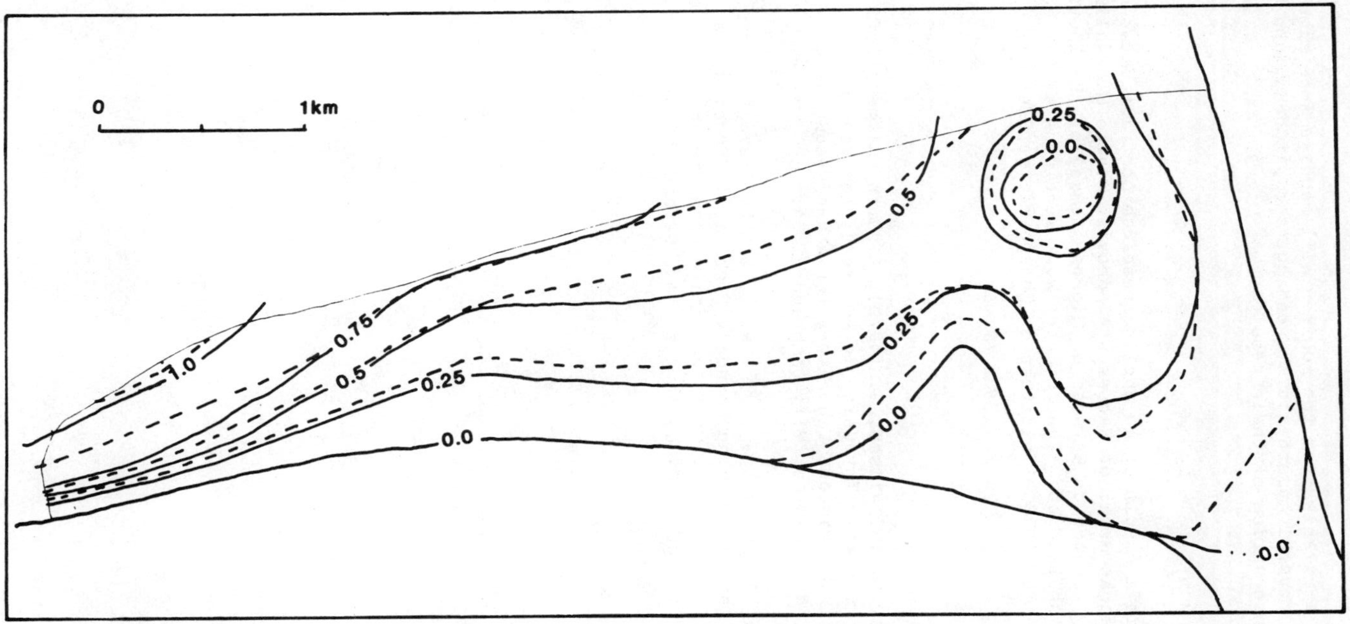


FIGURE 5 - Comparison between field data (——) and numerical simulation (- - -)

ter safely available from the aquifer will require more simulations, for several hypothesis of recharge and withdraw, and the carefull observation of their results.

ACKNOWLEDGEMENTS

This work is a result of an applied research contract between DGRAH (Direcção Geral de Recursos e Aproveitamentos Hidráulicos) and CEHIDRO (Centro de Estudos de Hidrosistemas) of the Instituto Superior Técnico. The computer work were done at the computer center of INESC (Instituto de Engenharia de Sistemas e Computadores).

REFERENCES

Sá da Costa, A. and Wilson, J. (1979), A numerical model of Seawater Intrusion in Aquifers, Tecnical Report 247 R.M. Parsons Laboratory for Water Resources and Hydrodynamics, MIT, Cambridge, USA.

Sá da Costa, A. and Wilson, J. (1980), Coastal Seawater Intrusion: A Moving Boundary Problem, in Proceedings of the 4th Int. Conf. on Finite elements in Water Resources, ed. by Wang et al., Oxford, Mississippi, USA.

# NUMERICAL SIMULATION OF INFILTRATION INTO POROUS MEDIA AND RESPONSE OF THE WATER TABLE

F. Stauffer

Federal Institute of Technology, Zurich, Switzerland

## INTRODUCTION

Two-dimensional unsteady saturated - unsaturated flow in layered porous media induced by local infiltration is considered. Special attention is given to the influence of the hysteresis phenomenon in the water characteristics of the soil on the flow. Consideration of the hysteretic behaviour is restricted to the drainage - imbibition cycle. Processes of this type frequently occur during the natural or artificial recharge of ground water or during irrigation. In order to simulate such flows, a Finite-Element - Galerkin - model is proposed. The model is used to simulate laboratory experiments. The simulation results are compared with measured data.

## GOVERNING EQUATIONS

The flow condition is assumed to satisfy the generalized Darcy law for saturated and partially saturated flow:

$$q = -k \cdot (p/\varrho \cdot g + z) \tag{1}$$

Here, q is the specific flux, k is the hydraulic conductivity, z is the vertical coordinate, p is the fluid pressure, g is the gravitational constant and $\varrho$ is the density of the fluid, which is assumed to be constant. Frictional effects due to air flow in the unsaturated zone and evaporation are neglected.

Under these conditions and taking into consideration the flow continuity, the unsaturated flow is governed by the Richards equation:

$$n \cdot \frac{\partial S}{\partial t} = \nabla \cdot (k \cdot \nabla (p/\varrho \cdot g + z)) \tag{2}$$

where n is the porosity, S is the fluid saturation and t is the time. For saturated conditions, the time derivative vanishes.

The problem of the simulation of two-dimensional infiltration into soils involves the solution of the Richards equation (2), together with the soil characteristics $p_c(S)$ and $k(S)$ including initial and boundary conditions, where $p_c$ is the capillary pressure with $p_c = -p$ for $p < 0$. Since one is interested in simultaneous saturated - unsaturated flow and since the soil may be layered, it is advantageous to use the fluid pressure $p(x,z)$ or the pressure height $h(x,z) = p/\varrho \cdot g$ as the independent variable. x is the horizontal coordinate.

For the representation of the soil characteristics $p_c(S)$ and $k(S)$, simple algebraic functions are chosen (Riedi and Stauffer, 1982). The boundary drying and wetting curves of the relation between capillary pressure $p_c$ and saturation S are described by

$$p_c = p_b \cdot \left(\frac{S-S_r}{1-S_r}\right)^{-1/\lambda} \cdot \left(\frac{S_m-S}{S_m-S_r}\right)^{\delta} \tag{3}$$

for $p_c \geqslant 0$
and $S = S_m$ for $p > 0$. The parameters $p_b$, $S_r$, $S_m$, $\lambda$ and $\delta$ are constant for the boundary curve considered. The primary wetting curves are approximated by modifying the approach of Mualem (1974). It is

$$S\,(p_c = 0 \searrow p_{c1} \nearrow p_c) = S_w(p_c) + [S_m - S_w(p_c)] \cdot \frac{[S_d(p_{c1}) - S_w(p_{c1})]}{[S_m - S_w(p_{c1})]} \tag{4}$$

The approach describes the saturation of the imbibition cycle after a drainage from the capillary pressure $p_c = 0$ to $p_c = p_{c1}$. $S_w$ denotes the boundary wetting curve, $S_d$ the boundary drying curve and $S_m$ is equal to $S_d$ ($p_c = 0$). To describe the relation $k(s)$, the approach of Brooks and Corey (1966) is used, or

$$k(S) = k_s \cdot \left(\frac{S-S_r}{1-S_r}\right)^{3+2/\lambda} \tag{5}$$

The parameters $\lambda$ and $S_r$ correspond to the function $p_c(S)$. $k_s$ is the hydraulic conductivity at saturation.

## FINITE ELEMENT FORMULATION

The procedure for the numerical integration of the differential equation (2) is based on the model of Neuman (1972). It uses a Galerkin - type Finite Element method. Quadrilateral elements are employed which are composed of four triangles with linear shape functions. The integration over the solution domain leads to an ordinary differential equation system

$$[A] \cdot \{h_i\} + [F] \cdot \left\{\frac{\partial h_i}{\partial t}\right\} + \{B\} = 0 \tag{6}$$

with [ ] being m x m matrices and { } being vectors of length m. m is the number of nodes. $h_i$ is the representation of the pressure height at node i. For the integration over time, a fully implicit backward difference scheme is used which results in the following linear equation system for the unknown vector $\{h_i(t+\Delta t)\}$:

$$([A(t+\Delta t/2)] + 1/\Delta t \cdot [F(t+\Delta t/2)] \cdot \{h_i(t+\Delta t)\} =$$
$$= \{B(t+\Delta t/2)\} + 1/\Delta t \cdot [F(t+\Delta t/2)] \cdot \{h_i(t)\} \quad (7)$$

$\Delta t$ is the length of the time step. Since the coefficients in [A], {B} and [F] are dependent on the variable h which is unknown at the time level $t+\Delta t/2$, they have to be estimated and improved by an iterative process. For each iteration, the value of $h_i(t+\Delta t)$ is used to obtain an improved estimate of $h_i(t+\Delta t/2)$ with

$$h_i(t+\Delta t/2) = 0.5 \cdot (h_i(t) + h_i(t+\Delta t)) \quad (8)$$

With this estimate, the coefficients are reevaluated. The iteration is terminated if the maximum absolute change in the values of $h_i$ between two successive iterations is smaller than a given measure. Up to this point, the general procedure corresponds to the approach of Neuman.

For the evaluation of the coefficients in the matrix [F], the values of the derivative $\partial S/\partial p_c$ have to be determined for the given values of $h_i$. The hysteresis in the drainage - imbibition cycle of the water characteristic is approximately considered using eq. (4). The resulting expression is stated in the form

$$\frac{\partial S}{\partial p_c}(p_c) = \frac{\partial S_W}{\partial p_c}(p_c) \cdot f(p_{cl}) \quad (9)$$

It is, therefore, possible to establish $f(p_{cl})$ before the modelling procedure and to determine the primary wetting curve by using the corresponding boundary wetting curve and a constant coefficient f. The existing procedure with single valued water characteristics can, therefore, easily be generalized to the drainage - imbibition cycle with minimal additional storage requirement and computational effort. If the hysteresis is neglected, f = 1.

Infiltration processes into soils often result in sharp fronts of the pressure and water content profiles. The front region is characterized by rapid temporal changes in the value of the variable h, so that small time steps have to be used. Since this change occurs primarily in the actual front region, a domain - restriction method is applied in order to reduce the computational effort. According to this method, the entire equation system (7) is solved only for the first iteration. For the following iterations, the variable $h_i$ at a given point which undergoes a small change of the value during the time step, is regarded as a constant, provided that this condition is also satisfied for all other points of the associated elements. This procedure considerable reduces the computational effort.

For the simulation of infiltration processes into relatively dry soils, very small time steps have to be used in the beginning stage of the process to get stable solutions. In the later stages, larger steps can be used. During the programme execution, the length of the time step is automatically adjusted. $\Delta t$ is increased by a given factor, when the iterative solution process of the preceding time step has rapidly converged as measured by the number of iterations necessary for the fulfilment of the convergence criterion. Otherwise, $\Delta t$ is reduced.

## SIMULATIONS

Simulations were performed for fluid infiltration through a finite strip with constant flow rate into a two-dimensional layered porous medium. The data of the initial and boundary conditions as well as of the water characteristics were taken from laboratory investigations at a sand packing (Stauffer and Job, 1982). The packing consisted of three horizontal layers of two different medium sands (material 1 and 2, figure 1).

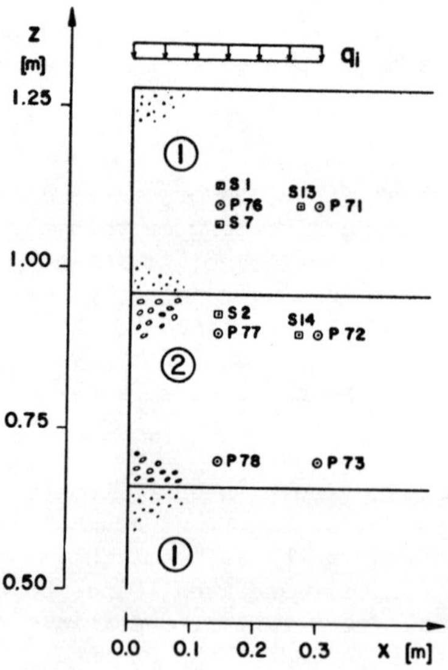


Fig. 1

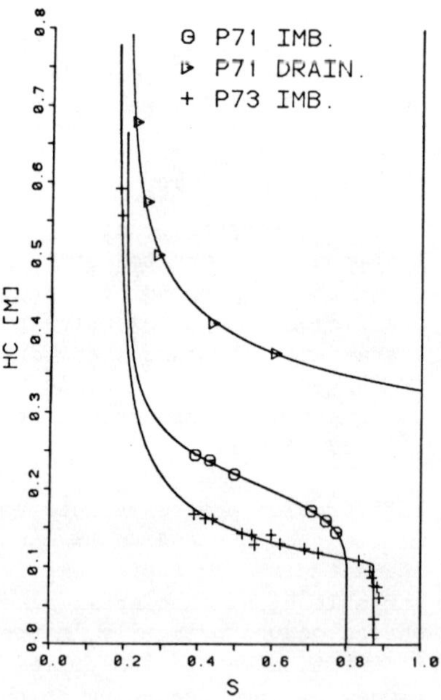


Fig. 2

The values of the hydraulic conductivity $k_s$ and of the porosity n were

| | $k_s$ [mm · $s^{-1}$] | n |
|---|---|---|
| mat. 1 | 0.25 | 0.37 |
| mat. 2 | 0.73 | 0.39 |

The boundary wetting and drying curves of the water characteristics of the two materials were determined at points P71 and P73 (fig. 1). The parameters were

| | $\frac{p_b}{\varrho \cdot g}$ [m] | $S_r$ | $\lambda$ | $\delta$ | $S_m$ |
|---|---|---|---|---|---|
| mat. 1 wetting | 0.204 | 0.201 | 5.99 | 0.131 | 0.803 |
| mat. 1 drying | 0.330 | 0.201 | 5.03 | 0.0 | 1.0 |
| mat. 2 wetting | 0.102 | 0.174 | 2.43 | 0.010 | 0.875 |

Figure 2 shows the capillary pressure height $h_c = p_c / \varrho{\cdot}g$ as a function of the saturation S.

The initial pressure distribution was quasi-static and was controlled by the level z of the water table. The distribution was obtained by lowering the water table from the top of the sand packing to the level z = 0.1 m. Since the sand packing was initially fully saturated, the resulting fluid retention profile corresponded to the first boundary drying curve. The flow rate during the unsteady flow was $q = 0.082\ \text{mm} \cdot \text{s}^{-1}$ along a strip of finite length (fig. 1). The outflow level (fig. 3) was fixed at z = 0.1 m.

The grid used for the simulation is shown in fig. 3. During the early stage, the domain was restricted to $0 \leqslant x \leqslant 0.7$ m.

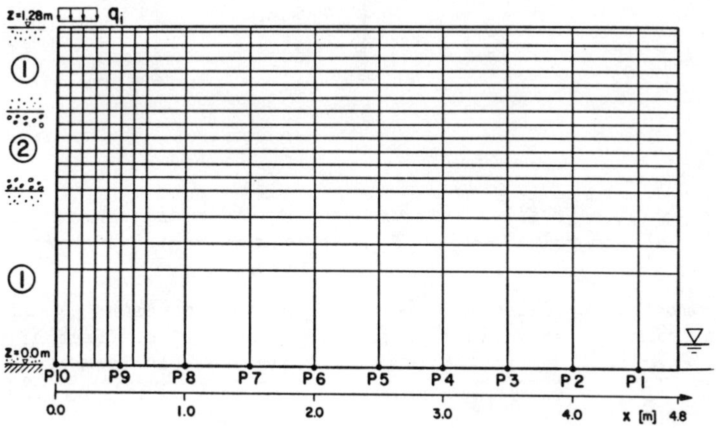


Fig. 3

Since the initial water content was about residual in the region $z \geqslant 0.66$ m, the pressure distribution in this region was replaced by a uniform distribution $p/\varrho{\cdot}g = -0.5$ m in order to reduce the computational effort. Consequently, the hysteresis was neglected in this region. This modification is justified in infiltration problems, because the saturation behind the front is much higher than residual. In the region with $z < 0.66$ m, the initial pressure distribution was assumed to be hydrostatic.

Simulations were performed for the first 75 min. after the start of the experiment. Time series of the saturation S at different stations are shown in figs. 4 and 5, together with measured data. The values of the saturation at the stations P71, P72, P76 and P77 were determined from the pressure data by means of eq. (3).

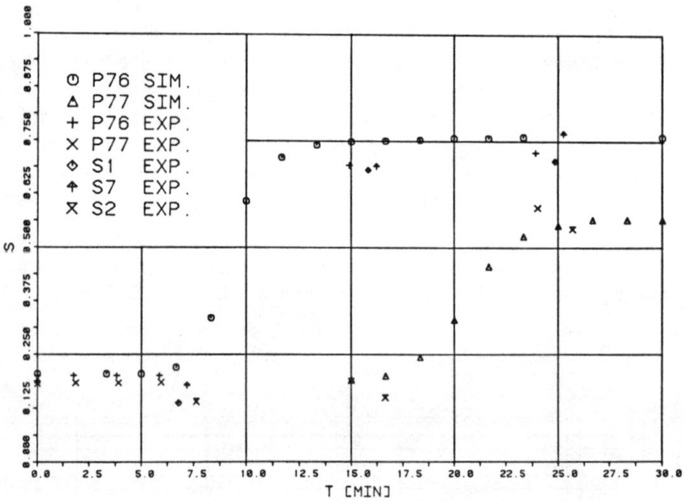


Fig. 4

The saturation profiles S(z) of the simulation and the experiment at the coordinate x = 0.14 m for t = 75 min. are shown in fig. 6.

The computed approximate position of the wetting front at different times is shown in fig. 7, demonstrating the effect of the soil layering. The front is approximately characterized by

$$h_f = p_c\,(\bar{S}) \,/\, \varrho \cdot g \tag{10}$$

with $\bar{S} = [\,S(p_{cmin}) - S\,(p_{cmax})\,]\,/2$. $p_{cmin}$ and $p_{cmax}$ indicate the capillary pressure before and after the passage of the front.

Time series of the measured pressure heights at the lower boundary (fig. 3) are shown in fig. 8. Simulations were performed with considering (fig. 9) and without considering (fig. 10) the hys-

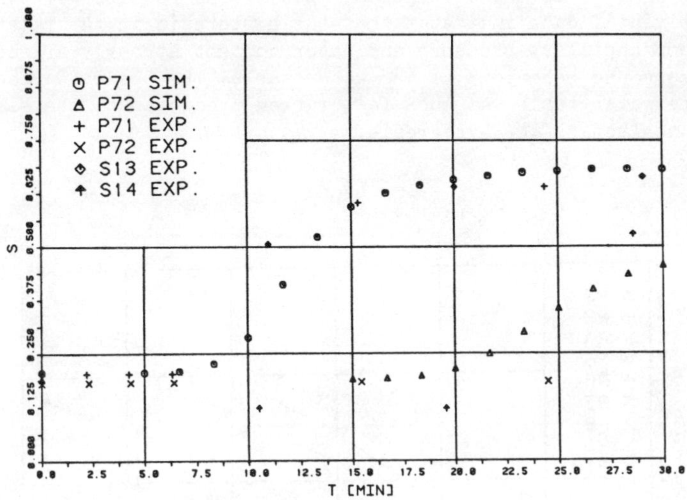


Fig. 5

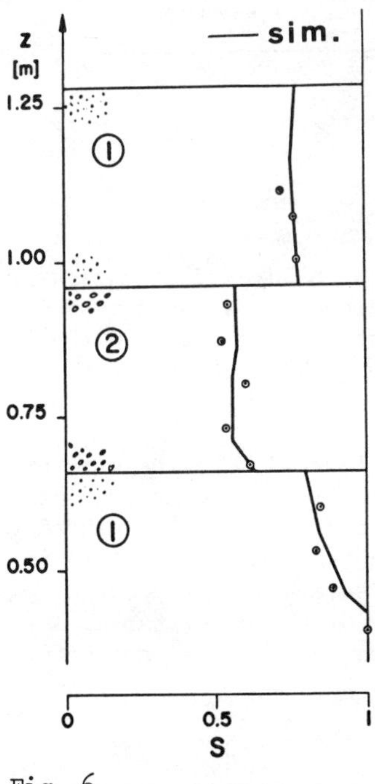


Fig. 6

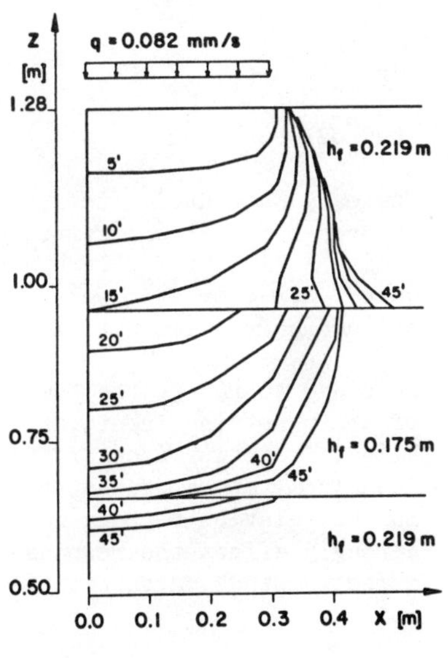


Fig. 7

teresis phenomenon. In the latter case, the drainage characteristic $S_d(p_c)$ is used. The comparison of the results with the experimental data indicates that the hysteresis in the relation between capillary pressure and water content strongly affects the response behaviour of the water table. The observed reaction of the water table was much faster than predicted by the simulation neglecting the hysteresis.

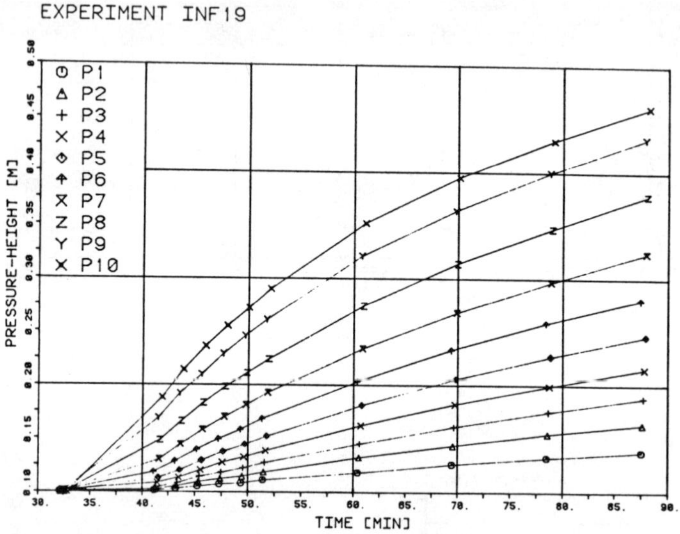


Fig. 8

CONCLUSIONS

Unsteady saturated - unsaturated flow in a layered sand packing induced by local infiltration was simulated by a two-dimensional Finite - Element - Galerkin model. Hysteresis effects in the water characteristic were approximately taken into account for the drainage - imbibition cycle.

In order to reduce the computational effort due to the existence of sharp wetting fronts, a domain - restriction method was applied. The comparison of the simulation results with measured data from laboratory experiments indicates that the hysteresis in the relation between capillary pressure and water content may strongly affect the response behaviour of the water table. The effect depends on the initial condition and the soil characteristics.

This work was supported by ETH Project No. 0.330.076.98/5.

SIMULATION INF19 WITH HYSTERESIS

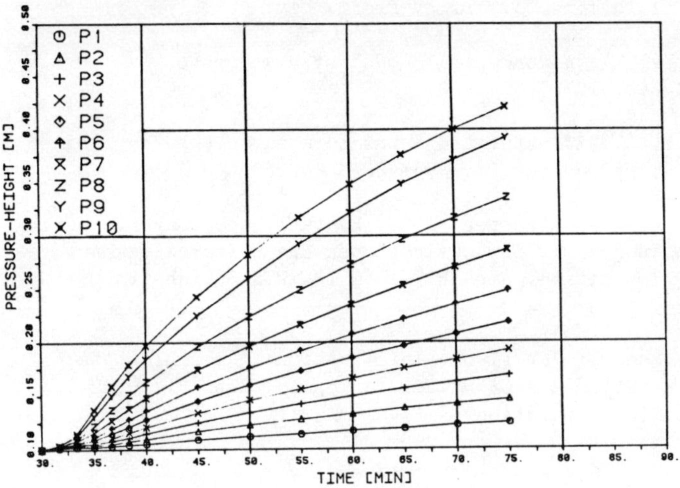


Fig. 9

SIMULATION INF19 WITHOUT HYSTERESIS

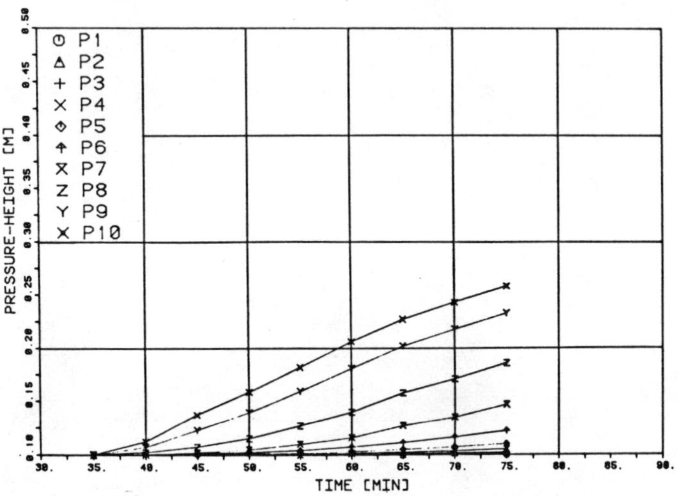


Fig. 10

REFERENCES

Brooks, R. and A.T. Corey (1966) Properties of porous media affecting fluid flow. J. Irrig. Drain. Div. 92: 61 - 88.

Mualem, Y. (1974) A conceptual model of hysteresis. Water Res. Res. 10: 514 - 520.

Neuman, S.P. (1973) Saturated - unsaturated seepage by finite elements. J. Hydraulics Div. 99: 2233 - 2250.

Riedi, G.A. and F. Stauffer (1982) Hydraulische Charakteristika von Sandpackungen mit Berücksichtigung des Hysteresiszyklus Drainage - Imbibition. Rep. R17 - 82 IHW ETH Zürich (in preparation).

Stauffer, F. and D. Job (1982) Infiltration in geschichtete Böden und Reaktion des Grundwasserspiegels. Rep. R18 - 82 IHW ETH Zürich (in preparation).

# COMPARISON OF THREE MODELS OF A DUAL AQUIFER SYSTEM

Swaminathan, K. and Elango, K.
Department of Civil Engineering
Indian Institute of Technology
Madras 600 036, India.

## INTRODUCTION

This paper deals with modelling of a dual aquifer system having a configuration as shown in Fig.1 with contiguous unconfined and confined regions. Recharging of the aquifer occurs over the unconfined region and the aquifer is exploited by pumping through wells located in the confined region. The other end of the confined aquifer is in contact with a constant-head water body, such as sea. The confined region may in general be inhomogeneous, usually with the thickness varying with distance towards the sea.

The occurrences of aquifers, which can be idealised to such a dual aquifer configuration have been reported in literature. These include Linconshire-limestone aquifer in England as reported by Rushton and Redshaw (1979), an aquifer in Long Island, New York referred by Collins et al (1972), the coastal aquifer north of Madras and the Neyveli aquifer, India. Thus modelling of flow phenomenon in such aquifers may contribute to their efficient and safe exploitation.

Three types of models, namely an analytical model, a finite element model and a physical model of sand-box type, have been used in the study to assess their relative performances. The results presented in this paper pertain to the case of an inhomogeneous confined aquifer portion with one specific combination of magnitudes of recharge and pumpage.

## FEATURES OF MODELS

Major features of the three models are presented

concisely.

Analytical Model
The analytical model is an extension of the analysis of a dual aquifer system, with homogeneous confined aquifer portion as given by Chan et al (1977) to an inhomogeneous condition based on the perturbation method. The solution is written as the sum of a zeroth order solution corresponding to the homogeneous case, $s_o$ and a first order solution $s_1$ with the perturbation parameter $\alpha$ as the ratio of the incremental thickness to the thickness of the confined region at the interface with the unconfined region. Thus

$$s = s_o + \alpha\, s_1$$

Following a procedure similar to the derivation of the zeroth order solution by Chan et al (1977), the first order solution for a periodic recharge and a constant pumpage case is obtained using finite Fourier cosine transform and Fourier series representation.

A computer program is written to evaluate the resulting expressions.

Finite element model
The numerical model is based on finite element and Galerkin's weighted residual methods. A digital computer program, based on the finite element method for two-dimensional depth averaged, unconfined flow has been presented by Trösch (1975).

This program has been suitably modified to simulate contiguous unconfined-confined aquifer regions with the interface between these two regions being fixed in space and over time and to provide for spatially varying storage coefficient and temporally and spatially varying recharge as reported by Swaminathan (1982). Thus apart from the coefficient of permeability, the storage coefficient as well as the recharge can be treated as elementwise variable parameters.

The elements used are mixed, curved, isoparametric quadrilaterals, providing for an arbitrary combination of linear, quadratic and cubic sides. Provisions are available for considering inhomogeneous, anisotropic media, handling of point-wise pumpage and Dirichlet and Neumann boundary conditions, including prescription of temporally varying heads at selected nodes.

The additional data required are for the following
- (i) vector TOP containing the top level at each node, and is of length NP, the number of nodes;
- (ii) vector BOTTOM containing the bottom level at each node and also is of length NP;
- (iii) vector SS, containing the value of storage coefficient of each element, and is of length NEL, number of elements and
- (iv) matrix SZFL, containing the values of recharge rate at every interval of time of simulation, and is of size NEL·NTIM, in which NTIM is the total number of time intervals for simulation.

Apart from adopting these additional informations properly, certain other changes also were necessitated on availability of system subroutines, overflow limit etc.

The finite element discretization for the problem considered is shown in Fig.3.

<u>Physical model</u>
A sand model is used as the physical model of the dual aquifer system. The results from this model are used as the basis for verification and validation of the performances of the analytical and the numerical models. The physical model is illustrated in Fig.2. The confined region has length of 30 cm and has thickness varying from 25 cm at the interface to 40 cm at the constant-head boundary. The unconfined region is 72 cm long and 66 cm high. The model width is 64 cm.

Polythene tubes are kept at bottom with sensing ends at points coinciding with the nodes in the finite element model. There are two manometric banks wherein these tubes terminate. Clean river sand of effective size 0.25 mm is used. A recharge tray above the unconfined region can be operated to supply a time varying recharge rate. A single well representing an eye of pumping is kept at the centre of the confined region. The value of the constant water level at the boundary of confined region is kept at 41.5 cm by suitable arrangement.

The data is recorded photographically with 35 sets of simultaneous exposures of both the manometer banks and later digitized using a microfilm reader, yielding a total of about 4800 values for this configuration.

Calibration Procedure

The calibration phase seeks to determine the values of the hydrogeological parameters for the analytical and numerical models, that reproduce the experimentally determined vauues for the space-time distribution of piezometric levels as 'best' as possible.

The three experimental runs are carried out. In the first run, recharge rate R is 11 cc/sec for half the cycle and the constant pumpage is 1.25 cc/s, the second run has R = 11 cc/sec but Q = 2.5 cc/sec and in the third run R = 14 cc/sec and Q = 2.5 cc/sec.

The criterion for choice of optimal parameters is based on least square method. Thus the results for selected 30 nodes at four time levels namely beginning, one quarter period, half-period and three quarter period of a total of 20 minutes have been used as basis for evaluation of the root-mean square deviations. Repeated evaluations with different sets of values for the parameters are used for the selection of the best estimates of the parameters.

RESULTS

Typical results for the case with recharge of 11 cc/sec and pumpage of 2.5 cc/sec are herein presented. Four types of plots are made to bring out the temporal and spatial variations of piezometric levels over the entire aquifer domain considered and described in the following:

i) Temporal variations: The unsteady variations of piezometric head during a cycle, for two representative points namely one corresponding to location at x = -72 cm and y = 0 in the unconfined region and the other at point 46 corresponding to x = 155 cm and y = 0 cm in the confined aquifer region, near the eye of pumping are presented in Fig.4 from the results of the three models. The aquifer bottom level is chosen as the datum for representing the piezometric levels.

ii) Spatial variations: The spatial variation over the entire aquifer domain at the quarter period, just prior to the commencement of recharge are shown in Fig.5 as isopiestic lines drawn at 1 cm interval. The estimated location of the dividing stream line indicating the region of influence of the well, is also shown.

iii) Space time variations: The combined picture of space time variations in the aquifer is brought out in Fig.6 along a longitudinal section passing through the well. The time instants correspond to two extreme states namely the beginning and end of recharge.

iv) Time variations of inflow, outflows and storage: The results from the finite element model are used in this plot as shown in Fig.7. In this plot, apart from the time variations of the input functions namely recharge and pumpage the outflow from the aquifer through the constant head boundary and the cumulative changes in storage volume in the total system are presented.

## DISCUSSION

From the graphs presented the satisfactory performance of the three models is seen. From the temporal variations plot it is seen that the range of the piezometric head, between the maximum and minimum values are approximately same for the three models. The trend depicted is in consonance with the recharge input.

The analytical model shows a lower value in the unconfined region compared to the other models, the difference being about 5 percent. This may be due to the linearization of equation for the unconfined portion. In the finite element model an iterated value is used for representing the transmissibility which depends on the flow thickness. This difference is seen also in the isopiestic contours plotted.

## CONCLUSION

The models studied are found to be effective in flow simulation for a complex aquifer system. The finite element model can be used to estimate the critical pumpages for different rates of recharge and this enables the estimation of safe extraction rates from the dual-aquifer, systems without flow reversal as in the case of a coastal aquifer.

## REFERENCES

Chan, Y.K., Mullineaux, N. and Reed, J.R. (1977) Analytical Solution for Drawdown in an Unconfined-confined Rectangular Aquifer. Journal of Hydrology, 34, 287-296.

Collins, M.A., Gelhar, L.W. and Wilson, J.L. (1972) Heleshaw Model of Long Island Aquifer System. Journal of Hyd. Dn., Proc. ASCE, 98, HY9:1701-1714.

Rushton, K.R. and Redshaw, S.C. (1979) Numerical Analysis by Analog and Digital Methods - Seepage and Groundwater Flow. John Wiley and Sons Inc., London.

Swaminathan, K. (1982), Studies on a Dual Aquifer System. Ph.D. thesis submitted to Indian Institute of Technology, Madras, India.

Trösch, J. (1975) Numerische Simulation Dupuit'scher Grundwasserströmungen. Mitteilungen der Versuchanstalt für Wasserbau, Hydrologie und Glaziologie an der Eidg. Tech. Hochschule, Zürich, Vol1 and 2: No.14 and 15.

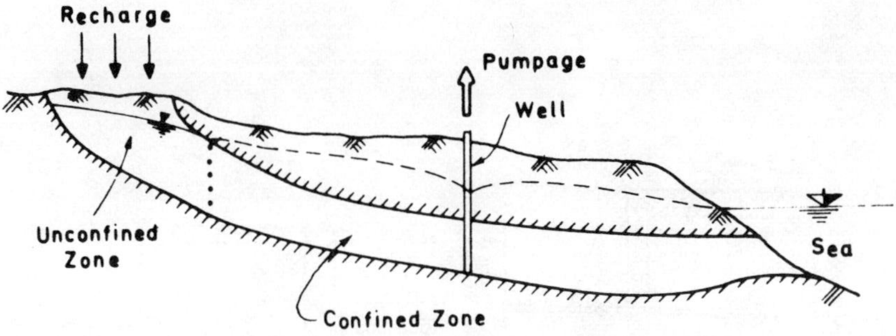


A TYPICAL COASTAL AQUIFER

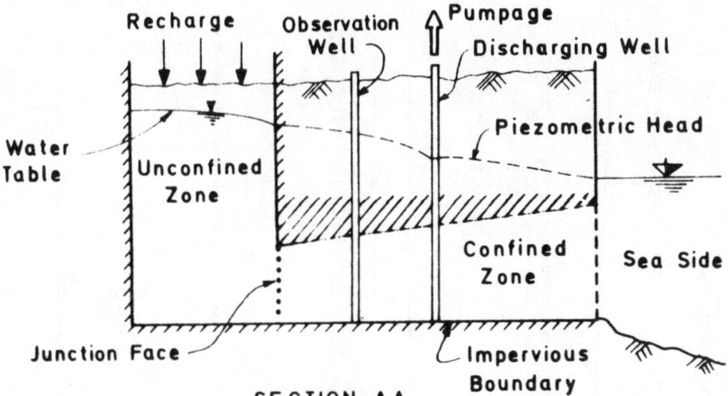


SECTION-AA

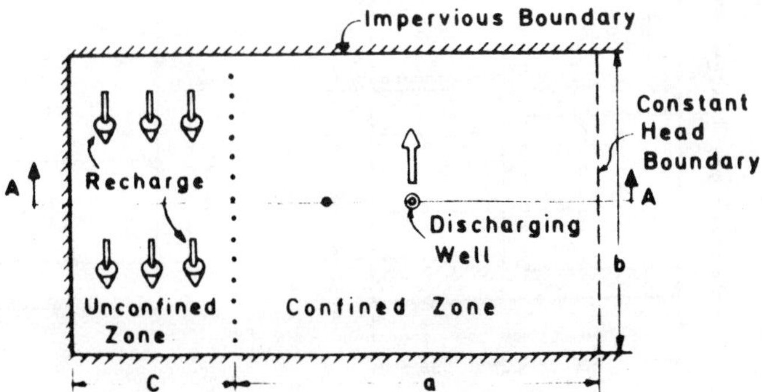


FIG 1 PLAN VIEW OF THE DUAL AQUIFER SYSTEM
(SIMPLIFIED VERSION)

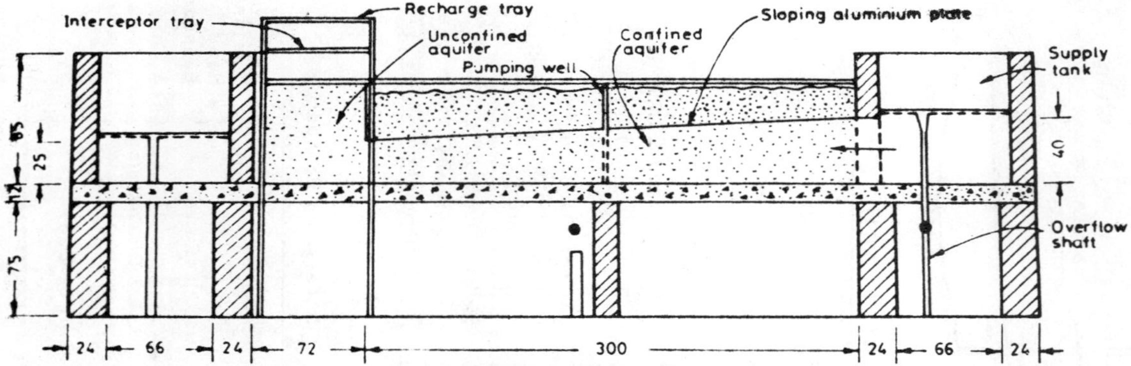


FIG. 2 .SECTIONAL ELEVATION OF INHOMOGENEOUS AQUIFER MODEL.

SCALE 1:30 ALL DIMENSIONS IN cm

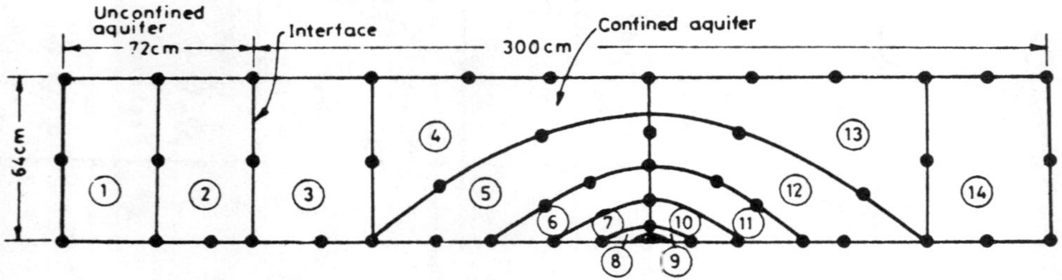


FIG. 3 .FEM DISCRETISATION FOR SAND MODEL.

SCALE 1:20

• - Nodal points (69)

- Half well

⑤ - Element number

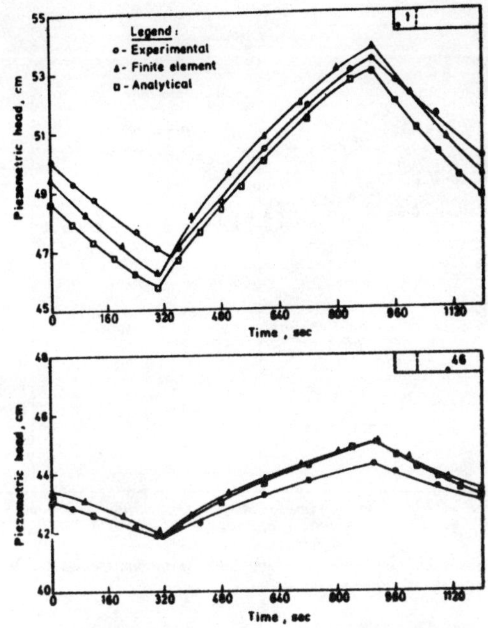


Recharge : 11.0 cc/sec Pumpage : 2.5 cc/sec.

FIG. 4 . INHOMOGENEOUS CASE : TEMPORAL VARIATIONS OF PIEZO METRIC HEAD.

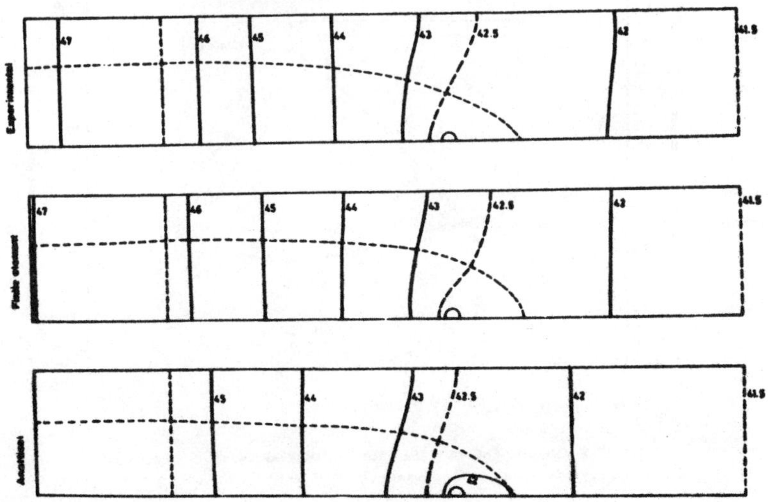


Recharge : 11.0 cc/sec Pumpage : 2.5 cc/sec

FIG. 5 . INHOMOGENEOUS CASE : RUN V ISO PIESTIC LINES - 300 sec

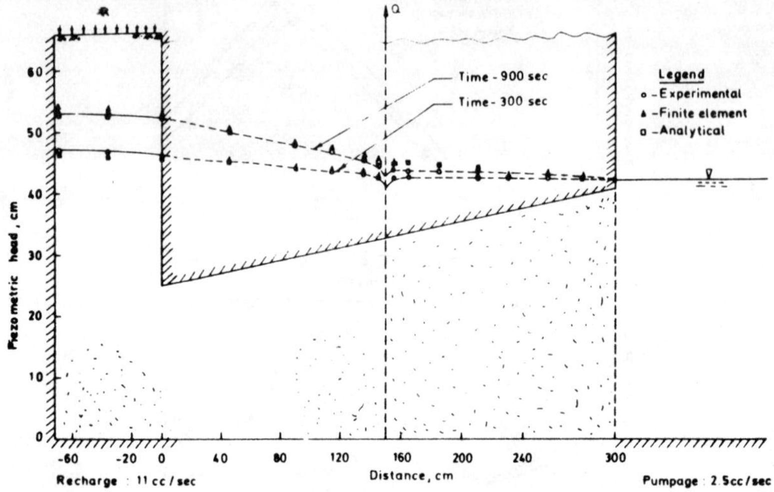


FIG. 6 . INHOMOGENEOUS CASE : RUN V SPACE TIME VARIATIONS OF HEAD ALONG THE LONGITUDINAL SECTION OF THE MODEL.

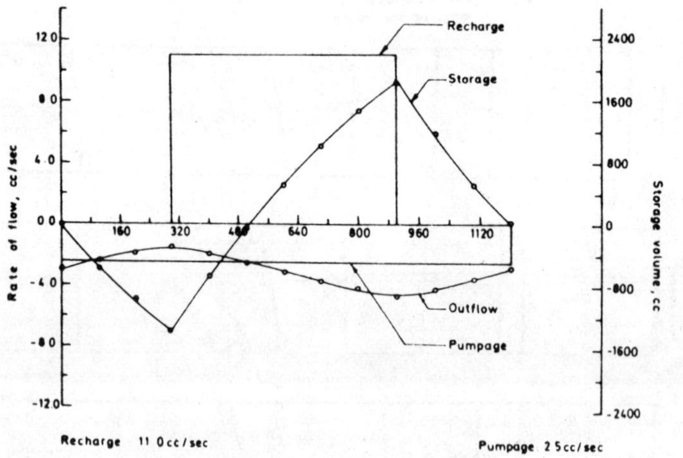


FIG 7 INHOMOGENEOUS CASE RUN V TIME VARIATIONS OF INFLOWS OUTFLOWS AND STORAGE

# 11. GROUNDWATER, SOIL AND INTERACTION

# A HYBRID MODEL TO SIMULATE LANDSUBSIDENCE DUE TO GROUNDWATER RECOVERY

Frans B.J. Barends
Delft Soil Mechanics Laboratory, Delft, Holland

## INTRODUCTION

Due to a gradual increase of groundwater recovery and gas/oil production in the past decennia landsubsidence becomes manifest as a serious environmental problem, in particular in urban areas and wetlands.
Acquired data are usually insufficient and mostly irrelevant to simulate the actual phenomenon properly. For example, values for the transmissivity obtained from pumping tests are only partly useful and data concerning the consolidation characteristics of compressible soil layers are rarely available.

Areas of recovery are generally stratified consisting of aquifers and aquitards. Subsidence, i.e. the surface settlement due to deformations in the system, can be best controlled or predicted by a quasi three-dimensional flow-consolidation model, since then it fits the quality of available and obtainable data. This approach is not new (Terzaghi, 1925). A new aspect presented here is the true coupling between the flow in the aquifers and the consolidation in the aquitards in a simple form.

For the various processes involved a different time scale exists.
Aquifer compaction will be completed in a short time, aquitard consolidation will take a much longer time, and subsequently phreatic changes in a semi-permeable top layer takes place at the last (Sellmeijer, 1976). In many approaches recommanded in literature this fact has been simply overlooked. The existing large difference between aquifers and aquitards with respect to hydro-mechanical properties allows to consider different time periods in which only the relevant processes dominate. Hence, the complete problem can be simplified drastically, by no means violating its general validity. More precise, three periods are to be regognised in the development of landsubsidence due to an operating well system. The first period is typified by the flexibility of both the aquifers and aquitards, in the second

period the consolidation of the aquitards dominates, while in the third period the consolidation reaches an end and a constant leak (phreatic response) through the aquitards is established.

In this respect the applicability of the well-function after Hantush (1964) for a leaky confined compressible aquifer has to be questioned, since it combines features of a different time scale. For flow towards a well see appendix.

Another aspect is stressed, to wit: efficiency. A practical procedure to simulate subsidence requires a special treatment of the critical part, which is the consolidation of the aquitards. Application of a numerical scheme for this phenomenon will unavoidably result in time consuming calculations. Therefore, it is preferred to adopt an analytical explicit formulation of the consolidation process parameterized in terms of the yet unknown piezometric heads in the adjacent aquifers. The seepage flow in the aquifers including the topical leakage from the consolidating aquitards can be generated numerically, thus no limitation concerning inhomogeneities and geometries. The coupling between both is established iteratively.
By this approach no integration in time is required, since at any instant the stage of the consolidation process is known explicitly. Long term simulation represents no specific difficulty, and relatively low computation costs. A multiple well system is easily incorporated. The surface subsidence is found by summation of the vertical settlement of the various aquitards.

## THE PHYSICAL SYSTEM

As has been outlined before it is the purpose to consider a practical method to evaluate landsubsidence due to recovery of groundwater. In this regard the coupled flow-consolidation process in a horizontally stratified subsoil consisting of aquifers and compressible aquitards is simplified to attain a convenient mathematical formulation including the essential phenomena involved, amenable to analytical (and conventional numerical) methods. Therefore only horizontal flow will be considered in the aquifers and vertical consolidation in the aquitards.
Vertical compaction of the aquifers and the pore water compressibility can be accounted for as well. Although surely horizontal deformation may occur, even in the same order as the vertical one (Bear, 1981), and although more or less complex methods exist to evaluate them, these deformations are not considered here.
On a regional scale this simplification is acceptable.

### Horizontal flow in an aquifer

The mass belance of pore water seeping through a horizontal rigid aquifer states:

$$\nabla \cdot u = -q_{io}/D \qquad (1)$$

where u is the mean horizontal filter velocity, D is the aquifer height, and $q_0$ represents the vertical leak into the adjacent aquitards. Introduction of Darcy's law:

$$u = -K\nabla H, \qquad (2)$$

yields for equation (1):

$$\nabla\cdot\nabla H = q_0/KD. \qquad (3)$$

Here, K is the hydraulic (horizontal) permeability of the aquifer and H is the piezometric head (a function of time and horizontal coordinates). Equation (3) is easily extended to include the vertical deformation of the aquifer (Cooper, 1966) and the compressibility of the pore water (Verruijt, 1969):

$$\nabla\cdot\nabla H = \partial H/c\partial t + q_0/KD, \qquad (4)$$

in which c represents the laterally confined consolidation coefficient to be determined by oedometer tests.

<u>Vertical consolidation in an aquitard</u>

Soil can be visualised as a skeleton of particles encompassing interconnected voids, which in general contains pore water. For the range of stress variation due to pumping wells the individual soil particles (and the pore water) can be considered incompressible. However, the bulk volume of the soil skeleton can change due to local deformations at the intergranular contact points and due to (irreversible) rearrangements of the particles into a new position. The strength of the soil depends on the structural frame of the solid particles. Shear can be resisted by the skeleton itself, but compression may be resisted by the skeleton and the pore water. In fully saturated soil a reduction of the volume is possible only if the pore water can escape from the pores. Thus, changes in the pore geometry can generate pore water motion. In the mechanical equilibrium both phases take part. This process is referred to as consolidation first highlighted by Terzaghi (1925).

In a similar manner the stress conditions for the aquitards will be affected by groundwater recovery. As a consequence they will consolidate and deform. This deformation causes the pore water to seep out into the adjacent aquifers altering the actual piezometric head. The total deformation of the aquitards will result in a surface displacement, the landsubsidence.

Since the compaction of the aquitards is a slow process, the subsidence reveals in particular a retarded transient character. This means, that <u>the observed subsidence consitutes only a part of the total effect already induced by the actual recovery.</u>

In the present approach vertical linear consolidation is assumed, which starts from the pore water mass balance:

$$\partial(n\rho w)/\partial z + \partial(n\rho)/\partial t = 0 \qquad (5)$$

in which $\rho$ is the water density (here considered constant), n is the porosity and w the water velocity. Introduction of the relation between n and the local vertical soil skeleton velocity v, according to:

$$\partial v/\partial z = \partial n/\partial t + \partial (nv)/\partial z, \quad (6)$$

yields:

$$\partial (q+v)/\partial z = 0, \quad (7)$$

in which: $q = n(w-v)$, represents the relative vertical filter velocity. Darcy's law is defined in terms of the relative pore water motion:

$$q = -K'\partial h/\partial z \ ; \ h = z + p/\rho g. \quad (8)$$

K' is the vertical hydraulic permeability of the aquitard, h is the piezometric head and p denotes the pore pressure. A linear soil deformation behaviour is adopted, expressed in terms of the effective stress: $\sigma' = \sigma - p$, according to:

$$\partial v/\partial z = -\alpha\partial\sigma'/\partial t = \alpha(p-\sigma)/\partial z, \quad (9)$$

in which $\sigma$ represents the total vertical stress. Substitution of (8) and (9) into equation (7) yields:

$$c'\partial^2 h/\partial z^2 = \partial(h-\sigma/\rho g)/\partial t. \quad (10)$$

The laterally confined consolidation coefficient c' is related to other material properties according to:

$$c' = K'/\rho g\alpha. \quad (11)$$

If the vertical total stress is stationary, equation (10) reduces to the familiar diffusion equation.

### The mathematical formulation

For a multi-layer system the flow-consolidation process is governed by field equation (4) for the aquifers and by field equation (8) for the aquitards. The coupling between the seepage flow in the aquifer and the consolidation in the adjacent aquitards is embodied at one hand in the boundary leakage $q_0$ being the momentary vertical filter velocity at the separations and at the other hand in the piezometric head at the separation, which acts as a (transient) pressure boundary condition for the consolidating aquitard.

The problem description is completed by additional conditions for the aquifer flow (the well system and the far boundaries), and by a compatible initial state. It is to be noted that usually the total stress does not vary ($d\sigma=0$), but this is not the case for a phreatic top layer (Sellmeijer, 1976). For the sake of simplicity the pore water has been considered incompressible. It is not essential for this discussion.

## AN ANALYTICAL SOLUTION FOR A PLANE FLOW SITUATION

The critical part of the subsidence development is embodied in the consolidation of the aquitard. An analytical solution is sought expressed in terms of the piezometric head of the adjacent aquifers. Since this piezometric head itself is affected by the consolidation, no method that excludes the flow in the aquifer will give correct results. A complete analytical solution can be found applying complex mathematical methods (Verruijt, 1969). To overcome this fact the flow and consolidation are decoupled; this approach is common practice. The corresponding piezometric head distribution might be acceptable, but the subsidence may deviate significantly. This can be illustrated on the base of an analytical solution of a two-layer system, an aquitard on top of an aquifer. A similar problem in the field of heat diffusion has been considered by Tang(1981). The exact solution permits to evaluate the validity of approximate solving methods for more complicated cases.

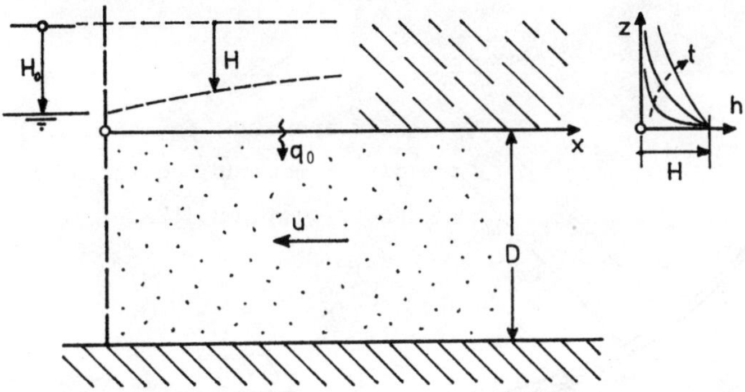


Fig. 1: Scheme of a two-layer system.

The flow in a two-layer system (Fig. 1) induced by a unit step disturbance $H_0$ at: $(x,z,t) = (0,0,0)$, is described by:

$$\partial^2 H/\partial x^2 = \partial H/c\partial t + q_0/KD \; ; \; x \geq 0 \; , \quad t \geq 0;$$

$$H(0,t) = H_0 \; ; \quad H(x,0) = 0 \; ; \quad H(\infty,t) = 0;$$

$$\partial^2 h/\partial z^2 = \partial h/c'\partial t \; ; \quad z \geq 0 \; , \quad t \geq 0;$$

$$h(0,t) = H(x,t) \; ; \; h(z,0) = 0 \; ; \quad h(\infty,t) = 0;$$

$$q_0 = -K'\partial h/\partial z \; ; \quad z \to 0.$$

Note, that: $d\sigma = 0$, and the compatible initial state. Applying the Laplace transformation technique the transform solution yields:

$$\bar{H} = (H_0/s) \exp[-x\sqrt{s/c + (K'/KD)\sqrt{s/c'}}]$$

$$\bar{h} = \bar{H} \exp[-z\sqrt{s/c'}].$$

Using the properties of the Poisson integral the inverse transform can be obtained. The result is:

$$h = H_0 \, B\,(\eta,\zeta,\varepsilon) \; ; \tag{12}$$

$$\eta = x/(2\sqrt{ct}) \; ; \quad \zeta = z/(2\sqrt{c't}) \; ; \quad \varepsilon = cK'/c'KD;$$

$$B = 2/\sqrt{\pi} \int_\eta^\infty \exp[-\xi^2] \operatorname{erfc}[\zeta + \eta^2\varepsilon\sqrt{c't}/2\xi^2]\, d\xi.$$

For this type of problem an approximate inversion technique

exists (the direct method, see Barends, 1981), which yields:

$$h = H_0 \exp[-\sqrt{2}\zeta - \eta\sqrt{2+2\varepsilon\sqrt{2c't}}] . \tag{13}$$

The uncoupled approach yields:

$$h = H_0 \,\mathrm{erf}[\zeta] \exp[-2\eta\sqrt{\varepsilon\sqrt{c't/\pi}}] , \quad \text{for } c \to \infty . \tag{14}$$

In Fig. 2 the solutions (12), (13) and (14) (for $c \to \infty$) are presented as a function of $\eta\sqrt{\varepsilon\sqrt{c't}}$ .
Obviously the direct method is a reasonable approximation.

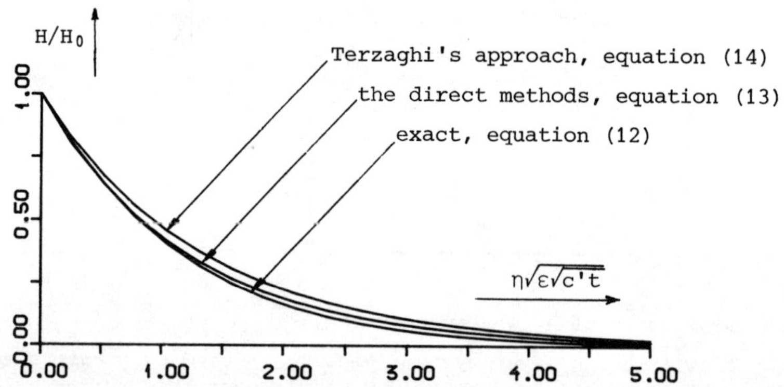


Fig. 2 Exact versus approximate solutions for the plane-symmetrical situation.

An exact analytical solution for an aquitard with finite height d is not found. In this case the direct method will give an acceptable result, to wit:

$$h = H\{\exp[-\sqrt{2}(\zeta-\delta)] - \exp[\sqrt{2}(\zeta-\delta)]\}/N; \tag{15}$$

$$H = H_0 \exp[-\eta\sqrt{2+2\varepsilon\mu\sqrt{2c't}}] ;$$

$$\delta = d/(2\sqrt{c't}) \ ; \ N = \exp[\sqrt{2}\delta] - \exp[-\sqrt{2}\delta] ;$$

$$\mu = (1+\exp[-2\sqrt{2}\delta])/(1-\exp[-2\sqrt{2}\delta]) .$$

Fig. 2 also presents this result, if a conformal variable is substituted. If $\varepsilon\mu\sqrt{2c't} < 1$ the compaction of the aquifer itself becomes unsignificant; in formula:

$$t < t_1 = \varepsilon^{-2}/2c' . \tag{16}$$

If $\delta$ tend to zero, or:

$$t > t_2 = d^2/4c', \tag{17}$$

equation (15) becomes:

$$H = H_0 \exp[-x/\lambda] \; ; \quad \lambda = \sqrt{KDd/K'}, \tag{18}$$

identical to the solution in the situation of a leaking aquifer (a constant leaking factor $\lambda$). Therefore, three periods can be distinguished, separated by $t_1$ and $t_2$. In the first period the compaction of the aquifer and the consolidation of the aquitard are significant, in the second period only the consolidation prevails, and in the third period a constant leak is established. For practical cases the ratio: $t_1/t_2$ , is about 1%, remarkable small!

## THE VERTICAL SETTLEMENT

The vertical strain in the equitard due to pore pressure changes in accordance with equation (9) is:

$$\varepsilon = -\alpha(\sigma'-\sigma'_0) = \alpha(p-p_0) = K'h/c'. \tag{19}$$

The corresponding vertical settlement S becomes:

$$S = \int_0^d (K'h/c')dz. \tag{20}$$

For the considered case of a two-layer system, employing equation (15), this integral can be evaluated:

$$S = HK'\sqrt{2t/c'}\; F(\sqrt{2}\delta);$$
$$F(a) = (1-\exp^2[-a])/(1-\exp[-2a]). \tag{21}$$

This expression is proportional to the piezometric head in the aquifer. It is stressed that the piezometric head in the aquifer is also a function of time.

For $\delta$ tending to zero the settlement becomes:

$$S = K'dH/2c' \; ; \; H = H_0 \exp[-x/\lambda], \tag{22}$$

which is fully compatible with the stationary stage (a linear pressure distribution over the aquitard).

The advantage of this method which includes the time-dependent behaviour, is that the effect of temporary withdrawal (pumping test, periodic drawdown) can be evaluated including its retarded transient character.

## A HYBRID MODEL

The geometry and homogeneity of the subsoil stratification varies naturally. The assumption of a uniform semi-infinite homogeneous situation as considered in the analytical solution previously described does not represent a realistic case, although it reflects the character of the phenomena quite clearly. Therefore, a numerical procedure is adopted allowing the geometry and homogeneity to vary.

For the considered two-layer system, the flow in the aquifer is treated numerically applying the finite element method, whereas

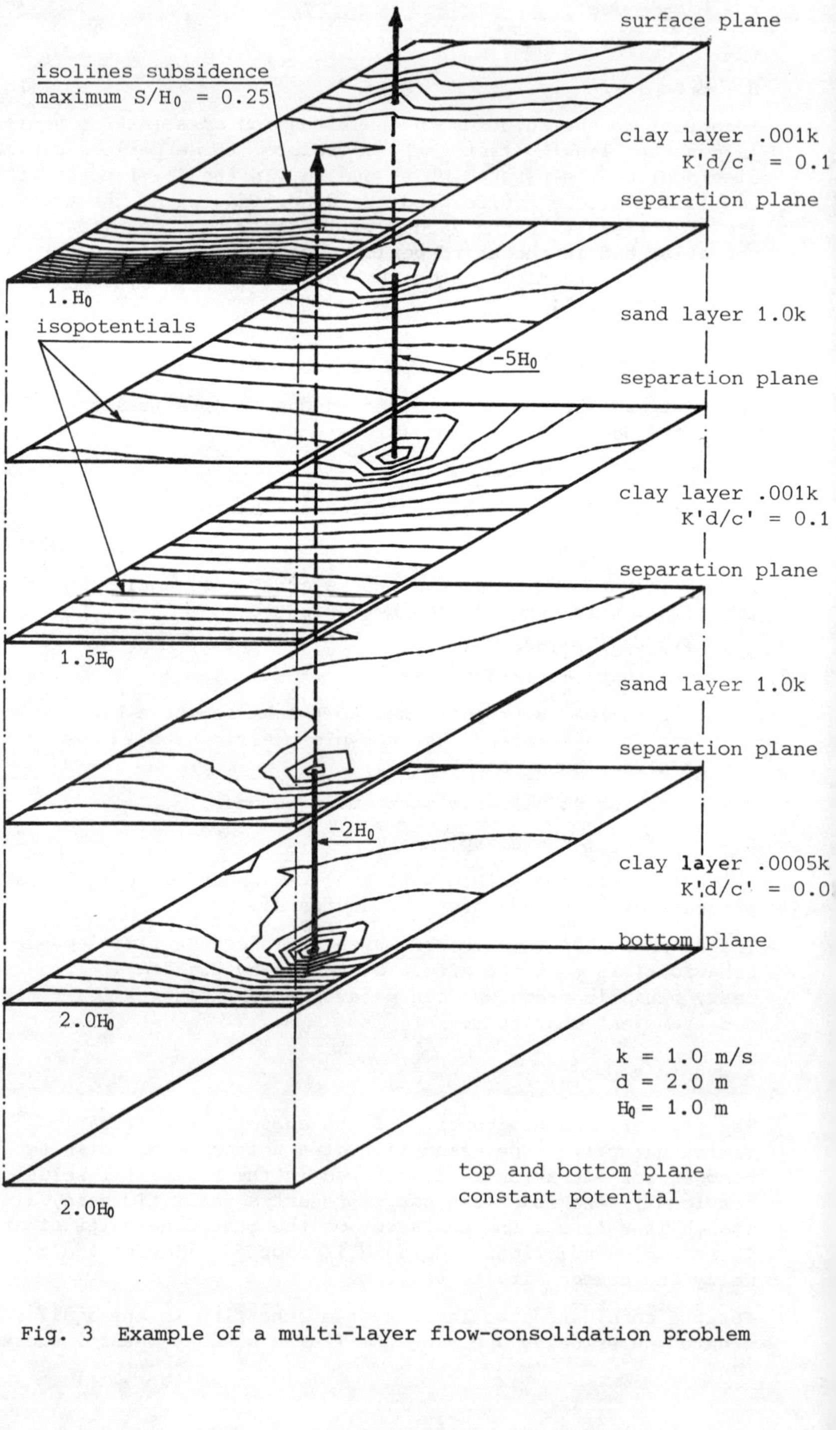


Fig. 3 Example of a multi-layer flow-consolidation problem

the consolidation of the aquitard is described by an analytical formula, equation (15). Obviously the material parameters and geometry may vary with respect to the horizontal position. The coupling between the numerical and analytical part is established by an iterative procedure.

For the considered case the flow in the aquifer is described by the following recursive equation:

$$\nabla \cdot KD\nabla H^{i} = K'H^{i-1}G(\sqrt{2}\delta)/\sqrt{2c't}; \qquad (23)$$

$$G(a) = (1+\exp[-2a])/(1-\exp[-2a]).$$

The corresponding settlement can be determined by equation (21) employing the calculated piezometric head. Since the flow in the aquifer is effected at a relative small scale, the iterative procedure converges fastly.

The essential advantage of this approach is the fact, that the time-dependence is explicitly formulated. At any instant the computation is the same. Hence, for large time values the evaluation of the subsidence does not require more computational effort.

For a multi-layer system the functions F and G, which actually constitute the core of the method, are different. A similar analysis can be worked out for such a case. The results are not mentioned here. An example of a multi-layer flow system is presented in Fig. 3. Two different wells are acting. The solution is generated by the SEEP code (Barends, 1980).

ACKNOWLEDGEMENT

The author is grateful to Dr. C.J. van Duyn for his assistance in finding the analytical part of the method presented.

LITERATURE

Barends, F.B.J. (1980) Nonlinearity in Groundwater Flow. Ph.D-thesis, Delft Univ.Techn., Delft, Holland, 128 pp.

Barends, F.B.J. (1981) Landsubsidence due to a Well in an Elastic Saturated Subsoil, Proc. Euromech. 143: Flow and Transport in Porous Media, A.A. Balkema Pub., Rotterdam. 1:11-18.

Bear, J.; M. Yavuz (1981) Mathematical Model for Regional Landsubsidence due to pumping, Water Res., 17(4):947-958.

Cooper, H.H. (1966) The Equation of Groundwater Flow in Fixed and Deforming Coordinates, J.Geophys.Res. 20(71):4785-4790.

Jacob, C.E. (1940) The Flow of Water in an Elastic Artesian Aquifer, Trans.Am.Geophys.Union, 21:574-586.

Hantush, M.S. (1964) Hydraulics of Wells, Adv.Hydro Sc.,1:281-431.

Sellmeijer, J.B. (1976) Nonsteady Groundwater Flow in Semi-confined Aquifers, Soils & Found.Jap.Soc.SMFE, 16(3):33-50.

Tang, D.H.; E.O. Frind; E.A. Sudicky (1981) Contaminant transport in Fractured Porous Media, Water Res.Res., 17(3):555-564.

Terzaghi, K. (1925) Erdbaumechanik auf bodenphysikalischer Grundlage, Franz Deuticke, Vienna.

Verruijt, A, (1969) Elastic Storage in Aquifers, chap.8 in: Flow through Porous Media (De Wiest, ed.), Acc.Press.London.

## Appendix

### The axi-symmetrical case

Axially symmetrical horizontal flow in a semi-confined compressible aquifer towards a well acting at a constant production Q, while a semi-permeable consolidating aquitard on top affects the actual flow, can be correctly described using the analysis previously applied for the plane-symmetrical case. The exact solution yields:

$$h = Q/(2\pi KD) \int_0^\infty B(\eta',\zeta,\varepsilon)\, d\theta \quad ; \quad \eta' = r \cosh(\theta)/2\sqrt{c't} \;, \qquad (24)$$

in which the function B is defined in equation (12). The parameter $\eta'$ includes the radial coordinate r.

The complete elaboration of this solution will be presented elsewhere. The result is mentioned here to emphasize the fact that also for the axi-symmetrical case similar conclusions hold to wit, compaction of the aquifer and a constant leak through a confining layer are phenomena with a different time scale.

Many approaches in literature just combine these two phenomena. This however makes sense physically, only if the following is valid:

$$dK'c/DKc' \ll 1 \; . \qquad (25)$$

In many real situations this does not hold. Subsequently the application of for example Hantush' well-function may lead to erroneous interpretations.

# FINITE ELEMENT ANALYSIS OF SOIL MOISTURE FLOW AND MOISTURE STRESS IN A DESSICATING SOIL

JOHN L. NIEBER

Texas A&M University

## INTRODUCTION

Numerous investigators have reported qualitative and quantitative observations of fractures in natural soils and the influence of these fractures on soil-water-air-plant relations. Bouma and Dekker (1978), Stirk (1954), Fox (1964), Johnson (1961), DeVos and Virgo (1969) and Virgo (1981) have reported field observations of fractures in natural clay soils. Surface connected fractures were reported to be up to 12 centimeters in width and 2 meters in depth.

Bouma et al. (1978), Kissel et al. (1973), Thomas and Phillips (1979), Stirk (1954), Blake et al. (1973), Richie et al. (1972), and Bouma et al. (1981) all acquired data regarding the movement (saturated or unsaturated) of water through fractures in soils. These investigators reported that fractures assist in improving the infiltration capacities and drainage rates for slowly permeable soils. It was also noted that with soil fractures large volumes of water bypass the soil medium and thereby reduce the opportunity for interaction between the soil and constituents dissolved in the water. This condition is potentially detrimental to pollution control as well as to nutrient conservation.

Theoretical studies of water movement in fractured soils have been reported by Hoogmoed and Bouma (1980), Bouma and DeLatt (1981) and Nieber (1981). Hoogmoed and Bouma (1980) demonstrated the importance of soil fractures on improving the infiltration capacities of slowly permeable soils. They demonstrated this using a one-dimensional formulation for infiltration into porous media. Nieber (1981) performed a similar analysis but used a two-dimensional formulation for soil moisture flow. Bouma and DeLatt (1981) demonstrated by water balance calculations that horizontal and vertical soil fractures tend to significantly decrease the availability of moisture to plants.

The fracturing of dessicating soils is similar to the fracturing of other solids such as rock, ice, metals, plastics and organic materials when placed under conditions where the material expands or contracts nonuniformly. Lachenbruch (1962) studied the thermally induced fracturing of permafrost in the Arctic. Kindle (1917, 1932), Longwell (1928), and Corte and Higashi (1964) have experimentally investigated fracturing of dessicating soils. Rao (1975) presented a theoretical analysis of moisture stress of a clay soil. Smith (1959) performed a detailed investigation of the shrinkage/swelling properties of Texas Blackland soils and the effect of these properties on the physical characteristics of natural soil horizons.

This paper will present an analysis of the transient soil moisture flow and transient stress development in soils having significant shrink/swell potential. The analysis is performed using the finite element method to solve the equations for soil moisture flow and the equations of linear elasticity. The application presented is for the soil moisture flow and stress development in the soil around a row crop.

SOIL MOISTURE FLOW ANALYSIS

Transient soil moisture flow is adequately described by the Richards equation given as,

$$C \frac{\partial \phi}{\partial t} = \nabla \cdot (K \nabla \phi) \qquad (1)$$

where, $\phi = h + e$ = hydraulic head [m],
$h$ = water pressure head [m],
$e$ = elevation head [m],
$C = C(h) = \frac{d\theta}{dh}$ = specific moisture capacity [1/m],
$K = K(\theta)$ = hydraulic conductivity [m/sec],
$\theta = \theta(h)$ = volumetric moisture content [$m^3/m^3$],
$t$ = time [sec], and
$\nabla$ = "del" operator.

The use of this equation implies the assumption that the soil and moisture are both incompressible. It also assumes that the flow of moisture occurs under isothermal conditions and that air within the soil is at the ambient atmospheric pressure. For a more exact analysis of soil moisture flow in a shrinking/swelling soil it would be necessary to avoid these assumptions by including additional governing equations. However, for the analysis given here the use of the Richards equation alone should provide an adequate approximation.

The boundary conditions for the soil moisture flow problem are either specified hydraulic head or specified moisture flux. The initial condition is just the initial spatial distribution of hydraulic head.

To solve the Richards equation along with the boundary conditions and initial condition the finite element method was used to transform the space derivatives and the finite difference method was used to transform the time derivative. The resulting set of nonlinear algebraic equations are solved by an under-relaxation iterative technique at each time step. The system of equations and the solution procedures are described in detail by Neuman (1973) and Nieber (1980).

Relations for the coefficients C(h) and K(θ) can be described by the equations

$$C(h) = \frac{nABh^{B-1}}{(A+h^B)^2}$$

and (2)

$$K(\theta) = K_0\left(\frac{\theta-\theta_r}{n}\right)^N$$

where, $n = \theta_0 - \theta_r$ = drainable porosity,
$\theta_0$ = "natural" saturation moisture content,
$\theta_r$ = residual moisture content,
$K_0$ = saturated hydraulic conductivity,
$A$ = soil moisture parameter dependent on the maximum size soil pore [$m^B$], and
$B,N$ = soil moisture parameters dependent on soil pore size distribution [1].

STRESS ANALYSIS

The stress analysis presented here was based on the assumption of plane strain and linear elastic conditions. The equations governing this problem are (1) equations of equilibrium, (2) an equation of compatibility, (3) equations of stress-strain relations, and (4) equations of boundary conditions. The equations of equilibrium are

$$\frac{\partial\sigma_x}{\partial x} + \frac{\partial\tau_{xy}}{\partial y} = 0$$

(3)

$$\frac{\partial\sigma_y}{\partial y} + \frac{\partial\tau_{xy}}{\partial x} = 0$$

where, $\sigma_x$ and $\sigma_y$ are the normal stresses [KPa] in the x- and y- directions respectively and $\tau_{xy}$ is the shear stress [KPa] in the xy plane. The equation of compatibility is

$$\nabla^2(\sigma_x + \sigma_y + \alpha E \Delta\theta) = 0 \tag{4}$$

where $\alpha$ is the moisture expansion coefficient [1], E is Young's elastic modulus [KPa], and $\Delta\theta$ is the change in moisture content from the initial moisture content.

The stress-strain relations for the plane strain problem are

$$\varepsilon_x = \frac{1}{E}(\sigma_x - \nu\sigma_y) + \alpha\Delta\theta$$

$$\varepsilon_y = \frac{1}{E}(\sigma_y - \nu\sigma_x) + \alpha\Delta\theta \tag{5}$$

$$\varepsilon_z = 0$$

$$\gamma_{xy} = \tau_{xy}/D$$

where $\varepsilon_x$, $\varepsilon_y$ and $\varepsilon_z$ are the respective strains [1] in the x, y and z directions, $\gamma_{xy}$ is the shear strain [1] in the xy plane, $\nu$ is Poisson's ratio [1], and $D(= E/2(1 + \nu))$ is the shear modulus of elasticity [KPa].

The boundary condition equations are simply the specification of either the stress ($\sigma_x$, $\sigma_y$, $\tau_{xy}$) or the displacement ($u_x$,$u_y$) at points along the boundary of the solution region.

The finite element method was used to obtain a solution to these equations. The program code for plane stress analysis contained in Segerlind (1976) was modified for plane strain analysis for use in the present application. The program uses constant strain (linear displacement) triangles. Constant strain triangles may not be the best choice for the analysis presented here because for many solid mechanics problems they do not provide satisfactory accuracy. However, without a detailed analysis it was assumed that constant strain triangles would provide sufficient accuracy for this application.

The solution from the soil moisture flow problem, in the form of soil moisture distribution at various times, was used as input to the stress analysis solution. Thus it was possible to simulate transient stress distributions caused by soil drying.

## PROBLEM DESCRIPTION

Johnson (1961) describes an agricultural field experiment concerning moisture conservation through the control of soil cracking pattern and crack size distribution. He hypothesized that cracking control should be feasible by reducing evapotranspiration on the soil area where cracking is desired.

Therefore, in his field experiment, Johnson planted wheat on 14 inch rows and after the wheat had emerged wheat was removed along 18 inch wide strips running perpendicular to the row direction. The result of the experiment was that single large cracks developed along the strips. Uncontrolled or "natural" cracks occurred in the areas between rows.

The application presented in this paper was inspired by the results reported by Johnson (1961). The problem chosen was to analyze the soil moisture and stress distribution that might occur between two rows of a crop during a drying period. An illustration of the problem analyzed is given in Figure 1. Indicated on this figure is the region analyzed. The region was chosen to be 0.177 meters wide and 0.60 meters long. The finite element grid for the region analyzed is illustrated in Figure 2. The boundary conditions for both moisture flow and stress are indicated in the illustration.

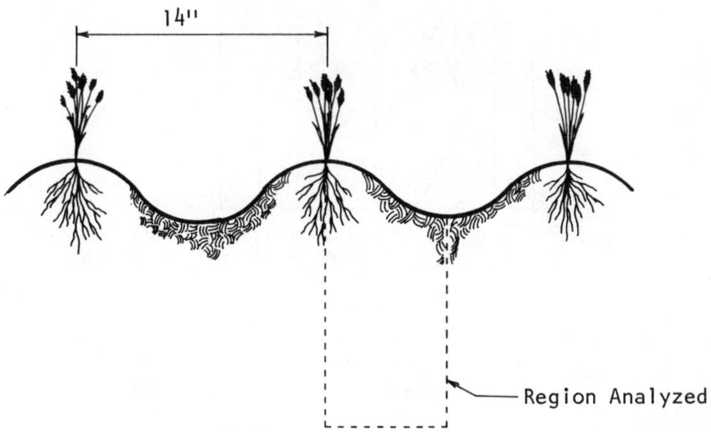


Figure 1. Illustration of a vertical cross-section of soil taken perpendicular to the rows in a furrowed field.

The soil moisture parameters for Equation (2) were assigned values of $A = 7.1\ m^{0.874}$, $B = 0.874$, $N = 15.02$, $\theta_0 = 0.42$, $\theta_r = 0.24$, and $K_0 = 5\text{x}10^{-7}$ m/sec. These parameters were chosen because they corresponded to the parameters measured for a soil classified as Norwood silt loam.

The soil elastic parameters $\alpha$, $\nu$ and E were not measured for the Norwood silt loam. However, the values reported by Corte and Higashi (1964) for a similar soil were adopted for use in this analysis. The value of $\nu$ was assigned to be 0.35. The values of E and $\alpha$ were reported by Corte and Higashi to be

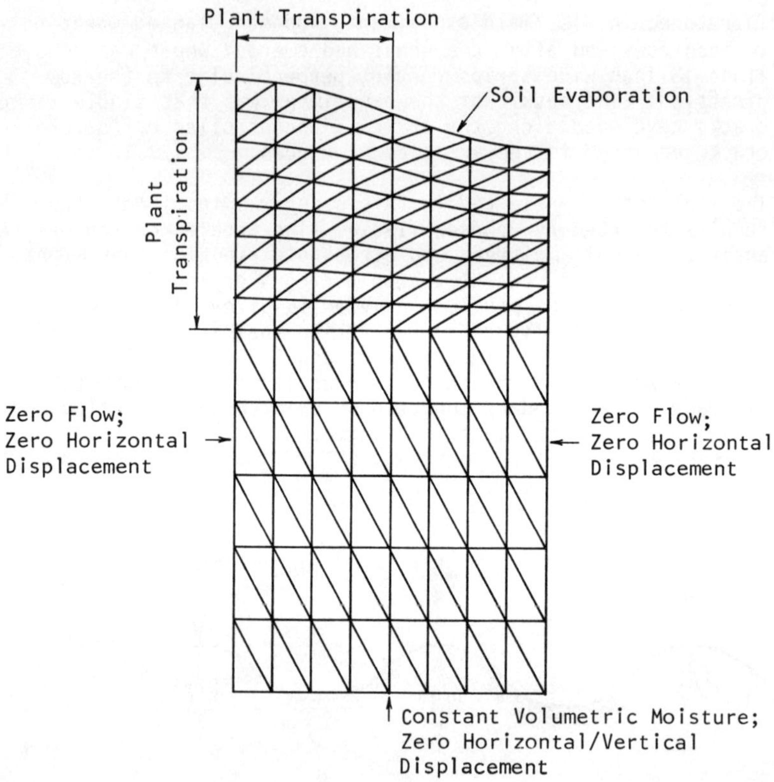


Figure 2. Finite element discretization of the region analyzed with boundary conditions for the moisture flow and stress indicated.

moisture dependent. The moisture dependency for E[KPa] was reported as $E = 4.80 \times 10^6 \exp(-18\ \theta)$ and this relation was used in the present analysis. Although $\alpha$ was reported to be moisture dependent it was used here as a constant at 0.40.

The transpiration rate from the plant and the evaporation rate from the bare soil surface were each assigned to vary diurnally from 6 mm/day at noon to 1 mm/day at midnight. Root uptake for transpiration was assumed to be greatest near the soil surface and decreased with depth. The moisture content at the 0.60 meter depth was maintained at 0.372. The initial moisture content of the region was 0.372.

## RESULTS AND DISCUSSION

The transient moisture content distributions and transient maximum principal stress distributions are presented in

Figures 3 and 4. The distributions correspond to times of 368 minutes, 1108 minutes, 1802 minutes and 3464 minutes following the beginning of drying. As time proceeds the drying front advances into the profile. Due to the constant water content condition at the base of the region it is expected that eventually a nearly steady-state condition will prevail. Absolute steady-state conditions will not be achieved due to the variations in evaporation demand.

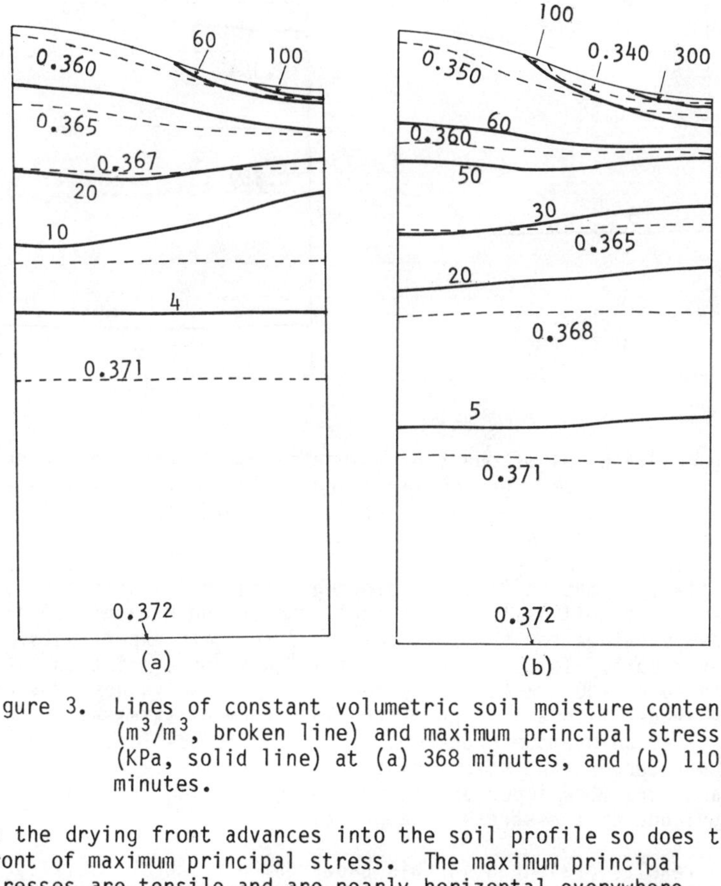


Figure 3. Lines of constant volumetric soil moisture content ($m^3/m^3$, broken line) and maximum principal stress (KPa, solid line) at (a) 368 minutes, and (b) 1108 minutes.

As the drying front advances into the soil profile so does the front of maximum principal stress. The maximum principal stresses are tensile and are nearly horizontal everywhere. The stress distributions presented in Figures 3 and 4 show a concentration of largest stresses toward the center of the furrow. This is the location where the most significant soil cracking is observed to occur (see Johnson, 1961). Large values of maximum principal stress do extend into the region of the root mass. However, the development of large cracks within the root mass region is less likely due to the added tensile strength provided to the soil mass by the roots.

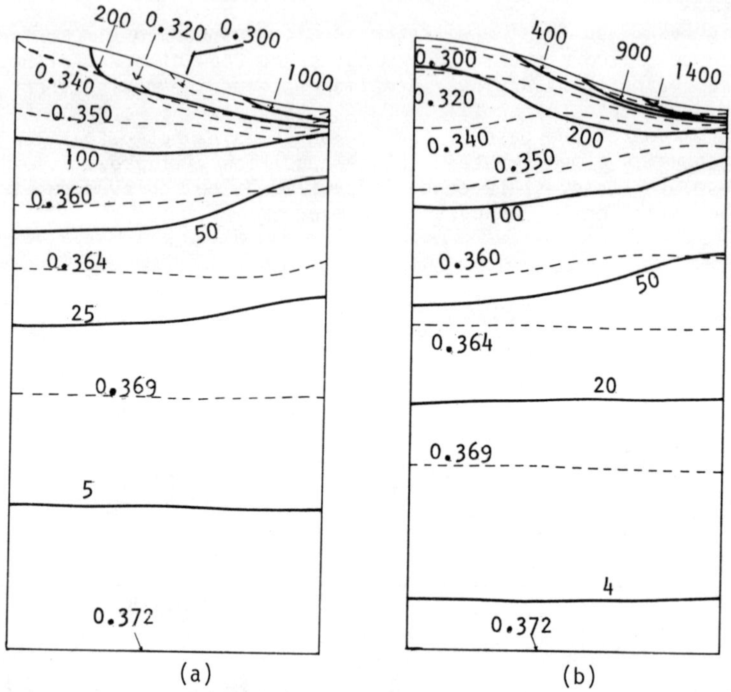


Figure 4. Lines of constant volumetric soil moisture content ($m^3/m^3$, broken line) and maximum principal stress (KPa, solid line) at (a) 1802 minutes, and (b) 3464 minutes.

Little information is available regarding the tensile stress at which a soil will fail (crack). Corte and Higashi (1964) provide values for the soil they used in their experiments. They reported failure stresses of 117 KPa for a moist condition to 22,100 KPa for a dry condition. These values provide the reader with a feeling of the magnitude of stresses required to promote cracking. They are not absolute values though because the stress distribution and the rate of stress change are more important in promoting cracking than the magnitude of the stress at a point.

The results presented in this paper demonstrate the utility of computer simulation techniques for analyzing the stress distributions within a soil undergoing volume changes. The conditions analyzed were not complex since the dynamics of root water extraction, wetting/drying cycles, and the dynamics of crack propagation were not considered. To provide useful interpretations of field observations of soil cracking in differing climatic and geographical regions it will be necessary to incorporate these conditions into the analysis.

## REFERENCES


Blake, G., E. Schlichting, and U. Zimmerman. 1973. Water recharge in a soil with shrinkage cracks, Soil Sci. Soc. Am. Proc., 37:669-672.

Bouma, J. and L. W. Dekker. 1978. A case study of infiltration into dry clay soil, I. Morphological observations, Geoderma, 20:27-40.

Bouma, J., L. W. Dekker, and J. H. M. Wosten. 1978. A case study of infiltration into dry clay soil, II. Physical measurements, Geoderma, 20:41-51.

Bouma, J. and P. J. M. DeLatt. 1981. Estimation of the moisture supply capacity of some swelling clay soils in the Netherlands, J. Hydrol., 49:247-259.

Bouma, J., J. W. van Hoorn, and G. H. Stoffelsen. 1981. Measuring the hydraulic conductivity of soil adjacent to tile drains in a heavy clay soil in the Netherlands, J. Hydrol., 50:371-381.

Corte, A. and A. Higashi. 1964. Experimental Research on Dessication Cracks in Soil, U. S. Army Corps of Engineers, Cold Regions Research and Engineering Laboratory, Research Report 66. 76 pp.

DeVos, J. H. and K. J. Virgo. 1969. Soil structure in Vertisols of the Blue Nile Clay Plains, Sudan, J. Soil Sci., 20:189-206.

Fox, W. E. 1964. Cracking characteristics and field capacity in a swelling soil, Soil Sci., 98:413.

Hoogmoed, W. B. and J. Bouma. 1980. A simulation model for predicting infiltration into cracked clay soil, Soil Sci. Soc. Am. J., 44:458-461.

Johnson, W. C. 1961. Controlled soil cracking as a possible means of moisture conservation on wheatlands of the Southwestern Great Plains, Agron. J., 54:323-325.

Kindle, E. M. 1917. Some factors affecting the development of mud-cracks, J. Geol., 25:135-144.

Kindle, E. M. 1932. A note on mud-cracks and associated joint structure, Am. J. Sci., 5:329-330.

Kissel, D. E., J. T. Ritchie, and E. Burnett. 1973. Chloride movement in undisturbed clay soil, Soil Sci. Soc. Am., Proc. 37:21-24.

Lachenbruch, A. H. 1962. Mechanics of Thermal Contraction Cracks and Ice-Wedge Polygons in Permafrost, The Geological Society of America, Waverly Press, Baltimore, 69 pp.

Longwell, C. R. 1928. Three common types of desert mud-cracks, Am. J. Sci., 15:136-145.

Neuman, S. P. 1973. Saturated-unsaturated seepage by finite elements, J. Hydraul. Div., ASCE, 99(HY12), 2233-2250.

Nieber, J. L. 1980. Evaluation of coefficient matrices for linear triangular elements involving unsaturated soil moisture flow. In: Wang, S. Y. et al. (eds.), Finite Elements in Water Resources, Proceedings of the Third International Conference on Finite Elements in Water Resources, 2.260-2.269.

Nieber, J. L. 1981. Simulation of infiltration into cracked soils, Presented at the 1981 Summer Meeting of the American Society of Agricultural Engineers, Paper No. 81-2026.

Rao, V. N. M. 1975. Stresses in a Maxwell viscoelastic cylinder due to transient temperature and moisture gradients, Trans. Am. Soc. Agri. Eng., 18:1165-1169.

Ritchie, J. T., D. E. Kissel, and E. Burnett. 1972. Water movement in undisturbed swelling clay soil, Soil Sci. Soc. Am., Proc., 36:874-879.

Segerlind, L. J. 1976. Applied Finite Element Analysis, John Wiley & Sons, New York, 422 pp.

Smith, R. M. 1959. Some Structural Relationships of Texas Blackland Soils with Special Attention to Shrinkage and Swelling, USDA-ARS 41-28, 51 pp.

Stirk, G. B. 1954. Some aspects of soil shrinkage and the effect of cracking upon water entry into soil, Aust. J. Agri. Res., 5:279-290.

Thomas, G. W. and R. E. Phillips. 1979. Consequences of water movement in macropores, J. Environ. Qual., 2:149-153.

Virgo, K. J. 1981. Observations of cracking in Somali Vertisols, Soil Sci., 131:60-61.

---

The research reported in this paper was supported by the Texas Agricultural Experiment Station, and by the Department of Agricultural Engineering at Texas A&M University.

# 12. FLOW IN ROCKY MATERIAL

# NUMERICAL MODELLING OF RUBBLE MOUND BREAKWATERS

A'Alim A. Hannoura, University of New Orleans; and Frans B. Barends, Delft Soil Mechanics Laboratory

## INTRODUCTION

Rubble-mound structures have been used for thousands of years for harbours and coastal lines protection against wave action. Perhaps the oldest use of rubble-mound breakwaters was that found in the ancient City of Alexandria at the Pharaoh Port. The course of progress in design procedures for rubble-mound has been from mere experience and judgement to physical modelling and the use of empirical equations. The purpose of this paper is to shed some lights on the numerous facets of the problem of wave interaction with rubble-mounds and the possibilities of using numerical modelling in the design of such structures.

## BACKGROUND

The application of numerical modelling techniques to the investigation of rubble-mound structures is, relatively, at an early stage. This is not due to the inefficiency of the modelling techniques but it is largely attributed to the lack of understanding of the physical processes associated with the problem of wave interaction with rubble-mounds. In the design of rubble-mound structures coastal engineers are usually interested in evaluating wave transmission through and reflection from the structure, and the stability of the armour layers due to wave action. Among the earliest research in the area of numerical and theoretical modelling of rubble-mound structures are models developed by Sollitt and Cross (1972); Nasser and McCorquodale (1975); and Madsen and White (1975). These early trials are marked by oversimplification of the real problem (Hannoura, 1978).

## DEFINITION OF THE PROBLEM

When a progressive wave collides against a rubble-mound breakwater, transmission and reflection of the wave take place on the sea-side of the breakwater. The resultant internal flow is, often, non-Darcy and one or two phase. Two phase flow may occur due to external and/or internal wave breaking (Fuhrboter, 1970; Nasser, 1974; and Sollitt and Cross, 1972). At the same time the breakwater is subjected to unsteady wave forces.

The flow domain is not rectangular. Sea-side and harbour-side slopes, usually, have multiple slopes and berms. The construction material varies in size from the armour layer where artificial large units are placed on the slope to the core zone where quarry run is used. Between the armour layer and the core zone, layers of different sizes and hydraulic characteristics are placed.

## FORMULATION OF THE PROBLEM

### The Methodology

The current formulation of the wave interaction with a rubble-mound breakwater problem is limited to the analysis of the internal flow, i.e., wave motion in the porous structure. At this stage, the problem is formulated in two phases: (a) the development of the hydrodynamics equations, i.e., continuity and momentum, for unsteady non-Darcy flow in rubble-mounds. The hydrodynamics equations are simplified to a set of one-dimensional unsteady equations and an implicit-explicit finite difference scheme is uded to obtain the instantaneous phreatic surface. A further simplification of the hydrodynamics equations is introduced in order to obtain a system of hyperbolic differential equations which are solvable by the method of characteristics. This will give the equations of the $\alpha$ and $\beta$ characteristics which are used to determine the maximum time step for the finite difference solution required to insure the stability and convergence of the solution. (b) In the second phase of the solution uses the finite element method to analyse the flow domain after every time step or after a predeterminate number of time steps. This phase is necessary to overcome the simplifications assumed in the first phase of the overall solution. The introduction of the finite element method will yield representative values of the actual pressure distribution and the hydraulic conductivity of the rubble-mound matrix. The solution domain is divided into a number of vertical sub-domains and then weighted averages for pressure and hydraulic conductivity are obtained and transferred to the one-dimensional finite difference solution. This hybrid model, i.e., the use of the finite difference method for time integration and the finite element for space integration, would require, only about 40% of the CUP time required for a complete finite element solution (Hannoura, 1980). Appendix "A" gives the numerical formulation.

### The Flow Coefficients

There are several flow coefficients involved in the determination of the unsteady free surface due to wave interaction with a rubble-mound structure. The accuracy of the numerical results are heavily dependent on the values assigned to these coefficients. The coefficients, in the problem considered, are: (a) added mass coefficient; (b) friction coefficient; and (c) air entrainment factor, i.e., the flow is influenced by air entrainment near the interface and its effect on the friction coefficient. The literature on these coefficients for unsteady two phase flow in rubble-mounds is very limited (Hannoura and Barends, 1981).

### The Boundary Conditions

Sea-side The finite difference solution assumes that the core of the breakwater is impervious so only the location of the wave front on the interface slope is needed. On the other hand, the finite element solution requires a full specification of the instantaneous wave potential on the slope.

Harbour-side Assuming zero one-dimensional velocity for the finite difference solution implies no transmission through the breakwater which is not true in all cases. The hybrid model could overcome this approximation by utilization of the free surface location at the core, from a previous period, as a boundary condition for the next period. Therefore the finite element solution has to be used to check the validity of the continuity through the breakwater. Such a process may be undesirable and a full finite element solution would be an alternative.

### The Initial Conditions

The numerical solution starts from rest, i.e., zero velocity and celerity. Then a regular periodical wave is assumed at the sea-side interface. Irregular waves can also be used.

## NUMERICAL OUTPUT

The hybrid model results are given in terms of instantaneous phreatic surface and pressure distribution. These results may be utilized for transmission, reflection and stability calculations.

## EXPERIMENTAL VERIFICATION

Initial verification of the model was limited to a comparison of maximum and minimum phreatic surfaces (Hannoura, 1978). This agreement was very satisfactory. Due to the limitation of experimental data on the nature of wave impact and instantaneous potential along the sea-side interface; long-period waves could not be modelled as well as short-period waves. These early experiments were conducted at the University of Windsor in a relatively small wave flume (0.46 x 0.61 x 13.16 m).

Recently, in a large scale test done by a team of the Delft Soil Mechanics Laboratory using the Delta Flume (5.0 x 7.0 x 240.0 m), a verification of internal pressure provided a very good agreement between experimental and numerical results.

## NUMERICAL MODELLING AND RUBBLE-MOUND BREAKWATER DESIGN

While physical modelling of rubble-mound breakwaters would remain as an indispensable stage in the testing of these structures, numerical modelling could be used to save time and cost in the design process. Numerical modelling of different design alternatives would provide the design engineer with estimates of transmission and reflection coefficients, and slope stability information for different sea conditions. From these numerical results one design alternative would evolve. At this stage physical modelling could be applied and modifications of the final design would be evaluated both numerically and experimentally.

A complete modelling of the problem requires the use of an additional numerical model to simulate the dynamic phase of the problem. Such an analysis could be achieved through the use of computer codes developed by Sweet et al (1980).

## CONCLUSIONS AND FUTURE WORK

The use of numerical models in the investigation of some recent failure cases of rubble-mound structures has shown the potential for using such techniques not only after the fact but also for design. However, in order to use numerical models successfully an understanding of the physical flow processes in rubble-mounds and the structural aspects of construction material, e.g., artifical armour units are needed. Advances in computation techniques would, eventually, enable the modelling of wave motion, i.e., transformation, along the sea-side interface (Vinje and Brevig, 1981).

## APPENDIX "A" ON THE NUMERICAL FORMULATION

Applying the one-dimensional momentum equation to the internal flow, the following svstem of equations of motion and continuity for unsteady non-Darcy flow in a rubble-mound structure is obtained:

$$C_m \frac{\partial u}{\partial t} + \frac{u}{m} \frac{\partial u}{\partial x} + 2cm \frac{\partial c}{\partial x} C_p = -gFmu \qquad (1)$$

and

$$\frac{\partial c}{\partial t} + \frac{u}{m}\frac{\partial c}{\partial x} + \frac{c}{2m}\frac{\partial u}{\partial x} = 0 \qquad (2)$$

in which $C_m$ = inertia coefficient, u=horizontal Darcy velocity, m=porosity, c=wave celertiy=$\sqrt{g(H_o + \eta)}$, $C_p$ = pressure distribution correction factor, t=time, y=acceleration due to gravity, F=non-Darcy friction term, $H_o$ = mean water level, $\eta$= internal wave amplitude, and x is the horizontal axis.

Characteristic formulation

Considering the total derivatives of the unknown quantities u and c as:

$$du = \frac{\partial u}{\partial t}\, dt + \frac{\partial u}{\partial x}\, dx \qquad (3)$$

and

$$dc = \frac{\partial x}{\partial t}\, dt + \frac{\partial c}{\partial x}\, dx \qquad (4)$$

The following system of equations for unsteady non-Darcy flow in a porous media, assuming $C_p$ = constant is obtained:

$$\begin{Bmatrix} C_m & u/m & 0 & 2cC_p m \\ 0 & c/2m & 1 & u/m \\ dt & dx & 0 & 0 \\ 0 & 0 & dt & dx \end{Bmatrix} \begin{Bmatrix} \frac{\partial u}{\partial t} \\ \frac{\partial u}{\partial x} \\ \frac{\partial c}{\partial t} \\ \frac{\partial c}{\partial x} \end{Bmatrix} = \begin{Bmatrix} -gFm \\ 0 \\ du \\ dc \end{Bmatrix} \qquad (5)$$

From equation (5), the characteristic directions can be written as:

$$\alpha = \frac{dx}{dt}\Big|_{+} = (1 + \frac{1}{Cm})\frac{u}{2m} + \sqrt{(\frac{u}{m})^2(\frac{C_m + 3}{4C_m} - 1) + c^2\frac{C_p}{C_m}} \qquad (6)$$

and

$$\beta = \frac{dx}{dt}\Big|_{-} = (1 + \frac{1}{C_m})\frac{u}{2m} - \sqrt{(\frac{u}{m})^2(\frac{C_m + 3}{4C_m} - 1) + c^2\frac{C_p}{C_m}} \qquad (7)$$

The characteristic solution proceeds by solving the governing equations along the characteristic directions $\alpha$ and $\beta$.

An explicit finite difference scheme is used to discretize equations (1) and (2) in order to obtain u and c at the internal grid points, i.e.

$$u(i,j+1) = u(i,j) - \frac{\Delta t}{2\Delta x C_m} \frac{u(i,j)}{m}[u(i,j+1) - u(i-1,j)] + mc(i,j)c(i,j)[C_p(i+1,j) - C_p(i-1,j)] + 2mC_p(i,j)c(i,j)[c(i+1,j) - c(i-1,j)] + 2gF(i,j)mu(i,j)\Delta x \quad (8)$$

and

$$c(i,j+1) = c(i,j) - \frac{\Delta t}{2m\Delta x} u(i,j)[c(i+1,j) - c(i-1,j)] + \frac{c(i,j)}{2}[u(i+1,j) - u(i-1,j)] \quad (9)$$

in which $F(i,j) = a+b|u(i,j)|$; a and b are the non-Darcy friction coefficients. The values for $F(i,j)$ and $C_p(i,j)$ are obtained from depth integration using the finite element solution.

Equations (6) and (7) are used to control the discretization of the x-t plane.

## Finite element formulation

The momentary two-dimensional internal flow can be described by:

$$\frac{\partial}{\partial x}\left(K(|\nabla\psi|)\frac{\partial\psi}{\partial x}\right) + \frac{}{y}\left(K(|\nabla\psi|)\frac{\partial\psi}{\partial y}\right) = 0 \quad (10)$$

in which $K(|\nabla\psi|)$ = non-linear hydraulic conductivity and $\psi$= piezometric head. McCorquodale (1970) gives the corresponding functional form of equation (10) as:

$$\chi = \iint_A K(|\nabla\psi|)(\psi_x^2 + \psi_y^2) + G(|\nabla\psi|)\,dydx \quad (11)$$

where $G(|\nabla\psi|)$ is a continuity function. The solution of equation (10) is now reduced to finding the distribution of the piezometric function, $\psi$, which minimizes $\chi$, equation (11), subjected to certain boundary conditions. The integral of equation (11) is given by:

$$\sum_{e}\sum_{j} K(|\nabla\psi|)^e S^e_{ij}\psi_j = 0 \qquad (12)$$

in which $S^e_{ij} = (b_i b_j + c_i c_j)|A^e$ = the element stiffness matrix

$$b_i = y_i - y_k$$

$$c_i = x_k - x_j$$

and i,j and k are the element node numbers.

Equation (12) gives a set of non-linear simultaneous equations in the unknown values of $\psi$.

REFERENCES


Fuhrboter, A.
1970. Air Entrainment and Energy Dissipation in Breakers, 12th. Conf. on Coastal Eng., Proc., ASCE, Vol. 1, Chapter 24.

Hannoura, A.; and Barends, F.
1981. Non-Darcy Flow; A State of the Art, "Flow and Transport in Porous Media", A. Verruijt and F. Barends, Editors, A.A. Balkema, Rotterdam, The Netherlands.

Hannoura, A.
1980. A Hybrid Finite Element Model Applied to Unsteady Flow Problems, Proc. of the 3rd. Int. Conf. on F.E. in Water Res., S.Y. Wang et al, Editor, U.S.A.

Hannoura, A.
1978. Numerical and Experimental Modelling of Unsteady Flow in Rockfill Embankments, Ph.D. Thesis, Univ. of Windsor, Canada.

Madsen, O.; and White, S.
1975. Reflection and Transmission Characteristics of Porous Rubble-Mound Breakwaters, MIT, Dept. of Civ. Eng., Ralph M. Parson Lab., Report No. 207, U.S.A.

McCorquodale, J.
1970. Variational Approach to Non-Darcy Flow, J. Hyd. Div., ASCE, Vol. 96, No. HY11.

Nasser, M.; and McCorquodale, J.
1975. Wave Motion in Rockfill, J. Waterways, Harbor and Coastal Engineering Div., ASE, Vol. 101, No. WW2.

Nasser, M.
1974. Theoretical and Experimental Analyses of Wave Motion in Rockfill Structures, Ph.D. Thesis, Univ. of Windsor, Canada.

Sollitt, C.; and Cross, R.
1972. Wave Reflection and Transmission at Permeable Breakwaters, MIT, Dept. of Civ. Eng., Ralph M. Parson Lab., Report No. 147.

Sweet, J.; Barends, F.; Engels, C.; and Kogel, H.
1980. A Method for Dynamic Soil-Structure Interaction Problems, 1st. Int. Conf. on Num. Methods for Non-Linear Problems, Swansea, U.K.

Vinje, T.; and Brevig, P.
1981. Breaking Waves on Finite Water Depths, Dept. of Marine Tech., NIT, Univ. of Trondheim, Norway.

# FINITE ELEMENT SIMULATION OF CONTAMINANT TRANSPORT IN FRACTURED ROCK NEAR KARLSHAMN, SWEDEN

Allen M. Shapiro and Johan Andersson

Dept. of Water Resources Engineering
Royal Institute of Technology
100 44 Stockholm, Sweden

## ABSTRACT

A finite element model predicting the steady-state movement of fluid in fractured rock is calibrated with hydrologic information collected from investigations of fractured rock near Karlshamn, Sweden. Alternative representations of the hydraulic conductivity field and finite element mesh lead to similar results for water table elevations, but provide contrasting information in investigations of hypothetically proposed problems of contaminant transport. Such results indicate the reliability of making predictions about phenomena involving the transport of contaminants that are based solely upon hydrologic data such as water level measurements.

## INTRODUCTION

Even under the best circumstances, field information can at most be considered sparse in dealing with the majority of problems in subsurface hydrology. Transferring what little information there is, to a model that is to be used for predictive purposes can lead to the same physical system being described by alternative means. However, once a model has been established the adjustment of internal parameters can under most circumstances provide adequate correlations with field information. Yet when extending the use of such models, care must be exercised for problems which lie outside the situation for which the model was calibrated.

As an illustration of these points, finite element models predicting fluid movement and contaminant transport in fractured rock are formulated from information taken from the vicinity of Karlshamn, Sweden. The area under investigation represents a unique situation in collecting data from fractured formations since a series of underground caverns exist which have been evacuated of water. This has allowed an intimate inspection of fracture zones and their orientation, and thus alleviates a major problem encountered in the modelling of most fractured formations. Yet even with this wealth of information problems in the model conceptualization remain acute. Establishing the conductivity of the fractures and conceptualizing them for modelling purposes still require attention.

In this investigation two alternatives of describing the fractured rock near Karlshamn, which are drawn from the same underlying physical interpretation, are calibrated with existing water table elevations. However, in extending these models to the study of contaminant movement one is not able to assess their accuracy or establish the magnitude of additional model parameters in either representation. The contrasting information provided by the two models serves as an example of extrapolating the use of a given model beyond its calibrated ability.

## GEOHYDROLOGY

The city of Karlshamn lies on the southeastern coast of Sweden with the Baltic Sea as its eastern boundary. The study area (see Figure 1) which is situated just south of the city acts as a land bridge of approximately 350 meters in width, connecting the Swedish mainland to the north and the Sternö peninsula to the south (Axelsson and Carlsson, 1980). The primary geologic characteristic of the area is the diabase dike which intrudes into the coastal gneiss. The dike is almost vertically oriented with a north-south direction, and is approximately 200 meters in width. Randomly oriented cooling fractures exist in the diabase dike, and as a result of changing stress orientations major directional fracture zones have also been formed. These fracture zones are comprised of several planar fractures which intersect, not necessarily at the same point, but densely located within a given area. The zones of fracturing which have been noted to extend in both the diabase and the gneiss constitute the major means of fluid movement in the rock formation. The other randomly oriented fractures in the area act to a lesser extent while the rock itself is regarded as being impervious. Further geologic details of the study area are discussed in Larsson, et al. (1977).

Specific information about the fracture zones in the study area (see Figure 1) has been collected from a visual inspection of the underground caverns. The bottom elevation of the caverns extends to a depth of thirty meters below sea level in some areas and currently, the caverns are being pumped to maintain a water table elevation which keeps them dry. Other water table measurements have been established from bore holes that have been drilled in the vicinity of the cavern area. The magnitude and distribution of the pumpage from the caverns and

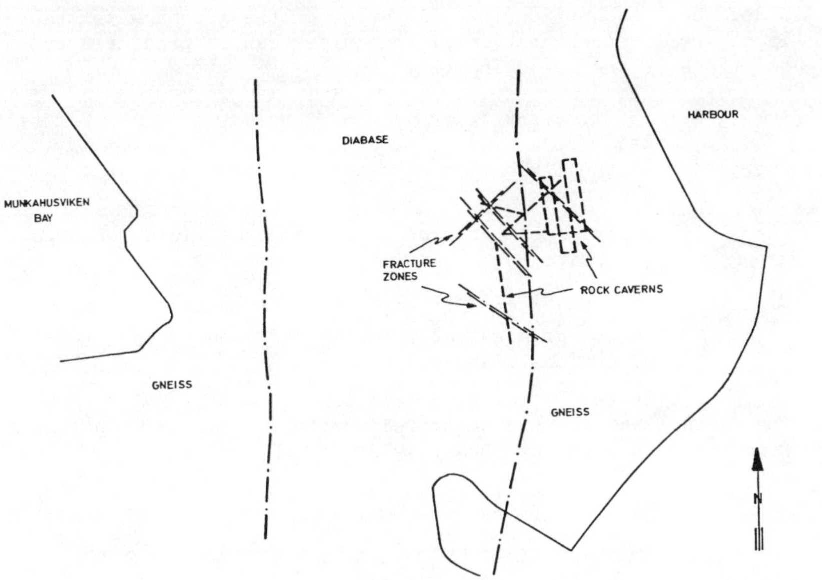


Figure 1 Geologic description of the vicinity near Karlshamn, Sweden

details of water table measurements are reported in Andersson, et al. (1981). Also provided in Andersson, et al. (1981) are the net infiltration from rainfall, its distribution, and a discussion of the physical boundaries to the study area.

## MODEL CONCEPTUALIZATION

The information obtained from the geologic inspection of the underground caverns is not sufficient in formulating a model which can describe ground water and contaminant movement in and around the cavern area. Information needs to be extrapolated into the surrounding region. Thus, the fracture zones noted in Figure 1 are extended along their orientations and additional fractures are added in keeping with the general pattern recognized from the mapping conducted in the caverns. The rock is also assumed to be applicable to a continuum representation similar to that used in the analysis of porous media. The make-up of the fracture zones and the highly interconnected nature of the cooling and stress fractures make this assumption possible. Furthermore, the ground water elevations are assumed to represent steady-state values and the infiltration rate is taken to be independent of seasonal fluctuations.

Under these assumptions the elevation of the ground water table in an areal representation of an aquifer can be predicted by solutions to the following equation:

$$\underset{\sim}{\nabla}\cdot\underset{\sim}{q} - \sum_{k=1}^{N} Q_k\, \delta(x-x_k)\, \delta(y-y_k) - S = 0 \tag{1}$$

where $\underset{\sim}{q}$ is the fluid flux over the saturated aquifer thickness,

$Q_k$ denotes point sources (or sinks) of fluid recharge (or pumpage),

$S$ is the net infiltration from rainfall,

$\underset{\sim}{\nabla}$ is the gradient operator in the areal space dimensions x and y, and

$\delta$ is the Dirac delta function.

The fluid flux over the saturated aquifer thickness in phreatic formations is given by (Bear, 1972),

$$\underset{\sim}{q} = -\tfrac{1}{2}\, \underset{\approx}{K}\cdot\underset{\sim}{\nabla}(h^2) \tag{2}$$

where $h$ is the average hydraulic head measured from the base of the aquifer, and

$\underset{\approx}{K}$ is the hydraulic conductivity tensor.

The corresponding equation of mass conservation for a chemically inert species in an areal representation of steady-state flow in a phreatic aquifer is the following:

$$\theta h \frac{\partial c}{\partial t} + \underset{\sim}{\nabla}\cdot(\underset{\sim}{q}c) - \underset{\sim}{\nabla}\cdot(\underset{\approx}{D}\cdot\underset{\sim}{\nabla}c) - \sum_{k=1}^{N} Q_k\, \bar{\bar{c}}_k\, \delta(x-x_k)\, \delta(y-y_k)$$

$$- S\,\bar{c} = 0 \tag{3}$$

where $\theta$ is porosity,

$c$ is species concentration,

$\bar{\bar{c}}_k$ is the concentration of injected fluid (or in the case of pumpage $\bar{\bar{c}}_k = c$),

$\bar{c}$ is the concentration of infiltrating water, and

$\underset{\approx}{D}$ is the dispersion tensor over the saturated aquifer thickness.

The dispersion tensor is defined for an areal representation of an aquifer by (Pinder and Gray, 1979),

$$D_{ij} = (D_d + \alpha_T\, |\underset{\sim}{q}|)\, \delta_{ij} + (\alpha_L - \alpha_T)\, \frac{q_i q_j}{|\underset{\sim}{q}|} \tag{4}$$

with $\alpha_L$ and $\alpha_T$ denoting the coefficients of longitudinal and transverse dispersivity, respectively, and $D_d$ the coefficient of molecular diffusion.

Spatially and temporally discretized representations of the previous equations are used to obtain the values of hydraulic head and species concentration. A Galerkin finite element formulation is employed in the discretization of the spatial domain while a finite difference approximation is applied to the time derivative of equation (3). The models of Pinder and Voss (1979) and Pinder and Gray (1979) are employed for this purpose. Quadrilateral elements having linear basis functions along each side are used in conjunction with a fully-implicit time derivative. In addition, an asymmetric weighting function is applied to the convective term of equation (3) as a means of eliminating oscillations in the species concentration solution.

Alternative representations of the geologic conditions in Karlshamn are to be expressed by different formulations of the finite element mesh. Two finite element grids are generated, each with the intent of simulating the existing ground water table and analyzing a hypothetical problem of contaminant movement near the cavern area. The representation of the fracture zones is to be expressed by the means with which the parameter values are assigned. For example, in Figure 2 the fracture zones are to be identified by large values of hydraulic conductivity placed at nodal locations. This provides an implicitly defined fracture zone in the vicinity of a row of nodes due to the interpolation between nodal values. Alternatively, the finite element mesh displayed in Figure 3 indicates the fracture zones by thin elements where values of hydraulic conductivity are defined on an element-wise basis. In both instances the hydraulic conductivity of the fracture zones is to be represented by directionally oriented values.

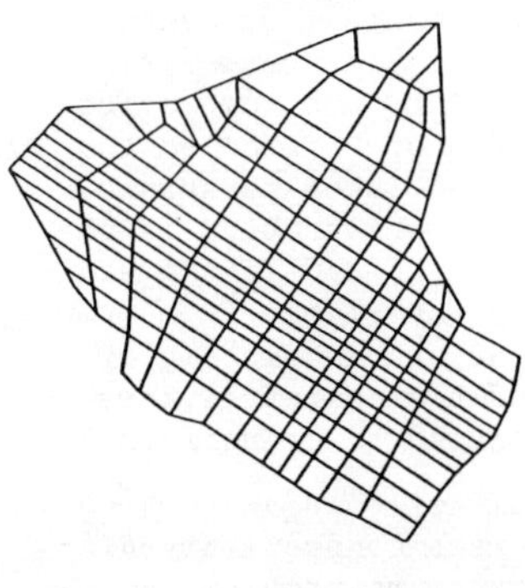

Figure 2 Mesh I - Finite element mesh of the study area

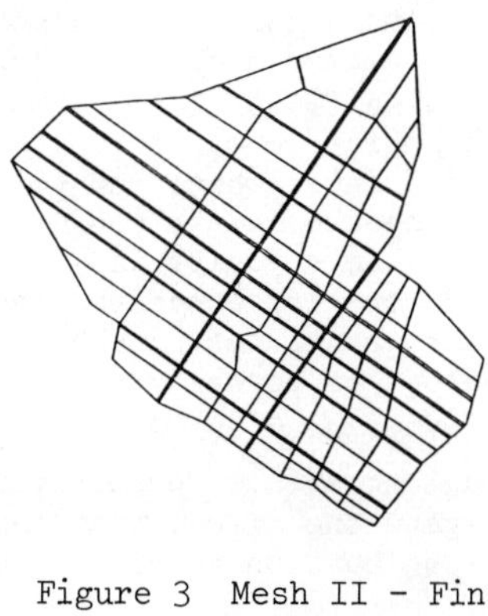

Figure 3 Mesh II - Finite element mesh of the study area

The distribution of hydraulic conductivity for each mesh is obtained by calibrating solutions to equation (1) with actual water level measurements. From these results, transient species concentration distributions are analyzed from the placing of a point source of known concentration in the vicinity of the cavern area.

## RESULTS

The hydraulic conductivity distribution employed in reproducing field values of hydraulic head showed large differences for the rock and fracture values that were used for each finite element mesh. Values employed for the finite element mesh displayed in Figure 2 were $1.5 \times 10^{-10}$ m/s for the rock and $1.5 \times 10^{-7}$ m/s for nodal values of fracture conductivity. The resulting hydraulic head solution is shown in Figure 4. For the finite element mesh pictured in Figure 3, element values of hydraulic conductivity range between $1.0 \times 10^{-5}$ m/s for the fracture zones and $7.5 \times 10^{-9}$ m/s for the rock, with the result being the hydraulic head distribution of Figure 5. The values stated above all lie within the range noted by other authors who have studied similar types of rock in the Karlshamn area (see e.g. Axelsson and Carlsson, 1980 and Ekman and Gentzschien, 1980).

The difference in the values used for each mesh signifies the non-unique nature of defining hydraulic conductivity when values over some area of the domain are not supplied. One can also imagine that alternative spatial distributions of hydraulic conductivity other than those used could have reproduced the results shown in Figures 4 and 5.

Hydraulic head values in the vicinity of the rock caverns are matched almost exactly between the two models. However, variations do exist in other areas. In magnitude these variations are not significant (approximately 1 - 2 m) but some effect will be noticable on the contaminant transport problem that is proposed. This concern is addressed in a later paragraph. Also of concern in the contaminant transport problem is the fact that the porosity and dipersivities needed in the evaluation of species concentration are not available from pump tests and tracer tests. Therefore, the two models employ the same values of porosity ($\theta = 0.01$) and the dispersivities ($\alpha_L = 100$ m, $\alpha_T = 10$ m), which are assumed to be uniform over the modelled area. The ultimate result of this assumption will investigate the effect that the different conductivity distributions have on the contaminant transport problem.

Results after 1.14 yrs and 4 yrs are displayed in Figures 6 and 7 for each of the finite element configurations. After 1.14 yrs there is a noticable difference in the extent of the

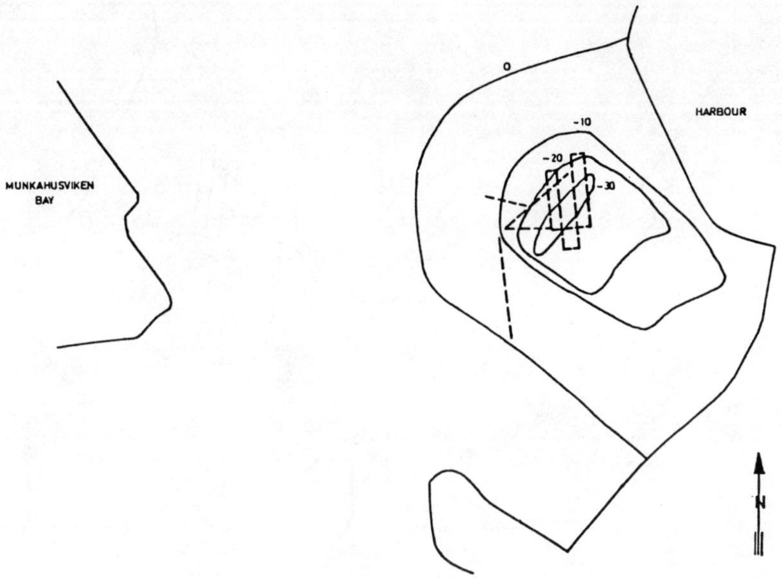


Figure 4  Steady-state water table elevations generated from Mesh I

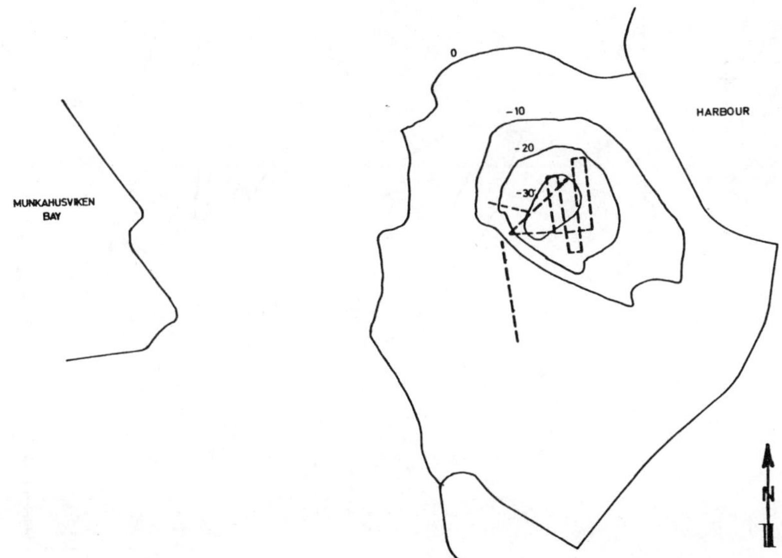


Figure 5  Steady-state water table elevations generated from Mesh II

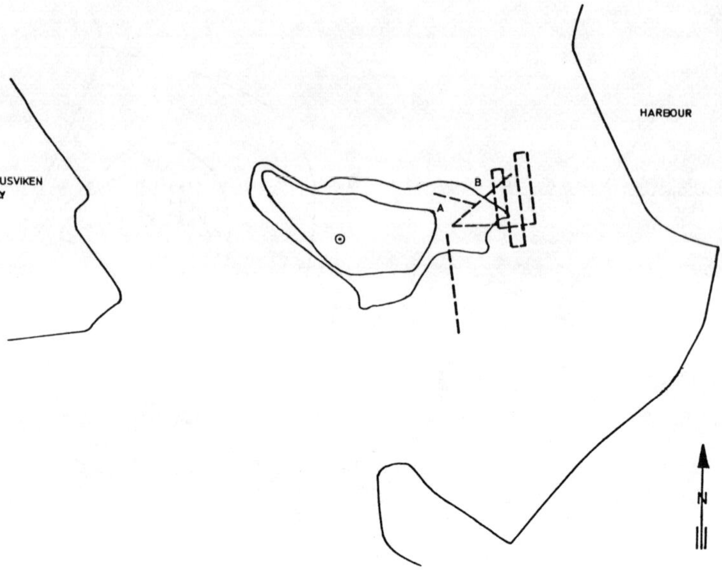


Figure 6  Five-percent iso-concentration contours at 1.14 yrs (A) and 4 yrs (B) generated from Mesh I

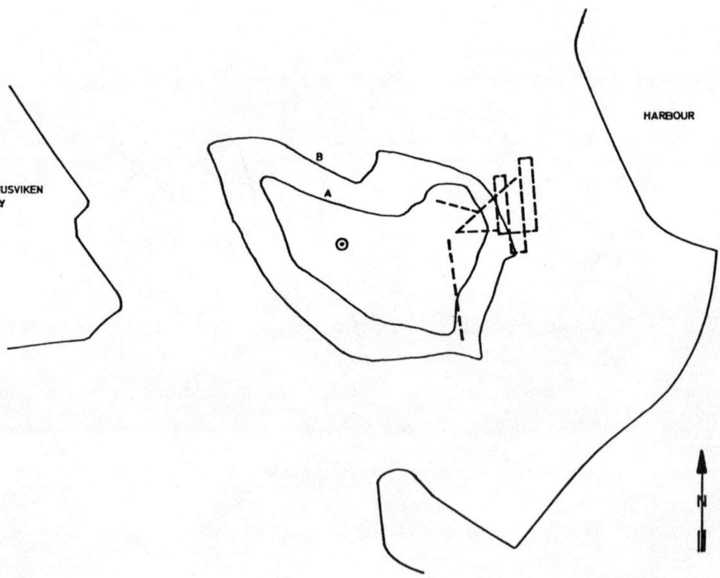


Figure 7  Five-percent iso-concentration contours at 1.14 yrs (A) and 4 yrs (B) generated from Mesh II

contaminant plume. In Figure 7 (mesh II) the five percent iso-concentration contour has already encroached upon the cavern area. A similar point is not reached using mesh I until almost three years of elapsed time. Because the hydraulic head distribution between the two models is essentially the same in the cavern area, the difference is attributed to the grid formulation and the choice of hydraulic conductivity values. The diffenence which occurs in the southeastern region of the modelled area is attributed to the differences in the hydraulic head distributions shown in Figures 4 and 5.

After four years the difference between the modelled results diminishes in the vicinity of the cavern area. This is the result of the stagnation point which is present and the fact that the flow lines in the two models are essentially the same. Thus, as a steady-state is approached the lines of iso-concentration will become similar in comparing each of the finite element representations.

## CONCLUSIONS

In modelling fractured rock the means by which one transfers physical information to a conceptual model has an effect on the results predicted, especially when the model is extended beyond the situation for which it was calibrated. This is illustrated here with the extension of a steady-state water balance to the study of contaminant transport. A similar result could also be expected if conditions of transient flow were considered.

In investigating the problem of contaminant transport, alternative hydraulic conductivity distributions yielded similar calibrated results for water table elevations but conflicting information with regard to contaminant movement. Since the same values for other model parameters were used in each of the model representations, one can argue that alternative methods of defining the fracture zones should also lead to different values of the other model parameters between each of the meshes. This undoubtedly is true, however, without field information one is unable to assess the uniqueness of the values of porosity and dispersivities between the two models.

## REFERENCES

Andersson, J., K. Cederwall and A. Shapiro, (1981) Rock cavern at Sutudden, Karlshamn (In Swedish), Dept. of Hydraulics Report No. 11, Royal Institute of Technology, Stockholm.

Axelsson, C. and L. Carlsson, (1980) Model calculations of groundwater conditions on the Sternö Peninsula, Sixth Nordic Hydrologic Conference, Uppsala University, Sweden.

Bear, J., (1972) Dynamics of Fluids in Porous Media, American Elsevier, New York.

Ekman, L. and B. Gentzschein, (1980) Additional work and summary of geohydrologic information in the Sternö area, Karlshamn (In Swedish), KBS Technical Report TR 80-01, Stockholm, Sweden.

Larsson, I., T. Lundgren and U. Wiklander, (1977) The Blekinge coastal gneiss, geology and hydrogeology (In Swedish), KBS Technical Report 25, Stockholm, Sweden.

Pinder, G. and W. Gray, (1979) Galerkin finite element models for aquifer simulation, Princeton University Water Resources Program, Research Report 76-WR-5, Princeton, New Jersey.

Pinder, G. and C. Voss, (1979) AQUIFEM, A finite element model for aquifer evaluation, Dept. of Water Resources Engineering Report No. 7911, Royal Institute of Technology, Stockholm.

# 13. TRANSPORT PROCESSES IN GROUNDWATER FLOW

# FINITE ELEMENT SOLUTIONS OF COUPLED GROUNDWATER FLOW AND TRANSPORT EQUATIONS UNDER TRANSIENT CONDITIONS INCLUDING THE EFFECT OF THE SELECTED TIME STEP SIZES

S. C. Anand and A. Pandit

Clemson University

## INTRODUCTION

The subject of mass transport of hazardous and/or potentially hazardous materials in the groundwater environment is one that has received wide attention, both in the recent past and in the present, as is evidenced by the large number of papers in the literature on this subject. Pinder and Frind (1972) were one of the first investigators to use the Galerkin's procedure in conjunction with the finite element technique to model the transport of salt water in aquifers. Pickens and Lennox (1976) developed models to simulate the transient two-dimensional movement of conservative wastes in steady-state, saturated groundwater flow systems. Huyakorn and Taylor (1976), on the other hand, compared three different formulations that could be used to solve the steady-state saltwater encroachment problem. They also illustrated that by using the 'subregion analysis' one could overcome the instability problems arising from high velocities near the seawater face for an idealized rectangular confined aquifer. Frind (1980) extended the solution of the transient saltwater intrusion in coastal aquifers by considering the presence of an aquitard above a fully confined aquifer that extends out under the sea. In most of these studies, either the results were presented for fixed values of the field parameters or the density variations due to the presence of contaminants were neglected, an assumption that is not valid as shown by the authors (1980).

The steady-state, coupled, two-dimensional finite element model developed by the authors (1980) is extended in this paper to incorporate time-dependent effects and a complete analysis to study the saltwater encroachment in coastal aquifers under transient conditions is conducted for a wide range of the field parameters. In particular, the influence of the selected time step sizes on computational effort is investigated.

## MATHEMATICAL FORMULATION

The mathematical model is formulated in terms of two equations, a flow equation and a time-dependent transport equation, which are coupled by the density and velocity terms. A brief description and necessary derivations of these equations are given as follows:

### Flow Equation

The flow equation in terms of the stream function, $\psi$, is given by

$$\frac{\partial^2\psi}{\partial x^2} \quad \frac{\partial^2\psi}{\partial y^2} = -\frac{1}{a}\frac{\partial C}{\partial x} , \qquad (1)$$

in which $\psi$ and C are the dimensionless stream function and saltwater concentration, respectively, and $a$ is the discharge parameter (Anand et al., 1980).

### Transport Equation

The general form of the convective-dispersion equation including the effects of hydrodynamic dispersion can be written as

$$D\left(\frac{\partial^2 C}{\partial x^2} + \frac{\partial^2 C}{\partial y^2}\right) - u\frac{\partial C}{\partial x} - v\frac{\partial C}{\partial y} = \frac{\partial C}{\partial t} \qquad (2)$$

where, $D(L^2t^{-1})$ is the coefficient of hydrodynamic dispersion, u and v are the horizontal and vertical components of the pore velocity and t is the time.

Using the dimensionless variables

$$x' = \frac{x}{d} , \; y' = \frac{y}{d} , \; u' = \frac{\partial\psi}{\partial y'} , \; v' = -\frac{\partial\psi}{\partial x'}$$
$$u' = \frac{u}{U} , \; v' = \frac{v}{U} , \; C' = \frac{C}{C_1} \qquad (3)$$

where U and d are the characteristic velocity and depth of the aquifer, respectively; and $C_1$ is the concentration of the seawater, and introducing another dimensionless variable

$$t' = \frac{tU}{d} , \qquad (4)$$

Eq. (2) can be nondimensionalized as

$$\frac{\partial^2 C'}{\partial x'^2} + \frac{\partial^2 C'}{\partial y'^2} = \frac{Ud}{D}\left[u'\frac{\partial C'}{\partial x'} + v'\frac{\partial C'}{\partial y'}\right] + \frac{Ud}{D}\frac{\partial C'}{\partial t'} \qquad (5)$$

or

$$\frac{\partial^2 C}{\partial x^2} + \frac{\partial^2 C}{\partial y^2} = Pe\left(\frac{\partial\psi}{\partial y}\frac{\partial C}{\partial x} - \frac{\partial\psi}{\partial x}\frac{\partial C}{\partial y}\right) + Pe\frac{\partial C}{\partial t} \qquad (6)$$

in which Pe = Ud/D is the Peclet number (primes in Eq. (6) have been removed for convenience).

## FINITE ELEMENT EQUATIONS BY GALERKIN'S METHOD

The finite element formulation of the flow equation for a triangular element with no internal or midside nodes, and assuming linear shape functions for the stream function, will remain exactly identical to that given by the authors (1980). The final form of the equations becomes

$$\iint_A \left( \frac{\partial\{N\}}{\partial x} \frac{\partial[N]}{\partial x} + \frac{\partial\{N\}}{\partial y} \frac{\partial[N]}{\partial y} \right) dA \; \{\psi\} = \int_\ell \{N\} \frac{\partial\psi}{\partial x} c_x \, d\ell + \int_\ell \{N\} \frac{\partial\psi}{\partial y} c_y \, d\ell + \frac{1}{\alpha} \iint_A \{N\} \frac{\partial[N]}{\partial x} dA \; \{C\} \; . \tag{7}$$

In Eq. (7), $[N] = [N_1 \; N_2 \; N_3]$ are the linear shape functions, $\{\psi\} = [\psi_1 \; \psi_2 \; \psi_3]^T$ and $\{C\} = [C_1 \; C_2 \; C_3]^T$ are the stream functions and concentrations at the nodal points 1,2 and 3 of the triangular element, respectively. In addition, $c_x$ and $c_y$ are the direction cosines at the boundaries.

Similarly, the transport equation, Eq. (6), can be easily reduced for a triangular element to the form

$$\left[ \iint_A \left( \frac{\partial\{N\}}{\partial x} \frac{\partial[N]}{\partial x} + \frac{\partial\{N\}}{\partial y} \frac{\partial[N]}{\partial y} \right) dA + Pe \iint_A \left( \{N\} \frac{\partial[N]}{\partial y} \{\psi\} \frac{\partial[N]}{\partial x} - \{N\} \frac{\partial[N]}{\partial x} \{\psi\} \frac{\partial[N]}{\partial y} \right) dA \right] \{C\} + \left[ Pe \iint_A \{N\}[N] \, dA \right] \frac{\partial\{C\}}{\partial t} = \int_\ell \{N\} \left( \frac{\partial C}{\partial x} c_x + \frac{\partial C}{\partial y} c_y \right) d\ell \tag{8}$$

which is exactly similar to the finite element form of the transport equation derived for the steady-state case by the authors (1980) except for the additional second term on the left hand side associated with the concentration rate.

The flow and transport equations, Eqs. (7) and (8), respectively, for an element can also be written symbolically in the matrix form as

$$[k]\{\psi\} = \{f\}, \quad \text{and} \tag{9}$$

$$[r]\{C\} + [s] \left\{ \frac{\partial C}{\partial t} \right\} = \{q\} \; , \tag{10}$$

in which matrices [k], [r], [s] and vectors {f}, {q} and $\{\frac{\partial C}{\partial t}\}$ are of the size (3 x 3) and (3 x 1), respectively.

The corresponding equations for the total system can then be expressed as

$$[K]\{\psi\} = \{f\}, \quad \text{and} \tag{11}$$

$$[R]\{C\} + [S] \left\{ \frac{\partial C}{\partial t} \right\} = \{Q\} \tag{12}$$

in which [K], [R] and [S] are (n x n) matrices and $\{\psi\}$, {F}, {Q}, {C} and $\{\partial C/\partial t\}$ are vectors of size n, where n represents the total number of nodes in the finite element mesh. Equation (12) will be further modified by utilizing any one of the temporal finite difference schemes.

### Temporal Finite Difference Approximation

Three weighted average approximation finite difference schemes, namely the explicit, the implicit, and Crank-Nicolson Method, can be employed for the solution of the transient Eq. (12).

Using the implicit method, that has been reported to give stable results, Eq. (12) transforms to

$$([R] + (\frac{1}{\Delta t})[S])\{C\}_{t+\Delta t} = (\frac{1}{\Delta t})[S]\{C\}_t + \{Q\}_{t+\Delta t} \qquad (13)$$

where $\Delta t$ is the size of the time step.

Equations (11) and (13) represent a coupled, nonlinear, transient system and are solved by an iterative technique, once the initial and boundary conditions are known.

SOLUTION PROCEDURE

Using the concentration value defined at each nodal point by the initial condition, Eq. (11) is solved for the stream function $\psi_o$, at time $t_o$. These values of the stream function, $\psi_o$, are then used in Eq. (13) to solve for concentrations, $C_1^o$, at time $\Delta t$. Concentrations, $C_1^o$, are now used in Eq. (11) to obtain new values of the stream function, defined as $\psi_1^1$, which are utilized in Eq. (13), with the initial conditions, to yield an updated set of concentration values, $C_1^1$. If the difference between $C_1^1$ and $C_1^o$ is found to be less than a small prescribed value, $\varepsilon$, the iterative process for this time step is stopped and one proceeds to the second time step. Otherwise, $C_1^1$ are utilized once again in Eq. (11) to calculate new values of the stream function, $\psi_1^2$, that are in turn used in Eq. (13) to obtain $C_1^2$. This iterative procedure is continued until convergence is reached, which yields the values for the stream function and concentration after the first time step.

To obtain a solution for any time step m, the iterative procedure described above is repeated except that concentrations found in step (m-1) are used in the first term on the right hand side of Eq. (13).

PROBLEM DESCRITION

The problem under consideration is a study of the time dependent saltwater encroachment in a homogeneous, isotropic, confined, rectangular, coastal aquifer. The analysis is carried out for region sizes with varying length to depth ratios. A region size with length to depth ratio of 2:1 is shown in Fig. 1 along with the required boundary conditions.

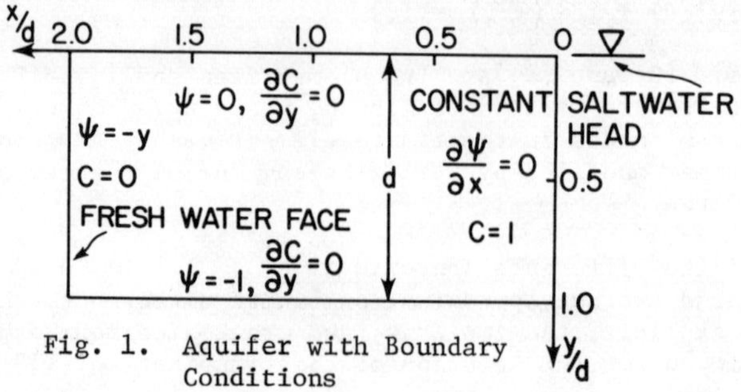


Fig. 1. Aquifer with Boundary Conditions

| Boundary Conditions For Stream Function | For Saltwater Concentration | Initial Conditions |
|---|---|---|
| $\psi\ (\ell, y, t) = -y$ | $C(\ell, y, t) = 0$ | $C(0, y, 0) = 1$ |
| $\psi\ (x, 0, t) = 0$ | $\frac{\partial C}{\partial y}(x, 0, t) = 0$ | $C(x, y, 0) = 0$ |
| $\frac{\partial \psi}{\partial x}\ (0, y, t) = 0$ | $C(0, y, t) = 1$ | |
| $\psi\ (x, 1, t) = -1$ | $\frac{\partial C}{\partial y}(x, 1, t) = 0$ | |

## RESULTS AND DISCUSSION

### Solutions for Various Region Sizes with Uniform Time Steps

Considerations for region size selection as well as the choice of the appropriate mesh size have been discussed in detail by the authors in an earlier paper (1980). Region sizes with length to depth ratios of 10 to 1, 4 to 1 and 3 to 1 were found to be appropriate for low, intermediate, and high values of the field parameters, respectively. Three combinations of Pe and $a$, with respective region sizes, are defined as Cases 1, 2 and 3, and are shown in Table 1. In an effort to obtain transient solutions for various combinations of the field parameters, using the aforementioned region sizes, analyses were performed with various uniform time step sizes, $\Delta t$, ranging from 0.04 to 0.32, as shown in Table 1. In each case, enough time increments are utilized so that the total elapsed time is reasonably close to T*, the time required to obtain the steady state solution.

It can be observed from Table 1 that, for Pe = 1 and $a = 0.01$, 192 iterations are required for 160 time increments of size 0.04 each, whereas 136 iterations are needed for 80 time steps of size 0.08 yielding an average of 1.2 and 1.7 iterations per time step, respectively. This shows that the average number of iterations per time step increases with an increase in the value of $\Delta t$. A similar observation can also be made for both Pe = 100 and 200. However, this does not imply that the total number of iterations to obtain the final solution will also increase with increasing $\Delta t$. Infact a comparison, for each case, leads to the conclusion that $\Delta t = 0.08$ yields the most economical solution for all three combinations of the field parameters. The respective CPU times are also given in Table 1.

Although one expects the finite element models with larger number of nodal points to require more CPU time per iteration due to greater number of unknowns involved in the solutions, that is not the case for the three problems discussed in Table 1. It may be noted that for $\Delta t = 0.08$ the average CPU time of 1.38 seconds utilized per iteration for Case 1 is less than the average CPU time of 1.72 seconds for Case 2, even though the number of nodal points used in the respective models are 331 and 141. A similar observation can be made by comparing the number of nodal points and the average CPU time per iteration for Cases 1 and 3, respectively. This anomaly may be

Table 1. Various Parameters and Results for Problems Solved With Uniform Time Steps

| Case | Field Parameters | | Region Size | | T* | Time Step $\Delta t$ | Total Time Steps | Total Elapsed Time | CPU** Time (Secs.) | Total No. of Itera-tion | Avg. CPU Time Per Iteration (Secs.) | No. of Iterations Per Time Step |
|---|---|---|---|---|---|---|---|---|---|---|---|---|
| | Pe | $\alpha$ | Length-Depth Ratio | No. of Nodal Points | | | | | | | | |
| 1 | 1 | 0.01 | 10:1 | 331 | 6.5 | 0.04 | 160 | 6.4 | 286 | 192 | 1.49 | 1.20 |
| | | | | | | 0.08 | 80 | 6.4 | 188 | 136 | 1.38 | 1.70 |
| | | | | | | 0.16 | Solution does not converge | | | | | |
| 2 | 100 | 0.1 | 4:1 | 141 | 3.2 | 0.04 | 80 | 3.2 | 203 | 105 | 1.93 | 1.31 |
| | | | | | | 0.08 | 40 | 3.2 | 141 | 82 | 1.72 | 2.05 |
| | | | | | | 0.16 | 20 | 3.2 | 155 | 97 | 1.60 | 4.85 |
| | | | | | | 0.32 | Solution does not converge | | | | | |
| 3 | 200 | 0.2 | 3:1 | 279 | 3.2 | 0.04 | 80 | 3.2 | 375 | 115 | 3.26 | 1.44 |
| | | | | | | 0.08 | 40 | 3.2 | 273 | 91 | 3.00 | 2.27 |
| | | | | | | 0.16 | 20 | 3.2 | 270 | 90 | 3.00 | 4.50 |
| | | | | | | 0.32 | Solution does not converge | | | | | |

*Approximate time required to attain steady state solution.

**On IBM 370/3033

explained by the fact that pivoting in the iteration process is required more often for higher values of Peclet numbers, thus necessitating additional computational effort per iteration. It should also be mentioned here that the iterative process does not converge to yield a solution if the selected time step sizes are too large. As can be seen in Table 1, no solutions are possible if $\Delta t = 0.16$ for Case 1 and equal to 0.32 for Cases 2 and 3.

It should be emphasized that all parameters, including time, in these analyses are nondimensionalized as defined in Eq. (3). Therefore, any two nondimensional time steps of equal magnitude for different field parameters could have vastly different sizes in real time.

The gradual movement of the saltwater front for various cases considered in Table 1 is presented in Fig. 2 for a uniform time step size $\Delta t = 0.08$.

## Analysis With Variable Time Steps

If one appreciates the fact that the movement of the saltwater front is far more rapid in the beginning than near the steady-state, as observed in the previous section, it appears logical to use a time step scheme in which the size of $\Delta t$ corresponds to the encroachment rate of the saltwater. Thus, several schemes of gradually increasing time step sizes are utilized to determine the progress of the seawater for the three cases discussed in the previous section. These schemes, entitled A to H and shown in Table 2, are designed in such a manner that the value of $\Delta t$ is doubled after every five (an arbitrarily chosen number) steps until the total elapsed time is reasonably close to $T^*$, the approximate time required to reach the steady-state solution. Schemes A to D, for which the total elapsed time is close to $T^* = 6.5$, are used for $Pe = 1$, while Schemes E to H, in which the total elapsed time is close to $T^* = 3.2$, are used for $Pe = 100$ and $Pe = 200$. It may be noted that in Schemes E to H, the size of $\Delta t$ is not incremented to 0.32 once it attains a value of 0.16. This is done because the iterative process does not converge to a solution for $Pe = 100$ and $Pe = 200$ when $\Delta t = 0.32$. Transient solutions were obtained for the three combinations of the field parameters using the various time step schemes described above. The details of these solutions are summarized in Table 3.

The optimum solution scheme for any case is the one that requires the minimum total number of iterations to reach the steady-state. It can be seen from Table 3 that all of the Schemes A through D require approximately the same total number of iterations and computer time for Case 1. Similarly, for Case 2, Schemes E through H are equally good as far as computer time requirements are concerned. However, for Case 3, the large size assigned to the initial time step in Scheme H leads to considerable savings of computational effort as compared to Schemes E and F.

It may also be observed that, for Case 1, the number of total time steps required to obtain the final solution decreases

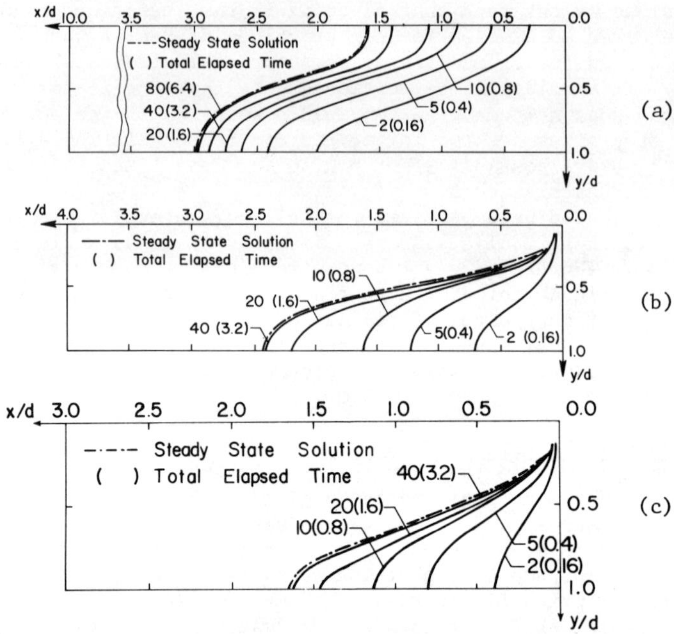


Fig. 2. 0.5 Isochlores at Various Time Steps with $\Delta t = 0.08$
(a) $Pe = 1$, $\alpha = 0.01$, (b) $Pe = 100$, $\alpha = 0.1$, (c) $Pe = 200$, $\alpha = 0.2$

from 35 to 20 for Schemes A to D, respectively, due to the relatively large value of the initial time step. However, a corresponding reduction of the required computer time does not occur. While 85 seconds of CPU time are required for Scheme A, 99 seconds are needed for Scheme D. A similar increase in terms of the total number of iterations, from 59 to 78, can also be observed. This phenomenon may be explained by realizing the fact that many more iterations are necessary during the initial time steps if the size of the selected $\Delta t$ is large, thus more than offsetting the savings accrued through the use of fewer time steps.

It is again noted that there is necessarily no direct relationship between the number of nodal points used in a finite element model and the average CPU time needed per iteration, due to the fact that the pivoting requirements for various combinations of the field parameters are different. It has been observed in the present analysis that pivoting is more frequently required for higher values of Peclet numbers, thus, requiring larger CPU time per iteration for these cases. This may be readily observed by comparing the respective values of the number of nodal points and the average CPU time per iteration for Cases 1 and 2 as well as Cases 1 and 3 in Table 3.

Table 2. Variable Time Step Schemes

| | Scheme A | | | Scheme B | | | Scheme C | | | Scheme D | | |
|---|---|---|---|---|---|---|---|---|---|---|---|---|
| | Time Step Nos. | $\Delta t$ | Total Elapsed Time, t | Time Step Nos. | $\Delta t$ | Total Elapsed Time, t | Time Step Nos. | $\Delta t$ | Total Elapsed Time, t | Time Step Nos. | $\Delta t$ | Total Elapsed Time, t |
| Pe = 1, $\alpha$ = 0.01 | 1–5 | 0.01 | 0.05 | 1–5 | 0.02 | 0.10 | 1–5 | 0.04 | 0.20 | 1–5 | 0.08 | 0.40 |
| | 6–10 | 0.02 | 0.15 | 6–10 | 0.04 | 0.30 | 6–10 | 0.08 | 0.60 | 6–10 | 0.16 | 1.20 |
| | 11–15 | 0.04 | 0.35 | 11–15 | 0.08 | 0.70 | 11–15 | 0.16 | 1.40 | 11–15 | 0.32 | 2.80 |
| | 16–20 | 0.08 | 0.75 | 16–20 | 0.16 | 1.50 | 16–20 | 0.32 | 3.00 | 16–20 | 0.64 | 6.00 |
| | 21–25 | 0.16 | 1.55 | 21–25 | 0.32 | 3.10 | 21–25 | 0.64 | 6.20 | | | |
| | 26–30 | 0.32 | 3.15 | 26–30 | 0.64 | 6.30 | | | | | | |
| | 31–35 | 0.64 | 6.35 | | | | | | | | | |
| | Scheme E | | | Scheme F | | | Scheme G | | | Scheme H | | |
| Pe = 100, $\alpha$ = 0.1<br>Pe = 200, $\alpha$ = 0.2 | 1–5 | 0.01 | 0.05 | 1–5 | 0.02 | 0.10 | 1–5 | 0.04 | 0.20 | 1–5 | 0.08 | 0.40 |
| | 6–10 | 0.02 | 0.15 | 6–10 | 0.04 | 0.30 | 6–10 | 0.08 | 0.60 | 6–22 | 0.16 | 3.12 |
| | 11–15 | 0.04 | 0.35 | 11–15 | 0.08 | 0.70 | 11–26 | 0.16 | 3.16 | | | |
| | 16–20 | 0.08 | 0.75 | 16–30 | 0.16 | 3.10 | | | | | | |
| | 21–35 | 0.16 | 3.15 | | | | | | | | | |

Table 3. Various Parameters and Results for Problems Solved with Variable Time Steps

| Case | Field Parameters | | Region Size | | T* | Variable Time Steps Scheme | Total Time Steps | Total Elapsed Time | CPU** Time (Secs.) | Total No. of Iterations | Avg. CPU Time Per Iteration (Secs.) |
|---|---|---|---|---|---|---|---|---|---|---|---|
| | Pe | $\alpha$ | Length-Depth Ratio | No. of Nodal Points | | | | | | | |
| 1 | 1 | 0.01 | 10:1 | 331 | 6.5 | A | 35 | 6.35 | 85 | 59 | 1.44 |
| | | | | | | B | 30 | 6.30 | 81 | 57 | 1.42 |
| | | | | | | C | 25 | 6.20 | 86 | 64 | 1.34 |
| | | | | | | D | 20 | 6.00 | 99 | 78 | 1.27 |
| 2 | 100 | 0.1 | 4:1 | 141 | 3.2 | E | 35 | 3.15 | 154 | 87 | 1.77 |
| | | | | | | F | 30 | 3.10 | 151 | 86 | 1.75 |
| | | | | | | G | 26 | 3.16 | 144 | 85 | 1.70 |
| | | | | | | H | 22 | 3.12 | 169 | 97 | 1.74 |
| 3 | 200 | 0.2 | 3:1 | 279 | 3.2 | E | 35 | 3.15 | 343 | 121 | 2.83 |
| | | | | | | F | 30 | 3.10 | 317 | 113 | 2.80 |
| | | | | | | G | 26 | 3.16 | 277 | 96 | 2.89 |
| | | | | | | H | 22 | 3.12 | 245 | 88 | 2.76 |

*Approximate time required to attain steady state solution.

**On IBM 370/3033.

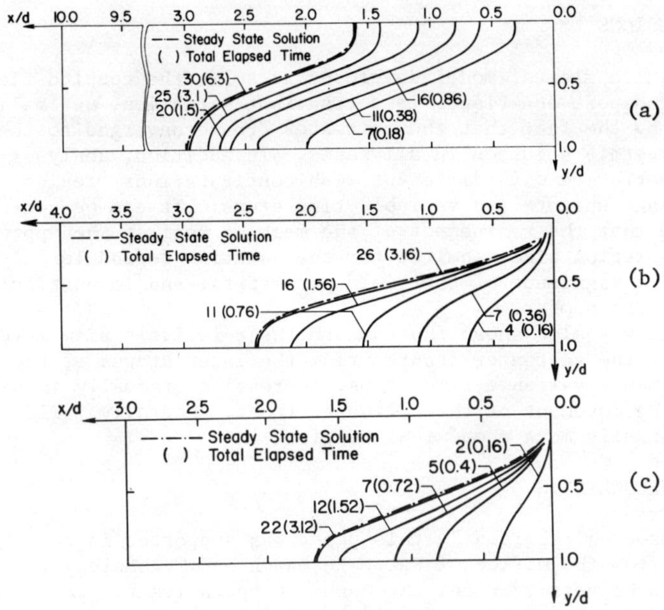


Fig. 3. 0.5 Isochlores at Various Times. (a) Case 1, Scheme B; (b) Case 2, Scheme G; (c) Case 3, Scheme H.

Comparison of The Results Using Uniform and Variable Time Steps

The transient progress of the saltwater front for all three cases, using Schemes B, G and H are shown in Fig. 3. In order to provide a visual comparison with the results obtained by using uniform time steps, 0.5 isochlors have been plotted in this figure, at times that are reasonably close to the ones shown in Fig. 2.

A comparison of the results for all three cases using uniform and variable time steps, respectively, indicates that the positions of the isochlors for any given elapsed time are virtually identical. This indicates that the size of the time step utilized does not have an appreciable influence on the accuracy of the solution. On the other hand, a comparison of the total number of iterations required to attain the final solution for Case 1 shows that while 136 iterations are needed (see Table 1) with the most efficient uniform time step of 0.08, only 57 iterations are required (see Table 3) if the best available variable time step Scheme B is used. This results in over 100 percent reduction in terms of the needed CPU time for the solution (from 188 seconds to 81 seconds). A somewhat smaller reduction in CPU time of 18.5 percent occurs for Case 3 with variable time steps, whereas computational time, as well as total number of iterations, for Case 2 remain virtually the same for both the uniform and variable time steps.

## CONCLUSIONS

The finite element model developed to solve the coupled flow and transport equations, in a transient situation, was validated by the fact that the saltwater front converged to the steady-state solution in all cases. In addition, analyses were performed with different mesh configurations, region sizes and uniform and variable time steps. It can be concluded that the refinement of the mesh as well as the appropriate region size required for the solution depend largely upon the magnitude of the field parameters, and in particular the Peclet number.

It was also shown that due to the relatively slow movement of the saltwater front during the later stages of the encroachment, variable time steps, increasing gradually in size with the movement of the saltwater front, generally yield computationally more economical results.

## ACKNOWLEDGEMENTS

The research reported in this paper was supported in part by a grant from the Office of Water Research and Technology to Clemson University under the Annual Cooperative Program. This financial support is gratefully acnowledged.

## REFERENCES

Anand, S.C., Pandit, A., and Sill, B.L. (1980) Some Considerations in Finite Element Solutions to Coupled Groundwater Flow and Convective-Dispersion Equations. Proc. 3rd. Int. Conf. on Finite Elements in Water Res., Univ. of Mississippi, Miss.: 3.97-3.107.

Frind, E.O. (1980) Seawater Intrusion in Continuous Coastal Aquifer-Aquitard Systems. Proc. 3rd. Int. Conf. on Finite Elements in Water Res., Univ. of Mississippi, University, Miss.: 2:177-2:198.

Huyakorn, P. and Taylor, C. (1976) Finite Element Models for Coupled Groundwater Flow and Convective Dispersion. Proc. Ist. Int. Conf. on Finite Elements in Water Res., Princeton University, Princeton, NJ.: 1:131-1:151.

Pickens, J.F. and Lennox, W.C. (1976) Numerical Simulation of Waste Movement in Steady Groundwater Sistems, Water Res. Research, 12,2:171-180.

Pinder, G.F. and Frind, E.O. (1972) Application of Galerkin's Procedures to Aquifer Analysis, Water Res., Research, 8,1:108-120.

# NUMERICAL ESTIMATION OF THE SOLUTION OF THE DIFFUSION EQUATION WITH FIXED DIRICHLET-TYPE BOUNDARY CONDITIONS

JERZY CHOŁAST,ANDRZEJ NIEMIEC

Academy of Agriculture,Inst.of Applied Mathem.
ul Grunwaldzka 53, Wrocław, Poland

## §1 Introduction

Since 1956 the problem of infiltration has been modeled by differential equation with boundary conditions of the Dirichlet type.Till today full analitical solution of this problem is not know. To obtain quantative analysis of this problem the asymptotic methods are used. The first qualitative results find in 1974 Peletier and Atkinson /1/ . In that paper the algorithm to obtain upper and lower limit function was presented.In this paper the authors generalised the results of Peletier and Atkinson which constitied the ground for constructing a computer program. This program can be used for all diffusivity functions greater or equal to zero, and for funkction given in the form of the table of their values. This is very usefull in applications. If a certain condition for diffusivity function is satisfied, then we obtain the condition for the front of viscosivity .

§2 Formulation of the problem

We assumed, that the porous medium is isotropic, homogenous, the processes are isotermic, the chemical concentration does not exist. We also assumed that the influence of gravitation can be neglected. This last assumption is satisfied in heavy soil.Above assumptions leads us to the following initial-boundary problem:

1. $\Theta_t = (D(\Theta)\Theta_x)_x$ for $x,t>0$

1´. $\Theta(x,0) = \Theta_0$ =const for $x>0$

$\Theta(0,t) = \Theta_1$ =const for $t>0$

where $\Theta$ humidity for unit volume of soil $/m^3 H_2O m^{-3}$ of soil /, $D(\cdot)$ -diffusivity of soil relativly to the water $/m^2 s^{-1}$ / ,x-depth of the soil, t-time, the lower indexes t and x denotes the partial differentiation.Searching to the solution of the problem 1. with the conditions 1´ in the automorphic form our problem can be formulated as ordinary nonlinear differential equation:

2. $-s f' = (D^*(f) f)'$ $s>0$ a.e.,

with boundary conditions:

2´ $f(0) = 1$

$\int_0^\infty f(s)ds = \lim_{s\to 0+} D(f)f'$

where $s = x/\sqrt{2t}$ , $f(s) = (\Theta(x,t) - \Theta_0)/(\Theta_1 - \Theta_0)$ ,

$D^*(f) = D((\Theta_1 - \Theta_0) f + \Theta_0)$ and "`" denotes differentiation related to variable s.

Atkinson and Peletier proved, that the problem 2-2′ is equivalent to the integral equation:

3. $$u(x) = 2\int_x^1 \frac{D^*(v)}{v + \int_0^v D^*(w)/u(w)dw} dv \ , \quad 0 \leq x \leq 1 \ ,$$

where $u(x) = [f^{-1}(x)]^2$ / $f^{-1}$ converse function for f /. In the following part of this paper we write for convention $f^{-1}(s) = s(f)$ . The equation 3 can be writen as:

3′. $$u(f) = 2\int_f^1 \frac{D^*(v)}{v + \int_0^v D^*(w)/u(w)dw} dv$$

It can be shown, that the solution of equations 3-3′ is limited by two sequences of functions $\{v_n\}$ and $\{w_n\}$:

4. $$v_n(f) \leq u^2(f) \leq w_m(f) \quad n,m = 0,1,2,\ldots$$

where :

5. $$v_0(f) = \left[\int_f^1 D^*(v)dv\right]^2 / \int_0^1 D^*(v)\,dv,$$
$$w_0(f) = 2\int_f^1 D^*(v)/v\,dv \quad ,$$

and

6. $$v_{n+1}(f) = \mathcal{L}\, v_n(f)$$
$$w_{m+1}(f) = \mathcal{L}\, w_m(f), \quad n,m = 0,1,2,\ldots \quad ,$$

here $\mathcal{L}$ is the integral operator given by:

7. $$\mathcal{L}\, g(x) = 2\int_x^1 \frac{D^*(z)}{z + \int_0^z D^*(y)/g(y)dy} dz \quad , 0 \leq x \leq 1$$

defined for functions g motincreasing and greater or equal to 0.

The proof of formula 4. is given by the assumption,

that the integral $\int_0^1 D^*(f)/f\,df$ is infinite. The authors proofed the formula 4. without this assumption.

It is easy to see, that the humidity along whole soil profil is given by:

8. $$x(\Theta,t_o) = 2t_o\, u((\Theta-\Theta_o)/(\Theta_1-\Theta_o))$$

for $0 \leq \Theta_o \leq \Theta \leq \Theta_1 \leq 1$, where $x=x(\Theta,t_o)$ is converse function to $\Theta=\Theta(x,t_o)$, for given $t=t_o$ .

From 4. we obtain the estimation for $x(\Theta,t_o)$:

9. $$\sqrt{2t_o}\sqrt{v_n\left(\frac{\Theta-\Theta_o}{\Theta_1-\Theta_o}\right)} \leq x(\Theta,t_o) \leq \sqrt{2t_o}\sqrt{w_m\left(\frac{\Theta-\Theta_o}{\Theta_1-\Theta_o}\right)}$$

$n,m=0,1,\ldots$ .

The velocity of flow on the boundary of soil profil at $t=t_o$ is given by:

10. $$i(t_o) = D^*(\Theta)\Theta_x\big|_{x=0} = -(\Theta_1-\Theta_o)\int_0^1\sqrt{u(f)}\,df\; 1/\sqrt{2t_o}$$

During the diffusion process the amount of water, wich goes to the soil is given by:

11. $$M(t_o) = \int_0^t i(t)\,dt = (\Theta_1-\Theta_o)\int_0^1\sqrt{u(f)}\,df\,(2t_o)^{1/2}$$

If the integral $\int_0^\infty D^*(f)/f\,df < \infty$ , then the front of wiscosity is finite, and the distance of it is given as:

12. $$x_o(t_o) = x(0,t_o) = \sqrt{2t_o}\sqrt{\int_0^1 u(f)\,df}\ ,$$

and can be estimated by formula:

13. $$\sqrt{2t_o}\sqrt{\int_0^1 v_n(f)\,df} \leq x_o(t_o) \leq \sqrt{2t_o}\sqrt{\int_0^1 w_m(f)\,df}\ ,$$

$n,m=0,1,2,\ldots$ .

The computer program was written in ALGOL 1900 / for ICL 1900 / .

## §3 Numerical results.

In the tabels are given the numerical results for diffusivity functions $D^*(f) = \beta \exp(\alpha f)$, $D^*(f) = f$, and $D^*(f) = f^{\alpha}/(1-f)^{\beta}$. This functions are good known in literature. For the function $D^*(f) = 1/2\, f\,(1-f)$ the exact solution is known, $f(s) = 1-s$. The numerical error for this function is less then 1%/for 20 iterations and the partition of $[0,1]$ for 200 points /

The diffusivity function $D^*(f) = 0$ iff $f < 1/2$, and 1 iff $f \geq 1/2$ was considered as a test of application of this method for not-continous functions. The numerical results shows, that the solution of the problem 2-2′ could not be continous function.

## §4 Colloraries

1. The presented method can be used for all diffusivity functions $D^*(f)$, integrable . $L^1(0,1)$ in particular for the function $D^*(f)$ given in the form of the table of their values.
2. This method permitted to find the upper and lower estimation of the solution of the problem 1-1′ for all $t, x > 0$.
3. If the integral 7. is finite , then this method gives us the estimation of the front of viscosivity.
4. By the integration of the lower and upper estimation we obtain the estimator of the intensivity of infiltration on the boundary.

## Tabble 1

Number of iterations=20 ,Partition of 0,1 interval=200

$D\,f = \exp(\alpha f)$

| F | D(F) | V(NMAX,F) | W(NMAX,F) |
|---|---|---|---|
| 0.0000 | 0.00674 | 0.00000 | 0.00000 |
| 0.0500 | 0.00865 | 7.14645 | 7.18217 |
| 0.1000 | 0.01111 | 7.00865 | 7.04502 |
| 0.1500 | 0.01426 | 6.90323 | 6.94011 |
| 0.2000 | 0.01832 | 6.80602 | 6.84338 |
| 0.2500 | 0.02352 | 6.70825 | 6.74610 |
| 0.3000 | 0.03020 | 6.60476 | 6.64313 |
| 0.3500 | 0.03877 | 6.49146 | 6.53041 |
| 0.4000 | 0.04979 | 6.36438 | 6.40397 |
| 0.4500 | 0.06393 | 6.21920 | 6.25953 |
| 0.5000 | 0.08208 | 6.05091 | 6.09210 |
| 0.5500 | 0.10540 | 5.85351 | 5.89571 |
| 0.6000 | 0.13534 | 5.61963 | 5.66300 |
| 0.6500 | 0.17377 | 5.34097 | 5.38481 |
| 0.7000 | 0.22313 | 5.00326 | 5.04959 |
| 0.7500 | 0.28650 | 4.59451 | 4.64264 |
| 0.8000 | 0.36788 | 4.09572 | 4.14503 |
| 0.8500 | 0.47237 | 3.48053 | 3.53219 |
| 0.9000 | 0.60653 | 2.71975 | 2.77176 |
| 0.9500 | 0.77880 | 1.77313 | 1.82067 |
| 1.0000 | 1.00000 | 0.00000 | 0.00000 |

$D(f) = f^{\alpha}/(1-f)^{\beta}$

| F | D(F) | V(NMAX,F) | W(NMAX,F) |
|---|---|---|---|
| 0,0050 | 0.07160 | 85.87127 | 85.87127 |
| 0,0500 | 0.25420 | 85.86745 | 85.86745 |
| 0,1000 | 0.41152 | 85.86484 | 85.86484 |
| 0,1500 | 0.58143 | 85.86253 | 85.86253 |
| 0,2000 | 0.78125 | 85.86027 | 85.86027 |
| 0,2500 | 1.02640 | 85.85795 | 85.85795 |
| 0,3000 | 1.33603 | 85.85546 | 85.85546 |
| 0,3500 | 1.73680 | 85.85273 | 85.85273 |
| 0,4000 | 2.26805 | 85.84964 | 85.84965 |
| 0,4500 | 2.99020 | 85.84607 | 85.84607 |
| 0,5000 | 4.00000 | 85.84183 | 85.84183 |
| 0,5500 | 5.45946 | 85.83664 | 85.83665 |
| 0,6000 | 7.65466 | 85.83009 | 85.83010 |
| 0,6500 | 11.12466 | 85.82149 | 85.82149 |
| 0,7000 | 16.97250 | 85.80963 | 85.80963 |
| 0,7500 | 27.71281 | 85.79218 | 85.79219 |
| 0,8000 | 50.00000 | 85.76416 | 85.76416 |
| 0,8500 | 105.79894 | 85.71263 | 85.71263 |
| 0,9000 | 300.00000 | 85.59310 | 85.59311 |
| 0,9500 | 1743.55958 | 85.12581 | 85.12581 |
| 0,9950 | 564269.44203 | 60.81657 | 60.81658 |
| 1,0000 | 1.00000 | 0.00000 | 0.00000 |

$D(f) = f$

| F | D(F) | V(NMAX,F) | W(NMAX,F) |
|---|---|---|---|
| 0.0050 | 0.00500 | 1.14248 | 1.15149 |
| 0.0500 | 0.05000 | 1.10283 | 1.11215 |
| 0.1000 | 0.10000 | 1.05795 | 1.06763 |
| 0.1500 | 0.15000 | 1.01215 | 1.02221 |
| 0.2000 | 0.20000 | 0.96538 | 0.97584 |
| 0.2500 | 0.25000 | 0.91760 | 0.92847 |
| 0.3000 | 0.30000 | 0.86874 | 0.88004 |
| 0.3500 | 0.35000 | 0.81873 | 0.83049 |
| 0.4000 | 0.40000 | 0.76752 | 0.77973 |
| 0.4500 | 0.45000 | 0.71501 | 0.72770 |
| 0.5000 | 0.50000 | 0.66112 | 0.67430 |
| 0.5500 | 0.55000 | 0.60578 | 0.61944 |
| 0.6000 | 0.60000 | 0.54886 | 0.56300 |
| 0.6500 | 0.65000 | 0.49028 | 0.50487 |
| 0.7000 | 0.70000 | 0.42991 | 0.44490 |
| 0.7500 | 0.75000 | 0.36766 | 0.38293 |
| 0.8000 | 0.80000 | 0.30343 | 0.31878 |
| 0.8500 | 0.85000 | 0.23717 | 0.25224 |
| 0.9000 | 0.90000 | 0.16900 | 0.18347 |
| 0.9500 | 0.95000 | 0.09945 | 0.11099 |
| 0.9950 | 0.99500 | 0.03759 | 0.04345 |
| 1.0000 | 1.00000 | 0.00000 | 0.00000 |

$D(f) = \frac{1}{2} f\,(1-f)$

| F | D(F) | V(NMAX,F) | W(NMAX,F) |
|---|---|---|---|
| 0.0050 | 0.00499 | 0.99495 | 1.00303 |
| 0.0500 | 0.04875 | 0.94995 | 0.95840 |
| 0.1000 | 0.09500 | 0.89996 | 0.90883 |
| 0.1500 | 0.13875 | 0.84997 | 0.85928 |
| 0.2000 | 0.18000 | 0.80000 | 0.80975 |
| 0.2500 | 0.21875 | 0.75006 | 0.76025 |
| 0.3000 | 0.25500 | 0.70009 | 0.71077 |
| 0.3500 | 0.28875 | 0.65017 | 0.66133 |
| 0.4000 | 0.32000 | 0.60027 | 0.61192 |
| 0.4500 | 0.34875 | 0.55041 | 0.56254 |
| 0.5000 | 0.37500 | 0.50059 | 0.51320 |
| 0.5500 | 0.39875 | 0.45083 | 0.46389 |
| 0.6000 | 0.42000 | 0.40114 | 0.41464 |
| 0.6500 | 0.43875 | 0.35156 | 0.36542 |
| 0.7000 | 0.45500 | 0.30211 | 0.31626 |
| 0.7500 | 0.46875 | 0.25286 | 0.26715 |
| 0.8000 | 0.48000 | 0.20391 | 0.21810 |
| 0.8500 | 0.48875 | 0.15541 | 0.16912 |
| 0.9000 | 0.49500 | 0.10767 | 0.12020 |
| 0.9500 | 0.49875 | 0.06138 | 0.07136 |
| 0.9950 | 0.49999 | 0.02261 | 0.02752 |
| 1.0000 | 0.50000 | 0.00000 | 0.00000 |

$D(f) = 0$ iff $x < 0.5$, $1$ iff $x > 0.5$

| F | D(F) | V(NMAX,F) | W(NMAX,F) |
|---|---|---|---|
| 0.0050 | 0.00000 | 0.88522 | 0.89612 |
| 0.0500 | 0.00000 | 0.88522 | 0.89612 |
| 0.1000 | 0.00000 | 0.88522 | 0.89612 |
| 0.1500 | 0.00000 | 0.88522 | 0.89612 |
| 0.2000 | 0.00000 | 0.88522 | 0.89612 |
| 0.2500 | 0.00000 | 0.88522 | 0.89612 |
| 0.3000 | 0.00000 | 0.88522 | 0.89612 |
| 0.3500 | 0.00000 | 0.88522 | 0.89612 |
| 0.4000 | 0.00000 | 0.88522 | 0.89612 |
| 0.4500 | 0.00000 | 0.88522 | 0.89612 |
| 0.5000 | 1.00000 | 0.87959 | 0.89056 |
| 0.5500 | 1.00000 | 0.77175 | 0.78397 |
| 0.6000 | 1.00000 | 0.67233 | 0.68563 |
| 0.6500 | 1.00000 | 0.57930 | 0.59351 |
| 0.7000 | 1.00000 | 0.49124 | 0.50617 |
| 0.7500 | 1.00000 | 0.40779 | 0.42253 |
| 0.8000 | 1.00000 | 0.32610 | 0.34175 |
| 0.8500 | 1.00000 | 0.24775 | 0.26313 |
| 0.9000 | 1.00000 | 0.17130 | 0.18610 |
| 0.9500 | 1.00000 | 0.09856 | 0.11016 |
| 0.9950 | 1.00000 | 0.03664 | 0.04245 |
| 1.0000 | 1.00000 | 0.00000 | 0.00000 |

Literature

1. F.W.Atkinson.L.A.Peletier, Similarity solution of the nonlinear diffusion equation, Arch.Rat.Mech.Anal vol 54, 1974, pp.373-392

2. W.R.Gardner, Solutions of the flow equation for drying of soils and other porous media, Soil Sci., vol 23, 1959, pp183-187.

3. M.Kutilek, Aplikovana Hydropedologie, Ceske Vysoke Uceni v Praze, 1975.

4. J.Parlange, Theory of water-movment in soils-one dimensional infiltration, Soil Sci., vol 111, nr 3, 1971, pp 170-174

5. J.R.Philip, The theory of infiltration , 1, The infiltration equation and its solution, Soil Sci., vol 83, 1957, pp 345-357.

6. P.K.Toliakis, C.D.Tzimopoulos, D.K.Toliakas, A simple solution for horisontal absorbtion of water by soils for exponential soil water diffusivity , Water Resources Research, vol 16, nr 4, 1980, pp821-825.

# THE PRINCIPAL DIRECTION TECHNIQUE: A NEW APPROACH TO GROUNDWATER CONTAMINANT TRANSPORT MODELING

E. O. Frind

Department of Earth Sciences, University of Waterloo, Waterloo, Ontario, Canada

## ABSTRACT

The problem of contaminant transport in long thin hydrogeologic systems is investigated and a new simulation technique based on Galerkin finite elements, but formulated in terms of the principal directions of transport and structured as an alternating direction solution scheme, is applied. It is shown that the new technique is both more accurate and more efficient than the conventional finite element technique. The accuracy is unaffected by an extreme geometric aspect ratio or a large contrast between the physical parameters in different directions.

## INTRODUCTION

Contamination of groundwater systems by hazardous waste materials originating at surface sources is now a widespread problem. Because a contaminant plume can travel for considerable distances and thus degrade valuable groundwater resources, the understanding and prediction of plume formation is of vital importance.

Pinder (1973) first applied the Galerkin finite element technique to the problem of contaminant travel in the horizontal plane. Other contributions by various researchers have since followed. Unfortunately, cross-sectional problems have been found to be more difficult than areal problems because of the large length-to-thickness ratio of most hydrogeologic systems and the possibly large contrast between the longitudinal and the transverse transport processes. The classical Galerkin finite element technique may in such situations be particularly vulnerable to numerical dispersion; also computing cost may be high.

We will, therefore, focus here on the problem of simulating advective-dispersive transport in long thin systems. We will attempt to take advantage of the physical characteristics of such systems in developing a simulation scheme having acceptable accuracy and efficiency.

## CONCEPTUAL BASIS

Figure 1 shows an idealized cross section of a shallow unconfined aquifer containing a contaminant plume originating at a surface source. Actual systems of this type have been described by Kimmel and Braids (1980), and MacFarlane et al (1982). Assuming steady-state flow and a conservative contaminant of low density, the relevant physical transport processes are advection, hydrodynamic dispersion in the direction of flow, and hydrodynamic dispersion in the transverse direction, the latter often being on the order of molecular diffusion.

Conceptually, a simulation environment that is favourable to the attainment of optimum accuracy can be created by formulating the problem in a natural coordinate system defined by the principal directions of transport, namely the direction of flow and the direction orthogonal to the flow. The solution domain in this case will consist of the bundle of streamtubes containing the plume. Conditions for optimum accuracy arise because advective transport, which is the major cause of numerical dispersion, will be reduced to a one-dimensional process which in this form is more readily controllable.

The finite element solution can be structured as an alternating direction scheme similar to that developed by Peaceman and Rachford (1955). In this way, the time-consuming direct solution of large matrices is avoided and computing cost is held to a reasonable level, while the ability to handle irregular domains is retained. The simulation marches through time taking pairs of alternating steps, each solving implicitly in one direction only. The resulting decoupling of the spatial coordinates at a given time level may further enhance conditions for optimum accuracy because it supports the containment of numerical dispersion.

A similar alternating direction technique based on stream-

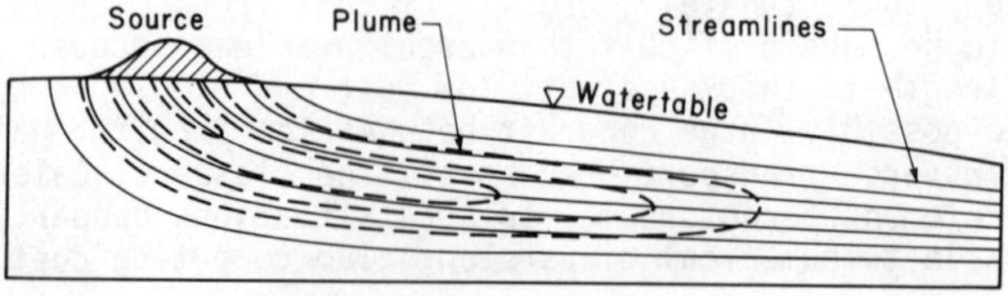


Figure 1. Idealized Flow System with Plume.

function coordinates, but using finite differences, was developed by Shamir and Harleman (1966, 1967).

THEORY

We will consider a quadrilateral flow domain defined in orthogonal curvilinear coordinates x and y (Figure 2), where x follows the streamlines. The flow field is assumed to be steady and the geometric distortion of the domain is small locally. Further assuming that longitudinal dispersion is dominated by mechanical dispersion, while transverse dispersion consists mainly of molecular diffusion, we can write the advective-dispersive transport equation in principal direction coordinates as:

$$\frac{\partial}{\partial x}\left(\alpha v \frac{\partial u}{\partial x}\right) + \frac{\partial}{\partial y}\left(D \frac{\partial u}{\partial y}\right) - v \frac{\partial u}{\partial x} - \frac{\partial u}{\partial t} = 0 \qquad (1)$$

where $u = u(x,y,t)$ is the concentration, $\alpha$ is the longitudinal dispersivity (L), v is the fluid velocity (L/T), D is the effective molecular diffusivity ($L^2/T$), and t is time. The coefficients can be spatially variable. The velocity however can be expressed in terms of the constant discharge $Q = 2b\theta v$ within a streamtube of width 2b (Fig. 2), where $\theta$ is the porosity.

The boundary condition at the inflow boundary x = 0, in the presence of a source, can be of either the first type

$$u(0,y) = u_o(y) \qquad (2)$$

or the third type

$$vu - v\alpha \frac{\partial u}{\partial x} = \frac{q_o(y)}{\theta} u_o(y) \qquad (3)$$

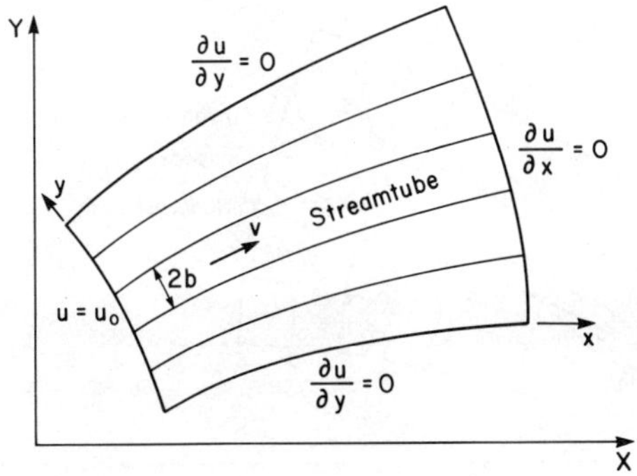


Figure 2. Deformed Quadrilateral Flow Domain.

where $u_0(y)$ is the specified source function, and $q_0(y)$ is the fluid flux outside the domain. The boundary conditions along the remaining three sides, assuming that the contaminant plume is entirely contained within the domain, is of the second type

$$\frac{\partial u}{\partial \underset{\sim}{n}} = 0 \tag{4}$$

where $\underset{\sim}{n}$ is the normal to the boundary.

Equation (1) can be written in the form:

$$(L_x + L_y)u = \frac{u^{n+1} - u^n}{\Delta t} \tag{5}$$

where $L_x$ and $L_y$ represent the differential operators, and the time derivative has been replaced by a finite difference approximation, with n indicating the time level, and $\Delta t = t^{n+1} - t^n$. Equation (5) can be split into two steps in the following manner.

$$\underline{\text{Step 1}} \quad {}^*L_x\, u^{n+\frac{1}{2}} + L_y\, u^n = \frac{u^{n+\frac{1}{2}} - u^n}{\Delta t/2} \tag{6a}$$

$$\underline{\text{Step 2}} \quad L_x\, u^{n+\frac{1}{2}} + {}^*L_y\, u^{n+1} = \frac{u^{n+1} - u^{n+\frac{1}{2}}}{\Delta t/2} \tag{6b}$$

where the * symbol indicates the term containing the unknowns. In each equation, the two spatial derivatives together are centered in time with respect to the temporal derivative. When averaged over the two half time steps, the resulting expression is centered at time n+1/2, giving second-order accuracy in time. Because $L_x$ contains the advective operator, the procedure must commence with Step 1.

The domain can be subdivided into deformed quadrilateral

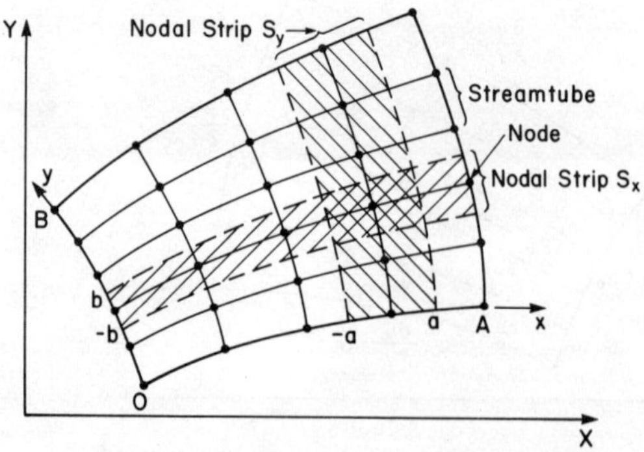


Figure 3. Finite Element Discretization with Nodal Strips.

elements that join at the N nodes of the grid (Figure 3) and that have straight sides. The formulation of the corresponding finite element equations can be broadly based on the standard Galerkin approach (Pinder and Gray, 1977) provided that the formulation is consistent with the structure of the alternating direction scheme. This scheme allows a spatial connection of points at any given time level in the implicitly-solved direction only, and provides no direct spatial linkage between point values of the time derivative. These relationships must be maintained in the selection of both the weighting functions and the interpolation functions in the Galerkin scheme.

The required spatial decoupling in the explicit direction can be achieved by dividing the domain into nodal strips oriented along the implicit direction for each step. Figure 3 shows typical strips $S_x$ and $S_y$ for Steps 1 and 2 respectively. The strips extend half-way across the element on either side of a node, occupying, respectively, the intervals [-b,b] and [-a,a], each with respect to the strip axis. Basis functions consistent with the spatially decoupled formulation are as shown in Figures 4(a) and 4(b), for Steps 1 and 2 respectively. The basis functions associated with node j can be expressed as:

<u>Step 1</u>

$$w_j(x) = \left.\begin{aligned} &\frac{x - x_{j-1}}{x_j - x_{j-1}} && x_{j-1} \leq x \leq x_j \\ &= \frac{x_{j+1} - x}{x_{j+1} - x_j} && x_j \leq x \leq x_{j+1} \end{aligned}\right\} \tag{7a}$$

<u>Step 2</u>

$$w_j(y) = \left.\begin{aligned} &\frac{y - y_{j-1}}{y_j - y_{j-1}} && y_{j-1} \leq y \leq y_j \\ &= \frac{y_{j+1} - y}{y_{j+1} - y_j} && y_j \leq y \leq y_{j+1} \end{aligned}\right\} \tag{7b}$$

Corresponding interpolation functions that can be used as trial solutions in the Galerkin equations are:

<u>Step 1</u>

$$u \simeq \hat{u} = \sum_{j=1}^{N} u_j(t)\, w_j(x) \tag{8a}$$

<u>Step 2</u>

$$u \simeq \hat{u} = \sum_{j=1}^{N} u_j(t)\, w_j(y) \tag{8b}$$

where it should be noted that the nodal index j increases in the implicit direction always, that is, in the x-direction for

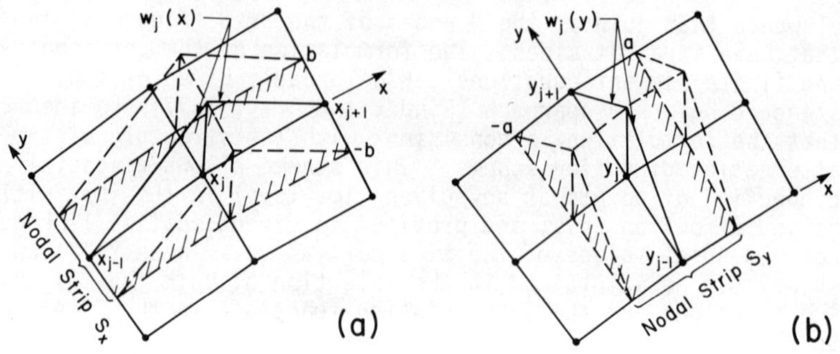


Figure 4. Basis Functions $w_j(x)$ and $w_j(y)$.

Step 1 and in the y-direction for Step 2. These functions permit spatial interpolation in the implicitly-solved direction only.

The Galerkin equations corresponding to equations (6) can be written as:

$$\int_x \int_y \{L_x u^{n+\frac{1}{2}} + L_y u^n - \frac{u^{n+\frac{1}{2}} - u^n}{\Delta t/2}\} w_i(x)\, dy\, dx = 0 \qquad i = 1,\ldots,N \tag{9a}$$

$$\int_y \int_x \{L_x u^{n+\frac{1}{2}} + L_y u^{n+1} - \frac{u^{n+1} - u^{n+\frac{1}{2}}}{\Delta t/2}\} w_i(y)\, dx\, dy = 0 \qquad i = 1,\ldots,N \tag{9b}$$

or, upon substitution of eqn. (3):

$$\int_x \int_y \{[\frac{\partial}{\partial x}(\alpha v \frac{\partial u}{\partial x}) - v \frac{\partial u}{\partial x}]^{n+\frac{1}{2}} + [\frac{\partial}{\partial y}(D \frac{\partial u}{\partial y})]^n - \frac{u^{n+\frac{1}{2}} - u^n}{\Delta t/2}\} \cdot w_i(x)\, dy\, dx = 0, \quad i=1,\ldots,N \tag{10a}$$

$$\int_y \int_x \{[\frac{\partial}{\partial x}(\alpha v \frac{\partial u}{\partial x}) - v \frac{\partial u}{\partial x}]^{n+\frac{1}{2}} + [\frac{\partial}{\partial y}(D \frac{\partial u}{\partial y})]^{n+1} - \frac{u^{n+1} - u^{n+\frac{1}{2}}}{\Delta t/2}\} \cdot w_i(y)\, dx\, dy = 0, \quad i=1,\ldots,N \tag{10b}$$

It is understood that, because of the form of the interpolation functions eqn. (8), $u \simeq \hat{u} = \hat{u}(x,t)$ in eqn. (10a) and $u \simeq \hat{u} = \hat{u}(y,t)$ in eqn. (10b). Integrating across the strip width intervals [-b,b] and [-a,a], for strips $S_x$ and $S_y$ respectively, we obtain:

$$\int_x 2b \left\{\frac{\partial}{\partial x}\left(\alpha v \frac{\partial u}{\partial x}\right) - v \frac{\partial u}{\partial x}\right\}^{n+\frac{1}{2}} w_i(x)\, dx$$

$$+ \int_x \left\{\left[D \frac{\partial u}{\partial y}\right]_b - \left[D \frac{\partial u}{\partial y}\right]_{-b}\right\}^n w_i(x)\, dx$$

$$- \int_x 2b \frac{u^{n+\frac{1}{2}} - u^n}{\Delta t/2} w_i(x)\, dx = 0 \qquad i=1,\ldots,N \qquad (11a)$$

$$\int_y \left\{\left[\alpha v \frac{\partial u}{\partial x}\right]_a - \left[\alpha v \frac{\partial u}{\partial x}\right]_{-a} - [vu]_a + [vu]_{-a}\right\}^{n+\frac{1}{2}} w_i(y)\, dy$$

$$+ \int_y 2a \left\{\frac{\partial}{\partial y}\left(D \frac{\partial u}{\partial y}\right)\right\}^{n+1} w_i(y)\, dy$$

$$- \int_y 2a \frac{u^{n+1} - u^{n+\frac{1}{2}}}{\Delta t/2} w_i(y)\, dy = 0 \qquad i=1,\ldots,N \qquad (11b)$$

Upon substitution of $v = Q/2b\theta$, the first integral in eqn. (11a) is found to be independent of the streamtube width. The second integral in (11a) and the first integral in (11b) represent the mass flux crossing the lateral boundaries of the strips $S_x$ and $S_y$ respectively. The integration across the strips conserves mass locally in the transverse direction.

The second-derivative terms in both equations may be integrated by parts to yield:

$$\int_x \left\{- \frac{\alpha Q}{\theta} \frac{\partial u}{\partial x} \frac{\partial w_i(x)}{\partial x} - \frac{Q}{\theta} \frac{\partial u}{\partial x} w_i(x)\right\}^{n+\frac{1}{2}} dx$$

$$+ \left[\frac{\alpha Q}{\theta} \frac{\partial u}{\partial x} w_i(x)\right]_{x=0}^{A}$$

$$+ \int_x \left\{\left[D \frac{\partial u}{\partial y}\right]_b - \left[D \frac{\partial u}{\partial y}\right]_{-b}\right\}^n w_i(x)\, dx$$

$$- \int_x 2b \frac{u^{n+\frac{1}{2}} - u^n}{\Delta t/2} w_i(x)\, dx = 0 \qquad i=1,\ldots,N \qquad (12a)$$

$$\int_y \{[\alpha v \frac{\partial u}{\partial x}]_a - [\alpha v \frac{\partial u}{\partial x}]_{-a} - [vu]_a + [vu]_{-a}\}^{n+\frac{1}{2}} w_i(y)\, dy$$

$$+ \int_y \{-2a\, D \frac{\partial u}{\partial y} \frac{\partial w_i(y)}{\partial y}\}^{n+1} dy$$

$$+ [2a\, D \frac{\partial u}{\partial y} w_i(y)]_{y=0}^{B}$$

$$- \int_y 2a \frac{u^{n+1} - u^{n+\frac{1}{2}}}{\Delta t/2} w_i(y)\, dy = 0 \qquad i=1,\ldots,N \qquad (12b)$$

The natural boundary terms arising out of the integration by parts vanish at the boundaries x = A, y = 0, and y = B, because of boundary condition (4).

The trial solutions, eqns. (8a) and (8b), may be substituted into the implicit terms in eqns. (12a) and (12b) respectively. This substitution maintains the required spatial linkage between point values and permits spatial interpolation between nodes that are linked implicitly. Because the time derivative at a point depends only on values at successive time levels and not on values at neighbouring points at the same time level, it can be represented directly by point values without spatial interpolation. The resulting equations can be arranged in the form of coefficients of the nodal values of the unknown and the time derivative, as follows:

$$\sum_{j=1}^{N} u_j^{n+\frac{1}{2}} \int_x \{\frac{\alpha Q}{\theta} \frac{\partial w_j}{\partial x} \frac{\partial w_i}{\partial x} + \frac{Q}{\theta} \frac{\partial w_j}{\partial x} w_i\}\, dx - [\frac{\alpha Q}{\theta} \frac{\partial u}{\partial x}]_{x=0}$$

$$- \int_x \{[D \frac{\partial u}{\partial y}]_b - [D \frac{\partial u}{\partial y}]_{-b}\}^n w_i(x)\, dx$$

$$+ \sum_{j=1}^{N} \frac{u_j^{n+\frac{1}{2}} - u_j^n}{\Delta t/2} \int_x 2b\, w_i(x)\, dx = 0 \qquad i=1,\ldots,N \qquad (13a)$$

$$- \int_y \{[\alpha v \frac{\partial u}{\partial x}]_a - [\alpha v \frac{\partial u}{\partial x}]_{-a} - [vu]_a + [vu]_{-a}\}^{n+\frac{1}{2}} w_i(y)\, dy$$

$$+ \sum_{j=1}^{N} u_j^{n+1} \int_y 2a\, D \frac{\partial w_j}{\partial y} \frac{\partial w_i}{\partial y}\, dy$$

$$+ \sum_{j=1}^{N} \frac{u_j^{n+1} - u_j^{n+\frac{1}{2}}}{\Delta t/2} \int_y 2a\, w_i(y)\, dy = 0 \qquad i=1,\ldots,N \qquad (13b)$$

The transverse flux terms in both equations can be evaluated from known nodal values of u, using a finite difference approximation for the transverse gradients. Because the strip boundaries are located halfway between nodes, the gradient approximation is second-order accurate.

The coefficients of the unknowns $u_i$, which are all in the form of one-dimensional integrals, can be easily evaluated by direct integration piecewise over the elements. The resulting equations can be conveniently written in matrix form as:

$$[R_x]\{u\}^{n+\frac{1}{2}} = \{F_y\}^{n} + \{F_o\} \tag{14a}$$

$$[R_y]\{u\}^{n+1} = \{F_x\}^{n+\frac{1}{2}} + \{F_o\} \tag{14b}$$

where $[R_x]$ and $[R_y]$ are tridiagonal coefficient matrices, $\{F_y\}$ and $\{F_x\}$ are the flux vectors, and $\{F_o\}$ contains the dispersive flux boundary term.

For the case of elements of constant size 2a x 2b, the $i^{th}$ row in matrices $[R_x]$ and $[R_y]$ becomes:

$$R_{x_i} = \left[\left(-\frac{\alpha Q}{2a\theta} - \frac{Q}{2\theta}\right)_{i-1} \quad \left(\frac{\alpha Q}{a\theta} + \frac{8ab}{\Delta t}\right)_i \quad \left(-\frac{\alpha Q}{2a\theta} + \frac{Q}{2\theta}\right)_{i+1}\right] \tag{15a}$$

$$R_{y_i} = \left[\quad \left(-\frac{aD}{b}\right)_{i-1} \quad \left(\frac{2aD}{b} + \frac{8ab}{\Delta t}\right)_i \quad \left(-\frac{aD}{b}\right)_{i+1} \quad\right] \tag{15b}$$

The corresponding flux terms in $\{F_y\}$ and $\{F_x\}$, written in the form of vector products, are:

$$F^n_{y_i} = \left[\quad \left(\frac{aD}{b}\right)_{i-1} \quad \left(-\frac{2aD}{b} + \frac{8ab}{\Delta t}\right)_i \quad \left(\frac{aD}{b}\right)_{i+1} \quad\right] \begin{Bmatrix} u_{i-1} \\ u_i \\ u_{i+1} \end{Bmatrix}^n \tag{16a}$$

$$F^{n+\frac{1}{2}}_{x_i} = \left[\left(\frac{\alpha Q}{2a\theta} + \frac{Q}{2\theta}\right)_{i-1} \quad \left(-\frac{\alpha Q}{a\theta} + \frac{8ab}{\Delta t}\right)_i \quad \left(\frac{\alpha Q}{2a\theta} - \frac{Q}{2\theta}\right)_{i+1}\right] \begin{Bmatrix} u_{i-1} \\ u_i \\ u_{i+1} \end{Bmatrix}^{n+\frac{1}{2}} \tag{16b}$$

Equations (14) must be modified further to take account of the boundary condition at x = 0. In the case of a first-type boundary, eqn. (14a) is partitioned in the usual way, with the rows corresponding to the fixed values eliminated and the linking part transferred to the right-hand side. In the case of a third-type boundary, boundary condition (3) can be substituted into the boundary term in eqn. (13a), which then yields a contribution to both the coefficient matrix $[R_x]$ and the boundary flux vector $\{F_o\}$ in eqn. (14a). In equation (14b), the third-type boundary flux at x = 0 appears in the form of a transverse flux.

Upon substitution of the boundary conditions, equations (14) can be solved sequentially for the unknown nodal values $u^{n+\frac{1}{2}}$ and $u^{n+1}$. The solution is efficiently performed by means of the Thomas algorithm for tridiagonal matrices.

The Principal Direction equations can also be developed using a general weighted residual (non-Galerkin) approach with two-dimensional interpolation functions (Frind and Pinder, 1982).

STABILITY

For the case of a uniform system, the numerical stability of the Principal Direction Technique can be easily analyzed using the classical von Neuman approach (Peaceman, 1977). Assuming a constant grid spacing $\Delta x$, $\Delta y$, and constant dispersion coefficients $D_x \equiv \alpha v$ and $D_y \equiv D$, we can define the stability parameters

$$\nu = \frac{v\,\Delta t}{\Delta x} \tag{17}$$

$$\rho_1 = \frac{D_x\,\Delta t}{\Delta x^2} \tag{18a}$$

$$\rho_2 = \frac{D_y\,\Delta t}{\Delta y^2} \tag{18b}$$

The error at a node with coordinates $x = i\Delta x$, $y = j\Delta y$, and time level n, can be expressed in the form of a Fourier expansion (Peaceman, 1977) as

$$\varepsilon_{ijn} = \gamma^n \exp(\hat{\imath}\, p_x\, i\Delta x) \exp(\hat{\imath}\, p_y\, j\Delta y) \tag{19}$$

where $\gamma$ is the amplitude, $\hat{\imath} = \sqrt{-1}$, and $p_x$, $p_y$ are indices in the expansion. Upon substitution of eqn. (19) into equations (14), the error amplification factors $\gamma_1$ and $\gamma_2$, for Steps 1 and 2 respectively, are found to be:

$$\gamma_1 = \frac{1 - 2\rho_2\,\beta_y}{[(1 + 2\rho_1\,\beta_x)^2 + (\frac{\nu}{2})^2\,\beta_v]^{1/2}} \tag{20a}$$

$$\gamma_2 = \frac{[(1 - 2\rho_1 \beta_x)^2 + (\frac{\nu}{2})^2 \beta_v]^{1/2}}{1 + 2\rho_2 \beta_y} \tag{20b}$$

where

$$\beta_v = \sin^2 p_x \Delta x$$

$$\beta_x = \sin^2 p_x \Delta x/2$$

$$\beta_y = \sin^2 p_y \Delta y/2$$

The amplification factor for a double step is

$$\gamma_1\gamma_2 = \frac{(1 - 2\rho_2 \beta_y)[(1 - 2\rho_1 \beta_x)^2 + (\frac{\nu}{2})^2 \beta_v]^{1/2}}{(1 + 2\rho_2 \beta_y)[(1 + 2\rho_1 \beta_x)^2 + (\frac{\nu}{2})^2 \beta_v]^{1/2}} \tag{21}$$

and the condition for stability is that

$$|\gamma_1||\gamma_2| \leq 1 \tag{22}$$

When $\nu = 0$, the purely dispersive case of Peaceman and Rachford (1955) is obtained; for this case each step itself is conditionally stable, while the double step is unconditionally stable.

For the advective-dispersive case, we let the advective parameter assume its maximum value of $\nu = 1$ (see following section), and we let $\beta_v$, $\beta_x$, and $\beta_y$ take on their limiting values. Upon analyzing the resulting forms of equations (20) and (21), we find the following:

1. Step 1 is stable when $\rho_1 < \infty$ or $\rho_2 \leq 1$.
2. Step 2 is unstable.
3. The combination of Steps 1 and 2 is unconditionally stable.

The numerical stability of the technique is thus assured.

## CONTROL OF NUMERICAL DISPERSION

In the Principal Direction Technique, the control of numerical dispersion is straightforward because each of the two solutions is in one-dimensional form. In this form, the accuracy of the solution is related in specific ways to the values of the dimensionless parameters $\nu$, $\rho_1$, and $\rho_2$. Once these relationships are established, spatial and temporal discretizations consistent with a given level of accuracy can be selected on the basis of these parameters.

The parameter $\nu$ represents the distance travelled by advection during one time step, relative to the grid spacing. Because this distance should not exceed one space increment, we have the criterion

$$\nu \leq 1 \qquad (23)$$

Violation of this constraint is known to result in smearing of the front.

The parameters $\rho_1$ and $\rho_2$ can be interpreted as measures of the spread of the front in the respective directions, relative to the spatial discretization. In the x-direction, transport is advective-dispersive; for this type of process, inadequate spatial discretization is known to result in oscillatory behaviour. The solution accuracy is therefore related to the value of $\rho_1$. A similar relationship exists for $\rho_2$, except that it can be expected to be less critical because transport in the y-direction is purely diffusive.

We will use the system shown in Figure 5 to demonstrate the sensitivity of the solution with respect to $\rho_1$ and $\rho_2$. The physical parameters are:

fluid velocity, $v = 0.1$ m/day

longitudinal dispersivity, $\alpha = 1.0$ m

molecular diffusivity, $D = 8 \times 10^{-5}$ m$^2$/day

The boundary conditions and the source function are shown in the figure. An analytical solution of the advective-dispersive equation in two dimensions developed by Cleary (1978) will be used as a basis of comparison. Table 1 shows four discretization schemes for the numerical solution and the corresponding values of $\nu$, $\rho_1$ and $\rho_2$.

The sensitivity with respect to $\rho_1$ is expressed by the longitudinal profiles at y = 0, shown in Figure 6 for Cases 1 to 3. The profiles indicate that very good accuracy is obtained with

Table 1. Discretization Schemes for Sensitivity Analysis.

| Case | $\Delta x$ m | $\Delta y$ m | $\Delta t$ days | $\nu$ | $\rho_1$ | $\rho_2$ |
|---|---|---|---|---|---|---|
| 1 | 2.0 | 0.2 | 20 | 1.0 | 0.5 | 0.04 |
| 2 | 4.0 | 0.2 | 40 | 1.0 | 0.25 | 0.04 |
| 3 | 8.0 | 0.2 | 80 | 1.0 | 0.125 | 0.04 |
| 4 | 2.0 | 0.4 | 20 | 1.0 | 0.5 | 0.01 |

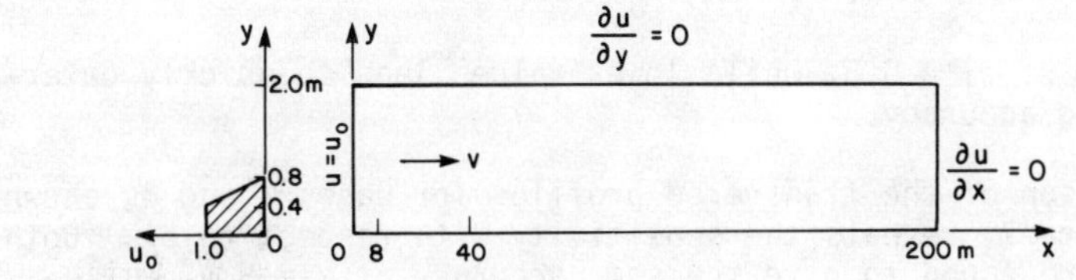


Figure 5. Rectangular Flow Domain and Source Function.

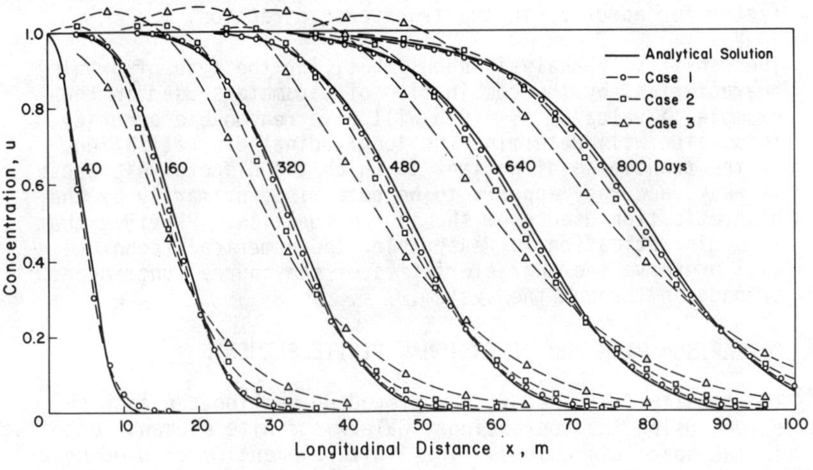


Figure 6. Longitudinal Profiles, Sensitivity Analysis.

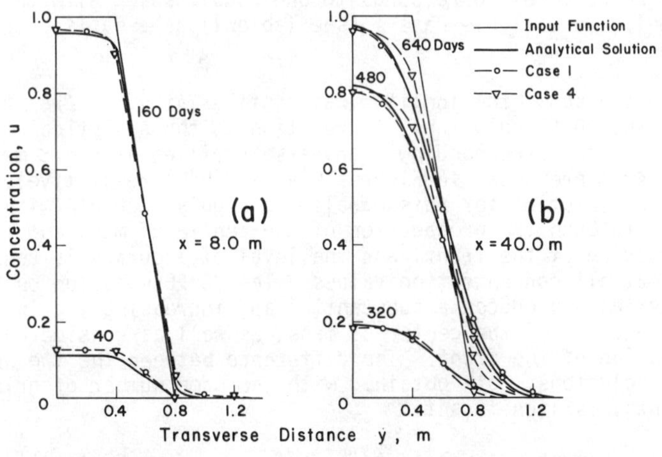


Figure 7. Transverse Profiles, Sensitivity Analysis.

a value of $\rho_1 = 0.5$, while lower values lead to quickly deteriorating accuracy.

Comparison of the transverse profiles for Cases 1 and 4, shown in Figure 7, reveals the sensitivity with respect to $\rho_2$. Both cases are found to give the same accuracy at x = 8 m, while Case 1 produces only slightly better accuracy than Case 4 at x = 40 m. Thus sensitivity with respect to $\rho_2$ is low. Comparison with the source function shows that the discretization used over the source function is probably the determining factor for accuracy in the transverse direction.

The sensitivity analysis shows that, for the type of problem characterized by the combination of parameters used in the example, a value of $\rho_1 = 0.5$ will give reasonable accuracy. This value will determine the longitudinal discretization. In the transverse direction, in which the dispersive process is weak, accuracy appears to be determined primarily by the discretization used over the source function. Provided that this discretization is reasonable, the numerical technique will preserve the characteristics of the source function as it propagates through the system.

## COMPARISON WITH TWO-DIMENSIONAL FINITE ELEMENTS

The problem discussed in the preceding section can also be solved using the conventional Galerkin finite element technique in two space dimensions. The finite element model used here (Frind, 1980) has rectangular elements, bilinear basis functions, exact integration, and a Gauss solver. The spatial discretization scheme is identical for both models. The time discretization is such that one fully implicit solution with the 2D FE model corresponds to one double sweep with the PD model. We will use Case 1 (see Table 1) as a basis of comparison.

Figure 8 shows the longitudinal profiles at y = 0 from the PD and the 2D FE solutions, in relation to the analytical solution. The corresponding transverse profiles at x = 8 m and x = 40 m are shown in Figures 9(a) and 9(b) respectively. The PD solution for this problem is highly accurate with respect to both the propagation of the centre of mass and the spreading of the front, and the level of accuracy is consistent at all concentration values. The 2D FE solution on the other hand produces a substantial and increasing lag in the propagation of the center of mass, as well as considerable smearing of the front. The difference between the two numerical solutions, both obtained with the same number of grid points, is significant.

In an attempt to explain this difference, the FE solution was repeated with a shorter time step of $\Delta t$ = 10 days; this resulted in a decrease of only about 30% in the error. Further experiments with linear triangles substituted for the rectangles in the 2D FE model produced results that were virtually identical to those obtained with rectangles. These experiments indicate that the difference in the accuracy of the two methods cannot be due to any advantage the PD method might have in the time domain on account of its double sweep. However, it can be due to the spatially decoupled form of PD

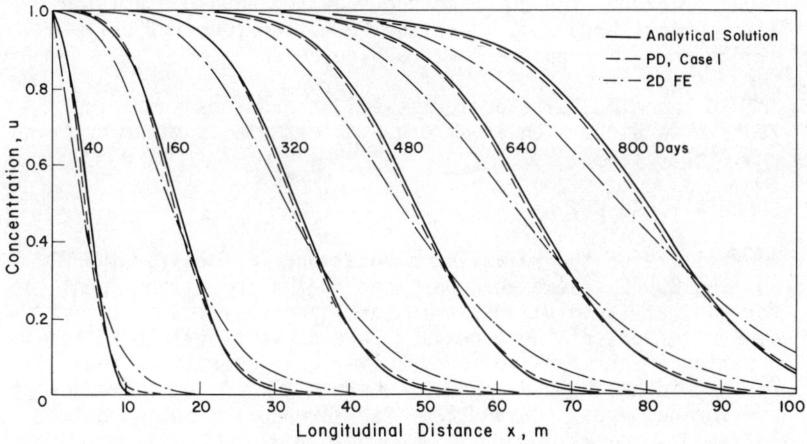


Figure 8. Longitudinal Profiles, Model Comparison.

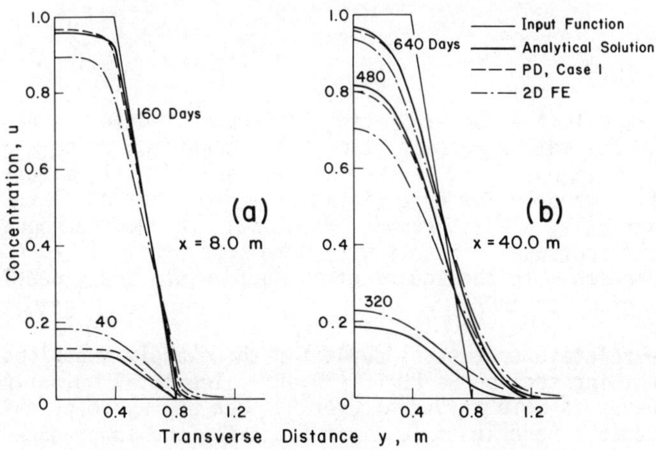


Figure 9. Transverse Profiles, Model Comparison.

which contrasts with the standard FE form in which all nodes are bidirectionally connected.

The spatially decoupled form also eliminates the problem of numerical dominance of one term over another within the same matrix that may occur in the 2D FE method in the presence of sharply contrasting physical parameters. Furthermore, roundoff error occuring during the matrix solution is sharply reduced because of the smaller size of the matrices. In obtaining the solutions it was found that, as a result of roundoff error, the 2D FE method required double precision arithmetic (IBM 3033), while the PD method required only single precision on the same machine.

The PD technique also performs well in problems where the transverse dispersion is strong. An example is given by Frind and Pinder (1982).

COMPUTATIONAL EFFORT

An estimate of the relative computational effort for the PD and the 2D FE schemes can be based on the operations count for the matrix solutions, which are generally the most time-consuming part of the process. In a conventional 2D FE model, assuming a Gaussian-type band solver, the operations count for a grid having n x m nodes (where $n > m$) is on the order of $nm^3$ for decomposition and $2nm^2$ for forward and backward solution. The decomposition is normally performed only when the time step changes. In the PD model, which uses the Thomas algorithm, the operations count for each double sweep constituting one time step is approximately 12 nm. The relative computational effort for the two techniques is therefore

$$\frac{PD}{FE} = \frac{12}{m(\frac{m}{k} + 2)} \tag{24}$$

where k is the number of time steps executed in the FE scheme with one matrix decomposition. The condition most favourable for FE is when k is large; in this case PD will be more efficient only for grid widths of $m > 6$. For small k, PD will practically always be more efficient. In PD, the operations count increases linearly with grid size, while in 2D FE, it increases with the square of the grid width and linearly with the grid length.

The relative operations count for the example described in the preceding section is PD/FE = 0.480. The total computing cost however is more favourable for PD than expressed by this ratio because of the lower level of precision and lower core storage requirements. CPU time for the PD solution using 100 x 10 elements was 0.284 seconds per time step (IBM 3033 under VM, Fortran G).

## CONCLUSION

Although further work is needed, it is evident that the Principal Direction Technique shows good potential in the analysis of contaminant transport in long thin hydrogeologic systems under steady-state flow conditions. Compared with the conventional Galerkin finite element technique in two dimensions, it has the following advantages:

1. High accuracy in simulating the propagation of a plume, both with respect to the center-of-mass translation and the front spreading.
2. Capability to handle systems with large geometric aspect ratios and large contrasts in the physical parameters.
3. Low computing cost.

The high degree of accuracy obtained appears to be a result of the more effective control of numerical dispersion in one-dimensional form that is achieved by the choice of the principal directions of transport as natural coordinates, and the spatial decoupling of the coordinate directions. It may also be related to the attainment of local mass conservation in the PD scheme, as opposed to global mass conservation in the conventional finite element scheme. Accuracy is unaffected by large contrasts in the magnitudes of the directional parameters because these parameters are computationally separated.

Provided that further tests with curvilinear systems confirm the present findings, the Principal Direction Technique can become an effective simulation tool that may permit systems to be studied in greater detail, and results to be interpreted with greater confidence, than is now possible.

## ACKNOWLEDGEMENTS

The author is indebted to G. F. Pinder and V. V. Nguyen who both contributed to this work in important ways through discussion and helpful suggestions at critical times. Thanks also go to E. A. Sudicky for weeding out errors in the manuscript. Most of the work was done while on leave at Princeton University. Financial support was provided by the U.S. National Science Foundation, through Grant No. CME 7905887, and by the Natural Sciences and Engineering Research Council of Canada, through Grant No. A8368.

## REFERENCES

Cleary, R. W. (1978), Analytical Models for Groundwater Pollution and Hydrology. Report 78-WR-15, Dept. of Civil Engineering, Princeton University, Princeton N.J.

Frind, E. O. (1980), Seawater Intrusion in Continuous Coastal Aquifer-Aquitard Systems. Proceedings of the Third International Conference on Finite Elements in Water Resources, The University of Mississippi, University, Miss.

Frind, E. O. and G. F. Pinder (1982), The Principal Direction Technique for Solution of the Advection-Dispersion Equation. Proceedings, Tenth IMACS World Congress, Concordia University, Montreal, Canada.

Kimmel, A. E. and O. C. Braids (1980), Leachate Plumes in Groundwater from Babylon and Islip Landfills, Long Island, New York. U.S. Geological Survey Professional Paper 1085.

MacFarlane, D. S., J. A. Cherry, R. W. Gillham and E. A. Sudicky (1982), Hydrogeological Studies of a Sandy Aquifer at an Abandoned Landfill: 1. Groundwater Flow and Plume Delineation. J. of Hydrology, accepted for publication.

Peaceman, D. W. and H. H. Rachford (1955), The Numerical Solution of Parabolic and Elliptic Differential Equations. SIAM J. 3: 28-41.

Peaceman, D. W. (1977), Fundamentals of Rcscrvoir Simulation. Developments in Petroleum Science 6, Elsevier Publ. Co., New York.

Pinder, G. F. (1973), A Galerkin-Finite Element Simulation of Groundwater Contamination on Long Island, New York. Water Res. Research, 9, 3: 1657-1669.

Pinder, G. F. and W. G. Gray (1977), Finite Element Simulation in Surface and Subsurface Hydrology. Academic Press, New York.

Shamir, U. Y. and D. R. F. Harleman (1966), Numerical and Analytical Solutions of Dispersion Problems in Homogeneous and Layered Aquifers. Report No. 89, Dept. of Civil Engineering, Mass. Institute of Technology, Cambridge, Mass.

Shamir, U. Y. and D. R. F. Harleman (1967), Numerical Solutions for Dispersion in Porous Mediums. Water Res. Research, 3, 2: 557-581.

# NUMERICAL SOLUTION OF HEAT AND WATER TRANSPORT IN AN UNSATURATED FREEZING SOIL

Ulrich Hornung
Institut für Numerische und instrumentelle Mathematik
der Universität Münster

## COUPLED HEAT AND WATER FLOW IN A POROUS MEDIUM

In an isotropic and homogeneous porous medium we consider the water content $\Theta[g/m^3]$ and the temperature $\tau[^oK]$ as functions of time t[sec] and space x[m]. The vector q $[g/m^2/sec]$ describing the rate of water flow is given by

$$q = - D \nabla \Theta \tag{1}$$

according to Darcy's law, where $D[m^2/sec]$ denotes the diffusivity. Here we neglect thermomigration, i.e. we assume that there is no water flow induced by thermal gradients (for a discussion of this effect see Person (1981)). The continuity equation for mass is

$$\partial_t \Theta = - \nabla q \,. \tag{2}$$

The flow rate r $[J/m^2/sec]$ of heat is the sum

$$r = r_1 + r_2 \tag{3a}$$

of the flow rate by diffusion

$$r_1 = - L \nabla \tau \tag{3b}$$

according to Fourier's law, where $L[J/m/^oK/sec]$ denotes the conductivity of the medium, and the flow rate by convection

$$r_2 = c \tau q \,, \tag{3c}$$

where $c[J/^oK/g]$ is the specific heat of water. The continuity equation for the enthalpy $E[J/m^3]$ reads

$$\partial_t E = - \nabla r \, . \tag{4}$$

In the system of equations (1) ... (4) the functions D, L and E depend of $\Theta$ and $\tau$. Denoting the enthalpies of the porous medium and of the water by $E_o$ and $E_1$, resp., we have

$$E = E_o + E_1 \, , \tag{5a}$$

$$E_o = c_o \cdot \rho_o \cdot \tau , \tag{5b}$$

where $c_o$ [J/°K/g] is the specific heat of the skeleton, $\rho_o$ [g/m³] its density, and

$$E_1 = \begin{cases} \Theta \cdot c_1 \cdot \tau & \tau \leq 0 \, , \\ \Theta \cdot (\ell_1 + c_2) \cdot \tau & \tau > 0 \, , \end{cases} \tag{5c}$$

where $c_1$, $c_2$ [J/°K/g] are the specific heats of ice and water, resp., and $\ell_1$ [J/g] is the latent heat of phase change. A typical situation is represented in figures 1, 2 and 3.

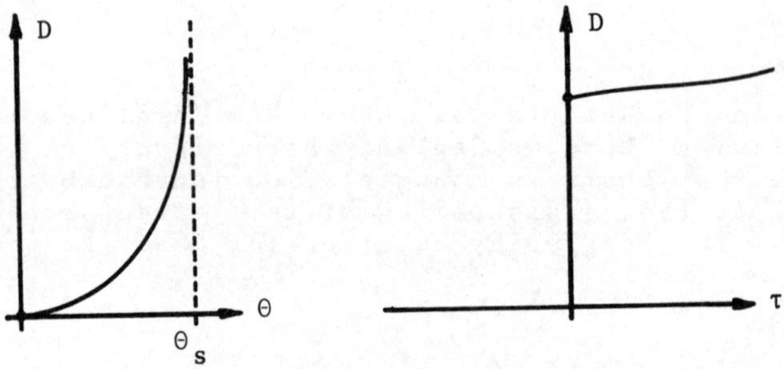


Figure 1

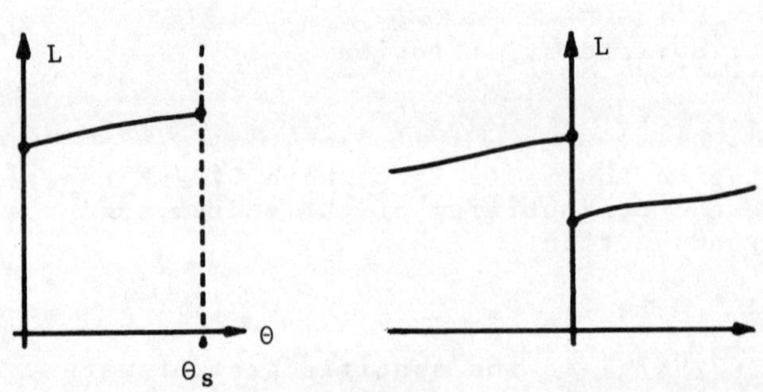


Figure 2

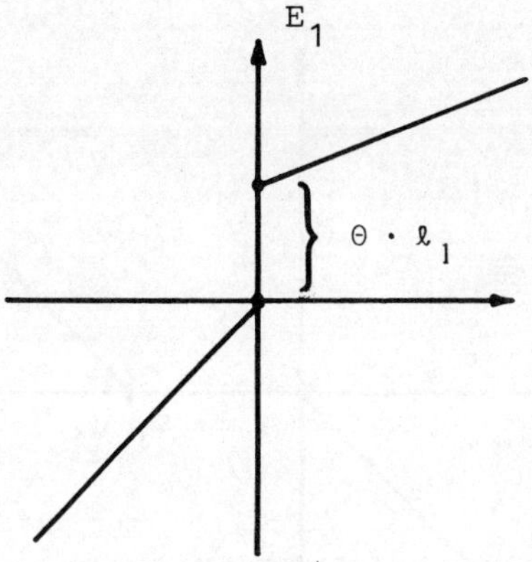


Figure 3

The diffusivity vanishes for $\tau < 0$, since there is no water flow in the frozen part of the soil. The conductivity L is larger in the solid than in the liquid phase, therefore there is a jump at zero temperature.

Models that are similar to the above have been treated by Jame/Norum (1980) and Thomas et al. (1980).

THE INITIAL BOUNDARY VALUE PROBLEM

The formulation of a mathematical model is simplified using the variables $u = \Theta$ and $v = E$. Then the temperature $w = \tau$ becomes a function $w = F(u,v)$ of u and v, and the same holds for $D(u,v)$ and $L(u,v)$. From (5) we obtain

$$w = \begin{cases} \dfrac{v}{c_o \cdot \rho_o + \ell_1 \cdot u} & , \quad v < 0 , \\ 0 & , \quad 0 \leq v \leq \ell_1 \cdot u , \\ \dfrac{v - \ell_1 \cdot u}{c_o \cdot \rho_o + c_2 \cdot u} & , \quad \ell_1 \cdot u < v \quad . \end{cases} \tag{6}$$

The situation is illustrated in figures 4, 5 and 6.

Now the model described in the preceding paragraph leads to the system of equations

$$\begin{aligned} q &= - D \cdot \nabla u \\ \partial_t u &= - \nabla q \\ \partial_t v &= \nabla(L \cdot \nabla w - c \cdot w \cdot q) \quad , \end{aligned} \tag{7}$$

which holds in a bounded region $\Omega \subset \mathbb{R}^n$ with n=1,2, or 3.

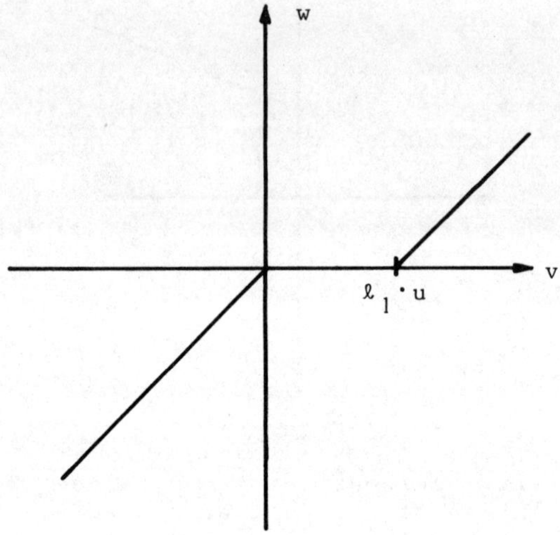


Figure 4

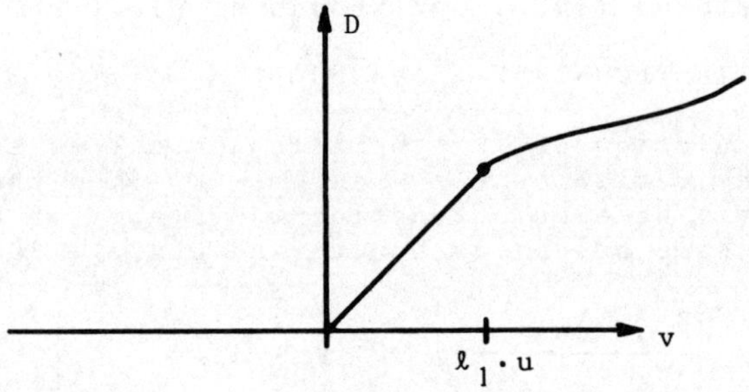


Figure 5

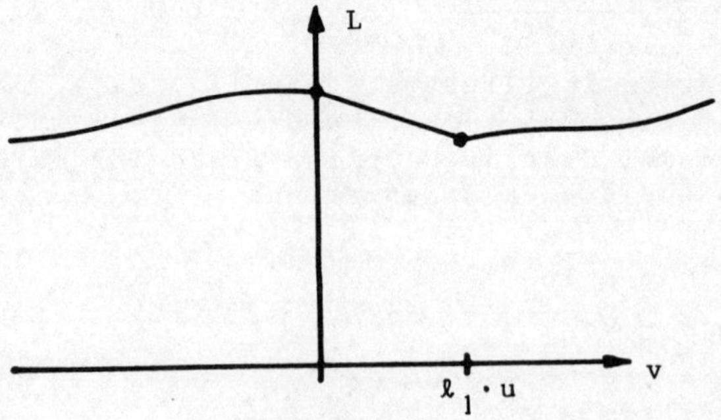


Figure 6

We assume that the normal water flow $q_\nu$ is prescribed on the boundary $\Gamma$ of $\Omega$, and that the values of w are given on $\Gamma$. Then we have

$$\begin{cases} q_\nu = g \\ w = f \end{cases} \tag{8}$$

on $\Gamma$. If some distributions of u and v are known at the beginning, we have

$$\begin{cases} u = u_o \\ v = v_o \end{cases} \tag{9}$$

for $t = 0$.

The initial boundary value problem (6) ... (9), which we call (P), has to be understood in the distributional sense. To be precise, we define the sets

$$Q = [0,T] \times \Omega$$

$$Q_1 = \{(t,x) \in Q : v(t,x) > 0\} ,$$

$$R = [0,T] \times \Gamma ,$$

and

$$R_1 = \{(t,x) \in R : v(t,x) > 0\} .$$

Now a triplet (u,v,w) is called a <u>weak solution of problem (P)</u>, if the following holds:

$$u,v \in L^2(Q), \; w \in L^2(0,T,H^1(\Omega)), \quad w \mid R = f , \tag{10a}$$

there is a $q_1 \in (L^2(Q))^n$ such that

$$\int_Q q_1 \cdot \varphi dxdt = - \int_Q u \nabla \varphi \; dxdt \tag{10b}$$

for all $\varphi \in C_o^\infty(Q)$ with supp $\varphi \subset Q_1$ ,

$$\left.\begin{aligned} q &= - D(u,v) q_1 \\ w &= F(u,v) \end{aligned}\right\} \text{ a.e. on } Q , \tag{10c}$$

$$\int_Q [u \cdot \partial_t \varphi + q * \nabla\varphi] dxdt + \int_\Omega u_o \cdot \varphi \, dx = \int_{R_1} g \cdot \varphi \, d\Gamma dt , \tag{10d}$$

and

$$\int_Q [v \cdot \partial_t \varphi + (L \nabla w - c\, w\, q) * \nabla \varphi] dxdt$$

$$+ \int_\Omega v_o \cdot \varphi \, dx \;=\; 0$$

for all $\varphi \in C^\infty(Q)$ with $\varphi(T,\cdot) = 0$ and $\varphi(t,x) = 0$ for $x \in \Gamma$.

## MATHEMATICAL TREATMENT OF SPECIAL CASES

Up to now there is no mathematical theory of the full problem (P) described above. In this paragraph we give a short review on existence, unicity, and stability theorems for some special situations and we mention known results of numerical analysis.

(a) The theory of water flow, i.e. the theory of equations (1) and (2) is relativily well understood. Fundamental facts from hydrology are discussed in Bear (1979), Hálek/Svec (1979) and Kovacs (1981). Unsteady saturated/unsaturated flow is dealt with in Hornung (1982). A study of degenerate problems of this type is given by DiBenedetto/Showalter (1982).

(b) The theory of thermal flow, i.e. the theory of equations (3) and (4), is very well developped. Analytical techniques are handled in Crank (1975), a general existence theory is given by Ladyzenskaya et al. (1968), numerical methods are discussed in Zienkiewicz (1981).

A survey on theoretical results on Stefan problems is given in Nitsche (1979). Cannon/DiBenedetto (1980) gave an existence theorem in a very general situation; Niezgodka/Pawlow (1982) proved uniqueness. A comparison of various numerical techniques can be found in Furzeland (1980); some new aspects are discussed in Wood et al. (1981).

(c) There is no theory on system (7), even if we specialize on problems without phase change. If in addition the convective term is neglected, we can identify $v = w$, and the system simplifies to

$$\begin{aligned} \partial_t u &= \nabla(D \nabla u) , \\ \partial_t v &= \nabla(L \nabla v) . \end{aligned} \tag{11}$$

Now the method of Cannon et al. (1976) applies, and

existence, uniqueness, and stability are obtained.

## A NUMERICAL METHOD

In the system (7) there is a strongly nonlinear coupling between the equations for water and heat flow. At a point of phase change there is a rapid change of the enthalpy v, and consequently the functions D and L are discontinuous. Therefore the type of the equation for water flow depends on the solution of the heat flow equation.

For these reasons the numerical method is fully implicit with respect to the discretization of time. Let u and v be the functions that have been calculated for time t. Then the approximations $u^*$ and $v^*$ for time $t^* = t + k$, $k > 0$ being the time step size, are obtained by solving the system of equations

$$\begin{aligned} q^* &= - D^* \cdot \nabla u^* , \\ u^* - u &= - k \cdot \nabla q^* , \\ v^* - v &= k \cdot \nabla (L^* \cdot \nabla w^* - c\, w^* q) . \end{aligned} \tag{12}$$

These are nonlinear equations which are solved by the following iterative procedure: Let $u^o$ and $v^o$ be some initial guesses for $u^*$ and $v^*$ (these starting values can be obtained by extrapolation from the previous time steps). Then for $m = 0,1,2,...$ the system

$$\begin{aligned} q^{m+1} &= - D^m \nabla u^{m+1} \\ u^{m+1} - u &= - k \nabla q^{m+1} \\ v^{m+1} - v &= k \nabla (L^m \nabla w^{m+1} - c\, w^{m+1} q^{m+1}) \end{aligned} \tag{13}$$

is solved for $u^{m+1}$ and $v^{m+1}$. Here $u^{m+1}$ is calculated first as the solution of a linear problem, then $q^{m+1}$ is used to find $v^{m+1}$. To this end the nonlinear function $w = F(u,v)$ is linearized with respect to v, i.e. we replace equation (6) by

$$w^{m+1} = \begin{cases} \dfrac{v^{m+1}}{c_o \cdot \rho_o + \ell_1 \cdot u^{m+1}} , & v^m < 0 \\ 0 , & 0 \le v^m \le \ell_1 \cdot u^{m+1} \\ \dfrac{v^{m+1} - \ell_1 \cdot u^{m+1}}{c_o \cdot \rho_o + c_2 \cdot u^{m+1}} , & \ell_1 \cdot u^{m+1} < v^m \end{cases} \tag{14}$$

Now a linear problem has to be solved in order to obtain $v^{m+1}$. This iteration method can be looked at as a Newton method applied to the nonlinearity $w = F(u,v)$, while there is an ordinary updating of the functions D and L. The iteration is stopped as soon as two subsequent functions $v^m$ and $v^{m+1}$ differ by less than some prescribed tolerance.

Standard finite element techniques are applied to discretize the space variable. Since the solution of the problem has very low continuity properties triangles with the piecewise linear elements are appropriate. Some care must be used to deal with the internal boundary at the freezing front. Since we use the enthalpy method to treat the jump condition with respect to the temperature, the latent heat is taken into account. On the other hand there is no normal water flow at front, i.e. we have to introduce zero Neumann conditions at the boundary $\Omega \cap \partial\Omega_1$.

In an artificial test case the numerical method described in this paper worked very satisfactory. Due to the implicit character of the scheme there are no stability restrictions on the time step size.

## MODIFICATIONS AND APPLICATIONS

Recently a similar problem has been successfully treated by DiBenedetto (1982), namely a two-phase Stefan problem in which the Navier-Stokes-equations govern the water flow in the liquid phase. Problem (P) becomes more complicated for saturated/unsaturated flow, since then another degeneracy is introduced. Some authors have studied heaving of the soil, an effect not yet fully understood, in which some electro-chemical mechanism induces an additional water flow towards the frozen zone, cf. Menot (1979) and Fremont (1982).

An application of the model described in this paper is the simulation of the operation of a heat pump. If for a longer period during the winter heat is extracted from the soil, the capillary water freezes in the neighbourhood of the pipes. Experimental and numerical results on this are contained in Müller (1979) and Dunand (1982).

## ACKNOWLEDGEMENT

This study was partially supported by the Deutsche Forschungsgemeinschaft (Grant Ho 782/2).

## REFERENCES


Bear, J.
1979. Hydraulic of Groundwater. McGraw-Hill, New York.

Cannon, J.R., E. DiBenedetto
1980. On the existence of weak solutions to an n-dimensional Stefan problem with nonlinear boundary conditions. SIAM J. Math. Anal. 11, 632-645.

Cannon, J.R.,Ford, W.T., Lair, A.V.
1976. Quasilinear parabolic systems. J. Diff. Equat. 20, 441-472.

Crank, J.
1975. The Mathematics of Diffusion. Clarendon Press, Oxford.

DiBenedetto, E.
1982. The Stefan Problem with Convection. In: R. Gorenflo, K.-H. Hoffmann (Eds.) Variational Methods and Ill-Posed Problems, Springer, Berlin, Heidelberg, New York.

DiBenedetto, E., Showalter, R.E.
1982. A free-boundary problem for a degenerate parabolic system. To appear.

Dunand, A.
1982. Thèse, Grenoble.

Fremond, M.
1982. Frost action in soils. To appear.

Furzeland, R.M.
1980. A comparative study of numerical methods for mooving boundary problems. J. Inst. Math. Appl. 26, 411-430.

Hálek, V., Švec, J.
1979. Groundwater Hydraulics. Elsevier, Amsterdam, Oxford, New York.

Hornung, U.
1982. A unilateral boundary value problem for unsteady waterflow in porous media. In: R. Gorenflo, K.-H. Hoffmann (Eds.) Variational Methods and Ill-Posed Problems, Springer, Berlin, Heidelberg, New York.

Jame, Y.W., Norum, D.I.
1980. Heat and transfer in a freezing unsaturated porous medium. Water Res. Research 16, 811-819.

Kovács, G.
1981. Seepage Hydraulics. Elsevier, Amsterdam, Oxford, New York.

Ladyzenskaja, O.A., Solonnikov, V.A., Ural'ceva, N.N.
1968. Linear and Quasilinear Equations of Parabolic Type. AMS, Providence.

Menot, J.M.
1979. Drainage et gel des milieux poreux partiellement satures. Thèse, Paris.

Müller, P.
1979. Ermittlung des spezifischen Energieverbrauchs verschiedener Wärmepumpensysteme und Untersuchung der Auswirkungen des Betriebes von Wärmepumpensystemen auf die Umwelt. In: H. Ehringer, G. Hoyaux (Eds.) Energy. Meeting on Heat Pumps Research, Development, and Application, ECSC-EEC-EAEC, Brüssel, 43-66.

Niezgódka, M., Pawlow, I.
1982. A generalized Stefan problem in several space variables. Appl. Maths. & Optimization. To appear.

Nitsche, J.A.
1979. A Finite Element Method for Parabolic Free Boundary Problems. Seminar on Free Boundary Problems, Pavia.

Person, J.-P.
1981. Caractérisation des propriétés thermohydrauliques d'un sol non saturé. Application au transport on au stockage d'énergie par systèmes en terrés. Thèse, Grenoble.

Thomas, H.R., Morgan, K., Lewis, R.W.
1980. A fully nonlinear analysis of heat and mass transfer problems in porous bodies. Inst. J. Numer. Meth. Engin. 15, 1381-1393.

Wood, A.S., Ritchie, S.I.M., Bell, G.E.
1981. An efficient implementation of the enthalpy method. Int. J. Numer. Meth. Engin. 17, 301-305.

Zienkiewicz, O.C.
1981. Finite elements methods in thermal problems. In: R.W. Lewis, K. Morgan, O.C. Zienkiewicz (Eds.) Numerical Methods in Heat Transfer, Wiley, Chichester/New York/Brisbane/Toronto, 1-25.

# FE MODEL OF STOCHASTIC GROUND WATER POLLUTION

M. Majumdar and S. Majumdar

Assistant Professors
Indian Institute of Technology, Kharagpur, India

## INTRODUCTION

Waste disposal in soil, excessive use of fertilisers in agricultural lands or seepage from septic tanks and underground sewer lines are major sources of ground water pollution. The problem of salt water intrusion in coastal aquifers is another common problem wherein the extent of salt water intrusion has to be known to decide upon the safe pumping rates. In general these problems are treated as dispersion problems the main objective being to find out the concentration profiles along with the extent of its spread, which in turn helps to decide upon the permissible quantities of the fertilizers to be applied, the quantities of waste disposal that can be safely disposed or the safe pumping rate in a coastal aquifer.

## THEORY

The dispersion problem may be expressed by the following mathematical model (5) :
Dispersion equation:

$$\nabla.(K\rho.(\nabla c/\rho)) - \nabla.(V c) = \partial c/\partial t \qquad ..(1)$$

Continuity equation: $\nabla.(\rho V) = -\partial\rho/\partial t \qquad ..(2)$

Darcy's law: $V = -k/n\ (\frac{p}{\rho g} + y) \qquad ..(3)$

State equations of the mixture:

$$\rho = f(c);\ \mu = g(c);\ D = D(c) \qquad ..(4)$$

in which c = pollutant concentration
D = molecular diffusion coefficient

$K$ = dispersion tensor
$k$ = hydraulic conductivity tensor
$p$ = pressure
$t$ = time
$V$ = pore velocity
$n$ = porosity
$\mu$ = dynamic viscosity
$\rho$ = density of the mixture.

Analytical solutions are available only for the simplest cases and numerical solution is the only viable alternative for most practical problems. However, for the numerical solution it is difficult to obtain proper or acceptable values for the coefficients in these equations. In order to overcome some of these limitations,a semistatistical approach has been adopted in the present work to simulate the dispersion through granular media. This way evaluation of the coefficients can be bypassed.

In the present study, first the macroscopic, flow field through a concerned region has been established with the help of the finite element method(FEM), and then,the dispersion of the pollutant from buried source has been simulated by the Monte Carlo procedure. The latter approach has been used by several investigations in the past(2,3,6,8) to simulate dispersion problem. It should be realized that though the ground water flow may be steady state in nature the propagation of pollutant may be time-dependent because of random nature and intermittent injection of the pollutant. The present approach brings out this aspect clearly. Indeed the most important feature of this investigation is that the tortuosity of the flow is replaced by the concept of "bifurcation density". The spreading of a pollutant in a porous media is mainly due to a continuous branching process of the pollutant, and branching is said to occur only when a particle divides the flow from an incoming capillary into two or more branches. A flow may have a tortuous path but may not deviate from the main velocity distribution. In such cases the dispersion will be restricted to a very small region.

The problem investigated here is that of an unconfined ground water flow wherein the pollutant source is considered to be a buried point source at a specified location. The analysis is divided into four parts as follows:
(1) Finding the velocity profile in the region of interest using FEM;
(2) Determination of bifurcation density and

variance of average flow direction;
(3) Monte Carlo simulation to depict dispersion, and
(4) Determination of concentration profile.

## 1. Velocity Distribution

The unconfined flow problem defined in Fig.1 is solved for potential function using FEM. The governing flow equation in isotropic homogeneous porous media is

$$\nabla^2 \phi = 0 \qquad ..(5)$$

$$\text{and} \quad V = -k \, \nabla \phi \qquad ..(6)$$

in which $\phi = (\frac{p}{\rho g} + y)$, is the potential function.

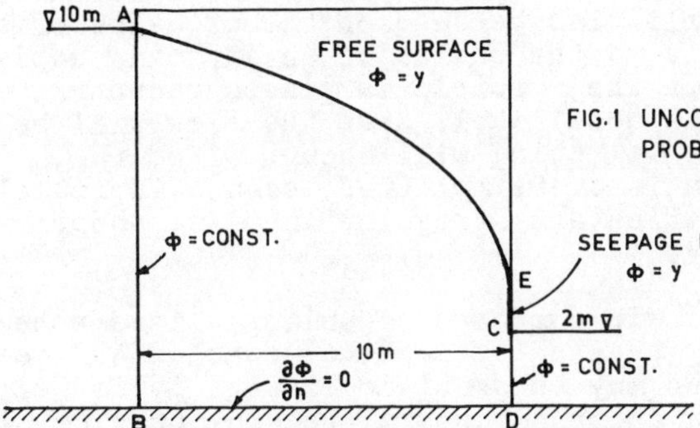


FIG.1 UNCONFINED FLOW PROBLEM

The boundary conditions are shown in the figure. The variational integral for the given equation is

$$I = \int_{\forall} k(\nabla \phi)^2 \, d\forall - 2u^* \int_S \phi \, ds \qquad ..(7)$$

in which $u^* = -k \, \partial\phi/\partial n$. The first integral is taken over the entire volume and the second integral over the entire boundary. The region is divided into triangular elements and a linear shape function $\phi = N^T \Phi$ is used. In terms of $\Phi$, the velocity V is given as

$$V = B^T \Phi$$

$$I^e = \int_{\forall} \Phi^T BB^T \, d\forall - 2u^* \int_S N^T \Phi \, dS \qquad ..(8)$$

where e stands for element. The application of minimization condition results in

$$G\Phi - U = 0 \qquad ..(9)$$

with $G = \int_{\forall} BB^T \, d\forall$ and $U = u^* \int_S N \, dS$.
Eq.(9) is solved for the potential function $\phi$ from which velocities in each element are calculated. For the location of free surface initially a profile is chosen and a solution for $\phi$ is obtained. If on the free surface $\phi = y$ is not satisfied a new surface $y = \phi$ is selected modifying all the elements accordingly. The process is repeated until the solution satisfied the necessary boundary condition at the

free surface.

## 2. Bifurcation Density and Variance

Finite element analysis of a subsurface flow yields macroscopic information of the flow field. But actual local velocity at any point in the media is dependent on the microscopic arrangement of capillaries around the point of interest. The ensemble average of the capillaries coincide with the macroscopic flow direction.
More often the flow out of a capillary is split into two or more capillaries because of the presence of a particle directly at the outlet of an incoming capillary (Fig.2). If the particle is placed eccentrically, then it is more likely that the flow will be one sided and no splitting will occur. Of course, this explanation is oversimplified because in actual practice capillaries are irregular in size, shape and connectivity.

In case the media is composed of uniform sized spherical particles, then an estimate of some of the geometric properties may be obtained easily. It is well known that cubic and rhombohedral packing provide two extreme cases of porosity .476 and .26 respectively(1). But for the sake of generality let us assume random packing. If particle diameter is d and porosity is n, then approximate number of particles per unit volume is

$$N = 6(1 - n)/\pi d^3 \quad .. (10)$$

However, only a fraction of these particles will lie close to the average velocity direction. If the centre of a particle lies within a distance of one capillary diameter, it may be assumed that branching has occurred. Thus the number of particles per unit length within a diameter d in the average flow direction is

$$m = \pi(\epsilon d)^2 N/4 = 3(1 - n)\epsilon^2/2d \quad .. (11)$$

where $\epsilon d$ is capillary diameter.

It may be intuitively shown that for rhombohedral packing, the capillary diameter is .154d and for cubic packing it is .414d. Also minimum number of particles r are encountered if they are placed in a row, where $r = 1/d$ .. (12)
Then the maximum probability of a particle being considered hit is

$$w = m/r = (3/2)(1 - n)\epsilon^2 \quad .. (13)$$

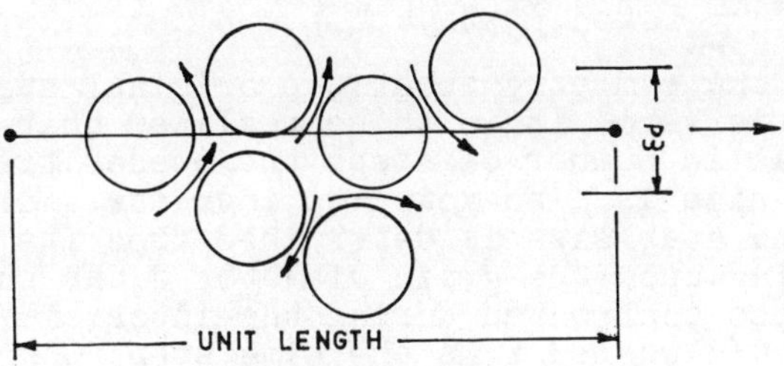


FIG. 2 BIFURCATION DENSITY

Thus the probability, w, of bifurcation or the 'bifurcation density' is mainly dependent on the square of capillary diameter ratio. In a two dimensional case it may be assumed that as the fluid travels a distance of one diameter it has the probabilities, w, of branching and (1-w) of continuing in the average direction. On this basis the transition matrix of Markov process is generated.

When a pollutant particle starts from its original position given by state 0 (i.e. average flow direction), after one step the probability of going to state +1 or -1 (i.e. bifurcating in plus and minus transverse direction) is $\beta = w/2$ and the probability of staying at state 0 is $\alpha = 1-w$. States $\pm 1$ correspond to step 1/2 d away from 0. On this line the one step transition matrix may be written as

$$\mathbb{P} = \begin{bmatrix} -N \\ \vdots \\ -1 \\ 0 \\ +1 \\ \vdots \\ +N \end{bmatrix} \overset{\begin{matrix} -N \ldots\ldots -1 & 0 & +1 \ldots\ldots +N \end{matrix}}{\begin{bmatrix} \beta & \alpha & & & & \\ & & & & & \\ & \alpha & \beta & \alpha & & \\ & & \alpha & \beta & \alpha & \\ & & & & & \\ & & & & \alpha & \beta \end{bmatrix}} \quad .. \quad (14)$$

It follows naturally that at the end of N steps, the N-step transition matrix $\mathbb{P}^N = \underbrace{\mathbb{P} \ldots \mathbb{P}}_{N \text{ times}} = \mathbb{P}^{N-1} . \mathbb{P}$

Using this matrix the probability $p_{ij}$ and the corresponding state value $x_j$ are used to calculate the variance at the end of N steps.

$$\sigma_N^2 = \sum_{-N}^{+N} p_{oj}\, x_j^2 - \mu^2 \qquad ..(15)$$

Due to the symmetry about the original state 0 the mean $\mu$ is zero. It is to be realized that with the increase in number of steps the needed transition matrix size is also going to increase. For every cell the step size is determined from its velocity V. Depending upon the grain diameter d, the number of particles confronted during the interval $V.\delta t$ is $N = V\delta t/d$ where $\delta t$ is the time step. Consequently variance corresponding to N steps $\sigma^2{}_N$ is required. It is obvious that for every cell, the value of V is different leading to distinct values for variance.

## 3. Monte Carlo Simulation

Dispersion from the pollutant source is simulated using the Monte Carlo procedure. A large number of the pollutant particles are discharged at the source point and their paths are traced. Position of a particle at the end of (i+1)th step, $P_{i+1}$ is dependent on its position $P_i$ at the end of ith step, a convection component $V\ \delta t$ and a random component $s.\ \delta t$.

$$\vec{P}_{i+1} = \vec{P}_i + \vec{V}\delta t + \vec{s}\,\delta t \qquad .. (16)$$

The value of step size $|\vec{s}|$ is chosen as follows. In the region of dominant mechanical dispersion the effect of molecular diffusion is insignificant and hence can be neglected. In this study dispersion is considered to be occuring mainly due to mechanical dispersion. In a porous medium the mechanical dispersion occurs due to pore geometry which in turn depends upon the particle size distribution, pore size distribution and the packing. Due to the tortuosity of the medium the microscopic velocity is different from the general flow direction, $\vec{V}$, at every point. The spread of the pollutant occurs due to the bifurcation of path. But as the average flow is in the $\vec{V}$ direction the probability of transport in the $\vec{V}$ direction is greater than the probability of transport in any other direction .It is suggested that the random step size is normally distributed with mean corresponding to the average flow direction and variance $\sigma^2$ which needs to be determined from the bifurcation characteristics of the soil as discussed earlier. Thus normally distributed random numbers $N(0,\sigma)$ were generated and selected at random to determine the random step size as follows.

$$\vec{s} = |\vec{V} \sin R| \{\sin(\theta - R)i + \cos(\theta - R)j\} \qquad .. (17)$$

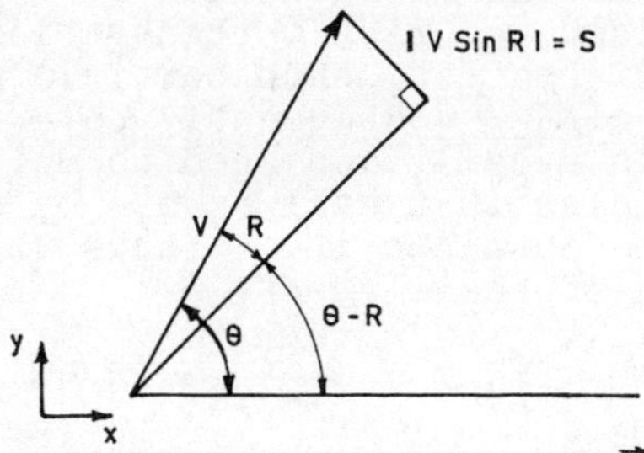


FIG. 3 DEFINITION SKETCH FOR $\vec{S}$

where, as seen from Fig.3, $\theta = \tan^{-1}(V_x/V_y)$, $V_x$ and $V_y$ being the x and y component of the average velocity vector $\vec{V}$; and R is the random number selected from the population $N(0, \sigma)$.

## 4. Concentration Profiles

The results can be analysed from two different view points. In steady flow, we may assume that a number of particles all starting at the same time from the pollution source occupy successive positions. This way the picture of movement of the tracer cloud in the flow region with time, t, assuming the steady flow in the region, is unaffected. The second picture is equally significant. Considering the injection of the pollutant as continuous, we get a steady state picture of the concentration profile in the region. This can be easily obtained by superposition of all the time steps simultaneously. The volume occupancy of the pollutant is considered negligible as compared to the main flow. Each particle may be thought of as a small value of concentrated pollutant drifting in flow. Let q be the quantity injected per unit time. If T is the time elapsed from the beginning of the process, a certain part of the concentrated pollutant qT remains in the region, while the rest flows out of the region. The average time of stay in the region reflects this.

Let $m_{ij}$ be the number of particles in jth cell at ith time step. Since a total number of $M_p$ particles are discharged from the source point, at any instant i, $M_{pi} = \Sigma_j\, m_{ij}$. For a given time i, $m_{ij}$ $(j=1,2,\ldots,n)$ represents the position of pollutant cloud at that instant. But as in most practical cases discharge is continuous, and if summation over time i is taken, i.e. $\sum_{i=1}^{T} m_{ij}$, T being the expected duration of the pollutant in the region in which steady state picture for each cell is obtained. If q is the rate of pollutant discharge with concentration c and $M_p$ particles are discharged, each particle represents a concentration $qcT/M_p$. Consequently to arrive at the concentration in a cell, one need to multiply the number of particles in a cell $\Sigma_i m_{ij}$ by $qcT/M_p$. The intensity can be obtained if this value is divided by the area of the element, or

$$c_j = \frac{qcT}{M_p.A_j} \Sigma_i m_{ij}$$

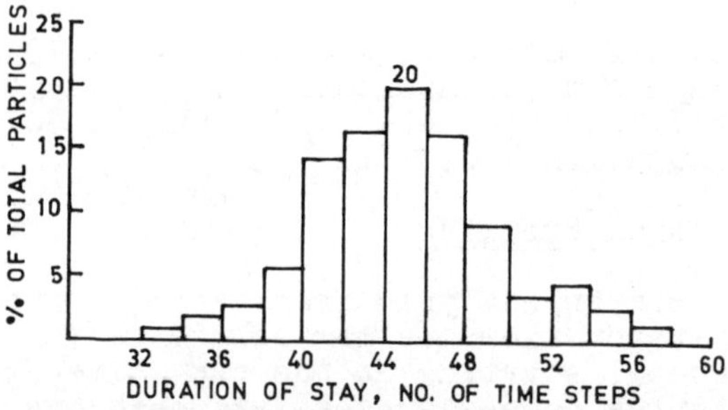


FIG.4 DISTRIBUTION OF LENGTH OF STAY OF POLLUTANT PARTICLES IN THE REGION

## RESULTS AND DISCUSSION

The method described above is illustrated through the unconfined flow problem of Fig.1. The value of $\epsilon$, n and k are taken as .186, .3 and 10.m/day respectively, while the soil particle size of 1 mm is used. First the velocities in various cells are calculated using the finite element method. Then the Monte Carlo simulation is carried out for a time step size of .05 day. The pollutant is discharged from a position of (1.5, 6.0)m as indicated in Fig.5. For every position $P_i$ the corresponding cell position is located and the same cell velocity V is used in the simulation expression. The concentration in each cell is calculated as described earlier. The distribution of length of stay of particles in the region is shown in Fig.4. The average stay of pollutant particles is about 45 time steps (2.25 days). Positions of pollutant cloud at the end of 20 and 40 time steps i.e. one day and two day cloud are shown in Fig.5 while the steady state concentration profile in the region is given in Fig.6. The figures indicate that dispersion is not very extensive. For uniform sized particles, the range of capillary diameter and porosity, $.154 \leq \epsilon \leq .414$ and $.26 \leq n \leq .476$ respectively fix the possible range of dispersion. However, when nonspherical particles are involved the situation becomes complex and the approach suggested here has to be modified.

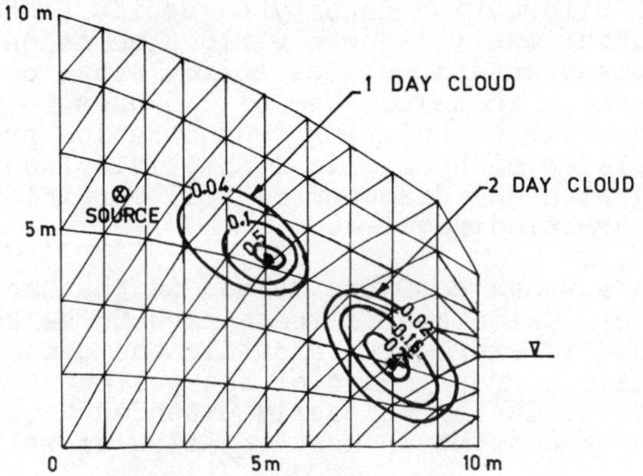


FIG.5 MOVEMENT OF POLLUTANT CLOUD

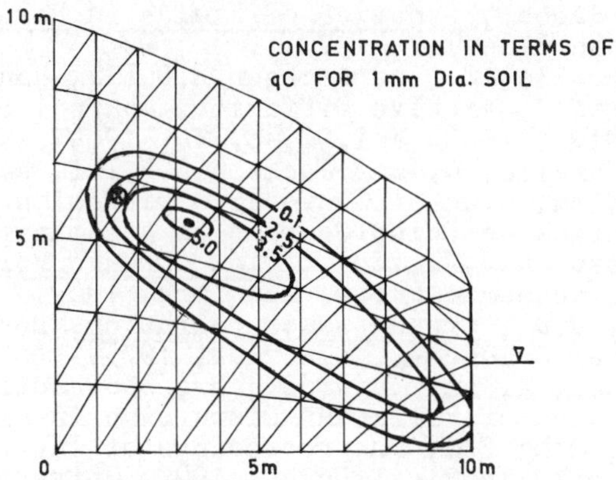


FIG. 6 STEADY STATE CONCENTRATION PROFILE

## SUMMARY AND CONCLUSIONS

This investigation introduces a new stochastic approach to the solution of dispersion in porous media. The main features include (1) use of FEM for solution of velocity profile in cells, (2) introduction of bifurcation density to depict probability of branching and (3) Monte Carlo simulation to obtain transient as well as steady state concentration profiles in cells. Use of FEM adds to the convenience while calculating concentration profiles. The knowledge of hydraulic conductivity and porosity along with the diameter of soil particles is the only information required.

The main advantage of the method is that no dispertion or diffusion coefficients need to be known. The method is extremely versatile and can account for transient propagation of the pollutant in a steady flow. It can be easily extended to simulate pollutant transport in unsteady flow as well.

## REFERENCES


1. Bear, Jacob.,"Dynamics of Fluids in Porous Media, American Elsevier, 1972.
2. Bugliarello, G. and Jackson,E.D.III, Random Walk Study of Convective Diffusion, Journal of Engg. Mech.Div., ASCE, Vol.90,No.EM4, 1964, 49-77.
3. Burgliarello, G. and Jackson,E.D.III, Random Walk Simulation of Convective Diffusion from Instantaneous Pcint Sources in a Laminar Field.
4. De Wiest, R.J.M.,(Ed)., Flow through Porous Media. Academic Press, New York, 1969.
5. Fried, J.J., Ground Water Pollution, Development in Water Science-4, Elsevier, 1975.
6. Majumdar, M. and Majumdar, S., Stochastic Simulation of Pollutant Dispersion in Flow Analysed by FEM, in Proceedings of Int. Conf. Comp.Appl. in Civil Engg., 1979, University of Roorkee, Roorkee, India, IX-19-IX-24.
7. Martin, M.C. and Carey, G.F., Introduction of Finite Element Analysis, McGraw Hill, 1973.
8. Makano, Y. and Murrmann, R.P., A Statistical Method for Analysis of Diffusion in Soils, Soil Sci.Soc. Am. Proc., Vol.35, 1971, 397-402.
9. Zienkiewicz, O.C. and Cheung, Y.K., The Finite Element Method in Structural and Continuum Mechanics, McGraw Hill, 1968.

# 14. NUMERICAL TECHNIQUES FOR THE SOLUTION OF THE TRANSPORT EQUATION

# ON THE NUMERICAL SOLUTION OF THE DIFFUSION-CONVECTION EQUATION

J.F. Botha, B.M. Herbst and S.W. Schoombie

University of the Orange Free State, Bloemfontein

## INTRODUCTION

The numerical solution of the diffusion-convection equation

$$D_t u(x,t) = aD_{xx}u(x,t) - bD_x u(x,t) \qquad 1$$

plays an important part in modelling various phenomena associated with water resources research. If the problem is linear and diffusion flow dominates, any suitable numerical method will suffice for the solution of Equation 1. However, when convection forms a strong component of the flow or if the equation is non-linear, some rather special techniques have to be used.

In this paper two methods recently developed for this purpose will be reviewed. The first method referred to as the moving finite element method, by its originators [Herbst, Mitchell and Schoombie (1982)], will be discussed in the following section. This is followed by a discussion of the alternating direction collocation method of Botha and Celia (1981) and the paper concluded with a discussion of a few numerical results.

## THE MOVING FINITE ELEMENT METHOD

The solution of Equation 1, subject to suitable boundary and initial conditions, represents in many cases of physical interest a travelling wave with a very steep front. For a stationary problem such a steep front can usually be accommodated by refining the mesh in the vicinity of the front. In the case of Equation 1 the front will always be moving. The same approach applied to this problem would, therefore, require a mesh which moves with the front. The following discussion will show that such a mesh can indeed be constructed.

Consider the normalised time-dependent mesh

$$\Delta(t) : 0 = x_0 < x_1(t) < \text{----} < x_{n-1}(t) < x_n = 1 \qquad 2$$

where $x_0$ and $x_n$ are fixed for all values of t. Let

$$v_i(x,t) = \begin{cases} v((x - x_i)/h_i) & x \in I_i \\ v((x - x_{i+1})/h_{i+1}) & x \in I_{i+1} \end{cases}$$

be a set of piecewise linear basis functions defined over each of the intervals $I_j = [x_{j-1}, x_j]$ with length $h_j = (x_j - x_{j-1})$. These basis functions can now be used to construct a time-dependent trial solution

$$U(x,t) = \sum_{i=0}^{n} U_i(t)v_i(x,t) \qquad 3$$

for Equation 1, such that

$$D_tU(x,t) = \sum_{i=0}^{n} [D_tU(t)v_i + c_iDx_i(t)]$$

where $\quad c_i = D_{x_i}U = - v_iD_xU$

The use of Equation 3 as a trial function when solving Equation 1 by the Galerkin method, leads in general to a system of 2(n-1) unknowns. [The (n-1) function parameters $U_j$ and (n-1) nodes $x_j$]. This problem can be circumvented if there exists a set of 2(n-1) basis functions defined over $\Delta(t)$. This demand on the basis functions is clearly met by the set of cubic Hermite polynomials, henceforth denoted by $S_i(x)$ and $T_i(x)$. The use of these polynomials as weighting functions allows one to reduce Equation 1 to the so-called Petrov-Galerkin equations

$$\sum_i [(S_j,v_i)DU_i(t) + (S_j,c_i)Dx_i(t)] - (S_j,AU) = 0$$

$$\sum_i [(T_j,v_i)DU_i(t) + (T_j,c_i)Dx_i(t)] - (T_j,AU) = 0$$

where $\quad AU = aD_{xx}U - bD_xU$

Evaluation of the inner products in these equations yields in general two sets of non-linear ordinary differential equations which can be solved.

The first set of differential equations can be shown to be a finite difference replacement of Equation 1, while the second set represents an equidistributing principle which can be expressed in the form [see Herbst *et al* (1982)]

$$h_i^2D_{xx}u(x_i^-,t) = h^2_{i+1}D_{xx}u(x_i^+,t) + O(h^3) \qquad 4$$

As this principle holds for all time t, Equation 4 ensures that the nodes of the mesh are always concentrated where the second derivative is at its maximum.

## THE ALTERNATING DIRECTION COLLOCATION METHOD

This method has been recently developed by Botha and Celia (1981) for solving three-dimensional parabolic equations of the form

$$D_t u(\underline{x},t) = a(\underline{x},t,u,\nabla u)\nabla^2 u(\underline{x},t) + f(\underline{x},t,u,\nabla u) \qquad 5$$

but can of course also be applied to Equation 1.

The basic philosophy behind the method is to approximate the solution $u(\underline{x},t)$ over the interval $[t_n,t_{n+1}]$ in terms of the linear Lagrange polynomials $\ell_n(t)$, the exact function values $u(\underline{x},t_n) = u(t_n)$ and the error function $e_i(t)$ as

$$D^q u(t) = D^q \ell_n(t)u(t_n) + D^q \ell_{n+1}(t)u(t_{n+1}) + D^q e_1(t)$$

Substitution of this expression in Equation 5 yields, in general, a semi-discretized equation which when collocated at the time $t_{n+\frac{1}{2}} = t_n + \Delta t/2$ ($\Delta t = t_{n+1} - t_n$) leads to an elliptic equation of the form

$$[u(t_{n+1}) - u(t_n)] / \Delta t = a(t_{n+\frac{1}{2}})\nabla^2[u(t_n) + u(t_{n+1})]/2 + f(t_{n+\frac{1}{2}}) + E(t_{n+\frac{1}{2}}) \qquad 6$$

Equation 6 can be discretized in various ways. One method suggested by Botha and Celia is to expand $u(t_{n+1})$ in terms of a Taylor series. This yields an approximation $O(\Delta t)$ which can be useful when solving a linear equation of the form of Equation 5.

A much more useful second order correct in time approximation can, however, be derived from Equation 6 if one expands the coefficients $a(t_{n+\frac{1}{2}})$ and $f(t_{n+\frac{1}{2}})$ rather than $u(t_{n+1})$. This leads to an approximation of the form

$$(u^{n+1} - u^n)/\Delta t = a^{n+\frac{1}{2}}\nabla^2[u^{n+1} + u^n] + f^{n+\frac{1}{2}}$$

where, using w to denote either a or f,

$$w^{n+\frac{1}{2}} = w(\underline{x},\{u^{n+1} + u^n\}/2,\{\nabla u^{n+1} + \nabla u^n\}/2)$$

and the subscripts denote an approximation to the exact value. This equation can also be expressed in terms of the difference $\Delta u^{n+1} = u^{n+1} - u^n$ and a parameter $\lambda$ as

$$\Delta u^{n+1} - (\lambda\Delta t)\nabla^2(\Delta u^{n+1}) = [(a^{n+\frac{1}{2}}/2) - \lambda]\nabla^2 u^{n+1} + [(a^{n+\frac{1}{2}}/2) + \lambda]\nabla^2 u^n + f^{n+\frac{1}{2}} \qquad 7$$

Although $\lambda$ can in general be any constant, a detailed analysis of the fully discretized version of (7), based on piecewise polynomials and orthogonal collocation, shows that there exists an optimal

$$\lambda = \max_{\underline{x},t}[a(\underline{x},u(t_n),\nabla u(t_n))/2]$$

PRACTICAL APPLICATIONS

The efficiency of the moving mesh finite element method in tracing sharp fronts can be judged from the results in figure 1, which represents the solution of Equation 1 at the time steps t = 0, 0.4, 1.2 subject to the initial condition

$$u(x,0) = 1 - x + \delta x(1 - x) \qquad \delta > 0$$

and boundary conditions

$$u(0,t) = 1, \ u(1,t) = 0$$

with $a = 10^{-3}$, b =1, $\delta$ = 0,05 and n = 10. Notice in particular how closely the nodes follow the region where the second derivative is large.

The initial placing of the mesh nodes must of course satisfy the equidistributing principle which can be conveniently expressed in the form

$$\int_{x_{i-1}}^{x_i} |D_{xx}u(x,0)|^{1/2}\, dx = (1/n)\int_0^1 |D_{xx}u(0,x)|^{1/2}\, dx$$

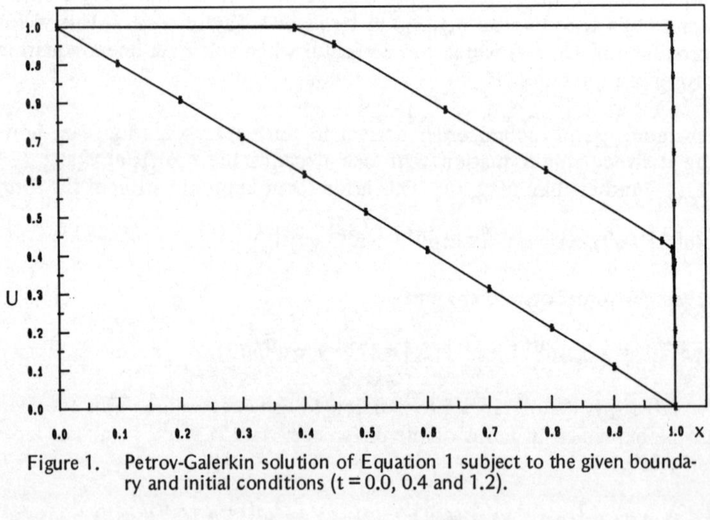


Figure 1. Petrov-Galerkin solution of Equation 1 subject to the given boundary and initial conditions (t = 0.0, 0.4 and 1.2).

Results obtained thus far has proved that the alternating direction collocation method is one of the most efficient methods for solving linear and mildly non-linear equations. It was, therefore, decided to apply it to the highly non-linear infiltration equation (expressed in terms of the soil moisture content $\Theta$, moisture diffusivity $M(\Theta)$ and unsaturated hydraulic conductivity $K(\Theta)$,

$$D_t\Theta(z,t) = D_z\,[M(\Theta)D_z\,\Theta(z,t)] - D_z K(\Theta)$$

The initial and boundary conditions were chosen so as to represent a soil which is initially at a constant soil moisture content and then subjected to a rainfall intensity of Q (mm/s). These conditions can be expressed mathematically as

$$\Theta(z,0) = \Theta_0$$

$$\Theta(L) = 1$$

$$= MD_z \Theta = Q - K(\Theta) \qquad 8$$

The value of L was fixed at 1 m, while values of $K(\Theta)$ and $M(\Theta)$ were determined from the moisture retention curve of a sand given in Hillel (1977), using the method of Jackson (1972).

The most accurate collocation solution of Equation 7 can usually be obtained by using a $C^1$-continuous Hermite piecewise polynomial (see e.g. Percell and Wheeler (1980)]. For a high intensity rainfall and dry soil this may cause some problems as the boundary condition, Equation 8, would then lead to a discontinuity in $D_z\ \Theta(0,t)$. For a value of $\Theta_0$ = 0,02 and Q = 1.306 $10^{-2}$ (mm/s) this discontinuity is so severe that the method failed altogether. However, by using the more physical assumption that the intensity increases continuously from zero over a period of time, this problem can usually be overcome successfully, especially if one uses basis functions defined over two or more intervals.

## SUMMARY

It is shown in this paper that the moving mesh finite element method is one of the most accurate methods available for solving diffusion-convection equations involving steep fronts. Difficulties may, however, be experienced when applying the alternating direction collocation method to highly non-linear problems with boundary conditions resulting in severe discontinuities in the function or its first derivative.

## REFERENCES

Botha, J.F. and Celia, M.A. (1981). Alternating Direction Approximations for a Parabolic Partial Differential Equation. Submitted for Publication.

Herbst, B.M., Mitchell, A.R. and Schoombie, S.W. (1982). A Moving Petrov-Galerkin Method for Transport Equations. To Appear in Int. J. Numer. Meth. Engng.

Hillel, D. (1977). Computer Simulation of Soil Water Dynamics. International Development Research Centre, Ottawa.

Jackson, R.A. (1972). On the Calculation of Hydraulic Conductivity. Soil Sci. Soc. Amer. Proc. 38 : 380-383.

Percell, P. and Wheeler, M.F. (1980). A $C^1$-finite Element Collocation Method for Elliptic Equations. SIAM J. Numer. Anal. 17 : 605-622.

# TRANSPORT SIMULATION USING THREE DIMENSIONAL ALTERNATING DIRECTION COLLOCATION

Michael A. Celia and George F. Pinder

Department of Civil Engineering, Princeton University

## INTRODUCTION

The purpose of this paper is to extend our earlier work on alternating direction collocation methods (ADC) to include three dimensional nonlinear transport simulation. Earlier papers (Celia et al, 1980; Hayes et al, 1981) described the two dimensional linear formulation and the theoretical underpinnings of the methodology. We will not revisit these matters again in this work but rather focus on the generalized, multidimensional non-linear formulation.

## THEORETICAL FORMULATION

Let us consider the following generalized expression

$$\frac{\partial u}{\partial t} + B_i(\underset{\sim}{x},u)\,\frac{\partial u}{\partial x_i} - C_i(\underset{\sim}{x},u)\,\frac{\partial^2 u}{\partial x_i^2} = F(\underset{\sim}{x},t,u) \tag{1}$$

where $\underset{\sim}{x}$ designates a three dimensional spatial coordinate system, $t$ is time, and $B_i$, $C_i$ and $F$ are functions of the spatial coordinates, time and the dependent variable $u$. The discretization of the spatial domain into finite elements generally results in the formation of irregular subspaces. Orthogonal collocation is most readily applied to rectangles. Thus a transformation of the irregular subspace (element) to a rectangle or more specifically a square, is in order. The appropriate relationship is

$$x_i = x_i(\xi_j) \qquad i,j=1,2,3 \tag{2}$$

where $\xi_j$ are the local coordinates normally associated with quadrilateral finite element transformations (see Fig. 1). Combining Equations (1) and (2) and employing the chain rule of differentiation, we obtain Eq. (1) in terms of the local coordinates $\xi_i$ :

$$\frac{\partial u}{\partial t} + b_i[\underset{\sim}{x}(\xi),u]\frac{\partial u}{\partial \xi_i} - c_i[\underset{\sim}{x}(\xi),u]\frac{\partial^2 u}{\partial \xi_i^2} = f[\underset{\sim}{x}(\xi),t,u,\frac{\partial^2 u}{\partial \xi_i \partial \xi_j}] \qquad i,j = 1,2,3; \quad j \neq i \tag{3}$$

Note that the cross-derivative has been transferred to the right hand side of Eq. (3).

The immediate objective is to approximate Eq. (3) using the collocation method. To begin, a trial function is selected which satisfies the boundary conditions imposed on Eq. (1). For a second-order equation such as Eq. (1), Hermite cubics with $C^1$ continuity are suitable. We write the trial function as

$$u(\underset{\sim}{\xi},t) \approx \hat{u}(\underset{\sim}{\xi},t) = \sum_{q=1}^{Q} U_q(t)\ \phi_q[\underset{\sim}{\xi}] \tag{4}$$

where it is understood that $\phi_q$ are Hermite cubics and that $U_q$ represents the appropriate undetermined coefficients. Let us further specify that the $\phi_q(\xi)$ are generated as tensor products of one dimensional Hermite polynomials defined in the three coordinate directions $\xi_i$. One can now write the basis function $\phi_q(\underset{\sim}{\xi})$ as

$$\phi_q(\xi_1,\xi_2,\xi_3) = \alpha_r(\xi_1)\beta_s(\xi_2)\gamma_t(\xi_3) \tag{5}$$

where $\alpha_r(\xi_1),\beta_s(\xi_2)$, and $\gamma_t(\xi_3)$ are the three one-dimensional Hermite polynomials defined at node $r,s,t$.

The collocation method requires the evaluation of the residual obtained through combination of Eq. (4) and (3) at each of $P$ collocation points, that is

$$\left[\frac{\partial \hat{u}}{\partial t} + b_i\frac{\partial \hat{u}}{\partial \xi_i} - c_i\frac{\partial^2 \hat{u}}{\partial \xi_i^2} - f\right]_{\underset{\sim}{\xi}_p} = 0 \qquad p=1,2,\ldots,P \tag{6}$$

where we have omitted, for clarity, the specific functional relationships for $\hat{u}$ and $f$. The actual locations of the eight collocation points in each element will be the same with respect to the local $\xi_i$ coordinates. Let us now employ a finite difference approximation for the time derivative in Eq. (6). There results

$$\left[\frac{\hat{u}^{n+1}-\hat{u}^{n}}{\Delta t} + b_i\frac{\partial \hat{u}^{n+\theta}}{\partial \xi_i} - c_i\frac{\partial^2 \hat{u}^{n+\theta}}{\partial \xi_i^2} - f^{n+\theta}\right]_{\underset{\sim}{\xi}_p} = 0 \qquad p=1,2,\ldots,P \tag{7}$$

where $\theta$ is a parameter $(0 \le \theta \le 1)$ which locates the space derivative in the time domain, that is $\hat{u}^{n+\theta} = \theta\hat{u}^{n+1} + (1-\theta)\hat{u}^{n}$. To minimize roundoff error, it is advantageous to introduce the relationship

$$\delta^n = \hat{u}^{n+1} - \hat{u}^n \tag{8}$$

which describes the change in $\hat{u}$ over one time step. Equation (7) now becomes

$$\left[\delta^n + \Delta t\theta b_i \frac{\partial\delta^n}{\partial\xi_i} - \Delta t\theta c_i \frac{\partial^2\delta^n}{\partial\xi_i^2} = \Delta t\left\{f^{n+\theta} - b_i \frac{\partial u^n}{\partial\xi_i} + c_i \frac{\partial^2 u^n}{\partial\xi_i^2}\right\}\right]_{\underset{\sim}{\xi}_p} \qquad p=1,2,\ldots,P \tag{9}$$

This equation is the general statement of collocation in three space dimensions with a variably-weighted finite difference scheme in time.

Because the ADC method requires a spatially invariant coefficient matrix, Eq. (9) is now modified. Specifically, the operators $\lambda_i^1 \frac{\partial\delta^n}{\partial\xi_i} + \lambda_i^2 \frac{\partial^2\delta^n}{\partial\xi_i^2}$ are added to each side of Eq. (9), and the result rearranged to yield

$$\left[\delta^{n,m+1} + \Delta t\,\lambda_i^1 \frac{\partial\delta^{n,m+1}}{\partial\xi_i} - \Delta t\,\lambda_i^2 \frac{\partial^2\delta^{n,m+1}}{\partial\xi_i^2} = \Delta t\left\{f^{n+\theta} - b_i \frac{\partial\hat{u}^n}{\partial\xi_i} + c_i \frac{\partial^2\hat{u}^n}{\partial\xi_i^2} + (\lambda_i^1 - \theta b_i)\frac{\partial\delta^{n,m}}{\partial\xi_i} - (\lambda_i^2 - \theta c_i)\frac{\partial^2\delta^{n,m}}{\partial\xi_i^2}\right\}\right]_{\underset{\sim}{\xi}_p} \qquad p=1,2,\ldots,P \tag{10}$$

where $m$ is the iteration level. Because this scheme is iterative, we denote it as iterative alternating direction collocation IADC. To decompose Eq. (10) into a series of one-dimensional problems, it is necessary to obtain factors. This requires that we "*complete the square*" by adding the term:

$$\hat{a}(\delta) = (a_1a_2 + a_2a_3 + a_3a_1 + a_1a_2a_3)\delta \tag{11a}$$

to each side of Eq. (10) where

$$a_i = \Delta t(\lambda_i^1 \frac{\partial}{\partial\xi_i} - \lambda_i^2 \frac{\partial^2}{\partial\xi_i^2}) \tag{11b}$$

Upon substitution of Eq. (4) and (5) into the resulting equation, we obtain

$$\sum_{q=1}^{Q} \delta_q \Bigg[ (\alpha_r + \Delta t\ \lambda_1^1 \frac{d\alpha_r}{d\xi_1} - \Delta t\ \lambda_1^2 \frac{d^2\alpha_r}{d\xi_1^2})(\beta_s + \Delta t\ \lambda_2^1 \frac{d\beta_s}{d\xi_2} - \Delta t\ \lambda_2^2 \frac{d^2\beta_s}{d\xi_2^2})$$

$$\cdot(\gamma_t + \Delta t\ \lambda_3^1 \frac{d\gamma_t}{d\xi_3} - \Delta t\ \lambda_3^2 \frac{d^2\gamma_t}{d\xi_3^2}) = \Delta t \Big\{ f^{n+\theta} - b_i \frac{\partial \hat{u}^n}{\partial \xi_i} + c_i \frac{\partial^2 \hat{u}^n}{\partial \xi_i^2}$$

$$+ (\lambda_i^1 - \theta b_i) \frac{\partial \delta^{n,m}}{\partial \xi_i} - (\lambda_i^2 - \theta c_i) \frac{\partial^2 \delta^{n,m}}{\partial\ \xi_i^2} + \frac{\hat{a}}{\Delta t} (\delta^{n,m}) \Big\} \Bigg]_{\underset{\sim}{\xi}_p}$$

$$p=1,2,\ldots,P \tag{12}$$

where $Q$ is the number of nodes and the interpretation of $q$ will be apparent after examination of Equation (13a) to follow. The objective now is to decompose Eq. (12) into three steps such that, for each step, it is necessary to solve equations in only one space dimension.

The following algorithm satisfies these constraints.

$$\left[ \sum_{r=1}^{R} \delta^{(1)}_{r,s,t} (\alpha_r + \Delta t\ \lambda_1^1 \frac{d\alpha_r}{d\xi_1} - \Delta t\ \lambda_1^2 \frac{d^2\alpha_r}{d\xi_1^2}) = \text{RHS} \right]_{\xi_p} \quad \begin{array}{l} p=1,2,\ldots,P_R \\ s=1,2,\ldots,S \\ t=1,2,\ldots,T \end{array} \tag{13a}$$

where $n = R \cdot S \cdot T$, RHS denotes the right hand side of Eq. (12), and $P_R$ is the number of collocation points in the $\xi_1$ coordinate direction

$$\left[ \sum_{s=1}^{S} \delta^{(2)}_{r,s,t} (\beta_s + \Delta t\ \lambda_2^1 \frac{d\beta_s}{d\xi_2} - \Delta t\ \lambda_2^2 \frac{d^2\beta_s}{d\xi_2^2} = \delta^{(1)}_{r,s,t} \right]_{\xi_p} \quad \begin{array}{l} p=1,2,\ldots,P_S \\ r=1,2,\ldots,R \\ t=1,2,\ldots,T. \end{array} \tag{13b}$$

$$\left[\sum_{t=1}^{T}\delta_{r,s,t}\left(\gamma_t + \Delta t\lambda_3^1 \frac{d\gamma_t}{d\xi_3} - \Delta t\lambda_3^2 \frac{d^2\gamma_t}{d\xi_3^2}\right) = \delta^{(2)}_{r,s,t}\right]_{\xi_p}$$

$p=1,2,\ldots,P_T$
$s=1,2,\ldots,S$
$r=1,2,\ldots,R$ .

(13c)

In the development of Equation (13), we have assumed, for simplicity in presentation, that the domain is a cube and the elements are rectangular parallelpipeds (bricks). Moreover, we have not formally introduced the Hermite polynomial expansion with its associated eight undetermined coefficients per node. The principal complexity associated with the introduction of the deformed geometry is the transformation from $\underset{\sim}{x}$ to $\underset{\sim}{\xi}$ . Provided the nodes are numbered such that $\xi_i$ is geometrically consistent from element to element, the precedure is a straightforward extension of the two-dimensional work presented in Frind and Pinder [1979].

Consider now the impact of introducing the Hermite expansion. For each $\xi_1$ sweep one must accommodate 8R unknowns. Thus each $\xi_1$ sweep must generate 8R equations determined from 8R collocation points. One should visualize one $\xi_1$ sweep as composed of four passes; each pass generates 2R equations in 2R unknowns. The selection of unknowns is easily determined, provided one is consistent in each coordinate direction. For example, in the first pass in the $\xi_1$ direction (see Figure 2), one would select unknowns associated with $\alpha_r$, $\frac{d\alpha_r}{d\xi_1}$, $\beta_s$ and $\gamma_t$ ; in the second, one would select those identified with $\alpha_r$ , $\frac{d\alpha_r}{d\xi_1}$, $\beta_s$ and $\frac{d\gamma_t}{d\xi_3}$ and finally, in the fourth, those identified with $\alpha_r$ , $\frac{d\alpha_r}{d\xi_1}$ , $\frac{d\beta_s}{d\xi_2}$ and $\frac{d\gamma_t}{d\xi_3}$. One next increments 4 and repeats the $\xi_1$ sweep.

## EXAMPLE APPLICATION

To demonstrate the application of three-dimensional ADC the response of a "pie-shaped" section of an aquifer surrounding a partially penetrating well was simulated. The finite element net is illustrated in plan and cross-section in figure 3. The flux specified at the well is 2.0, the hydraulic conductivity is 1.0 and the storage coefficient is 0.17.

Because we are employing iterative ADC, it is important to determine the rate of convergence of the method. Figure 4 illustrates the rate of convergence of the iterative scheme where $\omega$ is a relaxation parameter used in the sense $\delta^{k+1} = \omega\delta^{k} + (1-\omega)\delta^{k-1}$. While it is apparent the initial rate of convergence increases as the parameter approaches unity, the error (defined as the supremum of the solution at k versus the solution at convergence) overshoots the solution and with the larger value of $\omega$ the solution thereafter converges more slowly than when under-relaxed.

As would be expected, the number of iterations required to achieve a specified error tolerance depends upon the size of the time step. This relationship is illustrated quantitatively in figure 5. It is evident that one would normally achieve optimal efficiency by choosing the largest time step compatible with a specified time truncation error because the number of iterations required for convergence increases arithmetically while the time step increases approximately exponentially.

Experiments were conducted to examine the error associated with the geometric transformation inherent in the deformed (non rectangular) areal elements. The angle $\phi$ illustrated in figure 3 was varied over the range $10° \le \phi \le 60°$ which resulted in considerable geometric distortion. The response of the error to this range of values was negligible.

In summary, the iterative ADC method appears to be an efficient and robust approach to the solution of multidimensional parabolic partial differential equations such as encountered in groundwater hydrology.

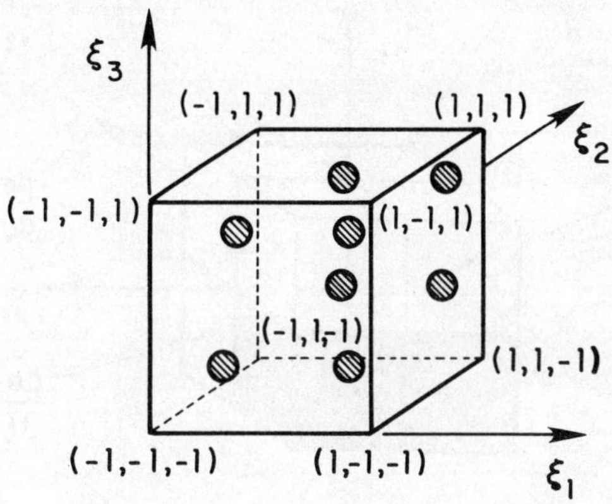


Figure 1: Local $\xi_i$ co-ordinate system. Orthogonal collocation points are indicated diagrammatically by circles.

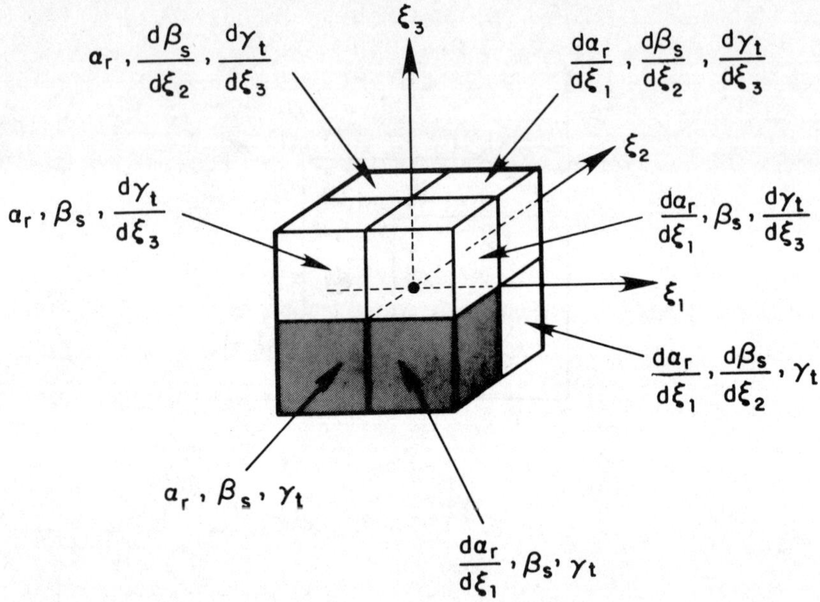


Figure 2: Basis function components associated with passes along $\xi_1, \xi_2, \xi_3$. The first pass along $\xi_1$ would consider $\alpha_r, \beta_s, \gamma_t$ and $d\alpha_r/d\xi_1$, $\beta_s$, $\gamma_t$, the shaded cubes. Each small cube is associated with one collocation point and each element contains eight cubes.

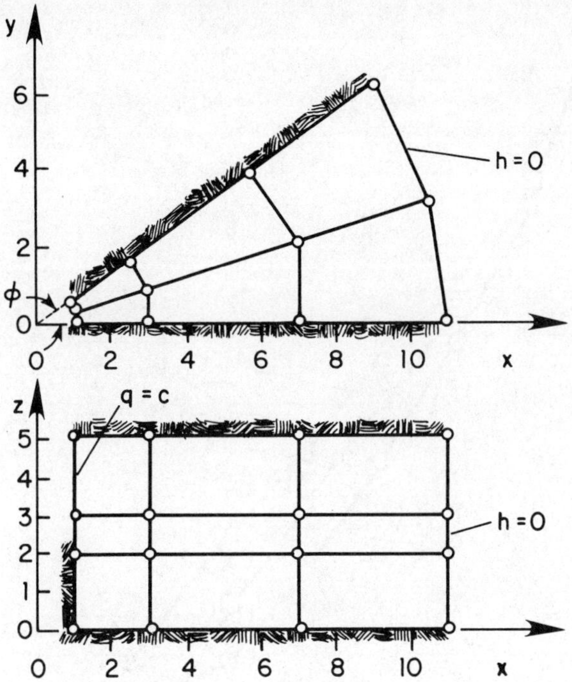


Figure 3: Collocation finite-element net: h is head, q is specified normal gradient.

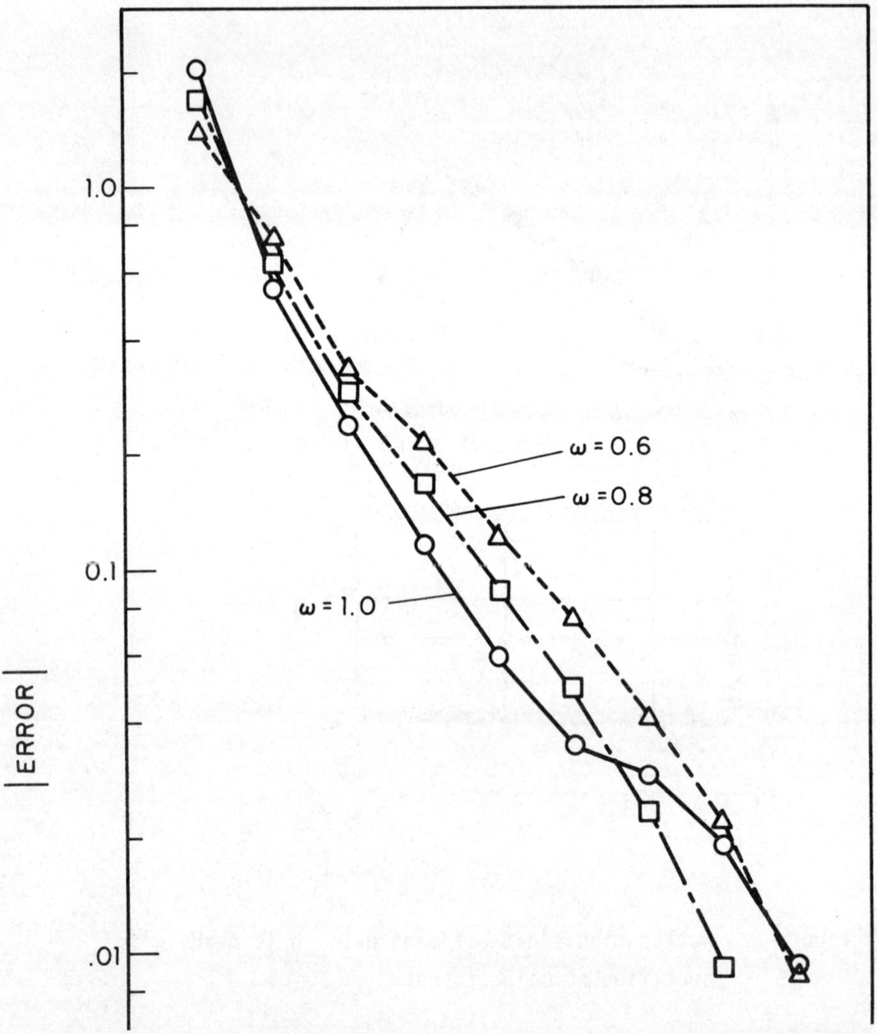


Figure 4: Rate of convergence of iterative ADC for problem of figure 3.

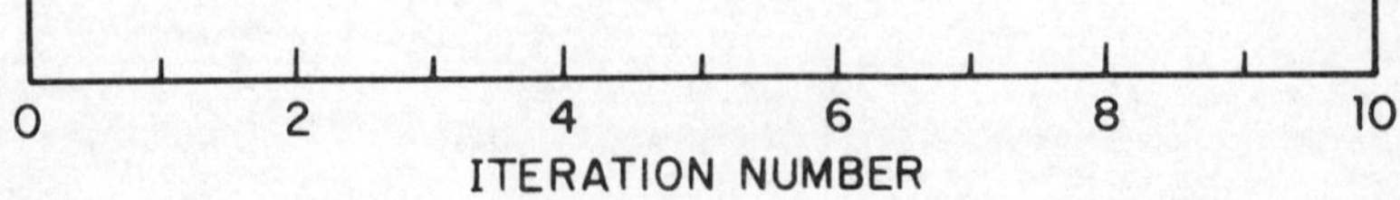

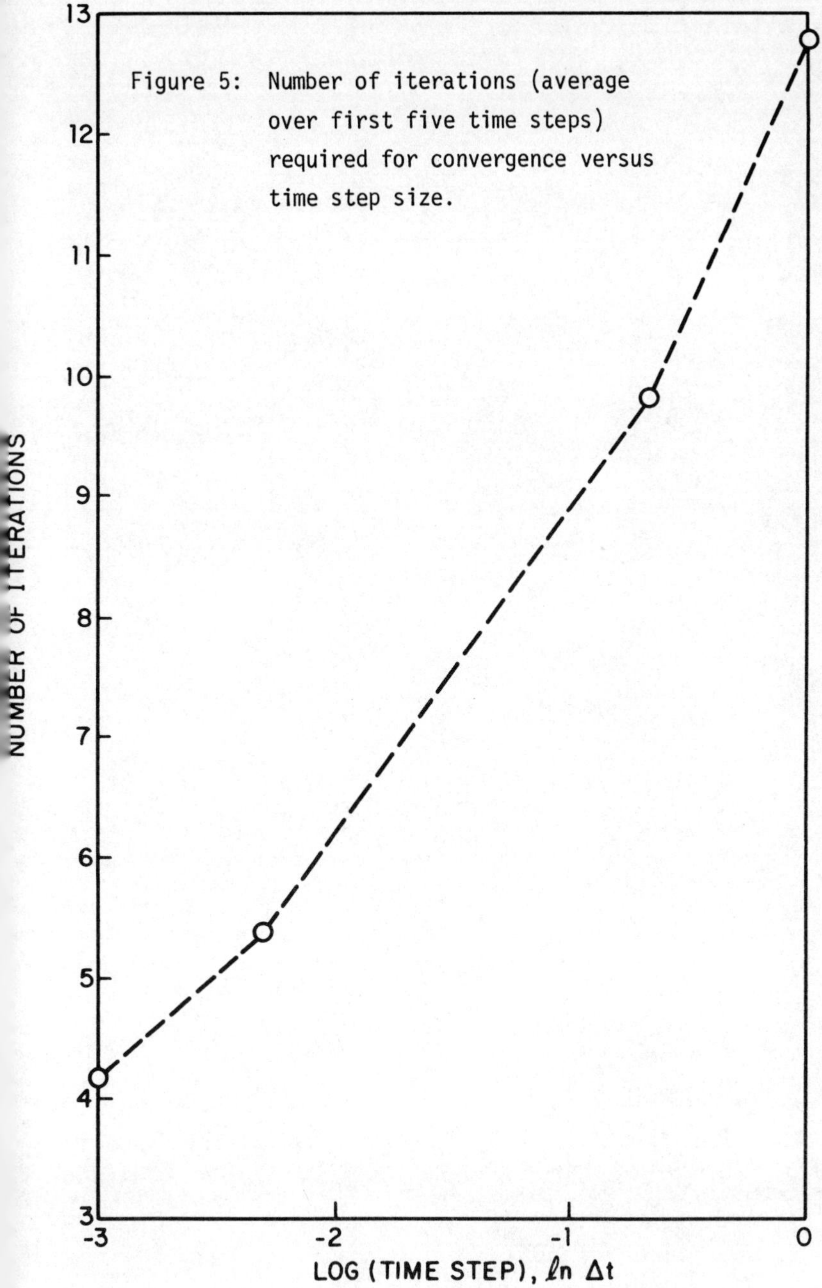


Figure 5: Number of iterations (average over first five time steps) required for convergence versus time step size.

# THE DISCRETE ELEMENT METHOD FOR DIFFERENT UPWIND SCHEMES AND CURVING BOUNDARIES

J. Häuser, D. Eppel
Institut für Physik, GKSS-Research Center
2054 Geesthacht, Germany

F. Tanzer
1. Phys. Institut, Universität Giessen
6300 Giessen, Germany

For the assessment of environmental impact on the aquatic life simulation of transport of discharged pollutants in tidal rivers, bays, and estuaries is essential. Hence,simulation models should be capable to predict the propagation of plumes caused by discharges of industrial sites,municipalities, and power plants. It is for that purpose that UTRANS (Unified Transport) model [1] has been developed. The present version of the model allows the calculation of passive transport phenomena for any type of pollutant both conservative and nonconservative (e.g., temperature). Flow field and water levels are either calculated by harmonic functions or by coupling a hydrodynamic model. The calculated examples of this paper use the simple, but highly accurate 1D model FLUSS [2]. FLUSS calculates cross-section averaged velocities from measured water-levels. In addition, work is in progress to couple UTRANS to Eraslan's SEDONE [3] model, which calculates the 1D flow-field solving the 1D Navier-stokes equation, incorporating three-dimensional effects. In the near future the coupling of Thacker's [4] 2D, irregular grid finite-difference storm surge model will be investigated in order to simulate the complete transport process for any 2D multiconnected irregularly shaped region of interest.

Although transport of matter and energy in river flow problems and oceanic flow is governed by the same fundamental equations the nature of plume modeling in rivers is aggravated by the problem of irregular geometry. It is well known from infrared measurements that thermal plumes in the Elbe river are closely following the shorlines having considerable longitudinal extension (in particular for tidal rivers with severe reversing flow conditions). The width of such a plumes varies, in general, only between 200 - 400 m. Hence, this type of transport problems is geometry dominated. Thus, an inaccurate approximation of the shoreline geometry may lead to substantial errors near the boundary. Since that is the region of the plume the calculated pollutant distribution may suffer from these inaccuracies.

The present paper discusses a transient, 2D far-field model for impact analysis of pollutant discharges. The model presented performs numerical simulation of short time (within tidal cycle) transient pollutant distributions. The contributions of vertical variations of velocity component and pollutant concentration are incorporated in the development of the depth-averaged integral form of the respective conservation principles. Upon spatially integrating the conservation equations over the enclosure surfaces of a DE, a system of weakly coupled, nonlinear ordinary-differential equations is obtained. Since the spatial integration results in a non-stiff mathematical system, the present version of UTRANS utilizes either a Runge-Kutta method with Fehlberg's scheme [5, 6], where six function evaluations are used to provide a local error estimate along with step-size control, or the standard Runde-Kutta-Gill method without step-size control. For systems of ordinary differential equations which arise from convection type problems (that is, good stability properties along the imaginary axis are demanded) it can be proved that the classical fourth order Runge-Kutta methods possess the maximally stable amplification factor [7]. Hence, these methods are preferred against predictor-corrector schemes or extrapolation procedures.

While the time integration of the resulting system of ordinary-differential equations poses no severe problems, the spatial integration over the enclosure surfaces of a DE gives rise to the question of accuracy of the obtained numerical results. Accuracy is a matter of both the exact representation of a solution area having curving shorelines and appropriate mathematical methods. Therefore the use of irregular grids for the representation of the curving boundaries is a significant improvement over conventional models using uniform grids, which forces the boundaries to be represented in a stair-step fashion. It is, however, near the shore-line that the calculated values for pollutant concentration are most needed. Hence, resorting to high convective order schemes [8, 12] or flux-correcting transport techniques [13,16] is of no value if only a rough geometrical representation of the solution area is utilized. However, not all of the high-order convective schemes can be used since the number of necessary neighboring points may not be available when irregular grids are used. In most river simulation problems the assumption of a stationary source (e.g., thermal discharge of a power plant) is valid. Hence, most of the time Raithby's criterion [17] is fulfilled and second-upwind (donor-cell) [18], using the local Peclet number, should give the desired results. The following upwind schemes are discussed:

(i) second upwind (donor-cell method) for approximation of convective terms,

(ii) second upwind along with central differencing depending upon the grid Peclet number,

(iii) second order donor-sell method [19],

(iv) quadratic interpolation upstream differencing,

(v) quadratic interpolation upstream differencing with cross terms.

First, these methods are employed to a solution domain using an equidistant mesh. The flow field is assumed to be uniform and a constant depth of 1 m is used. Neumann boundary conditions $\frac{\partial c}{\partial n} = 0$ (c denotes any concentration) are utilized for solid boundaries while Dirichlet conditions are specified for open boundaries for the test problem.

Second, and much more important for the behavior (that is, convection stability) of the above mentioned numerical schemes, is their performance when applied to calculation of temperature distributions in the tidal river Weser. Calculations are performed for Oct. 4, 1979 in the vicinity of the discharge of the power plant KKU (length of solution area ca. 15 km). Here the region of interest is irregular with curved boundaries and a highly irregular topography of the river-bottom. To obtain satisfactory results it is necessary to accurately reproduce the topography in the numerical model (see also [20]). Numerous field-measured data, including various temperature profiles at different cross-sections were recorded for that day [21]. Hence, detailed comparisons between measured data and computed results were possible. Thus, a thorough evaluation of the mathematical methods employed was performed (a detailed discussion of Weser results is also given in [22]).

Although additional work on UTRANS remains to be completed results from the application of the Weser and Elbe examples as well as comprehensive comparisons with field-measured data demonstrate that the model behaves properly.

REFERENCES

[ 1] Häuser, J., Eppel, D., and F. Tanzer (in preparation): A Tutorial Introduction to the Unified Transport Model UTRANS.

[ 2] Fiedler, H., Müller, A., and D. Nolte, 1981: FLUSS - Ein eindimensionales Modell des Wärme- und Stofftransportes in Flüssen. Teil I: Grundlagen. GKSS 81/E/12, Geesthacht, 52 pp.

[ 3] Hetrick, D.M., Eraslan, A.H., and M.R. Patterson, 1979: SEDONE: A Computer Code for Simulating Tidal-Transient, One-Dimensional Hydrodynamic Conditions and Three-Layer, Variable-Size Sediment Concentrations in Controlled Rivers and Estuaries. NUREG/CR-0430 ORNL/NUREG ITM-256, 295 pp.

[ 4] Thacker, W.C., 1978: Irregular-Grid Finite-Difference Techniques for Storm Surge Calculations for Curving Coastlines. Marine Forecasting, pp. 261 - 283.

[ 5] Fehlberg, E., 1970: Klassische Runge-Kutta Formeln vierter und niedriger Ordnung mit Schrittweiten-Kontrolle und ihre Anwendung auf Wärmeleitungsprobleme. Computing 6, pp. 61 - 71.

[ 6] Shampine, L.F. and C.A. Allen, 1973: Numerical Computing. W.B. Saunders. 258 pp.

[ 7] Praagman, N., 1979: Numerical Solution of the Shallow Water Equations by a Finite Element Method. T.H. Delft, Mathematics Department. 149 pp.

[ 8] Fischer, K., 1978: Convective Difference Schemes and Hermite Interpolation. Intl. J. for Numerical Methods in Eng., Vol. 12, pp. 931 - 940.

[ 9] Glass, J. and W. Rodi, 1981: A Higher Order Numerical Scheme for Scalar Transport, SFB 80/T/193, Universität Karlsruhe, 87 pp.

[10] Leonhard, B.P., 1979: A Stable and Accurate Convective Modelling Procedure Based on Quadratic Upstream Interpolation. Computer Math. in Applied Mech. and Eng., 19, pp. 55 - 98.

[11] Leonhard, B.P., 1981: A Stable, Accurate, Economical, and Comprehendible Algorithm for the Navier-Stokes and Scalar Transport Equations. Numerical Methods in Laminar and Turbulent Flow, Pineridge Press, pp. 543 - 553.

[12] Häuser, J., Eppel, D. and F. Tanzer, 1981: Solution of Fluid Flow Problems by a Direct Spline Interpolation Method. Numerical Methods in Laminar and Turbulent Flow, Pineridge Press, pp. 1165 - 1185.

[13] Boris, J.P. and D.L. Book, 1976: Solution of Continuity Equations by the Method of Flux-Corrected Transport. Methods in Computational Physics, Vol. 16, pp. 85 - 128.

[14] Book, D.L. (Ed.), 1981: Finite Difference Techniques for Vectorized Fluid Dynamics Calculations. Springer Verlag, 226 pp.

[15] Zalesak, S.T., 1979: Fully Multidimensional Flux-Corrected Transport Algorithms for Fluids. J. Comp. Phys., 31, pp. 335 - 362.

[16] Eppel, D., Häuser, J. and F. Tanzer, in preparation: Numerical Comparisons of Several Standard FD-Methods with FCT-Algorithms.

[17] Raithby, G.D., 1976: A Critical Evaluation of Upstream Differencing applied to Problems Involving Fluid Flow. Computer Methods in Applied Mechanics and Engineering, 9, pp. 75 - 103.

[18] Roache, P.J., 1976: Computational Fluid Dynamics. Hermosa Publishers. 446 pp.

[19] Hirt, C.W., 1968: Heuristic Stability Theory for Finite Difference Equations. J. of Comp. Phys., 2, pp.339-355.

[20] Holz, K.P., Nitsche, G. and Volkhard Wetzel, 1981: Verifikation eines numerischen Tidemodells, Wasserwirtschaft, 71, pp. 289 - 294.

[21] Institut für Physik der GKSS, 1980: Gewässeranalytische Untersuchungen auf der Unterweser im Herbst 1979. GKSS/E/27, Geesthacht, 239 pp.

[22] Häuser, J., Eppel, D., Müller, A., Nehlsen, A. and F. Tanzer, 1981: A Thermal Impact Assessment Model with Measured Field Data Applied to the Tidal River Weser. III. Waste Heat Management and Utilization Conference, Hemisphere Publishing Corporation, Washington DC, 20 pp.

# FINITE ELEMENT TECHNIQUES IN TRANSPORT PHENOMENA

J.C. Heinrich
E. Envia

Department of Aerospace and Mechanical Engineering
University of Arizona
Tucson, Arizona 85721 USA

## INTRODUCTION

A wide spectrum of processes involving transport of physical quantities of a variety of types can be successfully modeled by assuming that the transport takes place mainly through the basic processes of diffusion and convection. This assumption leads to governing equations or systems of equations of the form

$$\frac{D\phi}{Dt} = \underset{\sim}{\nabla} K \underset{\sim}{\nabla} \phi + S \tag{1}$$

where $\phi$ represents the transported quantity (in the present context $\phi$ may be heat, velocity, contaminent, etc); $\frac{D\phi}{Dt}$ is the total derivative of $\phi$; $\underset{\sim}{\nabla} \equiv \left(\frac{\partial}{\partial x_i}\right)^T$ is the "del" operator; K is the diffusion coefficient and S represents sinks and sources.

Numerical solutions of these equations are necessary in many practical situations, and in recent years, finite element methods of approximation have become increasingly popular among scientists and engineers. There are two main reasons for finite element methods to be attractive. For one, it exhibits optimal global approximation properties in self-adjoint elliptic problems, and even though these may not be useful in non-self-adjoint or hyperbolic equations, semi-discrete Galerkin type approximations of space-time problems show that high accuracy can be attained and strongly encourage the use of finite element methods in parabolic and hyperbolic problems. On the other hand, finite element formulations provide great flexibility when some generality is expected of the numerical model, to be able to handle diverse geometries, different types of boundary conditions, variable physical properties and even schemes of different orders of approximation.

It appears, however, that all too often the desire of generality leads to misuse of proposed models in situations that lie out of their intended range of applicability without an adequate supporting analysis. This invariably leads to inaccurate or incorrect results, and is very easy to overlook in Equation (1), where special cases range from the extreme of changing the type of operator to be discretized to the more subtle complications (but maybe even more difficult to solve) of dealing with different time and/or length scales in the same problem.

Despite the fact that significant progress has been made in the finite element modeling of transport processes, much work still remains to be done, especially in the areas of highly convective flows, hyperbolic equations and the modeling of problems in three spatial dimensions. The most serious difficulty with regard to finite element solutions of practical flow situations would appear to be its yet unsatisfactory efficiency. We also find, however, situations where new schemes are needed that will incorporate the physics of the problems better.

In the next section, we present the model equations to be discussed. In section 3, we give a summarized account of the work that has been done so far in the area of finite element modeling of convective transport. We must point out that this section will discuss only a limited number of contributions from today's enormous amount of literature on the subject, and, therefore, we are forced to leave out of this discussion a number of noteworthy contributions in the area. In section 4, we present some preliminary results of some aspects of our own work in this area and, finally, in section 5, some conclusions and recommendations based on our own experience are given.

## 2. TRANSPORT EQUATIONS

Throughout this paper, we will consider almost exclusively equations where convection is important. We consider, first, the linear convective diffusion equation, which we write in two space dimensions as

$$\frac{\partial\phi}{\partial t} + u\frac{\partial\phi}{\partial x} + v\frac{\partial\phi}{\partial y} = \frac{\partial}{\partial x}\left(K\frac{\partial\phi}{\partial x}\right) + \frac{\partial}{\partial y}\left(K\frac{\partial\phi}{\partial y}\right) + S, \tag{2}$$

where u and v are the components of a known velocity field independent of $\phi$. Equation (2) is assumed to be valid over a simply connected region $\Omega$ with boundary $\Gamma$ where for simplicity, we consider only two types of boundary conditions

$$\phi = \bar{\Phi} \quad \text{on} \quad \Gamma_1, \text{ and} \tag{3}$$

$$-K\left(\frac{\partial\phi}{\partial x}n_x + \frac{\partial\phi}{\partial y}ny\right) = 0 \quad \text{on} \quad \Gamma_2 = \Gamma - \Gamma_1, \tag{4}$$

and initial condition

$$\phi(x,y,0) = \phi_o(x,y). \tag{5}$$

Here, $n_x$ and $n_y$ are components of the unit outward normal to $\Gamma$.

In one space dimension, Equation (2) reduces to

$$\frac{\partial\phi}{\partial t} + u\frac{\partial\phi}{\partial x} = \frac{\partial}{\partial x}\left(K\frac{\partial\phi}{\partial x}\right) + S, \tag{6}$$

with appropriate forms of the initial and boundary conditions. Equation (6) will be used extensively in section 4.

The second class of transport problems considered here is given by the non-linear two-dimensional equations for viscous incompressible flows. They are:

Continuity: $\frac{\partial u}{\partial x} + \frac{\partial v}{\partial y} = 0$ (7)

Momentum conservation in the x-direction:

$$\frac{\partial u}{\partial t} + u\frac{\partial u}{\partial x} + v\frac{\partial u}{\partial y} = -\frac{\partial p}{\partial x} + \nu[2\frac{\partial^2 u}{\partial x^2} + \frac{\partial}{\partial y}(\frac{\partial u}{\partial y} + \frac{\partial v}{\partial x})] \quad (8)$$

Momentum conservation in the y-direction:

$$\frac{\partial v}{\partial t} + u\frac{\partial v}{\partial x} + v\frac{\partial v}{\partial y} = -\frac{\partial p}{\partial y} + \nu[\frac{\partial}{\partial x}(\frac{\partial u}{\partial y} + \frac{\partial v}{\partial x}) + 2\frac{\partial^2 v}{\partial y^2}] \quad (9)$$

The boundary conditions associated with these equations will be those of prescribed velocities in part of the boundary, and prescribed surface stresses in the remaining of it. A prescribed velocity field is given as initial condition. With Equations (3) and (4), we associate the one-dimensional Burgers equation

$$\frac{\partial u}{\partial t} + u\frac{\partial u}{\partial x} = \nu\frac{\partial^2 u}{\partial x^2} \quad (10)$$

with appropriate boundary and initial conditions. In the above equations, p is the modified fluid pressure and $\nu$ the kinematic viscosity.

For discretization purposes, the above equations are rewritten in a weighted residuals averaged form in the space variables. For example, Equation (1) becomes

$$\int_\Omega [W(\frac{\partial \phi}{\partial t} + u\frac{\partial \phi}{\partial x} + v\frac{\partial \phi}{\partial y}) + K(\frac{\partial W}{\partial x}\frac{\partial \phi}{\partial x} + \frac{\partial W}{\partial y}\frac{\partial \phi}{\partial y})]d\Omega = \int_\Omega WSd\Omega, \quad (11)$$

where Green's theorem has been used to reduce the order of differentiation as is standard in finite element methods. The function $\phi$ is then approximated by

$$\phi(x,y,t) \cong \sum_{i=1}^{m} N^i(x,y)\cdot\phi_i(t), \quad (12)$$

and we seek solutions over the m-dimensional space spanned by the basis or shape function $N^i(x,y)$ which will be the standard piece-wise bilinear functions defined over quadrilateral elements. We will allow for the weighting functions $W^i(x,y)$ to be possibly different than the shape functions in a general Petrov Galerkin form. Substitution of Equation (12) and the prescribed weighting functions into Equation (11) leads to a system (in general non-linear) of ordinary differential equations

$$\underset{\sim}{M}\,\dot{\underset{\sim}{\phi}} + \underset{\sim}{K}\,\underset{\sim}{\phi} = \underset{\sim}{F} \quad (13)$$

where

$$m_{ij} = \int_\Omega W^i N^j d\Omega, \quad k_{ij} = \int_\Omega [W^i(u\frac{\partial N^j}{\partial x} + v\frac{\partial N^j}{\partial y}) + K(\frac{\partial W^i}{\partial x}\frac{\partial N^j}{\partial x} + \frac{\partial W^i}{\partial y}\frac{\partial N^j}{\partial x})]d\Omega,$$

$$f_i = \int_\Omega W^i S d\Omega \quad \text{and} \quad \phi_i = \phi(x_i, y_i, t).$$

Time marching schemes of several types can then be used to discretize the time variable, resulting in a system of algebraic equations in the unknown values $\phi(x_i, y_i, t^m) \equiv \phi_i^m$.

In one-dimensional problems, piecewise linear basis functions will be considered. Obviously, other types of elements can and have been used in the discretization of the above equations, especially the linear and quadratic triangles and the biquadratic Lagrangian rectangle, however, the most natural element for convection problems appear to be the linear element in one-dimension and the bilinear element in two-dimensions. Our discussions will be centered almost exclusively in these elements.

## 3. FINITE ELEMENT ALGORITHMS

Early attempts to obtain finite element approximations to Equation (1) immediately encountered difficulties in situations where convection dominates in the problem, and it became evident that something special had to be done for finite element algorithms to produce meaningful solutions.

In the context of Equation (2), Adey and Brebbia [1] proposed an algorithm based on a mixed Euler-Lagrange formulation, in which, at each time step, diffusion is allowed to take place and the convection is calculated afterwards, assuming a fixed fluid mass in each element and a known velocity field. This algorithm has been successfully used in linear problems with quadratic triangular elements.

Along similar lines, an algorithm has been proposed more recently by Varoḡlu and Finn [2]. They used finite elements in space and time, and discretized the equation in such a way that (e.g., in one space dimension), a combination of linear triangles and bilinear rectangles is used where the element sides follow the characteristic directions. Extension of this algorithm to treat problems in two space dimensions is given in Reference [3].

In Zienkiewicz, et.al. [4], it was proposed seeking an algorithm that would follow the idea of upwinding in finite difference methods. This led to the first Petrov-Galerkin type schemes, for the particular case of steady state solutions of Equation (2) in Christie, et.al. [5] for one space dimension, and in Heinrich, et.al. [6] for two-dimensional steady state situations. The basic idea in these methods is to break the symmetry that results from standard Galerkin approximations of the convective term by using appropriate test functions which, in general, are different from the basic functions. In references [5] and [6], test functions were introduced which were obtained from the basic piecewise linear or bilinear functions by addition of a quadratic perturbation. Appropriate choice of the magnitude of this perturbation then led to a successful class of algorithms for steady-state convection and diffusion. The combination of test and trial functions mentioned above is not unique, and a variety of algorithms involving different types of elements and weighting functions have been proposed in the literature, which has been reviewed in Heinrich and Zienkiewicz [7]. The basic concepts in most of these algorithms are, however, contained in References [5], [6].

Still in the context of steady state convection-diffusion problems, the idea of adding an anisotropic balancing dissipation was put forward in Kelly, et.al. [8], Hughes and Brooks [9] and Griffiths and Mitchell [10]. It essentially developed from the fact that in one dimensional problems, the Petrov-Galerkin algorithm mentioned above is equivalent to a standard Galerkin scheme applied to a modified equation with an additional diffusion term. In two (or three) dimensional problems, consider the constant coefficients equation

$$-\underset{\sim}{\nabla}^T \underset{\sim}{K} \underset{\sim}{\nabla} \phi + \underset{\sim}{u}^T \underset{\sim}{\nabla} \phi = 0 \tag{14}$$

where $\underset{\sim}{u}$ denotes the velocity vector field. The diffusivity tensor $\underset{\sim}{K}$ for isotropic diffusion in a Cartesian coordinate system becomes

$$\underset{\sim}{K} = k \underset{\sim}{I} \tag{15}$$

where $\underset{\sim}{I}$ is the identity matrix and k the diffusivity constant. We then seek to solve a modified equation given by

$$\underset{\sim}{\nabla}^T (\underset{\sim}{K} + \overline{\underset{\sim}{K}}) \underset{\sim}{\nabla} \phi - \underset{\sim}{u}^T \underset{\sim}{\nabla} \phi = 0 \tag{16}$$

where $\overline{\underset{\sim}{K}}$ is a highly anisotropic tensor with principal direction oriented parallel to the known velocity vector $\underset{\sim}{u}$. A transformation is, therefore, required to coordinates (t,m) parallel and perpendicular to $\underset{\sim}{u}$. In that system we have

$$\overline{\underset{\sim}{K}}(t,m) = \overline{k} \begin{bmatrix} 1 & 0 \\ 0 & 0 \end{bmatrix}$$

where $\overline{k}$ is calculated as in the one dimensional problem and is given by

$$\overline{k} = \frac{1}{2} |\underset{\sim}{u}| h \alpha, \tag{17}$$

Here h is the local grid size and $\alpha$ a parameter. For the one dimensional constant coefficients case, it has been shown that the value

$$\alpha = \coth \frac{|\underset{\sim}{u}| h}{2k} - \frac{2k}{|\underset{\sim}{u}| h} \tag{18}$$

renders the exact solution to the equation at nodal points. The matrix $\overline{K}$ in the (x,y) coordinate system is then obtained by

$$\overline{K}_{(x,y)} = T^T \, \overline{K}_{(t,m)} \, T \tag{19}$$

$$\text{where } T = \begin{bmatrix} \cos\theta & \sin\theta \\ -\sin\theta & \cos\theta \end{bmatrix} \tag{20}$$

and $\theta$ is the angle that the velocity vector, $\underset{\sim}{u}$, defines with the positive x-coordinate axis.

This scheme is especially attractive because diffusion is added only in the direction of flow, and the amount of diffusion to be added can be locally calculated with great accuracy from the one dimensional case. It is also important to point out that this scheme can be formulated as a Petrov-Galerkin scheme in more than one way.

Extensions of the above Petrov-Galerkin and added dissipation algorithms to the solution of the non-linear Navier-Stokes Equations (7)-(9) have been performed in Heinrich and Zienkiewicz [11], Hughes, et.al. [12] and others. However, extensions of the Petrov-Galerkin methods turned out unsatisfactory (arguments towards explaining why are given in the next section), and although successful applications of the added dissipation concept to the transient Navier-Stokes equations have been performed (e.g. Brooks and Hughes [13]), our level of understanding in this area is still limited and further theoretical analysis is needed in this type of non-linear system.

For transient linear problems, governed by Equation (2), a large amount of work is found in the published literature, however, as far as application of Petrov-Galerkin methods is concerned, most of them are restricted to straightforward use of available elements and test functions developed for the steady state case, and with some exceptions (e.g., Griffiths [14], Griffiths and Mitchell [15]), they lack an appropriate supporting analysis. Not surprisingly, the results indicate little or no advantage in their use and one analysis performed by Shapiro and Pinder [16] showed at least one such scheme to be divergent. In the next section, we present preliminary results of an analysis directed towards better understanding the use of Petrov-Galerkin methods in Equation (2) for time dependent problems. We should also mention that application of a Petrov-Galerkin method to a class of degenerate parabolic equations was successful in producing the first direct solutions ever of these equations, (see Bergman and Heinrich [17]).

Finally, for the particular case of linear hyperbolic equations, obtained as a limiting case of Equation (2) when the diffusivity K is zero, (that is, transport by convection only), a special Petrov-Galerkin algorithm was constructed by Donea, et.al. [18], using biquadratic trial functions, a modification of the added diffusion algorithm was proposed by Hughes and Brooks [9], which shows superior performance. In the same context, a series of Petrov-Galerkin type algorithms was constructed by Morton and Parrot [19] for the one dimensional case, which deserves special attention. These schemes are based on standard difference discretizations of the time variable such as the forward Euler method, the Theta method and the Leap-Frog method. The test functions in the spatial variable were chosen so that, in conjunction with linear trial functions, the final scheme has the property that waves are propagated without distortion in the special case when the characteristics pass through the nodes at each time level. The resulting algorithms showed a marked improvement in the stability and accuracy over the standard schemes. These methods have since been further extended by Griffiths [20].

In this section, we have presented an account of the basic finite element algorithms that have been developed, which incorporate special features to deal with convective transport problems. A detailed review of the literature is not possible due to space limitations, and is not practicable today without a careful selection of the growing literature in the area. It should be clear from the above discussion that much more progress remains to be done, and active research in the area is currently being carried on among others by most of the aforementioned authors.

## 4. ONE DIMENSIONAL ANALYSIS

Error bounds and convergence proofs for finite element algorithms are normally obtained in terms of Sobolev or energy norms, which are more natural for integral formulations of the weighted residuals type. However, in practice, when meshes of a size bounded away from zero are used, these norms are sometimes too permissive, in the sense that only the globally averaged properties are satisfied, and no local accuracy need be achieved. This is a source of difficulty in many finite element algorithms, and for this reason we have chosen to look at a more basic local analysis of some algorithms for convective transport based on Taylor series expansions, although this analysis is certainly limited, we will see that it can still give us valuable information about the schemes. We will restrict ourselves here to the one dimensional Equation (6), the fully implicit backward time marching scheme and piecewise linear trial functions. Boundary conditions will not be included, since our aim is to determine the behavior of the schemes at general internal nodes. Obviously, the boundary condition effects will have to be included in a final analysis. To establish the ideas, first consider the steady-state equation

$$-(K + \bar{K}) \frac{d^2\phi}{dx^2} + u\frac{d\phi}{dx} = 0, \tag{21}$$

where $\bar{K}$ is an added diffusion which we write in the form given in Equation (17), the mesh size, h, will be assumed uniform.

The Galerkin discretization of Equation (21) using piecewise linear trial functions yields, for a typical node, a difference equation of the form

$$(1 + (\alpha - 1)\frac{\gamma}{2})\phi_{i+1} - 2(1 + \alpha\frac{\gamma}{2})\phi_i + (1 + (\alpha + 1)\frac{\gamma}{2})\phi_{i-1} = 0 \tag{22}$$

$$\text{where } \gamma = \frac{uh}{K} \tag{23}$$

Expanding the above difference equation in Taylor series, and using the original differential equation to replace higher order derivatives in terms of $\phi''$, we obtain

$$(22) = K\,\phi_i'' - u\phi_i' + \frac{K}{\gamma}\left[(\frac{2}{\gamma} + \alpha)(\cosh\gamma - 1) - \sinh\gamma\right]\phi_i'' \tag{24}$$

The last term to the right hand side of Equation (24) is the exact truncation error, and can be rewritten as

$$f(\alpha,\gamma)\phi_i'' = \frac{K}{\gamma}\left(\left(\frac{2}{\gamma}+\alpha\right)\tanh\frac{\gamma}{2} - 1\right)\sinh\gamma\ \phi_i''. \tag{25}$$

The function $f(0, \gamma)$ in Equation (25) can be easily shown to be negative for all values of $\gamma \neq 0$. We conclude, therefore, that the difference Equation (22) with $\alpha = 0$, derived from the Galerkin approximation to Equation (21), where $\bar{K} = 0$, solves a modified underdiffused equation. This justifies the use of added diffusion to improve the numerical solution. Moreover, if we require Equation (25) to be identically zero, we find that this happens for $\alpha = \coth\frac{\gamma}{2} - \frac{2}{\gamma}$, which is identical to Equation (18), so that for linear steady state conditions, the Petrov-Galerkin and added diffusion schemes are in firm setting.

A similar analysis can be performed for the transient Equation (6), although this time we cannot expect exact answers. The consistent mass, backward implicit Galerkin approximation to Equation (6) yields a typical difference equation of the form

$$\frac{h}{6\Delta t}\left[\phi_{i+1}^{M+1} + 4\phi_i^{M+1} + \phi_{i-1}^{M+1}\right] - \frac{K}{h}\left[\phi_{i+1}^{M+1} - 2\phi_i^{M+1} + \phi_{i-1}^{M+1}\right] +$$

$$\frac{u}{2}\left[\phi_{i+1}^{M+1} - \phi_{i-1}^{M+1}\right] = \frac{h}{6\Delta t}\left[\phi_{i+1}^{M} + 4\phi_i^{M} + \phi_{i-1}^{M}\right] \tag{26}$$

Proceeding as in the previous steady state case, we can reduce all higher order derivatives to terms involving $\phi''$, $\dot{\phi}'$, $\dot{\phi}''$,..., and can write

$$(26) = (\dot{\phi}-K\phi''+u\phi')_i^{M+1} + \frac{K}{\gamma}\left(1-\frac{2}{\gamma}\tanh\frac{\gamma}{2}\right)\sinh\gamma(\phi'')_i^{M+1} -$$

$$\left(\frac{\Delta t}{2} + \frac{K}{u^2\gamma}\left(1-\frac{2}{\gamma}\tanh\frac{\gamma}{2}\right)\right)\sinh\gamma(\ddot{\phi})_i^{M+1} + \left[\frac{\Delta t^2u^2}{6} +\right.$$

$$\left.\frac{2K^2}{u^2\gamma}\left(1-\frac{2}{\gamma}\tanh\frac{\gamma}{2}\right)\sinh\gamma\right](\dot{\phi}'')_i^{M+1} + \text{HOT} \tag{27}$$

where we have replaced $\dot{\phi}'$ in terms of $\ddot{\phi}$ and $\dot{\phi}''$ and the higher order terms depend on $\Delta t^2$, $h^4$ and strictly higher order derivatives which cannot be reduced. The coefficients of $\phi''$, $\ddot{\phi}$ and $\dot{\phi}''$ in Equation (27) are, therefore, exact, and we see that Equation (26) solves a modified differential equation (of infinite order) which is, again, underdiffused, by exactly the same amount as in the steady state case.

An added diffusion discretization of Equation (6) yields a difference equation which can be shown to be equal to

$$(\dot{\phi}-K\phi''+u\phi')_i^{M+1} + \frac{K}{\gamma}\left[1-\left(\frac{2}{\gamma}+\alpha\right)\tanh\frac{\gamma}{2}\right]\sinh\gamma(\phi'')_i^{M+1} + \left[\frac{u\Delta t}{2} + \frac{K}{u\gamma}\right.$$

$$\left.\left[1-\left(\frac{2}{\gamma}+\alpha\right)\tanh\frac{\gamma}{2}\right] + \frac{\alpha h}{2}\right]\sinh\gamma(\dot{\phi}')_i^{M+1} + \left[-\frac{K\Delta t}{2} + \frac{u^2\Delta t^2}{6} + \frac{K^2}{u^2\gamma}\right.$$

$$\left.\left(1-\left(\frac{2}{\gamma}+\alpha\right)\tanh\frac{\gamma}{2}\right)\sinh\gamma + \frac{\alpha Kh}{2u}\right](\dot{\phi}'')_i^{M+1} + \text{HOT} \tag{28}$$

Choosing $\alpha$ from Equation (18) and using $\ddot{\phi}$ instead of $\dot{\phi}'$ thus reduces to

$$(\dot{\phi}-K\phi''+u\phi')_i^{M+1} + \left(-\frac{\Delta t}{2} - \frac{\alpha h}{2u}\right)(\ddot{\phi})_i^{M+1}\left(\frac{u^2\Delta t^2}{6} + \frac{\alpha Kh}{u}\right)(\dot{\phi}'')_i^{M+1} + \text{HOT} \tag{29}$$

where the diffusion term has been eleminated, but an extra error has been added to the time discretization term $\ddot{\phi}$ and to the dispersion term $\phi''$. In fact, a spatial dependence has been added to the time discretization error. On the other hand, if an equivalent Petrov-Galerkin discretization is performed, where the mass matrix is now modified by the weighting function with $\alpha = \coth\frac{\gamma}{2} - \frac{2}{\gamma}$, we obtain, for the right hand side

$$(\dot{\phi}-K\phi''+M\phi')_i^{M+1} - \frac{\Delta t}{2}(\ddot{\phi})_i^{M+1} + \left(\frac{u^2\Delta t^2}{6} + \frac{\alpha Kh}{2u} - \frac{\alpha uh\Delta t}{4}\right)(\dot{\phi}'')_i^{M+1} + \text{HOT} \quad (30)$$

where no additional error is introduced in the time discretization but two additional terms appear in the coefficient of the dispersion term. If a lumped mass formulation is used in conjunction with Petrov-Galerkin or added dissipation, the difference equations for $\alpha = \coth\frac{\gamma}{2} - \frac{2}{\gamma}$ are equal to

$$(\dot{\phi}-K\phi''+u\phi')_i^{M+1} + \left(-\frac{\Delta t}{2} - \frac{\alpha h}{2u}\right)(\ddot{\phi})_i^{M+1} + \left(-\frac{h^2}{6} + \frac{u^2\Delta t^2}{6} + \frac{\alpha Kh}{u}\right)(\phi'')_i^{M+1} + \text{HOT} \quad (31)$$

which differs from Equation (29) by an additional dispersion term.

From the above results, some conclusions can be drawn:

1) Even though we are adding the correct amount of dissipation to the operator when $\alpha = \coth\frac{\gamma}{2} - \frac{2}{\gamma}$, we are also adversely affecting other terms in the operator and a second corrective factor should be introduced in transient equations to better approximate the original differential equation. 2) The use of non-consistent schemes to treat the mass matrix will not improve the scheme, this is consistent with reported numerical results.

The limiting cases, $u = 0$ and $K = 0$, are of special interest, and it is easier to treat them separately. For the pure diffusion case $u = 0$, the consistent mass scheme yields

$$(\dot{\phi}-K\phi'')_i^{M+1} + \left(\frac{h^2}{12} - \frac{K\Delta t}{2}\right)(\dot{\phi}'')_i^{M+1} + \text{HOT} \quad (32)$$

The lumped mass equations on the other hand give

$$(\dot{\phi}-K\phi'')_i^{M+1} + \left(-\frac{h^2}{12} - \frac{K\Delta t}{2}\right)(\dot{\phi}'')_i^{M+1} + \text{HOT} \quad (33)$$

Obviously, addition of dissipation is not an issue in this case. This also suggests that mass lumping is not appropriate for this kind of problem and gives us an optimal time step to use $\Delta t = \frac{h^2}{6K}$, which makes the dispersion term vanish in Equation (32). The use of this time step shows significant improvement in numerical results. In the pure convection, hyperbolic case $K = 0$, some interesting behaviors can be observed. We can look at five different cases; for each of which the difference equations equal:

i) Consistent mass Galerkin

$$(\dot{\phi}+u\phi')_i^{M+1} - \frac{\Delta t u^2}{2}(\phi'')_i^{M+1} + \frac{\Delta t^2 u^2}{6}(\dot{\phi}'')_i^{M+1} + \text{HOT} \tag{34}$$

ii) Lumped mass Galerkin

$$(\dot{\phi}+u\phi')_i^{M+1} - \frac{\Delta t u^2}{2}(\phi'')_i^{M+1} + \left(\frac{\Delta t^2 u^2}{6} - \frac{h^2}{6}\right)(\dot{\phi}'')_i^{M+1} + \text{HOT} \tag{35}$$

iii) Consistent mass Galerkin with added diffusion

$$(\dot{\phi}+u\phi')_i^{M+1} - \left(\frac{\Delta t u^2}{2} + \frac{\alpha u h}{2}\right)(\phi'')_i^{M+1} + \frac{\Delta t^2 u^2}{6}(\dot{\phi}'')_i^{M+1} + \text{HOT} \tag{36}$$

iv) Lumped mass Galerkin with added diffusion

$$(\dot{\phi}+u\phi')_i^{M+1} - \left(\frac{\Delta t u^2}{2} + \frac{\alpha u h}{2}\right)(\phi'')_i^{M+1} + \left(\frac{\Delta t^2 u^2}{6} - \frac{h^2}{6}\right)(\dot{\phi}'')_i^{M+1} + \text{HOT} \tag{37}$$

v) Consistent mass Petrov-Galerkin (of reference [6])

$$(\dot{\phi}+u\phi')_i^{M+1} - \frac{\Delta t u^2}{2}(\phi'')_i^{M+1} + \left(\frac{\Delta t^2 u^2}{6} - \frac{\alpha h u \Delta t}{4}\right)(\dot{\phi}'')_i^{M+1} + \text{HOT} \tag{38}$$

From where we can, again, draw some conclusions:

1) The over diffusive character of the backward implicit time stepping scheme is immediately evident from Equations (34)-(38)

2) Perhaps surprising is the fact that mass lumping would appear to have a positive effect in this case, by adding a coefficient to the dispersion term which makes this term vanish when the Courant number $C = \frac{u\Delta t}{h}$ equals one.

3) From Equation (38), it follows that the Petrov Galerkin method considered here does not affect the diffusion error, on the other hand, $\alpha$ can be chosen so as to eliminate the dispersion error. It is clear now why applications of this Petrov-Galerkin method to equation (6), when the diffusivity K vanishes, did not produce improvement in the solutions, when used in conjunction with the backward implicit time marching scheme.

Numerical experiments have been carried out to relate to the above theoretical results. In general, we find an excellent agreement between what is predicted and the numerical results in the two limiting cases of $u = 0$ and $K = 0$, the use of the suggested diffusions and time steps resulting in the expected changes in the solution. Experiments have also been made where an additional dispersion term $\beta\,\dot{\phi}''$ is introduced in the original equation and the parameter $\beta$ chosen so as to eliminate the $\dot{\phi}''$ term in the truncation error, which significantly improved the solutions. For the full equation, however, when both the velocity u and diffusion K are important, our results turn out to be less sensitive to modification, and we have not, so far, been able to observe the kind of improvement that we would expect, as in the limiting cases. This could be due to the fact that this is a more complex equation where, over different time scales, a different term dominates the behavior of the solution (see e.g. Lick [21]), and it could be a sign that we have to introduce more physical insight in the numerical model. These aspects, as well as algorithms using other time stepping schemes, are currently under investigation.

We now consider non-linear problems. We use as a model the simple non-linear Burger's equation in steady state, that is

$$-\nu\frac{d^2u}{dx^2} + u\frac{du}{dx} = 0 \tag{39}$$

If a Petrov-Galerkin scheme of the type proposed in Reference [6] is applied to Equation (39), the resulting non-linear difference equation for a typical interior node $x_i$ is

$$\frac{\nu}{h^2}(u_{i-1}-2u_i+u_{i+1})+\frac{1}{6h}\Big[(1-\frac{3\alpha}{2})u^2_{i+1}+u_{i+1}u_i+3\alpha u^2_i-u_iu_{i-1} - (1+\frac{3\alpha}{2})u^2_{i-1}\Big]= 0 \tag{40}$$

Restricting our attention, for the moment, only to the term in $\alpha$, we find

$$-\frac{\alpha}{4}(u^2_{i-1}-2u^2_i+u^2_{i+1})= -\alpha(\frac{1}{2}(u'_i)^2-\frac{h}{2}u_iu''_i-\frac{h^3}{8}(u''_i)^2+ \text{HOT}) \tag{41}$$

Equation (41) suggests that this Petrov-Galerkin method is not quite appropriate for this non-linear problem. In fact, it is very difficult to try to interpret the terms involved, and although the numerical evidence shows that it does stabilize the answer for a suitable range of $\alpha$, it also shows poor accuracy, and extensions to two dimensions run into convergence difficulties as can be surmised from References [11] and [12]. In an effort to use an added diffusion scheme, we can try and calculate using the Galerkin Scheme [$\alpha = 0$ in Equation (40)], an approximation to the viscosity to be added to the system. Expanding in a non-linear series and collecting terms of order $h^2$, we get on the right hand side

$$(\nu u''_i+u_iu'_i)+\frac{h^2}{3}(\frac{1}{2}u_iu'''_i+u'_iu''_i)-\frac{h^2}{12}\nu u^{iv}_i+0(h'') \tag{42}$$

the coefficient of $h^2$ can be manipulated and finally written in the form $\frac{h^2}{12}(2u'_i+\frac{u_i}{\nu})u''_i$. Since the coefficient of $u''_i$ is variable, and we would like to cancel this term by adding a viscous term of the form $-\frac{d}{dx}(\bar{\nu}\frac{du}{dx})$, the added viscosity $\bar{\nu}$ turns out to be

$$\bar{\nu} = \frac{h^2y^2}{12\nu} \tag{43}$$

Numerical experience with this added diffusion scheme, however, shows that $\bar{\nu}$, as givey by Equation (43), is too rough an approximation to the necessary added viscosity. This results in inaccurate approximations to the solution for high values of the Reynolds number. Calculating the next term in the non-linear expansion reveals that the first two terms at least are identical to the ones in the linear case, this strongly suggests that the right additional viscosity to use is $\bar{\nu} = \frac{1}{2}|u|h\alpha$ with $\alpha = \coth\frac{\gamma}{2} - \frac{2}{\gamma}$ as before. Our numerical experience so far, indeed, supports the above statement and indicates that for the steady state Navier-Stokes Equations, the added diffusion algorithm explained in section 3 and References [8] and [9] is justified. In the transient case, more work is still needed to understand the effect of the added diffusion in the solution. So far, the successful calculations performed by

Brooks and Hughes [13] are extremely encouraging. Nevertheless it appears to be inevitable that some kind of compromise between added diffusion and mesh refinement has to be reached for adequate solutions of viscous flow problems.

## 5. CONCLUSIONS

Finite element modeling of convective transport is an area of increasing research activity and great practical potential. In a relatively short period of time, many good ideas and a large number of algorithms have been contributed, of which a small sample has been mentioned in this paper. Of the many proposed finite element schemes, we feel that those of the Petrov-Galerkin and added diffusion type (which, at least, in the linear case, can be shown to be equivalent), are the most appealing, especially if application to three-dimensional situations are to be performed. So far, steady state linear equations can be adequately modeled in most cases, while transients remain somewhat evasive and still new ideas will be needed to improve the accuracy of solutions. Moreover, the concept of anisotropic balancing diffusion has been useful in greatly improving our capabilities to solve the non-linear Navier-Stokes equations, although a more solid theoretical basis is needed in this case. In the case of the Navier Stokes equations, the problems are compounded by the fact that solutions are still too costly and improvements of the efficiency of finite element algorithms constitutes an important research area.

In this paper, we have also attempted to point out the fact that some finite element algorithms have been, and continue to be, inadequately used in situations out of their intended range of applicability, without an appropriate analysis, with the result of poor performance. In the particular case of Petrov-Galerkin methods, we have presented here a preliminary attempt at establishing the local properties of these algorithms in one space dimension, through a Taylor series analysis, for one particular set of schemes has been presented, and it explains, with good precision, the results of numerical experiments reported in the literature. We conclude that: 1) Petrov-Galerkin schemes are optimal for linear steady-state equations, 2) Each variation of a particular algorithm needs to be analyzed individually, 3) Petrov-Galerkin schemes of the type considered here do not seem to be appropriate for use in non-linear Burgers or Navier-Stokes equations, however, the added diffusion scheme seems to be optimal in this case, 4) Modifications aimed only at correcting the diffusive behavior of the discretized equation are not sufficient in transient equations and additional sophistication is needed for accurate modeling of the time history, and, 5) The present analysis is insufficient and needs to be complemented with other approaches to include aspects of wave propagation, both in the differential equation and the discretized form.

Presently, work is in progress to analyze other algorithms, in particular, those based in the Crank-Nicolson second order in time scheme and the forward explicit time marching scheme, and to try to combine the present approach with other methods of analysis. Also, improved numerical schemes obtained in the context of one dimensional problems are not always easy to generalize to higher space dimensions, especially for time dependent problems. In this context, we believe that their formulation as Petrov-Galerkin, or added diffusion methods, should give a better insight into finding these extensions.

REFERENCES

1. Adey, R. and C. Brebbia (1974) "Finite Element Solution for Effluent Dispersion. Numerical Methods in Fluid Dynamics, (C. Brebbia and J.J. Connor, editors) Pentch Press, London, 325-354.
2. Varoḡlu, E and W.D. Finn (1980) "Finite Elements Incorporating Characteristics for One Dimensional Diffusion-Convection Equations", J. Comp. Phys. 34, 371-389.
3. Variḡlu, E. (1980) "A Finite Element Model Incorporating Characteristics for the Convection Diffusion Equation", Third International Conference on Finite Element Methods in Water Resources, (S.Y. Wang, C.V. Alonso, C.A. Brebbia, W.G. Gray and G.F. Pinder, editors) University of Mississippi, 3.3-3.19.
4. Zienkiewicz, O.C., R.H. Gallagher and P. Hood (1976) "Newtonian and Non-Newtonian Viscous Incompressible Flow. Temperature Induced Flows. Finite Element Solutions". The Mathematics of Finite Elements and Applications, Vol. 2 (J. Whiteman editor) Academic Press, London, 235-267.
5. Christie, I., D.F. Griffiths, A.R. Mitchell and O.C. Zienkiewicz (1976) "Finite Element Methods for Second Order Differential Equations with Significant First Derivatives", Int. J. Num. Meth. Engng. 10, 1389-1396.
6. Heinrich, J.C., P.S. Huyakorn, O.C. Zienkiewicz and A.R. Mitchell (1977) "An 'Upwind' Finite Element Scheme for Two-Dimensional Convective-Transport Equation", Int. J. Num. Meth. Engng. 11, 131-143.
7. Heinrich, J.C., and O.C. Zienkiewicz (1979) "The Finite Element Method and 'Upwinding' Techniques in the Numerical Solution of Convection Dominated Flow Problems", Finite Element Methods for Convection Dominated Flows (T.J.R. Hughes, editor) ASME publication, AMD Vol. 34, 105-136.
8. Kelly, D.W., S. Nakazawa, O.C. Zienkiewicz and J.C. Heinrich (1980) "A Note on Upwinding and Anisotropic Balancing Dissipation in Finite Element Approximations to Convective Diffusion Problems", Int. J. Num. Meth. Engng. 15, 1705-1711.
9. Hughes, T.J.R., and A. Brooks (1979) "A Multi-Dimensional Upwind Scheme with No Cross-Wind Diffusion", Finite Element Methods for Convection Dominated Flows (T.J.R. Hughes, editor) ASME publication AMD Vol. 34, 19-35.
10. Griffiths, D.F., and A.R. Mitchell (1979) "On Generating Upwind Finite Element Methods", Finite Element Methods for Convection Dominated Flows (T.J.R. Hughes, editor) ASME publication AMD Vol. 34, 91-104.

11. Heinrich, J.C., and O.C. Zienkiewicz (1979) "Solution of Non-Linear Second Order Differential Equations with Significant First Derivatives by a Petrov-Galerkin Finite Element Method", Numerical Analysis of Singular Perturbation Problems (P. Hemker and J. Miller, editors) Academic Press, London, 251-273.
12. Hughes, T.J.R., W.K. Liu and A. Brooks, (1979) "Finite Element Analysis of Incompressible Viscous Flows by the Penalty Function Formulation", J. Comp. Physics 30, 1-60.
13. Brooks, A. and T.J.R. Hughes (to appear 1982) "Streamline Upwind/Petrov-Galerkin Formulations for Convection Dominated Flow with Particular Emphasis on the Incompressible Navier-Stokes Equations", Comp. Meth. Appl. Mech. Engng.
14. Griffiths, D.F. (1979) "Towards Time Stepping Algorithms for Convective Diffusion", Numerical Analysis of Singular Perturbation Problems (P. Hemker and J. Miller, editors) Academic Press, London, 199-216.
15. Mitchell, A.R., and D.F. Griffiths (1979) "Semi-Discrete Galerkin Methods for Time Dependent Conduction Convection Methods", The Mathematics of Finite Elements and Applications III (J. Whiteman, editor) Academic Press, London, 19-34.
16. Shapiro, A., and G.F. Pinder (1981) "Analysis of Upstream Weighted Collocation Approximation to the Transport Equation", J. Comp. Physics 39, 46-71.
17. Bergman, L.A., and J.C. Heinrich (1981) "Petrov-Galerkin Finite Element Solution for the First Passage Probability and Moments of First Passage Time of the Randomly Accelerated Free Particle", Comp. Meth. Appl. Mech. Engng. 27, 345-362.
18. Donea, J., S. Giuliani and H. Laval (1979) "Accurate Explicit Finite Element Schemes for Convective Conductive Heat Transfer Problems", Finite Element Methods for Convective Dominated Flows (T.J.R. Hughes, editor) ASME publication AMD Vol. 34, 149-165.
19. Morton, K.W. and A.D. Parrot (1978) "Generalized Galerkin Methods for First Order Hyperbolic Equations", University of Reading, Numerical Analysis report 4/78.
20. Griffiths, D.F.(1982) "A Petrov-Galerkin Method for Hyperbolic Equations", The Mathematics of Finite Elements and Applications IV (J. Whiteman, editor) Academic Press, London.
21. Lick, W.J. (1967) "Wave Propagation in Real Gasses", Ad vances in Applied Mechanics 10, 1-72.

# EULERIAN-LAGRANGIAN METHODS FOR ADVECTION-DISPERSION

Shlomo P. Neuman and Shaul Sorek

University of Arizona

## ABSTRACT

Three different Eulerian-Lagrangian schemes are used to solve the one-dimensional advection-dispersion equation. Advection is formally decoupled from dispersion in a manner analogous to Neuman (1981) which does not leave room for ambiguity. The resulting advection problem is solved by a novel approach called the method of reverse streaklines, and by the more conventional method of continuous particle tracking. Dispersion is handled by implicit finite elements on a fixed grid, using linear and quadratic basis functions. The results are compared with a third Eulerian-Lagrangian method in which the concentration function remains undecomposed. Preliminary results suggest that the first two methods may, after further improvement, work well for a wide range of Peclet numbers. The third method suffers from numerical dispersion and overshoot when Peclet numbers are large.

## INTRODUCTION

The advection-dispersion equation is widely used to describe Fickian transport of tracers and pollutants in the atmosphere, oceans, lakes, rivers, and the subsurface. The nature of this equation can be conveniently characterized by the dimensionless Peclet number

$$Pe = vL/D_m \tag{1}$$

where v is modulus of velocity, L is a characteristic length, and $D_m$ is molecular diffusion. For example, in the case where an inert chemical species is spreading due to molecular diffusion in a one-dimensional velocity field, the governing equation can be written as

$$\frac{\partial c}{\partial t} = \frac{\partial^2 c}{\partial x^2} - Pe \frac{\partial c}{\partial x} \tag{2}$$

where c is concentration, x is spatial coordinate defined relative to L, and t is time. When Pe is small, dispersion dominates, and the equation is parabolic in character. When Pe is large, advection dominates, and the character of the equation changes to hyperbolic. In nonuniform and nonsteady flow fields, the character of the advection-dispersion equation may vary in space and time, being predominantly parabolic in some regions and predominantly hyperbolic in others.

Most conventional numerical methods for solving the advection-dispersion equation can be classified into two major categories, Eulerian or Lagrangian, depending on the emphasis that they place on the parabolic or hyperbolic nature of the problem. In the Eulerian method, the equation is discretized by a finite difference or finite element grid fixed in space. Lagrangian methods utilize either a deforming grid, or a fixed grid in deforming coordinates. A review of these methods, together with a discussion of their relative advantages and disadvantages, can be found in Neuman (1981).

A third approach that has been gaining popularity in recent years is the mixed Eulerian-Lagrangian method. This method combines the simplicity of the fixed Eulerian grid with the computational power of the Lagrangian approach, the latter being especially effective in regions with high Peclet numbers. Neuman (1981) has reviewed existing Eulerian-Lagrangian methods and proposed a new technique consisting of two steps: 1,

Formal decomposition of the concentration field into two parts, one controlled by pure advection, the other essentially by dispersion, and 2, solution of the resulting advection problem by the method of characteristics on a fixed space time grid, coupled with an adaptive explicit-implicit finite element solution of the dispersion problem on another fixed grid. The formal decomposition proposed by Neuman is rigorous and differs from previous such attempts in that it leaves no room for ambiguity. However, the particular numerical implementation of this idea described in Neuman's paper suffers from some numerical dispersion. The purpose of the present work is to test two alternative numerical schemes in which the advection part of the problem is handled with the aid of moving particles. These two schemes are compared with the results of another Eulerian-Lagrangian method in which the concentration function is left in its original form, without decomposition. All three methods perform well in regions of low Pe, but the methods based on moving particles are shown to be superior when Pe is large.

## DECOMPOSITION OF CONCENTRATION FIELD

Consider the one-dimensional advection-dispersion equation

$$[1 + s(x,t)] \frac{\partial c}{\partial t} = \frac{\partial}{\partial x} [D^*(x,t) \frac{\partial c}{\partial x} - v(x,t)c] \\ -\lambda(x,t)c + q(x,t) \quad \text{on} \quad (X_L, X_R)x(0,\tau] \tag{3}$$

where c, x, t are the same as in (2), D* is dispersion coefficient, v is velocity, $\lambda$ is radioactive decay coefficient, q is source term, s is retardation factor, $x_L$ and $x_R$ are left and right boundary points, respectively, and $(0,\tau]$ is time interval of interest. The parameters of (3) satisfy $D^* \geq 0$, $\lambda \geq 0$, and $s > 0$. The equation is to be solved for c, subject to the initial and boundary conditions

$$c(x,0) = c_o(x) \quad \text{on} \quad (x_L, x_R) \tag{4}$$

$$-D^* \frac{\partial c}{\partial x} + vc + \alpha_i(c - C_i) = Q_i$$

$$\text{on } x_i x(0,\tau],\ i=L,R \tag{5}$$

Here $c_o(x)$, $C_i(t)$, and $Q_i(t)$ are prescribed functions, and $\alpha_i(t)$ controls the type of boundary condition prevailing at $x_i\ x(0,\tau]$: If $\alpha_i \to \infty$ , equation (5) is a prescribed concentration condition; if $\alpha_i = 0$, it is a prescribed mass flux condition; otherwise, it is a mixed condition.

Using the hydrodynamic derivative

$$\frac{D}{Dt} = \frac{\partial}{\partial t} + v \frac{\partial}{\partial x} \tag{6}$$

We can rewrite (3) in the form

$$\frac{Dc}{Dt} + s \frac{\partial c}{\partial t} = \frac{\partial}{\partial x} [D^* \frac{\partial c}{\partial x}] - fc + q \quad \text{on } (x_L, x_R)x(0,\tau] \tag{7}$$

where $f(x,t) = \partial v/\partial x + \lambda$. Here c no longer represents concentration at a point in space-time, but rather the concentration of a fluid particle moving at the velocity v. The pathline of this particle is described by the hydrodynamic derivative of x, which leads to the characteristic equation

$$\frac{Dx}{v} = Dt \qquad \text{on} \quad (x_L, x_R)x(0,\tau] \tag{8}$$

Neuman (1981) has shown that (7)-(8) and (4)-(5) can be replaced by two sets of equations, one in terms of $\overline{c}(x,t)$, the other in terms of $\mathring{c}(x,t)$, where

$$c = \overline{c} + \mathring{c} \tag{9}$$

One way to perform such a decomposition is to let $\overline{c}$ satisfy the homogeneous first order partial differential equation

$$\frac{D\overline{c}}{Dt} = \frac{\partial \overline{c}}{\partial t} + v \frac{\partial \overline{c}}{\partial x} = 0 \quad \text{on} \quad (x_L, x_R)x(0,\tau] \tag{10}$$

subject to the initial condition

$$\bar{c}(x,0) = \beta(x)c_o(x) \qquad \text{on} \qquad (x_L, x_R) \tag{11}$$

and the Couchy condition

$$v\bar{c} + \alpha_i(\bar{c} - \beta_i C_i) = \alpha_i Q_i$$

$$\text{on } x_1 x(0,\tau]; \quad i=L,R \tag{12}$$

Here $x_i$ is an inflow boundary at which fluid particles are entering the system; conditions at outflow boundaries have no effect on $\bar{c}$ and are thus irrelevant. The functions $\beta(x)$, $\beta_i(t)$, and $\alpha_i(t)$ are prescribed to vary between 0 and 1 (more will be said about their choice later). Clearly, the $\bar{c}$ value of a given fluid particle remains constant as the latter is advected through the flow field. The function $\mathring{c}$ must then satisfy the equation

$$\frac{D\mathring{c}}{Dt} + s\frac{\partial \mathring{c}}{\partial t} = \frac{\partial}{\partial x}\left[D^* \frac{\partial \mathring{c}}{\partial x}\right] - f\mathring{c} + q + \bar{g}$$

$$\text{on} \quad (x_L, x_R)x(0,\tau] \tag{13}$$

where

$$\bar{g}(x,t) = \frac{\partial}{\partial x}\left[D^* \frac{\partial \bar{c}}{\partial x}\right] - f\bar{c} - s\frac{\partial \bar{c}}{\partial t}$$

subject to the initial and boundary conditions

$$\mathring{c}(x,0) = [1-\beta(x)]c_o(x) \qquad \text{on} \qquad (x_L, x_R) \tag{14}$$

$$-D^* \frac{\partial \mathring{c}}{\partial x} + v\mathring{c} + \alpha_i[\mathring{c} - (1-\beta_i)C_i] = (1-\gamma_i)Q_i + \bar{h}$$

$$\text{on } x_i x(0,\tau]; \quad i=L,R \tag{15}$$

where $\bar{h}(x,t) = D^*\partial\bar{c}/\partial x$. Note that $\bar{g}$ in (13) and $\bar{h}$ in (15) act as source functions depending on $\bar{c}$.

We see that the hyperbolic-parabolic advection-dispersion problem defined by (3)-(5) can be formally decoupled into a purely hyperbolic "advection problem" defined in terms of $\bar{c}$ by (10)-(12), and another predominantly parabolic "dispersion problem" defined in terms of $\overset{\circ}{c}$ by (13)-(15). The approach is to first solve the advection problem for $\bar{c}$, as the latter is independent of $\overset{\circ}{c}$, and then solve the residual dispersion problem for $\overset{\circ}{c}$, with $\bar{g}$ and $\bar{h}$ as source terms generated by the first problem. Alternative ways to decompose c into $\bar{c}$ and $\overset{\circ}{c}$ have been discussed by Neuman (1981).

NUMERICAL APPROACH

In Neuman's (1981) original paper on this subject, equations (10)-(12) were solved on a fixed space-time grid by the method of characteristics, and (13)-(15) were solved on a fixed but different space-time grid by finite elements. The use of such conjugate space-time grids required projecting results from one grid onto another, which led to some numerical dispersion. The purpose of the present work is to examine the possibility of eliminating such numerical dispersion by solving the advection problem with moving particles instead of a fixed grid. We will discuss two manners of moving the particles through the flow field.

The first approach, called the "Method of Reverse Streaklines", appears to be novel. The second approach is similar in principle, if not in detail, to the discrete particle technique used by Garder et al. (1964), Redell and Sunada (1970), and Konikow and Bredehoeft (1978) in what these authors call the "Method of Characteristics". However, our approach differs from theirs in that it is based on a rigorous decomposition of c into $\bar{c}$ and $\overset{\circ}{c}$, which in turn enables us to solve for $\overset{\circ}{c}$ by implicit or adaptive explicit-implicit finite elements (the above authors use explicit finite differences which may be relatively time consuming). Furthermore, our manner of projecting $\bar{c}$ from the particles onto the finite element grid points, and $\overset{\circ}{c}$ in the reverse direction, is different from theirs.

A third method examined in this paper is one that solves (7) in its original form without decomposing c into $\bar{c}$ and $\overset{\circ}{c}$, in a manner similar in principle to that of Hinstrup et al. (1977), but using finite elements instead of the finite differences they have used.

## ADVECTION BY REVERSE STREAKLINES

In order to solve the dispersion problem (13)-(15) for $\overset{\circ}{c}$, one must know the source terms $\bar{g}$ and $\bar{h}$ which, in turn, depend on $\bar{c}$. Thus, the advection problem must be solved first. The purpose of the reverse streaklines method is to provide accurate values of $\bar{c}$ at the nodal points of the finite element grid used for solving the dispersion problem at each time step. Let the vector of desired $\bar{c}$ values at the nodes, at time $t_k$, be designated by $\underline{\bar{c}}^k$. In order to compute $\bar{c}_n^k$ at node n, it is sufficient to identify the $\bar{c}(x,0)$ value at a point $x = {}^k x_n$ where ${}^k x_n$ is the location at time t=0 of a fluid particle reaching the nodal position $x_n$ at time $t_k > 0$. In other words,

$$\bar{c}_n^k = \bar{c}({}^k x_n, 0) \tag{16}$$

where

$${}^k x_n = x_n - \int_0^{t_k} v\, Dt; \quad {}^k x_n \in (x_L, x_R) \tag{17}$$

the integral in (17) being taken with respect to that particle which happens to reach $x_n$ at time $t_k$. If ${}^k x_n$ falls outside the flow region interior such that ${}^k x_n \notin (x_L, x_R)$, then (16) and (17) must be replaced by

$$\bar{c}_n^k = \left.\frac{\alpha_i \beta_i C_i + \gamma_i Q_i}{v + \alpha_i}\right|_{t=t_i}; \quad i = L,R \tag{18}$$

where $t_i$ is the time at which the particle has left the i-th boundary point, i.e.,

$$x_i = x_n - \int_{t_i}^{t_k} v\, Dt \tag{19}$$

Later in the text we will show how $\bar{c}^k_{n}$, n=1,2,..., N, where N is total number of finite element grid points, is used to compute $\overset{\circ}{c}^k_n$ for each n. Once this is accomplished, one has the option of redefining $\bar{c}(x,0)$ and $\overset{\circ}{c}(x,0)$ according to

$$\begin{aligned} \bar{c}({}^k x_n, 0) &= \beta(x_n, t_k) c^k_n \\ \overset{\circ}{c}({}^k x_n, 0) &= [1-\beta(x_n, t_k)] c^k_n \end{aligned} \tag{20}$$

where $c^k_n = \bar{c}^k_n + \overset{\circ}{c}^k_n$. In particular, if $\beta \equiv 1$, then $\overset{\circ}{c}(x,0) \equiv 0$. As will be seen below, this means that $\overset{\circ}{c}$ is zero at the beginning of each time step.

Clearly, a particle reaching $x_n$ at $t_k$ lies upstream, or to the rear, of another particle arriving at $t_{k-1}$. If we were dealing with a two- or three-dimensional velocity field, the locus of all particles reaching $x_n$ during the time interval $(0,\tau]$ would form a reverse streakline, and hence the name of our method. In the special case where flow takes place at steady state, streaklines coincide with streamlines, and (17) gives rise to the recursive relationship

$$ {}^k x_n = {}^{k-1} x_n - \int_{t_{k-1}}^{t_k} v \, Dt; \quad {}^k x_n \ \varepsilon (x_L, x_R) \tag{21}$$

When this happens, our approach can be appropriately termed the "Method of Reverse Characteristics."

The chief advantage of this method is that since $\bar{c}(x,0)$ and the boundary conditions can be specified with arbitrary accuracy, $\bar{c}^k_n$ in (16) and (18) may be computed as accurately as desired. Furthermore, since the moving particles coincide with the finite element grid points at each time step, there is no need to interpolate in order to project their $\bar{c}$ values onto these grid points. A possible disadvantage is that the reverse streaklines or characteristics, whose number is equal to the number of nodes, may not cover the entire flow field: one can easily visualize a two- or three-dimensional flow system in which point sources remain unidentified because they are by-passed by the streaklines. In steady state flow fields, this could be remedied by designing the finite elements so that their sides coincide with streamlines passing through all point sources. Another disadvantage is that in time-varying velocity fields, the recursive formula (21) does not apply, and the flow path of each particle must be traced over the entire time period, $(0,t_k]$, at every discrete time level, $t_k$.

ADVECTION BY CONTINUOUS PARTICLE TRACKING

In the method of reverse streaklines, a new set of particles is introduced at each time step. A more conventional approach is to introduce particles at strategic points in the space-time domain of interest and continuously track their positions along the pathlines until the particles leave the flow region. In our work, any existing particle, p, located at point $x_p$ at time $t_k$, is assigned a $\bar{c}$ value equal to $\bar{c}_p^k \equiv \bar{c}(x_p, t_k)$ for the duration of the time interval $(t_k, t_{k+1})$. Any new particle, r, introduced at $(x_r, t_k)\varepsilon(x_L, x_R)x(0,\tau]$ is assigned a $\bar{c}$ value equal to $\bar{c}_r^k \equiv \bar{c}(x_r, t_k)$. A new particle introduced along a boundary, $x_r = x_i$, $(x_i, t_k)\varepsilon x_i x(0,\tau]$, i=L,R, is assigned the $\bar{c}$ value

$$\bar{c}_r^k = \frac{\alpha_i \beta_i C_i + \gamma_i Q_i}{v + \alpha_i} \Bigg|_{x_i = x_r,\ t=t_k} \tag{22}$$

for the duration of the time interval.

At the end of the time step, each particle, p, reaches a new position

$$x_p^{k+1} = x_p^k + \int_{t_k}^{t_{k+1}} v \, Dt \tag{23}$$

At this stage, we use linear interpolation between the $\bar{c}$ values of particles in the immediate vicinity of each finite element grid point, n, in order to compute $\bar{c}_n^{k+1} \equiv \bar{c}(x_n, t_{k+1})$. The latter values are used to evaluate $\bar{g}$ and $\bar{h}$ in the finite element equations for $\overset{\circ}{c}$, in the manner described below.

Suppose that $\bar{c}$ and $\overset{\circ}{c}$ at the end of a time step, $(t_{k-1}, t_k)$, are known. Then prior to starting the next time step, $(t_k, t_{k+1})$, one has the option to redefine $\bar{c}$ and $\overset{\circ}{c}$ according to

$$\begin{aligned} \bar{c}(x,t_k) &= \beta(x,t_k)c(x,t_k) \\ \overset{\circ}{c}(x,t_k) &= [1-\beta(x,t_k)]c(x,t_k) \end{aligned} \qquad (24)$$

The same can be done for boundary points by specifying $\beta_i(t_k)$ and $\gamma_i(t_k)$ in (12). In the examples presented below, $\beta(x,t_k) \equiv 1$, so that $\bar{c}$ at the beginning of each time step becomes identical to the total concentration, c, and $\overset{\circ}{c}$ becomes identically equal to zero.

## DISPERSION BY FINITE ELEMENTS

Our method of solving (13)-(15) for $\overset{\circ}{c}$ is similar in principle to that outlined by Neuman (1981). The approach is to identify each grid location, $x_n$, in the finite element mesh with a particle having reached $x_n$ at time $t_{k+1}$. We approximate the functions $\bar{c}(x,t)$ and $\overset{\circ}{c}(x,t)$ by $\bar{c}^N(x,t)$ and $\overset{\circ}{c}^N(x,t)$, respectively, according to

$$\begin{aligned} \bar{c}(x,t) &\simeq \bar{c}^N(x,t) \equiv \sum_{n=1}^{N} \bar{c}_n(t)\,\xi_n(x) \\ \overset{\circ}{c}(x,t) &\simeq \overset{\circ}{c}^N(x,t) \equiv \sum_{n=1}^{N} \overset{\circ}{c}_n(t)\,\xi_n(x) \end{aligned} \qquad (25)$$

Here N is total number of finite element grid points, $\bar{c}_n$ and $\overset{\circ}{c}_n$ are nodal values of $\bar{c}$ and $\overset{\circ}{c}$, respectively, and $\xi_n(x)$ are basis functions satisfying $\xi_n(x_m) = \delta_{nm}$, $\delta_{nm}$ being Kronecker delta (i.e., $\delta_{nm} = 1$ if n=m and $\delta_{nm} = 0$ if $n \neq m$).

Two kinds of basis functions are used, linear (chapeau) and quadratic. Let $\xi_n^e$ (x) be the value of $\xi_n$ (x) within element e, so that

$$\xi_n = \bigcup_e \xi_n^e \qquad (26)$$

where the union sign is taken over the two elements in the immediate vicinity of n. In the case of linear interpolation, consider e to represent $(x_n, x_{n+1})$. Then

$$\xi_n^e(x) = \frac{1}{2}(1-\eta)$$

$$\xi_{n+1}^e(x) = \frac{1}{2}(1+\eta)$$

where

$$\eta = \frac{x - x_{n+1}}{x_{n+1}-x_n}$$

In the case of quadratic interpolation, e may represent $(x_n, x_{n+2})$ where $x_{n+1}$ is located in the middle of the interval. Then

$$\left.\begin{aligned} \xi_n^e(x) &= \frac{\mu}{2}(\mu - 1) \\ \xi_{n+1}^e(x) &= 1 - \mu^2 \\ \xi_{n+2}^e(x) &= \frac{\mu}{2}(\mu+1) \end{aligned}\right\} \qquad \mu = \frac{x - x_{n+1}}{x_{n+1}}$$

Application of the Galerkin orthogonalization process to (13) gives

$$\int_{x_L}^{x_R} \left\{ \frac{D\overset{\circ}{c}}{Dt} + s\,\frac{\partial \overset{\circ}{c}}{\partial t} - \frac{\partial}{\partial c}\left[D^* \frac{\partial \overset{\circ}{c}{}^N}{\partial x}\right] + f\overset{\circ}{c}{}^N - q - \overline{g}^N \right\}\xi_n \, dx = 0$$

$$n = 1,2,\ldots, N \qquad (29)$$

where

$$\overline{g}^N = \frac{\partial}{\partial x}\left[D^* \frac{\partial \overline{c}^N}{\partial x}\right] - f\overline{c}^N - s\,\frac{\partial \overline{c}}{\partial t}$$

and the time derivatives are treated for the moment as known functions. Integration by parts yields

$$\int_{x_L}^{x_R} \{ [\frac{D\overset{\circ}{c}}{Dt} + s\,\frac{\partial(\overset{\circ}{c} + \bar{c})}{\partial t} + f(\bar{c}^N + \overset{\circ}{c}^N) - q]\xi_n + D^* \frac{\partial(\bar{c}^N + \overset{\circ}{c}^N)}{\partial x} \frac{\partial \xi_n}{\partial x} \}dx + \delta_{Ln}\, D^* \frac{\partial(\bar{c}^N + \overset{\circ}{c}^N)}{\partial x}\Bigg|_{x = x_L} - \delta_{Rn}\, D^* \frac{\partial(\bar{c}^N + \overset{\circ}{c}^N)}{\partial x}\Bigg|_{x = x_R} = 0$$

$$n=1,2,\ldots,N \qquad (30)$$

Where L and R are the nodes at $x_L$ and $x_R$, respectively. As will be seen below, by regarding each node as a moving particle, the dispersion problem for $\overset{\circ}{c}$ becomes purely parabolic. It has been shown by Neuman and Narasimhan (1977) that for such parabolic problems, there is an advantage in adapting the lumped-mass finite element approach, in which the first two integral terms on the left side of (30) are approximated by

$$\int_{x_L}^{x_R} \frac{D\overset{\circ}{c}}{Dt}\,\xi_n\,dx \simeq \frac{D\overset{\circ}{c}_n}{Dt} \int_{x_L}^{x_R} \xi_n\,dx$$

$$\int_{x_L}^{x_R} s\,\frac{\partial(\bar{c}+\overset{\circ}{c})}{\partial t}\,\xi_n\,dx \simeq \frac{\partial(\bar{c}_n+\overset{\circ}{c}_n)}{\partial t} \int_{x_L}^{x_R} s\,\xi_n\,dx \qquad (31)$$

Note that this is analogous to what one does in most conventional finite difference schemes. Substituting (15), (25), and (31) into (30) leads to the following system of differential equations,

$$(\underline{\underline{A}} + \underline{\underline{B}} + \underline{\underline{F}})\underline{\overset{\circ}{c}} + \underline{\underline{W}}\,\frac{D\underline{\overset{\circ}{c}}}{Dt} + \underline{\underline{S}}\,\frac{\partial \underline{\overset{\circ}{c}}}{\partial t} = \underline{Q} \tag{32}$$

Here $\underline{\underline{A}}$ is a symmetric positive-definite "dispersion matrix" of order N whose (n,m)th term is defined as

$$A_{nm} = \int_{x_L}^{x_R} D^* \frac{\partial \xi_n}{\partial x} \frac{\partial \xi_m}{\partial x}\, dx \tag{33}$$

$\underline{\underline{B}}$ is a diagonal "boundary matrix" of order N, whose terms are

$$\begin{aligned} B_{nm} &= (\delta_{Ln} - \delta_{Rn})\,(v\big|_{x=x_n} + \alpha_n) \qquad && \alpha_n < \infty \\ B_{nn} &= 0 && \alpha_n \to \infty \end{aligned} \tag{34}$$

$\underline{\underline{F}}$ is a square symmetric matrix of order N defined by

$$F_{nm} = \int_{x_L}^{x_R} f\ \xi_n \xi_m\, dx \tag{35}$$

$\underline{\overset{\circ}{c}}$ is the N-dimensional vector of $\overset{\circ}{c}_n$ values. $\underline{\underline{W}}$ is a diagonal "capacity" matrix of order N such that

$$W_{nn} = \int_{x_L}^{x_R} \xi_n\, dx \tag{36}$$

$\underline{\underline{S}}$ is a diagonal "retardation matrix" of order N having the terms

$$S_{nn} = \int_{x_L}^{x_R} s\ \xi_n\, dx \tag{37}$$

Finally, $\underline{Q}$ is a "source vector" defined as

$$Q_n = \Psi_n + \int_{x_L}^{x_R} q\, \xi_n\, dx - \sum_{n=1}^{N} (A_{nm} + F_{nm})\, \bar{c}_m - S_{nn} \frac{\partial \bar{c}_n}{\partial t} \quad (38)$$

where

$$\Psi_n = (\delta_{Rn} - \delta_{Ln})[\alpha_n(1-\beta_n)C_n + (1-\gamma_n)Q_n];\ \alpha_n < \infty \quad (39)$$

Note that when $\alpha_n \to \infty$, $\overset{\circ}{c}_n$ is known by virtue of (15), and the n-th row of (32) is not needed to compute $\underline{\overset{\circ}{c}}$. Thus, $\psi_n$ need not be evaluated a priori, but it can be obtained a posteriori from (32) after $\underline{\overset{\circ}{c}}$ has been computed, if so desired.

The differential matrix equation (32) is integrated in time by finite differences. If this integration would take place at grid points fixed in space, the hydrodynamic derivative $D\underline{\overset{\circ}{c}}/Dt$ would have to be expressed as the sum of a partial derivative $\partial\underline{\overset{\circ}{c}}/\partial t$ and an advective term. The problem is avoided by viewing each node as a particle reaching $x_n$ and $t_{k+1}$ and defining

$$\frac{D\overset{\circ}{c}_n}{Dt} = \frac{\overset{\circ}{c}{}^{k+1}_n - {}^{k}\overset{\circ}{c}_n}{\Delta t} \quad (40)$$

where $\overset{\circ}{c}{}^{k+1}_n = \overset{\circ}{c}(x_n, t_{k+1})$

$${}^{k}\overset{\circ}{c}_n = \overset{\circ}{c}({}^{k}x_n, t_k)$$

$${}^{k}x_n = x_n - \int_{t_k}^{t_{k+1}} v\, Dt$$

With this, the backward difference equivalent of (32) becomes

$$[\underline{\underline{A}} + \underline{\underline{B}} + \underline{\underline{F}} + \frac{1}{\Delta t}(\underline{\underline{W}} + \underline{\underline{S}})]\underline{\overset{\circ}{c}}{}^{k+1} = \underline{Q}^{k+1} + \frac{1}{\Delta t}(\underline{\underline{W}}\,{}^{k}\underline{\overset{\circ}{c}} + \underline{\underline{S}}\,\underline{\overset{\circ}{c}}{}^{k}) \quad (41)$$

and

$$Q_n^{k+1} = \Psi_n + \int_{x_L}^{x_R} q\, \xi_n\, dx - \sum_{n=1}^{N} (A_{nm} + F_{nm})\bar{c}_m^{\,k+1} - S_{nn} \frac{\bar{c}_n^{\,k+1} - \bar{c}_n^{\,k}}{\Delta t} \quad (42)$$

Schemes other than backward difference, such as the adaptive explicit-implicit method described in Neuman (1981), are also possible but have not been explored in the present study.

After solving (41) for $\overset{\circ}{\underline{c}}{}^{k+1}$, one can compute the total concentration at the grid points according to $\underline{c}^{k+1} = \overset{\circ}{\underline{c}}{}^{k+1} + \bar{\underline{c}}^{k+1}$. To obtain the total concentration, $c_p^{k+1}$, at a moving particle situated in the interior of an element, one first computes

$$\overset{\circ}{c}{}^{N}(x_p, t_{k+1}) = \sum_{n=1}^{N} \overset{\circ}{c}{}_n^{k+1} \; \xi_n(x_p) \tag{43}$$

and then calculates

$$c_p^{k+1} = \bar{c}_p^{k+1} + \overset{\circ}{c}{}^{N}(x_p, t_{k+1}) \tag{44}$$

This completes the time step.

Another approach is to express the total concentration, c, as

$$c(x,t) \simeq c^N(x,t) = \sum_{n=1}^{N} c_n(t)\, \xi_n(x) \tag{45}$$

and replace (41)-(42) with

$$[\underline{\underline{A}} + \underline{\underline{B}} + \underline{\underline{F}} + \frac{1}{\Delta t}(\underline{\underline{W}} + \underline{\underline{S}})]\underline{c}^{k+1} = \underline{Q}^{*k+1} + \frac{1}{\Delta t}(\underline{\underline{W}}{}^{k}\underline{c} + \underline{\underline{S}}{}^{k}\underline{c}) \tag{46}$$

where

$$Q_n^{*k+1} = \Psi_n + \int_{x_L}^{x_R} q \;\; \xi_n \, dx + B_{nn} \, \bar{c}_n^{k+1} \tag{47}$$

and the n-th component of ${}^{k}\underline{c}$ is given by ${}^{k}c_n = c({}^{k}x_n, t_k)$. The procedure for the (k+1)th time step here is to first solve the advection problem for $\bar{c}^{k+1}$, then interpolate to determine ${}^{k}\underline{c}$, solve (46) for $\underline{c}^{k+1}$ in the same manner as one would solve (41) for $\overset{\circ}{\underline{c}}{}^{k+1}$, determine $\overset{\circ}{\underline{c}}{}^{k+1}$ according to

$$\overset{\circ}{\underline{c}}{}^{k+1} = \underline{c}^{k+1} - \bar{\underline{c}}^{k+1} \tag{48}$$

and finally compute $c_p^{k+1}$ at all particles in the interior of elements with the aid of (44).

Another option, utilized in some of our examples, is to set $\beta \equiv \beta_i \equiv \alpha_i \equiv 0$ so that $\bar{c} \equiv 0$ and $\mathring{c} \equiv c$. This means that c remains undecomposed and there is no need to solve a separate advection problem. This approach is similar in principle to that of Hinstrup et al. (1977).

EXAMPLES

Only a small number of the optional solution strategies presented in the previous paragraphs have been tested as of this writing. The following four examples show how some of these Eulerian-Lagrangian strategies perform under a relatively wide range of conditions. We expect to obtain improved results in the near future.

Example 1 concerns the problem of solving

$$\frac{\partial c}{\partial t} = D^* \frac{\partial^2 c}{\partial x^2} - \mathbf{v} \frac{\partial c}{\partial x} \qquad \text{on } (0,x_R)x(0,\tau] \tag{49}$$

subject to

$$\begin{aligned} c(x,0) &= 0 && \text{on} && (0,x_R) \\ c(0,t) &= 1 && \text{on} && (0,\tau] \\ c(x_{R,}\, t) &= 0 && \text{on} && (0,\tau] \end{aligned} \tag{50}$$

The physical and grid parameters, in an arbitrary system of consistent units, are $D^* = 0.01$, $v = 0.05$, $x_R = 2.5$, $\Delta x = 0.05$, and $\Delta t = 0.5$. The problem is strongly dispersion-dominated with a Peclet number $Pe = v\Delta x/D^* = 0.25$. One would therefore expect to obtain good results without decomposing c into $\bar{c}$ and $\mathring{c}$ as explained in the last paragraph of the preceding section. The results for t = 10, obtained by solving for c directly (i.e., without decomposition) with linear as well as quadratic basis functions, are identical and shown in Fig. 1. As expected, they compare very well with the analytical solution,

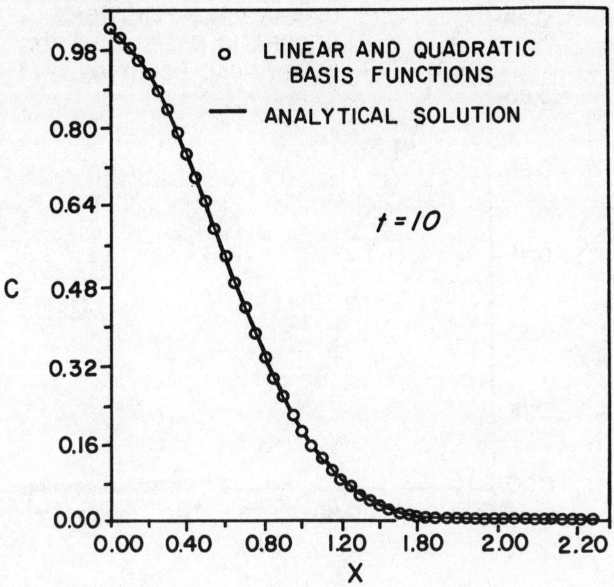


Figure 1. Results of Example 1 without decomposing c.

$$c(x,t) = \frac{1}{2}\operatorname{erfc}\left(\frac{x-vt}{2\sqrt{D^*t}}\right) + \frac{1}{2}\exp\left(\frac{vx}{D^*}\right)\operatorname{erfc}\left(\frac{x+vt}{2\sqrt{D^*t}}\right) \qquad (51)$$

which is valid for $x_R \to \infty$ .

Fig. 2 shows results obtained with linear and quadratic basis functions when the method of reverse streaklines is used with $\beta(x) \equiv \beta_i \equiv 1$ ($\gamma_i$ is irrelevant here) but without the use of (20). Both results compare well with the analytical solution except for a slight kink just ahead of the dispersing front. The kink of the quadratic solution is smaller than that of the linear solution. It appears to result from a discontinuity in the $\overset{\circ}{c}$ function near the front, as illustrated in Fig. 3. Fig. 4 shows that one way to eliminate the problem is to reduce $\Delta x$, from 0.05 to 0.025. Another possibility is to smooth the kink out with cubic splines, the result of which is shown in Fig. 5. We are currently investigating the potential utility of other options such as (20), (45)-(48), etc.

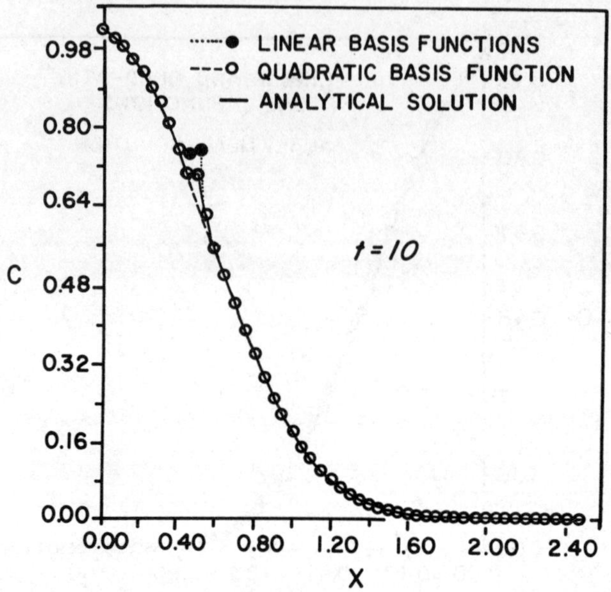


Figure 2. Results of Example 1 by reverse streaklines.

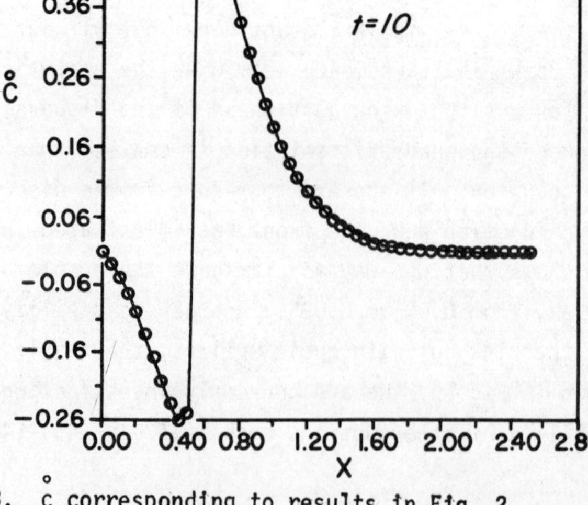


Figure 3. $\overset{\circ}{c}$ corresponding to results in Fig. 2.

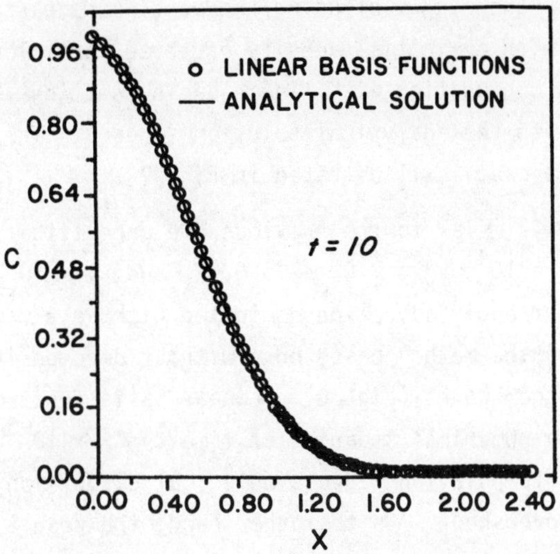


Figure 4. Result of Example 1 by reverse streaklines with $\Delta x$ halved.

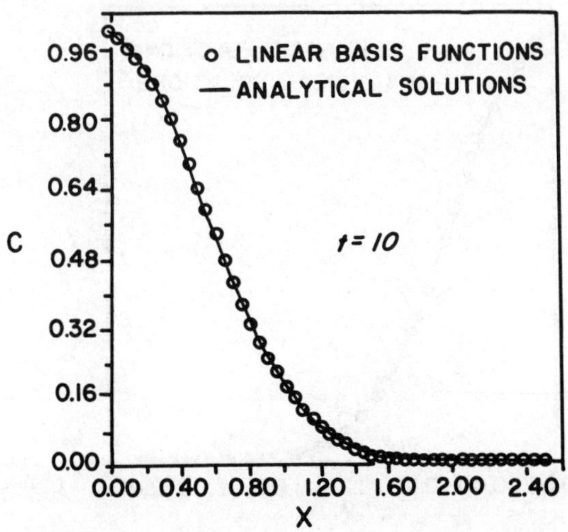


Figure 5. Result of Example 1 by reverse streaklines using cubic spline smoothing.

Fig. 6 shows the result of using the continuous particle tracking method with an initial density of three particles per element, in conjunction with linear basis functions. Clearly, agreement with the analytical solution is good. The variation of $\mathring{c}$ in this case is illustrated in Fig. 7.

Example 2 is similar to the previous one except that now $D^* = 1.0$, $v = 10^4$, $x_R = 1.0$, $\Delta x = 0.02$, and $t = 10^{-6}$. The problem is strongly advection dominated with $Pe = v\Delta x/d^* = 200$. In this case the method based on c without decomposition into $\bar{c}$ and $\mathring{c}$ no longer works (Fig. 8): Linear basis functions lead to considerable numerical smearing at time $t = 5 \times 10^{-5}$, whereas quadratic interpolation reduces numerical dispersion at the expense of overshoot. On the other hand, the result obtained with reverse streaklines and linear basis functions, illustrated in Fig. 9, compares quite well with the analytical solution; the $\mathring{c}$ function for this case appears in Fig. 10. Continuous particle tracking with linear basis functions and

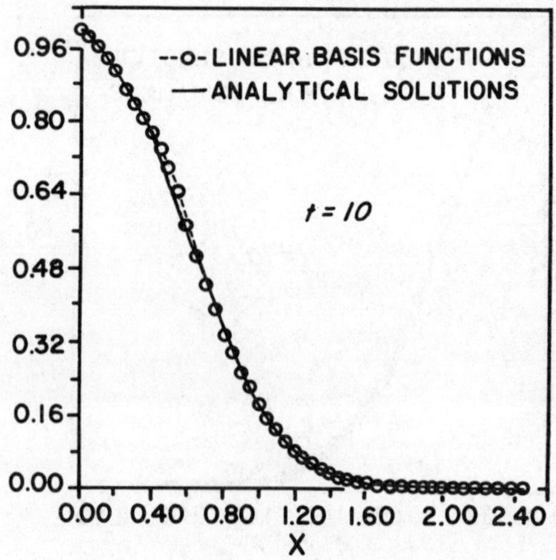


Figure 6. Result of Example 1 by continuous particle tracking.

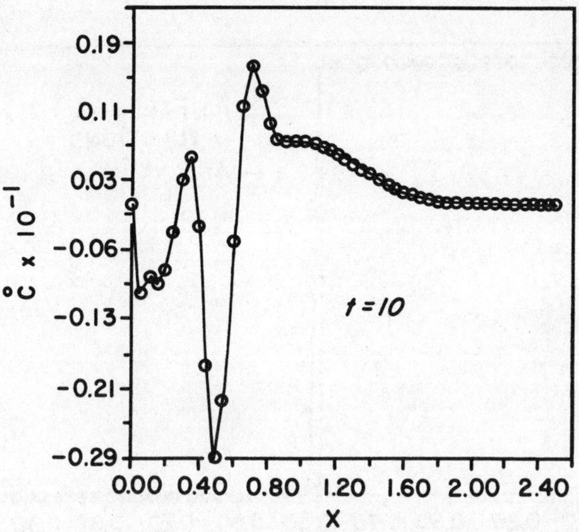


Figure 7. $\overset{\circ}{c}$ corresponding to result in Fig. 6.

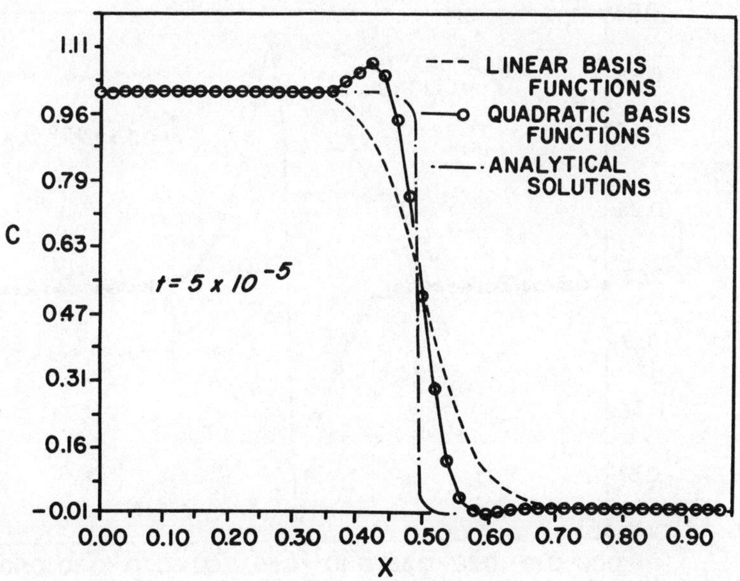


Figure 8. Results of Example 2 without decomposing c.

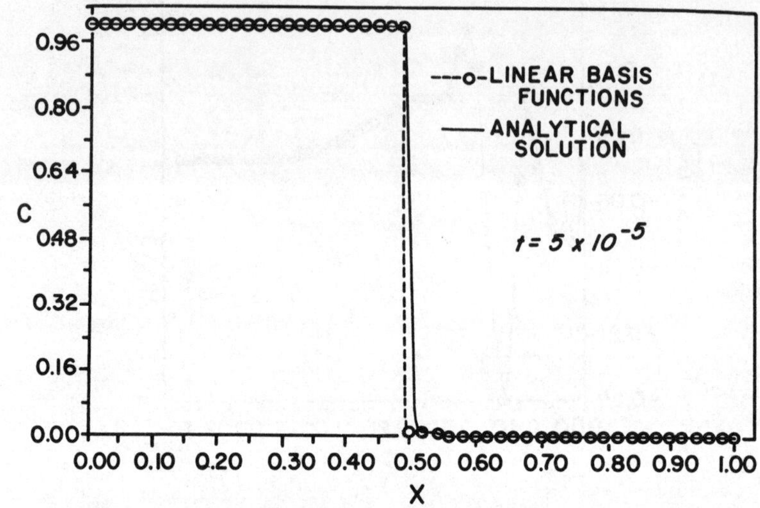


Figure 9. Result of Example 2 by reverse streaklines.

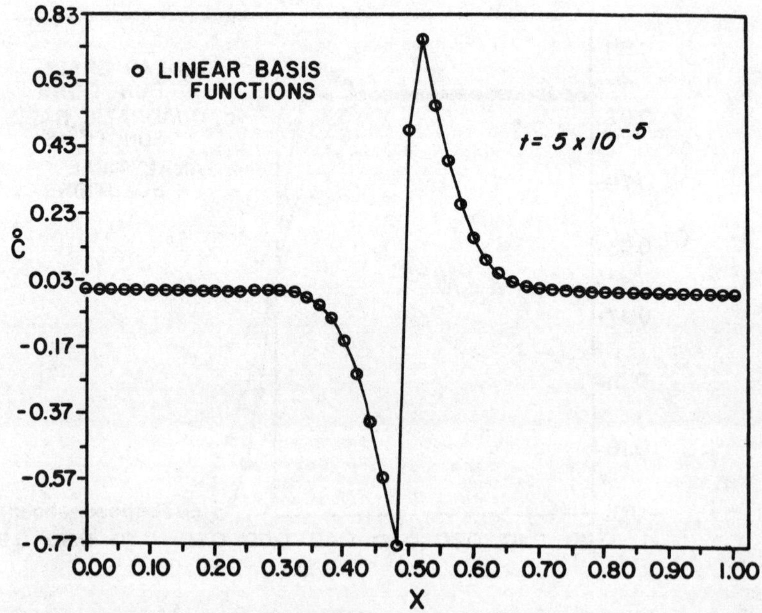


Figure 10. $\overset{\circ}{c}$ corresponding to result in Fig. 9.

an initial density of three particles per element leads to the result in Fig. 11, which is good except for a slight shift to the right. Fig. 12 shows the behavior of $\mathring{c}$ in this case.

Example 3 concerns the advection and dispersion of a rectangular wave. The problem is defined by (49) subject to

$$\begin{aligned} c(x,0) &= 1 \qquad \text{when } 0.1 \leq x \leq 0.2 \\ c(x,0) &= 0 \qquad \text{when } 0 \leq x < 0.1;\ 0.2 < x \leq 1.0 \\ c(0,t) &= c(1,t) = 0 \end{aligned} \tag{52}$$

and its analytical solution is

$$c(x,t) = \frac{1}{2}\operatorname{erf}\left(\frac{b-x+vt}{2\sqrt{D^{*}t}}\right) + \frac{1}{2}\operatorname{erf}\left(\frac{b+x-vt}{2\sqrt{D^{*}t}}\right) \tag{53}$$

where $2b = 0.1$ is the width of the rectangle. The result at $t = 6\text{x}10^{-4}$ for $D^{*} = 1.0$, $v = 10^{3}$, $\Delta x = 0.01$, and $\Delta t = 6\text{x}10^{-6}$, when c remains undecomposed, shows numerical dispersion as in the previous example. When the method of reverse streaklines is applied in the same manner as before, the result is

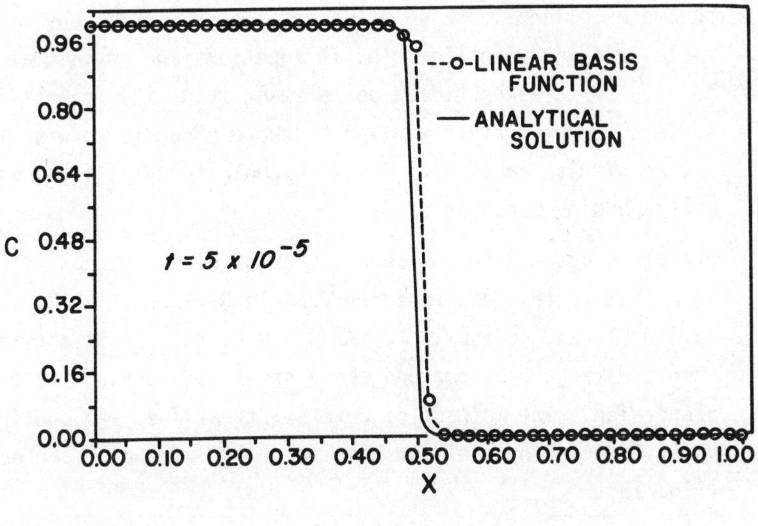


Figure 11. Result of Example 2 by continuous particle tracking.

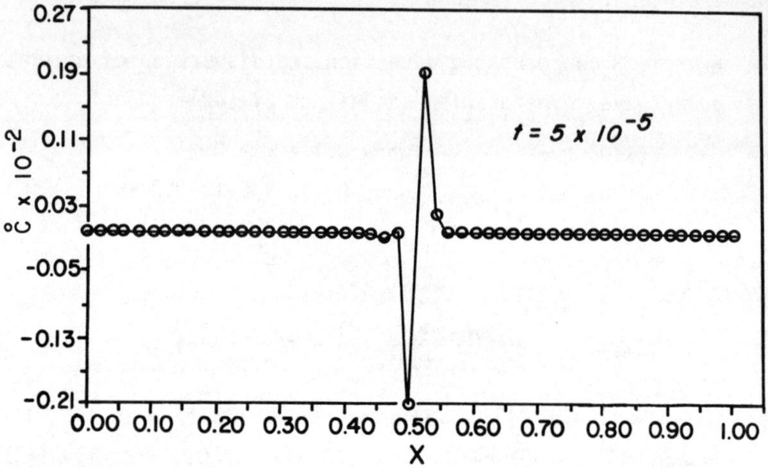


Figure 12. $\overset{\circ}{c}$ corresponding to result in Fig. 11.

poor unless $\Delta x$ is reduced down to 0.0067, in which case they become excellent. The solution obtained by continuous particle tracking with linear basis functions and an initial density of three particles per element (Fig. 13) is better but suffers from two kinks similar to those encountered with the method of reverse streaklines in Example 1. Fig. 14 shows the corresponding behavior of $\overset{\circ}{c}$.

Example 4 deals with the pure advection of a rectangular wave identical to that described in Example 3. Here $D^* = 0$, $v = 1$, $x = 0.01$, and $\Delta t = 6\times10^{-3}$. At $t = 0.6$, the result obtained when c remains undecomposed again shows considerable numerical dispersion. The methods of reverse streaklines and continuous particle tracking yield results which are virtually perfect (Fig. 15).

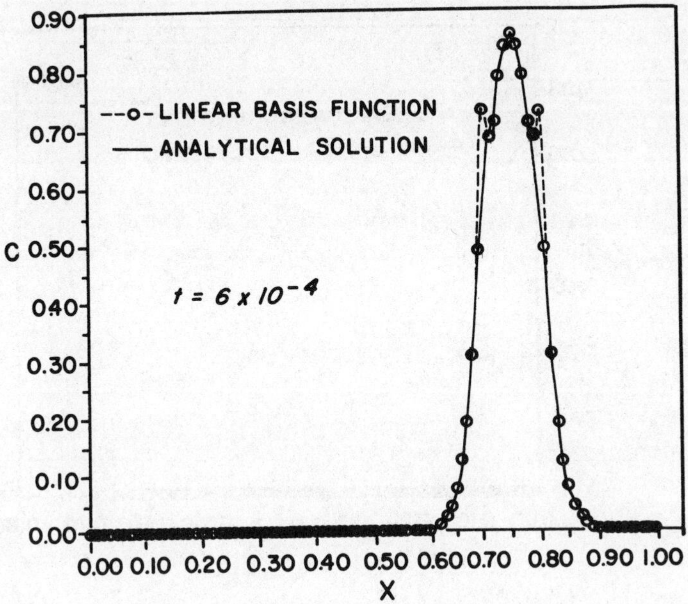


Figure 13. Result of Example 3 by continuous particle tracking.

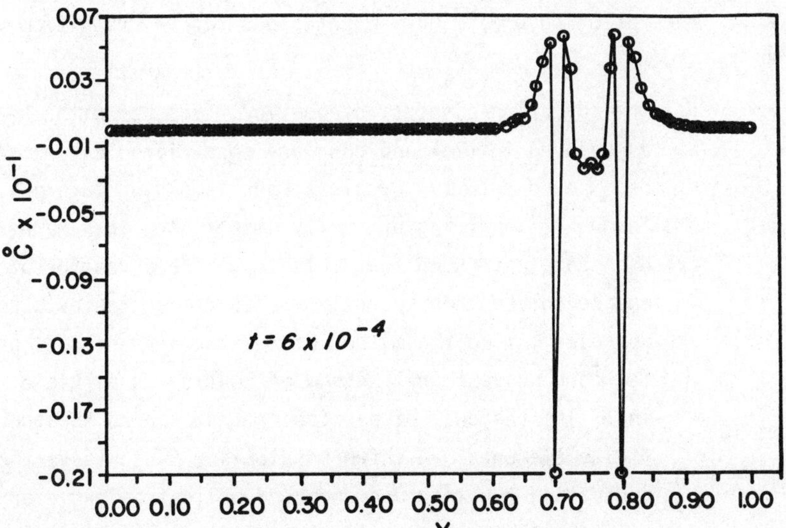


Figure 14. $\overset{\circ}{c}$ corresponding to result in Fig. 13.

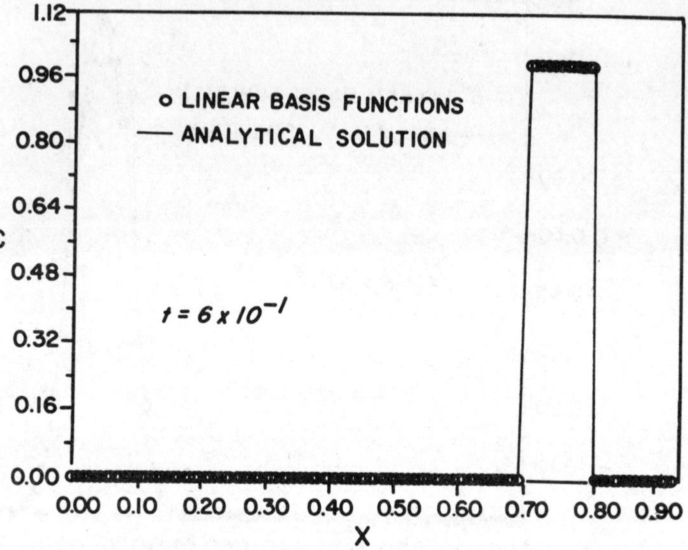


Figure 15. Results of Example 4 by reverse streaklines and continuous particle tracking.

CONCLUSIONS

The following preliminary conclusions can be drawn from this study:

1. The advection-dispersion equation, together with the associated initial and boundary conditions, can be formally decomposed into two problems, one involving pure advection, the other involving primarily dispersion, in a manner which does not leave room for ambiguity. The advection problem can be solved independently at each time step by a new approach called the method of reverse streaklines, or by the more conventional method of continuous particle tracking. The residual dispersion problem can be treated by finite elements. A third possibility is to leave the concentration function undecomposed while treating every finite element grid point as a moving particle during each time step. All three methods are Eulerian-Lagrangian because they use a finite element grid fixed in space,

together with particles moving through the flow field.

2. The third method in which the concentration function remains undecomposed works very well for problems which are strongly dispersion-dominated. When advection becomes important, this method suffers from numerical dispersion and overshoot.

3. The method of reverse streaklines works extremely well for advection-dominated problems. When the effect of dispersion becomes important, the method as applied to date exhibits irregularities. Strategies for the elimination of these irregularities are under investigation.

4. The method of continuous particle tracking works very well for advection-dominated problems. When dispersion becomes important, the method as applied to date works well in some cases and exhibits irregularities in others. Strategies for the elimination of these irregularities are under investigation.

## REFERENCES

Garder, A. O., Peaceman, D. W., and Pozzi, A. L., Jr.

1964 Numerical Calculation of Multidimensional Miscible Displacement by the Method of Characteristics, Soc. Petrol. Engin. Jour., 4(1), 26-36.

Hinstrup, P., Kej, A., and Kroszynski, U.

1977 A High Accuracy Two-dimensional Transport-dispersion Model for Environmental Applications, Proc. 18th Intern. Assoc. Hydraul. Res. Congress, Baden-Baden, Germany, 13, 129-137.

Konikow, L. F. and Bredehoeft, J. D.

1978 Computer Model of Two-dimensional Solute Transport and Dispersion in Ground Water, U. S. Govt. Printing Office, Washington, D. C., pp. 90.

Neuman, S. P.

1981 A Eulerian-Lagrangian Numerical Scheme for the Dispersion-convection Equation Using Conjugate Space-time grids, Jour. Comp. Phys., 41(2), 270-294.

Neuman, S. P., and Narasimhan, T. N.

1977 Mixed Explicit-implicit Iterative Finite Element Scheme for Diffusion Type Problems: I. Theory, Intern. Jour. Num. Meth. Engin., 11, 309-323.

Reddell, D. L., and Sunada, D. K.

1970 Numerical Simulation of Dispersion in Groundwater Aquifers, Colorado State Univ. Hydrology Paper 41, pp. 79.

# NUMERICAL SOLUTION OF DIFFUSIVE-CONVECTIVE DIFFERENTIAL PROBLEMS BY GENERAL FINITE DIFFERENCE FORMS

M. Reali, R. Rangogni and V. Pennati

ENEL - DSR - CRIS , Milano, Italy

## ABSTRACT

General finite difference methods can be applied directly with either regular or irregular cartesian networks and are, consequently, suitable for handling differential problems defined on irregular space domains.

Specific applications for the numerical solution of one and two dimensional diffusive-convective differential problems are introduced.

An important characteristics of these methods is the use of a general finite difference approximation, for the first space derivative, one order higher than for the second space derivative.

Numerical checks for a few chosen problems with known analytic solution indicate that the present general methods are both efficient and accurate ones.

## INTRODUCTION

Finite difference Methods (FDM), Collatz (1966), are oftentimes considered to be inadequate when differential problems on irregular domains have to be solved.

Assumed difficulties are, however, generally related to the use of regular cartesian networks which are inherently unsuitable for a proper handling of boundary conditions on irregular domains.

In Reali et al. (1980) we have shown how compact analytic expressions of finite difference forms (FDF) for the partial derivatives of a function u of several independent variables x, y, ..., can be obtained in a straightforward way for any grid point (either internal or on the boundary) of a general

(irregular) cartesian network.
In this paper, we highlight some features of numerical differentiation for general cartesian networks and apply general FDF for the numerical solution of two one-dimensional and two two-dimensional diffusive-convective differential problems.
These types of problems are of considerable general interest in water resources research, in particular for the mathematical modelling of waste heat discharge from thermal power plants, and for the control of various polluting processes.
A salient feature of the novel numerical methods we employ is the use of a four-point FDF of third order for the convective (first space) derivative while the first (time) derivative and the second (space) derivatives are represented to second order.
Checks between known analytic solutions and numerical solutions prove the novel methods to be very promising ones, simple and sufficiently accurate.

## NUMERICAL DIFFERENTIATION ON IRREGULAR NETWORKS

The problem of numerical differentiation on irregular cartesian networks has been considered from a general and practical viewpoint in Reali et al. (1980, 1981) where closed form analytic expressions for the analogues of the derivatives of a function $u = u(x, y)$ of two independent cartesian variables $x, y$ have been obtained.
General FDF of given order of approximation are obtained as solution of suitable systems of truncated Taylor expansions of function $u$. The procedure can be extended in a straightforward way to more dimensional spaces.
For any grid point $P_i$ (either inside or on the boundary of the domain of definition of function $u$) a given derivative such as

$$\left(\frac{\partial u}{\partial x}\right)_{P_i},\ \left(\frac{\partial u}{\partial y}\right)_{P_i},\ \left(\frac{\partial^2 u}{\partial x\,\partial y}\right)_{P_i},\ \left(\frac{\partial^2 u}{\partial x^2}\right)_{P_i},\ \left(\frac{\partial^2 u}{\partial y^2}\right)_{P_i},$$

is approximated by a suitable linear combination of values of $u$ at point $P_i$ and at chosen grid points $P_{i,k}$ about point $P_i$.

The analogues $\left[\frac{\partial u}{\partial x}\right]_{P_i}$, and $\left[\frac{\partial^2 u}{\partial x^2}\right]_{P_i}$ of the first and second derivatives $\left(\frac{\partial u}{\partial x}\right)_{P_i}$ and $\left(\frac{\partial^2 u}{\partial x^2}\right)_{P_i}$, would be given by expressions

$$\left[\frac{\partial u}{\partial x}\right]_{P_i} = \sum_k \alpha_{i,k}\, u_{i,k} \qquad (1)$$

$$\left[\frac{\partial^2 u}{\partial x^2}\right]_{P_i} = \sum_k \beta_{i,k}\, u_{i,k} \qquad (2)$$

where $\alpha_{i,k}$ and $\beta_{i,k}$ are numerical coefficients depending only on the coordinates of grid points $P_i$ and $P_{i,k}$ belonging to the grid line through $P_i$ parallel to the x-axis, $u_{i,k}$ are grid values of function u at points $P_i$ and $P_{i,k}$ and the number of terms in the sums depends on the required order of approximation. With four terms, in particular, the first and the second derivative analogues would be, respectively, third and second order correct.
Similarly, normal boundary derivatives $(\frac{\partial u}{\partial n})_{P_i}$ and mixed derivatives $(\frac{\partial^2 u}{\partial x\, \partial y})_{P_i}$ would be represented, to the required order of approximation, by general expressions

$$\left[\frac{\partial u}{\partial n}\right]_{P_i} = \sum_k \gamma_{i,k}\, u_{i,k} \qquad (3)$$

$$\left[\frac{\partial^2 u}{\partial x\, \partial y}\right]_{P_i} = \sum_k \delta_{i,k}\, u_{i,k} \qquad (4)$$

which require grid points $P_{i,k}$ generally not aligned with grid point $P_i$.

## NUMERICAL APPLICATIONS

We demonstrate the applicability of the general FDF above introduced by solving numerically two one and two two-dimensional diffusive convective problems.
It is required to find a function u $(\underline{x}, t)$ satisfying (in adimensional space and time variables)

$$\frac{\partial u}{\partial t} + \lambda \cdot \nabla u = \nabla^2 u \qquad \underline{x} \in D,\ t \geq 0 \qquad (5)$$

$$u(\underline{x} \in \Gamma, t) = f(t) \qquad (6)$$

$$u(\underline{x}, 0) = g(\underline{x}) \qquad (7)$$

In a given space domain D with boundary $\Gamma$ .

In the present formulation, diffusivity has been taken to be unity.
The problems we have considered have known analytic solutions by means of which we have been able to check our numerical solutions.
The general FDM here employed require the following steps.

1) Regular space-time networks with constant space and time increments $\Delta x$ and $\Delta t$ are used for the one dimensional problems. Irregular space-time networks with non constant space increments $\Delta x$ and $\Delta y$ and constant time increments $\Delta t$ are used for the two dimensional problems.
2) A linear system of N equations for finding the N values $u_{i,n+1}$ $(i=1, 2, \ldots N;\ n = 0, 1, 2, \ldots)$ at all N internal grid points $P_i$ at each time instant $t = t_{n+1} = (n+1)\,\Delta t$ is set up by discretising the given diffusive-convection equation (5) according to a generalized Crank-Nicolson scheme (Reali et al. (1980, 1981), the initial and boundary conditions (6) and (7) being introduced as proper given data. Accordingly, the time derivative $\frac{\partial u}{\partial t}$ is approximated by a second order analogue

$$\left[\frac{\partial u}{\partial t}\right]_{u,n+\frac{1}{2}} = \frac{u_{i,n+1} - u_{i,n}}{\Delta t} \quad ; \tag{8}$$

the first space derivative $\frac{\partial u}{\partial x}$ is approximated by a third order analogue

$$\left[\frac{\partial u}{\partial x}\right]_{i,n+\frac{1}{2}} = \frac{1}{2}\left\{\left[\frac{\partial u}{\partial x}\right]_{i,n} + \left[\frac{\partial u}{\partial x}\right]_{i,n+1}\right\} \tag{9}$$

by means of four point general FDF;
the second space derivative $\frac{\partial^2 u}{\partial x^2}$ is approximated by a second order analogue

$$\left[\frac{\partial^2 u}{\partial x^2}\right]_{i,n+\frac{1}{2}} = \frac{1}{2}\left\{\left[\frac{\partial^2 u}{\partial x^2}\right]_{i,n} + \left[\frac{\partial^2 u}{\partial x^2}\right]_{i,n+1}\right\} \tag{10}$$

by means of four point general FDF.

3) The algebraic system of equations discretizing the initial differential problem (5), (6), (7) is solved directly to obtain the required values $u_{i,n+1}$ $(n=0, 1, 2, \ldots)$.

The numerical solutions obtained have been checked on the basis of the standard error norms $\|\cdot\|_\infty$ and $\|\cdot\|_2$,

$$\|\underline{s}(t_n)\|_\infty = \operatorname{Max} |u_{i,n} - \tilde{u}_{i,n}| \quad (1 \leqq i \leqq N) \tag{11}$$

$$\|\underline{s}(t_n)\|_2 = \left[\sum_{i=1}^{N} |u_{i,n} - \tilde{u}_{i,n}|^2\right]^{1/2} \tag{12}$$

in which $u_{i,n}$ and $\tilde{u}_{i,n}$ represent, respectively, the exact and the numerical solution at grid point $P_i$ at time instant $t_n$. The i-component of the error vector $\underline{s}(t_n)$ at time instant $t_n$ is $(u_{i,n} - \tilde{u}_{i,n})$ and N represents the number of grid points at which function u is evaluated.

## ONE DIMENSIONAL PROBLEMS

A salient feature of our numerical solution method is represented by the use of a third order approximation for the convective derivative. A second order approximation for this derivative has also been considered to allow comparison with other solution methods.

### Problem 1

In this problem $\lambda = (b, 0)$, $(b = 1.0,\ 2.0)$, $\bar{D} = \{x : 0 \leq x \leq 1\}$, and the analytic solution is

$$u(x,t) = \exp\left(\frac{b}{2}x - \frac{b^2}{4}t\right).$$

Numerical results are given in table I.

TABLE I - Time dependence of error norms $\|\cdot\|_\infty$ and $\|\cdot\|_2$ for problem 1 $(0 \leqq x \leqq 1,\ \Delta x = 0.1\ \ \Delta t = 0.1)$

| Time steps | Norms | b = 1 | | b = 2 | |
|---|---|---|---|---|---|
| | | second order | third order | second order | third order |
| 5 | $\|\cdot\|_2$ | $5.6 \cdot 10^{-5}$ | $4.3 \cdot 10^{-6}$ | $9.9 \cdot 10^{-4}$ | $2.5 \cdot 10^{-4}$ |
| | $\|\cdot\|_\infty$ | $2.4 \cdot 10^{-5}$ | $1.8 \cdot 10^{-6}$ | $4.4 \cdot 10^{-4}$ | $1.1 \cdot 10^{-4}$ |
| 10 | $\|\cdot\|_2$ | $4.9 \cdot 10^{-5}$ | $3.8 \cdot 10^{-4}$ | $6.1 \cdot 10^{-4}$ | $1.5 \cdot 10^{-4}$ |
| | $\|\cdot\|_\infty$ | $2.1 \cdot 10^{-5}$ | $1.6 \cdot 10^{-6}$ | $2.7 \cdot 10^{-4}$ | $6.6 \cdot 10^{-5}$ |
| 100 | $\|\cdot\|_2$ | $5.2 \cdot 10^{-6}$ | $3.8 \cdot 10^{-7}$ | $7.5 \cdot 10^{-8}$ | $1.8 \cdot 10^{-8}$ |
| | $\|\cdot\|_\infty$ | $2.3 \cdot 10^{-6}$ | $1.7 \cdot 10^{-7}$ | $3.3 \cdot 10^{-8}$ | $8.2 \cdot 10^{-9}$ |

Problem 2

In this problem $\lambda = (0.4,\ 0)$, $\bar{D} = \{x : 0 \leqslant x \leqslant 5\}$ and the analytic solution is $u(x, t) = \exp(0.4\,x) - 5\,x + 2\,t$. Numerical results are given in table II.

TABLE II - Time dependence of error norms $\|\cdot\|_\infty$ and $\|\cdot\|_2$ for problem 2 $(0 \leqslant x \leqslant 5,\ \Delta x = 0.5)$

| Time steps | Norms | $\Delta t = 1$ | | $\Delta t = 10$ | |
|---|---|---|---|---|---|
| | | second order | third order | second order | third order |
| 5 | $\|\cdot\|_2$ | $9.3 \cdot 10^{-3}$ | $2.2 \cdot 10^{-4}$ | $1.1 \cdot 10^{-2}$ | $2.8 \cdot 10^{-4}$ |
| | $\|\cdot\|_\infty$ | $4.2 \cdot 10^{-3}$ | $1.3 \cdot 10^{-4}$ | $4.8 \cdot 10^{-3}$ | $1.9 \cdot 10^{-4}$ |
| 10 | $\|\cdot\|_2$ | $1.0 \cdot 10^{-2}$ | $2.2 \cdot 10^{-4}$ | $1.1 \cdot 10^{-2}$ | $2.1 \cdot 10^{-4}$ |
| | $\|\cdot\|_\infty$ | $4.6 \cdot 10^{-3}$ | $1.3 \cdot 10^{-4}$ | $4.7 \cdot 10^{-3}$ | $1.2 \cdot 10^{-4}$ |
| 50 | $\|\cdot\|_2$ | $1.0 \cdot 10^{-2}$ | $2.3 \cdot 10^{-4}$ | $1.1 \cdot 10^{-2}$ | $2.2 \cdot 10^{-4}$ |
| | $\|\cdot\|_\infty$ | $4.7 \cdot 10^{-3}$ | $1.3 \cdot 10^{-4}$ | $4.7 \cdot 10^{-3}$ | $1.3 \cdot 10^{-4}$ |
| 100 | $\|\cdot\|_2$ | $1.0 \cdot 10^{-2}$ | $2.3 \cdot 10^{-4}$ | $1.1 \cdot 10^{-2}$ | $2.2 \cdot 10^{-4}$ |
| | $\|\cdot\|_\infty$ | $4.7 \cdot 10^{-3}$ | $1.3 \cdot 10^{-4}$ | $4.7 \cdot 10^{-3}$ | $1.3 \cdot 10^{-4}$ |

TWO DIMENSIONAL PROBLEMS

We have considered two problems defined in the same domain D, with a computational network G placed on it, as shown in Fig. 1.

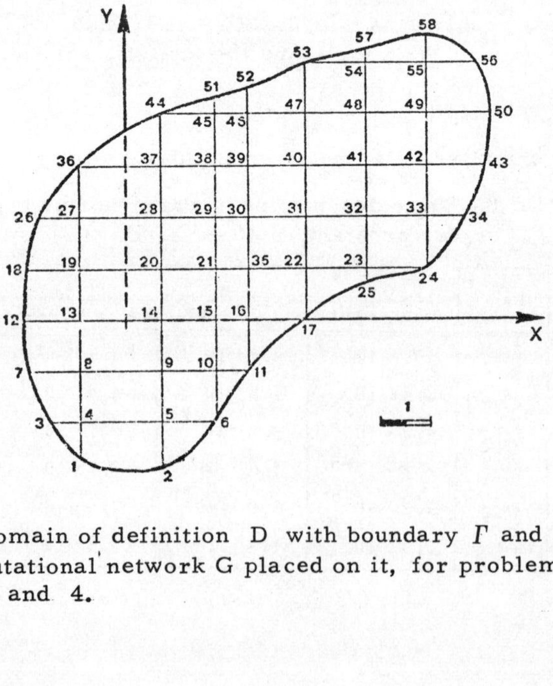


FIG. 1 - Domain of definition D with boundary $\Gamma$ and computational network G placed on it, for problems 3 and 4.

Problem 3

In this problem $\lambda = (1, 1)$, and the analytic solution is $u(x, y, t) = 2(t-x) + \exp(x) - \exp(y)$. Two computational networks have been employed. The one shown in Fig. 1 with N=35 internal grid points and a more refined one with N = 62 internal grid points.

Numerical results are given in table III.

TABLE III - Time dependence of error norms $\|\cdot\|_\infty$ and $\|\cdot\|_2$ for problem 3.

| Time $t_n = n \cdot \Delta t$ | | $u_{max}$ | N = 35 | | N = 62 | |
|---|---|---|---|---|---|---|
| n | $\Delta t$ | | $\|\cdot\|_\infty$ | $\|\cdot\|_2$ | $\|\cdot\|_\infty$ | $\|\cdot\|_2$ |
| 20 | 0.5 | $1431 \cdot 10^3$ | 32.30 | 63.43 | 9.18 | 24.53 |
| 100 | 1 | $1611 \cdot 10^3$ | 32.30 | 63.43 | -- | -- |
| 5 | 20 | $1611 \cdot 10^3$ | 48.23 | 91.23 | 14.18 | 37.40 |
| 25 | 10 | $1911 \cdot 10^3$ | 32.14 | 63.64 | 9.37 | 25.43 |
| 50 | 10 | $2411 \cdot 10^3$ | 32.32 | 63.30 | -- | -- |

Problem 4

In this problem $\lambda = (-1, -1)$ and the analytic solution is $u(x, y, t) = \exp(0.2x + 0.3y + 0.63t)$.

The computational network shown in Fig. 1 has been used.

Numerical results are given in table IV.

TABLE IV - Time dependence of error norms $\|\cdot\|_\infty$ and $\|\cdot\|_2$ for problem 4.

| Time $t_n = n \cdot \Delta t$ | | $u_{max}$ | N = 35 | | N = 62 | |
|---|---|---|---|---|---|---|
| n | $\Delta t$ | | $\|\cdot\|_\infty$ | $\|\cdot\|_2$ | $\|\cdot\|_\infty$ | $\|\cdot\|_2$ |
| 20 | 0.5 | $9897 \cdot 10^3$ | 17.74 | 57.94 | 17.35 | 72.14 |
| 20 | 1. | $5499 \cdot 10^6$ | $3695 \cdot 10^4$ | $1204 \cdot 10^5$ | $3682 \cdot 10^4$ | $1583 \cdot 10^5$ |

## CONCLUSION

The methods of solution of one and two dimensional diffusive-convective problems here described have good characteristics of simplicity, efficiency and accuracy.
In particular, the direct numerical solutions obtained with relatively course grids for two dimensional problems would seem to suggest the present methods as useful ones for applications in water resources research.
It is to be noticed that the present general FDM considerably extend previous standard FDM in as much as they can easily handle directly differential problems defined on irregular domains.

## REFERENCES

Collatz L. (1966) The numerical treatment of differential equations. Springer Verlag, Berlin.

Reali M. and Rangogni R. (1980) High order finite difference forms for any grid by a general Taylor expansion technique. Internal report no. 2792 (1979) and 2792/1 (1980) ENEL-DSR-CRIS, Via Ornato 90/14 - Milano

Reali M. and Rangogni R. (1980) On the derivation of high order finite difference forms for general networks. Green Series Report no. 385 (1980) ENEL-DSR-Via G. B. Martini, 3 - Rome

Reali M., Rangogni R. and Pennati V. (1981) Accurate finite difference solution of the heat conduction equation for a general two-dimensional domain. Numerical Methods in Thermal Problems - R. W. Lewis, K. Morgan and B.A. Schrefler Eds., Pineridge Press. Swansea, 1981

Reali M., Rangogni R. and Pennati V. (1981) Compact analytic expressions of two-dimensional finite difference forms. Internal report no. 2990 (1981) ENEL-DSR-CRIS, Via Ornato 90/14 - Milano

# PETROV-GALERKIN METHODS FOR TWO DIMENSIONAL FLOW CALCULATIONS

R. Wait and M. Parsaei

University of Liverpool, Liverpool, England.

## Introduction

The relative merits of different finite element methods for convective flow calculations appear likely to remain a subject for debate for some considerable time. The purpose of this paper is not an attempt to resolve this debate. But it is to be hoped, that it does provide a little additional insight into the problem.

Various authors have pointed out that finite element solutions to convective flow problems can be more accurate than the corresponding finite difference solutions (Donéa and Giuliana, 1981) assuming a consistent mass formulation is used. However, there are considerable variations in the quality of results reported for mass lumped formulations (Gresho *et al*, 1978; Kawahara *et al*, 1982, for example).

In this short note we attempt to show how it is possible to use quadratic elements and construct a lumped scheme that is sufficiently accurate for many types of flow calculations. It does not contain excessive numerical dissipation and it does not support spurious wiggles.

## Petrov-Galerkin in One-Dimension

Initially we shall consider finite element solutions of the simple advection equation

$$u_t + au_x = 0 \qquad \begin{cases} t > 0 \\ x \in R \end{cases} \qquad (1)$$

using Petrov-Galerkin methods based on quadratic shape functions. If we write the semi-discrete approximation in the form

$$U(x,t) = \sum_i U_i(t)\ \phi_i(t)$$

where the trial functions $\phi_i$ are piecewise quadratic, it follows that the nodal values satisfy

$$\int_R (U_t + a\, U_x)\, \psi_i \, dx = 0 \qquad \forall_i \, . \tag{2}$$

Following Griffiths (1982) the test functions $\psi_i$ are taken to be perturbations of the corresponding $\phi_i$ such that

$$\psi_i = \phi_i + \sigma_i \tag{3a}$$

where

$$\int_{\mathrm{supp}(\sigma_i)} \sigma_i = 0 \, . \tag{3b}$$

In addition, as the purpose of lumping is to reduce the computation time for each time-step, it was considered worthwhile to allow only simple perturbations that were easy to compute. Thus in each element it was assumed that the perturbations could be expressed in terms of local shape functions. If in an element $\varepsilon$ we denote the shape functions by $\phi_i^\varepsilon$ then the test functions are written as

$$\psi_i^\varepsilon = \phi_i^\varepsilon + \sum_k \alpha_{ki}\, \phi_k^\varepsilon \, .$$

If the unperturbated mass matrix is

$$\left\{M_{ki}^\varepsilon\right\} = \left\{\int_\varepsilon \phi_k^e\, \phi_i^e\right\}$$

it follows that the perturbed form satisfies

$$\bar{M}_{ki}^\varepsilon = M_{ki}^\varepsilon + \sum_j \alpha_{jk}\, M_{ji}^\varepsilon$$

and thus no additional integrations are necessary. Similar modifications occur in the convection matrix

$$\left\{C_{ki}^\varepsilon\right\} = \left\{\int_\varepsilon \frac{d\phi_i^\varepsilon}{dx}\, \phi_k\right\} .$$

This approach also ensures that the structure of the matrices is preserved and so no additional storage is required.

Using this approach it is possible to specify five perturbations corresponding to each integer node and two to each mid-element node, these are all illustrated in figure 1. The object is then to manipulate the coefficients in order to achieve the desired local accuracy in a manner analogous to that of Griffiths (1982) for linear elements.

It is well known (Gresho <u>et al</u>, 1978; Hedstrom, 1979) that quadratic elements lead to a secondary spurious wave motion travelling five times as fast as the true solution and in the opposite direction. The size of the spurious motion depends to a large extent on the shape of the initial data. It is frequently possible to ignore these secondary effects but if a consistent mass approximation is used they can persist and lead to inaccuracies.

## The Effects of Lumping

If we wish to consider the order of accuracy of the solution, it is necessary either to eliminate the low order effects of the spurious wave by smoothing as suggested by Hedstrom (1979), or to define accuracy in terms of approximation by the primary mode alone. In any fully discrete approximation it would be necessary to carry out the Hedstrom-smoothing at each time-step. In the subsequent analysis we assume that this can be done but we do not attempt it here.

Mass lumping can also have a significant effect on the propogation of the spurious solution. Assume that the true solution is a wave with speed $\omega$, then the consistent Crank-Nicolson type scheme leads to an approximation containing undamped components with speeds approximating $\omega$ and $-5\omega$. If a mass lumped Euler scheme

$$\underset{\sim}{U}^{n+1} = M\underset{\sim}{U} + (a\Delta t)\, C\underset{\sim}{U}^{n} \quad . \tag{4}$$

is used; it follows that the spurious mode is damped by the factor $\frac{2}{5}$ at each time-step. It is possible to construct examples in which this dramatic damping can be observed in the results. The primary wave is only first order accurate and itself contains considerable dispersion and so the results are not very accurate. In fact the method is unstable in this form.

In order to analyse the accuracy of the schemes. we consider the effect on a single Fourier component such that

$$U_j^n = e^{i\omega x_j}$$

and

$$U_j^{n+1} = e^{i\bar{a}\omega\Delta t}\, e^{i\omega x_j}$$

for some complex valued $\bar{a}$. It is then possible to define the order of the method as p such that

$$e^{i\bar{a}\omega\Delta t} - e^{ia\omega\Delta t} = 0(\Delta t^{p}),$$

the order of dissipation d such that

$$|e^{i\bar{a}\omega\Delta t}|^{2} = 1 - 0(\Delta t^{d})$$

and to analyse the phase error in terms of

$$a - \mathrm{Re}(\bar{a}) \ .$$

If we write the perturbation as

$$\sigma_i = \sum_{j=1}^{5} \alpha_i^{(j)} \sigma_i^{(j)} \qquad \text{for the integer nodes}$$

and

$$\sigma_i = \sum_{j=1}^{2} \beta_i^{(j)} \sigma_i^{(j)} \qquad \text{for the half nodes} \quad ,$$

it follows that the resulting formulae can be expressed as in table I. If we impose the additional condition

$$\alpha^{(2)} = 20r\,\alpha^{(5)}$$

in order to simplify the algebra, it follows that the stability condition is

$$r \leqslant .21 \quad .$$

Whereas the standard Galerkin method, with all the parameters zero, leads to an unconditionally unstable formula if lumped as shown in (4).

Table I. Accuracy of Petrov-Galerkin formulae

| Order accuracy | Order dissipation | conditions on parameters |
|---|---|---|
| 1 | 2 | (I) $\begin{cases}\beta^{(2)} = 0 \\ \alpha^{(3)} = 0\end{cases}$ or (II) $\begin{cases}\beta^{(2)} = 0 \\ \beta^{(1)} = -1\end{cases}$ |
| 2 | 4 | (III) $\begin{cases}\text{(I) and} \\ \alpha^{(1)} = \Big[4\alpha^{(5)}(\beta^{(1)} + 1 - 20r^2) \\ \quad -r^2(4\alpha^{(2)} - \beta^{(1)} - 3)\Big]/(1+\beta^{(1)})\end{cases}$ <br> or <br> (IV) $\begin{cases}\text{(II) and} \\ \alpha^{(3)} = r(-\frac{2}{15}\alpha^{(2)} + \frac{1}{15} - \frac{8}{3}\alpha^{(5)} r)\end{cases}$ |
| 3 | 4 | (V) $\begin{cases}\text{(IV) and} \\ \alpha^{(4)} = -\frac{5}{2r}\alpha^{(1)}(1 + 4r^2) \\ \quad - \frac{5}{3r}\alpha^{(2)}(2 - r^2) \\ \quad + \frac{5}{6r}(2 + 8r^2) \\ \quad + \frac{10}{3}\alpha^{(5)}(1 + 10r^2)\end{cases}$ |
| 4 | 6 | (V) and <br> $\alpha^{(1)} = \frac{1}{15}(38r^2 + 1)$ <br> $+ \frac{2\alpha^{(2)}}{15}(7r^2 - 1)$ <br> $+ \frac{4\alpha^{(5)}}{3} r(14r^2 + 1)$ |

In all formulae $r \equiv \frac{a\Delta t}{\Delta x}$ .

Problems in Two Dimensions

In order to solve problems of the form

$$u_t + au_x + bu_y = 0 ,$$

we assume that a tensor product form of finite element approximation is being used. This implies that we employ the full 9-node biquadratic elements rather than 8-node elements. If tensor product basis functions are used with the Galerkin equations

$$\iint_R (U_t + a\, U_x + b\, U_y)\, \psi_{ij}\, dxdy = 0 \qquad \forall_{ij} \tag{5}$$

then the discretized equations (4) with Euler time discretization, are replaced by

$$\underset{\sim}{U}^{n+1} = \left\{ M_1 \otimes M_2 + \Delta t (aC_1 \otimes M_2 + bM_1 \otimes C_2) \right\} \underset{\sim}{U}^n . \tag{6}$$

Here, i = 1, 2 refer to the x, y directions respectively and $M_i$ and $C_i$ are the corresponding univariate mass and convection matrices. In order to simplify the analysis of the perturbation terms, the Galerkin equations (6) are replaced by

$$\underset{\sim}{U}^{n+1} = (M_1 + a\Delta t\ C_1) \otimes (M_2 + b\Delta t\ C_2)\ \underset{\sim}{U}^n \tag{7}$$

thus the formulation used in Alternating Direction Galerkin (for example, Mitchell and Wait; 1977) is adopted even though the equations are explicit and the alternating direction mode of solution is not necessary.

Once (7) is assumed then it is possible to use the parameters calculated for one-dimensional problems directly in the two dimensional formulation. If the three spurious modes are ignored then the accuracy results follow directly for the two-dimensional approximation.

If a and b vary from element to element, then the formulation (7) is assumed to hold only for element matrices and not for the global formulation. In such cases it is understood that the values referred to in Table I are local values only.

The results of a numerical test are shown in figure 2, many more results of tests on one-and-two-dimensional problems are available in Wait and Parsaei (1982).

REFERENCES

Donéa, J and Giuliani, S. (1981) "A simple method to operate high-order accurate convection operators for explicit schemes based on linear finite elements", Int. J. Numer. Methods in Fluids, 1, 63 - 79

Gresho, P.M., Lee, R.L. and Sani, R.L. (1978) "Advection-Dominated flows, with emphasis on the consequences of mass lumping" in Finite Elements in Fluids III (ed. Gallagher et al) Wiley, 335 - 350.

Griffiths, D.F. (1982) "A Petrov-Galerkin method for hyperbolic equations", in Mathematics of Finite Elements and its Applications IV, (ed. Whiteman) Academic, to appear.

Kawahara, M., Hirano, H. and Tsubota, K. (1982) "Selective lumping finite element method for shallow water flow", Int. J. Numer. Mathods in Fluids, 2, 89 - 112.

Hedstrom, G.W. (1979) "The Galerkin method based on Hermite cubics", S.I.A.M. J. Numer. Anal., 16, 385 - 393.

Mitchell, A.R. and Wait, R. (1977) "The finite element method for partial differential equations", Wiley.

Wait, R. and Parsaei (1982) "Lumped Petrov-Galerkin methods for hyperbolic equations", to appear.

Figure 1.

Perturbations $\sigma_i^{(1)}$, .. , $\sigma_i^{(5)}$ at an integer node and $\sigma_i^{(1)}$, $\sigma_i^{(2)}$ at a half node.

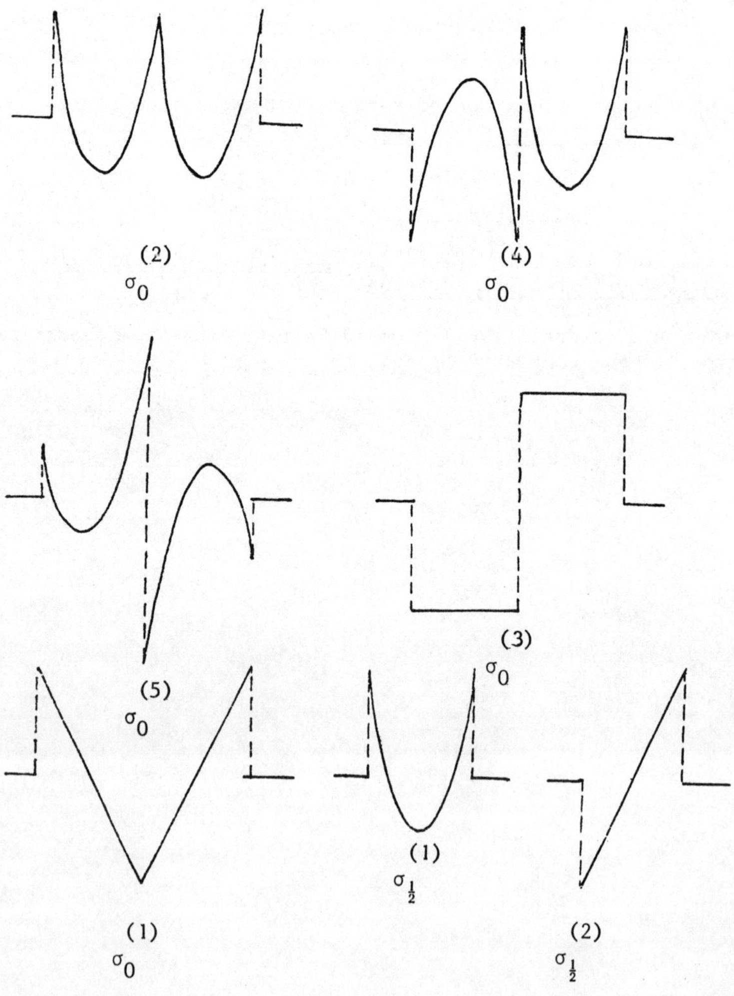

Figure 2.

A square unit pulse moves such that a = 2b, 4th order tensor product approximation with $\alpha^{(2)} = -1$, $a\frac{\Delta t}{\Delta x} = .2$ . It is shown after 1, 40, 80 and 120 time steps. The contours are .3, .5, .7, .9 (and 1.1).

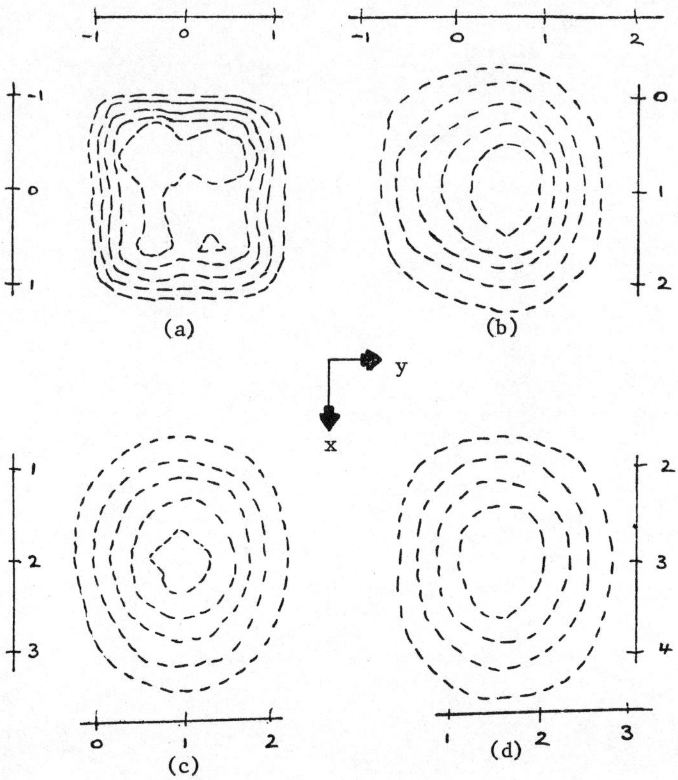

# NUMERICAL SOLUTION OF THREE-DIMENSIONAL AND TIME-DEPENDENT ADVECTION-DIFFUSION EQUATIONS BY COLLOCATION METHODS

Hans Wengle

Institute for Fluid Dynamics, LRT
Hochschule der Bundeswehr München
8014 Neubiberg, Federal Republic of Germany

## 1. Introduction

Collocation is a general principle to provide from continuous mathematical models the corresponding discretized systems of equations. The method is a subclass of the weighted residual methods and it is defined by requiring the residual to be zero at certain selected points, the so-called collocation points. The residual is defined as the difference between the evaluation of the continuous form and the discretized form of the differential equation to be solved.

Two different collocation methods had and still have considerable success in solving problems in fluid mechanics and chemical engineering. The first method is the so-called pseudo-spectral method, publications on which first appeared in the early 1970's by Orszag (1971), Kreiss and Oliger (1972), Gazdag (1973), and Fornberg (1975). The pseudo-spectral,and related spectral methods, were originally developed for problems in turbulent fluid flows; for references see the review of Orszag and Israeli (1974). The second method, orthogonal collocation, was systematically used for the first time by Frazer, Jones and Skan (1937). Later, the method was further developed and applied to problems in transport phenomena and chemical reactor theory by Villadsen and Stewart (1967). For theoretical aspects associated with orthogonal collocation methods, the reader is referred to the book of Villadsen and Michelsen (1978) and the reviews of Finlayson (1975, 1980).

Based on their success and the large amount of experience in application documented in the recent literature, the two numerical techniques have been selected for the solution of advection-diffusion problems, in particular the solution of models for the advective and turbulent-diffusive transport of an air-pollutant in the atmospheric boundary layer. To predict the dynamic behavior of trace contaminants in given three-dimensional and time-dependent flow fields, a transport equation must be solved of the general form (1), where c is the mean concentration of a chemically inert contaminant, u, v, and w are the given components of the mean wind field, and $K_x$, $K_y$, and $K_z$ are given eddy-diffusivities:

$$\frac{\partial c}{\partial t} + u\,\frac{\partial c}{\partial x} + v\,\frac{\partial c}{\partial y} + w\,\frac{\partial c}{\partial z} = \frac{\partial}{\partial x}(K_x\,\frac{\partial c}{\partial x}) + \frac{\partial}{\partial y}(K_y\,\frac{\partial c}{\partial y}) + \frac{\partial}{\partial z}(K_z\,\frac{\partial c}{\partial z}) \quad (1)$$

The concentration field may be three-dimensional and time-varying due to the three-dimensional and time-dependent flow field and, in general, the eddy-diffusivities are also functions of position and time. In a wind field not varying in time, the concentration field may still be time-dependent due to unsteady source emission. In many cases, the turbulent transport in the main wind direction (say x-direction) may be neglected in comparison to the advective transport in the same direction. To complete the mathematical description of the unsteady diffusion of an inert contaminant from an elevated point source, we define proper initial and boundary conditions. For example, we assume that the plume is confined between ground (z=0) and an elevated stable layer at z=H and that the point source is located at $(x_S=0, y_S, z_S)$ and begins emitting at time t=0 with constant source strength S into an atmosphere without background concentration:

| | | |
|---|---|---|
| t=0: | $c=0$ | (2a) |
| z=0 and H: | $(K_Z \partial c/\partial z)=0$ | (2b) |
| y=0 and $y_{max}$: | $c=0$ | (2c) |
| x=0: | $c=(S/u(0,y_S,z_S,t))\delta(y-y_S)\delta(z-z_S)/\Delta y_S \Delta z_S$ | (2d) |
| $x=x_{max}$: | $\partial^2 c/\partial x^2=0$ | (2e) |

If the turbulent transport in x-direction is neglected, the last boundary condition (2e) is not required. The source condition (2d) is derived from a mass flux rate integral over an area $\Delta y_S \Delta z_S$ containing the point source.

Two-dimensional and time-dependent line-source diffusion problems were solved by Christensen and Prahm (1976), Berkowitz and Prahm (1978), Wengle and Seinfeld (1978), and Lee (1981, 1982) using pseudo-spectral methods and by Wengle, Van den Bosch and Seinfeld (1978) using orthogonal collocation. The more general three-dimensional point-source diffusion problem was solved by Fleischer and Worley (1978) using global orthogonal collocation and by Wengle (1981) using orthogonal collocation on multiple elements.

Our object in the current work is to develop a standard approach for solving rather general advection-diffusion equations of the form (1), and to demonstrate that, with a systematic modular programming technique, the two collocation methods mentioned above are not more difficult to apply than classical finite-differece methods. In this paper, we first compare the two numerical solution techniques in general terms and then, we concentrate on two problem areas which are (a) the treatment of non-periodic boundary conditions using pseudo-spectral methods and (b) the problem of how to avoid an unfavorable structure of the coefficient matrix of the resulting set of collocation equations using orthogonal collocation. For application we selected the point-source diffusion as an important problem in air pollution modeling.

## 2. COMPARISON OF PSEUDO-SPECTRAL AND ORTHOGONAL COLLOCATION METHODS IN GENERAL TERMS

Both methods are based on the expansion of the dependent variable in series of functions. They differ in

- the choice of the expansion functions,
- the selection of the collocation points and
- the way the resulting discretized system is solved.

In the pseudo-spectral method we choose function expansions in such a way that the coefficients of the expansion and the spatial derivatives may be evaluated directly by a Fast-Fourier-Transform algorithm (FFT). It has been shown by Orszag (1971a)that the derivatives can be evaluated in this way accurately, and without introducing significant phase-errors. If we select equidistant collocation points, we may use cosine-, sine-, or the more general Fourier-series; if we prefer non-equidistant collocation points, the choice of Chebyshev polynomials as basic functions and of the collocation points $x_j=0.5L(1+\cos(\pi j/N))$ $(j=0,1,..,N)$ still allows us to evaluate the expansion coefficients and the spatial derivatives by FFT's.

In orthogonal collocation the basic functions must be orthogonal polynomials and the collocation points are selected to be the roots of certain orthogonal polynomials (Jacobi polynomials) which leads to high accuracy, hence the name orthogonal collocation. It is convenient to select the basic functions in such a way that the coefficients of the expansion are the values of the dependent variable at the collocation points; Lagrange polynomials satisfy this requirement. Differentiating the expansions term by term leads to expressions for the derivatives, and the values of the derivatives of the Lagrange polynomials in these expansions need to be evaluated only once at the beginning of the program (as long as the collocation points are fixed). If a single expansion is used to represent the dependent variable over the entire domain in each coordiante direction the method is called global collocation; if each coordinate direction is subdivided first into smaller regions, and a separate expansion is used for each subregion, the method is called orthogonal collocation on multiple elements or orthogonal collocation on finite elements or global spline collocation.

Figure 1 shows schematically the basic steps in the application of the two methods to the numerical solution of a two-dimensional and time-dependent line-source diffusion problem:
Pseudo-spectral method: because of the zero-gradient boundary conditions (2b), a cosine-series has been chosen in the vertical direction and, in horizontal direction, we used a Chebyshev polynomial expansion to handle properly the "inflow" boundary condition (2d) which requires the numerical realization of a Dirac-delta function. For the time step, the Adams-Bashforth or the Runge-Kutta type integration schemes are appropriate.
Orthogonal collocation: it is also possible to collocate in time direction but usually, no internal collocation points are used

in that direction and therefore, the time step is equivalent to an implizit finite-difference Euler step. The resulting system of algebraic equations has a full coefficient matrix and therefore, the global collocation scheme will require reasonably low computational effort only if the number of internal collocation points is relatively small, i.e. the coefficient matrix is small. To improve the efficiency of the method we propose (in section 4) to use the multiple element approach and an alternating direction implizit (ADI) solution of the collocation equations.

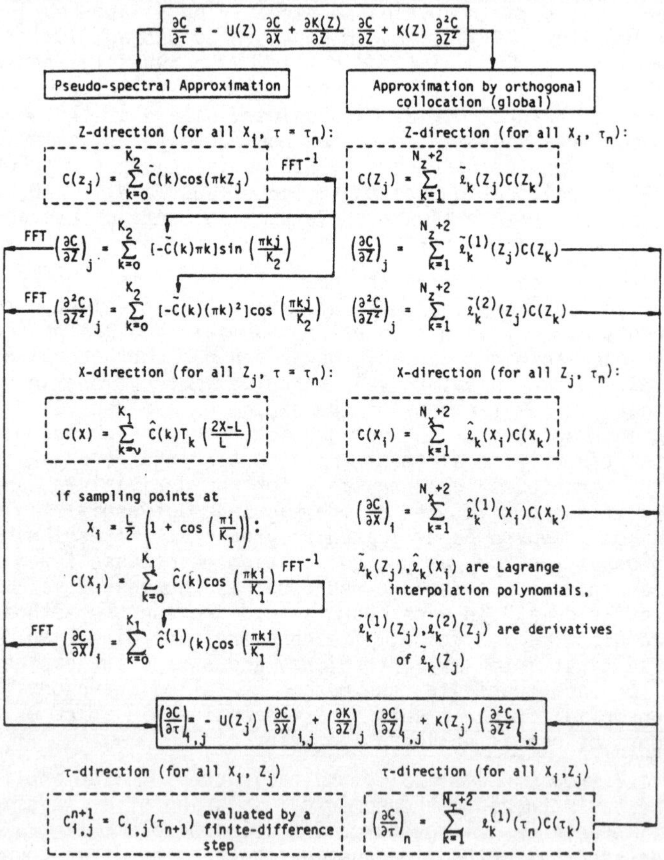


Figure 1 Basic steps in the numerical solution of a two-dimensional and time-dependent advection-diffusion equation by (global) pseudo-spectral and global orthogonal collocation methods

## 3. TREATMENT OF NON-PERIODIC BOUNDARY-CONDITIONS USING PSEUDO-SPECTRAL METHODS

Periodic boundary conditions are the natural boundary conditions for Fourier expansions.However, most practical problems require non-periodic boundary conditions to be satisfied, and the corresponding discontinuities of a function and/or of its derivatives at the boundaries create a strong overshooting of a finite Fourier expansion (Gibbs phenomenon).To avoid the resulting slow convergence rate, we may choose a Chebyshev polynomial expansion; then, the convergence rate depends only on the smoothness of the function to be approximated and not on the nature of the boundary condition. In that case we must use collocation points $x_j=0.5L\cdot(1+\cos(\pi j/N))$ $(j=0,1,..,N)$ to be able to evaluate the expansions by FFT's. This is indeed a very good choice for boundary-layer type problems. If we prefer (or if we have to use) equidistant collocation points, we may also choose different expansion functions which automatically satisfy the non-periodic boundary conditions. For example, zero-gradient boundary conditions (2b) can be satisfied using cosine-series, and zero-function-value boundary conditions (2c) can be satisfied using sine-series. A more general principle is to subtract (from the function to be approximated) a simple function $F_p$ (typically a polynomial but not necessarily only these) so that the difference $F_r$ has as much smoothness as we wish. For example, we choose $F_p(x)$ to be a Nth degree polynomial and, in x-direction, we represent F(x) as

$$F(x) = F_p(x) + F_r(x) \tag{3}$$

$$F_p(x) = \sum_{i=0}^{N} a_i x^i \tag{4}$$

The unknown coefficients $a_i$ are found from the periodicity conditions for the desired periodic function $F_r(x)$ and its derivatives $F_r^{(n)}(x)$ at both ends (x=0 and L) of the domain of interest [0,L]:

$$F_r^{(n)}(0) = F_r^{(n)}(L) \qquad 0 \leq n \leq N-1 \tag{5}$$

The expressions for the coefficients $a_i$ derived from (5) involve the function values and up to the (N-1)th derivatives of F(x) at x=0 and x=L which must either be given or must be approximated (e.g. by polynomial interpolation). If these coefficients are good, the residual function $F_r$ will appear to the FFT algorithm as a sufficiently periodic function. This will result into an increased convergence rate for the direct evaluation of the derivatives of $F_r$. The derivatives of the polynomial $F_p$ can be evaluated analytically and therefore, the derivatives of the original non-periodic function F(x) can be evaluated.

In figure 2 we present a schematic outline of the reduction-to-periodicity approach applied to the numerical solution of the advection-diffusion equation (1). Note that the diffusive terms in (1) have been written differently in figure 2 to provide a clear separation between first and second derivatives of c.

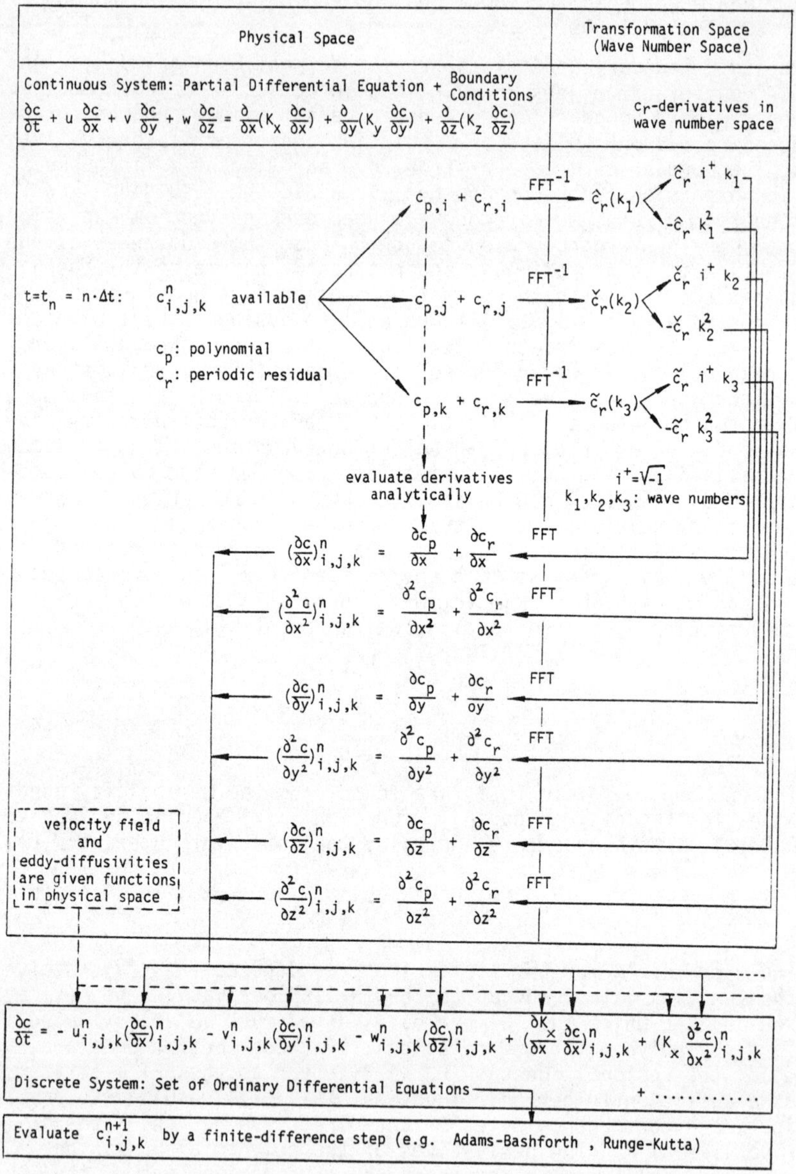


Figure 2 Pseudo-spectral approximation of a three-dimensional and time-dependent advection-diffusion problem with non-periodic boundary conditions using the reduction-to-periodicity principle

## 4. NUMERICAL SOLUTION BY ORTHOGONAL COLLOCATION ON MULTIPLE ELEMENTS

In contrast to global collocation (see figure 1) we first subdivide the computational domain into smaller subregions (fig. 3a). Then, we normalize the independent variables such that they vary from 0 to 1 within each of the subelements. To provide a finite set of algebraic equations, we follow the steps of figure 3b, individually for each subelement. Unfortunately, the number of unknowns will be larger than the number of equations resulting from solving the differential equation at all internal collocation points and from the boundary conditions. Therefore, additional equations must be provided by requiring continuous first derivatives normal to the boundaries between the subelements. As already mentioned, in time direction we use no internal collocation points which is equivalent to an implizit Euler step. If we neglect turbulent diffusion in x-direction, this direction also will be a so-called "marching"-direction, and we use at most one internal collocation point. From this it follows that we may use the result at the end of a $\Delta X(ix)$-element as the initial condition for the calculations in the $\Delta X(ix+1)$-element.

As a result of using the multiple element approach we are able to use more subelements in those regions where steep gradients are to be expected and to evaluate the collocation equations faster than when global collocation is used. Now, the coefficient matrix has a banded structure with a large number of zero entries even within the band. However, the band width will still be relatively large in comparison to the size of the matrix and, for a time-dependent problem, the computing time will be fairly large if we solve the collocation equations directly.

To improve the ratio of computing time to physical time, it is of advantage to use an alternating direction implizit- (ADI-) approach to solve the collocation equations. To advance the differential equation over one time element $\Delta T$, two intermediate values $c^*$ and $c^{**}$ are evaluated applying the scheme of equations (6) to (8), given by Brian (1961). The system of collocation equations is solved in three steps: in each step, the unknowns from partial derivatives with respect to one particular coordinate direction are treated implizitly while the unknowns from derivatives with respect to all the other coordinates are taken from the last time step or from an intermediate step. From a physical point of view, we therefore advance the system first by advection in x-direction, then by advection and turbulent diffusion in z-direction and finally, by advection and turbulent diffusion in y-direction. From a computational point of view, we solve in each of the three steps a series of sets of algebraic equations, each of which has a small coefficient matrix with block-diagonal structure. The blocks are small and have all the same simple structure; the block size is determined by the number of internal collocation points on a line in a particular coordinate direction within a subelement.

a.)

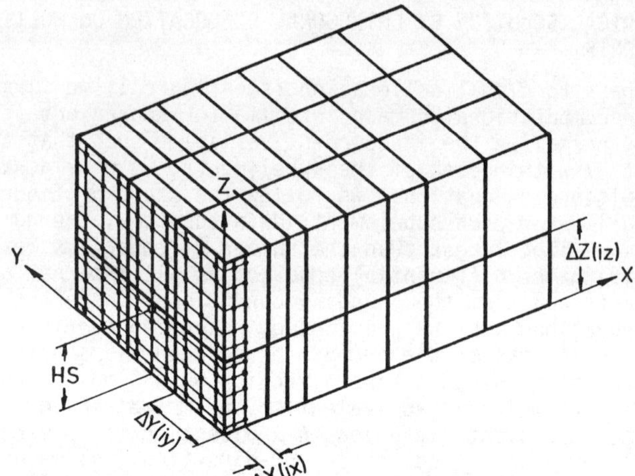


b.)

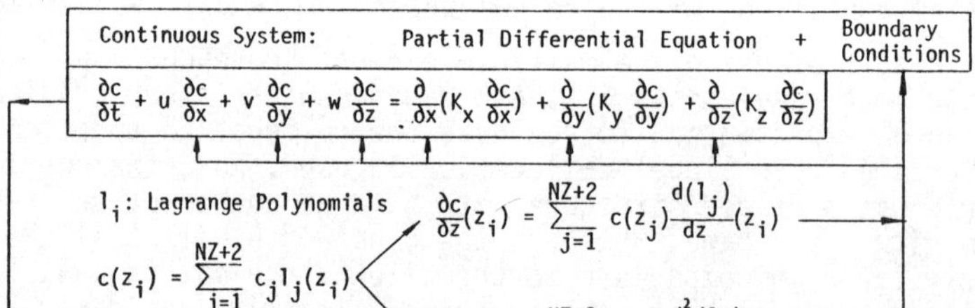


Figure 3 Numerical solution by orthogonal collocation on multiple elements:
a.) subdivision of computational domain
b.) basic steps to provide discretized system

$$\frac{c^{*} - c^{n}}{T(n)/2} = -u\frac{\partial c^{*}}{\partial x} - v\frac{\partial c^{n}}{\partial y} - w\frac{\partial c^{n}}{\partial z} + K_y\frac{\partial^2 c^{n}}{\partial y^2} + K_z\frac{\partial^2 c^{n}}{\partial z^2} \; \frac{\partial K_y}{\partial y}\frac{\partial c^{n}}{\partial y} + \frac{\partial K_z}{\partial z}\frac{\partial c^{n}}{\partial z} \tag{6}$$

$$\frac{c^{**} - c^{n}}{T(n)/2} = -u\frac{\partial c^{*}}{\partial x} - v\frac{\partial c^{n}}{\partial y} - w\frac{\partial c^{**}}{\partial z} + K_y\frac{\partial^2 c^{n}}{\partial y^2} + K_z\frac{\partial^2 c^{**}}{\partial z^2} + \frac{\partial K_y}{\partial y}\frac{\partial c^{n}}{\partial y} + \frac{\partial K_z}{\partial z}\frac{\partial c^{**}}{\partial z} \tag{7}$$

$$\frac{c^{n+1} - c^{n}}{T(n)/2} = -u\frac{\partial c^{*}}{\partial x} - v\frac{\partial c^{n+1}}{\partial y} - w\frac{\partial c^{**}}{\partial z} + K_y\frac{\partial^2 c^{n+1}}{\partial y^2} + K_z\frac{\partial^2 c^{**}}{\partial z^2} + \frac{\partial K_y}{\partial y}\frac{\partial c^{n+1}}{\partial y} + \frac{\partial K_z}{\partial z}\frac{\partial c^{**}}{\partial z} \tag{8}$$

## 5. CONCLUDING REMARKS

Within the general framework of collocation methods, the pseudo-spectral method, as applied hitherto, can be considered as a global collocation method. The key subroutine used by all pseudo-spectral methods is a FFT algorithm and therefore, the proper treatment of non-periodic boundary conditions is of crucial importance. It is not necessary (and not desirable) to introduce into the differential equation an artificial sink term to force the solution to be a periodic function, as proposed in the first papers of Christensen and Prahm (1976). The Gibbs phenomenon can be avoided by applying a general reduction-to-periodicity principle (see section 3). It must be emphasized that the decomposition (3) is exact, even for arbitrary values of the coefficients $a_i$ since they only serve to define a simple function $F_p$ to be subtracted from the original non-periodic function F to generate a sufficiently smooth and periodic residual function $F_r$ the derivatives of which can be evaluated efficiently and accurately by FFT's. More details can be found in Gottlieb and Orszag (1977), Roache (1978) and Wengle and Seinfeld (1978).

The numerical solution method "orthogonal collocation on multiple elements" is also a flexible and efficient method, as long as the number of collocation points is relatively small. If the collocation equations are solved directly, the resulting program is rather complicated, whereas an ADI solution approach generates a very simple structure of the program. However, the ADI procedure needs slightly more internal collocation points to reach comparable accuracy. The key subroutine here is a solver for sets of algebraic equations with block-diagonal structure of the coefficient matrix. A ratio of computing time to physical time of 1/6 has been reached on a Burroughs B7800 computer, see Wengle (1981). With respect to the cost of model calculations, a further improvement in that ratio is desirable.

REFERENCES


Berkowicz, R. and Prahm, L.P. (1978) Pseudospectral simulation of dry deposition from a point source.
Atmospheric Environment 12, 379-387.

Brian, P.L.T. (1961) A finite-difference method of higher order accuracy for the solution of 3-dim. transient heat conduction problems. A.I.C.H.E. J. 7, 367-370.

Christensen, O. and Prahm, L.P. (1976) A pseudospectral model for dispersion of atmospheric pollutants.
J.Appl.Meteor. 15, 1284-1294.

Finlayson, B.A. (1975) Weighted residual methods and their relation to finite-element methods in flow problems,
in: Gallagher, R.H., Oden, J.T., Taylor, C. and Zienkiewicz,O.C. Finite Elements in Fluids, Vol.2, John Wiley and Sons, London.

Finlayson, B.A. (1980) Orthogonal collocation on finite elements - progress and potential.
Mathematics and Computers in Simulation 22, 11-17.

Fleischer, M.T. and Worley Jr., F.L. (1978) Orthogonal collocation - application to diffusion from point sources.
Atmospheric Environment 12, 1349-1357.

Frazer, R.A., Jones, W.P. and Skan, S.W. (1937) Approximations to functions and to the solutions of differential equations.
Great Britain Aero.Res.Conf., London, Reprint No.1799, Reprinted in Grat Britain Air Ministry Aero.Res.Comm.Tech.Rep. 1, 516-549.

Fornberg, B. (1975) On a Fourier method for the integration of hyperbolic equations. SIAM J.Num.Anal. 12, 509-528.

Gazdag, J. (1973) Numerical convective schemes based on accurate computation of space derivatives. J.Comp.Phys. 13, 100-113.

Gottlieb, D. and Orszag, S.A. (1977) Numerical Analysis of Spectral Methods: Theory and Applications.
Society for Industrial and Applied Mathematics,Philadelphia,USA.

Kreiss, H.O. and Oliger, J. (1972) Comparison of accurate methods for the integration of hyperbolic equations. Tellus 24, 199-215.

Lee, H.N. (1981) An alternate pseudospectral model for pollutant transport, diffusion and deposition in the atmosphere.
Atmospheric Environment 15, 1017-1024.

Lee, H.N. (1982) Numerical modeling of pollutant transport and dispersion using pseudospectral method.
Third joint Conf. on Appl. of Air Pollution Meteor., Jan 11-15, San Antonia, Texas, USA.

Orszag, S.A. (1971) Numerical Simulation of incompressible flows within simple boundaries. I. Galerkin (spectral) representations. Studies in Applied Mathematics 4, 293-327.

Orszag, S.A. (1971a) Numerical simulation of incompressible flows within simple boundaries: accuracy.
J.Fluid Mech. 49, 75-112.

Orszag, S.A. and Israeli, M. (1974) Numerical simulation of viscous incompressible flows.
Annual Reviews of Fluid Mechanics 6, 281-318.

Roache, P. (1978) A pseudo-spectral FFT technique for non-periodic problems. J.Comp.Phys. 27, 204-220.

Villadsen, J.V. and Stewart, W.E. (1967) Solution of boundary-value problems by orthogonal collocation.
Chem.Engng.Sci. 21, 1483-1501.

Villadsen, J.V. and Michelsen, M.L. (1978) Solution of Differential Equation Models by Polynomial Approximations.
Prentice-Hall Inc., Englewood Cliffs, New Jersey.

Wengle, H. and Seinfeld, J.H. (1978) Pseudospectral solution of atmospheric diffusion problems.
J.Comp.Phys. 26, 87-106.

Wengle, H., Van den Bosch, B. and Seinfeld, J.H. (1978) Solution of atmospheric diffusion problems by pseudo-spectral and orthogonal collocation methods.
Atmospheric Environment 12, 1021-1032.

Wengle, H. (1981) Numerical solution of three-dimensional and time-dependent advection-diffusion equations by "orthogonal collocation on multiple elements".
12th Int.Tech.Meeting on Air Pollution Modeling and its Applications, August 25-28, Palo Alto, USA,
Plenum Publishing Corp., New York.

# 15. SIMULATION OF TRANSPORT PROCESSES

# THERMAL DISCHARGE ANALYSIS IN ST-LAWRENCE RIVER BY FINITE ELEMENTS : A CASE STUDY

J.F. Cochet, G. Dhatt et G. Hubert

Université de Technologie de Compiègne, Compiègne, France

## INTRODUCTION

The finite element method has become a strong analytical tool for solving complex flow problems in large rivers, estuaries and oceans. When coupled with ever developing interactive graphics, it becomes an indispensable aid to the decision making process for a design engineer. In this article, we present results of an environmental impact case study dealing with problem of nuclear thermal discharge. The finite element method is employed to obtain velocity and temperature profiles in the St-Lawrence river flow (near Gentilly, Quebec) to include stationary and tidal phenomena. We discuss various aspects of model calibration such as choice of boundary conditions, initial conditions, bed slope and Chezy's friction coefficient. The numerical results are compared, wherever possible, with those obtained by the Lasalle reduced model or with measured in-situ values [1, 2].

## MATHEMATICAL MODEL [3, 4, 5]

In this study, we employ the depth averaged two-dimensional model representing the flow and thermal discharge phenomenon (vertical accelleration is neglected giving hydrostatic pressure). The fluid is assumed to be well mixed along the depth and the density is assumed constant. The decoupled flow and transport-diffusion equations are given by :

<u>Flow equations</u>

$$\frac{\partial u}{\partial t} + u\,\frac{\partial u}{\partial x} + v\,\frac{\partial u}{\partial y} = F_x \qquad (1\text{-a})$$

$$\frac{\partial v}{\partial t} + u\,\frac{\partial v}{\partial x} + v\,\frac{\partial v}{\partial y} = F_y \qquad (1\text{-b})$$

$$\frac{\partial h}{\partial t} + \frac{\partial (H + h)u}{\partial x} + \frac{\partial (H + h)v}{\partial y} = 0 \qquad (1\text{-c})$$

where u, v are the depth averaged velocities (Fig. 1)

$$u = \frac{1}{H + h} \int_{-H}^{h} u \, dz \qquad v = \frac{1}{H + h} \int_{-H}^{h} v \, dz \qquad (2)$$

H(x,y) and h(x,y) define bed and free surface profiles respectively as shown in Fig. 1. $F_x$ and $F_y$ are the summation of following force components :

- Coriolis force

$$F_x = - fv \qquad F_y = fu \qquad (3\text{-a})$$

- Hydrostatic force

$$F_x = - g \frac{\partial h}{\partial x} \qquad F_y = - g \frac{\partial h}{\partial y} \qquad (3\text{-b})$$

- Bed friction force

$$F_x = - g \frac{u\sqrt{u^2 + v^2}}{C^2 (H + h)} \qquad F_y = - g \frac{v\sqrt{u^2 + v^2}}{C^2 (H + h)} \qquad (3\text{-c})$$

- Surface friction force (wind force)

$$F_x = A \frac{u'\sqrt{u'^2 + v'^2}}{H + h} \qquad F_y = A \frac{v'\sqrt{u'^2 + v'^2}}{H + h} \qquad (3\text{-d})$$

where u', v' are the wind velocity components, A is wind-fluid friction coefficient

- Viscosity force (artificial or turbulent)

$$F_x = \nu \, \Delta \, u \qquad F_y = \nu \, \Delta \, v$$

where $\nu$ is the viscosity coefficient, normally introduced for numerical stability and to represent flows with vortex. $\Delta$ is the Laplacian operator.

<u>Transport-diffusion equation</u>

$$\frac{\partial T}{\partial t} + u \frac{\partial T}{\partial x} + v \frac{\partial T}{\partial y} - \frac{1}{(H + h)} \left(\frac{\partial}{\partial x} D_x (H + h) \frac{\partial T}{\partial x} + \frac{\partial}{\partial y} D_y (H + h) \frac{\partial T}{\partial y}\right)$$

$$= (S - P) \; / \; (H + h) \qquad (4)$$

where T is the depth averaged temperature

$$T(x,y) = \frac{1}{(H + h)} \int_{-H}^{h} T(x,y,z)\, dz \tag{5}$$

$D_x$ and $D_y$ are the dispersion coefficients. The values of $D^x$ along the flow are normally chosen superior to the values along transverse direction. S is the source term (the thermal discharge input $m^3\ °C/sec - m^2$) and P is the heat loss to environment ($P = K \cdot (T - T_0)$ where $K = m3\ °C/°C - m^2 - sec$ and $T_0$ is the ambiant air temperature).

FINITE ELEMENT MODEL [4, 6]

The integral formulation based on weighted residue method is given by :

$$W = \int_A [\delta u(\text{Eq. 1-a}) + \delta v(\text{Eq. 1-b}) + \delta h(\text{Eq. 1-c}) + \delta T(\text{Eq. 4})]\, dA = 0 \tag{6}$$

where $\delta u$, $\delta v$, $\delta h$, $\delta T$ are Galerkin type weighting functions and u, v, h, T are the unknows of the problem. A given domain A is represented by a set of sub-domains $A^e$ called "finite elements" to give :

$$W = \sum_{\text{element}} W^e = 0 \tag{7}$$

Over each element, we chose a finite element nodal approximation such that :

$$u(x,y,t) = \langle N(x,y)\rangle \{U_n(t)\} \qquad \delta u = \langle N\rangle \{\delta U_n\}$$

$$v = \langle N\rangle \{V_n\} \qquad \delta v = \langle N\rangle \{\delta V_n\} \tag{8}$$

(8)

$$T = \langle N\rangle \{T_n\} \qquad \delta T = \langle N\rangle \{\delta T_n\}$$

$$h = \langle N_h\rangle \{h_n\} \qquad \delta T = \langle N\rangle \{\delta h_n\}$$

where $\langle N\rangle$ and $\langle N_h\rangle$ are approximation functions for elements of Fig. 2. The choice of approximation of Eq. (8) leads to a matrix representation of Eq. (7) :

$$[M] \left\{\frac{du}{dt}\right\} + [K]\{u\} = \{F\} \tag{9-a}$$

$$[\underline{M}] \left\{\frac{dT}{dt}\right\} + [\underline{K}]\{T\} = \{\underline{F}\} \tag{9-b}$$

where $\{u\}$ represents all the nodal unknowns ($U_n$, $V_n$, $h_n$) and $\{T\}$ the nodal temperature unknowns. The nonlinear Eq. (9-a) is solved by the Newton-Raphson type implicit Euler method and the linear unsymmetrical Eq. (9-b) is solved by the implicit Euler method. The stationary flow is obtained as a limiting case of the solution of Eq. (9-a). The stationary temperature values are obtained directly from Eq. (9-b) by the neglecting the $[M]\{\partial T/\partial t\}$ term.

We employ either a 9-4 nodes quadrilateral or a 6-3 nodes triangle. Due to facilities in mesh refinements and automatique mesh generation, we have employed systematically the triangular elements in this study.

## APPLICATION TO THE ST-LAWRENCE SEA-WAY

The geometrical domaine is 11 kms long and about 4-5 kms wide which is situated near Gentilly, Quebec. The depth profile varies rapidly along the cross-section i.e. from 13 meters to zero meter at certain parts due to vegetation zones at the middle of the domaine which divide the water in northern and southern arms. A typical cross section is given in Fig. 3. The bed slope along the flow is taken as $5 \times 10^{-5}$.

### Velocity profiles

We chose the Chezy's coefficient C as variable depending on the depth of the profile. The water level boundary conditions (not velocity values) are applied along upstream and downstream limites which are obtained form in-situ values. The model is calibrated by adjusting C and water-levels for a flow of 12000 $m^3$/sec. The results are compared with those obtained from measured values and from Lasalle (1) reduced model. The calibrated model is then applied to 7000 and 16000 $m^3$/sec flows. The excellent comparison for velocity profiles with measured values demonstrated the exacteness of our calibration. An interesting aspect of this study is that the present model is free from any oscillations in velocity and water level values for stationary flow. One may mention that it is necessary to use artificial viscosity terms for proper convergence which require no-slip conditions along land boundaries. Typical velocity profile for 12000 $m^3$/sec flow is given in Fig. 3. A comparative flow distribution along northern and southern parts is given below :

| Q $m^3$/s | our model | | measured | |
|---|---|---|---|---|
| | north | south | north | south |
| 7000 | 91 % | 9 % | 93 % | 7 % |
| 12000 | 82 % | 18 % | 79 % | 21 % |
| 16000 | 73 % | 27 % | 73 % | 27 % |

## Thermal discharge

We studied the influence of different thermal discharges in the northern and in the southern branches on the eventual heated zone. At the point of discharge, we employed an equivalent boundary condition on the temperature.

Along upstream, the temperature is imposed zero and no temperature is imposed along down stream. Along land boundaries, we impose no heat loss conditions. A detailed comparative study has been performed by chosing different values of $D_x$ (along flow) and $D_y$ (transverse to flow). The influence of heat loss coefficient to environment (K) has been studied as well. The details of the parametric study may be found in the recent CENTREAU report [7]. Typical temperature profiles for a thermal discharge are shown in Fig. 4. The influence of different values of $D_y$ keeping $D_x$ constant is resumed in the following table

| Thermal Discharge m3 °C/sec. | $D_y$ | Heated zone S $km^2$, Length L, width b<br>2°C S | L | b | 1°C S | L | b | 0.5°C S | L | b |
|---|---|---|---|---|---|---|---|---|---|---|
| North | | | | | | | | | | |
| 2000 | 10 | - | - | - | - | - | - | 0.20 | 0.4 | 0 |
| 2000 | 1.0 | - | - | - | 0.05 | - | - | 0.26 | 0.6 | 0 |
| 2000 | 0.1 | - | - | - | 0.05 | - | - | 0.28 | 0.65 | 0 |
| South | | | | | | | | | | |
| 1200 | 1.0 | 0.20 | 0.25 | - | 0.49 | 1.10 | 0.2 | 1.14 | 4.2 | 0.4 |
| 1200 | 0.1 | 0.20 | 0.25 | - | 0.37 | 1.00 | 0.1 | 0.95 | 2.8 | 0.4 |

$Q = 12000\ m^3/sec.,\ D_x = 50,\ K_{EXT} = 100$ Watt °C/$m^2$

Various numerical experimentations for studying thermal discharge phenomenon permit us to make following remarks :

- Due to large flow, the thermal discharge in the northern branch results in very little effect on water temperature. However, one observes significant influence on water temperature from the southern discharge due to a relatively smaller water flow (1/5 of northern part)

- Any thermal discharge should be made in the centre of main stream for reducing heated surface

- For discharge in the main flow zone, the parameters $D_x$, $D_y$, K have little influence on the temperature distribution. However the influence of these parameters becomes important if thermal discharge is made in the zones where river velocity is small or there is a separation of water into the northern or southern branches.

## CONCLUSION

We have applied the present model to study stationary and non-stationary flows in the St-Lawrence sea-way. Our experience [7] has shown that this model may be reliably employed to predict velocity profiles in diverse types of flow conditions. We have shown as well the reliability of transport-diffusion model without using any upwinding for estimating temperature profiles. We recommend to employ water-level conditions along upstream and downstream boundaries. For thermal flow, the temperature values should be employed only along upstream boundary. It is found that $D_x$ values along flow direction has relatively little influence on precision except that it is necessary for numerical stability (10 in the present study). However $D_y$ along transverse direction may be chosen as small as necessary. We observed that $D_y$ has as well little influence on thermal transport.

Presently work is under way to study 3-dimensional flow using layered model (3). At the same we are developing penality formulation for 2-D flows which may result in reduced computational effort. Preliminary results of this work are given in reference [8].

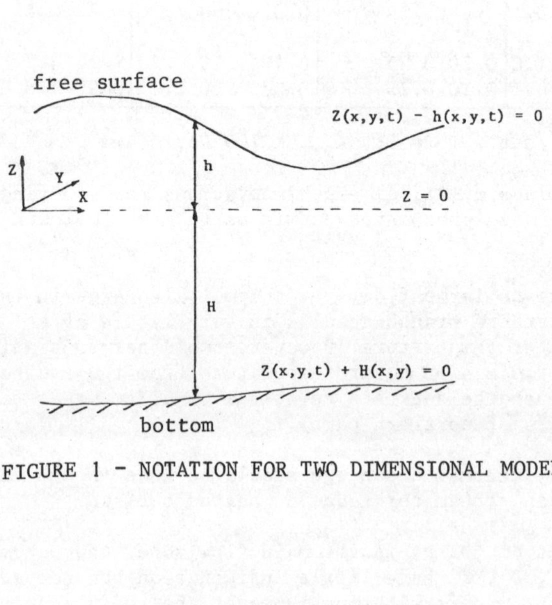


FIGURE 1 - NOTATION FOR TWO DIMENSIONAL MODEL

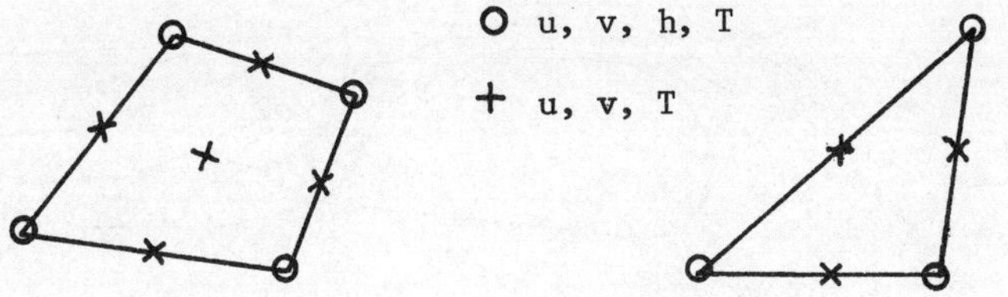


9-4 nodes quadrilateral 6-3 nodes triangle

FIGURE 2 - QUADRILATERAL AND TRIANGULAR ELEMENTS

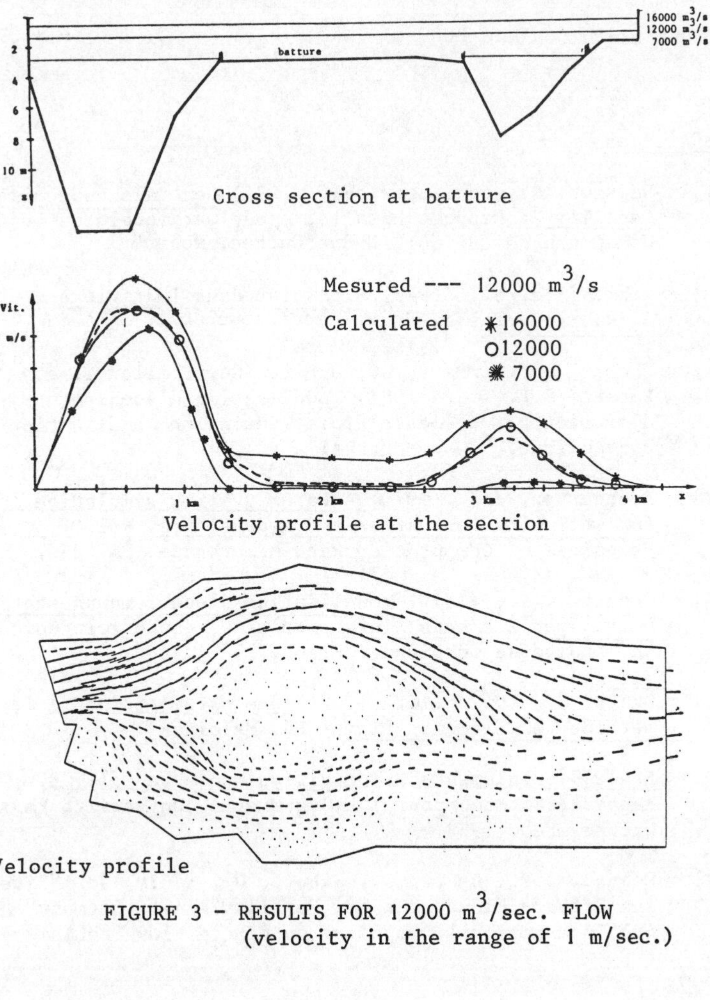


Velocity profile

FIGURE 3 - RESULTS FOR 12000 $m^3$/sec. FLOW
(velocity in the range of 1 m/sec.)

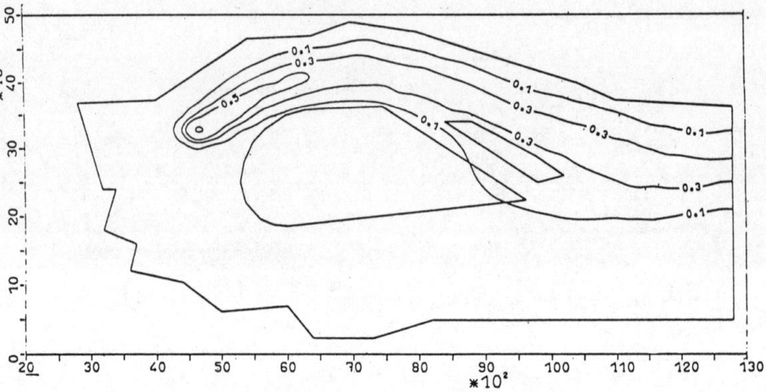


FIGURE 4 - DISCHARGE 2000 $m^3$ °C/sec. ; $D_x$ = 50. ; $b_y$ = 1. ; Kext = 100. Watt/°C $m^2$

References


[1] - Hausser, R., Boulanger, F., (1974) Centrale nucléaire de Gentilly 2. Rapport Lasalle, étude sur modèle du rejet d'eau chaude LHL 601, Hydro Québec, Montréal

[2] - Robitaille, S., (1975) Diffusion dans la rivière St-Maurice. Thèse, Univ. Laval, Québec

[3] - Cochet, J.F., Dhatt, G., Hubert, G., Ouellet Y. et Robert, J.L. (mars 1982)"Modélisation d'écoulements et de diffusion par éléments finis", Séminaire modélisation hydraulique, SCGC, Montréal.

[4] - Leendertse, J.J., (1970) A water quality simulation model for well mixed estuaries and coastal sea, Vol. 1, Principal of computation Rand Memorandum, RM 6230.

[5] - Cochet, J.F., (1979) Modélisation d'écoulements stationnaires par éléments finis, Thèse de docteur-ingénieur, Université de Compiègne, France.

[6] - Dhatt, G., Touzot, G., (1981) Une présentation de la méthode des éléments finis, Ed. Maloine, Paris.

[7] - Etude d'écoulements et de diffusion de Gentilly 3 , Rapport CENTREAU pour l'Hydro-Québec, Université Laval, Québec, Feb. 1982.

[8] - Cochet, J.F., Dhatt, G., Hubert, G., (July 1982)"River and Estuary Flows by a New Penality Finite Element", 4th International Symposium on FEM in Flow Problem.

# WATER QUALITY MODELING WITH FINITE ELEMENTS AND INTERACTIVE GRAPHICS

Peter N. French

Department of Environmental Engineering, Cornell University, Ithaca, New York 14853

## INTRODUCTION

This year marks the tenth anniversary of the passage into law by the United States Congress of the Federal Water Pollution Control Act Ammendments of 1972. Within this law, responsibilities were given to various State and/or Federal agencies for developing and/or reviewing classifications and waste load allocations for all streams and all waste dischargers. Similarly, the public was charged with responding to those classifications and allocations through required public meetings and hearings.

The method used by the appropriate agencies to determine waste load allocations for each discharge usually involves a computerized water quality simulation model. Given the necessary stream standards which have to be maintained, and a flow or set of conditions which characterizes the critical water quality period, the simulation model can be used to determine the maximum allowable loadings of heat, bacteria, biochemical oxygen demand and other pollutants.

Since both temperature and dissolved oxygen vary diurnally, the critical waste load for a system becomes linked not only to a flow and a point in space, but also to a time of day. Small changes from either the critical time of day, the critical point in space, or the critical flow (or other conditions) should produce improvements in water quality, and thus result in a smaller chance of contravening stream standards.

These variations over flow, space and time create a problem for water quality management agencies. The problem arises because it is difficult to take conventional results from batch processed simulation models and transmit them to interested parties in a meaningful way. As a result, water quality is expressed in static terms: certain wasteloads do or do not contravene standards. Because of this problem in transmitting results, it is difficult for people to ask, or agencies to answer, questions about the "gray areas" of water quality. For example, even though a particular waste load results in the contravention of water quality standards, how extensive (in space and time) is that violation? A violation of small magnitude may be ecologically acceptable for one or two hours, but may not be acceptable for 10 or 12 hours. In effect, such questions are a challenge to have the regulatory agencies illustrate to the public the need for, and the bases behind, water quality improvement. For the most part, the agencies involved lack the mechanism for meeting this challenge successfully.

One available technology, capable of improving not only the modeling process itself but also the comprehension of results from analytical models, is computer graphics. Several previous papers have described the coupling of resource planning models with both vector graphics (French et al. 1980, Johnson and Loucks 1980) and color graphics (Taylor et al. 1981a,b) as methods for improving the data definition process and understanding model output.

This paper discusses a water quality simulation model which has been implemented within the framework of an interactive computer graphics package. The capabilities, requirements, and limitations of the model are discussed. Examples of applications to a real system are presented. Based upon this study, some recommendations are made for future research in that particular basin.

## WATER QUALITY SIMULATION MODEL

Although different algorithms are used, the water quality simulation model is based upon the mathematical representation of the U.S. Environmental Protection Agency's QUAL-II model. This model formulation was selected because:

a) it can account for the diurnal nature of water quality,

b) it is more complex than just a Streeter-Phelps approach in that nutrient loading and uptake by photosynthesis is considered,

c) data input requirements are not excessive, and

d) those input data which are required could be entered quickly and easily in an interactive environment.

Certain assumptions are made to reduce the dimensionality of the Fickian diffusion equation. Specifically, points along the stream are assumed to be mixed transversally and concentrations

(C) and velocities (U) used refer to depth-averaged values. These simplifications allow a longitudinal dispersion coefficient (D) to replace both longitudinal and vertical diffusivities, thus resulting in the following one dimensional equation:

$$A dx \frac{\partial C}{\partial t} = \frac{\partial}{\partial x}\left(AD\frac{\partial C}{\partial x}\right) dx - \frac{\partial}{\partial x}\left(AUC\right) dx + A dx \frac{dC}{dt} + S \tag{1}$$

where A is the cross sectional area of a portion of the stream having a length dx. The loading term, S, is composed of point and non-point loads, and can include benthic contributions. The source/sink term, Adx(dC/dt), reflects settling, growth and decay (respiration) of biological species, and oxidation-reduction reactions. Because it is extremely general, Equation 1 can apply to a wide variety of substances. The most complex substance modeled, algae, will serve as an example of the utilization of the equation.

## Algal Growth Term

When modeling the growth of algae in aquatic systems, the source/sink term in Equation 1 can be formulated in many ways. Thomann et al. (1975) used light, available phosphorus, and a combination of ammonia and nitrate. In their model "Water Quality for River-Reservoir Systems", the Hydrologic Engineering Center, U.S. Army Corps of Engineers (1975), used phosphate, ammonia and nitrate, carbon dioxide, and light.

The term used in this model to represent algal growth is based upon a multiplicative Michaelis-Menten representation of the nutrients used for the substrate. The nutrients used include light, nitrate, and phosphate. At any depth, z, the algal growth rate, $\mu(z)$, can be defined as:

$$\mu(z) = \hat{\mu}(T) \cdot \frac{L(z)}{L(z) + K_\ell} \cdot \frac{N(z)}{N(z) + K_n} \cdot \frac{P(z)}{P(z) + K_p} \tag{2}$$

where L, N, and P represent the amount of light and the concentrations of nitrate and phosphate, respectively. Similarly, $K_i$ ($i=\ell,n,p$) represent the half saturation concentrations for each of the nutrients, and $\hat{\mu}$ is the maximum specific growth rate at 20 degrees C, corrected to ambient temperatures.

The use of multiple nutrients for defining the algal growth rate introduces nonlinearities into Equation 1. Before discretizing the differential equations, a quasilinearization procedure described by Lee (1968) is applied to differential equations involving algae, nitrate or phosphate. Light nonlinearity is removed by assuming that diurnal variations in the algal concentration do not appreciably change the amount of light penetrating the water.

Since the quasilinearization results in an approximation to the correct solution, an iterative process is used until some desired degree of convergence is attained. Using ε as a convergence criterion and C as the concentration of constituent i at location j and time t, iterations n continue until

$$\left| (C^t_{ij})_{n+1} - (C^t_{ij})_n \right| \leq \varepsilon^t_{i,n+1} \tag{3}$$

The convergence criterion is determined by the user's selection of a convergence fraction, f, which ranges from 0.0 to 1.0. After each solution, a convergence criterion at the current iteration is obtained for each constituent involved by:

$$\varepsilon^t_{i,n+1} = (1.0 - f)\ \underset{j}{\mathrm{Min}}(C^t_{ij})_{n+1} \tag{4}$$

If the solution does not converge the equations are solved again, with all C(n+1) from the current iteration becoming C(n) in the next iteration. When convergence is attained, the program increments to the next time step.

Equation Discretization

Since the stream paths are drawn in and each site (node) is located interactively by the user, the resulting river system invariably consists of non-uniform reach lengths (elements). This, and the improved accuracy over difference methods for parabolic equations, make spatial discretization by the Galerkin finite element procedure the preferred method. The type of shape function used, either linear or quadratic, is interactively selected by the user.

Regardless of the approach used, many models assume that all hydraulic parameters - flow, dispersion, velocity, cross-sectional area - and rate constants apply to the element, or stream segment, and that they are constant throughout that element. Although this greatly simplifies the integration, this assumption may not be physically realistic for all cases. A more reasonable approach is to assume that parameters, like concentrations, do vary within an element, and to integrate accordingly (Gallagher 1975). The resulting set of linear equations may be represented as

$$[A]\{\dot{C}_i\} + [B]\{C_i\} \pm \sum_{j \varepsilon J} [D_j]\{C_j\} = \{f\} \tag{5}$$

where [A] is a matrix of elemental volumes, [B] contains the transport components, [D] defines the growth, settling, or decay components, {f} is the loading vector, and J is the set of all constituents which, as sources or sinks, interact with constituent i. A backward differencing scheme is used to represent the time derivative.

## Data Preparation

Just as Equation 1 is general in terms of the types of constituents which it can represent, so too is the overall model general in terms of the river basin configurations which can be simulated. Although the original QUAL-II model failed to maintain a mass balance at junction nodes (Grenney et al. 1978), the model described herein incorporates each junction mass balance into the solution. Thus, dendritic streams are easily simulated.

Loadings to the system can either be at nodes, or distributed along elements (according to the shape function) to represent non-point source runoff. Withdrawals at nodes also are permitted, with the transport matrix, [B], being modified so that the concentration of the water withdrawn matches that at the withdrawal node, and so that a mass balance is maintained. All loads to the system are defined in terms of concentrations and flows, with the actual values being defined interactively.

One of the user's options in defining model operating characteristics is whether dispersion will be considered, or whether the transport term will be strictly convective. In either case, upstream boundary conditions are expressed as $C(0,t) = Co(t)$.

If dispersion is to be considered, use of a first order boundary condition typically requires that $dC/dx = 0$ when evaluated over the last element in the system. For a quadratic shape function this presents no problem. However, with a linear element this condition forces the concentration at the last two sites to be equal. Thus the last reach either must be very short, or else must be far enough downstream so that a steady state situation has been attained. Because neither of these requirements can be guaranteed for every situation, a new type of condition utilizing the last two elements was developed. This new boundary condition requires that the change in concentration per unit length of element be the same for the last two elements, or

$$\lfloor N' \rfloor_{ij} \{C\}_{\substack{i\\j}} = \lfloor N' \rfloor_{jk} \{C\}_{\substack{j\\k}} \qquad (6)$$

## Model Solution

After creation of the global matrix and insertion of the appropriate boundary conditions, the solution is computed. The equation solver is based upon Gupta and Tanji's (1978) algorithm for sparse, asymmetric matrices. Solution results are stored at up to 24 hourly time steps in a disk file, thus permitting rapid data retrieval for displays of diurnal variations. Figure 1 illustrates diurnal variations for light, temperature, phosphate, algae and the nitrogen series at a site in a hypothetical river basin.

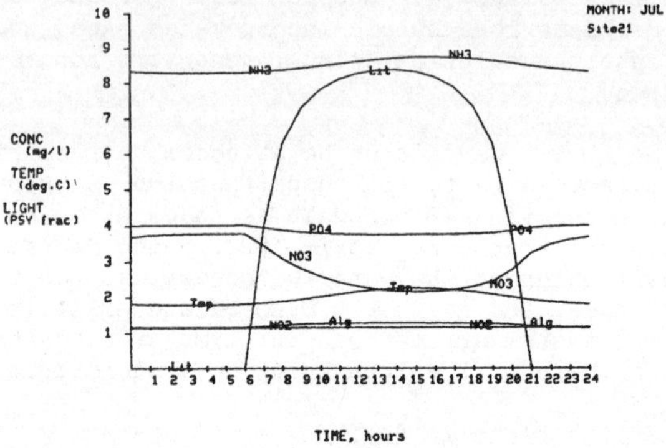


Figure 1.-- Simulation results from a hypothetical basin showing diurnal variation in light (photosynthetic fraction x 10); temperature (x 0.1); $NH_3$, $NO_2$, $NO_3$, and $PO_4$ (all x 100); and algae.

## APPLICATIONS

The Interactive Water Quality Planning Package was used to simulate conditions in the Han River in South Korea. The Han is one of four major rivers in Korea, and serves both as water supply and effluent recipient for the capital city of Seoul. Seoul has one waste treatment plant, located upstream at the mouth of Chung Kae Stream (site J7, Figure 2). This plant treats approximately 20% of Seoul's wastewater with an efficiency of 50%. The remaining untreated wastewater reaches the Han River through a number of small tributaries. Rather than modeling each of the tributaries, their contributions were considered as point discharges to the Han River.

Although input data requirements for the simulation model are modest, data necessary for the simulation of the Han River are scarce. No data are available for algae or for non-point source contributions. Point source data are questionable, since most of the wastewater generated by the city of Seoul is both untreated and unregulated. Data available for calibration purposes essentially limited simulations to dissolved oxygen and BOD.

Data on dissolved oxygen and BOD, provided by Dr. Yong Chung of Yonsei University, served as initial estimates for boundary conditions and point discharges (Chung 1981). Concentrations of non-point loadings were estimated to be approximately the same as headwater concentrations. A decay rate of 1.0 per day was used for BOD (Chung 1981), and Churchill et al's. (1962) equation was used to determine reaeration rates (Chung 1975).

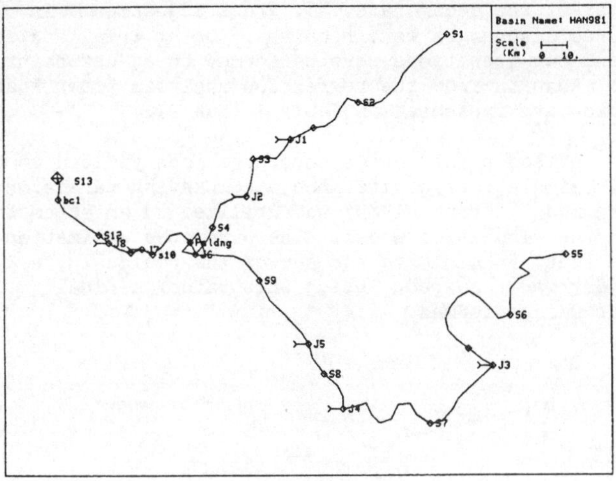


Figure 2.-- Vector representation of the Han River.

| | BOD | | | Dissolved Oxygen | | |
|---|---|---|---|---|---|---|
| | Run 1 | Run 2 | Run 3 | Run 1 | Run 2 | Run 3 |
| Slope | 0.34 | 0.76 | 0.75 | 0.53 | 0.93 | 1.00 |
| Intercept | 0.20 | 0.84 | 0.71 | 4.37 | -0.05 | -0.83 |
| Corr. coef. | 0.64 | 0.85 | 0.87 | 0.51 | 0.44 | 0.42 |

Table 1. Linear regression results for BOD and dissolved oxygen using: original, estimated point source loading data (Run 1); modified point source loading data (Run 2); and modified point source loading data and "optimal" K1 (Run 3).

Results of the BOD and dissolved oxygen simulations, using quadratic shape functions, are shown in Figure 3 together with observed data points (KNIEP 1980). The simulation and observations were taken for September, a typically average flow month. Using the accompanying statistical package (French 1980), linear regression analyses were performed on observed versus simulated data. The results (see Table 1, Run 1) demonstrated that the model was far from being calibrated.

The magnitude of the discrepancies suggested that, rather than refinements of the decay rate, K1, gross adjustments in the point source discharges were required. Using trial and error simulation, two iterations were performed to adjust point loadings. Results from the regression analyses after load modification are presented in Table 1 (Run 2).

After the revised point source concentrations yielded more reasonable simulation results, the accompanying parameter estimation model (French 1980) was utilized in an attempt to calibrate the simulation model. The parameter estimation model suggested that, to minimize the sum of the residuals, a K1 of 0.68 per day was required. Using this value, a final simulation was performed.

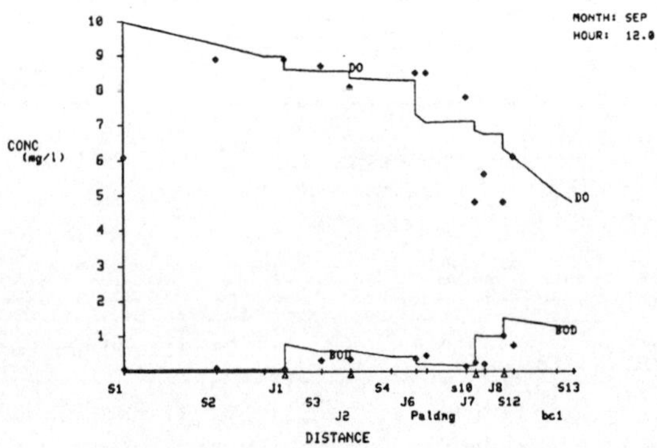


Figure 3.-- Simulation results for dissolved oxygen and BOD (x0.1) using original, estimated point source loading data and K1 = 1.0/day.

A visual comparison of the simulated results and the observed data (Figure 4) show a close correlation for BOD throughout the Han River. Results from the regression analyses are given in Table 1 (Run 3). The model results for dissolved oxygen are good in the reaches upstream of Seoul, but seriously overestimate DO in the downstream reaches. One possible cause for this, although not verifiable, is urban drainage from Seoul.

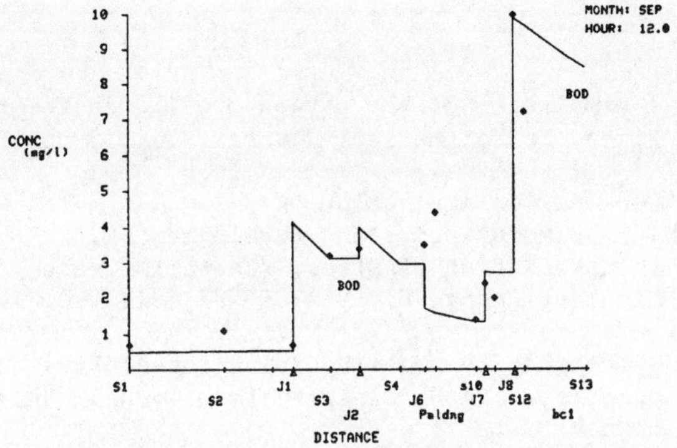


Figure 4.-- BOD concentration versus distance downstream after simulation with the "optimal" K1 value of 0.68/day

## CONCLUSIONS

As originally conceived, this paper intended to illustrate how a set of numerical and statistical techniques, combined with interactive computer graphics, could be used to quickly and easily simulate water quality and display results to the user effectively. Although this was accomplished, the results were unexpected. Instead of being able to calibrate and verify the simulation model, and thus use it to predict future conditions, the results pointed up both deficiencies and discrepancies in the data base. Specifically, photosynthesis could not be simulated, nonpoint loads had to be estimated, and point loads had to be adjusted from between -80% and +50% of the original data. Based upon a visual comparison of simulated and observed data, the adjustment of point sources was able to effect a greater improvement in model results than was the use of a more sophisticated parameter estimation model.

The results from the simulations performed point to the need for a more comprehensive water quality sampling program for the Han River which will more accurately quantify both point and nonpoint source discharges. Until such data are available, reliable estimates of future water quality conditions cannot be made.

## ACKNOWLEDGEMENT

I wish to express my appreciation to Dr. Yong Chung of Yonsei University, Seoul, Korea, for providing geophysical and water quality data for the Han River.

## REFERENCES

Chung, Y.
1975. A simulation of the oxygen profile in the Han River, Yonsei Medical Journal, 16, 1, pp. 29-39.

Chung, Y.
1981. Personal communication.

Churchill, M.A.; Elmore, H.L.; and Buckingham, R.A.
1962. The prediction of stream reaeration rates, J. Sanitary Engng. Div., ASCE, 88, SA4, pp. 1-46.

French, P.N.
1980. Water quality modeling using interactive computer graphics, Ph.D. dissertation, Cornell University, Ithaca, NY.

French, P.N.; Johnson, L.E.; Loucks, D.P.; and Greenberg, D.P.
1980. Water resources planning using computer graphics, J. Water Resources Planning and Management Division, ASCE, 6, WR1, pp. 21-42.

Gallagher, R.H.
1975. Finite Element Analysis Fundamentals, Prentice-Hall, NJ.

Grenney, W.J.; Teuscher, M.C.; and Dixon, L.S.
1978. Characteristics of the solution algorithms for the QUAL-II river model, J. Water Pollution Control Federation, 50, pp. 151-157.

Gupta, S.K. and Tanji, K.K.
1977. Computer program for solution of large, sparse, unsymmetric systems of linear equations, Int. J. Num. Meth. Engng., 11, pp. 1251-1259.

Hydrologic Engineering Center, U.S. Army Corps of Engineers.
1975. Water quality for river-reservoir systems, Davis, CA.

Johnson, L.E. and Loucks, D.P.
1980. Interactive multiobjective planning using computer graphics, Computers and Operations Research, 7, 1-2, (Great Britain).

KNIEP (Korean National Institute of Environmental Preservation).
1980. The water quality survey data for rivers of Korea.

Lee, E.S.
1968. Quasilinearization and invariant imbedding, Academic Press, NY.

Taylor, M.R.; French, P.N.; and Loucks, D.P.
1981a. Interactive graphics for flood plain management. Proc., 1981 Water Forum, ASCE, San Francisco, CA.

Taylor, M.R.; French, P.N.; Loucks, D.P.; and Greenberg, D.P.
1981b. Interactive color graphics for watershed planning. Proc., First international conference on computing in civil engineering. ASCE, New York, NY.

Thomann, R.V.; Di Toro, D.M.; and O'Connor, D.J.
1974. Preliminary model of Potamic Estuary Phytoplankton, J. Env. Engng. Div., ASCE, 100, EE3.

# TIDAL CURRENT AND SUBSTANCE DISPERSION BY FINITE ELEMENT METHOD USING FINITE DIFFERENCE TECHNIQUE

Kazuo Murakami and Masayuki Morikawa

Port and Harbour Research Institute, Ministry of Transport, JAPAN

## INTRODUCTION

Recently tide and water quality simulations by finite element method are carried out very frequently. This is because the expression of configuration by arbitary triangular or trapezoidal elements in finite element method is more accurate than the expression by square grids in finite difference method. This means that the accuracy of numerical simulation rises up by use of finite element method. Tidal computation by finite element approximation does not always satisfy continuity condition in each element completely. The reason is that the approximation with the method of weighted residual seeks to the minimum error from the true solution in some average sense. Therefore, there is any possibility to include the small deviation in each element in continuity condition. Michael Gee and MacArthur(1978) investigated the evaluation of continuity errors. They conclude that the internal continuity errors can be reduced to acceptable levels by increasing network detail, particularly in area of large curvature of the velocity field. There is a serious problem, however, when the solution applies to diffusion equation based on mass conservative law. When some conditions, which are no source, no sink or decay, initial concentration and the same boundary condition, are imposed to diffusion problem, the concentration should not change the value due to convection or diffusion. However, the concentration may change if the flow field does not satisfy continuity equation in each element. That is, unreasonable results such as source or decay are brought due to wrong continuity condition.

To solve the problem, the computation of tide by continuity equation uses the finite difference technique in stead of finite element technique, that is a change of water surface elevation is calculated with the difference between volumes of inflow and outflow in each element. The same technique applies to diffusion equation in order to represent the convective term.

## FINITE ELEMENT METHOD

Governing equation of tide and tidal current are expressed as follows;

$$H_{,t} + q_{x,x} + q_{y,y} = q_I \tag{1}$$

$$q_{x,t} + (\bar{u}q_x)_{,x} + (\bar{u}q_y)_{,y} - fq_y + (F_p - F_{xx})_{,x} - F_{yx,y} + \frac{1}{\rho_o}(\tau_x^s - \tau_x^b) - \frac{1}{\rho_o}P^s H_{,x} - g\zeta h_{,x} = 0 \tag{2}$$

$$q_{y,t} + (\bar{v}q_x)_{,x} + (\bar{v}q_y)_{,y} + fq_x - F_{xy,x} + (F_p - F_{yy})_{,y} + \frac{1}{\rho_o}(\tau_y^s - \tau_y^b) - \frac{1}{\rho_o}P^s H_{,y} - g\zeta h_{,y} = 0 \tag{3}$$

in which,
$x, y$ = cartesian coordinates,
$\bar{u}, \bar{v}$ = ensemble averaged velocities in $x$ and $y$ directions,
$q_x, q_y$ = specific discharges per unit width in $x$ and $y$ directions,
$f$ = coriolis parameter = $2\omega \sin \phi$ ,
$\omega$ = phase velocity of earth rotation,
$\phi$ = latitude,
$\rho_o$ = average density,
$P^s$ = surface pressure,
$\zeta$ = surface displacement(positive upwards),
$h$ = depth,
$H$ = total depth = $h + \zeta$,
$F_p$ = specific pressure force measure,
$F_{xx}, F_{xy}, F_{yy}$ = internal specific force measure,
$\tau_x^s, \tau_y^s$ = surface shear stresses, and
$\tau_x^b, \tau_y^b$ = bottom shear stresses.

Boundary conditions on force boundary(open sea) are expressed as

$$F_{nn} = F_{nn}^* \tag{4}$$

$$F_{ns} = F_{ns}^* \tag{5}$$

and boundary conditions on discharge boundary(land or river) are expressed as

$$q_n = q_n^* \tag{6}$$

$$q_s = q_s^* \tag{7}$$

In which, the superscript * signifies a prescribed value.

Generally, $q_n^*$ is zero on land boundary. Basic equations (1) to (3) are solved by nemerical simulation with boundary conditions (4) to (7). Concerning with this problem, finite difference method (e.g. Leendertse(1967)) has been used to solve a tidal computation. In the last ten years, finite element method is developed to apply to fluid dynamics (e.g. Oden et al.(1975)). A Galerkin method, which is one technique of the weighted residual method, is used to solve Eqs. (1) to (7). The Galerkin technique, in which the weighting function and the trial function are same, is the most popular method to solve some fluid problems. Using the Weak Formulation, Eqs. (1) to (3) are written as Eqs. (8) to (10).

$$\int_\Omega \{H_{,t} + q_{x,x} + q_{y,y} - q_I\}\Delta H dA = 0 \tag{8}$$

$$\begin{aligned}\int_\Omega [\{q_{x,t} + (\bar{u}q_x)_{,x} + (\bar{u}q_y)_{,y} - fq_y + \frac{1}{\rho_0}(\tau_x^s - \tau_x^b) \\ - \frac{1}{\rho_0}p^s H_{,x} - g\zeta h_{,x}\}\Delta q_x - \{(F_p - F_{xx})\Delta q_{x,x} \\ - F_{yx}\Delta q_{x,y}\}]dA - \int_{S_q}\{\alpha_{nx}(F_{xx} - F_p) + \alpha_{ny}F_{yx}\}\Delta q_x dS \\ - \int_{S_F} F_{nx}^*\Delta q_x dS + \int_{S_q}\{\alpha_{nx}q_n - \alpha_{ny}q_s - q_x^*\}\Delta q_x dS = 0\end{aligned} \tag{9}$$

$$\begin{aligned}\int_\Omega [\{q_{y,t} + (\bar{v}q_x)_{,x} + (\bar{v}q_y)_{,y} + fq_x + \frac{1}{\rho_0}(\tau_y^s - \tau_y^b) \\ - \frac{1}{\rho_0}p^s H_{,y} - g\zeta h_{,y}\}\Delta q_y - \{F_{xy}\Delta q_{y,x} + (F_p - F_{xx})\Delta q_{y,y}\}]dA \\ - \int_{S_q}\{\alpha_{nx}F_{xy} + \alpha_{ny}(F_{yy} - F_p)\}\Delta q_y dS - \int_{S_F} F_{ny}^*\Delta q_y dS \\ \int_{S_q}\{\alpha_{ny}q_n + \alpha_{nx}q_s - q_y^*\}\Delta q_y dS = 0\end{aligned} \tag{10}$$

where, $\Delta H$, $\Delta q_x$, $\Delta q_y$ are weighting functions. Wang and Connor (1975) solved these equations by the Split-Time Method. In this scheme, velocities and surface elevations are obtained at alternating time levels. That is, there is half-time step difference between time levels of velocity and surface elevation. Surface elevations are obtained from Eq. (8) at time $t=(n+(1/2))T$, and then velocities are obtained from Eqs. (9) and (10) at time $t=(n+1)T$. The similar method has been used to tidal computation by finite difference method.

<u>Continuity Equation</u>

According to the method of Wang and Connor, a change of surface elevation is solved by means of finite element method. Finite element method solves matrics equations considering the whole area. Therefore, finite element method satisfies a continuity condition for whole area completely, but the method does not always satisfy the condition for each element. By this reason, a calculation of surface elevation by continuity equation uses the finite difference technique in stead of finite element method. A change of surface elevation is defined as the difference of mass between inflow and outflow in some area.

$$(\zeta_i^{n+(1/2)} - \zeta_i^{n-(1/2)})A_i = \Delta t[Q_{inflow} - Q_{outflow}] \tag{11}$$

in which, $\zeta_i$ = surface elevation at node $i$,
$\Delta t$ = time step,
$A_i$ = area which represents $\zeta_i$,
$Q_{inflow}$ = volume of inflow into the area, and
$Q_{outflow}$ = volume of outflow from the area.

in which, the area $A_i$ is shown in Figure 1. The volume of inflow and outflow in the region is obtained by velocities at nodes in each element as shown in Figure 2. The volume of inflow or outflow is estimated with normal component velocities to lines LM and MN. Normal component velocity $\bar{v}'_L$ is expressed by coordinate transformation as

$$\begin{pmatrix} \bar{u}'_L \\ \bar{v}'_L \end{pmatrix} = \begin{pmatrix} \cos\theta_{LM} & \sin\theta_{LM} \\ -\sin\theta_{LM} & \cos\theta_{LM} \end{pmatrix} \begin{pmatrix} \bar{u}_L \\ \bar{v}_L \end{pmatrix} \tag{12}$$

in which,

$$\theta_{LM} = \tan^{-1}\frac{(y_M - y_L)}{(x_M - x_L)} \tag{13}$$

therefore, the volume entering to the region from $LM$ is expressed as

$$Q_{LM} = \frac{1}{2}(\bar{v}'_L H_L + \bar{v}'_M H_M)\,\overline{LM} \tag{14}$$

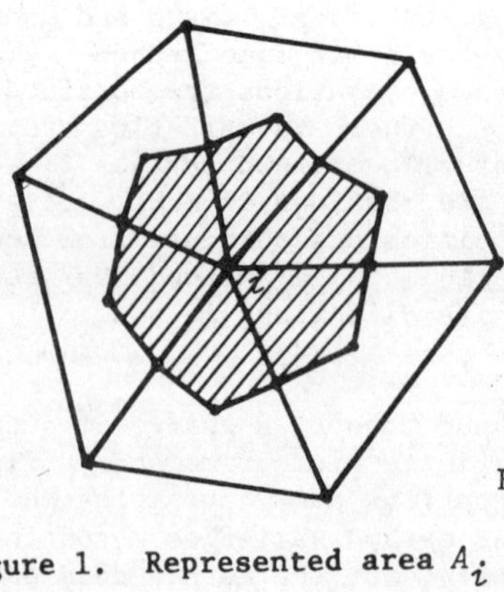


**Figure 1. Represented area $A_i$**

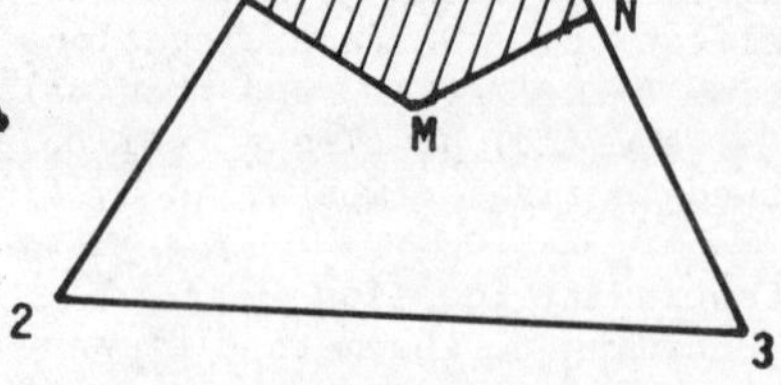


Figure 2. Computational points L,M,N of velocity in order to estimate the volumes $Q_{inflow}$ and $Q_{outflow}$.

in which $\overline{LM}$ is the distance from points $L$ to $M$. Similarly, the volume from $MN$ is expressed as

$$Q_{MN} = \frac{1}{2} (\bar{v}''_M H_M + \bar{v}''_N H_N) \overline{MN} \quad (15)$$

In this way, $Q_{inflow}$ and $Q_{outflow}$ are calculated in all neighbouring finite elements, and then surface elevation is obtained by the continuity equation (11). From the above consideration, the continuity condition in any region is satisfied completely.

Law of Mass Conservation

Diffusion equation is based on law of mass conservation.

$$\frac{\partial C}{\partial t} + \frac{\partial}{\partial x}(\bar{u}C) + \frac{\partial}{\partial y}(\bar{v}C) = \frac{\partial}{\partial x}(E_{xx}\frac{\partial C}{\partial x} + E_{xy}\frac{\partial C}{\partial y}) + \frac{\partial}{\partial y}(E_{xy}\frac{\partial C}{\partial x} + E_{yy}\frac{\partial C}{\partial y}) + R \quad (16)$$

in which,

$C$ = layer integrated concentration $(C=cH)$,

$E_{xx}, E_{xy}, E_{yy}$ = dispersion coefficients,

$R$ = source or sink.

The flow field, which is used to solve a diffusion equation, must satisfy the continuity condition. Generally, flow field which is obtained by numerical simulation is substituted to Eq. (16) to solve diffusion equation. On occasion of that the expression of convective terms in Eq. (16) should be adopted the same technique in tidal computation. Convective terms and diffusive terms are considered separately in Eq.(16). Then the change of concentration in a region is expressed as following equations.

$$\int_{\Omega}\{(\frac{\partial C}{\partial t} - \ddot{R})\Delta W - Q_x\frac{\partial \Delta W}{\partial x} - Q_y\frac{\partial \Delta W}{\partial y}\}dA + \int_{S_q} Q_n^* \Delta W dS = 0 \quad (17)$$

and

$$\frac{\partial C}{\partial t} A_i = \sum_{j}^{j=N} ( c_{in} Q_{inflow} - c_{out} Q_{outflow} )_j \quad (18)$$

in which,

$$Q_x = - E_{xx}H\frac{\partial c}{\partial x} - E_{xy}H\frac{\partial c}{\partial y} \quad (19)$$

$$Q_y = - E_{xy}H\frac{\partial c}{\partial x} - E_{yy}H\frac{\partial c}{\partial y} \quad (20)$$

$c_{in}$ = concentration of inflow,

$c_{out}$ = concentration of outflow,

$N$ = number of elements neighbouring node $i$.

In this way, the law of mass conservation of diffusion equation agrees with the continuity equation of tide.

## APPLICATIONS

The finite element method using the finite difference technique is applied to tidal computation and substance dispersion calculations. Comparisons between finite element method and the finite element method using the finite difference technique are carried out by means of these calculation results.

### Results of Tidal Computation

As an example, tidal computation in Tokyo Bay is carried out by both techniques. Figure 3 represents computed velocity vectors by finite element method according to Wang & Connor. Figure 4 represents computed velocity vectors by finite element method using finite difference technique. There is no difference between these results. Figures 5 and 6 show the comparisons of tide at Tokyo Port and tidal velocity at Uraga Channel. These are considered as almost the same results. The fact means that the computational result using the finite difference technique is reasonable, and the problem of continuity condition in finite element method is not serious in an average sense.

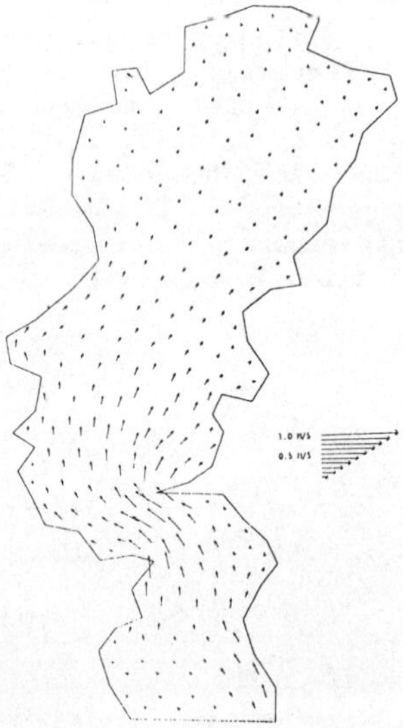

Figure 3. Computed vectors by finite element method.

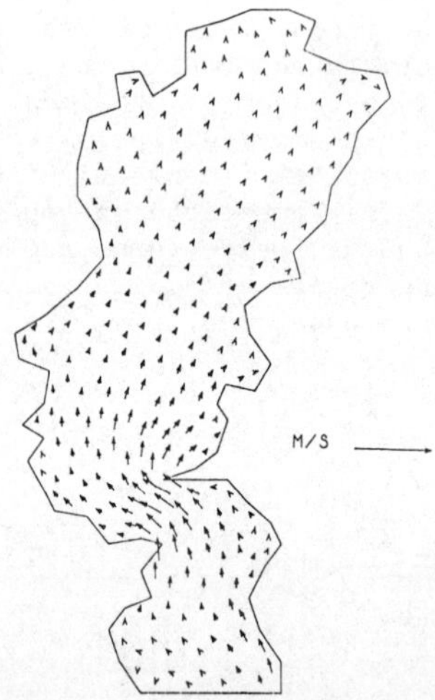


Figure 4. Computed vectors by finite element method using the finite difference technique.

### Computational Results of Substance Dispersion

Comparison between the numerical results of substance dispersion using the finite difference technique and finite element method is carried out in uniform flow. As Heinrich

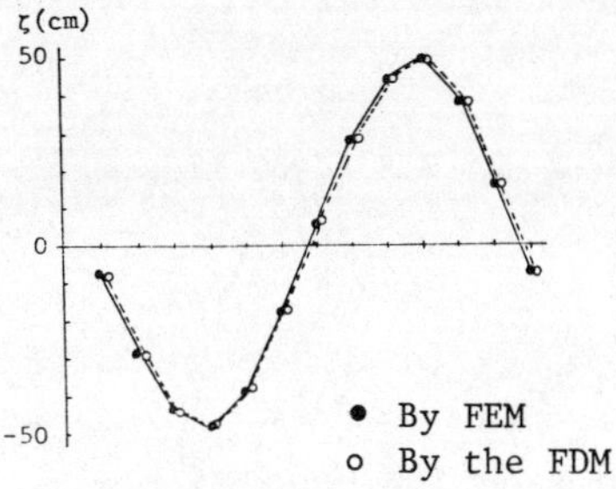


Figure 5. Computed surface elevations at Tokyo Port.

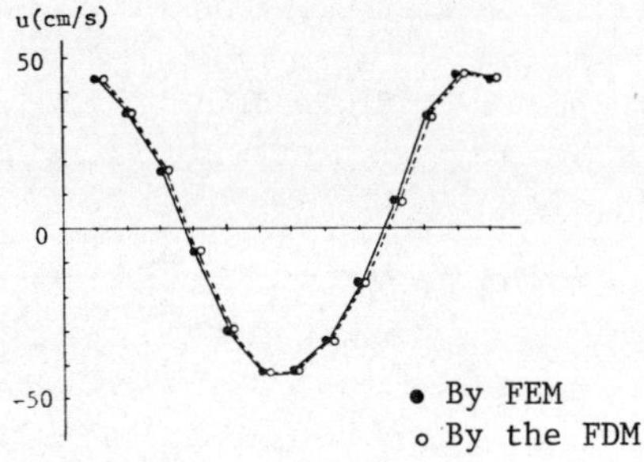


Figure 6. Computed tidal velocities at the Bay Mouth.

et al.(1977) pointed out, if Peclet Number $\nu = VL/K$ is greater than 2, the solution becomes unstable. In which, $V$ is velocity, $L$ is mesh size, and $K$ is diffusion coefficient. The unstability phenomenon is appeared in finite difference and finite element methods. In order to avoid the phenomenon, upstream finite difference is adopted. A problem in upstream finite difference is to create numerical diffusion, which is pointed out by Bella et al.(1970) as psudo-diffusion. Figure 7 shows the concentration distributions solved by finite element method due to Christodoulou et al.(1976). Peclet numbers are 25 in Run A, and 1.46 in Run B. Unreasonable negative concentrations are appeared in Run A. On the contrary of that the concentration distribution of Run B agrees well with analytical solution drawn with a solid line in Figure 7. These results mean that the numerical diffusion is not generated by finite element method. While the concentration distribution of Run C using the finite difference technique is shown in the same figure. Diffusion coefficients are 102 $m^2/s$ in Run B and 30 $m^2/s$ in Run C. Numerical diffusion is created in case of the result using the finite difference technique. It is considered that the psudo-diffusion coefficient is 72 $m^2/s$ according to Bella et al.

$$D_p = \frac{V}{2} ( L - V\Delta t ) \tag{21}$$

in which, $V = 0.1$ m/s, $L = 1500$m, $\Delta t = 600$ sec. From above mentioned considerations, the solution by finite element method is better than the one by using the finite difference technique, if the continuity condition is satisfied in each element and Peclet number is less than the critical number. In order to solve these problems, upwind finite element scheme (e.g. Heinrich et al.(1977)) or flux-corrected transport formulation(e.g. Boris et al.(1973)) is considered for solving the continuity equation.

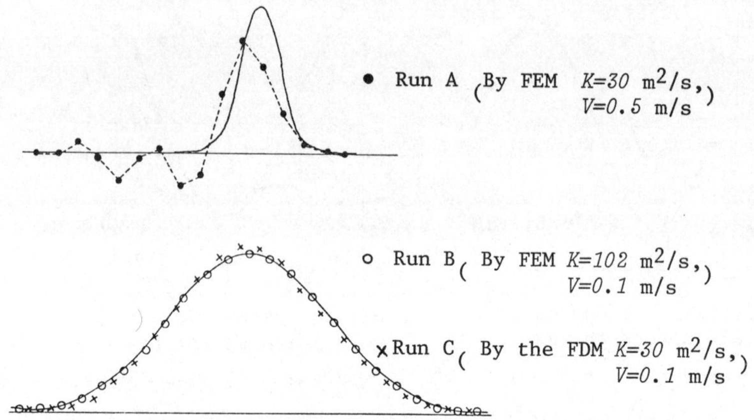


Figure 7. Computed concentration distributions in uniform flow. (Solid lines show analytical solutions)

Numerical Results of Substance Dispersion in Tokyo Bay

As previously mentioned, the finite difference technique produces the numerical diffusion. On the other hand, finite element method does not satisfy the continuity condition in each element. Figure 8 shows the result of concentration by finite element method on the conditions of that there is no source, no sink or decay, and initial concentration is the same as boundary concentration. Diffusive transport flux and convective transport flux should be zero, if there is no concentration gradient and continuity condition is satisfied. As shown in Figure 8, however, there are high and low concentration areas. On the contrary, the same concentrations in every nodes are obtained by the finite element method using the finite difference technique. The fact means that the finite element method using the finite difference technique is regarded as appropriate method. In Tokyo Bay, a computation of substance dispersion is conducted using the finite difference technique as shown in Figure 9. In the figure, the concentration contour lines are computational results after 100 days from initial condition which is 1 ppm in all region. Figure 10 shows the COD(Chemical Oxygen Demand) concentration distribution measured by field survey in Tokyo Bay. The diffusive substance is assumed to be conservative material in case of numerical simulation. Therefore, there is some difference between computed results and field measurement results. But, the patterns of substance dispersion are considered as similar in both results. In order to obtain more accurate result, it is necessary to consider the non-conservative property of organic material in the simulation model.

## CONCLUDING REMARKS

In uniform flow field and Tokyo Bay, tidal current and substance dispersion computations by the finite element method using the finite difference technique are carried out. These considerations afford the following conclusions.

(i) In flow problem, finite element method does not satisfy the continuity condition in each element.

(ii) Using the finite difference technique, the continuity condition is satisfied in each element. The result of tidal computation is almost the same as one obtained by finite element method.

(iii) Numerical diffusion is produced by the finite

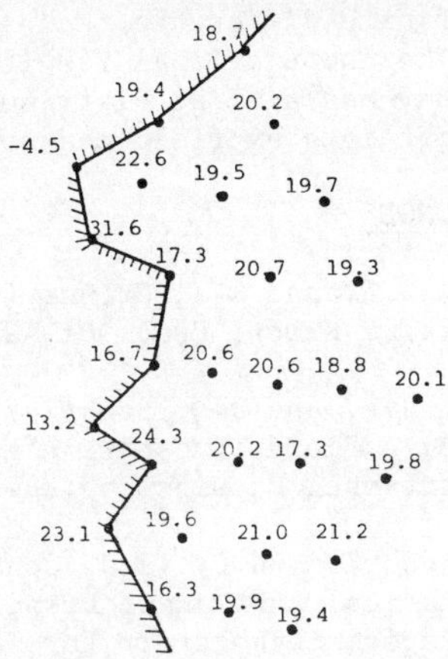


Figure 8. Computed concentration distribution by finite element method. (Initial condition is 20 ppm)

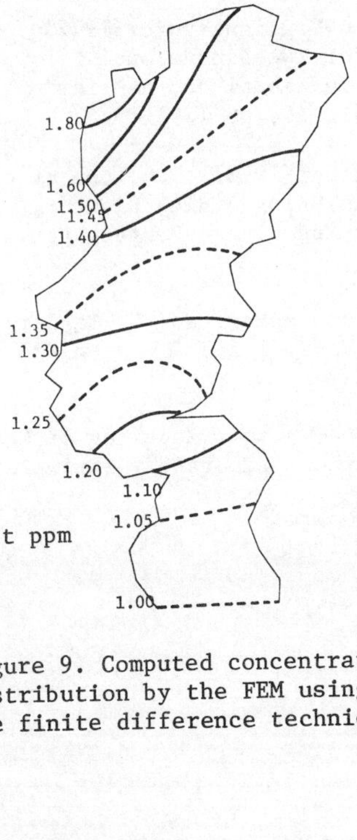


Figure 9. Computed concentration distribution by the FEM using the finite difference technique.

Figure 10. Measured concentration distribution in the Tokyo Bay.

element method using the finite difference technique. Finite element method does not produce the numerical diffusion.

(iv) Unreasonable source or decay does not appear in substance dispersion results by the finite element method using the finite difference technique.

(v) The computational result of substance dispersion in Tokyo Bay has similar distribution pattern to COD distribution obtained by field measurement.

## REFERENCES

Bella, D.A. and W.J. Grenney(1970) Finite-Difference Convective Error, Proc. of ASCE, Vol. 96, SA6, pp 1361-1375.

Boris, J.P. and D.L. Book(1973) Flux-Corrected Transport I. SHASTA, A Fluid Transport Algorithm That Works, Journal of Computational Physics, pp 38-69.

Christodoulou, G.C., J.J. Connor, and B.R. Pearce(1976) Mathematical Modeling of Dispersion in Stratified Waters, TR219, R.M. Parsons Laboratory for Water Resources and Hydrodynamics, M.I.T.

Leendertse, J.J.(1967) Aspects of a Computational Model for Long-Period Water-Wave Propagation, Memorandum, RM-5294-PR, Rand Corporation.

Heinrich, J.C., P.S. Huyakorn, and O.C. Zienkiewicz(1977) An 'Upwind' Finite Element Scheme for Two-Dimensional Convective Transport Equation, International Journal for Numerical Method in Engineering, Vol. 11, pp 1361-1375.

Michael Gee, D. and Robert C. MacArthur(1978) Development of Generalized Free Surface Flow Models Using Finite Element Technique, Finite Elements in Water Resources, Pentech Press, London, pp 2.61-2.77.

Oden, J.T., O.C. Zienkiewicz, R.H. Gallagher, and C. Taylor (1975) Finite Elements in Fluids, Vols. I and II, John Wiley & Sons.

Wang, J.D. and J.J. Connor(1975) Mathematical Modeling of Near Coastal Circulation, TR200, R.M. Parsons Laboratory for Water Resources and Hydrodynamics, M.I.T.

# SIMULATION ANALYSIS FOR DIFFUSION OF DISCHARGED WARM WATER IN THE BAY BY FINITE ELEMENTS

Toyoki Tanaka
Engineering Research Laboratories, Toray Industries, Inc., Otsu 520 Japan
and Yoshio Ono, Toshikazu Ishise
Toray Engineering Co., Ltd., Japan
and Kisabro Nakata
Ministry of International Trade and Industry, Japan

## INTRODUCTION

Due to a power plant siting, large amount of warm cooling water from it continuously enter a coastal bay.
Since such cooling water emission probably changes the coastal environment, environmental impact assessment (EIA) is necessary before a plant siting.
The prediction of cooling water plays an important role on EIA.
In this paper, the behaviour of discharged warm water in the bay is studied under the influence of the natural heat exchange between water surface and the atmosphere.
A mathematical model is developed in which thermal diffusion problems above mentioned are solved using the finite element method under some practical assumptions.
In general, for predicting the cooling water distribution in the coastal bay area exactly, it is necessary to seize the phenomena as three dimensional density current. However we are interested in the macroscopic thermal dynamics in the bay area, therefore we introduced the horizontally two dimensional model assuming on vertical profile of water temperature and velocity.
The similar investigations were studied by Wada (1974), Spurr (1974) or Macqueen (1972) using another type of mathematical model.

Tachibana bay in Shikoku area, where the cooling water emission from a power plant enter, is an example in this study.
Observations on three dimensional temperature distribution in the bay, were executed twice in 1979 by Ministry of International Trade and Industry, Japan. From these observations, we assumed the vertical profile of water temperature according to Wada et al. (1974).
For the purpose of inspecting the validity of the model, the comparison is made between computations and observations.

The numerical pattern of water temperature in the bay is simulated well compared with that of the actual measurement.

## BASIC EQUATION OF THERMAL DIFFUSION

Three dimensional thermal diffusion equation is given as follows.

$$C_p\left(\frac{\partial T}{\partial t} + U_0\frac{\partial T}{\partial x} + V_0\frac{\partial T}{\partial y} + W_0\frac{\partial T}{\partial z}\right)$$

$$= \frac{\partial}{\partial x}\left(Kx\frac{\partial T}{\partial x}\right) + \frac{\partial}{\partial y}\left(Ky\frac{\partial T}{\partial y}\right) + \frac{\partial}{\partial z}\left(Kz\frac{\partial T}{\partial z}\right) + Q - \beta T \tag{1}$$

where $C_p$: the specific heat ($Kcal/{}^\circ C \cdot m^3$)

x, y, z: cartesian coordinate (m)
$U_0, V_0, W_0$: convective velocities corresponding to x, y and z direction, respectively (m)
Kx,Ky,Kz: thermal diffusion coefficient corresponding to x, y and z direction, respectively ($kcal/{}^\circ C \cdot m \cdot sec$)
Q: heat loading ($Kcal/m^3 \cdot sec$)
$\beta$: heat loss coefficient ($Kcal/{}^\circ C \cdot m^3 \cdot sec$)

We integrate equation(1) from sea surface ($-\eta$) to the warm water layer bottom(Hw).
Introducing the assumption of vertical water temperature distribution function according to Wada et al. (1974) given as follows

$$T(x, y, z, t) = T_0 + (Ts - T_0)\, g(z) \tag{2}$$

where
T : water temperature
$T_0$: natural environmental water temperature
Ts: water surface temperature $Ts = Ts(x, y, t)$
g(z): the vertical temperature distribution function

$$g(z) = \exp\left\{-2\left(\frac{z+\eta}{Hw+\eta}\right)^2\right\} \tag{3}$$

$\eta$: tidal sea level
Hw: the thickness of warm water layer

and introducing the vertical distribution function of current velocity components given by

$$\begin{aligned} U_0 &= Us \cdot g(z) \\ V_0 &= Vs \cdot g(z) \\ W_0 &= 0 \end{aligned} \tag{4}$$

where suffix s means surface,
then we can get the following vertically integrated two dimensional thermal diffusion equation,

$$\frac{\partial Ts}{\partial t} + \frac{A}{B} Us \frac{\partial Ts}{\partial x} + \frac{A}{B} Vs \frac{\partial Ts}{\partial y} = Kx \frac{\partial^2 Ts}{\partial x^2} + Ky \frac{\partial^2 Ts}{\partial y^2} + \frac{\bar{Q} - \bar{\alpha} Ts}{Cp} \qquad (5)$$

where

$$\bar{\alpha} = \beta + \frac{0.9049Kz}{(Hw + \eta)^2}$$

$$\bar{Q} = 1.672Q + \{\frac{0.9049Kz}{(Hw + \eta)^2} + 0.672\beta\}To \qquad (6)$$

and if we put $s = \frac{Z + \eta}{Hw + \eta}$

then

$$A \equiv \int_{-\eta}^{Hw} g(z)^2 dz = (Hw + \eta) \int_0^1 \exp(-4s^2) ds$$

$$B \equiv \int_{-\eta}^{Hw} g(z)\, dz = (Hw + \eta) \int_0^1 \exp(-2s^2) ds \qquad (7)$$

$$\frac{A}{B} = 0.7375,$$

In shallower area than Hw, we assumed that the temperature profile above mentioned was held to sea bottom (Hz), and that the sea bottom was regarded as adiabatic plane.
Then we can newly obtain the following parameters in equation (5).

$$\bar{\alpha} = \beta$$

$$\bar{Q} = \frac{(Hz + \eta)}{B} Q + \{1 - \frac{(Hz + \eta)}{B}\} \beta To$$

if we put $h = \frac{Hz + \eta}{Hw + \eta} \qquad (0 < h < 1)$

then

$$A = h(Hz + \eta)\int_0^h \exp(-4s^2)ds$$
$$B = h(Hz + \eta)\int_0^h \exp(-2s^2)ds \qquad (8)$$

BOUNDARY CONDITION

Natural boundary condition of thermal diffusion equation is given as follows.

$$Kx\frac{\partial T}{\partial x}nx + Ky\frac{\partial T}{\partial y}ny + Kz\frac{\partial T}{\partial z}nz + \delta(T - Te) + q = 0 \qquad (9)$$

where

Kx, Ky, Kz: thermal diffusion coefficient (Kcal/°C·m·sec)
nx, ny, nz: direction cosine of normal vector out to boundary (°C)
Te: temperature out of boundary (°C)
q: heat flux (Kcal/$m^2$·sec)
$\delta$: heat conduction coefficient (Kcal/°C·$m^2$·sec)

Integrating equation (9), we can get following equation,

$$Kx\frac{\partial Ts}{\partial x}nx + Ky\frac{\partial Ts}{\partial y}ny + \bar{q} + \bar{\gamma}Ts = 0 \qquad (10)$$

where

$$\bar{q} = 1.672(q - \delta Te) + \{\frac{0.9049Kz}{(Hw + \eta)^2} + 0.672\delta\}To$$
$$\bar{\gamma} = \delta - \frac{0.9049Kz}{(Hw + \eta)^2} \qquad (11)$$

In shallower area, considering bottom boundary condition, we can newly get following parameters in equation (10).

$$\bar{q} = \frac{(Hz + \eta)}{B}(q - \delta Te) - \delta\{1 - \frac{(Hz + \eta)}{B}\}To \qquad (12)$$

$$\bar{\gamma} = \delta$$

$$\left.\frac{dg(z)}{dz}\right|_{Hz} = 0 \qquad \text{(adiabatic condition)} \tag{13}$$

FINITE ELEMENTS FORMULATION

The Galerkin's method of weighted residual now is used. Equation (5) and (10) can be written in the discretized form using Galerkin's techniques.

$$\int_v Wi\,(\frac{\partial Ts}{\partial t} + \frac{A}{B}\,Us\frac{\partial Ts}{\partial x} + \frac{A}{B}\,Vs\,\frac{\partial Ts}{\partial y}$$

$$- Kx\,\frac{\partial Ts^2}{\partial x^2} - Ky\,\frac{\partial Ts^2}{\partial y^2} + \frac{\bar{\alpha}Ts-\bar{Q}}{Cp})\,dv \tag{14}$$

$$+\int_s Wi\,(Kx\,\frac{\partial Ts}{\partial x}\,n_x + Ky\,\frac{\partial Ts}{\partial y}\,n_y + \bar{q} + \bar{\gamma}Ts)\,ds = 0$$

Integration equation (14) by parts, using Green's theorem, the following matrix equation can be obtained,

$$[M]Ts + [R + K + S]Ts + F = 0 \tag{15}$$

where

$$[M] = \int WiWjdV$$

$$[R] = \frac{A}{B}Us\int Wi\frac{\partial Wj}{\partial x}\,dV + \frac{A}{B}Vs\int Wi\frac{\partial Wj}{\partial y}\,dV$$

$$[K] = Kx\int \frac{\partial Wi}{\partial x}\frac{\partial Wj}{\partial y}\,dV + Ky\int \frac{\partial Wi}{\partial x}\frac{\partial Wj}{\partial y}\,dV \tag{16}$$

$$[S] = \frac{\bar{\alpha}}{Cp}\int WiWjdV + \bar{\gamma}\int WiWjds$$

$$F = \frac{\bar{Q}}{Cp}\int WidV + \bar{q}\int Wids$$

where [ ] means matrix representation.

## HEAT EXCHANGE PROCESS BETWEEN SEA SURFACE AND THE ATMOSPHERE

Heat exchange process between sea surface and the atmosphere is formulated according to Wada(1974), who proposed the heat loss coefficients (β) as a linear function of water temperature, difference between temperatures of water and atmosphere, and wind velocity given as follows.

$$\beta \doteqdot G(Tw,\ Tw - Ta) + H(Tw)\cdot W \tag{17}$$

where

Tw: water temperature
Ta: atmosphere temperature
W: wind velocity

Then heat budget through sea surface (ΔQ) that is corresponding to the last two terms of equation (1), is given as follows.

$$\begin{aligned} \Delta Q &= Qs-(Qb+Qe+Qc) \\ &= Qo-\beta\cdot Tw \end{aligned} \tag{18}$$

where

Qs: natural heat input from the sun
Qb: heat loss by long wave back radiation
Qe: heat loss by evaporation
Qc: sensible heat loss; heat loss by conduction and convection
Qo: heat addition term

If we introduce the concept of natural equilibrium temperature $T_M$, then we get,

$$\Delta Q = 0 \tag{19}$$

accordingly, the next equation is obtained.

$$Qo = \beta\cdot T_M \tag{20}$$

## CONDITIONS AND RESULTS OF NUMERICAL SIMULATION

We applied this numerical model to the region of discharged warm cooling water diffusion from the power plant and industries located in the innermost of the practical Tachibana Bay in Shikoku area, Japan.

TIDAL CURRENT SIMULATION

Current velocity field and tidal sea levels are computed using FEM technique by Tanaka et al. (1978) of barotropic tidal current model.
Figure 1 shows the FEM idealization of Tachibana Bay.
Figure 2 shows the averaged current pattern involved tidal residual current in the bay.

THERMAL DIFFUSION SIMULATION

We can calculate the parameter $\beta$ and Qo using the method mentioned above by meteorological condition in observation duration in the bay. The results are as follow.

summer : $Tw = 28.0°C$, $Ta = 27.0°C$, $W = 1.2$ m/sec
$T_M = 26.3°C$, and $Hw = 6.4$ m
$\beta = 0.0117$ (Kcal/°C·m$^2$·sec) and
$Qo = 0.307$ (Kcal/m$^2$·sec)

winter : $Tw = 12.4°C$, $Ta = 4.5°C$, $W = 4.0$ m/sec
$T_M = 13.4°C$ and $Hw = 2.5$ m
$\beta = 0.0125$ (Kcal/°C·m$^2$·sec) and
$Qo = 0.167$ (Kcal/m$^2$·sec)

In computation, from the view point of setting the averaged natural environmental water temperature $T_0$, we used $T_0$ as a linear function of water depth from correlation relation between observed results and water depth, and set $T_0$ at each element. The open boundary condition on water temperature is given using observation results.
As for thermal emission from land, we considered the following two typed sources using observation results.

(a) from power plant and industry
(b) from rivers

Thermal diffusion coefficient is determined from observed results.

$Kx = Ky = 40$ Kcal/°C·m·sec, $Kz = 0$.

The correlations in the cases of summer and winter between computation and observation are shown in Figure 3 (a), (b). The obtained correlation is above statistical significant level.
The results of numerical simulation are shown in Figure 4 (a), (b) as horizontal temperature distribution pattern in the bay.

## CONCLUSION

A thermal diffusion model is established for the prediction of two dimensional discharged cooling water distribution in the coastal bay.
To improve the accuracy of the model, we pay attention to the physical point of view rather than to the numerical point of view.
An intrinsic physical assumption sometimes brings a practical solution for a complicated phenomenon.
We studied under the assumption of two dimensional flow and diffusion, vertical profile of temperature and velocity, the influence of water depth geometry, and heat balance between water surface and the atmosphere.
There are fortunately large amount of field data in Tachibana Bay to introduce the assumption.
As the result of this study, we can get a reasonable model to agree well with actual phenomena.

## REFERENCES


J. F. Macqueen, "Heat Discharges, Part 2: A Numerical Model." Water Pollution Research Laboratory Tech. Paper No. 13 (1972).

G. Spurr and R. A. Scriven, "British Experience of Physical Behavior of Heated Effluents in the Atmosphere and in Various Types of Aquatic Systems", IAEA : Tech. Rep. Ser. No.155, (1974)

A. Wada, et al., "Study on Adaptability of Prediction Method of Simulation Analysis for Diffusion of Discharged Warm Water in the Sea," CRIEPI, Japan Tech. Rep. 73011 (1974)

T. Tanaka and Y. Ono, "Finite Element Analysis of Typhoon Surge in Ise Bay," Proc. of U.S. - Japan Seminar on Interdisciplinary Finite Element Analysis, Cornell Univ. (1978)

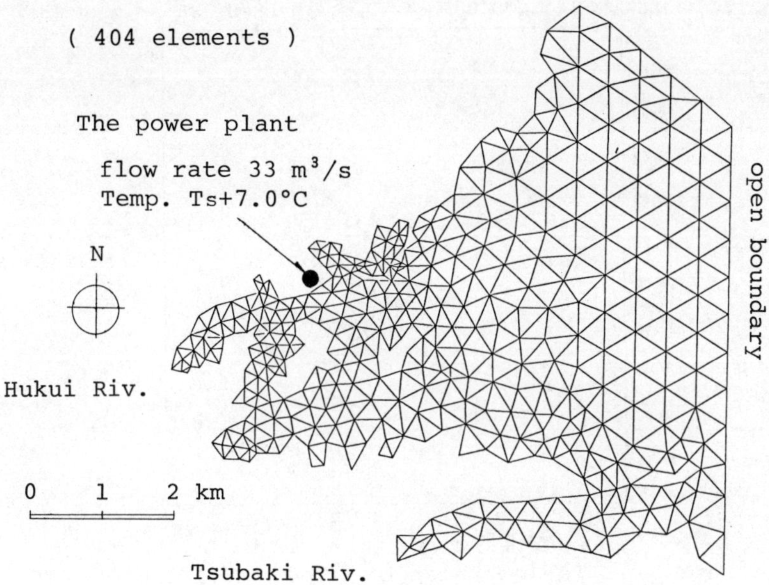


Figure 1. FEM idealization of Tachibana Bay

Figure 2. The averaged current pattern of the bay

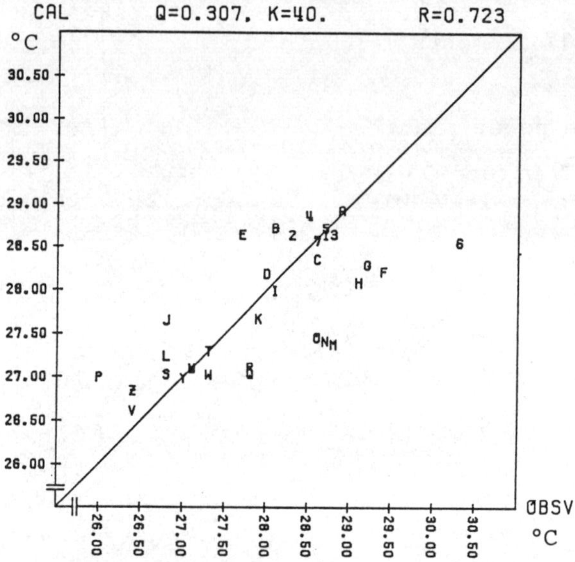


Figure 3 (a). The correlation between computation and observation (summer season)

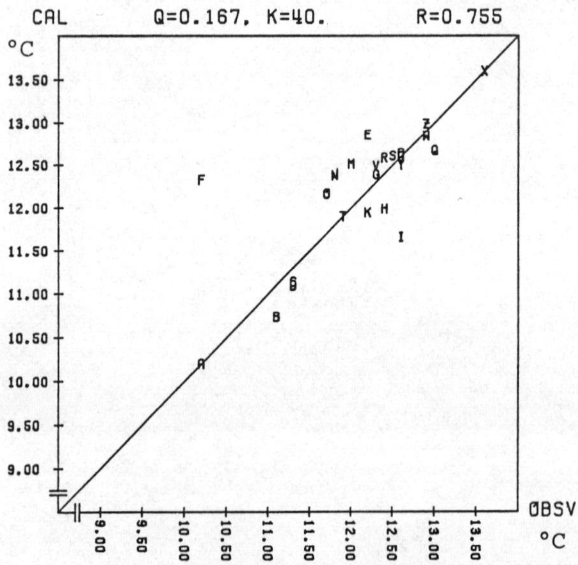


Figure 3 (b). The correlation between computation and observation (winter season)

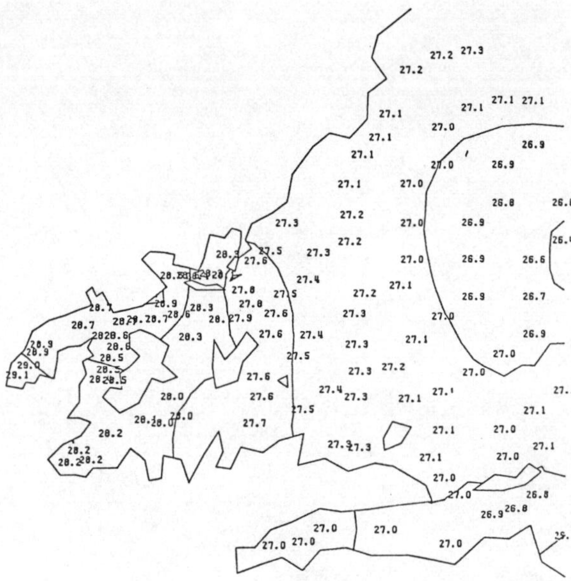

Figure 4 (a). The temperature distribution pattern in the bay (summer season) °C

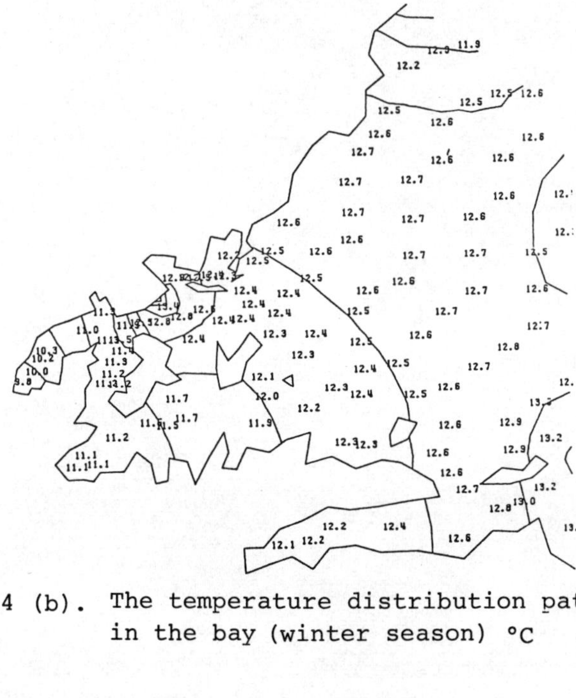

Figure 4 (b). The temperature distribution pattern in the bay (winter season) °C

# 16. NUMERICAL MODELS OF SEDIMENT TRANSPORT

# HYDROTRANSPORTATION AS A POINT PROCESS

Marek Erlich, Andrzej Niemiec
Academy of Agriculture, Wrocław, Poland

## §1 Introduction

In his baic paper /1/ H.Einstein considered the probabilistic model of the motion of the bed-load material.He assumed, that the motion of the grain has diphase character:the moving period and the rest period. Einstei assumed,that the time of the moving period is very short compared to the duration of the rest period.This assumption leads us to the description of hydrotransportation process as a simple Poisson process. But in orginal Einstein paper, the transition from the motion of one particle to the motion of the group of particles is not clear.Einstein assumed, that the motion of the group of particles form the process of the same character.An interesting generalisation of the Einstein model was presented by Shen and Cheong /7/. They suggested for the description of the hydrotransportation process a nonhomogenous Poisson process. Yang and Sayre/8/ obtained from experimental data gamma-distribution for the length of the rest period. In our paper /2/ we considered the motion of river bed-load material as a Markov chain. We found the estimators of transition probability in one step. In that model we assumed, that the motion of one particle in 3-dimensional space form 3 independent Markov chains : forward/backward, left/right and top/bottom.

We assumed, that if the distance between two grains is large, then the motion of them is independent, and the influence of the motion of one grain to another grain is expressed in the definition of one-step transition probability. This model was constructed under the assumption of the homogenity of transported material.

In this paper we proposed a more general model of the motion of the group of particles in any space. We assumed only diphase character of one particle motion.

§2 General model of the motion of one grain

Let $(\Omega, \mathcal{F}, P)$ be a probabilistic space/ $\mathcal{F}$ - the $\sigma$-field of measurable sets, P-probabilistic measure /, and:

1. $N: \Omega \times T \longrightarrow \mathfrak{X}$ , where $\mathfrak{X} \subseteq R$ /real numbers/ $T=[0,T]$ -time . For more convenient notation we assumed that $\mathfrak{X} \subseteq R$, but in general $\mathfrak{X} \subseteq R^k$.

The process $N(t)=N(t,\omega)$ is called the step process iff there exists for all $t \in T$ a sequence of points $s_1 < s_2 < \ldots < s_n \leq t$ , $s_i \in T, i=1,\ldots,n$, such that in the intervals $[0,s_1),[s_1,s_2),\ldots,[s_n,t]$ , $N(t)$ is constant .This process is right-continous and stochastically continous, ie $P(N(t+)-N(t)>0) = 0$. According to Einstein assumption the motion of one grain is a step process, and the moments $s_1,\ldots,s_n$ are motion moments.

The step process can be analogously defined when $\mathfrak{X} \subset R^3$, $\mathfrak{X} = [0,a] \times [0,h] \times R_+$ , where a= river width, h = river depth ,and $R_+$ is the river length. Without a loss of generality we assumed that is a separable Borel set with diameter $diam(\mathfrak{X}) = \infty$ ,/$diam(A) = \sup(r(x,y), x,y \in A)$, r-metric in A /., with natural topology of $\mathbb{R}$ /3/ .

The following properities of N(t) can be showed:

i If $n(t)$ = number of. jumps of $N(t)$ to the moment t , then $n(t)$ is a stochastically continous Poisson process and:

2. $P(n(t)=k)=(\lambda(t))^k/k! \exp(-\lambda(t))$,
$\lambda(t) = E\,n(t)$ .

ii Let $\eta_1,\eta_2,\ldots,\eta_k$ be the length of jumps of $N(t)$ and $\tau_1,\tau_2,\ldots,\tau_k$ are periods between jumps, then $\tau_1,\ldots,\tau_k$ are independent random variables ,and :

3. $N(t) = \sum_{k=0}^{n(t)} \eta_k$

iii If $E\eta_i > 0, E\tau_i > 0$ and $D^2\eta_i < \infty, D^2\tau_i < \infty$ $i=1,2,\ldots$ then:

4. $P(\lim_{t\to\infty} N(t)=\infty) = 1$ .

iv If $\tau_i$ are independent, idetnically distributed random variables /iidrv/ with exponential distribution function and $\eta_i$ are iidrv with distribution function $F(x)$/ i.e. $F(x)=P(\eta_i < x)$ / then the function $\lambda(t) = E\, n(t)$ fullyfiles integral equation:

5. $$\lambda(t) = 1\!\!1_{[0,\infty)} \int_0^t \lambda\,(t-s)dF(s)$$

From properities ii and iv and Wald equality it follows that there exists a function $n(t)$ for wich:

6. $$n(t) = E\,N(t) = \lambda(t)E\,\eta_i\ ,$$

assuming the equal distribution function of $\eta_i$.

It is easy to see that :

7. $$P(N(t) < x) = 1\!\!1_{[0,\infty)} {}^{*}\sum \lambda(t)^n/n!\ \exp(-\lambda(t))\,\mathcal{G}_n(x)$$

where $\mathcal{G}_n$ is the distribution function of the sum $\eta_1+\eta_2+\ldots+\eta_n$ .From this formula we can obtain Einstein's formula for the average velocity of the grain.

§3 <u>The motion of the group of particles</u>

We will consider the hydrotransportation process of the group of M nonhomogenous particles. Let $N^i(t)$ be a process of motion of one particle / $i=1,\ldots,M$ / which started from certain level $x_i$ / i.e $N(0)=x_i$, $x_i \neq x_j$, $i\neq j$/

If the measures $\mu^i_\eta$ of $\eta^i_k$ and the measures $\mu^i_\tau$ of $\tau^i_k$ / the lendth of the jump and the duration of the rest period respectivelly / are nonatomic / i.e. measure $\mu$ is nonatomic iff $\mu(\{x\}) = 0$ / then for all $t,s \in T$

9. $P(N^i(t) = N^j(s)) = 0$ , $i \neq j, i,j=1,\ldots,M$

wich states that the motion of particlas are independent. When the measures $\mu$ are atomic measures with atoms $c_i$ then the grains do not meet only when $c_i/c_j \in Q$ /the set of the rational number /.When the fractions $c_i/c_j \in Q$ then the grains can meet one another periodically , with the period greater then the least common multiplier.

Let us consider a set $A=\{x_1,\ldots,x_M\}$ of initial points of bed-load material, and a set $A_t = \{x: x=N^i(t),$ $N^i(0)=x_i, x_i \in A\}$ wich is the image of A after time $t \in T$ In these terms oure problem is:knowing A /the set of initial values / we should describe $A_t$ /the image of A -evaluation of A /$\mathcal{E}$.

We will describe the simplest case. Let us assume that all particles moving in the simple Poisson processwith a parameter $\lambda_i$, and the length of jump of each grain equal to $a_i$ /i=1,...,M /,$a_i/a_j \in Q$ if$i \neq j$ Let $EA_t=\{v: v=EN^i(t), N^i(0)=x_i, x_i \in A\}$ are the average image of A in the sense of average value of all trajectories. In the case described:

10. $$EA_t = \{x : x = a_j \lambda_j t + x_j;\ A \ni x_j\}$$

Futhermore the average diameter of $A_t$ is equal to:

11. $$E\,\mathrm{diam}(A_t) = \max_j (a_j \lambda_j t + x_j) - \min_i (a_i \lambda_i t + x_i).$$

When $t \to \infty$ then $E\,\mathrm{diam}(A_t) \to \infty$.

When we consider a motion of M homogenous grains in then $\bar{N}(t) = (N^1(t), \ldots, N^M(t)) \in \mathfrak{X}^M$, and $\mu(s) = P(N(t+) - -N(t) < s)$ is the product measure on $\mathfrak{X}^M$. When $\mu(\cdot)$ is non-atomic measure, then we obtained the expression:

12. $$P(N^i(t) \in [x, x+dx], N^j(t) \in [y, y+dy] \mid N^i(t) = x, N^j(t) = y) = \begin{cases} 0 \text{ iff } x = y \\ P(N^i(t) \in [x, x+dx] \mid N^i(t) = x) \cdot P(N^j(t) \in [y, y+dy] \mid N^j(t) = y), \end{cases}$$

wich is analogous to the expression 7. in the paper /2/ If $\mu$ is the atomic measure, then 12. is invalid , and we cannot obtain a proper expression in a simple way.

Mathematically, a more interesting case is, when $A \subset \mathfrak{X}$ is an uncoutable set / e.g. $A = [0,1]$ /. In general, the set $A_t$ must not be a borel set in $\mathfrak{X}$. The authors showed, that when $\Omega$ is constructed in special way, then $A_t$ is a borel set in $\mathfrak{X}$, and is also measurable respectivelly Lebesgue measure.

## Literature

/1/ Einstein,H.A. Bed-load transportation as a probability problem,Sedimentation/Einstein/ H.W.Shen ed Fort-Collins,Colorado 1972 .

/2/ Erlich,M.; Niemiec,A. Multidimensional stochastic model of bed-load transportation in river 9/ in polish / Archiwum Hydrotechniki 1/1981.

/3/ Gichmann,N7 Skorochod,A.W. Introduction to the theory of stochastic processes /in polish/, Warszawa,1970.

/4/ Grandel,J. Double stochastic point processes, Lecture Notes in Mathematics,Springer-Verlag 1976

/5/ Grandel,J.Point processes and random measures, Adv.Appl.Prob,1/1977.

/6/ Shen,H.W.;Cheong,H.F. Dispersion of contaminated sediment bed-load material,J.of Hydr.Div.ASCE vol 93,HY 11,nov 1973

/7/ Shen,H.W.;Cheong,H.F. Stochastic sediment bed-load model load models,Appl.of Stochastic processes in sediment transport,H.W.Shen,H.Kikkawa ed, WRP Colorado 1980.

/8/ Yang,C.T.;Sayre,W.W. Stochastic model for sand dispersion,Proc ASCE HY feb 1971.

# FINITE ELEMENT MODELLING OF SEDIMENT TRANSPORT IN AN ESTUARY

A. Fritz and K.-P. Holz

Hannover University

## INTRODUCTION

Shallow waters with intertidal flats and gradual changes of bathymetry due to sedimentation and erosion processes are some of the main characteristics of the German north sea coast. Owing to the morphological instability in many areas, extensive dredging work is necessary to maintain navigation channels in both the coastal zone and in tidal rivers. Attempts to obtain a better insight into the sediment transport processes in these areas have been made in the past, and include field observations and investigations using movable-bed hydraulic models. Due to the considerable expense of field-investigations and problems of hydraulic scaling in physical models, the latter methods are gradually being supplemented by numerical methods.

Numerical models for the simulation of sediment transport phenomena may be classified according to the scale concerned. For small-scale problems, such as the transport of sediment over dunes and ripples, two-dimensional vertical models for flow, turbulence and sediment transport are commonly used (Puls,1981). As soon as larger scale bed forms such as trenches are involved, effects of turbulence are neglected and a vertical integrated flow model is adopted (Kerssens,v.Rijn,1977; Alonso,Wang,1980). The vertical distribution of sediment over the depth is normally retained. Because of excessive computing costs it is impracticable to extend this approach to horizontal two-dimensional models. This problem has been avoided, for example, by Lepetit et al.(1979) who made a further simplification using a vertical integrated transport model to compute scour around a coastal structure. However, when the area under investigation covers several kilometers in width, such as in an estuary, and when time-dependent tidal flow is required, the numerical solution of the transport equation might still be too costly. For this reason Sündermann et al.(1977), as well as Amos et al.(1980), only computed net transport rates, as obtained by summation.

In this paper a similar approach has been applied to predict bathymetric changes for a period of two years in the Jade estuary. An attempt is also made to test the reliability of the method by simulating the siltation in a one-dimensional trench. The numerical results are compared to experimental and field data.

## ASSUMPTIONS AND EQUATIONS

The objective of this paper is to simulate the sediment transport in the Jade estuary. Figure 1 shows its original bathymetry and the area (dotted) to be modelled. The tidal range in this estuary is about 3.3 m and the flow conditions are well mixed. Since the transport of material is dominated by tidal flow, the effect of wave-induced currents on erosion and sedimentation processes is not considered. The bed material is non-cohesive sand with a grainsize diameter of $d_{50}$= 0.2 mm.

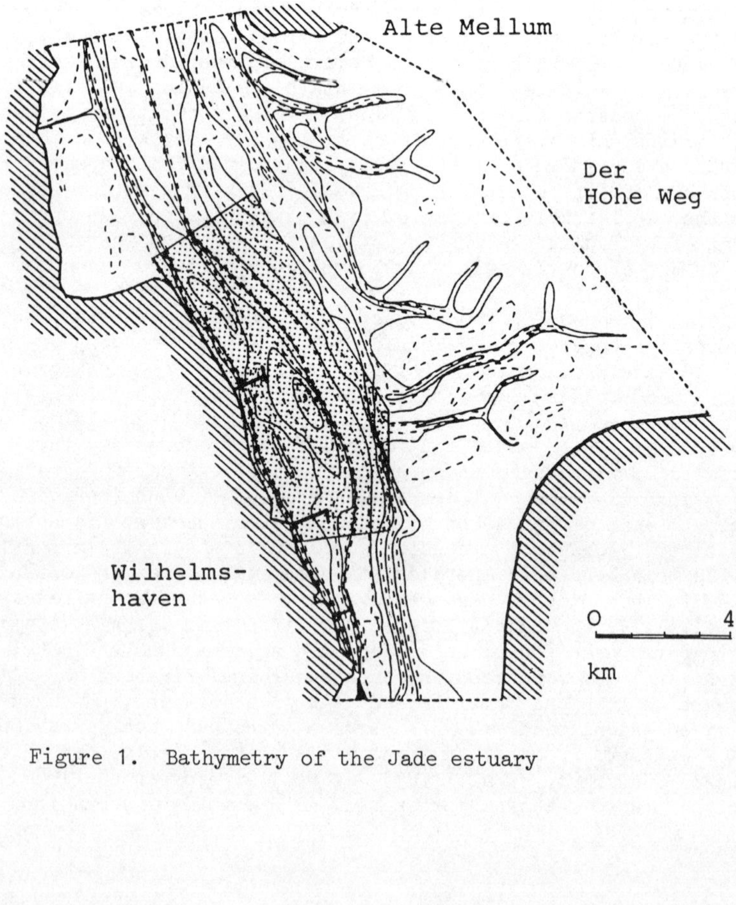


Figure 1. Bathymetry of the Jade estuary

The computation of the flow-field starts from the vertically integrated equations for shallow water waves :

$$\int_A \{q_{i,t} + (v_j q_i),_{ij} - \Omega\varepsilon_{ij}q_j + g(a+h)h,_i$$
$$+ \frac{\lambda}{a+h}\sqrt{v_j v_j}\, q_i - A_H q_{i,jj}\}\, dA \quad = 0 \qquad (1)$$

$$\int_A \{h,_t + q_{i,i}\}\, dA \quad = 0 \qquad (2)$$

In the tensor notation used, $q_i$ denotes the vertically averaged flow-rate, $v_i$, the corresponding velocities, h, the free surface elevation, a, the water depth, and $\Omega$, the Coriolis parameter. The influence of bottom friction as well as that of turbulence are parametrized with the friction parameter, $\lambda$, and the eddy viscosity, $A_H$. Integration is performed over the horizontal plane, A.

For the computation of large-scale sediment transport in which the horizontal reach is much larger than the water depth, an instantaneous entrainment of material is assumed until an equilibrium load, depending on actual flow conditions, is reached. This equilibrium load is given by Zanke (1982), based on experimental investigations for steady flow conditions :

$$T = T_{bed} + T_{sus(pended)} \qquad (3a)$$

$$T_{bed} = T^o_{bed} \cdot (v_j v_j - v_c^2)^2 \cdot f_{bed}(d, w_s, \nu, \rho') \qquad (3b)$$

$$T_{sus} = T^o_{sus} \cdot (v_j v_j - v_c^2) \cdot (v_j v_j - v_a^2) \cdot (a+h) \cdot f_{sus}(d, w_s, \nu, \rho') \qquad (3c)$$

Herein $T^o_{bed}$ and $T^o_{sus}$ are constants, $v_c$ and $v_a$ are the critical velocities for the transport of bed load and suspended load, respectively. $f_{bed}$ and $f_{sus}$ are functions of grain parameters such as d ( grain diameter ), $w_s$ ( settling velocity ), $\nu$ ( viscosity ), and $\rho'$ ( density difference between grain material and the ambient fluid ). It is assumed that the total load T is transported in the direction of the main flow, and therefore it can be split up into vectorial components $T_i$ according to the velocity components $v_i$.

For refined morphological modelling, bed load and suspended load should be accounted for independently. The concentration distribution of suspended material over the water depth may be determined from the suspended load $T_{sus}$, using an empirical relationship, given by Rouse (1938) :

$$C(x_3) = C_o \cdot \left\{ \frac{H-x_3}{x_3} \cdot \frac{z_o}{H-z_o} \right\}^{\alpha} \tag{4}$$

$$\alpha = w_s / \kappa v_*$$

Herein $v_*$ denotes the shear velocity, $\kappa$ ($\kappa$=0.4) is the von Kármán constant, and H is the actual water depth.

CONTINUITY OF MATERIAL

Accretion and erosion at the flow bed are computed from the continuity equation :

$$a_{,t} + T_{i,i} = 0 \tag{5}$$

which may be directly applied for bed load. The computation of the suspended load is more complicated, since the concentration distribution varies with the flow field. The relation can be described by a differential equation. However, for far-field simulations, the numerical solution would be very time-consuming. For this reason a simple bookkeeping strategy is proposed here.

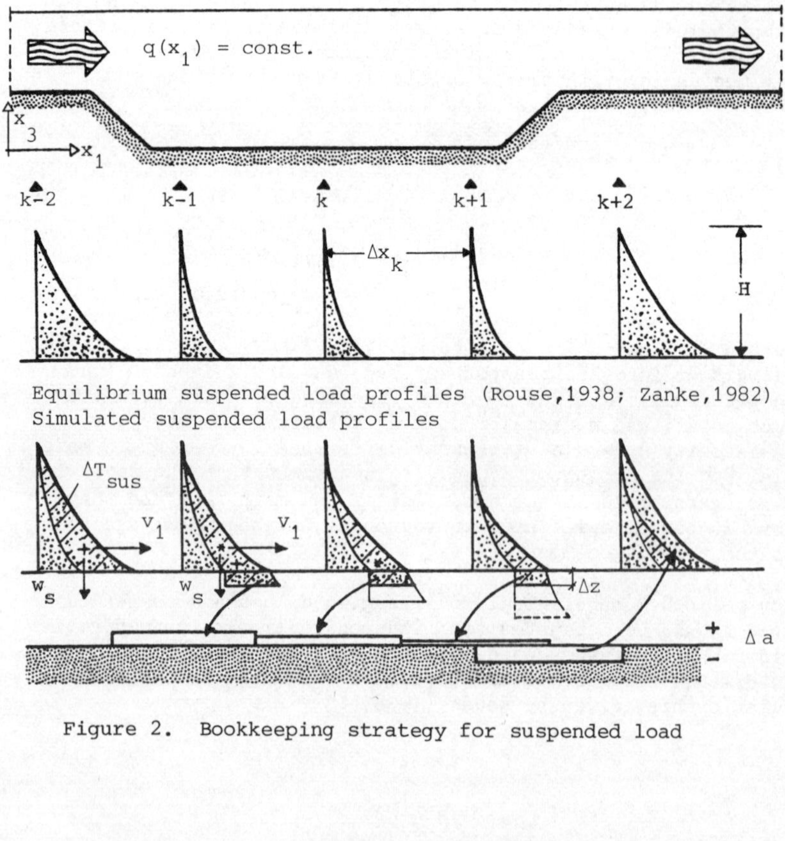


Figure 2. Bookkeeping strategy for suspended load

Considering equation 4 and assuming $z_o$ to be the thickness of the bed load layer, the shape of the concentration distribution is known at all nodes, as shown in figure 2. The constant $C_o$ can easily be determined from $T_{sus}$, which represents the integral of equation 4, taken over the water-depth.

The rate of erosion is determined on an element level by comparing the total load (T) at equilibrium for the actual velocities at the element nodes. The bed level variation $\Delta a$ is obtained from equation 5. The rate of accretion is obtained from the same relation, which is applied for the bed load directly.

For suspended load, however, a more complex treatment is required. On an element level, the equilibrium load $T_{sus}$ is computed at the nodes. If the upstream load $T_{sus}(k-1)$ is larger than the downstream load $T_{sus}(k)$, as shown in figure 2, oversaturation prevails within this element, which corresponds to a residual mass :

$$\Delta T_{sus} = T_{sus}(k) - T_{sus}(k-1) \tag{6}$$

A certain amount of this quantity will settle out within the element. In the bookkeeping strategy this process is simulated as follows :

- Compute the concentration distribution $C_{\Delta T}(x_3)$ corresponding to $\Delta T_{sus}$.
- Compute the mean settling path $\Delta z$ for the particles in suspension $\quad \Delta z = w_s \cdot \Delta t$ .
  The time interval follows from the horizontal velocity component $\quad \Delta t = \Delta x(k-1) / v_x$ .
- Add $C_{\Delta T}(x3)$ with a vertical shift of $\Delta z$ to the concentration distribution at the downstream node.
- Compute $T_{sus}$ from $C_{\Delta T}(x3)$ by integration over $0 \leqq x_3 \leqq \Delta z$ and then calculate the accretion within this element according to equation 5.
- Store the computed concentration profile $C_k(x_3)$ for $x_3 \geqq 0$ at node k.

For practical numerical computation, it turns out that a stepwise constant approximation of the actual concentration distribution is much easier to formulate. This method was applied in the following examples.

The simulation of sedimentation and erosion leads to a time-dependent problem, which is treated as a cyclic three-step procedure. In the first step, the flow field is computed. In the second step, the transport rates and continuity of sediment are calculated. In the third step, an extrapolation for the morphological changes over a time interval $\Delta T$ is made. For the extrapolated bathymetry, the flow field is then recomputed, and the whole cycle repeated.

## SILTATION IN A TRENCH

The strategy outlined above was tested against experimental data for the flow across a trench, as given by Kerssens et al. (1977). In this experiment, a trench having the geometry shown in figure 3 was formed in a 0.170 mm sand bed in a laboratory flume. The experiment was run with an upstream inflow rate of 0.195 $m^3$/m s. Sediment was supplied at a constant rate of 0.041 kg/m s, and most of the load was carried in suspension. Due to hydraulic sorting, the $d_{50}$ of the suspended load was about 0.110 mm, with a fall velocity of 0.01 m/s .

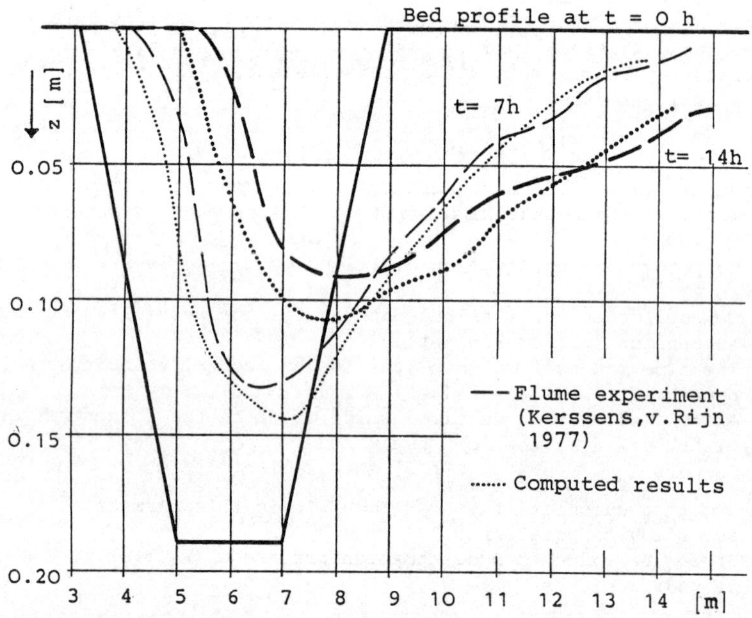


Figure 3. Evolution of bed profiles in a trench

The bed profile was calculated after 7 and 14 hours of continuous flow. The profiles are shown in figure 3, together with the experimental data from Kerssens et al.. The agreement between the observed and simulated bed profiles is satisfactory. A further comparison was made with numerical results obtained by Kerssens et al. (1977) and Alonso et al. (1980), using a vertical two-dimensional transport model. The comparison with the computed bed profiles from Kerssens and Alonso is acceptable.

The results given in figure 3 were obtained by approximating the concentration distribution stepwise within 4 equal intervals over the depth. Although the results were not greatly effected by an increase of intervals over the depth, the minimum number required is two. For the computation, 16 iterations were found to give satisfactory results for the 7 hrs simulation, whilst 35 iterations were necessary for the 14 hrs period.

## SEDIMENT TRANSPORT IN THE JADE ESTUARY

The encouraging results obtained using the simple bookkeeping strategy for the trench problem suggested the use of the method for modelling sediment transport in an estuary. For this purpose, the Jade estuary was chosen, since it has no fresh-water inflow, the interior part of the estuary is well protected against incoming waves, and the morphology is quite homogeneous. In addition, the mean grain-size is about $d_{50}$ = 0.2 mm, which is rather coarse. The corresponding settling velocity is about $w_s$ = 0.029 m/s .

The morphological calculation was performed for the area indicated in figure 1. The whole area was discretized into elements, and the flow-field computed over one tidal cycle. The flow model was based on an explicit finite element model as described by Holz et al.(1980), which includes the simulation of intertidal flats. As boundary conditions, the water-level history for a mean annual tide were prescribed along the northern and southern boundaries. After a careful calibration of the flow model, the velocities obtained were used as input for the sediment transport model. This model could be further simplified for the intended large-scale computations. Assuming a concentration distribution according to equation 4, the vertical coordinate of its centerpoint, $z_s$, may be calculated. This is compared to the settling velocity $w_s$, the maximum horizontal velocity in an element and the mean segment length $\Delta x$. If

$$z_s < w_s \cdot \Delta t \qquad \text{and} \qquad \Delta t = \Delta x \,/\, v_{max}$$

it is assumed, that no material is transported across the element. Since this condition was fulfilled on average for the whole finite element flow grid, the continuity equation 5 could be solved directly using equation 3 for the total load. This morphological computation was repeated every 10 minutes over a tidal cycle and the resulting bathymetry-changes extrapolated to a period of two years. The results were compared to morphological changes based upon echo soundings in the field for the same time interval.

Figure 4 shows a comparison between computed and observed regions of accretion of more than 1.00 m over the original bathymetry for the dotted area in figure 1. The agreement with field data is fairly good, except for the shaded dead-water zone behind the head of the jetty. This is an open pile supported bridge which could not be modelled numerically.

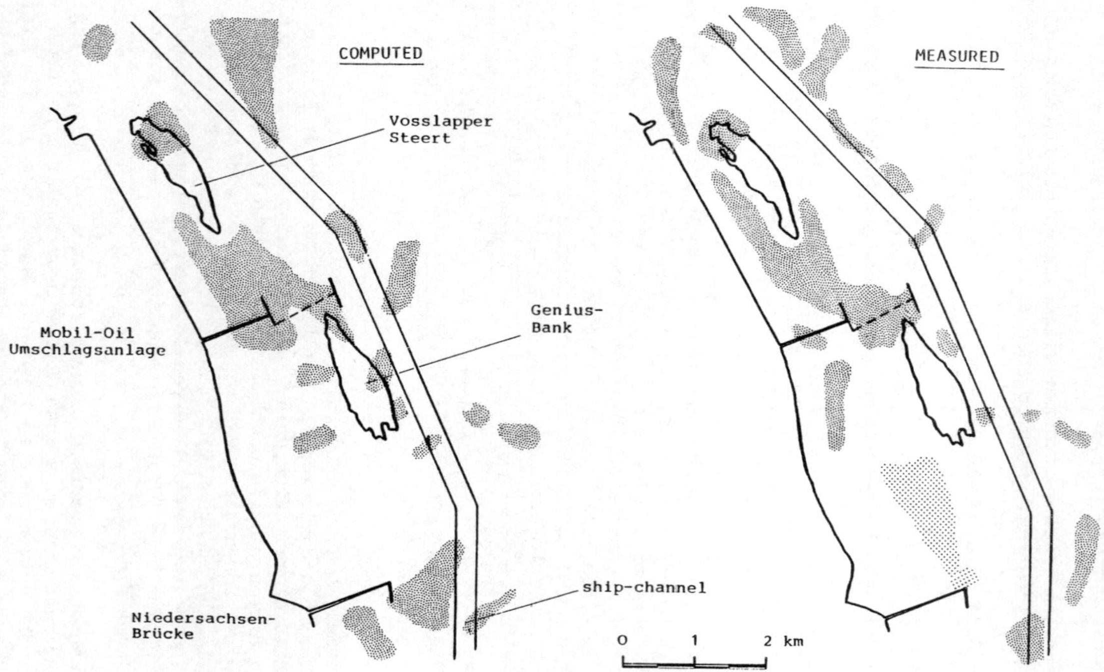


Figure 4. Regions of sedimentation in the Jade estuary, 1974-76

## CONCLUSIONS

A first step towards modelling far-field sediment transport in estuaries was made by testing and applying a simple bookkeeping strategy for the simulation of morphological changes. The proposed strategy was shown to successfully describe the siltation in a trench in both, qualitative and quantitative terms. Although fair agreement was obtained between observed and predicted morphological changes in the Jade estuary, in qualitative terms, a certain degree of caution is required, concerning quantitative changes. In order to obtain better quantitative predictions, more detailed and precise field data are necessary. Better predictions would also be expected by development of the flow model to give more reliable estimates of slack-water velocities, by including the influence of waves, and by refining the transport model. This is especially important if finer bed material is involved.

The computations have shown, however, that a simple model involving a limited amount of computational expense is of great value for the far-field modelling of sedimentation processes in estuaries.

## ACKNOWLEDGEMENTS

The authors gratefully acknowledge the support given by the Studienstiftung des deutschen Volkes to the first author, who is currently preparing a doctoral thesis on the work described.

## REFERENCES

Alonso, C.V., Wang, S.Y. (1980) Simulation of Local Scour and Fill in Sand Bed Streams. Proceedings. Third International Conf. on Finite Elements in Water Resources, Mississippi, :6.57-6.68.

Amos, G.L., Greenberg, D.A. (1980) The Simulation of Suspended Particle Matter in the Monos Bay, Bay of Fundy - A Region of Potential Tidal Power Development. Canadian Coastal Conference, Burlington, Ont., Canada.

Holz, K.-P., Nitsche, G. (1980) Tidal Wave Analysis for Estuaries with Intertidal Flats. Proceedings. Third International Conf. on Finite Elements in Water Resources, Mississippi, :5.113-5.126.

Kerssens, P.J.M., van Rijn, L.C. (1977) Model for Non-steady Suspended Sediment Transport. Delft Hydraulics Laboratory, Publication no. 191.

Lepetit, J.P., Hauguel, A. (1979) A Numerical Model for Sediment Transport. In Nihoul, J.C.J. (ed.), Marine Forecasting, Elsevier, Amsterdam, :453-463.

Puls, W. (1981) Numerical Simulation of Bed Form Mechanics. Mitteilungen des Inst. f. Meereskunde der Universität Hamburg, Nr. XXIV.

Rouse, H. (1938) Transportation of Sediment. Chapter XIV of Fluid Mechanics for Hydraulic Engineers. Mc Graw-Hill Book Co.

Sündermann, J., Krohn, J. (1977) Numerical Simulation of Tidal Caused Sand Transport in Coastal Waters. Proceedings. Seventeenth Congress of the International Association for Hydraulic Research, 1:173-181.

Zanke, U. (1982) Grundlagen der Sedimentbewegung. Springer-Verl. Berlin, Heidelberg, New York.

# MATHEMATICAL MODELLING OF TIME DEPENDENT WAVE ATTENUATION AND DISCRETE SOLID BODY TRANSPORT IN GRAVITY DRIVEN PARTIALLY FILLED PIPE FLOWS

J. A. Swaffield and Sarah Bridge,
Brunel University Drainage Research Group
L. S. Galowin, U.S. Department of Commerce,
National Bureau of Standards

## SUMMARY

The method of characteristics is applied to solve the unsteady partially filled pipe flow equations applying to the flow in building drainage systems. Wave attenuation, together with the propagation of steep fronted waves and the transport of discrete solids, is modelled and comparisons drawn with experimental investigations undertaken with representative pipe size and flow loading values.

## UNSTEADY FLOW IN PARTIALLY FILLED BUILDING DRAINAGE PIPE NETWORKS

Building drainage pipe networks operate under conditions of unsteady free surface flow. As such the effects of flow attenuation, namely a reduction in flow depth and velocity along the pipe length, must be considered in system design. Two main effects are identifiable, namely an under-estimation of the carrying capacity of any drain and an increased probability of solid deposition as the flow depth decreases downstream. Both these effects are accentuated if water conserving criteria are included in the design of appliances served by the system. There is therefore a need to develop a predictive method that will accurately provide flow depth and velocity profiles along a system that includes multiple flow entry junctions and the range of pipe geometry associated with drainage systems. Such a solution should also be capable of dealing with a range of boundary conditions, including moving boundaries such as steep-fronted waves and solids in transport.

Although there are a number of analysis methods available to solve the appropriate unsteady flow equations, as demonstrated by Price (1974), with reference to flood routing, the technique chosen was the method of characteristics that has been widely used in the analysis of unsteady flow, Wylie and

Streeter (1978). This technique has advantages in its ability to deal simply with moving boundary conditions and the ability of the solution technique to allow boundary equations to be developed and tested in semi isolation.

Preliminary work was reported by Swaffield, Bridge and Galowin (1981), based on joint research at Brunel University, U.K., and N.B.S., Washington. The current paper outlines the mathematical models utilised to predict wave attenuation and represent wave and solid motion. Experimental verification is also presented based on unsteady flow conditions compatible with conditions in drainage pipe networks.

BASIS FOR A MODEL INCORPORATING THE METHOD OF CHARACTERISTICS

Figure 1 illustrates the derivation of the equations of motion and continuity for unsteady partially filled pipe flow, namely:

$$g \frac{\partial h}{\partial x} + g\,(S - S_o) + V \frac{\partial V}{\partial x} + \frac{\partial V}{\partial t} = 0 \qquad (1)$$

$$VT \frac{\partial h}{\partial x} + T \frac{\partial h}{\partial t} + A \frac{\partial V}{\partial x} = 0 \qquad (2)$$

where the unsteady friction is modelled by the local time equivalent steady flow parameters and Manning's loss coefficient.

Equations 1 and 2 may be combined linearly to yield a total differential equation:-

$$\frac{dV}{dt} + \frac{g}{c}\frac{dh}{dt} + g\,(S - S_o) = 0 \qquad (3)$$

provided that this relation is confined by the condition

$$\frac{dx}{dt} = V \pm c \qquad (4)$$

where $c = \sqrt{gA/_T}$, the local wave propagation velocity.

The restraint provided by equation 4 confines the solution to two characteristic lines in an x - t plane, Figure 1, the characteristic gradients depending on the flow condition as defined by $V >$ or $<$ c.

In common with other applications of the method of characteristics, the solution requires initial values of velocity and depth along the pipe at time zero, together with suitable boundary equations to define either the pipe terminations or any moving boundary within the pipe length.

These boundary conditions will be dealt with separately for wave attenuation and steep fronted wave and solid motion.

## WAVE ATTENUATION IN LONG PIPES

Experimental verification of the model described was undertaken at Brunel University using a 14 m long, 100 mm diameter glass drain capable of gradient adjustment over a wide range. Surge profiles at pipe entry were initially generated by a discharge valve controlling the output of a pressure chamber supplying water to the pipe via a vertical riser and vertical swept 90° junction. Figure 2 compares maximum depth readings at seven points along the pipe, measured by means of pressure transducers, to predicted values assuming normal depth entry flow conditions. The under-estimation of maximum depth early in the pipe length was thought to be due to this entry boundary condition, and alternative simulations based on calculated critical and measured depth-flow rate at pipe entry were investigated.

Normally linear interpolation techniques are employed to yield the base conditions for each time step, i.e. points R and S in Figure 1. If the initial steady flow depth is constrained at pipe entry, this assumption is not valid over the first few computing sections and therefore a suitable approximation was to allow the flow velocity and wave speed to vary with the square root of distance along the pipe during the calculation of the initial flow conditions only.

Ideal entry conditions at the swept riser junction should be critical depth. However, due to pipe curvature, the measured depth consistently exceeded the appropriate critical value. Figure 2 also compares the maximum depth profiles along the test pipe for both alternative entry boundary conditions to the experimental results, indicating an improved model.

The tranquil entry flow conditions above were representative of flow in drains relatively remote from any vertical stack connection. To extend the simulation to include entry flow with high kinetic energy, flow was introduced into the 100 mm diameter test pipe from above via a 40 mm vertical tube supplied with water from a variable height tank.

Figure 3 compares the maximum depth profiles along the test pipe to a simulation based on entry flow kinetic energy as the determinant of the upstream boundary flow depth. Although reasonable agreement was achieved on maximum depth predictions, close study of the individual depth time profiles at each of the seven pressure tappings along the 14 m pipe, indicated a time lag in the predicted peak depth at locations between 2 and 7 m from pipe entry. This effect was considered to be due to the local formation of a steep fronted wave that attenuated within this distance and led to a further investigation of the practicality of inclusion of such a wave propagation in the model.

STEEP-FRONTED WAVES IN UNSTEADY PARTIALLY FILLED PIPE FLOW

The steep-fronted wave is a region of rapidly varying flow which is not accurately modelled by the gradually varied equations employed previously. A steep-fronted wave theoretically exists when the relative change in the Froude number across the discontinuity spans the value of unity (the absolute value must be less than unity on one side and either greater or less than unity on the other). This type of wave is caused by an increase or decrease in flow either upstream or downstream. However, the condition of interest here is an increase in upstream flow, causing a rapidly varying positive wave to propagate down the pipe.

The wave may be considered as a moving boundary with the five necessary equations to determine wave speed and flow conditions upstream and downstream being derived by bringing the system to rest by superposition of a reversed wave speed, Figure 4, as follows:-
(i) the continuity equation applied across the wave

$$(V_1 - V_W)A_1 = (V_2 - V_W)A_2 \tag{5}$$

(ii) the force equation applied across the wave

$$\rho g A_1 \bar{h}_1 - \rho g A_2 \bar{h}_2 = \rho A_2 (V_2 - V_W)^2 - \rho A_1 (V_1 - V_W)^2 \tag{6}$$

The first two expressions are applicable to rapidly varied flow and are valid across the wave, the remaining expressions applying to the flow either side of the wave.

(iii) $C^+$ characteristic P'R associated with the supercritical flow

$$V_{P'} - V_R + \frac{g}{c_R}(h_{P'} - h_R) + g(S_R - S_o)\Delta t = 0 \tag{7}$$

(iv) $C^-$ characteristic P'C, also associated with supercritical flow

$$V_{P'} - V_C - \frac{g}{c_C}(h_{P'} - h_C) + g(S_C - S_o)\Delta t = 0 \tag{8}$$

(v) $C^-$ characteristic P'S associated with downstream subcritical flow

$$V_{P'} - V_S - \frac{g}{c_S}(h_{P'} - h_S) + g(S_S - S_o)\Delta t = 0 \tag{9}$$

Solution of these boundary equations at each time step allows the wave speed to be calculated and the progress of the wave to be traced through the x - t grid.

In order to begin the calculation, an initial value of the wavespeed at the inception of the discontinuity must be found. Fox (1977) assumes the velocity of the wave is known from a previous step of integration and offers no method of calculating an initial value. The solution adopted is to locate the inception of the steep-fronted wave (the point at which the flow becomes supercritical) and take the depth at that node as the higher depth of the discontinuity. The lower depth is found by extrapolating backwards from the next two nodes downstream. These two depths are used to find the wavespeed at the moment the steep-fronted wave is generated. Figure 5 illustrates the effect of including a steep-fronted simulation.

Many authors have considered the problem of simulating steep-fronted waves in open channels and whilst recognising that the equations of gradually varied flow are not theoretically applicable, have found them more reliable than the rapidly varied equations due to the twin problems of recognising the inception of the steep-fronted wave and then calculating its wavespeed. Martin and DeFazio (1969), Chaudry and Contractor (1973) and Terzidis and Strelkoff (1970) found that many rapidly varied wave forms can be adequately simulated using the differential equations of gradually varied flow. The authors have experienced difficulty in experimentally verifying this method of simulating steep-fronted waves and feel it is not necessarily applicable in partially filled building drainage pipe networks.

## SOLID TRANSPORT IN UNSTEADY PARTIALLY FILLED PIPEFLOW

Discrete solids may be treated in a similar manner to the steep-fronted wave example. Figure 6 illustrates the technique for developing the necessary boundary equations:-

(i) the continuity of flow equation with the body brought to rest by superposition of its reverse velocity

$$(V_1 - V_S)A_1 = (V_2 - V_S)A_2 \qquad (10)$$

(ii) the $C^+$ characteristic on the upstream face, for both sub and supercritical flow conditions, identical to equation 7.

(iii) the $C^-$ characteristic on the downstream face, in subcritical flow PS, Figure 6, where P is the predicted solid position based on force calculations at the preceding timestep. Note the solid track is represented by PZ at slope $1/V_S$. For supercritical downstream flow the $C^-$ characteristic is represented by PS', equation 9. For this equation to be true in both sub and supercritical flow, S' must lie downstream of Z, or $V_S > V_2 - c$, to avoid intersection of characteristics. If $V_S < V_2 - c$, then flow into the downstream pipe is unaffected by upstream conditions and a critical flow depth would be expected, thus $V_2 = c$ and hence for $V_S \nless 0$ it follows that $V_S > V_2 - c$ at all times. The corollary to this is that for stationary solids the downstream depth tends to the critical value appropriate to the leakage flow past the solid.

(iv) leakage flow past the solid provides the final boundary expression. Laboratory tests with tethered solids yield a relationship of the form $Q = K\,(SE - SE_O)^2$ where $SE_O$ is the upstream specific energy necessary to generate flow past the solid and K is a curve fit coefficient. The value of SE is taken as $h_1 + V_1^2/2g$, but in the moving solid case $V_1$ is interpreted as the relative water to solid velocity $(V_1 - V_S)$

$$Q = (V_1 - V_S)A_1 = K\left(h_1 + \frac{(V_1 - V_S)^2}{2g} - SE_O\right)^2 \qquad (11)$$

As the value of $V_S$ is assumed constant from the previous time step calculation, a quartic in $V_1$ may be formed and calculation of the upstream and downstream flow condition achieved.

The forces acting on the solid may be combined in an equation of motion

$$F_1 - F_2 + (mg - F_B)\sin\alpha - \mu(mg - F_B)\cos\alpha = m\frac{dV_S}{dt} \qquad (12)$$

that yields a value of solid velocity across the next time step and hence predicts the future solid position and allows the solution to cycle.

Figure 7 illustrates a comparison between the model described and transport of a single solid from rest along a 14 m length of 100 mm diameter clear UPVC pipe. Solid velocity records were obtained by using six pairs of light interruption photocell sensors linked to an Apple II data logger. The solid employed was a water filled 80 mm x 37.5 mm diameter closed rigid plastic cylinder of s.g. 1.10. Buoyancy forces were calculated on the basis of a linear depth profile along the solid between its upstream and downstream faces.

Reasonable agreement is represented over the majority of the solid transport distance, although this deteriorates slightly towards the end of the pipe. Observations suggest that the solid motion becomes more sensitive to its passage over pipe section couplings, spaced at 2 m intervals, at the lower solid velocities. This is the equivalent to an enhanced local friction factor not included in the model.

Depth measurements were possible visually due to the duration of transport and were relatively constant as a ratio of the cylinder diameter. Again the model predictions are supported by the experimental observations.

## CONCLUSIONS

The agreement achieved by use of the proposed models is sufficient to justify further development to include the effects of changes in pipe geometry and multiple inflows.

The objective of the work will remain the introduction of improved design methods. However, it is also hoped that the techniques developed will form a base for future consideration of the effects of both design and legislative code changes.

## ACKNOWLEDGMENTS

The work reported has been supported by the National Bureau of Standards and undertaken jointly at N.B.S. and Brunel University, Uxbridge, U.K.

## NOTATION

| | |
|---|---|
| A | Flow area |
| c | Wave propagation speed = $\sqrt{g\ A/T}$ |
| F | Hydrostatic forces, Froude N°, = V/c |
| $F_B$ | Buoyancy forces, $F_F$ Friction force |
| g | Acceleration due to gravity |
| h | Depth, $\bar{h}$ Centroid depth |
| K | Empirical constant, solid leakage flow |
| L | Pipe length |
| m | Hydraulic mean depth or solid mass |
| n | Manning's loss coefficient |
| S | Slope of energy grade line = $n^2 V^2 / m^{1.33}$ |
| $S_o$ | Pipe slope, also $\alpha$ |
| SE | Flow specific energy, $h + V^2/2g$ |
| T | Free surface width |
| t | Time co-ordinate |
| V | Flow mean velocity, suffix indicates position |
| $V_s$ | Solid velocity. $V_w$ Steep fronted wave velocity |
| x | Distance co-ordinate |
| $\Delta t$ | Computational time step |
| $\mu$ | Solid to wall sliding friction factor |
| $\rho$ | water density |

| | |
|---|---|
| Chaudry, Y.M. and Contractor, D.N. (1973) | Application of the implicit method to surges in open channels. Wat. Res.Res. 9, 6, 1973. |
| Fox, J. (1977) | Hydraulic analysis of unsteady flow in pipe networks. Macmillan Press, London. |
| Martin, C.S. and DeFazio, F.G. (1969) | Open channel surge simulation by digital computer. J. Hyd. Div., ASCE, 95, HY6, 1969 |
| Price, R.K. (1974) | Comparison of four numerical methods for flood routing. Journal ASCE. |
| Swaffield, J.A., Bridge, S. and Galowin, L.S. (1981) | Wave attenuation in long drainage pipes, a numerical solution of the unsteady partially filled pipe flow equations. CIB W62 Conf. Technische Fachhochschule, Berlin |
| Terzidis, G. and (1970) Strelkoff, T. | Computation of open-channel surges and shocks. J. Hyd. Div., ASCE, 96, HY12, 1970. |
| Wylie E.B. and (1978) Streeter, V.L. | Fluid Transients, McGraw Hill. |


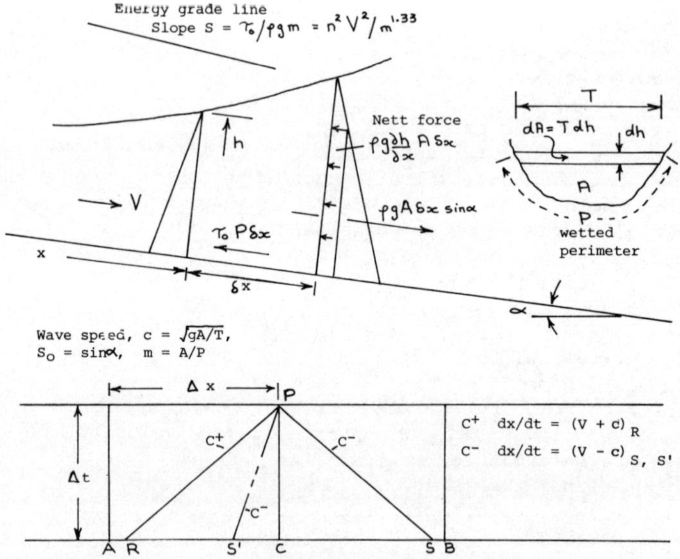


Figure 1: Summary of method of characteristics solution of the unsteady partially filled pipe flow equations

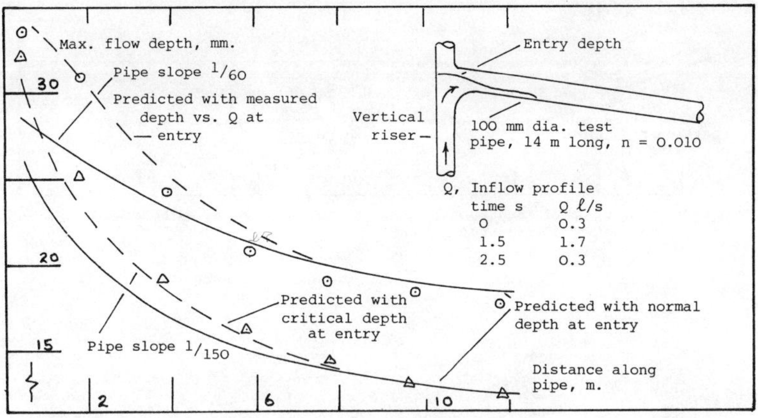


Figure 2: Comparison of observed and predicted attenuation of a pressure vessel generated wave, various entry boundary conditions

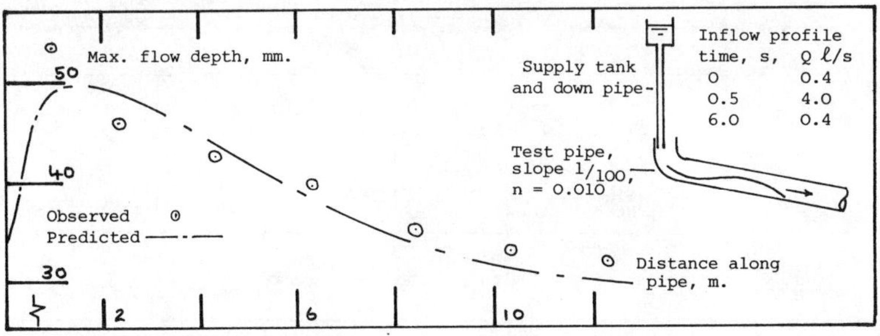


Figure 3: Comparison of observed and predicted attenuation of cistern generated wave, entry conditions based on inflow kinetic energy

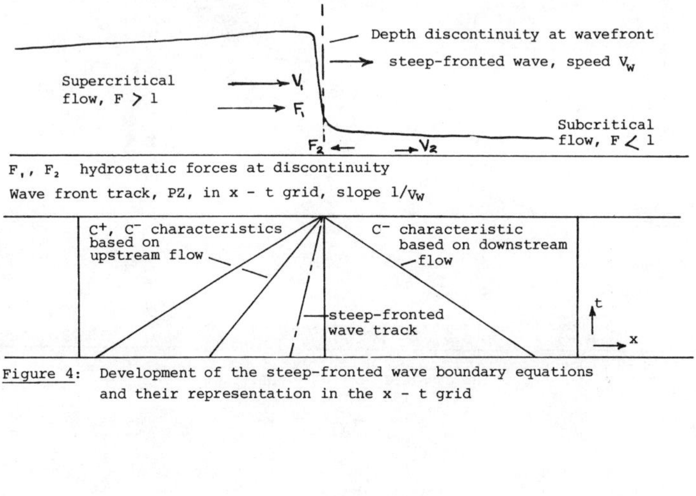


Figure 4: Development of the steep-fronted wave boundary equations and their representation in the x - t grid

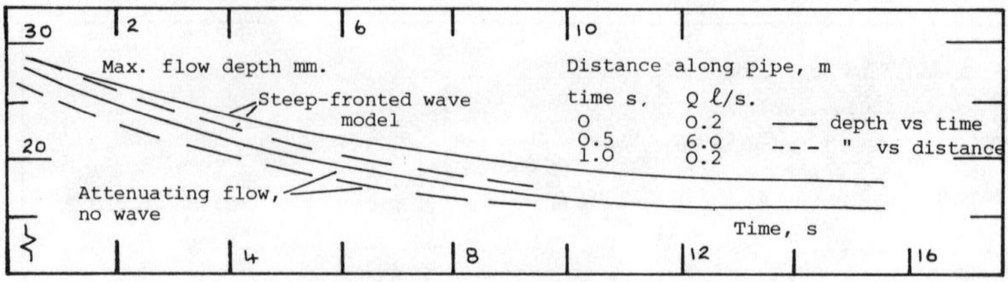


Figure 5: Comparison of theoretical flow attenuation with and without the inclusion of steep-fronted wave propagation.

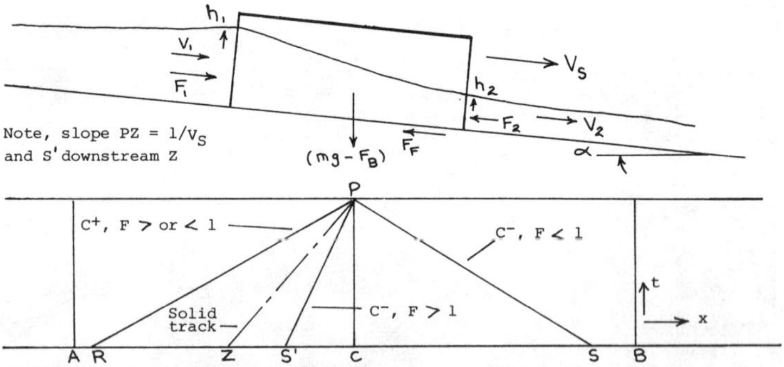


Figure 6: Development of the moving solid boundary equations

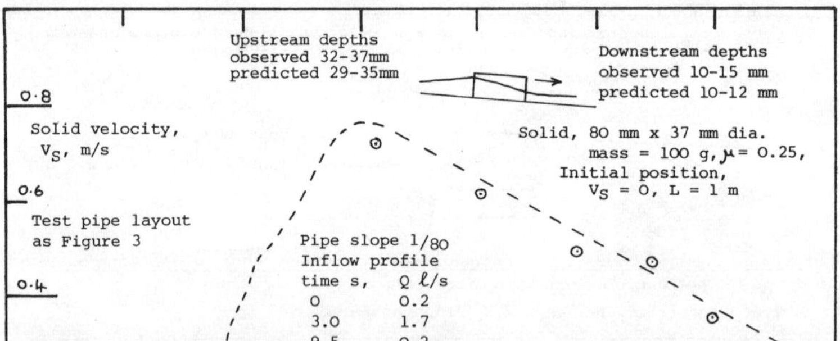


Figure 7: Comparison of observed and predicted solid transport from rest utilizing the developed boundary equations

| | |
|---|---|
| Chaudry, Y.M. and Contractor, D.N. (1973) | Application of the implicit method to surges in open channels. Wat. Res.Res. 9, 6, 1973. |
| Fox, J. (1977) | Hydraulic analysis of unsteady flow in pipe networks. Macmillan Press, London. |
| Martin, C.S. and DeFazio, F.G. (1969) | Open channel surge simulation by digital computer. J. Hyd. Div., ASCE, 95, HY6, 1969 |
| Price, R.K. (1974) | Comparison of four numerical methods for flood routing. Journal ASCE. |
| Swaffield, J.A., Bridge, S. and Galowin, L.S. (1981) | Wave attenuation in long drainage pipes, a numerical solution of the unsteady partially filled pipe flow equations. CIB W62 Conf. Technische Fachhochschule, Berlin |
| Terzidis, G. and (1970) Strelkoff, T. | Computation of open-channel surges and shocks. J. Hyd. Div., ASCE, 96, HY12, 1970. |
| Wylie E.B. and (1978) Streeter, V.L. | Fluid Transients, McGraw Hill. |


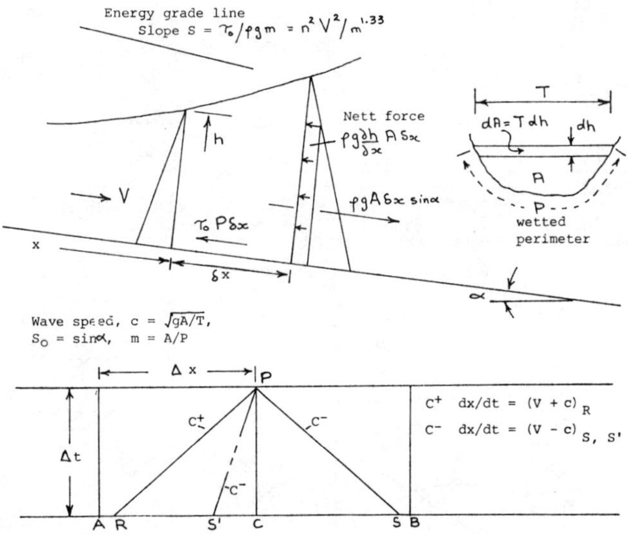


Figure 1: Summary of method of characteristics solution of the unsteady partially filled pipe flow equations

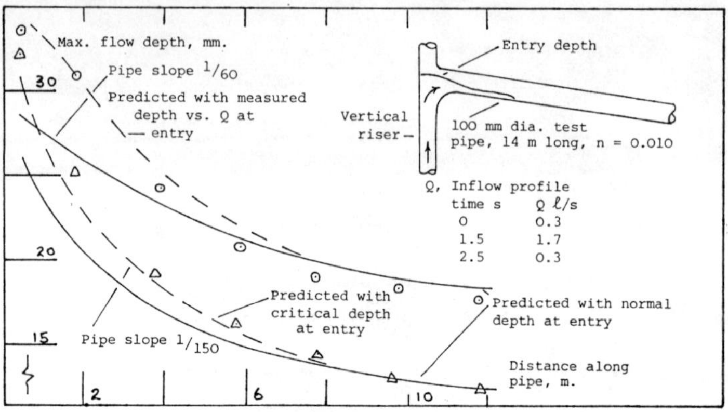


Figure 2: Comparison of observed and predicted attenuation of a pressure vessel generated wave, various entry boundary conditions

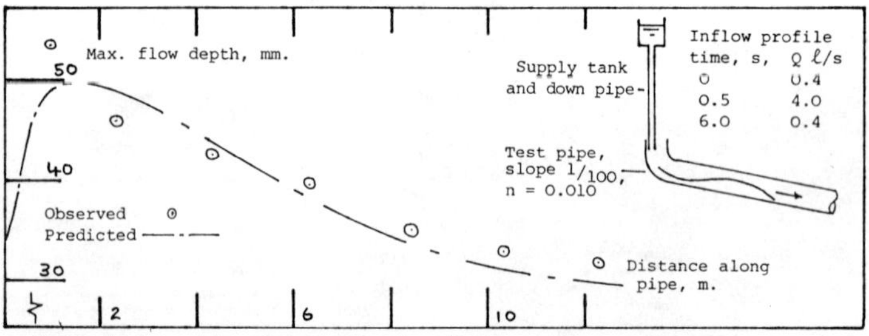


Figure 3: Comparison of observed and predicted attenuation of cistern generated wave, entry conditions based on inflow kinetic energy

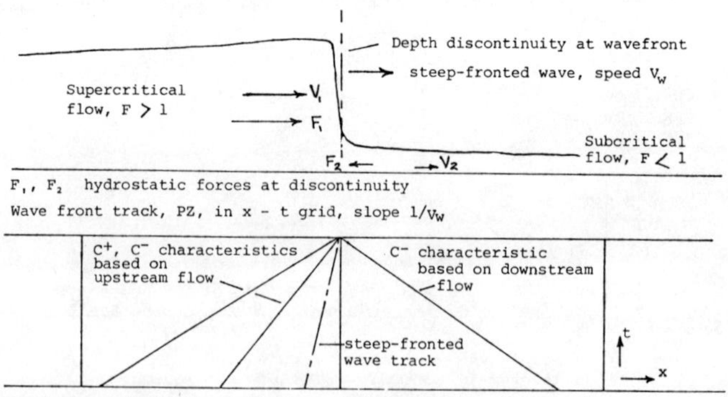


Figure 4: Development of the steep-fronted wave boundary equations and their representation in the x - t grid

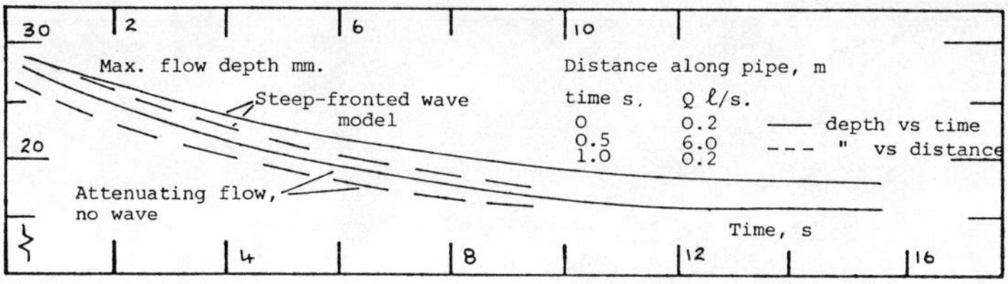


Figure 5: Comparison of theoretical flow attenuation with and without the inclusion of steep-fronted wave propagation.

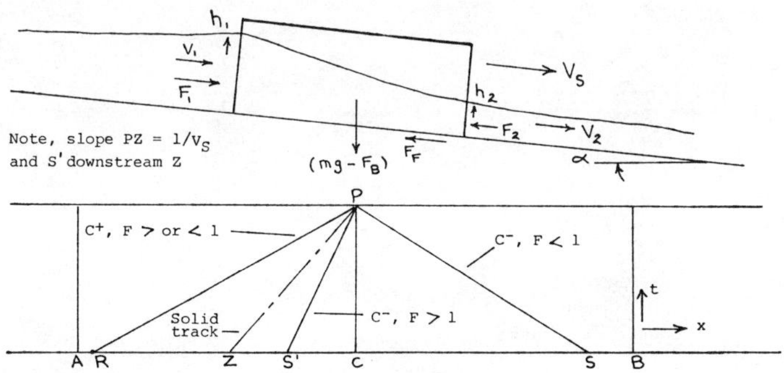


Figure 6: Development of the moving solid boundary equations

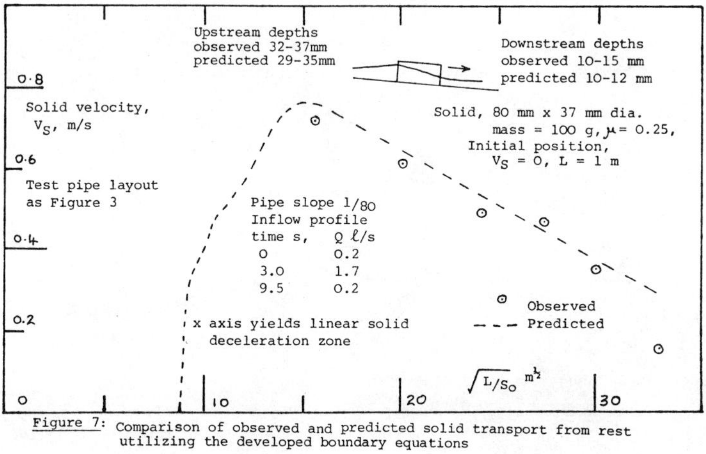


Figure 7: Comparison of observed and predicted solid transport from rest utilizing the developed boundary equations

# COMPUTER SIMULATION OF SEDIMENTATION PROCESSES

S. Y. Wang

The University of Mississippi, University, Mississippi, USA

## INTRODUCTION

This paper is to report a newly developed project to study the sedimentation processes in the Mississippi Sound, in which the quantity of sediment movement is one of the world's largest. The marine environment, water resources, pollution control, and industrial development in the region are greatly affected by these sedimentation processes. Similar phenomena have occured all over the world during the history of the earth. Tremendous efforts have been made by scientists of many different countries for many decades. A very brief survey on historical development of both analytical and experimental studies on sedimentation phenomena is to be presented first, followed by a more detailed review on computer simulation methodology. And, finally, the current state of the art as well as the finite element model developed are outlined.

## SURVEY ON SEDIMENTATION STUDIES

The dynamics of the detailed sedimentation processes has not yet established satisfactorily. This is due to the fact that not only the interaction between the turbulent flows in which sediment particles are transported and particles of bed material is extremely difficult to describe; but the distribution of size, shape, and specific weight of particles as well as the characteristics of turbulent flow are of stochastic nature. Difficulties encountered in analytical study of sedimentation processes had prompted early investigators to obtain approximate understanding by direct field observation as well as controlled experimentation in the laboratory. Among these early contributions, DuBoy's (1879) tractive-force formula was introduced and, then, modified by Kramer (1932). Also developed are classical impact theory concerning the critical bottom velocity by Einstein (1902), Indri (1941), and many others. Based upon these theories, numerous bed load empirical formulae were developed including Chang (1952), Meyer-Peter and Muller

(1948), Colby (1964), Engelund-Hansen (1966), Inglis (1947), among others. The dimensional analysis had been also applied in attempting to develop universal bed load formulae in terms of non-dimensional parameters. Some typical contributors are Bar and Herbertson (1966, 1968), Garde and Albertson (1961), and Bogardi (1964).

Empirical relations to predict capacity of sediment transport in suspension had been established in parallel to the bed load studies in the past few decades. Rouse (1939), Vanoni and Brooks (1906, 1957), Frankl (1955), Bogardi (1949), Sanoyan (1968), and many others had made significant contributions in this area. In addition, total load (a combination of bed and suspended loads) formulae have also been developed. In many cases, when the suspended load is negligible, they are identical to those of the bed load (Vanoni). It is regrettable that the length limitation does not allow the inclusion of numerous other contributions of great significance into this short paper. Readers are suggested to refer to the excellent books of Raudkivi (1967), Rossinsky (1970), Graf (1971), Bogardi (1978), Chien (1956), and Vanoni (1975) for more extensive reviews.

Although contributions resulting from field observations and laboratory experiments have provided the scientists with physical insight into some basic characteristics of sedimentation processes as well as the engineers with tools of prediction needed in projects of hydraulic engineering; one should be aware of the fact that each empirical formula can only be applied to a rather special situation, and that the development of this formula by empirical means is both costly and time-consuming. Those latter drawbacks are especially crucial when the hydraulic systems investigated become more and more complicated. Therefore, scientists in the field are continuously developing more and more cost-effective research methodology to advance the understanding of the basic characteristics of sediment transport phenomena. With the spectacular advent of digital computers and numerical methods, the numerical modeling method has become popular and promising.

## NUMERICAL MODELING

Early mathematical modeling were due to Helmholtz's (1888) original finding and Baschin's (1899) proof that a wave formed on a sand bed is similar to that on the interface of two fluids with different density. The simplified inviscid and potential flow models were solved by Exner (1925), and Kennedy (1961) et al. for flows in alluvial channels. Mathematical models of sediment routing and morphological transients were also developed by Cunge (1973), Mahmood (1974), Simons (1975), Lopez (1978), and others. In most of these analytical and numerical studies, empirical relations between the hydrodynamic properties and sediment transport capacity of the flow were used to complete the mathematical models. It not only was, but also is a necessity, because there is no pure analytical model in existence, which can adequately describe various kinds of sediment movement in a flow at the present. This fact is the reason that the contribution of empirical investigations has been briefly discussed in this paper.

Numerous studies on suspended sediment movement in streams or coastal waters had been also reported recently by Smith/O'Connor (1977), Cunge (1973), and Kerssens (1979), just to cite a few. One- and two-dimensional diffusion or convection-diffusion equation were solved by finite difference schemes in these and other studies. Most two-dimensional cases are based on the vertically integrated equation, also referred to as the depth-averaging scheme, which cannot be applied to problems having high variation of flow properties over depth. But, nevertheless, it reduces the computer storage and computing time a great deal. Therefore, this type of model is still very popular. One of the difficulties associated with the finite difference method is the inconvenience in handling the irregular boundary geometry. Unfortunately, the environment in which the sediment is transported is usually highly irregular. In order to develop more realistic model, a coordinate transformation, either analytical or numerical, has been used to reduce the complicated geometry to a simpler configuration, which can be handled by the finite difference schemes. The numerical transformation scheme of Thompson (1973) developed for the aerodynamics has found its application in hydraulics recently. The other alternative to alleviate the irregular geometry problem is the application of the Finite Element Method.

## FINITE ELEMENT MODELING

Finite Element Method has been applied to model structural and solid mechanics problems with great success. Its application in fluid flow simulation without sediment transport has been reported only in the last two decades. Although adequate modeling of hydrodynamic phenomenon of fluid flow is a prerequisit for modeling sediment transport in the flow, the limited length of this paper does not allow a thorough review on the historical development as well as the state-of-the-art of the finite element modeling of various flows in natural environment such as rivers, lakes, coastal water, etc. A brief survey on modeling the free surface flows by the finite element method is given by Wang (1978), Connor and Brebbia (1976), Pinder and Gray (1977), et al, however. Therefore, the brief discussion on finite element modeling of sediment transport in flows is presented below.

One-dimensional diffusion-convection equation was solved by Price (1968), by several techniques including a variational approach of Galerkin type, which introduced an alternative to the conventional analytical, finite-differencing, and characteristic approaches. Cavendish (1969) reported some solutions to two-dimensional steady diffusion problems, and Culham (1971) applied the same technique to the nonlinear, unsteady problem of one spatial dimension. Shortly after Price's contribution, one- and two-dimensional diffusion-convection problems were solved by Guymon (1970) using finite elements. And then, Smith, Farraday, and O'Connor (1973) discussed the limitations of the previously developed variational approaches based on exponential type of transformation, such as the one used by Guymon and adopted a more general finite elements based on either Rayleigh-Ritz or Galerkin approach.

Meanwhile, Ariathurai and Krone (1976) reported a finite element model for cohesive sediment transport in estuaries based on the Method of Weighted Residual. Approximate solutions of the convection-diffusion equation in two horizontal dimensions were in reasonable agreement with data collected from flume tests. It should be noted that empirical relations between sediment particle movement and fluid properties had also been used in their model. Another two-dimensional model in a horizontal plane was developed by Leimkuhler, et al, using the depth-averaging scheme. It further demonstrated the versatility of the finite element method in expression of irregular geometry, complicated boundary conditions, and inhomogeneous material properties. Their model had been verified using data from the Massachusetts Bay. One should also note the fact that empirical information is very important in determining sediment transport coefficients, such as dispersion coefficients, decay factors, etc. These quantities are essential to the accuracy of the mathematical model. Most of the contributions discussed above are dealing primarily with the suspended sediments.

More recently, Alonso, Wang, and Su (1980, 1982) had been successful in simulating errosion or scouring and deposition or back-filling of bed materials in flows in alluvial channels with or without obstructions. Both suspended load and bed load have been considered. Results have been verified with laboratory testing data provided by the Delft Hydraylics Laboratory of The Netherlands.

## SEDIMENT TRANSPORT MODEL

A complete numerical simulation model to represent the sediment transport in a flow over loose boundary requires not only the accurate solution of hydrodynamic characteristics of the turbulent flow in which sediments are transported, but also the development of adequate mathematical theory governing the dynamic interactions between solid sediment particles and the flow. Since the scientific accomplishments in these two areas just mentioned are far from satisfactory, the numerical modeling methodology needs further improvement. The model used in this study is briefly described below.

The turbulent flow in "instantaneously frozen" boundaries is governed by the continuity and momentum equations. For computing efficiency, the vertically integrated equations are used. They are:

$$\frac{\partial(hU)}{\partial t} + \frac{\partial(hU^2)}{\partial x} + \frac{\partial(hUV)}{\partial y} + gh\frac{\partial\eta}{\partial x} - \frac{1}{\rho}\frac{\partial(hT_{xx})}{\partial x} + \frac{1}{\rho}\frac{\partial(hT_{xy})}{\partial y}$$

$$-\frac{1}{\rho}(\tau_x^s - \tau_x^b) = 0 \qquad (1)$$

$$\frac{\partial(hV)}{\partial t} + \frac{\partial(hUV)}{\partial x} + \frac{\partial(hV^2)}{\partial y} + gh\frac{\partial\eta}{\partial y} - \frac{1}{\rho}\frac{\partial(hT_{yx})}{\partial x} + \frac{1}{\rho}\frac{\partial(hT_{yy})}{\partial y}$$

$$-\frac{1}{\rho}(\tau_y^s - \tau_y^b) = 0 \qquad (2)$$

$$p = p_o + \rho g(\eta - z) \qquad (3)$$

$$\frac{\partial h}{\partial t} + \frac{\partial(hU)}{\partial x} + \frac{\partial(hV)}{\partial y} = 0 \qquad (4)$$

where p and $\rho$ are pressure and density of water respectively, h is hydraulic depth, $\eta$ is bed elevation, g is gravitational acceleration, U and V are vertically averaged velocity components, $T_{ij}$ are effective shear stresses on vertical planes, $\tau_i^s$ and $\tau_i^b$ are shear stresses on free surface and bed surface respectively.

One of the turbulence closure schemes should be chosen to complete the hydrodynamic model. For example, if one wants to adopt the simplest scheme, the eddy-viscosity approximation, the effective shear stress terms in equation (2) and (3) should be replaced by:

$$-\varepsilon h\left(\frac{\partial^2 U}{\partial x^2} + \frac{\partial^2 U}{\partial y^2}\right) \quad \text{and,} \quad -\varepsilon h\left(\frac{\partial^2 V}{\partial x^2} + \frac{\partial^2 V}{\partial y^2}\right) \qquad (5)$$

where $\varepsilon$ is the so called effective eddy viscosity.

In addition, the shear stresses on the free surface are usually neglected, unless otherwise specified, and the bed shear stresses are related to the roughness of the bed materials as well as flow properties by empirical means. One of the popular selections is given below:

$$(\tau_x^b \; ; \; \tau_y^b) = \rho f (U^2 + V^2)^{1/2} \, (U \; ; \; V) \tag{6}$$

where:

$$f = gC^{-2} \quad \text{or} \quad f = gn^2 \, (1.49)^{-2} h^{-1/3} \tag{7}$$

in which C and n are Chezy and Manning's coefficients respectively.

As discussed in the previous sections of this paper, there are several different ways to model the sediment transport in turbulent flows including: (i) to consider all sediments as suspended particles, (ii) to consider the total load as a combination of suspended load and bed load, and (iii) to use the empirical total load formula similar to that of the bed load. After having tried all three different approaches, it has been experienced that the third approach is by far the simplest. Therefore, the third approach is outlined here. One may refer to Alonso and Wang (1980) for other alternatives.

One of the total load formulae may be selected to predict the sediment transport capacities is of the form:

$$q_i = c_g \rho V_i^m \tag{8}$$

where $q_i$ is the bed material transport function in the direction of depth-averaged velocity component $V_i$, $c_g$ and m are the coefficient and exponent respectively. For a particular hydraulic system, bed geometry, and material characteristics, $c_g$ and m are estimated using experimental information. Then, the bed material continuity equation below

$$q_{s,x} + q_{s,y} + (1 - \lambda) z_{b,t} = 0 \tag{9}$$

can be used to calculate a new bed elevation $z_b$. The mathematical model has been completely developed.

This set of differential equations is solved by the finite element method for several basic cases. Results of two typical cases are shown below, with Figure 1 presenting the local scour and deposition around a spur dike in a channel with sand bed and Figure 2 presenting similar simulated bed deformations with time at a transition section of a channel having a sand bed.

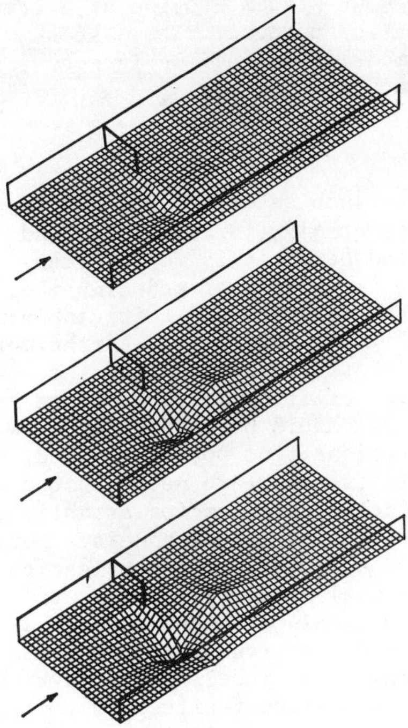

Fig. 1. Local Scour and Deposition near a Spur Dike

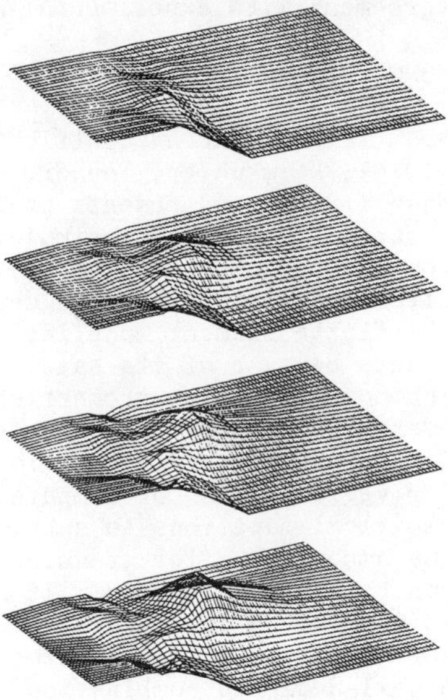

Fig. 2. Local Scour and Deposition at a Transition Section of Alluvial Channel

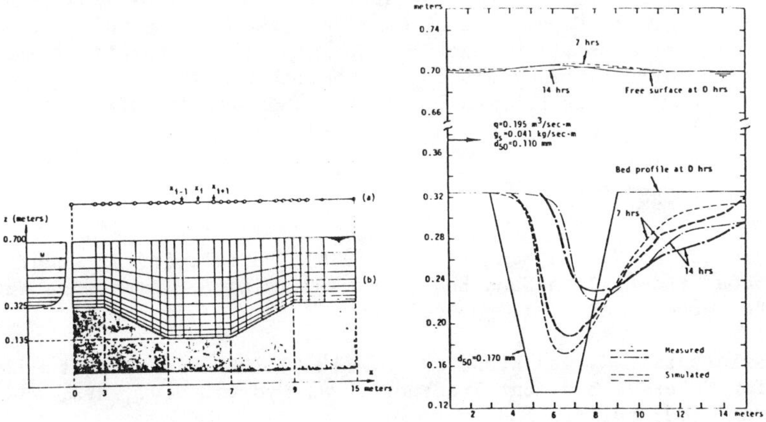


Fig. 3. F.E. Simulation of Back-filling of a Trench (Alonso & Wang, 1980)

A similar model has been applied to a vertically two-dimensional problem of back-filling and down-stream erosion of a trench built on a sand bed for verification purpose. It is seen clearly that the computational model developed is in good agreement with experimental data.

## CONCLUDING REMARKS

From several recent contributions such as Ariathurai and Krone (1976), Keimkuhler, Connor, Wang, et al. (1975), Alonso and Wang (1980), and resents presented here, it is fully demonstrated that computer simulation is capable of modeling the complicated phenomena of sediment transport and this viable alternative saves a great deal of money and time. Furthermore, the finite element modeling is a better approach than the others because of its saitiability of irregular boundaries and inhomogeous median properties. It should be emphasized, however, that experimental investigations from both field and laboratory are still very important. They are needed not only for verifications of computer models, but also for establishing empirical functions to supplement the numerical models. But, the amount of effort required for experimental investigations can be greatly reduced; if numerical modeling is also utilized to simulate a large variety of physical phenomena through a series of parametric studies of sediment transport problems. A well-designed combination of numerical modeling and physical experimentation is going to be the most cost-effective tool in scientific research.

## ACKNOWLEDGEMENTS

The work reported herein is partially supported by NOAA Office of Sea Grant, Department of Commerce, the Mississippi-Alabama Sea Grant Consortium, the USDA Sedimentation Laboratory, and The University of Mississippi. Contributions of Dr. C.V. Alonso of Sedimentation Lab and Dr. T. Y. Su of Jackson State University to this paper are also acknowledged.

## REFERENCES

Alonso, C. V. and Wang, S. Y. (1980), "Simulation of Local Scour and Fill in Sand Bed Streams," Finite Elements in Water Resources, Vol. III, pp. 6.57-68.

Ariathurai, R. and Krone, R. B. (1976), "Finite Element Model for Cohesive Sediment Transport," J. Hydrau. Div., HY3, ASCE, Vol. 102, pp. 323-338.

Barr, D. I. H. and J. G. Herbertson (1968), "Similitude Theory Applied to Correlation of Flume Sediment Transport Data," Water Resources Research, Vol. 4, No. 2, pp. 307-316.

Baschin, O. (1899), "Die Entstehung Wellenahnlicher Oberflachenformen," 2. Gesellschft, Erdkunde, Bd. XXXIV.

Bogardi, J. (1964), "Dimensional Analysis and Similarity," Lectures Delivered in Utah State University C.E., 260.

Bogardi, J. (1978), Sediment Transport in Alluvial Streams, Akademiai Kiado, Budapest.

Bogardi, J. (1949), "The Movement of Suspended Sediment Between Zahony and Razompuszta in the Tisza River," Vizugyi Kozlemenyek No. 3-4.

Cavendish, J. C., H. S. Price, and R. S. Varga (1969), "Galerkin Methods for the Numerical Solution of Boundary Value Problems," Soc. Petrol. Eng. J., 16, 204-220.

Chang, Y. L. (1947), "Laboratory Investigation of Flume Traction and Transportation," j. ASCE Hydrau. Div. Vol. 76, HY11.

Chien, N. (1956), "The Present Status of Research on Sediment Transport," Transactions, ASCE, Volume 121, pp. 833-868.

Colby, B. R. (1964), "Discharge of Sands and Mean Velocity and Relationships in Sands-Bed Streams," Paper 462-A, U.S. Geological Survey, Washington, D.C.

Connor, J. J. and Brebbia, C. A. (1976), Finite Element Techniques for Fluid Flow, Newnes-Butterworth, London.

Culham, W. E., and R. S. Varga (1971), "Numerical Methods for Time Dependent Nonlinear Boundary Value Problems," Soc. Petrol. Eng. J., 248(4), 374-388.

Cunge, J. A. and Perdreau, N. (1973), "Mobile Bed Fluvial Mathematical Models," La Houille Blanche, No. 7.

DuBoys, P. (1879), "Le Rhone et les Rivieres a Lit Affouillable," Annales des Ponts et Chaussees.

Einstein, H. A. (1943), "Flow on Movable Bed," Proc. Second Hydraulic Conf., Univ. of Iowa, Bulletin 27.

Engelund, F and Hansen, E. (1966), "Investigations of Flow in Alluvial Streams," Acta Polytech. Scand., Ci-35.

Exner, F. M. (1925), "Uber die Weschselwirkung Zwischen Wassar und Geshiebe in Flussen," Sitzber. Akad. Wiss. Wien, pt. 11a, Bd. 134.

Frankl, F. I. (1955), "Energy Equations of Suspended-Sediment Transportation," DAN SSR, Volume 192.

Garde, R. J. and M. L. Albertson (1961), "Bed-Load Transport in Alluvial Channels," La Huille Blanche, N. 3, Mai-June.

Graf, W. H., Hydraulics of Sediment Transport, McGraw-Hill, New York.

Guymon, G. L. (1970), "A Finite Element Solution of the One-Dimensional Diffusion-Convection Equation," Water Resour. Res., 6(1), 204-210.

Guymon, G. L., V. H. Scott, and L. R. Herrmann (1970), "A General Numerical Solution of the Two-Dimensional Diffusion-Convection Equation by the Finite Element Method," Water Resour. Res., 6(6), 1611-1617.

Helmholtz, H. (1888), "Uber Atmosphaerische Bewegungen," Sitzber, Konigl. Preuss, Akad. Wissl, Berlis.

Indri, E. (1941), "Problem del Transporto Solido ed i Risultati di Recenti Ricerche," Societa Editrice Riviste Industrie Elettriche.

Inglis, C. C. (1947), "Meanders and Their Bearing on River Training," Inst. of Civil Engineers, Maritime and Waterways Engineers Division, London, England, pp. 3-54.

Kennedy, J. F. (1961), "Stationary Waves and Anti-Dunes in Alluvial Channels" Report No. KH-R-2, W. M. Keck Laboratory, Cal. Tech.

Kerssens, P. M. J. and Van Rijn, L. C. (1979), "Model for Suspended Sediment Transport," J. Hydrau. Div., ASCE, V. 105, 401-776.

Kramer, H. (1932), Modellgeschiebe und Schleppkraft, Berlin.

Leimkuhler, W., Connor, J., Wang, J., et. al., (1975), "Two-Dimensional Finite Element Dispersion Model," Proc. of Symp. on Modeling Tech. Vol II.

Lopez, S. J. L. (1978), "Mathematical Modeling of Sediment Deposition in Reservoirs," Hydrology Papers No. 95 CSU, Ft. Collins, Colorado.

Mahmood, K. (1974), "Mathematical Modeling of Morphological Transients in Sandbed Canals," J. of IAHR, Vol. 2.

Meyer-Peter, E. and R. Muller (1948), "Formulas for Bed-Load Transport," Report on Second Meeting of IAHR, Stockholm, Sweden, pp. 39-64.

O'Connor, B. A. (1971), "Mathematical Model for Sediment Distribution," Proc. 14th Congr. Int. Ass. Hydraul. Res., 4.

Pinder, G. F. and Gray, W. G. (1977), Finite Element Simulation in Surface and Subsurface Hydrology, Academic Press, NY.

Price, H. S., J. C. Cavendish, and R. S. Varga (1968), "Numerical Methods of Higher-Order Accuracy for Diffusion-Convection Equations," Soc. Petrol. Eng. J., 243(3), 293-303.

Raudkivi, A. J. (1967), Loose Boundary Hydraulics, Pergamon Press, Oxford.

Rossinsky, K. I. (1970), Sediment Transport in Open Channels, Institut Vodnykh Problem, Izd, "NAUKA," Moscow.

Rouse, H. (1939), "Experiments on the Mechanics of Sediment Suspension," Proc. Fifth Int. Congress of Applied Mechanics.

Sanoyan, V. G. (1968), "Theoretical Determination of the Sediment Transporting Capacity of Watercourses ...," Symposium on Current Problems in River Training and Sediment Movement, 3, Budapest.

Simons, D. B., Richardson, E. V., and Mahmood, K. (1975), "One-Dimensional Modeling in Alluvial River Flow," Proceedings on Unsteady Flow in Open Channel, WRP.

Smith, I. M., Farraday, R. V., and O'Connor, B. A. (1973), "Raleigh-Ritz and Galerkin Finite Elements for Diffusion-Convection Problems," Water Resou. Res. Vol. 9, No. 3, pp. 593-606.

Smith, T. J. and O'Connor, B. A. (1977), "A Two-Dimensional Model for Suspended Sediment Transport," IAHR Proc. 7th Cong. Vol. 1, pp. 79-87.

Thompson, J. F., Jr., Shanks, S. P., and Wu, J. C. (1973), "Numerical Solutions of 3-D Navier-Stokes Equations in Integro-Differential Form: Flow about a Finite Body," Proc. AIAA Comp. Fluid Dynamics Conf.

Vanoni, V. A. and N. H. Brooks (1957), "Laboratory Studies of the Roughness and Suspended Load of Alluvial Streams," M.R.D. Sediment Series No. 11.

Vanoni, V. A. (1975), editor, Sedimentation Engineering, ASCE-Manuals and Reports on Engineering Practice No. 54.

Wang, S. Y. and Alonso, C. V. (1978), "Numerical Modeling of Three Dimensional Viscous Flows in an Open Channel," Applied Numerical Modeling, Vol. 2, IV-2, Pentech Press, England.

Wang, S. Y. (1982), "Computational Simulation of Sedimentation Processes," to be published in the Proceedings of the Int'l. Conf. on Computational Methods and Experimental Measurements,

Washington, D.C.

Wang, S. Y. (1982), "Finite Element Modeling of Sediment Transport, The State of the Art," to be published in the Proceedings of 4th Int'l. Symposium on Finite Element Methods in Flow Problems, July, 1982.

# 17. FLOW INDUCED FORCES IN STRUCTURES

# A NEW HYDRODYNAMIC MODEL FOR RESTRICTED FLUID DOMAINS.

Arcilla, A.S.

Visiting Professor, Politech. Univ. of Barcelona

## 1. INTRODUCTION

The problem to be solved is the prediction of the motion of a floating rigid-body in a restricted fluid domain. It encompasses a wide variety of practical cases: entrance manoeuvres of ships in harbours, tow-out of platforms, etc. To cover all these cases the following requirements will be imposed to the mathematical model:

- Arbitrary geometry of rigid-body and fluid domain.
- Motion of the body composed of two parts: average accelerated motion and zero-mean oscillatory motion.

The last requirement means that standard manoeuvres (starting from rest or with a curved trajectory) under wave action can be considered. This departs from usual models in which average accelerated motions are considered only with simple fluid domain geometries (FDG's), while complex geometries are only considered for bodies whose average position is stationary, using a 3D sink-source technique or similar means.

The aim of this paper is to formulate a model with the above requirements. It will be organised as follows:

- Brief review of currently used mathematical models.
- The proposed formulation for the problem.
- Conclusions.

## 2. REVIEW OF THE MOST FREQUENT APPROACHES TO THE PROBLEM.

Models to predict the motion of a rigid body are based on Newton's second law, the problem being the evaluation of the forces acting on the body. There are two basic approaches to obtain these forces: the first is the well-known method of the

hydrodynamic coefficients, in which forces are expressed as a combination of variables significant to the problem, affected by some coefficients called "hydrodynamic". The method is described in (Norrbin, 1971) and (Abkowitz, 1964).
The second approach is the construction of a mathematical model starting from a set of simplifying hypothesis. Models can be constructed using as basic independent variable the frequency w or the time t. Models in the frequency domain are used in connection with spectral descriptions of a sea-state, therefore providing valuable probabilistic information in a simple way (e.g. Price et al., 1974). Furthermore, they obviate the integration of the equations of motion (Cummins, 1962). As a drawback, these models are only valid for linear systems. On the other hand, models in the time domain, though using a linear composition of responses, can accept any exciting force and are, therefore, more suitable for non-linear problems (Van Oortmerssen, 1976). They are based on the impulse response function technique (see (Price, 1974) and (Ogilvie, 1964),e.g.)
The problem we are trying to solve is non-linear and only under very particular conditions a linear formulation is acceptable. This suggests the use of a model in the time domain. Furthermore, the requirements imposed at the beginning imply that the FDG encountered by the body varies with time (e.g. imagine a ship carrying out an entrance manoeuvre). Therefore an approach based on hydrodynamic coefficients would require a continuous evaluation of these coefficients. This could be performed by dividing the total period of time ellapsed in the manoeuvre into time steps; each time step is associated to a constant FDG. A set of coefficients would, then, be evaluated for each time step. This means an overwhelming number of model tests for many of these coefficients do not have a clear physical meaning, particularly in restricted fluid domains (Tuck, 1978). Because of these reasons a mathematical model, starting from basic hypothesis, is the approach selected in this paper.
Let us now focus our attention on the hydrodynamic forces acting on the body. They have two components: "viscous" forces, associated to the viscous (rotational) motion of the fluid, and "potential" forces, associated to the irrotational motion of the fluid. A detailed model will be constructed in this paper only for the latter. It is recognised the need to develop a similar model in which due consideration is paid to viscous forces, but here they will be evaluated with a standard velocity squared law (e.g. Remery et al., 1973)
Models to evaluate potential forces with arbitrary FDG have been developed, as far as the author knows, only for very specific types of motion of the body, the most frequent being

that of bodies whose average position remains stationary (Faltinsen et al., 1974), (Van Oortmerssen, 1976). The objective of this paper is to develop a formulation able to cope with the arbitrary FDG and average accelerated motion of the body (superposed to a zero-mean oscillatory motion). Some of the difficulties to be encountered are:

i) Formulating the conservation laws for the fluid motion in an accelerated, body-bound, system of axes. Alternatively, using a fixed system of axes in which the immersed surface of the body varies at each time step.

ii) Composing the incident wave trains to the waves generated by the motions of the body.

iii) Establishing the boundary condition at the immersed surface of the body on its exact position, rather than on its mean position as is usual. This is due to the existance of a non-stationary mean immersed surface (Timman and Newman, 1962).

## 3. THE PROPOSED FORMULATION OF THE PROBLEM

The model will simulate the motion of a rigid body in a restricted fluid domain. It will be based on Newton's law of dynamics, which, in a moving system of axes, may be written as (Abkowitz, 1964)

$$F_k = L_{kj}\,\ddot{x}_j + N_{kjl}\,\dot{x}_j\,\dot{x}_l \qquad k=1-6,\ j=1-6,\ l=1-6. \tag{1}$$

in which Fk: total force (k=1,2,3) and moment (k=4,5,6) components acting on the body.

$\ddot{x}_j, \dot{x}_j, x_j$ : linear (k=1,2,3) and angular (k=4,5,6) components of acceleration, velocity and displacement respectively.

Lkj, Nkjl: matrices of coefficients depending on the mass distribution of the body.

The total force on the body, Fk, has two distinct origins: part of it, Ek, is due to the "external disturbances" (wind, waves, current, tug boats, etc.) The rest comes from the hydromechanic reaction on the body, HMk. To calculate Fk the following simplifying hypothesis will be introduced: The exciting force, Ek, will be evaluated considering only the average motion of the body. The hydromechanic reaction will be evaluated considering the actual motion but supposing the only disturbance in the fluid domain is that due to the motion of the body. Fk is then obtained composing linearly Ek and HMk:

$$Fk(t) = Ek(t) + HMk(t) \tag{2}$$

A similar hypothesis is often used in offshore engineering to calculate the wave action on a structure (Faltinsen, 1979).

In this paper only HMk(t) will be evaluated. However, it will be done for an arbitrary motion of the body, resulting from an

arbitrary exciting force Ek(t).
HMk(t) is composed of a hydrostatic part,HSk, and a hydrodynamic part, HDk. Only the latter will be considered in detail, the former presenting no problem (it may be found, e.g. in Newman, 1977).
HDk(t) will be evaluated assuming the fluid is incompressible. Therefore the only possible motions of the fluid particles are pure translation and pure rotation (Lighthill, 1979). The former may be described in terms of a potential function; the associated hydrodynamic force on the body will be the potential reaction Pk(t). The latter, being associated to the viscous properties of the fluid, will make up the viscous reaction,Vk(t).
. The emphasis, as it has been said, will be on Pk(t).
The potential solution would be completely determined at each instant by the mass conservation law and appropiate boundary conditions, if a fixed boundary existed. However, the existance of a free-surface introduces a memory effect because wave trains generated by the motion of the body take some time to fade away. Furthermore, as the wave trains and the body move at different velocities, a diffraction/reflection problem appears at each time instant. This difficulty, together with the ones mentioned in the previous section, disappear if the following fictitious problem is solved instead of the real one:
The body is fixed and the fluid domain moves with ($-\ddot{x}_j$), j= 1-6. This situation (System 2 or S2) is obtained from the real one (System 1 or S1) by applying to every element of the system the additional accelerations ($-\ddot{x}_j$). In S2, the body-bound axes may be used conveniently for they are now fixed in space. To use this approach it is necessary, however, to evaluate the "corrector" stresses on the body, arising as a consequence of the additional accelerations ($-\ddot{x}_j$). Therefore:

$$F_{S2} = F_{S1} + \text{Corrector Stresses.} \tag{3}$$

in which:

$F_{S2}, F_{S1}$ : potential stresses in S2 and S1 respectively.

Let us now evaluate the corrector stresses associated to the additional accelerations of translation ($-\ddot{x}_j$), j=1,2,3.
These accelerations (in the fluid domain) must be generated by a force field per unit volume: $-\rho\,\ddot{x}_j$ , $j=1,2,3$
This force field must be generated by a pressure gradient:

$$-(-\rho\ddot{x}_j)=\rho\,\ddot{x}_j \quad \text{in the j - direction j = 1,2,3} \tag{4}$$

The resulting pressure field is:

$$\rho\,\ddot{x}_j\,x_j \quad \text{for a fixed j=1,2,3} \tag{5}$$

Therefore, the generalised force (corrector stress) in mode k(k=1-6) due to acceleration in mode j(j=1-3), CSkj, is:

$$CS_{kj}=-\iint_S \rho\,\ddot{x}_j x_j n_k\,dS=-\rho\ddot{x}_j\iint_S x_j n_k\,dS=-\rho\ddot{x}_j\,a_{jk} \tag{6}$$

in which:
S:immersed surface of the body, varying with time.
$x_j$ : j -coordinate of immersed surface points.
$n_k$:generalised components (k=1-6) of a unit vector, normal to S, positive towards the fluid.
Observe that $\ddot{x}_j$ is constant in the fluid domain and $a_{jk}$ is a function only of the immersed surface geometry.
The CSkj so evaluated are associated to a purely translational fluid motion. Therefore they only affect (correct) the potential reaction Pk(t). Furthermore, they are easy to obtain because the fluid motion is irrotational and, the subsequent hydrodynamic pressure, normal to S.
This is not the case for the corrector stresses due to the additional accelerations of rotation ($-\ddot{x}_j$), j=4,5,6. Now the hydrodynamic pressure will have two components, normal and tangential to S respectively. Neglecting the latter, the corrector stresses may be obtained following the procedure used before. If the accelerations are split up into radial and tangential components, the following corrector stresses for the viscous reaction are obtained (S.-Arcilla, 1980):

$$CS_{kj} = -\rho \ddot{x}_j b_{jk} - \rho (\dot{x}_j)^2 c_{jk} \tag{7}$$

in which:
$b_{jk} = \iint_S r^2 \theta n_k \, dS$ , is the coefficient for the tangential accelerations.
$c_{jk} = \iint_S r^2 n_k \, dS$ , ditto for the radial accelerations
$(r, \theta)$ :cylindrical coordinates of considered point with respect to the axis of rotation.
Now, the potential reaction in S2 will be evaluated. The actual reaction will be subsequently obtained with equation (3).
To calculate Pk(t) the impulse response function technique will be used. This means that Pk(t) will be calculated in the time domain by dividing the total time period into time steps and evaluating $Pk^i(t)$ in each time step $I_i$(i denotes time step).
Due to the free-surface effect there is an instantaneous potential reaction, $PI_k^i$, and a differed potential reaction, $PD_k^i$, (Cummins, 1962). They are obtained from the corresponding potential functions: $\Psi^i$, $\chi^i$.
Time steps are selected so that each of them is associated to a constant FDG. Therefore, selection depends on the average motion; the zero-mean oscillatory motion may be neglected for this purpose.
To evaluate $\Psi^i$ and $\chi^i$ the motion of the body is decomposed into the six possible modes $X_j$, j=1-6 (3 translations and 3 rotations). The motion in each mode may be obtained as a sequence of impul-

sive motions. Therefore, in each time step the body performs an impulsive motion in each mode:

$$\dot{x}_j(t) = \text{constant} \qquad t \in I_i,\ j = 1-6 \tag{8}$$

Hence, the only corrector stresses between S2 and S1, in Ii, arise from the radial accelerations. They only affect the viscous reaction. Thus, it is equivalent to calculate $Pk^i$ in S1 or S2. It will be done in S1 in which only the immersed surface S moves (in S2 it is the rest of the boundary that moves, complicating the evaluation of $\Psi^i$ and $\chi^i$).

The instantaneous potential $\Psi^i$ is obtained as follows:

$$\Psi^i = \Psi^i_j \ \dot{x}_j \qquad j = 1-6 \tag{9}$$

in which: $\Psi^i_j$ is the potential function for a displacement with unit velocity in mode j in Ii. It varies with time through the superscript i, but, in each time step, since the FDG is constant, $\Psi^i_j$ is a function: $\Psi^i_j(x_1, x_2, x_3)$.

Similarly, $\chi^i$ is obtained as:

$$\chi^i = \chi^i_j \ \dot{x}_j \ \Delta\tau_i \qquad \text{(i is fixed)} \tag{10}$$

in which: $\chi^i_j$ represents the time evolution of $\Psi^i_j$ (due to the free-surface effect) beginning at the end point of Ii, $\tau_i$.

$(\dot{x}_j \Delta\tau_i)$ : is the displacement in mode j in Ii.

Therefore $\chi^i_j$ are functions: $\chi^i_j(x_1, x_2, x_3; (t-\tau_i))$.

$\Psi^i_j$ is obtained by distributing fundamental sources over all the boundaries: S(immersed surface),B(bottom), F(free-surface), C(control surface, sufficiently far away from the body).

Fundamental sources are elementary solutions of Laplace's equation and may be written as: $m/4\pi r$

in which: r is the distance from the source to the considered point and m is the source (or sink) intensity.

The boundary is discretised into plane elements of 3 or 4 sides. Let k denote the element number in Ii: k=1-Ni. The source intensities will be assumed constant in each element: $m_k$ ,k=1-Ni. There are, therefore, Ni unknowns. Ni equations are obtained imposing the boundary conditions at the centre-points of each element.

$\Psi^i_j$ may then be written as:

$$\Psi^i_j = \sum_{k=1}^{N_i} \iint_{\Delta S_k} \frac{m^i_{jk}}{4\pi r}\, ds \tag{11}$$

in which: $\Delta$Sk: surface area of panel (element) k. $m^i_{jk}$:source intensity of panel k for the calculation of $\Psi^i_j$.

The boundary conditions for $\Psi^i_j$ are the usual ones for potential flow (Van Oortmerssen, 1976. e.g.) but for the condition on C:

$$S: \quad \frac{\partial \Psi^i_j}{\partial n} = n_j \tag{12}$$

$$B: \quad \frac{\partial \Psi^i_j}{\partial n} = 0 \tag{13}$$

$$F: \quad \Psi^i_j = 0 \tag{14}$$

$$C: \quad \Psi^i_j = 0 \tag{15}$$

n: unit vector, normal to S, with components nj.

j = 1-6

The condition on C for this special potential problem, in which $\psi_j^i$ describes the flow due to an impulsive displacement, may be obtained considering that the disturbance is limited in magnitude and duration. Under these conditions the energy flux $\frac{dE}{dt}$ through a control surface (assumed to be a cylinder of radius r, centered at the disturbance point) is (Stoker, 1957):

$$\frac{dE}{dt} = -\frac{1}{r}K \quad , \quad K \text{ being a bounded constant} \tag{16}$$

When C is far enough from the body (r is large), and considering the fluid was initially at rest, it is concluded that condition (15) is appropiate to the problem.

The calculation of $\chi_j^i$ follows a similar pattern. It is the first attempt, as far as the author knows, to evaluate the differed potential in the time domain from the boundary conditions.

$\chi_j^i$ is obtained by distributing fundamental sources over the same boundary as before. The same discretisation into plane elements will be assumed. The only difference is that, since $\chi_j^i$ is a function of time, the source intensities will also be a function of time. However, they will be spatially constant inside every panel as before. Thus:

$$\chi_j^i(x_1, x_2, x_3; [t-\tau_i]) = \sum_{k=1}^{N_i} \iint_{\Delta S_k} \frac{n_{jk}^i(t-\tau_i)}{4\pi r} dS \tag{17}$$

(The nomenclature is the same as before)

There are Ni unknown functions: $n_{jk}^i(t-\tau_i)$ . With boundary conditions imposed at the centre-point of every panel the problem may be solved as follows. Boundary conditions on S( $\frac{\partial \chi_j^i}{\partial n} = 0$ ), B( $\frac{\partial \chi_j^i}{\partial n} = 0$ ) and C( $\chi_j^i = 0$)provide a system of ($N_i - N_{Fi}$)linear equations:

$$n_{jl}^i(t-\tau_i) A_{lk} = n_{jm}^i(t-\tau_i) B_{mk} \quad , \quad j=1-6 \tag{18}$$

in which: Alk, Bmk are the appropiate matrices of coefficients.
$l=1-(N_i-N_{Fi})$ ; $k=1-(N_i-N_{Fi})$ ; $m=1--N_{Fi}$

$N_{Fi}$: number of panels in the free-surface in Ii.

Therefore, with these conditions (the standard conditions for potential flow but for the one on C) the source intensities on S,B,C are expressed as a function of the intensities on F. Hence, when these last ones are calculated the problem is solved. Let us now impose the boundary condition on F:

$$\left(\frac{\partial^2 \chi_j^i}{\partial t^2} + g\frac{\partial \chi_j^i}{\partial x_3} = 0\right) \quad \text{at the centre of every panel in F.}$$

This condition provides $N_{Fi}$ second order linear differential equations. $2N_{Fi}$ initial conditions are required. They are as follows:

-Initially the fluid is at rest: $\chi_j^i((t-\tau_i)=0) = 0$ (19)

Using equation (18) it is easy to show that it is enough to impose this condition to the $N_{Fi}$ unknowns in F. Like this, $N_{Fi}$ initial conditions are obtained:

$$n^i_{jm}((t-\tau_i)=0)=0 \qquad m=1-N_{Fi}\,,\ j=1-6 \tag{20}$$

-The free-surface elevations at $(t-\tau_i)=0$ obtained from $\Psi^i_j$ and $\chi^i_j$ must coincide. Therefore (Van Oortmerssen, 1976, e.g.):

$$\frac{\partial \chi^i_j(t-\tau_i)}{\partial t} = -g\,\frac{\partial \Psi^i_j}{\partial x_3} \qquad (t-\tau_i)=0\ ,\ \text{at F} \tag{21}$$

This, when imposed at the centre of every panel in F, provides $N_{Fi}$ initial conditions,whose structure is:

$$\dot{n}^i_{jm}((t-\tau_i)=0) = r^i_{jm} \qquad m=1-N_{Fi}\,,\ j=1-6 \tag{22}$$

in which: $r^i_{jm}$ is the appropiate vector of coefficients.
This defines the source intensities in F. Subsequently the sources in S,B,C may be calculated and the problem is solved. Observe that boundary conditions on F, for $\Psi^i_j$ and $\chi^i_j$ , are linearised. Therefore, only the linear potential is being calculated. The hydrodynamic pressure, however, will be obtained with the non-linear Bernoulli equation, integrated over the exact immersed surface in each time step (this is possible because the impulse-response-function technique has been applied to the potential reaction instead of to the potential function as usual). This means that two second order effects from the linear potential will be considered. The linear contribution from the quadratic potential could be introduced following the method proposed in (Lighthill, 1979).
The hydrodynamic pressure is given by Bernoulli's equation for unsteady irrotational flow:

$$p^i(t) = -\rho\left(\frac{\partial \phi^i}{\partial t} + \frac{1}{2}\,\frac{\partial \phi^i}{\partial x_j}\,\frac{\partial \phi^i}{\partial x_j}\right) \qquad j=1,2,3 \tag{23}$$

in which $\phi^i$ is $\Psi^i$ or $\chi^i$.
The potential reaction is then:

$$P^i_k(t) = -\iint_{S^i} p^i(t)\, n_k\, ds \qquad k=1-6 \tag{24}$$

in which: $S^i$ is the immersed surface in Ii

$n_k$ are the components of a unit vector normal to $S^i$ .

Substituting the expressions of $\Psi^i$ and $\chi^i$, $PI^i_k$ and $PD^i_k$ are obtained. They have the following structure:

$$PI^i_k(t) = -\left[\ddot{x}_j M^i_{kj} + \dot{x}_j B^i_{kj} + (\dot{x}_j)^2 R^i_{kj}\right] \tag{25}$$

$$PD^i_k(t-\tau_i) = -\left[\dot{x}_j \Delta\tau_i\, K^i_{kj}(t-\tau_i) + (\dot{x}_j \Delta\tau_i)^2 S^i_{kj}(t-\tau_i)\right] \tag{26}$$

in which: i is fixed; j=1-6 ; k=1-6.
The values of the coefficients $M^i_{kj}$ , $B^i_{kj}$ , $R^i_{kj}$ and the retardation functions $K^i_{kj}$ ,$S^i_{kj}$ may be found in (S.-Arcilla, 1980).
The total potential reaction, $PR^i_k(t)$, in Ii is then:

$$PR^i_k(t) = PI^i_k(t) + \sum_{l=1}^{i-1} PD^l_k(t-\tau_l) \tag{27}$$

Introducing these expressions into (1) and (2), together with equations (6) and (7) and standard expressions for the viscous and hydrostatic reactions, the following structure is obtained for the equations of motion:

$$E_k^i(t) = A_{kj}^i \ddot{x}_j + D_{kjl}^i \dot{x}_j \dot{x}_l + P_{kj}^i \dot{x}_j + C_{kj}^i x_j + \int_k^i(t) \quad (28)$$

in which: k=1-6 ; j=1-6 ; l=1-6

The memory function $\int_k^i(t)$ comes from the retardation functions $K_{kj}^i$ and $S_{kj}^i$ of first and second order respectively. These equations (28), together with 12 initial conditions: $x_j(t=\tau_i)$ and $\dot{x}_j(t=\tau_i)$ ; j=1-6, complete the proposed formulation for the problem.

## 4. CONCLUSIONS

A formulation to predict the motion with 6 d.o.f. of a rigid body in a restricted fluid domain has been proposed. The problem has been solved in the time domain, the emphasis being placed on the potential reaction.

It is recognised the need to develop a similar model for the viscous reaction. Due attention should also be paid to the "external exciting force" and to the numerical propagation of errors, which is now being studied for the proposed model.

## 5. REFERENCES


Abkowitz, M.A., 1964. Lectures on ship hydrodynamics. Rep.Nº Hy-5.Hydro-og Aerodynamisk Lab.Lyngby, Denmark.

Arcilla, A.S.- 1980. Dynamics and kinematics of marine structures. Tech. Report Nº17. Ministry Public Works. Madrid. Spain (in Spanish)

Cummins, W.E., 1962. The impulse response function and ship motions. DTMB. Rep. 1661. Washington D.C.

Faltinsen, O.M., Michelsen, F.C. 1974. Motions of large structures in waves at Zero Froude number. Int. Symp. Dynamics Marine Vehicles and Structures in Waves. London.

Faltinsen, O.M. 1979. Sea loads and motions of marine structures. IInd. WEGEMT School. Trondheim.

Lighthill, J., 1979. Waves and hydrodynamic loading. BOSS 79. London.

Newman, J.N., 1977. Marine hydrodynamics. M.I.T. Press.

Norrbin, N.H., 1971. Theory and observations on the use of a mathematical model for ship manoeuvring in deep and confined waters. Publ.Nº68 SS PA, Göteborg.

Ogilvie, T.F.,1964. Recent progress towards the understanding and prediction of ship motions. 5th Symp. Nav. Hydrodyn. Bergen.

Price, W.G. and Bishop, R.E.D., 1974. Probabilistic theory of ship dynamics. Chapman and Hall.

Remery, G.F.M. and Van Oortmerssen, G.,1973. The mean wave, wind and current forces on offshore structures. O.T.C. paper Nº1741.

Stoker, J.J., 1957. Water waves. Interscience Publishers. New York.

Timman, R. and Newman, J.N., 1962. The coupled damping coefficients of a symmetric ship. J. Ship Research March.

Tuck, E.O., 1978. Hydrodynamic problems of ships in restricted waters. Ann. Rev. Fluid Mech. 10.

Van Oortmerssen, G. 1976. The motions of a moored ship in waves. Ph. D. Thesis. Wageningen.

# WAVE ACTION ON FLOATING STRUCTURES AND WAVE PROPAGATION USING FINITE ELEMENTS METHOD.

A. LEJEUNE - J. MARCHAL - Th. HOFFAIT - M. SAHLOUL - S. GRILLI

Laboratoires d'Hydraulique et des Constructions Hydrauliques
Institut du Génie Civil - Université de Liège - Belgique

## SUMMARY

Two problems are solved in this paper. The first study is calculating wave action forces on floating structures by means of a potential theory. The finite elements method is used to research water particles' velocities and accelerations. Heaving and rolling movements are studied by developping equations for the bidimensional problem. The theory of the wave sollicitations computation on a fixed body in the bidimensional and three dimensional cases are also presented.

To test programm performances, we have realized a complete computation of the rectangular caisson movements and we have compared these results with scale model data in a wave tank : the concordance between numerical and experimental heave and roll transfer functions was excellent, except at the rolling natural frequency and when the viscosity has an influence. In the second example we have studied the comportment of a cylindrical caisson in forced heaving motion using 249 knots. Numerical and experimental results are in good concordance for $C_M$ curves but for $C_A$ curves differences appear because we have neglected viscosity effects and because measured accuracy of the damping force was not sufficient.

In the second study we resolve the problem of wave propagation around structures : Berkhoff diffraction - refraction is used to study this problem. Using a bidimensional grid of finite elements method we were able to reproduce the waves' field in the three dimensional case.
We give a example.

## INTRODUCTION

In this paper, two problems concerning sea flows are described using both finite elements methods. In the first one, wave actions and floating structures movements are computed and in the second wave propagations is studied.

## PART A. - WAVE ACTIONS ON FLOATING STRUCTURES

### 1. Wave actions

Following assumptions are made for this study; wave theory is linear Stokes theory and ratio between floating structure diameter D and wave length L is defined as $0,2 < D/L < 1$. In that case viscous effects can be neglected as it could be very easily demonstrated.
The complex motion of the structure on sea is analysed and studied only by separating it into components. The three translation components are surging along the longitudinal X axis swaying along lateral Y axis, and heaving along the vertical Z axis. The three rotational components about these axes are respectively called rolling, pitching and yawing (figure 1).

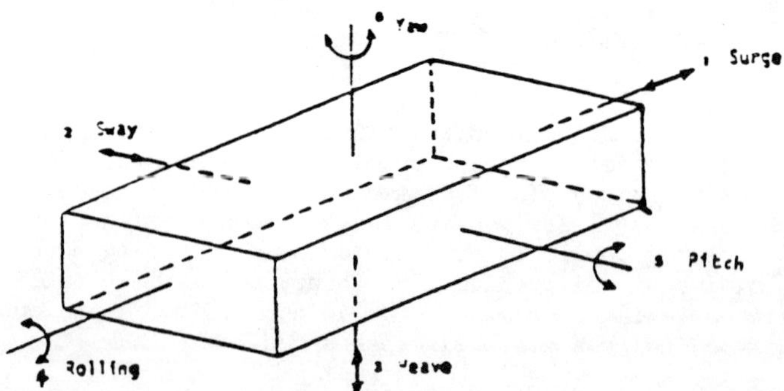


Figure 1

Only heaving and rolling will be presented : i.e. the most important movements as far as stability is concerned.

### 2. Potential Theory

#### A. Potential Flow

To calculate forces due to wave action, we need velocity and acceleration of water particles. To do that, we use velocity potential theory.

Using ideal fluid, the potential flow around a floating structure in sea is a linear combination of the following potentials.

1. $\phi_I$ is the potential of the incident wave, which is spreading out in the whole sea surface and then coming from the infinite. We use the linear Stokes potential defined above. The integration of dynamic pressures on the hull surface deduced from $\phi_I$, gives the Froude Krilov force.
2. $\phi_d$ is the potential of the diffracted wave, around the fixed structure. The integration of dynamic pressures, deduced from $\phi_I+\phi_d$, gives the total hydrodynamic force due to the

wave action

3. $\phi_R$ is the potential due to the structure in still water. To define those potential, we have to determinate the added mass and damping coefficients of the movements. In fact, the structure has 6 degrees of freedom i.e. 6 own movements independent of wave characteristics. So the potential will be a sum of 6 potentials

$$\Phi_R = \sum_{i=1}^{6} V_i \cdot \Phi_{Ri}$$

where $V_i$ is the amplitude of velocity (translation or rotation).

$\phi_{Ri}$ is the potential of an unit velocity following the i degree of freedom

In this condition, the total potential $\phi_t$ could be written as follow

$$\Phi_t = \Phi_I + \Phi_D + \sum_{i=1}^{6} V_i \cdot \Phi_{Ri}$$

B. Boundary Conditions

Boundary conditions are the following :

1. on the sea bottom : $\frac{\partial \Phi}{\partial n} = 0$

2. at the free surface : $\frac{\partial^2 \Phi}{\partial t^2} + g \frac{\partial y(x,t)}{\partial t} = 0$

and than we obtain the condition at the linearized free surface : $\frac{\partial^2 \Phi}{\partial t^2} + g \frac{\partial \Phi}{\partial y} = 0$

and if we pose $\phi = \phi e^{i\omega t}$ $\quad \frac{\partial \phi}{\partial y} - \frac{\omega^2}{g} \phi = 0$

3. on the structure : normal fluid velocity on the structure is equal to the velocity along the hull. This condition is presented in different forms

diffraction $\frac{\partial(\Phi_I + \Phi_D)}{\partial n} = 0$ on the hull

and structure movement $\frac{\partial \Phi_{Ri}}{\partial n} = \vec{u} \cdot \vec{V}$ on the hull

4. Radiation conditions: the radiation imposes that at boundary of fluid space, diffraction potential $\phi_D$ and radiation potential $\phi_R$ tend to zero.

$$\Phi_D \to 0 \qquad r \to \infty \qquad \Phi_R \to 0$$

and using polar coordinates, it becomes : $\frac{\partial \Phi_D}{\partial r} + \frac{1}{c} \frac{\partial \Phi_D}{\partial t} = 0$

In fact the space fluid is limited, and we impose that at its boundaries, radiation waves are only composed by divergent cylindrical waves coming from inside.
In that case, wave celerity C is deined by $c = \frac{gT}{2\pi}\, th(\frac{2\pi d}{L})$

and if $\Phi = \Phi_I + \Phi_D$ the above equation is changing in

$$\frac{\partial\Phi}{\partial r} + \frac{1}{c}\frac{\partial\Phi}{\partial t} = \frac{\partial\Phi_I}{\partial r} + \frac{1}{c}\frac{\partial\Phi_I}{\partial t}$$

3. Finite Elements Method

The finite elements method is used to solve the following flow differential equation $A(\phi) = 0$
The fluid space is divided in small elements in which potential $\phi$ is approached by the relation

$$\Phi_a = \underline{N}\ \underline{\phi}'$$

where $\phi'$ is the vector of potential values at the element knots
$\underline{N}$ is the vector of ponderation functions only depending on element knots coordinates.
The studied phenomena being periodical, $\phi'$ must be written (using complex notation) as follow :

$$\Phi' = \phi\ e^{j\omega t}$$

where $\underline{\phi}$ is the vector of potential amplitude at the element knots
then

$$\Phi_a = \underline{N}(x,y)\underline{\phi}\ e^{j\omega t}$$

The discretized problem formulation is obtained by the Galerkin method.

A body motions in still water : $\phi_{Ri}$

A. Heaving

We shall develop here the equations for the bidimensional problem, the three dimensional case being the same but heavier for the statement. The contributions to the second member of equation (1) will be the following :

free surface : $\int_{\Gamma_1} \underline{N}^T(-\frac{1}{g})\underline{N}\ \underline{\phi} e^{i\omega t}(-\omega^2)d\Gamma_1$

where $\Gamma_1$ is the boundary curve at the free surface

bottom : the condition $\frac{\partial\Phi}{\partial n} = 0$
gives no contribution. In this case we shall only stop the finite element mesh

body surface : The body surface has a periodical motion that can be written in the form

$$y = y_m.e^{i\omega t}$$

its speed will be

$$\vec{V} = \dot{y} = y_m.i.\omega.e^{i\omega t} = y_m.\omega.e^{i(\omega t+\frac{\pi}{2})}$$

where $\theta$ is the angle between OX axe and normal outside the fluid on the boundary between body and fluid.
The contribution will thus have the following form

$$\int_{\Gamma_3} \underline{N}^T \sin\theta y_m i\omega e^{i\omega t} d\Gamma_3$$

radiation boundary :

$$\int_{\Gamma_4} \underline{N}^T(-\frac{1}{c})\underline{N}\ \underline{\phi} e^{i\omega t} d\Gamma_4$$

symetry condition : The condition $\frac{\partial\Phi}{\partial n} = 0$ gives also no contribution to the second member, and we obtain the following system

$$\int_S (\frac{\partial\underline{N}^T}{\partial x}, \frac{\partial\underline{N}^T}{\partial y}) \begin{pmatrix} \frac{\partial\underline{N}}{\partial x} \\ \frac{\partial\underline{N}}{\partial y} \end{pmatrix} \underline{\phi}\ e^{i\omega t}\ d_S = \int_{\Gamma_1} \underline{N}^T(-\frac{1}{g})\underline{N}\ \underline{\phi} e^{i\omega t}(-\omega^2) d\Gamma_1$$

$$+ \int_{\Gamma_3} \underline{N}^T \sin\theta y_m i\omega e^{i\omega t} d\Gamma_3 + \int_{\Gamma_4} \underline{N}^T(-\frac{1}{c})\underline{N}\ \underline{\phi} e^{i\omega t} d\Gamma_4$$

In matricial form, the system is

$$(-i\omega\underline{\underline{K}}_2+\omega^2\underline{\underline{K}}_3-\underline{\underline{K}}_1)e^{i\omega t}\underline{\phi} + i\omega y_m\underline{P}_1 e^{i\omega t} = \underline{0}$$

symplifying the time harmonic term, we obtain a quasistatic complex system :

$$(\underline{\underline{K}}_1+i\omega\underline{\underline{K}}_2-\omega^2\underline{\underline{K}}_3)\ \underline{\phi} = y_m i\omega\underline{P}_1$$

with

$$\underline{\underline{K}}_1 = \int_S (\frac{\partial\underline{N}^T}{\partial x}, \frac{\partial\underline{N}^T}{\partial y}) \begin{pmatrix} \frac{\partial\underline{N}}{\partial x} \\ \frac{\partial\underline{N}}{\partial y} \end{pmatrix} dS$$

coming from the continuity condition on the field

$$\underline{\underline{K}}_2 = \frac{1}{c}\int_{S_4} \underline{N}^T\ \underline{N}\ dS_4$$

representing the radiation condition

$$\underline{\underline{K}}_3 = \frac{1}{g}\int_{\Gamma_1} \underline{N}^T\ \underline{N}\ d\Gamma_1$$

explaining the free surface condition

$$\underline{P}_1 = \int_{\Gamma_3} \underline{N}^T\ \sin\theta\ d\Gamma_3$$

introducing the sollicitation due to body motion.

If, in equation of the discretized problem, we uncouple the nodal values vector in their real and imaginery parts ($\phi=\phi_1+i\ \phi_2$) we obtain the expression :

$$\left[(\underline{\underline{K}}_1-\omega^2\underline{\underline{K}}_3)+i\omega\underline{\underline{K}}_2\right](\underline{\phi}_1+i\underline{\phi}_2) = iy_m\omega\underline{P}_1$$

If we separate real and imaginary parts, we obtain :

$$\begin{bmatrix} (\underline{\underline{K}}_1-\omega^2\underline{\underline{K}}_3) & -\omega\underline{\underline{K}}_2 \\ -\omega\underline{\underline{K}}_2 & -(\underline{\underline{K}}_1-\omega^2\underline{\underline{K}}_3) \end{bmatrix} \begin{bmatrix} \underline{\phi}_1 \\ \underline{\phi}_2 \end{bmatrix} = \begin{bmatrix} \underline{0} \\ -y_m\omega\underline{P}_1 \end{bmatrix}$$

It is this system that must be resolved by classical algorithms of the finite elements method.
With the $\phi_1$ and $\phi_2$ well known values, we can compute the forces and pressions in all the space and then also heaving hydrodynamic coefficients.

B. Rolling
The studied motion takes the following form $\alpha = \alpha_m e^{i\omega t}$

that gives for the body surface condition the following contribution

$$\frac{\partial\Phi}{\partial n} = \vec{n}.\vec{V} = \alpha_m . i\omega(x \sin\theta - y \cos\theta)e^{i\omega t}$$

where x and y are the coordonates of the body considered point in rotation.

The other limiting conditions are the same that in the case of heaving and we obtain the same system with for second member :

$$\begin{bmatrix} 0 \\ -\alpha_m.\omega.\underline{P}_2 \end{bmatrix}$$

with

$$\underline{P}_2 = \int_{\Gamma_3} \underline{N}^T (x \sin\theta - y \cos\theta) d\Gamma_3$$

C. Wave sollicitations on a fixed body

The computation of this problem takes not the same way in the bi or three dimensional case.
Of course, the radiation condition is given by

$$\frac{\partial\Phi}{\partial n} + \frac{1}{c}\frac{\partial\Phi}{\partial t} = \frac{\partial\Phi_I}{\partial n} + \frac{1}{c}\frac{\partial\Phi_I}{\partial t}$$

In the bi-dimensional case, the wave $(\phi_I)$ follows of course the condition

$$\frac{\partial\Phi_I}{\partial n} + \frac{1}{c}\frac{\partial\Phi_I}{\partial t} = 0$$

and we obtain the system

$$\begin{bmatrix} (\underline{\underline{K}}_1-\omega^2\underline{\underline{K}}_3) & -\omega\underline{\underline{K}}_2 \\ -\omega\underline{\underline{K}}_2 & -(\underline{\underline{K}}_1-\omega^2\underline{\underline{K}}_3) \end{bmatrix} \begin{bmatrix} \phi_1 \\ \phi_2 \end{bmatrix} = \begin{bmatrix} 0 \\ 0 \end{bmatrix}$$

The system solution is meanwhile not $\phi_1=\phi_2=0$ because we must

also artificially impose the $\phi_1+i\phi_2=\phi_I$ values at the boundaries corresponding at the incident wave.
In the three dimensional case, the complete discretized system will be

$$(\underline{\underline{K}}_1-\omega^2\underline{\underline{K}}_3) + i\omega\underline{\underline{K}}_2\ \underline{\phi} = \underline{P}_4$$

or

$$\begin{bmatrix} (\underline{\underline{K}}_1-\omega^2\underline{\underline{K}}_3) & -\omega\underline{\underline{K}}_2 \\ -\omega\underline{\underline{K}}_2 & -(\underline{\underline{K}}_1-\omega^2\underline{\underline{K}}_3) \end{bmatrix} \begin{bmatrix} \underline{\phi}_1 \\ \underline{\phi}_2 \end{bmatrix} = \begin{bmatrix} \mathrm{Re}\ \underline{P}_4 \\ \mathrm{Im}\ \underline{P}_4 \end{bmatrix}$$

with

$$\mathrm{Re}\ \underline{P}_4 = \int_{S_4} \underline{N}^T f_1(x,y,z)dS_4$$

$$\mathrm{Im}\ \underline{P}_4 = \int_{S_4} \underline{N}^T f_2(x,y,z)dS_4$$

$f = f_1 + i\ f_2$ where the f function represents the contribution

$$\frac{\partial\Phi_I}{\partial n} + \frac{1}{c}\frac{\partial\Phi_I}{\partial t}$$

due to the wave sollicitation

. COMPUTATION EXAMPLES

First example

To test the programme performances, we have realized a complete computation of a rectangular caisson and we have compared these results with scale model data in a wave tank.
The used discretization is represented in figure 2.
It has 238 elements with 6 or 8 knots, so in total 705 knots.
The obtanied results are reproduced in figures 3 and 4. For heaving the Y motions amplitudes are divided by the solliciting wave amplitude H while the rolling angle $\alpha_A$ by the maximum angle of the tangent at the free surface wave.

$$\alpha_H = \frac{\pi H}{L}$$

The periods variation is represented in the diagram by the reduced frequency (which is a non-dimensional coefficient). Its value is

$$\overline{\omega} = \frac{\omega^2 b}{g}$$

where $\omega$ is the wave frequency
b is half the breadth of the caisson

The experimental study has been realized in the wave tank of the Hydrodynamic Laboratorium of Liège University.
The concordance between numerical and experimental results is excellent, except at the rolling natural frequency where the computed amplitudes are too big and when the viscosity has an

influence.
Nevertheless the computed rolling natural period is the same that its experimental value : it is very important for the design of structures for which we always must avoid resonance.

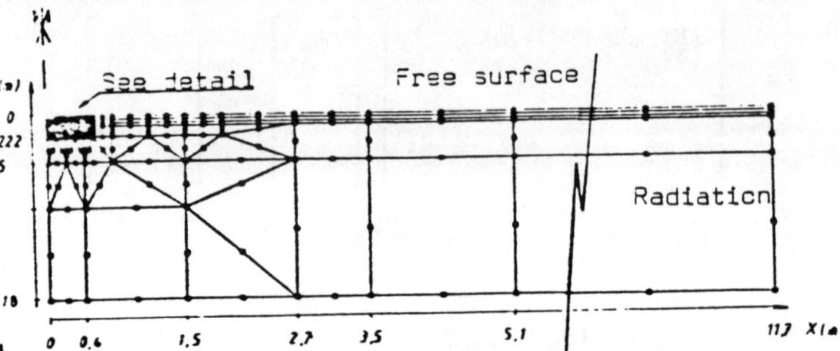


Figure 2.

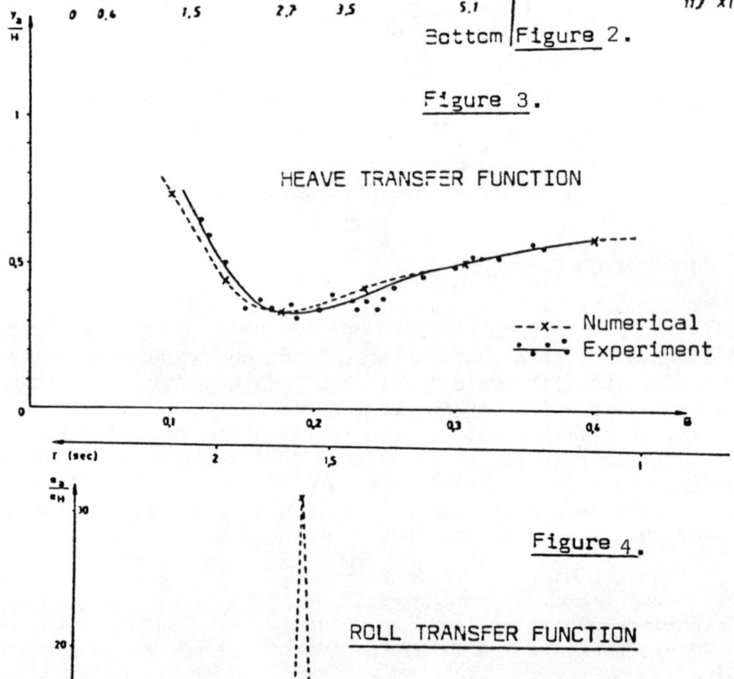


Figure 3.

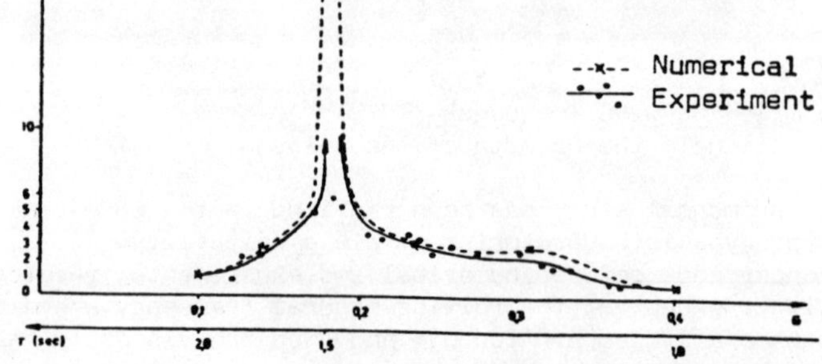


Figure 4.

Second example

We have realized the study of a cylindrical caisson in forced heaving motion.

The used discretization has 142 elements with 8 knots, so in total 249 knots. It is a tridimensional discretization representing a quarter of the fluid space (figure 5)

The obtained results are reproduced in figures where $C_M$ and $C_A$ are respectively the added mass and damping coefficients.

The concordance between numerical and experimental results is good for $C_m$, not for $C_A$ curve. That's because we have neglected viscosity in the computed results and because the damping force is small in this case so that measures accuracy is not sufficient.

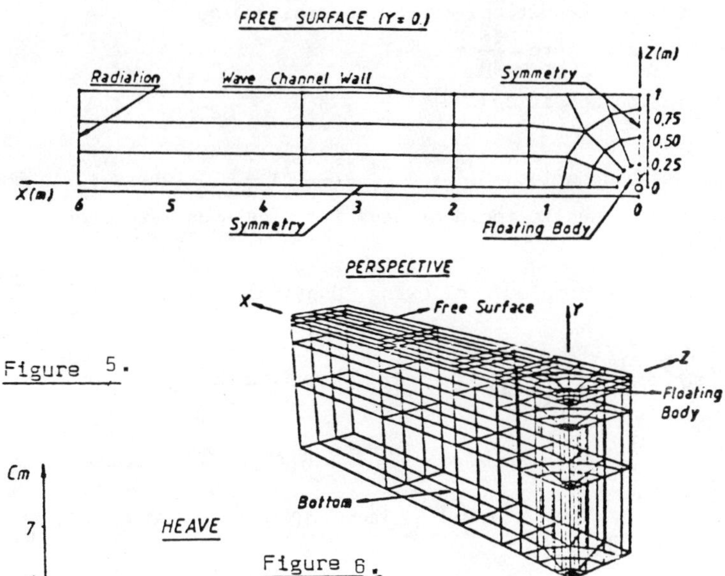


Figure 5.

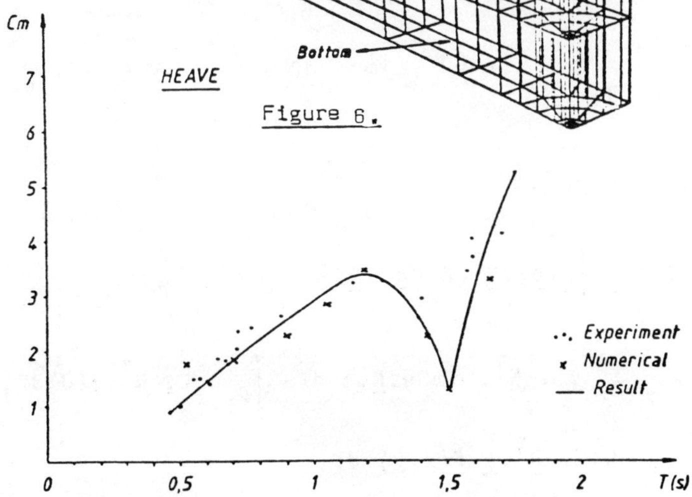


Figure 6.

## PART B. STUDY OF WAVE PROPAGATION AROUND STRUCTURES IN THE SEA

### 1. Model used

Berkhoff diffraction-refraction model is used to study the wave propagation and is well adjusted to a large range of depths in the sea.
The equation is the following one

$$\nabla (C . Cg . \nabla\phi) + \frac{\omega^2 Cg}{C} \phi = 0$$

where the wave celerity $c = \frac{g}{k} \tanh (kd)^{1/2}$

and the group wave velocity $C_g = C\frac{1}{2} (1 + \frac{2dk}{\sinh 2dk})$

d is the depth and k the wave number.
The boundary conditions are defined as follow

- on the structure $\frac{\partial\phi}{\partial n} = 0$
- radiation condition $\frac{\partial\phi}{\partial n} + \frac{i\omega}{c} \phi = 0$
- on sea boundaries : $\frac{\partial\phi}{\partial n} + \frac{1}{c}\frac{\partial\phi}{\partial t} = \frac{\partial\phi_I}{\partial n} + \frac{1}{c}\frac{\partial\phi_I}{\partial t}$ where $\phi_I$ is the potential due to incident wave (for instance wave Stokes theory)

To solve the problem, following functional is used

$$\pi(\phi) = \frac{1}{2} \int_S \left[\nabla\phi . C . Cg . \nabla\phi - \frac{\omega^2 Cg}{C} \phi . \phi\right] dS$$

$$+ \frac{1}{2} \int_{\Gamma_2+\Gamma_3} \phi . C . Cg . i\omega\phi \; d\Gamma - \int_{\Gamma_I} \phi . C . Cg . (\overline{\frac{\partial\phi}{\partial n}}) d\Gamma$$

### 2. Finite Element Method

Like in the part A, classical method is used $\phi = N(x,y)^T \phi$ and

$$\pi(\phi) = \frac{1}{2} \int_S \left[\underline{\phi} . \nabla\underline{N}^T . C . Cg . \nabla\underline{N} . \underline{\phi} - \underline{\phi} . N^T . \frac{\omega^2 Cg}{C} . \underline{N} . \underline{\phi}\right] dS$$

$$+ \frac{1}{2} \int_{\Gamma_2+\Gamma_3} \underline{\phi} . \underline{N}^T . i\omega Cg . \underline{N} . \underline{\phi} \; d\Gamma - \int_{\Gamma_1} C . Cg . \underline{N}^T . \phi(0) d\Gamma_1$$

$$- \int_{\Gamma_3} C . Cg . \underline{N}^T . \underline{\phi} \; f(x,y) \; d\Gamma_3$$

Finally the matrix form is given by

$$\begin{bmatrix} (\underline{\underline{K}}_1 - \omega^2 \underline{\underline{K}}_2) & -\omega \underline{\underline{K}}_3 \\ -\omega \underline{\underline{K}}_3 & -(\underline{\underline{K}}_1 - \omega^2 \underline{\underline{K}}_2) \end{bmatrix} \begin{bmatrix} \underline{\phi}_1 \\ \underline{\phi}_2 \end{bmatrix} = \begin{bmatrix} \Re e\ \underline{P} \\ -\mathrm{Im}\ \underline{P} \end{bmatrix}$$

and characteristics of wave (i.e. celerity, amplitude, wave length) are computed in the field.

3. Example

This example reproduces the waves' field of a 250 s period and 10 m amplitude tsumani around a circular island of 4 km radius. The used mesh is composed by 54 bi-dimensional elements of second degree.

The cylindrical discretization is extended up to a distance of 60 km. The boundary conditions are described in the theorical part of this study.

Obtained results of wave amplitudes are reproduced in the figure 7 (diagramm.) These results are in accordance with Berkoff values.

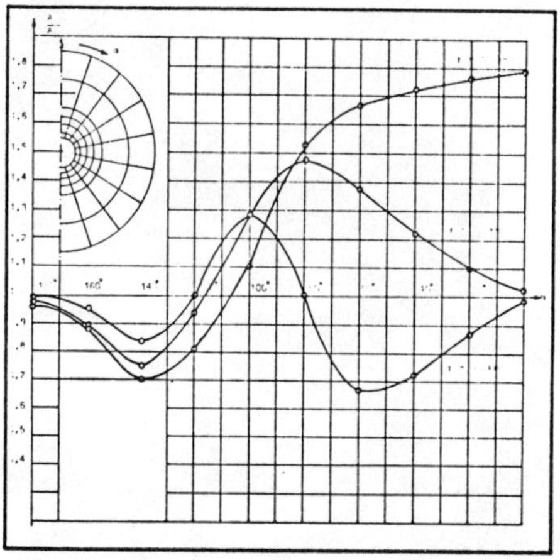

Figure 7

# NUMERICAL SOLUTION OF SOME LINEAR EVOLUTION PROBLEMS IN NAVAL HYDRODYNAMICS.

Christian Licht

Ecole Nationale Supérieure des Techniques Avancées
Institut Français du Pétrole

## 1. INTRODUCTION

Naval Hydrodynamics deals with the motion of bodies in the sea. So we have to solve coupled-systems of ordinary differential equations and boundary-value problems. These problems involve unbounded domains and free boundary. We avoid the last (but not the least) difficulty by the hypothesis of linearisation near a steady state. A method to attack these problems is to consider the whole system fluid + body. We introduce the space $\mathcal{H}$ of possible states $u$ with finite mechanical energy, on which energy generally defines an hibert-norm. Then the problem can be expressed in the form of a linear evolution equation $\frac{du}{dt} = Au$, $u(o) = u^o$, with $A$ infinitesimal generator of a unitary group $U(t)$ on $\mathcal{H}$. This (which implies existence and uniqueness of $u$) agrees with the physical properties of conservation of energy and reversibility. Because of certain stationary conditions of the problem ($\Delta\phi = 0$ if $\phi$ is the potential flow in the case of an incompressible fluid) occuring in the definition of $\mathcal{H}$, we do not make use of an internal approximation of $\mathcal{H}$ but of the Trotter's theory on approximation of semi-groups, in order to justify the convergence of our numerical method. This one involves, in fact, an implicit scheme of time discretisation of the equations, a truncation of the fluid domain with an absorbing boundary condition and finite elements in the truncated domain. §§ 2,3 develop these ideas through an important practical example (immersed elastic structure). Numerical results, comments and prospects about an other example (rigid floating body) are given in § 4.

## 2. THE THEORETICAL PROBLEM

### 2.1 Statement of the problem

We consider the linear problem connected with the transient vibrations near equilibrium of an elastic structure $\mathcal{B}$ in the sea. The fluid is heavy, perfect and incompressible. We assume the flow is irrotational. $O(x_1,x_2,x_3)$ is a right-handed coordinate system. G, the fluid domain at equilibrium, is bounded by $S_I$ (the immersed surface of $\mathcal{B}$), $S_F$ (the still free surface), $S_B$ (the bottom) and $\mathcal{B}$ by $S_I$, $S_E$ (the emerged surface of $\mathcal{B}$), $S_0$ (along which the structure cannot move).

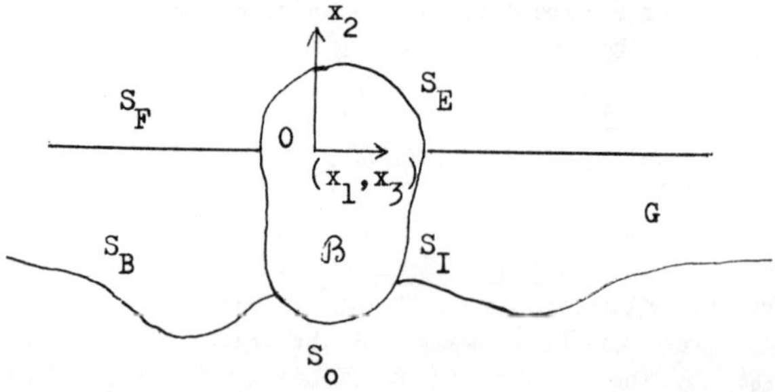


Figure 1

The mechanical state of the system fluid-structure can be described by four parameters $(\phi, \eta, \vec{s}, \vec{v})$ : $\phi$ the potential of the flow, $\eta$ the elevation of the free surface, $\vec{s}$ the displacement field in $\mathcal{B}$, $\vec{v}$ the velocity field in $\mathcal{B}$.

Let $\rho_f$ (resp. $\rho_s$ ) the density of the fluid (resp. of the structure), $\vec{g}$ the gravity acceleration, $\sigma(\vec{s})$ (resp. $\varepsilon(\vec{s})$) the stress (resp. strain) tensor in $\mathcal{B}$. $\vec{n}$ is the normal pointing out of the fluid, t the time. The well-known equations of the fluid motion are :

$$\Delta\phi = 0 \qquad \text{in } G \qquad (1)$$

$$\phi_t = -g\eta \qquad \text{on } S_F \qquad (2)$$

$$\eta_t = \partial_n\phi \qquad \text{on } S_F \qquad (3)$$

$$\partial_n\phi = \vec{v}.\vec{n} \qquad \text{on } S_I \qquad (4)$$

$$\partial_n\phi = 0 \qquad \text{on } S_B \qquad (5)$$

The equations of the vibrations of the structure are :

$$\sum_{j=1}^{3} \partial_j \sigma_{ij}(\vec{s}) - \rho_s \partial_t v_i = 0 \qquad \text{in } \mathcal{B} \qquad (6)$$

$$\vec{s} = 0 \qquad \text{on } S_o \qquad (7)$$

$$\sigma(\vec{s})\,\vec{n} = 0 \qquad \text{on } S_E \tag{8}$$

$$\sigma(\vec{s})\,\vec{n} = -\,\rho_f(\phi_t - \vec{s}.\vec{g})\,\vec{n} \qquad \text{on } S_I \tag{9}$$

For the sake of simplicity (see [5]) we make Tong hypothesis and replace equation (9) by

$$\sigma(\vec{s})\,\vec{n} = -\,\rho_f(\phi_t - \vec{s}.\vec{n}\ \vec{g}.\vec{n})\,\vec{n} \qquad \text{on } S_I \tag{10}$$

So can be introduced a symmetric bilinear form of potential energy:

$$a(\vec{s},\vec{\varphi}) \equiv \int_{\mathcal{B}} \sum_{i,j} \sigma_{ij}(\vec{s})\,\varepsilon_{ij}(\vec{\varphi}) - \rho_f \int_{S_I} \vec{g}.\vec{n}\ \ \vec{s}.\vec{n}\ \ \vec{\varphi}.\vec{n} \tag{11}$$

If the initial state $u^o$ is given, the problem (P) is to determine the state $u = (\phi, \eta, \vec{s}, \vec{v})$ at any time.

## 2.2 Resolution of the problem

**2.2.1 Functional frame :** The mechanical energy of the system is proportional to

$$E(u) = \rho_f \int_G |\nabla\phi|^2 + \rho_f g \int_{S_F} \eta^2 + a(\vec{s},\vec{s}) + \int_{\mathcal{B}} \rho_s\,|\vec{v}|^2 \tag{12}$$

E(u) defines the square of an hilbert-norm on the space

$$\mathcal{F} = H_D(G) \times L^2(S_F) \times V(\mathcal{B}) \times L^2(\mathcal{B})^3 \tag{13}$$

$H_D(G)$ is the closure of the Sobolev space $H^1(G)$ for $\left(\int_G |\nabla.|^2\right)^{1/2}$

$V(\mathcal{B}) = \{ \vec{s} \in H^1(\mathcal{B})^3 \ :\ \vec{s}\,|_{S_0} = 0 \}$

(we suppose for the sake of simplicity that a is coercive on $V(\mathcal{B})$)

Taking in account the stationary conditions (1),(5),(7), we consider the space $\mathcal{h}$ of possible states with finite energy :

$$\mathcal{h} = \{ u \in \mathcal{F} \ :\ \int_G \nabla\phi.\nabla\varphi = 0 \qquad \forall \varphi \in {}_oH_1^o(G) \} \tag{14}$$

with

$$H_1^o(G) = \{ \varphi \in H^1(G)\ :\varphi|_{S_F} = 0 \}$$

$${}_oH_1^o(G) = \{ \varphi \in H_1^o(G)\ :\varphi|_{S_I} = 0 \}$$

**2.2.2 Weak formulation of (P) :** We construct an operator A with domain D(A) in $\mathcal{h}$ so that

$$\frac{du}{dt} = Au \quad , \quad u(o) = u^o \tag{15}$$

may be a weak formulation of (P).

For $\eta \in L^2(S_F)$, if the elliptic problem

$$\begin{cases} \int_G \nabla\psi.\nabla\varphi = 0 & \forall \varphi \in H_1^o(G) \\ \psi|_{S_F} = \eta \end{cases} \tag{16}$$

has a (unique !) solution in $H^1(G)$, we put $\eta \in D(B), \psi = B\eta$ . **Moreover** if $\eta \in D(B)$, there is at most one solution $(\vec{w}, \gamma)$ in $L^2(\mathcal{B})^3 \times \overset{o}{H}_1(G)$ of :

$$\int_{\mathcal{B}} \rho_s \vec{w}.\vec{\varphi} + \rho_f \int_{S_I} \gamma \vec{\varphi}.\vec{n} = \rho_f \int_{S_I} g\, B\eta \vec{\varphi}.\vec{n} - a(\vec{s},\vec{\varphi}) \qquad \forall \vec{\varphi} \in V(\mathcal{B}) \tag{17}$$

So we can verify (see [5] ) that A can be defined by

$$D(A) = \{u \in h : \partial_n \phi|_{S_F} \in L^2(S_F),\ \vec{v} \in V(\mathcal{B}),\ \partial_n \phi|_{S_I} = \vec{v}.\vec{n},$$

$$\eta \in D(B), \exists (\vec{w}, \gamma) \in L^2(\mathcal{B})^3 \times \overset{o}{H}_1(G) \tag{18}$$

$$\text{solution of equation (17)} \}$$

$$Au = (- g\, B\eta + \gamma,\ \partial_n \phi|_{S_F},\ \vec{s},\ \vec{w}) \tag{19}$$

2.2.3 Resolution of (P) : We can prove (see [5] ) that A is shew-adjoint, thus it is the infinitesimal generator of a continuous unitary group U(t) of operators on $h$ and if $u^o \in D(A)$, equation (15) has a unique solution $C^1(\mathbb{R}, h) \cap C^o(\mathbb{R}, D(A))$ :

$$u(t) = U(t)\, u^o \tag{20}$$

It is noticeable that in the course of the proof we solve

$$u - \lambda Au = f\ ,\quad f \in h\ ,\ \lambda \in \mathbb{R}\ . \tag{21}$$

u is obtained very easily from the solutions of "split" elliptic problems in $\mathcal{B}$ and in G. So we see the practical possibility of using an implicit scheme of time discretisation.

## 3. NUMERICAL ANALYSIS OF THE PROBLEM.

### 3.1 Presentation of a numerical method.

Because of the conditions (14) satisfied by the elements u of $h$ (i.e. $\Delta\phi = 0$ in G, $\partial_n \phi|_{S_B} = 0$), we cannot, except in very particular cases, use an internal approximation of $h$ . We will utilize Trotter's theory of approximation of semi-groups with variable spaces (see [7] ) that we briefly recall.

Let $h_m$ a sequence of Hilbert space (not necessarily include in $h$) and $P_m$ a linear (not necessarily surjective) map from $h$ to $h_m$ so that :

i) $\| P_m \| \leq 1 \quad \forall m \in \mathbb{N}$

ii) $\lim\limits_{m\to\infty} \| P_m x \|_{h_m} = \| x \|_h \quad \forall x \in h$

And let the evolution equations $\frac{du}{dt} = Au$ in $h$ and $\frac{du_m}{dt} = A_m u_m$ in $h_m$ with A (resp. $A_m$) infinitesimal generator of U(t) on $h$ (resp. $U_m(t)$ on $h_m$).

If $\exists \lambda \in \mathbb{R}^{+}$ :

$$\lim_{m\to\infty} \| P_m (I - \lambda A)^{-1} f - (I - \lambda A_m)^{-1} P_m f \|_{h_m} , \quad \forall f \in h$$

we have $\lim_{m\to\infty} \| P_m U(t) u^o - U_m(t) P_m u^o \|_{h_m} = 0$

For each $u^o$, the convergence is uniform on $[0,T]$ , $T< +\infty$, and it is the same for the implicit scheme :

$$u_m^{k+1} - u_m^k = \Delta t \; A_m \; u_m^{k+1} \qquad k = 0,1,\dots,K \qquad \Delta t = T/K.$$

Here the construction of $P_m$ takes into account the two difficulties in view of an immediate use of the implicit scheme in $h$ : firstly truncation of the fluid domain, secondly use of finite elements. For sufficiently large n, we note $G(n), S_F(n)$, $S_B(n)$ the troncature of $G, S_F, S_B$ by the vertical plane $|x_1| = n$, $|x_3| = n$ ; $S(n)$ is the surface of truncation. We consider the problem $(P_n)$ posed in $G(n)$ with the same equations than (P). On $S(n)$ we add a condition for the flow-potential in order that $(P_n)$ is well-posed :

$$\phi_t + \nu \partial_n \phi = 0 \quad , \quad \nu \in [0, +\infty] \quad \text{on } S(n) \qquad (22)$$

For each $\nu$, this condition has an obvious physical interpretation. As in 2.2, $(P_n)$ can be expressed in the form of a linear evolution equation $\frac{du_n}{dt} = A_n u_n$ in an Hilbert space $h_n$, with $A_n$ maximal-dissipative if $\nu \in ]0, +\infty[$ or skew-adjoint if $\nu \in \{0, +\infty\}$. Finally, we discretise $h_n$ in $h_n^h$ by use of finite element in $G(n)$ and build a linear (dissipative or conservative) operator $A_n^h$ on $h_n^h$.

For example in the (technically) simplest case $\nu = 0$, $h_n^h$ and $P_n^h$ are as follows. $X_n^h$ is a subspace of finite elements of $\{\varphi \in H^1(G(n)) : \varphi|_{S(n)} = 0\}$ , $Y_n^h$ is the space of the traces on $S_F(n)$ of the elements of $X_n^h$ , $V_n^h$ is a subspace of finite elements of $V(\mathcal{B})$. We put

$$\mathcal{F}_n^h = X_n^h \times Y_n^h \times V_n^h \times V_n^h$$

$$h_n^h = \{ u \in \mathcal{F}_n^h : \int_{G(n)} \nabla\phi . \nabla w = 0 \quad \forall w \in X_n^h \text{ and } w|_{S_I \cup S_{B(n)}} = 0 \}$$

For each u in $h_n^h$, $u_n^h = P_n^h u$ is the projection of u (considered as an element of $\mathcal{F}$ ) on $\mathcal{F}_n^h$ (considered as a subspace of $\mathfrak{F}$ with a suitable extension by 0 on $G \setminus G(n)$ and $S_F \setminus S_F(n)$ for the two first components). The elements which enter in the construction of A can be defined by "variational formulation", thus the construction of $A_n^h$ is easy (see [4] and [5]). The calculous of $(I - \lambda A_n^h)^{-1} f_n^h$ (see the next section) calls into play approximations of solutions of the elliptic problems in $\mathcal{B}$ and G connected with equation (21). And the convergence of $(I - \lambda A_n^h)^{-1}$ results from the combination of properties of continuous dependance as a function of the domain and properties of the finite element method.

Thus by use of implicit scheme in $\mathcal{A}_n^h$, we can define an effective numerical method whose convergence is proved when first $n \to \infty$ and next the time-step $\Delta t$ and the space discretisation parameter h go independently to 0. It is also noticeable that we can obtain an error estimation for the approximation of the truncated problem $(P_n)$.

### 3.2 The algorithm

We describe it in the case $\nu > 0$ (modifications in the other cases are obvious) $X_n^h$ (resp. $V_n^h$ ) is now a subspace of finite elements of $H^1(G(n))$ (resp. $V(\mathcal{B})$).

a) preliminary calculous

We form the matrix R of the bilinear form $\mathcal{R}$ on $X_n^h$ :

$$\mathcal{R}(\phi,\varphi) = \int_{G(n)} \nabla\phi . \nabla\varphi + \frac{1}{g\Delta t^2} \int_{S_F(n)} \phi\varphi + \frac{1}{\nu\Delta t} \int_{S(n)} \phi\varphi$$

We build the matrix of the operator $W = V_n^h \to X_n^h$ :

$$\mathcal{R}( W (\vec{v}), \varphi ) = \int_{S_I} \vec{v}.\vec{n}\ \varphi \qquad \forall \varphi \in X_n^h$$

We form the matrix M of the bilinear form $\mathcal{M}$ on $V_n^h$ :

$$\mathcal{M}(\vec{v},\vec{w}) = \int_{\mathcal{B}} \rho_s\ \vec{v}.\vec{w} + \int_{S_I} W(\vec{v})\ \vec{w}.\vec{n} + \Delta t^2\ a\ (\vec{v},\vec{w})$$

b) initialisation

c) the K steps : $u^k = (\phi^k, \eta^k, \vec{s}^{\,k}, \vec{v}^k) \to u^{k+1}$

- calculous of $\psi$:

$$\mathcal{R}(\psi,\varphi) = \frac{1}{g\Delta t^2} \int_{S_{F(n)}} (\phi^k - g\Delta t\, \eta^k)\varphi + \frac{1}{\nu\Delta t} \int_{S(n)} \phi^k\varphi \qquad \forall \varphi \in X_n^h$$

- <u>calculous of $u^{k+1}$</u>

$$\cdot \mathcal{M}(\vec{v}^{k+1},\vec{w}) = \rho_f \int_{S_I} (\phi^k - \psi)\, \vec{w}.\vec{n} + \int_{\mathcal{B}} \rho_s\, \vec{v}^k.\vec{w} - \Delta t\, a(\vec{s}^k,\vec{w}) \qquad \forall w \in V_n^h$$

$$\cdot\ \vec{s}^{k+1} = \vec{s}^k + \Delta t\, \vec{v}^{k+1}$$

$$\cdot\ \phi^{k+1} = \psi + W(\vec{v}^{k+1})$$

$$\cdot\ \eta^{k+1} = -\frac{1}{g\Delta t}(\phi^{k+1} - \phi^k)$$

The computer implementation of this algorithm is easy. The matrix R and M which are symmetric positive-definite are factorised by Cholesky's method. And at each step we have only to compute the second members by combinaison of informations <u>issued only from the previous step</u>. Thus this method is fast and the required storage is essentially this of the matrix R and M.

<u>3.3 Remark</u>

It seems also possible to reduce the fluid domain of calculous by using quasi localized finite elements (see [4] ) : i.e. finite element in G(R) and harmonic functions in G / G(R) with nul normal derivative on $S_B \setminus S_B(R)$. We can explicit a base if $S_B \setminus S_B(R)$ is flat.

## 4. NUMERICAL RESULTS AND COMMENTS

Here are given numerical results obtained by the previous method in the two-dimensional problem of the motion of a <u>rigid</u> floating body. We refer to [2] for a theoretical study and to [4] for a numerical study and other numerical results. At time $t = 0$, the fluid is at rest and the body is kept without velocity from $y_o$ about static equilibrium. Figure 3 shows the vertical motion of the body with different boundary conditions 20 m away. By using previously truncation at farther distances, we are sure to possess a good approximation for the first 8 s. of the motion in the open sea. We notice that $\nu = C = (gd)^{1/2}$ (d is the depth here equal to 5 m.) gives the result. Physically, it is well-known that a front of surface-waves is outgoing practically with a finite velocity C. So before the time $\theta$ when the front reaches the surface of truncation, all the boundary conditions give coherent results ; after $\theta$ reflections are different and for $\nu = C$ the absorption of energy is maximal (see figure 2). We have very often noticed that a good choice of $\nu$ allows us to divide by 2 or 3 the distance of truncation with regard to Dirichlet's or Neumann's approximation.

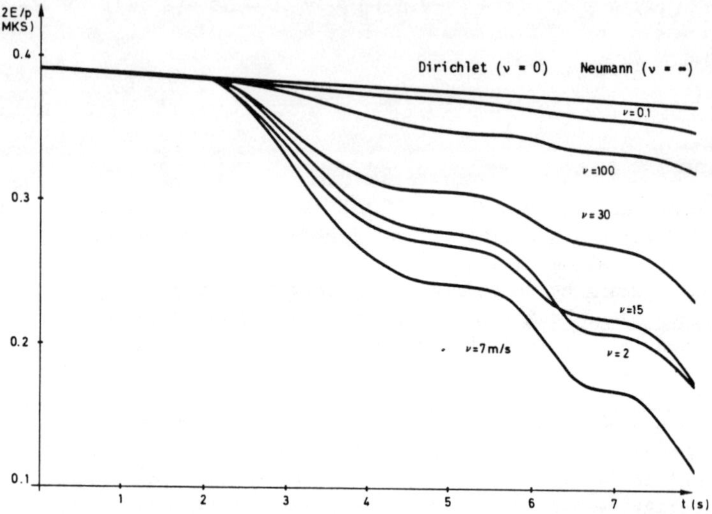


Figure 2 : Energy of the truncated and discretised system
( $\phi_t + \nu \partial_n \phi = 0$ 20 m away)

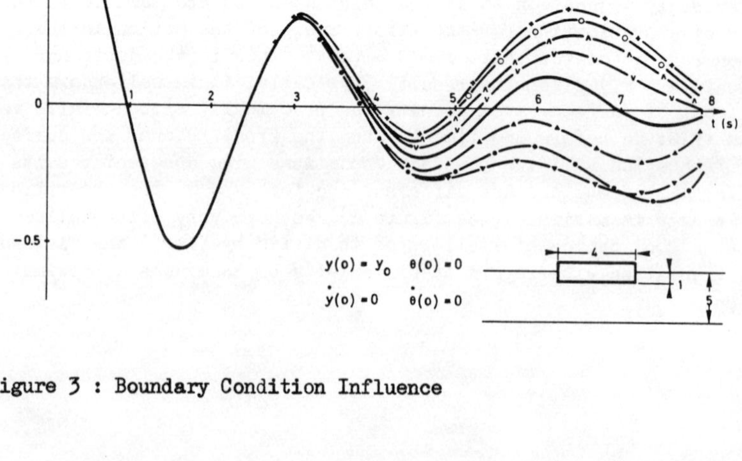


Figure 3 : Boundary Condition Influence

It should be very interesting to justify theoretically these remarks, to obtain an error estimation of the process of truncation and to develop even more efficient absorbing boundary conditions. In the case of a compressible fluid, the problem is hyperbolic (see [5] ), and it is possible to use the technics of [6] . If the fluid is incompressible, surface waves are dispersive, so the situation is more complex and studies are presently being carried out....
At last we compare our results with those obtained by very different methods : [1] makes use of a transient Green function and integrodifferential equations, [3] uses a Fourier-Laplace transform with respect to time. These results deal with infinite depth, so we have used a depth d = 10 R where R is the radious of the considered circular cylinder.

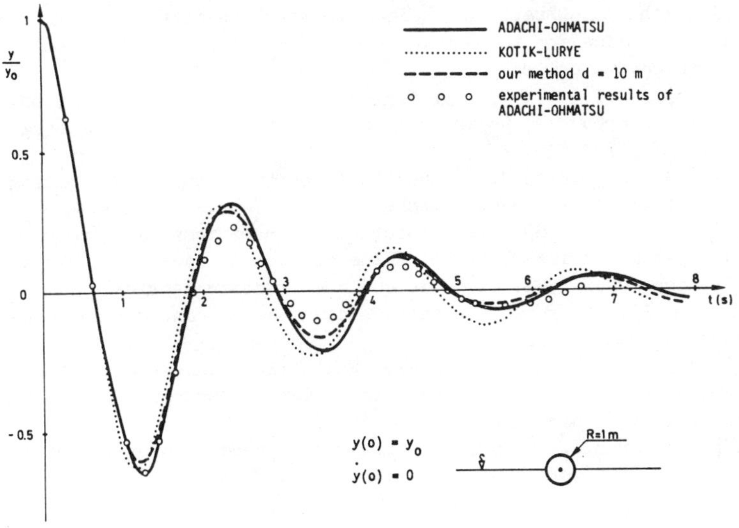


Figure 4 : Heave motion

## 5. CONCLUDING REMARKS.

This paper shows the theoretical and numerical interest of the theory of semi-groups in order to study some transient problems. One can find in [5] other problems in Naval Hydrodynamics (for example underwater acoustic) which enter in the scope of these technics. In particular this method can deal with totally arbitrary shape of the fluid domain and more complex situations (external excitation of the fluid-body system ...). It seems also appropriate to many transient linear problems of Physics or Engineering which involve unbounded domains.
Implementation of the numerical method is easy. The computing times are reasonable and use of an appropriate boundary condition allows to reduce the domain of calculous. It should be very interesting to develop even more efficient absorbing boundary conditions especially if we want to follow phenomena for a long time.

## References

[1] Adachi H., Ohmatsu S., (1980) On the Time-dependent potential and its application to wave problem. $13^{th}$ O.N.R. Symposium on Naval Hydrodynamics, Tokyo.

[2] Beale J.T., (1977) Eigenfunction expansion for object floating in an open sea. Communications on Pure and Applied Mathematics, XXX : 283-313.

[3] Kotik J., Lurye J., (1968) Heave oscillations of a floating cylinder or sphere. Schiffstechnik, 15 : 37-38.

[4] Licht C., (1980) Etude théorique et numérique de l'évolution d'un système fluide-flotteur. Thèse de docteur ingénieur, Nantes.

[5] Licht C., (1982) Etude de quelques modèles décrivant les vibrations d'une structure élastique dans la mer. Rapport de recherche E.N.S.T.A., Paris.

[6] Majda A., Enquist B., (1979) Radiation boundary conditions for acoustic and elastic wave calculations. Communications on Pure and Applied Mathematics, XXXII : 313-357.

[7] Trotter H.F., (1958) Approximation of semi-groups of operators. Pacific Journal of Mathematics, 8 : 887-919.

# CALCULATION OF THE VALUE OF THE HYDRODYNAMIC FORCES ACTING ON A CULVERT TAINTER VALVE BY MEANS OF THE FINITE ELEMENTS METHOD

Viliam Strauss

Institute of Hydrology and Hydraulics of the Slovak Academy of Sciences, Bratislava, Czechoslovakia

Complicated cases of flow in hydrotechnical structures is possible to solve only by the use of simplifying assumptions. The step of idealization influences mainly the results of quantitative solution. In spite of this the results of such solutions provide often valuable knowledge which enables reduction of research on a physical model. In some cases we replace the real flow by the irrotational flow of ideal liquid. It makes possible the formation of a mathematical model on the basis of the theory of potential flow. It is known that this theory provides good results by the solution of tasks where the losses of mechanical energy are relatively small so that it is possible to neglect them. Usual are the solutions of the ground water flow, spillways, flow beneath the gate, etc. In the last time mathematical models of the flow in hydrotechnical structures occur which even approximately do not fulfil the conditions of the potential flow because the turbulent zones cuase large losses. In spite of this it was shown that einther the idealization of such phenomena provides a piece of knowledge quickly and relatively simply. This knowledge enables a mutual comparison of several alternatives and their evaluation so that the research on a physical model may be reduced to a selected solution.

By the hydraulic research of culvert tainter valve of the Danuble hydroelectric plant lock of Gabčíkovo we used - except of the physical models - a mathematical model, too.

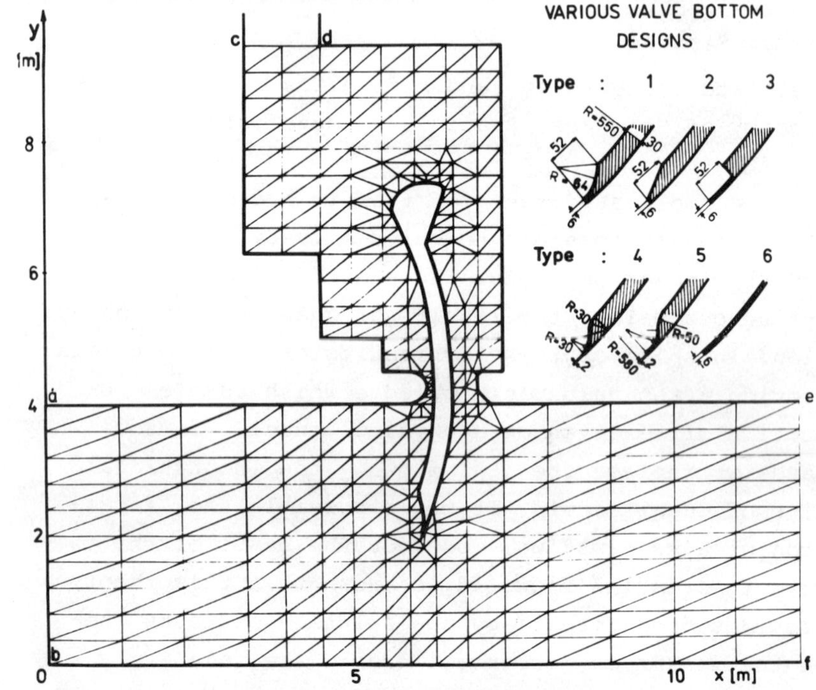


Figure 1. Net for the calculation of poten - tial flow by the method of finite elements and types of the valve lip

It was based on the potential theory of flow. We used this although we knew that the flow around the partially opened valve does not - even appoximately - fulfil the conditions of the irrotational flow. The results were used for a mutual comparison of alternatives of the lip`s different shape. In reality the valve dams up the culvert area of 4 x 4 m and its radius is 580 cm. The valve is placed in the shaft in such a way that in its intermediate position it is possible to draw the water above its upper part to the zone of lower pressure behind the valve /fig.1/. In the time of filling and emptying the lock the tainter valves should be strained equally as the so called "reversed" valves, i. e. pulled trunnion arms.

The solution of the potential flow for a certain zone means to find the solution of the Laplace equation of velocity potential which has the following form the two-dimensional flow:

$$\frac{\partial \varphi^2}{\partial x^2} + \frac{\partial \varphi^2}{\partial y^2} = 0 \qquad /1/$$

Knowing the function $\varphi = \varphi(x,y)$ we can determine the fields of velocity for the whole zone because in every point there is:

$$V_x = \frac{\partial \varphi}{\partial x}; \quad V_y = \frac{\partial \varphi}{\partial y}; \qquad /2/$$

and for next the pressures and flow pattern, i. e. the net of streamlines and equipotential, too.

The equation 1 has the analytical solution only for some simple cases of flow. For a general case only a numerial calculation is possible. We used the solution by means of the finite elements method. In the last years its exploitation for potential flow was worked up in detail on the Faculty of Civil Engineering of the Technical University of Prague / Kazda, J. 1978 /.

The net of triangular elements for the half opening of the valve is illustrated in fig. 1. The solution of the velocity potential in the net knots and the components of velocity in the elements was performed by means of a computer with the help of the FEFI 0 programme. Basically it was elaborated for ground water potential flow, showed by the equa-

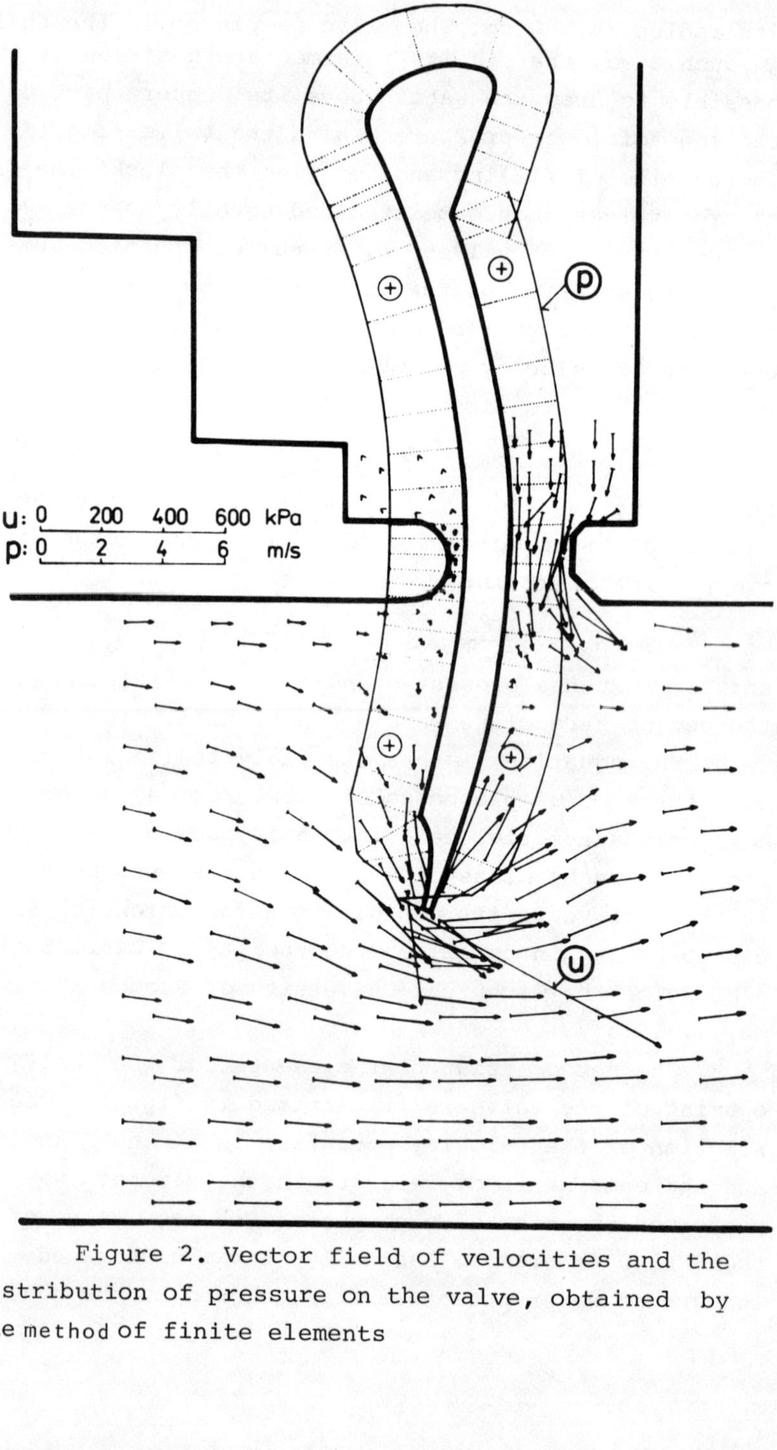


Figure 2. Vector field of velocities and the distribution of pressure on the valve, obtained by the method of finite elements

tion where $k_x(x,y)$ a $k_y(x,y)$ are the coefficients of filtration in the direction of the axis, $H$ is the piezometric head $H = h + p/\varrho g$

When $k_x = k_y = 1$ enclosed into the computation, it is possibleby means of the FEFI O programme - to solve the problem with velocity potential, too. Then the value $H$ is given as the sum of geodetic, pressure and velocity head.

The boundary conditions of the solution are:

- the shape of the culvert, culvert shaft and the double skin plate of the valve itself /Neumann´s boundary condition/

- the known potentials in definite points of the net /Dirichlet´s boundary condition/. According to fig. 1 on the boundaries of the zone ab and ef we suppose vertical equipotential lines and on the boundary cd horizontal equipotential line is supposed. The potentials on these boundaries were given according to the water level which was determinated experimentally on a model /Strauss, V. 1978/.

On the described mathematical model we wanted to calculate how the velocity and pressure are changed according to the shape of the valve lip by otherwise equal conditions. Fig. 1. shows the six searched types of the lip.

For all alternatives of lip the field of velocities and the pressure distribution on the double skin plate /fig. 2./ were solved. The flow pattern illustrated by the velocity vectors does not show substantial differencies for the individual types. As shown in table 1. the maximum local velocities under the lip of the valve are rather different.

Table 1

| Type of lip | Net element | $v_x$ $m.s^{-1}$ | $v_y$ $m.s^{-1}$ | Velocity in centre of gravity of the element |
|---|---|---|---|---|
| 1 | 296 | 7,95 | -4,16 | 8,90 |
| 2 | 296 | 8,56 | -4,00 | 9,34 |
| 3 | 296 | 9,005 | -4,64 | 10,20 |
| 4 | 296 | 15,19 | -8,215 | 17,27 |
| 5 | 296 | 8,656 | -4,108 | 9,58 |
| 6 | 294 | 10,02 | -5,424 | 11,40 |

Pressures in the lip zones of all searched lip types are illustrated in fig. 3. The maximum values of the depression which occurs behind the valve are drawn in table 2. /100 kPa = 10 m $H_2O$/

Table 2

| Type of lip | Point of net | $\bar{v}$ $/m.s^{-1}/$ | y /m/ | H /m/ | Pressure p /kPa/ |
|---|---|---|---|---|---|
| 1 | 193 | 5,34 | 2,02 | 26,496 | - 56,30 |
| 2 | 192 | 4,98 | 2,00 | 25,859 | - 23,16 |
| | 193 | 6,24 | 2,02 | 26,368 | -167,98 |
| 3 | 192 | 5,17 | 2,00 | 25,926 | - 42,46 |
| | 193 | 5,13 | 2,02 | 25,515 | - 42,79 |
| 4 | 192 | 6,98 | 2,00 | 25,293 | -275,44 |
| | 193 | 8,91 | 2,01 | 26,309 | -595,10 |
| 5 | 192 | 5,01 | 2,00 | 25,873 | - 26,28 |
| | 193 | 6,34 | 2,02 | 26,389 | -180,02 |
| 6 | 196 | 5,73 | 2,00 | 26,328 | -113,82 |
| | 197 | 6,81 | 2,02 | 25,938 | -250,05 |

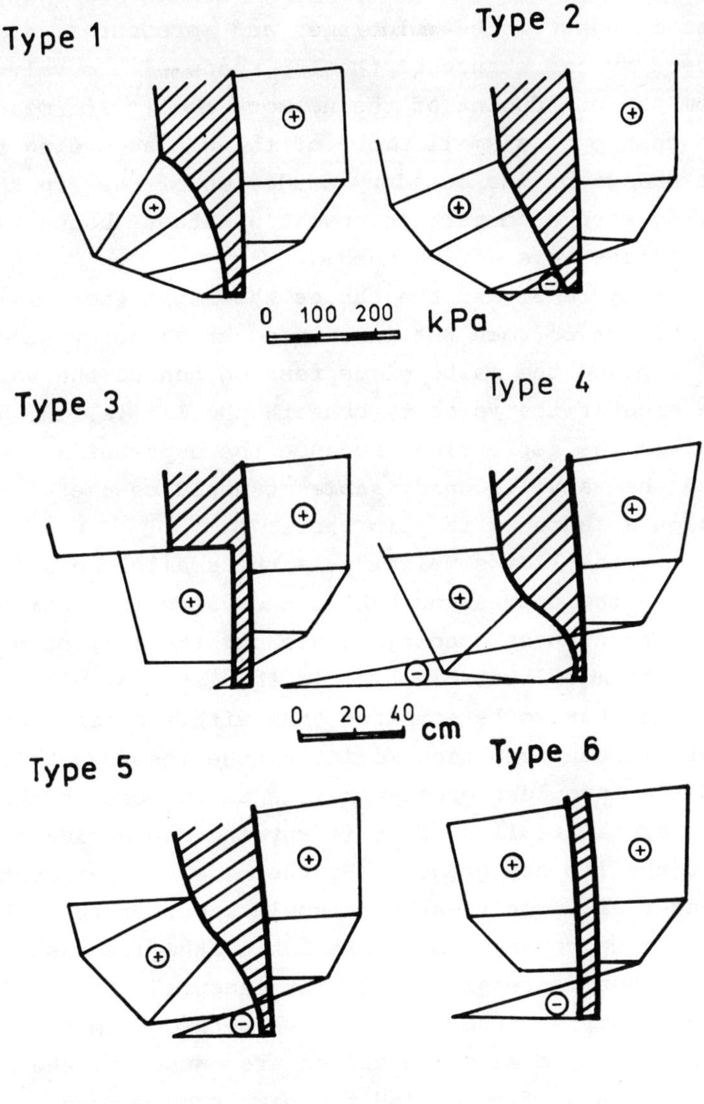


Figure 3. Distribution of pressures in the zone of reversed valve lip

The aim of the caltulacions was to find out what changes of the velocities and pressures are caused by the change of the lip shape of the valve. From the standpoint of the network model it meant the change of a small range of the net according to the change of the boundary condition of the lip shape. To make a comparison possible, other boundary conditions were not changed.

Figure 3. and the tables show that there is a connection between the maximum point velocity under the lip and the value of depression behind the valve. The greater the velocity beneath the lip is, the greater is the depression. Because the depression behind the valve is unadvisable it would be useful to design a shape of the lip where:

- the flow´s velocity is the smallest one
- the depression behind the valve is minimum.

When these conditions are our starting points then the most suitable type is the lst one.

It has to be stressed that with the consideration of idealized flow /omitting the losses/ the calculation provides greater values of velocities than they are in reality. That is why the depressions behind the lip are greater. By the model experiments using real velocities which could occur in the culvert, a decrease of pressure behind the lip was observed but depressions were not measured at all. It is evident that the differencies between the calculation and the measured values are caused by the high turbulence of flow behind the partly submerged valve. The mutual comparison of the calculation of several alternatives together with the results of the model experiments - in spite of the differencies - enable to design an optimal shape of the lower part of the valve.

References:

Kazda, J. /1978/ Teorie přehrad - aplikace programových systémů /Theorie of the Dams, Application of Programming Systems/, Publishing house of the Technical University of Prague

Strauss, V. /1978/ Výskum hradiaceho segmentu v obtokoch plavebných komôr stupňa Gabčíkovo /Hydraulic Model Investigation of the Culvert Tainter Valves, Lock Gabčíkovo, Danube River/, Research report of the Institute of Hydrology and Hydraulics, Slovak Academy of Sciences, Bratislava

# SIMULATION OF FREE JET TRAJECTORIES FOR THE DESIGN OF AERATION DEVICES ON HYDRAULIC STRUCTURES

C.Y. Wei and F.G. DeFazio

Harza Engineering Company, Chicago

## INTRODUCTION

With the demand for energy growing rapidly worldwide, more and more high head hydroelectric projects are being proposed and constructed. Control of the flood flows through these projects requires design of high velocity spillways and/or outlet works. The flow through a high velocity spillway chute can easily exceed 40 meters per second and protection of concrete chute surface from damages due to cavitation is essential to provide satisfactory operation. One method of protection is to incorporate aeration devices in the spillway design to minimize such damages. A commonly used aeration device is an aeration ramp which deflects the flow from the concrete chute surface. This deflection creates a subatmospheric cavity downstream from the ramp which draws air into the flow. The air entrained is primarily in the boundary layer and protects the concrete surface by reducing and/or eliminating the formation of cavitation bubbles.

To achieve a proper degree of protection against cavitation, it is necessary to ensure that enough air is entrained into the jet. Experience to date indicates that the amount of air entrained into a jet is proportional to the velocity of the jet and the length of the cavity formed underneath the jet. It is therefore important to determine the jet trajectories for a given ramp geometry over the range of operating flow conditions. Pan, Shao et al. (1980) developed a calculation method using conformal transformations with the aid of physical models to estimate the cavity length. The procedure did not include the effect of subatmospheric pressure in the cavity and applies to straight chute surfaces only.

Since many spillway chute surfaces are curved and the cavities behind aeration ramps are subjected to various degree of subatmospheric pressures, it is important that the mathematical model used to predict cavity lengths is two-dimensional, incorporates the curved geometry of the chute surface, and the sub-

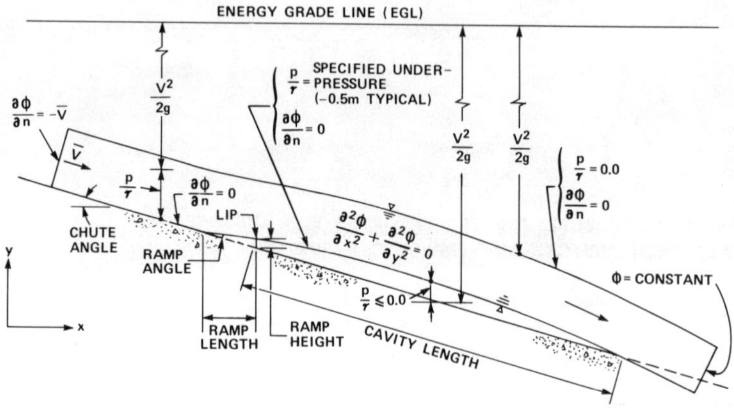


Figure 1.--Definition sketch for the aeration ramp trajectory problem.

atmospheric pressure along the lower free surface. The finite element method is a suitable method that can be used to include these additional requirements provide that the inherent non-linear boundary conditions which result from the unknown upper and lower free surfaces can also be successfully treated.

This paper describes the assumptions and a procedure used to analyze the free jet problem. The ideal potential flow theory is applied to simulate the high velocity flow passing an aeration ramp and the governing Laplace equation and associated boundary conditions are solved numerically by the finite element method.

## The Governing Equation and Boundary Conditions

The fluid flow is assumed incompressible, and irrotational. Only the steady state condition is considered in this study, and the flow is governed by the Laplace equation

$$\frac{\partial^2 \phi}{\partial x^2} + \frac{\partial^2 \phi}{\partial y^2} = 0 \tag{1}$$

in which $\phi$ is the velocity potential function, and x and y are respectively the Cartesian coordinates in the horizontal and vertical directions. The velocity potential function is defined such that

$$u = -\frac{\partial \phi}{\partial x}; \text{ and } v = -\frac{\partial \phi}{\partial y} \tag{2}$$

where u and v are the x and y velocity components.

As shown in Figure 1, the upper free surface is subject to a constant atmospheric pressure and the lower free surface is subject to a subatmospheric under-pressure induced by the entrainment of the air into the jet, that is

$$\frac{p}{\gamma} \leqslant 0.0 \tag{3}$$

where p is pressure and $\gamma$ is the specific weight of water. When the pressure is atmospheric, $p/\gamma = 0.0$, and the under-pressure normally varies from approximately atmospheric to about -1.0 meter of water. In the analysis, the jet is assumed to be solid or free of air bubbles and the effect of air entrainment on the trajectory is accounted for through Equation 3.

Along the free surfaces, the chute, and the ramp surfaces, the streamline tangency condition must be satisfied, that is the normal velocity components must be zero.

$$\frac{\partial \phi}{\partial n} = 0 \tag{4}$$

The incoming flow through the upstream boundary is assumed to be uniformly distributed such that

$$\frac{\partial \phi}{\partial n} = -\bar{v} \tag{5}$$

where $\bar{v}$ is the incoming velocity of the jet, and n is the unit outward normal.

The downstream boundary of the flow region is considered as a constant potential line. Similar to the free surfaces, the shape and the location of this boundary is not exactly known a priori. In the study, an arbitrary value of potential is assigned to this boundary, i.e.

$$\phi = \text{constant} \tag{6}$$

The Bernoulli equation is applied throughout the entire flow region such that

$$H = \frac{v^2}{2g} + \frac{p}{\gamma} + y \tag{7}$$

where H is the total head or the energy grade line (EGL) calculated at the ramp, v is velocity and y is elevation above a datum. The total head is assumed constant throughout the entire flow region being analyzed.

<u>Finite Element Procedure</u>

The design of an aeration ramp involves calculations of free jet trajectories with relatively high incoming jet velocities. The Froude numbers (defined as $\bar{v}/\sqrt{gd}$ where g is the gravitational acceleration and d is the depth of the incoming flow) range from about 2 to 12 are normally considered for the design of a ramp, and the curvature of the free surfaces of the jet are relatively mild. The free surface profiles can therefore be adequately represented by piecewise linear sections. Hence, elements with straight sides are used in this study. For lower Froude number cases where the jet profiles exhibit stronger curvature, the selection of an element which can accommodate curved sides may be desirable.

The basic finite element procedure applied to solve the ideal fluid flow problem defined previously is given by Chan et al. (1973). The governing partial differential equation is converted into finite element equations through the variational principle. The flow region to be analyzed is subdivided into eight-node quadrilateral elements, and each element consists of four triangular subelements. Each triangular subelement has three corner nodes and three mid-side nodes to facilitate quadratic representation of velocity potential in each element. Hence, the velocity components within an element vary linearly. The resulting system of linear equations are solved simultaneously for nodal velocity potentials by the Gaussian elimination method.

Free Surface and Mesh Adjustments

Free jet problems which involve two free surfaces are very difficult to solve because the locations of both free surfaces are not known initially. There are two boundary conditions which need to be satisfied on the free surfaces as given by Equations 3 and 4. Chan's procedure (Chan et al. 1973) for the determination of a free surface is applied individually to the upper and lower free surfaces. The free surfaces can be adjusted according to the following steps:

(1) An initial free surface location is assumed.
(2) Based on the constant pressure condition and the Bernoulli equation, the velocity potential is calculated for each free surface node starting from the downstream end where an arbitrary value of the velocity potential is assigned.
(3) Steps 1 and 2 are repeated for the second free surface.
(4) Form the system matrix for the nodal velocity potentials through the finite element procedure. The system matrix is modified to account for the boundary conditions.
(5) The system of equations are solved for nodal velocity potentials, and velocity components are calculated for all free surface nodes.
(6) The new free surface velocity vectors are applied to find the new free surfaces with spline fits.

These six steps constitute a basic computation cycle. At the end of the fifth step, the solution adequacy is checked and the computation can be terminated when the calculated downstream outflow satisfies the continuity criterion and the total head based on the re-calculated nodal velocity at each free surface node satisfies the EGL specified to a desired degree of accuracy. It has been indicated that satisfying the pressure condition on a free surface that may be improperly located, and then adjusting the surface to satisfy the streamline tangency condition is the most suitable method for the high velocity free surface flow problems (Chan et al. 1973, Larock et al. 1976). In step 5, the free surface nodal velocity components, including the velocity at the lip of the ramp, are calculated from the appropriate triangular subelements having one side in common with the free surface.

During the course of the free surface adjustments, the portion of the flow domain bounded by the upper and lower free surfaces may shift considerably in the direction of adjustments especially near the downstream boundary. It is therefore important that the finite element mesh is flexible, can be regenerated, and that the original intended element size distribution established for local accuracy requirements is maintained. For each free surface node, a fixed reference node located on the opposite free surface and on the same transverse mesh line is assigned. When the free surface node is given a new location, all the nodes on the same transverse mesh line are also adjusted proportionally with respect to the fixed reference node. A similar procedure is also applied to the opposite free surface.

As defined in Equation 6, the downstream boundary is a constant potential line. The velocity distribution along the boundary is not necessarily uniform, and the upper and lower free surfaces are not parallel to each other. Hence, the boundary would exhibit some curvature. The downstream boundary must junction with the free surfaces orthogonally. This constant potential boundary should also be adjusted after the upper and lower free surfaces are adjusted in step 6. A spline interpolation scheme can be applied to minimize the transverse velocity components using the upper junction node as a fixed reference point. The slope of the boundary at this fixed point should be normal to the slope defined by the last two corner nodes on the upper free surface. Thus, the lower junction node is also adjusted with downstream boundary. However, a simple circular arc which passes through the fixed reference node on the upper free surface and re-locates the lower junction node on the lower free surface has been found to be more direct method for adjusting the downstream boundary. The center of the circular arc is located at the intersection of the two straight lines defined by the last two corner nodes on each free surface.

## Solution Strategy

Experience indicates that, in general, when the calculated downstream outflow through the constant potential boundary satisfies the continuity condition with respect to the upstream inflow, the remaining boundary conditions are also satisfactorily satisfied. It has also been observed that the accuracy of the final solution is very sensitive to the initial mesh location, and reasonable solution can be obtained within 15 computation cycles when the difference between inflow and outflow, $\Delta Q$, is less than $\pm$ 0.01% of the inflow and the difference between the actual given total head and the computed total head at each free surface node is less than ± 0.1% of the upstream depth.

The lower boundary of the initial finite element mesh is constructed in accordance with the trajectory of a projectile issuing from the lip of the ramp using the average velocity calculated at the ramp and the specified under-pressure. The transverse lines are then constructed normal to this initial lower free surface. To alter the initial position of the mesh covering the free jet portion of the flow region, it is only necessary to alter the exit angle of the projectile in the direction desired during the initial generation of the mesh. Hence, the correct initial mesh can be obtained by selecting a proper exit angle at the ramp which determines the lower boundary profile as illustrated in the following section.

## FREE OVERFALLS

The free overfall problem has been studied by many researchers (Rouse 1943, Markland 1965, Strelkoff et al. 1970, Chow et al. 1979). Markland (1965) applied relaxation method in the complex-potential plane and obtained solutions for values of approach Froude number equal to, and greater than, unity. Strelkoff and Moayeri (1970) also obtained solutions through an integral equation derived from conformal mapping and singularity distributions. The results of these researchers were compared with the finite element solutions. Three runs with 15 computation cycles per run were made for each case. The initial mesh for each run were set up using the projectile equation and a slightly different exit angle at the brink. The initial exit angle of the final run was estimated by interpolation from the initial exit angles used in the earlier runs with respect to the values of $\Delta Q$ calculated at the downstream boundary. After the initial mesh was generated, the exit angle was set equal to the ramp slope for the subsequent free surface adjustments.

In the analysis, zero under-pressure was assumed and the upstream boundary was located at a distance equal to 3.5 times the incoming flow depth from the brink so that the results can be compared with the normalized experimental data given by Rouse. The initial finite element mesh and the final solution in flow net form for $F_r$ = 1.0 case are shown in Figures 2a and 2b respectively. Figure 2c shows the free surfaces obtained for $F_r$ = 1.0, 2.0, 4.0, and 8.0 cases, and these results agree exceptionally well with Markland's solution except the free surface in the approach channel for $F_r$ = 1.0. In Markland's analysis, the calculation was extended to ten depths instead of 3.5 depths from the brink and a lower approach free surface was obtained for $F_r$ = 1.0. However, when the length of the approach channel in the finite element analysis was extended to seven depths, the calculated free surface in the approach channel matched well with Markland's solution. For $F_r$ = 1.0 case, the upper free surface compares very well with Rouse's data while

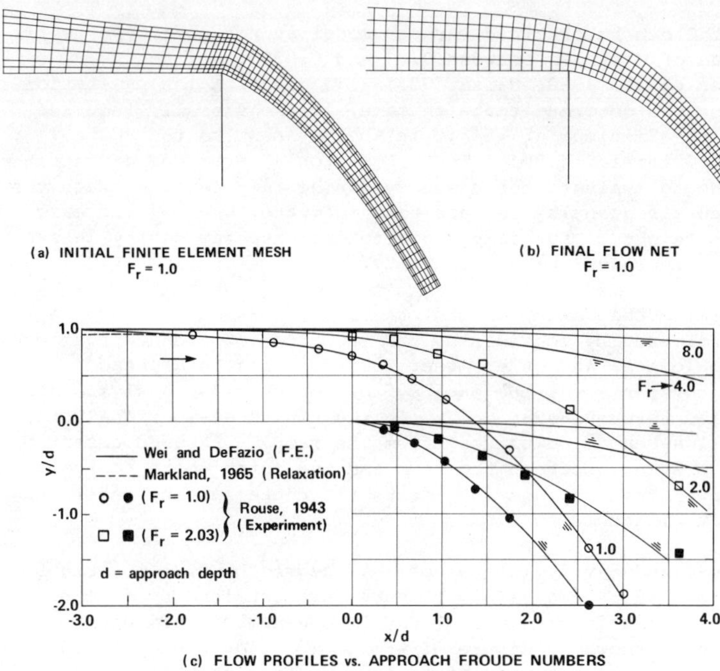


Figure 2.--F.E. solutions of the free overfalls.

slightly lower lower free surface was measured, and this observation also holds true for other cases tested for $F_r$ = 1.74, 2.03, 2.48, and 3.26. The lower measured lower free surface is probably due to the fact that the actual velocity is smaller in the boundary layer near the bottom of the test flume. Rouse's data also tend to show slightly higher upper free surfaces near the brink. The computed brink depth for $F_r$ = 1.0 was 0.710 of the approach depth which is slightly lower than the theoretical value of 0.715 given by Rouse, and the calculated vertical thickness of the jet was also very close to 2/3 of the approach depth as given by Rouse.

## TRAJECTORY ANALYSES FOR THE DESIGN OF AERATION RAMPS

When designing an aeration ramp, it is important to make certain that adequate amount of air is entrained into the jet through the cavity for the flow under consideration. Hamilton (1978) proposed that the quantity of air demanded by a jet can be expressed by the following empirical equation:

$$q_a = CV_L L \tag{8}$$

where $q_a$ is the volume of air demanded by the jet per unit time and unit width of the chute, C is a constant, $V_L$ is the average jet velocity over the ramp and L is the cavity length. The constant C can be obtained through model and/or prototype tests. The value of C varies between 0.01 and 0.035 (Hamilton 1978, Pan, Shao et al. 1980, Harza 1981). To minimize the cavitation damage on the concrete chute surface, it is generally required to provide a minimum of 8% air in the jet next to the chute surface (Quintela, 1980). Thus, Equation 10 provides a practical means to evaluate the adequacy of the ramp design. When the estimated air quantity is less than adequate, one may increase the ramp height or the ramp angle to increase the cavity length and hence increase the air quantity.

To determine the cavity length, the total head at the ramp is first estimated by considering the entrance loss, chute surface friction losses, and other minor losses. This total head is assumed constant over the entire flow region being analyzed and is applied through Equation 7. In the finite element analysis, the spillway chute downstream from the ramp is assumed absent, and the distance between the ramp and the point where the computed lower free surface intersects the chute surface determines the cavity length.

The normalized cavity lengths of a 17 degree by 0.75 m (height) ramp on a 15.5 degree straight chute were obtained by finite elements for various values of the Froude numbers and under-pressures as shown in Figure 3. For comparison, model (1:50) and prototype observations of cavity length with total energy heads of 35 to 40 meters were made and these results are also included in Figure 3. It indicates that as the under-pressure becomes greater the cavity length is reduced because the jet trajectory is lowered by the increased pressure differential between the upper and lower free surface. This effect is more pronounced at the higher Froude numbers because the jet is thinner and its mass is effected more by the pressure differential. The under-pressure is also dependent on the flow path of the air from atmosphere to the cavity. A restricted flow path such as through a conduit to the cavity will result in a larger pressure differential than an unrestricted opening. Therefore, the actual cavity length will occur where air demand will equal the air supply for the same under-pressure. Since the prototype flows are very turbulent with considerable spray, a range of estimated cavity lengths is shown for each flow tested. The prototype under-pressures estimated from the finite element solution of -0.05 m to -0.5 m were confirmed by piezometric readings taken at the prototype ramp.

For preliminary design purpose, at a selected ramp site, the cavity lengths can be computed for different ramp lengths and ramp angles for the flow condition under consideration. Typical design curves are shown in Figure 4 for ramps on a 15.5 degree chute as well as ramps on a 21.8 degree chute with a 55 m radius upstream concave curve. An under-pressure of -0.5 m was used. The chart can be used to size the ramp in accordance with the cavity length and air quantity required for the flow condition

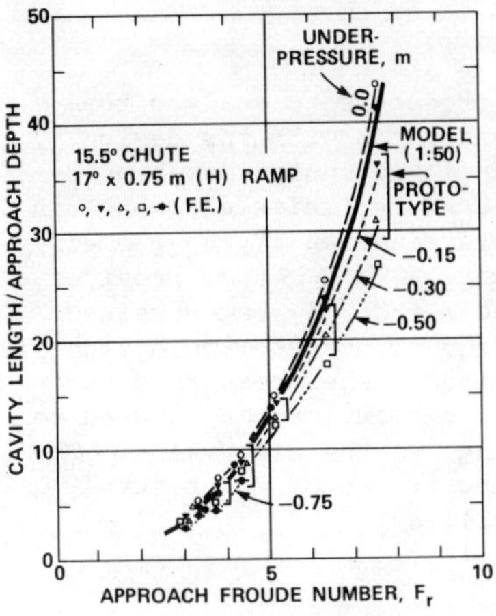


Figure 3.--Normalized cavity length vs. Froude number.

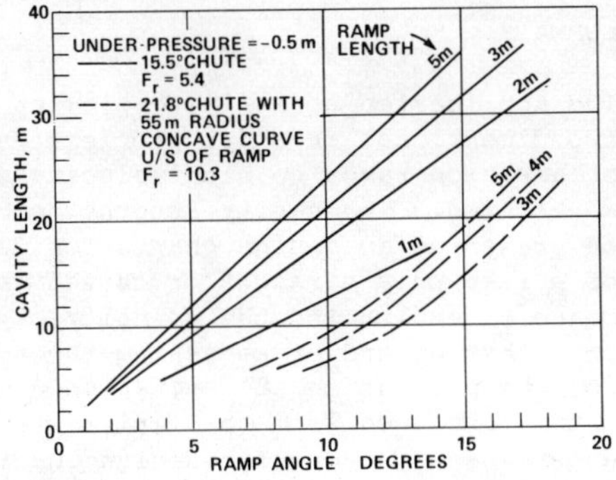


Figure 4.--Design curves for aeration ramps.

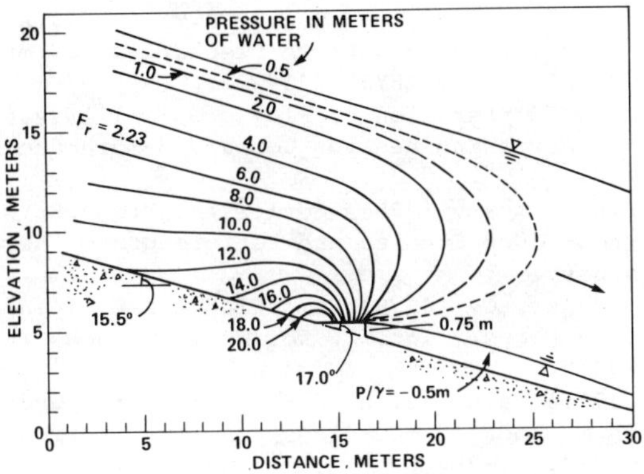


Figure 5.--Free surfaces and pressure distribution for $F_r$ = 2.23 flow.

being considered. It can be seen that the relationship between cavity length and ramp angle is not linear for curved chute surface (Figure 4).

The computed upper and lower free surfaces for the 17 degree ramp with a Froude number of 2.23 and an under-pressure of -0.5 m are shown in Figure 5. The calculated pressure distribution at the ramp which is useful for the design of the side walls is also shown in Figure 5.

## SUMMARY

The application of a finite element procedure to analyze the two dimensional free jet problems associated with the design of aeration ramps on high velocity spillway chutes has been demonstrated. The cavity lengths computed by finite elements can be presented as design charts for determination of the geometry of a ramp on a straight or curved chute. This method provides a means of estimating the quantity of air that may be entrained into the jet and of assessing the adequacy of the ramp design for the prevention of cavitation damages. The pressure distribution obtained from the finite element solution can be used to determine the hydraulic design loadings on the side walls. The procedure is also useful for analyzing trajectories of jets issuing from the flip bucket of a spillway.

## REFERENCES

Chan, S.T.K., Larock, B.E., and Hermann, L.R. (1973) Free-Surface Ideal Fluid Flows by Finite Elements. J. of the Hyd. Div., ASCE, 99, HY6, 959-974.

Chow, W.L., and Han, T. (1979) Inviscid Solution for the Problem of Free Overfall. J. of Applied Mech., 46, 1-5.

Diersch, H.J., Schirmer, A., and Busch, K.F. (1977) Analysis of Flows with Initially Unknown Discharge. J. of the Hyd. Div., ASCE, 103, HY3, 213-232.

Hamilton, W.S. (1978) Aeration of Flow down Spillway. An Unpublished Harza Engineering Company Memorandum, February 27.

Harza Engineering Company (1981) Guri Final Stage-Existing Chute 3: Report on 1980 Studies and Performance of Aeration Ramps, January.

Larock, B.E., Taylor, C. (1976) Computing Three-Dimensional Free Surface Flows. Int. J. for Num. Methods in Engrg., 10, 1143-1152.

Markland, E. (1965) Calculation of Flow at a Free Overfall by Relaxation Method. Proceedings, the Institution of Civil Engineers, 31, 71-78.

Pan, S.B., Shao, Y.Y., Shi, Q.S., and Dong, X.L. (1980) The Self-Aeration Capacity of the Water Jet over the Aeration Ramp. J. of Hyd. Engrg. (Shuili Xuebao, PRC), No. 5, 13-22.

Rouse, H. (1943) Discussion on "Energy Loss at the Base of a Free Overfall" by Moore, Trans., ASCE, 108, 1383-1387.

Strelkoff, T., and Moayeri, M.S. (1970) Pattern of Potential Flow in a Free Overfall. J. of the Hyd. Div., ASCE, 96, HY4, 879-901.

Quintela, A.C. (1980) Flow Aeration to Prevent Cavitation Erosion. Water Power & Dam Construction, January, 17-22.

# 18. PARAMETER ESTIMATION AND OPTIMIZATION

# A GENERAL APPROACH TO WATER QUALITY

Eroğlu,V., Yücel M.,
İ.T.Ü. C.Eng.Dept., I.D.M.M.A.C.Eng.Dept.

## 1. OPTIMIZATION OF THE TREATMENT PERFORMANCES

A river has been divided into N number of reaches starting from the first polluting source. Where the tributaries joined the river, the junction point was taken at the of the section, (Fig 1)

The oxygen deficit in any section of the river at any time has been calculated by the following formulae of streeter-phelps, (1).

$$D = \frac{k}{r-k} . L_o . (e^{-kt} - e^{-rt}) . D_o . e^{-rt} \tag{1}$$

where:
D : dissolved oxygen deficit at t=t , mg/1
D : dissolved oxygen deficit at t=0 , mg/1
$L_o$: $BOD_5$ at t=0 , mg/1
k : oxygen consumtion coefficient, $day^{-1}$
r : oxygen take up coefficient, from the air, $day^{-1}$

We need to know the location and the value of the maximum deficit, ($t_c$ , $D_c$). Starting from the equation 1

$$t_c = \frac{1}{r-k} . \ln \{ \frac{r}{k} . (1 - \frac{(r-k)}{k} . \frac{D_o}{L_o}) \} \tag{2}$$

$$D_c = \frac{k}{r} . L_o . e^{-rte}$$

are written.
We assume that the rate of flow is constant through a section. The value of the dissolved oxygen can be expressed as.

$$O = DO - D \tag{4}$$

where

O: dissolved oxygen concentration, mg/l
DO: dissolved oxygen saturation value, mg/l

Time of flow to the section where the minimum dissolved oxygen concentration exists and the oxygen deficit at that section can be found by (equation 2) and (equation 3) respectively.

The mathematical formulation of the optimization problem with equal degrees of treatment can be done in the following form:

Function to be optimized $\sum_{i=1}^{N} C_i(\varepsilon_i)$ (5)

Constraint $DO_i \geqslant SO_i \quad i=1,2,\ \ldots,\ N$ (6)

Requirement for the minimum degree of treatment: $\varepsilon_i > 0{,}35 \quad i=1,2,\ \ldots,\ N$ (7)

$DO_i$ : Minimum dissolved oxygen at section i, mg/l
$SO_i$ : Standart " " " " " mg/l
$_i$ : treatment efficiency

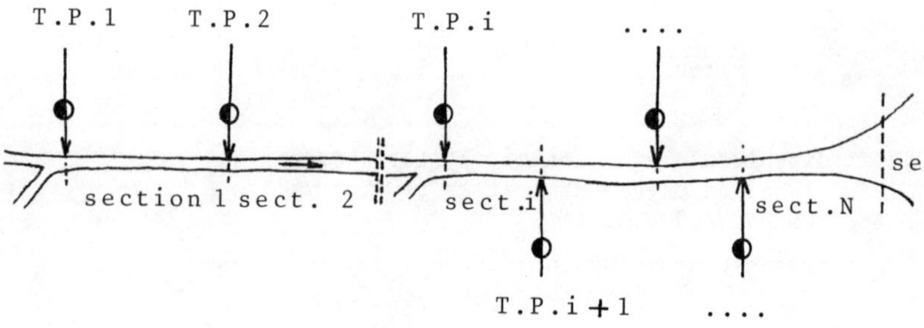


Figure 1. Reaches on a river

Requirement for the equal degrees of treatment

$$\varepsilon_1 = \varepsilon_2 = \ldots\ldots\ldots\ldots = \varepsilon_n \qquad (8)$$

The method developed to solve this problem is outlined as follows: The initial construction costs and the operational costs of the plant increase with the increasing degree of treatment. Threfore minimum degree of treatment meeting the requiremts for the quality standards will give the optimum solution .

At least a primary mechanical treatment should be applied to keep the floating and settling matters at a certain amount and to remove them at the discharge points. The plants constructed for that purpose have a $BOD_5$ removal efficiency of $\varepsilon_i \geqslant 0,35$, then we may assume the same value for $\varepsilon_i$.

Starting from the fact that $\varepsilon_i$=0,35 for the first treatment plant at the upper end of the river, we calculate the dissolved oxygen concentration in the upper, critical and lower ends of the section. When we assume $\varepsilon_1$=0,35, we face with two cases in the first section of the river, either the standart is met or not. In the second case we increase the $\varepsilon$ with an amount of $\Delta\varepsilon$ and continue in the same way till the standart is met. We take the increased last value of the first treatment plant as the initial value of the second plant. We again control the standart. If it is met we proceed with the third plant and if it is not met we return to the first plant and again increase its $\varepsilon$ as $\Delta\varepsilon$ and so on. This procedure will continue till the standards are met in all sections.

The $\varepsilon_1 = \varepsilon_2 \, ........ = \varepsilon_n$ efficiencies are the optimum values rendering the total cost minimum in the case of equal degree of treatment. Since the optimum $\varepsilon_i$ value was found, the minimum total cost can be calculated by the equation 5.

It is very difficult to do the computations manually in the case of many outfalls, a computer programm was developed for the solution of the problem (2).

From the accuracy point of the calculation we should choose the $\Delta\varepsilon$ values as little as necessary. It is convenient to choose $\Delta\varepsilon$ between 0,01 and 0,05. For running time economy in the computers, we may first start the programm by choosing a value of 0,05 to 0,10 and obtain an approximate optimum value. We subtract a $\Delta\varepsilon$ from that value and use the obtained value as the first efficiency of the first plant. For example we may take $\Delta\varepsilon = 0,10$ at the beginning and find the first optimum performance as $\varepsilon_i$=0,60 and then reach to an accurate value between 0,60 to 0,50. By taking $\varepsilon_1$=0,50 and $\Delta\varepsilon = 0,01$ we may obtain the accurate optimum treatment performance

REFFERENCE

1. THOMANN, R.V., "Systems Analysis and Water Quality Management", Mc Graw-Hill Book Company, New York, 1972
2. EROĞLU, V."Nehir Havzalarındaki Tasfiye Tesisi Verimlerinin Optimizasyonu", İ.T.Ü. İnşaat Fakültesi Matbaası, 1980
3. LIEBMAN, J.C., LYNN, W.R., "The Optimal Allocation of Stream Dissolved Oxygen", Water Resources Res, V.2, No.3, Third Quarter, (1966), pp. 581-591

# AQUIFER PARAMETER IDENTIFICATION USING STOCHASTIC APPROXIMATION

S.K.Goyal

Associate Professor in Civil Engineering
Punjab Engineering College, Chandigarh, India.

## ABSTRACT

Identification of diffusivity distribution in an inhomogenous aquifer from groundwater level fluctuations using stochastic approximation is presented. Equation governing the radial ground water flow is suitably modified into a state-space formulation using an explicit finite difference scheme. The observation dynamics is represented in terms of the previous drawdowns at the considered point and the two neighbouring points and the noise in modelling and observations. Taking expectation of the observation dynamics, the parameters are represented explicitly in terms of the second order moments of the observed drawdons. Stochastic approximation is used to estimate these moments recursively which in turn yield the diffusivity parameter. The method is conceptually simple, takes into account the uncertainties in modelling and observations and provides direct and realistic estimates of diffusivity distribution. The algorithm is successfully applied to recover diffusivity values from the simulated drawdown data for a radially heterogenous reservoir.

## INTRODUCTION

Identification of aquifer parameters from ground water levels has been attempted by a variety of methods, such as - regression analysis (Jahns,1966), gradient searching (Chen et al, 1974), quasilinearisation (Yeh,1975), linear programming (Kleinecke,1971), use of flatness or smoothness criterion (Emsellem and de Marsily,1971; Nutbrown,1975), algebraic method using spline interpolation (Sagar,1973), energy dissipation approach (Nelson,1968) or the use of groundwater hydrographs and maps (Headworth,1972). These methods either require a considerable computational effort or yield physically unrealistic estimates or require a large amount of water level data. Moreover these methods do not account for the uncertainties in modelling and observations.

Chen and Seinfeld (1972) attempted the parameter identification for a diffusion equation in the presence of measurement errors, using a complex nonlinear filter where the parameters are considered as additional state variables. But the procedure requires a considerable computational effort.

The short-comings in the existing methods for the identification of aquifer parameters can be overcome using stochastic approximation (Kubrusly,1978) technique.

FLOW EQUATIONS AND STATE-SPACE FORMULATION

Problem under study relates to the unsteady radial flow in a nonleaky, confined and heterogenous reservoir described as

$$\frac{1}{r}\frac{\partial}{\partial r}[T(r).r\frac{\partial s}{\partial r}] = S\frac{\partial s}{\partial t}, \tag{1}$$

where r is the radial distance from the pumped well; t is the time since pumping started; s is the drawdown; T(r) is the transmissivity at a distance r and S is the storage coefficient assumed as constant over the aquifer region.

Equation 1 can be rewritten as

$$\frac{1}{r}\frac{\partial}{\partial r}[D(r).r\frac{\partial s}{\partial r}] = \frac{\partial s}{\partial t}, \tag{2}$$

where D(r)=T(r)/S and is termed as diffusivity at the radial distance r.

Simplifying Equation 2,

$$\frac{\partial s}{\partial t} = D(r)\frac{\partial^2 s}{\partial r^2} + \frac{\partial s}{\partial r}[\frac{D(r)}{r} + \frac{\partial D(r)}{\partial r}] \tag{3}$$

Let $$\frac{D(r)}{r} + \frac{\partial D(r)}{\partial r} = \alpha_1(r) \tag{4}$$

and $$D(r) = \alpha_2(r) \tag{5}$$

$$\therefore \quad \frac{\partial s}{\partial t} = \alpha_1(r)\frac{\partial s}{\partial r} + \alpha_2(r)\frac{\partial^2 s}{\partial r^2} \tag{6}$$

Applying explicit central and forward differences and rearranging, Equation 6 becomes

$$\begin{aligned} s_{i,k+1} = & [\alpha_2(r)\frac{\Delta t}{(\Delta r)^2}]s_{i-1,k} + [1-\alpha_1(r)\frac{\Delta t}{\Delta r} \\ & -2\alpha_2(r)\frac{\Delta t}{(\Delta r)^2}]\,s_{i,k} + [\alpha_1(r)\frac{\Delta t}{\Delta r} \\ & +\alpha_2(r)\frac{\Delta t}{(\Delta r)^2}]s_{i+1,k}, \end{aligned} \tag{7}$$

where $i \in \{2,\ldots,P\}$; P is the number of nodepoints (pumped well being the node 1)

$k \varepsilon \{1,\ldots,N\}$; N is the number of discrete points in time

$\Delta t$ = time internal

$\Delta r$ = space interval along the radial direction.

$$\text{Let} \quad \alpha_2(r)\frac{\Delta t}{(\Delta r)^2} = p_i^2 \tag{8}$$

$$\text{and} \quad \alpha_1(r)\frac{\Delta t}{\Delta r} = p_i^1 \tag{9}$$

$$\therefore \quad s_{i,k+1} = a_i^1 s_{i-1,k} + a_i^2 s_{i,k} + a_i^3 s_{i+1,k}, \tag{10}$$

where

$$a_i^1 = p_i^2 \tag{11}$$

$$a_i^2 = (1-p_i^1-2p_i^2) \tag{12}$$

$$a_i^3 = (p_i^1+p_i^2) \tag{13}$$

From Equations 11 to 13,

$$a_i^1 + a_i^2 + a_i^3 = 1 \tag{14}$$

Equation 10 has been obtained using explicit finite difference scheme and as such it is tractable only if the space and time intervals satisfy the stability criterion. According to Rushton (1973), the stability criterion for a heterogenous radial reservoir becomes

$$\frac{t.D_{max}}{(\Delta r)^2} \leq 0.25, \tag{15}$$

where $D_{max}$ is the maximum value of the diffusivity in the aquifer region.

Further Equation 10 represents the drawdown in terms of the previous drawdowns when there is no error in modelling the flow. Accounting for the errors in modelling, Equation 10 gets modified as

$$s_{i,k+1} = a_i^1 s_{i-1,k} + a_i^2 s_{i,k} + a_i^3 s_{i+1,k} + w(k),\ldots \tag{16}$$

where w(k) is the space-invariant modelling noise which is white gaussian with zero mean, and covariance Q.

The parameters $a_i^1$, $a_i^2$ and $a_i^3$ in Equation 16 are determinable provided the true values of the draw-downs are known. The water level observations are subject to error and are represented as

$$y_i(k+1) = s_{i,k+1}+v(k+1) \tag{17}$$

where $y_i(k+1)$ is the measurement at the node i, at the time k+1 and v(k+1) is the space-invariant measurement noise which is assumed white gaussian with zero mean, and covariance R. Substituting for $s_{i,k+1}$ from Equation 16 in Equation 17, substituting back for s values in terms of y values and using Equation 14,

$$y_i(k+1) = \underline{y}_i'(k)\underline{a}_i - v(k)+w(k)+v(k+1), \tag{18}$$

where $\underline{a}_i = [a_i^1 \quad a_i^2 \quad a_i^3]'$

$\underline{y}_i(k) = [y_{i-1}(k) \quad y_i(k) \quad y_{i+1}(k)]$

and ′ represents the transpose of a matrix.
The parameters $\underline{a}_i$ are to be determined from Equation 18 for the given observed drawdowns.

PARAMETER EXPLICITATION

The parameters can be expressed explicitly in terms of second order moments of the observation process as follows. Premultiplying both sides of Equation 18 by $\underline{y}_i(k)$ and taking expectation,

$$E[\underline{y}_i(k)y_i(k+1)]= E[\underline{y}_i(k)\underline{y}_i'(k)]\underline{a}_i - \underline{R}, \tag{19}$$

where $\underline{R} = R\,[1 \quad 1 \quad 1]$

Let $E[\underline{y}_i(k).y_i(k+1)]+\underline{R}=P_i^1 \qquad (20)$

and $E[\underline{y}_i(k)\ \underline{y}_i'(k)] = P_i^o \qquad (21)$

$\therefore \quad \underline{a}_i = [P_i^o]^{-1} . P_i^1 \qquad (22)$

Equation 22 describes the parameters explicitly in terms of the correlations $P_i^1$ and $P_i^o$, provided the matrix $P_i^o$ is invertible. $P_i^o$ and $P_i^1$ are not known completely but since these are second order moments of the noisy observation process and as such these can be determined using the recursive estimation technique of stochastic approximation (Kubrusky,1978).

STOCHASTIC APPROXIMATION ALGORITHM

Stochastic approximation is an extension of the gradient method to stochastic systems. In general,it is a class of recursive algorithms for solving equations of the form

$$g(X) = E\{f(y,X)\} = 0 \tag{23}$$

where y is the observed value of the dependent variable and X is the unknown parameter vector.

Solution of Equation 23 by stochastic approximation has the following structure (Ljung,1977),

$$\hat{X}(k+1) = \hat{X}(k) + \gamma(k+1) . f[y(k+1), \hat{X}(k)], \tag{24}$$

where $\hat{X}(k)$ is the estimate of the unknown parameter vector X at the time k; y(k+1) is the one-step ahead prediction for the measurement sequence and $\gamma(k+1)$ is a decreasing scaler sequence which assures convergence of $\hat{X}(k+1)$ to X.

The correlation matrix $P_i^o$ and the $P_i^1$ can be evaluated recursively (Kubrusly,1978) as an extension of the general structure as follows.

$$P_i^o(k+1) = P_i^o(k) + \mu(k)[\underline{y}_i(k)\underline{y}_i'(k) - P_i^o(k)] \tag{25}$$

$$P_i^1(k+1) = P_i^1(k) + \lambda(k)[\underline{y}_i(k) y_i(k+1) + \underline{R} - P_i^1(k)] \tag{26}$$

$\mu(k)$ and $\lambda(k)$ are real sequences which assure convergence. These may be taken among the class

$$\mu(k) = \frac{1}{(\theta k+\psi)^{\nu}} : 0 < \theta \leq 1, \psi > 1, \quad 0.5 < \nu \leq 1.$$

$$\lambda(k) = \frac{1}{(\theta k+\psi)^{\nu}} : 0 < \theta \leq 1, \psi \geq 1, \quad 0.5 < \nu \leq 1.$$

According to Equation 22,

$$\underline{a}_i(k) = [P_i^o(k)]^{-1} . P_i^1(k) \tag{27}$$

$P_i^o(k)$ and $P_i^1(k)$ are evaluated from Equations 25 and 26 and thus the parameter vector $\underline{a}_i(k)$ is known from Equation 27.

Equation 27 requires inversion of the correlation matrix $P_i^o(k)$ at each time step. The time-aonsuming matrix inversion can be avoided by suitably modifying Equation 27.

From Equation 25, using the matrix inversion lemma (Mendel, 1976),

$$[P_i^o(k+1)]^{-1} = \frac{1}{1-\mu(k)}\left\{[P_i^o(k)]^{-1} - \frac{\mu(k)[P_i^o(k)]^{-1}\underline{y}_i(k)\underline{y}_i'(k)[P_i^o(k)]^{-1}}{1-\mu(k)+\mu(k)\underline{y}_i'[P_i^o(k)]^{-1}\underline{y}_i(k)}\right\} \tag{28}$$

Using Equations 26 and 28, it can be proved that

$$\underline{a}_i(k+1) = \frac{1-\lambda(k)}{1-\mu(k)} \{1-\mu(k)[P_i^o(k+1)]^{-1} \underline{y}_i(k).$$
$$\underline{y}_i'(k)\}.\underline{a}_i(k)+\lambda(k)[P_i^o(k+1)]^{-1}.$$
$$[y_i(k+1)\underline{y}_i(k) + \underline{R}] \tag{29}$$

Equation 29 provides asymptotically unbiased estimates of the parameters $\underline{a}_i$, as $P_i^o(k+1)$ can be calculated according to Equation 28. Knowing $a_i^1(k)$ and $a_i^2(k)$ and $a_i^3(k)$, the parameters $p_i^1(k)$ and $p_i^2(k)$ are evaluated using least squares. From $p_i^2(k)$, the diffusivity D(r) can be determined using Equations 8 and 5.

## APPLICATION TO DATA

Water level data for a radially heterogenous reservoir are simulated using an explicit finite difference scheme, and ass assuming, the number of nodes 6, time interval = 1 day, space interval = 1000 m, initial head $h_o$=10 throughout, head at the well=$h_o e^{-(k-1)\Delta t}$, head at the outer boundary =head at the node 5; diffusivities at the nodes 1 to 6 = $2.0 \times 10^5$, $1.6 \times 10^5$, $1.2 \times 10^5$, $0.7 \times 10^5$, $0.4 \times 10^5$ and $0.1 \times 10^5$ $m^2$/day respectively. Heads and thus the drawdowns at various nodes with time are simulated.

The stability criterion (Equation 15) is checked for accepta-bility, assuming a maximum diffusivity of $2.0 \times 10^5$ $m^2$/day which gives a stability factor of 0.2. Stochastic approxima-tion algorithm (Equations 28 and 29) is applied to recover the diffusivities at the nodes 3 and 4. Initial values of $\underline{a}_i$ in Equation 29 were prescribed, based on an initial diffusi-vity of $0.5 \times 10^5$ $m^2$/day, as,

$$\underline{a}_i(1) = [0.05 \qquad 0.88 \qquad 0.07]'$$

Initial covariance $P_i^o(1)$ was assumed as an identity matrix. Measurement noise covaraince R was taken as zero. Convergence factors were chosen as

$$\mu(k+1) = \frac{1}{k+1.5},$$

and

$$\lambda(k+1) = \frac{1}{k+2.0}.$$

The algorithm updates the estimates with each additional set of observations. Diffusivity estimates at the end of success-ive sets of observations are given in Table 1.

Table 1 Diffusivity estimates with successive sets of observations

| Node | 3 | 4 |
|---|---|---|
| Initial estimate ( x $10^5$ $m^2$/day ) | 0.5000 | 0.5000 |
| Diffusivity estimate ( x $10^5$ $m^2$/day) after processing observation set | | |
| 1 | 0.6369 | 0.5026 |
| 2 | 0.7712 | 0.5083 |
| 3 | 0.8977 | 0.5187 |
| 4 | 1.0200 | 0.5355 |
| 5 | 1.1263 | 0.5597 |
| 6 | 1.2089 | 0.5916 |
| 7 | 1.2559 | 0.6311 |
| 8 | 1.2579 | 0.6775 |
| True value (x$10^5$ $m^2$/day) | 1.2000 | 0.7000 |

## DISCUSSION OF RESULTS AND CONCLUSIONS

It is apparent from Table 1, that the estimate for node 3 increases with each observation but tends to stabilise as more data are processed and the final estimate differs from the true value by 4.8 percent. The estimate for node 4 has been progressively increasing with each observation and the final estimate differs from the true value by 3.2 percent. It is expected that the estimate will stabilise as more data are processed. The algorithm has been applied to two nodes, but since the operation of the algorithm is pointwise, it can be extended to other nodes also.

It can be concluded that the stochastic approximation technique provides direct and realistic estimates of the diffusivity distribution in a radially heterogenous aquifer with a small number of observations. The estimates are updated as more and more data are processed till these stabilise. The computational effort, in terms of computer time and memory, is short. The technique also accounts for the uncertainties in modelling and observations.

## REFERENCES

Chen,W.H., and Seinfeld, J.H.(1972) Estimation of Spatially Varying Parameters in Partial Differential Equations. Int. J. Control, 15, 3 : 487-495.

Chen, W.H., Gavalas, G.R., Seinfeld, J.H., and Wasserman, M.L. (1974) A New Algorithm for Automatic History Matching. Soc. Pet. Eng. J., 14 : 593 - 608.

Emsellem, Y., and de Marsily, G. (1971) An Automatic Solution for the Inverse Problem. Water Resources Research, 7, 5 : 1264 - 1283.

Headworth,H.G. (1972) The Analysis of Water Level Fluctuations in the Chalk of Hampshire. J. Inst. Water Engrs., 26 : 107 - 124.

Jahns, H.O. (1966) A Rapid Method for Obtaining a Two-dimensional Reservoir Description from Well Pressure Response Data. Soc. Pet. Eng. J., 6, 4 : 315 - 327.

Kleinecke, D. (1971) Use of Linear Programming for Estimating Geohydrologic Parameters of Groundwater Basins. Water Resources Research, 7, 2 : 367 - 374.

Kubrusly, C.S. (1978) Linear DPS Identification in a Stochastic Environment. Comm. GSC - 10/78, Elect. Engg. Deptt., Catholic University, Rio de Janeiro.

Ljung, L.(1977) Analysis of Recursive Stochastic Algorithms. I.E.E.E. Trans. Automatic Control, AC-22 : 551 - 575.

Mendel,J.M. (1976) Discrete Techniques of Parameter Estimation. Marcel Dekker Inc., New York.

Nelson,R.W. (1968) In Place Determination of Permeability Distribution for Heterogenous Porous Media through Analysis of Energy Dissipation. Soc. Pet. Eng. J., 8 : 33 - 42.

Nutbrown, D.A. (1975) Identification of Parameters in a Linear Equation of Groundwater Flow. Water Resources Research,11 : 581 - 588.

Rushton, K.R. (1973) Discrete Time-steps in Digital Computer Analysis of Aquifers Containing Pumped Wells. J. Hydrology, 18 : 1 - 19.

Sagar, B. (1973) Caliberation and Validation of Aquifer Models. Tech. Report 17, Deptt. of Hydrology and Water Resources, Univ. of Arizona, Tucson, Arizona.

Yeh, W.W.G. (1975) Optimal Identification of Parameters in an Inhomogenous Medium with Quadratic Programming. Soc. Pet. Eng. J., 15:371-375.

# IDENTIFICATION OF SOIL PARAMETERS FOR AN INFILTRATION PROBLEM

U. Hornung and W. Messing

Institut für Numerische und instrumentelle Mathematik, University of Münster, West Germany

## INTRODUCTION

The numerical simulation of water flows in porous media is considered. Prerequisit for a valid model is a satisfactory measurement of the soil physical properties. In the case of unsaturated or saturated/unsaturated water flows the properties of the soil are represented by a pair of nonlinear functions. Measurement of these functions is up to now done by a sequence of laboratory experiments, each resulting in some information about a small part of these functions. The purpose of this paper is to study the possibility of identifying the functions from a rather simple experiment.

## FORMULATION OF THE PROBLEM

### General

The identification of material properties by inverse procedures has been examined in different fields during the last years. In Chavent and Cohen (1978) a procedure is proposed to identify a nonlinear coefficient in a degenerate parabolic equation, which describes displacement of oil by water. Chen and Liu (1981) examined the determination of a spatially varying thermal conductivity. Richter (1981) gave a relation of the inverse problem for an elliptic equation to the solution of a hyperbolic problem. Error estimates for the numerical solution of an inverse problem are given in the paper of Falk (1980).

In the field of groundwater flow Neuman (1980) proposed a strategy to identify the spatial distribution of aquifer properties from observed water levels at different points. An inverse problem for a nonlinear diffusion equation, e.g. the horizontal unsaturated flow equation, was considered by Cannon and DuChateau (1980). These authors solved the identification problem using additional measurements of boundary data. In DuChateau (1981) an algorithm is proposed, to determine the coefficient from a two-parameter family. The problem of identifying the soil physical properties for unsaturated water flow from infiltration experiments in the field will be discussed in Haverkamp (1982).

## Basic equations

Unsaturated/saturated water flows in an isotropic, homogeneous vertical soil column are modelled by the equation

$$\frac{\partial \Theta^*}{\partial t^*} = \frac{\partial}{\partial z^*} \left(K^* \cdot \frac{\partial}{\partial z^*} (\psi - z^*)\right) , \tag{1}$$

where $\Theta^*[cm^3/cm^3]$ is the volumetric water content, $K^*[cm/sec]$ is the conductivity, $\psi[cm]$ the pressure head, $z^*[cm]$ with $0 \leq z^* \leq L^*$ the depth, and $t^*[sec]$ the time. $\Theta^*$ and $K^*$ describe the soil physical properties. Both are nonlinear and monotonic functions of the pressure head $\psi$. They are bounded by $\Theta_r \leq \Theta^* \leq \Theta_s$ and $0 < K^* \leq K_s$, where $\Theta_r$ and $\Theta_s$ are the residual and saturated water content respectively and $K_s$ is the conductivity at saturation. $\Theta^*$ and $K^*$ are constant in the saturated part of the flow region, such that equation (1) is elliptic in this part, otherwise it is parabolic.

In order to get a dimensionless formulation, the following transformation is performed

$$\begin{cases} T^* = L^* \cdot (\Theta_s - \Theta_r)/K_s , \\ z = z^*/L^* , \quad t = t^*/T^* , \\ h(t,z) = - \psi(t^*,z^*)/L^* , \\ K(h) = K^*(\psi)/K_s , \quad \Theta(h) = (\Theta^*(\psi)-\Theta_r)/(\Theta_s-\Theta_r) . \end{cases} \tag{2}$$

Thus equation (1) becomes

$$\frac{\partial \Theta}{\partial t} = - \frac{\partial}{\partial z} \left(K \cdot \frac{\partial}{\partial z} (h + z)\right) . \tag{3}$$

Since the function $-\Theta(h)$ is monotone, equation (3) is an elliptic/parabolic partial differential equation for $t > 0$ and $0 < z < 1$.

## The inverse problem

The objective is the identification of the coefficients $\Theta(h)$ and $K(h)$ for $0 \leq h \leq 1$ by means of the following experiment.
Starting with the steady state

$$h_o(z) = 1 - z \qquad \text{for } 0 \leq z \leq 1 \qquad (4)$$

at $t = 0$, a constant flow rate

$$K \cdot \frac{\partial}{\partial z}(h + z)\Big|_{z=0} \equiv q_o > 0 \quad \text{for } t > 0 \qquad (5)$$

is supplied at the top of the soil column. Simultaneously a constant pressure head

$$h\Big|_{z=1} = 0 \qquad \text{for } t > 0 \qquad (6)$$

is maintained at the bottom. Equations (3), (4), (5), (6) and the coefficients $\Theta$ and $K$ make up a complete flow problem, which can be solved using numerical methods. If in addition the outflow at the bottom

$$q_1(t) = K \cdot \frac{\partial}{\partial z}(h + z)\Big|_{z=1} \qquad (7)$$

is measured, the question arises, whether it is possible to identify the coefficients $K(h)$ and $\Theta(h)$ by means of the response $q_1(t)$ related to the input $q_o$.

## NUMERICAL PROCEDURE FOR THE DIRECT PROBLEM

The flow problem given by equations (3), (4), (5) and (6) is solved numerically based upon the implicit-time discretisation

$$-\frac{\Theta(H^{j+1})-\Theta(H^j)}{\Delta t^j} = \delta_z\left(\frac{K^j+K^{j+1}}{2} \cdot \delta_z(H^{j+1}+z)\right), \qquad (8)$$

where $\delta_z$ means the usual finite difference discretisation. Because of the terms $\Theta(H^{j+1})$ and $K^{j+1} = K(H^{j+1})$ equations (8) become nonlinear. The necessity to solve systems of nonlinear equations is compensated by the advantage to preserve the water balance of the system (cf. Hornung and Messing (1981, 1982)). Equations (8) are solved by an iterative procedure, which is the Newton-algorithm with respect to the nonlinearity $\Theta(h)$. Strictly speaking for $\nu = 0,1,2,...$ the linear system

$$-\left(\Theta^{j+1,\nu} + \frac{d\Theta}{dh} \cdot (H^{j+1,\nu+1} - H^{j+1,\nu}) - \Theta^j\right) = \Delta t^j \cdot \delta_z\left(\frac{K^j+K^{j+1,\nu}}{2} \cdot \delta_z(H^{j+1,\nu+1}+z)\right) \qquad (9)$$

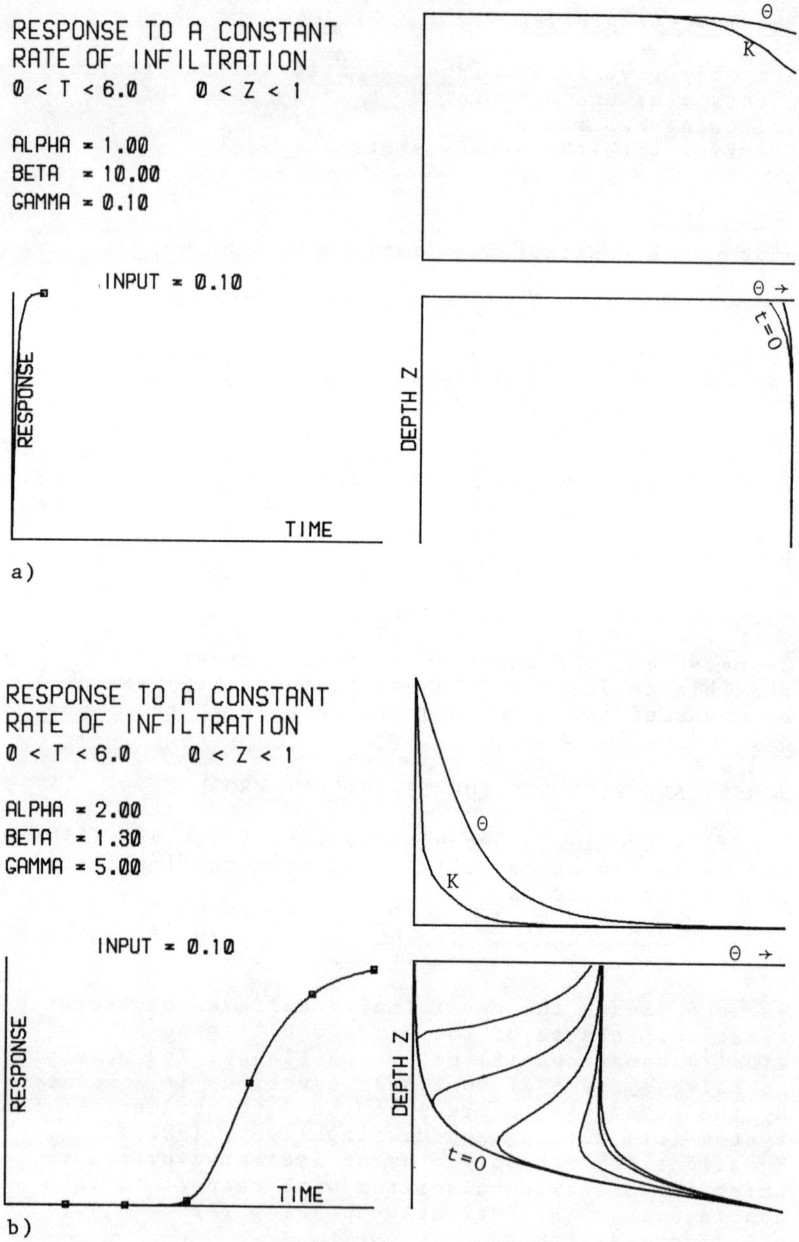


Figure 1: Coefficients Θ(h) and K(h) for $0 \leq h \leq 1$, water content profiles during simulation, and input/response relationship for two triplets of parameters α, β, γ.

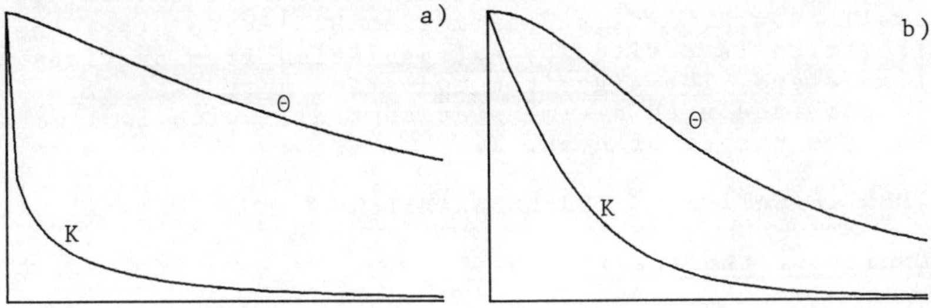


Figure 2: Coefficients Θ(h) and K(h) for $0 \leq h \leq 1$ corresponding to the triplets a) $\alpha=1$, $\beta=1.3$, $\gamma=1.05$ and b) $\alpha=2$, $\beta=2$, $\gamma=1$.

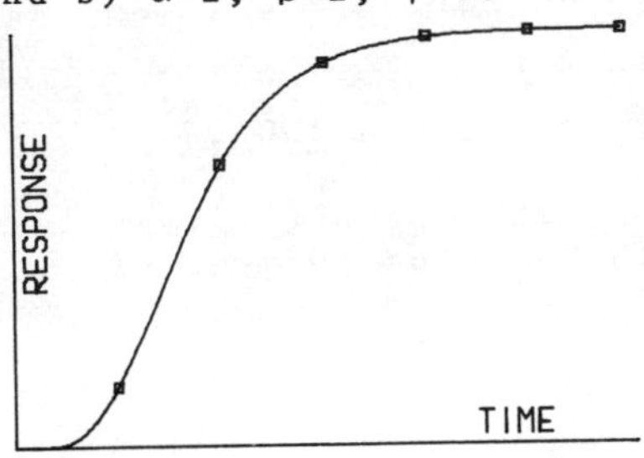


Figure 3: Input/response relationship corresponding to the two different soil physics in fig. 2.

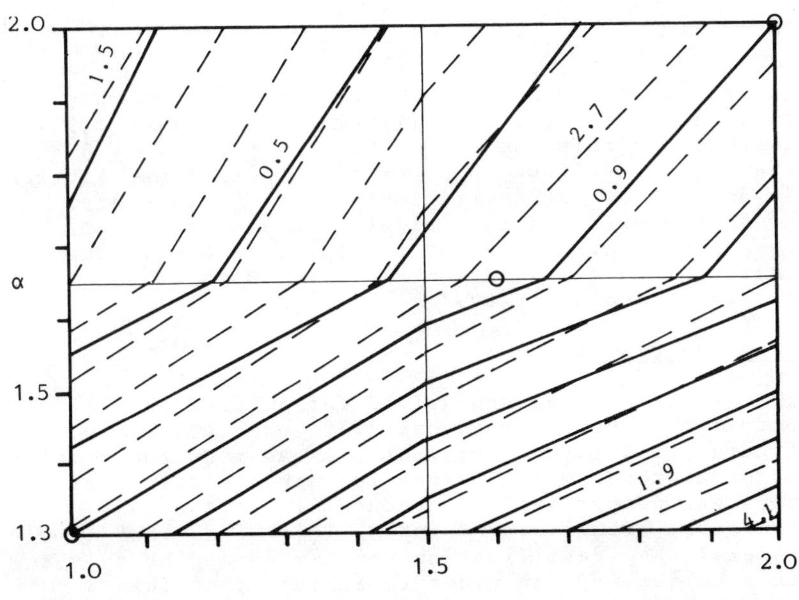


Figure 4: Isolines of $t_{10}$ (solid lines) and $t_{90}$ (dashed lines) with $\Delta t_x = 0.2$ for $\gamma=1.02$. Points marked by O are related to the same response.

is solved until the difference $|H^{j+1,\nu+1} - H^{j+1,\nu}|$ is small enough. A first estimation $H^{j+1,0}$ to start the iteration is derived by extrapolation from previous time steps. The algorithm converges very fast if it is combined with a time step control, which is coupled to the change of H and K.

## IDENTIFICATION OF THREE PARAMETERS

### Choice of the coefficients

Following a suggestion of van Genuchten (1980) the 3-parameter family

$$\Theta(h) = \begin{cases} (1 + (\alpha h)^{\beta})^{-\gamma} & \text{for } h > 0 , \\ 1 & \text{for } h \leq 0 , \end{cases} \tag{10}$$

with $\alpha, \gamma > 0$, $\beta > 1$ is chosen. The conductivity K is related to the function $\Theta(h)$ and the formula of Mualem (1976)

$$\begin{cases} K(h) = k(\Theta(h)) \quad \text{with} \\ k(\Theta) = \Theta^{1/2}\left( \int_0^{\Theta} \frac{d\tau}{h(\tau)} \Big/ \int_0^{1} \frac{d\tau}{h(\tau)} \right)^2 \end{cases} \tag{11}$$

is used to calculate K. The function $h(\Theta)$ is known from equation (10). The triplet of parameters $\alpha,\beta,\gamma$ is chosen within the bounds $0.2 \leq \alpha \leq 2$, $1.3 \leq \beta \leq 10$ and $0.1 \leq \gamma \leq 5$ due to numerical limitations. Figures 1a and b show the coefficients $\Theta$ and K, the water content profiles during simulation and the input/response relationship for two triplets $\alpha,\beta,\gamma$. The input $q_o = 0.1$ was chosen, this is 10% of the saturated conductivity.

### Negative results

The direct problem was solved for a large set of parameters and the responses were plotted, see e.g. figure 1. It turned out that one can find two triplets of parameters, which belong to complete different soil physical properties and produce almost the same response, cf. figures 2 and 3. This means that in general the identification problem cannot be solved in a unique way. In order to exclude that this phenomenon is purely local, which means that the response is insensitive to changes of parameters in some region, a more detailed analysis was carried out. Three characteristic points of the response curve, namely the times $t_{10}$, $t_{50}$, and $t_{90}$, which are defined as the moments at which 10%, 50%, or 90% of the input has been flowing out, were recorded and the approximate isolines plotted, see figure 4. The interpretation of figure 4 is that within the part of the parameter region which is shown the response changes considerably but is the same for the mentioned triplets. Moreover the isolines for $t_{10}$ and $t_{90}$, which connect these two triplets, take approximately the same path. The consequence of this coincidence is that for all parameters on the path the response is almost the same. This was verified by choosing some more points on this path and comparing the response to the others.

IDENTIFICATION OF TWO PARAMETERS

Since it is impossible to identify the coefficients within the 3-parameter family (10), the number of the parameters was reduced. The restriction

$$\gamma = 1 - 1/\beta \tag{12}$$

has been introduced by van Genuchten (1980) in order to get a closed-form expression for equation (11). He succeeded in fitting the curves given by equations (10), (11), and (12) to measured data. The bounds for the two parameters where now chosen as $0.5 \leq \alpha \leq 4$ and $1.1 \leq \beta \leq 10$. The isolines for $t_{10}$ and $t_{90}$ within this parameter region are shown in figure 5. For $\beta \geq 1.5$ and $\alpha \leq 1.5$ the isolines look very similar, which means that the response is almost insensitive to changes of $\beta$ and identification is impossible. Moreover within the parameter region $1.5 \leq \alpha \leq 2.5$ and $2 \leq \beta \leq 5$ a path is found on which $t_{10}$ and $t_{90}$ coincide. The three points $(\alpha,\beta) = (1.75,5)$, $(1.93,3.5)$, and $(2.25,2)$ correspond to different soil physical properties and result in the same response.

The only situation in which parameter identification can be possible is that the isolines of $t_{10}$ and $t_{90}$ are transversal. From figure 5 it is clear that transversality holds for the case $t_{10} > 1$, i.e. the infiltration process must be slow enough if the input-output relationship is used to identify the soil. The infiltration experiment gives sufficient information for the solution of the inverse problem only if the water takes long enough to move from the top to the bottom of the soil column. In this case the inverse problem has a locally unique solution. The condition mentioned can be written as $t_{10}^{*} > T^{*} = L^{*} \cdot (\Theta_s - \Theta_r)/K_s$.

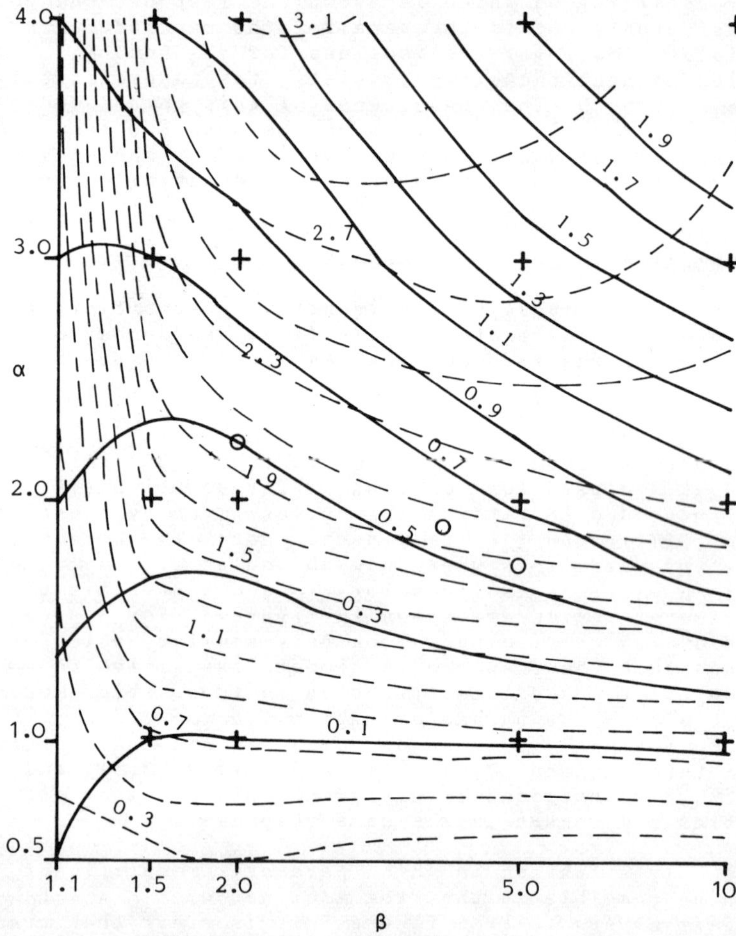


Figure 5: Isolines of $t_{10}$ (solid lines) and $t_{90}$ (dashed lines) with $\Delta t_x = 0.2$ in the case of the 2-parameter family. Points marked by O are related to the same response.

Further studies have to be done to investigate whether a combination of infiltration and defiltration experiments yield more information for the inverse problem.

## CONCLUSIONS

Identification of soil physical parameters from the input-output relationship in an infiltration experiment is possible only under certain circumstances. It is shown that there is no way to determine the properties of the soil from a three parameter family. A necessary condition when using two parameters is that the infiltration process is slow, i.e. that the response time is long enough. Otherwise the experiment is too simple to give information which characterizes the soil in a unique manner.

## ACKNOWLEDGEMENT

This study was supported by the Deutsche Forschungsgemeinschaft (Grant Ho 782/2).

## REFERENCES

Cannon, J.R., DuChateau, P.
1980. An inverse problem for a nonlinear diffusion equation. SIAM J. Appl. Math. 39, 272-289.

Chavent, G., Cohen, G.
1978. Numerical approximation and identification in a 1-d parabolic degenerated non-linear diffusion and transport equation. In: J. Stoer (ED.) Optimization Techniques, Part 1, LN in Control and Information Sciences, Springer, 282-293.

Chen, Y.M., Liu, J.Q.
1981. A numerical algorithm for remote sensing of thermal conductivity. J. of Comp. Physics 43, 315-326.

DuChateau, P.
1981. Monotonicity and uniqueness results in identifying an unknown coefficient in a nonlinear diffusion equation. SIAM J. Appl. Math. 41, 310-323.

Falk, R.S.
1980. Error estimates for the approximate identification of a constant coefficient from boundary flux data. Numer. Funct. Anal. & Opt. 2, 121-154.

Haverkamp, R.
1982. These, Grenoble. To appear.

Hornung, U., Messing, W.
1981. Simulation of two-dimensional saturated/unsaturated flows with an exact water balance. In:A.Verruijt & F.B.J. Barends (Eds.) Flow and Transport in Porous Media, A.A. Balkema, Rotterdam, 91-96.

Hornung, U., Messing, W.
1982. Truncation errors in the numerical solution of Richards' equation. To appear in Advances in Water Resources.

Mualem, Y.
1976. A new model for predicting the hydraulic conductivity of unsaturated porous media. Water Res. Res. 12, 513-522.

Neuman, S.P.
1980. Adjoint-state finite element equations for parameter estimation. In: Wang, Alonso, Brebbia, Gray, Pinder (Eds.), Finite Elements in Water Resources. University of Mississippi, 2.66-2.75.

Richter, G.R.
1981. An inverse problem for the steady state diffusion equation. SIAM J. Appl. Math. 41, 210-221.

van Genuchten, M.Th.
1980. A closed-form equation for predicting the hydraulic conductivity of unsaturated soils. Soil Sc. Soc. of Am. Journal 44, 892-898.

# OPTIMAL OPERATION OF WELLS IN A MULTILAYERED LEAKY AQUIFER

Takeshi Kawatani

Faculty of Engineering, Kobe University, Japan

## STATEMENT OF PROBLEM

The problem dealt with herein is to determine the optimal operation of production wells in a multilayered groundwater reservoir. The aquifer is assumed to be divided by interlying a relatively thin aquitard. Namely, the upper aquifer is regarded as a phreatic leaky aquifer while the lower aquifer as a confined leaky aquifer (see Fig.1). Hereafter, the water-table aquifer and the confined aquifer are referred to as I-layer and III-layer respectively. The aquitard is called II-layer. It is also assumed that some wells penetrate the I-layer only, some have screens only in the III-layer and some completely penetrate both I- and III-layers.

By the way, the overdevelopment of groundwater may cause troubles like land subsidence or intrusion of seawater into reservoir. Then it becomes necessary to regulate either the quantity of water pumped from wells or the water levels in wells. The control of quantity may be easier than that of water level. However, the problems associated with the usage of groundwater, e.g., land subsidence, are more directly related to the decline of water table or of piezometric head than the quatity of water pumped from wells. Therefore, in this study, the optimal operation of production wells will be determined in terms of controlling the water levels.

## GOVERNING EQUATIONS AND FINITE ELEMENT FORMULATION

The equation governing the steady saturated groundwater flow with pumpage is derived from the equation of continuity and the Darcy law and written in the form

$$L(h) = \frac{\partial}{\partial x_i} \left(K_{ij} \frac{\partial h}{\partial x_j}\right) - q_o = 0 \tag{1}$$

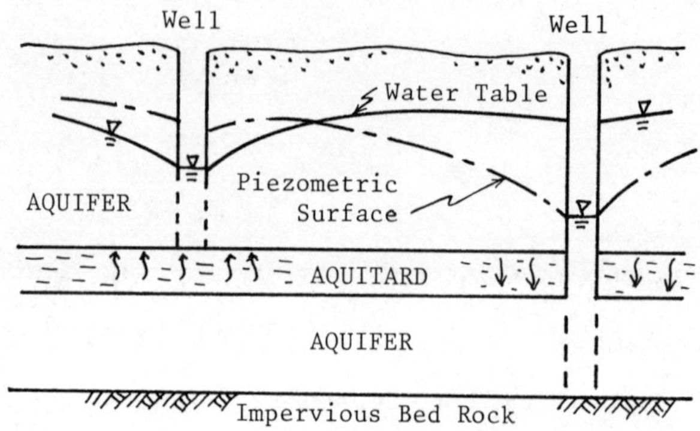


Figure 1 Illustration of a multilayered groundwater reservoir for analysis.

where h is the piezometric head, $q_o$ is the pumping rate per unit volume of porous medium and $K_{ij}$ is the hydraulic conductivities.

Equation (1) will be transformed into finite element form by using the Galerkin weighted residual method (Zienkiewicz 1971). The piezometric head h is approximated by a finite series of the form

$$h(x_i) = h_n I_n(x_i) \qquad (i = 1,2,3) \tag{2}$$

where $I_n(x_i)$ are the interpolation functions and $h_n$ are the unknown coefficients which represent solutions of Equation (1) at specified node within the solution domain. Substituting Equation (2) into Equation (1) and integrating the weighted residual equation by parts, one obtains

$$\left( \int_V K_{ij} \frac{\partial I_n}{\partial x_i} \frac{\partial I_m}{\partial x_j} dv \right) h_m = - \int_s q_i l_i I_n ds - \int_V q_o I_n dv \tag{3}$$

where V stands for the flow domain bounded by the surface s, $l_i$ is the direction cosines of the normal to s at any point and $q_i$ is the components of the specific discharge, i.e.,

$$q_i = - K_{ij} \frac{\partial h}{\partial x_j} \tag{4}$$

Usually, the vertical scale of aquifer is much smaller than the horizontal scale. Hence, the vertical component of the hydraulic gradient is negligible except in the very vicinity of well and the Dupuit approximation (Bear 1972) is applicable. In other words, water flows horizontally in the aquifers. On the other hand, water flows in the vertical direction in the aquitard where the vertical hydraulic gradient is predominant. Therefore, the interpolation function $I_n(x_i)$ becomes a function of x and y in the I- and III-layers and a function of z only in the II-layer.

The flow domain is segmented by a system of columns whose base areas are denoted by $A_e$. Moreover, each column is divided into three subdomains $e_I$, $e_{II}$ and $e_{III}$ corresponding to the phreatic aquifer, aquitard and confined aquifer, respectively as shown in Fig.2. In practice, by virtue of the Dupuit approximation, the interpolation functions defined within $e_I$ and $e_{III}$ become functions of x and y. Thus, the former function is defined in the upper plane of each volume element shown in Fig.3, while the latter is defined in the base plane of the volume element. The interpolation function for $e_{II}$ is a function of z only and defined along the vertical side of the volume element in Fig.3.

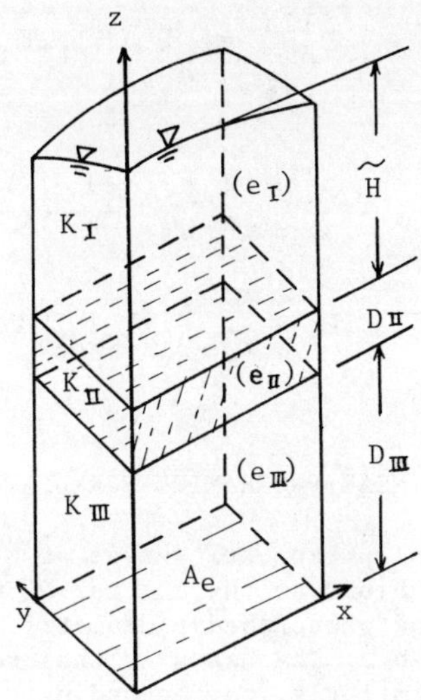

Figure 2 Segmentation of flow domain

Let $N_n$ represent the interpolation function defined within each subregion e. Then, taking the Dupuit approximation into account, Equation (3) is written in the form

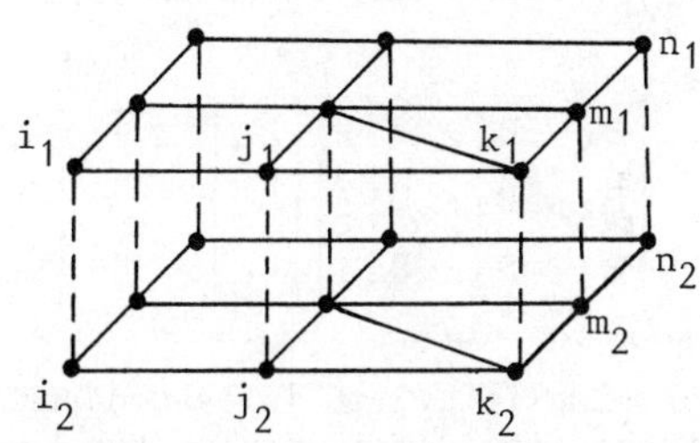

Figure 3 Volume elements

$$[E]\{h\} = -\{Q\}-\{P\} \qquad (5)$$

where $[E] = [E^I]+[E^{II}]+[E^{III}]$

$$E^I_{nm} = \sum_{e_I} (K_I H) \int_{A_e} \left\{\frac{\partial N_n}{\partial x}\frac{\partial N_m}{\partial x} + \frac{\partial N_n}{\partial y}\frac{\partial N_m}{\partial y}\right\} dxdy$$

$$E^{II}_{nm} = \sum_{e_{II}} (K_{II} A_e) \int_{D_{II}} \left\{\frac{\partial N_n}{\partial z}\frac{\partial N_m}{\partial z}\right\} dz = \pm\sum_{e_{II}} \left\{\frac{K_{II} A_e}{D_{II}}\right\}$$

$$\left\{\begin{array}{l} + \text{ for } n = m \\ - \text{ for } n \neq m \end{array}\right.$$

$$E_{nm}^{III} = \sum_{e_{III}} (K_{III} D_{III}) \int_{A_e} \{\frac{\partial N_n}{\partial x}\frac{\partial N_m}{\partial x} + \frac{\partial N_n}{\partial y}\frac{\partial N_m}{\partial y}\} dxdy$$

$$Q_n = \sum_e \int_s (q_i l_i N_n) ds$$

$$P_n = \sum_e \int_v (q_o N_n) dv$$

LINEAR-PROGRAMING FORMULATION

The piezometric heads at the nodes corresponding to the wells subject to the control of water level are represented by $\{h_c\}$. The prescribed piezometric heads on the boundary are denoted by $\{h_b\}$. The heads at the ordinary nodes are designated by $\{h_o\}$. Similarly, corresponding to $\{h_c\}$, $\{h_o\}$ and $\{h_b\}$, vectors $\{Q^o\}$ and $\{P\}$ are divided into three parts, i.e., $\{Q_c\}$, $\{Q_o\}$, $\{Q_b\}$ and $\{P_c\}$, $\{P_o\}$, $\{P_b\}$, respectively. Futhermore, the coefficient matrix $[E]$ is also partitioned into submatrices, e.g.,$[B_{11}]$, $[B_{12}]$, etc. Then, Equation (5) is rewritten as

$$\begin{bmatrix} B_{11} & B_{12} & B_{13} \\ B_{21} & B_{22} & B_{23} \\ B_{31} & B_{32} & B_{33} \end{bmatrix} \begin{Bmatrix} h_c \\ h_o \\ h_b \end{Bmatrix} = - \begin{Bmatrix} Q_c \\ Q_o \\ Q_b \end{Bmatrix} - \begin{Bmatrix} P_c \\ P_o \\ P_b \end{Bmatrix} \qquad (6)$$

For simplicity, it is assumed that no flow crosses the boundary of the flow domain except the boundary along which the piezometric heads are prescribed. In this case, the vector $\{Q\}$ does not need to be taken into account. It is also assumed that there in no well other than the controlled wells. Then it follows that $\{P\} = \{0\}$ and $\{P_b\} = \{0\}$. Eliminating $\{h_o\}$ between the first and second equations of Equation (6), one obtains

$$\{P_c\} = [D]\{h_c\} + \{R\} \qquad (7)$$

where
$$[D] = - [B_{11}] + [B_{12}][B_{22}]^{-1}[B_{21}]$$

$$\{R\} = [F]\{h_b\}$$

$$[F] = - [B_{13}] + [B_{12}][B_{22}]^{-1}[B_{23}]$$

Thus, the relationship between the pumping rates and the water levels in the controlled wells has been obtained (Ueda 1979).

Now consider the well fully penetrating both aquifers, i.e., the I- and III-layers. Let $h_u$ and $h_l$ denote the piezometric heads in the I- and III-layers respectively at the nodes corresponding to the well penetrating both aquifers. Then, at these nodes, the condition of $h_u = h_l$ must be satisfied in Equation (7). In algorithm. the l-th row and column of [D] are added to its u-th row and column, respectively. Thus, [D] which is an (M,M) matrix is reduced to a matrix $[\tilde{D}]$ of (M-1,M-1). Moreover, the l-th components of vectors {P} and {R} are also added to their u-th components respectively. The vectors {P} and {R} are reduced to $\{\tilde{P}\}$ and $\{\tilde{R}\}$ which consist of (M-1) components. When there are more than two wells penetrating two aquifers, the same procedure as mentioned above is carried out repeatedly. Thus, Equation (7) becomes

$$\{\tilde{P}\} = [\tilde{D}]\{h_c\} + \{\tilde{R}\} \tag{8}$$

Let $\{\hat{h}\}$ be the lowest limits of water levels in the controlled wells, and write $\{h_c\}$ as

$$\{h_c\} = \{H\} + \{\hat{h}\}$$

Then, Equation (8) is rewritten as

$$\{\tilde{P}\} = [\tilde{D}]\{H\} + \{\hat{R}\} \tag{9}$$

where

$$\{\hat{R}\} = [\tilde{D}]\{\hat{h}\} + \{\tilde{R}\}$$

OPTIMIZATION

Let the objective function G be the total quantity of water pumped from the production wells, which will be maximized under given constraints. That is,

$$\max G = \max \sum_{\alpha=1}^{k} \tilde{P}_\alpha \tag{10}$$

On the other hand, the constraints in this study are assumed to be

$$H_\alpha \geqq 0 \tag{11}$$

and

$$\tilde{P}_\alpha = \tilde{D}_{\alpha\beta} H_\beta + \hat{R}_\alpha \geq 0 \tag{12}$$

The latter constraint states that the productivity of individual well should be non-negative.

The simplex method (Dantzig 1963) is employed to find out the optimal water level in each production well under the given conditions.

EXAMPLE

The plane view of the segmented groungwater reservoir analyzed in this study is shown in Fig.4 and the vertical section of the reservoir along y = 600 m is illustrated in Fig.5. In these figures, the coordinate system, the dimensions of the reservoir, the hydraulic conductivities and the boundary conditions are also presented. It is assumed that there are five production wells which are numbered as shown in Fig.4. Wells No.1 and No.2 are pumped from the water-table aquifer while wells No.3 and No.4 are pumped only from the confined aquifer. Well No.5 is fully penetrating both aquifers.

The object is to maximize the total amount of water pumped from the five wells. Meanwhile, the lowest limits of water levels in the wells are given as follows

| Well | No.1 | No.2 | No.3 | No.4 | No.5 |
|---|---|---|---|---|---|
| $\hat{h}$ (m) | 62.5 | 45.0 | 62.5 | 55.0 | 50.0 |

Furthermore, a certain amount of water should be ensured to the downstream area. For this, the constraint on the piezometric head along x = 1800 m is assigned as

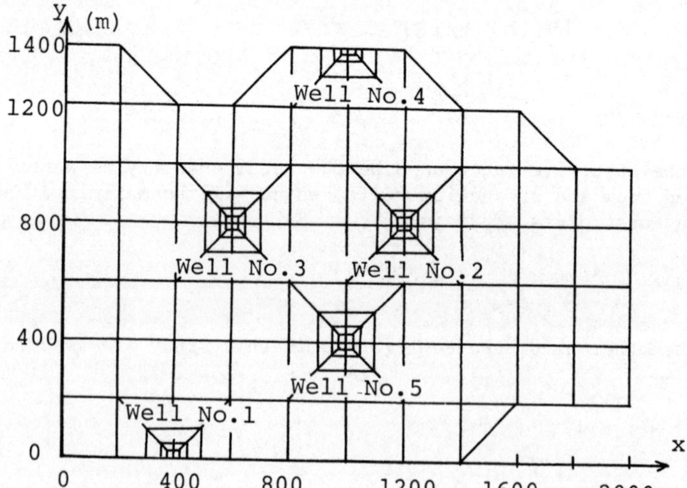


Figure 4 Plane view of the segmented flow domain analyzed herein.

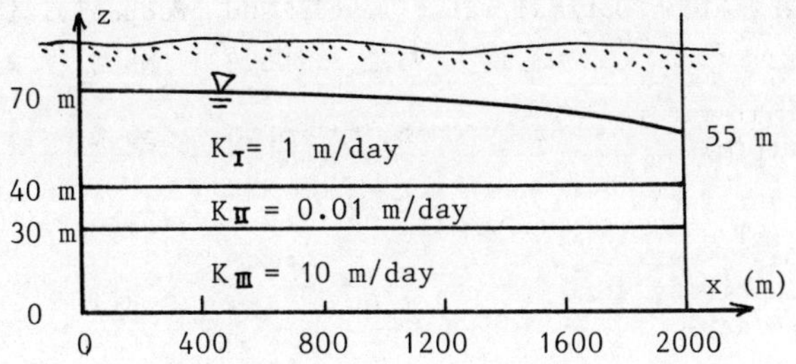


Figure 5 Vertical section of the groundwater reservoir analyzed herein.

$h \geqq 56$ m along x = 1800 m

and in addition to the above constraint the condition of $\{\tilde{P}\}$ = {0} along x = 1800 m should be satisfied because the quantity of water pumped from "virtual" wells along x = 1800 m must be equal to zero.

The algorithm of this analysis is as follows;

(i) Estimation of water table and determination of the coefficient matrix; Prescribing the water levels in the wells as the boundary condition which are presumed from the constraints on them, the water table is tentatively determined. Then, the coefficient matrix [E] in Equation (5) is obtained.

(ii) Linear-programing formulation; Partitioning the matrix [E] into the submatrices [B] as shown in Equation (6), the matrix $[\tilde{D}]$ and the vector $\{\hat{R}\}$ in Equation (9) are obtained through the manipulations described in Equations (7) and (8).

(iii) Optimization; Given the constraints on the water levels and on the the amount of water pumped from each well, the optimal solutions to maximize the total quantity of water are determined by employing the simplex method.

(iv) Modification of the water table; Compare the optimal water levels determined in (iii) with those presumed in (i). When the difference between them exceeds a certain norm, the procedures mentioned in (i) to (iv) are repeated after modifying the water table until the difference becomes small.

Thus, the optimal water levels, the water table, the piezometric surface and the pumping rates etc. are determined.

## Results

The optimal solutions of the water levels in the production wells and the productivity of each well are listed in the following table. The results reveal that it is impossible to pump the water from Well No.2 under the given constraints. The total amount of water is 3360 $m^3$/day.

Table 1 Optimal water levels and productivities

| Well | No.1 | No.2 | No.3 | No.4 | No.5 |
|---|---|---|---|---|---|
| Optimal water level (m) | 62.5 | 58.76 | 62.5 | 55.0 | 55.68 |
| (Constraint) (m) | (62.5) | (45.0) | (62.5) | (55.0) | (50.0) |
| Productivity ($m^3$/day) | 73 | 0 | 403 | 967 | 1917 |

The piezometric head along x = 1800 m lies between 56 m and 56.37 m which satisfies the constraint assigned to ensure a certain amount of water to the downstream area. The water table in the phreatic aquifer and the piezometric surface of the confined leaky aquifer are shown in Fig.6 and Fig.7, respectively. The water table and the piezometric surface in the x-z plane through the well are shown in Fig.8(a), (b) and (c).

The results indicate that the method described herein can provide a useful tool to perform the efficient usage of the ground water in a layered aquifer.

## REFERENCES

Bear, J. (1972) Dynamics of Fluids in Porous Media. pp.764, American Elsevier Publishing Company, Inc., New York.

Dantzig, G.B. (1963) Linear Programing and Extensions. Princeton Univ. Press, Princeton.

Ueda, T., K. Jinno and Y. Chyono (1979) On the Optimal Well Discharge in Ground Water Area. Proceedings of JSCE, No.283, 33-43. (in Japanese)

Zienkiewicz, O.C. (1971) The Finite Element Method in Engineering Science. pp.521, McGraw-Hill, London.

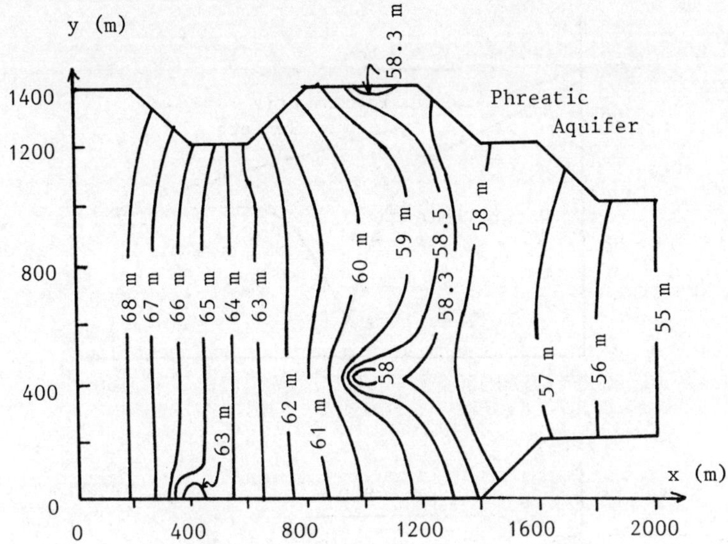


Figure 6 Water table in the phreatic aquifer.

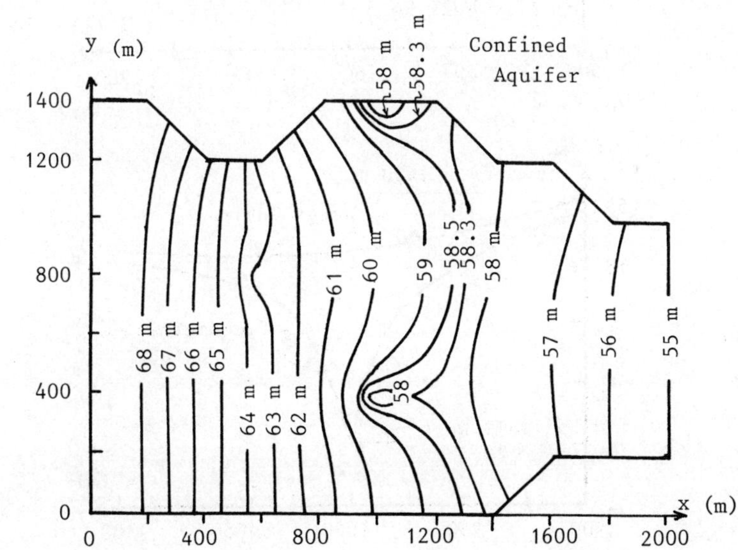


Figure 7 Piezometric surface of the confined aquifer.

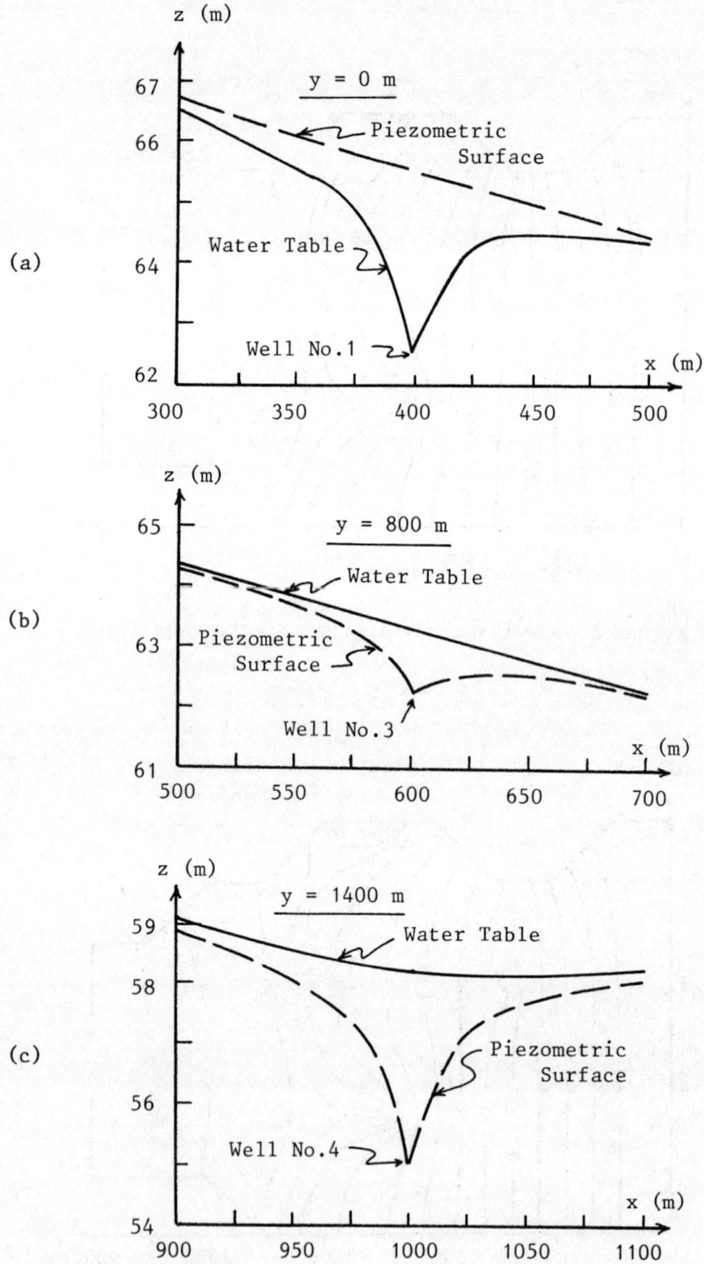


Figure 8 Water table and piezometric surface in the x-z plane through the well.

# 19. BOUNDARY ELEMENT METHOD

# SOME APPLICATIONS OF THE BOUNDARY ELEMENT METHOD FOR POTENTIAL PROBLEMS

C.A. Brebbia and L. Wrobel

Computational Mechanics Centre, Southampton, UK, and
Federal University of Rio de Janeiro, Brazil.

## 1. INTRODUCTION

The boundary element method is now firmly established as an important alternative technique to the prevailing numerical methods of analysis in continuum mechanics [1][2]. One of the most important types of applications is for the solution of a range of problems such as temperature diffusion, some types of fluid flow motion, flow in porous media and many others which can be written in function of a potential and whose governing equation is the Laplacian type. All these potential cases can generally be efficiently and economically analysed using boundary elements.

The technique consists in the transformation of the partial differential equation describing the behaviour of the unknown inside and on the boundary of the domain into an integral equation relating only boundary values, and then finding out the numerical solution of this equation. If values at internal points are required, they can be calculated afterwards from the boundary data. Since all numerical approximations take place only at the boundary, the dimensionality of the problem is reduced by one and a smaller system of equations obtained in comparison with those achieved through differential methods.

Integral equations for linear problems can be based on Fredholm's [3] type formulation. Alternatively, they can be formulated through the application of Green's third identity [4]. More recently, it has been demonstrated that the same integral relationships can be obtained through weighted residual considerations [1]. In this way, it becomes easier to relate and combine the Boundary Element Method with other numerical techniques, such as the Finite Element Method, as well as to extend it for the analysis of problems governed by more complex partial differential equations, including non-linearities. In what follows the weighted residual formulation will be used in preference to the others as it is more general and more in line with the approximate methods familiar to engineers and other analysts.

In 1963, Jaswon [5] and Symm [6] presented a numerical technique to solve Fredholm boundary integral equations. The technique consists of discretizing the boundary into a series of small segments (elements), assuming that the source density remains constant within each segment. By using the method of collocation, the discretized equation is applied to a number of particular points (nodes) in each element, and the influence coefficients computed approximately using Simpson's rule. Exception is made for the singular coefficients resulting from the self-influence of each element, which are computed either analytically (for Dirichlet type problems) or by the summation of the off-diagonal coefficients plus the free term (for Neumann's problems). This produces a system of linear algebraic equations which can then be solved.

Applying such techniques, Jaswon and Symm obtained accurate solutions for simple two-dimensional Neumann and Dirichlet problems. They also proposed a more general numerical formulation for solving Cauchy boundary value problems through the application of Green's third identity, which yields a boundary integral equation where boundary values and normal derivatives of the physical variable play the role of the fictitious source densities. Results using this formulation were reported by Symm [6] and Jaswon and Ponter [7].

Hess and Smith [8] developed a parallel work for the solution of Neumann type boundary value problems, more specifically the problem of potential flow around arbitrary bodies. Applying basically the same (indirect) technique they computed the quantities of interest (potential and velocity) from the source density distribution. They extended the method to analyse a variety of body shapes: two-dimensional, axisymmetric and fully three-dimensional results were presented. The influence coefficients were all computed analytically for the two and three-dimensional cases. For axisymmetric cases, the influence coefficients were computed numerically using Simpson's rule but the number of sub-elements was scaled in such a way that the further the element lies from the actual node, the fewer the number of sub-elements used in the calculation. The singular-self-influence-coefficients were computed analytically by means of series expansions.

Harrington et al [9] applied the technique to solve some two dimensional electrical engineering problems with the more general impedance boundary condition, which is of the Robin type, i.e. it prescribes a linear relation between the potential and its normal derivative. They also proposed a piecewise variation for the source density. Mautz and Harrington [10] solved axisymmetric electrical engineering problems with Dirichlet boundary conditions, again employing the indirect formulation and assuming the source density to remain constant with each element. Some of their numerical considerations were later discussed by Jaswon and Symm [11].

This paper explains how different orders of boundary elements, such as constant or high order, can be applied to solve potential type problems. The methodology is first presented for the Poisson's equation and then extended to anisotropic media and non-linear materials. Several examples are presented to illustrate the technique and results are compared against analytical solutions such as those obtained using finite elements.

## 2. POTENTIAL PROBLEMS

Let us first consider a potential function u over a domain $\Omega$, where the following governing equation is to be satisfied,

$$\nabla^2 u - b = 0 \quad \text{in } \Omega \tag{1}$$

The boundary conditions corresponding to this problem are of two types: essential conditions, such as $u = \bar{u}$ on $\Gamma_1$ and natural conditions such as $q = \partial u/\partial n = \bar{q}$ on $\Gamma_2$. The total boundary is $\Gamma = \Gamma_1 + \Gamma_2$.

For our numerical solution u will be approximated and we can minimize the error thus introduced by weighting the governing equation by a new function $u^*$. This gives

$$\int_\Omega (\nabla^2 u - b)u^* d\Omega = \int_{\Gamma_2} (q-\bar{q})u^* d\Gamma - \int_{\Gamma_2} (u-\bar{u})q^* d\Gamma \tag{2}$$

where $q^* = \partial u^*/\partial n$.

After integrating by parts twice the terms in the Laplacian equation (2) becomes:

$$\int_\Omega (\nabla^2 u^*)u d\Omega - \int_\Omega b u^* d\Omega = -\int_{\Gamma_2} \bar{q}u^* d\Gamma - \int_{\Gamma_2} q u^* d\Gamma + \int_{\Gamma_2} u q^* d\Gamma + \int_{\Gamma_1} \bar{u} q^* d\Gamma \tag{3}$$

The function $u^*$ is now assumed to be the fundamental solution of the equation, representing a concentrated potential at a point i, i.e. the solution of:

$$\nabla^2 u^* + \Delta_i = 0 \tag{4}$$

where $\Delta_i$ is the Dirac delta function. Hence equation (3) can be written as:

$$u_i + \int_{\Gamma_2} uq^* d\Gamma + \int_{\Gamma_1} \bar{u}q^* d\Gamma + \int_{\Omega} bu^* d\Omega$$

$$= \int_{\Gamma_2} \bar{q}u^* d\Gamma + \int_{\Gamma_1} qu^* d\Gamma \tag{5}$$

It can be noted that for an isotropic three-dimensional medium the fundamental solution of equation (4) is:

$$u^* = \frac{1}{4\pi r} \tag{6}$$

where r is the distance from the point of application of the concentrated potential to the point under consideration. For two dimensions the solution is:

$$u^* = \frac{1}{2\pi} \ln\left(\frac{1}{r}\right) \tag{7}$$

Other fundamental solutions for potential problems are given by Brebbia and Walker [2].

Equation (5) is valid for any point inside the domain, but in order to formulate the problem as a boundary technique one needs to take this point to the boundary. For this case equation (5) can be written as:

$$c_i u_i + \int_{\Omega} uq^* d\Omega + \int_{\Omega} bu^* d\Omega = \int_{\Gamma} qu^* d\Gamma \tag{8}$$

Note that we have written the integrals for the whole $\Gamma = \Gamma_1 + \Gamma_2$ boundary but that, depending on which part of the boundary we consider, the $\bar{u}$ or $\bar{q}$ values will be known. We can introduce these boundary conditions at a later stage.

The value of $c_i$ is: = 1 for an internal point, = 0 for an external point, = $\frac{1}{2}$ for a boundary point on a smooth boundary.

Boundary Elements

Equation (8) can now be applied on the boundary of the domain under consideration. This boundary can be divided into

N elements. The points where the unknown values are considered are called 'nodes' and are similar to those of finite elements. The main difference is that now elements and nodes are defined only on the Γ boundary. The functions u and q over each boundary element are given by:

$$u = \underset{\sim}{\Phi}^T \underset{\sim}{u}$$

$$q = \underset{\sim}{\Psi}^T \underset{\sim}{q} \tag{9}$$

and equation (8) is discretized as follows:

$$c_i u_i + \sum_{j=1}^{N} \int_{\Gamma_j} u q^* d\Gamma + \sum_{k=1}^{M} \int_{\Omega_k} b u^* d\Omega = \sum_{j=1}^{N} \int_{\Gamma_j} q u^* d\Gamma \tag{10}$$

Note that M internal elements or cells need to be defined to compute the integrals in Ω but these elements do not introduce any further unknown and hence the problem is still a boundary problem .

We can substitute the u and q values given by equation (9) into (10) and carry out the integrations (usually numerically). This gives for each node, after assembling, the following equation,

$$c_i u_i + \sum_{j=1}^{n} \hat{H}_{ij} u_j + B_i = \sum_{j=1}^{n} G_{ij} q_j \tag{11}$$

or

$$\sum_{j=1}^{n} H_{ij} u_j + B_i + \sum_{j=1}^{n} G_{ij} q_j \tag{12}$$

where the $B_i$ term is the result of having integrated the domain term and $H_{ij} = \hat{H}_{ij}$ for $i \neq j$ and $H_{ij} = \hat{H}_{ij} + c_i$ for $i = j$. n is the total number of nodes.

The whole set can be written in matrix form as follows,

$$\underset{\sim\sim}{HU} + \underset{\sim}{B} = \underset{\sim\sim}{GQ} \tag{13}$$

Note that $n_1$ values of u and $n_2$ values of q are known on the boundary, hence equation (13) can be reordered in such a way that all the unknowns are on the left hand side, i.e.

$$\underset{\sim\sim}{AX} = \underset{\sim}{F} \tag{14}$$

where $\underset{\sim}{X}$ is the vector of unknowns u's and q's.

When the surface is not smooth at the point 'i' the $c_i = \frac{1}{2}$ value is no longer valid. However, we can always calculate the diagonal terms of $\underset{\sim}{H}$ by the fact that when a uniform potential is applied over the whole boundary, the normal derivatives must be zero. Hence equation (13) in the absence of the vector $\underset{\sim}{B}$ and for a closed body, becomes,

$$\underset{\sim}{H}\underset{\sim}{U} = \underset{\sim}{0} \tag{15}$$

where $\underset{\sim}{U}$ is a uniform potential. Thus the sum of all the elements of $\underset{\sim}{H}$ in any row ought to be zero and the value of the coefficient on the diagonal can be easily calculated once the off-diagonal coefficients are all known, i.e.

$$H_{ii} = - \sum_{\substack{j=1 \\ (i \neq j)}}^{n} H_{ij} \tag{16}$$

Once the values of the boundary unknowns are found, the values of u and q anywhere can be calculated using equation (8). The internal fluxes are obtained by computing the derivatives of equation (8).

<u>Internal Source Treatment</u>

In many practical cases the boundary-value problems for Poisson's equation may be reduced to similar problems for Laplace's equation by subtracting out a particular solution independent of the boundary conditions.

Another alternative way of including internal sources into the formulation is by transforming the corresponding domain integral in (8) into equivalent boundary integrals, for the cases when the function b is harmonic in $\Omega$, i.e. $\nabla^2 b = 0$. If we compute a Galerkin-type function $v^*$ such that $\nabla^2 v^* = u^*$, we can write Green's second identity in the form,

$$\int_\Omega (b\ \nabla^2 v^* - v^* \nabla^2 b)\, d\Omega = \int_\Gamma \left\{ b \frac{\partial v^*}{\partial n} - v^* \frac{\partial b}{\partial n} \right\} d\Gamma \tag{17}$$

Hence,

$$\int_\Omega b u^* d\Omega = \int_\Gamma \left\{ b \frac{\partial v^*}{\partial n} - v^* \frac{\partial b}{\partial n} \right\} d\Gamma \tag{18}$$

One such function $v^*$ for two dimensional problems is given in [12] as,

$$v^* = \frac{r^2}{8\pi}\left\{\ln\left(\frac{1}{r}\right) + 1\right\} \tag{19}$$

It can easily be seen that,

$$\nabla^2 v^* = \frac{1}{r}\frac{\partial}{\partial r}\left(r\,\frac{\partial v^*}{\partial r}\right) = \frac{1}{2\pi}\ln\frac{1}{r} = u^* \tag{20}$$

Notice that as a special case we could have concentrated source at an interior point '$\ell$'. For this case b becomes,

$$b = Q_\ell\ \Delta_\ell \tag{21}$$

where $\Delta_\ell$ is a Dirac delta function. For the case of several such sources equation (8) can be generalized to give,

$$c_i u_i + \int_\Gamma u q^* d\Gamma + \int_\Omega b u^* d\Omega + \sum (Q_\ell u_\ell^*) = \int_\Gamma q u^* d\Gamma \tag{22}$$

These concentrated sources are very simple to handle as they do not require any special integration.

## 3. ORTHOTROPY AND ANISOTROPY

Let us now assume that the medium over which the problem is defined is orthotropic (figure 1 ). The governing equation in the directions of orthotropy can be written as,

$$k_1 \frac{\partial^2 u}{\partial y_1^2} + k_2 \frac{\partial^2 u}{\partial y_2^2} = 0 \tag{23}$$

for the two-dimensional case, being $k_i$ the medium property coefficient in the direction of orthotropy i.

The fundamental solution for this equation is [2],

$$u^* = \frac{1}{(k_1 k_2)^{\frac{1}{2}}}\ln\frac{1}{r} \tag{24}$$

where r is the reduced distance between source and observation points, i.e.

$$r(\xi,x) = \left\{\frac{1}{k_1}\,[y_1(\xi) - y_1(x)]^2 + \frac{1}{k_2}\,[y_2(\xi) - y_2(x)]2\right\}^{\frac{1}{2}}$$

Applying the divergence theorem to the terms of equation

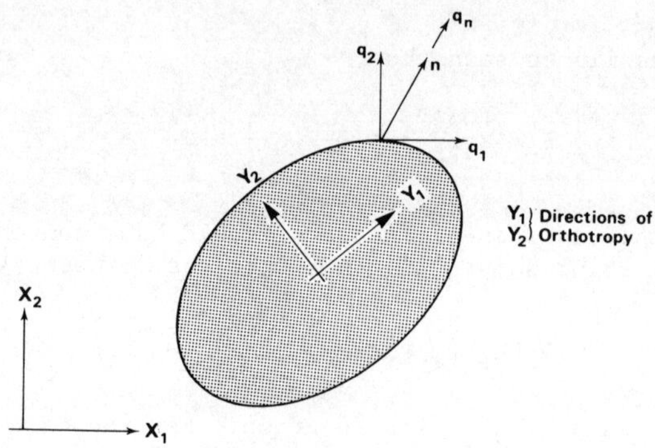


Figure 1 Orthotropic Medium

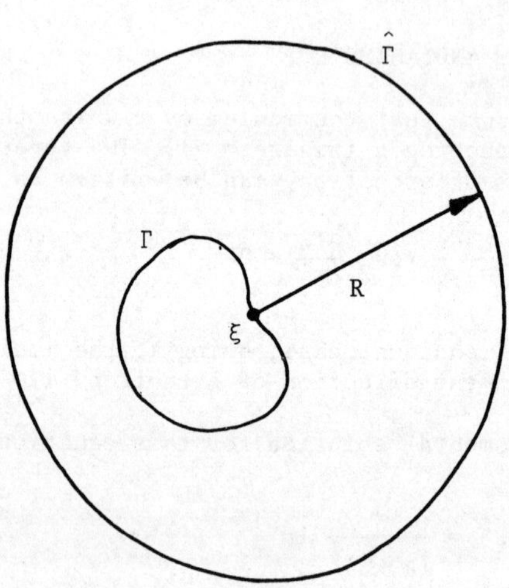


Figure 2 Infinite Region with Cavity

(23) yields,

$$\int_\Omega \left\{k_1 \frac{\partial^2 u}{\partial y_1^2} + k_2 \frac{\partial^2 u}{\partial y_2^2}\right\} d\Omega = \int_\Gamma \left\{k_1 \frac{\partial u}{\partial y_1} n_{y1} + k_2 \frac{\partial u}{\partial y_2} n_{y2}\right\} d\Gamma \qquad (25)$$

where $n_{y1}$ and $n_{y2}$ are the direction cosines of the outward normal to the surface (figure 1). The term enclosed in brackets in the right-hand side integral is the normal boundary flux q. Analogously, we can define,

$$q^* = k_1 \frac{\partial u^*}{\partial y_1} n_{y1} + k_2 \frac{\partial u^*}{\partial y_2} n_{y2} \qquad (26)$$

The problem can then be solved in the same way as for isotropic problems, i.e. by transforming the governing equation (23) plus boundary conditions into a boundary integral equation similar to (22).

For fully anisotropic media the governing equation becomes,

$$k_{11} \frac{\partial^2 u}{\partial x_1^2} + 2k_{12} \frac{\partial^2 u}{\partial x_1 \partial x_2} + k_{22} \frac{\partial^2 u}{\partial x_2^2} = 0 \qquad (27)$$

for two-dimensional problems; the coefficients $k_{ij}$ defining the medium properties. This equation has the following fundamental solution [13];

$$u^* = \frac{1}{|k_{ij}|^{\frac{1}{2}}} \ln \frac{1}{r} \qquad (28)$$

where $|k_{ij}|$ is the determinant of the medium property coefficients matrix and r is

$$r(\xi,x) = \left\{\frac{1}{k_{11}} [x_1(\xi) - x_1(x)]^2 + \frac{2}{k_{12}} [x_1(\xi) - x_1(x)][x_2(\xi) - x_2(x)] + \frac{1}{k_{22}} [x_2(\xi) - x_2(x)]^2\right\}^{\frac{1}{2}}$$

The normal boundary flux q is now given by

$$q = \left\{k_{11} \frac{\partial u}{\partial x_1} + k_{12} \frac{\partial u}{\partial x_2}\right\} n_{x1} + \left\{k_{12} \frac{\partial u}{\partial x_1} + k_{22} \frac{\partial u}{\partial x_2}\right\} n_{x2} \qquad (29)$$

Analogously, we have

$$q^* = \left[k_{11}\frac{\partial u^*}{\partial x_1} + k_{12}\frac{\partial u^*}{\partial x_2}\right] n_{x1} + \left[k_{12}\frac{\partial u^*}{\partial x_1} + k_{22}\frac{\partial u^*}{\partial x_2}\right] n_{x2} \quad (30)$$

## 4. TREATMENT OF INFINITE REGIONS

Although the boundary integral equation (22) has been derived considering that the domain $\Omega$ is bounded,all concepts presented thus far are also valid for infinite regions, provided that certain regularity conditions concerning the behaviour of the functions are satisfied.

Let $\hat{\Gamma}$ be the surface of a circle (or a sphere if the problem is three dimensional) of radius R surrounding the surface $\Gamma$ and centred at the point $\xi$ (figure 2). A boundary integral equation similar to (22) for the finite domain $\Omega$ enclosed by the actual surface $\Gamma$ and the fictitious surface $\hat{\Gamma}$ can be written as,

$$c_i u_i + \int_\Gamma u q^* d\Gamma + \int_{\hat{\Gamma}} u q^* d\Gamma = \int_\Gamma q u^* d\Gamma + \int_{\hat{\Gamma}} q u^* d\Gamma \quad (31)$$

If we let the radius $R \to \infty$, equation (31) will only be valid for points in $\Gamma$ (and $\Omega$) if

$$\lim_{R \to \infty} \int_{\hat{\Gamma}} \{qu^* - uq^*\}\, d\Gamma = 0 \quad (32)$$

For three dimensional problems, since

$$d\Gamma = |J|\, d\eta_1 d\eta_2 \quad ; \quad |J| = O(R^2)$$

$$u^* = O(R^{-1})$$

$$q^* = O(R^{-2})$$

where $|J|$ is the Jacobian and O( ) represents the asymptotic bebehaviour of the functions as $R \to \infty$, the condition of equation (32) is satisfied if the function u behaves at most as $O(R^{-1})$, such that its derivative $q = O(R^{-2})$. These are the regularity conditions at infinity and they ensure that each term in the integral in equation (32) behaves at most as $O(R^{-1})$, i.e. they approach zero as $R \to \infty$.

For two dimensional problems, we have that the function $u^*$ behaves as the logarithm of R and its derivative $q^* = O(R^{-1})$ as $R \to \infty$. The regularity conditions at infinity for this case imply that u behaves at most as ln R such that $q = O(R^{-1})$. Note that now the terms in the integral in equation (32) do not approach zero separately as $R \to \infty$ since the Jacobian is $O(R)$, but they cancel each other thus fulfilling condition (32).

Therefore, applying condition (32) into equation (31) yields,

$$c\,u + \int_{\Gamma} uq^{*}d\Gamma = \int_{\Gamma} qu^{*}d\Gamma \qquad (33)$$

that is, the boundary integral equation obtained for points on the internal surface $\Gamma$ of the infinite regular region is of the same form as equation (22) for finite regions.

For unbounded domains, it is also possible to evaluate the diagonal coefficients of matrix $\underset{\sim}{H}$ through constant potential conditions but care must be taken because of the regularity condition at infinity are violated as the function u is now assumed to be constant everywhere in $\Omega$. Since it can easily be shown that,

$$\lim_{R \to \infty} \int_{\hat{\Gamma}} q^{*}d\Gamma = -1 \qquad (34)$$

for both two and three dimensional problems, we have instead of (16) the following relationship

$$H_{ii} = - \sum_{\substack{j=1 \\ (j\neq 1)}}^{n} H_{ij} + 1 \qquad i = 1,2 \ldots N. \qquad (35)$$

## 5. NONLINEAR MATERIALS [14][15]

In many practical applications in potential problems, the conductivity is a function of the potential, i.e. $k = k(u)$. In this case the governing equation becomes,

$$\frac{\partial}{\partial x_1}\left(k\frac{\partial u}{\partial x_1}\right) + \frac{\partial}{\partial x_2}\left(k\frac{\partial u}{\partial x_2}\right) + \frac{\partial}{\partial x_3}\left(k\frac{\partial u}{\partial x_3}\right) + b = 0 \qquad (36)$$

where b is the source term. The boundary conditions for this equation are

$$\begin{aligned} u &= \bar{u} && \text{on } \Gamma_1 \\ q &= k\frac{\partial u}{\partial n} = \bar{q} && \text{on } \Gamma_2 \end{aligned} \qquad (37)$$

Part of the boundary can also have convective conditions of the type

$$q = h(u_f - u) \qquad \text{on } \Gamma_3 \tag{38}$$

where $u_f$ is the potential or temperature in the surrounding medium and h is a heat transfer coefficient for instance.

Equation (36) is non-linear as $k = k(u)$, and in order to solve it using boundary elements we need to define a source term which will contain all non-linearities and apply a recursive procedure to find the correct solution. This technique implies using domain cell as explained in the section on Poisson's equation. Another way of solving the problem which is much more elegant and efficient is using Kirchoff's transformation to render the non-linear problem into a linear one

We need to start by defining a new variable $\psi(u)$ such that

$$\vec{\nabla}\psi = \frac{d\psi}{du}\, du \tag{39}$$

Comparing the right hand side of (39) with equation (36) we now define $\psi$ in such a way that,

$$\frac{d\psi}{du} = k(u) \tag{40}$$

or in integral form,

$$\psi = \int_{u_o}^{u} k(u)\, du \tag{41}$$

This equation is called Kirchoff's transformation, where $u_o$ is an arbitrary reference value. It follows from (41) that

$$\frac{\partial\psi}{\partial x_i} = k\,\frac{\partial u}{\partial x_i} \qquad i = 1,2,3 \tag{42}$$

Equation (36) can now be written as,

$$\nabla^2\psi + b = 0 \tag{43}$$

Hence we have arrived at the same form as the homogeneous isotropic Poisson's equation but written in terms of a new variable $\psi$. Solutions for the cases of constant conductivity may now be used replacing u by $\psi$. The problem is still linear provided that the boundary conditions are of the essential ($\Gamma_1$) or natural ($\Gamma_2$) type. If the boundary conditions are of a mixed type - i.e. conditions on $\Gamma_3$ - the Kirchoff transformation process introduces non-linearities.

The conditions on $\Gamma_1$ and $\Gamma_2$ corresponding to equation (43) are

$$\psi = \bar{\psi} \qquad \text{on } \Gamma_1$$
$$\frac{\partial \psi}{\partial n} = \bar{q} \qquad \text{on } \Gamma_2 \tag{44}$$

These equations are related to the original ones by

$$\psi = \bar{\psi} = \int_{u_o}^{\bar{u}} k \, du \qquad \text{on } \Gamma_1$$
$$\frac{\partial \psi}{\partial n} = k \frac{\partial u}{\partial n} = \bar{q} \qquad \text{on } \Gamma_2 \tag{45}$$

The third type of boundary condition gives,

$$\frac{\partial \psi}{\partial n} + h \{K^{-1}(\psi) - u_f\} = 0 \tag{46}$$

which is a non-linear equation. $K^{-1}$ is the inverse Kirchoff transform, i.e.

$$K^{-1} (\psi) = u \tag{47}$$

Transforming the problem into a boundary integral equation does not present now any special difficulties. The resulting equations will not present any domain integrals other than those due to the source terms b and can be written as,

$$c(\xi) \psi(\xi) + \int_{\Gamma_{1+2}} \psi(x) q^*(\xi,x) d\Gamma + \int_{\Gamma_3} h K^{-1}(\xi) u^*(\xi,x) d\Gamma$$
$$= \int_{\Gamma_{1+2}} q(x) u^*(\xi,x) d\Gamma + \int_{\Omega} b(x) u^*(\xi,x) d\Omega + \int_{\Gamma_3} h u_f u^*(\xi,x) d\Gamma \tag{48}$$

These equations can be discretized in the usual way. In the

presence of $\Gamma_3$ type boundary, the system of equations is non-linear and the equations need to be solved iteratively.

The dependence of the conductivity on the potential can be of different types,

i) <u>Exponential Law</u> In this case,

$$k = k_o \exp\left\{\beta\left(\frac{u-u_o}{u_o}\right)\right\} \tag{49}$$

where $u_o$, $\beta$ and $k_o$ are material constants. For this case,

$$u = \bar{u} = \int_{u_o}^{\bar{u}} k_o \exp\left\{\beta\left(\frac{u-u_o}{u_o}\right)\right\} du =$$

$$= k_o \frac{u_o}{\beta} \exp\left\{\beta\left(\frac{\bar{u}-u_o}{u_o}\right)\right\} - k_o \frac{u_o}{\beta} \tag{50}$$

and the potential can be written as,

$$u = u_o + \frac{u_o}{\beta} \ln(\psi\beta/(k_o u_o) + 1) \tag{51}$$

Notice that equations (59) and (51) are valid for all cases, i.e. $\beta \gtrless 0$, $(u-u_o) \gtrless 0$.

ii) <u>Power Law</u> Another possibility is to define,

$$k = k_o \left\{1 + \beta\left(\frac{u-u_o}{u_o}\right)\right\}^n \tag{52}$$

where n is given. The boundary conditions in this case are,

$$\psi = \bar{\psi} = \frac{k_o u_o}{(n+1)\beta} \left\{1 + \beta\left(\frac{\bar{u}-u_o}{u_o}\right)\right\}^{n+1} - \frac{k_o u_o}{(n+1)\beta} \tag{53}$$

with the following functions for temperatures at any point.

$$u = u_o + \frac{u_o}{\beta} \left\{1 + \psi \frac{(n+1)}{k_o u_o}\right\}^{\frac{1}{n+1}} - \frac{u_o}{\beta} \tag{54}$$

For the special case of linear function, i.e. n = 1, we have

$$\bar{\psi} = \frac{k_o u_o}{2\beta}\left\{1 + \left[\frac{\bar{u}-u_o}{u_o}\right]\right\}^2 - \frac{k_o u_o}{2\beta} \qquad (55)$$

and

$$u = u_o + \frac{u_o}{\beta}\left\{1 + \frac{2\beta\psi}{k_o u_o}\right\}^{\frac{1}{2}} - \frac{u_o}{\beta} \qquad (56)$$

In this case, equation (56) has two roots, a positive and a negative one. From the physical point of view the positive root is the only valid one, as the other will imply negative conductivity which has no physical meaning. Equations (55) and (56) are valid for all cases, i.e. $\beta \gtrless 0$, $(u-u_o) \gtrless 0$ provided $k(u) > 0$.

## 6. APPLICATIONS

<u>Example 1</u>

Let us first consider the problem of a circular cavity of unit radius in an infinite two-dimensional region with Neumann boundary conditions, i.e. a constant radial influx of 31.21 J/(m² s) specified along the cavity surface (figure 3).

The solution will have a logarithmic potential behaviour at infinity. The exact solution of this problem is,

$$u = -31.32 \ln R \qquad (a)$$

which shows the expected behaviour.

Results for the temperature at points on the boundary $\Gamma$ and inside the domain $\Omega$ are given in table 1 and for the radial flux at points in $\Omega$ in table 2 compared to the exact solution. Taking symmetry into account, only one quarter of the cavity surface was subdivided using constant elements.

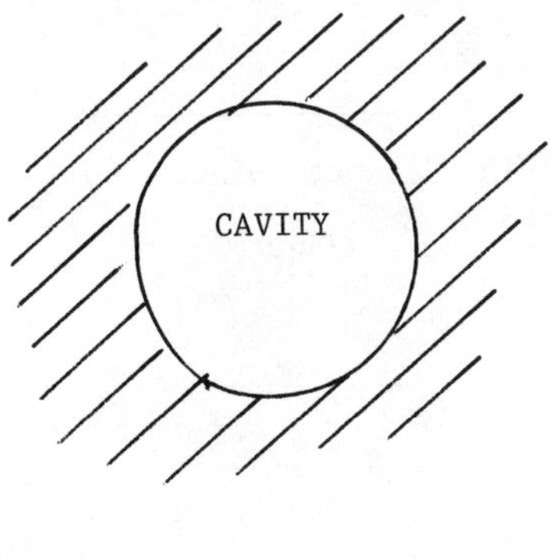


Figure 3 Circular cavity in an infinite two-dimensional region

| R | BEM (N=4) | BEM (N=8) | EXACT |
|---|---|---|---|
| 1.0 | 0.48 | 0.12 | 0.00 |
| 1.5 | 12.57 | 12.63 | 12.65 |
| 2.0 | 21.49 | 21.60 | 21.63 |
| 3.0 | 34.07 | 3.423 | 34.28 |
| 5.0 | 49.91 | 50.15 | 50.22 |
| 10.0 | 71.40 | 71.75 | 71.86 |
| 100.0 | 142.81 | 143.50 | 143.72 |
| 1000.0 | 214.21 | 215.24 | 215.58 |

Table 1 Temperature (-) at points on infinite region

| R | BEM (N=4) | BEM (N=8) | EXACT |
|---|---|---|---|
| 1.5 | 20.68 | 20.77 | 20.81 |
| 2.0 | 15.51 | 15.58 | 15.61 |
| 3.0 | 10.34 | 10.39 | 10.40 |
| 5.0 | 6.20 | 6.23 | 6.24 |
| 10.0 | 3.10 | 3.12 | 3.12 |
| 100.0 | 0.31 | 0.31 | 0.31 |
| 1000.0 | 0.03 | 0.03 | 0.03 |

Table 2 Radial flux at points on infinite region

Example 2

The previously discussed formulation for infinite regions can be readily applied to practical problems such as potential fluid flow past obstacles.

As an example, we study a two-dimensional potential flow with uniform onset velocity V in the $x_1$ direction around a NACA 0018 aerofoil, whose shape is shown in figure 4. For the solution of this problem, we employ a stream function $\psi$ defined as

$$q_i = \frac{\partial\psi}{\partial x_2} \qquad ; \qquad q_2 = - \frac{\partial\psi}{\partial x_1} \qquad \text{(a)}$$

It is now convenient to separate the stream function $\psi$ into two parts,

$$\psi = \psi_1 + \psi_2 \qquad \text{(b)}$$

where $\psi_1 = Vx_2$ defines the steady onset flow and $\psi_2$ is a perturbation stream function. Since the perturbation decays at infinity, we require that $\psi_2 = 0(R^{-1})$ at most as $R \to \infty$. Furthermore, as $\nabla^2\psi = 0$, we also have that $\nabla^2\psi_2 = 0$ and the problem can now be solved in terms of the perturbation $\psi_2$.

Considering the surface of the aerofoil as the streamline $\psi = 0$, the boundary conditions of the problem are

$$\psi_2 = - \psi_1 = - Vx_2 \qquad \text{on } \Gamma \qquad \text{(c)}$$

As the problem is antisymmetric with respect to the $x_1$-axis, only one half of the aerofoil needed to be analysed. The constant boundary elements discretization employed for the solution is shown in figure 4. Results for the tangential velocity presented in the same figure are in good agreement with results given in [16].

Example 3

This example studies the steady-state heat conduction problem of a semi-infinite medium bounded internally by two parallel and equal cylinders, as shown in figure 5. The interface $\hat{\Gamma}$ is at zero temperature, the temperature at infinity is also zero and the surfaces of the cylinders are isotherms.

If the depth d is much greater than the cylinder radius a, this problem can be seen as an approximation to the more practical one of two equally loaded electricity cables laid direct in the ground in horizontal formation. Of interest in this kind of problem is the determination of the external

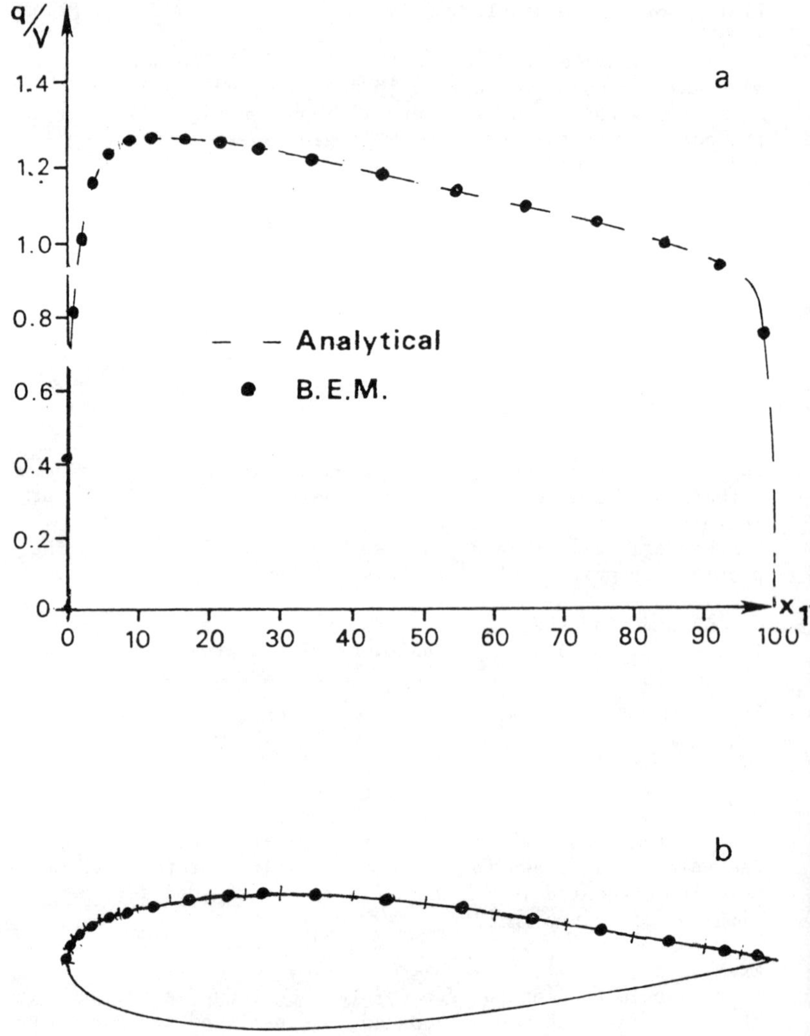


Figure 4 NACA 0018 aerofoil: a) Results; b) Discretization

thermal resistance of each cable.

The thermal resistance G per unit length between the surface $\Gamma$ at a temperature c and the surface $\bar{\Gamma}$ at zero temperature, through a medium with thermal conductivity k, is given by

$$G = -\frac{c}{k \int_{\Gamma} q \, d\Gamma} \tag{a}$$

Results for the non-dimensional ratio Gk are presented in Table 3 for several values of d/a, for a unit cylinder surface temperature c. Two different cases were considered, i.e. when the cables are touching (b = 0) and when the cable spacing equals one diameter (b = a). These results were obtained by sub-dividing the surface of one cylinder into 32 elements and considering symmetry with respect to the $x_2$-axis. Also shown in the table are the results obtained through an approximate analytical solution [17]; the agreement between both solutions is very good.

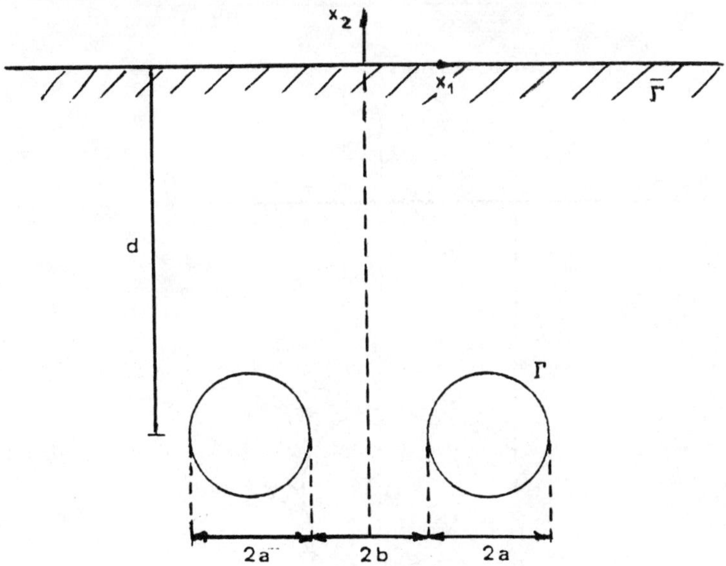


Figure 5 Semi-infinite region bounded internally by two equal cylinders

| b | d/a | BEM | ANALYTICAL |
|---|---|---|---|
| b = 0 | 10 | 0.810 | 0.810 |
| | 25 | 1.102 | 1.102 |
| | 50 | 1.322 | 1.322 |
| | 100 | 1.543 | 1.543 |
| b = a | 10 | 0.726 | 0.724 |
| | 25 | 1.016 | 1.014 |
| | 50 | 1.236 | 1.235 |
| | 100 | 1.457 | 1.456 |

Table 3 Values of ratio Gk

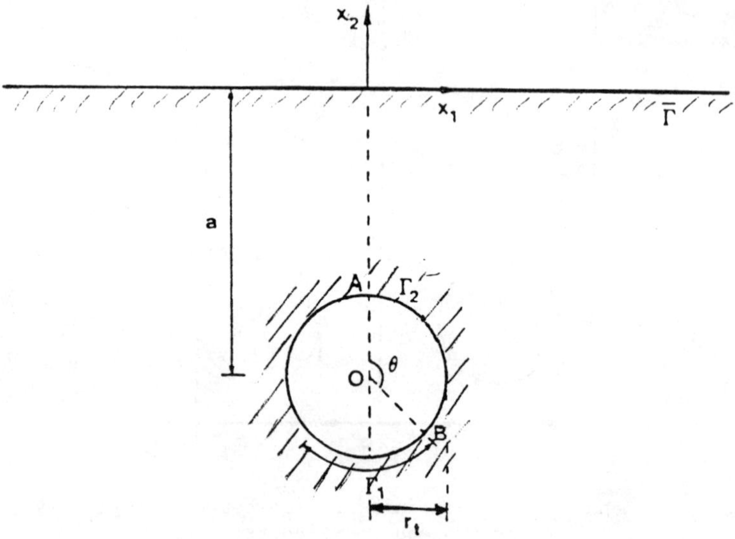


Figure 6 Groundwater flow round a tunnel

Example 4

The problem represented in figure 6 that is, a two-dimensional problem of groundwater flow round a tunnel with permeable invert was also studied with linear elements. If the medium is homogeneous and isotropic the problem is reduced to that of Laplace's equation for the groundwater pressure u. The boundary conditions of the problem are,

$$u = d \qquad \text{on } \bar{\Gamma}$$
$$u = 0 \qquad \text{on } \Gamma_1 \qquad \text{(a)}$$
$$q = -\cos\theta \qquad \text{on } \Gamma_2$$

where d is the depth of the river and $\theta$ is the angle measured from the vertical (see figure 6). The surface $\Gamma_1$ is the permeable invert of the tunnel where we assume that water flows in freely and $\Gamma_2$ is the impermeable part of the tunnel lining, where the condition of no flow across the surface holds. Notice that for a point at infinity we have the condition $u = d - x_2$.

The problem can be reformulated by subtracting out the solution at infinity. The groundwater pressure u is divided into two parts,

$$u = u_1 + u_2 \qquad \text{(b)}$$

where $u_1 = d - x_2$ satisfies the infinity condition. Then we have that $u_2$ tends to zero at infinity and furthermore, that $\nabla^2 u_2 = 0$ such that the problem can now be solved in terms of $u_2$. The boundary conditions for $u_2$ become then,

$$u_2 = 0 \qquad \text{on } \bar{\Gamma}$$
$$u_2 = -(d+a-r_t \cos\theta) \qquad \text{on } \Gamma_1 \qquad \text{(c)}$$
$$q_2 = 0 \qquad \text{on } \Gamma_2$$

where $r_t$ is the radius of the tunnel and a is the distance of the centre of the tunnel below the bottom of the river. The numerical values adopted for the parameters are d = 60, a = 30, $r_t = 3.5$ and $\hat{AOB} = 3/4$.

The boundary elements analysis of the problem employed the fundamental solution for semi-infinite regions discussed in reference [2] thus no discretization of the interface $\bar{\Gamma}$ was necessary.

Results for the function $u_2$ at some boundary points are presented in table 4 for three different discretizations

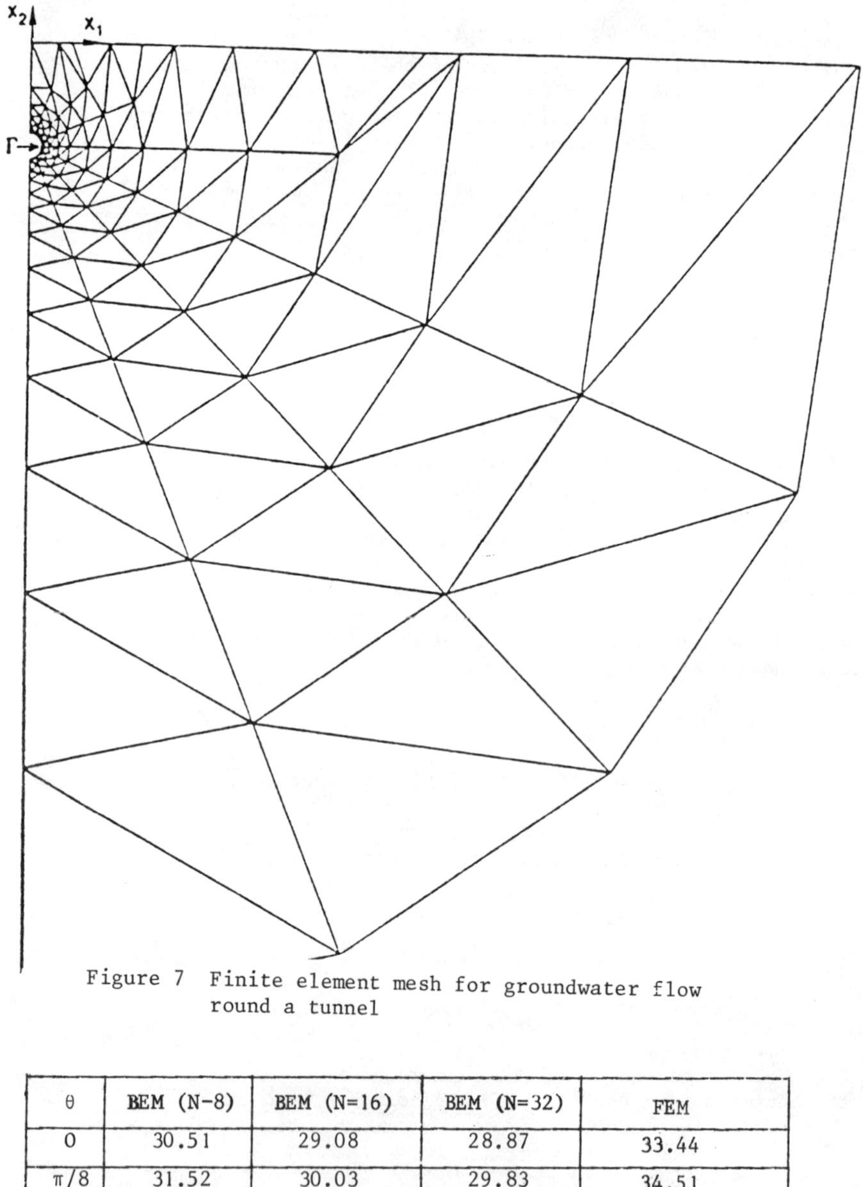


Figure 7 Finite element mesh for groundwater flow round a tunnel

| θ | BEM (N-8) | BEM (N=16) | BEM (N=32) | FEM |
|---|---|---|---|---|
| 0 | 30.51 | 29.08 | 28.87 | 33.44 |
| π/8 | 31.52 | 30.03 | 29.83 | 34.51 |
| π/4 | 34.60 | 32.87 | 32.73 | 37.76 |
| 3π/8 | 39.96 | 38.05 | 37.85 | 43.46 |
| π/2 | 48.24 | 45.96 | 45.72 | 52.34 |
| 5π/8 | 61.43 | 58.42 | 58.10 | 66.46 |

Table 4 Values of $u_2$ (-) on tunnel surface

using linear boundary elements and compared to a finite element solution [18] obtained by discretizing the whole semi-infinite region into 152 triangular finite elements plus some infinite elements (see figure 7). The discrepancy between both solutions is due to the coarseness of the finite element mesh around the tunnel (see figure 7), which does not take into account properly the discontinuity on the radial flux at the point B.

Example 5

Skerget [19] analysed a heat transfer problem in an engine valve as shown in figure 8. The problem is axisymmetric with prescribed mixed-type boundary conditions, which render the problem non-linear. The problem was solved using a direct iterative technique with prescribed accuracy $\max|u^{j+1}-u^{j}| \leq 1$, which was achieved in only seven cycles. Here we take the following numerical values $k_o = 80$, $u_o = 300$, $\beta = 0.5$. Temperatures obtained at some internal points are shown in table 5 where they are compared against those resulting from an analysis with a constant diffusivity equal to $k_o$.

| R | Z | $k = k_o$ | $k = k(T)$ |
|---|---|---|---|
| 0.100 | 0.004 | 772.8 | 761.2 |
| 0.100 | 0.008 | 766.6 | 758.2 |
| 0.100 | 0.012 | 760.5 | 755.1 |
| 0.100 | 0.016 | 754.5 | 752.2 |
| 0.112 | 0.012 | 758.6 | 754.6 |
| 0.012 | 0.08 | 611.5 | 655.0 |
| 0.012 | 0.1 | 562.7 | 614.8 |
| 0.012 | 0.11 | 542.5 | 596.7 |

Table 5 Temperatures at internal points for engine valve

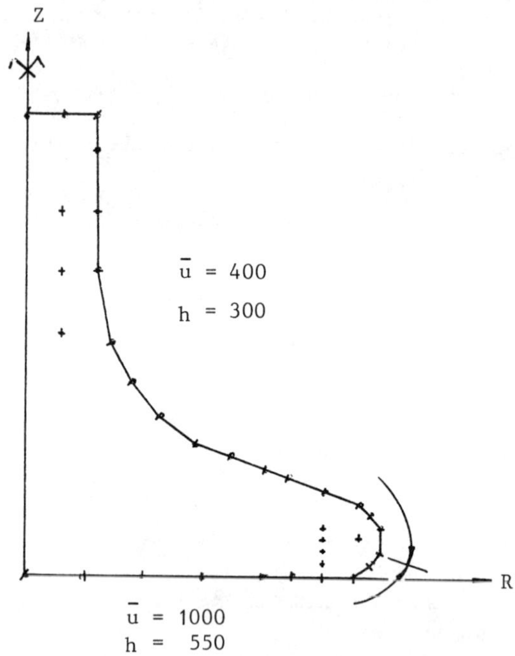


Figure 8 Boundary element discretization with prescribed boundary conditions for engine valve (26 quadratic elements, 53 nodes)

## 7. CONCLUSIONS

The present paper shows some applications of boundary element methods to solve potential problems of the type that can occur in fluid flow. The case of Laplace's or Poisson's equations is discussed in detail, together with their extension to anisotropic problems and infinite regions. The paper also extends previous work on material nonlinearity and demonstrates how many of these problems can be linearized by using the Kirchoff's transformation.

REFERENCES


1. BREBBIA, C.A. "The Boundary Element Method for Engineers" Pentech Press, London and Halstead Press, New York, 1978. Second Edition, 1980.

2. BREBBIA, C.A. and S. WALKER "Boundary Element Techniques in Engineering", Butterworths, London, 1980.

3. FREDHOLM, I. "Sur une Classe d'equations fonctionelles", Acta Meth., 27, 365-390, 1903.

4. KELLOGG, O.D. "Foundations of Potential Theory". Springer-Verlag, Berlin, 1929.

5. JASWON, M.A. "Integral Equation Methods in Potential Theory I", Proc. Royal Society A, 275, 23-32, 1963.

6. SYMM, G.T. "Integral Equation Methods in Potential Theory II", Proc. Royal Society A, 275, 33-46, 1963.

7. JASWON, M.A. and A.R. PONTER "An Integral Equation Solution of the Torsion Problem", Proc. Royal Society A, 273, 237-246, 1963.

8. HESS, J.L. and A.M.O. SMITH "Calculation of Potential Flow about Arbitrary Bodies" Progress in Aeronautical Sciences, Vol. 8, D. Küchemann (ed.) Pergamon Press, London, 1967.

9. HARRINGTON, R.F., K. PONTOPPIDAN, P. ABRAHAMSEN and N.C. ALBERTSEN, "Computation of Laplacian Potentials by an equivalent-source method", Proc. IEE, Vol. 116, No. 10, 1715-1720, 1969.

10. MAUTZ, J.R. and R.F. HARRINGTON "Computation of Rotationally Symmetric Laplacian Potentials", Proc. IEE, Vol. 117, No. 4, 850-852, 1970.

11. JASWON, M.A. and G.T. SYMM "Integral Equation Methods in Potential Theory and Elastostatics", Academic Press, London, 1977.

12. FAIRWEATHER, G., F.J. RIZZO, D.J. SHIPPY and Y.S. WU "On the Numerical Solution of Two-Dimensional Potential Problems by an Improved Boundary Integral Equation Method" Journal of Computational Physics, Vol. 31, 96-112, 1979.

13. CHANG, Y.P., C.S. YANG and D.J. CHEN "The Use of Fundamental Green's Functions for the Solution of Heat Conduction in Anisotropic Media". Int. Journal of Heat and Mass Transfer, Vol. 16, 1905-1918, 1973.

14. BIALECKI, R. and A.J. NOWAK "Boundary Value Problems for Nonlinear Applied Mathematical Modelling, Vol. 5, December 1981.

15. SVELEK, P. and C. BREBBIA "Nonlinear Potential Problems", Chapter 1 in "Progress in Boundary Element Methods. Vol.2". Pentech Press, London, 1982

16. ABBOTT, I.H. and A.E. VON DOENHOFF "Theory of Wing Sections", Dover, New York, 1959.

17. GOLDENBERG, H. "External Thermal Resistance of Two Buried Cables", Proc. IEE, Vol. 116, No. 5, 822-826, 1969.

18. WOOD, W.L. "On the Finite Element Solution of the Exterior Boundary-Value Problem", Int. Journal Num. Methods Engng, Vol. 10, 885-891, 1976.

19. SKERGET, P. Internal Report, Computational Mechanics Centre, Southampton, 1982.

# BOUNDARY ELEMENT METHOD: PROCESSING OF THE SOURCE TERM OF THE POISSON EQUATION BY MEANS OF BOUNDARY INTEGRALS ONLY

Di Monaco, A. and Rangogni, R.

ENEL - DSR - Centro Ricerca Idraulica e Strutturale - Milano (Italy)

SUMMARY

The most important advantage of the Boundary Element Method is that no internal discretization of the domain is required. However this advantage is generally lost when source terms are present in the governing differential equation. This paper shows, with reference to the 3D Poisson equation, that, in the case of a harmonic source term, it is possible to eliminate the volume integral from the formulation. In fact the volume integral can be transformed into a boundary integral by using the Green formula. The procedure is applied to the solution of a simple problem and the results are compared with the analytical solution.

## INTRODUCTION

Several numerical techniques have been used to solve numerically elliptic boundary values problems, the most common of which are the finite element method (FEM) and the finite difference method (FDM).
A more recent method of increasing popularity is the boundary element method (BEM) which, though being of narrower applicability, has shown to be very efficient in the solution of certain type of problems.
This last method is based on the transformation of elliptic (or parabolic) differential equations into integral equations, in which only the values of the unknown function u, and its normal derivative $\partial u/\partial n$ on the boundary of the domain, appear.
The unknowns of the new problem become the function u and/or its normal derivate on the boundary, according to the assigned data.

Once these values have been determined, it is subsequently possible to compute approximate values of the function u in an arbitrary internal point of the domain by a simple integration on the boundary.
To this purpose it is necessary to discretize the boundary of the domain with elementary surfaces (or arcs in two dimensions) and to approximate the unknown function u and its normal derivative $\partial u/\partial n$ by means of trial functions.
Several approximations have been used by the researchers: constant, linear and quadratic polynomials (see Symm 1963, Jaswon 1977, Brebbia 1978, Di Monaco et al. 1980).
The main feature which makes this method advantageous with respect to other ones is, therefore, that no internal discretization of the computational domain is required.
However, the advantage of avoiding the internal discretization of the domain is generally lost when source terms are present in the differential equation governing the problem. Such terms occur, for example, in the Poisson equation.
In this case, as it is usually found in the current technical literature (see for example Brebbia 1978), there appear in the mathematical formulation volume integrals that cannot generally be processed analytically. It is therefore necessary to have recourse to numerical procedures that require a supporting grid composed of simple sub-regions within the definition domain.
In this paper we present a procedure which enables to process, under certain conditions, the source term in the Poisson equation by means of boundary integrals only.
This formulation is applied to three-dimensional problems and was found independently from Fairweather et al. (1979) who applied it to two-dimensional problems.
Our main goal, however, is to draw attention on the advantages of this procedure, which allows to avoid internal discretization of the domain in presence of source terms.

## MATHEMATICAL FORMULATION

Let us suppose D is a bounded domain in the three-dimensional space $\mathbb{R}^3$ with piecewise smooth boundary S.
It is known (Smirnov, 1964) that applying Greens third identity to any sufficiently smooth function u, defined on $\overline{D}$ (closure of D), for any P = (X, Y, Z) $\in$ D, the following equation can be deduced.

$$u(P)\,\eta(P) = \int_S \left(u \frac{\partial}{\partial n}\frac{1}{r} - \frac{1}{r}\frac{\partial u}{\partial n}\right) dS - \int_D \frac{1}{r}\nabla^2 u\, dD, \qquad (1)$$

where $\nabla^2$ denotes the Laplace differential operator, n denotes the inward normal unit vector on S, r is the distance between the fixed point P and the integration point Q = (x, y, z) and finally,

$$\eta(P) = \int_S \frac{\partial}{\partial n} \frac{1}{r} \, dS \qquad (2)$$

represents the solid angle in P. In particular if S has an unique tangent plane in P (when $P \in S$) then we have

$$\eta(P) = \begin{cases} 4\pi & \text{if } P \in D \\ 2\pi & \text{if } P \in S \\ 0 & \text{if } P \notin \overline{D} \end{cases} \qquad (3)$$

If u is assigned on S (Dirichlet problem) then the unknown is $\frac{\partial u}{\partial n}$ while if $\frac{\partial u}{\partial n}$ is prescribed on S (Neumann problem) the unknown is the function u. When we have a mixed problem the unknowns are, on one part of S, the function u and, on the other part of S, the function $\frac{\partial u}{\partial n}$. When $\nabla^2 u = 0$ in D (Laplace equation), then the integral on D vanished. On the contrary, if $\nabla^2 u = f$ (Poisson equation) we need also to compute the volume integral in equation (1) and, in general, it is impossible to compute it analytically. On the other hand the numerical evaluation of the volume integral requires an internal mesh in D and this implies a supplementary work in the preparation of the input data. However it is possible, in many cases, to turn the volume integral in Equation (1) into a surface integral. To this purpose we recall the Green's formula applied to two generic, sufficiently regular functions g and h

$$\int_D (g\nabla^2 h - h\nabla^2 g) \, dD = - \int_S (g\frac{\partial h}{\partial n} - h\frac{\partial g}{\partial n}) \, dS \qquad (4)$$

Substituting g with f in Equation (4) and supposing that $\nabla^2 f = 0$, we obtain:

$$\int_D f\nabla^2 h \, dD = - \int_S (f\frac{\partial h}{\partial n} - h\frac{\partial f}{\partial n}) \, dS \qquad (5)$$

If we now are able to find a function h such that $\nabla^2 h = r^{-1}$, then the volume integral in Equation (5) becomes exactly the volume integral in Equation (1). Such a function really exists, and for 3D problems is given by:

$$h(x, y, z) = \frac{1}{2} r = \frac{1}{2} \left[ (x - X)^2 + (y - Y)^2 + (z - Z)^2 \right]^{1/2} \qquad (6)$$

Using Equations (6) and (5), Equation (1) becomes:

$$u(P)\,\eta(P) = \int_S (u\frac{\partial}{\partial n}\frac{1}{r} - \frac{1}{r}\frac{\partial u}{\partial n}) \, dS + \int_S (f\frac{\partial h}{\partial n} - h\frac{\partial f}{\partial n}) \, dS \qquad (7)$$

which implies only surface integrals. Equation (7) is the basis for the application of BEM to Poisson equation $\nabla^2 u = f$ in the hypothesis that f is a harmonic function. In practical problems it is frequent that the source term f is a contant f = k, and in this case it is certainly verified that $\nabla^2 f = 0$.
Moreover it is also $\frac{\partial f}{\partial n} = 0$ and Equation (7) becomes

$$u(P)\,\eta(P) = \int_S \left(u\frac{\partial}{\partial n}\frac{1}{r} - \frac{1}{r}\frac{\partial u}{\partial n}\right) dS + \int_S k\,\frac{\partial h}{\partial n}\, dS \tag{8}$$

Equation (8) has been used to solve the problem presented in the following section. To complete, we recall (Fairweather 1979) that in two-dimensions the function h, which allows a similar treatment, is

$$\bar{h} = \frac{1}{4} r^2 (\lg r - 1) \tag{9}$$

where r denotes again the distance (in the plane) between the point P and the integration point Q.
The numerical application of Equation (8) requires, as usual, to discretize the boundary surface S into a number M of simpler surface elements and to approximate u and $\frac{\partial u}{\partial n}$ on each element with suitable functions.
For instance, in the simplest case when both u and $\partial u/\partial n$ are taken as constant on each surface element $S_j$, denoting with $u_j$ and $v_j$, respectively, these constant values, Equation (8) becomes

$$u(P)\,\eta(P) = \sum_{j=1}^{N} \left[G_j(P) u_j + H_j(P) v_j + k F_j(P)\right] \tag{10}$$

where N is the total number of free parameters (in this case N = M), and

$$\begin{aligned} G_j(P) &= \int_{S_j} \frac{\partial}{\partial n}\frac{1}{r}\, dS \\ H_j(P) &= \int_{S_j} \frac{1}{r}\, dS \\ F_j(P) &= \int_{S_j} \frac{\partial h}{\partial n}\, dS \end{aligned} \tag{11}$$

If higher orders of approximation are used Equation (11) must be suitably modified, but formally Equation (10) always holds. If we now choose P on S we can write so many equations of type (10) as many unknown values $u_j$ or $v_j$ we have.
The solution of the system obtained in this way gives us the unknown values. Then, by means of Equation (10), an approximate evaluation of the solution u at any point of D can subsequently be obtained.

## PARTICULAR FEATURES OF THE COMPUTER PROGRAM USED FOR THE NUMERICAL EXAMPLE

Against the advantage of not requiring an internal discretization of the domain D, the BEM presents the disadvantage that the matrix of the algebraic system of equations, which it leads to, is not symmetric and in general is fully populated. From this point of view BEM does not appear to be very competitive with respect to the already mentioned FDM and FEM. We note, however, with regard to this point, that it is also possible to insert in the above formulation a "least square" procedure, which leads to a symmetric matrix.
Besides, this approach brings about some advantages in the treatment of problems defined on non-homogeneous domains or, more in general, in the treatment through the domain subdivision into sub-regions. The aforementioned advantages consist essentially in the possibility of building up the matrix of each subdomain separately and to assemble them in a way perfectly similar to that used in FEM. This introduces some degree of sparsity in the matrix and allows the use of well known and efficient techniques for the solution of the algebraic system, such as the band algorithm or rather the frontal solution. In this particular approach BEM can be seen as a generalization of FEM. The above procedure allows also to deal, without further complications, with the presence of sharp corners on the boundary.
The formulation of BEM, coupled to a least square method, is presented in Di Monaco et al. (1982 a) and is the base of the program used to solve the numerical example shown in the next section.
We think useful to close this section with an outline of the technique followed - and implemented in our computer program - in dealing with axialsymmetric problems. The problem, posed in an orthogonal cartesian coordinates system $(x, y, z)$, is first of all transformed in a cylindrical coordinates system $(r, \varphi, z)$ and than reduced to two dimensions $(r, z)$, through an integration with respect to $\varphi$ between 0 and $2\pi$. In this way we get a formulation which implies only integrals over the plane curve generating the surface **S**.
Practical calculations require an evaluation of elliptic integrals of first and second kind which, in turn, require suitable approsimations and techniques for dealing with singularities. For further details about these techniques, refer to Di Monaco ed al. (1982 b).

## NUMERICAL EXAMPLE

We show the results of a simple axialsymmetric problem solved by the method presented in the previous sections just to demonstrate its applicability and asses its accuracy.

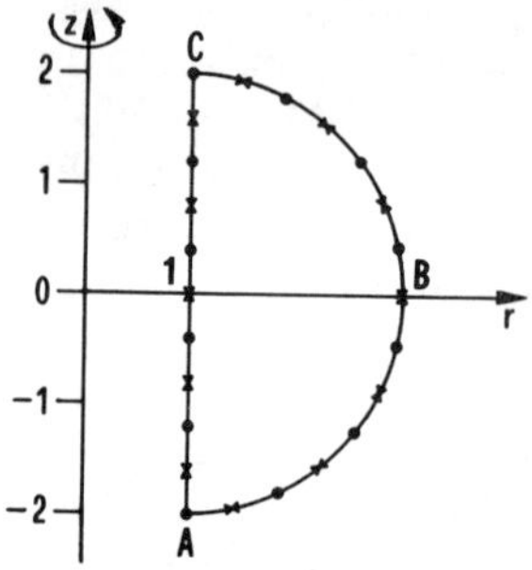


Figure 1 - Section in the axial plane (r, z) of the annular domain chosen for the numerical example.

We consider the following Poisson-Dirichlet problem defined on the domain D shown in fig. 1

$$\begin{aligned} &\nabla^2 u\,(x, y, z) = 6 \quad (x, y, z) \in D \\ &u\,(x, y, z) = x^2 + y^2 + z^2 \quad (x, y, z) \in S \end{aligned} \tag{12}$$

The analytical solution of Equations (12) is

$$u\,(x, y, z) = x^2 + y^2 + z^2 \quad (x, y, z) \in D \tag{13}$$

The curved part of the boundary in fig. 1 was divided into 7 isoparametric second order elements and the straight part into 5 straight second order elements.
The numerical solution represented by the values of the function $\hat{v}$ (approximation of $\partial u/\partial n$) in the 24 boundary nodes, was checked against the exact solution $v = \partial u/\partial n$, deduced from Equation (13) and a percentage error was defined on the boundary as follows

$$E\,(\%) = \frac{\hat{v} - v}{v_{max}} \cdot 100 \tag{14}$$

In fig. 2 the diagrams of the computed function $\partial u/\partial n$ and percentage error E (%) along the rectified boundary, are shown.

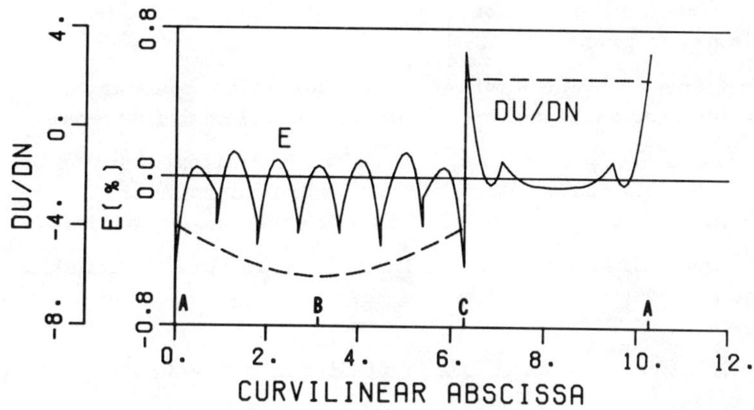


Figure 2 - Diagrams of computed $\partial u/\partial n$ and E (%) along the rectified boundary.

## CONCLUSION

We have shown a procedure for dealing with the source term in the 3D Poisson equation by means of boundary integrals only.

Such technique, whenever it can be applied, has the great a-dvantage of avoiding internal discretization of the domain.

The main purpose of this paper was to draw attention on such a procedure which, in our opinion, though already proposed in the literature for 2D problems, does not seem to have been taken into proper consideration by the researchers.

## REFERENCES


Brebbia, C.A. (1978) The Boundary Element Method for Engineers. Pentech Press, London.

Brebbia, C.A. (Ed) (1980) New Developments in Boundary Element Methods. Proc. of the second Int. Seminar on BEM held at the University of Southampton, March 1980.

Brebbia, C.A. (1981) Boundary Element Methods. Proc. of the third Int. Seminar, Irvine, California, July 1981.

Di Monaco, A. and Rangogni, R.(1980) Il Boundary Element Method ad elementi curvi del secondo ordine nel problema piano di Laplace. L'Energia Elettrica, 57, 12 : 565 - 574.

Di Monaco, A. and Rangogni, R. (1982a) A least square procedure coupled to the Boundary Element Method for elliptic problems. In preparation.

Di Monaco, A.and Rangogni, R. (1982b) The Boundary Element Method for axial symmetric elliptic problems. In preparation

Fairweather, G. et al (1979) On the Numerical Solution of two-dimensional problems by an improved Boundary Integral Equation Method, Journal of Comp. Physics 31, 96-112

Jaswon, M.A. and Symm, G.T. (1977) Integral Equation Methods in potential Theory and Elastostatics. Accademic Press, London.

Smirnov V.I. (1964) A course of Higher Mathematics, Vol II, Pergamon Press, Oxford.

Symm, G.T. (1963) Integral Equations Methods in Potential theory. Proc. Roy. Soc. A, 275, part. II, 33-46.

# FINITE ELEMENT ANALYSIS OF MOMENTUM TRANSFER IN RIVERS OF COMPLEX CROSS SECTION

Miodrag Radojković, Marko Ivetić

Institute of Hydraulic Engineering
Faculty of Civil Engineering, Belgrade, Yugoslavia

## SUMMARY

In the most of the current one-dimensional mathematical models of open channel flow lateral momentum transfer due to turbulence and secondary currents is neglected when subdivision of compound channel cross-section is applied to compute cross sectional roughness parameter in function of local roughness and/or geometric characteristics along the cross section.

On the contrary, many investigatiors have shown experimentaly that neglecting of lateral momentum exchange between typical parts of compound cross section (main channel and flood plain) may result in serious errors in computation of discharges in each subsection, specially in flood plains where errors of more than 100% were reported.

The aim of the paper is description of the possibility for computing lateral momentum transfer in standard onedimensional models by introduction of simple turbulence model based on mixing length theory. The proposed procedure is verified by comparison with experimental results of other investigators.

## INTRODUCTION

In the last decade open channel hydraulics is significantly improved by develomepent of highly sophisticated mathematical models capable to solve many problems with 2D and 3D resolution [6]. Nevertheless, one dimensional models are still the most widely used engineering tool and any reasonable improvement of them deserves attention.

An example where possible improvements of one dimensional computations can be introduced, is flow in rivers of compound cross section usually analysed by subdivision of section into a number of nearly homogeneous subareas and supposing shear free water boundaries between them [1]. Total discharge in cross section is obtained as sum of discharges for each subarea computed by some known discharge-resistance formula. But, it has been well established that supposition of shear free water boundaries is not always valid, specially in rivers consisting of main channel and one or more flood plains [3].

The problem of introducing this type of boundary shear in river computations was attacked by many researchers with laboratory experiments in channels of idealised geometries. Although some empirical formulas were recomanded [5] their general validity in natural conditions, where geometry of the problem differs significantly from one investigated, is questionable.

In the paper, the authors suggested a computational approach to solve the problem by mathematical modelling of the flow pattern in compound river cross section by finite element method. Results of available experimental research were used to verify the model.

Proposed model can be used for prediction of velocity distribution and other flow parameters for the conditions where recorded data on flow are not available.

## PROBLEM DESCRIPTION

Momentum conservation equation for one dimensional steady flow in a natural watercourse of variable cross section reads:

$$\frac{\partial \bar{\alpha}\ \bar{V}^2\bar{A}}{\partial x_1} + g\bar{A}\ \frac{\partial z}{\partial x_1} + \frac{1}{2}\ \bar{C}_\tau \bar{P}\bar{V}|\bar{V}| = 0 \qquad (1)$$

Friction coefficient for the whole section $\bar{C}_\tau$ and velocity distribution coefficient $\bar{\alpha}$ have to be determined in terms of section geometry and local friction coefficient distribution along the bottom. This is ussualy done in a way that a small element is taken from the section and the following suppositions, in analysis of momentum balanse for the element, applied:

1. vertical boundaries of each element are shear free (no net loss or gain of momentum of one element from another);
2. level slope is constant across the width of river;
3. ratio of inertial forces to gravity and pressure forces have nearly the same value for the whole section and for each element.

Then the balance of forces acting on each element (Fig.1) is combined with ballance of forces acting on the whole cross section applying the condition that total discharge in the cross section is sum of discharges for each element to obtain:

$$\bar{C}_\tau = \frac{\bar{A}^2\bar{R}}{\left[\int_0^{\bar{B}} \frac{H\sqrt{R}}{\sqrt{C_\tau}}\, dx_2\right]^2} \qquad (2)$$

$$\alpha = \frac{\alpha_v \, \bar{C}_\tau}{\bar{A} \, \bar{R}} \int_o^{\bar{B}} \frac{HR}{C_\tau} \, dx_2 \qquad (3)$$

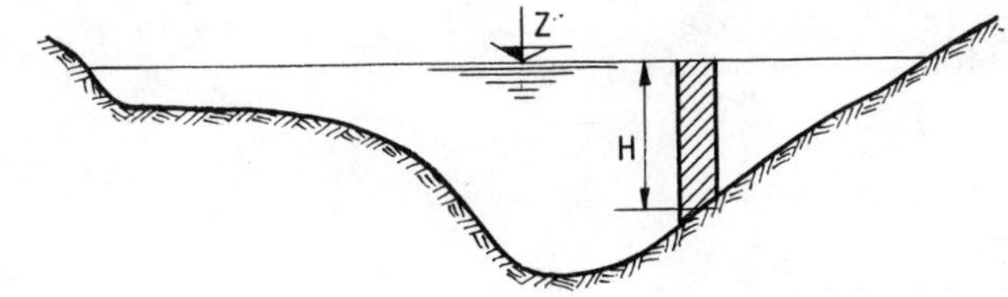


Fig. 1. Element of river section used for deriving momentum equation in the direction of mean flow

Validity of any supposition given above is questionable but only the first one will be discussed in the sequel.

Problem of shear-stress distribution in complex cross sections was studied by many authors experimentally in laboratory flumes consisting of main channel and one or more flood plains as typical example of complex cross section (Fig.2).

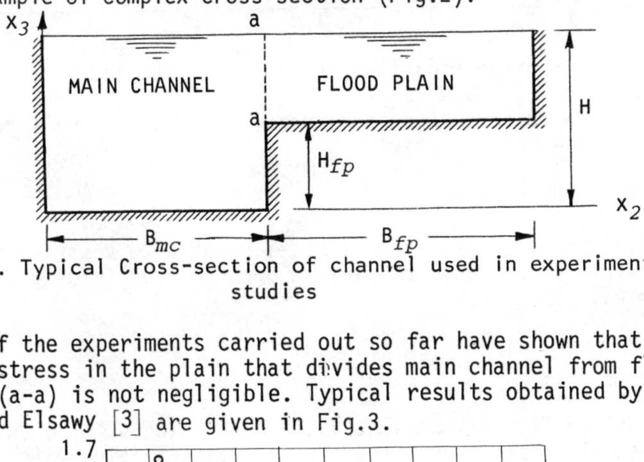


Fig. 2. Typical Cross-section of channel used in experimental studies

Most of the experiments carried out so far have shown that the shear stress in the plain that divides main channel from flood plain (a-a) is not negligible. Typical results obtained by Myers and Elsawy [3] are given in Fig.3.

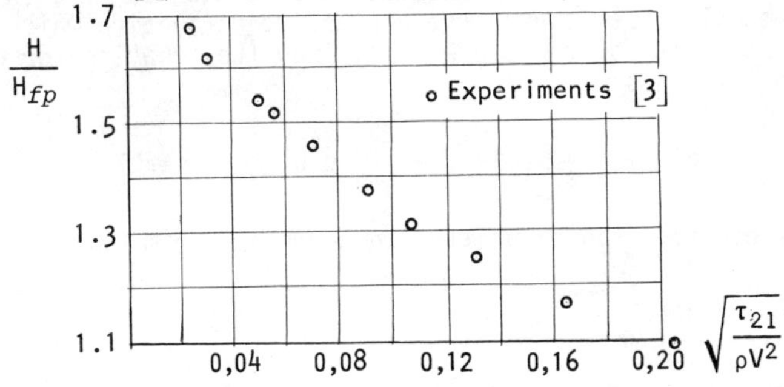


Fig.3. Stress measured in dividing plane

As the consequence, discharge capacity of the main channel is decreased and of the flood plain increased compared to the discharge capacity that would have occured if there had been no shear stress along the dividing plane. If this shear stress could be computed the real discharge capacity for each part od river section would be possible to evaluate as was done in [4] for above mentioned experiments (Fig. 4).

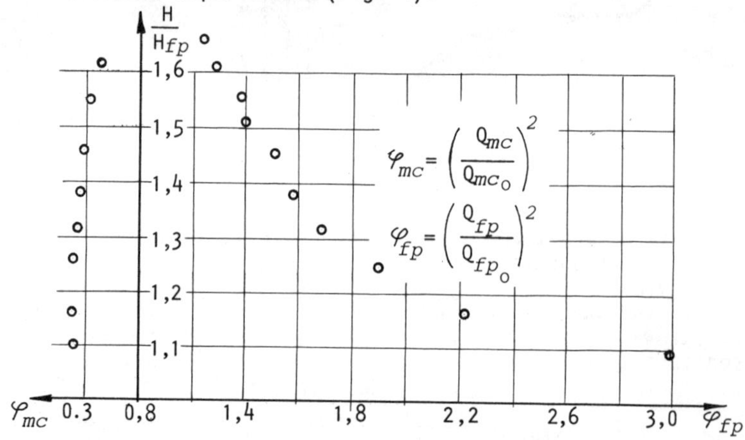


Fig. 4. Discharge capacities of main channel and flood plain

The method to compute this boundary shear stress will be discussed in the next chapter.

## MATHEMATICAL MODEL

Derivation of a mathematical model capable to describe incompressible steady uniform flow in open channel is originated from Reynolds equations:

$$\frac{\partial\, u_i u_j}{\partial x_i} = f_j - \frac{1}{\rho}\frac{\partial p}{\partial x_j} + \frac{\mu}{\rho}\frac{\partial}{\partial x_i}\left(\frac{\partial u_j}{\partial x_i}\right) - \frac{\partial}{\partial x_i}\overline{u_i' u_j'} \tag{4}$$

In uniform flow all derivatives in direction of flow ($x_1$) are set equal to zero, except the pressure derivative which is substituted with bottom slope ($I_o$). Then the equation for direction $x_1$ reads:

$$\frac{\partial\, u_i u_1}{\partial x_i} = gI_o + \frac{\mu}{\rho}\frac{\partial}{\partial x_i}\left(\frac{\partial u_1}{\partial x_i}\right) - \frac{\partial}{\partial x_i}\overline{u_i' u_1'} \qquad i=2,3 \tag{5}$$

This equation can be written in condensed form as:

$$\frac{\partial \tau_{i1}/\rho}{\partial x_i} = g\, I_o \tag{6}$$

where:

$$\frac{\tau_{i1}}{\rho} = \overline{u_i' u_1'} - \frac{\mu}{\rho}\frac{\partial u_1}{\partial x_i} + u_i\, u_1 \qquad (7)$$

$\tau_{i1}$ can be regarded as "effective shear stress" which accounts for:

- Reynolds stresses $(\overline{u_i' u_1'})$
- viscous stresses $(\frac{\mu}{\rho}\frac{\partial u_1}{\partial x_i})$
- effect of cecondary currents $(u_i u_1)$

This stress appears in the ballance of momentum for cross section element given in Fig. 1 and can be computed if the velocity distribution in the cross section is known.

For that purpose Reynolds stresses have to be modelled to close the problem description and then all terms included in $\tau_{i1}$ can be computed.

According to the state of the art of turbulence modelling this problem can be expected to be efficiently solved by some sophisticated model such as K -ε [6], but,as it would require at least 5 equations to solve,it seemed appropriate to attack the problem in the first stage of research in a simpler way. One of the possibilities was introduced by Soviet authors [7]:

$$\frac{\tau_{i1}}{\rho} = \ell_m\, u_1 \frac{\partial u_1}{\partial x_i} \qquad (8)$$

where $\ell_m$ is some kind of mixing length (not the same as one defined by Prandtl). Then, after substitution of (8) in (6) one obtaines:

$$\frac{\partial}{\partial x_2}\, \ell_m u_1 \frac{\partial u_1}{\partial x_2} + \frac{\partial}{\partial x_3}\, \ell_m\, u_1 \frac{\partial u_1}{\partial x_3} = g\ I_o \qquad (9)$$

This is the elliptic nonlinear partial differential equation that can be easily solved by FEM for given boundary conditions.

To generalize the solution, equation (9) can be written in dimensionless form after applying uniform flow conditions:

$$g\ I_o = \frac{1}{2}\, \overline{C}_\tau\, \frac{\overline{V}}{R}$$

as

$$\frac{\partial}{\partial \xi_2}\, \ell_* \omega_1 \frac{\partial \omega_1}{\partial \xi_2} + \frac{\partial}{\partial \xi_3}\, \ell_* \omega_1 \frac{\partial \omega_1}{\partial \xi_3} = \frac{1}{2}\, \overline{C}_\tau\, \frac{\overline{B}}{R}$$

where:

$$\omega_1 = \frac{u_1}{\overline{V}} \qquad \xi_i = \frac{x_1}{B} \qquad \ell_* = \frac{\ell(\xi_2,\xi_3)}{\overline{B}}$$

Boundary condition along the solid boundary is defined by standard procedure which suppose logarithmic velocity distribution in the wall region [6] that can be written as:

$$\omega_W = \sqrt{C_\tau}(1{,}768 \ln Re\sqrt{C_\tau} + 3.542) \tag{12}$$

where referent quantities in Reynolds number are distance of the point where $\omega_W$ acts from wall and cross section average velocity respectively. Following practical recommendation that $30 < Re\sqrt{C_\tau} < 100$ [6] it is adopted that:

$$\omega_W = 10\sqrt{C_\tau} \tag{13}$$

Boundary condition at free surface is prescribed as zero velocity gradient.

Equation (9) can be now solved for given distribution of mixing length across the section. Any distribution should satisfy obvious condition:

$$\int_\Omega \omega \, d\Omega = 1 \tag{14}$$

that should be incorporated in computation procedure.

Velocity distribution coefficient is computed as secondary result from the definition:

$$\alpha = \int_\Omega \omega^2 d\Omega \tag{15}$$

EXPERIMENTAL VERIFICATION

Proposed procedure is verified by experimental of Crory and Elsawy |2| carried out in the laboratory flume with the characteristics:

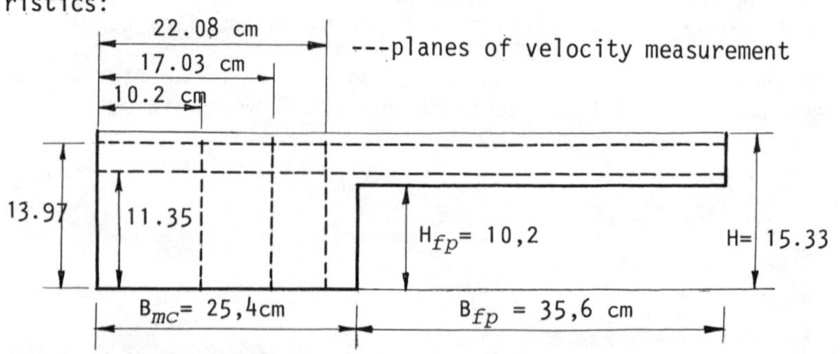


Fig. 5. Geometric parameters of the laboratory flume [2]

In one of the experiments, selected from the series, with $B_{mc}$ = = 25.4 cm and $B_{fp}$= 35.6 cm, for flow depth H = 15.33 cm measured discharge was Q = 13.94 ℓ/s giving $\bar{C}_\tau$ = 0.0063. Velocity distribution measurements were done in several horizontal and vertical plains as given in Fig.5 by Laser Doppler Annemometry.

Mathematical model described in the pervious chapter was applied to compute velocity distribution in the cross section by simple FEM program with rectangular mesh given in Fig.6.

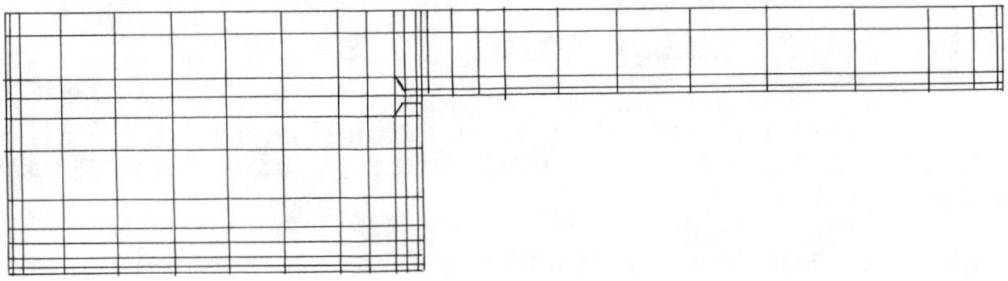

Fig. 6. Finite element mesh used in computations

Distribution of mixing length in the cross section is given in Fig. 7.

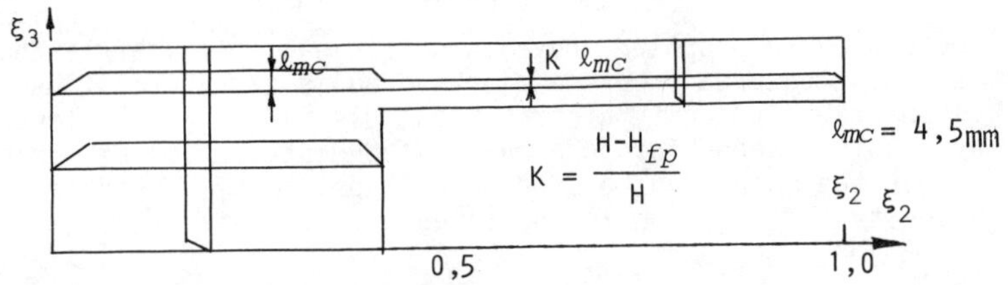


Fig. 7. Mixing length distribution

Computed velocities are compared with measured ones in horizontal plains (Fig.8) and vertical planes (Fig.9).

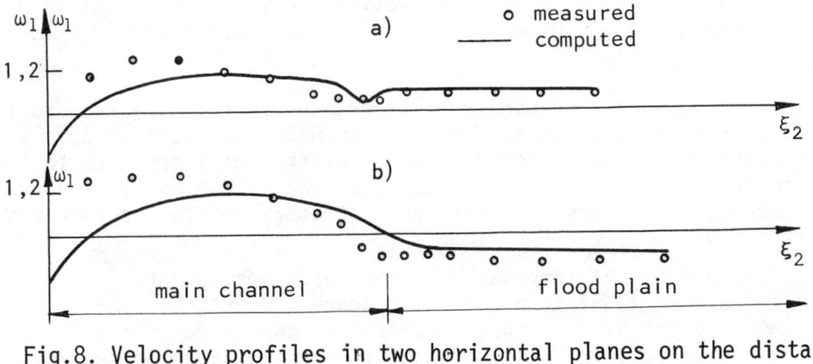


Fig.8. Velocity profiles in two horizontal planes on the distance a) $\xi_3$= 0,229, and b) $\xi_3$= 0,186 from the main channel bed

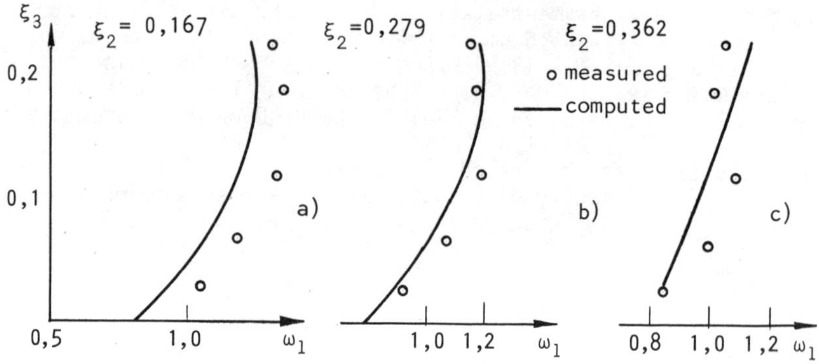


Fig. 9. Velocity profiles in three vertical planes on the distance a) $\xi_2$= 0,167, b) $\xi_2$= 0,279 and c) $\xi_2$= 0,362 from the main channel wall

Boundary shear in plane dividing main channel from flood plain was evaluated indirectly from the bottom shear distribution measured by Preston tube. Experimentally determined value of average shear in vertical plane a-a $\left(\frac{\tau_{21}}{\rho V^2}\right)_c = 8.5\times10^{-3}$, compares fairly good with computed one $\left(\frac{\tau_{21}}{\rho V}\right)_c = 8\times10^{-3}$.

The procedure proposed above is sufficient to calibrate the model when measured cross section hydraulic characteristics for given stage (velocity distribution, discharge, level slope) are available . Then, for once calibrated distribution of mixing length the model can be used to extrapolate $\overline{C}_\tau$, $\overline{\alpha}$ and relative velocity distribution in the range of interest.

## CONCLUSION

Proposed method for analysis of shear stress distribution in rivers of complex cross section,although applied very crude closure model which coupled the effects of turbulence with ones of secondary currents, have given satisfactory prediction of velocity distribution and effective shear stresses in the investigated cross section.

Therefore, in the opinion of authors, contrary to practice applied so far to investigate flow in compound cross section solely by experiments, only computational experimental approach to the problem is promising to produce results that can be of value to practical engineers. The following strategy for future research is proposed:

- developement of computation method with appropriate turbulence model to predict secondary currents
- measurement of all three velocity components (mean and fluctuating values) in cross section of representative geometries that will be used to verify developed model

- field measurements of velocity components in rivers to verify real life applicability of the method

## NOTATION

(symbols with overbar denote quantities defined for the whole cross section).

$A$ - cross section area
$B$ - cross section width
$C_\tau$ - local bottom friction coefficient
$\overline{C}_\tau$ - friction coefficient for cross section
$H$ - water depth
$I_o$ - bottom slope
$\overline{P}$ - wetted perimeter of cross section
$Q_{mc}$ - discharge capacity of main channel
$Q_{mco}$ - discharge capacity of main channel that would be computed without shear in deviding plane
$Q_{fp}$ - discharge capacity of flood plain
$Q_{fpo}$ - discharge capacity of mean channel that would be computed without shear in deviding plane
$R$ - local hydraulic radius
$\overline{R}$ - cross section hydraulic radius
$Re$ - Reynolds number
$V$ - average velocity in cross section
$f_j$ - gravitational force component
$g$ - gravity constant
$\ell_m$ - mixing length
$\ell_*$ - dimensionless mixing length
$p$ - presure
$u_i$ - mean velocity component
$u_i$ - fluctuation component
$\alpha$ - velocity distribution coefficient for one cross section element
$\overline{\alpha}$ - velocity distribution coefficient for cross section
$\tau_{i1}$ - effective stress component
$\rho$ - water density
$\mu$ - dynamic viscosity
$\omega_i$ - dimensionless mean velocity component
$\omega_w$ - dimensionless velocity at wall
$\xi_i$ - dimensionless cartesian coordinates
$\overline{u_i u_j}$ - Reynolds stress component
$\Omega$ - dimensionless cross sectional area

REFERENCES

[1] Abbot,M.B. (1979) Computational Hydraulics. Pitman Publishing, London.

[2] Crory,P.M.,Elsawy,E.M. (1980) An Experimental Investigation into the Interaction between a Rivers Deep Section and its Flood Plain. International Symposium on River Engineering, Belgrade 1980.

[3] Myers,W.R.C.,Elsawy,E.M. (1975) Boundary Shear in Channels with Flood Plain, Journal ASCE, HY 7.

[4] Radojković, M. (1976) Mathematical Modelling of Rivers with Flood Plains. Symposium Rivers 76, Fort Collins.

[5] Rajaratnam,N.,Ahmadi,R.(1981) Hydraulics of Channels with Flood Plains, Journal od Hydraulic Research No 1.

[6] Rodi,W. (1980) Turbulence Models and Their Application in Hydraulics,IAHR Publication, Delft.

[7] Sherenkow,I.A. (1978) Two Dimensional Problems of Subcritical Flow in Open Channels. "Energy", Moscow.